INDEX OF TABLES

COLLEGE PHYSICS

SIXTH EDITION

COLLEGE PHYSICS

SIXTH EDITION

FRANCIS W. SEARS
Late Professor Emeritus
Dartmouth College

MARK W. ZEMANSKY
Late Professor Emeritus
City College of the City University of New York

HUGH D. YOUNG
Professor of Physics
Carnegie-Mellon University

ADDISON-WESLEY PUBLISHING COMPANY

Reading, Massachusetts ▲ Menlo Park, California ▲ Don Mills, Ontario
Wokingham, England ▲ Amsterdam ▲ Sydney ▲ Singapore
Tokyo ▲ Mexico City ▲ Bogotá ▲ Santiago ▲ San Juan

This book is in the Addison-Wesley Series in Physics

Bruce Spatz
Sponsoring Editor
Loretta Bailey
Art Editor
Ann Delacy
Manufacturing Supervisor
Marion E. Howe
Production and Copy Editor

Martha K. Morong
Production Manager
Oxford Illustrators, Ltd.
Illustrator
Vanessa Piñeiro-Robbins
Text and Cover Designer
Keith Wollman
Developmental Editor

Library of Congress Cataloging in Publication Data

Sears, Francis Weston
    College physics.

    Includes index.
    1. Physics. I. Zemansky, Mark Waldo
II. Young, Hugh D. III. Title.
QC23.S369 1985   530   84-16835
ISBN 0-201-07836-3

BCDEFGHIJ-RN-898765

P R E F A C E

In this new edition of *College Physics,* the basic goals are unchanged from earlier editions. We aim to provide a broad yet rigorous introduction to physics at the beginning college level for students whose mathematical preparation includes algebra and trigonometry; no calculus is used. Our primary emphasis is on helping students to develop problem-solving ability in applying a broad range of physical principles.

Three principal objectives have guided this revision: to make the book more uniform in texture and difficulty level; to add interest and relevance to the examples and problems; and to broaden the scope of applications of basic physics, especially in the life-science areas. Here are some of the more important changes.

LEVEL AND CONTENT

Great care has been taken to maintain the rigor and precision of exposition that has characterized earlier editions of this book. However, several sections that have proved to be too demanding for this book's audience have been completely rewritten and simplified. These include the material on

thermodynamics,

Gauss's law,

dielectrics,

magnetic fields,

induced electromotive force,

geometrical optics.

Some sections that were excessively detailed have been condensed or in a few cases eliminated. The treatments of magnetic materials, polarization of light, and illumination have been considerably condensed, and these topics no longer appear as separate chapters. The discussion of color has been omitted, and the number of chapters has been reduced from 48 to 44. A significant *addition* to the opening chapter is a completely new section on order-of-magnitude estimates.

PROBLEM-SOLVING STRATEGIES

Specific problem-solving strategies have been introduced explicitly wherever they are appropriate. A typical example occurs in Section 29–2 in connection with applications of Kirchhoff's rules; procedures for applying these rules, including the various hazards of algebraic sign errors, are introduced carefully and are applied to several examples. Similar specific problem-solving strategies are introduced in the following sections: 2–6, 3–4, 4–5, 6–5, 6–6, 8–2, 18–1, 25–5, 27–4, and 38–3. These and other systematic problem-solving procedures are discussed with nearly every worked-out example.

NEW EXAMPLES AND PROBLEMS

The examples and end-of-chapter problem collections have been carefully reviewed with an eye toward filling gaps in problem coverage and adding straightforward "confidence-builder" problems. An increased effort has been made throughout to relate physics to everyday experience. About 200 new problems and 50 new examples, most of them involving human activity, have been added, and many other problems and examples have been revised. There are now about 1500 problems and about 700 questions at the ends of the chapters.

NEW CHAPTER OVERVIEW

Each chapter now begins with a new introductory paragraph giving the student an overview of the content of the chapter and its relation to other chapters. In this way the unity of the fabric of physics and the continuity of its various threads can be continually emphasized.

LIFE-SCIENCE APPLICATIONS

Many new topics having relevance for the life sciences have been added, including

diffusion and osmosis,

radiation dosimetry,

medical applications of x-rays,

CAT scanners,

the centrifuge and sedimentation,

ballistocardiography,

blood-pressure measurements,

electron microscopy,

temperature control in warm-blooded animals.

TOPICS OF CURRENT INTEREST

Several other new topics of current interest have been added or expanded, including

solar energy,

internal-combustion engines,

heat pumps,

fiber optics,

the twin paradox,

practical household wiring,

electrical safety,

band theory of solids,

the new definition of the meter,

superconductivity,

semiconductors

semiconductor devices,

recent developments in fundamental particle theory.

DESIGN AND FORMAT

The successful design features of the previous edition of the book have been retained. We use a single-column, two-color format; the second color is used to add clarity to figures through color-coding of vectors, graphs, materials, and other significant features. Color is also used to identify important equations and relations and to set off examples.

To help tie physics to everyday experience and to enhance the book's visual appeal to the student, about 50 carefully chosen photographs have been added.

UNITS AND NOTATION

We have moved closer to 100% SI units. English units are retained in a few problems and examples in the first half, but SI units are used exclusively in the second half. The joule is used as the standard unit of energy of all forms, including heat. As usual, boldface symbols are used for vector quantities, and in addition boldface $+$, $-$, and $=$ signs are used to remind the student at every opportunity of the crucial distinctions between operations with vectors and those with numbers.

FLEXIBILITY

Our book is adaptable to a wide variety of course outlines. Many instructors will find that it contains too much material for a one-year course, and the format has been designed to facilitate tailoring the book to individual needs by omitting certain chapters or sections. For example, any or all of the chapters on relativity, hydrostatics, hydrodynamics, acoustics, electromagnetic waves, optical instruments, and several others can be omitted without loss of continuity. In addition, some sections that are unusually challenging or somewhat out of the main stream have been identified with an asterisk preceding the section title. These too may be omitted without loss of continuity.

Conversely, however, many topics that were regarded a few years ago as of peripheral importance and were purged from introductory courses have now come to the fore again in the life sciences, earth and space sciences, and environmental problems. An instructor who wishes to stress these kinds of applications will find this text a useful source for discussion of the appropriate principles.

In any case, it should be emphasized that instructors should not feel constrained to work straight through the book from cover to cover. Many chapters are, of course, inherently sequential in nature, but within this general limitation instructors should be encouraged to select from among the contents those chapters that fit their needs, omitting material that is not relevant for the objectives of a particular course.

REVIEWERS

With the death of Professor Zemansky in December 1981, I have taken full responsibility for this new edition. I wish to thank my many colleagues and students who have contributed suggestions, in particular these reviewers who provided critical and constructive comments throughout the writing process. Their valuable suggestions are greatly appreciated.

Carl A. Babski (Miami Dade Community College)

Daria Bouadana (New York City Technical College)

M.A. Hakeem (Southern Illinois University)

Jean P. Krisch (University of Michigan, Ann Arbor)

Fred T. Pregger (Trenton State College)

Barney Sandler (New York City Technical College)

James W. Simmons (Emory University)

Robert L. Wild (University of California, Riverside)

Robert M. Wood (University of Georgia)

We began the preparation of the sixth edition of *College Physics* by surveying a number of those who had taught from the fifth edition, and their responses were of great help. I would like to thank those who participated in this survey:

Harvey H. Bird (Fairleigh Dickinson University)

William Briscoe (George Washington University)

Howard L. Brooks (DePauw University)

Albert C. Claus (Loyola University of Chicago)

Francis J. Curry (Kutztown University)

S. Deser (Brandeis University)

Warren Deshotels (Marquette University)

Mary Ann Fisher (College of Southern Idaho)

Lewis Ford (Texas A&M University)

I.E. Dean Hamdan (Montclair State College)

Ernest Henninger (DePauw University)

Richard H. Hodson (Montclair State College)

Edwin R. Jones, Jr. (University of South Carolina)

Hans-Otto Meyer (Indiana University)

Willis Roy Minton (John C. Calhoun State Community College)

Henry F. Muyskens (Olympic College)

Frank R. Pomilla (York College—CUNY)

Francesc S. Roig (University of California, Santa Barbara)

Julius H. Taylor (Morgan State University)

In addition, David Dahlbacka contributed numerous suggestions for examples and problems, especially in the first half of the book.

ACKNOWLEDGEMENTS

A special debt of gratitude is owed to Professor Robert Kraemer of Carnegie-Mellon for many stimulating discussions about physics pedagogy and for major contributions to the high-energy physics material. An equally important debt of a different kind is owed to the late Dr. Jon Van Zaig for his support and encouragement when they were most needed. Finally and most important, I offer my gratitude to my wife Alice and our children Gretchen and Rebecca for their love, support, and emotional sustenance during a very difficult period in my life. May all men be blessed with love such as theirs.

As usual, communications from readers are most welcome, especially when they concern any errors or deficiencies that may be found in this edition.

Pittsburgh, Pennsylvania   H.D.Y.
November 1984

CONTENTS

IMPULSE AND MOMENTUM

EQUILIBRIUM OF A RIGID BODY

ROTATION

ELASTICITY

PERIODIC MOTION

FLUID STATICS

FLUID DYNAMICS

TEMPERATURE AND EXPANSION

15

QUANTITY OF HEAT

16

HEAT TRANSFER

17

THERMAL PROPERTIES OF MATTER

18

THE FIRST LAW OF THERMODYNAMICS

19

THE SECOND LAW OF THERMODYNAMICS

20

MOLECULAR PROPERTIES OF MATTER

21

MECHANICAL WAVES

22

VIBRATING BODIES

1

UNITS, PHYSICAL QUANTITIES, AND VECTORS

The study of physics is an adventure. It is challenging, sometimes frustrating, occasionally painful, and often richly rewarding and satisfying. It appeals to the emotions and the esthetic sense as well as to the intellect. The student of physics can experience anew the joy of discovery felt by such scientific giants as Galileo, Newton, Maxwell, and Einstein, whose achievements form the foundation for our present understanding of the physical world. A real feeling of discovery comes to a student who learns the value of physics in solving practical problems, in gaining insight into everyday phenomena, and most important as a monument to the achievements of the human intellect in its quest for understanding of the world we all live in.

We begin our study with an introduction to several important matters of language, including units for physical quantities, precision of numbers, and vector algebra. This language is needed in every subsequent chapter of the book, so it is important to learn it thoroughly at the beginning.

1–1

INTRODUCTION

Physics is an *empirical* study. Everything we know about the physical world and about the principles that govern its behavior has been learned through *observations* of the phenomena of nature. The ultimate test of any physical theory is its agreement with observations and measurements of physical phenomena.

Thus physics is inherently a science of *measurement*. Lord Kelvin (1824–1907), one of the pioneers in investigating energy relations in heat and thermal phenomena, stated this principle eloquently:

I often say that when you can measure what you are speaking about, and express it in numbers, you know something about it; but when you cannot express it in numbers, your knowledge is of a meagre and unsatisfactory kind; it may be the beginning of knowledge, but you have scarcely, in your thoughts, advanced to the stage of **science,** *whatever the matter may be.*

Any number or set of numbers used for a quantitative description of a physical phenomenon is called a *physical quantity.* To *define* a physical quantity we must either specify a procedure for *measuring* the quantity, or specify a way to *calculate* the quantity from other quantities that can be measured. For example, we can define *distance* and *time* by describing procedures for measuring them, and then define the *speed* of a moving body as the distance traveled divided by the time of travel.

A definition in terms of a procedure for measuring the defined quantity is called an *operational definition*. Certain fundamental quantities can be defined *only* with operational definitions. The fundamental quantities of mechanics are usually taken to be mass, length, and time. For other areas of physics other fundamental quantities, including temperature and electric charge, will be introduced later.

1–2

STANDARDS AND UNITS

In measuring a quantity, we always compare it with some established reference standard. To say that a rope is 30 meters long is to say that it is 30 times as long as an object whose length has been *defined* to be one meter. Such a standard is called a *unit* of the quantity; thus the *meter* is a unit of distance, and the *second* is a unit of time.

For precision in making measurements, it is essential to have precise and reproduceable definitions of the units of measurement. When the metric system was established in 1791 by the Paris Academy of Sciences, the meter was originally defined as one ten-millionth of the distance from the equator to the North Pole, and the second as the time for a pendulum one meter long to swing from one side to the other. In later years these definitions have been modified and greatly refined.

Since 1889 the definitions of the basic units have been established by an international organization called the General Conference on Weights and Measures, to which all the major countries of the world send representatives. The system of units defined by this organization, based on the metric system, has been known officially since 1960 as the *International System*, abbreviated SI because of the French equivalent, *Système International*.

Until 1960 the standard of *time* was based on the mean solar day, the time interval between successive arrivals of the sun at its highest point, averaged over a year. The present atomic standard was adopted in 1967. The two lowest energy states of the cesium atom have slightly different energies, depending on whether the spin of the outermost electron is parallel or antiparallel to the nuclear spin. Electromagnetic radiation (microwaves) of precisely the proper frequency causes transitions from one of these states to the other. One second is now defined as the time required for 9,192,631,770 cycles of this radiation.

In 1960 the meter was defined by an atomic standard, in terms of the wavelength of the orange-red light emitted by atoms of krypton (^{86}Kr) in a glow discharge tube; one meter was defined as 1,650,763.73 of these wavelengths. In November 1983 the standard was changed again, in a more radical way. The new definition of the meter is the distance light travels in 1/299,792,458 second. This has the effect of *defining* the speed of light to be precisely 299,792,458 m·s^{-1}, and defining the meter to be consistent with this number and with the above definition of the second. The reason for this change is that at present the speed of light and time intervals can both be measured much more precisely than can distances.

The standard of *mass* is the mass of a cylinder of platinum–iridium alloy, designated as *one kilogram,* kept at the International Bureau of Weights and Measures at Sèvres, near Paris. An atomic standard of mass has not yet been adopted because it is not yet possible to measure masses on an atomic scale with as great precision as on a macroscopic scale.

Once the fundamental units are defined, it is easy to introduce larger and smaller units for the same physical quantities. In the metric (SI) system these additional units are always related to the fundamental ones by multiples of 10 or $\frac{1}{10}$. Thus one kilometer (1 km) is 1000 meters, one centimeter (1 cm) is $\frac{1}{100}$ meter, and so on. The multiplicative factors are most conveniently expressed in exponential notation; thus $1000 = 10^3$, $\frac{1}{1000} = 10^{-3}$, and so on. The names of the additional units are always derived by adding a prefix to the name of the fundamental unit. For example, the prefix "kilo-", abbreviated k, always means a unit larger by a factor of 1000; thus

$$1 \text{ kilometer} \quad = 1 \text{ km} \quad = 10^3 \text{ meters} \quad = 10^3 \text{ m},$$

$$1 \text{ kilogram} \quad = 1 \text{ kg} \quad = 10^3 \text{ grams} \quad = 10^3 \text{ g},$$

$$1 \text{ kilowatt} \quad = 1 \text{ kW} \quad = 10^3 \text{ watts} \quad = 10^3 \text{ W}.$$

Table 1–1 lists the standard SI prefixes with their meanings and abbreviations. We note that most of these are multiples of 10^3.

TABLE 1–1	PREFIXES FOR POWERS OF TEN											
Power of ten	10^{-18}	10^{-15}	10^{-12}	10^{-9}	10^{-6}	10^{-3}	10^{-2}	10^3	10^6	10^9	10^{12}	10^{15}	10^{18}
Prefix	atto-	femto-	pico-	nano-	micro-	milli-	centi-	kilo-	mega-	giga-	tera-	peta-	exa-
Abbreviation	a	f	p	n	μ	m	c	k	M	G	T	P	E

In pronunciation of unit names with prefixes, there is *always* an accent on the *first* syllable; some examples are *kil*-o-gram, *kil*-o-meter, *cen*-ti-meter, and *mic*-ro-meter. Accenting "kilometer" on the *second* syllable is not correct!

Following are several examples of the use of multiples of 10 and their prefixes, and some additional time units.

$$1 \text{ nanometer} = 1 \text{ nm} = 10^{-9} \text{ m} \quad \text{(used by optical designers)},$$
$$1 \text{ micrometer} = 1 \ \mu\text{m} = 10^{-6} \text{ m} \quad \text{(used commonly in biology)},$$
$$1 \text{ millimeter} = 1 \text{ mm} = 10^{-3} \text{ m}$$
$$1 \text{ centimeter} = 1 \text{ cm} = 10^{-2} \text{ m} \bigg\} \text{(used most often)}$$
$$1 \text{ kilometer} = 1 \text{ km} = 10^{3} \text{ m} \quad \text{(a common European unit}$$
$$\text{of distance).}$$

$$1 \text{ microgram} = 1 \ \mu\text{g} = 10^{-9} \text{ kg},$$
$$1 \text{ milligram} = 1 \text{ mg} = 10^{-6} \text{ kg},$$
$$1 \text{ gram} = 1 \text{ g} = 10^{-3} \text{ kg}.$$

$$1 \text{ nanosecond} = 1 \text{ ns} = 10^{-9} \text{ s},$$
$$1 \text{ microsecond} = 1 \ \mu\text{s} = 10^{-6} \text{ s},$$
$$1 \text{ millisecond} = 1 \text{ ms} = 10^{-3} \text{ s}.$$

$$1 \text{ minute} = 1 \text{ min} = 60 \text{ s},$$
$$1 \text{ hour} = 1 \text{ hr} = 3600 \text{ s},$$
$$1 \text{ day} = 1 \text{ day} = 86{,}400 \text{ s}.$$

EXAMPLE   The staff of a science museum built a globe 13 m in diameter to represent the earth, which has a diameter of about 13,000 km. A visitor suggested placing a dot on the globe to represent the city where the museum is located. If the city is approximately circular in shape, with a diameter of 5 km, how large should the dot be? How does this compare with the diameter of a human hair (about 20 μm)?

Solution   First, since 1 km = 1000 m,

$$13{,}000 \text{ km} = 13{,}000 \times 1000 \text{ m} = 13{,}000{,}000 \text{ m}.$$

The actual earth is 1,000,000 (one million) times as large in diameter as the globe, so the diameter of the dot should be (1/1,000,000) × 5 km. Also, 5 km = 5000 m = 5,000,000 mm, so the diameter of the dot is (1/1,000,000) × 5,000,000 mm = 5 mm. To compare this with the diameter of a hair, note that 5 mm = 5 × 1000 μm = 5000 μm. Thus the dot is 5000/20 times, or about 250 times, as large as the diameter of a hair. ◀

We mention the British system of units, used only in the United States and the British Commonwealth and being replaced by SI in the latter. British units are officially *defined* in terms of SI units, as follows:

Length: 1 inch = 2.54 cm   (exactly),

   Mass: 1 pound-mass = 0.45359237 kg   (exactly).

The fundamental British unit of time is the second, with the same definition as in the SI. In the British system, the *pound* is a unit of *force*, and is a force equal to the weight of one pound-mass, under specified conditions. In this book the pound-mass is not used, and the pound is *always* a unit of force.

In physics, British units are used only in mechanics and thermodynamics; there is *no* British system of electrical units. For illustrative purposes, a few problems in each chapter in the first half of this book use British units, but the last half of the book uses only SI units.

1–3

UNIT CONSISTENCY AND CONVERSIONS

Relations among physical quantities are often expressed by equations in which the quantities are represented by algebraic symbols. An algebraic symbol always denotes both a *number* and a *unit*. For example, s might represent a distance of 10 m, t a time of 5 s, and v a speed of 2 m/s or $2 \text{ m} \cdot \text{s}^{-1}$. (In this book negative exponents are usually used with units, to avoid use of the fraction bar.)

An equation must always be *dimensionally consistent;* this means that two terms may be added or equated only if they have the same units. For example, if a body moving with constant speed v travels a distance s in a time t, these quantities are related by the equation

$$s = vt. \tag{1–1}$$

If s is measured in meters, then the product vt must also have units of meters. Using the above numbers as an example, we may write

$$10 \text{ m} = (2 \text{ m} \cdot \text{s}^{-1})(5 \text{ s}).$$

The unit (s^{-1}) or $(1/\text{s})$ cancels the unit (s) on the right side; thus the right side *does* have units of meters (m), as it must. This illustrates the fact that in calculations, units are treated just like algebraic symbols with respect to multiplication, division, and cancellation.

In all calculations needed for problems, units should *always* be written with all numbers and treated as in the above example. This provides a very useful check on calculations; if at some stage in a calculation an expression is found to have inconsistent units, an error has been made somewhere. In all worked-out numerical examples throughout this book, units will be carried through the calculations, and the student is strongly urged to follow this practice in solving problems.

The algebraic properties of units provide a convenient procedure for converting a quantity from one unit to another. Equality is sometimes used to represent the same physical quantity expressed in two different units. For example, to say that 1 min = 60 s does not mean that the number 1 is equal to the number 60, but rather than 1 min represents the same physical time interval as 60 s. Thus one may multiply by 1 min and then divide by 60 s, or multiply by the quantity (1 min/60 s), without changing the physical meaning. To find the number of seconds in 3 min, we write

$$3 \text{ min} = (3 \text{ min})\left(\frac{60 \text{ s}}{1 \text{ min}}\right) = 180 \text{ s}.$$

EXAMPLE 1   American women in the age group 19 to 22 years have an average height of 5 ft, 4 in. What is this height in centimeters? In meters?

Solution   We first express the height in inches:

$$5 \text{ ft} = \left(\frac{12 \text{ in.}}{1 \text{ ft}}\right) 5 \text{ ft} = 60 \text{ in.}$$

$$5 \text{ ft, 4 in.} = 5 \text{ ft} + 4 \text{ in.} = 60 \text{ in.} + 4 \text{ in.} = 64 \text{ in.}$$

Then

$$64 \text{ in.} = \left(\frac{2.54 \text{ cm}}{1 \text{ in.}}\right) 64 \text{ in.} = 163 \text{ cm.}$$

(The product has been rounded to the nearest centimeter.) Finally,

$$163 \text{ cm} = \left(\frac{1 \text{ m}}{100 \text{ cm}}\right) 163 \text{ cm} = 1.63 \text{ m.} \qquad \blacktriangleleft$$

EXAMPLE 2   An American woman drives a car in Germany at $50 \text{ km} \cdot \text{h}^{-1}$ (50 kilometers per hour). Express this speed in meters per second.

Solution

$$50 \text{ km} \cdot \text{h}^{-1} = (50 \text{ km} \cdot \text{h}^{-1})\left(\frac{1000 \text{ m}}{1 \text{ km}}\right)\left(\frac{1 \text{ h}}{3600 \text{ s}}\right) = 13.89 \text{ m} \cdot \text{s}^{-1}. \qquad \blacktriangleleft$$

1–4

PRECISION AND SIGNIFICANT FIGURES

Measurements can never be made with absolute precision; physical quantities obtained from experimental observations always have some uncertainty. A distance measured with an ordinary ruler or meter stick is usually precise only to the nearest millimeter, while a precision micrometer caliper can measure distances dependably to 0.01 mm or even less. The precision of a number is often indicated by following it with the symbol ± and a second number indicating the maximum likely error. If the diameter of a steel rod is given as 56.47 ± 0.02 mm, this means that the true value is unlikely to be less than 56.45 mm or greater than 56.49 mm. The terms *likely* and *unlikely* can also be given more precise meaning by use of statistical concepts, but these are beyond the scope of this book.

Precision can also be expressed in terms of the maximum likely *fractional* error or *percent* error. A resistor labeled "47 ohms, 10%" probably has a true resistance differing from 47 ohms by no more than 10% of 47 ohms or about 5 ohms; that is, the true value is between about 42 and 52 ohms. In the steel rod example above, the fractional uncertainty is (0.02 mm)/(56.47 mm) or about 0.00035; the percent uncertainty is (0.00035)(100%) or about 0.035%.

The expressions "parts per million," "parts per billion," and so on, are in common use. For example, the above uncertainty of 0.00035 or 0.035% would be stated as 350 parts per million, often abbreviated 350 ppm. Concentrations of substances in solutions or mixtures are often described in this way, such as a concentration of a pollutant in drinking water of 10 parts per million.

When numbers having uncertainties or errors are used to compute other numbers, these too will be uncertain. It is especially important to understand this when a number obtained from measurements is to be compared with a value obtained from a theoretical prediction. Suppose a student wants to verify the value of π, the ratio of circumference to diameter of a circle. The correct value, to ten digits, is 3.141592654. He draws a circle and measures its diameter and circumference to the nearest millimeter, obtaining the values 135 mm and 424 mm, respectively. He punches these into his pocket calculator and obtains the quotient 3.140740741. Does this agree with the true value or not?

To answer this question we must first recognize that at least the last six digits in the student's result are meaningless because they imply a greater precision in the result than is possible with his measurements. The number of meaningful digits in a number is called the number of *significant figures;* in general, no numerical result can have more significant figures than do the numbers from which it is computed. Thus the student's value of π has only *three significant figures* and should be stated simply as 3.14, or conceivably as 3.141 (rounded to four figures). Within the limit of three significant figures, the student's value of π *does* agree with the true value.

In the examples and problems in this book, it is usually assumed that the numerical values given are precise to three or at most four significant figures, and thus the answers should be stated to at most four significant figures. The student will usually do the arithmetic with a pocket calculator having a display with five to ten digits. To write down a ten-digit result from a calculation with numbers having three significant figures is not only unnecessary; it is genuinely wrong because it misrepresents the precision of the results. Such results must *always* be rounded to keep only the correct number of significant figures or, in doubtful cases, at most one more.

In calculations with very large or very small numbers, significant-figure considerations are simplified greatly by use of powers-of-ten notation, sometimes called *scientific* notation. The distance from the earth to the sun is about 149,000,000,000 m, but to write the number in this form gives no indication of the number of significant figures. Certainly not all 12 are significant! Instead, we move the decimal point 11 places to the left (corresponding to dividing by 10^{11}) and multiply by 10^{11}. That is,

$$149,000,000,000 \text{ m} = 1.49 \times 10^{11} \text{ m}.$$

In this form it is clear that the number of significant figures is three.

Similar considerations are applicable when very large or very small numbers have to be multiplied or divided. For example, the energy E corresponding to the mass m of an electron is given by the equation

$$E = mc^2, \tag{1–2}$$

where c is the speed of light. The appropriate numbers are $m = 9.11 \times 10^{-31}$ kg and $c = 3.00 \times 10^8$ m·s^{-1}. We find

$$E = (9.11 \times 10^{-31} \text{ kg})(3.00 \times 10^8 \text{ m·s}^{-1})^2$$
$$= (9.11)(3.00)^2(10^{-31})(10^8)^2 \text{ kg·m}^2\text{·s}^{-2}$$
$$= (82.0)(10^{[-31+(2\times8)]}) \text{ kg·m}^2\text{·s}^{-2}.$$

Many pocket calculators use scientific notation and thus save the trouble of adding exponents; but the student should be able to do such calculations by hand when necessary. Incidentally, it should be noted that the value of c has three significant figures even though two of them are zeros. To greater precision, $c = 2.997925 \times 10^8$ m·s^{-1}; thus it would *not* be correct to write $c = 3.000 \times 10^8$ m·s^{-1}.

1–5

ESTIMATES AND ORDERS OF MAGNITUDE

Knowing the precision of numbers that represent physical quantities is important, as indicated in the preceding section. But there are situations where even a very crude estimate of a quantity may provide useful information. We may know how to calculate a certain quantity if we are given the necessary input data, but the data may be unavailable or may have to be guessed at. Or the calculation may be too complicated to carry out exactly. In any case, the result of an approximate calculation will also be a guess, but there are situations where such a guess is useful, even if it is wrong or uncertain by a factor of two or ten or more. Such calculations are often called "order-of-magnitude" calculations or estimates.

EXAMPLE   You are writing an international espionage novel in which the hero escapes across the border with a billion dollars' worth of gold in his suitcase. Is this possible? Would it fit? Would it be too heavy to carry?

Solution   The news reports say that gold is selling for around $400 an ounce. On a particular day it may be $200 or $600, but never mind. We also know that an ounce is about 30 grams. Actually, an ordinary (avoirdupois) ounce is 28.35 g; an ounce of gold is a troy ounce, which is 9.45% more. Again, never mind. Ten dollars' worth of gold has a mass somewhere around one gram, so a billion (10^9) dollars' worth is a hundred million (10^8) grams, or a hundred thousand (10^5) kilograms. This corresponds to a weight in English units of around 200,000 lb, or a hundred tons. Whether the precise number is fifty tons or two hundred doesn't matter; either way, our hero is not about to carry it across the border in a suitcase.

We can also estimate the *volume* of this gold. If its density were the same as that of water (1 g·cm^{-3}), the volume would be 10^8 cm^3 or 100 m^3. But gold is a heavy metal; we might guess its density as ten times that of water. It's actually 19.3 times as dense as water. But guessing ten, we find a volume of 10 m^3. Visualize ten cubical stacks of gold bricks, each one meter on a side, and ask whether it would fit in a suitcase!

◀

Problems 1–15 through 1–28 at the end of this chapter are of the estimating or "order-of-magnitude" variety. Some are silly, and most require a lot of guesswork for the needed input data. Don't try to look up a lot of data; make the best guesses you can. Even when they are wrong by a factor of ten, the results can be useful and interesting.

1–6

VECTORS AND VECTOR ADDITION

Some physical quantities are described completely by a single number with a unit; examples are time, temperature, mass, density, and quantity of electric charge. Many other quantities, however, have a *directional* quality that cannot be described by a single number. A familiar example is *velocity;* to describe the motion of a body we must say not only *how fast* it is moving but also *in what direction. Force* is another example; when we push or pull on a body, we are said to exert a force on it. To describe a force, we need to describe the *direction* in which it acts, as well as its *magnitude,* which is a quantitative description of "how much" or "how hard" the force pushes or pulls, in terms of a standard unit of force. The SI unit of force is the *newton* (N), and the British unit of force is the pound (lb), sometimes called the pound-force:

$$1 \text{ lb} = 4.448 \text{ N}.$$

A physical quantity described by a single number (possibly with an algebraic sign, as for electric charge) is called a *scalar* quantity, and a quantity having both magnitude (the "how much" aspect of the quantity) and direction is called a *vector* quantity. Calculations with scalar quantities involve the operations of ordinary arithmetic, but calculations with vector quantities are somewhat different. Because vector quantities play an essential role in all areas of physics, we turn now to a discussion of their nature and the operation of *vector addition.*

We begin with a vector quantity called *displacement.* When a particle, which we represent as a point, moves from one location in space to another, it is said to undergo a displacement. In Fig. 1–1a, the change of position from point P_1 to point P_2 is represented graphically by the directed line segment P_1P_2, with an arrowhead to represent the sense of the motion. Displacement is a vector quantity because to describe it completely we must state not only how far the particle moves but also *in what direction;* a displacement of 3 km north is not the same as a displacement of 3 km southeast.

Usually it is most convenient to represent a displacement by a single letter, such as A in Fig. 1–1a. To distinguish symbols representing vector quantities from those for scalar quantities, vector symbols are always printed using boldface type, as shown. In handwriting, vector symbols are usually underlined or written with an arrow above, as shown in the figure, to indicate that they represent vector quantities.

The vector from point P_3 to point P_4 in Fig. 1–1b has the same length and direction as that from P_1 to P_2, so these two displacements are *equal,* even though they start at different points. By definition, two vector quantities are equal if they have the same magnitude (length) and direction, no matter where they are located in space. The vector B, however, is *not* equal to A because its direction is opposite to that of A. To represent the relationship of the two, we define the *negative* of a vector quantity as a vector quantity having the same magnitude but opposite direction to that of the original vector. The negative of vector quantity A is denoted as $-A$, and we use a boldface "minus" to emphasize the vector nature of

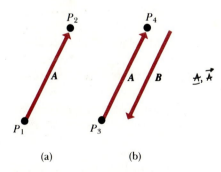

1–1 (a) Vector A is the displacement from point P_1 to point P_2. (b) The displacement from P_3 to P_4 is equal to that from P_1 to P_2, but displacement B is the negative of displacement A.

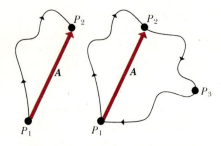

1–2 A displacement is always a straight line segment, directed from the starting point to the endpoint, even if the actual path is curved. When a point ends at the same place it started, the displacement is zero.

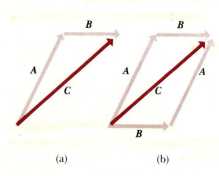

(a)   (b)

1–3 (a) Vector **C** is the vector sum of vectors **A** and **B**. (b) The order in vector addition is immaterial.

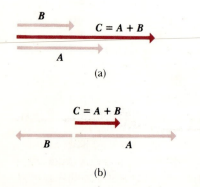

(a)

(b)

1–4 Vector sum of (a) two parallel vectors, and of (b) two antiparallel vectors.

the quantities. Thus the relation between **A** and **B** may be written as **A = −B** or **B = −A.** Note that a boldface "equals" sign is also used to emphasize that equality of two vector quantities is not the same relationship as equality of scalar quantities.

==Displacement is always a straight line segment, directed from the starting point to the endpoint,== even though the path taken by the particle may be curved. Thus in Fig. 1–2, when the point moves from P_1 to P_2 along the curved path shown, the displacement is the vector **A** shown. Also, when it continues on to P_3 and then returns to P_1, the displacement for the entire trip is zero.

The *magnitude* of a vector quantity (its length in the case of displacement vectors) is represented by the same letter used for the vector but in light italic type rather than boldface italic. An alternative notation is the vector symbol with vertical bars on both sides. Thus

$$\text{Magnitude of } \boldsymbol{A} = A = |\boldsymbol{A}|. \tag{1–3}$$

By definition, the magnitude of a vector quantity is a scalar (a single number) and is always positive.

When a particle undergoes a displacement, **A,** followed by a second displacement **B,** as shown in Fig. 1–3a, the total effect is the same as though it had started at the same initial point and undergone a single displacement **C,** as shown. We call displacement **C** the *vector sum* of displacements **A** and **B**; the relationship is expressed symbolically as

$$\boldsymbol{C} = \boldsymbol{A} + \boldsymbol{B}. \tag{1–4}$$

The boldface "plus" sign emphasizes that obtaining the sum of (adding) two vector quantities is not the same operation as adding two scalar quantities.

If the displacements are made in the reverse order, as in Fig. 1–3b, with **B** first and **A** second, the result is the same, as the figure shows. Thus we may write

$$\boldsymbol{C} = \boldsymbol{B} + \boldsymbol{A} \quad \text{and} \quad \boldsymbol{A} + \boldsymbol{B} = \boldsymbol{B} + \boldsymbol{A}. \tag{1–5}$$

Thus vector addition is *commutative;* the order of the terms is immaterial. Formally, we have also used the *transitive* property of vector equality: Two vector quantities each equal to a third are also equal to each other.

Figure 1–3b also suggests an alternative graphical representation of the vector sum; when vectors **A** and **B** are both drawn from a common point, vector **C** is the diagonal of a parallelogram constructed with **A** and **B** as two adjacent sides.

The vector sum is often called the *resultant;* thus the vector sum **C** of vectors **A** and **B** can be called the resultant displacement.

Figure 1–4 illustrates a special case in which two vectors are parallel, as in (a), or antiparallel, as in (b). If they are parallel, the magnitude of the vector sum **C** equals the sum of the magnitudes of **A** and **B.** If they are antiparallel, the magnitude of the vector sum equals the *difference* of the magnitudes of **A** and **B.** The vectors in Fig. 1–4 have been displaced slightly sidewise to show them more clearly, but they actually lie along the same geometric line.

When more than two vectors are to be added, we may first find the vector sum of any two, add this vectorially to the third, and so on. This

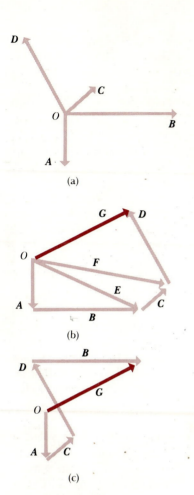

(a)

(b)

(c)

1–5 Polygon method of vector addition.

process is illustrated in Fig. 1–5, which shows in part (a) four vectors *A, B, C,* and *D.* In Fig. 1–5b, vectors *A* and *B* are first added, giving a vector sum *E*; vectors *E* and *C* are then added by the same process, to obtain the vector sum *F*; finally, *F* and *D* are added, to obtain the vector sum

$$G = A + B + C + D.$$

Evidently the vectors *E* and *F* need not have been drawn; we need only draw the given vectors in succession, with the tail of each at the head of the one preceding it, and complete the polygon by a vector *G* from the *tail* of the first to the *head* of the last vector. The order in which the vectors are drawn makes no difference, as shown in Fig. 1–5c; the reader may wish to try other possibilities.

Diagrams for addition of displacement vectors need not be drawn actual size; it is often convenient to use a scale in which the distance on the diagram is *proportional* to the actual distance, such as 1 cm for 5 km, just as on a map. Furthermore, such scales can also be used with other vector quantities whose units are not distance units. For example, in a diagram for force vectors (discussed in Section 2–2), one might use a scale in which a vector 1 cm long represents a force of magnitude 5 N. Then a 20-N force is represented by a vector 4 cm long with the appropriate direction. Thus sums of vector quantities can always be obtained by drawing scale diagrams and measuring the appropriate distances and angles. This is seldom done in practice because there are better methods available. Some of these are discussed in the next section.

A displacement or other vector quantity may be multiplied by a scalar quantity. The displacement 2*A* is by definition a displacement in the same direction as that of *A* but twice the magnitude. In some cases the scalar quantity is a physical quantity having units. The reader may be familiar with the relationship *F = ma*; force (a vector quantity) is equal to the product of mass (a scalar quantity) and acceleration (a vector quantity). The magnitude of the force is equal to the mass multiplied by the magnitude of acceleration, and the unit of the magnitude of force is the product of the unit of mass and that of the magnitude of acceleration.

The special case of multiplication by −1 has already been mentioned; $(-1)A = -A$ is by definition a vector having the same magnitude as *A* but opposite direction. This provides the basis for defining vector *subtraction*. The difference *A − B* of two vectors is defined to be the vector sum of *A* and *−B*:

$$A - B = A + (-B). \tag{1–6}$$

The boldface **+,−,** and **=** signs remind us of the vector nature of these operations.

1–7

COMPONENTS OF VECTORS

Addition and subtraction of vectors are greatly facilitated by the use of *components*. To define components we use a rectangular (cartesian) coordinate axis system as in Fig. 1–6. Any vector lying in the *xy*-plane can be represented as the sum of a vector parallel to the *x*-axis and one parallel

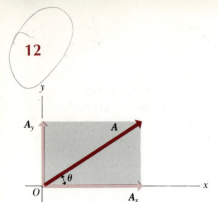

1–6 Vectors A_x and A_y are the rectangular components of A in the directions of the x- and y-axes.

to the y-axis. These two vectors are labeled A_x and A_y in the figure; they are called the *rectangular component vectors of vector* A. The relationship is expressed formally as

$$A = A_x + A_y. \tag{1–7}$$

Each component vector lies along a coordinate-axis direction, by definition, so a single number is sufficient to describe each component vector. Thus the number A_x is, apart from a possible negative sign, the *magnitude* of the component vector A_x. We further specify that A_x is positive when A_x points in the *positive* axis direction and negative when in the opposite direction. The two numbers A_x and A_y are called simply the *components* of A.

If the magnitude A of the vector A and its direction, given by angle θ in Fig. 1–6, are known, the components can be calculated. From the definitions of the trigonometric functions,

$$\frac{A_x}{A} = \cos \theta \quad \text{and} \quad \frac{A_y}{A} = \sin \theta;$$

$$A_x = A \cos \theta \quad \text{and} \quad A_y = A \sin \theta. \tag{1–8}$$

The components of a vector are themselves numbers, and they may be positive or negative. In Fig. 1–7, the component B_x is negative, since its direction is opposite to that of the positive x-axis, and also since the cosine of an angle in the second quadrant is negative; B_y is positive, but both C_x and C_y are negative.

Either the magnitude and direction or the components of a vector quantity can be used to describe it, and either provides a *complete* description. Equations (1–8) show how to obtain the components if the magnitude and direction are given; conversely, if the components are given, the magnitude and direction can be obtained. Applying the Pythagorean theorem to Fig. 1–6, we find

$$A = \sqrt{A_x{}^2 + A_y{}^2}. \tag{1–9}$$

Also, from the definition of the tangent of an angle,

$$\tan \theta = \frac{A_y}{A_x} \quad \text{and} \quad \theta = \arctan \frac{A_y}{A_x}. \tag{1–10}$$

There is a slight complication in applying Eq. (1–10). Suppose $A_x = 2$ m and $A_y = -2$ m. Then $\tan \theta = -1$. But there are two angles having a tangent of -1: 135° and 315° (or $-45°$). To decide which is correct we must look at the individual components; because A_x is positive and A_y negative, the angle must be in the fourth quadrant; thus $\theta = 315°$ (or $-45°$) is the correct value.

The use of components provides an efficient means of calculating the vector sum of several vectors. In principle such sums may always be found using diagrams such as those in Figs. 1–3 and 1–5; but this requires either making and measuring a scale diagram (which is difficult to do with precision) or the trigonometric solution of oblique triangles (which can become quite complicated). The component method, by contrast, requires only right triangles and simple computations and can be carried out with great precision.

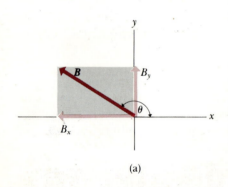

(a)

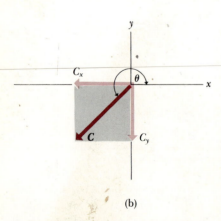

(b)

1–7 Components of a vector may be positive or negative numbers.

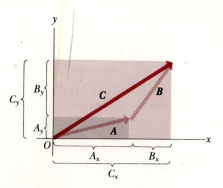

1–8 C_x is the x-component of the vector sum C of vectors A and B, and is equal to the sum of the x-components of A and B. The y-components are similarly related.

Figure 1–8 shows two vectors A and B and their vector sum (resultant) C, along with the x- and y-components of all three vectors. It can be seen from the diagram that the x-component C_x of the vector sum is simply the sum $(A_x + B_x)$ of the x-components of the vectors being added, and similarly for the y-components. Thus

$$C_x = A_x + B_x,$$
$$C_y = A_y + B_y. \tag{1–11}$$

Thus once the components of A and B are known, perhaps by use of Eqs. (1–8), the components of the vector sum can be computed. Then, if needed, the magnitude and direction of C can be obtained from Eqs. (1–9) and (1–10).

This discussion of the sum of two vectors can easily be extended to any number. Let R be the vector sum of $A, B, C, D, E, \ldots$ Then

$$R_x = A_x + B_x + C_x + D_x + E_x + \ldots,$$
$$R_y = A_y + B_y + C_y + D_y + E_y + \ldots. \tag{1–12}$$

EXAMPLE The pilot of a private plane flies 20.0 km in a direction 60° north of east, then 30.0 km straight east, then 10.0 km straight north. How far and in what direction is she from the starting point?

Solution The situation is shown in Fig. 1–9. We have chosen the x-axis as east and the y-axis as north, the usual choice for maps. Let A be the first displacement, B the second, C the third, and R the vector sum or resultant displacement. The components of A are

$$A_x = (20.0 \text{ km})(\cos 60°) = 10.0 \text{ km},$$
$$A_y = (20.0 \text{ km})(\sin 60°) = 17.3 \text{ km}.$$

The components of all the displacements and the calculations can be arranged systematically, as in Table 1–2.

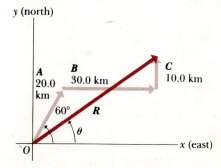

1–9 Three successive displacements $A, B,$ and C, and the resultant or vector sum displacement $R = A + B + C$.

TABLE 1–2				
Distance	**Angle**	**x-component**	**y-component**
$A = 20.0$ km	60°	10.0 km	17.3 km
$B = 30.0$ km	0°	30.0 km	0
$C = 10.0$ km	90°	0	10.0 km
		$R_x = 40.0$ km	$R_y = 27.3$ km

$$R = \sqrt{(40.0 \text{ km})^2 + (27.3 \text{ km})^2} = 48.4 \text{ km};$$

$$\theta = \arctan \frac{27.3 \text{ km}}{40.0 \text{ km}} = 34.3°.$$

Alternatively, find θ first, as above, then use Eqs. (1–8) to find R:

$$R = \frac{R_x}{\cos \theta} = \frac{40.0 \text{ km}}{\cos 34.3°} = 48.4 \text{ km},$$

or

$$R = \frac{R_y}{\sin \theta} = \frac{27.3 \text{ km}}{\sin 34.3°} = 48.4 \text{ km}.$$ ◀

The above discussion has been confined to vectors lying in the *xy*-plane, but it is easily generalized to vectors that do not all lie in a single plane. We introduce a *z*-axis perpendicular to the *xy*-plane; then in general a vector A has components A_x, A_y, and A_z in the three coordinate directions. The magnitude A is given by

$$A = \sqrt{A_x^2 + A_y^2 + A_z^2}. \tag{1–13}$$

QUESTIONS

1–1 What are the units of the number π?

1–2 The rate of climb of a mountain trail was described in the guidebook as 150 meters per kilometer. How can this be expressed as a number with no units?

1–3 Suppose you are asked to compute the cosine of 3 meters. Is this possible?

1–4 Hydrologists describe the rate of volume flow of rivers in "second-feet." Is this unit technically correct? If not, what would be a correct unit?

1–5 A highway contractor stated that in building a bridge deck he had poured 200 yards of concrete. What do you think he meant?

1–6 Does a vector having zero length have a direction?

1–7 What is your weight in newtons?

1–8 What is your height in centimeters?

1–9 What physical phenomena (other than a pendulum or cesium clock) could be used to define a time standard?

1–10 Could some atomic quantity be used for a definition of a unit of mass? What advantages or disadvantages would this have compared to the one-kilogram platinum cylinder kept at Sèvres?

1–11 How could you measure the thickness of a sheet of paper with an ordinary ruler?

1–12 Can you find two vectors having different lengths that have a vector sum of zero? What length restrictions are required for three vectors to have a vector sum of zero?

1–13 What is the displacement when a bicyclist travels from the north side of a circular race track of radius 500 m to the south side? When she makes one complete circle around the track?

1–14 What are the units of volume? Suppose a student tells you a cylinder of radius r and height h has volume given by $\pi r^3 h$. Explain why this cannot be right.

1–15 An angle (measured in radians) is a number with no units, since it is a ratio of two lengths. Think of other geometrical or physical quantities that are unitless.

1–16 Can you find a vector quantity that has components different from zero but a magnitude of zero?

1–17 One sometimes speaks of the "direction of time," evolving from past to future. Does this mean that time is a vector quantity?

PROBLEMS

1–1 Starting with the definition 1 in. = 2.54 cm, compute the number of kilometers in one mile, to five significant figures.

1–2 The density of water is $1 \text{ g} \cdot \text{cm}^{-3}$. What is this value in kilograms per cubic meter?

1–3 Convert the following speeds, as indicated.

a) $60 \text{ mi} \cdot \text{hr}^{-1}$ to feet per second

b) $100 \text{ km} \cdot \text{hr}^{-1}$ to meters per second

1–4 What is the mass in kilograms of a person weighing 170 lb?

1–5 Compute the number of seconds in a day, and in a year (365 da).

1–6 If one deutschmark (the West German unit of currency) is worth 40 cents and gasoline costs 1.30 deutschmarks per liter, what is its cost in dollars per gallon? Use the conversion factors in Appendix F.

1–7 The gasoline consumption of a small car is 17.0 km/l. How many miles per gallon is this? Use the conversion factors in Appendix F.

1–8 The speed limit on a highway in Lower Slobbovia was given as 150,000 furlongs per fortnight. How many miles per hour is this? (One furlong is 1/8 mile, and a fortnight is 14 days.)

1–9 The piston displacement of a certain automobile engine is given as 2.0 liters. Using only the facts that 1

liter = 1000 cm³ and 1 in. = 2.54 cm, express this volume in cubic inches.

1–10 What is the percent error in each of the following approximations to π?

  a) 22/7   b) 355/113

1–11 What is the percent error in the approximate statement 1 yr = $\pi \times 10^7$ s?

1–12 Estimate the percent error in measuring

  a) a distance of about 50 cm with a meter stick;

  b) a mass of about 1 g with a chemical balance;

  c) a time interval of about 4 min with a stopwatch.

1–13 The mass of the earth is 5.98×10^{24} kg, and its radius is 6.38×10^6 m. Compute the density of the earth, using powers-of-ten notation and the correct number of significant figures.

1–14 An angle is given, to one significant figure, as 5°, meaning that its value is between 4.5° and 5.5°. Find the corresponding range of possible values of the cosine of the angle. Is this a case where there are more significant figures in the result than in the input data?

1–15 A box of typewriter paper is $11'' \times 17'' \times 9''$; it is marked "10 M." Does that mean it contains ten thousand sheets or ten million?

1–16 What total volume of air does a person breathe in a lifetime? How does that compare with the volume of the World Trade Center?

1–17 The example in Section 1–5 showed that it is impossible to carry a billion dollars in gold in a suitcase. Could you carry a *million* dollars' worth?

1–18 Could the water-supply needs of Los Angeles be met by hauling water in by truck? By railroad?

1–19 How many kernels of corn are there in a 1-liter whiskey bottle?

1–20 How many hairs do you have on your head?

1–21 What is the total length of film in a two-hour feature movie? What is the diameter of the reel if it is all on a single reel?

1–22 How many times does a human heart beat during a lifetime?

1–23 How much would it cost to paper the United States with dollar bills?

1–24 How many drops of water are in the oceans?

1–25 How many dollar bills would have to be stacked up to reach the moon? Would that be cheaper than building and launching a spacecraft?

1–26 If you filled your room with rocks, what would they weigh? What if you filled it with crumpled newspapers?

1–27 If all the gasoline burned in cars and trucks in the United States in one year were placed in a cubical tank, how big would it be?

1–28 How many cars can pass through a two-lane tunnel through a mountain, in one hour?

1–29 Two points P_1 and P_2 are described by their x- and y-coordinates, (x_1, y_1) and (x_2, y_2), respectively. Show that the components of the displacement A from P_1 to P_2 are $A_x = x_2 - x_1$ and $A_y = y_2 - y_1$. Also derive expressions for the magnitude and direction of this displacement.

1–30 When two vectors A and B are drawn from a common point, the angle between them is θ. Show that the magnitude of their vector sum is given by

$$\sqrt{A^2 + B^2 + 2AB \cos \theta}.$$

1–31 Find the magnitude and direction of the vector represented by each of the following pairs of components.

  a) $A_x = 3$ cm, $A_y = -4$ cm

  b) $A_x = -5$ m, $A_y = -12$ m

  c) $A_x = -2$ km, $A_y = 3$ km

1–32 A postal employee drives a delivery truck 1 mile north, then 2 miles east, then 3 miles northwest (at 45° between north and west). Determine the resultant displacement

  a) by drawing a scale diagram, and

  b) by using components.

1–33 A bug starts at the center of a 12-inch phonograph record and crawls along a straight radial line to the edge. While this is happening the record turns through an angle of 45°. Draw a sketch of the situation and describe the magnitude and direction of the bug's displacement.

1–34 A spelunker is surveying a cave. He follows a passage 100 m straight east, then 50 m in a direction 30° west of north, then 150 m at 45° west of south. After a fourth unmeasured displacement he finds himself back where he started. Using a scale drawing, determine the fourth displacement (magnitude and direction).

1–35 Find graphically the magnitude and direction of the vector sum of the three forces in Fig. 1–10,

  a) using the polygon method and a scale drawing, and

  b) using components. Estimate the percent error in the graphical method.

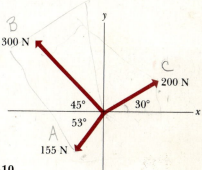

Figure 1–10

1-36 Find graphically the vector sum *A* + *B* and the vector difference *A* − *B* in Fig. 1–11.

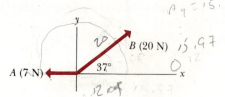

A_x = 12
A_y = 15.

y

B (20 N)   15.97

37°

A (7 N)   O
                                        x

12 of 15.97

Figure 1-11

1-37 Find the vector sum *A* + *B* and the vector difference *A* − *B* in Fig. 1–11, using components.

1-38 Vector *A* is 2 in. long and is 60° above the x-axis in the first quadrant. Vector *B* is 2 in. long and is 60° below the x-axis in the fourth quadrant. Find graphically

a) the vector sum *A* + *B*, and

b) the vector differences *A* − *B* and *B* − *A*.

1-39 Obtain the vector sum and differences requested in Problem 1–38, using the method of components.

1-40 Vector *A* has components $A_x = 2$ cm, $A_y = 3$ cm, and vector *B* has components $B_x = 4$ cm, $B_y = -2$ cm. Find

a) the components of the vector sum *A* + *B*,

b) the magnitude and direction of *A* + *B*,

c) the components of the vector difference *A* − *B*,

d) the magnitude and direction of *A* − *B*.

1-41 A disoriented physics professor drives 5 mi east, then 4 mi south, then 2 mi west. Find the magnitude and direction of the resultant displacement.

1-42 A sailor in a small sailboat encounters shifting winds. He sails 2 km east, then 4 km southeast, then an additional distance in an unknown direction. His final position is 5 km directly east of the starting point. Find the magnitude and direction of the third leg of the voyage.

1-43 Vector *M*, of magnitude 5 cm, is at 36.9° counterclockwise from the +x-axis. It is added to vector *N*, and the resultant is a vector of magnitude 5 cm, at 53.1° counterclockwise from the +x-axis. Find

a) the components of *N*, and

b) the magnitude and direction of *N*.

1-44 A vector *A* of length 10 units makes an angle of 30° with a vector *B* of length 6 units. Find the magnitude of the vector difference *A* − *B* and the angle it makes with vector *A*

a) by the triangle method, and

b) by the method of rectangular resolution.

1-45 Two vectors *A* and *B* have the same magnitude. Under what circumstances does the vector sum *A* + *B* have the same magnitude as *A* and *B*? When does the vector difference *A* − *B* have this magnitude?

2

EQUILIBRIUM
OF A PARTICLE

Our study of physics begins in the area of *mechanics,* the study of forces and motion and their relationships. Mechanics is based on three principles called *Newton's laws of motion.* These principles are classified as *natural laws;* they are fundamental statements about the behavior of the physical world. They are *empirical* laws, deduced from experiment and observation. They cannot be derived from anything more fundamental.

These laws were clearly stated for the first time by Sir Isaac Newton (1642–1727) and were published in 1686 in his *Philosophiae Naturalis Principia Mathematica* (The Mathematical Principles of Natural Science). Many men had preceded Newton in his field, however; the most outstanding was Galileo Galilei (1564–1642) who, in his studies of accelerated motion, had laid much of the groundwork for Newton's three laws. In this chapter we study particles (and bodies that can be represented as particles) in *equilibrium,* that is, at rest or in uniform motion. This study uses Newton's first and third laws; the second law is introduced in Chapter 4.

2–1

FORCE

Force is a central concept in all of physics. When we push or pull on a body, we are said to exert a *force* on it. Forces can also be exerted by inanimate objects: a stretched spring exerts forces on the bodies to which its ends are attached; compressed air exerts a force on the walls of its container; a locomotive exerts a force on the train it is pulling or pushing. The force of gravitational attraction exerted on every physical body by the earth is called the *weight* of the body. Gravitational forces (and electrical and magnetic forces also) can act through empty space without contact. A force on an object resulting from direct contact with another object is called a *contact force;* viewed on an atomic scale, contact forces

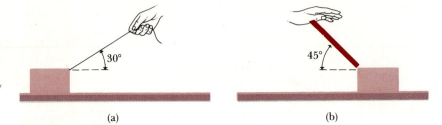

2–1 Force may be exerted on the box by either (a) pulling it, or (b) pushing it.

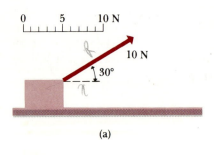

(a)

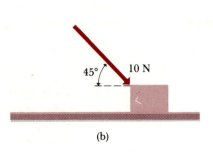

(b)

2–2 Force diagram for the forces acting on the box in Fig. 2–1.

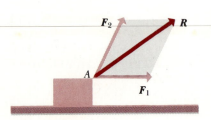

2–3 A force represented by the vector R, equal to the vector sum of F_1 and F_2, produces the same effect as the forces F_1 and F_2 acting simultaneously.

arise chiefly from electrical attraction and repulsion of the electrons and nuclei making up the atoms of material.

As mentioned in Section 1–6, force is a *vector* quantity. To describe a force, we need to describe the *direction* in which it acts, as well as its *magnitude,* which is a quantitative description of "how much" or "how hard" the force pushes or pulls, in terms of a standard unit of force. In Chapter 4 we shall see how a unit of force can be defined in terms of the units of mass, length, and time. As mentioned in Section 1–6, the SI unit of force is the *newton,* abbreviated N. In the British system the unit of force is the *pound-force,* equal to about 4.448 N. In the cgs system of units, the unit of force is the *dyne,* equal to 10^{-5} N.

An instrument commonly used to measure forces is the spring balance, which consists of a coil spring, enclosed in a case for protection, carrying at one end a pointer that moves over a scale. A force exerted on the balance changes the length of the spring, and the change can be read on the scale.

One way to calibrate such a balance is to make a number of bodies each having a weight of exactly 1 N. Then, when two, three, or more of these are suspended simultaneously from the balance, the total force stretching the spring is 2 N, 3 N, and so on, and the corresponding positions of the pointer can be labeled 2 N, 3 N, etc. The calibrated spring balance can then be used to measure the magnitude of an unknown force.

Suppose we slide a box along the floor by pulling it with a string or pushing it with a stick, as in Fig. 2–1. Our point of view is that the motion of the box is caused not by the *objects* that push or pull on it, but by the *forces* these objects exert. The forces in the two cases can be represented as in Fig. 2–2; the labels indicate the magnitude and direction of the force, and the length of the arrow, to some chosen scale, also shows the magnitude.

Some vector quantities, of which force is one, are not *completely* specified by their magnitude and direction alone. The effect of a force depends also on the position of the point at which the force is applied. For example, when one pushes horizontally against a door, the effectiveness of a given force in setting the door in motion depends on the distance of the point of application from the line of the hinges, about which the door rotates. We return to this consideration in Chapter 8.

Now consider the following physical problem. Two forces, represented by the vectors F_1 and F_2 in Fig. 2–3, are applied simultaneously at the same point A of a body. Is it possible to produce the same effect by

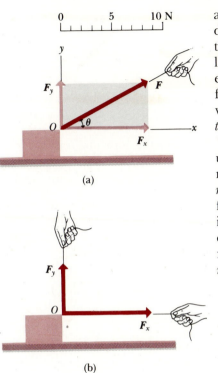

(a)

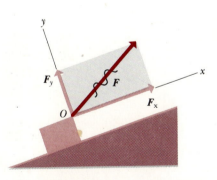

(b)

2–4   The inclined force **F** may be replaced by its rectangular components F_x and F_y. $F_x = F \cos \theta$, $F_y = F \sin \theta$.

applying a *single* force at *A*, and if so, what should be its magnitude and direction? The question can be answered only by experiment; investigation shows that a single force, represented in magnitude, direction, and line of action by the *vector sum* **R** of the original forces, is in all respects equivalent to them. This single force is called the *resultant* of the original forces. Hence the mathematical process of *vector addition* of two force vectors corresponds to the physical operation of finding the *resultant of two forces*, simultaneously applied at a given point.

The fact that forces can be combined by vector addition is of the utmost importance, as we shall see in the following chapters. Furthermore, this fact also permits a force to be represented by means of *components*, as we have done with displacements in Section 1–7. In Fig. 2–4a, force **F** is exerted on a body at point *O*. The rectangular components of **F** in the directions *Ox* and *Oy* are F_x and F_y; and it is found that simultaneous application of the forces F_x and F_y, as in Fig. 2–4b, is equivalent in all respects to the effect of the original force. *Any force can be replaced by its rectangular components, acting at the same point.*

As a numerical example, let

$$F = 10.0 \text{ N}, \qquad \theta = 30°.$$

Then,

$$F_x = F \cos \theta = (10.0 \text{ N})(0.866) = 8.66 \text{ N},$$

$$F_y = F \sin \theta = (10.0 \text{ N})(0.500) = 5.00 \text{ N},$$

and the effect of the original 10.0-N force is equivalent to the simultaneous application of a horizontal force, to the right, of 8.66 N, and a lifting force of 5.00 N.

The axes used to obtain rectangular components of a vector need not be vertical and horizontal. For example, Fig. 2–5 shows a block being pulled up an inclined plane by a force **F**, represented by its components F_x and F_y, parallel and perpendicular to the sloping surface of the plane.

We shall often have occasion to find the vector sum of several forces acting on a body, which again may be called their *resultant*. The Greek letter Σ is often used in a shorthand notation for this sum. Thus if the forces are labeled F_1, F_2, F_3, and so on, and their resultant is **R,** then the operation

$$\boldsymbol{R = F_1 + F_2 + F_3 + \cdots}$$

is often abbreviated

$$R = \sum F, \qquad (2\text{–}1)$$

where ΣF is read "sum of the forces." (The Greek letter Σ, equivalent to the Roman S, is an abbreviation for "sum.") This means, of course, the vector sum. In terms of components, we may write

$$R_x = \sum F_x, \qquad R_y = \sum F_y, \qquad (2\text{–}2)$$

in which ΣF_x is read as the sum of the x-components of the forces. We also note that a boldface $\boldsymbol{\Sigma}$ is used in Eq. (2–1) as a reminder that the sum is a vector sum, and a lightface Σ is used in Eqs. (2–2) because components are ordinary numbers.

2–5 F_x and F_y are the rectangular components of **F**, parallel and perpendicular to the sloping surface of the inclined plane. When a vector is represented in terms of its components, the original vector should be crossed out to avoid duplication in the diagram.

Finally, these can be combined to form the resultant **R,** whose magnitude is

$$R = \sqrt{R_x^2 + R_y^2},$$

since R_x and R_y are perpendicular to each other.

The angle α between R and the x-axis can now be found from any one of its trigonometric functions. For example, $\tan \alpha = R_y/R_x$. As with the individual vectors, the components R_x and R_y may be positive or negative, and the angle α may be in any of the four quadrants.

EXAMPLE 1   In Fig. 2–6a, three forces F_1, F_2, F_3 act at point O; find their resultant, or vector sum.

Solution   First we choose a set of coordinate axes. We can always choose one axis to be parallel to one of the forces; this usually simplifies the calculations. In Fig. 2–6b, the x-axis coincides with F_1. We first resolve each force into x- and y-components. According to the usual conventions of analytic geometry, x-components toward the right are considered positive and those toward the left, negative. Upward y-components are positive and downward y-components are negative.

Force F_1 lies along the x-axis and need not be resolved. The components of F_2 are

$$F_{2x} = F_2 \cos \theta, \qquad F_{2y} = F_2 \sin \theta.$$

Both of these are positive, and F_{2x} has been slightly displaced upward to show it more clearly. The magnitudes of the components of F_3 are

$$F_3 \cos \phi, \qquad F_3 \sin \phi.$$

However, both the components of F_3 are negative because the corresponding component vectors point in the negative axis directions. Hence

$$F_{3x} = -F_3 \cos \phi, \qquad F_{3y} = -F_3 \sin \phi.$$

We now imagine F_2 and F_3 to be removed and replaced by their rectangular components. To indicate this, the vectors F_2 and F_3 are crossed out lightly. All of the x-components can now be combined into a single force R_x whose magnitude R_x equals the algebraic sum of the magnitudes of the x-components, and all of the y-components can be combined into a single force R_y of magnitude R_y. Thus

$$R_x = F_1 + F_2 \cos \theta - F_3 \cos \phi,$$

and

$$R_y = F_2 \sin \theta - F_3 \sin \phi. \qquad \blacktriangleleft$$

EXAMPLE 2   In Fig. 2–6, let $F_1 = 120$ N, $F_2 = 200$ N, $F_3 = 150$ N, $\theta = 60°$, and $\phi = 45°$. Find the x- and y-components of the resultant or vector sum R, and find its magnitude and direction.

Solution   The computations can be arranged systematically as in Table 2–1.

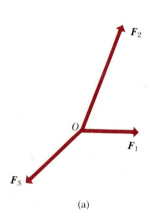

(a)

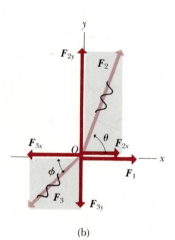

(b)

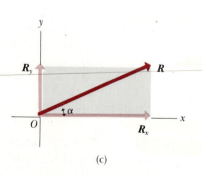

(c)

2–6   Vector R, the resultant of F_1, F_2, and F_3, is obtained by the method of rectangular resolution. The rectangular components of R are $R_x = \Sigma F_x$ and $R_y = \Sigma F_y$.

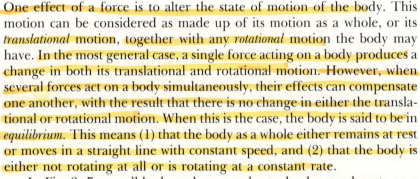

TABLE 2–1			
Force	**Angle**	**x-component**	**y-component**
$F_1 = 120$ N	0	$+120$ N	0
$F_2 = 200$ N	$60°$	$(200 \cos\theta)\ +100$ N	$(200 \sin\theta)\ +173$ N
$F_3 = 150$ N	$45°$	$(-150 \cos\theta)\ -106$ N	$(-150 \sin\theta)\ -106$ N
		$R_x = \Sigma F_x = +144$ N	$R_y = \Sigma F_y = +67$ N

$$R = \sqrt{(114\ \text{N})^2 + (67\ \text{N})^2} = 132\ \text{N},$$

$$\alpha = \arctan\frac{67\ \text{N}}{114\ \text{N}} = \arctan 0.588 = 30.4°.$$

◀

2–2

EQUILIBRIUM. NEWTON'S FIRST LAW

One effect of a force is to alter the state of motion of the body. This motion can be considered as made up of its motion as a whole, or its *translational* motion, together with any *rotational* motion the body may have. In the most general case, a single force acting on a body produces a change in both its translational and rotational motion. However, when several forces act on a body simultaneously, their effects can compensate one another, with the result that there is no change in either the translational or rotational motion. When this is the case, the body is said to be in *equilibrium.* This means (1) that the body as a whole either remains at rest or moves in a straight line with constant speed, and (2) that the body is either not rotating at all or is rotating at a constant rate.

In Fig. 2–7 a small body such as a rock or a hockey puck rests on a horizontal surface having negligible friction, such as an air-hockey table or a slab of wet ice. If a single force F_1 acts on the body, as in Fig. 2–7a, and if the body is originally at rest, it immediately starts to move. If it is in motion at the start, the effect of the force is to change the motion, either in direction or speed, or both. In either case, the body is not in equilibrium.

Now suppose we apply a second force F_2, as in Fig. 2–7b, equal in magnitude to F_1 but opposite in direction. Experiment then shows that the body is in equilibrium; if it is initially at rest it remains at rest, and if it is initially moving it continues to move in the same direction with constant speed. In this case the two forces are negatives of each other, $F_2 = -F_1$, and so their vector sum is zero.

$$R = F_1 + F_2 = 0.$$

For brevity, we shall often speak of two forces as simply being "equal and opposite," meaning that their magnitudes are equal and that one is the negative of the other.

This discussion can be generalized immediately to a body with any number of forces acting on it. We have noted that forces may be combined according to the rules of vector addition. When a body is in equi-

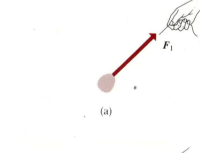

(a)

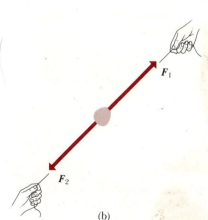

(b)

2–7 A small body acted on by two forces is in equilibrium if the forces are equal in magnitude and opposite in direction.

librium, the vector sum, or resultant, of all the forces acting on it must be zero. Each component of the resultant must therefore be zero. Hence, for a body in equilibrium,

$$R = \sum F = 0,$$

or

$$\sum F_x = 0, \qquad \sum F_y = 0. \tag{2-3}$$

When Eqs. (2–3) are satisfied, the body is in equilibrium, provided that it may be represented as a point or that all the forces acting on it are applied at the same point. If the body is extended in space and has forces applied at various points, then it may have a tendency to *rotate*. In that case there is a second condition for equilibrium, representing the condition that there must be no tendency for rotation. We return to this second condition in Chapter 8, but in the present chapter we are concerned only with equilibrium of pointlike bodies.

The condition for equilibrium represented by Eqs. (2–3) is called *Newton's first law of motion*. Newton did not state his first law in exactly these words. His original statement (translated from the Latin in which the *Principia* was written) reads:

Every body continues in its state of rest, or of uniform motion in a straight line, unless it is compelled to change that state by forces impressed on it.

Although rotational motion was not explicitly mentioned by Newton, it is clear from his work that he fully understood the conditions that the forces must satisfy when the rotation is zero or is constant.

2–3

DISCUSSION OF NEWTON'S FIRST LAW

Newton's first law of motion is not as self-evident as it may seem. This law asserts that in the absence of any applied force a body either remains at rest or moves uniformly in a straight line. It follows that once a body has been set in motion, no force is needed to keep it moving.

This assertion appears to be contradicted by everyday experience. Suppose we exert a force with the hand to push a book along a horizontal tabletop. After the book has left our hand, and we are no longer exerting a force on it, it does *not* continue to move indefinitely, but slows down and eventually comes to rest. If we wish to keep it moving uniformly we must continue to exert some forward force on it. But this force is required only because a frictional force is exerted on the sliding book by the tabletop, in a direction opposite to the motion of the book. The smoother the surfaces of the book and table, the smaller the frictional force and the smaller the force we must exert to keep the book moving. The first law asserts that if the frictional force could be eliminated completely, no forward force at all would be required to keep the book moving, once it had been set in motion. More than that, however, the law implies that if the *resultant* force on the book is zero, as it is when the frictional force is balanced by an equal forward force, the book also continues to move uniformly. In other words, *zero resultant force is equivalent to no force at all.*

Furthermore, the first law defines by implication what is known as an *inertial frame of reference.* To understand what is meant by this term we must recognize that the motion of a given body can be described only with respect to, or relative to, some other body. Its motion relative to one body may be very different from that relative to another. Thus a passenger in an aircraft that is making its take-off run may be at rest relative to the aircraft but be moving faster and faster relative to the earth.

A *frame of reference* (or a *reference system*) means a set of coordinate axes attached to (or moving with) some specified body or bodies. Suppose we consider a frame of reference attached to the aircraft referred to above. Everyone knows that during the take-off run, while the aircraft is going faster and faster, a passenger feels the back of his seat pushing him forward, although he remains at rest relative to a frame of reference attached to the aircraft. Newton's first law does not therefore correctly describe the situation; a forward force *does* act on the passenger, but nevertheless (relative to the aircraft) he remains at rest.

Suppose, on the other hand, that the passenger is standing in the aisle on roller skates. He then starts to move *backward,* relative to the aircraft, when the take-off run begins, even though no backward force acts on him. Newton's first law again does not correctly describe the facts.

We can now define an inertial frame of reference as one relative to which a body *does* remain at rest or move uniformly in a straight line when no force (or no resultant force) acts on it. That is, *an inertial frame of reference is one in which Newton's first law correctly describes the motion of a body on which zero resultant force acts.*

An aircraft gaining speed during take-off is evidently *not* an inertial system. For many purposes a reference system attached to the earth can be considered an inertial system, although it is not precisely so, due to effects related to the earth's rotation and other motions. But if a frame of reference *is* inertial, then a second frame moving uniformly (i.e., with constant speed in a straight line, without rotation) relative to it is *also* inertial, since any body in uniform motion as seen by an observer in the first frame will also appear to an observer in the second frame to be in uniform motion. The speed and direction appear different to the two observers, but both observe that Newton's first law is obeyed.

Thus there is no single, unique inertial frame of reference—there are infinitely many; but the motion of any one relative to any other is uniform in the above sense. It follows that the concepts of "absolute rest" and "absolute motion" have no physical meaning. Experiments show that a frame of reference at rest relative to the so-called fixed stars is inertial, within the limits of experimental error, and so every frame in uniform motion relative to this one is also inertial. An aircraft in steady flight over the earth is just as suitable an inertial frame of reference as the earth itself.

Finally, Newton's first law contains a qualitative definition of the concept of *force,* or at least of one aspect of the force concept, as "that which changes the state of motion of a body." (Of course, forces also produce other effects, such as changing the length of a coil spring.)

When a body at rest relative to the earth is observed to start moving, or when a moving body speeds up, slows down, or changes its direction, we can conclude that a resultant force is acting on it. This effect of a force can be used to define a *unit* of force, and we shall show in Chapter 4 how this is done.

2–4

NEWTON'S THIRD LAW OF MOTION

Any individual force is but one aspect of a mutual interaction between *two* bodies. It is found that *whenever one body exerts a force on another, the second always exerts on the first a force that is equal in magnitude, is opposite in direction, and has the same line of action.** A single isolated force is therefore an impossibility.

The two forces involved in every interaction between two bodies are often called an "action" and a "reaction," but this does not imply any difference in their nature, or that one force is the "cause" and the other its "effect." *Either* force may be considered the "action," and the other the "reaction" to it.

This property of forces was stated by Newton in his *third law of motion.* In his words,

> To every action there is always opposed an equal reaction: or, the mutual actions of two bodies upon each other are always equal, and directed to contrary parts.

As an example, suppose that a man pulls on one end of a rope attached to a block, as in Fig. 2–8. The weight of the block and the force exerted on it by the surface are not shown. The block may or may not be in equilibrium. The resulting action–reaction pairs of forces are indicated in the figure. (The lines of action of all the forces lie along the rope; the force vectors have been offset from this line to show them more clearly.) Vector F_1 represents the force exerted *on* the rope *by* the man. Its reaction is the equal and opposite force F_1' exerted *on* the man *by* the rope. Vector F_2 represents the force exerted on the block by the rope. The reaction to it is the equal and opposite force F_2', exerted on the rope by the block:

$$F_1' = -F_1, \qquad F_2' = -F_2. \tag{2–4}$$

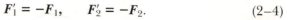

It is very important to realize that the forces F_1 and F_2', although they are opposite in direction and have the same line of action, do *not* constitute an action–reaction pair. For one thing, both of these forces act on the *same* body (the rope) while an action and its reaction necessarily act on *different* bodies. Furthermore, the forces F_1 and F_2' are not necessarily equal in magnitude. If the block and rope are moving to the right with increasing speed, the rope is not in equilibrium and F_1 is greater in magnitude than F_2'. Only in the special case when the rope remains at rest or moves with constant speed are the forces F_1 and F_2' equal in

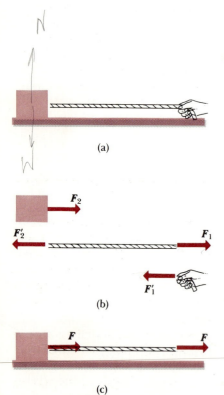

(a)

(b)

(c)

2–8 (a) A man pulls on a rope attached to a block. The forces that the rope exerts on the block and on the man are shown. (b) Separate diagrams showing the forces acting on the block, on the rope, and on the man. (c) If the rope is in equilibrium so that $F_2' = F_1$, the rope can be considered to transmit a force from the man to the block, and vice versa.

---

*There are instances, such as the motion of electrically charged particles, where this law is not obeyed, but it is correct for all the macroscopic forces we shall encounter in mechanics.

magnitude, but this is an example of Newton's *first* law, not his *third*. Even when the speed of the rope is changing, however, the action–reaction forces F_1 and F_1' are equal in magnitude to *each other*, and the action–reaction forces F_2 and F_2' are equal in magnitude to *each other*, although then F_1 is not equal to F_2'.

In the special case when the rope is in equilibrium, and when no forces act on it except those at its ends, F_2' equals $-F_1$ by Newton's *first* law. Since F_2 *always* equals $-F_2'$ by Newton's *third* law, then in this special case F_2 also equals F_1 and the force exerted on the block by the rope is equal to the force exerted on the rope by the man. The rope can therefore be considered to "transmit" to the block, without change, the force exerted on it by the man, as in Fig. 2–8c. This point of view is often useful, but it is important to remember that it applies only under the restricted conditions as stated.

A body such as the rope in Fig. 2–8, which is subjected to pulls at its ends, is said to be in *tension*. The tension at any point equals the force exerted at that point. Thus in Fig. 2–8b the tension at the right-hand end of the rope equals the magnitude of F_1 (or of F_1') and the tension at the left-hand end equals the magnitude of F_2 (or of F_2'). If the rope is in equilibrium and if no forces act except at its ends, the tension is the same at both ends. If, for example, the magnitudes of F_1 and F_2 are each 50 N, the tension in the rope is 50 N (*not* 100 N).

One statement made above needs emphasis: The two forces in an action–reaction pair *never* act on the same body. Remembering this general principle can often help clear up confusion regarding action–reaction pairs.

2–5

IDEALIZED MODELS

The phenomena of nature are seldom simple; often a phenomenon involves several interrelated principles, and the relationships can be extremely complex. Often one must make simplifying assumptions and approximations in order to facilitate analysis of such a phenomenon and to focus attention on its most significant aspects. This process of simplifying and idealizing is called making a *model*, and models play an essential role in applications of principles of physics.

Here is an example. A baseball is thrown into the air, and we want to calculate where it lands and with what velocity. The ball has a somewhat irregular surface; it is spinning as well as moving through the air. It is affected by the force of gravity, which decreases slightly as the ball ascends. Additional effects are caused by air resistance (friction) and by buoyancy in the air. There may be variable air currents that complicate these effects further.

Clearly, the analysis is hopelessly complicated if we try to include all these effects; we need to make a simplified model. If we approximate the ball as a uniform sphere and assume it moves through still air, we can calculate the air resistance and buoyancy forces, although these are still complex. The buoyant force is usually much smaller than the force of gravity, and if the ball moves slowly enough the friction force is also small, so we may wish to omit these completely. We may feel that the shape and spinning motion are irrelevant for an overall description of motion; if so, we can pretend that the ball is a *particle* (a point mass without any size, shape, or spin). Finally, we can assume that the gravitational force is constant during the motion. Then at last the problem is simple enough to permit detailed analysis. Such problems are considered in Chapter 5.

Because of the approximations and simplifications, one should not expect this analysis to predict *precisely* how the ball will move; but of course an approximate description is a lot better than no description at all, which may be the alternative if the problem is so complex as to defy analysis.

The concept of an idealized model plays an important role in equilibrium problems. In general the forces acting on a body *do not* all have lines of action passing through a single point, and as a result the body undergoes both translational and rotational motion. However, if in a particular problem the rotational motion is irrelevant, then the body may be represented as a particle or point. In such a case, all the forces are considered to act at this point.

2–6

EQUILIBRIUM OF A PARTICLE

In the remainder of this chapter, we shall consider several examples and problems involving the equilibrium of particles. It is surprising how many situations of interest and importance in engineering, in the life and earth sciences, and in everyday life involve the equilibrium of particles. In the analysis of such problems, the following systematic problem-solving procedure is strongly recommended.

1. Make a simple sketch of the apparatus or structure, showing dimensions and angles.

2. Choose some object (a knot in a rope, for example) as the particle in equilibrium. Draw a separate diagram of this object and show by arrows (use a colored pencil) *all* of the forces exerted *on* it by other bodies. This is called the *force diagram* or *free-body diagram*. When a system is composed of several particles, it may be necessary to construct a separate free-body diagram for each one. Do *not* show, in the free-body diagram of a chosen particle, any of the forces exerted *by* it. These forces (which are the reactions to the forces acting *on* the chosen particle) all act *on other bodies* and appear in the free-body diagrams of those bodies.

3. Construct a set of rectangular axes and resolve into rectangular components all forces acting on the particle. Cross out lightly those forces that have been resolved.

4. Set the algebraic sum of all x-forces (or force components) equal to zero, and the algebraic sum of all y-forces (or components) equal to zero. This provides two independent equations, which can be solved simultaneously for two unknown quantities (which may be forces, angles, distances, etc.).

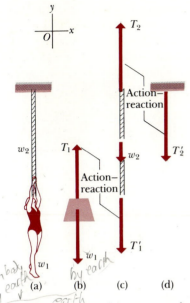

2–9 (a) Gymnast hanging at rest from vertical rope. (b) The gymnast is isolated and all forces acting on her are shown. (c) Forces on the rope. (d) Downward force on the ceiling. Lines connect action–reaction pairs.

A remark about notation is in order. Thus far, vector quantities have been denoted by boldface italic symbols, and this usage will be continued throughout this book. However, it is sometimes convenient in diagrams to label a vector by using a lightface italic symbol to represent the *magnitude* of the quantity, and to describe its direction (if that is not obvious in the diagram) separately, perhaps by use of an angle. When both light and bold symbols appear on a diagram, confusion can be avoided by recalling that a bold symbol always represents the vector quantity itself, while a light symbol represents either the magnitude of a vector quantity or a component.

A force that will be encountered in many problems is the *weight* of a body, that is, the force of gravitational attraction exerted on the body by the earth. This is but one aspect of a mutual interaction between the earth and the body. At the same time, the body attracts the earth with a force equal in magnitude and opposite in direction to the force the earth exerts on the body. Thus if a body weighs 10 N, the earth pulls down on it with a force of 10 N, and the body pulls *up* on the earth with a force of 10 N, another example of an action–reaction pair.

EXAMPLE 1   A gymnast has just begun climbing up a rope hanging from a gymnasium ceiling, as in Fig. 2–9a. She stops, suspended from the lower end of the rope by her hands. Her weight is 600 N, and the weight of the rope is 100 N. Analyze the forces on the gymnast, and on the rope.

Solution   Figure 2–9b is a free-body diagram for the gymnast. The forces acting on her are her weight w_1 and the upward force T_1 exerted on her by the rope. If we take the x-axis horizontal and the y-axis vertical, there are no x-components of force. The y-components are those associated with the forces T_1 and w_1. Force T_1 acts in the positive y-direction, and its y-component is just the magnitude T_1, a positive (scalar) quantity. But w_1 acts in the negative y-direction, and its y-component is the *negative* of the magnitude w_1. Thus the algebraic sum of y-components is $T_1 - w_1$. Then from the equilibrium condition, $\Sigma F_y = 0$, we have

$$\Sigma F_y = T_1 - w_1 = 0,$$

$$T_1 = w_1. \quad \text{(First law)}$$

Let us emphasize again that the forces w_1 and T_1 are *not* an action–reaction pair, although they are equal in magnitude, opposite in direction, and have the same line of action. The weight w_1 is a force of attraction exerted on the body by the earth. Its reaction is an equal and opposite force of attraction exerted on the earth by the body. This reaction is one of the set of forces acting *on the earth*, and therefore it does not appear in the free-body diagram of the suspended block.

The reaction to the force T_1 is a downward force, T_1', of equal magnitude:

$$T_1 = T_1'. \quad \text{(Third law)}$$

The force T_1' is shown in part (c), which is the free-body diagram of the rope. The other forces on the rope are its own weight w_2 and the upward force T_2 exerted on its upper end by the ceiling. The y-component of T_2

is positive, while those of w_2 and T_1' are negative. Since the rope is also in equilibrium.

$$\sum F_y = T_2 - w_2 - T_1' = 0$$

$$T_2 = w_2 + T_1'. \qquad \text{(First law)}$$

The reaction to T_2 is the downward force T_2' in part (d), exerted on the ceiling by the rope:

$$T_2 = T_2'. \qquad \text{(Third law)}$$

With the numbers given, the gymnast's weight w_1 is 600 N, and the weight of the rope w_2 is 100 N. Then

$$T_1 = w_1 = 600 \text{ N},$$

$$T_1' = T_1 = 600 \text{ N},$$

$$T_2 = w_2 + T_1' = 100 \text{ N} + 600 \text{ N} = 700 \text{ N},$$

$$T_2' = T_2 = 700 \text{ N}.$$

If the weight of the rope were so small as to be negligible, then in effect no forces would act on it except at its ends. The forces T_2 and T_2' would then each equal 600 N and, as explained earlier, the rope could be considered to transmit a 600-N force from one end to the other without change. We could then consider the upward pull of the rope on the gymnast as an "action" and the downward pull on the ceiling as its "reaction." The tension in the rope would then be 600 N. ◀

EXAMPLE 2 In Fig. 2–10, a hanging lamp of weight w hangs from a cord, which is knotted at point O to two other cords, one fastened to the ceiling, the other to the wall. We wish to find the tensions in these three cords, assuming the weights of the cords to be negligible.

Solution To use the conditions of equilibrium to determine an unknown force, we must consider some body that is in equilibrium and on which the desired force acts. The suspended lamp is one such body; as shown in the preceding example, the tension T_1 in the vertical cord supporting the lamp is equal in magnitude to the weight of the lamp. The other cords do not exert forces on the lamp (because they are not attached directly to it) but they do act on the knot at O. Hence we consider the *knot* as a particle in equilibrium, considering the weight of the knot itself as negligible.

Free-body diagrams for the lamp and the knot are shown in Fig. 2–10b, where T_1, T_2, and T_3 are the *magnitudes* of the forces shown; the directions of these forces are indicated by the vectors on the diagram. An x-y coordinate axis system is also shown, and the force of magnitude T_3 has been resolved into its x- and y-components.

Considering the lamp first, we note that there are no x-components of force. The y-component of force exerted by the cord is in the positive y-direction and is just T_1. The y-component of the weight is, however, in the negative y-direction and is $-w$. The equilibrium condition for the lamp, that the algebraic sum of y-components of force must be zero, is

$$T_1 + (-w) = 0.$$

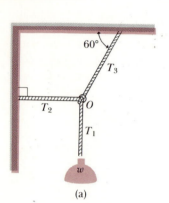

(a)

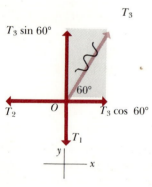

$T_3 \sin 60°$

T_2 $T_3 \cos 60°$

T_1

y x

T_1

w

(b)

2–10   (a) A lamp of weight w is suspended from a cord knotted at O to two other cords. (b) Free-body diagrams for the lamp and for the knot, showing the components of force acting on each.

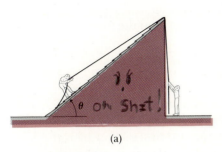

(a)

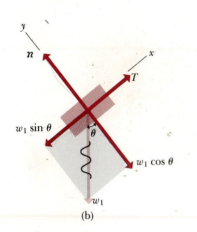

(b)

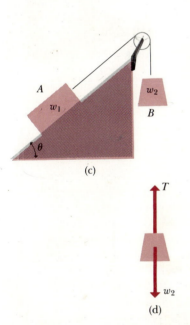

(c)

(d)

2–11 Forces on a body in equilibrium on a frictionless inclined plane. (a) The physical situation; (b) free-body diagram for body A; (c) idealized model of situation; (d) free-body diagram for body B.

Hence,

$$T_1 = w.$$

The tension in the vertical cord equals the weight of the lamp, as we have already noted.

We now consider the knot at O. There are both x- and y-components of force acting on it, so there are two separate equilibrium conditions. The algebraic sum of the x-components must be zero, and the algebraic sum of y-components must separately be zero. (Note that x- and y-components are *never* added together in a single equation.) We find

$$\sum F_x = 0: \qquad T_3 \cos 60° - T_2 = 0,$$

$$\sum F_y = 0: \qquad T_3 \sin 60° - T_1 = 0.$$

Because $T_1 = w$, the second equation can be rewritten as

$$T_3 = \frac{T_1}{\sin 60°} = \frac{w}{\sin 60°} = 1.155 \, w.$$

This result can now be used in the first equation:

$$T_2 = T_3 \cos 60° = (1.155 \, w) \cos 60° = 0.577 \, w.$$

Thus all three tensions can be expressed as multiples of the weight of the lamp, which is assumed to be known. To summarize,

$$T_1 = w,$$

$$T_2 = 0.577 \, w,$$

$$T_3 = 1.155 \, w.$$

Thus if the weight of the lamp is $w = 50$ N, then

$$T_1 = 50 \text{ N},$$

$$T_2 = (0.577)(50 \text{ N}) = 28.9 \text{ N},$$

$$T_3 = (1.155)(50 \text{ N}) = 57.7 \text{ N}.$$

We note that T_3 is greater than the weight of the lamp; if this seems strange, one should note that it must be large enough so that its vertical component is equal to w in magnitude, so the force itself must have somewhat larger magnitude than this. ◀

EXAMPLE 3 A steeplejack of weight w_1 is trying to find a leak in an icy, perfectly frictionless roof with slope angle θ. He is secured by a rope passing over the top and down the vertical side of the building. The other end of the rope is held by his helper (weight w_2) on the ground. The helper finds that to hold the man on the roof he must put his full weight on the rope. If we know the weight w_1 of the steeplejack, what is the weight w_2 of the helper?

Solution We represent the situation by the idealized model shown in Fig. 2–11c, where A is the steeplejack and B the helper. Free-body diagrams for the two bodies are shown in parts (b) and (d) of the figure. Here, as in preceding examples, we label and refer to forces by using light italic symbols, which in equations represent the *magnitudes* of these forces. The forces on body B are its weight w_2 and the force T exerted on it by the cord. Because it is in equilibrium,

$$T = w_2.$$

Body A is acted on by its weight w_1, the force T exerted on it by the rope, and the force n exerted on it by the plane. We use the same symbol T for the force exerted on each body by the rope because, as discussed above, these forces are equivalent to an action–reaction pair and have the same magnitude. The pulley, which corresponds to the frictionless top edge of the roof, changes the directions of the forces exerted by the rope but not their magnitudes; and the two forces at the ends have the same magnitude even though their directions are different. If there is no friction, the force n can have no component along the plane that would resist the motion of the body on the plane; hence it must be perpendicular or *normal* to the surface of the plane.

It is simplest to choose x- and y-axes parallel and perpendicular to the surface of the plane, because then only the weight w_1 needs to be resolved into components. The conditions of equilibrium for body A give

$$\sum F_x = T - w_1 \sin \theta = 0,$$

$$\sum F_y = n - w_1 \cos \theta = 0.$$

Thus if $w_1 = 800$ N and $\theta = 30°$, we find

$$w_2 = T = w_1 \sin \theta = (800 \text{ N})(0.500) = 400 \text{ N},$$

and

$$n = w_1 \cos \theta = (800 \text{ N})(0.866) = 693 \text{ N}.$$

Note carefully that, *in the absence of friction*, the same weight w_2 of 400 N is required whether the system remains at rest or moves with constant speed in *either* direction. This is *not* the case when friction is present.   ◀

2–7

FRICTION

Whenever the surface of one body slides over that of another, *each* body exerts a frictional force on the other, parallel to the surfaces. The force *on* each body is opposite to the direction of its motion relative to the other. Thus when a block slides from left to right along the surface of a table, a frictional force to the left acts on the block and an equal force toward the right acts on the table. Frictional forces may also act when there is no relative motion. When we push horizontally on a heavy packing case resting on the floor, the case may not move because of an equal and opposite frictional force exerted on the case by the floor.

In Fig. 2–12a, a box is at rest on a horizontal surface, in equilibrium under the action of its weight w and the upward force P exerted on it by the surface. The lines of action w and P have been displaced slightly to show these forces more clearly.

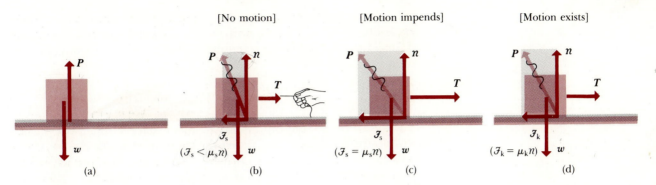

[No motion]   [Motion impends]   [Motion exists]

2–12 The magnitude of the friction force $\mathcal{F}_s$ is less than or equal to $\mu_s n$ when there is no relative motion and is equal to $\mu_k n$ when motion exists.

Suppose now that a rope is attached to the box as in Fig. 2–12b and the tension T in the rope is gradually increased. Provided the tension is not too great, the box remains at rest. The force P exerted on the box by the surface is inclined toward the left as shown, since its horizontal component opposes T. The component of P parallel to the surface is called the *force of static friction*, $\mathcal{F}_s$. The other component, n, is the *normal* force exerted on the block by the surface. From the conditions of equilibrium, the force of static friction $\mathcal{F}_s$ is equal in magnitude and opposite in direction to the force T, and the normal force n is equal and opposite to the weight w.

As the force T is increased further, a limiting value is reached at which the box suddenly "breaks loose" and starts to slide. In other words, there is a certain *maximum* value that the force of static friction $\mathcal{F}_s$ can have. Figure 2–12c is the force diagram when T is just below its critical value. If the force T exceeds this value, the box is no longer in equilibrium.

Experiment shows that, for a given pair of surfaces, the magnitude of the maximum value of $\mathcal{F}_s$ depends on that of the normal force n; when the normal force increases, a greater force is required to make the box slide. In general the details of this relationship can be quite complex, but in some cases it is approximately true that the maximum value of $\mathcal{F}_s$ is *directly proportional* to n, with proportionality factor μ_s, called the *coefficient of static friction*. The actual force of static friction can have any magnitude between zero (when there is no applied force parallel to the surface) and a maximum value given by $\mu_s n$. Thus,

$$\mathcal{F}_s \leq \mu_s n. \tag{2–5}$$

The equality sign holds only when the applied force T, parallel to the surface, has such a value that motion is about to start (Fig. 2–12c). When T is less than this value (Fig. 2–12b), the inequality sign holds and the magnitude of the friction force must be computed from the conditions of equilibrium.

As soon as sliding begins, it is found that the friction force *decreases*. This new friction force, for a given pair of surfaces, also depends on the magnitude of the normal force, and again it is convenient, though not always precise, to represent the relation as a proportionality. The proportionality factor, μ_k, is called the *coefficient of sliding friction* or *kinetic friction*. Thus, when the box is in motion, the force of sliding or kinetic friction is given by

$$\mathcal{F}_k = \mu_k \mathcal{n}. \qquad\qquad (2\text{--}6)$$

This is shown in Fig. 2–12d.

The force P exerted on the box by the surface is called the *contact force*. The frictional force $\mathcal{F}$ is always the component of the contact force *parallel* to the surface, and the normal force $\mathcal{n}$ is always the component *normal* (perpendicular) to the surface. We use special script symbols for these quantities to emphasize their special role in representing the contact force. In addition, use of the script letter $\mathcal{n}$ prevents our confusing this symbol for normal force with the abbreviation for newton, N. In many problems to be considered later, the normal force is *not* equal in magnitude to the weight of the body, because the surface is not horizontal or because there are additional components of force normal to the surface. Nevertheless, $\mathcal{n}$ always denotes the normal component of contact force exerted by the surface on the body. By definition, $\mathcal{n}$ and $\mathcal{F}$ for a given body are always perpendicular. Therefore Eqs. (2–5) and (2–6) are not *vector* equations, but relations between the *magnitudes* $\mathcal{n}$ and $\mathcal{F}$ of these forces.

Usually in problems involving a contact force, it is convenient not to represent that contact force as a single vector, such as P in Fig. 2–12, but to represent it immediately in terms of the components $\mathcal{n}$ and $\mathcal{F}$, which we may call the normal and frictional components of the contact force. We shall follow this pattern most of the time in solving problems involving contact forces; the symbol P will usually not appear.

The coefficients of friction depend primarily on the nature of the surfaces in contact, being relatively large if the surfaces are rough, and small if they are smooth. The coefficient of sliding friction varies somewhat with the relative velocity, but for simplicity we shall assume it to be independent of velocity. It is also nearly independent of the contact area. Equations (2–5) and (2–6) are useful empirical relations but do not represent *fundamental* physical laws like Newton's laws. Typical numerical values are given in Table 2–2.

TABLE 2–2 COEFFICIENTS OF FRICTION

Materials	Static, μ_s	Kinetic, μ_k
Steel on steel	0.74	0.57
Aluminum on steel	0.61	0.47
Copper on steel	0.53	0.36
Brass on steel	0.51	0.44
Zinc on cast iron	0.85	0.21
Copper on cast iron	1.05	0.29
Glass on glass	0.94	0.4
Copper on glass	0.68	0.53
Teflon on Teflon	0.04	0.04
Teflon on steel	0.04	0.04

Table 2–2 lists coefficients of friction only for solids. Liquids and gases show frictional effects also, but the simple equation $\mathcal{F} = \mu n$ does not hold. It will be shown in Chapter 13 that there exists a property of liquids and gases called *viscosity*, which determines the friction force between two surfaces sliding over each other with a layer of liquid or gas between them. Gases have the lowest viscosities of all materials at normal temperatures, and therefore, to reduce friction to a value close to zero, it is convenient to have an object slide on a layer of gas.

This is the principle of the *hovercraft*, a type of vehicle containing powerful blowers that maintain a cushion of air between the vehicle and the ground so that it literally floats on a layer of air. A familiar laboratory example of the principle is the linear air track, conceived originally by H. V. Neher, and R. B. Leighton at the California Institute of Technology, and developed further by John Stull of Alfred University. Elastic bumpers are provided for the inverted V-shaped riders. (See Fig. 2–13a.) When one rider only is placed on the air track and is given a push, it will hit the stationary bumper and then proceed to move back and forth many times before coming to rest. The frictional force is velocity-dependent, but at typical speeds the effective coefficient of friction is of the order of 0.001.

A similar device is the frictionless air table shown in Fig. 2–13b, developed by Harold A. Daw of New Mexico State University. The pucks, made of plastic, are supported by over 1600 tiny air jets about an inch apart. Two-dimensional collisions, both elastic and inelastic, may be demonstrated on this table. The same principle is used in "air-hockey" games.

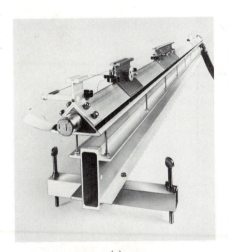

(a)

(b)

2–13 (a) The Ealing-Stull linear air track. Inverted Y-shaped sliders ride on a layer of air streaming through many fine holes in the inverted V-shaped surface. (b) The Ealing-Daw two-dimensional air table. Plastic pucks slide on a cushion of air issuing from more than a thousand minute holes in the tabletop. (Photograph courtesy of the Ealing Corporation.)

EXAMPLE 1 Suppose the object shown in Fig. 2–12 is a 500-N crate that a delivery company has just unloaded in your driveway. You find that to get it moving toward your garage you have to push with a horizontal force of magnitude 200 N, but that once it is moving you can keep it moving with constant speed by pushing with 100 N force. What are the coefficients of static and kinetic friction?

Solution From Fig. 2–12c and the data above, we have

$$\sum F_y = n - w = n - 500 \text{ N} = 0, \qquad n = 500 \text{ N},$$

$$\sum F_x = T - \mathcal{F}_s = 8 \text{ N} - \mathcal{F}_s = 0, \qquad \mathcal{F}_s = 200 \text{ N},$$

$$\mathcal{F}_s = \mu_s n \quad \text{(motion impends)}.$$

Hence we have

$$\mu_s = \frac{\mathcal{F}_s}{n} = \frac{200 \text{ N}}{500 \text{ N}} = 0.40.$$

From Fig. 2–12d, we have

$$\sum F_y = n - w = n - 500 \text{ N} = 0, \qquad n = 500 \text{ N},$$

$$\sum F_x = T - \mathcal{F}_k = 4 \text{ N} - \mathcal{F}_k = 0, \qquad \mathcal{F}_k = 100 \text{ N},$$

$$\mathcal{F}_k = \mu_k n \quad \text{(motion exists)}.$$

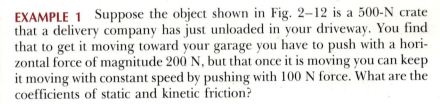

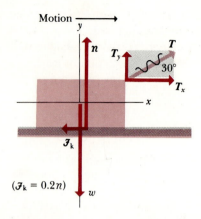

2–14 Forces on a box being dragged to the right on a level surface at constant speed.

Hence

$$\mu_k = \frac{\mathcal{F}_k}{n} = \frac{100\text{ N}}{500\text{ N}} = 0.20. \qquad \blacktriangleleft$$

EXAMPLE 2 In Example 1, what is the friction force if the box is at rest on the surface and a horizontal force of 50 N is exerted on it?

Solution We have

$$\sum F_x = T - \mathcal{F}_s = 50\text{ N} - \mathcal{F}_s = 0, \qquad \text{(First Law)}$$

$$\mathcal{F}_s = 50\text{ N}.$$

Note that in this case $\mathcal{F}_s < \mu_s n$. $\quad \mu_s = \frac{50\text{N}}{500\text{N}} = 0.1$

$$50 < (0.1)(500)$$

EXAMPLE 3 In Example 1 you try to move the crate by tying a rope around it and pulling upward on the rope at an angle of 30° to the horizontal. What tension in the rope is required to make the crate move with constant speed?

Solution Figure 2–14 is a free-body diagram showing the forces on the crate. We note that the normal force n is *not* equal in magnitude to the weight of the crate because the force exerted by the rope has an additional vertical component. From the equilibrium conditions,

$$\sum F_x = T\cos 30° - 0.2\,n = 0,$$

$$\sum F_y = T\sin 30° + n - 500\text{ N} = 0.$$

These are two simultaneous equations for the two unknown quantities T and n. To solve them we can eliminate one unknown and solve for the other, as follows. Rearrange the second equation to the form

$$n = 500\text{ N} - T\sin 30°.$$

Substitute this expression for n back into the first equation, obtaining

$$T\cos 30° - 0.2(500\text{ N} - T\sin 30°) = 0.$$

Finally, solve this equation for T, then substitute the result back into either of the original equations to obtain n. The results are

$$T = 103.5\text{ N}, \qquad n = 448.2\text{ N}. \qquad \blacktriangleleft$$

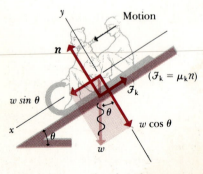

2–15 Forces on a toboggan sliding down a hill (with friction) at constant speed.

EXAMPLE 4 A toboggan loaded with vacationing students slides down a long, snow-covered slope having a coefficient of kinetic friction μ_k. The slope has just the right angle to make the toboggan slide with constant speed. Find the slope angle.

Solution We represent the situation by a block sliding down a plane inclined at an angle θ, as shown in Fig. 2–15. The forces on the block (toboggan) are its weight w and the normal and frictional components of the force exerted on it by the plane. We take axes perpendicular and parallel to the surface of the plane and represent the weight in terms of its components in these two directions, as shown. The equilibrium conditions in component form are then

$$\sum F_x = w \sin \theta - \mathcal{F}_k = w \sin \theta - \mu_k n = 0,$$
$$\sum F_y = n - w \cos \theta = 0.$$

Hence

$$\mu_k n = w \sin \theta, \qquad n = w \cos \theta.$$

Dividing the first of these equations by the second, we find

$$\mu_k = \frac{\sin \theta}{\cos \theta} = \tan \theta.$$

It follows that a block, regardless of its weight, slides down an inclined plane with constant speed if the tangent of the slope angle of the plane equals the coefficient of kinetic friction. Measurement of this angle then provides a simple experimental method for determining the coefficient of kinetic friction.

FIY ! BE FREE

QUESTIONS

2–1 Can a body be in equilibrium when only one force acts on it?

2–2 A helium balloon hovers in midair, neither ascending nor descending. Is it in equilibrium? What forces act on it?

2–3 If the two ends of a rope in equilibrium are pulled with forces of equal magnitude and opposite direction, why is the total tension in the rope not zero?

2–4 A horse is hitched to a wagon. Since the wagon pulls back on the horse just as hard as the horse pulls on the wagon, why doesn't the wagon remain in equilibrium, no matter how hard the horse pulls?

2–5 A clothesline is hung between two poles, and then a shirt is hung near the center. No matter how tightly the line is stretched, it always sags a little at the center. Explain why.

2–6 A man sits in a chair that is suspended from a rope. The rope passes over a pulley suspended from the ceiling and the man holds the other end of the rope in his hands. What is the tension in the rope, and what force does the chair exert on the man?

2–7 How can pushing *down* on a bicycle pedal make the bicycle move *forward*?

2–8 A car is driven up a steep hill at constant speed. Dis-

cuss all the forces acting on the car; in particular, what pushes it up the hill?

2–9 Can the coefficient of friction ever be greater than unity? If so, give an example; if not, explain why not.

2–10 A block rests on an inclined plane with enough friction to prevent sliding down. To start the block moving, is it easier to push it up the plane, down the plane, or sideways? Why?

2–11 In pushing a box up a ramp, is it better to push horizontally or to push parallel to the ramp?

2–12 In stopping a car on an icy road, is it better to push the brake pedal hard enough to "lock" the wheels and make them slide, or to push gently so the wheels continue to roll? Why?

2–13 When one stands with bare feet in a wet bathtub, the grip feels fairly secure, and yet a catastrophic slip is quite possible. Discuss this situation with respect to the two coefficients of friction.

2–14 The horrible squeak made by a piece of chalk held against a blackboard at the wrong angle results from alternate sticking and slipping of the chalk against the blackboard. Interpret this phenomenon in terms of the two coefficients of friction. Can you think of other examples of "slip-stick" behavior?

36 EQUILIBRIUM OF A PARTICLE

PROBLEMS

2–1

a) Find graphically the horizontal and vertical components of a 40-N force with a direction 50° above the horizontal to the right. Let 1 cm = 5 N.

b) Check your results by calculating the components.

2–2 A raccoon pushes a tin box along the floor as in Fig. 2–1b by applying a force of 40 N making an angle of 30° with the horizontal. Using a scale of 1 cm = 5 N, find the horizontal and vertical components of the force by the graphical method. Check your results by calculating the components.

2–3 A man is dragging a trunk up the loading ramp of a mover's truck. The ramp has a slope angle of 20°, and the man pulls upward at an angle of 30° to the ramp.

a) How large a force F is necessary in order that the component F_x parallel to the plane shall be 16 N?

b) How large will the component F_y then be?

Solve graphically, letting 1 cm = 2 N, and by calculating components.

2–4 The three forces shown in Fig. 1–10 act on a body located at the origin.

a) Find the x- and y-components of each of the three forces.

b) Use the method of rectangular resolution to find the resultant of the forces.

c) Find the magnitude and direction of a fourth force that must be added to make the resultant force zero. Indicate the fourth force by a diagram.

2–5 Use the method of rectangular resolution to find the resultant of the following set of forces and the angle it makes with the positive x-axis: 200 N, along the x-axis toward the right; 300 N, 60° above the x-axis to the right; 100 N, 45° above the x-axis to the left; 200 N, along the negative y-axis.

2–6 The resultant of four forces is 1000 N in the direction 30° west of north. Three of the forces are 400 N, 60° north of east; 200 N, south; and 400 N, 53° west of south. Find the rectangular components of the fourth force.

2–7 Two forces, F_1 and F_2, act at a point. The magnitude of F_1 is 8 N and its direction is 60° above the x-axis in the first quadrant. The magnitude of F_2 is 5 N and its direction is 53° below the x-axis in the fourth quadrant.

a) What are the horizontal and vertical components of the resultant force?

b) What is the magnitude of the resultant?

c) What is the magnitude of the vector difference $F_1 - F_2$?

2–8 Two forces, F_1 and F_2, act upon a body in such a manner that the resultant force R has a magnitude equal to that of F_1 and makes an angle of 90° with F_1. Let $F_1 = R = 10$ N. Find the magnitude of the second force, and its direction (relative to F_1).

2–9 Two men and a boy want to push a crate in the direction marked x in Fig. 2–16. The two men push with forces F_1 and F_2, whose magnitudes and directions are indicated in the figure. Find the magnitude and direction of the smallest force that the boy should exert.

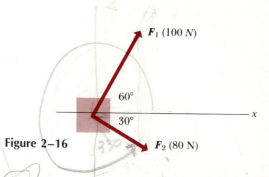

Figure 2–16

2–10 Two men pull horizontally on ropes attached to a post; the angle between the ropes is 45°. If man A exerts a force of 300 N and man B a force of 200 N, find the magnitude of the resultant force and the angle it makes with A's pull. Solve

a) graphically, using a scale diagram, and

b) analytically, using components.

Let 1 cm = 50 N in (a).

2–11 Imagine that you are holding a book weighing 4 N at rest on the palm of your hand. Complete the following sentences. Remember that the two forces in an action–reaction pair *never* act on the same body!

a) A downward force of magnitude 4 N is exerted on the book by _____.

b) An upward force of magnitude _____ is exerted on _____ by the hand.

c) Is the upward force (b) the reaction to the downward force (a)?

d) The reaction to force (a) is a force of magnitude _____, exerted on _____ by _____. Its direction is _____.

e) The reaction to force (b) is a force of magnitude _____, exerted on _____ by _____. Its direction is _____.

f) That the forces (a) and (b) are equal and opposite is an example of Newton's _____ law.

g) That forces (b) and (e) are equal and opposite is an example of Newton's _____ law.

Suppose now that you exert an upward force of magnitude 5 N on the book.

h) Does the book remain in equilibrium?

i) Is the force exerted on the book by the hand equal and opposite to the force exerted on the book by the earth?

j) Is the force exerted on the book by the earth equal and opposite to the force exerted on the earth by the book?

k) Is the force exerted on the book by the hand equal and opposite to the force exerted on the hand by the book?

Finally, suppose that you snatch your hand away while the book is moving upward.

l) How many forces then act on the book?

m) Is the book in equilibrium?

n) What balances the downward force exerted on the book by the earth?

2–12 A bottle is given a push along a tabletop, and slides off the edge of the table.

a) What forces are exerted on it while it is falling from the table to the floor?

b) What is the reaction to each force, that is, on what body and by what body is the reaction exerted? Neglect air resistance.

2–13 Two 10-N weights are suspended at opposite ends of a rope that passes over a light frictionless pulley. The pulley is attached to a chain that goes to the ceiling.

a) What is the tension in the rope? _10N_

b) What is the tension in the chain? _20N_

2–14 In Fig. 2–17, a man lifts a weight w by pulling down on a rope with a force F. The upper pulley is attached to the ceiling by a chain, and the lower pulley is attached to

the weight by another chain. If $w = 400$ N, find the tension in each chain and the force F if the weight is lifted at constant speed. Assume the weights of the rope, pulleys, and chains to be negligible.

2–15 In Fig. 2–10a, let the weight of the hanging lamp be 80 N.

a) Find the tension in each cord.

b) If the 60° angle is changed to 45°, find the tension in each cord.

2–16 A picture frame is hung against a wall, suspended by two wires attached to its upper corners. If the two wires make the same angle with the vertical, what must this angle be for the tension in each wire to equal the weight of the frame?

2–17 Figure 2–18 shows a technique called rappelling, used by mountaineers for descending vertical rock faces. The climber sits in a rope seat, and the rappel rope passes through a friction device attached to this seat. Suppose the rock is perfectly smooth, and the climber's feet push horizontally into the rock; if the climber's weight is 800 N, find the tension in the rope and the force his feet exert on the rock face.

Figure 2–18

2–18 A man proposes to pull his car out of a mudhole on a country road by stretching a rope between the front of his car and a solid object directly in front of the car, and then pushing sideways at the midpoint of the rope. Suppose that the rope is 50 m long, and that a sideways force of 300 N is required to displace the midpoint sideways 5 m. Determine the tension in the rope and the forward component of force on the car.

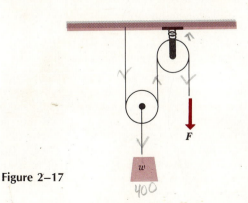

Figure 2–17

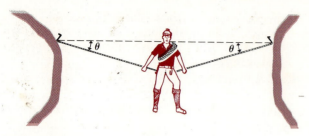

Figure 2–19

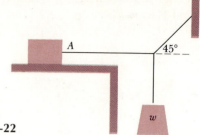

Figure 2–22

2–19 Figure 2–19 illustrates a mountaineering technique called a Tyrolean traverse. A rope is stretched tightly between two points, and the climber slides across the rope. The climber's weight is 800 N, and the breaking strength of the rope (typically nylon, 11 mm diameter) is 20,000 N.

a) If the angle θ is 15°, find the tension in the rope.

b) What is the smallest value the angle θ can have if the rope is not to break?

2–20 Find the tension in each cord in Fig. 2–20 if the weight of the suspended body is 200 N.

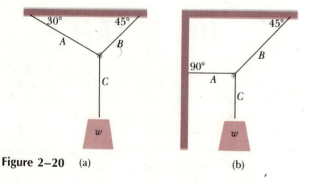

Figure 2–20   (a)   (b)

2–21 In Fig. 2–21, find the weight of the suspended body if the tension in the diagonal string is 20 N.

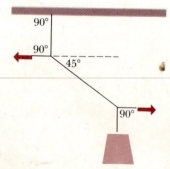

Figure 2–21

2–22

a) Block A in Fig. 2–22 weighs 100 N. The coefficient of static friction between the block and the surface on which it rests is 0.30. The weight w is 20 N and the

system is in equilibrium. Find the friction force exerted on block A.

b) Find the maximum weight w for which the system will remain in equilibrium.

2–23 A flexible rope of weight w hangs between two hooks at the same height, as shown in Fig. 2–23. At each end the rope makes an angle θ with the horizontal.

a) What is the magnitude and direction of the force **F** exerted by the rope on the hook at the left?

b) What is the tension T in the rope at its lowest point?

Figure 2–23

2–24 A 30-N tool box is pulled at constant speed up a frictionless icy sloping roof by a weight of 10 N hanging from a rope attached to the box and passing over a frictionless pulley at the top of the roof. (See Fig. 2–11c.) Find

a) the slope angle of the roof,

b) the tension in the rope, and

c) the normal force exerted on the box by the roof.

2–25

a) A large rock rests upon a rough horizontal surface. A bulldozer pushes on the rock with a horizontal force that is slowly increased, starting with zero force. Draw a graph with T along the x-axis and the friction force $\mathcal{F}$ along the y-axis, starting at $T = 0$ and showing the region of no motion, the point where motion impends, and the region where motion exists.

b) A rock of weight w rests on a rough horizontal plank. The slope angle of the plank θ is gradually increased until the rock starts to slip. Draw two graphs, both with θ along the x-axis. In one graph show the ratio of the normal force to the weight n/w as a function of θ. In the second graph, show the ratio of the friction force to the weight $\mathcal{F}/w$. Indicate the region of no motion, the point where motion impends, and the region where motion exists.

2–26 A box of strawberries weighing 20 N rests on a horizontal surface. The coefficient of static friction between box and surface is 0.40 and the coefficient of sliding friction is 0.20.

a) How large is the friction force exerted on the box?

b) How great will the friction force be if a horizontal force of 5 N is exerted on the box?

c) What is the minimum force that will start the box in motion?

d) What is the minimum force that will keep the box in motion once it has been started?

e) If the horizontal force is 10 N, how great is the friction force?

2–27 A dictionary is pulled to the right at constant velocity by a 10-N force acting 30° upward at above the horizontal. The coefficient of sliding friction between the book and the surface is 0.5. What is the weight of the book?

2–28 A man pushes a board having a weight of 20 N through a circular saw by pushing down on it with a stick at an angle of 60°, as shown in Fig. 2–24. The coefficient of friction between board and saw table is 0.3.

a) What force must he exert to make the board move with constant speed?

b) Suppose the coefficient of friction is greater than 0.3; show that if it is greater than a certain critical value the man cannot push the board, no matter how hard he pushes. Find the critical value of μ_k.

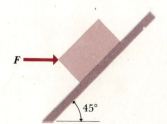

Figure 2–24

2–29 A man pushes a 200-N box of books up a 45° ramp by pushing horizontally on it, as shown in Fig. 2–25. The coefficient of kinetic friction is 0.3.

a) What force must he exert to make the box move with constant speed?

Figure 2–25

b) Suppose μ_k is greater than 0.3; show that if it exceeds a certain critical value, the man cannot push the box up the ramp, no matter how hard he pushes. Find the critical value of μ_k.

2–30 A block weighing 100 N is placed on an inclined plane of slope angle 30° and is connected to a second hanging block of weight w by a cord passing over a small frictionless pulley, as in Fig. 2–10. The coefficient of static friction is 0.40 and the coefficient of sliding friction is 0.30.

a) Find the weight w for which the 100-N block moves up the plane at constant speed.

b) Find the weight w for which it moves down the plane at constant speed.

c) For what range of values of w will the block remain at rest?

2–31 A safe weighing 2000 N is to be lowered at constant speed down skids 4 m long, from a truck 2 m high.

a) If the coefficient of sliding friction between safe and skids is 0.30, will the safe need to be pulled down or held back?

b) How great a force parallel to the skids is needed?

2–32

a) If a force of 86 N parallel to the surface of a 20° inclined plane will push a 120-N block up the plane at constant speed, what force parallel to the plane will push it down at constant speed?

b) What is the coefficient of sliding friction?

2–33 Block A in Fig. 2–26 weighs 4 N and block B weighs 8 N. The coefficient of sliding friction between all surfaces is 0.25. Find the force P necessary to drag block B to the left at constant speed

a) if A rests on B and moves with it,

b) if A is held at rest, and

c) if A and B are connected by a light flexible cord passing around a fixed frictionless pulley.

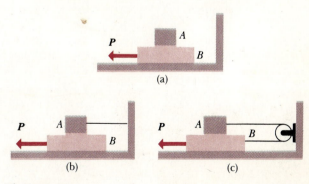

(a)

(b)   (c)

Figure 2–26

3

MOTION ON A STRAIGHT LINE

Mechanics deals with the relations of force, matter, and motion. The preceding chapters have been concerned with forces, and we are now ready to discuss mathematical methods for describing motion. This branch of mechanics is called *kinematics*.

Motion may be defined as a continuous change of position. We shall consider first a description of motion of a single point. Such a model is adequate whenever rotation and similar complications are absent, or when the body is small enough to be considered a point.

The simplest case is motion of a point along a straight line. In later chapters we shall see that more general motions in space can always be represented by means of their projections onto the three coordinate axes. Thus, while displacement is in general a vector quantity, as discussed in Chapter 1, we consider first situations in which only one component of displacement is different from zero.

3–1

AVERAGE VELOCITY

Let us consider a hockey player skating the puck down the center line of the ice toward the opposition's net. The puck moves along a straight line, which we shall use as the x-axis of our coordinate system, as shown in Fig. 3–1a. The puck's distance from the origin O at center court is given by the coordinate x, which varies with time. At time t_1 the puck is at point P, with coordinate x_1, and at time t_2 it is at point Q, where its coordinate is x_2. The displacement during the time interval from t_1 to t_2 is the vector from P to Q; the x-component of this vector is $(x_2 - x_1)$, and all other components are zero.

It is convenient to represent the quantity $x_2 - x_1$, the *change* in x, by means of a notation using the Greek letter Δ (capital delta) to designate a change in any quantity. Thus we write

$$\Delta x = x_2 - x_1, \tag{3–1}$$

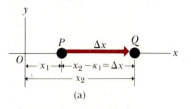

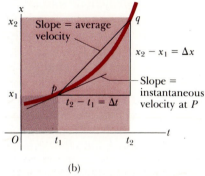

3–1 (a) Hockey puck moving on the x-axis. (b) Coordinate–time graph of the motion. The average velocity between t_1 and t_2 equals the slope of the chord pq. The instantaneous velocity at P equals the slope of the tangent at p.

in which Δx is not a product but is to be interpreted as a single symbol representing the change in the quantity x. Similarly, we denote the time interval from t_1 to t_2 as $\Delta t = t_2 - t_1$.

The *average velocity* of the particle is defined as the ratio of the displacement Δx to the time interval Δt. We represent average velocity by the letter v with a bar to signify average value. Thus

$$\bar{v} = \frac{x_2 - x_1}{t_2 - t_1} = \frac{\Delta x}{\Delta t}. \tag{3–2}$$

Strictly speaking, this defines the x-component of the average velocity, which is really a vector quantity. However in this chapter all vectors have *only* x-components, so it is not essential to distinguish between a vector and its x-component. In the following chapters we shall return to this distinction in the more general context of two-dimensional motion.

In Fig. 3–1b the coordinate x is graphed as a function of time. It is important to understand that the curve in this figure does *not* represent the path of the particle in space; as Fig. 3–1a shows, the path is a straight line. Rather, the graph provides a convenient representation of the change of position with time. The points on the coordinate–time graph corresponding to points P and Q are labeled p and q.

In Fig. 3–1b the average velocity is represented by the *slope* of the chord pq, that is, the ratio of vertical to horizontal intervals on the right triangle of which pq is the hypotenuse, evaluated using the scales and units with which x and t are plotted. That is, $\bar{v}$ is the ratio of $x_2 - x_1$ (or Δx) to $t_2 - t_1$ (or Δt).

In this discussion we have not specified whether the speed of the hockey puck is or is not constant during the time interval $\Delta t = t_2 - t_1$. It may have started from rest, reached a maximum speed, and then slowed down. But that does not matter; to calculate the average velocity we need only the total displacement $\Delta x = x_2 - x_1$ and the total time interval $\Delta t = t_2 - t_1$.

3–2

INSTANTANEOUS VELOCITY

Implicit in the above paragraph is the idea that when the velocity of a moving point varies, it is possible to define a velocity at one specific instant of time or at one specific point in the path. Such a velocity is called *instantaneous velocity*, and it needs to be defined carefully.

Suppose we wish to find the instantaneous velocity of the hockey puck in Fig. 3–1 at the point P. The average velocity between points P and Q is associated with the entire displacement Δx, and with the entire time interval Δt. Imagine the second point Q to be taken closer and closer to the first point P, and let the average velocity be computed over these shorter and shorter displacements and time intervals. The instantaneous velocity at the first point can then be defined as the value that the average velocity approaches when the second point is taken closer and closer to the first. Although the displacement then becomes extremely small, the time interval by which it is divided also becomes small, and the quotient is not necessarily a small quantity.

In mathematical language the instantaneous velocity is called the *limit* (abbreviated lim) of $\Delta x/\Delta t$ as Δt approaches zero. Thus instantaneous velocity v is given by

$$v = \lim_{\Delta t \to 0} \frac{\Delta x}{\Delta t}. \qquad (3\text{–}3)$$

Since Δt is always taken as positive, v has the same algebraic sign as Δx. Hence a positive value of v means motion in the direction of increasing x (toward the right in Fig. 3–1a), and a negative v means motion in the direction of decreasing x (toward the left in Fig. 3–1a).

Thus the sign of v always has a definite meaning with reference to the positive coordinate direction. Note, however, that a particle may have positive x and negative v; that is the case if in Fig. 3–1a the particle is located to the right of O but is moving toward the left. Similarly, a particle to the left of O but moving toward the right has negative x but positive v. Finally, a particle to the left of O and moving toward the left has both x and v negative; in this case "decreasing x" means that x is negative and is becoming *more negative* (decreasing in the algebraic sense).

Strictly speaking, Eq. (3–3) defines the x-component of the instantaneous velocity, which in more general motion must be treated as a vector quantity. However, in one-dimensional motion along a straight line, it is customary to call v the instantaneous velocity. When the term velocity is used without a qualifying adjective, one nearly always means *instantaneous* rather than *average* velocity.

As point Q approaches point P in Fig. 3–1a, point q approaches point p in Fig. 3–1b. In the limit, the slope of the chord pq equals the slope of the tangent to the curve at point p, due allowance being made for the scales to which x and t are plotted. *The instantaneous velocity at any point of a coordinate–time graph therefore equals the slope of the tangent to the graph at that point.* If the tangent slopes upward to the right, its slope is positive, the velocity is positive, and the motion is toward the right. If the tangent slopes downward to the right, the velocity is negative. At a point where the tangent is horizontal, its slope is zero and the velocity is zero.

If we express distance in meters and time in seconds, velocity is expressed in meters per second ($\text{m} \cdot \text{s}^{-1}$). Other common units of velocity are feet per second ($\text{ft} \cdot \text{s}^{-1}$), centimeters per second ($\text{cm} \cdot \text{s}^{-1}$), miles per hour ($\text{mi} \cdot \text{hr}^{-1}$), and kilometers per hour ($\text{km} \cdot \text{hr}^{-1}$).

EXAMPLE A cheetah is crouched in ambush 20 m to the north of an observer's blind (see Fig. 3–2). At time $t = 0$, the cheetah charges an antelope in a clearing 50 m north of the observer. The observer estimates that during the first 2 s of the attack, the cheetah's motion is described by the equation $x = a + bt^2$, where $a = 20$ m and $b = 5 \text{ m} \cdot \text{s}^{-2}$.

a) Find the displacement of the cheetah in the time interval between $t_1 = 1$ s and $t_2 = 2$ s.

At time $t_1 = 1$ s, the cheetah's position is

$$x_1 = 20 \text{ m} + (5 \text{ m} \cdot \text{s}^{-2})\,(1 \text{ s})^2 = 25 \text{ m}.$$

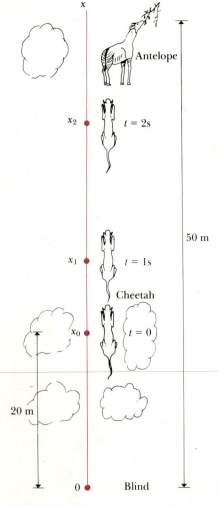

3–2 A cheetah attacking an antelope from ambush.

At time $t_2 = 2$ s,

$$x_2 = 20 \text{ m} + (5 \text{ m} \cdot \text{s}^{-2}) (2 \text{ s})^2 = 40 \text{ m}.$$

The displacement is therefore

$$x_2 - x_1 = 40 \text{ m} - 25 \text{ m} = 15 \text{ m}.$$

b) Find the average velocity in this time interval:

$$\bar{v} = \frac{x_2 - x_1}{t_2 - t_1} = \frac{15 \text{ m}}{1 \text{ s}} = 15 \text{ m} \cdot \text{s}^{-1}.$$

c) Find the instantaneous velocity at time $t_1 = 1$ s. The position at time $t = 1 \text{ s} + \Delta t$ is

$$x = 20 \text{ m} + (5 \text{ m} \cdot \text{s}^{-2}) (1 \text{ s} + \Delta t)^2$$
$$= 25 \text{ m} + (10 \text{ m} \cdot \text{s}^{-1})\Delta t + (5 \text{ m} \cdot \text{s}^{-2}) (\Delta t)^2.$$

The displacement during the interval Δt is

$$\Delta x = 25 \text{ m} + (10 \text{ m} \cdot \text{s}^{-1})\Delta t + (5 \text{ m} \cdot \text{s}^{-2}) (\Delta t)^2 - 25 \text{ m}$$
$$= (10 \text{ m} \cdot \text{s}^{-1})\Delta t + (5 \text{ m} \cdot \text{s}^{-2}) (\Delta t)^2.$$

The average velocity during Δt is

$$\bar{v} = \frac{\Delta x}{\Delta t} = 10 \text{ m} \cdot \text{s}^{-1} + (5 \text{ m} \cdot \text{s}^{-2}) \Delta t.$$

For the instantaneous velocity at $t = 2$ s, we let Δt approach zero in the expression for $\bar{v}$: $v = 10 \text{ m} \cdot \text{s}^{-1}$. This corresponds to the slope of the tangent at point p in Fig. 3–1b. ◀

The term *speed* has two different meanings. It may mean the *magnitude* of the instantaneous velocity; two automobiles traveling at $50 \text{ mi} \cdot \text{hr}^{-1}$, one north and the other south, both have a speed of $50 \text{ mi} \cdot \text{hr}^{-1}$. In another sense, the *average speed* of a body means the total length of path covered, divided by the elapsed time. Thus, if an automobile travels 90 mi in 3 hr, its average speed is $30 \text{ mi} \cdot \text{hr}^{-1}$, even if the trip starts and ends at the same point. The average *velocity*, in the latter case, would be zero, since the total displacement is zero.

3–3

AVERAGE AND INSTANTANEOUS ACCELERATION

When the velocity of a moving body changes with time, the body is said to have an *acceleration*. Just as velocity is a quantitative description of rate of change of position with time, so acceleration is a quantitative description of rate of change of velocity with time.

Considering again the motion of a particle along the *x*-axis, we suppose that at time t_1 the particle is at point P and has velocity v_1, and that at a later time t_2 it is at point Q and has the velocity v_2.

The *average acceleration* $\bar{a}$ of the particle as it moves from P to Q is defined as the ratio of the change in velocity to the elapsed time.

$$\bar{a} = \frac{v_2 - v_1}{t_2 - t_1} = \frac{\Delta v}{\Delta t}, \qquad (3\text{--}4)$$

where t_1 and t_2 are the times corresponding to the velocities v_1 and v_2. Again, strictly speaking v_1 and v_2 are values of the *x-component* of instantaneous velocity, and Eq. (3–4) defines the x-component of average acceleration.

EXAMPLE 1   An astronaut has left the space shuttle on a tether line to test a new personal maneuvering device. Her on-board partner measures her velocity before and after certain maneuvers, as follows:

a) $v_1 = 0.8 \text{ m} \cdot \text{s}^{-1}$, $v_2 = 1.2 \text{ m} \cdot \text{s}^{-1}$;

b) $v_1 = 1.6 \text{ m} \cdot \text{s}^{-1}$, $v_2 = 1.2 \text{ m} \cdot \text{s}^{-1}$;

c) $v_1 = -0.4 \text{ m} \cdot \text{s}^{-1}$, $v_2 = -1.0 \text{ m} \cdot \text{s}^{-1}$;

d) $v_1 = -1.6 \text{ m} \cdot \text{s}^{-1}$, $v_2 = -0.8 \text{ m} \cdot \text{s}^{-1}$.

If $t_1 = 2$ s and $t_2 = 4$ s, find the average acceleration for each set of data.

Solution

a) $\bar{a} = \dfrac{1.2 \text{ m} \cdot \text{s}^{-1} - 0.8 \text{ m} \cdot \text{s}^{-1}}{4 \text{ s} - 2 \text{ s}} = 0.2 \text{ m} \cdot \text{s}^{-2}$.

b) $\bar{a} = \dfrac{1.2 \text{ m} \cdot \text{s}^{-1} - 1.6 \text{ m} \cdot \text{s}^{-1}}{4 \text{ s} - 2 \text{ s}} = -0.2 \text{ m} \cdot \text{s}^{-2}$.

c) $\bar{a} = \dfrac{-1.0 \text{ m} \cdot \text{s}^{-1} - (-0.4 \text{ m} \cdot \text{s}^{-1})}{4 \text{ s} - 2 \text{ s}} = -0.3 \text{ m} \cdot \text{s}^{-2}$.

d) $\bar{a} = \dfrac{-0.8 \text{ m} \cdot \text{s}^{-1} - (-1.6 \text{ m} \cdot \text{s}^{-1})}{4 \text{ s} - 2 \text{ s}} = +0.4 \text{ m} \cdot \text{s}^{-2}$.

We note in particular that when the point moves in the negative direction with increasing speed (c), the acceleration is negative, and that when it moves in the negative direction with decreasing speed (d), the acceleration is positive.   ◀

We can now define *instantaneous acceleration*, following the same procedure used for instantaneous velocity. Consider this situation: A sport car driver has just entered the final straightaway at the Grand Prix. He reaches point P at time t_1, moving with velocity v_1, and crosses the finish line at point Q at time t_2 with velocity v_2, as shown in Fig. 3–3a. Figure 3–3b is a graph of instantaneous velocity v plotted as a function of time, points p and q corresponding to positions P and Q in Fig. 3–3a. The average acceleration is represented by the slope of the chord pq, computed using the appropriate scales and units of the graph.

The *instantaneous acceleration* of a body, that is, its acceleration at some one instant of time or at some one point of its path, is defined in the same way as instantaneous velocity. Let the second point Q in Fig.

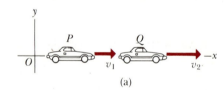

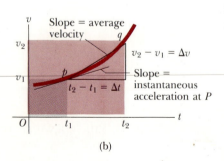

3–3   (a) Car moving on the x-axis. (b) Velocity–time graph of the motion. The average acceleration between t_1 and t_2 equals the slope of the chord pq. The instantaneous acceleration at P equals the slope of the tangent at p.

3–3a be taken closer and closer to the first point P, and let the average acceleration be computed over shorter and shorter intervals of time. The instantaneous acceleration at the first point is defined as the limiting value of the average acceleration when the second point is taken closer and closer to the first:

$$a = \lim_{\Delta t \to 0} \frac{\Delta v}{\Delta t}. \tag{3–5}$$

Instantaneous acceleration plays an important part in the laws of mechanics; average acceleration is less frequently used. From now on when the term "acceleration" is used we shall understand it to mean instantaneous acceleration.

When a particle moves in a curved path, the *direction* of its velocity changes, and this change in direction also gives rise to an acceleration, as will be explained in Chapter 5.

As point Q approaches point P in Fig. 3–3a, point q approaches point p in Fig. 3–3b, and the slope of the chord pq approaches the slope of the tangent to the velocity–time graph at point p. *The instantaneous acceleration at any point of the graph therefore equals the slope of the tangent to the graph at that point.*

If we express velocity in meters per second and time in seconds, acceleration is expressed in meters per second, per second ($m \cdot s^{-1} \cdot s^{-1}$). This is usually written as $m \cdot s^{-2}$, and is read "meters per second squared." Other common units of acceleration are feet per second squared ($ft \cdot s^{-2}$) and centimeters per second squared ($cm \cdot s^{-2}$).

EXAMPLE 2   Suppose the velocity of the car in Fig. 3–3 is given by the equation

$$v = m + nt^2,$$

where $m = 10 \ m \cdot s^{-1}$ and $n = 2 \ m \cdot s^{-3}$.

a) Find the change in velocity of the car in the time interval between $t_1 = 2 \ s$ and $t_2 = 5 \ s$.

   At time $t_1 = 2 \ s$,

$$v_1 = 10 \ m \cdot s^{-1} + (2 \ m \cdot s^{-3}) \ (2 \ s)^2$$
$$= 18 \ m \cdot s^{-1}.$$

At time $t_2 = 5 \ s$,

$$v_2 = 10 \ m \cdot s^{-1} + (2 \ m \cdot s^{-3}) \ (5 \ s)^2$$
$$= 60 \ m \cdot s^{-1}.$$

The change in velocity is therefore

$$v_2 - v_1 = 60 \ m \cdot s^{-1} - 18 \ m \cdot s^{-1}$$
$$= 42 \ m \cdot s^{-1}.$$

b) Find the average acceleration in this time interval.

$$\bar{a} = \frac{v_2 - v_1}{t_2 - t_1} = \frac{42 \ m \cdot s^{-1}}{3 \ s} = 14 \ m \cdot s^{-2}.$$

This corresponds to the slope of the chord pq in Fig. 3–3b.

c) Find the instantaneous acceleration at time $t_1 = 2$ s.
At time $t = 2$ s $+ \Delta t$,

$$v = 10 \text{ m} \cdot \text{s}^{-1} + (2 \text{ m} \cdot \text{s}^{-3}) (2 \text{ s} + \Delta t)^2$$
$$= 18 \text{ m} \cdot \text{s}^{-1} + (8 \text{ m} \cdot \text{s}^{-2})\Delta t + (2 \text{ m} \cdot \text{s}^{-3}) (\Delta t)^2.$$

The change in velocity during Δt is

$$\Delta v = 18 \text{ m} \cdot \text{s}^{-1} + (8 \text{ m} \cdot \text{s}^{-2})\Delta t + (2 \text{ m} \cdot \text{s}^{-3}) (\Delta t)^2 - 18 \text{ m} \cdot \text{s}^{-1}$$
$$= (8 \text{ m} \cdot \text{s}^{-2})\Delta t + (2 \text{ m} \cdot \text{s}^{-3}) (\Delta t)^2.$$

The average acceleration during Δt is

$$\bar{a} = \frac{\Delta v}{\Delta t} = 8 \text{ m} \cdot \text{s}^{-2} + (2 \text{ m} \cdot \text{s}^{-3})\Delta t.$$

The instantaneous acceleration at $t = 2$ s, obtained by letting Δt approach zero, is $a = 8 \text{ m} \cdot \text{s}^{-2}$. This corresponds to the slope of the tangent at point p in Fig. 3–3b. ◄

A few remarks about the *sign* of acceleration may be helpful. When the acceleration and velocity of a body have the same sign, the body is speeding up. If both are positive, it moves in the positive direction with increasing speed. If both are negative, the body moves in the negative direction with a velocity that becomes more and more negative with time, and again the body's speed increases.

When v and a have opposite signs, the body is slowing down. If v is positive and a negative, the body moves in the positive direction with decreasing speed. If v is negative and a positive, the body moves in the negative direction with a velocity that is becoming less negative, and again the body slows down.

The term *deceleration* is sometimes used, either for a negative value of a or for a decrease in speed. Because of the possible ambiguity, this term is best avoided; it is not used in this book. Instead, we recommend careful attention to the interpretation of the algebraic sign of a in relation to that of v.

3–4

MOTION WITH CONSTANT ACCELERATION

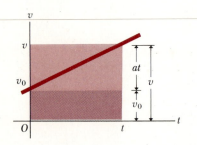

3–4 Velocity–time graph for rectilinear motion with constant acceleration.

The simplest kind of accelerated motion is straight-line motion in which the acceleration is constant, that is, in which the velocity changes at the same rate throughout the motion. The velocity–time graph is then a straight line, as in Fig. 3–4, the velocity increasing by equal amounts in equal intervals of time. The slope of a chord between any two points on the line is the same as the slope of a tangent at any point, and the average and instantaneous accelerations are equal. Hence in Eq. (3–4) the average acceleration $\bar{a}$ can be replaced by the constant acceleration a, and we have

$$a = \frac{v_2 - v_1}{t_2 - t_1}. \qquad (3\text{–}6)$$

Now let $t_1 = 0$ and let t_2 be any arbitrary later time t. Let v_0 represent the velocity when $t = 0$ (called the *initial* velocity), and let v be the velocity at the later time t. Then the preceding equation becomes

$$a = \frac{v - v_0}{t - 0},$$

or

$$v = v_0 + at. \tag{3–7}$$

This equation can be interpreted as follows: The acceleration a is the constant rate of change of velocity, or the change per unit time. The term at is the product of the change in velocity per unit time, a, and the duration of the time interval, t; therefore it equals the *total* change in velocity. The velocity v at the time t then equals the velocity v_0 at the time $t = 0$, plus the change in velocity at. Graphically, the ordinate v at time t, in Fig. 3–4, can be considered as the sum of two segments: one of length v_0 equal to the initial velocity, the other of length at equal to the change in velocity in time t.

To find the *displacement* of a particle moving with constant acceleration, we make use of the fact that when the acceleration is constant and the velocity–time graph is a straight line, as in Fig. 3–4, the average velocity in any time interval equals the average of the velocities at the beginning and the end of the interval. Hence the average velocity between zero and t is

$$\bar{v} = \frac{v_0 + v}{2}. \tag{3–8}$$

(This is *not* true in general, when the acceleration is not constant and the velocity–time graph is curved, as in Fig. 3–3b.)

By definition, the average velocity is

$$\bar{v} = \frac{x_2 - x_1}{t_2 - t_1}.$$

Now let $t_1 = 0$ and let t_2 be any arbitrary time t. Let x_0 represent the position when $t = 0$ (the *initial* position) and let x be the position at time t. Then the preceding equation becomes

$$x - x_0 = \bar{v}t. \tag{3–9}$$

Substituting the expression for $\bar{v}$ in Eq. (3–8) into Eq. (3–9), we obtain

$$x - x_0 = \left(\frac{v_0 + v}{2}\right)t. \tag{3–10}$$

Two more very useful equations can be obtained from Eqs. (3–7) and (3–10), first by eliminating v and then by eliminating t. When we substitute into Eq. (3–10) the expression for v in Eq. (3–7), we find

$$x - x_0 = \left(\frac{v_0 + v_0 + at}{2}\right)t,$$

or

$$x = x_0 + v_0 t + \tfrac{1}{2}at^2. \tag{3–11}$$

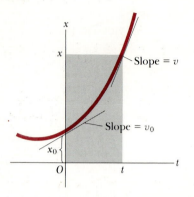

3–5 Coordinate–time graph for motion with constant acceleration.

When Eq. (3–7) is solved for t and the result substituted in Eq. (3–10), we have

$$x - x_0 = \left(\frac{v_0 + v}{2}\right)\left(\frac{v - v_0}{a}\right) = \frac{v^2 - v_0^2}{2a},$$

or, finally,

$$v^2 = v_0^2 + 2a(x - x_0). \tag{3–12}$$

Equations (3–7), (3–11), and (3–12) are the *equations of motion with constant acceleration*.

The curve in Fig. 3–5 is the coordinate–time graph for motion with constant acceleration. That is, it is a graph of Eq. (3–11); the curve is a *parabola*. The slope of the tangent at $t = 0$ equals the initial velocity v_0, and the slope of the tangent at time t equals the velocity v at that time. It is evident that the slope continually increases, and measurements would show that the *rate* of increase is constant, that is, that the acceleration is constant.

EXAMPLE   A motorcyclist passing through a small Iowa town accelerates as he passes the signpost at $x = 0$ marking the city limits. His acceleration is constant, $a = 4$ m·s^{-2}. At time $t = 0$ he is at $x = 5$ m and has velocity $v = 3$ m·s^{-1}.

a) Find his position and velocity at time $t = 2$ s.

b) Where is he when his velocity is 5 m·s^{-1}?

Solution   With reference to Eqs. (3–7), (3–11), and (3–12), we have $x_0 = 5$ m, $v_0 = 3$ m·s^{-1}, $a = 4$ m·s^{-2}.

a) From Eq. (3–11),

$$x = x_0 + v_0 t + \tfrac{1}{2}at^2$$
$$= 5 \text{ m} + (3 \text{ m·s}^{-1})(2 \text{ s}) + \tfrac{1}{2}(4 \text{ m·s}^{-2})(2 \text{ s})^2$$
$$= 19 \text{ m}.$$

From Eq. (3–7),

$$v = v_0 + at$$
$$= 3 \text{ m·s}^{-1} + (4 \text{ m·s}^{-2})(2 \text{ s})$$
$$= 11 \text{ m·s}^{-1}.$$

b) From Eq. (3–12),

$$v^2 = v_0^2 + 2a(x - x_0),$$
$$(5 \text{ m·s}^{-1})^2 = (3 \text{ m·s}^{-1})^2 + 2(4 \text{ m·s}^{-2})(x - 5 \text{ m}),$$
$$x = 7 \text{ m}.$$

Alternatively, we may use Eq. (3–7) to find first the *time* when $v = 5$ m·s^{-1}:

$$5 \text{ m·s}^{-1} = 3 \text{ m·s}^{-1} + (4 \text{ m·s}^{-2})(t),$$
$$t = \tfrac{1}{2} \text{ s}.$$

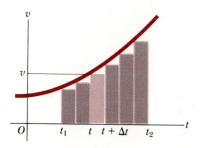

3–6   The area under a velocity–time graph equals the displacement.

Then from Eq. (3–11),

$$x = 5 \text{ m} + (3 \text{ m·s}^{-1})\,(\tfrac{1}{2}\text{ s}) + \tfrac{1}{2}(4 \text{ m·s}^{-2})\,(\tfrac{1}{2}\text{ s})^2$$
$$= 7 \text{ m.}$$

A special case of motion with constant acceleration is that in which the acceleration is zero. The *velocity* is then constant and the equations of motion become simply

$$v = v_0 = \text{constant,}$$

$$x = x_0 + vt.$$

When the acceleration is not constant, the preceding equations are not applicable. Even so, some of the same graphical considerations are useful; velocity is still the slope of the displacement-versus-time graph, and acceleration the slope of the velocity-versus-time graph. Figure 3–6 shows a motion in which the acceleration is not constant. We divide the total time into a large number of intervals, calling a typical one Δt. Let the velocity during that interval be v. The velocity changes slightly during Δt, but if the interval is very small, the change will also be very small. The displacement during that interval, neglecting the variation of v, is given by

$$\Delta x = v\,\Delta t.$$

This corresponds graphically to the *area* of the shaded strip with height v and width Δt, that is, the *area* under the curve corresponding to the interval Δt. Since the total displacement in any interval (say, t_1 to t_2) is the sum of the displacements in the small subintervals, this is given graphically by the *total* area under the curve between the vertical lines t_1 and t_2. This area may be obtained in various ways; the most obvious one is to draw the curve on graph paper and count squares, but there are various more subtle ways.

As an example of the application of this method, we consider again the case of constant acceleration, for which the velocity–time graph is Fig. 3–4. The total displacement $(x - x_0)$ from $t = 0$ to a later time t is given by the total area under the line, which can be obtained by computing the rectangular and triangular areas and adding (see Fig. 3–7). Carrying out this program, we find:

$$x - x_0 = v_0 t + \tfrac{1}{2}(t)(at) = v_0 t + \tfrac{1}{2}at^2,$$

in agreement with Eq. (3–11).

In problems with constant acceleration, several points are worth keeping in mind:

1. It is essential to decide at the beginning of a problem where the origin of coordinates is and which axis direction is positive; a diagram showing these choices is always helpful.

2. Once the positive axis direction has been chosen, the positive directions for velocity and acceleration are also determined. It would be inconsistent to define x as positive to the right of the origin and velocities toward the left as positive.

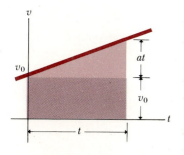

3–7   The area under a velocity–time graph equals the displacement $x - x_0$.

3. It often helps to express questions in prose. *When* does the particle arrive at a certain point? (I.e., at what value of *t*?) *Where* is the particle when its velocity has a certain value? (I.e., what is the value of *x*?) And so on. These questions can then be translated into equations that can be solved to give the required results.

3–5

FREELY FALLING BODIES

The most common example of motion with (nearly) constant acceleration is that of a body falling toward the earth. The motion of a falling body has attracted the attention of philosophers and scientists since ancient times. Aristotle asserted in his writings that heavy objects fall faster than light objects, in proportion to their weight. Galileo argued that the motion of a falling body should be nearly independent of its weight. According to legend, he tested his reasoning experimentally by dropping cannonballs and bullets from the Leaning Tower of Pisa, although there is no reference to such experiments in his own writing.

In any event, the motion of falling bodies has been studied with great care and precision by scientists in more recent times. It is found that when air resistance can be neglected all bodies, regardless of their size or weight, fall with the same acceleration at a particular point on the earth's surface; and if the distance fallen is small compared to the radius of the earth, the acceleration remains constant throughout the fall. In the following discussion the effect of air resistance and the decrease in acceleration with altitude will be neglected. This idealized motion is spoken of as "free fall," although the term includes rising as well as falling motion.

The acceleration of a freely falling body is called *acceleration due to gravity*, or the *acceleration of gravity*, and is denoted by the letter *g*. At or near the earth's surface its magnitude is approximately $32 \text{ ft} \cdot \text{s}^{-2}$, $9.8 \text{ m} \cdot \text{s}^{-2}$, or $980 \text{ cm} \cdot \text{s}^{-2}$. More precise values, and small variations with latitude and elevation, will be considered later. On the surface of the moon, the acceleration of gravity is due to the attractive force exerted on a body by the moon rather than Earth. On the moon, $g = 1.67 \text{ m} \cdot \text{s}^{-2} = 5.47 \text{ ft} \cdot \text{s}^{-2}$. Near the surface of the *sun*, $g = 274 \text{ m} \cdot \text{s}^{-2}$!

EXAMPLE 1   A bullet is dropped from the Leaning Tower of Pisa, from rest, and falls freely. Compute its position and velocity after 1, 2, 3, and 4 s. Take the origin 0 at the elevation of the starting point, the *y*-axis vertical, and the upward direction as positive.

Solution   The initial coordinate y_0 and the initial velocity v_0 are both zero. The acceleration is downward, in the negative *y*-direction, so $a = -g = -32 \text{ ft} \cdot \text{s}^{-2}$.

From Eqs. (3–11) and (3–7),

$$y = v_0 t + \tfrac{1}{2}at^2 = 0 - \tfrac{1}{2}gt^2 = (-16 \text{ ft} \cdot \text{s}^{-2})t^2,$$

$$v = v_0 + at = 0 - gt = (-32 \text{ ft} \cdot \text{s}^{-2})t.$$

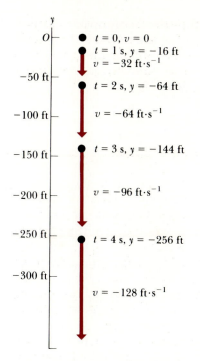

3–8 Position and velocity of a freely falling body.

When $t = 1$ s, $y = (-16 \text{ ft·s}^{-2})(1\text{ s})^2 = -16$ ft, and $v = (-32 \text{ ft·s}^{-2})(1\text{ s}) = -32 \text{ ft·s}^{-1}$. The body is therefore 16 ft below the origin (y is negative) and has a downward velocity (v is negative) of magnitude 32 ft·s^{-1}.

The position and velocity at 2, 3, and 4 s are found in the same way. The results are illustrated in Fig. 3–8; the student should check the numerical values. ◄

EXAMPLE 2 A ball is thrown (nearly) vertically upward from the cornice of a tall building, leaving the thrower's hand with a speed of 48 ft·s^{-1} and just missing the cornice on the way down, as in Fig. 3–9. The downward path is shown displaced to the right for clarity. Find (a) the position and velocity of the ball, 1 s and 4 s after leaving the thrower's hand; (b) the velocity when the ball is 20 ft above its starting point; (c) the maximum height reached and the time at which it is reached. Take the origin at the elevation at which the ball leaves the thrower's hand, the y-axis vertical and positive upward.

Solution The initial position y_0 is zero. The initial velocity v_0 is $+48$ ft·s^{-1}, and the acceleration is -32 ft·s^{-2}. The velocity at any time is

$$v = v_0 + at = 48 \text{ ft·s}^{-1} + (-32 \text{ ft·s}^{-2})t. \qquad (3\text{–}13)$$

The position at any time is

$$y = v_0 t + \tfrac{1}{2}at^2 = (48 \text{ ft·s}^{-1})t + \tfrac{1}{2}(-32 \text{ ft·s}^{-2})t^2. \qquad (3\text{–}14)$$

The velocity at any position is

$$v^2 = v_0{}^2 + 2ay = (48 \text{ ft·s}^{-1})^2 + 2(-32 \text{ ft·s}^{-2})y. \qquad (3\text{–}15)$$

a) When $t = 1$ s, Eqs. (3–13) and (3–14) give

$$y = +32 \text{ ft}, \qquad v = +16 \text{ ft·s}^{-1}.$$

The ball is 32 ft above the origin (y is positive) and it has an upward velocity (v is positive) of 16 ft·s^{-1} (less than the initial velocity, as expected).

When $t = 4$ s, again from Eqs. (3–13) and (3–14),

$$y = -64 \text{ ft}, \qquad v = -80 \text{ ft·s}^{-1}.$$

The ball has passed its highest point and is 64 ft *below* the origin (y is negative). It has a *downward* velocity (v is negative) of magnitude 80 ft·s^{-1}. Note that it is not necessary to find the highest point reached, or the time at which it was reached. The equations of motion give the position and velocity at *any* time, whether the ball is on the way up or the way down.

b) When the ball is 20 ft above the origin

$$y = +20 \text{ ft}$$

and, from Eq. (3–15),

$$v^2 = 1024 \text{ ft}^2\text{·s}^{-2}, \qquad v = \pm 32 \text{ ft·s}^{-1}.$$

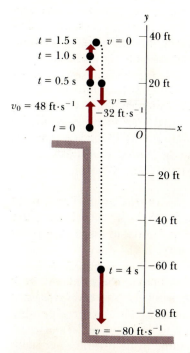

3–9 Position and velocity of a body thrown vertically upward.

The ball passes this point *twice*, once on the way up and again on the way down. The velocity on the way up is $+32 \text{ ft} \cdot \text{s}^{-1}$, and on the way down it is $-32 \text{ ft} \cdot \text{s}^{-1}$.

c) At the highest point, $v = 0$. Hence, from Eq. (3–15),

$$0 = (48 \text{ ft} \cdot \text{s}^{-1})^2 - (64 \text{ ft} \cdot \text{s}^{-2})y,$$

and

$$y = +36 \text{ ft}.$$

The time can now be found from Eq. (3–13), setting $v = 0$:

$$0 = 48 \text{ ft} \cdot \text{s}^{-1} + (-32 \text{ ft} \cdot \text{s}^{-2})t,$$

$$t = 1.5 \text{ s}.$$

Alternatively, to find the maximum height we may ask first *when* the maximum height is reached. That is, at what t is $v = 0$? As just shown, $v = 0$ when $t = 1.5$ s; substituting this value of t back into Eq. (3–14), we find

$$y = (48 \text{ ft} \cdot \text{s}^{-1})(1.5 \text{ s}) + \tfrac{1}{2}(-32 \text{ ft} \cdot \text{s}^{-2})(1.5 \text{ s})^2$$
$$= 36 \text{ ft}.$$

Although at the highest point the velocity is instantaneously zero, the *acceleration* at this point is still $-32 \text{ ft} \cdot \text{s}^{-2}$. The ball stops for an instant, but its velocity is continuously changing; the acceleration is constant throughout. ◀

Figure 3–10 is a "multiflash" photograph of a freely falling golf ball. This photograph was taken with aid of a stroboscopic light source, developed originally by Dr. Harold E. Edgerton of the Massachusetts Institute of Technology. This source produces a series of intense flashes of light. The interval between successive flashes is controllable at will, and the duration of each flash is so short (a few millionths of a second) that there is no blur in the image of even a rapidly moving body. The camera shutter is left open during the entire motion, and as each flash occurs, the position of the ball at that instant is recorded on the photographic film.

The equally spaced light flashes subdivide the motion into equal time intervals Δt. Since the time intervals are all equal, the velocity of the ball between any two flashes is directly proportional to the separation of its corresponding images in the photograph. If the velocity were constant, the images would be equally spaced. The increasing separation of the images during the fall shows that the velocity is continually increasing or the motion is accelerated. By comparing two successive displacements of the ball, we can find the *change* in velocity in the corresponding time interval. Careful measurement, preferably on an enlarged print, shows that this change in velocity is the same in each time interval. In other words, the motion is one of *constant* acceleration.

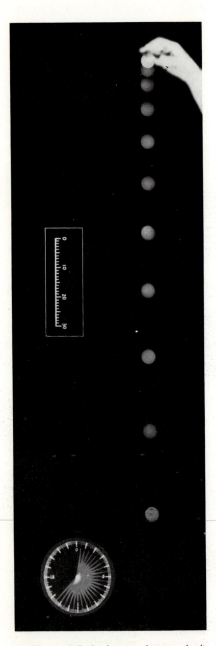

3–10 Multiflash photograph (retouched) of freely falling golf ball.

3–6

RELATIVE VELOCITY

The velocity of a body, like its position, must be described with reference to the origin of a coordinate system. Ordinarily this origin is taken to be fixed in some other body, but this second body may be in motion relative to a third, and so on. Thus when we speak of "the velocity of an automobile," we usually mean its velocity relative to the earth. But the earth is in motion relative to the sun, the sun is in motion relative to some other star, and so on.

Suppose a long train of flatcars is moving to the right along a straight level track, as in Fig. 3–11, and that a daring automobile driver is driving to the right along the flatcars. In Fig. 3–11, v_{FE} represents the velocity of the flatcars F relative to the earth E, and v_{AF} the velocity of the automobile A relative to the flatcars F. The velocity of the automobile relative to the earth, v_{AE}, is evidently equal to the *sum* of the relative velocities v_{AF} and v_{FE}:

$$v_{AE} = v_{AF} + v_{FE}. \tag{3–16}$$

Thus if the flatcars are traveling relative to the earth at $30 \ \text{km} \cdot \text{hr}^{-1} (= v_{FE})$ and the automobile is traveling relative to the flatcars at $40 \ \text{km} \cdot \text{hr}^{-1} (= v_{AF})$, the velocity of the automobile relative to the earth (v_{AE}) is $70 \ \text{km} \cdot \text{hr}^{-1}$.

We note that in straight-line motion the velocity v_{AE} is the *algebraic* sum of v_{AF} and v_{FE}. Thus if the automobile were traveling to the *left* with a velocity of $40 \ \text{km} \cdot \text{hr}^{-1}$ relative to the flatcars, $v_{AF} = -40 \ \text{km} \cdot \text{hr}^{-1}$ and the velocity of the automobile relative to the earth would be $-10 \ \text{km} \cdot \text{hr}^{-1}$; that is, it would be traveling to the left, relative to the earth.

EXAMPLE 1   A man is driving a car at $65 \ \text{km} \cdot \text{hr}^{-1}$ on a straight level road where the speed limit is $40 \ \text{km} \cdot \text{hr}^{-1}$. He is spotted by a motorcycle officer, who accelerates in pursuit. By the time the driver sees the motorcycle's flashing blue light, the motorcycle is traveling at $80 \ \text{km} \cdot \text{hr}^{-1}$. What is the motorcycle's velocity relative to the car?

Solution   Let the car be C, the motorcycle M, and the earth E. Then

$$v_{CE} = 65 \ \text{km} \cdot \text{hr}^{-1} \qquad \text{and} \qquad v_{ME} = 80 \ \text{km} \cdot \text{hr}^{-1},$$

and we wish to find v_{MC}.

3–11   Vector v_{FE} is the velocity of the flatcars F relative to the earth E, and v_{AF} is the velocity of automobile A relative to the flatcars.

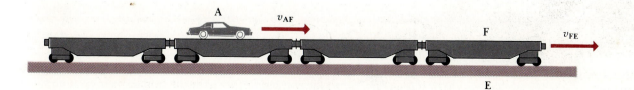

From the rule for combining velocities,

$$v_{ME} = v_{MC} + v_{CE}.$$

Thus

$$80 \text{ km} \cdot \text{hr}^{-1} = v_{MC} + 65 \text{ km} \cdot \text{hr}^{-1},$$

$$v_{MC} = 15 \text{ km} \cdot \text{hr}^{-1},$$

and the officer is overtaking the driver at $15 \text{ km} \cdot \text{hr}^{-1}$. ◀

EXAMPLE 2   How would the relative velocity be altered if the motorcycle were ahead of the car?

Solution   Not at all. The relative *positions* of the bodies do not matter. The velocity of M relative to C is still $+15 \text{ km} \cdot \text{hr}^{-1}$, but he is now pulling ahead of C at this rate. ◀

In more general motion, where the velocities do not all lie along a straight line, vector addition must be used to combine velocities. We shall return to problems of this sort at the end of Chapter 5.

QUESTIONS

3–1 When a particle moves in a circular path, how many coordinates are required to describe its position, assuming that the radius of the circle is given?

3–2 What is meant by the statement that space is three-dimensional?

3–3 Does the speedometer of a car measure speed or velocity?

3–4 Some European countries have highway speed limits of $100 \text{ km} \cdot \text{hr}^{-1}$. What is the equivalent number of miles per hour?

3–5 A student claims that a speed of $60 \text{ mi} \cdot \text{hr}^{-1}$ is the same as $88 \text{ ft} \cdot \text{s}^{-1}$. Is this relationship exact or only approximate?

3–6 In a given time interval, is the total displacement of a particle equal to the product of the average velocity and the time interval, even when the velocity is not constant?

3–7 Under what conditions is average velocity equal to instantaneous velocity?

3–8 When one flies in an airplane at night in smooth air, there is no sensation of motion even though the plane may be moving at $500 \text{ mi} \cdot \text{hr}^{-1}$. Why is this?

3–9 An automobile is traveling north. Can it have a velocity toward the north and at the same time have an acceleration toward the south? Under what circumstances?

3–10 A ball is thrown straight up in the air. What is its acceleration at the instant it reaches its highest point?

3–11 Is the acceleration of a car greater when the accelerator is pushed to the floor or when the brake pedal is pushed hard?

3–12 Under constant acceleration, the average velocity of a particle is half the sum of its initial and final velocities. Is this still true if the acceleration is *not* constant?

3–13 A baseball is thrown straight up in the air. Is the acceleration greater while it is being thrown, or after it is thrown?

3–14 How could you measure the acceleration of an automobile using only instruments located within the automobile?

3–15 If the initial position and initial velocity of a vehicle are known, and a record is kept of the acceleration at each instant, can its position after a certain time be computed from this data? Explain how this might be done.

PROBLEMS

3–1 In 1954 Roger Bannister became the first human to run a mile in less than 4 minutes. Suppose that a runner on a straight track covers a distance of 1 mile in exactly 4 minutes. What was his average velocity in (a) $\text{mi} \cdot \text{hr}^{-1}$? (b) $\text{ft} \cdot \text{s}^{-1}$? (c) $\text{cm} \cdot \text{s}^{-1}$?

3–2 A body moves along a straight line, its distance from the origin at any instant being given by the equation $x = 8t - 3t^2$, where x is in centimeters and t is in seconds. Find the average velocity of the body in the interval from $t = 0$ to $t = 1$ s, and in the interval from $t = 0$ to $t = 4$ s.

3–3 A hobbyist is testing a new model rocket engine by using it to propel a cart along a model railroad track. He determines that its motion along the x-axis is described by the equation $x = (10 \text{ cm} \cdot \text{s}^{-2})t^2$. Compute the instantaneous velocity of the body at time $t = 3$ s. Let Δt first equal 0.1 s, then 0.01 s, and finally 0.001 s. What limiting value do the results seem to be approaching?

3–4 A test driver at Incredible Motors, Inc., is testing a new-model car having a speedometer calibrated to read $\text{m} \cdot \text{s}^{-1}$ rather than $\text{mi} \cdot \text{hr}^{-1}$. The following series of speedometer readings was obtained during a test run.

Time (s)	0	2	4	6	8	10	12	14	16
Velocity ($\text{m} \cdot \text{s}^{-1}$)	0	0	2	5	10	15	20	22	22

a) Compute the average acceleration during each 2-s interval. Is the acceleration constant? Is it constant during any part of the time?

b) Make a velocity–time graph of the data above, using scales of 1 cm = 1 s horizontally, and 1 cm = 2 $\text{m} \cdot \text{s}^{-1}$ vertically. Draw a smooth curve through the plotted points. What distance is represented by 1 cm²? What is the displacement in the first 8 s? What is the acceleration when $t = 8$ s? When $t = 13$ s? When $t = 15$ s?

3–5 The graph in Fig. 3–12 shows the velocity of a motorcycle police officer plotted as a function of time.

a) Find the instantaneous acceleration at $t = 3$ s, at $t = 7$ s, and at $t = 11$ s.

b) How far does the officer go in the first 5 s? The first 9 s? The first 13 s?

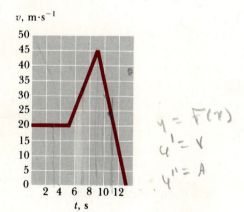

y = F(t)
u' = v
y'' = A

Figure 3–12

3–6 Figure 3–13 is a graph of the acceleration of a model railroad locomotive moving on the x-axis. Sketch the graphs of its velocity and coordinate as functions of time, if $x = v = 0$ when $t = 0$.

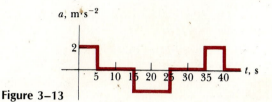

Figure 3–13

3–7 Figure 3–14 is a graph of the coordinate of a spider crawling along the x-axis. Sketch the graphs of its velocity and acceleration as functions of time.

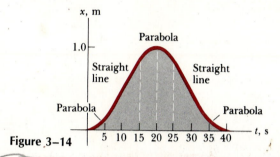

Figure 3–14

3–8 An astronaut has left Spacelab V to test a new space scooter for use in constructing Space Habitat I. His partner measures the following velocity changes, each taking place in a time interval of 10 s. What are the magnitude, the algebraic sign, and the direction of the average acceleration in each interval?

a) At the beginning of the interval the scooter is moving toward the right along the x-axis at 5 $\text{m} \cdot \text{s}^{-1}$, and at the end of the interval it is moving toward the right at 20 $\text{m} \cdot \text{s}^{-1}$.

b) At the beginning, toward the right at 20 $\text{m} \cdot \text{s}^{-1}$; at the end, toward the right at 5 $\text{m} \cdot \text{s}^{-1}$.

c) At the beginning, toward the left at 5 $\text{m} \cdot \text{s}^{-1}$; at the end, toward the left at 20 $\text{m} \cdot \text{s}^{-1}$.

d) At the beginning, toward the left at 20 $\text{m} \cdot \text{s}^{-1}$; at the end, toward the left at 5 $\text{m} \cdot \text{s}^{-1}$.

e) At the beginning, toward the right at 20 $\text{m} \cdot \text{s}^{-1}$; at the end, toward the left at 20 $\text{m} \cdot \text{s}^{-1}$.

f) At the beginning, toward the left at 20 $\text{m} \cdot \text{s}^{-1}$; at the end, toward the right at 20 $\text{m} \cdot \text{s}^{-1}$.

3–9 The makers of a certain automobile advertise that it will accelerate from 15 to 50 $\text{mi} \cdot \text{hr}^{-1}$ in 13 s. Compute

a) the acceleration in $\text{ft} \cdot \text{s}^{-2}$, and

b) the distance the car travels in this time,

assuming the acceleration to be constant.

3–10 An airplane taking off from a landing field has a run of 500 m. If it starts from rest, moves with constant acceleration, and makes the run in 30 s, with what velocity in $\text{m} \cdot \text{s}^{-1}$ did it take off?

3–11 An automobile starts from rest and acquires a velocity of 40 km·hr^{-1} in 15 s.

   a) Compute the acceleration in kilometers per hour per second, assuming it to be constant.

   b) If the automobile continues to gain velocity at the same rate, how many more seconds are needed for it to acquire a velocity of 60 km·hr^{-1}?

   c) Find the distances covered by the automobile in parts (a) and (b).

3–12 An antelope moving with constant acceleration covers the distance between two points 60 m apart in 6 s. Its velocity as it passes the second point is 15 m·s^{-1}.

   a) What is the acceleration? $a = ?$

   b) What is its velocity at the first point? $V_0 ?$

3–13 A ball is released from rest and rolls down an inclined plane, requiring 4 s to cover a distance of 100 cm.

   a) What is its acceleration in cm·s^{-2}?

   b) How many centimeters would it have fallen vertically in the same time?

3–14 The "reaction time" of the average automobile driver is about 0.7 s. (The reaction time is the interval between the perception of a signal to stop and the application of brakes.) If an automobile can accelerate at -5 m·s^{-2}, compute the total distance covered in coming to a stop after a signal is observed,

   a) from an initial velocity of 15 m·s^{-1},

   b) from an initial velocity of 30 m·s^{-1}.

3–15 At the instant the traffic lights turn green, an automobile that has been waiting at an intersection starts ahead with a constant acceleration of 2 m·s^{-2}. At the same instant a truck, traveling with a constant velocity of 10 m·s^{-1}, overtakes and passes the automobile.

   a) How far beyond its starting point will the automobile overtake the truck?

   b) How fast will it be traveling?

3–16 The engineer of a passenger train traveling at 30 m·s^{-1} sights a freight train whose caboose is 200 m ahead on the same track. The freight train is traveling in the same direction as the passenger train with a velocity of 10 m·s^{-1}. The engineer of the passenger train immediately applies the brakes, causing a constant acceleration of -1 m·s^{-2}, while the freight train continues with constant speed.

   a) Will there be a collision?

   b) If so, where will it take place?

3–17 A sled starts from rest at the top of a hill and slides down with constant acceleration. At some later time it is 80 ft from the top; two seconds after that it is 125 ft from the top, two seconds later 180 ft, and two seconds later 245 ft.

   a) What is the average velocity of the sled during each of the 2-s intervals after passing the 80-ft point?

   b) What is the acceleration of the sled?

   c) What was the velocity of the sled when it passed the 80-ft point?

   d) How long did it take to go from the top to the 80-ft point?

   e) How far did the sled go during the first second after passing the 80-ft point?

   f) How long does it take the sled to go from the 80-ft point to the midpoint between the 80-ft and the 125-ft mark?

   g) What is the velocity of the sled as it passes the midpoint in part (f)?

3–18 A subway train starts from rest at a station and accelerates at a rate of 2 m·s^{-2} for 10 s. It then runs at constant speed for 30 s, and slows down at -4 m·s^{-2} until it stops at the next station. Find the *total* distance covered.

3–19 A jet-propelled motorcycle starts from rest, moves in a straight line with constant acceleration, and covers a distance of 64 m in 4 s.

   a) What is the final velocity?

   b) How much time was required to cover half the total distance?

   c) What is the distance covered in one-half the total time?

   d) What is the velocity when half the total distance has been covered?

   e) What is the velocity after one-half the total time?

3–20 The speed of an automobile going north is reduced from 30 to 20 m·s^{-1} in a distance of 125 m. Find

   a) the magnitude and direction of the acceleration, assuming it to be constant;

   b) the elapsed time;

   c) the distance in which the car can be brought to rest from 20 m·s^{-1}, assuming the acceleration of part (a).

3–21 An automobile and a truck start from rest at the same instant, with the automobile initially at some distance behind the truck. The truck has a constant acceleration of 2 m·s^{-2} and the automobile an acceleration of 3 m·s^{-2}. The automobile overtakes the truck after the truck has moved 75 m.

   a) How long does it take the auto to overtake the truck?

   b) How far was the auto behind the truck initially?

   c) What is the velocity of each when they are abreast?

3–22

   a) With what velocity must a ball be thrown vertically upward in order to rise to a height of 20 m?

   b) How long will it be in the air?

3–23 A student throws a water balloon vertically downward from the top of a building; the balloon leaves the thrower's hand with a velocity of $10 \text{ m} \cdot \text{s}^{-1}$.

  a) What will be its velocity after falling for 2 s?

  b) How far will it fall in 2 s?

  c) What will be its velocity after falling 10 m?

  d) If it moved a distance of 1 m while in the thrower's hand, find its acceleration while in his hand.

  e) If the balloon was thrown from a point 40 m above the ground, in how many seconds will it strike the ground?

  f) What will the velocity of the ball be when it strikes the ground?

3–24 A hot-air balloonist, rising vertically with a velocity of $5 \text{ m} \cdot \text{s}^{-1}$, releases a sandbag at an instant when the balloon is 20 m above the ground.

  a) Compute the position and velocity of the sandbag at the following times after its release: $\frac{1}{4}$ s, $\frac{1}{2}$ s, 1 s, 2 s.

  b) How many seconds after its release will the bag strike the ground?

  c) With what velocity will it strike?

3–25 A stone is dropped from the top of a tall cliff, and 1 s later a second stone is thrown vertically downward with a velocity of $20 \text{ m} \cdot \text{s}^{-1}$. How far below the top of the cliff will the second stone overtake the first?

3–26 A flower pot falls off a window sill and falls past the window below. It takes 0.25 s to pass a window 3 m high. How far is the top of the window below the upper window sill?

3–27 A ball is thrown nearly vertically upward from a point near the cornice of a tall building. It just misses the cornice on the way down, and passes a point 160 ft below its starting point 5 s after it leaves the thrower's hand.

  a) What was the initial velocity of the ball? *48 f t · s⁻¹*

  b) How high did it rise above its starting point? *36 Ft.*

  c) What were the magnitude and direction of its velocity at the highest point? *zero*

  d) What were the magnitude and direction of its acceleration at the highest point? *32 f.s⁻² downward*

  e) What was the magnitude of its velocity as it passed a point 64 ft below the starting point? *80ft's⁻¹*

3–28 A juggler performs in a room whose ceiling is 3 m above the level of his hands. He throws a ball vertically upward so that it just reaches the ceiling.

  a) With what initial velocity does he throw the ball?

  b) What time is required for the ball to reach the ceiling?

He throws a second ball upward with the same initial velocity, at the instant that the first ball is at the ceiling.

  c) How long after the second ball is thrown do the two balls pass each other?

  d) When the balls pass each other, how far are they above the juggler's hands?

3–29 A celebrated jumping frog jumps vertically upward. It has a speed of $10 \text{ m} \cdot \text{s}^{-1}$ when it has reached one half its maximum height.

  a) How high does it rise?

  b) What are its velocity and acceleration 1 s after it jumps?

  c) 3 s after?

  d) What is the average velocity during the first half second?

3–30 A student determined to test the law of gravity for himself walks off a skyscraper 300 m high, stopwatch in hand, and starts his free fall (zero initial velocity). Five seconds later, Superman arrives at the scene and dives off the roof to save the student.

  a) What must Superman's initial velocity be in order that he catch the student just before the ground is reached?

  b) What must be the height of the skyscraper so that even Superman can't save him? (Assume that Superman's acceleration is that of any free falling body.)

3–31 A football is kicked vertically upward from the ground, and a student gazing out of the window sees it moving upward past him at $5 \text{ m} \cdot \text{s}^{-1}$. The window is 10 m above the ground.

  a) How high does the ball go above the ground?

  b) How long does it take to go from a height of 10 m to its highest point?

  c) Find its velocity and acceleration $\frac{1}{2}$ s after it left the ground.

3–32 A ball is thrown vertically upward from the ground with a velocity of $30 \text{ m} \cdot \text{s}^{-1}$.

  a) What time is required for the ball to rise to its highest point?

  b) How high does the ball rise?

  c) How long after it is thrown does the ball have a velocity of $10 \text{ m} \cdot \text{s}^{-1}$ upward?

  d) Of $10 \text{ m} \cdot \text{s}^{-1}$ downward?

  e) When is the displacement of the ball zero? *back where started From*

  f) When is the magnitude of the ball's velocity equal to half its velocity when thrown? *Time*

  g) When is the magnitude of the ball's displacement equal to half the greatest height to which it rises?

  h) What are the magnitude and direction of the acceleration while the ball is moving upward?

i) While moving downward?

j) When at the highest point?

3–33 A ball is released from rest at the top of an inclined plane 18 m long and reaches the bottom 3 s later. At the same instant that the first ball is released, a second ball is projected upward along the plane from its bottom with a certain initial velocity. The second ball is to travel part way up the plane, stop, and return to the bottom so that it arrives at the bottom simultaneously with the first ball. Both balls have the same constant acceleration.

a) Find the acceleration.

b) What must be the initial velocity of the second ball?

c) How far up the plane will it travel?

3–34 The rocket-driven sled Sonic Wind No. 2, used for investigating the physiological effects of large accelerations, runs on a straight, level track 3500 ft long. Starting from rest, it can reach a speed of $1000 \ \text{mi} \cdot \text{hr}^{-1}$ in 1.8 s.

a) Compute the acceleration, assuming it to be constant.

b) What is the ratio of this acceleration to that of a free falling body, g?

c) What is the distance covered?

d) A magazine article states that at the end of a certain run the speed of the sled was decreased from $632 \ \text{mi} \cdot \text{hr}^{-1}$ to zero in 1.4 s, and that during this time its passenger was subjected to more than 40 times the pull of gravity (that is, the acceleration was greater than 40 g). Are these figures consistent?

3–35 The first stage of a rocket to launch an earth satellite will, if fired vertically upward, attain a speed of $4000 \ \text{mi} \cdot \text{hr}^{-1}$ at a height of 26 mi above the earth's surface, at which point its fuel supply will be exhausted.

a) Assuming constant acceleration, find the time to reach a height of 26 mi.

b) How much higher would the rocket rise if it continued to "coast" vertically upward?

3–36 Suppose the acceleration of gravity were only $1.0 \ \text{m} \cdot \text{s}^{-2}$, instead of $10 \ \text{m} \cdot \text{s}^{-2}$.

a) Estimate the height to which you could jump vertically from a standing start.

b) How high could you throw a baseball?

c) Estimate the maximum height of a window from which you would care to jump to a concrete sidewalk

below. (Each story of an average building is about 4 m high.)

d) With what speed, in miles per hour, would you strike the sidewalk?

e) How many seconds would be required?

3–37 A spaceship takes a straight-line path from the earth to the moon, a distance of about 400,000 km. Suppose it accelerates at $10 \ \text{m} \cdot \text{s}^{-2}$ for the first 10 min of the trip, then travels at constant speed until the last 10 min, when it accelerates at $-10 \ \text{m} \cdot \text{s}^{-2}$, just coming to rest as it reaches the moon.

a) What is the maximum speed attained?

b) What total time is required for the trip?

c) What fraction of the total distance is traveled at constant speed?

3–38 A "moving sidewalk" in an airport terminal building moves $1 \ \text{m} \cdot \text{s}^{-1}$ and is 150 m long. If a man steps on at one end and walks $2 \ \text{m} \cdot \text{s}^{-1}$ relative to the moving sidewalk, how much time does he require to reach the opposite end if he walks

a) in the same direction the sidewalk is moving;

b) in the opposite direction?

3–39 Two piers A and B are located on a river, one mile apart. Two men must make round trips from pier A to pier B and return. One man is to row a boat at a velocity of $4 \ \text{mi} \cdot \text{hr}^{-1}$ relative to the water, and the other man is to walk on the shore at a velocity of $4 \ \text{mi} \cdot \text{hr}^{-1}$. The velocity of the river is $2 \ \text{mi} \cdot \text{hr}^{-1}$ in the direction from A to B. How long does it take each man to make the round trip?

3–40 The driver of a car wishes to pass a truck that is traveling at a constant speed of $20 \ \text{m} \cdot \text{s}^{-1}$ (about $50 \ \text{mi} \cdot \text{hr}^{-1}$). The car's maximum acceleration at this speed is $0.5 \ \text{m} \cdot \text{s}^{-2}$. Initially the vehicles are separated by 25 m, and the car pulls back into the truck's lane after it is 25 m ahead of the truck. The car is 5 m long and the truck 20 m.

a) How much time is required for the car to pass the truck?

b) What distance does the car travel during this time?

c) What is the final speed of the car, assuming its acceleration while passing the truck is constant?

NEWTON'S SECOND LAW. GRAVITATION

In the preceding chapters we have discussed separately the concepts of force and acceleration. In equilibrium problems we have made use of Newton's first law, which states that when the resultant force on a body is zero, the acceleration of the body is also zero. The next logical step is to ask how a body behaves when the resultant force on it is *not* zero. The answer to this question is contained in Newton's *second* law, which states that when the resultant force is not zero the body moves with accelerated motion, and that with a given force, the acceleration depends on a property of the body known as its *mass*.

This part of mechanics, which includes the study both of motion and of the forces that cause the motion, is called *dynamics*. In its broadest sense, dynamics includes nearly all of mechanics. Statics is concerned with special cases in which the acceleration is zero, and kinematics deals only with describing motion, not with its causes.

Unless otherwise stated, all velocities and accelerations are measured relative to an inertial reference system (Section 2–3), and in this chapter we shall consider only straight-line motion. Motion in a curved path will be discussed in the next chapter.

4–1

NEWTON'S SECOND LAW. MASS

We know from experience that an object at rest never starts to move by itself; a push or pull must be exerted on it by some other body. Similarly, a force is required to slow down or stop a body already in motion, and to make a moving body deviate from straight-line motion requires a sideways force. All these processes (speeding up, slowing down, or changing direction) involve a change in either the magnitude or direction of the velocity. Thus in each case the body *accelerates*, and an external force must act on it to produce the acceleration.

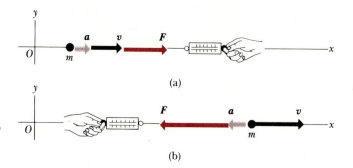

(a)

(b)

4–1 The acceleration **a** is proportional to the force **F** and is in the same direction as the force.

Let us consider several fundamental experiments. A small body, which we may model as a particle, rests on a level frictionless surface, and moves to the right along the x-axis of an inertial reference system, as in Fig. 4–1a. A constant horizontal force **F** is applied to the body. This force might be supplied by a spring balance, as described in Section 2–1, with the spring stretched a constant amount. We find that while the force is acting, the velocity of the body increases at a constant rate; that is, the body moves with *constant acceleration*. If the magnitude of the force is changed, the acceleration changes in the same proportion. Doubling the force doubles the acceleration, halving the force halves the acceleration, and so on. When the force is removed, the acceleration becomes zero; the body then moves with constant velocity, as discussed in Section 2–2.

Before the time of Galileo and Newton, it was generally believed that a force was necessary just to keep a body moving, even on a level friction-less surface or in "outer space." Galileo and Newton realized that *no force* is necessary to keep a body moving, once it has been set in motion, and that the effect of a force is not to *maintain* the velocity of a body, but to *change* its velocity. The *rate of change* of velocity, for a given body, is directly proportional to the force exerted on it.

In Fig. 4–1b, the velocity of the body is also toward the right, but the *force* is toward the left. Under these conditions the body moves more and more slowly to the right until it stops and then reverses its direction of motion and begins to move more and more rapidly toward the left, as long as the force continues to be applied. During this entire process the direction of the acceleration is toward the *left*, in the same direction as the force **F**. Hence we conclude not only that the *magnitude* of the acceleration is proportional to that of the force, but that the *direction* of the acceleration is the *same* as that of the force, regardless of the direction of the velocity.

To say that the rate of change of the velocity of a body is directly proportional to the force exerted on it is to say that the *ratio* of the force to the rate of change of velocity is a constant, regardless of the magnitude of the force. This constant ratio of force to rate of change of velocity is called the *mass m* of the body. Thus

$$m = \frac{F}{a},$$

or

$$F = ma. \tag{4–1}$$

The mass of a body can be thought of as the *force per unit of accelera-tion.* For example, if the acceleration of a certain body is found to be $5\ \text{m}\cdot\text{s}^{-2}$ when the force is 20 N, the mass of the body is

$$m = \frac{20\ \text{N}}{5\ \text{m}\cdot\text{s}^{-2}} = 4\ \text{N}\cdot\text{m}^{-1}\cdot\text{s}^2$$

and a force of 4 N must be exerted on the body for every $\text{m}\cdot\text{s}^{-2}$ of acceleration.

When a large force is needed to give a body a certain acceleration (i.e., speed it up, slow it down, or deviate it if it is in motion), the mass of the body is large; if only a small force is needed for the same accelera-tion, the mass is small. Thus the mass of a body is a quantitative measure of the property described in everyday language as *inertia.*

Another property of mass is demonstrated by measuring the masses of two bodies, using the procedure just described, and then fastening them together and measuring the mass of the composite body. If m_1 and m_2 are the individual masses, the mass of the composite body is always found to be $m_1 + m_2$. This very important result, which is reasonable but could not have been predicted in advance, shows that mass is an *additive quantity*, and that it is directly correlated with quantity of matter. In-deed, the concept of mass is one way to give the term "quantity of mat-ter" a precise meaning.

In the above discussion the particle moves along a straight line, the x-axis, and the force also lies along this direction. This is of course a special case; more generally, the force may also have a component in the y-direction, and the particle's motion need not be confined to a straight line. Furthermore, more than one force may act on the particle. Thus this formulation needs to be generalized to include motion in a plane or in space and the possibility of several forces acting simultaneously.

Experiment shows that when several forces act on a particle at the same time, the acceleration is the same as would be produced by a single force equal to the vector sum of these forces. This sum is usually most conveniently handled using the method of components. Thus when sev-eral forces act on a particle moving along the x-axis,

$$\sum F_x = ma. \tag{4–2}$$

Anticipating the more general discussion of velocity and acceleration to be given in Chapter 5, we remark briefly that when a particle moves in a plane, with position described by coordinates (x, y), the velocity is a vector quantity whose components v_x and v_y are equal to the rates of change of x and y, respectively, and the acceleration is a vector quantity with components a_x and a_y equal to the rates of change of v_x and v_y, respectively. Then a more general formulation of the relation of force to acceleration is

$$\sum F_x = ma_x, \qquad \sum F_y = ma_y. \tag{4–3}$$

This pair of equations is equivalent to the single vector equation

$$\sum \mathbf{F} = m\mathbf{a}, \tag{4–4}$$

where we write the left-hand side explicitly as $\Sigma\,\boldsymbol{F}$ to emphasize that the acceleration is determined by the resultant of all the forces acting on the particle.

Equation (4–4), or the equivalent pair of equations (4–3), is the mathematical statement of Newton's *second law of motion*. The acceleration of a body, the rate of change of its velocity, is equal to the resultant of all forces exerted on the particle, divided by its mass, and has the same direction as the resultant force.

The acceleration must always be measured relative to an inertial frame of reference, as defined in Section 2–3. Also, it is necessary to state the law for a *particle*, because when a force acts on an extended body the body may also *rotate*. In that case, different particles in the body have different accelerations. We return to this more general problem in Chapter 9.

In this chapter only straight-line motion is considered, and the line of motion is usually taken to be the *x*-axis. In these problems v_y and a_y are always zero. Individual forces may have *y*-components, but the *sum* of the *y*-components must always be zero. In Chapter 5 we consider more general motion in which v_y, a_y, and $\Sigma\,F_y$ need not be zero.

4–2

SYSTEMS OF UNITS

Nothing was said in the preceding discussion about the *units* to be used to measure force, mass, and acceleration. It is evident, however, from the equation $\Sigma\,\boldsymbol{F} = m\boldsymbol{a}$, that the units must be such that unit force imparts unit acceleration to unit mass; whatever system of units is chosen must meet this requirement.

Several systems of units are in common use in the United States; one is the meter-kilogram-second (mks) system, which also forms the basis of the Système Internationale (SI) introduced in Chapter 1. Other systems are the British or foot-pound-second system and the now obsolete centimeter-gram-second (cgs) system. This book uses principally SI units, not only because this system is convenient and widely used in scientific work, but also because there seems little doubt that SI will eventually be adopted worldwide for commerce and industry as well as scientific work. The United States and Britain are the only two major countries that do not use the metric system exclusively, and Britain is now in the process of converting.

In the mks (SI) system, the mass of the standard kilogram is the unit of mass, and the unit of acceleration is $1\ \mathrm{m\cdot s^{-2}}$. The unit force in this system is then *that force that gives a standard kilogram an acceleration of* $1\ \mathrm{m\cdot s^{-2}}$. This force is called *one newton* (1 N). One newton is approximately equal to one quarter of a pound-force (more precisely, 1 N = 0.22481 lb). Thus in the mks system,

$$F\ (\mathrm{N}) = m\ (\mathrm{kg}) \times a\ (\mathrm{m\cdot s^{-2}}).$$

The mass unit in the *centimeter-gram-second* (cgs) system is one gram, equal to $\frac{1}{1000}$ kg, and the unit of acceleration is $1\ \mathrm{cm\cdot s^{-2}}$. The unit force in this system is then *that force that gives a body whose mass is one gram, an acceleration of* $1\ \mathrm{cm\cdot s^{-2}}$. This force is called *one dyne* (1 dyn). Since

$1 \text{ kg} = 10^3 \text{ g}$ and $1 \text{ m} \cdot \text{s}^{-2} = 10^2 \text{ cm} \cdot \text{s}^{-2}$, it follows that $1 \text{ N} = 10^5 \text{ dyn}$. In the cgs system,

$$F \text{ (dyn)} = m \text{ (g)} \times a \text{ (cm} \cdot \text{s}^{-2}).$$

In the mks and cgs systems, we first selected units of mass and acceleration, and defined the unit of force in terms of these. In the *British engineering system,* we first select a unit of *force* (1 lb) and a unit of acceleration (1 ft $\cdot$ s^{-2}), and then define the unit of mass as *the mass of a body whose acceleration is* 1 ft $\cdot$ s^{-2} *when the resultant force on the body is* 1 lb. This unit of mass is called *one slug.* (The origin of the name is obscure; it may have arisen from the concept of mass as inertia or *sluggishness.*) Then in the British system,

$$F \text{ (lb)} = m \text{ (slugs)} \times a \text{ (ft} \cdot \text{s}^{-2}).$$

The pound is used in everyday life as a unit of quantity of matter (e.g., a pound of butter) but properly speaking it is a unit of *force* or *weight.* Thus a pound of butter is that quantity that has a weight of 1 lb. A useful fact in converting between mks and British units is that an object with a *mass* of 1 kg has a *weight* of about 2.2 lb (more precisely, 2.2046 lb).

The units of force, mass, and acceleration in the three systems are summarized in Table 4–1.

TABLE 4–1			
System of units	**Force**	**Mass**	**Acceleration**
mks	newton (N)	kilogram (kg)	$\text{m} \cdot \text{s}^{-2}$
cgs	dyne (dyn)	gram (g)	$\text{cm} \cdot \text{s}^{-2}$
engineering	pound (lb)	slug	$\text{ft} \cdot \text{s}^{-2}$

It should be emphasized that because of the above definitions there is an equivalence between force units and those of mass times acceleration. For example, $1 \text{ N} = 1 \text{ kg} \cdot \text{m} \cdot \text{s}^{-2}$; and in checking equations for consistency of units it is always appropriate to replace "N" wherever it appears by "kg $\cdot$ m $\cdot$ s^{-2}." Similarly, $1 \text{ lb} = 1 \text{ slug} \cdot \text{ft} \cdot \text{s}^{-2}$, and this conversion can be used in unit checks.

EXAMPLE 1   A constant horizontal force of 2 N is applied to a body of mass 4 kg, resting on a level frictionless surface. The acceleration of the body is

$$a = \frac{F}{m} = \frac{2 \text{ N}}{4 \text{ kg}} = 0.5 \text{ N} \cdot \text{kg}^{-1} = 0.5 \text{ (kg} \cdot \text{m} \cdot \text{s}^{-2}) \cdot \text{kg}^{-1}$$
$$= 0.5 \text{ m} \cdot \text{s}^{-2}.$$

Since the force is constant, the acceleration is constant also. Hence if the initial position and velocity of the body are known, the velocity and position at any later time can be found from the equations of motion with constant acceleration.   ◀

EXAMPLE 2   A body of mass 0.2 kg is given an initial velocity of 0.4 m $\cdot$ s^{-1} toward the right along a level laboratory tabletop. The body is observed to slide a distance of 1.0 m along the table before coming to rest. What was the magnitude and direction of the friction force $\mathscr{F}$ acting on it?

Solution We assume that the friction force is constant. The acceleration is then constant also; from the equations of motion with constant acceleration, we have

$$v^2 = v_0{}^2 + 2ax, \qquad 0 = (0.4 \text{ m} \cdot \text{s}^{-1})^2 + (2a)(1.0 \text{ m}),$$
$$a = -0.08 \text{ m} \cdot \text{s}^{-2}.$$

The negative sign means that the acceleration is toward the *left* (although the velocity is toward the right). The friction force on the body is

$$\mathscr{F} = ma = (0.2 \text{ kg})(-0.08 \text{ m} \cdot \text{s}^{-2}) = -0.016 \text{ N},$$

and is toward the left also. (A force of equal magnitude, but directed toward the right, is exerted *on* the table *by* the sliding body.) ◄

EXAMPLE 3 The acceleration of a certain body is found to be 5 ft·s⁻² when the resultant force on the body is 20 lb. The mass of the body is

$$m = \frac{F}{a} = \frac{20 \text{ lb}}{5 \text{ ft} \cdot \text{s}^{-2}} = 4 \ (\text{slug} \cdot \text{ft} \cdot \text{s}^{-2}) \cdot \text{ft}^{-1} \cdot \text{s}^2$$

$$= 4 \text{ slugs.} \qquad ◄$$

4–3

NEWTON'S LAW OF UNIVERSAL GRAVITATION

We have already encountered several times the force of gravitational attraction between a body and the earth. We now wish to study this phenomenon of gravitation in more detail.

    The law of universal gravitation was discovered by Newton, and was published by him in 1686. Newton's law of gravitation may be stated:

> *Every particle of matter in the universe attracts every other particle with a force that is directly proportional to the product of the masses of the particles and inversely proportional to the square of the distance between them.*

Thus

$$F_{\text{g}} = G \frac{m_1 m_2}{r^2}, \tag{4–5}$$

where F_{g} is the gravitational force on either particle, m_1 and m_2 are their masses, r is the distance between them, and G is a universal constant called the *gravitational constant*, whose numerical value depends on the units in which force, mass, and length are expressed.

    The gravitational forces acting on the particles form an action–reaction pair. Although the masses of the particles may be different, forces of *equal* magnitude act on each, and the action line of both forces lies along the line joining the particles.

    Newton's law of gravitation refers to the force between two *particles*. It can also be shown, however, that *the gravitational force exerted on or by a homogeneous sphere is the same as if the entire mass of the sphere were concen-*

trated in a point at its center. Thus if the earth were a homogeneous sphere, of mass m_E, the force exerted by it on a small body of mass m, at a distance r from its center, would be

$$F_g = G\frac{mm_E}{r^2},$$

provided the body lies outside the earth, i.e., that r is greater than the radius of the earth. A force of the same magnitude would be exerted *on* the earth by the body.

At points *inside* the earth these statements need to be modified. If one could drill a hole to the center of the earth and measure the force of gravity on a body at various distances from the center, the force would be found to *decrease* as the center is approached, rather than increasing as $1/r^2$. Qualitatively, it is easy to see why this should be so; as the body enters the interior of the earth (or other spherical body), some of the earth's mass is on the side of the body opposite from the center of the earth and pulls the body in the opposite direction. Exactly at the center of the earth, the gravitational force on the body is zero!

The magnitude of the gravitational constant G can be found experimentally by measuring the force of gravitational attraction between two bodies of known masses m_1 and m_2, at a known separation. For bodies of moderate size the force is extremely small, but it can be measured with an instrument invented by the Rev. John Michell and first used for this purpose by Sir Henry Cavendish in 1798. The same type of instrument was also used by Coulomb for studying forces of electrical and magnetic attraction and repulsion.

The Cavendish balance consists of a light, rigid T-shaped member (Fig. 4–2) supported by a fine vertical fiber such as a quartz thread or a thin metallic ribbon. Two small spheres, each of mass m_1, are mounted at the ends of the horizontal portion of the T, and a small mirror M, fastened to the vertical portion, reflects a beam of light onto a scale. To use the balance, two large spheres, each of mass m_2, are brought up to the positions shown. The forces of gravitational attraction between the large and small spheres twist the system through a small angle, thereby moving the reflected light beam along the scale.

By use of an extremely fine fiber, the deflection of the mirror may be made sufficiently large so that the gravitational force can be measured quite accurately. The gravitational constant, measured in this way, is found to be

$$G = 6.670 \times 10^{-11} \text{ N} \cdot \text{m}^2 \cdot \text{kg}^{-2}.$$

EXAMPLE 1   The mass m_1 of one of the small spheres of a Cavendish balance is 0.001 kg, the mass m_2 of one of the large spheres is 0.5 kg, and the center-to-center distance between the spheres is 0.05 m. The gravitational force on each sphere is

$$F_g = 6.67 \times 10^{-11} \text{ N} \cdot \text{m}^2 \cdot \text{kg}^{-2} \frac{(0.001 \text{ kg})(0.5 \text{ kg})}{(0.05 \text{ m})^2}$$

$$= 1.33 \times 10^{-11} \text{ N},$$

or about one hundred-billionth of a newton!   ◄

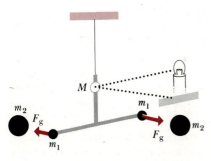

4–2   Principle of the Cavendish balance.

EXAMPLE 2   Suppose the spheres in Example 1 are placed 0.05 m from each other at a point in space far removed from all other bodies. What is the acceleration of each, relative to an inertial system?

Solution   The acceleration a_1 of the smaller sphere is

$$a_1 = \frac{F_g}{m_1} = \frac{1.33 \times 10^{-11} \text{ N}}{1.0 \times 10^{-3} \text{ kg}} = 1.33 \times 10^{-8} \text{ m} \cdot \text{s}^{-2}.$$

The acceleration a_2 of the larger sphere is

$$a_2 = \frac{F_g}{m_2} = \frac{1.33 \times 10^{-11} \text{ N}}{0.5 \text{ kg}} = 2.67 \times 10^{-11} \text{ m} \cdot \text{s}^{-2}.$$

In this case, the accelerations are *not* constant because the gravitational force increases as the spheres approach each other.   ◄

The three laws of motion and the law of gravitation are the principal elements in the *newtonian synthesis,* by which Newton was able to understand the motions of bodies on earth and the motions of the planets around the sun on the basis of the same fundamental physical laws. According to legend, the law of gravitation came to Newton as he watched an apple fall out of a tree and wondered what force accelerated it toward the earth. But to arrive at the $1/r^2$ form of the law of gravitation he had to do something much subtler; he had to take the observed motions of the planets, made initially by Tycho Brahe and systematized by Kepler, and show that Kepler's rules demanded a $1/r^2$ force law. This analysis is discussed in detail in Section 5–8; meanwhile we invite the reader to peek ahead at the last two paragraphs of that section.

4–4

MASS AND WEIGHT

The *weight* of a body can now be defined more generally than in the preceding chapters as *the resultant gravitational force exerted on the body by all other bodies in the universe.* At or near the surface of the earth, the force of the earth's attraction is so much greater than that of any other body that for practical purposes all other gravitational forces can be neglected and the weight can be considered as arising solely from the gravitational attraction of the earth. Similarly, at the surface of the moon, or of another planet, the weight of a body results almost entirely from the gravitational attraction of the moon or the planet. Thus if the earth were a homogeneous sphere of radius R and mass m_E, the weight w of a small body of mass m at or near its surface would be

$$w = F_g = \frac{Gmm_E}{R^2}. \tag{4–6}$$

The weight of a given body varies by a few tenths of a percent from point to point on the earth's surface, partly because of local deposits of ore, oil, or other substances whose density differs from the average, and partly because the earth is not a perfect sphere but is flattened somewhat at the poles. Equation (4–6) also shows that the weight of a given body decreases inversely with the square of its distance from the earth's center; at a radial distance of two earth radii, for example, it has decreased to one quarter of its value at the earth's surface.

The *apparent* weight of a body at the surface of the earth differs slightly in magnitude and direction from the earth's force of gravitational attraction because of the rotation of the earth about its axis. For the present, we shall ignore this effect and assume that the earth is an inertial reference system. Then when a body is allowed to fall freely, the force accelerating it is its weight w, and the acceleration produced by this force is the acceleration due to gravity, g. The general relation

$$\sum F = ma$$

therefore becomes, for the special case of a freely falling body,

$$w = mg. \tag{4–7}$$

Since, according to Eq. (4–6),

$$w = mg = G\frac{mm_E}{R^2},$$

it follows that

$$g = \frac{Gm_E}{R^2}, \tag{4–8}$$

showing that the acceleration due to gravity is the same for *all* bodies (since m cancels out) and is very nearly constant (since G and m_E are constants and R varies only slightly from point to point on the earth).

The weight of a body is a *force* and must be expressed in terms of the unit of force in the particular system of units one is using. Thus in the mks system the unit of weight is 1 N, in the cgs system it is 1 dyn, and in the engineering system it is 1 lb. Equation (4–7) indicates the relation between the mass and weight of a body in any consistent set of units.

For example, the weight of a standard kilogram, at a point where $g = 9.80$ m·s^{-2}, is

$$w = mg = (1 \text{ kg})(9.80 \text{ m}\cdot\text{s}^{-2}) = 9.80 \text{ N}.$$

At a second point where $g = 9.78$ m·s^{-2}, the weight is

$$w = 9.78 \text{ N}.$$

Thus, unlike the mass of a body, which is a constant, the weight varies from one point to another. On the moon, where $g = 1.67$ m·s^{-2}, the weight is 1.67 N, but the mass is still 1 kg.

When Eq. (4–8) is solved for m_E, we find

$$m_E = \frac{R^2 g}{G},$$

where R is the earth's radius. All of the quantities on the right are known, so the mass of the earth, m_E, can be calculated. Taking $R = 6370$ km $= 6.37 \times 10^6$ m, and $g = 9.80$ m·s^{-2}, we find

$$m_E = 5.96 \times 10^{24} \text{ kg}.$$

The volume of the earth is

$$V = \frac{4}{3} \pi R^3 = 1.08 \times 10^{21} \, m^3.$$

The mass of a body divided by its volume is known as its average *density*. (The density of water is $1 \, \text{g} \cdot \text{cm}^{-3} = 1000 \, \text{kg} \cdot \text{m}^{-3}$.) The average density of the earth is therefore

$$\frac{m_E}{V} = 5520 \, \text{kg} \cdot \text{m}^{-3} = 5.52 \, \text{g} \cdot \text{cm}^{-3}.$$

The density of most rock near the earth's surface, such as granites and gneisses, is about $3 \, \text{g} \cdot \text{cm}^{-3} = 3000 \, \text{kg} \cdot \text{m}^{-3}$, so the interior of the earth must have much higher density. Some basaltic rock found on the surface has a density of about $5 \, \text{g} \cdot \text{cm}^{-3}$.

The mass of a body can be measured in several different ways. One way is to use the relationship $m = F/a$; one may apply a measured force to the body, measure its acceleration, and obtain the mass as the ratio of force to acceleration. This method is used frequently in measuring masses of atomic and subatomic particles.

Another method uses a comparison technique, finding by trial some other body whose mass is equal to that of the given body but which is already known. Often it is easier to compare weights than masses. At a given point on the earth's surface, all bodies fall freely with the same acceleration g. Since the weight w of a body equals the product of its mass m and the acceleration g, it follows that if, at the same point, the weights of the two bodies are equal, their *masses* are equal also. The *equal-arm balance* is an instrument by means of which one can determine very precisely when the weights of two bodies are equal, and hence when their masses are equal.

Thus the property of matter called *mass* makes itself evident in two very different ways. The force of gravitational attraction between two particles is said to be proportional to the product of their masses; in this sense mass can be considered as *that property of matter by virtue of which every particle exerts a force of attraction on every other particle.* We may call this property *gravitational mass.* On the other hand, Newton's second law is concerned with the fact that a force (not necessarily gravitational) must be exerted on a particle in order to accelerate it, i.e., to change its velocity, either in magnitude or direction. This property can be called *inertial mass.* It is not at all obvious that the gravitational mass of a particle should be the same as its inertial mass, but experiment shows that the two *are* in fact the same. That is, if we have to push twice as hard on body A as we do on body B to produce a given acceleration, then the force of gravitational attraction between body A and some third body C is twice as great as the gravitational attraction between body B and body C, at the same distance. When an equal-arm balance is used to compare masses, it is actually *gravitational* mass that is being measured. Therefore the property represented by m in Newton's second law can be operationally defined as the result obtained by the prescribed methods of using an equal-arm balance.

Finally, we note that the SI units for mass and weight are frequently misused in everyday conversation. Expressions such as "This box weighs

6 kg" are nearly universal. What is meant, of course, is that the *mass* of the box (determined indirectly by weighing!) is 6 kg. This usage is so common that there is probably no hope of straightening things out, but it is important to recognize that the term *weight* is often used when *mass* is meant.

4–5

APPLICATIONS OF NEWTON'S SECOND LAW

We now discuss a number of applications of Newton's second law to specific problems. In all these examples, and in the problems at the end of the chapter, it will be assumed that the acceleration due to gravity is $9.80 \text{ m}\cdot\text{s}^{-2}$ or $32.0 \text{ ft}\cdot\text{s}^{-2}$, unless otherwise specified.

These examples will also be used to illustrate a systematic problem-solving procedure that the student is urged to study carefully and apply in the solution of assigned problems. The procedure consists of the following steps:

1. Select a body for analysis. Newton's second law will be applied to this body.

2. Draw a free-body diagram, just as with the problems of Chapter 2. Be sure to include all the forces acting *on* the chosen body, but be equally careful *not* to include any force exerted *by* the body on some other body. Some of the forces may be unknown; label them with algebraic symbols. Usually one of the forces will be the body's weight. Often it is useful to label this immediately as *mg* rather than *w*. If a numerical value of mass is given, the corresponding numerical value of weight may be computed.

3. Show location of coordinate axes explicitly in the free-body diagram, and then determine components of forces with reference to these axes. When a force is represented in terms of its components, cross out the original force so as not to include it twice.

4. If more than one body is involved, the above steps must be carried out for each body, and in addition there may be geometrical relationships between the motions of two or more bodies. Express these in algebraic form.

5. Write the Newton's-second-law equations for each body, usually Eqs. (4–3), and solve to find the unknown quantities.

6. Check special cases or extreme values of quantities, where possible, and compare the results for these particular cases with your intuitive expectations. Ask, "Does this result make sense?"

EXAMPLE 1 A 10-kg block of ice rests on a horizontal surface. What constant horizontal force T is required to give it a velocity of $4 \text{ m}\cdot\text{s}^{-1}$ in 2 s? Assume that the block starts from rest, that the friction force between the block and the surface is constant and equal to 5 N, and that all forces act at the center of the block. (See Fig. 4–3.)

Solution The mass of the block is given. The *y*-component of its acceleration is zero. The *x*-component of acceleration can be found from the

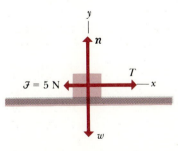

4–3 Force diagram for block in Example 1.

data on the velocity acquired in a given time. Since the forces are constant, the x-acceleration is constant and, from the equations of motion with constant acceleration,

$$a_x = \frac{v - v_0}{t} = \frac{4 \text{ m} \cdot \text{s}^{-1} - 0}{2 \text{ s}} = 2 \text{ m} \cdot \text{s}^{-2}.$$

The resultant of the x-forces is

$$\sum F_x = T - \mathcal{F},$$

and that of the y-forces is

$$\sum F_y = \mathcal{N} - w.$$

(Note that the *magnitude* of the friction force, $\mathcal{F} = 5$ N, is a positive quantity, but that the *component* of this force in the x-direction is negative, equal to $-\mathcal{F}$ or -5 N. Similarly, the magnitude of the weight is the positive quantity w, but the y-component of this force is $-w$.) Hence, from Newton's second law in component form,

$$T - \mathcal{F} = ma_x, \quad \mathcal{N} - w = ma_y = 0.$$

From the second equation, we find that

$$\mathcal{N} = w = mg = (10 \text{ kg})(9.80 \text{ m} \cdot \text{s}^{-2}) = 98.0 \text{ N},$$

and from the first,

$$T = \mathcal{F} + ma_x = 5 \text{ N} + (10 \text{ kg})(2 \text{ m} \cdot \text{s}^{-2}) = 25 \text{ N}. \quad \blacktriangleleft$$

EXAMPLE 2   An elevator and its load have a total mass of 800 kg. Find the tension T in the supporting cable when the elevator, originally moving downward at 10 m·s⁻¹, is brought to rest with constant acceleration in a distance of 25 m. (See Fig. 4–4.)

Solution   The weight of the elevator is

$$w = mg = (800 \text{ kg})(9.8 \text{ m} \cdot \text{s}^{-2}) = 7840 \text{ N}.$$

From the equations of motion with constant acceleration,

$$v^2 = v_0{}^2 + 2a(y - y_0), \quad a = \frac{v^2 - v_0{}^2}{2(y - y_0)}.$$

The initial velocity v_0 is -10 m·s⁻¹; the final velocity v is zero. If we take the origin at the point where the acceleration begins, then $y_0 = 0$ and $y = -25$ m. Hence

$$a = \frac{0 - (-10 \text{ m} \cdot \text{s}^{-1})^2}{2(-25 \text{ m})} = 2 \text{ m} \cdot \text{s}^{-2}.$$

The acceleration is therefore positive (upward). From the free-body diagram (Fig. 4–4) the resultant force is

$$\sum F = T - w = T - 7840 \text{ N}.$$

Since $\sum F = ma$,

$$T - 7840 \text{ N} = (800 \text{ kg})(2 \text{ m} \cdot \text{s}^{-2}) = 1600 \text{ N},$$
$$T = 9440 \text{ N}.$$

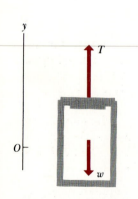

4–4   The resultant vertical force is $T - w$.

The tension must be *greater* than the weight by 1600 N to cause the upward acceleration while the elevator is stopping. ◀

EXAMPLE 3 In the above example, with what force do the feet of a passenger press downward on the floor, if the passenger's mass is 80 kg?

Solution We first find the force the floor exerts *on* the passenger's feet, which is the reaction force to the force required. The forces on the passenger are his weight $(80\ \text{kg})(9.8\ \text{m}\cdot\text{s}^{-2}) = 784\ \text{N}$ and the upward force F caused by the floor. Thus Newton's second law applied to the passenger, whose acceleration is the same as that of the elevator, gives

$$F - 784\ \text{N} = (80\ \text{kg})(2\ \text{m}\cdot\text{s}^{-2}) = 160\ \text{N},$$

and

$$F = 944\ \text{N}.$$

Thus the passenger pushes down on the floor with a force of 944 N while the elevator is stopping, and he *feels* a greater strain in his legs and feet than while standing at rest. If he stands on a bathroom scale calibrated in newtons, what will the scale read? ◀

EXAMPLE 4 A body (perhaps a child on a playground slide) slides down a frictionless plane inclined at an angle θ with the horizontal. What is its acceleration?

Solution The only forces acting on the body are its weight w and the normal force n exerted by the plane (Fig. 4–5b).
Take axes parallel and perpendicular to the surface of the plane and resolve the weight into x- and y-components. Then

$$\sum F_y = n - w\cos\theta,$$
$$\sum F_x = w\sin\theta.$$

But we know that $a_y = 0$, so from the equation $\sum F_y = ma_y$, we find that $n = w\cos\theta$. From the equation $\sum F_x = ma_x$, we have

$$w\sin\theta = ma_x,$$

and since $w = mg$,

$$a_x = g\sin\theta.$$ ◀

The mass does not appear in the final result, which means that any body, regardless of its mass, will slide down a frictionless inclined plane with an acceleration of $g\sin\theta$. In particular, when $\theta = 0$, $a_x = 0$, and when the plane is vertical, $\theta = 90°$ and $a_x = g$, as we should expect.

EXAMPLE 5 Consider the situation of Fig. 4–6 (which is the same figure as Fig. 2–8). Let the mass of the block be 4 kg and that of the rope be 0.5 kg. If the force F_1 is 9 N, what are the forces F_1', F_2, and F_2'? The surface on which the block moves is level and frictionless.

Solution We know from Newton's third law that $F_1 = F_1'$ and that $F_2 = F_2'$. Hence $F_1' = 9\ \text{N}$. The force F_2 could be computed by applying New-

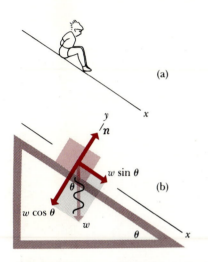

4–5 (a) A body on a frictionless inclined plane. (b) Free-body diagram.

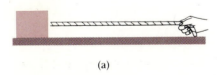

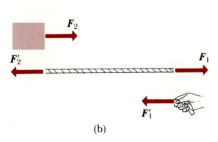

4–6 (a) A man pulls on a rope attached to a block. (b) Forces acting on the block, the rope, and the man's hand.

ton's second law to the block, if its acceleration were known, or the force F_2' could be computed by applying this law to the rope if its acceleration were known. The acceleration is not given, but it can be found by considering the block and rope together as a single system. The vertical forces on this system need not be considered. Since there is no friction, the resultant force acting *on* the system is the force F_1. (The forces F_2 and F_2' are not forces acting *on* the system but rather *internal* forces exerted by one part of the system on another. The force F_1' does not act on the system, but *on the man.*) Then, from Newton's second law,

$$\sum F = ma,$$

$$9 \text{ N} = (4 \text{ kg} + 0.5 \text{ kg})a,$$

$$a = 2 \text{ m} \cdot \text{s}^{-2}.$$

We can now apply Newton's second law to the block:

$$\sum F = ma,$$

$$F_2 = (4 \text{ kg})(2 \text{ m} \cdot \text{s}^{-2}) = 8 \text{ N}.$$

If we consider the rope alone, the resultant force on it is

$$\sum F = F_1 - F_2' = 9 \text{ N} - F_2',$$

and from the second law,

$$9 \text{ N} - F_2' = (0.5 \text{ kg})(2 \text{ m} \cdot \text{s}^{-2}) = 1 \text{ N},$$

$$F_2' = 8 \text{ N}.$$

In agreement with Newton's third law, which was tacitly used when the forces F_2 and F_2' were omitted in considering the system as a whole, we find that F_2 and F_2' are equal in magnitude. Note, however, that the forces F_1 and F_2' are *not* equal and opposite (the rope is not in equilibrium) and that these forces are *not* an action–reaction pair. ◄

EXAMPLE 6   In Fig. 4–7, a block of mass m_1 and weight $w_1 = m_1g$ moves on a level frictionless surface. It is connected by a light flexible cord passing over a small frictionless pulley to a second hanging block of mass m_2 and weight $w_2 = m_2g$. What is the acceleration of the system, and what is the tension in the cord connecting the two blocks?

Solution   The diagram shows the forces acting on each block. The forces exerted on the blocks by the cord can be considered an action–reaction pair, so we have used the same symbol T for each. For the block on the surface,

$$\sum F_x = T = m_1a_x,$$

$$\sum F_y = n - m_1g = m_1a_y = 0.$$

Applying Newton's second law to the hanging block, we obtain

$$\sum F_y = m_2g - T = m_2a_y.$$

We note that there is no necessity to use the same coordinate axes for the two bodies. It is convenient to take the $+y$ direction as downward for m_2, so both bodies move in positive axis directions.

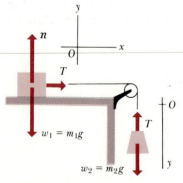

4–7   Force diagram for block on a frictionless horizontal surface and for the hanging block.

Now assuming that the string does not stretch, the speeds of the two bodies at any instant are equal, and so their accelerations have the same magnitude. That is, a_x for m_1 equals a_y for m_2. We can thus denote this common magnitude as simply a. Then the two equations containing accelerations are

$$T = m_1 a,$$

$$m_2 g - T = m_2 a.$$

These are two simultaneous equations for the two unknowns T and a. Adding the two equations eliminates T, giving

$$m_2 g = m_1 a + m_2 a = (m_1 + m_2)a$$

and

$$a = \frac{m_2}{m_1 + m_2} g.$$

Substituting this back into the first equation yields

$$T = \frac{m_1 m_2}{m_1 + m_2} g.$$

We see that the tension T is *not* equal to the weight $m_2 g$ of mass m_2, but is *less* by a factor of $m_1/(m_1 + m_2)$. Further thought shows that T *cannot* be equal to $m_2 g$; if it were, m_2 would be in equilibrium, and it isn't!

It is interesting to check some special cases. If $m_1 = 0$, we would expect that m_2 would fall freely and there would be no tension in the string. The equations do give $T = 0$ and $a = g$ for this case. Also, if $m_2 = 0$, we expect no tension and no acceleration; for this case the equations give $T = 0$ and $a = 0$. Thus in these two cases the general analytical results agree with intuitive expectations. ◄

EXAMPLE 7 A body of mass m is suspended from a spring balance attached to the roof of an elevator, as shown in Fig. 4–8. What is the reading of the balance if the elevator has an acceleration a, relative to the earth? Consider the earth's surface to be an inertial reference system.

Solution The forces on the body are its weight w (the gravitational force F_g exerted on it by the earth) and the upward force T exerted on it by the balance. The body is at rest relative to the elevator, and hence has an

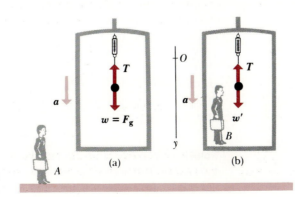

4–8 (a) To observer A, the body has a downward acceleration a, and he writes $w - T = ma$. (b) To observer B, the acceleration of the body is zero. He writes $w' = T$.

acceleration a relative to the earth. (We take the downward direction as positive.) The resultant force on the body is $w - T$, so, from Newton's second law,

$$w - T = ma, \qquad T = w - ma = m(g - a).$$

By Newton's third law, the body pulls down on the balance with a force equal and opposite to T, or equal to $(w - ma)$, so the balance reading equals $(w - ma)$.

If the same body were suspended in equilibrium from a balance attached to the earth, the balance reading would equal the weight w. To an observer riding in the accelerating elevator, the body *appears* to be in equilibrium and hence *appears* to be acted on by a downward force w' equal in magnitude to the balance reading, as in Fig. 4–8b. This *apparent* force w' can be called the *apparent weight* of the body. Then

$$w' = w - ma = m(g - a). \tag{4–9}$$

If the elevator is at rest, or moving vertically (either up or down) with constant velocity, $a = 0$ and the apparent weight equals the true weight. If the acceleration is downward, as in Fig. 4–8, so that a is positive, the apparent weight is less than the true weight; the body appears "lighter." If the acceleration is *upward*, a is negative, the apparent weight is greater than the true weight, and the body appears "heavier." These effects can be felt by taking a few steps in an elevator that is starting to move up or down after a stop. If the elevator falls freely, $a = g$, and since the true weight w also equals mg, the apparent weight w' is zero and the body *appears* "weightless." It is in this sense that an astronaut orbiting the earth in a space capsule is said to be "weightless."

The effect of this condition is exactly the same as though the body were in outer space with no gravitational force at all. The physiological effect of prolonged weightlessness is an interesting medical problem that is just beginning to be explored, with the advent of manned spacecraft. Gravity plays a role in blood distribution in the body, and one reaction to weightlessness is a decrease in volume of blood through increased excretion of water. In some cases astronauts returning to earth have experienced temporary impairment of sense of balance and a greater tendency toward motion sickness. ◀

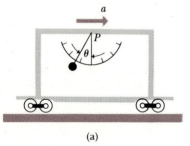

(a)

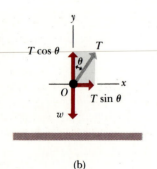

(b)

4–9 (a) A simple accelerometer. (b) The forces on the body are w and T.

EXAMPLE 8 Figure 4–9a represents a simple *accelerometer*. A small body is fastened at one end of a light rod pivoted freely at the point P. When the system has an acceleration a toward the right, the rod makes an angle θ with the vertical. In a practical instrument, some form of damping must be provided to keep the rod from swinging when the acceleration changes. The rod might hang in a tank of oil. The reader can make a primitive version of this device by tying a thread to the ceiling light in an automobile, tying a key or nearly any other small object to the other end, and using a protractor.

Solution As shown in the free-body diagram, Fig. 4–9b, two forces are exerted on the body: its weight $w = mg$ and the tension T in the rod. (We neglect the weight of the rod.) The resultant horizontal force is

$$\sum F_x = T \sin \theta,$$

and the resultant vertical force is

$$\sum F_y = T \cos \theta - mg.$$

The x-acceleration is the acceleration a of the system, and the y-acceleration is zero. Hence

$$T \sin \theta = ma, \qquad T \cos \theta = mg.$$

When the first equation is divided by the second, we get

$$a = g \tan \theta,$$

and the acceleration a is proportional to the tangent of the angle θ. ◀

Sometimes a frictional force opposing the motion of a body is not constant but increases with the body's speed. Two familiar examples are air resistance on a falling body and the viscous resisting force on a body falling through a liquid. For small speeds this force f is sometimes approximately proportional to speed v:

$$f = kv, \tag{4–10}$$

where k is a proportionality constant that depends on the dimensions of the body and the viscosity of the fluid. Taking the positive direction to be downward and neglecting any force associated with buoyancy in the fluid, we find that the net vertical component of force is $mg - kv$; Newton's second law gives

$$mg - kv = ma. \tag{4–11}$$

When the body first starts to move, $v = 0$, the resisting force is zero, and the initial acceleration is $a = g$. As the speed increases, the resisting force also increases, until finally it equals the weight in magnitude. At this time $mg - kv = 0$, the acceleration becomes zero, and there is no further increase in speed. The final speed, also called the *terminal* speed, is given by this equation; solving for v, we find

$$v_t = \frac{mg}{k}. \tag{4–12}$$

To find the precise functional relations of displacement and velocity to time requires methods of calculus, but Fig. 4–10 shows the general behavior of the position, velocity, and acceleration as functions of time.

In high-speed motion through air, the resisting force is approximately proportional to v^2 rather than to v; it is then called *air drag* or simply *drag*. Falling raindrops, airplanes, and cars moving at high speed all experience air drag. In this case Eq. (4–10) is replaced by $f = Kv^2$. We challenge the reader to show that the terminal speed is then given by

$$v_t = \sqrt{\frac{mg}{K}}, \tag{4–13}$$

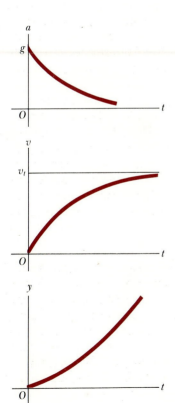

4–10 Graphs of acceleration, velocity, and position versus time for a body falling in a viscous fluid.

and that the units of the constant K are $\mathrm{N \cdot s^2 \cdot m^{-2}}$ or $\mathrm{kg \cdot m^{-1}}$.

EXAMPLE 9   For a human body falling through air, the numerical value of the constant K in Eq. (4–13) is found to be about $0.25 \text{ kg} \cdot \text{m}^{-1}$. If an 80-kg sky diver jumps out of an airplane and reaches terminal velocity before he opens his parachute, the terminal velocity is, from Eq. (4–13),

$$v_t = \sqrt{\frac{(80 \text{ kg})(9.8 \text{ m} \cdot \text{s}^{-2})}{0.25 \text{ kg} \cdot \text{m}^{-1}}}$$

$$= 56 \text{ m} \cdot \text{s}^{-1}. \qquad \blacktriangleleft$$

QUESTIONS

4–1 When a heavy weight is lifted by a string that is barely strong enough, the weight can be lifted by a steady pull, but if the string is jerked it will break. Why?

4–2 When a car accelerates, starting from rest, where is the force applied to the car to cause its acceleration? By what other body is the force exerted?

4–3 When a car stops suddenly, the passengers are thrown forward, away from their seats. What force causes this motion?

4–4 For medical reasons it is important for astronauts in outer space to determine their body mass at regular intervals. Devise a scheme for measuring body mass in a zero-gravity environment.

4–5 In SI units, suppose the fundamental units had been chosen to be force, length, and time instead of mass, length, and time. What would be the units of mass, in terms of the fundamental units?

4–6 A car traveling on a level road with coefficient of friction μ can never have an acceleration greater in magnitude than μg. Explain.

4–7 When a bullet is fired from a gun, what is the origin of the force that accelerates the bullet?

4–8 Why can a person dive into water from a height of 10 m without injury, while a person who jumps off the roof of a 10-m building and lands on a concrete street is likely to be seriously injured?

4–9 A passenger in a bus notices that a ball in the aisle that has been at rest suddenly starts to move toward the rear of the bus. Think of two different possible explanations, and devise a way to decide which is correct.

4–10 Two bodies on the two sides of an equal-arm balance exactly balance each other. If the balance is placed in an elevator and given an upward acceleration, do they still balance?

4–11 If a man in an elevator drops his briefcase but it does not fall to the floor, what can he conclude about the elevator's motion?

4–12 Suppose you are sealed inside a box on the planet Vulcan. After some time you notice that objects do not fall to the floor when released. How can you determine whether you and the box are in free fall or whether the Vulcans have somehow found a way to turn off the force of gravity?

4–13 Scales for weighing objects can be classified as those that use springs and those using standard weights to balance unknown weights. Which group would have greater accuracy when used in an accelerating elevator? When used on the moon? Does it matter whether you are trying to determine mass or weight?

4–14 Because of air resistance, two objects of unequal mass do *not* fall at precisely the same rate. If two bodies of identical shape but unequal mass are dropped simultaneously from the same height, which one reaches the ground first?

4–15 In discussing the Cavendish balance, a student claimed that the reason the small pivoted masses move toward the larger stationary masses rather than the larger toward the smaller is that the larger masses exert a stronger gravitational pull than the smaller. Please comment.

4–16 Since the earth is constantly attracted toward the sun by the gravitational interaction, why does it not fall into the sun and burn up?

4–17 "It's not the fall that hurts you; it's the sudden stop at the bottom." Translate this saying into the language of Newton's laws of motion.

4–18 Cavendish described his measurement of the gravitational constant G as "weighing the earth." This is an apt description not of the experiment itself but of the significance of the result. Why?

4–19 If action and reaction are always equal in magnitude and opposite in direction, why don't they always cancel each other and leave no net force for acceleration of the body?

4–20 A student wrote: "The only reason an apple falls downward to meet the earth instead of the earth falling upward to meet the apple is that the earth is much more massive and so exerts a much greater pull." Please comment.

PROBLEMS

In solving the problems of this chapter, the student is urged to use the systematic problem-solving procedures outlined in Section 4–5.

4–1

a) What is the mass of a body that weighs 1 N at a point where $g = 9.80$ m·s^{-2}?

b) What is the mass of a body that weighs 1 dyn at a point where $g = 980$ cm·s^{-2}?

4–2 The mass of a certain object is 10 kg.

a) What would its mass be if taken to the planet Mars?

b) Is the expression $F = ma$ valid on Mars?

c) Newton's second law is sometimes written in the form $F = wa/g$ instead of $F = ma$. Would this expression be valid on Mars?

4–3 A constant horizontal force of 40 N acts on a body on a *smooth horizontal* plane. The body starts from rest and is observed to move 100 m in 5 s.

a) What is the mass of the body?

b) If the force ceases to act at the end of 5 s, how far will the body move in the next 5 s?

c) What are the answers for (a) and (b) if the force is 10 lb and the body moves 250 ft in 5 s?

4–4 A .22 rifle bullet, traveling at 360 m·s^{-1}, strikes a block of soft wood, which it penetrates to a depth of 0.1 m. The mass of the bullet is 1.8 g. Assume a constant retarding force.

a) How much time was required for the bullet to stop?

b) What was the accelerating force, in newtons? in pounds?

4–5 A body of mass 15 kg rests on a frictionless horizontal plane and is acted on by a horizontal force of 30 N.

a) What acceleration is produced?

b) How far will the body travel in 10 s?

c) What will be its velocity at the end of 10 s?

4–6 A body of mass 0.5 kg is at rest at the origin, $x = 0$, on a horizontal frictionless surface. At time $t = 0$ a force of 0.001 N is applied to the body parallel to the x-axis, and 5 s later this force is removed.

a) What are the position and velocity of the body at $t = 5$ s?

b) If the same force is again applied at $t = 15$ s, what are the position and velocity of the body at $t = 20$ s?

4–7 A body of mass 10 kg is moving with a constant velocity of 5 m·s^{-1} on a horizontal surface. The coefficient of sliding friction between body and surface is 0.20.

a) What horizontal force is required to maintain the motion?

b) If the force is removed, how soon will the body come to rest?

4–8 A hockey puck leaves a player's stick with a velocity of 10 m·s^{-1} and slides 40 m before coming to rest. Find the coefficient of friction between the puck and the ice.

4–9 An electron (mass $= 9 \times 10^{-31}$ kg) leaves the cathode of a vacuum tube with zero initial velocity and travels in a straight line to the anode, which is 1 cm away. It reaches the anode with a velocity of 6×10^6 m·s^{-1}. If the accelerating force was constant, compute (a) the accelerating force, in newtons, (b) the time to reach the anode, (c) the acceleration. (The gravitational force on the electron may be neglected.)

4–10 The moon is 0.38×10^9 m from the earth and has a mass of 7.36×10^{22} kg. Find the gravitational force it exerts on a 1-kg body on earth; express your result also as a fraction of the body's weight (the earth's gravitational force on it.)

4–11 A communications satellite of mass 200 kg is in a circular orbit of radius 40,000 km around the earth (mass 5.97×10^{24} kg). What is the gravitational force on the satellite, and what fraction is this of its weight on earth?

4–12 A spaceship travels from earth directly toward the sun. At what distance from the center of the earth do the gravitational forces of the sun and the earth on the ship exactly cancel? Use the data in Appendix G.

4–13 In an experiment using the Cavendish balance to measure the gravitational constant G, it is found that a sphere of mass 0.8 kg attracts another sphere of mass 0.004 kg with a force of 13×10^{-11} N when the distance between the centers of the spheres is 0.04 m. The acceleration of gravity at the earth's surface is 9.80 m·s^{-2}, and the radius of the earth is 6400 km. Compute the mass of the earth from these data.

4–14 Two spheres, each of mass 6.4 kg, are fixed at points A and B (Fig. 4–11). Find the magnitude and direction of the initial acceleration of a sphere of mass 0.010 kg if released from rest at point P and acted on only by forces of gravitational attraction of the spheres at A and B.

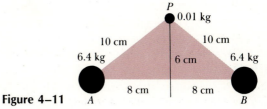

Figure 4–11

4–15 The mass of the moon is about one eighty-first, and its radius one fourth, that of the earth. What is the acceleration due to gravity on the surface of the moon?

4–16 What is the ratio of the gravitational pull of the sun on the moon to that of the earth on the moon? Use the data in Appendix G.

4–17 An elevator with mass 2000 kg rises with an acceleration of $1 \text{ m} \cdot \text{s}^{-2}$. What is the tension in the supporting cable?

4–18 A 4-kg bucket of water is accelerated upward by a cord whose breaking strength is 60 N. Find the maximum acceleration that can be given to the bucket without breaking the cord.

4–19 An elevator weighing 4000 lb rises with an acceleration of $4 \text{ ft} \cdot \text{s}^{-2}$. What is the tension in the supporting cable?

4–20 A 40-kg packing case is on the floor of a truck. The coefficient of static friction between the case and the truck floor is 0.30, and the coefficient of sliding friction is 0.20. Find the magnitude and direction of the friction force acting on the case

a) when the truck is accelerating at $2 \text{ m} \cdot \text{s}^{-2}$;

b) when it is accelerating at $-3 \text{ m} \cdot \text{s}^{-2}$.

4–21 A body hangs from a spring balance supported from the roof of an elevator.

a) If the elevator has an upward acceleration of $2.45 \text{ m} \cdot \text{s}^{-2}$ and the balance reads 50 N, what is the true weight of the body?

b) Under what circumstances will the balance read 30 N?

c) What will the balance read if the elevator cable breaks?

4–22 A transport plane is to take off from a level landing field with two gliders in tow, one behind the other. Each glider has a mass of 1200 kg, and the friction force or drag on each may be assumed constant and equal to 2000 N. The tension in the towrope between the transport plane and the first glider is not to exceed 10,000 N.

a) If a velocity of $40 \text{ m} \cdot \text{s}^{-1}$ is required for the take-off, what minimum length of runway is needed?

b) What is the tension in the towrope between the two gliders while the planes are accelerating for the take-off?

4–23

a) If the coefficient of friction between tires and dry pavement is 0.5, what is the shortest distance in which an automobile can be stopped when traveling at $60 \text{ mi} \cdot \text{hr}^{-1}$?

b) On wet pavement the coefficient of friction may be only 0.2. How fast should you drive on wet pavement

in order to be able to stop in the same distance as in part (a)?

4–24 A 5-kg block slides down a plane inclined at 30° to the horizontal. Find the acceleration of the block

a) if the plane is frictionless;

b) if the coefficient of kinetic friction is 0.2.

4–25 A 20-kg box rests on the flat floor of a truck. The coefficient of friction between box and floor is 0.1. The truck stops at a stop sign and then starts, with an acceleration of $2 \text{ m} \cdot \text{s}^{-2}$. If the box is 5 m from the rear of the truck when it starts, how much time elapses before it falls off the rear of the truck? How far does the truck travel in this time?

4–26 A 160-lb man stands on a bathroom scale in an elevator. As the elevator starts moving, the scale reads 200 lb.

a) Find the acceleration of the elevator (magnitude and direction).

b) What is the acceleration if the scale reads 120 lb?

c) If the scale reads zero, should the man worry? Explain.

4–27 A short commuter train consists of a locomotive and two cars. The mass of the locomotive is 1.00×10^5 kg and that of each car 5.00×10^4 kg. The train pulls away from a station with an acceleration of $0.5 \text{ m} \cdot \text{s}^{-2}$.

a) Find the tension in the coupler joining the locomotive to the first car, and in the coupler joining the two cars.

b) What total horizontal force must the locomotive wheels exert on the track?

4–28 A loaded elevator with very worn cables has a total mass of 2000 kg, and the cables can withstand a maximum tension of 24,000 N.

a) What is the maximum acceleration for the elevator if the cables are not to break?

b) What is the answer for part (a) if the elevator is taken to the moon, where $g = 1.67 \text{ m} \cdot \text{s}^{-2}$?

4–29 A packing crate rests on a ramp that makes an angle θ with the horizontal. The coefficient of kinetic friction is 0.50, and the coefficient of static friction is 0.75.

a) As the angle θ is increased, find the minimum angle at which the crate starts to slip.

b) At this angle, find the acceleration once the crate has begun to move.

c) How much time is required for the crate to slip 20 ft along the inclined plane?

4–30

a) What constant horizontal force is required to drag a 16-lb log along a horizontal surface with an acceleration of $4 \text{ ft} \cdot \text{s}^{-2}$ if the coefficient of sliding friction between log and surface is 0.5?

b) What weight, hanging from a cord attached to the 16-lb log and passing over a small frictionless pulley, will produce this acceleration?

4–31 A block of mass 5 kg resting on a horizontal surface is connected by a cord passing over a light frictionless pulley to a hanging block of mass 5 kg. The coefficient of friction between the block and the horizontal surface is 0.5. Find

a) the tension in the cord, and

b) the acceleration of each block.

4–32 A block having a mass of 2 kg is projected up a long 30° incline with an initial velocity of $22 \text{ m} \cdot \text{s}^{-1}$. The coefficient of friction between the block and the plane is 0.3.

a) Find the friction force acting on the block as it moves up the plane.

b) How long does the block move up the plane?

c) How *far* does the block move up the plane?

d) How long does it take the block to slide from its position in part (c) to its starting point?

e) With what velocity does it arrive at this point?

f) If the mass of the block had been 5 kg instead of 2 kg, would the answers in the preceding parts be changed?

4–33 A 20-kg block on a level frictionless surface is attached by a cord passing over a small frictionless pulley to a hanging block originally at rest 1 m above the floor. The hanging block strikes the floor in 2 s after the system is released from rest. Find

a) the weight of the hanging block;

b) the tension in the string while both blocks were in motion.

4–34 Block A in Fig. 4–12 has a mass of 2 kg and block B 20 kg. The coefficient of friction between B and the horizontal surface is 0.1.

a) What is the mass of block C if the acceleration of B is $2 \text{ m} \cdot \text{s}^{-2}$ toward the right?

b) What is the tension in each cord when B has the acceleration stated above?

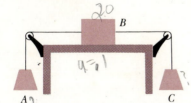

Figure 4–12

4–35 A block of mass 0.2 kg rests on top of a block of mass 0.8 kg. The combination is dragged along a horizontal surface at constant velocity by a hanging block of mass 0.2 kg as in Fig. 4–13a.

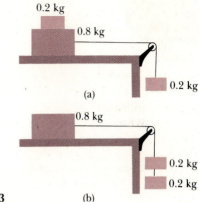

Figure 4–13 (a) (b)

a) The first 0.2-kg block is removed from the 0.8-kg block and attached to the hanging block, as in Fig. 4–13b. What is now the acceleration of the system?

b) What is the tension in the cord attached to the 0.8-kg block in part (b)?

4–36 Two 0.2-kg blocks hang at the ends of a light flexible cord passing over a small frictionless pulley, as in Fig. 4–14. A 0.1-kg block is placed on the right, and removed after 2 s.

a) How far will each block move in the first second after the 0.1-kg block is removed?

b) What was the tension in the cord before the 0.1-kg block was removed? After it was removed?

c) What was the tension in the cord supporting the pulley before the 0.1-kg block was removed? Neglect the weight of the pulley.

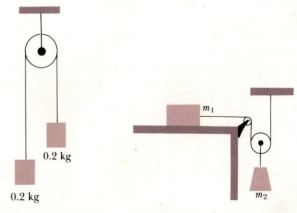

Figure 4–14   **Figure 4–15**

4–37 In terms of m_1, m_2, and g, find the acceleration of each block in Fig. 4–15 if $\mu_k = 0$.

4–38 The masses of bodies A and B in Fig. 4–16 are 20 kg and 10 kg, respectively. They are initially at rest on

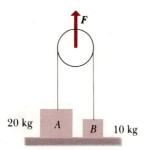

Figure 4–16

the floor and are connected by a weightless string passing over a weightless and frictionless pulley. An upward force F is applied to the pulley. Find the accelerations a_1 of body A and a_2 of body B when F is

a) 98 N,

b) 196 N,

c) 394 N,

d) 788 N.

4–39 The two blocks in Fig. 4–17 are connected by a heavy uniform rope of mass 4 kg. An upward force of 200 N is applied as shown.

a) What is the acceleration of the system?

b) What is the tension at the top of the heavy rope?

c) What is the tension at the midpoint of the rope?

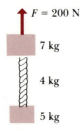

Figure 4–17

4–40 A man of mass 80 kg stands on a platform of mass 40 kg. He pulls a rope that is fastened to the platform and runs over a pulley on the ceiling. With what force does he have to pull in order to give himself and the platform an upward acceleration of $1 \text{ m} \cdot \text{s}^{-2}$?

4–41 What acceleration must the cart in Fig. 4–18 have in order that the block A will not fall? The coefficient of friction between the block and the cart is μ. How would the behavior of the block be described by an observer on the cart?

Figure 4–18

4–42 Which way will the accelerometer in Fig. 4–9 deflect under the following conditions?

a) The cart is moving to the right and speeding up.

b) The cart is moving to the right and slowing down.

c) The cart is moving to the left and speeding up.

d) The cart is moving to the left and slowing down.

e) The cart is at rest on a sloping surface.

f) The cart is given an upward velocity on a frictionless inclined plane. It first moves up, then stops, and then moves down. What is the deflection in each stage of the motion?

4–43 A body of mass $m = 5$ kg falls from rest in a viscous medium. The body is acted on by a net constant downward force of 20 N, and by a viscous retarding force proportional to its speed and equal to $5v$, where v is the speed in meters per second.

a) Find the initial acceleration, a_0.

b) Find the acceleration when the speed is $3 \text{ m} \cdot \text{s}^{-1}$.

c) Find the speed when the acceleration equals $0.1\, a_0$.

d) Find the terminal velocity, v_t.

4–44 A body falls from rest through a medium that exerts a resisting force that varies directly with the square of the velocity ($R = -kv^2$).

a) Draw a diagram showing the direction of motion and indicate, with the aid of vectors, all of the forces acting on the body.

b) Apply Newton's second law and infer from the resulting equation the general properties of the motion.

c) Show that the body acquires a terminal velocity, and calculate it.

4–45 A 20-kg monkey has a firm hold on a light rope that passes over a frictionless pulley and is attached to a 20-kg bunch of bananas, as shown in Fig. 4–19. The monkey

Figure 4–19

looks upward, sees the bananas, and starts to climb the rope to get at the bananas.

a) As he climbs, do the bananas move up, down, or remain at rest?

b) As he climbs, does the distance between him and the bananas decrease, increase, or remain constant?

c) The monkey releases his hold on the rope. What about the distance between him and the bananas while he is falling?

d) Before he reaches the ground, he grabs the rope to stop his fall. What do the bananas do?

4–46 An 80-kg man steps off of a platform 2.0 m above the ground. He keeps his legs straight as he falls, but at the moment his feet touch the ground his knees begin to bend, and his torso moves an additional 0.6 m before coming to rest.

a) What is his speed at the instant his feet touch the ground?

b) What is the acceleration of his torso as he slows down, assuming the acceleration is constant?

c) What force do his feet exert on the ground while he slows down? Express this force in newtons and also as a multiple of his weight.

MOTION IN A PLANE

Thus far our discussion of motion has been restricted to motion along a straight line. In this chapter we shall consider *plane motion*, that is, motion in a path that lies in a plane. Examples of such motion are the flight of a thrown or batted baseball, a projectile shot from a gun, a ball whirled at the end of a cord, the motion of the moon or of a satellite around the earth, and the motion of the planets around the sun.

In some of the problems to be considered the motion of the particle is known and we wish to determine its velocity and acceleration and the resultant force acting on it. An example is that of a ball attached to a cord and whirled in a circle at constant speed. What is the tension in the cord? In others, the force acting on a particle may be known at every point of space and we may wish to find the equation of motion of the particle. Examples are the orbit of a planet around the sun, or the path followed by a rocket. For these problems the kinematic language developed in Chapter 4 must be generalized; velocity and acceleration are in general *vector* quantities, subject to the methods of treatment introduced in Chapter 1.

5–1

AVERAGE AND INSTANTANEOUS VELOCITY

In Fig. 5–1a, a particle moves along a curved path in the *x-y* plane; points P and Q represent two positions of the particle. Its displacement as it moves from P to Q is the vector labeled Δs. The *x*- and *y*-components of Δs are the quantities Δx and Δy indicated on the figure.

Let Δt be the time interval during which the particle moves from P to Q. The average velocity during this interval is defined to be a vector quantity equal to the displacement divided by the time interval:

$$\text{Average velocity } \bar{v} = \Delta s/\Delta t. \tag{5–1}$$

That is, the average velocity is a vector quantity having the same direction as Δs and a magnitude equal to the magnitude of Δs divided by Δt.

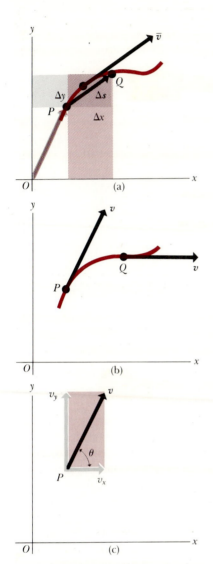

5–1 (a) The displacement Δs between points P and Q has components Δx and Δy. The average velocity v has the same direction as Δs. (b) The instantaneous velocity v at each point is always tangent to the path at that point. (c) Components of the instantaneous velocity at P are shown.

The magnitude of Δs is always the straight-line distance from P to Q regardless of the actual shape of the path taken by the particle. Thus the average velocity would be the same for any path that would take the particle from P to Q in the time interval Δt. Because the vector quantity $\bar{v}$ is associated with the entire displacement Δs, it has been drawn in Fig. 5–1a at a point intermediate between P and Q.

The *instantaneous* velocity v at the point P is defined in magnitude and direction as the *limit approached by the average velocity* when point Q is taken closer and closer to point P:

$$\text{Instantaneous velocity } v = \lim_{\Delta t \to 0} \frac{\Delta s}{\Delta t}. \tag{5–2}$$

As point Q approaches point P, the direction of the vector Δs approaches that of the tangent to the path at P, so that the instantaneous velocity vector at any point is *tangent* to the path at that point. The instantaneous velocities at points P and Q are shown in Fig. 5–1b.

It follows from Eq. (5–1) that the components $\bar{v}_x$ and $\bar{v}_y$ of average velocity are the corresponding components of displacement, divided by Δt. That is,

$$\bar{v}_x = \frac{\Delta x}{\Delta t}, \qquad \bar{v}_y = \frac{\Delta y}{\Delta t}.$$

Similarly, the instantaneous velocities of these projections, denoted by v_x and v_y, are equal respectively to the x- and y-components of the instantaneous velocity v, and we may write

$$v_x = \lim_{\Delta t \to 0} \frac{\Delta x}{\Delta t}, \qquad v_y = \lim_{\Delta t \to 0} \frac{\Delta y}{\Delta t}.$$

The magnitude of the instantaneous velocity is given by

$$|v| = \sqrt{v_x^2 + v_y^2},$$

and the angle θ in Fig. 5–1c by

$$\tan \theta = \frac{v_y}{v_x}.$$

Thus we may represent velocity, a vector quantity, in terms of its components or in terms of its magnitude and direction, just as with other vector quantities such as displacement and force. The *direction* of the instantaneous velocity of a particle at any point in its path is *always* in a direction tangent to the path at that point, no matter how complicated the motion.

5–2

AVERAGE AND INSTANTANEOUS ACCELERATION

In Fig. 5–2a the vectors v_1 and v_2 represent the instantaneous velocities, at points P and Q, of a particle moving in a curved path. The velocity v_2 necessarily differs in *direction* from the velocity v_1. The diagram has been constructed for a case in which it also differs in *magnitude*, although in some special cases the magnitude of the velocity may remain constant.

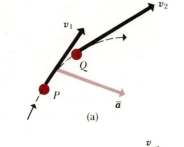

(a)

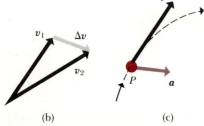

(b)   (c)

5–2 (a) The vector $\bar{a} = \Delta v/\Delta t$ represents the average acceleration between P and Q. (b) Construction for obtaining $\Delta v = v_2 - v_1$. (c) Instantaneous velocity v and instantaneous acceleration a at point P. Vector v is tangent to path; vector a points toward concave side of path.

The *average acceleration* $\bar{a}$ of the particle as it moves from P to Q is defined as the *vector change in velocity,* Δv, divided by the time interval Δt:

$$\text{Average acceleration } \bar{a} = \frac{\Delta v}{\Delta t}. \qquad (5\text{–}3)$$

Average acceleration is a vector quantity, in the same direction as the vector Δv.

The vector change in velocity, Δv, means the vector difference $v_2 - v_1$:

$$\Delta v = v_2 - v_1,$$

or

$$v_2 = v_1 + \Delta v.$$

Thus v_2 is the vector sum of v_1 and Δv. This relationship is shown in Fig. 5–2b.

The average acceleration vector, $\bar{a} = \Delta v/\Delta t$, is shown in Fig. 5–2a at a point midway between P and Q.

The *instantaneous acceleration* a at point P is defined in magnitude and direction as the limit approached by the average acceleration when point Q approaches point P and Δv and Δt both approach zero:

$$\text{Instantaneous acceleration } a = \lim_{\Delta t \to 0} \frac{\Delta v}{\Delta t}. \qquad (5\text{–}4)$$

The instantaneous acceleration vector at point P is shown in Fig. 5–2c. Note that it does *not* have the same direction as the velocity vector. Reference to the construction in Fig. 5–2c shows that the acceleration vector must always lie on the *concave* side of the curved path.

5–3

COMPONENTS OF ACCELERATION

Figure 5–3a again shows the motion of a particle referred to a rectangular coordinate system. The x- and y-components, a_x and a_y, of the instantaneous acceleration a of the particle are

$$a_x = \lim_{\Delta t \to 0} \frac{\Delta v_x}{\Delta t}, \qquad a_y = \lim_{\Delta t \to 0} \frac{\Delta v_y}{\Delta t}.$$

Thus, if the components a_x and a_y are known, the magnitude and direction of the acceleration a can be obtained, just as with velocity. When the acceleration is known, the force on the particle can be found, in magnitude and direction, from Newton's second law,

$$F = ma.$$

Conversely, if the force F is known at every point, the acceleration a and its components a_x and a_y can be found from Newton's second law.

The acceleration of a particle moving in a curved path can also be resolved into rectangular components $a_\perp$ and $a_\parallel$, in directions *normal* and

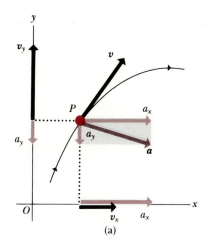

(a)

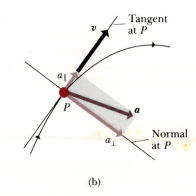

(b)

5–3 In (a) the acceleration **a** is resolved into rectangular components a_x and a_y. In (b) it is resolved into a normal component $a_\perp$ and a tangential component $a_\parallel$.

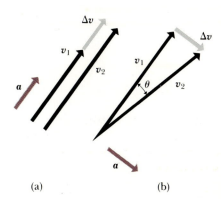

5–4 (a) When **a** is parallel to **v**, the magnitude of **v** increases but its direction does not change. (b) When **a** is perpendicular to **v**, the direction of **v** changes but its magnitude is constant.

parallel (or *tangential*) to the path, as shown in Fig. 5–3b. Unlike the rectangular components referred to a set of fixed axes, the normal and tangential components do not have fixed directions in space. They do, however, have a direct physical significance. The parallel component $a_\parallel$ arises from a change in the *magnitude* of the velocity vector **v**, while the normal component $a_\perp$ arises from a change in the *direction* of the velocity.

To understand these statements more fully, we consider the two special cases shown in Fig. 5–4. In Fig. 5–4a, the acceleration is at some instant *parallel* to the velocity v_1 at that instant. Then because **a** gives the rate of change of velocity, the change in **v** during a small time interval Δt is a vector Δv having the same direction as **a** and hence the same direction as v_1. Thus the velocity v_2 at the end of Δt, being equal to $v_1 + \Delta v$, is a vector having the same direction as v_1 but somewhat greater magnitude.

In Fig. 5–4b, the acceleration is perpendicular to the instantaneous velocity. Then in an interval Δt, the change Δv is a vector perpendicular to v_1, as shown. Again $v_2 = v_1 + \Delta v$, but in this case v_1 and v_2 have different directions. If the time interval is very small, the angle θ in the figure is also very small, Δv is very nearly perpendicular to *both* v_1 and v_2, and v_1 and v_2 have the same magnitude.

Thus when **a** is *parallel* to **v**, its effect is to change the magnitude of **v** but not its direction; when **a** is *perpendicular* to **v**, its effect is to change the direction of **v** but not its magnitude. In general **a** may have components both parallel and perpendicular to **v**, but the above statements are still valid for the individual components. In particular, a particle traveling along a curved path with constant speed has a nonzero acceleration, even though the magnitude of **v** does not change. In this case the acceleration is always perpendicular to **v** at each point. When a particle moves in a circle with constant speed, the acceleration is at each instant directed toward the center of the circle. This special case will be considered in detail in Section 5–5.

When a body moves in a curved path, a transverse force must be exerted on it to deviate it sidewise. The ratio of transverse (or normal) force to normal acceleration is the mass of the body, and this is also equal to the ratio of tangential force to tangential acceleration. That is, if $F_\perp$ and $F_\parallel$ are the normal and tangential components of force on a body moving in a curved path, Newton's second law (with the same value of m) applies to each of these components:

$$F_\perp = ma_\perp, \qquad F_\parallel = ma_\parallel.$$

If the force on a particle is *always* normal to the path, then $F_\parallel = 0, a_\parallel = 0$, and the particle has no tangential component of acceleration. The *magnitude* of the velocity then remains constant and the only effect of the force is to change the *direction* of motion, that is, to deviate the particle sidewise.

If the force has no normal component, then $F_\perp = 0, a_\perp = 0$, and there is no change in the *direction* of the velocity; that is, the particle moves in a straight line.

5-4

PROJECTILE MOTION

A *projectile* is any body that is given an initial velocity and then follows a path determined by the gravitational force acting on it and by air resistance. A batted baseball, a thrown football, an object dropped from an airplane, and a missile shot from a gun are all examples of projectiles. The path followed by a projectile is called its *trajectory*. The motion of a freely falling body discussed in Chapter 3 is a special case of projectile motion; in that case the trajectory is a straight line.

We shall consider only trajectories that are of short enough range so that the gravitational force may be considered constant in both magnitude and direction. Effects associated with air resistance and the rotation of the earth will not be considered. These simplifications form the basis of an idealized *model* of the physical problem; we neglect minor details in order to focus attention on the most important features of the problem.

The motion is most easily described and analyzed by use of a set of rectangular coordinate axes. We shall take the x-axis horizontal and the y-axis vertically upward. The x-component of the force on the projectile is then zero and the y-component is the weight of the projectile, $-mg$. Then from Newton's second law,

$$a_x = \frac{F_x}{m} = 0, \qquad a_y = \frac{F_y}{m} = \frac{-mg}{m} = -g.$$

That is, the horizontal component of acceleration is zero and the vertical component is downward and equal to that of a freely falling body. Since zero acceleration means constant velocity, the motion can be described as a combination of *horizontal motion with constant velocity* and *vertical motion with constant acceleration*.

The key to analysis of projectile motion is the fact that all the needed vector relationships, including Newton's second law and the definitions of velocity and acceleration, can be expressed in terms of separate equations for the x- and y-components of these vector quantities. It was seen in Chapter 4 that the single vector equation $F = ma$ is equivalent to the two component equations

$$F_x = ma_x \qquad \text{and} \qquad F_y = ma_y.$$

Similarly, each component of velocity is the rate of change of the corresponding coordinate, and each component of acceleration is the rate of change of the corresponding velocity component. In this sense the x and y motions are independent and may be analyzed separately. The actual motion is then the superposition of these separate motions.

Suppose that at time $t = 0$ our particle is at the point (x_0, y_0) and has velocity components v_{0x} and v_{0y}. As seen above, the components of acceleration are $a_x = 0, a_y = -g$. The variation of each coordinate with time is a case of motion with constant acceleration, and Eqs. (3–7) and (3–11) can be used directly. Substituting v_{0x} for v_0 and 0 for a_x, we find for x

$$v_x = v_{0x}, \tag{5–5}$$

$$x = x_0 + v_{0x}t. \tag{5–6}$$

Similarly, substituting v_{0y} for v_0 and $-g$ for a,

$$v_y = v_{0y} - gt, \tag{5–7}$$

$$y = y_0 + v_{0y}t - \tfrac{1}{2}gt^2. \tag{5–8}$$

Usually it is convenient to take the initial position as the origin; in this case, $x_0 = y_0 = 0$. This might be, for example, the position of a ball at the instant it leaves the thrower's hand, or the position of a bullet at the instant it leaves the gun barrel.

Figure 5–5 shows the path of a projectile that passes through the origin at time $t = 0$. The position, velocity, and velocity components of the projectile are shown at a series of times separated by equal intervals. As the figure shows, v_x does not change, but v_y changes by equal amounts in successive intervals, corresponding to constant y-acceleration.

The initial velocity v_0 is represented by its magnitude v_0 (the initial speed) and the angle θ_0 it makes with the positive x-axis. In terms of these quantities, the *components* v_{0x} and v_{0y} of initial velocity are

$$v_{0x} = v_0 \cos \theta_0,$$
$$v_{0y} = v_0 \sin \theta_0. \tag{5–9}$$

Using these relations in Eqs. (5–5) through (5–8) and setting $x_0 = y_0 = 0$, we obtain

$$x = (v_0 \cos \theta_0)t, \tag{5–10}$$

$$y = (v_0 \sin \theta_0)t - \tfrac{1}{2}gt^2, \tag{5–11}$$

$$v_x = v_0 \cos \theta_0, \tag{5–12}$$

$$v_y = v_0 \sin \theta_0 - gt. \tag{5–13}$$

These equations describe the position and velocity of the projectile in Fig. 5–5 at any time t.

Figure 5–6 shows parabolic trajectories of a bouncing golf ball.

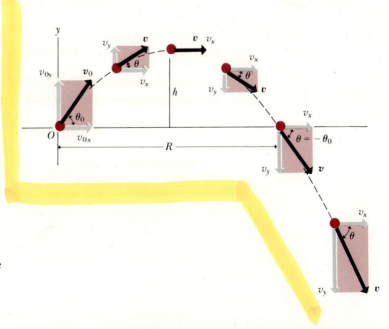

5–5 Trajectory of a body projected with an initial velocity v_0 at an angle of departure θ_0. The distance R is the horizontal range, and h is the maximum height. Time intervals between successive positions are equal.

5–6 Stroboscopic photograph of a bouncing golf ball, showing parabolic trajectories after each bounce. Successive images are separated by equal time intervals, as in Fig. 5–5. Each peak in the trajectories is lower than the preceding one because of energy loss during the "bounce" or collision with the horizontal surface. (Dr. Harold Edgerton, M.I.T., Cambridge, Massachusetts.)

A variety of additional information can be obtained from Eqs. (5–10) through (5–13). For example, the distance r of the projectile from the origin at any time is given by

$$r = \sqrt{x^2 + y^2}. \tag{5–14}$$

The projectile's speed (the magnitude of its resultant velocity) is

$$v = \sqrt{v_x^2 + v_y^2}. \tag{5–15}$$

The *direction* of the velocity, in terms of the angle θ it makes with the positive x-axis, is given by

$$\tan \theta = \frac{v_y}{v_x}. \tag{5–16}$$

The velocity vector v is tangent to the trajectory at each point, so its direction is the same as that of the trajectory.

Equations (5–10) and (5–11) give the position of the particle in terms of the parameter t. The equation in terms of x and y can be obtained by eliminating t. We find $t = x/v_0 \cos \theta_0$ and

$$y = (\tan \theta_0)x - \frac{g}{2v_0^2 \cos^2 \theta_0}x^2. \tag{5–17}$$

The quantities v_0, $\tan \theta_0$, $\cos \theta_0$, and g are constants, so the equation has the form

$$y = ax - bx^2,$$

where a and b are constants. This is the equation of a *parabola*.

EXAMPLE 1 A motorcycle stunt rider rides off the edge of a cliff with a horizontal velocity of magnitude $5 \text{ m} \cdot \text{s}^{-1}$. Find the rider's position and velocity after $\frac{1}{4}$ s (see Fig. 5–7).

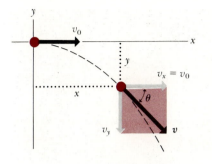

5–7   Trajectory of a body projected horizontally.

Solution   In this case, the initial angle θ_0 is zero. The initial vertical velocity component is therefore zero. The horizontal velocity component equals the initial velocity and is constant.

The x- and y-coordinates, when $t = \frac{1}{4}$ s, are

$$x = v_x t = (5 \text{ m}\cdot\text{s}^{-1})(\tfrac{1}{4}\text{ s}) = 1.25 \text{ m},$$

$$y = -\tfrac{1}{2}gt^2 = -\tfrac{1}{2}(9.8 \text{ m}\cdot\text{s}^{-2})(\tfrac{1}{4}\text{ s})^2 = -0.306 \text{ m}.$$

The distance from the origin at this time is

$$r = \sqrt{x^2 + y^2} = \sqrt{(1.25 \text{ m})^2 + (-0.306 \text{ m})^2} = 1.29 \text{ m}.$$

The components of velocity are

$$v_x = v_0 = 5 \text{ m}\cdot\text{s}^{-1},$$

$$v_y = -gt = (-9.8 \text{ m}\cdot\text{s}^{-2})(\tfrac{1}{4}\text{ s}) = -2.45 \text{ m}\cdot\text{s}^{-1}.$$

The resultant velocity has magnitude

$$v = \sqrt{v_x{}^2 + v_y{}^2} = 5.57 \text{ m}\cdot\text{s}^{-1}.$$

The angle θ is

$$\theta = \arctan\frac{v_y}{v_x} = \arctan\frac{-2.45 \text{ m}\cdot\text{s}^{-1}}{5.00 \text{ m}\cdot\text{s}^{-1}} = -26.1°.$$

That is, the velocity is $26.1°$ *below* the horizontal.   ◀

EXAMPLE 2   In Fig. 5–5, suppose the projectile is a home-run ball hit by Superman with an initial speed $v_0 = 160 \text{ ft}\cdot\text{s}^{-1}$ at an initial angle $\theta_0 = 53.1°$. Then

$$v_{0x} = v_0 \cos\theta_0 = (160 \text{ ft}\cdot\text{s}^{-1})(0.60) = 96 \text{ ft}\cdot\text{s}^{-1},$$

$$v_{0y} = v_0 \sin\theta_0 = (160 \text{ ft}\cdot\text{s}^{-1})(0.80) = 128 \text{ ft}\cdot\text{s}^{-1}.$$

a) Find the position of the ball, and the magnitude and direction of its velocity, when $t = 2.0$ s.

$$x = (96 \text{ ft}\cdot\text{s}^{-1})(2.0 \text{ s}) = 192 \text{ ft},$$

$$y = (128 \text{ ft}\cdot\text{s}^{-1})(2.0 \text{ s}) - \tfrac{1}{2}(32 \text{ ft}\cdot\text{s}^{-2})(2.0 \text{ s})^2 = 192 \text{ ft}.$$

$$v_x = 96 \text{ ft}\cdot\text{s}^{-1},$$

$$v_y = 128 \text{ ft}\cdot\text{s}^{-1} - (32 \text{ ft}\cdot\text{s}^{-2})(2.0 \text{ s}) = 64 \text{ ft}\cdot\text{s}^{-1},$$

$$v = \sqrt{v_x{}^2 + v_y{}^2} = 115.4 \text{ ft}\cdot\text{s}^{-1},$$

$$\theta = \arctan\frac{64 \text{ ft}\cdot\text{s}^{-1}}{96 \text{ ft}\cdot\text{s}^{-1}} = \arctan 0.667 = 33.7°.$$

b) Find the time at which the ball reaches the highest point of its flight, and find the height of this point.

   At the highest point, the vertical velocity v_y is zero. If t_1 is the time at which this point is reached,

$$v_y = 0 = 128 \text{ ft}\cdot\text{s}^{-1} - (32 \text{ ft}\cdot\text{s}^{-2})t_1,$$

$$t_1 = 4 \text{ s}.$$

The height h of the point is the value of y when $t = 4$ s:

$$h = (128 \text{ ft} \cdot \text{s}^{-1})(4 \text{ s}) - \tfrac{1}{2}(32 \text{ ft} \cdot \text{s}^{-2})(4 \text{ s})^2 = 256 \text{ ft}.$$

c) Find the *horizontal range R*, that is, the horizontal distance from the starting point to the point where the ball returns to earth, where $y = 0$. Let t_2 be the time at which this point is reached. Then

$$y = 0 = (128 \text{ ft} \cdot \text{s}^{-1})t_2 - \tfrac{1}{2}(32 \text{ ft} \cdot \text{s}^{-2})t_2{}^2.$$

This is a quadratic equation for t_2; it has two roots,

$$t_2 = 0 \quad \text{and} \quad t_2 = 8 \text{ s},$$

corresponding to the two times at which $y = 0$. The value $t_2 = 0$ is of course the time the ball left the ground; $t_2 = 8$ s is the time of its return. We note that this is just twice the time to reach the highest point. The time of descent therefore equals the time to rise.

The horizontal range R is the value of x when $t = 8$ s:

$$R = v_x t_2 = (96 \text{ ft} \cdot \text{s}^{-1})(8 \text{ s}) = 768 \text{ ft}.$$

The vertical component of velocity at this point is

$$v_y = 128 \text{ ft} \cdot \text{s}^{-1} - (32 \text{ ft} \cdot \text{s}^{-2})(8 \text{ s}) = -128 \text{ ft} \cdot \text{s}^{-1}.$$

That is, the vertical velocity has the same magnitude as the initial vertical velocity, but the opposite direction. Since v_x is constant, the angle *below* the horizontal at this point equals the initial angle θ_0. (In this analysis we have neglected the fact that the ball is struck a few feet above ground level rather than exactly at $y = 0$.)

d) If the ball did not hit the ground, it would continue to travel on *below* its original level. The ball park might be located atop a flat-topped hill that drops off steeply on one side. Then negative values of y, corresponding to times greater than 8 s, are possible. We challenge the reader to compute the position and velocity at a time 10 s after the start, corresponding to the last position shown in Fig. 5–5. The results are

$$x = 960 \text{ ft}, \qquad y = -320 \text{ ft},$$

$$v_x = 96 \text{ ft} \cdot \text{s}^{-1}, \qquad v_y = -192 \text{ ft} \cdot \text{s}^{-1}. \qquad \blacktriangleleft$$

EXAMPLE 3   Figure 5–8 shows an interesting example of the properties of projectile motion. A boy shoots an arrow from ground level at an apple hanging in a tree. At the same instant he releases the arrow, the apple falls from the tree and drops straight down, starting from rest. The arrow's path curves just enough to hit the apple regardless of its initial velocity. To show that this must happen, we note that the initial elevation of the apple is $x \tan \theta_0$, and that in time t it falls a distance $\tfrac{1}{2}gt^2$. Its elevation at the instant of collision is therefore

$$y = x \tan \theta_0 - \tfrac{1}{2}gt^2.$$

In this same time the arrow travels the distance x with constant x-component of velocity $v_0 \cos \theta$, so $x = v_0 \cos \theta_0 t$. Solving this for t and substitut-

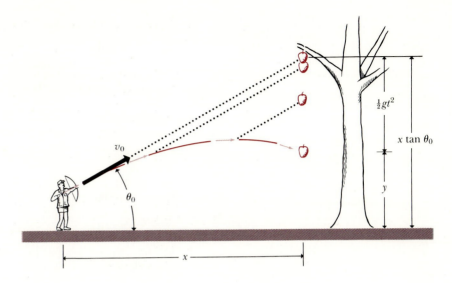

5–8 Trajectory of an arrow shot directly at a freely falling apple.

ing in the preceding equation, we obtain

$$y = x \tan \theta_0 - \frac{1}{2}g\left(\frac{x}{v_0 \cos \theta_0}\right)^2.$$

But this is the same expression as Eq. (5–17) for the path of the arrow. Thus at the instant the arrow reaches the line along which the apple is falling, the two heights are the same, and the two meet at this point.

◀

EXAMPLE 4   For a projectile launched with speed v_0 at initial angle θ_0, derive general expressions for the maximum height h and range R shown in Fig. 5–5. For a given v_0, what value of θ_0 gives maximum range?

Solution   We follow the same pattern as in Example 2b. First, *when* does the projectile reach its maximum height? At this point $v_y = 0$, so the time t_1 at maximum height is given by

$$v_y = v_0 \sin \theta_0 - g t_1 = 0,$$

$$t_1 = \frac{v_0 \sin \theta_0}{g}.$$

Next, what is the value of y at this time? From Eq. (5–11),

$$h = v_0 \sin \theta_0 \left(\frac{v_0 \sin \theta_0}{g}\right) - \frac{1}{2}g\left(\frac{v_0 \sin \theta_0}{g}\right)^2$$

$$= \frac{v_0{}^2 \sin^2 \theta_0}{2g}. \tag{5–18}$$

As expected, the maximum value of h occurs when the projectile is launched straight up, for then $\theta_0 = 90°$, $\sin \theta_0 = 1$, and $h = v_0{}^2/2g$. If it is launched horizontally, $\theta_0 = 0$ and the maximum height is zero!

To find the range, we first find the time t_2 when the projectile returns to the ground. At that time $y = 0$ and, from Eq. (5–11),

$$t_2(v_0 \sin \theta_0 = \tfrac{1}{2}gt_2) = 0.$$

The two roots of this quadratic equation for t_2 are $t_2 = 0$ and $t_2 = 2v_0 \sin \theta_0/g$. The first is obviously the time the projectile *leaves* the ground; the range R is the value of x at the second time. From Eq. (5–10),

$$R = (v_0 \cos \theta_0)\left(\frac{2v_0 \sin \theta_0}{g}\right).$$

According to a familiar trigonometric identity, $2 \sin \theta_0 \cos \theta_0 = \sin 2\theta_0$, so

$$R = \frac{v_0{}^2 \sin 2\theta_0}{g}. \tag{5–19}$$

The maximum value of $\sin 2\theta_0$, namely, unity, occurs when $2\theta_0 = 90°$, or $\theta_0 = 45°$, and this angle gives the maximum range for a given initial speed. ◄

EXAMPLE 5   In some problems we want to know what the departure angle θ_0 should be for a given v_0 to produce a certain range R. Suppose a football player wants to throw a football at $20\ \mathrm{m \cdot s^{-1}}$ to a receiver $30\ \mathrm{m}$ away; at what angle should he throw it?

Solution   From Eq. (5–19),

$$
\begin{aligned}
\theta_0 &= \frac{1}{2}\ \mathrm{arcsin}\ \frac{Rg}{v_0{}^2}\\[2mm]
&= \frac{1}{2}\ \mathrm{arcsin}\ \frac{(30\ \mathrm{m})(9.80\ \mathrm{m \cdot s^{-2}})}{(20\ \mathrm{m \cdot s^{-1}})^2}\\[2mm]
&= \frac{1}{2}\ \mathrm{arcsin}\ 0.735.
\end{aligned}
$$

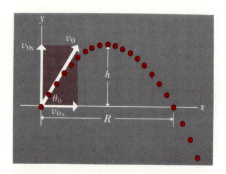

5–9   Trajectory of a body projected at an angle with the horizontal. (Reproduction of a multiflash photograph.)

There are *two* values of θ_0 between 0 and 90° satisfying this requirement, for arcsin 0.735 = 47.3° or 132.7°, giving $\theta_0 = 23.7°$ or 66.3°. Both these angles give the same range; the time of flight and the maximum height are greater for the higher-angle trajectory. Incidentally, the two possible values of θ_0 add to exactly 90°; is this a coincidence? ◄

Figure 5–9 is copied from a multiflash photograph of the trajectory of a ball; x- and y-axes and the initial velocity vector have been added. The horizontal distances between consecutive positions are all equal, showing that the horizontal velocity component is constant. The vertical distances first decrease and then increase, showing that the vertical motion is accelerated.

Figure 5–10 is made from a composite photograph of the three trajectories of a ball projected from a spring gun with departure angles of 30°, 45°, and 60°. The initial speed v_0 is approximately the same in all three cases. It will be seen that the horizontal ranges are (nearly) the same for the 30° and 60° angles and that both are less than the range when the angle is 45°.

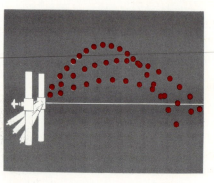

5–10   An angle of departure of 45° gives the maximum horizontal range. (Reproduction of a multiflash photograph.)

If the departure angle is *below* the horizontal as, for instance, in the motion of a ball after it rolls off a sloping roof, or the trajectory of a bomb released from a dive bomber, exactly the same principles apply. The horizontal velocity component remains constant and equal to $v_0 \cos \theta_0$. The vertical motion is the same as that of a body projected *downward* with an initial y-component of velocity $v_0 \sin \theta_0$, which is now a negative quantity.

5–5

CIRCULAR MOTION

It was shown in Section 5–3 that when a particle moves along a curved path it must have a component of acceleration perpendicular to the path, even if its speed is constant. For a *circular* path, there is a simple relation between the normal component of acceleration, the particle's speed, and the radius of the path. We now derive this relation for a particle moving in a circular path with constant speed. Such motion is called *uniform circular motion*.

Figure 5–11a represents a particle moving in a circular path of radius R with center at O. Vectors v_1 and v_2 represent its velocities at points P and Q. The vector change in velocity, Δv, is shown in Fig. 5–11b. The particle moves from P to Q in a time Δt.

The triangles OPQ and opq in Fig. 5–11 are *similar*, since both are isosceles triangles and the angles labeled $\Delta \theta$ are the same. Hence

$$\frac{\Delta v}{v_1} = \frac{\Delta s}{R} \qquad \text{or} \qquad \Delta v = \frac{v_1}{R} \Delta s.$$

The magnitude of the average normal acceleration $\bar{a}_\perp$ during Δt is therefore

$$\bar{a}_\perp = \frac{\Delta v}{\Delta t} = \frac{v_1}{R} \frac{\Delta s}{\Delta t}.$$

The *instantaneous* acceleration $a_\perp$ at point P is the limiting value of this expression, as point Q is taken closer and closer to point P:

$$a_\perp = \lim_{\Delta t \to 0} \frac{v_1}{R} \frac{\Delta s}{\Delta t} = \frac{v_1}{R} \lim_{\Delta t \to 0} \frac{\Delta s}{\Delta t}.$$

But the limiting value of $\Delta s/\Delta t$ is the speed v_1 at point P, and since P can be any point of the path, we can drop the subscript from v_1 and let v represent the speed at any point. Then

$$a_\perp = \frac{v^2}{R}. \tag{5–20}$$

The magnitude of the instantaneous normal acceleration is equal to the square of the speed divided by the radius. The direction is perpendicular to v and inward along the radius, toward the center of the circle. Because of this it is called a *central,* a *centripetal,* or a *radial* acceleration. (The term "centripetal" means "seeking a center.")

The unit of radial acceleration is the same as that of an acceleration resulting from a change in the *magnitude* of a velocity. Thus if a particle

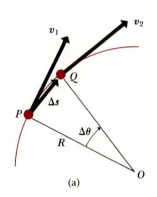

(a)

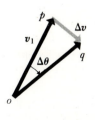

(b)

5–11 Construction for finding change in velocity, Δv, of a particle moving in a circle.

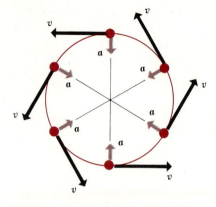

5–12 Velocity and acceleration vectors of a particle in uniform circular motion.

travels with a speed of 4 m·s^{-1} in a circle of radius 2 m, its radial acceleration is

$$a = \frac{(4 \text{ m·s}^{-1})^2}{2 \text{ m}} = 8 \text{ m·s}^{-2}.$$

Figure 5–12 shows the directions of the velocity and acceleration vectors at a number of points, for a particle revolving in a circle with a velocity of constant magnitude.

In the above discussion the particle's speed was assumed constant. If the speed varies, Eq. (5–20) still gives the normal component of acceleration, but in that case there is also a tangential component of acceleration, equal to the rate of change of speed:

$$a_\parallel = \lim_{\Delta t \to 0} \frac{\Delta v_\parallel}{\Delta t}.$$

If the speed is constant, there is no tangential component of acceleration and the acceleration is purely normal, resulting from the continuous change in *direction* of the velocity.

EXAMPLE 1   A car traveling at a constant speed of 20 m·s^{-1} rounds a curve of radius 100 m. What is its acceleration?

Solution   The magnitude of the acceleration is given by Eq. (5–20):

$$a_\perp = \frac{v^2}{R} = \frac{(20 \text{ m·s}^{-1})^2}{100 \text{ m}} = 4.0 \text{ m·s}^{-2}.$$

The direction of *a* at each instant is perpendicular to the velocity and directed toward the center of the circle.   ◄

EXAMPLE 2   In a carnival ride, the passengers travel in a circle of radius 5.0 m, making one complete circle in 4.0 s. What is the acceleration?

Solution   The speed is the circumference of the circle divided by the time for one revolution, called the *period* τ of the motion.

$$v = \frac{2\pi R}{\tau} = \frac{2\pi(5.0 \text{ m})}{4.0 \text{ s}} = 7.85 \text{ m·s}^{-1}.$$

The centripetal acceleration is

$$a_\perp = \frac{v^2}{R} = \frac{(7.85 \text{ m·s}^{-1})^2}{5.0 \text{ m}} = 12.3 \text{ m·s}^{-2}.$$

As in the preceding example, the direction of *a* is always toward the center of the circle. The magnitude of *a* is greater than *g*, the acceleration due to gravity, so this is quite a wild ride!   ◄

5–6

CENTRIPETAL FORCE

Newton's second law governs circular motion as well as all other motion of a particle. The acceleration toward the center of a particle in uniform circular motion must be caused by a *force* also directed toward the center.

Since the magnitude of the radial acceleration equals v^2/R, and its direction is toward the center, the magnitude of the net radial force on a particle of mass m must be

$$F = ma_\perp = m\frac{v^2}{R}. \qquad (5\text{--}21)$$

A familiar example of such a force occurs when one ties an object to a string and whirls it in a circle. The string must constantly pull in toward the center; if it breaks, then the inward force no longer acts, and the object flies off along a tangent to the circle.

More generally, there may be several forces acting on a body in uniform circular motion. In this case the *vector sum* of all forces on the body must have the magnitude given by Eq. (5–21) and must be directed toward the center of the circle. For such problems the problem-solving techniques developed in Section 4–5 are applicable, as the following examples will show.

The force in Eq. (5–21) is sometimes called *centripetal force*. This usage is unfortunate because it implies that this force is somehow different from ordinary forces, or that the fact of circular motion somehow generates an additional force; neither is correct. *Centripetal* refers to the *effect* of the force, namely to the fact that it causes circular motion in which the direction of the velocity changes but not its magnitude. In the equation $\Sigma F = ma$, the sum of forces must include only the real physical forces, pushes or pulls exerted by strings, rods, or other agencies; the quantity mv^2/R does *not* appear in ΣF but belongs on the ma side of the equation.

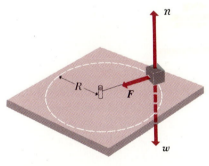

5–13 Body revolving uniformly in a circle on a horizontal frictionless surface.

EXAMPLE 1   A small body of mass 0.2 kg revolves uniformly in a circle on a horizontal frictionless surface such as an "air-hockey" table. It is attached by a cord 0.2 m long to a pin set in the surface. If the body makes two complete revolutions per second, find the force F exerted on it by the cord. (See Fig. 5–13.)

Solution   The circumference of the circle is

$$2\pi(0.2 \text{ m}) = 1.26 \text{ m}$$

so the velocity is $2.51 \text{ m} \cdot \text{s}^{-1}$. The magnitude of the centripetal acceleration is

$$a = \frac{v^2}{R} = \frac{(2.51 \text{ m} \cdot \text{s}^{-1})^2}{0.2 \text{ m}} = 31.6 \text{ m} \cdot \text{s}^{-2}.$$

Since the body has no vertical acceleration, the forces n and w are equal and opposite and the force F is the resultant force. Therefore

$$F = ma = (0.2 \text{ kg})(31.6 \text{ m} \cdot \text{s}^{-2})$$
$$= 6.32 \text{ N.} \qquad \blacktriangleleft$$

EXAMPLE 2   A child's indoor swing consists of a rope of length L anchored to the ceiling, with a seat at the lower end. The total mass of child and seat is m. They swing in a horizontal circle with constant speed v, as shown in Fig. 5–14; as they swing around, the rope makes a constant

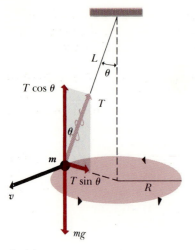

5–14 A child's indoor swing.

angle θ with the vertical. Assuming the time τ for one revolution (i.e., the period) is known, find the tension T in the rope and the angle θ.

Solution A free-body diagram for the system is shown in the figure. The forces on the system in the position shown are the weight mg and the tension T in the rope. The tension has a horizontal component $T \sin \theta$ and a vertical component $T \cos \theta$. (Note that the force diagram in Fig. 5–14 is exactly like that in Fig. 4–9b. The only difference is that in this case the acceleration a is the *radial* acceleration, v^2/R.) The system has no vertical acceleration, so the sum of the vertical components of force is zero. The horizontal force must equal the mass m times the radial (centripetal) acceleration. Thus the $\Sigma F = ma$ equations are

$$T \sin \theta = m\frac{v^2}{R}, \qquad T \cos \theta = mg.$$

When the first of these equations is divided by the second, the result is

$$\tan \theta = \frac{v^2}{Rg}. \tag{5–22}$$

Also, the radius R of the circle is given by $R = L \sin \theta$, and the speed is the circumference $2\pi L \sin \theta$ divided by the period τ:

$$v = \frac{2\pi L \sin \theta}{\tau}.$$

When these relations are used to eliminate R and v from Eq. (5–22), we obtain

$$\cos \theta = \frac{g\tau^2}{4\pi^2 L}, \tag{5–23}$$

or

$$\tau = 2\pi \sqrt{L(\cos \theta)/g}. \tag{5–24}$$

Once θ is determined, the tension T can be found from $T = mg/\cos \theta$. For a given length L, $\cos \theta$ decreases as the time is made shorter, and the angle θ increases. The angle never becomes 90°, however, since this requires that $\tau = 0$ or $v = \infty$. Equation (5–24) is similar in form to the expression for the time of swing of a simple pendulum, which will be derived in Chapter 11. Because of this similarity, the present system is called a *conical pendulum*. ◄

The reader may feel a temptation to add to the forces in Fig. 5–14 an extra outward force, to "keep the body out there" at angle θ, or to "keep it in equilibrium." Perhaps he has encountered the term *centrifugal force*; centrifugal means "fleeing from the center." This temptation must be resisted with all possible strength. First, the body doesn't stay "out there"; it is in constant motion around its circular path. Its velocity is constantly changing in direction, and the body is *not* in equilibrium. Second, if there *were* an additional outward force to balance the inward force, there would be no resultant inward force to cause the circular motion; in that case the body would move in a straight line, not a circle. What is sometimes called centrifugal force is, at least in an inertial frame

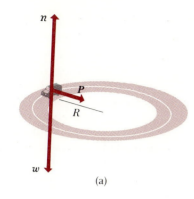

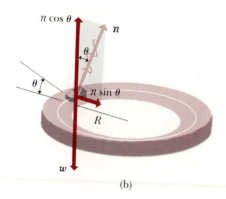

5–15 (a) Forces on a vehicle rounding a curve on a level track. (b) Forces when the track is banked.

of reference, not a force at all but a reflection of Newton's *first* law. In the absence of force a body moves in a straight line with constant speed, and for circular motion to occur there must be a net *inward* force.

EXAMPLE 3   The driver of a sports car is rounding a flat, unbanked curve of radius R. If the coefficient of friction between tires and road is μ_s, what is the maximum speed v at which he can take the curve without sliding?

Solution   Figure 5–15(a) is a free-body diagram for the situation. The acceleration v^2/R toward the center of the curve must be caused by the friction force $\mathcal{F}$, and there is no vertical acceleration. Thus we have

$$\mathcal{F} = m\frac{v^2}{R}, \qquad \mathcal{N} = mg.$$

The maximum friction force available is $\mathcal{F} = \mu_s \mathcal{N} = \mu_s mg$. Combining these relations to eliminate $\mathcal{F}$, we find

$$\mu_s mg = m\frac{v^2}{R},$$

or

$$v = \sqrt{\mu_s gR}.$$

If $\mu = 0.5$ and $R = 25$ m, then

$$v = \sqrt{(0.5)(9.8 \text{ m} \cdot \text{s}^{-2})(25 \text{ m})} = 11.1 \text{ m} \cdot \text{s}^{-1},$$

or about $25 \text{ mi} \cdot \text{hr}^{-1}$. Why did we use the coefficient of *static* friction for a moving car?   ◄

EXAMPLE 4   An engineer proposes to rebuild the curve in Example 3, banking it so that at a certain speed v *no* friction at all is needed for the car to make the curve. At what angle θ should it be banked?

Solution   The free-body diagram is shown in Fig. 5–15b. The normal force is no longer vertical but is perpendicular to the roadway, at an angle of θ with the vertical. Thus it has a vertical component $\mathcal{N} \cos \theta$ and a horizontal component $\mathcal{N} \sin \theta$, as shown. The horizontal component of $\mathcal{N}$ must now cause the acceleration v^2/R, and again there is no vertical acceleration. Thus

$$\mathcal{N} \sin \theta = \frac{mv^2}{R}, \qquad \mathcal{N} \cos \theta = mg.$$

Dividing the first of these equations by the second, we find

$$\tan \theta = \frac{v^2}{Rg}. \tag{5–25}$$

If $R = 25$ m and $v = 11.1 \text{ m} \cdot \text{s}^{-1}$ as in Example 3, then

$$\theta = \arctan \frac{(11.1 \text{ m} \cdot \text{s}^{-1})^2}{(25 \text{ m})(9.8 \text{ m} \cdot \text{s}^{-2})} = 26.6°.$$

Equation (5–25) shows that the tangent of the angle of banking is proportional to the square of the speed and inversely proportional to the radius. For a given radius no one angle is correct for all speeds. Hence in the design of highways and railroads, curves are banked for the *average speed* of the traffic over them. The same considerations apply to the correct banking angle of a plane when it makes a turn in level flight. ◀

5–7

MOTION IN A VERTICAL CIRCLE

Figure 5–16 represents a small body attached to a cord of length R and whirling in a *vertical* circle about a fixed point O to which the other end of the cord is attached. The motion, while circular, is not *uniform*, since the speed increases on the way down and decreases on the way up. The normal component of acceleration is still given by $a_\perp = v^2/R$, as discussed above, but v varies from point to point. In addition there is now a *tangential* component of acceleration.

The forces on the body at any point are its weight $w = mg$ and the tension T in the cord. We resolve the weight into a normal component, of magnitude $mg \cos \theta$, and a tangential component of magnitude $mg \sin \theta$, as in Fig. 5–16. The resultant tangential and normal forces are then

$$F_\| = mg \sin \theta \quad \text{and} \quad F_\perp = T - mg \cos \theta.$$

The tangential acceleration, from Newton's second law, is

$$a_\| = \frac{F_\|}{m} = g \sin \theta;$$

this is the same as the acceleration of a body sliding on a frictionless inclined plane with slope angle θ. The normal (radial) acceleration $a_\perp = v^2/R$ is

$$a_\perp = \frac{F_\perp}{m} = \frac{T - mg \cos \theta}{m} = \frac{v^2}{R};$$

to find the tension in the cord, we solve this for T:

$$T = m\left(\frac{v^2}{R} + g \cos \theta\right). \tag{5–26}$$

At the lowest point of the path, $\theta = 0$, $\sin \theta = 0$, and $\cos \theta = 1$. Hence at this point $F_\| = 0$, $a_\| = 0$, and the acceleration is purely radial (upward). The magnitude of the tension, from Eq. (5–26), is

$$T = m\left(\frac{v^2}{R} + g\right).$$

At the highest point, $\theta = 180°$, $\sin \theta = 0$, $\cos \theta = -1$, and the acceleration is again purely radial (downward). The tension is

$$T = m\left(\frac{v^2}{R} - g\right). \tag{5–27}$$

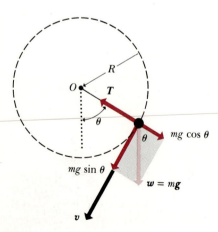

5–16 Forces on a body whirling in a vertical circle with center at O.

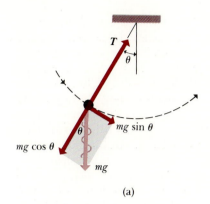

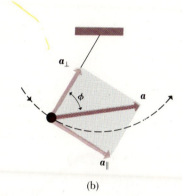

(a)

(b)

5-17 (a) Forces on a body swinging in a vertical circle. (b) The radial and tangential components of acceleration are combined to obtain the resultant acceleration *a*.

With motion of this sort, it is a familiar fact that when the speed at the highest point is less than some critical value v_c, the cord becomes slack and the path is no longer circular. To find this critical speed, we set $T = 0$ in Eq. (5-27):

$$0 = m\left(\frac{v_c^2}{R} - g\right), \quad v_c = \sqrt{Rg}.$$

If $R = 1$ m, then

$$v_c = \sqrt{(1 \text{ m})(9.8 \text{ m} \cdot \text{s}^{-2})} = 3.13 \text{ m} \cdot \text{s}^{-1}.$$

EXAMPLE In Fig. 5-17, a small body of mass $m = 0.10$ kg swings in a vertical circle at the end of a cord of length $R = 1.0$ m. If its speed $v = 2.0$ m·s^{-1} when the cord makes an angle $\theta = 30°$ with the vertical, find

a) the radial and tangential components of its acceleration at this instant,

b) the magnitude and direction of the resultant acceleration, and

c) the tension T in the cord.

Solution

a) The radial component of acceleration is

$$a_\perp = \frac{v^2}{R} = \frac{(2.0 \text{ m} \cdot \text{s}^{-1})^2}{1.0 \text{ m}} = 4.0 \text{ m} \cdot \text{s}^{-2}.$$

The tangential component of acceleration, due to the tangential force $mg \sin \theta$, is

$$a_\parallel = g \sin \theta = (9.8 \text{ m} \cdot \text{s}^{-2})(0.50) = 4.9 \text{ m} \cdot \text{s}^{-2}.$$

b) The magnitude of the resultant acceleration (see Fig. 5-17b) is

$$a = \sqrt{a_\perp^2 = a_\parallel^2} = 6.33 \text{ m} \cdot \text{s}^{-2}.$$

The angle ϕ is

$$\phi = \arctan\frac{a_\parallel}{a_\perp} = 50.8°.$$

c) The tension in the cord is given by $F_\perp = ma_\perp$: $T - mg \cos \theta = mv^2/R$, so

$$T = m\left(\frac{v^2}{R} + g \cos \theta\right) = 1.25 \text{ N}.$$

Note that the magnitude of the tangential acceleration is not constant but is proportional to the sine of the angle θ. Hence the equations of motion with constant acceleration *cannot* be used to find the speed at other points of the path. We shall show in the next chapter, however, how the speed at any point can be found from *energy* considerations.

◄

5–8

MOTION OF A SATELLITE

In discussing the trajectory of a projectile in Section 5–4, we assumed that the gravitational force on the projectile (its weight w) had the same direction and magnitude at all points of the trajectory. These conditions are approximately satisfied if the projectile remains near the earth's surface and if its trajectory is small compared to the earth's radius. Under these conditions (if air resistance is neglected) the trajectory is a *parabola*.

In reality, the gravitational force is directed toward the center of the earth and is inversely proportional to the square of the distance from the earth's center, so that it is *not* constant in either magnitude or direction. It can be shown that under an inverse square force directed toward a fixed point the trajectory must always be a circle, ellipse, parabola, or hyperbola.

Suppose that a very tall tower could be constructed as in Fig. 5–18, and that a projectile were launched from point A at the top of the tower in the "horizontal" direction AB. If the initial velocity is not too great, the trajectory will be like that numbered (1), which is a portion of an ellipse

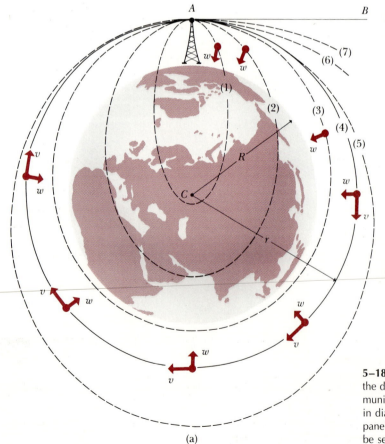

(a)

(b)

5–18 (a) Trajectories of a body projected from point A in the direction AB with different initial velocities. (b) A communications satellite before launch. The main body is 2.16 m in diameter and 2.82 m in height. Solar-energy collector panels on the sides and a parabolic antenna at the end can be seen. (Courtesy Hughes Aircraft Co.)

with the earth's center C at one focus. If the trajectory is so short that changes in the magnitude and direction of w can be neglected, the ellipse is approximately the same shape as a parabola.

Trajectories (2) to (7) illustrate the effect of increasing the initial velocity. Trajectory (2) is again a portion of an ellipse. Trajectory (3), which just misses the earth, is a *complete* ellipse, and the projectile has become an earth satellite. Its velocity when it returns to point A is the same as its initial velocity, and in the absence of retarding forces it will repeat its motion indefinitely. The earth's rotation will have moved the tower to a different point by the time the satellite returns to point A, but this does not affect the orbit.

Trajectory (4) is a special case in which the orbit is a circle. Trajectory (5) is again an ellipse, (6) is a parabola, and (7) is a hyperbola. Trajectories (6) and (7) are not closed orbits; for these the projectile never returns to its starting point, but "escapes" permanently from the earth's gravitational field.

For simplicity we shall consider only circular orbits; many manmade satellites have very nearly circular orbits. The force that provides the necessary centripetal acceleration is the gravitational attraction of the earth (mass m_E) for the satellite (mass m). If the radius of the circular orbit (measure from the center of the earth) is r, then the gravitational force is given by Gm_Em/r^2, as discussed in Section 4–4. This is to be equated to the product of the satellite mass m and its acceleration v^2/r. Thus for a circular satellite orbit, Newton's second law requires

$$\frac{Gm_Em}{r^2} = \frac{mv^2}{r}.$$

When this is solved for v, the result is

$$v = \sqrt{\frac{Gm_E}{r}}. \tag{5–28}$$

The relation shows that once the orbit radius is specified the satellite speed cannot be chosen arbitrarily but must be given by Eq. (5–28).

The orbit radius can also be related to the time required for one revolution, called the *period* τ of the satellite, or to the number of revolutions per unit time, called the *frequency* f. We first note that τ and f are related reciprocally; $\tau = 1/f$. Next, the speed is equal to the distance traveled in one revolution (the circumference $2\pi r$) divided by the time for one revolution. Thus

$$v = \frac{2\pi r}{\tau} = 2\pi r f. \tag{5–29}$$

An expression for τ can be obtained by solving this for τ and combining with Eq. (5–28):

$$\tau = \frac{2\pi r}{v} = 2\pi r \sqrt{\frac{r}{Gm_E}} = \frac{2\pi r^{3/2}}{\sqrt{Gm_E}}. \tag{5–30}$$

Equations (5–28) and (5–30) show that larger orbits correspond to longer periods and slower speeds.

EXAMPLE   An earth satellite revolves in a circular orbit at a height of 300 km above the earth's surface. Find its speed, its period, and its radial acceleration. The earth's radius is 6380 km = 6.38×10^6 m, and its mass is 5.98×10^{24} kg. (See Appendix G.)

Solution   The radius of the orbit is

$$r = 6380 \text{ km} + 300 \text{ km} = 6680 \text{ km} = 6.68 \times 10^6 \text{ m}.$$

From Eq. (5–28),

$$v = \sqrt{\frac{(6.67 \times 10^{-11} \text{ N} \cdot \text{m}^2 \cdot \text{kg}^{-2})(5.98 \times 10^{24} \text{ kg})}{6.68 \times 10^6 \text{ m}}}$$

$$= 7728 \text{ m} \cdot \text{s}^{-1}.$$

From Eq. (5–29),

$$\tau = \frac{2\pi(6.68 \times 10^6 \text{ m})}{7728 \text{ m} \cdot \text{s}^{-1}} = 5431 \text{ s} = 90.5 \text{ min}.$$

The radial acceleration is

$$a_\perp = \frac{v^2}{r} = \frac{(7728 \text{ m} \cdot \text{s}^{-1})^2}{6.68 \times 10^6 \text{ m}} = 8.94 \text{ m} \cdot \text{s}^{-2}.$$

This is somewhat less than the value of the free-fall acceleration g at the earth's surface and is in fact equal to the value of g at a height of 300 km above the earth's surface.   ◀

Our discussion has centered around the motion of manmade earth satellites, but it should be clear that it is also applicable to the circular motion of *any* body under its gravitational attraction to a stationary body. Other familiar examples are the motion of our moon and that of the moons of Jupiter and the motions of the planets around the sun in nearly circular orbits. The rings of Saturn, shown in Fig. 5–19, are composed of small particles in circular orbits around the planet.

Indeed, the study of planetary motion historically played a pivotal role in the development of physics. Johannes Kepler (1571–1630) spent several painstaking years analyzing the motions of the planets, basing his work on remarkably precise measurements made by the Danish astronomer Tycho Brahe (1546–1601) *without the aid of a telescope.* (The telescope was invented in 1608.) Kepler discovered that the orbits of the planets are (nearly circular) ellipses, and that the period of a planet in its orbit is proportional to the three-halves power of the orbit radius, as Eq. (5–30) shows.

But it remained for Newton (1642–1727) to show, with the aid of his laws of motion and law of gravitation, that this behavior of the planets could be understood on the basis of the very same physical principles he had developed to analyze *terrestrial* motion. From our historical perspective three hundred years later, there is absolutely no doubt that this *Newtonian synthesis,* as it has come to be called, is one of the greatest achievements in the entire history of science, certainly comparable in significance to the development of quantum mechanics, the theory of relativity, and the understanding of DNA, in our own century.

5–19 The rings of Saturn, photographed from Voyager 1 on November 1, 1980, from a range of 700,000 km. Individual particles move in circular orbits, with periods given by Eq. (5–30), using Saturn's mass. The outer portions of the rings take more time for one revolution than the inner portions. (Courtesy of Jet Propulsion Laboratory/NASA.)

*5–9

THE CENTRIFUGE

The centrifuge is an important laboratory tool for separating particles from a liquid having different density from that of the particles. Muddy water is a familiar example. In still water, the mud particles settle to the bottom, but the process may take several hours or days because the terminal velocities of the particles, given by Eq. (4–12), are extremely small.

The process may be hastened greatly by whirling the container in a circle at high speed. The particles move approximately in circles, but the resisting force of the fluid on a particle, given by Eq. (4–10), must now equal the mass m of the particle times the centripetal acceleration v^2/r. The terminal velocity is now given by

$$v_t = \frac{mv^2/r}{k}.$$

Thus the terminal velocity or *sedimentation rate* is increased by a factor v^2/gr, which may be many thousands.

Centrifuges are used to separate cream from milk, blood cells from plasma, silt from river water, macromolecules from solutions, and for many other applications. In one type of laboratory centrifuge the specimen is placed in a test tube in a rotor that holds it at an angle while rotating at high speed, driven by an electric motor.

EXAMPLE   A centrifuge rotor rotates at 9000 rev·min^{-1}, and the specimen is 10 cm from the axis. What is its acceleration? By what factor is the sedimentation rate increased?

Solution   The specimen moves in a circle of radius 0.10 m and circumference $2\pi(0.10 \text{ m}) = 0.628 \text{ m}$, 9000 times per minute or $9000/60 = 150$ times per second. Its speed is $(0.628 \text{ m})(150 \text{ s}^{-1}) = 94.2 \text{ m}\cdot\text{s}^{-1}$. Its acceleration is

$$a = \frac{v^2}{r} = \frac{(94.2 \text{ m}\cdot\text{s}^{-1})^2}{0.1 \text{ m}} = 8.88 \times 10^4 \text{ m}\cdot\text{s}^{-2}.$$

Thus the sedimentation rate is increased by a factor of

$$\frac{8.88 \times 10^4 \text{ m}\cdot\text{s}^{-2}}{9.8 \text{ m}\cdot\text{s}^{-2}} = 9063.$$

We have ignored the effects of buoyancy on particles in the liquid. This omission can be justified in detail, but we shall postpone that discussion until later.   ◀

*5–10

EFFECT OF THE EARTH'S ROTATION ON g

Because the earth rotates, it is not precisely an inertial frame of reference, and the apparent weight of an object on earth is not precisely equal to the earth's gravitational attraction. Figure 5–20 is a cutaway view of the earth, showing three observers, each holding a body of mass m hanging from a string. Each observer, unaware of the earth's rotation, *thinks* the body is in equilibrium under the action of the tension T in his string

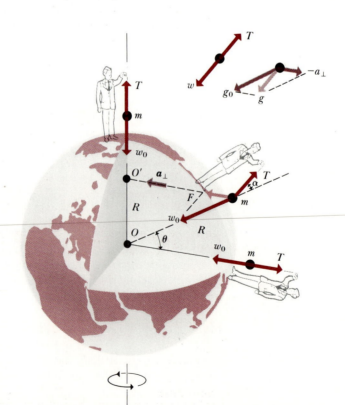

5–20   The resultant of the forces T and w_0 is equal to m times the centripetal acceleration $a_\perp = v^2/r$.

and an equal and opposite force w, which we shall call the *apparent weight*. This is different at different points on the earth, as we shall see.

If we assume the earth to be spherically symmetric, the earth's gravitational attraction, that is, the *true* weight, which we call w_0, has the same magnitude $F_g = Gmm_E/R^2$ at all points on the earth's surface. We also suppose that the center of the earth can be taken as the origin of an inertial coordinate system; this ignores the orbital motion of the earth, which is a much smaller effect than that of rotation about its axis. Then the body at the north pole really *is* in equilibrium in an inertial system, and the tension in that observer's string is equal to w_0. But the body at the equator is moving in a circle of radius R with speed v, and there must be a net inward force equal to the mass times the centripetal acceleration:

$$w_0 - T = mv^2/R. \tag{5–31}$$

Thus the magnitude of the apparent weight (equal to T at this location) is

$$w = w_0 - \frac{mv^2}{R}. \tag{5–32}$$

We also note that if the earth were not rotating, the body when released would have a free-fall acceleration g_0 given by $g_0 = w_0/m$, but that its actual acceleration relative to the observer at the equator is given by $g = w/m$. Dividing Eq. (5–32) by m and using these relations, we find

$$g = g_0 - \frac{v^2}{R}. \tag{5–33}$$

To evaluate v^2/R, we note that in one day (86,400 s) a point at the equator moves a distance equal to the earth's circumference, $2\pi R = 2\pi(6.38 \times 10^6$ m). Thus

$$v = \frac{2\pi(6.38 \times 10^6 \text{ m})}{86,400 \text{ s}} = 464 \text{ m} \cdot \text{s}^{-1},$$

and

$$\frac{v^2}{R} = \frac{(464 \text{ m} \cdot \text{s}^{-1})^2}{6.38 \times 10^6 \text{ m}} = 0.0337 \text{ m} \cdot \text{s}^{-2}.$$

Thus for a spherically symmetric earth the acceleration due to gravity should be about 0.03 m·s^{-2} less at the equator than at the poles.

Table 5–1 gives the values of g at various locations, showing variations of approximately this magnitude. There are also small variations

TABLE 5–1   VARIATIONS OF g WITH LATITUDE AND ELEVATION

Station	North latitude	Elevation, m	g, m·s^{-2}	g, ft·s^{-2}
Canal Zone	9°	0	9.78243	32.0944
Jamaica	18°	0	9.78591	32.1059
Bermuda	32°	0	9.79806	32.1548
Denver	40°	1638	9.79609	32.1393
Cambridge, Mass.	42°	0	9.80398	32.1652
Pittsburgh, Pa.	40.5°	235	9.80118	32.1561
Greenland	70°	0	9.82534	32.2353

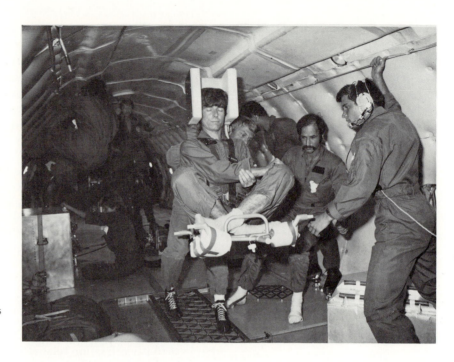

5–21 Astronauts conducting experiments in zero-gravity environment. (Courtesy of NASA, Johnson Space Center.)

due to lack of perfectly spherical symmetry of the earth, local variations in density, and differences in elevation.

At points intermediate between the equator and the poles, the difference in the magnitudes of g and g_0 is intermediate between zero and 0.0337 m·s^{-2}, and the *direction* of the apparent weight differs from the direction toward the center, by an angle α as shown.

A similar discussion can be applied to the phenomenon of "weightlessness" in satellites. A space vehicle in orbit is a freely falling body, with an acceleration $\boldsymbol{a}$ toward the earth's center equal to the value of the acceleration of gravity $\boldsymbol{g}$ at its orbit radius. The apparent weight $\boldsymbol{w}$ is given as before by

$$w = w_0 - ma = mg - a.$$

But, in this case,

$$g = a,$$

so

$$w = 0.$$

It is in this sense that an astronaut in the vehicle is said to be "weightless" or in a state of "zero g." (See Fig. 5–21.) The vehicle is like the freely falling elevator discussed in Section 4–5, Example 7, except that it has a large constant tangential speed along its orbit.

5–11

RELATIVE VELOCITY

The concept of relative velocity, introduced in Section 3–6 for motion along a straight line, is easily extended to include motion in a plane or in space. Suppose that, in the example preceding Eq. (3–16), the automobile is traveling at some instant not in the same direction as the train but perpendicular to this direction, *across* the flatcars at 40 km·hr^{-1}. If the flatcars are moving at 30 km·hr^{-1} relative to the earth, then the velocity of the automobile *relative to the earth* is the vector sum of these two velocities. That is, Eq. (3–16) is a special case of the more general *vector equation:*

$$v_{AE} = v_{AF} + v_{FE}. \tag{5–34}$$

Thus the automobile's velocity relative to the earth would be 50 km·hr^{-1} at an angle of 53° to the direction of the train's motion.

Equation (5–34) can be extended to include any number of relative velocities. For example, if a bug B crawls along the floor of the automobile with a velocity relative to the automobile of v_{BA}, his velocity relative to the earth is the vector sum of his velocity relative to the automobile and that of the velocity of the automobile relative to the earth:

$$v_{BE} = v_{BA} + v_{AE}.$$

When this is combined with Eq. (5–34), we find

$$v_{BE} = v_{BA} + v_{AF} + v_{FE}. \tag{5–35}$$

This equation illustrates the general rule for combining relative velocities.

1. Write each velocity with a double subscript in the *proper order,* meaning "velocity of (first subscript) relative to (second subscript)."

2. When relative velocities are added, the first letter of any subscript is to be the same as the last letter of the preceding subscript.

3. The first letter of the subscript of the first velocity in the sum, and the second letter of the subscript of the last velocity, are the subscripts, in that order, of the relative velocity represented by the sum. This somewhat lengthy statement should be clear when it is compared with Eq. (5–34) or (5–35).

Any of the relative velocities in an equation such as Eq. (5–34) can be transferred from one side of the equation to the other, with sign reversed. Thus Eq. (5–34) can be written

$$v_{AF} = v_{AE} - v_{FE}.$$

The velocity of the automobile relative to the flatcar equals the *vector difference* between the velocities of automobile and flatcar, each relative to the earth.

One more point should be noted. The velocity of body A relative to body B, v_{AB}, is the negative of the velocity of B relative to A, v_{BA}:

$$v_{AB} = -v_{BA}.$$

That is, v_{AB} is equal in magnitude and opposite in direction to v_{BA}. If the automobile is traveling to the *right* at 40 mi·hr^{-1}, relative to the flatcars, the flatcars are traveling to the *left* at 40 mi·hr^{-1}, relative to the automobile.

EXAMPLE 1 The compass of an aircraft indicates that it is headed due north, and its airspeed indicator shows that it is moving through the air

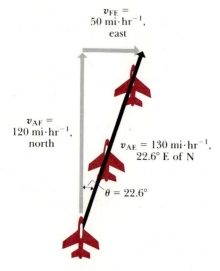

5–22 Vector diagram for aircraft flying north, wind blowing east, and resultant velocity vector of the plane.

at 120 mi·hr^{-1}. If there is a wind of 50 mi·hr^{-1} from west to east, what is the velocity of the aircraft relative to the earth?

Solution   Let subscript A refer to the aircraft, and subscript F to the moving air (which now corresponds to the flatcar in Fig. 3–11). Subscript E refers to the earth. We have given

$$v_{AF} = 120 \text{ mi·hr}^{-1}, \quad \text{due north,}$$
$$v_{FE} = 50 \text{ mi·hr}^{-1}, \quad \text{due east,}$$

and we wish to find the magnitude and direction of v_{AE}:

$$v_{AE} = v_{AF} + v_{FE}.$$

The three relative velocities are shown in Fig. 5–22. It follows from this diagram that

$$v_{AE} = 130 \text{ mi·hr}^{-1}, \quad \theta = \arctan\frac{50 \text{ mi·hr}^{-1}}{120 \text{ mi·hr}^{-1}} = 22.6° \text{ E of N.} \quad \blacktriangleleft$$

EXAMPLE 2   In what direction should the pilot head in order to travel due north? What will then be his velocity relative to the earth? (The magnitude of his airspeed, and the wind velocity, are the same as in the preceding example.)

Solution   We now have given

$$v_{AF} = 120 \text{ mi·hr}^{-1}, \quad \text{direction unknown,}$$
$$v_{FE} = 50 \text{ mi·hr}^{-1}, \quad \text{due east,}$$

and we wish to find v_{AE}, whose magnitude is unknown but whose direction is due north. (Note that both this and the preceding example require us to determine two unknown quantities. In the former example, these were the *magnitude and direction* of v_{AE}. In this example, the unknowns are the direction of v_{AF} and the *magnitude* of v_{AE}.)

The three relative velocities must still satisfy the *vector equation*

$$v_{AE} = v_{AF} + v_{FE}.$$

The appropriate vector diagram is shown in Fig. 5–23. We find

$$v_{AE} = \sqrt{(120 \text{ mi·hr}^{-1})^2 - (50 \text{ mi·hr}^{-1})^2} = 109 \text{ mi·hr}^{-1},$$
$$\theta = \arcsin\frac{50 \text{ mi·hr}^{-1}}{120 \text{ mi·hr}^{-1}} = 24.6°.$$

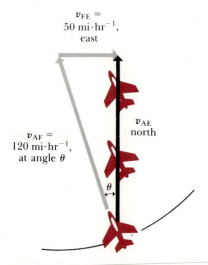

5–23   The resultant vector v_{AF} shows the direction in which the pilot should point the plane in order to have the plane travel due north.

That is, the pilot should head 24.6° W of N, and his ground speed will be 109 mi·hr^{-1}. $\quad \blacktriangleleft$

QUESTIONS

5–1 A simple pendulum (a mass swinging at the end of a string) swings back and forth in a circular arc. What is the direction of its acceleration at the ends of the swing? At the midpoint?

5–2 A football is thrown in a parabolic path. Is there any point at which the acceleration is parallel to the velocity? Perpendicular to the velocity?

5–3 When a rifle is fired at a distant target, the barrel is not lined up exactly on the target. Why not? Does the angle of correction depend on the distance of the target?

5–4 The acceleration of a falling body is measured in an elevator traveling upward at a constant speed of 9.8 m·s^{-1}. What result is obtained?

5–5 One can play catch with a softball in an airplane in level flight just as though the plane were at rest. Is this still possible when the plane is making a turn?

5–6 A package is dropped out of an airplane in level flight. If air resistance could be neglected, how would the motion of the package look to the pilot? To an observer on the ground?

5–7 No matter what the initial velocity, the motion of a projectile (neglecting air resistance) is *always* confined to a single plane. Why?

5–8 If a jumper can give himself the same initial speed regardless of the direction he jumps (forward or straight up), how is his maximum vertical jump (high jump) related to his maximum horizontal jump (broad jump)?

5–9 A manned space-flight projectile is launched in a parabolic trajectory. A man in the capsule feels weightless. Why? In what sense is he weightless?

5–10 The maximum range of a projectile occurs when it is aimed at 45°, if air resistance is neglected. Is this still true if air resistance is included? If not, is the optimum angle greater or less than 45°? Why?

5–11 A passenger in a car rounding a sharp curve is thrown toward the outside of the curve. What force throws him in this direction?

5–12 In uniform circular motion, what is the *average* velocity during one revolution? The average acceleration?

5–13 In uniform circular motion the acceleration is perpendicular to the velocity at every instant, even though both change continuously in direction. Is there any other motion having this property, or is uniform circular motion unique?

5–14 A certain centrifuge is claimed to operate at 100,000g. What does this mean?

5–15 If an artificial earth satellite is in an orbit with a period of exactly one day, how does its motion look to an observer on the rotating earth? (Such an orbit is said to be *geosynchronous;* most communications satellites are placed in such orbits.)

5–16 How could the motion of a satellite be used to determine the mass of the earth? Could a similar procedure be used to obtain the mass of the sun?

5–17 Raindrops hitting the side windows of a car in motion often leave diagonal streaks. Why? What about diagonal streaks on the windshield? Is the explanation the same or different?

5–18 In a rainstorm with a strong wind, what determines the best position in which to hold an umbrella?

5–19 In uninformed discussions of satellites, one hears questions such as "What keeps the satellite moving in its orbit?" and "What keeps the satellite up?" How do you answer these questions? Are your answers applicable to the moon?

PROBLEMS

5–1 A tourist throws a rock horizontally off the edge of a high cliff, with an initial speed of 10 m·s^{-1}.

a) Taking the y-axis to be upward, write equations for the x- and y-coordinates as functions of time.

b) Calculate the values of x and y at times t = 0, 0.5, 1.0, 1.5, and 2.0 s. Plot these points on graph paper and sketch the rock's trajectory.

c) Write equations for the x- and y-components of the rock's velocity.

d) Find the magnitude and direction of the velocity at time t = 1 s. Sketch the vector on the trajectory obtained in (b), at the appropriate point. Is the velocity tangent to the trajectory?

5–2 A golf ball is hit horizontally from a tee at the edge of a cliff. Its x- and y-coordinates are given as functions of time by

$$x = (20 \text{ m·s}^{-1})t, \qquad y = -(4.9 \text{ m·s}^{-2})t^2.$$

a) Compute the x- and y-coordinates at times t = 0, 1, 2, 3, and 4 s. Plot these positions on graph paper and sketch the trajectory.

b) Determine the ball's initial velocity and its acceleration.

c) Find the x- and y-components of velocity at time t = 2 s. Plot the velocity vector at the appropriate point on the trajectory obtained in (a). Is the velocity tangent to the trajectory?

5–3 A ball rolls off the edge of a tabletop 1 m above the floor and strikes the floor at a point 1.5 m horizontally from the edge of the table.

a) Find the time of flight.

b) Find the initial velocity.

c) Find the magnitude and direction of the velocity of the ball just before it strikes the floor. Draw a diagram to scale.

5–4 A block slides off a horizontal tabletop 1 m high with a speed of $3 \text{ m} \cdot \text{s}^{-1}$. Find

  a) the horizontal distance from the edge of the table at which the block strikes the floor, and

  b) the horizontal and vertical components of its velocity when it reaches the floor.

5–5 An airplane flying at $100 \text{ m} \cdot \text{s}^{-1}$ drops a box at an elevation of 2000 m.

  a) How much time is required for the box to reach the earth?

  b) How far does it travel horizontally while falling?

  c) Find the horizontal and vertical components of its velocity when it strikes the earth.

5–6 A block sliding toward the edge of a horizontal tabletop has a speed of $4 \text{ m} \cdot \text{s}^{-1}$ at a point 3 m from the edge. It slides off the edge of the table, which is 1 m high, and strikes the floor 1 m from the edge of the table. What was the coefficient of sliding friction between block and table?

5–7 A golf ball is driven horizontally from an elevated tee with a speed of $25 \text{ m} \cdot \text{s}^{-1}$. It strikes the fairway 2.5 s later.

  a) How far has it fallen vertically?

  b) How far has it traveled horizontally?

  c) Find the horizontal and vertical components of its velocity, and the magnitude and direction of its velocity, just before it strikes.

5–8 A baseball is thrown with an initial upward velocity component of $20 \text{ m} \cdot \text{s}^{-1}$ and a horizontal velocity component of $25 \text{ m} \cdot \text{s}^{-1}$.

  a) Find the position and velocity of the ball after 1 s, 2 s, 3 s, 4 s.

  b) How much time is required to reach the highest point of the trajectory?

  c) How high is this point?

  d) How much time (after it is thrown) is required for the ball to return to its original level?

  e) How far has it traveled horizontally during this time? Show your results in a neat sketch, large enough to show all features clearly. Label the value of t at each point.

5–9 A marksman fires a .22-caliber rifle horizontally at a target; the bullet has a muzzle velocity of $300 \text{ m} \cdot \text{s}^{-1}$. How much does the bullet drop in flight if the target is

  a) 50 m away,

  b) 150 m away?

5–10 A batted baseball leaves the bat at an angle of 30° above the horizontal and is caught by an outfielder 400 ft from the plate.

  a) What was the initial speed of the ball?

  b) How high did it rise?

  c) How long was it in the air?

5–11 Suppose the departure angle θ_0 in Fig. 5–8 is 15° and the distance x is 5 m. Where will the two objects meet if the initial speed of the first is

  a) $20 \text{ m} \cdot \text{s}^{-1}$,

  b) $5 \text{ m} \cdot \text{s}^{-1}$? Sketch both trajectories.

  c) Will they still meet if the departure angle is below horizontal?

5–12 If a baseball player can throw a ball a maximum distance of 60 m over the ground, what is the maximum vertical height to which he can throw it? Assume the ball to have the same initial speed in each case.

5–13 A player kicks a football at an angle of 37° with the horizontal and with an initial speed of $48 \text{ ft} \cdot \text{s}^{-1}$. A second player standing at a distance of 100 ft from the first in the direction of the kick starts running to meet the ball at the instant it is kicked. How fast must he run in order to catch the ball before it hits the ground?

5–14 A baseball leaves the bat at a height of 4 ft above the ground, traveling at an angle of 45° with the horizontal, and with a speed such that the horizontal range would be 400 ft. At a distance of 360 ft from home plate is a fence 30 ft high. Will the ball be a home run?

5–15 A projectile shot at an angle of 60° above the horizontal strikes a building 30 m away at a point 15 m above the point of projection.

  a) Find the speed of projection.

  b) Find the magnitude and direction of the velocity of the projectile when it strikes the building.

5–16 A boy throws a water-filled balloon at an angle of 53.1° with a speed of $10 \text{ m} \cdot \text{s}^{-1}$. A car is advancing toward the boy at a constant speed of $5 \text{ m} \cdot \text{s}^{-1}$. If the balloon is to hit the car, how far away should the car be when the balloon is thrown?

5–17 Prove that a projectile launched at angle θ_0 has the same range as one launched with the same speed at angle $(90° - \theta_0)$.

5–18 An airplane diving at an angle of 36.9° with the horizontal drops a bag of sand from an altitude of 800 m. The bag is observed to strike the ground 5 s after its release.

  a) What is the speed of the plane?

  b) How far does the bag travel horizontally during its fall?

  c) What are the horizontal and vertical components of its velocity just before it strikes the ground?

5–19 A 10-kg stone is dropped from a cliff in a high wind. The wind exerts a steady horizontal 50-N force on the stone as it falls. Is the path of the stone a straight line, a parabola, or some more complicated path? Explain.

5–20 A man is riding on a flatcar traveling with a constant speed of 30 ft·s⁻¹ (Fig. 5–24). He wishes to throw a ball through a stationary hoop 16 ft above the height of his hands in such a manner that the ball will move horizontally as it passes through the hoop. He throws the ball with a speed of 40 ft·s⁻¹ with respect to himself.

a) What must be the vertical component of the initial velocity of the ball?

b) How many seconds after he releases the ball will it pass through the hoop?

c) At what horizontal distance in front of the hoop must he release the ball?

16 ft

30 ft·s⁻¹

Figure 5–24

5–21 Find the acceleration of a point at the edge of a 12-in. record turning at $33\frac{1}{3}$ rev·min⁻¹.

5–22 An automobile travels around a circular track 2000 m in circumference at a constant speed of 30 m·s⁻¹.

a) Show in a scale diagram the velocity vectors of the automobile at the beginning and end of a time interval $\Delta t = 5$ s. Let 1 cm = 5 m·s⁻¹.

b) Find graphically the change in velocity, Δv, in this time interval.

c) Find the magnitude of the average acceleration during this interval, $\Delta v/\Delta t$.

d) What is the radial acceleration, v^2/R?

e) If the track is 10 m wide, what should be the elevation of the outer circumference above the inner circumference, for the speed above?

5–23 The radius of the earth's orbit around the sun (assumed circular) is 1.49×10^{11} m and the earth travels around this orbit in 365 days.

a) What is the magnitude of the orbital velocity of the earth, in meters per second?

b) What is the radial acceleration of the earth toward the sun, in meters per second squared?

5–24 A model of a helicopter rotor has four blades, each 2 m in length, and is rotated in a wind tunnel at 1500 rev·min⁻¹.

a) What is the linear speed of the blade tip, in meters per second?

b) What is the radial acceleration of the blade tip, expressed as a multiple of the acceleration of gravity, g?

c) A pressure-measuring device with a mass of 0.1 kg is mounted at the blade tip. Find the centripetal force on it, and compare with its weight.

5–25 A highway curve of radius 1600 ft is to be banked so that a car traveling 50 mi·hr⁻¹ will have no tendency to skid sideways. At what angle should it be banked?

5–26 A flat (unbanked) curve on a highway has a radius of 800 ft, and a car rounds the curve at a speed of 50 mi·hr⁻¹. What must be the minimum coefficient of friction to prevent sliding?

5–27 A stone of mass 1 kg is attached to one end of a string 1 m long, of breaking strength 500 N, and is whirled in a horizontal circle on a frictionless tabletop. The other end of the string is kept fixed. Find the maximum velocity the stone can attain without breaking the string.

5–28 An unbanked circular highway curve on level ground makes a turn of 90°. The highway carries traffic at 60 mi·hr⁻¹, and the centripetal force on a vehicle is not to exceed $\frac{1}{10}$ of its weight. What is the minimum length of the curve, in miles?

5–29 A coin placed on a 12-in. record will revolve with the record when it is brought up to a speed of $33\frac{1}{3}$ rev·min⁻¹, provided the coin is not more than 4 in. from the axis.

a) What is the coefficient of static friction between coin and record?

b) How far from the axis can the coin be placed, without slipping, if the turntable rotates at 45 rev·min⁻¹?

5–30

a) At how many revolutions per second must the apparatus of Fig. 5–25 rotate about the vertical axis in order that the cord shall make an angle of 45° with the vertical?

b) What is then the tension in the cord?

c) Find the angle θ that the cord makes with the vertical if the system is rotating at 1.5 rev·s⁻¹. (Set up the general equation relating the angle θ to the number

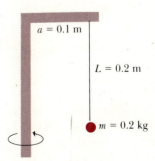

$a = 0.1$ m

$L = 0.2$ m

$m = 0.2$ kg

Figure 5–25

of revolutions per second, n, the lengths a and L, and the acceleration of gravity, g. Then find by trial the angle θ that satisfies this equation.)

5–31 The "Giant Swing" at a county fair consists of a vertical central shaft with a number of horizontal arms attached at its upper end. Each arm supports a seat suspended from a cable 5 m long, the upper end of the cable being fastened to the arm at a point 4 m from the central shaft.

   a) Find the time of one revolution of the swing if the cable supporting a seat makes an angle of 30° with the vertical.

   b) Does the angle depend on the weight of the passenger for a given rate of revolution?

5–32 The 4-kg block in Fig. 5–26 is attached to a vertical rod by means of two strings. When the system rotates about the axis of the rod, the strings are extended as shown in the diagram.

   a) How many revolutions per minute must the system make in order that the tension in the upper cord shall be 60 N?

   b) What is then the tension in the lower cord?

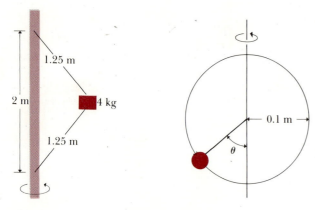

Figure 5–26   **Figure 5–27**

5–33 A bead can slide without friction on a circular hoop of radius 0.1 m in a vertical plane. The hoop rotates at a constant rate of 2 rev·s⁻¹ about a vertical diameter, as in Fig. 5–27.

   a) Find the angle θ at which the bead is in vertical equilibrium. (Of course it has a radial acceleration toward the axis.)

   b) Is it possible for the bead to "ride" at the same elevation as the center of the hoop?

   c) What will happen if the hoop rotates at 1 rev·s⁻¹?

5–34 A curve of 200-m radius on a level road is banked at the correct angle for a velocity of 15 m·s⁻¹. If an automobile rounds this curve at 30 m·s⁻¹, what is the minimum

coefficient of friction between tires and road so that the automobile will not skid? Assume all forces to act at the center of gravity.

5–35 A stunt pilot who has been diving vertically at a velocity of 400 mi·hr⁻¹ pulls out of the dive by changing his course to a circle in a vertical plane.

   a) What is the minimum radius of the circle in order that the acceleration at the lowest point shall not exceed "$7g$"?

   b) How much does a 180-lb pilot apparently weigh at the lowest point of the pullout?

5–36 A cord is tied to a pail of water and the pail is swung in a vertical circle of radius 1 m. What must be the minimum velocity of the pail at the highest point of the circle if no water is to spill from the pail?

5–37 The radius of a Ferris wheel is 5 m and it makes one revolution in 10 s.

   a) Find the difference between the apparent weight of a passenger at the highest and lowest points, expressed as a fraction of his weight. (That is, find the difference between the upward force exerted on the passenger by the seat at these two points.)

   b) What would the time of one revolution be if his apparent weight at the highest point were zero?

   c) What would then be his apparent weight at the lowest point?

   d) What would happen, at this rate of revolution, if his seat belt broke at the highest point, and if he did not hang onto his seat?

5–38 An earth satellite rotates in a circular orbit of radius 4400 mi (about 400 mi above the earth's surface) with an orbital speed of 17,000 mi·hr⁻¹.

   a) Find the time of one revolution.

   b) Find the acceleration of gravity at the orbit.

5–39 What is the period of revolution of a manmade satellite of mass m that is orbiting the earth in a circular path of radius 8000 km? (Mass of earth $= 5.98 \times 10^{24}$ kg.)

5–40 There exist several satellites moving in a circle in the earth's equatorial plane and at such a height above the earth's surface that they remain always above the same point. Find the height of such a satellite. (Such an orbit is said to be *geosynchronous*.)

5–41 It is desired to launch a satellite in a circular orbit 1000 m above the earth. What orbital speed must be imparted to the satellite?

5–42 The asteroid Toro was discovered in 1964. Its radius is about 5 km.

   a) Assuming the density of Toro is the same as that of earth, find its total mass, and find the acceleration due to gravity at its surface.

b) Suppose a body is to be placed in a circular orbit around Toro, with radius just slightly larger than the asteroid's radius. What is the speed of the body? Could you launch yourself into orbit around Toro by running?

5–43 What would be the length of a day if the rate of revolution of the earth were such that $g = 0$ at the equator?

5–44 The weight of a man as determined by a spring balance at the equator is 800 N. By how much does this differ from the true force of gravitational attraction at the same point?

5–45 A passenger on a ship traveling straight east with a speed of 18 knots observes that the stream of smoke from the ship's funnels makes an angle of 20° with the ship's wake. The wind is blowing from south to north. Assume that the smoke acquires a velocity (with respect to the earth) equal to the velocity of the wind, as soon as it leaves the funnels. Find the velocity of the wind.

5–46 An airplane pilot wishes to fly straight north. A wind of 60 mi·hr^{-1} is blowing toward the west. If the flying speed of the plane (its speed is still air) is 180 mi·hr^{-1}, in what direction should the pilot head? What is the speed of the plane over the ground? Illustrate with a vector diagram.

5–47 An airplane pilot sets a compass course straight west and maintains an air speed of 240 km·hr^{-1}. After flying for $\frac{1}{2}$ hr, he finds himself over a town that is 150 km west and 40 km south of his starting point.

a) Find the wind velocity, in magnitude and direction.

b) If the wind velocity were 120 km·hr^{-1} straight south, in what direction should the pilot set his course in order to travel straight west? Take the same air speed of 240 km·hr^{-1}.

5–48 When a train has a speed of 10 m·s^{-1} eastward, raindrops that are falling vertically with respect to the earth make traces that are inclined 30° to the vertical on the windows of the train.

a) What is the horizontal component of a drop's velocity with respect to the earth? With respect to the train?

b) What is the velocity of the raindrop with respect to the earth? With respect to the train?

5–49 A river flows straight north with a velocity of 2 m·s^{-1}. A man rows a boat across the river; his velocity relative to the water is 3 m·s^{-1} straight east.

a) What is his velocity relative to the earth?

b) If the river is 1000 m wide, how far north of his starting point will he reach the opposite bank?

c) How much time is required to cross the river?

5–50

a) In what direction should the rowboat in Problem 5–49 be headed in order to reach a point on the opposite bank directly east from the starting point?

b) What will be the velocity of the boat relative to the earth?

c) How much time is required to cross the river?

5–51 A motorboat is observed to travel 10 mi·hr^{-1} relative to the earth in the direction 37° north of east. If the velocity of the boat due to the wind is 2 mi·hr^{-1} eastward and that due to the current is 4 mi·hr^{-1} southward, what is the magnitude and direction of the velocity of the boat due to its own power?

CHAPTER

WORK
AND
ENERGY

Energy is one of the most important unifying concepts in all of physical science. Its importance stems from the principle of *conservation of energy*, which states that the total energy of all forms in an isolated system is constant. In this chapter we deal only with *mechanical* energy, that is, energy associated with the motion and position of mechanical systems. The concept of *work* is useful in analysis of changes in mechanical energy of a system when forces act on it. The relationship between work and mechanical energy is developed and used in a variety of problems, and the concept of *potential energy* is introduced as a convenient way to calculate work associated with certain kinds of forces.

CONSERVATION OF ENERGY

The concept of energy appears throughout every area of physics, and yet it is difficult to define in a general way just what energy *is*. Energy plays a central role in one of the fundamental natural laws called *conservation laws*, and looking at this role is as good a way as any to approach the question of what energy is.

A conservation law always concerns a transformation or an interaction that occurs within some physical system or in a system and its surroundings. Some quantities that describe the state or condition of the system and surroundings may change during the transformation or interaction, but there may be one or more quantities that remain constant or are *conserved*. A familiar example is conservation of *mass* in chemical reactions. It has been established by a very large amount of experimental evidence that the total mass of the reactants in a chemical reaction is always equal to the total mass of all the products of the reaction. That is, the total mass is always the same after the reaction occurs as before. This generalization is called the principle of *conservation of mass*, and it is obeyed in all chemical reactions.

Something similar happens in collisions between bodies. For a body of mass m moving with speed v we can define a quantity $\frac{1}{2}mv^2$, which we call the *kinetic energy* of the body. When two highly elastic or "springy" bodies (such as two hard steel ball bearings) collide, we find that the individual speeds change but that the total kinetic energy (the sum of the $\frac{1}{2}mv^2$ quantities for all the colliding bodies) is the same after the collision as before. We say that kinetic energy is *conserved* in such collisions. This result doesn't tell us what kinetic energy *is*, but only that it is useful in representing a conservation principle in certain kinds of interactions.

When two soft, deformable bodies, such as two balls of putty or chewing gum, collide, experiment shows that kinetic energy is *not* conserved. However, something else happens; the bodies become *warmer*. Furthermore, it turns out to be possible to work out a definite relationship between the temperature rise of the material and the loss of kinetic energy. We can define a new quantity, which we may call *internal energy*, that increases with temperature in a definite way, so that the *sum* of kinetic energy and internal energy *is* conserved in these collisions.

The significant discovery here is that it is *possible* to extend the principle of conservation of energy to a broader class of phenomena by defining a new form of energy. This is precisely how the principle has developed. Whenever an interaction has been studied in which it seems that the total energy in all known forms is *not* conserved, it has been found possible to define a new form of energy so that the *total* energy, including the new form, *is* conserved. These new forms have included energy associated with heat, with elastic deformations, with electric and magnetic fields, and in relativity theory even with mass itself. Conservation of energy has the status, along with a small number of partners, of a *universal* conservation principle; no exception to its validity has ever been found.

In this chapter we are concerned primarily with *mechanical* energy, that is, with energy associated with motion, position, and deformation of material bodies. In some interactions mechanical energy is conserved, and in others there is a conversion from mechanical energy to other forms, or the reverse, so that mechanical energy by itself is not conserved. In Chapters 15 and 18 we shall study in greater detail the relation of mechanical energy to heat, and in later chapters still other forms of energy will be considered.

6-2

WORK

In everyday life the term *work* is applied to any activity requiring the exertion of muscular or mental effort. In physics the term is used in a much more specific sense, which involves the application of a *force* to a body and *displacement* of that body. When a body moves a distance s along a straight line, while acted on by a constant force of magnitude F in the same direction as the motion, the work W done by the force is defined as

$$W = Fs. \tag{6-1}$$

This definition may seem quite arbitrary but, as we shall see in this chapter, work is directly related to quantities characterizing the *motion* of the body and is therefore a useful quantity.

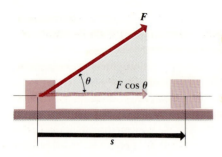

6–1 The work done by the force **F** in a displacement **s** is (F cos θ)·s.

More generally, the force need not have the same direction as the displacement. In Fig. 6–1, the force **F**, assumed constant, makes an angle θ with the displacement. The work W done by this force when its point of application undergoes a displacement **s**, is defined as the product of the magnitude of the displacement and the *component* of the force in the direction of the displacement.

The component of **F** in the direction of **s** is F cos θ. Then

$$W = (F \cos \theta)s. \tag{6-2}$$

An alternative interpretation of Eq. (6–2) is that s cos θ is the component of displacement in the direction of **F**. Thus the work may also be obtained by finding the component of displacement in the direction of **F** and multiplying by the magnitude of **F**.

Although it is calculated from two vector quantities, work itself is a *scalar* quantity. A 5-N force toward the east acting on a body that is displaced 6 m to the east does exactly the same amount of work as a 5-N force toward the north acting on a body displaced 6 m to the north. Work is, however, an *algebraic* quantity that can be positive or negative. When the component of the force is in the *same direction* as the displacement, the work W is *positive*. When it is *opposite* to the displacement, the work is *negative*. If the force is at *right angles* to the displacement, it has no component in the direction of the displacement and the work is *zero*.

Thus, when a body is lifted, the work done by the lifting force is positive; when a spring is stretched, the work done by the stretching force is positive; when a gas is compressed in a cylinder, the work done by the compressing force is positive. On the other hand, the work done by the gravitational force on a body being lifted is negative, since the (downward) gravitational force is opposite to the (upward) displacement. When a body slides on a fixed surface, the work done by the frictional force exerted *on the body* is negative, since this force is always opposite to the displacement of the body. No work is done by the frictional force acting *on the fixed surface* because there is no motion of this surface. Also, although it is considered "hard work" to hold a heavy object stationary at arm's length, no work is done in the technical sense because there is no motion! Even if one were to walk along a level floor while carrying the object, no work would be done, since the (vertical) supporting force has no component in the direction of the (horizontal) motion. Similarly, the work done by the normal force exerted on a body by a surface on which it moves is zero, and the work done by the centripetal force on a body moving in a circle is also zero.

The *unit* of work in any system is the unit of force multiplied by the unit of distance. In SI (mks) units the unit of force is the newton and the unit of distance is the meter; thus in this system the unit of work is one *newton meter* (1 N·m). This combination of units appears so frequently in mechanics that it is given a special name, the *joule* (abbreviated J).

$$1 \text{ joule} = (1 \text{ newton})(1 \text{ meter}) \quad \text{or} \quad 1 \text{ J} = 1 \text{ N·m}.$$

In the now-obsolete cgs system, the unit of work is one dyne-centimeter, also called one *erg*. Because 1 dyn = 10^{-5} N and 1 cm = 10^{-2} m,

$$1 \text{ erg} = 10^{-7} \text{ J}.$$

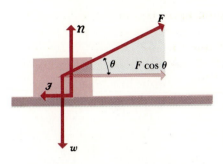

6–2 An object on a rough horizontal surface moving to the right under the action of a force **F** inclined at an angle θ.

In the British engineering system the unit of work is one *foot-pound* (1 ft-lb); no special name is given this unit. The following conversions are useful:

$$1 \text{ J} = 0.7376 \text{ ft} \cdot \text{lb}, \qquad 1 \text{ ft} \cdot \text{lb} = 1.356 \text{ J}.$$

When several external forces act on a body, it is useful to consider the work done by each separate force. Each of these may be computed from the definition of work in Eq. (6–2). Then, since work is a scalar quantity, the total work is the algebraic sum of the individual works.

EXAMPLE 1 Figure 6–2 shows a box being dragged along a horizontal surface by a workman who exerts a constant force **F** making a constant angle θ with the direction of motion. The other forces on the box are its weight **w**, the upward normal force **n** exerted by the surface, and the friction force **𝓕**. What is the work done by each force when the box moves a distance s along the surface to the right?

Solution The component of **F** in the direction of motion is $F \cos \theta$. The work of the force **F** is therefore

$$W_F = (F \cos \theta)s.$$

The forces **w** and **n** are both at right angles to the displacement. Hence,

$$W_w = 0, \qquad W_n = 0.$$

The friction force **𝓕** is opposite to the displacement, so the work done by the friction force is

$$W_{\mathcal{F}} = -\mathcal{F}s.$$

Since work is a scalar quantity, the total work W done by all forces on the box is the algebraic sum (*not* the vector sum) of the individual works:

$$\begin{aligned} W &= W_F + W_w + W_n + W_{\mathcal{F}} \\ &= (F \cos \theta)s + 0 + 0 - \mathcal{F}s \\ &= (F \cos \theta - \mathcal{F})s. \end{aligned}$$

But $(F \cos \theta - \mathcal{F})$ is the *resultant* force on the box. Hence *the total work done by all forces is equal to the work done by the resultant force.*

Suppose that $F = 50$ N, $\mathcal{F} = 15$ N, $\theta = 36.9°$, and $s = 20$ m. Then

$$\begin{aligned} W_F &= (F \cos \theta)s \\ &= (50 \text{ N})(0.800)(20 \text{ m}) = 800 \text{ N} \cdot \text{m}, \end{aligned}$$

$$\begin{aligned} W_{\mathcal{F}} &= -\mathcal{F}s \\ &= (-15 \text{ N})(20 \text{ m}) = -300 \text{ N} \cdot \text{m}, \end{aligned}$$

$$W = W_F + W_{\mathcal{F}} = 500 \text{ N} \cdot \text{m}.$$

As a check, the total work may be expressed as

$$\begin{aligned} W &= (F \cos \theta - \mathcal{F})s \\ &= (40 \text{ N} - 15 \text{ N})(20 \text{ m}) = 500 \text{ N} \cdot \text{m}. \end{aligned}$$ ◄

6–3

WORK DONE BY A VARYING FORCE

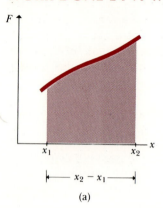

(a)

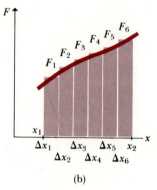

(b)

6–3 (a) Curve showing how F varies with x. (b) If the area is partitioned into small rectangles, their sum approximates the total work done during the displacement; the greater the number of rectangles used, the closer is the approximation.

In the preceding section we defined the work done by a *constant* force. In many important cases, however, the work is done by a force that varies in magnitude or direction during the displacement of the body on which it acts. Thus when a spring is stretched slowly, the force required to stretch it increases steadily as the spring elongates; when a body is projected vertically upward, the gravitational force exerted on it by the earth decreases inversely with the square of its distance from the earth's center.

Suppose a particle moves along a line under the action of a force directed along the line but varying with the particle's position. In Fig. 6–3a the force magnitude is shown as a function of the particle's coordinate x. To find the work done by this force, we divide the displacement into short segments Δx_1, Δx_2, etc., and in each segment approximate the varying force by one that is constant in each segment. The force then has approximately the value F_1 in segment Δx_1, F_2 in segment Δx_2, and so on. The work done in the first segment is then $F_1 \Delta x_1$, that in the second is $F_2 \Delta x_2$, and so on, and the *total* work is

$$W = F_1 \Delta x_1 + F_2 \Delta x_2 + \cdots \qquad (6\text{–}3)$$

But the products $F_1 \Delta x_1$, $F_2 \Delta x_2$, etc., are the *areas* of the various vertical strips, and the *total* work is evidently equal to the total area of these strips. As the subdivisions are made smaller and smaller, the total area of these strips becomes more and more nearly equal to the shaded area between the smooth curve and the horizontal axis, bounded by vertical lines at the ends of the displacement s as in Fig. 6–3b.

Hence if a graph is constructed of the magnitude of a variable force, as a function of displacement, the *work* done by the force can be determined by computing (or measuring) the *area* between the curve and the displacement axis.

As an example of this method, let us compute the work done when a spring is stretched. To keep a spring stretched at an elongation x beyond its unstretched length, a force F must be exerted at one end and a force equal in magnitude but opposite in direction at the other end, as shown in Fig. 6–4. If the elongation is not too great, F is directly proportional to x:

$$F = kx, \qquad (6\text{–}4)$$

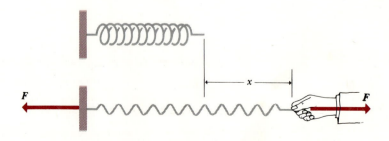

6–4 The force to stretch a spring is proportional to its elongation; $F = kx$.

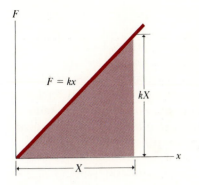

6–5 The work done in stretching a spring is equal to the area of the shaded triangle.

where k is a constant called the *force constant* or the *stiffness* of the spring. This direct proportion between force and elongation, for elongations that are not too great, was discovered by Robert Hooke in 1678 and is known as *Hooke's law*. It will be discussed in more detail in Chapter 10.

Suppose that forces equal in magnitude and opposite in direction are applied to the ends of the spring, and that the forces are gradually increased from zero. The left end is held stationary; the work of the force at this end is zero, but work is done by the varying force F at the moving end. This force is in the same direction as the displacement. In Fig. 6–5, the force F is plotted vertically and the displacement x of the moving end, representing the elongation or compression of the spring, is plotted horizontally. The work done by the force F when the elongation increases from zero to X is equal to the shaded triangular area. The area of a triangle equals one half the product of base and altitude. The base of the triangle is X, its altitude is kX, and its area is

$$\text{Area} = \tfrac{1}{2}(X)(kX) = \tfrac{1}{2}kX^2 = W. \tag{6–5}$$

The work is therefore equal to one half the product of the force constant k and the *square* of the final elongation X. When the elongation is doubled, the work increases by a factor of 4. Another view of this relationship is that the total work W is equal to the *average* force $\tfrac{1}{2}kX$ multiplied by the total displacement X.

During this process the spring also exerts a force on the hand, which moves, and we may ask what work the *spring* does on the *hand*. The displacement of the hand is the same as that of the end of the spring, but the force on it is the negative of the force on the spring because the two forces form an action–reaction pair. Thus the work done on the hand is the negative of the work done on the spring, namely $\tfrac{1}{2}kX^2$. In work calculations, it is always essential to specify on which body the work is being done!

EXAMPLE A woman weighing 600 N steps on a bathroom scale containing a heavy spring. The spring is compressed 1.0 cm under her weight. Find the force constant of the spring and the total work done on it during the compression.

Solution Whether the spring is stretched or compressed, the principle is the same. From Eq. (6–5),

$$k = \frac{F}{x} = \frac{600\text{ N}}{0.01\text{ m}} = 60,000\text{ N}\cdot\text{m}^{-1}.$$

Then from Eq. (6–5),

$$W = \tfrac{1}{2}kx^2 = \tfrac{1}{2}(60,000\text{ N}\cdot\text{m}^{-1})(0.01\text{ m})^2 = 3\text{ N}\cdot\text{m} = 3\text{ J}. \quad \blacktriangleleft$$

The reader should convince himself that the work done in changing the elongation or compression of a spring from x_1 to x_2 is

$$W = \tfrac{1}{2}kx_2{}^2 - \tfrac{1}{2}kx_1{}^2,$$

provided that $x = 0$ corresponds to zero elongation or compression, as above.

In some situations the *direction* of the force may change during the displacement, or the displacement may occur along a curved path. In these cases a further generalization is needed. In Fig. 6–3 and the accompanying discussion, F must be taken to be the *component* of force in a direction tangent to the path at each point. If a graph of this component as a function of the distance along the path can be constructed, then the work is still given by the area under the curve, as in Fig. 6–3b. Actually calculating work in such cases often requires methods of integral calculus, and such calculations will not be discussed here.

6–4

WORK AND KINETIC ENERGY

The work done on a body by a force is related very directly to the resulting change in the body's motion. To develop this relationship, we consider first a body of mass m moving along a straight line under the action of a constant resultant force of magnitude F directed along the line. The body's acceleration is given by Newton's second law, $F = ma$. Suppose the speed increases from v_1 to v_2 while the body undergoes a displacement $s = x_2 - x_1$. Then, from the analysis of motion with constant acceleration, and from Eq. (3–12), we have

$$v_2{}^2 = v_1{}^2 + 2as$$

$$a = \frac{v_2{}^2 - v_1{}^2}{2s}.$$

Hence

$$F = m\frac{v_2{}^2 - v_1{}^2}{2s},$$

and

$$Fs = \tfrac{1}{2}mv_2{}^2 - \tfrac{1}{2}mv_1{}^2. \qquad (6-6)$$

The product Fs is the work W done by the resultant force F. The quantity $\tfrac{1}{2}mv^2$, one half the product of the mass of the body and the square of its speed, is called its *kinetic energy, K*:

$$K = \tfrac{1}{2}mv^2. \qquad (6-7)$$

The first term on the right side of Eq. (6–6), containing the final speed v_2, is the final kinetic energy of the body, $K_2 = \tfrac{1}{2}mv_2{}^2$, and the second term is the initial kinetic energy, $K_1 = \tfrac{1}{2}mv_1{}^2$. The difference between these terms is the *change* in kinetic energy, and we have the important result that *the work done by the resultant external force on a body is equal to the change in kinetic energy of the body*:

$$W = K_2 - K_1 = \Delta K. \qquad (6-8)$$

Kinetic energy, like work, is a *scalar* quantity. The kinetic energy of a moving body depends only on its speed (the *magnitude* of its velocity) but not on the *direction* in which it is moving. The *change* in kinetic energy depends only on the work $W = Fs$ and not on the individual values of F and s. That is, the force F could have been large and the displacement s small, or the reverse might have been true. If the mass m and the speeds

v_1 and v_2 are known, the *work* of the resultant force can be found without any knowledge of the force F and the displacement s.

If the work W is *positive*, the final kinetic energy is greater than the initial kinetic energy and the kinetic energy *increases*. If the work is *negative*, the kinetic energy *decreases*. In the special case in which the work is *zero*, the kinetic energy remains *constant*.

In Eq. (6–8), W is the work done by the *resultant* force, that is, of the *vector sum* of *all* the forces acting on the body. Alternatively, one may calculate the work done by each separate force; W is then the *algebraic sum* of all these quantities of work. The example in Section 6–2 illustrates these two alternatives.

Although Eq. (6–8) was derived for the special case of a *constant* resultant force, it is true even when the force varies in any arbitrary way. The work done by *any* resultant force on a rigid body equals the change in kinetic energy of the body. To prove this statement, we divide the total displacement x into a large number of small segments Δx, just as in the calculation of work done by a varying force. The change of kinetic energy in segment Δx_1 is equal to the work $F_1 \Delta x_1$, and so on; the total kinetic energy is the sum of the changes in the individual segments and is thus equal to the total work done.

When SI units are used, m is measured in kilograms and v in meters per second. But from Eq. (6–8) kinetic energy must have the same units as work. To verify this we recall that $1 \text{ N} = 1 \text{ kg} \cdot \text{m} \cdot \text{s}^{-2}$. Thus

$$1 \text{ J} = 1 \text{ N} \cdot \text{m} = 1 \text{ kg} \cdot \text{m} \cdot \text{s}^{-2} \cdot \text{m} = 1 \text{ kg} \cdot \text{m}^2 \cdot \text{s}^{-2}.$$

The joule is thus the SI unit of *both* work and kinetic energy and, as we shall see later, of all kinds of energy. Similarly, in the obsolete cgs system, with m in grams and v in centimeters per second, we have

$$1 \text{ erg} = 1 \text{ dyn} \cdot \text{cm} = 1 \text{ g} \cdot \text{cm} \cdot \text{s}^{-2} \cdot \text{cm} = 1 \text{ g} \cdot \text{cm}^2 \cdot \text{s}^{-2}.$$

In the British engineering system,

$$1 \text{ ft} \cdot \text{lb} = 1 \text{ ft} \cdot \text{slug} \cdot \text{ft} \cdot \text{s}^{-2} = 1 \text{ slug} \cdot \text{ft}^2 \cdot \text{s}^{-2}.$$

EXAMPLE Refer again to the body in Fig. 6–2 and the numerical values given in the example of Section 6–2. The total work done by the external forces was shown to be 500 N·m. Hence the kinetic energy of the body increases by 500 N·m. To verify this, suppose the initial speed v_1 is $4 \text{ m} \cdot \text{s}^{-1}$ and the mass of the body is 10 kg. The initial kinetic energy is

$$K_1 = \tfrac{1}{2} m v_1{}^2 = \tfrac{1}{2}(10 \text{ kg})(4 \text{ m} \cdot \text{s}^{-1})^2 = 80 \text{ J}.$$

To find the final kinetic energy we first find the acceleration:

$$a = \frac{F}{m} = \frac{40 \text{ N} - 15 \text{ N}}{10 \text{ kg}} = 2.5 \text{ m} \cdot \text{s}^{-2}.$$

Then

$$v_2{}^2 = v_1{}^2 + 2as = (4 \text{ m} \cdot \text{s}^{-1})^2 + 2(2.5 \text{ m} \cdot \text{s}^{-2})(20 \text{ m}) = 116 \text{ m}^2 \cdot \text{s}^{-2},$$

$v_2 = 10.8 \text{ m} \cdot \text{s}^{-1}$, and

$$K_2 = \tfrac{1}{2}(10 \text{ kg})(116 \text{ m}^2 \cdot \text{s}^{-2}) = 580 \text{ J}.$$

The increase in kinetic energy is therefore 500 J.

To put things the other way around, if we are given $v_1 = 4 \text{ m} \cdot \text{s}^{-1}$ and $W = 500$ J, we may determine v_2 directly from the work–energy relation of Eq. (6–8), without having to find the acceleration at all. From Eq. (6–8), we write

$$\tfrac{1}{2}(10 \text{ kg})v_2^2 - \tfrac{1}{2}(10 \text{ kg})(4 \text{ m} \cdot \text{s}^{-1})^2 = 500 \text{ J}.$$

Solving this for v_2, we again obtain $v_2 = 10.8 \text{ m} \cdot \text{s}^{-1}$.   ◄

6–5

GRAVITATIONAL POTENTIAL ENERGY

When a gravitational force acts on a body undergoing a displacement, the force does work on the body, and we shall see that this work can be expressed conveniently in terms of the initial and final position of the body. In Fig. 6–6a, a body of mass m (and of weight $\boldsymbol{w} = m\boldsymbol{g}$) moves vertically from a height y_1 above an arbitrary reference plane to a height y_2. For the present, we consider only displacements near the earth's surface, so that variations of gravitational force with distance from the earth's center can be neglected. The downward gravitational force on the body is then constant and equal to $\boldsymbol{w}$. Let $\boldsymbol{P}$ represent the resultant of all other forces acting on the body. The direction of the gravitational force $\boldsymbol{w}$ is opposite to the upward displacement, and the work done by this force is

$$W_{\text{grav}} = -w(y_2 - y_1) = -(mgy_2 - mgy_1). \qquad (6\text{–}9)$$

The reader should check that this expression also has the correct sign when the body moves *downward* (so y_2 is less than y_1), and that it still gives a correct result when y_1, or y_2, or both, are negative, corresponding to a position *below* the reference plane.

Thus W_{grav} can be determined from the values of the quantity mgy at the beginning and end of the displacement. This quantity, the product of the weight mg and the height y above the origin of coordinates, is

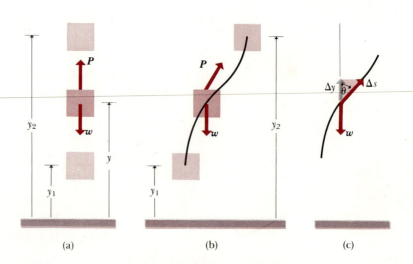

6–6 Work of the gravitational force $\boldsymbol{w}$ during the motion of an object from one point in a gravitational field to another.

(a)   (b)   (c)

called the *gravitational potential energy, U:*

$$U \text{ (gravitational)} = mgy. \tag{6–10}$$

Then the initial value of gravitational potential energy is $U_1 = mgy_1$ and the final value $U_2 = mgy_2$. The work W_{grav} done by the gravitational force during the displacement from y_1 to y_2 can then be written as

$$W_{\text{grav}} = U_1 - U_2 = -\Delta U.$$

Thus when the gravitational force does *positive* work, corresponding to downward displacement and decreasing y, the potential energy *decreases*; when the body moves upward the gravitational work is negative and the potential energy increases.

Now let W' represent the total work done by $\boldsymbol{P}$, that is, by all forces *other than* the gravitational force. The total work done by *all* forces is then $W = W_{\text{grav}} + W'$. Since the *total* work equals the change in kinetic energy,

$$W' + W_{\text{grav}} = K_2 - K_1 = \Delta K,$$

$$W' - (mgy_2 - mgy_1) = (\tfrac{1}{2}mv_2{}^2 - \tfrac{1}{2}mv_1{}^2). \tag{6–11}$$

The quantities $\tfrac{1}{2}mv_2{}^2$ and $\tfrac{1}{2}mv_1{}^2$ depend only on the final and initial *speeds*; the quantities mgy_2 and mgy_1 depend only on the final and initial *elevations*. Let us therefore rearrange this equation, transferring the quantities mgy_2 and mgy_1 from the "work" side of the equation to the "energy" side:

$$W' = (\tfrac{1}{2}mv_2{}^2 - \tfrac{1}{2}mv_1{}^2) + (mgy_2 - mgy_1) = \Delta K + \Delta U. \tag{6–12}$$

The left side of Eq. (6–12) contains only the work done by the force $\boldsymbol{P}$. The terms on the right depend only on the final and initial states of the body (its speed and elevation) and not specifically on the way in which it moved. As indicated, the first expression in parentheses on the right of Eq. (6–12) is the change in kinetic energy of the body, and the second is the change in its gravitational potential energy.

The sum of kinetic and potential energies is called the *total mechanical energy, E = K + U.* Equation (6–12) can also be written

$$W' = (\tfrac{1}{2}mv_2{}^2 + mgy_2) - (\tfrac{1}{2}mv_1{}^2 + mgy_1)$$

$$= (K_2 + U_2) - (K_1 + U_1) = E_2 - E_1 = \Delta E. \tag{6–13}$$

Hence, *the work done by all forces acting on the body,* **with the exception of the gravitational force,** *equals the change in the total mechanical energy of the body.* If the work W' is positive, the mechanical energy increases. If W' is negative, the mechanical energy decreases.

In the special case in which the *only* force on the body is the gravitational force, the work W' is zero. Equation (6–13) can then be written as $E_1 = E_2$, or

$$K_1 + U_1 = K_2 + U_2,$$

or

$$\tfrac{1}{2}mv_1{}^2 + mgy_1 = \tfrac{1}{2}mv_2{}^2 + mgy_2. \tag{6–14}$$

Under these conditions, then, the *total mechanical energy is constant*, or is *conserved*. This is a special case of the principle of *conservation of mechanical energy*.

EXAMPLE 1   A man holds a ball of mass $m = 0.2$ kg at rest in his hand. He then throws the ball vertically upward. In this process, his hand moves up 0.5 m before the ball leaves his hand with an upward velocity of 20 m·s^{-1}. Discuss the motion of the ball from the work–energy standpoint, assuming $g = 10$ m·s^{-2}.

Solution   First, consider the throwing process. Take the reference level $(y = 0)$ at the initial position of the ball. Then $K_1 = 0$, $U_1 = 0$. Take point 2 at the point where the ball leaves the thrower's hand. Then

$$U_2 = mgy_2 = (0.2 \text{ kg})(10 \text{ m·s}^{-2})(0.5 \text{ m}) = 1.0 \text{ J},$$
$$K_2 = \tfrac{1}{2}mv_2{}^2 = \tfrac{1}{2}(0.2 \text{ kg})(20 \text{ m·s}^{-1})^2 = 40 \text{ J}.$$

Let P represent the upward force exerted on the ball by the man in the throwing process. The work W' is then the work done by this force and is equal to the sum of the changes in kinetic and potential energy of the ball.

The kinetic energy of the ball increases by 40 J and its potential energy by 1 J. The work W' of the upward force P is therefore 41 J.

If the force P is constant, the work of this force is given by

$$W' = P(y_2 - y_1),$$

and the force P is then

$$P = \frac{W'}{y_2 - y_1} = \frac{41 \text{ J}}{0.5 \text{ m}} = 82 \text{ N}.$$

However, the *work* done by the force P is 41 J whether the force is constant or not.

Now consider the flight of the ball *after* it leaves the thrower's hand. In the absence of air resistance, the only force on the ball is then its weight $w = mg$. Hence the total mechanical energy of the ball remains constant. The calculations will be simplified if we take a new reference level at the point where the ball leaves the thrower's hand. Calling this point 1, we have

$$K_1 = 40 \text{ J}, \qquad U_1 = 0,$$
$$K + U = 40 \text{ J},$$

and the total mechanical energy at *any* point of the path equals 40 J.

Suppose we wish to find the speed of the ball at a height of 15 m above the reference level. Its potential energy at this elevation is $U = mgy = 30$ J. Its kinetic energy is therefore 10 J. To find its speed, we have

$$\tfrac{1}{2}mv^2 = K, \qquad v = \pm\sqrt{2K/m} = \pm 10 \text{ m·s}^{-1}.$$

The significance of the $\pm$ sign is that the ball passes this point *twice*, once on the way up and again on the way down. Its *potential* energy at this point is the same whether it is moving up or down. Hence its kinetic energy is the same and its *speed* is the same. The algebraic sign of the speed is + when the ball is moving up and − when it is moving down.

Next, let us find the height of the highest point reached. At this point, $v = 0$ and $K = 0$. Therefore $U = 40$ J, and the ball rises to a height h above the point where it leaves the thrower's hand, given by

$$mgh = (0.2 \text{ kg})(10 \text{ m} \cdot \text{s}^{-2})h = 40 \text{ J}.$$

Thus

$$h = 20 \text{ m}.$$

Finally, suppose we were asked to find the speed at a point 30 m above the reference level. The potential energy at this point would be 60 J. But the *total* energy is only 40 J, so the ball never reaches a height of 30 m. ◀

A body may travel from an initial elevation y_1 to a final elevation y_2 along a curved path, as shown in Fig. 6–6b. The work done by the gravitational force during this displacement is the same as when the body travels straight up, as in Fig. 6–6a. To prove this we divide the path into a large number of small segments Δs, of which a typical example is shown in Fig. 6–6c. According to the remark following Eq. (6–2), the work done during this displacement may be obtained by multiplying the component of displacement in the direction of the force by the magnitude of the force. As shown in Fig. 6–6c, the vertical component of displacement has magnitude $\Delta s \cos \theta = \Delta y$, but its direction is opposite to that of the force w. Thus the work done by w is

$$-mg \, \Delta s \cos \theta = -mg \, \Delta y$$

and is the same as though the body had been displaced straight up a distance Δy. Similarly, the *total* work done by the gravitational force depends only on the *total* vertical displacement $(y_2 - y_1)$ and is independent of any horizontal motion that may occur.

EXAMPLE 2 A child slides down a curved playground slide that is one quadrant of a circle of radius R, as in Fig. 6–7. If he starts from rest and there is no friction, find his speed at the bottom of the track. (The motion of this child is exactly the same as that of a child on a swing of length R, with the other end held at point 0.)

Solution The equations of motion with constant acceleration cannot be used, since the acceleration decreases during the motion. (The slope angle of the slide becomes smaller and smaller as the body descends.) However, if there is no friction, the only force on the child other than his weight is the normal force n exerted on him by the slide. The work done by this force is zero because at each point it is perpendicular to the small element of displacement near that point. Thus $W' = 0$ and mechanical energy is conserved. Take point 1 at the starting point and point 2 at the bottom of the slide. Take the reference level at point 2. Then $y_1 = R$, $y_2 = 0$, and

$$K_2 + U_2 = K_1 + U_1,$$
$$\tfrac{1}{2}mv_2{}^2 + 0 = 0 + mgR,$$
$$v_2 = \pm\sqrt{2gR}.$$

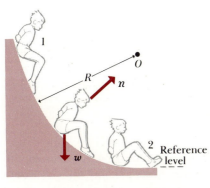

6–7 A child sliding down a frictionless curved slide.

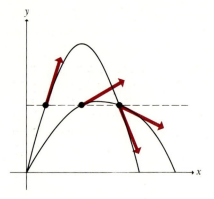

6–8 For the same initial speed, the speed is the same at all points at the same elevation.

The speed is therefore the same as if the child had fallen *vertically* through a height R. (What is now the significance of the $\pm$ sign?)

As a numerical example, let $R = 3$ m. Then

$$v = \pm\sqrt{2(9.8 \text{ m}\cdot\text{s}^{-2})(3\text{m})} = \pm 7.67 \text{ m}\cdot\text{s}^{-1}. \qquad \blacktriangleleft$$

EXAMPLE 3 Suppose a child of mass 25 kg slides down a slide of radius $R = 3$m, like that in Fig. 6–7, but his speed at the bottom is only 3 m$\cdot$s^{-1}. What work was done by the frictional force acting on the child?

Solution In this case, $W' = W_{\mathfrak{F}}$, and

$$
\begin{aligned}
W_{\mathfrak{F}} &= (\tfrac{1}{2}mv_2{}^2 - \tfrac{1}{2}mv_1{}^2) + (mgy_2 - mgy_1) \\
&= \tfrac{1}{2}(25 \text{ kg})(9 \text{ m}^2\cdot\text{s}^{-2}) - 0 + 0 - (25 \text{ kg})(9.8 \text{ m}\cdot\text{s}^{-2})(3 \text{ m}) \\
&= 112.5 \text{ J} - 735 \text{ J} = -622.5 \text{ J}
\end{aligned}
$$

The frictional work was therefore -622.5 J, and the total mechanical energy *decreased* by 622.5 J. The mechanical energy of a body is *not* conserved when friction forces act on it. $\blacktriangleleft$

EXAMPLE 4 In the absence of air resistance, the only force on a ball after it is thrown is its weight, and the mechanical energy of the ball is constant. Figure 6–8 shows two trajectories of a ball with the same initial speed (hence the same total energy) but with different angles of departure. At all points at the same elevation the potential energy is the same; hence the kinetic energy is the same and the speed is the same. $\blacktriangleleft$

EXAMPLE 5 A child of weight w sits on a swing of length l, as shown in Fig. 6–9. A *variable* horizontal force P that starts at zero and gradually increases is used to pull the child very slowly (so the kinetic energy is negligibly small) until the swing makes an angle θ with the vertical. Calculate the work done by the force P.

Solution The sum of the works W' done by all the forces other than the gravitational force must equal the change of total energy, that is, the change of kinetic energy plus the change of gravitational potential energy. Hence

$$W' = W_P + W_T = \Delta K + \Delta U = \Delta E.$$

Since T is perpendicular to the path of its point of application, $W_T = 0$; and since the swing was pulled very slowly at all times, the change of kinetic energy is also zero. Hence

$$W_P = \Delta U = w\,\Delta y,$$

where Δy is the distance that the child has been raised. From Fig. 6–9 Δy is seen to be $l(1 - \cos \theta)$. Therefore

$$W_P = wl(1 - \cos \theta). \qquad \blacktriangleleft$$

Thus far in this section it has been assumed that the changes in elevation are small enough so that the gravitational force on a body can be considered constant. But as discussed in Section 4–4, the gravitational

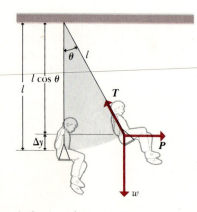

6–9 $\Delta y = l(1 - \cos \theta)$.

force is given in general by

$$F_g = \frac{Gmm_E}{r^2},$$

where m_E is the mass of the earth and r is the distance of the body from the earth's center. When the change in r is sufficiently large, we may no longer consider the gravitational force constant; it decreases as $1/r^2$. Calculation of the work done by this varying force requires methods of calculus; the result is that when the body moves from a distance r_1 from the earth's center to a distance r_2 the work W_{grav} done on the body by gravity is given not by $-mg(r_2 - r_1)$, as previously, but by

$$W_{grav} = Gmm_E\left(\frac{1}{r_2} - \frac{1}{r_1}\right). \tag{6–15}$$

If the gravitational force is the only force on the body, then this is equal to the change in kinetic energy from r_1 to r_2:

$$Gmm_E\left(\frac{1}{r_2} - \frac{1}{r_1}\right) = \tfrac{1}{2}mv_2{}^2 - \tfrac{1}{2}mv_1{}^2$$

or

$$\tfrac{1}{2}mv_1{}^2 - \frac{Gmm_E}{r_1} = \tfrac{1}{2}mv_2{}^2 - \frac{Gmm_E}{r_2}. \tag{6–16}$$

The quantity $-G(mm_E/r)$ is therefore the general expression for the gravitational potential energy of a body attracted by the earth:

$$U(\text{gravitational}) = -G\frac{mm_E}{r}. \tag{6–17}$$

The *total mechanical energy* of the body, the sum of its kinetic energy and potential energy, is

$$E = K + U = \tfrac{1}{2}mv^2 - G\frac{mm_E}{r}. \tag{6–18}$$

If the only force on the body is the gravitational force, the total mechanical energy remains constant, or is *conserved*.

EXAMPLE 6 In Jules Verne's story "From the Earth to the Moon" (written in 1865) three men were shot to the moon in a shell fired from a giant cannon sunk in the earth in Florida. What muzzle velocity would be needed (a) to raise a total mass m to a height above the earth equal to the earth's radius; (b) to escape from the earth completely? To simplify the calculation, neglect the gravitational pull of the moon.

Solution

a) In Eq. (6–16), let m be the total projectile mass, and let v_1 be the initial velocity. Then

$$r_1 = R, \qquad r_2 = 2R, \qquad v_2 = 0,$$

and

$$\tfrac{1}{2}mv_1{}^2 - G\frac{mm_E}{R} = 0 - G\frac{mm_E}{2R},$$

or

$$v_1{}^2 = \frac{Gm_E}{R}.$$

$$v_1 = \sqrt{\frac{(6.67 \times 10^{-11}\ \text{N}\cdot\text{m}^2\cdot\text{kg}^{-2})(5.98 \times 10^{24}\ \text{kg})}{6.38 \times 10^6\ \text{m}}}$$

$$= 7920\ \text{m}\cdot\text{s}^{-1} = 17{,}715\ \text{mi}\cdot\text{hr}^{-1}.$$

b) When v_1 is the escape velocity, $r_1 = R$, $r_2 = \infty$, $v_2 = 0$. Then

$$\tfrac{1}{2}mv_1{}^2 - G\frac{mm_E}{R} = 0 \qquad \text{or} \qquad v_1{}^2 = \frac{2Gm_E}{R}.$$

$$v_1 = \sqrt{\frac{2(6.67 \times 10^{-11}\ \text{N}\cdot\text{m}^2\cdot\text{kg}^{-2})(5.98 \times 10^{24}\ \text{kg})}{6.38 \times 10^6\ \text{m}}}$$

$$= 1.12 \times 10^4\ \text{m}\cdot\text{s}^{-1} = 25{,}050\ \text{mi}\cdot\text{hr}^{-1}.$$

We note that the speed of an earth satellite in a circular orbit of radius just slightly greater than R is, from Eq. (5–28), the same as the result of part (a), and that the escape velocity is larger than this by exactly a factor of $\sqrt{2}$. ◄

It may seem strange that the general expression for gravitational potential energy, Eq. (6–17), should contain a minus sign. The reason for this lies in the choice of a *reference state* or reference level in which the potential energy is considered zero. If we set $U = 0$ in Eq. (6–17) and solve for r, we find

$$r = \infty.$$

That is, the *gravitational potential energy of a body is zero when the body is at an infinite distance from the earth*. Since the potential energy *decreases* as the body approaches the earth, it must be *negative* at any finite distance from the earth. The *change* in potential energy of a body as it moves from one point to another is the same, whatever the choice of reference level, and it is only changes in potential energy that are significant.

Finally, we note that Eq. (6–15) can be rewritten as

$$W_{\text{grav}} = Gmm_E\left(\frac{r_1 - r_2}{r_1 r_2}\right).$$

If the particle stays close to the earth, then in the denominator we may replace r_1 and r_2 by R, the earth's radius, obtaining

$$W_{\text{grav}} = Gmm_E\left(\frac{r_1 - r_2}{R^2}\right).$$

But according to Eq. (4–8), $g = Gm_E/R^2$, so we finally obtain

$$W_{\text{grav}} = mg(r_1 - r_2),$$

which agrees with Eq. (6–9) with the y's replaced by r's. Thus Eq. (6–9) may be considered as a special case of the more general equation (6–15).

6–6

ELASTIC POTENTIAL ENERGY

The concept of potential energy is useful in calculations of work done by *elastic* forces, such as the spring discussed in Section 6–3. Figure 6–10 shows a body of mass m on a level surface. One end of a spring is attached to the body, and the other end is held stationary. We take our origin of coordinates ($x = 0$) as the position when the spring is neither stretched nor compressed. We now apply a force **P**, causing the body to accelerate. As soon as the spring begins to stretch, it begins to exert a force **F** on the body; we may call this an *elastic force*. If the force **P** is removed, the elastic force pulls the body back toward the original position, where the spring is unstretched. Hence **F** may be called a *restoring force*.

In Section 6–3 we spoke of the work done *on* the spring. Now, in the light of the relation between work done on a body and its change in kinetic energy, we shift our attention to the work that the spring does on the body of mass m; this is the negative of the work done *on* the spring. Thus in an elongation from $x = 0$ to a final value x, the work done on the mass *by* the spring is $-\frac{1}{2}kx^2$. Similarly, in a displacement from an initial elongation x_1 to a final elongation x_2, the elastic restoring force does an amount of work W_{el} given by

$$W_{el} = -\tfrac{1}{2}kx_2{}^2 - (-\tfrac{1}{2}kx_1{}^2).$$

The quantity $\frac{1}{2}kx^2$, one half the product of the force constant and the square of the coordinate of the body, is called the *elastic potential energy* of the system, U. (The symbol U is used for any form of potential energy.)

$$U \text{ (elastic)} = \tfrac{1}{2}kx^2. \tag{6–19}$$

The work done on the body by the elastic force can thus be expressed in terms of a change in potential energy, just as was done for gravitational work:

$$W_{el} = \tfrac{1}{2}kx_1{}^2 - \tfrac{1}{2}kx_2{}^2 = U_1 - U_2. \tag{6–20}$$

When x increases, the elastic work is negative and the potential energy increases; when x decreases, the elastic force does positive work and the potential energy decreases.

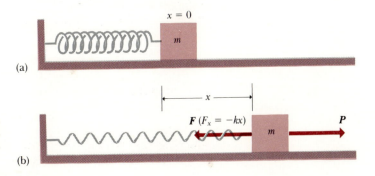

6–10 When an applied force **P** produces an extension x of a spring, an elastic restoring force **F** acts on the mass m. The x-component of **F** is $F_x = -kx$.

Let W' stand for the work done by the applied force $\mathbf{P}$. Setting the total work equal to the change in kinetic energy of the body, we have

$$W' + W_{\text{el}} = \Delta K,$$

$$W' - (\tfrac{1}{2}kx_2^2 - \tfrac{1}{2}kx_1^2) = (\tfrac{1}{2}mv_2^2 - \tfrac{1}{2}mv_1^2).$$

The quantities $\tfrac{1}{2}kx_2^2$ and $\tfrac{1}{2}kx_1^2$ depend only on the initial and final positions of the body and not specifically on the way in which it moved. Let us therefore transfer them from the "work" side of the equation to the "energy" side. Then

$$W' = (\tfrac{1}{2}mv_2^2 - \tfrac{1}{2}mv_1^2) + (\tfrac{1}{2}kx_2^2 - \tfrac{1}{2}kx_1^2). \tag{6-21}$$

Hence the work W' of the force $\mathbf{P}$ equals the sum of the change in the kinetic energy of the body and the change in its elastic potential energy.

Equation (6–21) can also be written

$$W' = (\tfrac{1}{2}mv_2^2 + \tfrac{1}{2}kx_2^2) - (\tfrac{1}{2}mv_1^2 + \tfrac{1}{2}kx_1^2) = (K_2 + U_2) - (K_1 + U_1)$$

$$= E_2 - E_1 = \Delta E. \tag{6-22}$$

The sum of the kinetic and potential energies of the body is its total mechanical energy; and *the work done by all forces acting on the body,* **with the exception of the elastic force,** *equals the change in the total mechanical energy of the body.*

If the work W' is positive, the mechanical energy increases. If W' is negative, it decreases. In the special case in which W' is zero, the mechanical energy remains constant or is *conserved.*

In that special case Eq. (6–22) may be rewritten

$$\tfrac{1}{2}mv_1^2 + \tfrac{1}{2}kx_1^2 = \tfrac{1}{2}mv_2^2 + \tfrac{1}{2}kx_2^2. \tag{6-23}$$

EXAMPLE 1 Let the force constant k of the spring in Fig. 6–10 be $24\ \text{N}\cdot\text{m}^{-1}$, and let the mass of the body be 4 kg. The body is initially at rest, and the spring is initially stretched 0.5 m. Then the body is released and moves back toward its equilibrium position. What is its speed when the spring is stretched 0.3 m?

Solution The equations of motion with constant acceleration cannot be used, since the spring force varies with position, but the speed can be found from energy considerations. The spring force is the only force acting on the body that does work; thus in Eq. (6–22), $W' = 0$. The total mechanical energy is constant, and Eq. (6–23) is applicable. We have $v_1 = 0$, $x_1 = 0.5$ m, $x_2 = 0.3$ m, and v_2 is to be found. From Eq. (6–23),

$$\tfrac{1}{2}(4\ \text{kg})(0)^2 + \tfrac{1}{2}(24\ \text{N}\cdot\text{m}^{-1})(0.5\ \text{m})^2$$

$$= \tfrac{1}{2}(4\ \text{kg})(v_2)^2 + \tfrac{1}{2}(24\ \text{N}\cdot\text{m}^{-1})(0.3\ \text{m})^2.$$

Evaluating the left side, we find that the total mechanical energy is 3 J, and solving for v_2 yields the result

$$v_2 = \pm 0.98\ \text{m}\cdot\text{s}^{-1}.$$

An alternative and perhaps more systematic way to set up this problem is to list the various energy quantities, some of which are unknown, as follows:

$$K_1 = \tfrac{1}{2}(4 \text{ kg})(0)^2 = 0,$$

$$U_1 = \tfrac{1}{2}(24 \text{ N} \cdot \text{m}^{-1})(0.5 \text{ m})^2 = 3 \text{ J},$$

$$K_2 = \tfrac{1}{2}(4 \text{ kg})(v_2)^2,$$

$$U_2 = \tfrac{1}{2}(24 \text{ N} \cdot \text{m}^{-1})(0.3 \text{ m})^2 = 1.08 \text{ J}.$$

Then from Eq. (6–23),

$$0 + 3 \text{ J} = (2 \text{ kg})v_2^2 + 1.08 \text{ J},$$

from which we again obtain $v_2 = \pm\ 0.98 \text{ m} \cdot \text{s}^{-1}$. What is the physical significance of the $\pm$ sign? ◀

EXAMPLE 2 . For the system of Example 1, suppose the body is initially at rest, and the spring is initially unstretched. Suppose that a constant force **P** of magnitude 10 N is exerted on the body, and that there is no friction. What will be the speed of the body when it has moved 0.5 m?

Solution The equations of motion with constant acceleration cannot be used, since the resultant force on the body varies as the spring is stretched. However, the speed can be found from energy considerations:

$$W' = \Delta K + \Delta U.$$

$$(10 \text{ N})(0.5 \text{ m}) = \tfrac{1}{2}(4 \text{ kg})v_2^2 - 0 + \tfrac{1}{2}(24 \text{ N} \cdot \text{m}^{-1})(0.50 \text{ m})^2 - 0,$$

$$v_2 = 1 \text{ m} \cdot \text{s}^{-1}. \qquad ◀$$

EXAMPLE 3 Suppose the force **P** is removed when the body has moved 0.5 m. *How much farther* does the body move before coming to rest?

Solution The elastic force is now the only force, and total mechanical energy is conserved. The kinetic energy is $\tfrac{1}{2}mv^2 = 2 \text{ J}$, and the potential energy is $\tfrac{1}{2}kx^2 = 3 \text{ J}$. The total energy is therefore 5 J (equal to the work of the force **P**). When the body comes to rest, its kinetic energy is zero and its potential energy is therefore 5 J. Hence

$$\tfrac{1}{2}kx_{\text{max}}^2 = \tfrac{1}{2}(24 \text{ N} \cdot \text{m}^{-1})(x_{\text{max}}^2) = 5 \text{ J}, \qquad x_{\text{max}} = 0.645 \text{ m}. \qquad ◀$$

EXAMPLE 4 A block of mass m, initially at rest, is dropped from a height h onto a spring whose force constant is k. Find the maximum distance y that the spring will be compressed. (See Fig. 6–11.)

Solution This is a process for which the principle of the conservation of mechanical energy holds. At the moment of release, the kinetic energy is zero. At the moment when maximum compression occurs, there is also no kinetic energy. Hence, the loss of gravitational potential energy of the block equals the gain of elastic potential energy of the spring. As shown in Fig. 6–11, the total fall of the block is $h + y$, whence

$$mg(h + y) = \tfrac{1}{2}ky^2, \qquad \text{or} \qquad y^2 - \frac{2mg}{k}y - \frac{2mgh}{k} = 0.$$

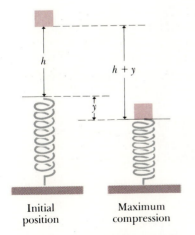

Initial position | Maximum compression

6–11 The total fall of the block is $h + y$.

Therefore, from the quadratic formula,

$$y = \frac{1}{2}\left[\frac{2mg}{k} \pm \sqrt{\left(\frac{2mg}{k}\right)^2 + \frac{8mgh}{k}} \right].$$

The positive root is the desired result; the negative root corresponds to the height to which the block and spring would rebound if they were fastened together after contact.   ◀

6–7

CONSERVATIVE AND DISSIPATIVE FORCES

We have seen that when an object acted on by a gravitational force (its weight) moves from any position above a zero reference level to any other position, the work done by the gravitational force is independent of the path and equal to the difference between the final and the initial values of a function called the *gravitational potential energy*. If the gravitational force is the only force on the object, the total mechanical energy (the sum of the kinetic and gravitational potential energies) is constant or *conserved*; the gravitational force is thus called a *conservative* force. When the object is ascending, the gravitational force does negative work, the kinetic energy decreases, and the potential energy increases. We may speak of a conversion of kinetic to potential energy. This conversion is reversible; when the object descends, the gravitational force does positive work, the kinetic energy increases, potential energy decreases, and potential energy has been reconverted into kinetic energy. The reversi-

6–12 Stroboscopic photograph of a pole-vaulter. The athlete's initial kinetic energy is partly converted to elastic potential energy in the flexed pole, then to gravitational potential energy as he rises and clears the bar, then back to kinetic energy as he drops on the other side. The elastic and gravitational forces are conservative. (Dr. Harold Edgerton, M.I.T., Cambridge, Massachusetts.)

bility of this conversion is an important aspect of the work done by a conservative force.

A similar discussion can be carried out for the elastic potential energy of a spring; again we may speak of a two-way conversion, kinetic to potential energy and the reverse, and again the process is completely reversible.

But now consider the work done by the frictional force acting on a body sliding on a rough surface. Such a force always does *negative* work on the body. If the body slides from one position to another and then back, the kinetic energy lost in the first displacement is not recovered in the second, because the friction force changes direction. There is no way to describe the work of the friction force in terms of a potential energy function. When the friction force acts alone, the total mechanical energy is *not* conserved. The friction force is therefore called a *nonconservative* or a *dissipative force*. *The mechanical energy of a body is conserved only when no dissipative forces act on it*.

We find that when friction forces act on a moving body, another form of energy is involved, associated with heat. The more general principle of conservation of energy includes this other form of energy, along with kinetic and potential energy; and when it is included, the *total* energy of any system remains constant. We shall study this general conservation principle more fully in a later chapter.

EXAMPLE   Example 3 in Section 6–5 illustrates the motion of a body acted on by a dissipative friction force. The initial mechanical energy of the body is its initial potential energy of 735 J. Its final mechanical energy is its final kinetic energy of 112.5 J. The frictional work $W_{\mathcal{F}}$ is -622.5 J. A quantity of heat equivalent to 622.5 J is developed as the body slides down the track. The sum of this energy and the final mechanical energy equals the initial mechanical energy, and the total energy of the system is conserved.   ◄

*6–8

INTERNAL WORK

Figure 6–13a shows a man standing on frictionless roller skates on a level surface, facing a rigid wall. Suppose that he sets himself in motion backward by pushing against the wall. The forces acting *on the man* are his

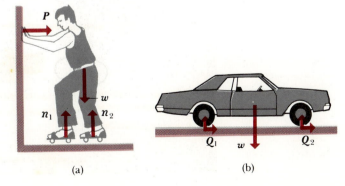

6–13   (a) External forces acting on a man who is pushing against a wall. The work done by these forces is zero. (b) External forces on an automobile. The work done by these forces is zero. In both cases, the work done by the internal force is responsible for the increase in kinetic energy.

(a)

(b)

weight w, the upward forces $\boldsymbol{n}_1$ and $\boldsymbol{n}_2$ exerted by the ground, and the horizontal force $\boldsymbol{P}$ (the reaction to the force with which the man pushes against the wall.) The works done by w and of $\boldsymbol{n}$ are zero because they are perpendicular to the motion. The force $\boldsymbol{P}$ is the unbalanced horizontal force that gives the system a horizontal acceleration. *The work done by $\boldsymbol{P}$, however, is zero because there is no motion of its point of application.* We are therefore confronted with a curious situation in which a force is responsible for acceleration, but its work is zero and is not equal to the increase in kinetic energy of the system!

The new feature in this situation is that the man is a composite system with several parts that can move in relation to each other and thus can do work on each other, even in the absence of any interaction with externally applied forces. Such work is called *internal work*. Although internal forces play no role in accelerating the composite system, their points of application can move so that work is done; thus the man's kinetic energy can change even though the *external* forces do no work.

The same principle holds in the case of an accelerated automobile. The portions of the rubber tires that are in contact with the rough roadway push back on the ground and the reactions to these forces, designated by $\boldsymbol{Q}_1$ and $\boldsymbol{Q}_2$ in Fig. 6–13b, are the external horizontal forces acting *on* the automobile, which are responsible for imparting the horizontal acceleration to the system. Since, however, the portions of the tires momentarily in contact with the road are at rest with respect to the road, the works done by the forces $\boldsymbol{Q}_1$ and $\boldsymbol{Q}_2$ are zero. As a result of the expanding gases in the cylinders of the automobile engine there are many internal forces, some of which do work; these internal forces are responsible for the increase in kinetic energy.

6–9 POWER

Time considerations are not involved in the definition of work. The same amount of work is done in raising a given weight through a given height whether the work is done in 1 s, or 1 hr, or 1 yr. In many instances, however, it is necessary to consider the *rate* at which work is done as well as the total amount of work accomplished. The rate at which work is done by a working agent is called the *power* developed by that agent.

If a quantity of work ΔW is done in a time interval Δt, the average power $\overline{P}$ is defined as

$$\text{Average power} = \frac{\text{work done}}{\text{time interval}},$$

$$\overline{P} = \frac{\Delta W}{\Delta t}.$$

If the rate at which work is done is not constant, this ratio may vary; in this case we may define an *instantaneous* power P by considering a *very small* time interval Δt:

$$P = \frac{\Delta W}{\Delta t} \quad (\Delta t \text{ very small}). \tag{6–24}$$

The SI unit of power is one joule per second $(1 \text{ J} \cdot \text{s}^{-1})$, which is called one *watt* (1 W). Since this is a rather small unit, the kilowatt (1 kW = 10^3 W) and the megawatt (1 MW = 10^6 W) are commonly used. The cgs power unit is one erg per second $(1 \text{ erg} \cdot \text{s}^{-1})$. No single term is assigned to this unit.

In the British system, where work is expressed in foot-pounds and time in seconds, the unit of power is one foot-pound per second. A larger unit called the *horsepower* (hp) is also used:

$$1 \text{ hp} = 550 \text{ ft} \cdot \text{lb} \cdot \text{s}^{-1}$$
$$= 33,000 \text{ ft-lb} \cdot \text{min}^{-1}.$$

That is, a 1-hp motor running at full load is doing 33,000 ft-lb of work every minute.

The watt is a familiar unit of *electrical* power; a 100-W light bulb converts electrical energy into light and heat at the rate of 100 joules per second. But there is nothing inherently electrical about the watt; the electric power consumption of a light bulb could be expressed in horsepower, and many automobile manufacturers are now rating their engines in kilowatts rather than horsepower.

From the relations between the newton, pound, meter, and foot, we can show that

$$1 \text{ hp} = 746 \text{ W} = 0.746 \text{ kW},$$

or about $\frac{3}{4}$ of a kilowatt, a useful figure to remember.

Since power is energy or work per unit time, the units of power may be used to define new units of work or energy. The kilowatt-hour is commonly used as a unit of electrical energy.

One kilowatt-hour is the work done in one hour by an agent working at the constant rate of one kilowatt.

Since such an agent does 1000 J of work each second, the work done in 1 hr is $3600 \times 1000 = 3,600,000$ J:

$$1 \text{ kWh} = 3.6 \times 10^6 \text{ J} = 3.6 \text{ MJ}.$$

Note that the horsepower-hour and the kilowatt-hour are units of *work* or *energy*, not power. Although energy is an abstract physical quantity, it nevertheless has a monetary value. A newton of force or a meter per second of velocity are not things that are bought and sold as such, but a kilowatt-hour of energy is a quantity offered for sale at a definite market rate. In the form of electrical energy, a kilowatt-hour can be purchased at a price varying from a few tenths of a cent to around ten cents, depending on the locality and the quantity purchased.

When a constant force F acts on a body that undergoes a displacement x in the direction of the force, the work done is

$$W = Fx,$$

and the average power is

$$\bar{P} = \frac{W}{t} = F\left(\frac{x}{t}\right). \tag{6–25}$$

But x/t is the average velocity, $\bar{v}$. Hence

$$\bar{P} = F\bar{v}$$

If the time interval is very small, Eq. (6–25) reduces to

$$P = F\left(\frac{\Delta x}{\Delta t}\right)$$

or

$$P = Fv, \tag{6–26}$$

where F and v are instantaneous values.

EXAMPLE 1 A jet airplane engine develops a thrust of 15,000 N (roughly 3000 lb). When the plane is flying at 300 m·s⁻¹ (roughly 600 mph), what horsepower is developed?

Solution

$$P = Fv = (1.5 \times 10^4 \text{ N})(300 \text{ m·s}^{-1}) = 4.5 \times 10^6 \text{ W}$$

$$= (4.5 \times 10^6 \text{ W})\left(\frac{1 \text{ hp}}{746 \text{ W}}\right) = 6030 \text{ hp.}$$

EXAMPLE 2 As part of a charity fund-raising drive, a Chicago marathon runner of mass 50 kg runs up the stairs to the top of the Sears tower, the tallest building in the United States (443 m), in 15 minutes. What is her average power output, in watts? In kilowatts? In horsepower?

Solution The total work is

$$W = mgh = (50 \text{ kg})(9.8 \text{ m·s}^{-2})(443 \text{ m}) = 2.17 \times 10^5 \text{ J.}$$

The time is 15 min = 900 s, so the average power is

$$\bar{P} = \frac{2.17 \times 10^5 \text{ J}}{900 \text{ s}} = 241 \text{ W} = 0.241 \text{ kW} = 0.323 \text{ hp.}$$

Alternatively, the average speed is (443 m)/(900 s) = 0.492 m·s⁻¹, so the average power is

$$\bar{P} = F\bar{v} = (mg)\bar{v} = (50 \text{ kg})(9.8 \text{ m·s}^{-2})(0.492 \text{ m·s}^{-1}) = 241 \text{ W.}$$

QUESTIONS

6–1 An elevator is hoisted by its cables at constant speed. Is the total work done on the elevator positive, negative, or zero?

6–2 A rope tied to a body is pulled, causing the body to accelerate. But according to Newton's third law the body pulls back on the rope with an equal and opposite force. Is the total work done then zero? If so, how can the body's kinetic energy change?

6–3 In Fig. 6–8, the projectile has the same initial kinetic energy in each case. Why does it not then rise to the same maximum height in each case?

6–4 Are there any cases where a frictional force can *increase* the mechanical energy of a system? If so, give examples.

6–5 An automobile jack is used to lift a heavy weight by exerting a force much smaller in magnitude than the weight. Does this mean that less work is done than if the weight had been lifted directly?

6–6 A compressed spring is clamped in its compressed position and is then dissolved in acid. What becomes of its potential energy?

6–7 A child standing on a playground swing can increase the amplitude of his motion by "pumping up," that is, by pulling back on the swing ropes at the appropriate points during the motion. Where does the added energy come from?

6–8 In a siphon, water is lifted above its original level during its flow from one container to another. Where does it get the needed potential energy?

6–9 A man bounces on a trampoline, going a little higher with each bounce. Explain how he increases his total mechanical energy.

6–10 Is it possible for the second hill on a roller-coaster track to be higher than the first? What would happen if it were higher?

6–11 Does the kinetic energy of a car change more when it speeds up from 10 to 15 m·s^{-1} or from 15 to 20 m·s^{-1}?

6–12 A car accelerates from an initial speed to a greater final speed while the engine develops constant power. Is the acceleration greater at the beginning of this process or at the end?

6–13 Time yourself while running up a flight of steps, and compute your maximum power, in horsepower. Are you stronger than a horse?

6–14 When a constant force is applied to a body moving with constant acceleration, is the power of the force constant? If not, how would the force have to vary with speed for the power to be constant?

6–15 An advertisement for a power saw states: "This power tool uses the energy of the motor to stop the blade rotation within 5 seconds after the switch is turned off." Is this an accurate statement? Please explain.

PROBLEMS

6–1 The locomotive of a freight train exerts a constant force of 60,000 N on the train while drawing it at 50 km·hr^{-1} on a level track. How much work does it do in a distance of 1 km?

6–2 A piano mover rolls a 200-kg piano at constant speed up a ramp 2 m long at an angle of 30° with the horizontal.

a) What force does he apply to the piano?

b) How much work does he do?

6–3 A fisherman reels in 20 m of line in pulling in a fish that exerts a constant resisting force of 30 N. How much work does the fisherman do?

6–4 A factory worker pushes a 40-kg box a distance of 5 m along a level floor, at constant speed, by pushing horizontally on it. The coefficient of friction between the box and the floor is 0.25.

a) What force is required?

b) How much work does the workman do?

6–5 In Problem 6–4, suppose the workman pushes forward and down, at an angle of 30° to the horizontal.

a) What force is required?

b) How much work is done by the workman? By the friction force?

6–6 The old oaken bucket that hangs in the well has a mass of 4 kg. We pull it up a distance of 5 m by pulling horizontally on a rope passing over a pulley at the top of the well.

a) How much work do we do in pulling the bucket up?

b) How much work is done by the gravitational force acting on the bucket?

6–7 A water skier is pulled by a tow rope behind a boat. He skis off to the side, and the rope makes an angle of 20° with his direction of motion. The tension in the rope is 120 N. How much work is done on the skier by the rope during a displacement of 500 m?

6–8 A ketchup bottle is pushed 2 m along a stationary horizontal surface by a horizontal force of 2 N. The opposing force of friction is 0.4 N.

a) How much work is done by the 2-N force?

b) What work is done by the friction force?

6–9

a) Compute the kinetic energy of a 1200-kg automobile traveling at 20 km·hr^{-1}.

b) How many times as great is the kinetic energy if the velocity is doubled?

6–10 Compute the kinetic energy, in joules, of a 2-g rifle bullet traveling at 500 m·s^{-1}.

6–11 An electron strikes the screen of a cathode-ray tube with a velocity of 10^7 m·s^{-1}. Compute its kinetic energy in joules. The mass of an electron is 9.11 × 10^{-31} kg.

6–12 A body of mass 2 kg is initially at rest on a horizontal frictionless plane. It is then pulled 4 m by a horizontal force of magnitude 25 N. Use the work–energy relation to find its final speed.

6–13 In Problem 6–12, suppose the body has an initial speed of 10 m·s^{-1} and is then pulled 4 m by a force of magnitude 25 N in the direction of the initial velocity. What is its final speed?

6–14 A body of mass 8 kg moves in a straight line on a horizontal frictionless surface. At one point in its path its speed is 4 m·s^{-1}, and after it has traveled 3 m its speed is 5 m·s^{-1} in the same direction. Use the work–energy relation to find the force acting on the body, assuming it is constant.

6–15 A force of magnitude 5 N acts on a 2-kg body moving initially in the direction of the force with a speed of 4 m·s^{-1}. Over what distance must the force act in order to change the body's speed to 6 m·s^{-1}?

6–16 A 5-kg block is lifted vertically at a constant velocity of 4 m·s^{-1} through a height of 12 m.

a) How great a force is required?

b) How much work is done? What becomes of this work?

6–17 A body moves a distance of 10 m under the action of a force that has the constant value of 5.5 N for the first 6 m and then decreases to a value of 2 N, as shown by the graph in Fig. 6–14.

a) How much work is done in the first 6 m of the motion?

b) How much work is done in the last 4 m?

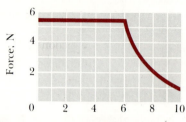

Figure 6–14

6–18 A block of mass 8 kg is pushed 5 m along a horizontal frictionless surface by a horizontal force of 40 N. The block starts from rest.

a) How much work is done? What becomes of this work?

b) Check your answer by computing the acceleration of the block, its final velocity, and its kinetic energy.

6–19 In the preceding problem, suppose the block had an initial speed of 4 m·s⁻¹, other quantities remaining the same.

a) How much work is done?

b) Check by computing the final velocity and the increase in kinetic energy.

6–20 What is the potential energy of an 800-kg elevator at the top of the Empire State Building 380 m above street level? Assume the potential energy at street level to be zero.

6–21 A 12-kg block is pushed 20 m up the sloping surface of a plane inclined at an angle of 37° to the horizontal, by a constant force F of 120 N acting parallel to the plane. The coefficient of friction between the block and plane is 0.25.

a) What is the work done by the force F?

b) Compute the increase in kinetic energy of the block.

c) Compute the increase in potential energy of the block.

d) Compute the work done by the friction force?

e) What can you say about the sum of (b), (c), and (d)?

6–22 A man of mass 80 kg sits on a platform suspended from a movable pulley and raises himself at constant speed by a rope passing over a fixed pulley (Fig. 6–15). Assuming no friction losses, find

a) the force he must exert,

b) the increase in his energy when he raises himself 1 m.

Answer part (b) by calculating his increase in potential energy, and also by computing the product of the force on the rope and the length of rope passing through his hands.

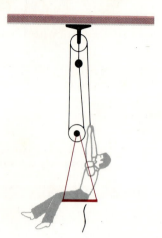

Figure 6–15

6–23 A barrel of mass 120 kg is suspended by a rope 10 m long.

a) What horizontal force is necessary to hold the barrel in a position displaced sideways 2 m from the vertical?

b) How much work is done in moving it to this position?

6–24 The system in Fig. 6–16 is released from rest with the 12-kg block 3 m above the floor. Use the principle of conservation of energy to find the velocity with which the block strikes the floor. Neglect friction and inertia of the pulley.

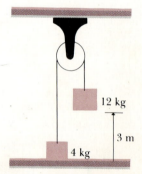

Figure 6–16

6–25 The force in newtons required to stretch a certain spring a distance x beyond its unstretched length is given by $F = 100x$.

a) What force will stretch the spring 0.1 m? 0.2 m? 0.4 m?

b) How much work is required to stretch the spring 0.1 m? 0.2 m? 0.4 m?

6–26 The scale of a certain spring balance reads from zero to 1200 N and is 0.1 m long.

a) What is the potential energy of the spring when it is stretched 0.1 m? 0.05 m?

b) When a 60-kg mass hangs from the spring?

6–27 A spring having a force constant of 200 N·m⁻¹ rests on a frictionless horizontal surface. One end is in contact with a stationary wall, and a 4-kg block is pushed against the other end, compressing the spring 0.1 m. The block is then released with no initial velocity.

a) What is the block's speed when the spring returns to its uncompressed length?

b) What is the block's speed when the spring is compressed 0.05 m?

6–28 You are asked to design spring bumpers for the walls of a parking garage. A freely rolling 1200-kg car moving at 0.5 m·s⁻¹ is to compress the spring no more than 0.05 m before stopping. What should be the force constant of the spring?

6–29 The spring of a spring gun has a force constant of 500 N·m⁻¹. It is compressed 0.05 m and a ball of mass 0.01 kg is placed in the barrel against the compressed spring.

a) Compute the speed with which the ball leaves the gun when released.

b) Determine the maximum speed if a constant resisting force of 10 N acts on the ball. (Note that in part (b) the maximum speed *does not* occur at the end of the barrel, because the ball actually slows down as it reaches the end. Why?)

6–30 A block of mass 1 kg is forced against a horizontal spring of negligible mass, compressing the spring an amount $x_1 = 0.2$ m. When released, the block moves on a horizontal tabletop a distance $x_2 = 1.0$ m before coming to rest. The spring constant k is 100 N·m⁻¹ (Fig. 6–17). What is the coefficient of friction, μ, between the block and the table?

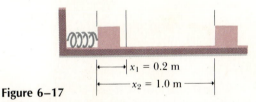

$x_1 = 0.2$ m
$x_2 = 1.0$ m

Figure 6–17

6–31 A 2-kg block is dropped from a height of 0.4 m onto a spring whose force constant k is 1960 N·m⁻¹. Find the maximum distance the spring will be compressed.

6–32 An 8-kg projectile is fired from a gun with a muzzle velocity of 300 m·s⁻¹ at an angle of departure of 45°. The angle is then increased to 90° and a similar projectile is fired with the same muzzle velocity.

a) Find the maximum height attained by each projectile.

b) Show that the total energy at the top of the trajectory is the same in the two cases.

c) Using the energy principle, find the height attained by a similar projectile if fired at an angle of 30°.

6–33 In a truck-loading station at a post office, a 2-kg package is released from rest at point A on a track that is

one quadrant of a circle of radius 1 m (Fig. 6–18). It slides down the track and reaches point B with a speed of 4 m·s⁻¹. From point B it slides on a level surface a distance of 3 m to point C, where it comes to rest.

a) What is the coefficient of sliding friction on the horizontal surface?

b) How much work is done by friction as the body slides down the circular arc from A to B?

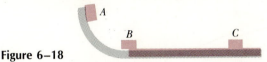

Figure 6–18

6–34 A small sphere of mass m is fastened to a weightless string of length 0.5 m to form a pendulum. The pendulum is swinging so as to make a maximum angle of 60° with the vertical.

a) What is the velocity of the sphere when it passes through the vertical position?

b) What is the instantaneous acceleration when the pendulum is at its maximum deflection?

6–35 A ball is tied to a cord and set in rotation in a vertical circle. Prove that the tension in the cord at the lowest point exceeds that at the highest point by six times the weight of the ball.

6–36 A car in an amusement-park ride rolls without friction around the loop-the-loop track shown in Fig. 6–19. It starts from rest at point A at a height $3R$ above the ground, where $R = 20$ m. When it reaches point B, at the end of a horizontal diameter of the loop, find

a) the radial acceleration experienced by the passengers,

b) the tangential acceleration, and

c) the resultant acceleration.

Show these accelerations in a diagram, approximately to scale.

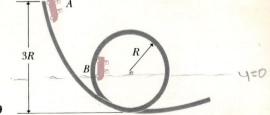

Figure 6–19

6–37 A meter stick, pivoted about a horizontal axis through its center, has a body of mass 2 kg attached to one end and a body of mass 1 kg attached to the other. The mass of the meter stick can be neglected. The system is released from rest with the stick horizontal. What is the velocity of each body as the stick swings through a vertical position?

6–38 A variable force **P** is maintained tangent to a frictionless cylindrical surface of radius a, as shown in Fig. 6–20. By slowly varying this force, a block of weight w is moved and the spring to which it is attached is stretched from position 1 to position 2. Calculate the work done by the force **P**.

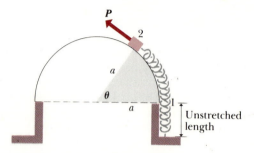

Figure 6–20

6–39 A 5-kg block is pushed up a frictionless plane inclined at 30° to the horizontal. It is pushed 2.0 m along the plane by a constant 100-N force parallel to the plane. If its speed at the bottom is 2 m·s^{-1}, what is its speed at the top?

6–40 In Problem 6–39, suppose the plane is not frictionless but has $\mu = 0.1$.

a) What is the speed of the block at the top?

b) What fraction of the work done by the force is dissipated by friction?

6–41 An 80-kg man jumps from a height of 2 m onto a platform mounted on springs. As the springs compress, the platform is pushed down a maximum distance of 0.2 m below its initial position, and it then rebounds.

a) What is the man's speed at the instant the platform is depressed 0.1 m?

b) If the man had just stepped gently onto the platform how far would it have been pushed down?

6–42 In Problem 6–36, if point A is at a height h (not equal to $3R$) above the bottom of the loop, what is the minimum value of h (in terms of R) so that the car moves around the loop without falling off at the top?

6–43 A ball of mass 0.5 kg is tied to a string of length 1.0 m, and the other end of the string is tied to a rigid support. The ball is held straight out horizontally from the point of support, with the string pulled taut, and is then released.

a) What is the speed of the ball at the lowest point of its motion?

b) What is the tension in the string at this point?

6–44 A skier starts at the top of a large frictionless spherical snowball, with very small initial velocity, and skis straight down the side. At what point does he lose contact with the snowball and fly off at a tangent? That is, at the instant he loses contact with the snowball, what angle does a radial line from the center to the skier make with the vertical?

6–45 In Section 6–5, Example 6, calculate the escape velocity, in meters per second, for an object at the surface of the earth. Why is this velocity independent of the object's mass?

6–46 Calculate the escape velocity for an object at the surface of the moon. Use the data in Appendix G.

6–47 There is a point on the line joining the centers of the earth and moon where the gravitational forces add to zero.

a) Determine the location of this point. The appropriate data are given in Appendix G.

b) With what velocity would a body have to be launched at the surface of the earth in order to just reach this point with zero velocity?

c) If a body is given a velocity just slightly greater than that found in (b), what is its velocity just as it reaches the surface of the moon?

6–48 Show that the escape velocity of a body at the surface of the earth is greater than the kinetic energy of the same body in a circular orbit around the earth (with orbit radius only slightly greater than earth's radius) by a factor of $\sqrt{2}$.

6–49 The asteroid Toro, discovered in 1964, has a radius of about 5 km and a mass of about 12×10^{15} kg. Calculate the escape velocity for a body at the surface of Toro. Could a person *jump* off Toro, i.e., jump with a velocity greater than the escape velocity?

6–50 If the potential energy of a 1-kg body, associated with the gravitational attraction of the earth and the moon, is zero far away from both bodies, what is it at the "zero-gravity" point described in Problem 6–47?

6–51 What average horsepower is developed by an 80-kg man while climbing in 10 s a flight of stairs that rises 6 m vertically? Express this power in watts and kilowatts.

6–52 The hammer of a pile driver has a mass of 500 kg and must be lifted a vertical distance of 2 m in 3 s. What horsepower engine is required?

6–53 A ski tow is to be operated on a 37° slope 300 m long. The rope is to move at 12 km·hr^{-1} and power must be provided for 80 riders at one time, with average mass 70 kg. Estimate the horsepower required to operate the tow.

6–54

a) If energy costs 5 cents per kWh, how much is one horsepower-hour worth?

b) How many ft-lb can be purchased for one cent?

6–55 Compute the monetary value of the kinetic energy of the projectile of a 14-in. naval gun, at the rate of 2 cents per kWh. The projectile has a mass of 600 kg and its muzzle velocity is 800 m·s^{-1}.

6–56 At 5 cents per kWh, what does it cost to operate a 10-hp motor for 8 hr?

6–57 The engine of an automobile develops 20 hp when the automobile is traveling at 50 km·hr^{-1}.

a) What is the resisting force acting on the automobile, in newtons?

b) If the resisting force is proportional to the velocity, what horsepower will drive the car at 25 km·hr^{-1}? At 100 km·hr^{-1}?

6–58 The engine of a motorboat delivers 40 hp to the propeller while the boat is moving at 10 m·s^{-1}. What would be the tension in the towline if the boat were being towed at the same speed?

6–59 A man whose mass is 70 kg walks up to the third floor of a building. This is a vertical height of 12 m above the street level.

a) How many joules of work has he done?

b) By how much has he increased his potential energy?

c) If he climbs the stairs in 20 s, what was his rate of working, in horsepower?

6–60 A pump is required to lift 800 kg (about 200 gallons) of water per minute from a well 10 m deep and eject it with a speed of 20 m·s^{-1}.

a) How much work is done per minute in lifting the water?

b) How much in giving it kinetic energy?

c) What horsepower engine is needed?

6–61 A 2000-kg elevator starts from rest and is pulled upward with a constant acceleration of 4 m·s^{-2}.

a) Find the tension in the supporting cable.

b) What is the velocity of the elevator after it has risen 15 m?

c) Find the kinetic energy of the elevator 3 s after it starts.

d) How much is its potential energy increased in the first 3 s?

e) What horsepower is required when the elevator is traveling 8 m·s^{-1}?

6–62 An automobile of mass 1200 kg has a speed of 30 m·s^{-1} on a horizontal road when the engine is developing 50 hp. What is its speed, with the same horsepower, if the road rises 1 m in 20 m? Assume all friction forces to be constant.

6–63

a) If 20 hp are required to drive a 1200-kg automobile at 50 km·hr^{-1} on a level road, what is the total retarding force due to friction, air resistance, etc.?

b) What power is necessary to drive the car at 50 km·hr^{-1} up a 10-percent grade (i.e., one rising 10 m vertically in 100 m horizontally)?

c) What power is necessary to drive the car at 50 km·hr^{-1} *down* a 2-percent grade?

d) Down what percent grade would the car coast at 50 km·hr^{-1}?

6–64 The total consumption of electrical energy in the United States is of the order of 10^{19} joules per year.

a) What is the average rate of energy consumption, in watts?

b) If the population of the United States is 200 million, what is the average rate of energy consumption per person?

6–65 The sun transfers energy to the earth by radiation, at a rate of approximately 1.4 kW per square meter of surface. If this energy could be collected and converted to electrical energy with 100-percent efficiency, how great an area would be required to collect the energy cited in the preceding problem?

6–66 The human heart is a powerful and extremely reliable pump. Each day it takes in and discharges over 7500 liters of blood. If the work done by the heart is equal to the work required to lift this amount of blood a height equal to that of the average American female (1.63 m), and if the density of blood is the same as that of water,

a) how much work does the heart do in a day, and

b) what is its power output, in watts?

IMPULSE
AND
MOMENTUM

In the preceding chapter the concepts of work and energy were developed from Newton's laws of motion. In this chapter we shall see how two similar concepts, those of *impulse* and *momentum*, also arise from these laws. These concepts are especially useful when sudden changes in motion of a body occur as a result of large forces acting for a short time, such as in impacts and collisions. The concept of momentum appears in all areas of physics, and the principle of conservation of momentum is one of the most fundamental of all physical laws.

7–1

IMPULSE AND MOMENTUM

Let us consider a particle of mass m moving along a straight line. For the moment we assume that the force F on this particle is constant and directed along the line of motion. If the particle's velocity at some initial time $t = 0$ is v_0, the velocity v at a later time t is given by

$$v = v_0 + at,$$

where the constant acceleration a is determined from the relation $F = ma$. When the above equation is multiplied through by m and ma is replaced by F, the result is

$$mv = mv_0 + Ft,$$

or

$$mv - mv_0 = Ft. \tag{7–1}$$

The right side of this equation, the product of the force and the time during which it acts, is called the *impulse* of the force, denoted by J. More generally, if a constant force F acts during a time interval from t_1 to t_2, the impulse of the force is defined to be

$$\text{Impulse} = J = F(t_2 - t_1). \tag{7–2}$$

The left side of Eq. (7–1) contains the product of mass and velocity of the particle at two different times. This product is also given a special name, *momentum*. It is also sometimes called *linear momentum* to distinguish it from a similar quantity called *angular momentum*, to be discussed later. We shall use the symbol p for momentum:

$$\text{Momentum} = p = mv. \tag{7–3}$$

In terms of these newly defined quantities, the meaning of Eq. (7–1) is that the impulse of the force from time zero to time t is equal to the change of momentum during that interval, that is, the momentum at the end of the interval minus that at the beginning. We see that there is nothing special about the two times zero and t, so we may also say that if the particle's velocity at time t_1 is v_1, and its velocity at time t_2 is v_2, then

$$F(t_2 - t_1) = mv_2 - mv_1. \tag{7–4}$$

Straight-line motion is, of course, a special case. Strictly speaking, the quantities F, v, and a should have been called the *components* along the straight line of the vector quantities $\mathbf{F}$, $\mathbf{v}$, and $\mathbf{a}$, respectively. The definitions of momentum are impulse and the relationship expressed by Eq. (7–4) are easily generalized for motion not confined to a straight line. We define impulse and momentum to be *vector quantities,* as follows:

$$\text{Impulse} = \mathbf{J} = \mathbf{F}(t_2 - t_1), \tag{7–5}$$

$$\text{Momentum} = \mathbf{p} = m\mathbf{v}. \tag{7–6}$$

The corresponding generalization of Eq. (7–4) is the vector equation

$$\mathbf{F}(t_2 - t_1) = m\mathbf{v}_2 - m\mathbf{v}_1. \tag{7–7}$$

This relation between impulse and momentum change has the same form as the relation between work and kinetic-energy change developed in Chapter 6. There are important differences, however. First, impulse is a product of a force and a *time* interval, whereas work is a product of a force and a *distance* and depends on the angle between force and displacement. Furthermore, both force and velocity are *vector* quantities, so impulse and momentum are also vector quantities, unlike work and kinetic energy, which are scalars. In motion along a straight line, where only one component of a vector is involved, the force and velocity may have components along this line that are either positive or negative.

EXAMPLE 1 A particle of mass 2 kg moves along the x-axis with an initial velocity of 3 m·s^{-1}. A force $F = -6$ N (i.e., in the negative x-direction) is applied for a period of 3 s. Find the final velocity.

Solution From Eq. (7–4),

$$(-6\text{N})(3\text{ s}) = (2\text{ kg})v_2 - (2\text{ kg})(3\text{ m·s}^{-1})$$

or

$$v_2 = -6 \text{ m·s}^{-1}.$$

The particle's final velocity is in the negative x-direction. ◄

The unit of impulse, in any system, equals the product of the units of force and time in that system. Thus, in SI (mks) units, the unit is one

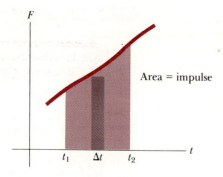

7–1 The area of the small rectangle is an approximation of the total change in momentum in the interval Δt.

newton second (1 N·s), in the cgs system it is one dyne second (1 dyn·s), and in the British system it is one pound second (1 lb·s).

The unit of momentum in SI (mks) units is one kilogram meter per second (1 kg·m·s^{-1}), in the cgs system it is one gram centimeter per second (1 g·cm·s^{-1}), and in the British system it is one slug foot per second (1 slug·ft·s^{-1}). Since

$$1 \text{ kg·m·s}^{-1} = (1 \text{ kg·m·s}^{-2})\text{s} = 1 \text{ N·s},$$

the units of momentum and impulse are equivalent.

The concept of impulse can be generalized to include forces that vary with time. We follow a procedure analogous to that used to compute work done by a variable force. If a variable force is applied during an interval from time t_1 to t_2, we divide this interval into a large number of subintervals. Calling a typical one of these Δt, we note that if it is sufficiently small we may regard the force as constant during this interval, and thus the impulse during Δt is $F\,\Delta t$. This also equals the momentum change in this interval. Graphically, $F\,\Delta t$ is represented by the area of the strip of width Δt under the curve that shows F as a function of t, as in Fig. 7–1. Then the total impulse between t_1 and t_2 is represented by the total area under the curve and is in turn equal to the total momentum change in the interval.

In cases where the force varies with time, we can always define an *average* force F_{av} such that the product $F_{av}(t_2 - t_1)$ represents the same area as the area under the curve in Fig. 7–1. This can be done even when the force varies in direction, so the statement

$$\boldsymbol{F}_{av}(t_2 - t_1) = \boldsymbol{J} = m\boldsymbol{v}_2 - m\boldsymbol{v}_1 \qquad (7\text{–}8)$$

is perfectly general, whether the actual force $\boldsymbol{F}$ is constant or not.

If the component in a given direction of the impulse of a force is *positive*, the corresponding component of momentum of the body on which the impulse acts *increases* algebraically; if negative, it decreases. If the impulse is zero, there is no change in momentum.

EXAMPLE 2   Consider the changes in momentum produced by the following forces:

a) A body moving on the x-axis is acted on for 2 s by a constant force of 10 N toward the right.

b) The body is acted on for 2 s by a constant force of 10 N toward the right and then for 2 s by a constant force of 20 N toward the left.

c) The body is acted on for 2 s by a constant force of 10 N toward the right, then for 1 s by a constant force of 20 N toward the left. The forces for each case are shown graphically in Fig. 7–2.

Solution

a) The impulse of the force is $(+10 \text{ N})(2 \text{ s}) = +20 \text{ N·s}$. Hence the momentum of *any* body on which the force acts increases by 20 kg·m·s^{-1}. This change is the same whatever the mass of the body and whatever the magnitude and direction of its initial velocity.

Suppose the mass of the body is 2 kg and that it is initially at rest. Its *final* momentum then equals its *change* in momentum and its final

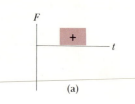

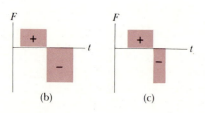

7–2   Graphic representation of the forces in Example 2.

velocity is 10 m·s^{-1} toward the right. (The reader should verify by computing the acceleration.)

Had the body been initially moving toward the right at 5 m·s^{-1}, its initial momentum would have been 10 kg·m·s^{-1} toward the right, its final momentum 30 kg·m·s^{-1}, and its final velocity 15 m·s^{-1} toward the right.

Had the body been moving initially toward the *left* at 5 m·s^{-1}, its initial momentum would have been -10 kg·m·s^{-1}, its final momentum $+10$ kg·m·s^{-1}, and its final velocity 5 m·s^{-1} toward the right. That is, the constant force of 10 N toward the right would first have brought the body to rest and then given it a velocity in the direction opposite to its initial velocity.

b) The impulse of this force is

$$(+10 \text{ N})(2 \text{ s}) - (20 \text{ N})(2 \text{ s}) = -20 \text{ N·s}.$$

The momentum of *any* body on which this impulse acts is changed by -20 kg·m·s^{-1}. The reader should examine various possibilities, as in the preceding example.

c) The impulse of this force is

$$(+10 \text{ N})(2 \text{ s}) - (20 \text{ N})(1 \text{ s}) = 0.$$

Hence the momentum of any body on which it acts is not changed. Of course, the momentum of the body is increased during the first 2 s but it is *decreased* by an equal amount in the next second. As an exercise, describe the motion of a body of mass 2 kg, moving initially to the left at 5 m·s^{-1}, and acted on by this force. It will help to construct a graph of velocity versus time. ◀

EXAMPLE 3 A ball of mass 0.4 kg is thrown against a brick wall. When it strikes the wall it is moving horizontally to the left at 30 m·s^{-1}, and it rebounds horizontally to the right at 20 m·s^{-1}. Find the impulse of the force exerted on the ball by the wall. If the ball is in contact with the wall for 0.01 s, find the average force on the ball during the impact.

Solution The initial momentum of the ball is

$$(0.4 \text{ kg})(-30 \text{ m·s}^{-1}) = -12 \text{ kg·m·s}^{-1}.$$

The final momentum is $+8$ kg·m·s^{-1}. The *change* in momentum is

$$mv_2 - mv_1 = 8 \text{ kg·m·s}^{-1} - (-12 \text{ kg·m·s}^{-1})$$
$$= 20 \text{ kg·m·s}^{-1} = J.$$

Hence, the impulse of the force exerted on the ball was 20 kg·m·s^{-1} = 20 N·s. Since the impulse is *positive*, the force must be toward the right.

The general nature of the force–time graph is shown by one of the curves in Fig. 7–3. The force is zero before impact, rises to a maximum, and decreases to zero when the ball leaves the wall. If the ball is relatively rigid, like a baseball, the time of collision is small and the maximum force is large, as in curve (a). If the ball is more yielding, like a tennis ball, the collision time is larger and the maximum force is less, as in curve (b). In any event, the *area* under the curve represents the impulse $J = 20$ N·s.

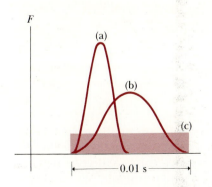

F

(a)

(b)

(c)

$\longmapsto$ 0.01 s $\longrightarrow$

7–3 Force–time graph for Example 3.

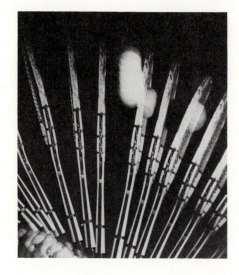

7–4 Multiple-flash stroboscopic photograph of a tennis racket hitting a ball during a serve. The exposure rate was 300 pictures per second. The ball is in contact with the racket for approximately 0.01 s. The ball flattens noticeably on both sides, and the frame of the racket bends and vibrates during and after the impact. (Dr. Harold Edgerton, M.I.T., Cambridge, Massachusetts.)

If the collision time is 0.01 s, then from Eq. (7–8),

$$F_{av}(0.01 \text{ s}) = 20 \text{ N} \cdot \text{s}, \qquad F_{av} = 2000 \text{ N}.$$

This average force is represented by the horizontal line (c) in Fig. 7–3. The actual force might vary with time as shown by curve (b) in the figure.
◀

Figure 7–4 is a stroboscopic photograph showing the impact of a tennis ball and racket during a serve.

7–2

CONSERVATION OF MOMENTUM

Let us consider now a system consisting of two bodies that interact with each other but not with anything else. Because each body exerts a force on the other, the momentum of each body changes. According to Newton's third law, the forces on the two bodies are always equal in magnitude and opposite in direction. Thus the two *impulses* in any given time interval are also equal and opposite. It follows that *in any time interval, the vector change in momentum of either particle is equal in magnitude and opposite in direction to the vector change in momentum of the other.*

This principle can be stated more simply by defining the *total momentum* of the system as the vector sum of the momenta of the separate bodies. If the change of momentum of one body is exactly the negative of that of the other, then the change in the *total* momentum must be zero. Thus *when two bodies interact only with each other, their total momentum is constant.*

A force that one part of a system exerts on another is called an *internal* force, and a force exerted on a part of a system by some body or agency outside the system is called an *external* force. When no external force acts on a system, or when the vector sum (resultant) of the external

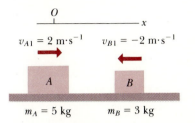

7–5 Momentum diagram for Example 1.

forces is zero, the total momentum of the system is constant, in both magnitude and direction. This is the *principle of conservation of linear momentum:*

> *When no resultant external force acts on a system, the total momentum of the system remains constant in magnitude and direction.*

The principle of conservation of momentum is one of the most fundamental and important principles of mechanics. Note that it is more general than the principle of conservation of mechanical energy; mechanical energy is conserved *only* when the internal forces are *conservative.* The principle of conservation of momentum holds whatever the nature of the internal forces, conservative or not.

EXAMPLE 1   Figure 7–5 shows two bodies moving toward each other on a frictionless surface or a level air track. There is no friction, and the resultant vertical force on each body is zero. Thus the resultant force on each body is the force exerted on it by the other body, and the total momentum of the system is constant, in magnitude and direction.

Suppose that, after the bodies collide, body B moves away with a final velocity of $+2$ m·s^{-1}. What is the final velocity of body A?

Solution   Let the final velocity of A be v_{A2}. Then we write an equation showing the equality of the total momentum before and after the collision:

$$(5 \text{ kg})(2 \text{ m·s}^{-1}) + (3 \text{ kg})(-2 \text{ m·s}^{-1}) = (5 \text{ kg})v_{A2} + (3 \text{ kg})(+2 \text{ m·s}^{-1}).$$

Solving this equation for v_{A2}, we find $v_{A2} = -0.4$ m·s^{-1}.   ◄

EXAMPLE 2   In Fig. 7–6, body A, having mass $m_A = 5$ kg, moves with initial velocity $v_{A1} = 2$ m·s^{-1} parallel to the x-axis and collides with body B, having mass $m_B = 3$ kg, initially at rest. After the collision, the velocity of A is found to be $v_{A2} = 1$ m·s^{-1} in a direction making an angle $\theta = 30°$ with the initial direction. What is the final velocity of B?

Solution   Again the total momentum of the system is the same before and after the collision. In this example, however, the velocities are not all along a single line, and the *vector* nature of momentum must be recognized. This is most easily accomplished by finding the components of each momentum in the x- and y-directions. Then momentum conservation requires that the sum of the x-components before the collision must equal that after the collision, and similarly for the y-components. Just as with force equilibrium problems, we write a separate equation for each component. For the x-components, we have

$$(5 \text{ kg})(2 \text{ m·s}^{-1}) + (3 \text{ kg})(0) = (5 \text{ kg})(1 \text{ m·s}^{-1})(\cos 30°) + (3 \text{ kg})v_{B2x}.$$

From this we find $v_{B2x} = 1.89$ m·s^{-1}.

Similarly, conservation of the y-component of total momentum gives

$$(5 \text{ kg})(0) + (3 \text{ kg})(0) = (5 \text{ kg})(1 \text{ m·s}^{-1})(\sin 30°) + (3 \text{ kg})v_{B2y},$$

from which $v_{B2y} = -0.83$ m·s^{-1}.

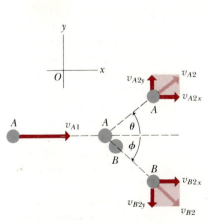

7–6 Velocity vector diagram for Example 2.

The magnitude of v_{B2} is

$$v_{B2} = \sqrt{(1.89 \text{ m} \cdot \text{s}^{-1})^2 + (-0.83 \text{ m} \cdot \text{s}^{-1})^2}$$
$$= 2.06 \text{ m} \cdot \text{s}^{-1},$$

and the angle of its direction from the positive x-axis is

$$\phi = \arctan \frac{-0.83 \text{ m} \cdot \text{s}^{-1}}{1.89 \text{ m} \cdot \text{s}^{-1}}$$
$$= -23.7°.$$

7–3

COLLISIONS

Suppose we are given the masses and initial velocities of two colliding bodies, and we wish to compute their velocities after the collision. If the collision is "head-on," as in the first example in Section 7–2, then momentum conservation provides the equation

$$m_A v_{A1} + m_B v_{B1} = m_A v_{A2} + m_B v_{B2} \qquad (7-9)$$

for the two unknowns v_{A2} and v_{B2}. This equation by itself does not provide enough information to determine two unknowns. If the collision is of the more general type of the second example in Section 7–2, then we have two equations, one each for the x- and y-components of momentum, but then we have four unknowns, the x- and y-components of each final velocity. Again we need more information to determine these unknowns. Hence momentum considerations alone are not sufficient to determine the final velocities.

If the interaction forces between the bodies are *conservative*, the total *kinetic energy* of the system is the same after the collision as before, and this provides an additional relation among the velocities. Such a collision is said to be *completely elastic, perfectly elastic*, or just plain *elastic*. A collision between two glass balls or two hard steel balls or two ivory billiard balls is almost completely elastic. As a model for elastic collisions, we may consider the situation shown in Fig. 7–7. When the bodies collide, the spring is momentarily compressed and some of the original kinetic energy is momentarily converted to elastic potential energy. The spring then expands, and when the bodies separate, this potential energy is reconverted to kinetic energy.

At the opposite extreme from a completely elastic collision is one in which the colliding bodies stick together and move *as a unit* after the collision. Such a collision is called a *completely inelastic* collision, or simply an *inelastic* collision. In Fig. 7–7, we could replace the spring with a ball of putty or chewing gum that squashes and sticks the two bodies together. Or we might keep the spring but add a coupling mechanism that locks the bodies together with the spring compressed at the instant the bodies reach a common velocity. A gumball striking a window shade, a bullet imbedding itself in a block of wood, or two cars colliding and locking bumpers are examples of inelastic collisions.

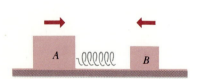

7–7 Body *A* (with spring) and body *B* approach each other on a frictionless surface.

7–4

INELASTIC COLLISIONS

For the special case of a completely inelastic collision between two bodies A and B, we have, from the definition of such a collision,

$$v_{A2} = v_{B2} = v_2.$$

When this is combined with the principle of conservation of momentum, we obtain

$$m_A v_{A1} + m_B v_{B1} = (m_A + m_B)v_2, \qquad (7–10)$$

and the final velocity can be computed if the initial velocities and the masses are known.

The total kinetic energy of a system after an inelastic collision is always less than before the collision. Suppose, for example, a body having mass m_A and initial velocity v_1 collides inelastically with a body of mass m_B initially at rest. After the collision the two bodies have a common velocity v_2 given by Eq. (7–10):

$$v_2 = \frac{m_A}{m_A + m_B}\, v_1. \qquad (7–11)$$

The kinetic energies K_1 and K_2 before and after the collision, respectively, are

$$K_1 = \tfrac{1}{2} m_A v_1{}^2,$$

$$K_2 = \tfrac{1}{2}(m_A + m_B)\left(\frac{m_A}{m_A + m_B}\right)^2 v_1{}^2,$$

so the ratio of final to initial kinetic energy is

$$\frac{K_2}{K_1} = \frac{m_A}{m_A + m_B}. \qquad (7–12)$$

The right side is always less than unity because the denominator of the fraction is always greater than the numerator. Thus in such a collision the final kinetic energy is always less than the initial. Even when the initial velocity of m_B is not zero, it is not difficult to prove that the kinetic energy after an inelastic collision is always less than before.

EXAMPLE 1   Suppose that the collision in Fig. 7–5 is completely inelastic, and that the masses and initial velocities are as shown. From momentum conservation,

$$(5 \text{ kg})(2 \text{ m} \cdot \text{s}^{-1}) + (3 \text{ kg})(-2 \text{ m} \cdot \text{s}^{-1}) = (5 \text{ kg} + 3 \text{ kg})v_2,$$

and

$$v_2 = 0.5 \text{ m} \cdot \text{s}^{-1}.$$

Since v_2 is positive, the system moves to the right after the collision.

The kinetic energy of body A before the collision is

$$\tfrac{1}{2} m_A v_{A1}{}^2 = \tfrac{1}{2}(5 \text{ kg})(2 \text{ m} \cdot \text{s}^{-1})^2 = 10 \text{ J},$$

and that of body B is

$$\tfrac{1}{2}m_B v_{B1}{}^2 = \tfrac{1}{2}(3 \text{ kg})(-2 \text{ m} \cdot \text{s}^{-1})^2 = 6 \text{ J}.$$

The total kinetic energy before collision is therefore 16 J. Note that the kinetic energy of body B is positive, although its velocity v_{B1} and its momentum mv_{B1} are both negative.

The kinetic energy after the collision is $\tfrac{1}{2}(m_A + m_B)v_2{}^2 = 1$ J. Hence, far from remaining constant, the final kinetic energy is only $\tfrac{1}{16}$ of the original, and $\tfrac{15}{16}$ is "lost" in the collision.

The energy is not really lost, of course; it is converted from mechanical energy to various other forms. If there is a ball of putty between the masses, it squashes irreversibly and becomes warmer. If the masses couple together like two freight cars, the energy goes into elastic waves that are eventually dissipated. If there is a spring between the bodies that is compressed when they are locked together, then the energy is stored as potential energy of the spring. Thus in all of these cases the *total* energy of the system is conserved, although the *kinetic* energy is not. However, in an isolated system, momentum is *always* conserved, whether the collision is elastic or not.   ◄

EXAMPLE 2   The *ballistic pendulum* is a device for measuring the velocity of a bullet. The bullet is allowed to make a completely inelastic collision with a body of much greater mass. The momentum of the system immediately after the collision equals the original momentum of the bullet, but since the *velocity* is very much smaller, it can be determined more easily. Although the ballistic pendulum has now been superseded by other devices, it is still an important laboratory experiment for illustrating the concepts of momentum and energy.

In Fig. 7–8, the pendulum, consisting perhaps of a large wooden block of mass M, hangs vertically by two cords. A bullet of mass m, traveling with a velocity v, strikes the pendulum and remains embedded in it. If the collision time is very small compared with the time of swing of the pendulum, the supporting cords remain practically vertical during this time. Hence, no external horizontal forces act on the system during the collision, and the horizontal momentum is conserved. Then if V represents the velocity of bullet and block immediately after the collision,

$$mv = (m + M)V, \qquad v = \frac{m + M}{m}V.$$

The kinetic energy of the system, immediately after the collision, is $K = \tfrac{1}{2}(m + M)V^2$.

The pendulum now swings to the right and upward until its kinetic energy is converted to gravitational potential energy. (Small frictional effects can be neglected.) Hence,

$$\tfrac{1}{2}(m + M)V^2 = (m + M)gy,$$
$$V = \sqrt{2gy},$$

and

$$v = \frac{m + M}{m}(\sqrt{2gy}).$$

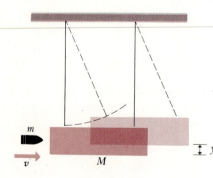

7–8   The ballistic pendulum.

By measuring m, M, and y, we can compute the original velocity v of the bullet.

Suppose $m = 5 \text{ g} = 0.005 \text{ kg}$, $M = 2 \text{ kg}$, and $h = 3 \text{ cm} = 0.03 \text{ m}$. Then, working backwards, we find the velocity V of the block just after impact:

$$V = \sqrt{2(9.8 \text{ m} \cdot \text{s}^{-2})(0.03 \text{ m})} = 0.767 \text{ m} \cdot \text{s}^{-1}.$$

Then we use momentum conservation to find the bullet's velocity v just before impact:

$$(0.005 \text{ kg})v = (2 \text{ kg} + 0.005 \text{ kg})(0.767 \text{ m} \cdot \text{s}^{-1}),$$
$$v = 307 \text{ m} \cdot \text{s}^{-1}.$$

We note that the kinetic energy just before impact is $\frac{1}{2}(0.005 \text{ kg})(307 \text{ m} \cdot \text{s}^{-1})^2 = 236 \text{ J}$, while just after impact it is $\frac{1}{2}(2.005 \text{ kg})(0.767 \text{ m} \cdot \text{s}^{-1})^2 = 0.589 \text{ J}$. Nearly all the kinetic energy disappears as the wood splinters and the bullet becomes hotter. ◀

7–5

ELASTIC COLLISIONS

Next let us consider a perfectly elastic collision between two bodies A and B. We first discuss a head-on collision in which all the initial and final velocities lie along the same line. Later we shall consider more general collisions.

Since both kinetic energy and momentum are conserved, we have

Conservation of kinetic energy

$$\tfrac{1}{2}m_A v_{A1}{}^2 + \tfrac{1}{2}m_B v_{B1}{}^2 = \tfrac{1}{2}m_A v_{A2}{}^2 + \tfrac{1}{2}m_B v_{B2}{}^2.$$

Conservation of momentum

$$m_A v_{A1} + m_B v_{B1} = m_A v_{A2} + m_B v_{B2}.$$

If the mass and initial velocities are known, these are two independent equations that can be solved simultaneously to find the two final velocities. The general solution of these equations is somewhat involved, and we will concentrate on a few important special cases.

Suppose mass m_B is initially at rest. We can then simplify the velocity notation by letting v be the initial velocity of A, and v_A and v_B the final velocities of A and B, respectively. Then the kinetic energy and momentum conservation equations are, respectively,

$$\tfrac{1}{2}m_A v^2 = \tfrac{1}{2}m_A v_A{}^2 + \tfrac{1}{2}m_B v_B{}^2, \tag{7–13}$$

$$m_A v = m_A v_A + m_B v_B. \tag{7–14}$$

Assuming the masses and v are known, we may solve for v_A and v_B. The simplest approach is somewhat indirect and uncovers an additional interesting feature of elastic collisions along the way.

We first rearrange Eqs. (7–13) and (7–14) as follows:

$$m_B v_B{}^2 = m_A(v^2 - v_A{}^2) = m_A(v - v_A)(v + v_A), \tag{7–15}$$

$$m_B v_B = m_A(v - v_A). \tag{7–16}$$

We now divide Eq. (7–15) by Eq. (7–16) to obtain

$$v_B = v + v_A. \tag{7-17}$$

We now substitute this back into Eq. (7–16) to eliminate v_B and then solve for v_A:

$$m_B(v + v_A) = m_A(v - v_A),$$

$$v_A = \frac{m_A - m_B}{m_A + m_B}\,v. \tag{7-18}$$

Finally, we substitute this result back into Eq. (7–17), to obtain

$$v_B = \frac{2m_A}{m_A + m_B}\,v. \tag{7-19}$$

A somewhat tedious calculation, but there's no substitute for persistence!

Now we can interpret the results. Suppose first that A is a table-tennis ball and B is a bowling ball. Then we expect A to bounce off with a velocity nearly equal to the original value but in the opposite direction, and B's velocity to be much smaller. Indeed, when m_A is much smaller than m_B, the fraction in Eq. (7–18) is approximately equal to (-1), and the fraction in Eq. (7–19) is much smaller than unity. We challenge the reader to make the corresponding reasonableness check for the opposite case where A is the bowling ball and B the table-tennis ball.

Another interesting case is that of equal masses. If $m_A = m_B$, then Eqs. (7–18) and (7–19) give $v_A = 0$, $v_B = v$. That is, the first body stops and the second leaves with the same velocity the first had before the collision, a phenomenon familiar to pool players.

Now comes the dividend. Equation (7–17) may be rewritten as

$$v = v_B - v_A. \tag{7-20}$$

Now $v_B - v_A$ is just the velocity of B relative to A before the collision, and v, apart from sign, is the velocity of B relative to A before the collision. Hence *the relative velocity has the same magnitude before and after the collision.* Although we have proved this only for one special case, it turns out to be a general property of *all* elastic collisions, even when both bodies are moving initially and the velocities do not all lie along the same line. The constancy of the magnitude of relative velocity provides an alternative definition of an elastic collision; in any collision where this condition is satisfied, the total kinetic energy is also conserved.

EXAMPLE 1 Suppose the collision shown in Fig. 7–5 is completely elastic. What are the velocities of A and B after the collision?

Solution From the principle of conservation of momentum,

$$(5\text{ kg})(2\text{ m}\cdot\text{s}^{-1}) + (3\text{ kg})(-2\text{ m}\cdot\text{s}^{-1}) = (5\text{ kg})v_{A2} + (3\text{ kg})v_{B2},$$

$$5v_{A2} + 3v_{B2} = 4\text{ m}\cdot\text{s}^{-1}.$$

Since the collision is completely elastic,

$$v_{B2} - v_{A2} = -(v_{B1} - v_{A1}) = -(-2\text{ m}\cdot\text{s}^{-1} - 2\text{ m}\cdot\text{s}^{-1}) = 4\text{ m}\cdot\text{s}^{-1}.$$

Solving these equations simultaneously, we obtain

$$v_{A2} = -1 \text{ m} \cdot \text{s}^{-1}, \quad v_{B2} = 3 \text{ m} \cdot \text{s}^{-1}.$$

Both bodies therefore reverse their directions of motion, A traveling to the left at $1 \text{ m} \cdot \text{s}^{-1}$ and B to the right at $3 \text{ m} \cdot \text{s}^{-1}$.

The total kinetic energy after the collision is

$$\tfrac{1}{2}(5 \text{ kg})(-1 \text{ m} \cdot \text{s}^{-1})^2 + \tfrac{1}{2}(3 \text{ kg})(3 \text{ m} \cdot \text{s}^{-1})^2 = 16 \text{ J},$$

which equals the total kinetic energy before the collision.   ◀

If an elastic collision is not head-on, the velocities do not all lie on a single line. Each final velocity then has two unknown components and there are four unknowns in all. Conservation of energy and of the x- and y-components of momentum give only three equations, so additional information is needed. This may take the form of the direction of one of the final velocities, perhaps obtained from geometrical considerations, or of one of the final velocity magnitudes.

EXAMPLE 2   In Fig. 7–6 suppose the 5-kg mass m_A has an initial velocity of $4 \text{ m} \cdot \text{s}^{-1}$ in the positive direction and a final velocity of $2 \text{ m} \cdot \text{s}^{-1}$ in an unknown direction. Find the final speed v_{B2} of the 3-kg mass and the angles θ and ϕ in the figure, assuming the collision is perfectly elastic.

Solution   Because the collision is elastic, the total final kinetic energy is the same as the initial kinetic energy. Thus

$$\tfrac{1}{2}(5 \text{ kg})(4 \text{ m} \cdot \text{s}^{-1})^2 = \tfrac{1}{2}(5 \text{ kg})(2 \text{ m} \cdot \text{s}^{-1})^2 + \tfrac{1}{2}(3 \text{ kg})v_{B2}{}^2,$$

from which

$$v_{B2} = 4.47 \text{ m} \cdot \text{s}^{-1}.$$

Conservation of the x- and y-components of total momentum gives, respectively,

$$(5 \text{ kg})(4 \text{ m} \cdot \text{s}^{-1}) = (5 \text{ kg})(2 \text{ m} \cdot \text{s}^{-1}) \cos\theta + (3 \text{ kg})(4.47 \text{ m} \cdot \text{s}^{-1}) \cos\phi,$$

$$0 = (5 \text{ kg})(2 \text{ m} \cdot \text{s}^{-1}) \sin\theta - (3 \text{ kg})(4.47 \text{ m} \cdot \text{s}^{-1}) \sin\phi.$$

These are two simultaneous equations for θ and ϕ; the simplest solution is to eliminate ϕ as follows: We solve the first equation for $\cos\phi$ and the second for $\sin\phi$; we then square each equation and add; since $\sin^2\phi + \cos^2\phi = 1$, this eliminates ϕ and leaves an equation which may be solved for $\cos\theta$ and hence for θ. This value may then be substituted back into either of the two equations and the result solved for ϕ. The details are left as an exercise; the results are

$$\phi = 36.9°, \quad \phi = 53.5°. \qquad ◀$$

The examples in this section and the preceding one show that collisions can be classified according to energy considerations. A collision in which kinetic energy is conserved is called *perfectly elastic* or simply *elastic*. A collision in which the total kinetic energy decreases and the two bodies have a common final velocity is called *completely inelastic* or simply *inelastic*. There are intermediate cases in which some kinetic energy is lost but not enough for the two bodies to reach a common final velocity; such collisions are called *semielastic*. There are also cases in which the final kinetic energy is *greater* than the initial value. This occurs, for example, when an explosive between the two bodies is detonated during the collision, blowing them apart with great speed. Such a collision might be called *superelastic*, although this term is not in common use. A special case of a superelastic collision is *recoil*, discussed in the next section.

Finally, momentum conservation can be applied even to systems that are not really isolated. It may happen that there are some external forces acting on the colliding bodies, but that the internal forces during the collision are so much larger than the external that the latter may be neglected during the actual collision. This is certainly the case, for example, when two cars collide at an icy intersection.

7–6

RECOIL

Figure 7–9 shows two blocks A and B, between which there is a compressed spring. When the system is released from rest, the spring exerts equal and opposite forces on the blocks until it has expanded to its natural unstressed length. It then drops to the surface, while the blocks continue to move. The original momentum of the system is zero, and if frictional forces can be neglected, the resultant external force on the system is zero. The momentum of the system therefore remains constant and equal to zero. Then if v_A and v_B are the velocities acquired by A and B, we have

$$m_A v_A + m_B v_B = 0, \qquad \frac{v_A}{v_B} = -\frac{m_B}{m_A}. \tag{7–21}$$

The velocities are of opposite sign and their magnitudes are inversely proportional to the corresponding masses.

The original kinetic energy of the system is also zero. The final kinetic energy is

$$K = \tfrac{1}{2}m_A v_A^2 + \tfrac{1}{2}m_B v_B^2.$$

The source of this energy is the original elastic potential energy of the system. The ratio of kinetic energies is

$$\frac{\tfrac{1}{2}m_A v_A^2}{\tfrac{1}{2}m_B v_B^2} = \frac{m_A}{m_B}\left(\frac{v_A}{v_B}\right)^2 = \frac{m_B}{m_A}. \tag{7–22}$$

Thus, although the final momenta are equal in magnitude, the final kinetic energies are inversely proportional to the corresponding masses, the body of smaller mass receiving the larger share of the original potential energy. The reason is that the change in *momentum* of a body equals

7–9 Conservation of momentum in recoil.

the *impulse* of the force acting on it, while the change in *kinetic energy* equals the *work* of the force. The forces on the two bodies are equal in magnitude and act for equal *times,* so they produce equal and opposite changes in momentum. The points of application of the forces, however, *do not* move through equal *distances* (except when $m_A = m_B$), since the acceleration, velocity, and displacement of the smaller body are greater than those of the larger. Hence more *work* is done on the body of smaller mass.

These considerations may be applied to the firing of a rifle. The initial momentum of the system is zero. When the rifle is fired, the bullet and the powder gases acquire a forward momentum, and the rifle (together with any system to which it is attached) acquires a rearward momentum of the same magnitude. Because of the relatively large mass of the rifle compared with that of the bullet and powder charge, the velocity and kinetic energy of the rifle are much *smaller* than those of the bullet and powder gases.

EXAMPLE A hunter holds a 3-kg rifle loosely in his hands and fires a bullet of mass 5 g with a muzzle velocity of 300 m·s^{-1}. What is the recoil velocity of the rifle? What are the kinetic energies?

Solution The total momentum is zero before and after firing. Thus

$$0 = (0.005 \text{ kg})(300 \text{ m·s}^{-1}) + (3 \text{ kg})v,$$

$$v = -0.5 \text{ m·s}^{-1}.$$

The negative sign means that the recoil is in the direction opposite to that of the bullet.

The kinetic energy of the bullet is

$$K_B = \tfrac{1}{2}(0.005 \text{ kg})(300 \text{ m·s}^{-1})^2 = 225 \text{ J},$$

and the kinetic energy of the rifle is

$$K_R = \tfrac{1}{2}(3 \text{ kg})(0.5 \text{ m·s}^{-1})^2 = 0.375 \text{ J}.$$

The bullet acquires much more kinetic energy than the rifle because the interaction force on it acts over a much longer distance, and hence does more work, than on the rifle. Finally, we note that

$$\frac{K_B}{K_R} = \frac{225 \text{ J}}{0.375 \text{ J}} = 600 = \frac{3 \text{ kg}}{0.005 \text{ kg}},$$

in accordance with Eq. (7–22). ◀

An interesting example of recoil is found in *ballistocardiography,* a technique developed in 1877 and still of some importance for the study of blood flow. The subject is placed on a horizontal cot suspended so that there are no horizontal forces on it; the total horizontal momentum is therefore constant. The left ventricle pumps blood into the aorta in pulses, ordinarily 60 to 80 per second (the pulse rate!). Each pulse contains of the order of 50 g of blood. The blood returning through the veins does not pulse, and hence there is a pulsating average flow velocity of blood in the body. Associated with this is a pulsating momentum, and so the remainder of the body must recoil with each pulse in order for the total horizontal momentum to remain constant. The total displacement of this recoil is only about 0.05 mm, but it can be measured with sensitive instruments. Such techniques have provided much useful information about the function of the heart and the circulatory system.

7–7

CENTER OF MASS

The principle of conservation of momentum for an isolated system can be restated in a useful way using the concept of *center of mass*. Suppose we have a collection of particles having masses m_1, m_2, and so on. Let the coordinates of m_1 be (x_1, y_1), those of m_2 be (x_2, y_2), and so on. The center of mass of the system is defined as the point having coordinates (X, Y) given by

$$X = \frac{m_1 x_1 + m_2 x_2 + m_3 x_3 + \cdots}{m_1 + m_2 + m_3 + \cdots},$$

$$Y = \frac{m_1 y_1 + m_2 y_2 + m_3 y_3 + \cdots}{m_1 + m_2 + m_3 + \cdots}, \tag{7–23}$$

Thus the center of mass represents a sort of weighted average position of the particles.

From this definition it follows directly that the *velocity* of the center of the mass is given by

$$V = \frac{m_1 v_1 + m_2 v_2 + m_3 v_3 + \cdots}{m_1 + m_2 + m_3 + \cdots}. \tag{7–24}$$

Let us denote the *total* mass $m_1 + m_2 + \cdots$ by M; then Eq. (7–24) may be rewritten as

$$MV = m_1 v_1 + m_2 v_2 + m_3 v_3 + \cdots = P. \tag{7–25}$$

The right side is simply the total momentum P of the system; hence we have proved that the total momentum is equal to the total mass times the velocity of the center of mass. It follows that for an isolated system, where the total momentum is constant, the velocity of the center of mass is also constant.

Proceeding one additional step, we note that the *rates of change* of the various velocities, that is, the accelerations, are related in the same way:

$$MA = m_1 a_1 + m_2 a_2 + m_3 a_3 + \cdots. \tag{7–26}$$

Now $m_1 a_1$ is equal to the vector sum of forces on the first particle, and so on, and the right side of Eq. (7–26) is equal to the vector sum of *all* the forces on *all* the particles. Just as in Section 7–2, we may classify each force as internal or external; because of Newton's third law, the internal forces all cancel in pairs, and what survives on the right side is the sum of the *external* forces only. Thus

$$\sum F_{\text{ext}} = MA.$$

That is, *the center of mass moves just as though all the mass were concentrated at that point and it were acted on by a resultant force equal to the sum of the external forces on the system.* For example, suppose we mark the center of mass of a hammer, which will be at some point partway down the handle. We throw the hammer spinning through the air. The motion appears complicated, but the center of mass follows a parabolic path just as though all the mass were concentrated at the center of mass. Or suppose a shell traveling in a parabolic trajectory explodes in flight, splitting into two fragments of equal mass. The fragments follow new parabolic paths, but the center of mass continues on the original trajectory just as though all the mass were still concentrated at that point. This phenomenon is shown in Fig. 7–10. A Fourth-of-July skyrocket exploding in air is a more spectacular example of the same idea.

This property of the center of mass is important in the analysis of the motion of rigid bodies, where the motion is described as a combination of motion of the center of mass and rotational motion about an axis through the center of mass. We return to this topic in Chapter 9.

In principle the position of the center of mass of a body can always be calculated by regarding it as a collection of particles and using Eqs. (7–23). In practice these calculations can become quite complex. Usually we shall deal with symmetric bodies in which the center of mass lies at the geometrical center. Other calculational techniques will be discussed in Chapter 8, in connection with the related concept of *center of gravity*.

EXAMPLE A 2-kg body and a 3-kg body are moving along the x-axis. At a particular instant the 2-kg body is 1 m from the origin and has a velocity of 3 m·s^{-1}, and the 3-kg body is 2 m from the origin and has a velocity of -1 m·s^{-1}. Find the position and velocity of the center of mass, and also find the total momentum.

Solution From Eq. (7–23),

$$X = \frac{(2 \text{ kg})(1 \text{ m}) + (3 \text{ kg})(2 \text{ m})}{2 \text{ kg} + 3 \text{ kg}} = 1.6 \text{ m},$$

and from Eq. (7–24),

$$V = \frac{(2 \text{ kg})(3 \text{ m·s}^{-1}) + (3 \text{ kg})(-1 \text{ m·s}^{-1})}{2 \text{ kg} + 3 \text{ kg}} = 0.60 \text{ m·s}^{-1}.$$

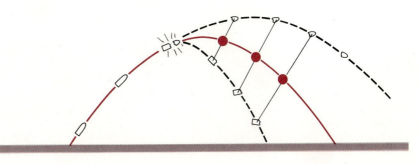

7–10 A shell explodes in flight. The fragments follow individual parabolic paths, but the center of mass continues on the same trajectory as the shell's path before exploding.

Strictly speaking, this is the x-component of velocity of the center of mass. The total momentum (strictly speaking, its x-component) is

$$P = (2 \text{ kg})(3 \text{ m} \cdot \text{s}^{-1}) + (3 \text{ kg})(-1 \text{ m} \cdot \text{s}^{-1}) = 3 \text{ kg} \cdot \text{m} \cdot \text{s}^{-1}).$$

Alternatively, from Eq. (7–25),

$$P = (5 \text{ kg})(0.6 \text{ m} \cdot \text{s}^{-1}) = 3 \text{ kg} \cdot \text{m} \cdot \text{s}^{-1}.$$ ◄

7–8

ROCKET PROPULSION

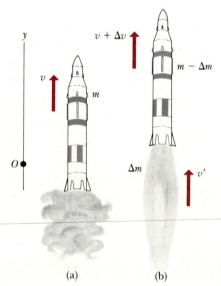

7–11 (a) Rocket at time t after take-off, with mass m and upward velocity v. Its momentum is mv. (b) At time $t + \Delta t$, the mass of the rocket (and unburned fuel) is $m - \Delta m$, its velocity is $v + \Delta v$, and its momentum is $(m - \Delta m)(v + \Delta v)$. The ejected gas has momentum $\Delta m(v - v_r)$.

A rocket is propelled by the ejection of a portion of its mass to the rear. The forward force on the rocket is the reaction to the backward force on the ejected material, and as more material is ejected, the mass of the rocket decreases. The analysis of this process is best handled by impulse–momentum considerations. In order not to bring in too many complicating forces, we shall consider a rocket fired vertically upward, and neglect air resistance and variations in g.

Figure 7–11a represents the rocket at a time t after take-off, when its mass is m and its upward velocity is v. The total momentum at this instant is thus mv. In a short time interval Δt, a mass Δm of gas is ejected from the rocket. Let v_r represent the downward speed of this gas *relative to the rocket*. The velocity v' of the gas relative to the earth is then

$$v' = v - v_r,$$

and its momentum is

$$\Delta m v' = \Delta m(v - v_r).$$

At the end of the time interval Δt, the mass of rocket and unburned fuel has decreased to $m - \Delta m$, and its velocity has increased to $v + \Delta v$. Its momentum is therefore

$$(m - \Delta m)(v + \Delta v).$$

Thus the *total* momentum at time $t + \Delta t$ is

$$(m - \Delta m)(v + \Delta v) + \Delta m(v - v_r).$$

Figure 7–11b represents rocket and ejected gas at this time.

We now make use of the impulse–momentum relation; the product of the resultant external force F on a system, and the time interval Δt during which it acts, is equal to the change in momentum of the system. If air resistance is neglected, the external force F on the rocket is its weight, $-mg$. (We take the upward direction as positive.) The change in momentum, in time Δt, is the difference between the momentum of the system at the end and at the beginning of the time interval. Hence

$$-mg \, \Delta t = (m - \Delta m)(v + \Delta v) + \Delta m(v - v_r) - mv$$
$$= m \, \Delta v - \Delta m v_r - \Delta m \, \Delta v.$$

When Δt is small, the term $\Delta m \, \Delta v$ may be dropped because it is a product of two small quantities and thus is much smaller than the other terms. Dropping this term, dividing by Δt, and rearranging, we obtain

$$m\left(\frac{\Delta v}{\Delta t}\right) = v_r\left(\frac{\Delta m}{\Delta t}\right) - mg. \tag{7-28}$$

The ratio $\Delta v/\Delta t$ is the acceleration of the rocket, so the left side of this equation (mass times acceleration) equals the resultant force on the rocket. The first term on the right equals the upward thrust on the rocket, and the resultant force equals the difference between this thrust and the weight of the rocket, mg. It will be seen that the upward thrust is proportional both to the relative velocity v_r of the ejected gas and to the mass of gas ejected per unit time, $\Delta m/\Delta t$.

The acceleration is

$$\frac{\Delta v}{\Delta t} = \frac{v_r}{m}\left(\frac{\Delta m}{\Delta t}\right) - g. \tag{7-29}$$

As the rocket rises, the value of g decreases, according to Newton's law of gravitation. (In "outer space," far from all other bodies, g becomes negligibly small.) If the values of v_r and $\Delta m/\Delta t$ remain approximately constant while the fuel is being consumed and m continually decreases, the acceleration *increases* until all the fuel is burned.

EXAMPLE   In the first second of its flight, a rocket ejects 1/60 of its mass with a velocity relative to the rocket of $2400 \text{ m} \cdot \text{s}^{-1}$. What is the acceleration of the rocket?

Solution   We have $\Delta m = m/60$, $\Delta t = 1$ s. So, from Eq. (7-29),

$$a = \frac{\Delta v}{\Delta t} = \frac{2400 \text{ m} \cdot \text{s}^{-1}}{(60)(1 \text{ s})} - 9.8 \text{ m} \cdot \text{s}^{-2} = 30.2 \text{ m} \cdot \text{s}^{-2}.$$

This is the initial value of acceleration. As mentioned above, if fuel burns at a constant rate (i.e., $\Delta m/\Delta t = $ constant), the acceleration increases with time because everything on the right side of Eq. (7-29) is constant except for m, which decreases with time. At time $t = 30$ s, when the mass has decreased to half its initial value, $a = 70.2 \text{ m} \cdot \text{s}^{-2}$. At about $t = 40$ s, the rocket reaches a speed of $2400 \text{ m} \cdot \text{s}^{-1}$, equal to the speed of the ejected mass, which is therefore at that instant at rest with respect to the earth. At later times, when v is greater than $2400 \text{ m} \cdot \text{s}^{-1}$, the ejected mass is actually moving upward, in the same direction as the rocket. Thus the velocity acquired by the rocket can be greater (and is often much greater) than the relative velocity v_r.   ◀

Figure 7–12 shows a spectacular example of rocket propulsion.

7–12   Launch of the Space Shuttle, a dramatic example of rocket propulsion. Successful launch and landing of the Space Shuttle are the first steps toward feasibility of a manned space station in orbit around the earth. (Courtesy NASA, Kennedy Space Center.)

QUESTIONS

7–1 Suppose you catch a baseball, and then someone invites you to catch a bullet with the same momentum or with the same kinetic energy. Which would you choose?

7–2 In splitting logs with a hammer and wedge, is a heavy hammer more effective than a lighter hammer? Why?

7–3 When a large, heavy truck collides with a passenger car, the occupants of the car are much more likely to be hurt than the truck driver. Why?

7–4 A glass dropped on the floor is more likely to break if the floor is concrete than if it is wood. Why?

7–5 "It ain't the fall that hurts you; it's the sudden stop at the bottom." Discuss.

7–6 When a person fires a rifle or shotgun, it is advisable to hold the butt firmly against the shoulder rather than a bit away from it, to minimize the impact on the shoulder. Why?

7–7 When a catcher in a baseball game catches a fast ball, he does not hold his arms rigid, but relaxes them so the mitt moves several inches while the ball is being caught. Why is this important? *increase collision time*

7–8 When rain falls from the sky, what becomes of its momentum as it hits the ground? Is your answer also valid for Newton's famous apple?

7–9 A man stands in the middle of a perfectly smooth, frictionless frozen lake. He can set himself in motion by throwing things, but suppose he has nothing to throw. Can he propel himself to shore *without* throwing anything?

7–10 A machine gun is fired at a steel plate. Is the average force on the plate from the bullet impact greater if the bullets bounce off or if they are squashed and stick to the plate?

7–11 How do Mexican jumping beans work? Do they violate conservation of momentum? Of energy?

7–12 Early critics of Robert Goddard, a pioneer in the use of rocket propulsion, claimed that rocket engines could not be used in outer space where there is no air for the rocket to push against. How would you answer such criticism? Would the criticism be valid for a jet engine in an ordinary airplane?

7–13 In a zero-gravity environment, can a rocket-propelled spaceship ever attain a speed greater than the relative speed with which the burnt fuel is exhausted?

PROBLEMS

7–1

a) What is the momentum of a 10,000-kg truck whose velocity is 20 m·s^{-1}?

What velocity must a 5,000-kg truck attain in order to have

b) the same momentum,

c) the same kinetic energy?

7–2 A baseball has a mass of about 0.2 kg.

a) If the velocity of a pitched ball is 30 m·s^{-1}, and after being batted it is 50 m·s^{-1} in the opposite direction, find the change in momentum of the ball and the impulse of the blow.

b) If the ball remains in contact with the bat for 0.002 s, find the average force of the blow.

7–3 A bullet having a mass of 0.05 kg, moving with a velocity of 400 m·s^{-1}, penetrates a distance of 0.1 m into a wooden block firmly attached to the earth. Assume the accelerating force constant. Compute

a) the acceleration of the bullet,

b) the accelerating force,

c) the time of acceleration,

d) the impulse of the collision. Compare the answer to part (d) with the initial momentum of the bullet.

7–4 A bullet emerges from the muzzle of a gun with a velocity of 300 m·s^{-1}. The resultant force on the bullet, while it is in the gun barrel, is given by

$$F = 400 - \frac{4 \times 10^5}{3}t,$$

where F is in newtons and t in seconds.

a) Construct a graph of F versus t.

b) Compute the time required for the bullet to travel the length of the barrel, assuming the force becomes zero just at the end of the barrel.

c) Find the impulse of the force.

d) Find the mass of the bullet.

7–5 A box, initially sliding on the floor of a room, is eventually brought to rest by friction. Is the momentum of the box conserved? If not, does this process contradict the principle of conservation of momentum? What becomes of the original momentum of the box?

7–6 A golf ball of mass 0.10 kg is initially at rest when it is struck by a club and given a speed of 50 m·s^{-1}. If the club and ball are in contact for 0.002 s, what average force acted on the ball? Is the effect of the ball's weight during the time of contact significant?

7–7 A tennis ball approaches a player's racket horizontally at 10 m·s^{-1}. After it is struck its velocity is horizontal, in the opposite direction, with magnitude 20 m·s^{-1}. The ball has mass 0.060 kg, and it is in contact with the racket for 0.01 s. What average force acted on the ball?

7–8 A steel ball of mass 0.5 kg is dropped from a height of 4 m onto a horizontal steel slab. The collision is elastic, and the ball rebounds to its original height.

a) Calculate the impulse delivered to the ball during impact.

b) If the ball is in contact with the slab for 0.002 s, find the average force on the ball during impact.

7–9 A baseball of mass 0.25 kg is struck by a bat. Just before impact, the ball is traveling horizontally at

$40 \text{ m} \cdot \text{s}^{-1}$, and it leaves the bat at an angle of 30° above horizontal with a speed of $60 \text{ m} \cdot \text{s}^{-1}$. If the ball and bat were in contact for 0.005 s, find the horizontal and vertical components of the average force on the ball.

7–10 Compare the damage to an automobile (and its occupants) in the following circumstances:

a) The automobile makes a completely inelastic head-on collision with an identical automobile traveling with the same speed in the opposite direction, and

b) it makes a completely inelastic head-on collision with a vertical rock cliff.

c) Which would be worse (for the occupants of a light car), to collide head-on with a truck traveling in the opposite direction with a momentum of equal magnitude, or to collide head-on with a truck having the same kinetic energy? (A reasonable criterion to use is the change in velocity of the car in each case.)

7–11

a) An empty freight car of mass 10,000 kg rolls at $2 \text{ m} \cdot \text{s}^{-1}$ along a level track and collides with a loaded car of mass 20,000 kg, standing at rest with brakes released. If the cars couple together, find their speed after the collision.

b) Find the decrease in kinetic energy as a result of the collision.

c) With what speed should the loaded car be rolling toward the empty car, in order that both shall be brought to rest by the collision?

7–12 When a bullet of mass 20 g strikes a ballistic pendulum of mass 10 kg, the center of gravity of the pendulum is observed to rise a vertical distance of 7 cm. The bullet remains embedded in the pendulum.

a) Calculate the original velocity of the bullet.

b) What fraction of the original kinetic energy of the bullet remains as kinetic energy of the system immediately after the collision?

c) What fraction of the original momentum remains as momentum of the system?

7–13 A bullet of mass 5 g is shot *through* a 1-kg wood block suspended on a string 2 m long. The center of gravity of the block is observed to rise a distance of 0.50 cm. Find the speed of the bullet as it emerges from the block if the initial speed is $300 \text{ m} \cdot \text{s}^{-1}$.

7–14 The pan of a spring balance has a mass of 200 g; when suspended from the coil spring, it is found to stretch the spring 10 cm. A lump of putty of mass 200 g is dropped from rest onto the pan from a height of 30 cm (Fig. 7–13). Find the maximum distance the pan moves downward.

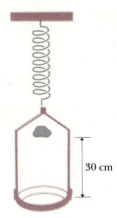

Figure 7–13

7–15 A bullet of mass 2 g, traveling in a horizontal direction with a velocity of $500 \text{ m} \cdot \text{s}^{-1}$, is fired into a wooden block of mass 1 kg, initially at rest on a level surface. The bullet passes through the block and emerges with its velocity reduced to $100 \text{ m} \cdot \text{s}^{-1}$. The block slides a distance of 20 cm along the surface from its initial position before coming to rest.

a) What was the coefficient of sliding friction between block and surface?

b) What was the decrease in kinetic energy of the bullet?

c) What was the kinetic energy of the block at the instant after the bullet passed through it?

7–16 A rifle bullet of mass 10 g is fired with a velocity of $800 \text{ m} \cdot \text{s}^{-1}$ into a ballistic pendulum of mass 5 kg, suspended from a cord 1 m long. Compute

a) the vertical height through which the pendulum rises,

b) the initial kinetic energy of the bullet,

c) the kinetic energy of bullet and pendulum after the bullet is embedded in the pendulum.

7–17 A 5-g bullet is fired horizontally into a 3-kg wooden block resting on a horizontal surface. The coefficient of sliding friction between block and surface is 0.20. The bullet remains embedded in the block, which is observed to slide 25 cm along the surface before stopping. What was the velocity of the bullet? 595 m/s

7–18 A rifle bullet of mass 0.01 kg strikes and embeds itself in a block of mass 0.99 kg which rests on a horizontal frictionless surface and is attached to a coil spring, as shown in Fig. 7–14. The impact compresses the spring

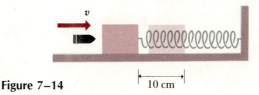

Figure 7–14   10 cm

elastic-bounces in stick together

10 cm. Calibration of the spring shows that a force of 1.0 N is required to compress the spring 1 cm.

a) Find the maximum potential energy of the spring.

b) Find the velocity of the block just after the impact.

c) What was the initial velocity of the bullet?

7–19 A 2000-kg automobile going eastward on Chestnut Street at 60 km·hr^{-1} collides with a 4000-kg truck which is going southward *across* Chestnut Street at 20 km·hr^{-1}. If they become coupled on collision, what is the magnitude and direction of their velocity immediately after colliding?

7–20 On a frictionless table, a 3-kg block moving 4 m·s^{-1} to the right collides with an 8-kg block moving 1.5 m·s^{-1} to the left.

a) If the two blocks stick together, what is the final velocity?

b) If the two blocks make a completely elastic head-on collision, what are their final velocities?

c) How much mechanical energy is dissipated in the collision of part (a)?

7–21 Two blocks of mass 300 g and 200 g are moving toward each other along a horizontal frictionless surface with velocities of 50 cm·s^{-1} and 100 cm·s^{-1}, respectively.

a) If the blocks collide and stick together, find their final velocity.

b) Find the loss of kinetic energy during the collision.

c) Find the final velocity of each block if the collision is completely elastic.

7–22

a) Prove that when a moving body makes a perfectly inelastic collision with a second body of equal mass, initially at rest, one-half of the original kinetic energy is "lost."

b) Prove that when a very heavy particle makes a perfectly elastic collision with a very light particle that is at rest, the light one goes off with twice the velocity of the heavy one.

7–23 A 10-kg block slides with a velocity of 20 cm·s^{-1} on a smooth level surface and makes a head-on collision with a 30-g block moving in the opposite direction with a velocity of 10 cm·s^{-1}. If the collision is perfectly elastic, find the velocity of each block after the collision.

7–24 A block of mass 200 g, sliding with a velocity of 12 cm·s^{-1} on a smooth, level surface, makes a perfectly elastic head-on collision with a block of mass m, initially at rest. After the collision, the velocity of the 200-g block is 4 cm·s^{-1} in the same direction as its initial velocity. Find

a) the mass m, and

b) its velocity after the collision.

7–25 A body of mass 600 g is initially at rest. It is struck by a second body of mass 400 g initially moving with a velocity of 125 cm·s^{-1} toward the right along the x-axis. After the collision, the 400-g body has a velocity of 100 cm·s^{-1} at an angle of 37° above the x-axis in the first quadrant. Both bodies move on a horizontal frictionless plane.

a) What is the magnitude and direction of the velocity of the 600-g body after the collision?

b) What is the loss of kinetic energy during the collision?

7–26 A small steel ball moving with speed v_0 in the positive x-direction makes a perfect elastic, noncentral collision with an identical ball originally at rest. After impact, the first ball moves with speed v_1 in the first quadrant at an angle θ_1 with the x-axis and the second with speed v_2 in the fourth quadrant at an angle θ_2 with the x-axis.

a) Write the equations expressing conservation of linear momentum in the x-direction, and in the y-direction.

b) Square these equations and add them.

c) At this point, introduce the fact that the collision is perfectly elastic.

d) Prove that $\theta_1 + \theta_2 = \pi/2$.

7–27 A jet of liquid of cross-sectional area A and density ρ moves with speed v_J in the positive x-direction and impinges against a perfectly smooth blade B, which deflects the stream at right angles but does not slow it down, as shown in Fig. 7–15.

a) If the blade is *stationary*, prove that the rate of arrival of mass at the blade is $\Delta m/\Delta t = \rho A v_J$.

b) If the impulse-momentum theorem is applied to a small mass Δm, prove that the x-component of the force acting on this mass for the time interval Δt is given by

$$F_x = -\frac{\Delta m}{\Delta t}v_J.$$

c) Prove that the *steady* force exerted *on* the blade in the x-direction is

$$F_x = \rho A v_J^2.$$

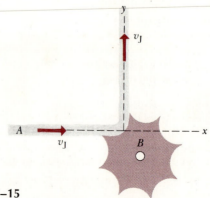

Figure 7–15

If the blade moves to the right with a speed v_B ($v_B < v_J$), derive the equations for

d) the rate of arrival of mass at the moving blade,

e) the force F_x on the blade, and

f) the power delivered to the blade.

7–28 A stone whose mass is 100 g rests on a horizontal frictionless surface. A bullet of mass 2.5 g, traveling horizontally at 400 m·s⁻¹, strikes the stone and rebounds horizontally at right angles to its original direction with a speed of 300 m·s⁻¹.

a) Compute the magnitude and direction of the velocity of the stone after it is struck.

b) Is the collision perfectly elastic?

7–29 A hockey puck B rests on a smooth ice surface and is struck by a second puck of equal mass, A, which was originally traveling at 30 m·s⁻¹ and which is deflected 30° from its original direction (Fig. 7–16). Puck B acquires a velocity at 45° with the original velocity of A.

a) Compute the speed of each puck after the collision.

b) Is the collision perfectly elastic? If not, what fraction of the original kinetic energy of puck A is "lost"?

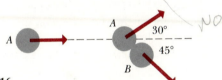

Figure 7–16

7–30 In Problem 7–29, suppose the collision is perfectly elastic, and A is deflected 30° from its initial direction. Find the final speed of each puck and the direction of B's velocity.

7–31 A neutron of mass m collides elastically with a nucleus of mass M, which is initially at rest. Show that if the neutron's initial kinetic energy is K_0, the maximum kinetic energy that it can *lose* during the collision is $4mMK_0/(M + m)^2$.

7–32 Imagine that a ball of mass 200 g rolls back and forth between opposite sides of a billiard table 1 m wide, with a velocity that remains constant in magnitude but reverses in direction at each collision with the cushions. The magnitude of the velocity is 4 m·s⁻¹.

a) What is the change in momentum of the ball at each collision?

b) How many collisions per unit time are made by the ball with one of the cushions?

c) What is the average time rate of change of momentum of the ball as a result of these collisions?

d) What is the average force exerted by the ball on a cushion?

e) Sketch a graph of the force exerted on the ball as a function of time, for a time interval of 5 s. [*Note:* This problem illustrates how one computes the average force exerted by the molecules of a gas on the walls of the containing vessel.]

7–33 A 20-kg projectile is fired at an angle of 60° above the horizontal and with a muzzle velocity of 400 m·s⁻¹. At the highest point of its trajectory the projectile explodes into two fragments of equal mass, one of which falls vertically with zero initial speed.

a) How far from the point of firing does the other fragment strike if the terrain is level?

b) How much energy was released during the explosion?

7–34 A railroad handcar is moving along straight frictionless tracks. In each of the following cases, the car initially has a total mass (car and contents) of 200 kg and is traveling with a velocity of 4 m·s⁻¹. Find the *final velocity* of the car in each of the three cases.

a) A 20-kg mass is thrown sideways out of the car with a velocity of 2 m·s⁻¹ relative to the car.

b) A 20-kg mass is thrown backwards out of the car with a velocity of 4 m·s⁻¹ relative to the car.

c) A 20-kg mass is thrown into the car with a velocity of 6 m·s⁻¹ relative to the ground and opposite in direction to the velocity of the car.

7–35 A bullet weighing 0.02 lb is fired with a muzzle velocity of 2700 ft·s⁻¹ from a rifle weighing 7.5 lb.

a) Compute the recoil velocity of the rifle, assuming it free to recoil.

b) Find the ratio of the kinetic energy of the bullet to that of the rifle.

7–36 Block A in Fig. 7–17 has a mass of 1 kg, and block B has a mass of 2 kg. The blocks are forced together, compressing a spring S between them, and the system is released from rest on a level frictionless surface. The spring is not fastened to either block and drops to the surface after it has expanded. Block B acquires a speed of 0.5 m·s⁻¹. How much potential energy was stored in the compressed spring?

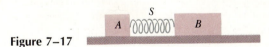

Figure 7–17

7–37 An open-topped freight car of mass 10,000 kg is coasting without friction along a level track. It is raining very hard, with the rain falling vertically downward. The car is originally empty and moving with a velocity of 1 m·s⁻¹. What is the velocity of the car after it has traveled long enough to collect 1000 kg of rain water?

7–38 An 80-kg man standing on ice throws a 0.2-kg ball horizontally with a speed of 30 m·s⁻¹.

a) With what speed and in what direction will the man begin to move?

b) If the man throws 4 such balls every 3 s, what is the average force acting on him? [*Hint:* Average force equals average rate of change of momentum.]

7–39 Find the average recoil force on a machine gun firing 120 shots per minute. The mass of each bullet is 10 g, and the muzzle velocity is 800 m·s^{-1}.

7–40 A rifleman, who together with his rifle has a mass of 100 kg, stands on roller skates and fires 10 shots horizontally from an automatic rifle. Each bullet has a mass of 10 g and a muzzle velocity of 800 m·s^{-1}.

a) If the rifleman moves back without friction, what is his velocity at the end of the 10 shots?

b) If the shots were fired in 10 s, what was the average force exerted on him?

c) Compare his kinetic energy with that of the 10 bullets.

7–41 A 1000-kg automobile is moving along a straight highway at 10 m·s^{-1}. Another car, with mass 2000 kg and speed 20 m·s^{-1} is 30 m ahead of the first.

a) Find the position of the center of mass of the two automobiles.

b) Find the total momentum from the above data.

c) Find the velocity of the center of mass.

d) Find the total momentum, using the velocity of the center of mass. Compare your result with that of part (b).

7–42 Find the position of the center of mass of the earth–moon system. Use the data in Appendix G.

7–43 A uniform steel rod 1 m in length is bent in a 90° angle at its midpoint. Determine the position of its center of mass. [*Hint:* The mass of each side of the angle may be assumed to be concentrated at its center.]

7–44 Three particles have the following masses and coordinates: (1) 2 kg, (3 m, 2 m); (2) 3 kg, (1 m, −4 m); (3) 4 kg, (−3 m, 5 m). Find the coordinates of the center of mass of the system.

7–45 A canoeist of mass 80 kg stands up in a 30-kg canoe of length 5 m. He walks from a point 1 m from one end to a point 1 m from the other end. If resistance to motion of the canoe in the water can be neglected, how far does the canoe move during this process?

7–46 A rocket burns 0.05 kg of fuel per second, ejecting it as a gas with a velocity relative to the rocket of 5000 m·s^{-1}.

a) What force does this gas exert on the rocket?

b) Would the rocket operate in free space?

c) If it would operate in free space, how would you steer it? Could you brake it?

EQUILIBRIUM OF A RIGID BODY

The discussion of equilibrium problems in Chapter 2 was concerned mostly with bodies that could be represented as points. It was mentioned, however, that for an extended body to be in equilibrium the body must also have no tendency to *rotate*. We now return to this more general problem of equilibrium of a rigid body and to the second condition for equilibrium, concerned with rotational equilibrium. In this context we introduce the concept of *moment* or *torque* of a force, which plays a central role in rigid-body equilibrium as well as in rotational motion of a rigid body, to be discussed in Chapter 9.

8–1

MOMENT OR TORQUE OF A FORCE

The tendency of a force to cause rotation depends on both the magnitude of the force and the position of the line along which it acts. For example, when one pushes a door open, a force applied near the knob is more effective than the same force applied near the hinge side. Similarly, in Fig. 8–1, the force F_1, if applied by itself to the body shown, would cause a *counterclockwise* rotation and a translation to the right, while force F_2 applied by itself would cause a *clockwise* rotation (and translation to the right), even though it has the same magnitude and direction as F_1.

It is useful to think of a body that is pivoted or hinged so it can rotate about some stationary axis. Suppose such a body is acted on by several forces, which act along lines lying in a plane perpendicular to the axis. In Fig. 8–2 this plane is the plane of the figure, and the axis of rotation is perpendicular to this plane at point O.

The tendency of force F_1 in Fig. 8–2 to cause a rotation about the axis through O is measured jointly by the magnitude of the force and the perpendicular distance l_1 between the line of action of the force and the axis. Clearly if $l_1 = 0$ there is *no* tendency to cause rotation. The role of distance l_1 is analogous to that of a wrench handle; a tight bolt is turned more easily by using a long-handled wrench than by a short-handled

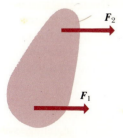

F_2

F_1

8–1

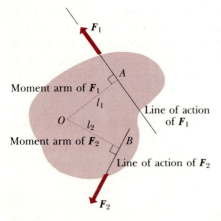

8–2 The moment or torque of a force about an axis is the product of the force and its moment arm.

one. Distance l_1 is called the *moment arm* of force F_1 about the axis O, and the product of F_1 and l_1 is called the *moment* or *torque* of the force about point O. These two terms are used interchangeably. We shall use the Greek letter Γ (capital "gamma") for moment or torque:

$$\Gamma = Fl. \tag{8–1}$$

In Fig. 8–2 the body is acted on by the forces F_1 and F_2, lying in the plane of the diagram. The moment arm of F_1 is the perpendicular distance OA of length l_1, and the moment arm of F_2 is the perpendicular distance OB of length l_2.

The effect of the force F_1 is to produce *counterclockwise* rotation about the axis, while that of F_2 is to produce *clockwise* rotation. To distinguish between these directions of rotation, we shall adopt the convention that *counterclockwise torques are positive, and clockwise torques are negative.* Hence the torque Γ_1 of the force F_1 about the axis through O is

$$\Gamma_1 = +F_1 l_1,$$

and the torque Γ_2 of F_2 is

$$\Gamma_2 = -F_2 l_2.$$

In SI (mks) units, where the unit of force is the newton and the unit of length the meter, the unit of torque is the newton-meter. If forces are expressed in pounds and lengths in feet, torques are expressed in pound-feet.

In problems involving torques, it is always advisable to include a symbol indicating the choice of positive direction of rotation; we shall use the symbol

$$\circlearrowright +$$

in examples in this chapter and in Chapter 9, where we deal with other rotational quantities as well as torque.

8–2

THE SECOND CONDITION FOR EQUILIBRIUM

We saw in Section 2–2 that, when a body is in equilibrium under the action of several coplanar forces, the vector sum of the forces must be zero. This is the *first condition for equilibrium,*

$$\sum F_x = 0, \quad \sum F_y = 0. \tag{8–2}$$

The second condition for equilibrium of a rigid body corresponds to the absence of any tendency of the body to rotate: *The sum of the torques of all forces acting on the body, calculated with respect to some specified axis, must be zero.* Furthermore, the body need not actually be pivoted about the chosen axis. In order to be in equilibrium the body must have no tendency to rotate about *any* axis, so the sum of torques must be zero, no matter what axis is chosen.

$$\sum \Gamma = 0 \ \textit{(about any arbitrary axis)}. \tag{8–3}$$

Although the choice of axis is arbitrary, one must, of course, use the *same* axis for all the torques.

For the present we regard the second condition for equilibrium of a rigid body as an *empirical* relation; we use it because it has been found experimentally to be correct. In Chapter 9 this condition will emerge as a special case of the general dynamical principle governing rotational motion of a rigid body, a principle that is derived from Newton's laws.

An additional observation that is often useful in solving problems is the following. *When a force is represented in terms of its components, the torque of that force with respect to a specified axis can be obtained by computing separately the torques of the components, each with its appropriate moment arm, and adding the results.* This is often easier than determining the moment arm of the original force. This technique is illustrated in some of the following examples.

The problem-solving techniques used for equilibrium of a point are equally useful in problems of equilibrium of a rigid body. These are discussed in Section 2–6, and we recommend that the reader review them. Briefly, the steps are as follows:

1. Sketch the physical situation, including dimensions.

2. Choose some body as the body in equilibrium, and draw a free-body diagram showing the forces acting on this body, and no others.

3. Draw coordinate axes and specify a positive sense of rotation, and represent forces in terms of their components with respect to the chosen axes.

4. Write equations expressing the equilibrium conditions, and solve to find unknown quantities.

EXAMPLE 1   A hanging flower basket B having weight w_2 is hung out over the edge of a balcony railing on a horizontal beam that rests on the balcony railing and is counterbalanced at its other end by a body of weight w_1, as shown in Fig. 8–3a. Find the weight w_1 needed, and the total upward force P exerted on the beam at point O. (This illustrates the principle of *cantilever* construction, by which decks, stairways, and the like can be built over empty space, supported entirely from one side.)

Solution   Figure 8–3b is a free-body diagram for the beam. The forces T_1 and T_2 are equal respectively to w_1 and w_2. The conditions of equilibrium, taking torques about an axis through O, perpendicular to the diagram, give

$$\sum F_y = P - T_1 - T_2 = 0, \quad \text{(First condition)}$$
$$\sum \Gamma_0 = T_1 l_1 - T_2 l_2 = 0. \quad \text{(Second condition)}$$

Let $l_1 = 1.2$ m, $l_2 = 1.6$ m, $w_2 = 15$ N. Then, from the equations above,

$$P = 35 \text{ N}, \quad T_1 = w_1 = 20 \text{ N}.$$

To illustrate that the total torque about *any* axis is zero, let us compute torques about an axis through point A:

$$\sum \Gamma_A = P l_1 - T_2 (l_1 + l_2)$$
$$= (35 \text{ N})(1.2 \text{ m}) - (15 \text{ N})(2.8 \text{ m}) = 0.$$

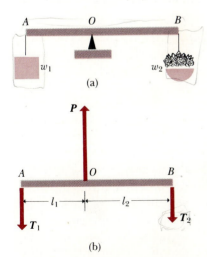

8–3   A flower basket cantilevered out over a balcony railing.

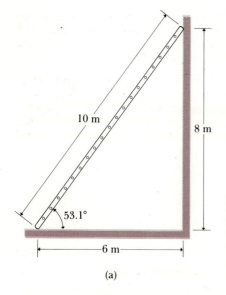

(a)

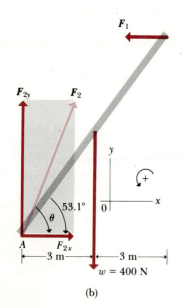

(b)

8–4   (a) A ladder in equilibrium leaning against a smooth frictionless wall. (b) Free-body diagram for the ladder. The force at the base is represented in terms of its *x*- and *y*-components.

The point about which torques are computed need not lie on the rod. To verify this, we challenge the reader to calculate the total torque about a point 0.4 m to the left of *A* and 0.4 m above it.   ◀

EXAMPLE 2   In Fig. 8–4a, a ladder 10 m long, of weight 400 N considered as concentrated at its center, leans in equilibrium against a vertical frictionless wall and makes an angle of 53.1° with the horizontal (forming a 3–4–5 right triangle). We wish to find the magnitudes and directions of the forces F_1 and F_2.

Solution   Figure 8–4b shows a free-body diagram. If the wall is frictionless, F_1 is horizontal. The direction of F_2 is unknown; except in special cases, its direction does *not* lie along the ladder. Instead of considering its magnitude and direction as unknowns, it is simpler to resolve the force F_2 into (unknown) *x*- and *y*-components and solve for these. The magnitude and direction of F_2 may then be computed. The first condition of equilibrium therefore provides the equations

$$\left.\begin{array}{l} \sum F_x = F_{2x} - F_1 = 0, \\ \sum F_y = F_{2y} - 400\ \text{N} = 0. \end{array}\right\} \quad \text{(First condition)}$$

In writing the second condition, torques may be computed about an axis through any point. The resulting equation is simplest if one selects a point through which the lines of action of two or more forces pass, since these forces then do not appear in the equation. Let us therefore take torques about an axis through point *A*.

$$\sum \Gamma_A = F_1(8\ \text{m}) - (400\ \text{N})(3\ \text{m}) = 0. \quad \text{(Second condition)}$$

From the second equation, $F_{2y} = 400$ N, and from the third,

$$F_1 = \frac{1200\ \text{N} \cdot \text{m}}{8\ \text{m}} = 150\ \text{N}.$$

Then from the first equation,

$$F_{2x} = 150\ \text{N}.$$

Hence

$$F_2 = \sqrt{(400\ \text{N})^2 + (150\ \text{N})^2} = 427.2\ \text{N},$$

$$\theta = \arctan\frac{400\ \text{N}}{150\ \text{N}} = 69.4°.   \quad ◀$$

EXAMPLE 3   Figure 8–5a shows a human arm lifting a dumbbell, and Fig. 8–5b shows a free-body diagram for the forearm, showing the forces involved. The forearm is in equilibrium under the action of the weight w, the tension T in the tendon connected to the biceps muscle, and the forces exerted on it by the upper arm, through the elbow joint. For clarity, the tendon force has been displaced away from the elbow farther than its actual position. We want to find the tendon tension and the components of force at the elbow, three unknown scalar quantities in all.

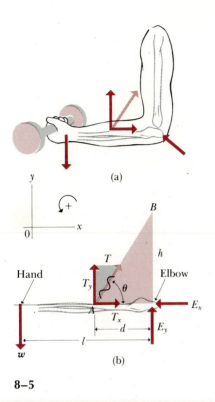

(a)

(b)

8–5

Solution   First we represent the tendon force in terms of its components T_x and T_y, using the given angle θ and the unknown magnitude T:

$$T_x = T \cos \theta, \qquad T_y = T \sin \theta.$$

We also represent the force at the elbow in terms of its components E_x and E_y. Next we note that if torques about the elbow joint are taken, the resulting torque equation does not contain the unknowns E_x and E_y, since their lines of action pass through this point, as also does T_x. The torque equation is then simply

$$lw - dT_y = 0.$$

From this we immediately find

$$T_y = \frac{lw}{d} \quad \text{and} \quad T = \frac{lw}{d \sin \theta}.$$

To find E_x and E_y we could now use the first conditions for equilibrium, $\Sigma F_x = 0$ and $\Sigma F_y = 0$. Instead, for added practice in using torques, we take torques about the point A where the tendon is attached:

$$(l - d)w + dE_y = 0 \quad \text{and} \quad E_y = -\frac{(l - d)w}{d}.$$

The negative sign of this result shows that our initial guess for the direction of E_y was wrong; it is actually vertically *downward*.

Finally, we take torques about point B in the figure:

$$lw - hE_x = 0 \quad \text{and} \quad E_x = \frac{lw}{h}.$$

We note that each stage of this calculation is simplified by choosing the point for calculating torques so as to eliminate one or more of the unknown quantities. In the last step, the force T has no torque about point B; thus when the torques of T_x and T_y are computed separately, they must add to zero. The reader is invited to verify this statement in detail.

As a specific example, suppose $w = 50$ N, $d = 0.1$ m, $l = 0.5$ m, and $\theta = 80°$. Since $\tan \theta = h/d$, we find

$$h = d \tan \theta = (0.1 \text{ m})(5.67) = 0.567 \text{ m}.$$

From the previous general results, we find

$$T = \frac{(0.5 \text{ m})(50 \text{ N})}{(0.1 \text{ m})(0.985)} = 254 \text{ N}.$$

$$E_y = -\frac{(0.5 \text{ m} - 0.1 \text{ m})(50 \text{ N})}{0.1 \text{ m}} = -200 \text{ N},$$

$$E_x = \frac{(0.5 \text{ m})(50 \text{ N})}{0.567 \text{ m}} = 44.1 \text{ N}.$$

The magnitude of the force at the elbow is

$$E = \sqrt{E_x^2 + E_y^2} = 205 \text{ N}.$$

As mentioned above, we have not explicitly used the first condition for equilibrium, that the vector sum of the forces be zero. As a check, the reader should verify that this condition *is* satisfied by the results. ◄

8–3

CENTER OF GRAVITY

In equilibrium problems where one of the forces acting on a body is its own weight, it is necessary to know the torque of this force about a specified point. It turns out that the torque can always be calculated correctly by assuming the entire force of gravity to be applied at a single point called the *center of gravity*. Furthermore, this point is identical with the *center of mass* introduced in Section 7–7.

Every particle of matter in a body is attracted by the earth, and the single force that we call the *weight* of the body is the resultant of all these forces of attraction. The direction of the force on each particle is toward the center of the earth, but the distance to the earth's center is so great that for all practical purposes the forces can be considered parallel to one another. Hence the weight of a body is the resultant of a large number of parallel forces.

Figure 8–6a shows a flat body of arbitrary shape in the *xy*-plane. Let the body be subdivided into a large number of small particles of weights w_1, w_2, etc., and let the coordinates of these particles be (x_1, y_1), (x_2, y_2), etc. The total weight W of the object is

$$W = w_1 + w_2 + \cdots = \sum w.$$

Each particle's weight also contributes to the total torque acting on the body. Computing torques about point O, we see that the torque associated with particle 1 is $w_1 x_1$, that for particle 2 is $w_2 x_2$, and so on, and that the total torque is

$$w_1 x_1 + w_2 x_2 + \cdots = \sum wx.$$

For convenience in the present discussion we are taking *clockwise* torques to be positive, the opposite choice from that used in preceding sections.

Along what line must the total weight W act in order for the total torque to equal the above expression? Suppose it acts along a line a distance X to the right of the origin. Then it must be true that $w_1 x_1 + w_2 x_2 + \cdots = WX$, or

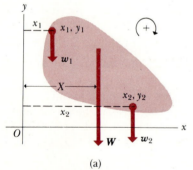

$$X = \frac{w_1 x_1 + w_2 x_2 + \cdots}{W} = \frac{\sum wx}{W} = \frac{\sum wx}{\sum w}. \qquad (8-5)$$

That is, the torque due to the weight of the object can always be obtained by assuming the total weight to be acting along a line determined by Eq. (8–5). Stated another way, in computing the torque due to the weight of an object, we must consider the weight as acting at a point whose *x*-coordinate X is given by Eq. (8–5). (See Fig. 8–6a.)

Now let the object and the reference axes be rotated 90° clockwise or, which amounts to the same thing, let us consider the gravitational forces to be rotated 90° counterclockwise, as in Fig. 8–6b. The total weight W is unaltered, but now it must act along a line a distance Y above the origin,

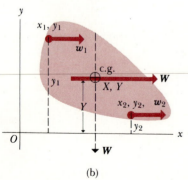

8–6 The body's weight **W** is the resultant of a large number of parallel forces. The line of action of **W** always passes through the center of gravity.

such that $WY = \Sigma wy$, or

$$Y = \frac{w_1 y_1 + w_2 y_2 + \cdots}{w_1 + w_2 + \cdots} = \frac{\Sigma \, wy}{\Sigma \, w} = \frac{\Sigma \, wy}{W}. \tag{8–6}$$

The point of intersection of the lines of action of W in the two parts of Fig. 8–6 has the coordinates X and Y and is called the *center of gravity* of an object. By considering some arbitrary orientation of the object, one can show that the line of action of W *always* passes through the center of gravity, and thus that the torque of W can always be obtained correctly by taking W to act at the center of gravity.

If we divide numerator and denominator in Eqs. (8–5) and (8–6) by g and use the fact that $w_1 = m_1 g$, and so on, the right sides of these equations become identical to those in the definitions of the center of mass in Section 7–7, Eqs. (7–23). Thus *the center of gravity of any body is identical to its center of mass.* We also note that the definition of center of gravity makes sense only in a uniform gravitational field, where g has the same value everywhere; otherwise the w's would depend on position. The center of *mass*, conversely, is defined independently of any gravitational effect.

If the centers of gravity of each of a number of bodies have been determined, the coordinates of the center of gravity of the combination can be computed from Eqs. (8–5) and (8–6), letting w_1, w_2, etc., be the weights of the bodies and x_1 and y_1, x_2 and y_2, etc., be the coordinates of the center of gravity of each.

Symmetry considerations are often useful in finding the position of the center of gravity. Thus the center of gravity of a homogeneous sphere, cube, circular disk, or rectangular plate is at its geometrical center. That of a cylinder or right circular cone is on the axis of symmetry, and so on.

EXAMPLE   Locate the center of gravity of the machine part in Fig. 8–7, consisting of a disk 4 cm in diameter and 2 cm long, and a rod 2 cm in diameter and 12 cm long, constructed of a homogeneous material.

Solution   By symmetry, the center of gravity lies on the axis and the center of gravity of each part is midway between its ends. The volume of the disk is 8π cm^3 and that of the rod is 12π cm^3. Since the weights of the two parts are proportional to their volumes,

$$\frac{w(\text{disk})}{w(\text{rod})} = \frac{w_1}{w_2} = \frac{8\pi}{12\pi} = \frac{2}{3}.$$

Take the origin O at the left face of the disk, on the axis. Then

$$x_1 = 1 \text{ cm}, \qquad x_2 = 8 \text{ cm},$$

and

$$X = \frac{w_1(1 \text{ cm}) + \tfrac{3}{2}w_1(8 \text{ cm})}{w_1 + \tfrac{3}{2}w_1} = 5.2 \text{ cm}.$$

The center of gravity is on the axis, 5.2 cm to the right of O.   ◄

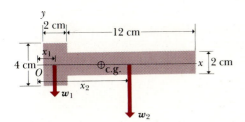

8–7

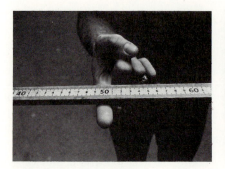

8–8 Locating the center of gravity of a body by balancing. The meter stick balances at its midpoint, but the hammer has a heavy head at one end. (Nancy Rodgers, Exploratorium.)

The center of gravity (center of mass) has several other important properties. First, when a body is suspended in a gravitational field from a single point, the center of gravity always hangs directly below the point of suspension; otherwise the body could not be in rotational equilibrium. This fact can be used for an experimental determination of the location of the center of gravity of an irregular body. Figure 8–8 shows examples of locating the center of gravity of a body by finding the balance point.

Second, a force applied to a body at its center of gravity does not tend to cause the body to rotate, while a force applied at any other point tends to cause both rotational and translational motion. Third, as discussed in Section 7–7, the total momentum of a body is always given by the product of its total mass and its center-of-mass velocity, and the acceleration of the center of mass is always the same as that of a point mass equal to the total mass of the body, acted on by the resultant external force. The second and third points will be considered in greater detail in Chapter 9, where we study the dynamics of rigid-body motion.

8–4

COUPLES

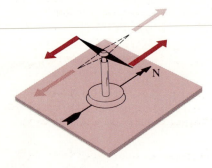

8–9 Forces on the poles of a compass needle.

It sometimes happens that the forces on a body include two forces of equal magnitude and opposite direction, having lines of action that are parallel but do not coincide. Such a pair of forces is called a *couple*. A common example is a compass needle in the earth's magnetic field, as shown in Fig. 8–9. The north and south poles of the needle are acted on by equal forces, one toward the north and the other toward the south. Except when the needle points in the N–S direction, the two forces do not have the same line of action.

Figure 8–10 shows a couple consisting of two forces, each of magnitude F, separated by a perpendicular distance l. The resultant R of the forces is

$$R = F - F = 0.$$

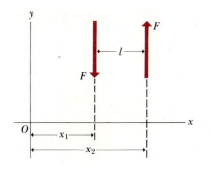

8–10 Two equal and opposite forces having different lines of action are called a *couple*. The torque of the couple is the same about all points and is equal to *Fl*.

The fact that the resultant force is zero means that a couple has no effect in producing translation (as a whole) of the body on which it acts. The only effect of a couple is to produce rotation.

The resultant torque of the couple in Fig. 8–10 about an arbitrary point O is

$$\sum \Gamma_0 = x_2 F - x_1 F$$
$$= (x_1 + l)F - x_1 F$$
$$= lF. \qquad (8-7)$$

Since the distances x_1 and x_2 do not appear in the result, we conclude that *the torque of a couple is the same about all points in the plane of the forces and is equal to the product of the magnitude of either force and the perpendicular distance between their lines of action.*

A plumber tightening a pipe fitting uses the principle of the couple. He applies a force to the end of the wrench handle and holds the pipe close to the fitting with the free hand, exerting an opposite force. Thus there is no net force sideways on the pipe, which would put a strain on the opposite end, but only a torque tending to screw the two parts together.

QUESTIONS

8–1 Mechanics sometimes extend the length of a wrench handle by slipping a section of pipe over the handle. Why is this a dangerous procedure?

8–2 Manuals for car-engine repair always specify the *torque* (moment) to be applied when tightening the cylinder-head bolts. Why is torque specified, rather than the *force* to be applied to the wrench handle?

8–3 Car tires are sometimes "balanced" on a machine that pivots the tire and wheel about the center, by laying weights around the wheel rim until it does not tip from the horizontal plane. Discuss this procedure in terms of the center of gravity.

8–4 Is it possible for a solid body to have no matter at its center of gravity? Consider, for example, a disk with a hole in the center, an empty bottle, and a hollow cylinder.

8–5 A man climbs a tall, old stepladder that has a tendency to sway. He feels much more unstable when standing near the top than when near the bottom. Why?

8–6 What determines whether a body in equilibrium is stable or unstable? As an example, discuss a cone resting on its base on a horizontal floor; balanced on its point; and lying on its side.

8–7 When a heavy load is placed in the back end of a truck, behind the rear wheels, the effect is to *raise* the front end of the truck. Why?

8–8 When a tall heavy object such as a refrigerator is

pushed across a rough floor, what determines whether it slides or tips?

8–9 Why is a tapered water glass with a narrow base easier to tip over than one with straight sides? Does it matter whether the glass is full or empty?

8–10 People sometimes prop open a door by wedging some object in the space between the hinged side of the door and the frame. Explain why this often results in the hinge screws being ripped out.

8–11 Discuss the action of a claw hammer in pulling out nails, in terms of moments.

8–12 In trying to lift an object that is much too heavy to be lifted directly, one often speaks of the importance of "leverage." Discuss this term in relation to moments, and explain how it is related to the concept of leverage in trading stock options and other securities.

8–13 Why is it easier to hold a 10-kg body in your hand at your side than to hold it with your arm extended horizontally?

8–14 In pioneer days when a Conestoga wagon was stuck in the mud, a man would grasp a wheel spoke and try to turn the wheel, rather than simply pushing on the wagon. Why?

8–15 Does the center of gravity of a solid body always lie within the body? If not, give a counterexample.

PROBLEMS

8–1 A heavy electric motor is to be carried by two men by placing it on a light board 2.0 m long. To lift the board with the motor on it, one man must lift at one end with a force of 600 N and the other at the opposite end with a force of 400 N. What is the weight of the motor, and where is it located?

8–2 In Problem 8–1, suppose the board is not light but weighs 200 N. What is the weight of the motor, and where is it located?

8–3 A child weighing 400 N sits on one end of a seesaw that is 3.0 m long and is pivoted 1.4 m from the child. If another child sitting at the opposite end just balances, what is his weight? Neglect the weight of the seesaw. What is the total force on the pivot?

8–4 A certain automobile has a wheelbase (distance between front and rear axles) of 3.0 m. If 60 percent of the weight rests on the front wheels, how far behind the front wheels is the center of gravity?

8–5 Two men are carrying a ladder 20 ft long, weighing 200 lb. If one man can lift a maximum of 80 lb and lifts at one end, at what point should the other man lift?

8–6 You are given (1) a meter stick through which a number of holes have been bored so that its center of gravity is not at its midpoint, (2) a knife edge, on which the meter stick can be pivoted, (3) a body whose weight is known to be w, and (4) a spool of thread. Using this equipment only, explain, with the aid of a diagram, how you would determine the weight of the meter stick.

8–7 A diving board 3 m long is supported at a point 1 m from the end, and a diver weighing 800 N stands at the free end, as shown in Fig. 8–11. Find the force at the support point and the force at the end that is held down, if the weight of the board is negligible.

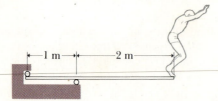

Figure 8–11

8–8 In Problem 8–7, suppose the board is of uniform cross section and weighs 400 N. Find the forces at the support.

8–9 A claw hammer is used to pull out a nail, as in Fig. 8–12. The handle is 0.3 m long, and the head pivots about a point 3 cm from the nail. If a force of 200 N is applied to the end of the handle, what force is applied to the nail? What force does the hammer head exert on the board into which the nail is driven?

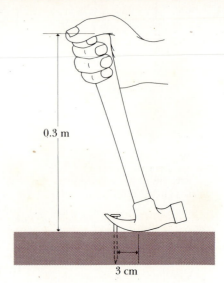

Figure 8–12

8–10 A trapdoor in the ceiling is hinged at one side; its total weight is 200 N. Find the net upward force needed to begin to open it, and the total force exerted on the door by the hinges, if

a) the upward force is applied at the center;

b) the upward force is applied at the center of the edge opposite the hinges.

8–11 A uniform plank 15 m long, weighing 400 N, rests symmetrically on two supports 8 m apart, as shown in Fig. 8–13. A boy weighing 640 N starts at point A and walks toward the right.

a) How far beyond point B can the boy walk before the plank tips?

b) How far from the right end of the plank should support B be placed in order that the boy walk just to the end of the plank without causing it to tip?

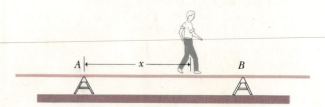

Figure 8–13

8–12 The beam in Fig. 8–14 weighs 200 N and its center of gravity is at its center. Find

a) the tension in the cable and

b) the horizontal and vertical components of the force exerted on the beam at the wall.

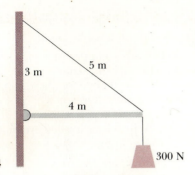

Figure 8–14

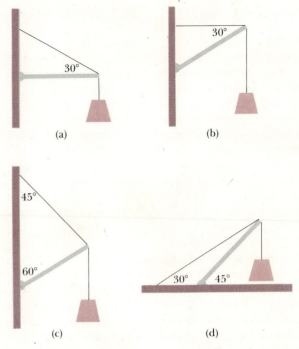

(a)

(b)

(c)

(d)

Figure 8–16

8–13 Find the tension in the cable *BD* in Fig. 8–15, and the horizontal and vertical components of the force exerted on the strut *AB* at pin *A*, using

a) the first and second conditions of equilibrium ($\Sigma F_x = 0, \Sigma F_y = 0, \Sigma \Gamma = 0$), taking torques about an axis through point *A* perpendicular to the plane of the diagram;

b) the second condition of equilibrium only, taking torques first about an axis through *A*, then about an axis through *B*, and finally about an axis through *D*. The weight of the strut can be neglected.

c) Represent the computed forces by vectors in a diagram drawn to scale, and show that the lines of action of the forces exerted on the strut at points *A*, *B*, and *C* intersect at a common point.

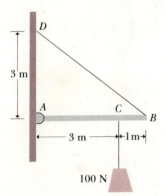

100 N

Figure 8–15

8–14 Find the tension *T* in each cable, and the magnitude and direction of the force exerted on the strut by the pivot, in each of the arrangements in Fig. 8–16. Let the weight of the suspended object in each case be 1000 N. Neglect the weight of the strut.

8–15 A horizontal boom 4 m long is hinged to a vertical wall at one end, and a 500-N body hangs from its outer end. The boom is supported by a guy wire from its outer end to a point on the wall directly above the boom.

a) If the tension in this wire is not to exceed 1000 N, what is the minimum height above the boom at which it may be fastened to the wall?

b) By how many newtons would the tension be increased if the wire were fastened 0.5 m below this point, the boom remaining horizontal? Neglect the weight of the boom.

8–16 Find the largest weight *w* that can be supported by the structure in Fig. 8–17 if the maximum tension the upper rope can withstand is 1000 N and the maximum compression the strut can withstand is 2000 N. The vertical rope is strong enough to carry any load required. Neglect the weight of the strut.

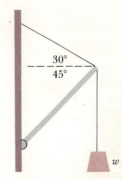

w

Figure 8–17

8–17 A ball of mass 1 kg is attached by a light rod 0.4 m in length to a second ball of mass 3 kg. Where is the center of gravity of this system?

8–18 In Problem 8–17, suppose the rod is uniform and has mass 2 kg. Where is the center of gravity of the system?

8–19 Three small objects of equal mass are located in the x,y-plane, at points having coordinates (0.1 m, 0), (0, 0.1 m), (0.1 m, 0.1 m). Determine the coordinates of the center of gravity. Can you use the symmetry of the situation to simplify your calculation?

8–20 A ladder 10 m long, weighing 600 N, leans against a vertical wall, making an angle of 30° with the vertical. A man weighing 800 N climbs up the ladder to a point 3 m above the ground. Where is the center of gravity of the system consisting of the man and the ladder?

8–21 In a molecule of carbon monoxide (CO), the nuclei of the two atoms are about 1.13×10^{-10} m apart. Where is the center of gravity of the molecule?

8–22 A solid cube 5 cm on a side is glued to a solid cube of the same material, 10 cm on a side, so that one corner and two edges of the two cubes coincide. Define a suitable coordinate system and determine the position of the center of gravity of the system.

8–23 A station wagon of mass 2000 kg has a wheelbase (distance between front and rear axles) of 3.0 m. Ordinarily 1100 kg rests on the front wheels and 900 kg on the rear wheels. A box of mass 200 kg is now placed on the tailgate, 1.0 m behind the rear axle. How much total weight now rests on the front wheels? On the rear wheels?

8–24 End A of the bar AB in Fig. 8–18 rests on a frictionless horizontal surface, while end B is hinged. A horizontal force P of 60 N is exerted on end A. Neglect the weight of the bar. What are the horizontal and vertical components of the force exerted by the bar on the hinge at B?

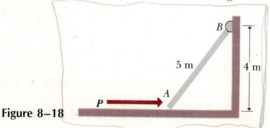

Figure 8–18

8–25 A single additional force is to be applied to the bar in Fig. 8–19 to maintain it in equilibrium in the position shown. The weight of the bar can be neglected.

a) What are the x- and y-components of the required force?

b) What is the tangent of the angle that the force must make with the bar?

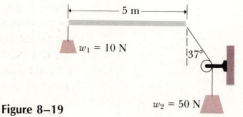

Figure 8–19

c) What is the magnitude of the required force?

d) Where should the force be applied?

8–26 A circular disk 0.5 m in diameter, pivoted about a horizontal axis through its center, has a cord wrapped around its rim. The cord passes over a frictionless pulley P and is attached to a body of weight 240 N. A uniform rod 2 m long is fastened to the disk, with one end at the center of the disk. The apparatus is in equilibrium, with the rod horizontal, as shown in Fig. 8–20.

a) What is the weight of the rod?

b) What is the new equilibrium direction of the rod when a second body weighing 20 N is suspended from the outer end of the rod, as shown by the broken line?

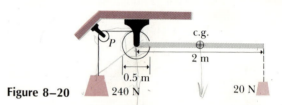

Figure 8–20

8–27 A roller whose diameter is 1.0 m weighs 360 N. What horizontal force is necessary to pull the roller over a brick 0.1 m high when the force is applied

a) at the center,

b) at the top of the roller?

8–28 The boom in Fig. 8–21 is uniform and weighs 2500 N.

a) Find the tension in the guy wire, and the horizontal and vertical components of the force exerted on the boom at its lower end.

b) Does the line of action of this force lie along the boom?

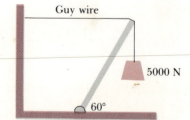

Figure 8–21

8–29 Two ladders, 4 m and 3 m long, respectively, are hinged at point A and tied together by a horizontal rope 0.6 m above the floor, as in Fig. 8–22. The ladders weigh 400 N and 300 N, respectively, and the center of gravity of each is at its center. If the floor is frictionless, find

a) the upward force at the bottom of each ladder,

b) the tension in the rope, and

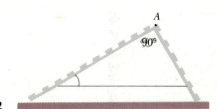

Figure 8–22

c) the force one ladder exerts on the other at point A.

d) If a load of 1000 N is now suspended from point A, find the tension in the rope.

8–30 A uniform ladder 10 m long rests against a vertical frictionless wall with its lower end 6 m from the wall. The ladder weighs 400 N. The coefficient of static friction between the foot of the ladder and the ground is 0.40. A man weighing 800 N climbs slowly up the ladder.

a) What is the maximum frictional force that the ground can exert on the ladder at its lower end?

b) What is the actual frictional force when the man has climbed 3 m along the ladder?

c) How far along the ladder can the man climb before the ladder starts to slip?

8–31 One end of a meter stick is placed against a vertical wall, as in Fig. 8–23. The other end is held by a light cord making an angle θ with the stick. The coefficient of static friction between the end of the meter stick and the wall is 0.30.

a) What is the maximum value the angle θ can have if the stick is to remain in equilibrium?   16.7

b) Let the angle θ be 10°. A body of the same weight as the meter stick is suspended from the stick as shown, at a distance x from the wall. What is the minimum value of x for which the stick will remain in equilibrium?   2A

c) When $\theta = 10°$, how large must the coefficient of static friction be so that the body can be attached at the left end of the stick without causing it to slip?   .53

Figure 8–23

8–32 One end of a post weighing 500 N rests on a rough horizontal surface with $\mu_s = 0.3$. The upper end is held by a rope fastened to the surface and making an angle of 37°

with the post, as in Fig. 8–24. A horizontal force F is exerted on the post as shown.

a) If the force F is applied at the midpoint of the post, what is the largest value it can have without causing the post to slip?

b) How large can the force be, without causing the post to slip, if its point of application is $\frac{7}{10}$ of the way from the ground to the top of the post?

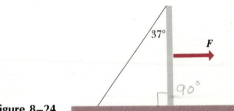

Figure 8–24

8–33 A door 1.0 m wide and 2.5 m high weighs 200 N and is supported by two hinges, one 0.5 m from the top and the other 0.5 m from the bottom. Each hinge supports half the total weight of the door. Assuming the door's center of gravity is at its center, find

a) the components of force exerted on the door by each hinge,

b) the magnitude and direction of the force exerted by each hinge.

8–34 A gate 4 m long and 2 m high weighs 400 N. Its center of gravity is at its center, and it is hinged at A and B. To relieve the strain on the top hinge, a wire CD is connected as shown in Fig. 8–25. The tension in CD is increased until the horizontal force at hinge A is zero.

a) What is the tension in the wire CD?

b) What is the magnitude of the horizontal component of force at hinge B?

c) What is the combined vertical force exerted by hinges A and B?

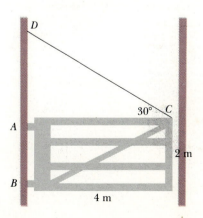

Figure 8–25

8–35 An engineer is designing a conveyor system for loading hay bales into a wagon. Each bale has a mass of 25 kg and is 0.75 m long, 0.25 m wide, and 0.5 m high. The coefficient of static friction between a bale and the conveyor belt is 0.40, and the belt moves with constant speed. The situation is shown in Fig. 8–26.

a) The angle θ of the conveyor is slowly increased. At some critical angle a bale will tip (if it doesn't slip first), and at some different critical angle it will slip (if it doesn't tip first). Find the two critical angles, and determine which happens first.

b) Would the outcome of (a) be different if the coefficient of friction were 0.60? If it were 0.50?

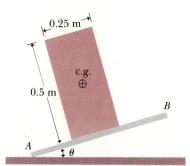

Figure 8–26

8–36 The hay bale of Problem 8–35 is dragged along a horizontal surface with constant speed by a force P, as shown in Fig. 8–27. The coefficient of kinetic friction is 0.40.

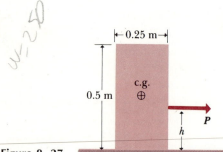

Figure 8–27

a) Find the magnitude of the force P.

b) Find the value of h for which the bale just begins to tip.

8–37 A garage door is mounted on an overhead rail, as in Fig. 8–28. The wheels at A and B have rusted so that they do not roll, but slide along the track. The coefficient of sliding friction is 0.5. The distance between the wheels is 2 m, and each is 0.5 m in from the vertical sides of the door. The door is symmetrical and weighs 800 N. It is pushed to the left at constant velocity by a horizontal force P.

a) If the distance h is 1.5 m, what is the vertical component of the force exerted on each wheel by the track?

b) Find the maximum value h can have without causing one wheel to leave the track.

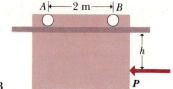

Figure 8–28

8–38 The objects in Fig. 8–29 are constructed of uniform wire bent into the shapes shown. Find the position of the center of gravity of each.

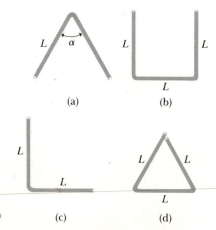

Figure 8–29

9

ROTATION

Up to this point our discussion of principles of dynamics has been concerned primarily with *particles;* the point mass serves as a *model* to represent a body whose size and shape are not relevant for the problem at hand. But there are many problems for which this model is inadequate; in the preceding chapter we discussed the equilibrium of a *rigid body.* This is also an idealization; real bodies always deform to some extent when forces are applied, while an idealized rigid body does not bend, stretch, or deform in any way. We found that equilibrium of a rigid body requires both the absence of any tendency for change in *translational* motion and the absence of any tendency for the body to change its *rotational* motion. Similarly, *motion* of a rigid body can always be represented as a combination of translational motion of some point in the body and rotational motion about an axis through that point. For simplicity we begin our discussion with rotation about a stationary axis. Just as with dynamics of a particle, we first develop kinematic language for describing rotational motion.

9–1

ANGULAR VELOCITY

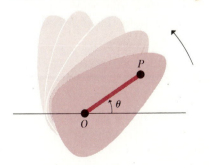

Figure 9–1 represents a rigid body of arbitrary shape rotating about a fixed axis through point O and perpendicular to the plane of the diagram. Line OP is a line fixed with respect to the body and rotating with it. The position of the entire body is evidently completely specified by the angle θ that the line OP makes with some reference line fixed in space, such as Ox. The motion of the body is therefore analogous to the rectilinear motion of a particle whose position is completely specified by a single coordinate such as x or y.

The equations of motion are greatly simplified if the angle θ is expressed in *radians.* One radian (1 rad) is the angle subtended at the center of a circle by an arc of length equal to the radius of the circle, as shown in Fig. 9–2a. Since the radius is contained 2π times

9–1 Body rotating about a fixed axis through point O.

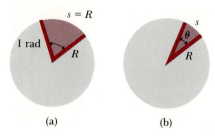

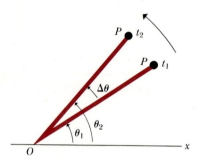

ROTATION

$s = R$

1 rad

R

(a)

s

θ

R

(b)

9–2 An angle θ in radians is defined as the ratio of the arc s to the radius R.

P t_2

P t_1

$\Delta\theta$

θ_2

θ_1

O x

9–3 Angular displacement $\Delta\theta$ of a rotating body.

($2\pi = 6.28 \ldots$) in the circumference, there are 2π or $6.28 \ldots$ rad in one complete revolution or $360°$. Hence,

$$1 \text{ rad} = \frac{360°}{2\pi} = 57.3 \text{ degrees},$$
$$360° = 2\pi \text{ rad} = 6.28 \text{ rad},$$
$$180° = \pi \text{ rad} = 3.14 \text{ rad}$$
$$90° = \pi/2 \text{ rad} = 1.57 \text{ rad}$$
$$60° = \pi/3 \text{ rad} = 1.05 \text{ rad},$$
$$45° = \pi/4 \text{ rad} = 0.79 \text{ rad},$$

and so on.

In general (Fig. 9–2b), if θ represents any arbitrary angle subtended by an arc of length s on the circumference of a circle of radius R, then θ (in radians) is equal to the length of the arc s divided by the radius R:

$$\theta = \frac{s}{R}, \qquad s = R\theta. \tag{9–1}$$

For example, if $\theta = 2\pi$ rad, then s is the circumference of the circle.

An angle in radians, being defined as the ratio of a length to a length, is a pure number. If $s = 1.5$ m and $R = 1$ m, the angle is usually described as $\theta = 1.5$ rad, but it would be equally correct to say simply $\theta = 1.5$.

In Fig. 9–3, a reference line OP in a rotating body makes an angle θ_1 with the reference line Ox, at a time t_1. At a later time t_2 the angle has increased to θ_2. The *average angular velocity* of the body, $\bar{\omega}$, in the time interval between t_1 and t_2, is defined as the ratio of the *angular displacement* $\theta_2 - \theta_1$, or $\Delta\theta$, to the elapsed time $t_2 - t_1$ or Δt:

$$\bar{\omega} = \frac{\Delta\theta}{\Delta t}.$$

The *instantaneous angular velocity* ω is defined as the limit approached by this ratio as Δt approaches zero:

$$\omega = \lim_{\Delta t \to 0} \frac{\Delta\theta}{\Delta t}. \tag{9–2}$$

Since the body is rigid, *all* lines in it rotate through the same angle in the same time, and *every point in a rotating rigid body has the same angular velocity*. If the angle θ is in radians, the unit of angular velocity is one radian per second ($1 \text{ rad} \cdot \text{s}^{-1}$). Other units, such as the revolution per minute ($\text{rev} \cdot \text{min}^{-1}$), are in common use. We note that $1 \text{ rev} \cdot \text{s}^{-1} = 2\pi \text{ rad} \cdot \text{s}^{-1}$.

9–2

ANGULAR ACCELERATION

If the angular velocity of a body changes, it is said to have an *angular acceleration*. If ω_1 and ω_2 are the instantaneous angular velocities at times t_1 and t_2, the *average angular acceleration* $\bar{\alpha}$ is defined as

$$\bar{\alpha} = \frac{\omega_2 - \omega_1}{t_2 - t_1} = \frac{\Delta\omega}{\Delta t},$$

and the *instantaneous angular acceleration* α is defined as the limit of this ratio when Δt approaches zero:

$$\alpha = \lim_{\Delta t \to 0} \frac{\Delta \omega}{\Delta t}. \tag{9–3}$$

The unit of angular acceleration is $1\ \text{rad} \cdot \text{s}^{-2}$ or $1\ \text{s}^{-2}$. Angular velocity and angular acceleration are exactly analogous to linear velocity and acceleration.

9–3

ROTATION WITH CONSTANT ANGULAR ACCELERATION

When the angular velocity of a body changes by equal amounts in equal intervals of time, the angular acceleration is constant. Under these circumstances the average and instantaneous angular accelerations are equal, whatever the duration of the time interval. One may therefore write

$$\alpha = \frac{\omega - \omega_0}{t - t_0},$$

where ω_0 is the angular velocity at some initial time t_0, ω is the angular velocity at some later time t, and α is the (constant) angular acceleration. It is usually convenient to take $t_0 = 0$, in which case this equation can be rearranged as follows:

$$\omega = \omega_0 + \alpha t. \tag{9–4}$$

Thus the angular velocity ω at any time t can be obtained if the initial value ω_0 and the angular acceleration α are known. Equation (9–4) also shows that the angular velocity at time t is the *sum* of the initial value ω_0 and the total amount αt by which the angular velocity changes during time t. This equation has precisely the same form as Eq. (3–7) for linear motion with constant acceleration, and it may be interpreted in the same way.

The angular *displacement* of a rotating body, the angle through which the body turns, corresponds to the linear displacement of a body moving along a straight line. The expression for the angular displacement can be found with the help of the average angular velocity. If the angular acceleration is constant, the angular velocity increases at a uniform rate and its *average* value during any time interval t equals half the sum of its values at the beginning and end of the interval. That is,

$$\bar{\omega} = \frac{\omega_0 + \omega}{2}.$$

If $\theta = \theta_0$ when $t = 0$, then $\theta = \theta_0 + \bar{\omega} t$ and hence

$$\theta = \theta_0 + \left(\frac{\omega_0 + \omega}{2} \right) t. \tag{9–5}$$

By combining Eqs. (9–4) and (9–5), we may obtain two more very useful equations. Substituting for ω in Eq. (9–5) the value of ω given by

Eq. (9–4), we have

$$\theta = \theta_0 + \frac{\omega_0 + (\omega_0 + \alpha t)}{2} \cdot t$$

or

$$\theta = \theta_0 + \omega_0 t + \tfrac{1}{2}\alpha t^2. \tag{9–6}$$

Also, substituting for t in Eq. (9–5) the value obtained by solving Eq. (9–4) for t, we have

$$\theta = \theta_0 + \left(\frac{\omega_0 + \omega}{2}\right)\left(\frac{\omega - \omega_0}{\alpha}\right) = \theta_0 + \frac{\omega^2 - \omega_0^2}{2\alpha},$$

or, finally,

$$\omega^2 = \omega_0^2 + 2\alpha(\theta - \theta_0). \tag{9–7}$$

Table 9–1 shows the similarity between the equations for motion with constant linear acceleration and those for motion with constant angular acceleration.

TABLE 9–1

Motion with constant linear acceleration	Motion with constant angular acceleration
$a = $ constant	$\alpha = $ constant
$v = v_0 + at$	$\omega = \omega_0 + \alpha t$
$x = x_0 + v_0 t + \tfrac{1}{2}at^2$	$\theta = \theta_0 + \omega_0 t + \tfrac{1}{2}\alpha t^2$
$x = x_0 + \dfrac{v_0 + v}{2} t$	$\theta = \theta_0 + \dfrac{\omega_0 + \omega}{2} t$
$v^2 = v_0^2 + 2a(x - x_0)$	$\omega^2 = \omega_0^2 + 2\alpha(\theta - \theta_0)$

EXAMPLE The angular velocity of a body is 4 rad·s^{-1} at time $t = 0$, and its angular acceleration is constant and equal to 2 rad·s^{-2}. A line OP in the body is horizontal at time $t = 0$.

a) What angle does this line make with the horizontal at time $t = 3$ s?

b) What is the angular velocity at this time?

Solution

a) We may define θ so that $\theta_0 = 0$. Then

$$\theta = \theta_0 + \omega_0 t + \tfrac{1}{2}\alpha t^2$$
$$= 0 + (4 \text{ rad·s}^{-1})(3 \text{ s}) + \tfrac{1}{2}(2 \text{ rad·s}^{-2})(3 \text{ s})^2$$
$$= 21 \text{ rad} = \frac{21}{2\pi} \text{ rev} = 3.34 \text{ rev.}$$

The body turns through three complete revolutions plus an additional 0.34 rev or (0.34 rev)(2π rad·rev^{-1}) = 2.14 rad = 123°. The line OP is thus turned through $123°$ and makes an angle of $57°$ with the horizontal.

b) $\omega = \omega_0 + \alpha t = 4 \text{ rad}\cdot\text{s}^{-1} + (2 \text{ rad}\cdot\text{s}^{-2})(3 \text{ s}) = 10 \text{ rad}\cdot\text{s}^{-1}.$

Alternatively, from Eq. (9–7),

$$\omega^2 = \omega_0{}^2 + 2\alpha(\theta - \theta_0)$$
$$= (4 \text{ rad}\cdot\text{s}^{-1})^2 + 2(2 \text{ rad}\cdot\text{s}^{-2})(21 \text{ rad})$$
$$= 100 \text{ rad}^2\cdot\text{s}^{-2},$$
$$\omega = 10 \text{ rad}\cdot\text{s}^{-1}. \qquad \blacktriangleleft$$

9–4

RELATION BETWEEN ANGULAR AND LINEAR VELOCITY AND ACCELERATION

In Section 5–5 we discussed the linear velocity and acceleration of a *particle* revolving in a circle. When a *rigid body* rotates about a fixed axis, every point in the body moves in a circle whose center is on the axis and that lies in a plane perpendicular to the axis. There are some useful and simple relations between the angular velocity and acceleration of the rotating body and the linear velocity and acceleration of points within it.

Let r be the distance from the axis to some point P in the body, so that the point moves in a circle of radius r, as in Fig. 9–4. When the radius makes an angle θ with the reference axis, the distance s to the point P, measured along the circular path, is

$$s = r\theta. \tag{9-8}$$

When the angle is very small, say $\Delta\theta$, the length of the arc Δs is also small, and the time interval Δt for the body to rotate through the angle $\Delta\theta$ is also small. We have

$$\Delta s = r\Delta\theta, \qquad \frac{\Delta s}{\Delta t} = r\frac{\Delta\theta}{\Delta t}.$$

That is, the average speed $\Delta s/\Delta t$ is r times the average angular velocity $\Delta\theta/\Delta t$, and in the limit, as $\Delta t \to 0$,

$$v = r\omega. \tag{9-9}$$

If the angular velocity about the given axis changes in magnitude (the axis remaining stationary) by $\Delta\omega$, the linear velocity in a direction tangent to a circle of radius r changes by Δv, where

$$\Delta v = r\Delta\omega.$$

If these changes take place in a short time interval Δt, then

$$\frac{\Delta v}{\Delta t} = r\frac{\Delta\omega}{\Delta t},$$

and, in the limit, as $\Delta t \to 0$,

$$a_\parallel = r\alpha, \tag{9-10}$$

where $a_\parallel$ is the *tangential* component of the linear acceleration of a point a distance r from the axis.

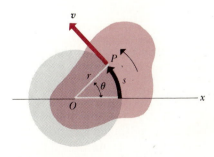

9–4 The distance s moved through by point P equals $r\theta$.

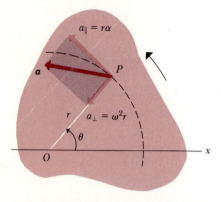

9–5 Nonuniform rotation about a fixed axis through point *O*. The tangential component of acceleration of point *P* equals $r\alpha$; the radial component equals $\omega^2 r$.

The *radial* component of acceleration v^2/r of the point *P*, Eq. (5–20), can also be expressed in terms of the angular velocity,

$$a_\perp = \frac{v^2}{r} = \omega^2 r. \tag{9–11}$$

This is true at each instant even when ω and v are not constant.

The tangential and radial components of acceleration of any arbitrary point *P* in a rotating body are shown in Fig. 9–5.

EXAMPLE A discus thrower turns with angular acceleration $\alpha = 50 \text{ rad} \cdot \text{s}^{-2}$. What are the radial and tangential components of acceleration of the discus at the instant when the angular velocity is $10 \text{ rad} \cdot \text{s}^{-1}$? The thrower's arm is 0.8 m long.

Solution

$$a_\perp = \omega^2 r = (10 \text{ s}^{-1})^2 (0.8 \text{ m}) = 80 \text{ m} \cdot \text{s}^{-2},$$

$$a_\parallel = r\alpha = (0.8 \text{ m})(50 \text{ s}^{-2}) = 40 \text{ m} \cdot \text{s}^{-2}.$$

$$a = \sqrt{a_\perp^2 + a_\parallel^2} = 89.4 \text{ m} \cdot \text{s}^{-2},$$

or about nine times the acceleration due to gravity.   ◀

9–5

KINETIC ENERGY OF ROTATION. MOMENT OF INERTIA

We have shown that the speed of a particle in a rigid body rotating about a fixed axis is

$$v = r\omega,$$

where *r* is the distance of the particle from the axis and ω is the angular velocity of the body. The kinetic energy of a particle of mass *m* is therefore

$$\tfrac{1}{2}mv^2 = \tfrac{1}{2}mr^2\omega^2.$$

The total kinetic energy of the body is the *sum* of the kinetic energies of all particles in the body,

$$K = \tfrac{1}{2}m_1 r_1^2\omega^2 + \tfrac{1}{2}m_2 r_2^2\omega^2 + \cdots$$

$$= \sum \tfrac{1}{2}mr^2\omega^2,$$

where the symbol Σ stands for *sum*, indicating that the quantity $\tfrac{1}{2}mr^2\omega^2$ is to be computed for each particle making up the body, and the results added. Since ω is the same for all particles in a rigid body,

$$K = \tfrac{1}{2}\left[\sum mr^2\right]\omega^2.$$

To obtain the sum $\Sigma\, mr^2$, the body is subdivided (in imagination) into a large number of particles, the mass of each particle is multiplied by the square of its distance from the axis, and these products are summed over

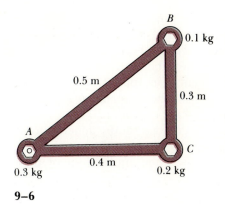

9–6

all particles. The result is called the *moment of inertia* of the body, about the axis of rotation, and is represented by I:

$$I = \sum mr^2. \qquad (9\text{–}12)$$

In SI (mks) units the unit of moment of inertia is one kilogram-meter^2 ($1 \text{ kg} \cdot \text{m}^2$).

The rotational kinetic energy of a rigid body can now be written as

$$K = \tfrac{1}{2}I\omega^2. \qquad (9\text{–}13)$$

The form of this expression is analogous to that of translational kinetic energy:

$$K = \tfrac{1}{2}mv^2.$$

That is, for rotation about a fixed axis, this moment of inertia I is analogous to mass m (or inertia), and angular velocity ω is analogous to linear velocity v.

EXAMPLE 1   An engineer is designing a one-piece machine part consisting of three connectors linked by molded struts, as in Fig. 9–6. The connectors can be considered particles connected by massless rods. What is the moment of inertia of this machine part

a) about an axis through point A, perpendicular to the plane of the diagram, and

b) about an axis coinciding with rod BC?

Solution

a) The particle at point A lies on the axis. Its distance *from* the axis is zero and it contributes nothing to the moment of inertia. Therefore,

$$I = \sum mr^2 = (0.1 \text{ kg})(0.5 \text{ m})^2 + (0.2 \text{ kg})(0.4 \text{ m})^2$$
$$= 0.057 \text{ kg} \cdot \text{m}^2.$$

b) The particles at B and C both lie on the axis. The moment of inertia is

$$I = \sum mr^2 = (0.3 \text{ kg})(0.4 \text{ m})^2 = 0.048 \text{ kg} \cdot \text{m}^2.$$

This illustrates the important fact that the moment of inertia of a body, unlike its mass, is *not* a unique property of the body but depends on the position of the axis about which it is computed.

c) If the body rotates about an axis through A and perpendicular to the plane of the diagram, with an angular velocity $\omega = 4 \text{ rad} \cdot \text{s}^{-1}$, what is the rotational kinetic energy?

$$K = \tfrac{1}{2}I\omega^2 = \tfrac{1}{2}(0.057 \text{ kg} \cdot \text{m}^2)(4 \text{ rad} \cdot \text{s}^{-1})^2 = 0.456 \text{ J}. \quad \blacktriangleleft$$

For a body that is *not* composed of discrete point masses but is a continuous distribution of matter, the summation expressed in the definition of moment of inertia, $I = \sum mr^2$, must be evaluated by the methods of calculus. The moments of inertia of a few simple but important bodies are listed in Fig. 9–7 for convenience.

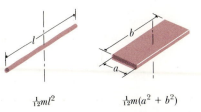

$\tfrac{1}{12}ml^2$

(a) Slender rod, axis through center

$\tfrac{1}{12}m(a^2 + b^2)$

(b) Rectangular plate, axis through center

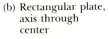

$\tfrac{1}{2}m(R_1^2 + R_2^2)$

(c) Hollow cylinder

$\tfrac{1}{2}mR^2$

(d) Solid cylinder

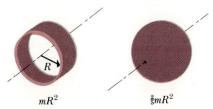

mR^2

(e) Thin-walled hollow cylinder

$\tfrac{2}{5}mR^2$

(f) Solid sphere

9–7   Moments of inertia. For each body, the axis is shown as a broken line.

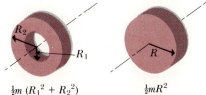

It should be noted that it is *not* correct in general to compute the moment of inertia of a body by assuming all the mass to be concentrated at the center of mass (center of gravity). For example, when a uniform thin rod of length L and mass M is pivoted about an axis through one end, perpendicular to the rod, the moment of inertia is $I = ML^2/3$ while, if we take the mass as concentrated at the center, distance $L/2$ from the axis, we obtain the *incorrect* result $I = M(L/2)^2 = ML^2/4$.

The concept of moment of inertia, together with the general work–energy principle, is very useful in problems involving the rotation of a rigid body. The following examples show some applications of these ideas.

EXAMPLE 2   A light flexible rope is wrapped several times around a solid cylinder of mass 50 kg and diameter 0.12 m, which rotates without friction about a stationary horizontal axis. The free end of the rope is pulled with a constant force of magnitude 9 N for a distance of 2 m. If the cylinder is initially at rest, find its final angular velocity and the final speed of the rope.

Solution   Because no energy is lost in friction, the final kinetic energy $\frac{1}{2}I\omega^2$ of the cylinder is equal to the work Fd done by the force, which is $(9\ \text{N})(2\ \text{m}) = 18\ \text{J}$. From Fig. 9–7, the moment of inertia is

$$I = \tfrac{1}{2}MR^2 = \tfrac{1}{2}(50\ \text{kg})(0.06\ \text{m})^2 = 0.09\ \text{kg} \cdot \text{m}^2.$$

The work–energy relation then gives

$$\tfrac{1}{2}(0.09\ \text{kg} \cdot \text{m}^2)\omega^2 = 18\ \text{J},$$
$$\omega = 20\ \text{rad} \cdot \text{s}^{-1}.$$

The final speed of the rope is equal to the tangential speed of the cylinder, which is given by Eq. (9–9):

$$v = r\omega = (0.06\ \text{m})(20\ \text{rad} \cdot \text{s}^{-1})$$
$$= 1.2\ \text{m} \cdot \text{s}^{-1}. \qquad \blacktriangleleft$$

EXAMPLE 3   A light flexible rope is wrapped several times around a solid cylinder of mass M and radius R, which rotates with no friction about a stationary horizontal axis, as in Fig. 9–8. The free end of the rope is tied to a mass m, which is released from rest a distance h above the floor. Find its speed and the angular velocity of the cylinder just as mass m strikes the floor.

Solution   Initially the system has no kinetic energy, but it has potential energy mgh. Just as mass m strikes the floor, the potential energy is zero, but both this mass and the cylinder have kinetic energy. The total kinetic energy is

$$K = \tfrac{1}{2}mv^2 + \tfrac{1}{2}I\omega^2. \qquad (9\text{–}14)$$

Now, according to Fig. 9–7, the moment of inertia of the cylinder is $I = \tfrac{1}{2}MR^2$. Furthermore, v and ω are related by $v = R\omega$, since the speed of mass m must be equal to the tangential velocity at the outer surface of

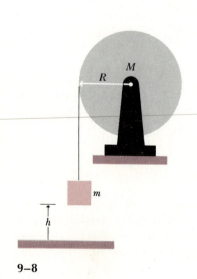

9–8

the cylinder. Using these relations and setting K equal to mgh, we obtain

$$mgh = \tfrac{1}{2}mv^2 + \tfrac{1}{2}(\tfrac{1}{2}MR^2)\left(\frac{v}{R}\right)^2 = \tfrac{1}{2}(m + \tfrac{1}{2}M)v^2,$$

$$v = \sqrt{\frac{2gh}{1 + M/2m}}.$$

When M is much larger than m, v is very small, as might be expected. When M is much smaller than m, v is nearly equal to the speed of a body in free fall with initial height h, namely, $\sqrt{2gh}$. ◄

9–6

TORQUE AND ANGULAR ACCELERATION

We are now ready to consider the *dynamics* of rotational motion of a rigid body about a stationary axis. As we shall see, the angular acceleration of the body is directly proportional to the sum of the torques with respect to the axis of rotation, and the proportionality is expressed simply by the moment of inertia.

To develop this relationship, we again imagine the body as made up of a large number of particles; a typical particle has mass m and distance r from the axis. We represent the total force acting on this particle in terms of a component $F_\perp$ along the radial direction and a component $F_\parallel$ tangent to the circle of radius r in which the particle moves during rotation. Then applying Newton's second law to this particle gives

$$F_\parallel = ma_\parallel. \tag{9–14}$$

Now $a_\parallel$ may be expressed in terms of the angular acceleration α, according to Eq. (9–10): $a_\parallel = r\alpha$. Using this relation and multiplying both sides of Eq. (9–14) by r, we obtain

$$F_\parallel r = mr^2\alpha. \tag{9–15}$$

We note that $F_\parallel r$ is just the torque of the force, and that mr^2 is the moment of inertia of the particle. (The component $F_\perp$ of force has no torque with respect to the axis.) Thus Eq. (9–15) may be rewritten as

$$\Gamma = I\alpha.$$

Now we may write an equation like this for every particle in the body, and add all these equations. The left side of the resulting equation is the sum of all the torques acting on all the particles, and the right side is the total moment of inertia multiplied by the angular acceleration, which is the same for every particle. Thus for the entire body

$$\sum \Gamma = I\alpha. \tag{9–16}$$

Finally, the sum $\sum \Gamma$ includes only the torques of the *external* forces, that is, the forces exerted on the body by agencies outside it. Torques corresponding to internal forces that the particles of the body exert on each other are not included because, according to Newton's third law, they add up to zero.

Equation (9–16) is the rotational analog of Newton's second law, $\Sigma F = ma$. In applying it to problems, we can use all of the same problem-solving strategies developed for Newton's second law, such as the use of free-body diagrams, defining coordinate systems (including indication of the positive sense of rotation), representing forces in terms of components, and all the rest. We strongly recommend that the reader review at this time the problem-solving strategies discussed at the beginning of Section 4–5.

EXAMPLE 1 A rope is wrapped several times around a uniform solid cylinder of radius 0.1 m and mass 50 kg, pivoted so it can rotate about its axis. What is the angular acceleration when the rope is pulled with a force of 20 N?

Solution The torque is $\Gamma = (0.1 \text{ m})(20 \text{ N}) = 2.0 \text{ m} \cdot \text{N}$ and the angular acceleration is

$$\alpha = \frac{\Gamma}{I} = \frac{2.0 \text{ m} \cdot \text{N}}{\frac{1}{2}(50 \text{ kg})(0.1 \text{ m})^2} = 8 \text{ rad} \cdot \text{s}^{-2}. \qquad \blacktriangleleft$$

EXAMPLE 2 In Example 3 at the end of Section 9–5, find the acceleration of mass m and the angular acceleration of the cylinder.

Solution We treat the two bodies separately. Figure 9–9 shows the forces acting on each body. We take the positive direction of mass m to be downward, and the positive sense of rotation for the cylinder to be counterclockwise. Applying Newton's second law to mass m yields the relation

$$mg - T = ma.$$

Applying Eq. (9–16) to the cylinder gives

$$RT = I\alpha = \tfrac{1}{2}MR^2\alpha.$$

Now the acceleration a of mass m must equal the tangential acceleration of a point on the surface of the cylinder, which, according to Eq. (9–10), is given by $a_{\parallel} = R\alpha$. We use this in the cylinder equation, divide by R, and substitute the resulting expression for T in the equation for m, obtaining

$$mg - \tfrac{1}{2}Ma = ma, \qquad a = \frac{mg}{m + M/2} = \frac{g}{1 + M/2m}.$$

The tension in the rope is *not* equal to the weight mg of mass m; if it were, m could not accelerate. From the above relations,

$$T = mg - ma = \frac{mg}{1 + 2m/M}.$$

When M is much larger than m, the tension is nearly equal to mg, and the acceleration is correspondingly much less than g. When M is zero, $T = 0$ and $a = g$; the mass then falls freely.

If mass m starts from rest at a height h above the floor, its velocity v when it strikes the ground is given by $v^2 - v_0^2 = 2ah$. In this case $v_0 = 0$

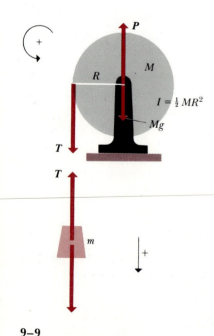

9–9

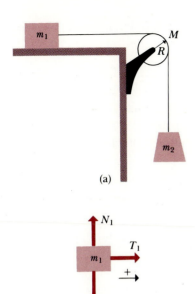

(a)

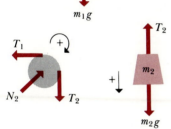

(b)

9–10

and

$$v = \sqrt{2ah} = \sqrt{\frac{2gh}{1 + M/2m}},$$

in agreement with the result obtained from energy considerations in Section 9–5. ◀

EXAMPLE 3 In Fig. 9–10a, mass m_1 slides without friction on the horizontal surface, the pulley is in the form of a thin cylindrical shell of mass M and radius R, and the string turns the pulley without slipping. Find the acceleration of each mass, the angular acceleration of the pulley, and the tension in each part of the string.

Solution Figure 9–10b shows free-body diagrams for the three bodies involved, and also shows a choice of positive directions for the various coordinates. We note that the two tensions T_1 and T_2 *cannot* be equal; if they were, the pulley could not have an angular acceleration. Hence, to label the tension in both parts of the string as simply T would be a serious error.

The equations of motion for masses m_1 and m_2 are

$$T_1 = m_1 a_1, \tag{9–17}$$

$$m_2 g - T_2 = m_2 a_2. \tag{9–18}$$

The unknown normal force N_2 acting on the axis of the pulley has no torque with respect to the axis of rotation, and the equation of motion of the pulley is

$$T_2 R - T_1 R = I\alpha = (MR^2)\alpha. \tag{9–19}$$

Assuming the string does not stretch or slip, we have the additional *kinematic* relations

$$a_1 = a_2 = R\alpha. \tag{9–20}$$

(The accelerations of m_1 and m_2 have different directions but the same magnitude.)

Equation (9–20) can be used to eliminate a_2 and α from Eqs. (9–17) through (9–19). The result is the following set of equations for the three unknowns T_1, T_2, and a_1:

$$T_1 = m_1 a_1,$$

$$m_2 g - T_2 = m_2 a_1,$$

$$T_2 - T_1 = M a_1.$$

These may be solved simultaneously; the simplest procedure is simply to add the three equations to eliminate T_1 and T_2, and then to solve for a_1. The result is

$$a_1 = \frac{m_2 g}{m_1 + m_2 + M}.$$

This result may then be substituted back into Eqs. (9–17) and (9–19) to find the tensions. The results are

$$T_1 = \frac{m_1 m_2 g}{m_1 + m_2 + M}, \qquad T_2 = \frac{(m_1 + M) m_2 g}{m_1 + m_2 + M}.$$

We note that if either m_1 or M is much larger than m_2, the accelerations are very small and T_2 is approximately $m_2 g$, while if m_2 is much larger than either m_1 or M, the acceleration is approximately g, as should be expected. ◀

9–7

WORK AND POWER IN ROTATIONAL MOTION

A force applied to a rotating body does *work* on the body; this work may be expressed in terms of the torque of the force and the angular displacement.

Suppose a force $\boldsymbol{F}$ acts as shown in Fig. 9–11 at the rim of a pivoted wheel of radius R while the wheel rotates through a small angle $\Delta\theta$. If this angle is small enough, the force may be regarded as constant during the correspondingly small time interval. By definition, the work of the force $\boldsymbol{F}$ is

$$\Delta W = F \, \Delta s.$$

But $\Delta s = R \, \Delta\theta$, so that

$$\Delta W = FR \, \Delta\theta.$$

Now FR is the *torque*, Γ, due to the force $\boldsymbol{F}$, so we have finally

$$\Delta W = \Gamma \, \Delta\theta. \tag{9–21}$$

If the torque is constant while the angle changes by a finite amount from θ_1 to θ_2,

$$W = \Gamma(\theta_2 - \theta_1) = \Gamma \, \Delta\theta. \tag{9–22}$$

That is, *the work done by a constant torque equals the product of the torque and the angular displacement.* If Γ is expressed in newton·meters, the work is in joules.

The force in Fig. 9–11 has no component along the *radial* direction. Such a component, if it existed, would do no work, since the displacement of the point of application has no radial component. Similarly, a radial component of force would make no contribution to the torque. Hence Eqs. (9–21) and (9–22) are still correct even when $\boldsymbol{F}$ does have a radial component. When both sides of Eq. (9–21) are divided by the small time interval Δt, during which the displacement occurs, we obtain

$$\frac{\Delta W}{\Delta t} = \Gamma\left(\frac{\Delta\theta}{\Delta t}\right).$$

But $\Delta W/\Delta t$ is the rate of doing work, or the *power*, and $\Delta\theta/\Delta t$ is the angular velocity. Hence

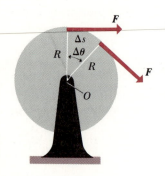

9–11

$$P = \Gamma\omega. \tag{9–23}$$

That is, *the instantaneous power developed by an agent exerting a torque equals the product of the torque and the instantaneous angular velocity.* This is the analog of $P = Fv$ for linear motion.

EXAMPLE The drive shaft of an automobile rotates at 3600 rpm and transmits 80 hp from the engine to the rear wheels. Compute the torque developed by the engine.

$$\omega = \frac{(3600 \text{ rev} \cdot \text{min}^{-1})(2\pi \text{ rad} \cdot \text{rev}^{-1})}{(60 \text{ s} \cdot \text{min}^{-1})}$$
$$= 120\pi \text{ rad} \cdot \text{s}^{-1},$$

$$80 \text{ hp} = (80 \text{ hp})(746 \text{ W} \cdot \text{hp}^{-1})$$
$$= 59{,}700 \text{ W},$$

$$\Gamma = \frac{P}{\omega} = \frac{59{,}700 \text{ W}}{120\pi \text{ rad} \cdot \text{s}^{-1}} = 158 \text{ m} \cdot \text{N}. \qquad \blacktriangleleft$$

*9–8

ROTATION ABOUT A MOVING AXIS

The foregoing analysis of the dynamics of rotational motion of a rigid body can be extended to some cases where the axis of rotation moves, that is, where both translational and rotational motion occur at once. Familiar examples of such motion include a ball rolling down a hill and a yo-yo unwinding at the end of a string. The key to this more general analysis, which we shall not derive in detail, is that Eq. (9–16) remains valid when the axis of rotation moves, provided the axis passes through the center of mass (center of gravity) of the body and does not change its direction.

Analysis of motion that includes both translation and rotation requires two separate equations of motion, one based on $\Sigma F = ma$ for the translational motion, the other based on $\Sigma \Gamma = I\alpha$ for the rotational motion. In addition, there is often a kinematic relation between the two motions, such as a wheel that rolls without slipping or a string that unwinds from a pulley while turning it.

EXAMPLE 1 A string is wrapped several times around a solid cylinder, and then the end of the string is held stationary while the cylinder is released from rest with no initial motion. Find the acceleration of the cylinder and the tension in the string.

Solution Figure 9–12 shows a free-body diagram for the situation, including the choice of positive coordinate directions. The equation for the translational motion of the center of mass is

$$Mg - T = Ma, \qquad (9\text{–}24)$$

and the equation for rotational motion about the axis through the center of mass is

$$TR = I\alpha = (\tfrac{1}{2}MR^2)\alpha. \qquad (9\text{–}25)$$

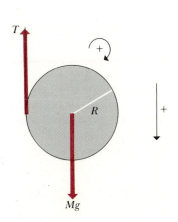

9–12

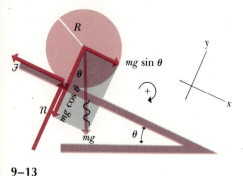

9–13

In addition, if the string unwinds without slipping there is the kinematic relation

$$a = R\alpha. \tag{9-26}$$

This may be used to eliminate α from Eq. (9–25), and then Eqs. (9–24) and (9–25) may be solved simultaneously for T and a. The results are

$$a = \tfrac{2}{3}g, \qquad T = \tfrac{1}{3}Mg. \qquad \blacktriangleleft$$

EXAMPLE 2   A solid sphere rolls without slipping down a plane inclined at angle θ to the horizontal. What is the acceleration of its center?

Solution   Figure 9–13 shows a free-body diagram, with positive coordinate directions indicated. The equations of motion for translational and rotational motion, respectively, are

$$mg \sin \theta - \mathcal{F} = ma,$$

$$\mathcal{F}R = I\alpha = (\tfrac{2}{5}mR^2)\alpha.$$

If the sphere rolls without slipping, then $a = R\alpha$. We use this to eliminate α and then solve for a and $\mathcal{F}$ to obtain

$$a = \tfrac{5}{7}g \sin \theta, \qquad \mathcal{F} = \tfrac{2}{7}mg \sin \theta.$$

We note that the acceleration is just $\tfrac{5}{7}$ as large as it would be if the sphere could slide without friction down the slope. Also, the friction force is essential to prevent slipping and thus to cause the angular acceleration of the sphere. An expression for the minimum coefficient of friction can be obtained by noting that the normal force is $n = mg \cos \theta$. To prevent slipping, the coefficient of (static) friction must be at least as great as

$$\mu_s = \frac{\mathcal{F}}{n} = \frac{\tfrac{2}{7}mg \sin \theta}{mg \cos \theta} = \tfrac{2}{7} \tan \theta.$$

If the plane is tilted only slightly, θ is small and only a small value of μ_s is needed to prevent slipping; but as the angle increases the required value of μ_s increases, as might be expected intuitively. $\blacktriangleleft$

*9–9

PARALLEL-AXIS THEOREM

If the moment of inertia I_{cm} of a body about an axis through its center of mass is known, then the moment of inertia I_P about any other axis parallel to the original one but displaced from it by distance d is easily obtained, by means of a relation called the *parallel-axis theorem*, which states

$$I_P = I_{cm} + Md^2. \tag{9-27}$$

To prove this theorem, we consider the body shown in Fig. 9–14. The origin of coordinates has been chosen to coincide with the center of

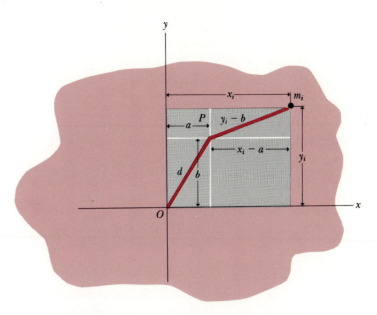

9–14

mass, and we wish to compute the moment of inertia about an axis through point P, perpendicular to the plane of the figure. Let m_i be a typical mass element having coordinates x_i and y_i. Then the moment of inertia about an axis through O perpendicular to the figure is

$$I_{\text{cm}} = \sum m_i(x_i^2 + y_i^2),$$

and the moment of inertia about the axis through P is

$$I_P = \sum m_i\left[(x_i - a)^2 + (y_i - b)^2\right].$$

We expand the squared terms and regroup, obtaining

$$I_P = \sum m_i(x_i^2 + y_i^2) - 2a \sum m_i x_i - 2b \sum m_i y_i + (a^2 + b^2) \sum m_i.$$

The first sum is recognized as I_{cm}; the second and third are zero because they represent the x- and y-coordinates of the center of mass, which are zero because we have taken our origin to be the center of mass. The final term is d^2 multiplied by the total mass, so the theorem is proved.

EXAMPLE 1 The moment of inertia of a thin rod about an axis through its midpoint, perpendicular to the rod, is $I = ML^2/12$ (as shown in Fig. 9–7). Find its moment of inertia about an axis perpendicular to the rod at one end.

Solution We have $I_{\text{cm}} = ML^2/12$ and $d = L/2$; Eq. (9–27) yields

$$I_P = \frac{ML^2}{12} + M\left(\frac{L}{2}\right)^2 = \frac{ML^2}{3}.$$ ◀

EXAMPLE 2 Find the moment of inertia of a thin uniform disk about an axis perpendicular to its plane at the edge.

Solution   From Fig. 9–7 we find $I_{cm} = MR^2/2$, and in this case $d = R$. Thus

$$I_P = \frac{MR^2}{2} + MR^2 = \frac{3MR^2}{2}. \qquad \blacktriangleleft$$

9–10

ANGULAR MOMENTUM AND ANGULAR IMPULSE

Figure 9–15a represents a small body of mass m moving in the plane of the diagram with a speed v and a momentum mv. We define its *angular momentum* about an axis through O perpendicular to the plane of the diagram, as the product of its linear momentum and the perpendicular distance from the axis to its line of motion. Angular momentum will be denoted by L. Thus, for a particle,

$$\text{Angular momentum} = L = mvr. \qquad (9\text{–}28)$$

This definition of angular momentum is analogous to the definition of moment or torque of a force, and indeed angular momentum is sometimes called *moment of momentum*.

Figure 9–15b shows a body of finite size rotating in the plane of the diagram about an axis through O. The velocity v of a small element of the body is related to the angular velocity ω of the body by $v = \omega r$. The angular momentum of the element is therefore $mvr = \omega mr^2$, and the total angular momentum of the body is $\Sigma \, \omega mr^2 = \omega \, \Sigma \, mr^2$. But $\Sigma \, mr^2$ is the moment of inertia of the body about its axis of rotation. Hence the angular momentum can be written as $I\omega$. Again denoting angular momentum by the symbol L, we have

$$L = I\omega. \qquad (9\text{–}29)$$

This is analogous to the definition of linear momentum mv for a particle.

When a constant torque Γ acts on a body of moment of inertia I, for a time interval from t_1 to t_2, the angular velocity changes from ω_1 to ω_2, according to the relation

$$\Gamma = I\alpha = I\left(\frac{\omega_2 - \omega_1}{t_2 - t_1}\right).$$

Rearranging this equation, we obtain

$$\Gamma(t_2 - t_1) = I\omega_2 - I\omega_1 = L_2 - L_1 = \Delta L. \qquad (9\text{–}30)$$

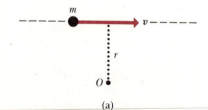

(a)

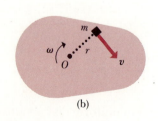

(b)

9–15   Angular momentum.

The product of the torque and the time interval during which it acts is called the *angular impulse* of the torque; we denote this quantity by J_θ.

$$\text{Angular impulse} = J_\theta = \Gamma(t_2 - t_1). \qquad (9\text{–}31)$$

This quantity is the rotational analog of the impulse of a force, defined in Section 7–1. Equation (9–30) states that *the angular impulse acting on a body equals the change of angular momentum of the body about the same axis.*

This relationship is easily generalized to a torque that varies with time, just as impulse was generalized in Section 7–1. Suppose Γ varies

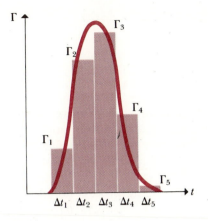

9–16

with time as in Fig. 9–16. We divide the total time interval into subinterval, as in the figure, drawing rectangles whose outline approximates the smooth curve. The larger the number of subintervals, the better the approximation. Equation (9–30) holds for each subinterval, so the *total* change in angular momentum is equal to the *total* angular impulse, represented graphically by the total area under the curve. Thus, when Γ varies with time,

$$\text{Angular impulse} = J_\theta = (\text{area under } \Gamma\text{-vs.-}t \text{ curve}). \qquad (9\text{–}32)$$

Thus the general relationship between angular impulse and angular momentum is

$$J_\theta = I\omega_2 - I\omega_1 = L_2 - L_1 = \Delta L, \qquad (9\text{–}33)$$

with J_θ given by Eq. (9–31) if Γ is constant, and by Eq. (9–32) otherwise. A torque that acts only during a short time interval, such as that shown in Fig. 9–16, is often called an *impulsive torque*.

9–11

CONSERVATION OF ANGULAR MOMENTUM

Here is an example of the usefulness of the concepts of angular impulse and angular momentum. Figure 9–17 shows two disks with moments of inertia I and I', rotating with initial constant angular velocities ω_0 and ω'_0, respectively. The disks are then pushed together by a force directed along the axis, without any torque on either disk. After a short time the disks reach a common final angular velocity ω.

During this time the larger disk exerts a torque Γ' on the smaller, and the smaller exerts a torque Γ on the larger. Both Γ and Γ' vary during the contact, becoming zero after the common final angular velocity is reached. At each instant the two torques are equal in magnitude and opposite in direction, because of Newton's third law. That is, at each instant $\Gamma = -\Gamma'$; the total impulses on the two bodies are thus related by $J_\theta = -J'_\theta$.

To see why this must be so, we consider a point of contact between the two bodies. The two forces that the bodies exert on each other at this common point have equal magnitude and opposite direction, according to Newton's third law. Because they act at the same point, they also have the same moment arm with respect to any axis. The two forces lie along the same line, and the torque of one is the negative of the torque of the other. The same thing is true of the pairs of forces at all the other contact points. Thus at any instant $\Gamma = -\Gamma'$, and the total impulses on the two bodies are related by $J_\theta = -J'_\theta$.

Now according to Eq. (9–33), the impulse on each disk equals the change of angular momentum of that disk:

$$J_\theta = I\omega - I\omega_0,$$

$$J'_\theta = I'\omega - I'\omega'_0.$$

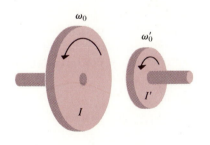

9–17 An impulsive torque acts when two rotating disks engage.

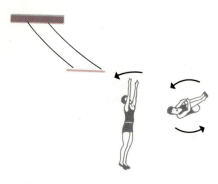

9–18 Conservation of angular momentum.

But because $J_\theta = -J'_\theta$, these are equal and opposite:

$$I\omega - I\omega_0 = -(I'\omega - I'\omega'_0)$$

or

$$I\omega_0 + I'\omega'_0 = (I + I')\omega. \tag{9–34}$$

The left side of Eq. (9–34) is the total angular momentum of the system (consisting of both disks) before contact, and the right side is the total angular momentum after contact. We have therefore derived the important result that the total angular momentum of the whole system is unaltered. When both disks are regarded as one system, then the torques Γ and Γ' are *internal* torques; during the engaging process no *external* torque acts. *When the resultant external torque on a system is zero, the angular momentum of the system remains constant;* hence any internal interaction between the parts of a system cannot alter its total angular momentum. This is the *principle of conservation of angular momentum,* and it ranks with the principles of conservation of linear momentum and conservation of energy as one of the most fundamental of physical laws.

A circus acrobat, a diver, or a skater performing a pirouette on the toe of one skate, all take advantage of the principle. Suppose an acrobat has just left a swing as in Fig. 9–18, with arms and legs extended and with a counterclockwise angular momentum. When he pulls his arms and legs in, his moment of intertia I becomes much smaller. Since his angular momentum $I\omega$ remains constant and I decreases, his angular velocity ω increases.

EXAMPLE 1   In the situation of Fig. 9–17, suppose the large disk has mass 2 kg, radius 0.2 m, and initial angular velocity 50 rad·s^{-1}, and the small disk has mass 4 kg, radius 0.1 m, and initial angular velocity 200 rad·s^{-1} (about 1900 rpm). Find the common final angular velocity after the disks are pushed into contact. Is kinetic energy conserved during this process?

Solution   The moments of inertia of the two disks are

$$I = \tfrac{1}{2}(2\text{ kg})(0.2\text{ m})^2 = 0.04\text{ kg·m}^2,$$

and

$$I' = \tfrac{1}{2}(4\text{ kg})(0.1\text{ m})^2 = 0.02\text{ kg·m}^2.$$

From conservation of angular momentum, we have

$$(0.04\text{ kg·m}^2)(50\text{ rad·s}^{-1}) + (0.02\text{ kg·m}^2)(200\text{ rad·s}^{-1})$$
$$= (0.04\text{ kg·m}^2 + 0.02\text{ kg·m}^2)\omega,$$
$$\omega = 100\text{ rad·s}^{-1}.$$

The initial kinetic energy is

$$K_0 = \tfrac{1}{2}(0.04\text{ kg·m}^2)(50\text{ rad·s}^{-1})^2 + \tfrac{1}{2}(0.02\text{ kg·m}^2)(200\text{ rad·s}^{-1})^2$$
$$= 450\text{ J}.$$

9–19 Conservation of angular momentum about a fixed axis.

The final kinetic energy is

$$K = \tfrac{1}{2}(0.04 \text{ kg} \cdot \text{m}^2 + 0.02 \text{ kg} \cdot \text{m}^2)(100 \text{ rad} \cdot \text{s}^{-1})^2 = 300 \text{ J}.$$

One-third of the kinetic energy was lost during this "angular collision," which is the rotational analog of an inelastic collision. We should not expect kinetic energy to be conserved, even though the resultant external force and torque are zero, since nonconservative (frictional) internal forces act during the contact.   ◀

EXAMPLE 2   A man stands at the center of a turntable, holding his arms extended horizontally, with a 5-kg mass in each hand, as in Fig. 9–19. He is set rotating about a vertical axis with an angular velocity of one revolution in 2 s. Find his new angular velocity if he drops his hands to his sides. The moment of inertia of the man may be assumed constant and equal to 6 kg·m². The original distance of the weights from the axis is 1 m, and their final distance is 0.2 m.

Solution   If friction in the turntable is neglected, no external torques act about a vertical axis and the angular momentum about this axis is constant. That is

$$I\omega = (I\omega)_0 = I_0\omega_0,$$

where I and ω are the final moment of inertia and angular velocity, and I_0 and ω_0 are the initial values of these quantities:

$$I = I_{\text{man}} + I_{\text{weights}},$$

$$I = 6 \text{ kg} \cdot \text{m}^2 + 2(5 \text{ kg})(0.2 \text{ m})^2 = 6.4 \text{ kg} \cdot \text{m}^2,$$

$$I_0 = 6 \text{ kg} \cdot \text{m}^2 + 2(5 \text{ kg})(1.0 \text{ m})^2 = 16 \text{ kg} \cdot \text{m}^2,$$

$$\omega_0 = 2\pi(\tfrac{1}{2}) \text{ rad} \cdot \text{s}^{-1},$$

$$\omega = \omega_0\left(\frac{I_0}{I}\right) = \pi \text{ rad} \cdot \text{s}^{-1}\frac{16 \text{ kg} \cdot \text{m}^2}{6.4 \text{ kg} \cdot \text{m}^2} = 2.5\pi \text{ rad} \cdot \text{s}^{-1} = 1.25 \text{ rev} \cdot \text{s}^{-1}.$$

That is, the angular velocity is more than doubled.

The initial kinetic energy is

$$K_0 = \tfrac{1}{2}(16 \text{ kg} \cdot \text{m}^2)(\pi \text{ rad} \cdot \text{s}^{-1})^2 = 79 \text{ J}.$$

The final kinetic energy is

$$K = \tfrac{1}{2}(6.4 \text{ kg} \cdot \text{m}^2)(2.5\pi \text{ rad} \cdot \text{s}^{-1})^2 = 197 \text{ J}.$$

Where did the extra energy come from?   ◀

EXAMPLE 3   A door 1.0 m wide, having a mass of 15 kg, is hinged at one side so it can rotate without friction about a vertical axis. A bullet having mass 10 g and speed 400 m·s⁻¹ is fired into the door, in a direction perpendicular to the plane of the door, and imbeds itself at the exact center of the door. Find the angular velocity of the door just after the bullet imbeds itself. Is kinetic energy conserved?

Solution   There is no external torque about the axis defined by the hinges, so angular momentum about this axis is conserved. The initial angular momentum of the bullet is given by Eq. (9–28):

$$L = mvr = (0.01 \text{ kg})(400 \text{ m}\cdot\text{s}^{-1})(0.5 \text{ m}) = 2.0 \text{ kg}\cdot\text{m}^2\cdot\text{s}^{-1}.$$

This is equal to the final angular momentum $I\omega$, where I is the total moment of inertia of door and bullet. To find I for the door alone, we note that dimensions parallel to the axis are irrelevant, and that its moment of inertia is the same as that of a rod 1 m long, pivoted at one end. As discussed in Example 1 of Section 9–11, this is given by

$$\frac{ML^2}{3} = \frac{(15 \text{ kg})(1.0 \text{ m})^2}{3} = 5.0 \text{ kg}\cdot\text{m}^2.$$

The moment of inertia of the bullet is

$$mr^2 = (0.01 \text{ kg})(0.5 \text{ m})^2 = 0.0025 \text{ kg}\cdot\text{m}^2.$$

Conservation of angular momentum requires that

$$2.0 \text{ kg}\cdot\text{m}^2\cdot\text{s}^{-1} = (5.0 \text{ kg}\cdot\text{m}^2 + 0.0025 \text{ kg}\cdot\text{m}^2)\omega,$$

$$\omega = 0.4 \text{ rad}\cdot\text{s}^{-1}.$$

The collision of bullet and door is inelastic, so we do not expect energy to be conserved. To check, we calculate initial and final kinetic energies:

$$K_i = \tfrac{1}{2}mv^2 = \tfrac{1}{2}(0.01 \text{ kg})(400 \text{ m}\cdot\text{s}^{-1})^2 = 800 \text{ J};$$

$$K_f = \tfrac{1}{2}I\omega^2 = \tfrac{1}{2}(5.0025 \text{ kg}\cdot\text{m}^2)(0.4 \text{ rad}\cdot\text{s}^{-1})^2 = 0.4 \text{ J}.$$

The final kinetic energy is only 1/2000 of the initial value.   ◀

*9–12

VECTOR REPRESENTATION OF ANGULAR QUANTITIES

9–20   Vector angular velocity of a rotating body.

A rotational quantity associated with an axis, such as angular velocity, angular momentum, or torque, can be represented by a *vector* along the axis. Thus angular velocity can be defined as a vector quantity with magnitude equal to the number of radians through which the body turns in unit time (our original definition) and direction along the axis of rotation.

A vector can have either of two opposite directions along a given line (corresponding to an arrowhead on either end of the arrow representing the vector); it is customary to define angular velocity as having the direction in which a righthand-thread screw would advance if turned with the body, as shown in Fig. 9–20. Alternatively, when the fingers of the right hand are wrapped around the axis with the fingers pointing in the direction of rotation, the thumb points in the direction of the vector angular velocity, as shown in Fig. 9–20. Similarly, a symmetric body turning about its axis of symmetry has a vector angular momentum directed along its axis, parallel to the angular velocity vector.

The definition of vector torque is perhaps less obvious, but a torque always tends to cause rotation about a certain axis, the axis about which the body would begin to rotate if initially at rest with only the torque under consideration acting on it. Thus the definition of torque can be similarly generalized, and it turns out that the relation of angular momentum to angular impulse is still valid for the generalized quantities. Thus for a constant vector torque $\boldsymbol{\Gamma}$,

$$\Delta \boldsymbol{L} = \boldsymbol{\Gamma} \, \Delta t. \tag{9–35}$$

These generalizations are not needed for the problems considered thus far in this chapter because when the axis of rotation maintains the same direction only one component of angular velocity is different from zero; such rotational motion is thus analogous to motion of a particle along a straight line. In problems where the direction of the axis changes, however, it is essential to consider the vector nature of the various angular quantities. The general formulation of the dynamics of rotational motion is somewhat complex and cannot be discussed in detail here, but following is an example of its application.

Figure 9–21 shows a familiar toy gyroscope. When the flywheel is set spinning by wrapping a string around its shaft and pulling, one possible motion when the shaft is supported at only one end is a steady circular motion of the axis in a horizontal plane. This is an interesting phenomenon, inasmuch as intuition suggests the end of the axis opposite O should drop if it is not supported. Vector angular momentum considerations permit an understanding of this behavior.

The forces acting on the gyroscope are its weight $\boldsymbol{w}$, acting downward at the center of gravity, and the upward force $\boldsymbol{P}$ at the pivot point O. These forces acting on a body at rest would tend to cause a rotation about an axis in the horizontal plane, perpendicular to the gyroscope

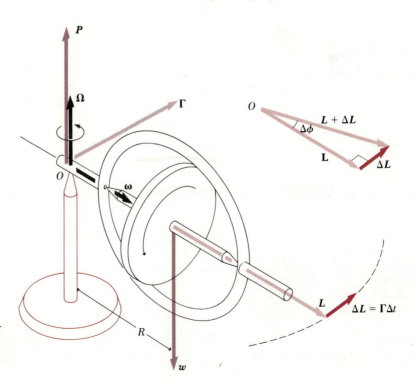

9–21 Vector $\Delta \boldsymbol{L}$ is the change in angular momentum produced in time Δt by the moment $\boldsymbol{\Gamma}$ of the force $\boldsymbol{w}$. Vectors $\Delta \boldsymbol{L}$ and $\boldsymbol{\Gamma}$ are in the same direction.

axis, and the associated torque Γ is in the direction shown. However, the body is *not* at rest, and the effect of this torque is not to *initiate* a rotational motion but to *change* a motion that already exists. At the instant shown the body has angular momentum described by the vector L, and the *change* ΔL in angular momentum in a short time interval is given by Eq. (9–34). Thus ΔL has the same direction as Γ. After time Δt the angular momentum is $L + \Delta L$ and, as the vector diagram shows, this means that the gyroscope axis has turned through a small angle $\Delta\phi$ given by $\Delta\phi = |\Delta L|/|L|$. Thus the motion of the axis is consistent with the torque–angular momentum relationship. This motion of the axis is called *precession.*

The rate at which the axis moves, $\Delta\phi/\Delta t$, is called the *precession angular velocity;* denoting this quantity by Ω, we find

$$\Omega = \frac{\Delta\phi}{\Delta t} = \frac{|\Delta L|/|L|}{\Delta t} = \frac{\Gamma}{L}. \tag{9–36}$$

Thus the precession angular velocity is *inversely* proportional to the angular velocity of spin about the axis. A rapidly spinning gyroscope precesses slowly, and as it slows down the precession angular velocity *increases!*

We may now return to the question of why the unsupported end of the axis does not fall. The reason is that the upward force P exerted on the gyroscope by the pivot is just equal in magnitude to its weight w, so that the resultant vertical *force* is zero and the vertical acceleration of the center of gravity is zero. In other words, the vertical component of its linear momentum remains zero, since there is no resultant vertical force. However, the resultant *torque* of these forces is *not* zero and the angular momentum changes.

If the top were not rotating, it would have no angular momentum L initially. Its angular momentum ΔL after a time Δt would be that acquired from the torque acting on it and would have the same direction as the torque. In other words, the top would rotate about an axis through O in the direction of the vector Γ. But if the top is originally rotating, the *change* in its angular momentum produced by the torque adds vectorially to the large angular momentum it already has, and since ΔL is horizontal and perpendicular to L, the result is a motion of precession, with both the angular momentum vector and the axis remaining horizontal.

QUESTIONS

9–1 What is the difference between tangential and radial acceleration, for a point on a rotating body?

9–2 A flywheel rotates with constant angular velocity. Does a point on its rim have a tangential acceleration? A radial acceleration? Are these accelerations constant? In magnitude? In direction?

9–3 A flywheel rotates with constant angular acceleration. Does a point on its rim have a tangential acceleration? A radial acceleration? Are these accelerations constant? In magnitude? In direction?

9–4 Can a single force applied to a body change both its translational and rotational motion?

9–5 How might you determine experimentally the moment of inertia of an irregularly shaped body?

9–6 Can you think of a body that has the same moment of inertia for all possible axes? For all axes passing through a certain point? What point?

9–7 A cylindrical body has mass M and radius R. Is it ever possible for its moment of inertia to be greater than MR^2?

9–8 In order to maximize the moment of inertia of a fly-wheel while minimizing its weight, what shape should it have?

9–9 In tightening cylinder-head bolts in an automobile engine, the critical quantity is the *torque* applied to the bolts. Why is this more important than the actual *force* applied to the wrench handle?

9–10 The flywheel of an automobile engine is included to increase the moment of inertia of the engine crankshaft. Why is this desirable?

9–11 A solid ball and a hollow ball with the same mass and radius roll down a slope. Which one reaches the bottom first?

9–12 A solid ball, a solid cylinder, and a hollow cylinder roll down a slope. Which reaches the bottom first? Last? Does it matter whether the radii are the same?

9–13 When an electric motor is turned on, it takes longer to come up to final speed if there is a grinding wheel attached to the shaft. Why?

9–14 Experienced cooks can tell whether an egg is raw or hard-boiled by rolling it down a slope (taking care to catch it at the bottom). How is this possible?

9–15 Consider the idea of an automobile powered by energy stored in a rotating flywheel, which can be "recharged" using an electric motor. What advantages and disadvantages would such a scheme have, compared to more conventional drive mechanisms? Could as much energy be stored as in a tank of gasoline? What factors would limit the maximum energy storage?

9–16 An electric grinder coasts for a minute or more after the power is turned off, while an electric drill coasts for only a few seconds. Why is there a difference?

9–17 Part of the kinetic energy of a moving automobile is in rotational motion of its wheels. When the brakes are applied hard on an icy street, the wheels "lock" and the car starts to slide. What becomes of the rotational kinetic energy?

PROBLEMS

9–1

a) What angle in radians is subtended by an arc 3 m in length, on the circumference of a circle whose radius is 2 m?

b) What angle in radians is subtended by an arc of length 78.54 cm on the circumference of a circle of diameter 100 cm? What is this angle in degrees?

c) The angle between two radii of a circle is 0.60 radian. What length of arc is intercepted on the circumference of a circle of radius 200 cm? of radius 200 ft?

9–2 Compute the angular velocity, in $rad \cdot s^{-1}$, of the crankshaft of an automobile engine that is rotating at 4800 $rev \cdot min^{-1}$.

9–3 A circular saw 0.6 m in diameter starts from rest and accelerates with constant angular acceleration to an angular velocity of 100 $rad \cdot s^{-1}$ in 20 s. Find the angular acceleration and the angle through which the saw has turned.

9–4 An electric motor is turned off, and its angular velocity decreases uniformly from 1000 $rev \cdot min^{-1}$ to 400 $rev \cdot min^{-1}$ in 5 s. Find the angular acceleration and the number of revolutions made by the motor in the 5-s interval. How many more seconds are required for the motor to come to rest?

9–5 A flywheel requires 3 s to rotate through 234 radians. Its angular velocity at the end of this time is 108 $rad \cdot s^{-1}$. Find its constant angular acceleration.

9–6 A flywheel whose angular acceleration is constant and equal to 2 $rad \cdot s^{-1}$ rotates through an angle of 100 radians in 5 s. How long had it been in motion at the beginning of the 5-s interval if it started from rest?

9–7 A record-player turntable starts from rest and turns with constant angular acceleration, reaching its final angular velocity of $33\frac{1}{3}$ $rev \cdot min^{-1}$ after 3 s. Find the components of acceleration of a point at the edge of a 12-inch record 2 s after the turntable is turned on. Also find the magnitude and direction of the resultant acceleration.

9–8 An electric fan blade 1.0 m in diameter is rotating about a fixed axis with an initial angular velocity of 2 $rev \cdot s^{-1}$. The angular acceleration is 3 $rev \cdot s^{-2}$.

a) Compute the angular velocity after 6 s.

b) Through what angle has the blade turned in this time interval?

c) What is the tangential velocity of a point on the tip of a blade at $t = 6$ s?

d) What is the resultant acceleration of a point on the tip of a blade at $t = 6$ s?

9–9 A grinding wheel having a diameter of 0.2 m starts from rest and accelerates with constant angular acceleration to an angular velocity of 900 $rev \cdot min^{-1}$ in 5 s.

a) Find the position at the end of 1 s of a point originally at the top of the wheel.

b) Compute and show in a diagram the magnitude and direction of the acceleration of this point at the end of 1 s.

9–10 A wheel rotates with a constant angular velocity of 10 $rad \cdot s^{-1}$.

a) Compute the radial acceleration of a point 0.5 m from the axis, from the relation $a_\perp = \omega^2 R$.

b) Find the tangential velocity of the point, and compute its radial acceleration from the relation $a_\perp = v^2/R$.

9–11 Find the required angular velocity of an ultracentrifuge, in rpm, in order that the radial acceleration of a point 1 cm from the axis shall equal $300{,}000g$ (i.e., 300,000 times the acceleration of gravity).

9–12 An automobile engine is idling at 500 rev·min⁻¹. When the accelerator is depressed, the angular velocity increases to 3000 rev·min⁻¹ in 5 s. Assume a constant angular acceleration.

a) What are the initial and final angular velocities, expressed in radians per second?

b) What was the angular acceleration, in radians per second squared?

c) How many revolutions did the engine make during the acceleration period?

d) The flywheel of the engine is 0.5 m in diameter. What is the linear speed of a point at its rim when the angular speed is 3000 rev·min⁻¹?

e) What was the tangential acceleration of the point during the acceleration period?

f) What is the radial acceleration of the point when the angular speed is 3000 rev·min⁻¹?

9–13

a) Prove that when a body starts from rest and rotates about a fixed axis with constant angular acceleration, the radial acceleration of a point in the body is directly proportional to its angular displacement.

b) Through what angle has the body turned at the instant when the resultant acceleration of a point makes an angle of 60° with the radial direction?

9–14 Small blocks, each of mass m, are clamped at the ends and at the center of a light rigid rod of length L. Compute the moment of inertia of the system about an axis perpendicular to the rod and passing through a point one quarter of the length from one end. Neglect the moment of inertia of the rod.

9–15 A thin rectangular sheet of steel is 0.3 m by 0.4 m and has mass 24 kg. Find the moment of inertia about an axis

a) through the center, parallel to the long sides;

b) through the center, parallel to the short sides;

c) through the center, perpendicular to the plane.

Do you notice anything strange about your results?

9–16 A bicycle wheel of radius 0.3 m has a rim of mass 1.0 kg and 50 spokes, each of mass 0.01 kg. What is its moment of inertia about its axis of rotation?

9–17 Find the moment of inertia of a rod 4 cm in diameter and 2 m long, of mass 8 kg,

a) about an axis perpendicular to the rod and passing through its center,

b) about an axis perpendicular to the rod and passing through one end,

c) about a longitudinal axis passing through the center of the rod.

9–18 The four bodies shown in Fig. 9–22 have equal masses m. Body A is a solid cylinder of radius R. Body B is a hollow thin cylinder of radius R. Body C is a solid square with length of side $2R$. Body D is the same size as C, but hollow (i.e., made up of four thin sticks). The bodies have axes of rotation perpendicular to the page and through the center of gravity of each body.

a) Which body has the smallest moment of inertia?

b) Which body has the largest moment of inertia?

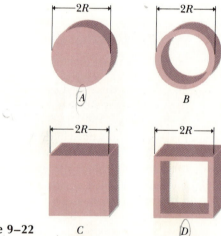

Figure 9–22

9–19 A flywheel consists of a solid disk 0.5 m in diameter and 0.02 m thick, and two projecting hubs 0.1 m in diameter and 0.1 m long. If the material of which it is constructed has a density of 6000 kg·m⁻³, find its moment of inertia about the axis of rotation.

9–20 A grinding wheel 0.2 m in diameter, of mass 3 kg, is rotating at 3600 rev·min⁻¹.

a) What is its kinetic energy?

b) How far would it have to drop in free fall to acquire the same kinetic energy?

9–21 A solid uniform disk of mass m and radius R is pivoted about a horizontal axis through its center, and a small body of mass m is attached to the rim of the disk. If the disk is released from rest with the small body at the end of a horizontal radius, find the angular velocity when the small body is at the bottom.

9–22 The flywheel of a gasoline engine is required to give up 300 J of kinetic energy while its angular velocity decreases from 600 rev·min^{-1} to 540 rev·min^{-1}. What moment of inertia is required?

9–23 The flywheel of a punch press has a moment of inertia of 25 kg·m^2 and it runs at 300 rev·min^{-1}. The flywheel supplies all the energy needed in a quick punching operation.

a) Find the speed in rev·min^{-1} to which the flywheel will be reduced by a sudden punching operation requiring 4000 J of work.

b) What must be the constant power supply to the flywheel in horsepower to bring it back to its initial speed in 5 s?

9–24 A magazine article described a passenger bus in Zurich, Switzerland, which derived its motive power from the energy stored in a large flywheel. The wheel was brought up to speed periodically, when the bus stopped at a station, by an electric motor, which could then be attached to the electric power lines. The flywheel was a solid cylinder of mass 1000 kg, diameter 1.8 m, and its top speed was 3000 rev·min^{-1}.

a) At this speed, what is the kinetic energy of the flywheel?

b) If the average power required to operate the bus is 25 hp, how long can it operate between stops?

9–25 A grindstone in the form of a solid cylinder has a radius of 0.5 m and a mass of 50 kg.

a) What torque will bring it from rest to an angular velocity of 300 rev·min^{-1} in 10 s?

b) What is its kinetic energy when it is rotating at 300 rev·min^{-1}?

9–26 The flywheel of a motor has a mass of 300 kg and a moment of inertia of 675 kg·m^2. The motor develops a constant torque of 2000 N·m and the flywheel starts from rest.

a) What is the angular acceleration of the flywheel?

b) What will be its angular velocity after making 4 revolutions?

c) How much work is done by the motor during the first 4 revolutions?

9–27 The flywheel of a stationary engine has a moment of inertia of 30 kg·m^2.

a) What constant torque is required to bring it up to an angular velocity of 900 rev·min^{-1} in 10 s, starting from rest?

b) What is its final kinetic energy?

9–28 A grindstone 1.0 m in diameter, of mass 50 kg, is rotating at 900 rev·min^{-1}. A tool is pressed normally against the rim with a force of 200 N, and the grindstone comes to rest in 10 s. Find the coefficient of friction between the tool and the grindstone. Neglect friction in the bearings.

9–29 A 60-kg grindstone is 1 m in diameter and has a moment of inertia of 3.75 kg·m^2. A tool is pressed down on the rim with a normal force of 50 N. The coefficient of sliding friction between the tool and stone is 0.6, and there is a constant friction torque of 5 N·m between the axle of the stone and its bearings.

a) How much force must be applied normally at the end of a crank handle 0.5 m long to bring the stone from rest to 120 rev·min^{-1} in 9 s?

b) After attaining a speed of 120 rev·min^{-1}, what must the normal force at the end of the handle become to maintain a constant speed of 120 rev·min^{-1}?

c) How long will it take the grindstone to come from 120 rev·min^{-1} to rest if it is acted on by the axle friction alone?

9–30 A constant torque of 20 N·m is exerted on a pivoted wheel for 10 s, during which time the angular velocity of the wheel increases from zero to 100 rev·min^{-1}. The external torque is then removed and the wheel is brought to rest by friction in its bearings in 100 s. Compute

a) the moment of inertia of the wheel,

b) the friction torque,

c) the total number of revolutions made by the wheel.

9–31 A cord is wrapped around the rim of a flywheel 0.5 m in radius, and a steady pull of 50 N is exerted on the cord, as in Fig. 9–23a. The wheel is mounted in frictionless bearings on a horizontal shaft through its center. The moment of inertia of the wheel is 4 kg·m^2.

a) Compute the angular acceleration of the wheel.

b) Show that the work done in unwinding 5 m of cord equals the gain in kinetic energy of the wheel.

c) If a mass having a weight of 50 N hangs from the cord, as in Fig. 9–23b, compute the angular acceleration of the wheel. Why is this not the same as in part (a)?

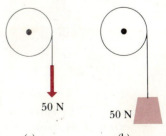

Figure 9–23　　(a)　　　　　(b)

50 N　　　　50 N

9–32 A flywheel 1.0 m in diameter is pivoted on a horizontal axis. A rope is wrapped around the outside of the flywheel, and a steady pull of 50 N is exerted on the rope. It is found that 10 m of rope are unwound in 4 s.

a) What was the angular acceleration of the flywheel?

b) What is its final angular velocity?

c) What is its final kinetic energy?

d) What is its moment of inertia?

9–33 A bucket of water of mass 20 kg is suspended by a rope wrapped around a windlass in the form of a solid cylinder 0.2 m in diameter, also of mass 20 kg. The bucket is released from rest at the top of a well and falls 20 m to the water.

a) What is the tension in the rope while the bucket is falling?

b) With what velocity does the bucket strike the water?

c) What was the time of fall? Neglect the weight of the rope.

9–34 A 5-kg block rests on a horizontal frictionless surface. A cord attached to the block passes over a pulley, whose diameter is 0.2 m, to a hanging block also of mass 5 kg. The system is released from rest, and the blocks are observed to move 4 m in 2 s.

a) What was the moment of inertia of the pulley?

b) What was the tension in each part of the cord?

9–35 Figure 9–24 represents an *Atwood's machine.* Find the linear accelerations of blocks A and B, the angular acceleration of the wheel C, and the tension in each side of the cord

a) if the surface of the wheel is frictionless;

b) if there is no slipping between the cord and the surface of the wheel.

Let the masses of blocks A and B be 4 kg and 2 kg, respectively, the moment of inertia of the wheel about its axis be 0.2 kg·m^2, and the radius of the wheel be 0.1 m.

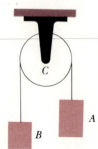

Figure 9–24

9–36 A block of mass $m = 5$ kg slides down a surface inclined 37° to the horizontal, as shown in Fig. 9–25. The coefficient of sliding friction is 0.25. A string attached to the block is wrapped around a flywheel on a fixed axis at

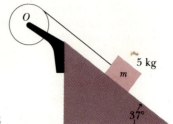

Figure 9–25

O. The flywheel has a mass $M = 20$ kg, an outer radius $R = 0.2$ m, and a moment of inertia with respect to the axis of 0.2 kg·m^2.

a) What is the acceleration of the block down the plane?

b) What is the tension in the string?

9–37 A solid cylinder rolls without slipping down a 30° slope. Find the acceleration, the frictional force, and the minimum coefficient of friction needed to prevent slipping.

9–38 A yo-yo is made from two uniform disks of radius a connected by a light axle of radius b. A string is wound several times around the axle; the inexperienced operator holds the end of the string stationary and releases the yo-yo with no initial velocity, and it drops as the string unwinds. Find the acceleration of the yo-yo and the tension in the string.

9–39 A lawn roller in the form of a hollow cylinder of mass M is pulled horizontally with a constant force F applied by a handle attached to the axle. If it rolls without slipping, find the acceleration and the frictional force.

9–40 Find the moment of inertia of a thin rod of mass M and length L, about an axis perpendicular to the rod, a distance $L/4$ from one end.

9–41 Find the moment of inertia of a thin square sheet of side length a and mass M, about an axis perpendicular to the sheet at a corner.

9–42 A uniform thin rod is bent into a square of side length a. If the total mass is M, find the moment of inertia about an axis through the center, perpendicular to the plane of the square.

9–43 What torque is developed by a 1-horsepower electric motor turning at 3600 rpm?

9–44

a) Compute the torque developed by an airplane engine whose output is 2000 hp at an angular velocity of 2400 rev·min^{-1}.

b) If a drum of negligible mass, 0.5 m in diameter, were attached to the motor shaft and the power output of the motor were used to raise a weight hanging from a rope wrapped around the drum, how large a weight could be lifted?

c) With what velocity would it rise?

9–45 A solid wood door 1.0 m wide and 2.0 m high is hinged along one side and has a total mass of 50 kg. Initially open and at rest, the door is struck with a hammer at its center; during the blow, an average force of 2000 N acts for 0.01 s. Find the angular velocity of the door after the impact.

9–46 The door of Problem 9–45 is struck at its center by a handful of sticky mud, of mass 0.5 kg, traveling 10 m·s⁻¹ just before impact. Find the final angular velocity of the door. Is the moment of inertia of the mud significant?

9–47 A target in a shooting gallery consists of a vertical square wooden board, 0.2 m on a side, of mass 2.0 kg, pivoted on an axis along its top edge. It is struck at the center by a bullet of mass 5 g, traveling 300 m·s⁻¹.

  a) What is the angular velocity of the board just after the bullet's impact?

  b) What maximum height above the equilibrium position does the center of the board reach before starting to swing down again?

  c) What bullet speed would be required for the board to swing all the way over after impact?

9–48 A large wooden turntable of radius 2.0 m and total mass 120 kg is rotating about a vertical axis through its center, with angular velocity 3.0 rad·s⁻¹. A bag of sand of mass 100 kg is dropped vertically onto it, at a point near the outer edge.

  a) Find the angular velocity of the turntable after the sandbag is dropped.

  b) Compute the kinetic energies before and after the sandbag is dropped. Why are they not equal?

9–49 Figure 9–26 shows part of a "fly-ball" governor, part of a speed controlling device used in old-fashioned steam engines. Each of the steel balls A and B has a mass of 0.5 kg, and they rotate about the vertical axis with an angular velocity of 4 rad·s⁻¹ at a distance of 15 cm from the axis. Collar C is now forced down until the balls are at a distance of 5 cm from the axis. How much work must be done to move the collar down?

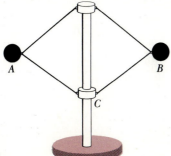

Figure 9–26

9–50 A man sits on an old-fashioned rotating piano stool holding a pair of dumbbells at a distance of 0.6 m from the

axis of rotation of the stool. He is given an angular velocity of 5 rad·s⁻¹, after which he pulls the dumbbells in until they are only 0.2 m distant from the axis. The moment of inertia of the man about the axis of rotation is 5 kg·m² and may be considered constant. Each dumbbell has a mass of 5 kg and may be considered a point mass. Neglect friction.

  a) What is the initial angular momentum of the system?

  b) What is the angular velocity of the system after the dumbbells are pulled in toward the axis?

  c) Compute the kinetic energy of the system before and after the dumbbells are pulled in. Account for the difference, if any.

9–51 A puck on a frictionless "air hockey" table has a mass 0.05 kg and is attached to a cord passing through a hole in the surface, as in Fig. 9–27. The puck is originally revolving at a distance of 0.2 m from the hole, with an angular velocity of 3 rad·s⁻¹. The cord is then pulled from below, shortening the radius of the circle in which the puck revolves to 0.1 m. The puck may be considered a point mass.

  a) What is the new angular velocity?

  b) Find the change in kinetic energy of the puck.

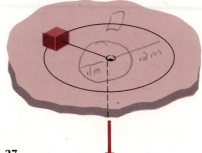

Figure 9–27

9–52 A uniform rod of mass 0.03 kg and 0.2 m long rotates in a horizontal plane about a fixed vertical axis through its center. Two small bodies, each of mass 0.02 kg, are mounted so that they can slide along the rod. They are initially held by catches at positions 0.05 m on each side of the center of the rod, and the system is rotating at 15 rev·min⁻¹. Without otherwise changing the system, the catches are released and the masses slide outward along the rod and fly off at the ends.

  a) What is the angular velocity of the system at the instant when the small masses reach the ends of the rod?

  b) What is the angular velocity of the rod after the small masses leave it?

9–53 A turntable rotates about a fixed vertical axis, making one revolution in 10 s. The moment of inertia of the turntable about this axis is 1200 kg·m². A man of mass 80 kg, initially standing at the center of the turntable, runs out along a radius. What is the angular velocity of the turntable when the man is 2 m from the center?

9–54 Disks *A* and *B* are mounted on a shaft *SS* and may be connected or disconnected by a clutch *C*, as in Fig. 9–28. The moment of inertia of disk *A* is one half that of disk *B*. With the clutch disconnected, *A* is brought up to an angular velocity ω_0. The accelerating torque is then removed from *A* and it is coupled to disk *B* by the clutch. (Bearing friction may be neglected.) It is found that 2000 J of heat are developed in the clutch when the connection is made. What was the original kinetic energy of disk *A*?

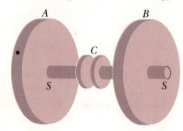

Figure 9–28

9–55 A man of mass 100 kg stands at the rim of a turntable of radius 2 m and moment of inertia 4000 kg·m², mounted on a vertical frictionless shaft at its center. The whole system is initially at rest. The man now walks along the outer edge of the turntable with a velocity of 1 m·s⁻¹, relative to the earth.

a) With what angular velocity and in what direction does the turntable rotate?

b) Through what angle will it have rotated when the man reaches his initial position on the turntable?

c) Through what angle will it have rotated when he reaches his initial position relative to the earth?

9–56 A man of mass 60 kg runs around the edge of a horizontal turntable mounted on a vertical frictionless axis through its center. The velocity of the man, relative to the earth, is 1 m·s⁻¹. The turntable is rotating in the opposite direction with an angular velocity of 0.2 rad·s⁻¹. The radius of the turntable is 2 m and its moment of inertia about the axis of rotation is 400 kg·m². Find the final angular velocity of the system if the man comes to rest, relative to the turntable.

9–57 Two flywheels, *A* and *B*, are mounted on shafts that can be connected or disengaged by a friction clutch *C* (see Fig. 9–28). The moment of inertia of wheel *A* is 8 kg·m². With the clutch disengaged, wheel *A* is brought up to an angular velocity of 600 rev·min⁻¹. Wheel *B* is initially at rest. The clutch is now engaged, until both wheels have the same angular velocity. The final angular velocity of the system is 400 rev·min⁻¹.

a) What was the moment of inertia of wheel *B*?

b) How much mechanical energy was lost in the process? (Neglect all bearing friction.)

9–58 The outstretched arms of a figure skater preparing for a spin can be considered a slender rod pivoting about an axis through its center; when her arms are brought in and wrapped about her body to execute the spin, they can be considered a thin-walled hollow cylinder. If her original angular velocity is 1 rev·s⁻¹, what is her final angular velocity? Her arms have a mass of 8 kg. When outstretched, they span 1.8 m; when wrapped, they form a cylinder of radius 25 cm. The moment of inertia of the remainder of her body is constant and equal to 5 kg·m².

9–59 The mass of the rotor of a toy gyroscope is 150 g and its moment of inertia about its axis is 1500 g·cm². The mass of the frame is 30 g. The gyroscope is supported on a single pivot, as in Fig. 9–29, with its center of gravity distant 4 cm horizontally from the pivot, and is precessing in a horizontal plane at the rate of one revolution in 6 s.

a) Find the upward force exerted by the pivot.

b) Find the angular velocity with which the rotor is spinning about its axis, expressed in rev·min⁻¹.

c) Copy the diagram, and show by vectors the angular momentum of the rotor and the torque acting on it.

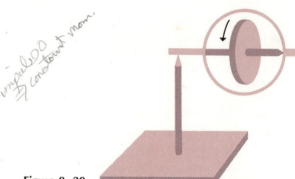

Figure 9–29

9–60 A demonstration gyro wheel is constructed by removing the tire from a bicycle wheel 1.0 m in diameter, wrapping lead wire around the rim, and taping it in place. The shaft projects 0.2 m at each side of the wheel and a man holds the ends of the shaft in his hands. The mass of the system is 5 kg and its entire mass may be assumed to be located at its rim. The shaft is horizontal and the wheel is spinning about the shaft at 5 rev·s⁻¹. Find the magnitude and direction of the force each hand exerts on the shaft under the following conditions:

a) The shaft is at rest.

b) The shaft is rotating in a horizontal plane about its center at 0.04 rev·s⁻¹.

c) The shaft is rotating in a horizontal plane about its center at 0.20 rev·s⁻¹.

d) At what rate must the shaft rotate in order that it may be supported at one end only?

10

ELASTICITY

The preceding two chapters have been concerned with equilibrium and motion of rigid bodies. The rigid body is an idealized *model* used to represent a body with a definite size and shape that does not deform appreciably when forces act on it. An ideal rigid body does not stretch, squeeze, or twist. Real materials always do deform to some extent when forces act on them, and in this chapter we consider the relations of the forces to the resulting deformations. Although these relations are determined ultimately by the forces acting between molecules of the materials, we concentrate in this chapter on quantities that can be measured directly, without attempting detailed understanding of the phenomena on a molecular level.

10–1

STRESS

Figure 10–1a shows a bar of uniform cross-sectional area A subjected to equal and opposite pulls F at its ends. The bar is said to be in *tension*. Consider a section through the bar at right angles to its length, as indicated by the broken line. Since every portion of the bar is in equilibrium, that portion at the right of the section must be pulling on the portion at the left with a force F, and vice versa. If the section is not too near the ends of the bar, these pulls are uniformly distributed over the cross-sectional area A, as indicated by the short arrows in Fig. 10–1b. We define the *stress* S at the section as the ratio of the force F to the area A:

$$\text{Stress} = \frac{F}{A}. \tag{10–1}$$

The stress is called a *tensile* stress, meaning that each portion *pulls* on the other, and it is also a *normal* stress because the distributed force is perpendicular to the area.

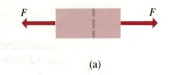

(a)

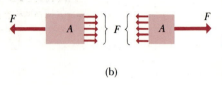

(b)

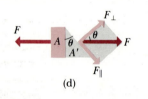

(c)

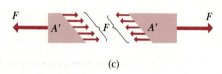

(d)

10–1   (a) A bar in tension. (b) The stress at a perpendicular section equals F/A. (c) and (d) The stress at an inclined section can be resolved into a normal stress $F_\perp/A'$, and a tangential or shear stress $F_\parallel \cdot A'$.

The SI unit of stress is the newton per square meter ($\text{N}\cdot\text{m}^{-2}$). This unit is also given the special name, the *pascal* (abbreviated Pa).

One pascal = 1 Pa = 1 $\text{N}\cdot\text{m}^{-2}$.

Other units of stress are the dyne per square centimeter ($\text{dyn}\cdot\text{cm}^{-2}$) and the pound per square foot ($\text{lb}\cdot\text{ft}^{-2}$). In the British system the hybrid unit, the pound per square inch ($\text{lb}\cdot\text{in}^{-2}$ or psi), is commonly used.

EXAMPLE   A human biceps (upper arm muscle) may exert a force of the order of 600 N on the bones to which it is attached. If the muscle has a cross-sectional area at its center of 50 cm^2 = 0.005 m^2, and the tendon attaching its lower end to the bones below the elbow joint has a cross section of 0.5 cm^2 = 5×10^{-5} m^2, find the tensile stress in each of these.

Solution   In each case the stress is the force per unit area. For the muscle,

$$\text{Tensile stress} = \frac{600 \text{ N}}{0.005 \text{ m}^2} = 120{,}000 \text{ N}\cdot\text{m}^{-2} = 1.2 \times 10^5 \text{ Pa}.$$

For the tendon,

$$\text{Tensile stress} = \frac{600 \text{ N}}{5 \times 10^{-5} \text{ m}^2} = 1.2 \times 10^7 \text{ Pa}.$$

The maximum force a muscle can exert depends on its cross-sectional area, but the maximum *stress* is nearly the same for a wide variety of muscles.   ◄

Returning to Fig. 10–1, we consider next a section through the bar at some arbitrary angle, as in Fig. 10–1c. The resultant force exerted on the portion at either side of this section, by the portion at the other, is equal and opposite to the force F at the end of the section. Now, however, the force is distributed over a larger area A' and is not at right angles to the area. If we represent the resultant of the distributed forces by a single vector of magnitude F, as in Fig. 10–1d, this vector can be resolved into a component $F_\perp$ normal to the area A', and a component $F_\parallel$ tangent to the area. The *normal* stress is defined, as before, as the ratio of the component $F_\perp$ to the area A'. The ratio of the component $F_\parallel$ to the area A' is called the *tangential* stress or, more commonly, the *shear* stress at the section:

$$\text{Normal stress} = \frac{F_\perp}{A'}$$

$$\text{Tangential (shear) stress} = \frac{F_\parallel}{A'}. \tag{10–2}$$

Stress is not a vector quantity since, unlike a force, we cannot assign to it a specific direction. The *force* acting on the portion of the body on a specified side of a section has a definite direction. Stress is one of a class of physical quantities called *tensors*.

A bar subjected to *pushes* at its ends, as in Fig. 10–2, is said to be in *compression*. The stress on the broken section, illustrated in part (b), is also a normal stress but is now a *compressive* stress, since each portion pushes

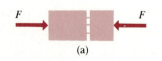

(a)

(b)

10–2   A bar in compression.

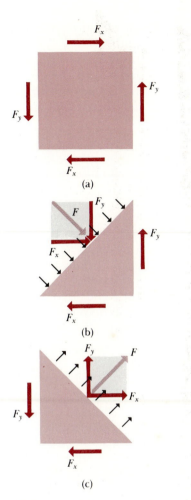

10–3 (a) A body in shear. The stress on one diagonal, part (b), is a pure compression; that on the other, part (c), is a pure tension.

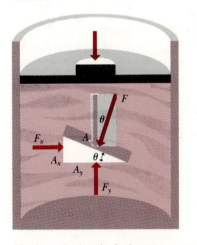

10–4 A fluid under hydrostatic pressure. The force on a surface in any direction is normal to the surface.

on the other. If we take a section in some arbitrary direction, it will be subject to both a tangential (shear) and a normal stress, the latter now being a compression.

As another example of a body under stress, consider the block of square cross section in Fig. 10–3a, acted on by two equal and opposite couples produced by the pairs of forces F_x and F_y distributed over its surfaces. The block is in equilibrium, and any portion of it is in equilibrium also. Thus the distributed forces over the diagonal face in part (b) must have a resultant F whose components are equal to F_x and F_y. The stress at this section is therefore a pure compression, although the stresses at the right face and the bottom are both shear stresses. Similarly, we see from Fig. 10–3c that the other diagonal face is in pure tension.

Consider next a fluid under pressure. The term "fluid" means a substance that can flow; hence the term applies to both liquids and gases. If there is a shear stress at any point in a fluid, the fluid slips sidewise so long as the stress is maintained. Hence in a fluid *at rest*, the shear stress must be zero everywhere. Figure 10–4 represents a fluid in a cylinder provided with a piston, on which is exerted a downward force. The triangle is a side view of a wedge-shaped portion of the fluid. If for the moment we neglect the weight of the fluid, the only forces on this portion are those exerted by the rest of the fluid, and since these forces can have no shear (or tangential) component, they must be normal to the surfaces of the wedge. Let F_x, F_y, and F represent the forces against the three faces. Since the fluid is in equilibrium, it follows that

$$F \sin \theta = F_x, \qquad F \cos \theta = F_y.$$

Also,

$$A \sin \theta = A_x, \qquad A \cos \theta = A_y.$$

Dividing the upper equations by the lower, we find

$$\frac{F}{A} = \frac{F_x}{A_x} = \frac{F_y}{A_y}.$$

Hence the force per unit area is the *same*, regardless of the direction of the section, and is always a compression. Any one of the preceding ratios defines the *hydrostatic pressure p* in the fluid,

$$p = \frac{F}{A}, \qquad F = pA. \tag{10–3}$$

Pressure has the same units as stress; commonly used units include 1 Pa $(= 1 \text{ N} \cdot \text{m}^{-2})$, 1 dyn $\cdot$ cm^{-2}, 1 lb $\cdot$ ft^{-2}, and 1 lb $\cdot$ in^{-2}. Also in common use is the *atmosphere*, abbreviated atm. One atmosphere is defined to be the average pressure of the earth's atmosphere at sea level.

One atmosphere = 1 atm = 1.013×10^5 Pa = 14.7 lb $\cdot$ in^{-2}.

Like other types of stress, pressure is not a vector quantity; no direction can be assigned to it. *The force against any area within (or bounding) a fluid at rest and under pressure is normal to the area, regardless of the orientation of the area.* This is what is meant by the common statement that "the pressure in a fluid is the same in all directions."

The stress within a solid can also be a hydrostatic pressure, provided the stress at all points of the surface of the solid is of this nature. That is, the force per unit area must be the same at *all* points of the surface, and the force must be normal to the surface and directed inward. This is *not* the case in Fig. 10–2, where forces are applied at the ends of the bar only, but it is automatically the case if a solid is immersed in a fluid under pressure.

10–2

STRAIN

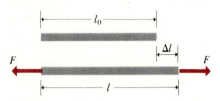

10–5   The longitudinal strain is defined as $\Delta l / l_0$.

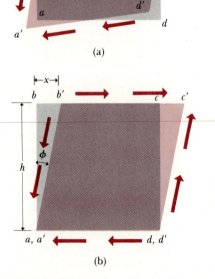

(a)

(b)

10–6   Change in shape of a block in shear. The shear strain is defined as x/h.

The term *strain* refers to the relative change in dimensions or shape of a body that is subjected to stress. Associated with each type of stress described in the preceding section is a corresponding type of strain.

Figure 10–5 shows a bar of natural length l_0 that elongates to a length $l = l_0 + \Delta l$ when equal and opposite pulls F are exerted at its ends. The elongation, Δl, of course, does not occur at the ends only; every element of the bar stretches in the same proportion as does the bar as a whole. The *tensile strain* in the bar is defined as the ratio of the increase in length to the original length:

$$\text{Tensile strain} = \frac{l - l_0}{l_0} = \frac{\Delta l}{l_0}. \qquad (10\text{–}4)$$

The *compressive strain* of a bar in compression is defined in the same way, as the ratio of the decrease in length to the original length.

Figure 10–6a illustrates the nature of the deformation when *shear* stresses act on the faces of a block, as in Fig. 10–3. The dotted outline *abcd* represents the unstressed block, and the solid lines $a'b'c'd'$ represent the block under stress. The centers of the stressed and unstressed block coincide in part (a). In part (b), the deformation is the same as in (a), but the edges *ad* and $a'd'$ coincide. The lengths of the faces under shear remain very nearly constant; all dimensions parallel to the diagonal *ac* increase in length, and those parallel to the diagonal *bd* decrease in length. Note that this is to be expected in view of the nature of the corresponding internal stresses (see Fig. 10–3). This type of strain is called a *shear strain*, and is defined as the ratio of the displacement x of corner b to the transverse dimension h:

$$\text{Shear strain} = \frac{x}{h} = \tan \phi. \qquad (10\text{–}5)$$

In practice x is nearly always much smaller than h, so $\tan \phi$ is very nearly equal to ϕ, and the strain is simply the angle ϕ (measured in radians, of course). Like all strains, shear strain is a pure number with no units because it is a ratio of two lengths.

The strain produced by a hydrostatic pressure, called a *volume strain*, is defined as the ratio of the change in volume, ΔV, to the original volume V. It also is a pure number:

$$\text{Volume strain} = \frac{\Delta V}{V}. \qquad (10\text{–}6)$$

10–3

ELASTICITY AND PLASTICITY

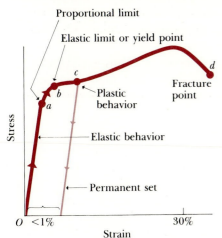

10–7 Typical stress–strain diagram for a ductile metal under tension.

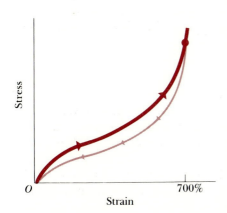

10–8 Typical stress–strain diagram for vulcanized rubber, showing elastic hysteresis.

We may now consider the relation between each of the three kinds of stress and its corresponding strain. When any stress is plotted against the appropriate strain, the resulting stress–strain diagram is found to have one of several different shapes, depending on the kind of material.

A typical stress–strain diagram for a ductile metal is shown in Fig. 10–7. The stress is a simple tensile stress and the strain is the percentage elongation. During the first portion of the curve (up to a strain of less than 1 percent), the stress and strain are proportional until the point *a*, the *proportional limit*, is reached. The fact that there is a region in which stress and strain are proportional is called *Hooke's law.* (Robert Hooke (1635–1703) was a contemporary of Newton.)

From *a* to *b*, stress and strain are not proportional, but nevertheless, if the load is removed at any point between *O* and *b*, the curve will be retraced and the material will return to its original length. In the region *Ob*, the material is said to be *elastic* or to exhibit *elastic behavior*, and the point *b* is called the *elastic limit*, or the *yield point*. Up to this point, the forces exerted by the material are *conservative;* when the material returns to its original shape, work done in producing the deformation is recovered. The deformation is said to be *reversible.*

If the material is loaded further, the strain increases rapidly, but when the load is removed at some point beyond *b*, say *c*, the material does not come back to its original length but traverses the thin line in Fig. 10–7. The length at zero stress is now *greater* than the original length, and the material is said to have a *permanent set.* Further increase of load beyond *c* produces a large increase in strain (even if the stress decreases) until a point *d* is reached at which *fracture* takes place. From *b* to *d*, the metal is said to undergo *plastic flow* or *plastic deformation.* A plastic deformation is *irreversible.* If large plastic deformation takes place between the elastic limit and the fracture point, the metal is said to be *ductile.* If, however, fracture occurs soon after the elastic limit is passed, the metal is said to be *brittle.*

Figure 10–8 shows a stress–strain curve for a typical sample of vulcanized rubber that has been stretched to over seven times its original length. During *no* portion of this curve is the stress proportional to the strain! The substance, however, is elastic, in the sense that when the load is removed the rubber is restored to its original length. On decreasing the load, the stress–strain curve is *not* retraced but follows the thin curve of Fig. 10–8.

The lack of coincidence of the curves for increasing and decreasing stress is known as *elastic hysteresis* (accent on the *third* syllable). (An analogous phenomenon observed with magnetic materials is called *magnetic hysteresis.*) When the stress–strain relation exhibits this behavior, the associated forces are *not* conservative, since the work done by the material in returning to its original shape is *less* than the work required to deform it. It can be shown that the area bounded by the two curves, that is, the area of the *hysteresis loop*, is proportional to the energy dissipated within the elastic or magnetic material.

The large elastic hysteresis of some types of rubber makes these materials very valuable as vibration absorbers. If a block of such material is placed between a piece of vibrating machinery and the floor, elastic hysteresis takes place during every cycle of vibration. Mechanical energy is converted to a form known as internal energy, which evidences itself by a rise in temperature. As a result, only a small amount of energy of vibration is transmitted to the floor.

The stress required to cause actual fracture of a material is called the *breaking stress* or the *ultimate strength*. Two materials, such as two steels, may have very similar elastic constants but vastly different breaking stresses.

10–4

ELASTIC MODULUS

The stress required to produce a given strain depends on the nature of the material under stress. The ratio of stress to strain, or the *stress per unit strain,* is called an *elastic modulus* of the material. The larger the elastic modulus, the greater the stress needed for a given strain.

Consider first longitudinal (tensile or compressive) stresses and strains. Experiment shows that up to the proportional limit, *a given longitudinal stress produces a strain of the same magnitude whether the stress is a tension or a compression.* Hence, the ratio of tensile stress to tensile strain, for a given material, equals the ratio of compressive stress to compressive strain. This ratio is called *Young's modulus* for the material, and will be denoted by Y:

$$Y = \frac{\text{tensile stress}}{\text{tensile strain}} = \frac{\text{compressive stress}}{\text{compressive strain}},$$

$$Y = \frac{F_\perp/A}{\Delta l / l_0} = \frac{l_0}{A}\frac{F_\perp}{\Delta l}. \tag{10–7}$$

If the proportional limit is not exceeded, the ratio of stress to strain is constant, and Hooke's law is therefore equivalent to the statement that *within the proportional limit, the elastic modulus of a given material is constant,* depending only on the nature of the material.

Since a strain is a pure number, the units of Young's modulus are the same as those of stress, namely, force per unit area. Some typical values are listed in Table 10–1.

When a material elongates under tensile stress, the dimensions *perpendicular* to the direction of stress become *shorter* by an amount proportional to the fractional change in length. If w is the original width and Δw the change in width, then it is found that

$$\frac{\Delta w}{w} = -\sigma\frac{\Delta l}{l}, \tag{10–8}$$

where σ is a dimensionless constant characteristic of the material, called *Poisson's ratio.* For many common materials σ has a value between 0.1 and 0.3. Similarly, a material under compressive stress "bulges" at the sides, and again the fractional change in width is given by Eq. (10–8).

TABLE 10–1 APPROXIMATE ELASTIC CONSTANTS

Material	Young's modulus, Y		Shear modulus, S		Bulk modulus, B		Poisson's ratio, σ
	Pa	lb·in^{-2}	Pa	lb·in^{-2}	Pa	lb·in^{-2}	
Aluminum	0.70×10^{11}	10×10^6	0.30×10^{11}	3.4×10^6	0.70×10^{11}	10×10^6	0.16
Brass	0.91×10^{11}	13×10^6	0.36×10^{11}	5.1×10^6	0.61×10^{11}	8.5×10^6	0.26
Copper	1.1×10^{11}	16×10^6	0.42×10^{11}	6.0×10^6	1.4×10^{11}	20×10^6	0.32
Glass	0.55×10^{11}	7.8×10^6	0.23×10^{11}	3.3×10^6	0.37×10^{11}	5.2×10^6	0.19
Iron	1.9×10^{11}	26×10^6	0.70×10^{11}	10×10^6	1.0×10^{11}	14×10^6	0.27
Lead	0.16×10^{11}	2.3×10^6	0.056×10^{11}	0.8×10^6	0.077×10^{11}	1.1×10^6	0.43
Nickel	2.1×10^{11}	30×10^6	0.77×10^{11}	11×10^6	2.6×10^{11}	34×10^6	0.36
Steel	2.0×10^{11}	29×10^6	0.84×10^{11}	12×10^6	1.6×10^{11}	23×10^6	0.19
Tungsten	3.6×10^{11}	51×10^6	1.5×10^{11}	21×10^6	2.0×10^{11}	29×10^6	0.20

The *shear modulus S* of a material, within the Hooke's law region, is defined as the ratio of a shear stress to the shear strain it produces:

$$S = \frac{\text{shear stress}}{\text{shear strain}}$$

$$= \frac{F_{\parallel}/A}{x/h} = \frac{h}{A}\frac{F_{\parallel}}{x} = \frac{F_{\parallel}/A}{\phi}. \tag{10–9}$$

(Refer to Fig. 10–6 for the meaning of x and of h.)

The shear modulus of a material is also expressed as force per unit area. For most materials it is one half to one third as great as Young's modulus. The shear modulus is also called the *modulus of rigidity* or the *torsion modulus*.

The more general definition of shear modulus is

$$S = \frac{\Delta F_{\parallel}/A}{\Delta x/h} = \frac{h}{A}\frac{\Delta F_{\parallel}}{\Delta x} = \frac{\Delta F_{\parallel}/A}{\Delta \phi}, \tag{10–10}$$

where Δx is the increase in x when the shearing force increases by $\Delta F_{\parallel}$.

The shear modulus has significance only for *solid* materials. A liquid or gas flows under the influence of a shear stress and cannot *permanently* support such a stress.

The modulus relating a hydrostatic pressure to the volume strain it produces is called the *bulk modulus*, and we shall represent it by B. The general definition of bulk modulus is the (negative) ratio of a change in pressure Δp to the volume strain $\Delta V/V_0$ (fractional change in volume) produced by it:

$$B = -\frac{\Delta p}{\Delta V/V_0} = -V_0\frac{\Delta p}{\Delta V}. \tag{10–11}$$

The minus sign is included in the definition of B because an *increase* of pressure always causes a *decrease* in volume. That is, if Δp is positive, ΔV is negative. By including a minus sign in its definition, we make the bulk modulus itself a positive quantity.

The change in volume of a *solid* or *liquid* under pressure is so small that the value of B is independent of the initial volume V_0. Provided the pressure is not too great, the ratio $\Delta p/\Delta V$ is constant also, the bulk modulus is constant, and we can replace Δp and ΔV by finite changes in pressure and volume. The volume of a *gas*, however, changes markedly with pressure, and the general definition of B must be used for gases.

The reciprocal of the bulk modulus is called the *compressibility k*. From its definition,

$$k = \frac{1}{B} = -\frac{\Delta V/V_0}{\Delta p} = -\frac{1}{V_0}\frac{\Delta V}{\Delta p}. \qquad (10\text{--}12)$$

The compressibility of a material thus equals the *fractional decrease in volume*, $-\Delta V/V_0$, *per unit increase Δp in pressure.*

The units of a bulk modulus are the same as those of pressure, and the units of compressibility are those of a *reciprocal pressure*. Thus the statement that the compressibility of water (see Table 10–2) is 46.4×10^{-6} atm^{-1} means that the volume decreases by 46.4 one-millionths of the original volume for each atmosphere increase in pressure.

TABLE 10–2   COMPRESSIBILITIES OF LIQUIDS

Liquid	Compressibility, k		
	Pa^{-1}	(lb·in^{-2})$^{-1}$	atm^{-1}
Carbon disulfide	93×10^{-11}	64×10^{-7}	94×10^{-6}
Ethyl alcohol	110×10^{-11}	76×10^{-7}	111×10^{-6}
Glycerine	21×10^{-11}	14×10^{-7}	21×10^{-6}
Mercury	3.7×10^{-11}	2.6×10^{-7}	3.8×10^{-6}
Water	45.8×10^{-11}	31.6×10^{-7}	46.4×10^{-6}

The various types of stress, strain, and elastic moduli are summarized in Table 10–3.

TABLE 10–3   STRESSES AND STRAINS

Type of stress	Stress	Strain	Elastic modulus	Name of modulus
Tension or compression	$\dfrac{F_\perp}{A}$	$\dfrac{\Delta l}{l_0}$	$Y = \dfrac{F_\perp/A}{\Delta l/l_0}$	Young's modulus
Shear	$\dfrac{F_\parallel}{A}$	$\tan\phi \approx \phi$	$S = \dfrac{F_\parallel/A}{\phi}$	Shear modulus
Hydrostatic pressure	$p\left(=\dfrac{F_\perp}{A}\right)$	$\dfrac{\Delta V}{V_0}$	$B = -\dfrac{p}{\Delta V/V_0}$	Bulk modulus

EXAMPLE 1 In an experiment to measure Young's modulus, a load of 500 kg, hanging from a steel wire of length 3 m and cross section 0.20 cm^2, was found to stretch the wire 0.4 cm above its no-load length.

What were the stress, the strain, and the value of Young's modulus for the steel of which the wire was composed?

Solution

$$\text{Stress} = \frac{F_\perp}{A} = \frac{(500 \text{ kg})(9.8 \text{ m} \cdot \text{s}^{-2})}{2.0 \times 10^{-5} \text{ m}^2}$$

$$= 2.45 \times 10^8 \text{ Pa};$$

$$\text{Strain} = \frac{\Delta l}{l_0} = \frac{0.004 \text{ m}}{3 \text{ m}} = 0.00133;$$

$$Y = \frac{\text{stress}}{\text{strain}} = \frac{2.45 \times 10^8 \text{ Pa}}{0.00133} = 1.84 \times 10^{11} \text{ Pa}. \qquad \blacktriangleleft$$

EXAMPLE 2 Suppose the object in Fig. 10–6 is a brass plate 1.0 m square and 0.5 cm thick. How large a force F must be exerted on each of its edges if the displacement x in Fig. 10–6b is 0.02 cm? The shear modulus of brass is 0.36×10^{11} Pa.

Solution The shear stress of each edge is

$$\text{Shear stress} = \frac{F_\parallel}{A} = \frac{F}{(1.0 \text{ m})(0.005 \text{ m})} = (200 \text{ m}^{-2})F.$$

The shear strain is

$$\text{Shear strain} = \frac{x}{h} = \frac{2 \times 10^{-4} \text{ m}}{1.0 \text{ m}} = 2.0 \times 10^{-4}.$$

$$\text{Shear modulus } S = \frac{\text{stress}}{\text{strain}} = 0.36 \times 10^{11} \text{ Pa} = \frac{(200 \text{ m}^{-2})F}{2.0 \times 10^{-4}},$$

$$F = 3.6 \times 10^4 \text{ N}. \qquad \blacktriangleleft$$

EXAMPLE 3 The volume of oil contained in a certain hydraulic press is $0.2 \text{ m}^3 = 200$ liters. Find the decrease in volume of the oil when subjected to a pressure increase of 2.04×10^7 Pa. The compressibility of the oil is $20 \times 10^{-6} \text{ atm}^{-1}$.

Solution We first convert the pressure to atmospheres:

$$2.04 \times 10^7 \text{ Pa} = 201 \text{ atm}.$$

From Eq. (10–12),

$$\Delta V = -kV \,\Delta p = -(20 \times 10^{-6} \text{ atm}^{-1})(0.2 \text{ m}^3)(201 \text{ atm})$$

$$= -8.04 \times 10^{-4} \text{ m}^3 = -0.804 \text{ liters}.$$

This represents a substantial compression of the oil—under the action of a very large pressure, nearly 3000 pounds per square inch. $\qquad \blacktriangleleft$

Although the three elastic moduli and Poisson's ratio have been discussed separately, they are not completely independent. For materials having no distinction between various directions (i.e., *isotropic* materials), only two of these are really independent. For example, the bulk and shear moduli may be expressed in terms of Young's modulus and Poisson's ratio:

$$B = \frac{Y}{3(1 - 2\sigma)}, \qquad S = \frac{Y}{2(1 + \sigma)}. \qquad (10\text{--}13)$$

For materials having directional properties, such as wood (which has a grain direction) and single crystals of materials, these relations do not hold, and the elastic behavior is more complex.

10–5

THE FORCE CONSTANT

The elastic moduli and Poisson's ratio are quantities that characterize the elastic properties of a material in a way that is independent of the size or shape of the particular specimen. They do not indicate directly how a particular rod, cable, or spring made of the material will distort under given forces.

In the particular case of tensile and compressive stresses and strains, we may solve Eq. (10–7) for $F_\perp$, obtaining

$$F_\perp = \frac{YA}{l_0}\Delta l;$$

Then if the quantity YA/l_0 is represented by a single letter k, and the elongation Δl is renamed x, we have

$$F_\perp = kx. \qquad (10\text{--}14)$$

We have already encountered this relationship in Sections 6–3 and 6–6 in connection with work done by a spring and the corresponding elastic potential energy. The elongation of a body in tension beyond its natural length is directly proportional to the stretching force. Or if the stress is compressive, the shortening of the body in compression is directly proportional to the compressing force. Hooke's law was first stated in this form, and later reformulated in terms of stress and strain.

When a helical spring (often called a *coil spring*) is stretched or compressed, the stress and strain in the wire are nearly pure *shear*, but the elongation or compression is still proportional to the stretching or compressing force, within certain limits of maximum force. That is, an equation of the form $F = kx$ is still valid; in this case the force constant k depends on the shear modulus of the material, its radius, the radius of the coils, and the number of coils. Of course, a coil spring can be compressed only if there are spaces between the coils!

In all these cases, the constant k, representing the ratio of force to elongation or compression, is called the *force constant* or *spring constant* of the spring. Its units are newtons per meter, dynes per centimeter, or pounds per foot. The reciprocal of the force constant, that is, the ratio of elongation or compression to force, is called the *compliance* of the spring.

QUESTIONS

10–1 When a wire is bent back and forth, it becomes hot. Why?

10–2 When a wire is stretched, the cross section decreases somewhat from the undeformed value. How does this affect the definition of tensile stress? Should the original or the decreased value be used?

10–3 Is the work required to stretch a metal rod proportional to the amount of stretch? Explain.

10–4 Why is concrete with steel reinforcing rods embedded in it stronger than plain concrete?

10–5 Is the shear modulus of steel greater or less than that of jelly? By roughly what factor?

10–6 Is the bulk modulus for steel greater or less than that of air? By roughly what factor?

10–7 Looking at the molecular structure of matter, discuss why gases are generally more compressible than liquids and solids.

10–8 Climbing ropes used by mountaineers are usually made of nylon. Would a steel cable of equal strength be just as good? What advantages and disadvantages would it have, compared to nylon?

10–9 How could you measure the force constants of the springs in an automobile?

10–10 In a nylon mountaineering rope, is a lot of mechanical hysteresis desirable or undesirable?

10–11 A spring scale for measuring weight has a spring and a scale that indicates how much the spring stretches under a given weight. Such scales are often illegal in commerce. (Cf. the inscription "no springs, honest weight" found on some grocery-store scales.) Why? (What property of a metal spring could affect its accuracy?)

10–12 Coil springs found in automobile suspension systems are sometimes designed *not* to obey Hooke's law. How can a spring be made so as to achieve this result? Why is it desirable?

10–13 Compare the mechanical properties of a steel cable, made by twisting many thin wires together, with those of a solid steel wire of the same diameter. What advantages does each have?

10–14 Electric power lines are sometimes made using wires with steel core and copper jacket, or strands of copper and steel twisted together. Why?

10–15 A spring is compressed, clamped in its compressed position, and then dissolved in acid. What becomes of the elastic potential energy?

10–16 When rubber mounting blocks are used to absorb machine vibrations through mechanical hysteresis, as discussed in Section 10–3, what becomes of the energy associated with the vibrations?

PROBLEMS

10–1 A steel rod 4 m long and 0.5 cm^2 in cross-sectional area is found to stretch 0.2 cm under a tension of 12,000 N. Under these conditions, determine

a) the stress;

b) the strain; and

c) Young's modulus for the material.

10–2 The elastic limit of a steel elevator cable is 40,000 lb·in^{-2}. Find the maximum upward acceleration that can be given a 2-ton elevator when supported by a cable whose cross section is $\frac{1}{2}$ in^2 if the stress is not to exceed $\frac{1}{4}$ of the elastic limit.

10–3 A nylon rope used by mountaineers elongates 1.5 m under the weight of an 80-kg climber. If the rope is 50 m in length and 9 mm in diameter, what is Young's modulus for this material? If Poisson's ratio for nylon is 0.2, find the change in diameter under this stress.

10–4 A relaxed biceps muscle requires a force of 25 N for an elongation of 5 cm, and the same muscle under maximum tension requires a force of 500 N for the same elongation. Regarding the muscle as a uniform cylinder of length 0.2 m and cross-sectional area 50 cm^2, find Young's modulus for the muscle tissue under each of these conditions.

10–5 A copper wire 4 m long and 1.0 mm in diameter was given the test below. A load of 20 N was originally hung from the wire to keep it taut. The position of the lower end of the wire was read on a scale.

Added load, N	Scale reading, cm
0	3.02
10	3.07
20	3.12
30	3.17
40	3.22
50	3.27
60	3.32
70	4.27

a) Make a graph of these values, plotting the increase in length horizontally and the added load vertically.

b) Calculate the value of Young's modulus.

c) What was the stress at the proportional limit?

10–6 A steel wire has the following properties:

Length = 5 m
Cross section = 0.05 cm^2
Young's modulus = 1.8×10^{11} Pa
Shear modulus = 0.6×10^{11} Pa
Proportional limit = 3.6×10^8 Pa
Breaking stress = 7.2×10^8 Pa

The wire is fastened at its upper end and hangs vertically.

a) How great a load can be supported without exceeding the proportional limit?

b) How much will the wire stretch under this load?

c) What is the maximum load that can be supported?

10–7

a) What is the maximum load that can be supported by an aluminum wire 0.1 cm in diameter without exceeding the proportional limit of 8×10^7 Pa?

b) If the wire was originally 5 m long, how much will it elongate under this load?

c) How much does the diameter change under this load?

10–8 A 5-kg mass hangs on a vertical steel wire 0.5 m long and 0.004 cm^2 in cross section. Hanging from the bottom of this mass is a similar steel wire that supports a 10-kg mass. For each wire, compute

a) the longitudinal strain, and

b) the elongation.

10–9 A copper rod of length 2 m and cross-sectional area 2.0 cm^2 is fastened end to end to a steel rod of length L and cross-sectional area 1.0 cm^2. The compound rod is subjected to equal and opposite pulls of magnitude 3×10^4 N at its ends.

a) Find the length L of the steel rod if the elongations of the two rods are equal.

b) What is the stress in each rod?

c) What is the strain in each rod?

10–10 A copper wire 8 m long and a steel wire 4 m long, each of cross section 0.5 cm^2, are fastened end to end and stretched with a tension of 500 N.

a) What is the change in length of each wire?

b) What is the elastic potential energy of the system?

10–11 A steel bar 0.2 cm square and 5 m long is stretched with a force of 400 N at each end. Find the stress, the strain, the total elongation, and the fractional change in thickness of the bar.

10–12 Two round rods, one of steel, the other of brass, are joined end to end. Each rod is 0.5 m long and 2 cm in diameter. The combination is subjected to tensile forces of 5000 N.

a) What is the strain in each rod?

b) What is the elongation of each rod?

c) What is the change in diameter of each rod?

10–13 In the Challenger Deep of the Marianas Trench, the depth of sea water is 10.9 km, and the pressure is 1.10×10^8 Pa (about 1.09×10^3 atm).

a) If a cubic meter of water is taken from the surface to this depth, what is the change in its volume?

b) What is the density of sea water at this depth? (At the surface, sea water a density of 1.03×10^3 kg·m^{-3}.)

10–14 A moonshiner produces pure ethanol, late at night, and stores it in a stainless steel tank in the form of a cylinder 20 cm in diameter with a tight-fitting piston at the top. The total volume of the tank is 200 l (= 0.2 m^3). In an attempt to squeeze a little more into the tank, he piles lead bricks on the piston, so the total mass of bricks and piston is 120 kg. What additional volume of ethanol can he squeeze into the tank?

10–15 A specimen of oil having an initial volume of 1000 cm^3 is subjected to a pressure increase of 12×10^5 Pa, and the volume is found to decrease by 0.3 cm^3. What is the bulk modulus for the material? The compressibility?

10–16 The compressibility of sodium is to be measured by observing the displacement of the piston in Fig. 10–4 when a force is applied. The sodium is immersed in an oil that fills the cylinder below the piston. Assume that the piston and walls of the cylinder are perfectly rigid, that there is no friction, and no oil leak. Compute the compressibility of the sodium in terms of the applied force F, the piston displacement x, the piston area A, the initial volume of the oil V_0, the initial volume of the sodium v_0, and the compressibility of the oil k_0.

10–17 Two strips of metal are riveted together at their ends by four rivets, each of diameter 0.5 cm. What is the maximum tension that can be exerted by the riveted strip if the shearing stress on the rivets is not to exceed 6×10^8 Pa? Assume each rivet to carry one-quarter of the load.

10–18 Find the mass per cubic meter of ocean water at a depth of 500 m where the pressure is about 5.0×10^6 Pa. The density at the surface is 1.03×10^3 kg·m^{-3}.

10–19 Compute the compressibility of steel, in reciprocal atmospheres, and compare with that of water. Which material is the more readily compressed?

10–20 A steel post 15 cm in diameter and 3 m long is placed vertically and is required to support a load of 10,000 kg.

a) What is the stress in the post?

b) What is the strain in the post?

c) What is the change in length of the post?

10–21 A hollow cylindrical steel column 5 m high shortens 0.03 cm under a compression load of 40,000 kg. If the inner radius of the cylinder is 0.80 of the outer one, what is the outer radius?

10–22 A bar of cross section A is subjected to equal and opposite tensile forces F at its ends. Consider a plane through the bar making an angle θ with a plane at right angles to the bar (Fig. 10–9).

a) What is the tensile (normal) stress at this plane, in terms of F, A, and θ?

b) What is the shear (tangential) stress at the plane, in terms of F, A, and θ?

c) For what value of θ is the tensile stress a maximum?

d) For what value of θ is the shear stress a maximum?

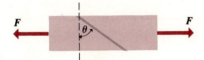

Figure 10–9

10–23 Suppose the block in Fig. 10–3 is rectangular instead of square but is in equilibrium under the action of shear stresses on those outside faces perpendicular to the plane of the diagram. (Note that then $F_x \neq F_y$.)

a) Show that the shear stress is the same on all outside faces perpendicular to the plane of the diagram.

b) Show that on all sections perpendicular to the plane of the diagram and making an angle of 45° with an end face, the stress is still a pure tension or compression.

10–24 In Fig. 10–6, suppose the body is a square steel plate, 10 cm on a side and 1 cm thick. Find the magnitude of force required on each of the four sides to cause a shear strain of 0.01.

10–25 In Fig. 10–6, suppose the body is a square aluminum plate 0.2 m on a side. It is desired that when forces are applied to the four edges, of equal magnitude 1.0×10^6 N each, the resulting shear strain is no greater than 0.01. What minimum thickness of plate is required?

11

PERIODIC MOTION

In previous chapters we have studied in detail the motion of bodies under the action of constant forces. In the present chapter we consider a class of problems in which the force and the resulting acceleration are *not* constant. Instead, in these problems a body has an *equilibrium* position; when displaced from this position, it experiences a force directed back toward the equilibrium position. But when the body returns to the equilibrium position, it is moving; so it overshoots, and the result is a back-and-forth motion past the equilibrium position. A familiar example is the pendulum, a mass suspended from a string. In the equilibrium position, the mass hangs straight down; when displaced from this position the mass does not simply return to the equilibrium position, but instead swings back and forth in a regular, repetitive manner.

Such a repetitive motion is said to be *oscillatory* or *periodic*. Many other examples might be cited: the balance wheel of a watch, the pistons in a gasoline engine, the strings in a musical instrument. The molecules of a solid body vibrate with oscillatory motion about their equilibrium positions in the crystal lattice, although of course this motion cannot be observed directly. The beating of the human heart is a periodic motion; Galileo is alleged to have used his own heartbeat for timing observations of motion. In many kinds of *wave motion*, the particles of the material in which the wave is traveling oscillate with period motion. An electric circuit in which there is an alternating current is described in terms of voltages, currents, and electric charges that oscillate with time. Thus a study of periodic motion will lay the foundation for future work in many different fields of physics.

11–1

BASIC CONCEPTS

In Chapter 10 we studied the ways an elastic body deforms when forces are applied. When the forces correspond to tensile or compressive stresses, the elongation or compression is directly proportional to the magnitude of the applied forces, provided the proportional limit is not

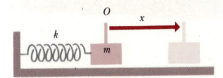

11–1 Model for periodic motion. If the spring obeys Hooke's law, the motion is simple harmonic motion.

exceeded. The proportionality is represented by the force constant k for each specific specimen of material, or spring, or similar device. To illustrate basic concepts, let us consider the system shown in Fig. 11–1. A body of mass m can move without friction along a straight line. It might be, for example, a glider on a linear track, as described in Section 2–7. The body is attached to one end of a spring, the other end of which is held stationary. We describe the position of the body with the coordinate x, taking the origin ($x = 0$) as the equilibrium position, where the spring is neither stretched nor compressed. When the body is displaced to the right, x is positive, the string is stretched, and it exerts a force on the body toward the left (the negative x-direction), toward the equilibrium position. When it is displaced to the left, x is negative, the spring is compressed, and it exerts a force on the body toward the right (the positive x-direction), again toward equilibrium. Thus the sign of the x-component of force *on the body* is always opposite to that of x itself. If, in addition, the spring obeys Hooke's law, then the force on the body is given by

$$F = -kx, \qquad (11-1)$$

where k is the *force constant* for the spring, discussed in Section 10–5. Equation (11–1) is valid for both positive and negative x.

Now suppose the body is displaced to the right a distance A and released. The spring exerts a restoring force; the body accelerates in the direction of this force, and moves toward the equilibrium position with increasing speed. The *rate* of increase (i.e., the acceleration) is not constant, since the accelerating *force* becomes smaller as the body approaches the equilibrium position.

When the body reaches the center, the restoring force has decreased to zero; but because of the velocity that has been acquired, the body "overshoots" the equilibrium position and continues to move toward the left. As soon as the equilibrium position is passed, the restoring force again comes into play, directed now toward the right. The body's speed thus decreases, at a rate that increases with increasing distance from O. It therefore comes to rest at some point to the left of O, and repeats its motion in the opposite direction.

As will be shown in the next section, the motion is confined to a range $\pm A$ on either side of the equilibrium position, and each back-and-forth movement takes place in the same interval of time. If there were no loss of energy by friction, the motion would continue indefinitely. This motion, under the influence of an elastic restoring force proportional to displacement and in the absence of all friction, is called *simple harmonic motion*, abbreviated SHM.

A *complete vibration* or *complete cycle* means one round trip, say from A to $-A$ and back to A, or from O to A to O to $-A$ and back to O.

The *periodic time*, or simply the *period* of the motion, represented by τ, is the time required for one complete vibration.

The *frequency, f,* is the number of complete vibrations per unit time. Evidently the frequency is the reciprocal of the period, or $\tau = 1/f$. The SI unit, one cycle per second, is called one *hertz* ($1 \text{ Hz} = 1 \text{ s}^{-1}$).

The *amplitude, A,* is the maximum displacement from equilibrium, that is, the maximum value of $|x|$. The total range of the motion is therefore $2A$.

11–2

ENERGY IN SIMPLE HARMONIC MOTION

The motion of a point mass m under a restoring force given by $-kx$, where k is the force constant and x the displacement from equilibrium, is an idealized model constituting the simplest possible example of periodic motion. In more complex examples the force may depend on displacement in a more complicated way, but when the force is *directly proportional* to displacement, the resulting motion is called *simple harmonic motion*. Since many more complex motions are *approximately* simple harmonic and may therefore be described approximately by this model, it is useful to analyze simple harmonic motion in some detail. We shall obtain expressions for the coordinate, velocity, and acceleration of a body moving with simple harmonic motion, just as we found those for a body moving with constant acceleration. It must be emphasized that the equations of motion with *constant* acceleration cannot be applied, since the acceleration is continuously changing.

Figure 11–2 represents the vibrating body of Fig. 11–1 at some instant when its displacement from the equilibrium position O is described by the coordinate x. The particle's mass is m, and the resultant force acting on it is simply the elastic restoring force $-kx$. From Newton's second law,

$$F = -kx = ma,$$

or

$$a = -\frac{k}{m}x. \tag{11–2}$$

Thus, an alternative description of the essential feature of simple harmonic motion is that *the acceleration at each instant is proportional to the negative of the displacement at that instant*. When x has its maximum positive value A, the acceleration has its maximum negative value $-kA/m$; and at the instant the particle passes the equilibrium position ($x = 0$), the acceleration is zero. Its velocity is, of course, *not* zero at this point.

The energy principle provides a convenient basis for analysis of some aspects of simple harmonic motion. The elastic restoring force is a *conservative* force, and the work done by this force can be represented in terms of a potential energy $U = \frac{1}{2}kx^2$, just as in Section 6–6. The kinetic energy is $K = \frac{1}{2}mv^2$ and, according to the principle of conservation of energy, the total energy $E = K + U$ is constant. That is,

$$E = \tfrac{1}{2}mv^2 + \tfrac{1}{2}kx^2 = \text{constant}. \tag{11–3}$$

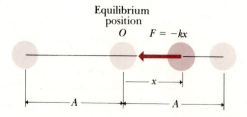

Equilibrium
position
O $F = -kx$

11–2   ⊢——— A ———⊦——— A ———⊣

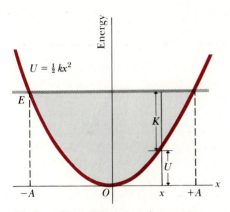

11–3 Relation between total energy E, potential energy U, and kinetic energy K, for a body oscillating with SHM.

The total energy E is also closely related to the amplitude A of the motion. When the particle reaches its maximum displacement $\pm A$, it stops and turns back toward equilibrium. At this instant, $v = 0$, so there is no kinetic energy; the (constant) total energy is thus equal to the potential energy at this point, which is $\frac{1}{2}kA^2 = E$. Thus,

$$\tfrac{1}{2}mv^2 + \tfrac{1}{2}kx^2 = \tfrac{1}{2}kA^2,$$

or

$$v = \pm\sqrt{\frac{k}{m}}\sqrt{A^2 - x^2}. \qquad (11\text{--}4)$$

This relation permits us to obtain the velocity (apart from a sign) for any given position, and Eq. (11–4) is analogous to the relation $v^2 - v_0^2 = 2a(x - x_0)$ for straight-line motion with constant acceleration.

The significance of Eq. (11–3) is shown by the graph in Fig. 11–3, in which energy is plotted vertically and the coordinate x horizontally. The curve represents the potential energy, $U = \frac{1}{2}kx^2$; this curve is a parabola. A horizontal line is drawn at a height equal to the total energy E. We see that the motion is restricted to values of x lying between the points at which the horizontal line intersects the parabola since, if x were outside this range, the potential energy would exceed the total energy, and that is impossible. The motion of the vibrating body is analogous to that of a particle released at a height E on a frictionless track shaped like the potential energy curve, and the motion is said to take place in a "potential energy well."

If a vertical line is constructed at any value of x within the permitted range, the length of the segment between the x-axis and the parabola represents the potential energy U at that value of x, and the length of the segment between the parabola and the horizontal line at height E represents the corresponding kinetic energy K. At the endpoints, therefore, the energy is all potential and at the midpoint it is all kinetic. The speed has its maximum value v_{max} at the midpoint:

$$\tfrac{1}{2}mv_{max}^2 = E, \qquad v_{max} = \sqrt{2E/m}. \qquad (11\text{--}5)$$

The fact that both the maximum potential energy and maximum kinetic energy are equal to the total energy (and thus to each other) may be used to relate the maximum velocity to the amplitude:

$$\tfrac{1}{2}kA^2 = \tfrac{1}{2}mv_{max}^2, \qquad v_{max} = \sqrt{\frac{k}{m}}A. \qquad (11\text{--}6)$$

11–3

EQUATIONS OF SIMPLE HARMONIC MOTION

The position–velocity relation given by Eq. (11–4) is very useful, but it does not tell us where the particle is at any given *time*. To have a complete description of the motion we need to know the position, velocity, and acceleration for any given time. The appropriate relations may be developed with the aid of the *circle of reference*, which we now introduce. In Fig. 11–4, point Q moves counterclockwise around a circle of radius A with

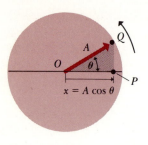

11–4 Coordinate of a body in simple harmonic motion.

constant angular velocity ω (measured in rad·s^{-1}). The vector from O to Q represents the position of point Q relative to O; θ is the angle this vector makes with the positive x-axis. This vector, whose horizontal component represents the actual motion, is called a *phasor;* this representation is also useful in many other areas of physics, including ac circuit analysis and optics. The radius of the circle of reference is always equal to the amplitude A of the motion.

Let P be a point on the horizontal diameter of the circle, directly below Q. Point P is called the projection of Q onto the diameter. Point Q is referred to as the *reference point,* and the circle in which it moves as the *reference circle.* As the reference point revolves, the point P moves back and forth along a horizontal line, keeping always directly below (or above) Q. We shall show that the motion of P is the same as that of a body moving under the influence of an elastic restoring force in the absence of friction, i.e., that it is *simple harmonic motion.*

The displacement of P at any time t is the distance OP or x; from the figure,

$$x = A \cos \theta.$$

If point Q is at the extreme right end of the diameter at time $t = 0$, then $\theta = 0$ at $t = 0$, and the angle θ at any time t is given by

$$\theta = \omega t.$$

Hence,

$$x = A \cos \omega t. \tag{11–7}$$

Now ω, the angular velocity of Q in radians per second, is related to f, the number of complete revolutions of Q per second, by

$$\omega = 2\pi f,$$

since there are 2π radians in one complete revolution. Furthermore, the point P makes one complete vibration for each revolution of Q. Hence, f is also the number of vibrations per second or the *frequency* of vibration of point P. Thus, Eq. (11–7) may also be written

$$x = A \cos 2\pi f t. \tag{11–8}$$

Equation (11–8) gives the displacement of point P at any time t after the start of the motion; note carefully that x represents the *displacement from the center of the path,* not the distance from the starting point.

The instantaneous velocity of P may be found with the aid of Fig. 11–5. The reference point Q moves with a tangential velocity given by Eq. (9–9):

$$v_\| = \omega A = 2\pi f A.$$

Since point P is always directly below or above the reference point, the velocity of P at each instant must equal the x-component of the velocity of Q. That is, from Fig. 11–5,

$$v = -v_\| \sin \theta = -\omega A \sin \theta;$$

$$v = -\omega A \sin \omega t, \tag{11–9}$$

or

$$v = -2\pi f A \sin 2\pi f t.$$

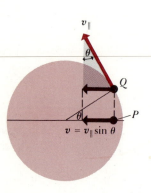

11–5 Velocity in simple harmonic motion.

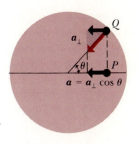

11–6 Acceleration in simple harmonic motion.

The minus sign is introduced because the direction of the velocity is toward the left. When Q is below the horizontal diameter, the velocity of P is toward the right, but since $\sin \theta$ is negative at such points, the minus sign is still needed. Equation (11–9) gives the velocity of point P at any time.

Finally, the acceleration of point P may be found by making use again of the fact that, since P is always directly below or above Q, its acceleration must equal the x-component of the acceleration of Q. Point Q, since it moves in a circular path with a constant angular velocity ω, has at each instant an acceleration toward the center given by Eq. (5–20) or (9–11):

$$a_\perp = \frac{v_\parallel^2}{A} = \omega^2 A = 4\pi^2 f^2 A.$$

From Fig. 11–6, the x-component of this acceleration is

$$a = -a_\perp \cos \theta;$$
$$a = -\omega^2 A \cos \omega t, \tag{11–10}$$

or

$$a = -4\pi^2 f^2 A \cos \omega t.$$

The minus sign is introduced because, in the position shown, the acceleration is toward the left. When Q is to the left of the center, the acceleration of P is toward the right but, since $\cos \theta$ is negative at such points, the minus sign is still required. Equation (11–10) gives the acceleration of P at any time.

Now comes the crucial step in showing that the motion of P is simple harmonic. We combine Eqs. (11–7) and (11–10), obtaining

$$a = -\omega^2 x. \tag{11–11}$$

Since ω is constant, the acceleration a at each instant equals a negative constant times the displacement x at that instant. But this is just the essential feature of simple harmonic motion, as given by Eq. (11–2). Hence the motion of P is indeed simple harmonic!

In order to make Eqs. (11–2) and (11–11) agree precisely, we must choose an angular velocity ω for the reference point Q such that

$$\omega^2 = \frac{k}{m}.$$

When that is done, the frequency of motion of Q, and thus of the actual particle, is given by

$$f = \frac{\omega}{2\pi} = \frac{1}{2\pi} \sqrt{\frac{k}{m}}. \tag{11–12}$$

This relation may be used to find the vibration frequency of a body of given mass when it is vibrating under the influence of an elastic restoring force of given force constant.

As Eqs. (11–7), (11–9), and (11–10) show, use of ω instead of f to describe the frequency of the motion simplifies these expressions by avoiding factors of 2π, which usually accompany f in such expressions.

Accordingly, it is common practice, for all the physical phenomena mentioned in Section 11–1, to use ω rather than f. In this context, ω is usually called the *angular frequency;* it is always 2π times the number of cycles per second. By convention, the hertz is *not* used as a unit of angular frequency, which is usually given the unit $\mathrm{rad \cdot s^{-1}}$ or simply $\mathrm{s^{-1}}$.

Since the period τ is the reciprocal of the frequency, Eq. (11–12) may also be written

$$\tau = 2\pi\sqrt{\frac{m}{k}}. \tag{11-13}$$

In SI units, m is expressed in kilograms and k in newtons per meter. The frequency f is then in vibrations per second, or hertz, and the period τ in seconds per vibration.

The general form of Eqs. (11–12) and (11–13) is as expected. When m is large, we expect the motion to be slow and ponderous, corresponding to small f and large τ. A large value of k means a very stiff spring, corresponding to large f and small τ. The most noteworthy feature of these equations, however, is that they *do not* contain the amplitude A of the motion; this may be surprising. If a given spring-mass system is given an initial displacement and released, and its frequency measured, and then it is stopped, given a *different* displacement, and released, the two frequencies are the same! To be sure, the maximum displacement, maximum speed, and maximum acceleration are all different in the two cases, but *not* the frequency. Thus an additional important characteristic of simple harmonic motion is that *the frequency is not dependent on the amplitude of the motion.*

It is helpful, in visualizing simple harmonic motion, to represent the coordinate, velocity, and acceleration of the vibrating body graphically. Graphs of these quantities against time are given in Fig. 11–7, which shows graphs of Eqs. (11–7), (11–9), and (11–10). Note that the velocity is maximum at times when the coordinate is zero, that is, at the center, while the velocity is zero when the coordinate is a maximum. The acceleration, on the other hand, is zero at the center and maximum at the ends of the path.

Throughout the above discussion we have assumed that the initial position of the particle (at time $t = 0$) is its maximum positive displacement A, but this is not an essential restriction. Different initial positions correspond to different initial positions of the reference point Q. For example, if at time $t = 0$ the phasor OQ makes an angle θ_0 with the positive x-axis, then the angle θ at time t is given not by $\theta = \omega t$ as before, but by

$$\theta = \theta_0 + \omega t. \tag{11-14}$$

The only change in the above discussion is to replace (ωt) in Eqs. (11–7), (11–9), and (11–10) by $(\omega t + \theta_0)$. These equations then become

$$x = A\cos(\omega t + \theta_0),$$
$$v = -\omega A\sin(\omega t + \theta_0),$$
$$a = -\omega^2 A\cos(\omega t + \theta_0) = -\omega^2 x. \tag{11-15}$$

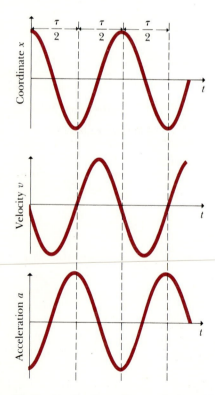

11–7 Graphs of coordinate, velocity, and acceleration of a body moving with simple harmonic motion.

The initial position x_0 and initial velocity v_0 (at time $t = 0$) are then given by

$$x_0 = A \cos \theta_0, \qquad v_0 = -\omega A \sin \theta_0.$$

The amplitude is then no longer equal to the initial displacement. This is reasonable: If at time $t = 0$ the particle has an initial displacement in the positive direction and also a positive velocity in that direction, then it will move *farther* in that direction before returning; hence A must be greater than x_0. The frequency and period relations, Eqs. (11–12) and (11–13), are unchanged.

EXAMPLE A spring is mounted as in Fig. 11–1. By attaching a spring balance to the free end and pulling sideways, we determine that the force is proportional to the displacement, a force of 4 N causing a displacement of 0.02 m. We attach a 2-kg body to the end, pull it aside a distance of 0.04 m, and release it.

a) Find the force constant of the spring.

$$k = \left| \frac{F}{x} \right| = \frac{4 \text{ N}}{0.02 \text{ m}} = 200 \text{ N} \cdot \text{m}^{-1}.$$

b) Find the period and frequency of vibration.

$$\tau = 2\pi \sqrt{\frac{m}{k}} = 2\pi \sqrt{\frac{2 \text{ kg}}{200 \text{ N} \cdot \text{m}^{-1}}} = \frac{\pi}{5} \text{s} = 0.628 \text{ s};$$

$$f = \frac{1}{\tau} = \frac{5}{\pi} \text{s}^{-1} = 1.59 \text{ s}^{-1} = 1.59 \text{ Hz};$$

$$\omega = 2\pi f = \sqrt{\frac{k}{m}} = 10 \text{ s}^{-1}.$$

c) Compute the maximum velocity attained by the vibrating body.

The maximum velocity occurs at the equilibrium position, where the coordinate is zero. Since

$$v = \pm\omega\sqrt{A^2 - x^2},$$

then when $x = 0$,

$$v = v_{max} = \pm\omega A.$$

Also,

$$A = 0.04 \text{ m},$$

$$v_{max} = \pm(10 \text{ s}^{-1})(0.04 \text{ m}) = \pm0.4 \text{ m} \cdot \text{s}^{-1}.$$

d) Compute the maximum acceleration.

From Eq. (11–10),

$$a = -\omega^2 x.$$

The maximum acceleration occurs at the ends of the path, where $x = \pm A$. Therefore

$$a_{max} = \mp\omega^2 A = \mp(10 \text{ s}^{-1})^2(0.04 \text{ m}) = \mp4.0 \text{ m} \cdot \text{s}^{-2}.$$

That is, $a = -4.0 \text{ m} \cdot \text{s}^{-2}$ when $x = +A = +0.04 \text{ m}$, and $a = +4.0 \text{ m} \cdot \text{s}^{-2}$ when $x = -A = -0.04 \text{ m}$.

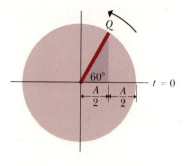

11–8

Alternatively, we may first find the maximum *force*. The force at any value of x is given by $F = -kx$, so the maximum force, occurring at $x = \pm A$, is $F_{max} = \mp kA$. Then the maximum acceleration is F_{max}/m:

$$F_{max} = \mp(200 \text{ N}\cdot\text{m}^{-1})(0.04 \text{ m}) = \mp 8 \text{ N},$$

$$a_{max} = \mp\frac{8 \text{ N}}{2 \text{ kg}} = \mp 4 \text{ m}\cdot\text{s}^{-2}.$$

e) Compute the velocity and acceleration when the body has moved halfway in toward the center from its initial position.

At this point, from Eq. (11–4),

$$x = \frac{A}{2} = 0.02 \text{ m},$$

$$v = -(10 \text{ s}^{-1})\sqrt{(0.04 \text{ m})^2 - (0.02 \text{ m})^2}$$

$$= -\left(\frac{2\sqrt{3}}{10}\right) \text{ m}\cdot\text{s}^{-1} = -0.346 \text{ m}\cdot\text{s}^{-1},$$

$$a = -\omega^2 x = -(10 \text{ s}^{-1})^2(0.02 \text{ m}) = -2.0 \text{ m}\cdot\text{s}^{-2}.$$

f) How much time is required for the body to move halfway in to the center from its initial position?

Note that the motion is neither one of constant velocity nor one of constant acceleration. The simplest method is to make use of the reference circle. While the body moves halfway in, the reference point revolves through an angle of 60° (Fig. 11–8). Since the reference point moves with constant angular velocity and in this example makes one complete revolution in $(\pi/5)$ s, the time to rotate through 60° is

$$\left(\frac{1}{6}\right)\left(\frac{\pi}{5}\right) \text{ s} = \frac{\pi}{30} \text{ s} = 0.105 \text{ s}.$$

The time may also be computed directly from the equation

$$x = A \cos \omega t,$$

$$A/2 = A \cos (10 \text{ s}^{-1})t,$$

$$\cos (10 \text{ s}^{-1})t = \frac{1}{2},$$

$$(10 \text{ s}^{-1})t = \arccos \frac{1}{2} = \frac{\pi}{3},$$

$$t = \frac{\pi}{30} \text{ s} = 0.105 \text{ s}. \quad \blacktriangleleft$$

11–4

MOTION OF A BODY SUSPENDED FROM A COIL SPRING

Figure 11–9a shows a coil spring of force constant k and no-load length l. When a body of mass m is attached to the spring as in part (b), it hangs in equilibrium with the spring extended by an amount Δl such that the

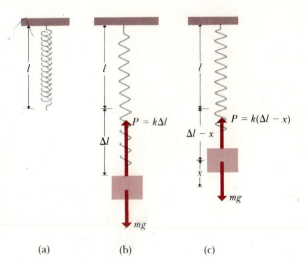

$P = k\Delta l$

Δl

$P = k(\Delta l - x)$

$\Delta l - x$

x

mg

mg

11–9 The restoring force on a body suspended by a spring is proportional to the coordinate measured from the equilibrium position.

(a)   (b)   (c)

upward force P exerted by the spring is equal to the weight of the body, mg. But $P = k\,\Delta l$, so

$$k\,\Delta l = mg.$$

Now suppose the body is at a distance x *above* its equilibrium position, as in Fig. 11–9c. The extension of the spring is now $\Delta l - x$, the upward force it exerts on the body is $k(\Delta l - x)$, and the resultant force F on the body is

$$F = k(\Delta l - x) - mg = -kx.$$

The resultant force is therefore proportional to the displacement of the body *from its equilibrium position;* and, if set in vertical motion, the body oscillates with an angular frequency

$$\omega = \sqrt{k/m}.$$

EXAMPLE A body of mass 5 kg is suspended by a spring, which stretches 0.1 m when the body is attached. It is then displaced downward an additional 0.05 m and released. Find the amplitude, period, and frequency of the resulting simple harmonic motion.

Solution Since the initial position is 0.05 m from equlibrium and there is no initial velocity, $A = 0.05$ m. To find the period we first find the force constant of the spring. The spring is stretched 0.1 m by a force of $(5\text{ kg})(9.8\text{ m}\cdot\text{s}^{-2})$, so

$$k = \frac{mg}{\Delta l} = \frac{(5\text{ kg})(9.8\text{ m}\cdot\text{s}^{-2})}{0.1\text{ m}} = 490\text{ N}\cdot\text{m}^{-1},$$

$$\tau = 2\pi\sqrt{\frac{m}{k}} = 2\pi\sqrt{\frac{5\text{ kg}}{490\text{ N}\cdot\text{m}^{-1}}} = 0.635\text{ s},$$

$$f = \frac{1}{\tau} = 1.57\text{ Hz.}$$ ◄

A similar analysis can be carried out for the case where a body is supported from below by a compressed spring, such as a bathroom scale containing a spring.

11–5

ANGULAR HARMONIC MOTION

Angular harmonic motion results when a body that is pivoted about an axis experiences a restoring torque proportional to the *angular* displacement from its equilibrium position. This type of vibration is very similar to linear harmonic motion. We have an angular displacement instead of a linear displacement, a restoring *torque* instead of a restoring force, and a moment of inertia instead of a mass. The corresponding equations may be written down immediately from the analogies between linear and angular quantities.

A restoring torque proportional to angular displacement is expressed by

$$\Gamma = -k'\theta, \tag{11–16}$$

where k' is a proportionality constant called the *torque constant*. The moment of inertia of the pivoted body corresponds to the mass of a body in linear motion. Hence, the angular frequency is now given by

$$\omega = \sqrt{\frac{k'}{I}}. \tag{11–17}$$

The corresponding period of oscillation is given by

$$\tau = \frac{2\pi}{\omega} = 2\pi\sqrt{\frac{I}{k'}}. \tag{11–18}$$

The form of this relation, the square root of an inertial quantity divided by a force quantity, is identical to Eq. (11–13).

Caution is required in the choice of symbols; ω must not be used for the angular velocity of the body (a quantity that varies with time) because it has already been used for the angular frequency of the motion (a constant for any given system). A natural choice for the body's angular velocity is Ω. Then in the equations in Sections 11–2 and 11–3 the appropriate formulas are obtained by replacing x by θ, v by Ω, and a by α.

The balance wheel of a watch is a familiar example of angular harmonic motion. If the hairspring behaves according to Eq. (11–16), the motion is *isochronous;* the period is constant even though the amplitude decreases somewhat as the mainspring unwinds.

11–6

THE SIMPLE PENDULUM

A simple pendulum consists of a point mass suspended by an inextensible weightless string in a uniform gravitational field. When pulled to one side of its equilibrium position and released, the pendulum bob vibrates about this position. We wish to analyze the motion of this idealized model, asking in particular whether it is simple harmonic.

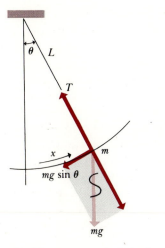

11–10   Forces on the bob of a simple pendulum.

The necessary condition for simple harmonic motion is that the restoring force F shall be directly proportional to the coordinate x and oppositely directed. The path of the bob is not a straight line, but the arc of a circle of radius L, where L is the length of the supporting cord. The coordinate x refers to distances measured *along this arc*. (See Fig. 11–10.) Hence, if $F = -kx$, the motion will be simple harmonic.

Figure 11–10 shows the forces on the bob at an instant when its coordinate is x. We choose axes tangent to the circle and along the radius, and resolve the weight into components. The restoring force F is

$$F = -mg \sin \theta. \qquad (11–19)$$

The restoring force is therefore proportional *not* to θ but to $\sin \theta$, so the motion is *not* simple harmonic. However, *if the angle θ is small,* $\sin \theta$ is very nearly equal to θ. For example, when $\theta = 0.1$ rad (about 6°), $\sin \theta = 0.0998$, a difference of only 0.2 percent. With this approximation, Eq. (11–19) becomes

$$F = -mg\theta = -mg\frac{x}{L},$$

or

$$F = -\frac{mg}{L}x. \qquad (11–20)$$

The restoring force is then proportional to the coordinate *for small displacements,* and the constant mg/L represents the force constant k. The angular frequency of a simple pendulum when its amplitude is small is therefore

$$\omega = \sqrt{\frac{k}{m}} = \sqrt{\frac{mg/L}{m}} = \sqrt{\frac{g}{L}}. \qquad (11–21)$$

The corresponding frequency and period relations are

$$f = \frac{\omega}{2\pi} = \frac{1}{2\pi}\sqrt{\frac{g}{L}}, \qquad (11–22)$$

$$\tau = \frac{2\pi}{\omega} = \frac{1}{f} = 2\pi\sqrt{\frac{L}{g}}. \qquad (11–23)$$

We note that these expressions do not contain the *mass* of the particle; this is because the restoring force, a component of the particle's weight, is proportional to m. Thus the mass appears on both sides of $F = ma$ and may be canceled out. For small oscillations the period of a pendulum for a given value of g is determined entirely by its length.

An elegant argument invented by Galileo 400 years ago also leads to this conclusion. Make a simple pendulum, said Galileo, measure its period, and then split it down the middle, string and all. Splitting it should not change the motion, so each half must swing with the same period as the original! Thus the period should not depend on the mass.

The form of the dependence on L and g in Eqs. (11–21) through (11–23) is generally what we should expect. It is a familiar fact that long pendulums have longer periods and smaller frequencies than shorter ones. Increasing g would increase the restoring force, which would cause the frequency to increase and the period to decrease.

We emphasize again that the motion of a pendulum is only *approximately* simple harmonic, and when the amplitude is not small the departures from simple harmonic motion can be substantial. What constitutes a "small" amplitude? It can be shown that the general equation for the time of swing, when the maximum angular displacement is α, is

$$\tau = 2\pi \sqrt{\frac{L}{g}}\left(1 + \frac{1^2}{2^2}\sin^2\frac{\alpha}{2} + \frac{1^2 \cdot 3^2}{2^2 \cdot 4^2}\sin^4\frac{\alpha}{2} + \cdots\right). \quad (11\text{--}24)$$

The time may be computed to any desired degree of precision by taking enough terms in the infinite series. When $\alpha = 15°$ (on either side of the central position), the true period differs from that given by the approximate Eq. (11–23) by less than 0.5%.

The usefulness of the pendulum as a timekeeper is based on the fact that the period is practically independent of the amplitude. Thus, as a clock runs down and the amplitude of the swings becomes slightly smaller, the clock will still keep very nearly correct time. This is also true of the vibrations of the quartz crystals used in digital clocks and watches; the motion is very nearly simply harmonic, so the period changes very little even though the amplitude decreases somewhat as the battery runs down.

The simple pendulum is also a precise and convenient method of measuring the acceleration of gravity g without actually resorting to free fall, since L and τ may readily be measured. More complicated pendulums find considerable application in the field of geophysics. Local deposits of ore or oil affect the local value of g, because their density differs from that of their surroundings. Precise measurements of this quantity over an area that is being prospected often furnish valuable information regarding the nature of underlying deposits.

11–7

THE PHYSICAL PENDULUM

A "physical" pendulum is any *real* pendulum, as contrasted with a simple pendulum in which all the mass is assumed to be concentrated at a point. In Fig. 11–11, a body of irregular shape is pivoted about a horizontal frictionless axis O and displaced from the vertical by an angle θ. The distance from the pivot to the center of gravity is h, the moment of inertia of the pendulum about an axis through the pivot is I, and the mass of the pendulum is m. The weight mg causes a restoring torque

$$\Gamma = -(mg)(h\sin\theta). \quad (11\text{--}25)$$

When released, the body oscillates about its equilibrium position; as in the case of the simple pendulum, the motion is not simple harmonic, since the torque Γ is proportional not to θ but to $\sin\theta$. However, if θ is small, we can again approximate $\sin\theta$ by θ, and the motion is approximately harmonic. With this approximation,

$$\Gamma \approx -(mgh)\theta,$$

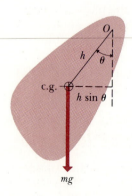

11–11 A physical pendulum.

and the effective torque constant is

$$k' = -\frac{\Gamma}{\theta} = mgh.$$

The angular frequency is

$$\omega = \sqrt{\frac{k'}{I}} = \sqrt{\frac{mgh}{I}}, \qquad (11\text{--}26)$$

and the period is

$$\tau = 2\pi \sqrt{\frac{I}{k'}} = 2\pi \sqrt{\frac{I}{mgh}}. \qquad (11\text{--}27)$$

EXAMPLE 1 Let the body in Fig. 11–11 be a meterstick pivoted at one end, and let L be the total length (1 m). Find its period of oscillation.

Solution From Example 1 of Section 9–9,

$$I = \tfrac{1}{3}mL^2, \qquad h = \frac{L}{2}, \qquad g = 9.8 \text{ m} \cdot \text{s}^{-2},$$

$$\tau = 2\pi \sqrt{\frac{\tfrac{1}{3}mL^2}{mgL/2}} = 2\pi \sqrt{\frac{2L}{3g}}$$

$$= 2\pi \sqrt{(2/3)(1 \text{ m})/9.8 \text{ m} \cdot \text{s}^{-2}} = 1.64 \text{ s.} \qquad \blacktriangleleft$$

EXAMPLE 2 Equation (11–27) may be solved for the moment of inertia I, giving

$$I = \frac{\tau^2 mgh}{4\pi^2}.$$

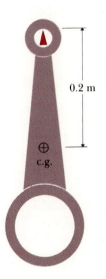

0.2 m

⊕
c.g.

The quantities on the right of the equation are all directly measurable. Hence the moment of inertia of a body of any complex shape may be found by suspending the body as a physical pendulum and measuring its period of vibration. The location of the center of gravity can be found by balancing. Since τ, m, g, and h are known, I can be computed. For example, Fig. 11–12 illustrates a connecting rod pivoted about a horizontal knife-edge. The connecting rod has a mass of 2 kg, and its center of gravity has been found by balancing to be 0.2 m below the knife-edge. When set into oscillation, it makes 100 complete vibrations in 120 s, so that $\tau = 120 \text{ s}/100 = 1.2 \text{ s}$. Therefore,

$$I = \frac{(1.2 \text{ s})^2 (2 \text{ kg})(9.8 \text{ m} \cdot \text{s}^{-2})(0.2 \text{ m})}{4\pi^2} = 0.143 \text{ kg} \cdot \text{m}^2. \qquad \blacktriangleleft$$

11–12

QUESTIONS

11–1 Think of several examples in everyday life of motion that is at least approximately simple harmonic. In what respects does each differ from SHM?

11–2 Does a tuning fork or similar tuning instrument undergo simple harmonic motion? Why is this a crucial question to musicians?

11–3 If a spring is cut in half, what is the force constant of each half? How would the frequency of SHM using a half-spring differ from that using the same mass and the entire spring?

11–4 The analysis of SHM in this chapter neglected the mass of the spring. How does the spring mass change the characteristics of the motion?

11–5 The system shown in Fig. 11–9 is mounted in an elevator, which accelerates upward with constant acceleration. Does the period increase, decrease, or remain the same?

11–6 A highly elastic "superball" bouncing on a hard floor has a motion that is approximately periodic. In what ways is the motion similar to SHM? In what ways is it different?

11–7 How could one determine the force constants of a car's springs by bouncing each end up and down?

11–8 Do the pistons in an automobile engine undergo simple harmonic motion?

11–9 Why is the "springiness" of a diving board adjusted for different dives and different weights of divers? How is the adjustment made?

11–10 In Fig. 11–1, suppose the stationary end of the spring is connected instead to another mass, equal to the original mass and free to slide along the same line. Could such a system undergo SHM? How would the period compare with that of the original system?

11–11 For the mass–spring system of Fig. 11–1, is there any point during the motion at which the mass is in equilibrium?

11–12 In any periodic motion, unavoidable friction always causes the amplitude to decrease with time. Does friction also affect the *period* of the motion? Give a qualitative argument to support your answer.

11–13 If a pendulum clock is taken to a mountaintop, does it gain or lose time, assuming it is correct at a lower elevation?

11–14 When the amplitude of a simple pendulum increases, should its period increase or decrease? Give a qualitative argument; do not rely on Eq. (11–24). Is your argument also valid for a physical pendulum?

11–15 A pendulum is mounted in an elevator that accelerates upward with constant acceleration. Does the period increase, decrease, or remain the same?

11–16 At what point in the motion of a simple pendulum is the string tension greatest? Least?

11–17 A child on a swing can increase his amplitude by "pumping up." Where does the extra energy come from? (The answer "from the child" is not sufficient!)

11–18 Could a standard of time be based on the period of a certain standard pendulum? What advantages and disadvantages would such a standard have, compared to the actual present-day standard discussed in Section 1–2?

PROBLEMS

11–1 A vibrating body goes through 5 complete vibrations in one second. Find the angular frequency and the period of the motion.

11–2 In Fig. 11–1, the mass is displaced 0.12 m from its equilibrium position and released with no initial velocity. After 2.0 s its displacement is found to be 0.12 m on the opposite side, and it has passed the equilibrium position once during this interval. Find

  a) the amplitude,

  b) the frequency,

  c) the period, and

  d) the angular frequency.

11–3 A harmonic oscillator is made using a body of mass 0.5 kg and a spring of unknown force constant. It is found to have a period of 0.20 s. Find the force constant of the spring.

11–4 A harmonic oscillator has a mass of 4 kg and a spring with force constant $100 \ \mathrm{N \cdot m^{-1}}$. Find the period, frequency, and angular frequency.

11–5 A body of unknown mass is attached to a spring of force constant $200 \ \mathrm{N \cdot m^{-1}}$, in the arrangement shown in Fig. 11–1. It is found to vibrate with a frequency of 3 Hz. Find the period, the angular frequency, and the mass.

11–6 A body of mass 0.25 kg is acted on by an elastic restoring force of force constant $k = 25 \ \mathrm{N \cdot m^{-1}}$.

  a) Construct the graph of elastic potential energy U as a function of displacement x, over a range of x from -0.3 m to $+0.3$ m. Let 1 cm = 0.1 J vertically, and 1 cm = 0.05 m horizontally.

  The body is set into oscillation with an initial potential energy of 0.6 J and an initial kinetic energy of 0.2 J. Answer the following questions by reference to the graph.

  b) What is the amplitude of oscillation.

  c) What is the potential energy when the displacement is one-half the amplitude?

  d) At what displacement are the kinetic and potential energies equal?

  e) What is the speed of the body at the midpoint of its path?

Find

   f) the period τ,

   g) the frequency f, and

   h) the angular frequency ω.

11–7 A body is vibrating with simple harmonic motion of amplitude 15 cm and frequency 4 Hz. Compute

   a) the maximum values of the acceleration and velocity,

   b) the acceleration and velocity when the coordinate is 9 cm, and

   c) the time required to move from the equilibrium position to a point 12 cm distant from it.

11–8 A body of mass 10 g moves with simple harmonic motion of amplitude 24 cm and period 4 s. The coordinate is +24 cm when $t = 0$. Compute

   a) the position of the body when $t = 0.5$ s,

   b) the magnitude and direction of the force acting on the body when $t = 0.5$ s,

   c) the minimum time required for the body to move from its initial position to the point where $x = -12$ cm, and

   d) the velocity of the body when $x = -12$ cm.

11–9 A force of 30 N stretches a spring 15 cm.

   a) What mass must be attached to the spring so that the system will oscillate with a period of $(\pi/4)$ s?

   b) If the amplitude of the motion is 5 cm, where is the body and in what direction is it moving $(\pi/12)$ s after it has passed the equilibrium position, moving to the right?

   c) What force does the spring exert on the body when it is 3 cm left of the equilibrium position, moving to the right?

11–10 A small block is executing simple harmonic motion in a horizontal plane with an amplitude of 10 cm. At a point 6 cm away from equilibrium, the velocity is 24 cm·s⁻¹.

   a) What is the period?

   b) What is the displacement when the velocity is ± 12 cm·s⁻¹?

   c) If a small object placed on the oscillating block is just on the verge of slipping at the endpoint of the path, what is the coefficient of friction?

11–11 The motion of the piston of an automobile engine is approximately simple harmonic.

   a) If the stroke of an engine (twice the amplitude) is 10 cm, and the angular velocity is 3600 rev·min⁻¹, compute the acceleration of the piston at the end of its stroke.

   b) If the piston has a mass of 0.5 kg, what resultant force must be exerted on it at this point?

   c) What is the velocity of the piston, in kilometers per hour, at the midpoint of its stroke?

11–12 A body of mass 2 kg is suspended from a spring of negligible mass and is found to stretch the spring 20 cm.

   a) What is the force constant of the spring?

   b) What is the period of oscillation of the body, if pulled down and released?

   c) What would be the period of a body of mass 4 kg hanging from the same spring?

11–13 The scale of a spring balance reading from zero to 180 N is 9 cm long. A body suspended from the balance is observed to oscillate vertically at 1.5 Hz. What is the mass of the body? Neglect the mass of the spring.

11–14 A body of mass 4 kg is attached to a coil spring and oscillates vertically in simple harmonic motion. The amplitude is 0.5 m, and at the highest point of the motion the spring has its natural unstretched length. Calculate the elastic potential energy of the spring, the kinetic energy of the body, its gravitational potential energy relative to the lowest point of the motion, and the sum of these three energies, when the body is

   a) at its lowest point,

   b) at its equilibrium position, and

   c) at its highest point.

11–15 A mass of 100 kg suspended from a wire whose unstretched length l_0 is 4 m is found to stretch the wire by 0.004 m. The cross-sectional area of the wire, which can be assumed constant, is 0.1 cm².

   a) If the load is pulled down a small additional distance and released, find the frequency at which it will vibrate.

   b) Compute Young's modulus for the wire.

11–16 A body of mass 5 kg hangs from a spring and oscillates with a period of 0.5 s. How much will the spring shorten when the body is removed?

11–17 Four passengers whose combined mass is 300 kg are observed to compress the springs of an automobile by 5 cm when they enter the automobile. If the total load supported by the springs is 900 kg, find the period of vibration of the loaded automobile.

11–18

   a) A block suspended from a spring vibrates with simple harmonic motion. At an instant when the displacement of the block is equal to one half the amplitude, what fraction of the total energy of the system is kinetic and what fraction is potential?

   b) When the block is in equilibrium, the length of the spring is an amount s greater than in the unstretched state. Prove that $\tau = 2\pi\sqrt{s/g}$.

11–19 Two springs with the same unstretched length but different force constants k_1 and k_2 are attached to a block of mass m on a level frictionless surface. Calculate the effective force constant in each of the three cases (a), (b), and (c) depicted in Fig. 11–13.

d) A body of mass m, suspended from a spring with a force constant k, vibrates with a frequency f_1. When the spring is cut in half and the same body is suspended from one of the halves, the frequency is f_2. What is the ratio f_2/f_1?

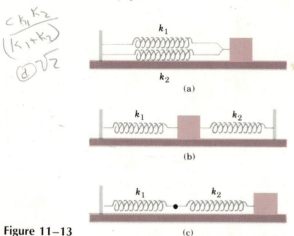

Figure 11–13

(a)

(b)

(c)

11–20 Two springs, each of unstretched length 0.2 m but having different force constants k_1 and k_2, are attached to opposite ends of a block of mass m on a level frictionless surface. The outer ends of the springs are now attached to two pins P_1 and P_2, 10 cm from the original positions of the ends of the springs. Let

$$k_1 = 1\ \text{N}\cdot\text{m}^{-1}, \qquad k_2 = 3\ \text{N}\cdot\text{m}^{-1}, \qquad m = 0.1\ \text{kg}.$$

(See Fig. 11–14.)

a) Find the length of each spring when the block is in its new equilibrium position, after the springs have been attached to the pins.

b) Find the period of vibration of the block if it is slightly displaced from its new equilibrium position and released.

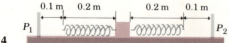

0.1 m   0.2 m   0.2 m   0.1 m

P_1 P_2

Figure 11–14

11–21 The block in Problem 11–20 and Fig. 11–14 is oscillating with an amplitude of 0.05 m. At the instant it passes through its equilibrium position, a lump of putty of mass 0.1 kg is dropped vertically onto the block and sticks to it.

a) Find the new period and amplitude.

b) Was there a loss of mechanical energy? If so, where did it go?

c) Would the answers be the same if the putty had been dropped on the block when it was at one end of its path?

11–22 A simple pendulum 4 m long swings with an amplitude of 0.2 m.

a) Compute the velocity of the pendulum at its lowest point.

b) Compute its acceleration at the ends of its path.

11–23 Find the length of a simple pendulum whose period is exactly 1 s at a point where $g = 9.80\ \text{m}\cdot\text{s}^{-2}$.

11–24 A pendulum clock that keeps correct time at a location where $g = 9.8000\ \text{m}\cdot\text{s}^{-2}$ is found to lose 10 seconds each day at a higher elevation. Find the value of g at this new location.

11–25 A certain simple pendulum has a period on earth of 2.0 s. What is its period on the surface of the moon, where $g = 1.7\ \text{m}\cdot\text{s}^{-2}$?

11–26 The balance wheel of a watch vibrates with an angular amplitude of π radians and with a period of 0.5 s.

a) Find its maximum angular velocity.

b) Find its angular velocity when its displacement is one-half its amplitude.

c) Find its angular acceleration when its displacement is $45°$.

11–27 A certain alarm clock ticks four times each second, each tick representing half a period. The balance wheel consists of a thin rim of radius 1.5 cm, connected to the balance staff by thin spokes of negligible mass. The total mass is 0.8 g.

a) What is the moment of inertia of the balance wheel?

b) What is the torque constant of the hairspring?

11–28 At one time it was proposed to define the second as the time required for a simple pendulum exactly one meter long to complete one side-to-side swing (one-half period).

a) By how much is this definition in error if $g = 9.8000\ \text{m}\cdot\text{s}^{-2}$?

b) For what value of g would this definition be exactly correct?

11–29 A thin uniform rod of length L is pivoted about an axis perpendicular to the rod at a distance $L/4$ from one end.

a) Find the moment of inertia about this axis.

b) Find the period of oscillation of the rod.

11–30 A thin uniform rod is pivoted about a perpendicular axis through the rod, and its period as a physical pendulum is measured. By trial and error a second pivot point is found so that the period is the same as for the first point. Show that in this case the period depends only on the distance L between the two points and is given by $\tau = 2\pi(L/g)^{1/2}$.

11–31 A monkey wrench is pivoted at one end and allowed to swing as a physical pendulum. The period is 0.9 s and the pivot is 20 cm from the center of gravity.

a) What is the ratio of moment of inertia to mass for the wrench, about an axis through the pivot?

b) If the wrench was initially displaced 0.1 rad from its equilibrium position, what is the angular velocity of the wrench as it passes through the equilibrium position?

11–32 A solid disk of radius $R = 12$ cm oscillates as a physical pendulum about an axis perpendicular to the plane of the disk at a distance r from its center. (See Fig. 11–15.)

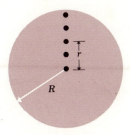

Figure 11–15

a) Calculate the period of oscillation (for small amplitudes) for the following values of r: 0, $R/4$, $R/2$, $3R/4$, R. (*Hint.* Use the parallel-axis theorem, Section 9–9.)

b) Let τ_0 represent the period when $r = R$, and τ the period at any other value of r. Construct a graph of the dimensionless ratio τ/τ_0 as a function of the dimensionless ratio r/R. (Note that the graph then describes the behavior of *any* solid disk, whatever its radius.)

11–33 It is desired to construct a pendulum of period 10 s.

a) What is the length of a *simple* pendulum having this period?

b) Suppose the pendulum must be mounted in a case not over 0.5 m high. Can you devise a pendulum, having a period of 10 s, that will satisfy this requirement?

11–34 A meter stick hangs from a horizontal axis at one end and oscillates as a physical pendulum. A body of small dimensions, and of mass equal to that of the meter stick, can be clamped to the stick at a distance h below the axis. Let τ represent the period of the system with the body attached, and τ_0 the period of the meter stick alone. Find the ratio τ/τ_0

a) when $h = 0.5$ m,

b) when $h = 1.0$ m.

c) Is there any value of h for which $\tau = \tau_0$? If so, find it and explain why the period is unchanged when h has this value.

CHAPTER 12

FLUID STATICS

A fluid is any substance that can flow; hence the term applies to both liquids and gases. Liquids and gases differ widely in their compressibilities; in most circumstances a gas is relatively easily compressed, while liquids are nearly incompressible. In this chapter we usually neglect the small volume changes that occur in a liquid under pressure.

Fluid statics is the study of fluids at rest (in equilibrium), and fluid dynamics is the study of fluids in motion. In this chapter we study equilibrium situations in fluids, including the concepts of density, pressure, buoyancy, and surface tension, as well as several practical instruments such as pumps and gauges.

12–1

DENSITY

The *density* of a homogeneous material is defined as its mass per unit volume. Units of density are one kilogram per cubic meter ($1 \text{ kg} \cdot \text{m}^{-3}$) in SI (mks) units, and one gram per cubic centimeter ($1 \text{ g} \cdot \text{cm}^{-3}$) in the cgs system. We shall represent density by the Greek letter ρ (rho):

$$\rho = \frac{m}{V}, \qquad m = \rho V. \qquad (12\text{–}1)$$

Typical values of density at room temperature are given in Table 12–1. The conversion factor

$$1 \text{ g} \cdot \text{cm}^{-3} = 1000 \text{ kg} \cdot \text{m}^{-3}$$

is useful in connection with this table.

The *specific gravity* of a material is the ratio of its density to that of water and is therefore a pure number. "Specific gravity" is an exceedingly poor term, since it has nothing to do with gravity. "Relative density" would describe the concept more precisely.

Density measurements are an important analytical technique in a wide variety of circumstances. The condition of an automobile storage

TABLE 12–1 DENSITIES			
Material	**Density** $g \cdot cm^{-3}$	**Material**	**Density** $g \cdot cm^{-3}$
Aluminum	2.7	Silver	10.5
Brass	8.6	Steel	7.8
Copper	8.9	Mercury	13.6
Gold	19.3	Ethyl alcohol	0.81
Ice	0.92	Benzene	0.90
Iron	7.8	Glycerin	1.26
Lead	11.3	Water	1.00
Platinum	21.4	Sea water	1.03

battery can be judged by measuring the density of the electrolyte, a sulfuric acid solution. As the battery discharges, the sulfuric acid (H_2SO_4) combines with lead in the battery plates to form insoluble lead sulfate ($PbSO_4$), decreasing the concentration of the solution. The density varies from about 1.30 $g \cdot cm^{-3}$ for a fully charged battery to 1.15 $g \cdot cm^{-3}$ for a discharged battery. Similarly, permanent-type antifreeze is usually a solution of ethylene glycol (density 1.12 $g \cdot cm^{-3}$) in water, with small quantities of additives to retard corrosion. The glycol concentration, which determines the freezing point of the solution, can be determined from a simple density measurement. Both these measurements are performed routinely in service stations with the aid of a simple hydrometer, which measures density by observation of the level at which a calibrated body floats in a sample of the solution. The hydrometer is discussed in Section 12–5.

Medical science finds many uses for density measurements; examples include tests on body fluids such as blood and urine. The density of normal human blood is about 1.04 to 1.06 $g \cdot cm^{-3}$. Since density increases with the concentration of red cells, an unusually low density may indicate anemia. Similarly, the usual density of urine is about 1.02 $g \cdot cm^{-3}$. Certain diseases lead to increased excretion of salts and a corresponding increase in urine density.

12–2

PRESSURE IN A FLUID

When the concept of hydrostatic pressure was introduced in Section 10–1, the *weight* of the fluid was neglected and the pressure was assumed the same at all points. It is a familiar fact, however, that atmospheric pressure is greater at sea level than at mountain altitudes and that the pressure in a lake or in the ocean increases with increasing depth below the surface. We therefore generalize the definition of pressure and define the pressure at *any point* as the ratio of the normal force ΔF exerted on a small area ΔA including the point, to the area ΔA:

$$p = \frac{\Delta F}{\Delta A}, \qquad \Delta F = p \, \Delta A. \tag{12–2}$$

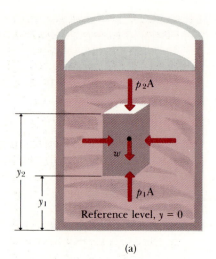

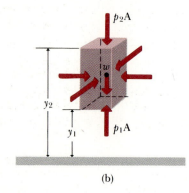

(a)

(b)

12–1 Forces on a rectangular element of fluid in equilibrium.

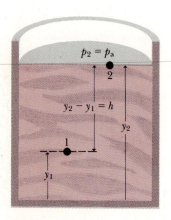

12–2

If the pressure is the same at all points of a finite plane surface of area A, these equations reduce to Eq. (10–3):

$$p = \frac{F}{A}, \qquad F = pA.$$

As noted in Section 10–1, the SI unit of pressure is the newton per square meter, also called the pascal:

$$1 \text{ Pa} = 1 \text{ N} \cdot \text{m}^{-2}.$$

Other commonly used units are the atmosphere (equal to normal atmospheric pressure at sea level) and the pound per square inch:

$$1 \text{ atm} = 1.013 \times 10^5 \text{ Pa} = 14.7 \text{ lb} \cdot \text{in}^{-2}.$$

We may derive a general relation between the pressure p at any point in a fluid in a gravitational field and the elevation y of the point. We consider the rectangular volume shown in Fig. 12–1. The bottom and top surfaces of the element are at elevations y_1 and y_2 above some reference level (where $y = 0$) and have area A. The height of this volume element is $y_2 - y_1$, its volume is $V = (y_2 - y_1)A$, its mass is $m = \rho V = \rho(y_2 - y_1)A$, and its weight is $w = mg = \rho g(y_2 - y_1)A$.

The total upward force at the bottom surface is $p_1 A$, and the total downward force at the top surface is $p_2 A$. The fluid in this volume is in equilibrium; thus the total vertical force, including the weight and the forces at the bottom and top surfaces, must be zero. This condition yields the equation

$$p_1 A - p_2 A - \rho g(y_2 - y_1)A = 0.$$

The area A may be divided out; the relation of pressure to elevation does not depend on the cross-sectional area of the volume. Thus we obtain

$$p_2 - p_1 = -\rho g(y_2 - y_1). \tag{12–3}$$

This equation shows that when point 1 is below point 2, $y_2 - y_1$ is positive but $p_2 - p_1$ is negative, that is, the pressure is greater at point 1 than at point 2. In other words, pressure increases with depth, which of course we already knew.

The horizontal forces on the opposite sides of the element also cancel out, so all the equilibrium conditions are satisfied.

Let us apply this equation to a liquid in an open container, such as that shown in Fig. 12–2. Take point 1 at any level in the fluid and let p represent the pressure at this point. Take point 2 at the top, where the pressure is atmospheric pressure, p_a. Then

$$p_a - p = -\rho g(y_2 - y_1),$$
$$p = p_a + \rho g h. \tag{12–4}$$

Note that the shape of the containing vessel does not affect the pressure, and that the pressure depends only on depth h. It also follows from Eq. (12–4) that if the pressure p_a is increased in any way, say by inserting a piston on the top surface and pressing down on it, the pressure p at any depth h must increase by exactly the same amount. This fact was recognized in 1653 by the French scientist Blaise Pascal (1623–1662) and is

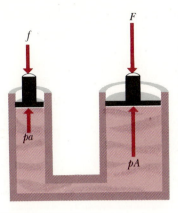

12–3

called "Pascal's law." It is often stated: *"Pressure applied to an enclosed fluid is transmitted undiminished to every portion of the fluid and the walls of the containing vessel."* We can see now that it is not an independent principle but a necessary consequence of the laws of mechanics.

Pascal's law is illustrated by the operation of a hydraulic press or jack, shown in Fig. 12–3. A piston of small cross-sectional area a is used to exert a small force f directly on a liquid such as oil. The pressure $p = f/a$ is transmitted through the connecting pipe to a larger cylinder equipped with a larger piston of area A. Since the pressure is the same in both cylinders,

$$p = \frac{f}{a} = \frac{F}{A} \quad \text{and} \quad F = \frac{A}{a} f.$$

It follows that the hydraulic press is a force-multiplying device with a multiplication factor equal to the ratio of the areas of the two pistons. Barber chairs, dentist chairs, car lifts and jacks, and hydraulic brakes are all devices that make use of the principle of the hydraulic press.

In deriving Eq. (12–4) we have assumed that the density ρ of the fluid is constant; this is a reasonable assumption for liquids, which are relatively incompressible, but it is *not* realistic for gases. In calculating the variation with altitude of the pressure in the earth's atmosphere, the variation in density must be included. This calculation requires methods of calculus and will not be attempted here. On the other hand, the density of gases is usually very much less than that of liquids, so over small changes of elevation the change in pressure is usually so small as to be negligible.

In many practical situations the significant quantity is the *difference* between the pressure in a container and atmospheric pressure. For example, if the pressure inside a tire is just equal to atmospheric pressure, the tire is flat. When we say the pressure in a tire is 32 pounds (actually $32 \ \text{lb} \cdot \text{in}^{-2}$), we mean it is *greater* than atmospheric pressure ($14.7 \ \text{lb} \cdot \text{in}^{-2}$) by this amount; the *total* pressure in the tire is then $46.7 \ \text{lb} \cdot \text{in}^{-2}$. The excess pressure above atmospheric pressure is usually called the *gauge pressure*, and the total pressure is sometimes called the *absolute pressure*. Engineers use the abbreviations psig and psia for "pounds per square inch gauge" and "pounds per square inch absolute."

EXAMPLE A household hot water heating system has an expansion tank in the attic, 12 m above the boiler. If the tank is open to the atmosphere, what is the gauge pressure in the boiler? The absolute pressure?

Solution From Eq. (12–4), the absolute pressure is

$$p = (1.01 \times 10^5 \ \text{Pa}) + (1000 \ \text{kg} \cdot \text{m}^{-3})(9.8 \ \text{m} \cdot \text{s}^{-2})(12 \ \text{m})$$
$$= 2.19 \times 10^5 \ \text{Pa} = 2.16 \ \text{atm} = 31.8 \ \text{lb} \cdot \text{in}^{-2}.$$

The gauge pressure is

$$p - p_a = 1.18 \times 10^5 \ \text{Pa} = 1.15 \ \text{atm} = 17.1 \ \text{lb} \cdot \text{in}^{-2}.$$

If there is a pressure gauge on the furnace, it is always calibrated to read gauge rather than absolute pressure.

We may also estimate the variation in atmospheric pressure over this height. The density of air at sea level and a temperature of 20°C is about $1.2 \ \text{kg} \cdot \text{m}^{-3}$, that is, about 0.12% of the density of water. If this density were constant over the 12-m height, then the atmospheric pressure at furnace level would be greater than that in the attic by an amount

$$\rho g h = (1.2 \ \text{kg} \cdot \text{m}^{-3})(9.8 \ \text{m} \cdot \text{s}^{-2})(12 \ \text{m}) = 141 \ \text{N} = 0.0014 \ \text{atm},$$

or a very small fraction of the pressures calculated above. Thus the variation in air pressure over this height is negligible.   ◄

12–3

PRESSURE GAUGES

The analysis in Section 12–2 can be applied directly to various devices used to measure pressure in a gas; these are usually called pressure *gauges*. The simplest type of pressure gauge is the open-tube manometer, shown in Fig. 12–4a. It consists of a U-shaped tube containing a liquid; one end of the tube is at the pressure p to be measured, while the other end is open to the atmosphere at pressure p_a.

The pressure at the bottom of the left column is $p + \rho g y_1$, while that at the bottom of the right column (the same point) is $p_a + \rho g y_2$, where ρ is the density of the manometric liquid. Since these pressures must be equal,

$$p + \rho g y_1 = p_a + \rho g y_2,$$

and

$$p - p_a = \rho g(y_2 - y_1) = \rho g h. \tag{12–5}$$

The pressure p is called the *absolute pressure*, whereas the difference $p - p_a$ between this and atmospheric pressure is called the *gauge pressure*. Thus the gauge pressure is proportional to the difference in height of the liquid columns.

The mercury barometer is a long glass tube that has been filled with mercury and then inverted in a dish of mercury, as shown in Fig. 12–4b. The space above the mercury column contains only mercury vapor, whose pressure, at room temperature, is so small that it may be neglected. From Eq. (12–3),

$$p_a = \rho g(y_2 - y_1) = \rho g h. \tag{12–6}$$

The SI unit of pressure is *one pascal* (1 Pa), equal to one newton per square meter ($1 \ \text{N} \cdot \text{m}^{-2}$). A related unit is *one bar*, defined as 10^5 Pa. Because mercury manometers and barometers are commonly used laboratory instruments, pressures are also sometimes described in terms of the height of the corresponding mercury column, as so many "inches of mercury" or "millimeters of mercury" (abbreviated mm Hg). The pressure due to a column of mercury one millimeter high has even been given a special name, *one Torr* (after Torricelli, inventor of the mercury barometer). Such units depend on the density of mercury, which varies with temperature, and on the value of g, which varies with location. For these and other reasons, they are gradually passing out of common use in favor of the pascal.

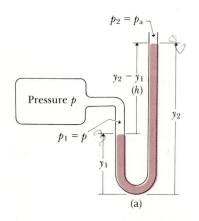

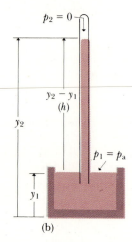

12–4 (a) The open-tube manometer. (b) The barometer.

One type of blood-pressure gauge commonly used by physicians, called a *sphygmomanometer,* includes a form of manometer similar to that shown in Fig. 12–4a. Blood-pressure readings, such as 130/80, refer to the maximum and minimum gauge pressures, measured in millimeters of mercury or Torrs. Because of height differences, the hydrostatic pressure varies at different points in the body; the standard reference point is the upper arm, level with the heart. Pressure is also affected by the viscous nature of blood flow, by valves throughout the vascular system, and by changes in diameters of blood vessels to regulate pressure.

EXAMPLE 1 Compute the atmospheric pressure on a day when the height of the barometer is 76.0 cm.

Solution The height of the mercury column depends on ρ and g as well as on the atmospheric pressure. As mentioned, ρ varies with temperature, and g with latitude and elevation above sea level. If we assume $g = 9.8 \text{ m} \cdot \text{s}^{-2}$ and $\rho = 13.6 \times 10^3 \text{ kg} \cdot \text{m}^{-3}$,

$$p_a = \rho g h = (13.6 \times 10^3 \text{ kg} \cdot \text{m}^{-3})(9.8 \text{ m} \cdot \text{s}^{-2})(0.76 \text{ m})$$
$$= 101{,}300 \text{ N} \cdot \text{m}^{-2} = 1.013 \times 10^5 \text{ Pa}.$$

In English units,

$$0.76 \text{ m} = 30 \text{ in.} = 2.5 \text{ ft},$$
$$\rho g = 850 \text{ lb} \cdot \text{ft}^{-3},$$
$$p_a = 2120 \text{ lb} \cdot \text{ft}^{-2} = 14.7 \text{ lb} \cdot \text{in}^{-2}. \qquad \blacktriangleleft$$

EXAMPLE 2 Blood pressure is ordinarily measured by placing the cuff around the upper arm, at the same level as the heart. If a person has a blood pressure of 130/80, measured in this way, what will be his blood pressure when he is standing and the cuff is placed on the calf of the leg, 1.0 m below the heart?

Solution The maximum (systolic) and minimum (diastolic) pressure are both greater by an amount $\rho g h$, where ρ is the density of blood and $h = 1.0$ m. Assuming blood has the same density of water, we find a pressure difference of $(1.0 \times 10^3 \text{ kg} \cdot \text{m}^{-3})(9.8 \text{ m} \cdot \text{s}^{-2})(1.0 \text{ m}) = 9.8 \times 10^3 \text{ Pa}$. Since $1.01 \times 10^5 \text{ Pa} = 760 \text{ mm Hg} = 1 \text{ atm}$, this is equal to 73 mm Hg, so the pressure at calf level would be 203/153. $\qquad \blacktriangleleft$

A pressure of $1.013 \times 10^5 \text{ Pa} = 14.7 \text{ lb} \cdot \text{in}^{-2}$ is called one *atmosphere* (1 atm). A pressure of exactly $10^5 \text{ N} \cdot \text{m}^{-2}$ is called one *bar*, and a pressure one one-thousandth as great is one *millibar*. Atmospheric pressures are of the order of 1000 millibars, and are now stated in terms of this unit by the National Weather Service.

The Bourdon-type pressure gauge is more convenient for more purposes than a liquid manometer. It consists of a flattened brass tube closed at one end and bent into a circular form. The closed end of the tube is connected by a gear and pinion to a pointer that moves over a scale. The open end of the tube is connected to the apparatus where the pressure is to be measured. When pressure is exerted within the flattened tube, it straightens slightly, just as a bent rubber hose straightens when water is admitted. The resulting motion of the closed end of the tube is transmitted to the pointer.

12–4

PUMPS

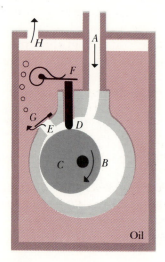

12–5 The rotary oil pump.

Many devices of modern life include glass or metal containers from which nearly all the air has been exhausted; examples include electric light bulbs, TV picture tubes, thermos bottles, and many others. Successful operation of these devices requires air pressures as low as 10^{-6} Pa, and special pumps are used to produce these "high" vacuums. Two commonly used types are the *rotary oil pump* for pressures as low as 10^{-2} Pa, and the *mercury or oil diffusion pump* for pressures as low as 10^{-6} Pa.

Figure 12–5 shows the principle of a common type of rotary oil pump. The container to be exhausted is connected to the tube A, which connects directly with the space marked B. As an eccentric cylinder C rotates in the direction shown, the point of contact between it and the inner walls of the stationary cylinder moves around in a clockwise direction, trapping some air in the space marked E. The sliding vane D is kept in contact with the rotating cylinder by the pressure of the rod F. When the air in E is compressed enough to increase the pressure slightly above atmospheric, the valve G opens and the air bubbles through the oil and leaves through an opening H in the upper plate. The cylinder is turned by a small electric motor.

For pressures below about 10^{-2} Pa, a diffusion pump is usually employed. In this pump a rapidly moving jet of mercury, or of a special oil of low vapor pressure, sweeps or pushes the air molecules away from the vessel to be exhausted. Air molecules from the vessel keep diffusing into the jet. A common type of diffusion pump is shown in Fig. 12–6. A

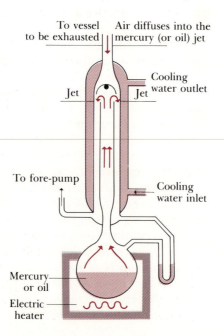

12–6 The diffusion pump.

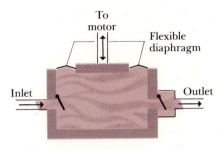

12–7 A diaphragm pump.

rotary oil pump is used to reduce the pressure within the diffusion pump to the low value necessary to ensure a well-defined jet to prevent air molecules from diffusing from the jet to the vessel. The air molecules that diffuse into the jet are removed by the oil pump (called in this case the *forepump*), and the mercury or oil condenses on the cool walls of the pump and returns to the well at the bottom.

Pumps are also used to compress gases and circulate liquids. Compressors for home refrigerators and air conditioners frequently use rotary pumps similar in principle to that shown in Fig. 12–5; reciprocating-piston pumps, equipped with pistons similar to those in gasoline engines, are also used for refrigerator compressors, and as air compressors. In these pumps, the essential principle is a varying volume and a valve arrangement to ensure one-way flow of the fluid being pumped.

In situations where it is important to prevent contamination of the fluid, the varying volume is often provided by a flexible diaphragm, as shown in Fig. 12–7. The center portion of the diaphragm moves up and down, driven by a motor. When it moves up, fluid is pushed into the chamber from the left by gravity or external air pressure; when it moves down, the fluid in the chamber is forced out the right side. Backflow is prevented by the two valves. Many automobile-engine fuel pumps operate on this principle, and the human heart comprises four such pumps, with the expansion and contraction of the four chambers by the heart muscles playing the role of the diaphragm in Fig. 12–7. Some artificial heart pumps use diaphragm pumps.

12–5

ARCHIMEDES' PRINCIPLE

Buoyancy is a familiar phenomenon; a body immersed in water seems to have less weight than when immersed in air, and a body whose average density is less than that of the fluid in which it is immersed can *float* in that fluid. Examples are the human body in water, or a helium-filled balloon in air.

Archimedes' principle states that *when a body is immersed in a fluid, the fluid exerts an upward force on the body equal to the weight of the fluid that is displaced by the body.* To prove this principle, we consider an arbitrary portion of fluid at rest. In Fig. 12–8, the irregular outline is the surface bounding this portion of fluid. The arrows represent the forces exerted by the surrounding fluid against small elements of the boundary surface.

Since the entire fluid is at rest, the x-component of the resultant of these surface forces is zero. The y-component of the resultant, F_y, must equal the weight mg of the fluid inside the arbitrary surface, and its line of action must pass through the center of gravity of this fluid.

Now suppose that the fluid inside the surface is removed and replaced by a solid body having exactly the same shape. The pressure at every point is exactly the same as before; the force exerted on the body by the surrounding fluid is unaltered and is therefore equal to the weight mg of fluid displaced. The line of action of this force again passes through the center of gravity of the displaced fluid.

In order for a balloon to float in air, its weight must be the same as the weight of an amount of air having the same volume as the balloon.

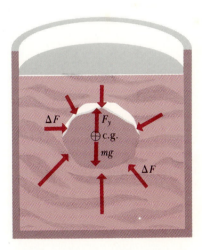

12–8 Archimedes' principle. The buoyant force F_y equals the weight of the displaced fluid.

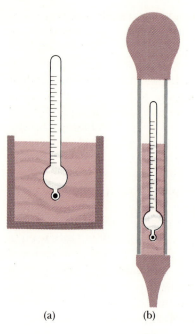

12–9 (a) A simple hydrometer. (b) Hydrometer used as a tester for battery acid or antifreeze.

That is, the average density of a floating balloon must be the same as that of the surrounding air. Similarly, when a submerged submarine is in equilibrium, its average density must be equal to that of the surrounding water.

A body whose average density is less than that of a liquid can float partially submerged at the free upper surface of the liquid. A familiar example is the hydrometer, shown in Fig. 12–9. The instrument sinks in the fluid until the weight of fluid it displaces is exactly equal to its own weight. In a fluid of greater density, less fluid is displaced for the same weight, so less of the instrument is immersed; i.e., in denser fluids it floats *higher*. This corresponds to the familiar fact that when swimming in sea water (density 1.03 g·cm^{-3}), one's body floats higher than in fresh water.

The hydrometer is weighted at its bottom end so the upright position is stable, and a scale in the stem at the top permits direct density readings. Figure 12–9b shows a hydrometer commonly used to measure density of battery acid or antifreeze. When the tube at the bottom is immersed in the fluid and the bulb is squeezed to expel air and then released (like a giant medicine dropper), the resulting pressure difference causes the fluid to rise into the outer tube, and the hydrometer floats in this sample of the fluid.

EXAMPLE 1   A block of brass a mass of 0.5 kg and a density of 8.0 × 10^3 kg·m^{-3}. It is suspended from a string. Find the tension in the string if the block is in air, and if it is completely immersed in water.

Solution   If the very small buoyant force of air can be neglected, the tension when the block is in air is just equal to its weight:

$$mg = (0.5 \text{ kg})(9.8 \text{ m·s}^{-2}) = 4.9 \text{ N}.$$

When immersed in water the block experiences an upward buoyant force equal to the weight of water displaced. To find this, we can first find the volume of brass:

$$V = \frac{m}{\rho} = \frac{0.5 \text{ kg}}{8.0 \times 10^3 \text{ kg·m}^{-3}} = 6.25 \times 10^{-5} \text{ m}^3.$$

Next, the weight of this volume of water is

$$w = mg = (\rho V)g = (1.0 \times 10^3 \text{ kg·m}^{-3})(6.25 \times 10^{-5} \text{ m}^3)(9.8 \text{ m·s}^{-2})$$
$$= 0.612 \text{ N}.$$

The tension in the string is the actual weight less the upward buoyant force, or 4.9 N − 0.612 N = 4.29 N.

A shortcut leading to the same result is as follows: The density of water is less than that of brass by a factor

$$\frac{1.0 \times 10^3 \text{ kg·m}^{-3}}{8.0 \times 10^3 \text{ kg·m}^{-3}} = \frac{1}{8}.$$

Therefore the weight of an amount of water of volume equal to that of the brass will be less than the weight of the brass by the same factor. Thus the weight of the displaced water is ($\frac{1}{8}$)(4.9 N) = 0.612 N. ◀

EXAMPLE 2   If the water container in the above example sits on a scale, how does the scale reading change when the brass is immersed in the water?

Solution   Taking the water and the brass together as a system, we note that its total weight does not change when the brass is immersed. Thus the sum of the two supporting forces, the string tension and the upward force of the scale and the container, must be the same in both cases. But the string tension was found to decrease by 0.612 N upon immersion of the brass, so the scale reading must *increase* by 0.612 N. Qualitatively we can see that the scale force must increase because when the brass is immersed, the water level in the container rises, increasing the pressure of water on the bottom of the container. ◀

An interesting medical application of buoyancy is found in the technique of *hydrotherapy*. A patient may be unable to lift a limb because of disease or injury in the associated muscles or joints. When the body is immersed in water it becomes, in effect, nearly weightless because its average density is only a little greater than that of water. As a result, the forces required to move the limb are greatly reduced, and therapeutic exercise becomes possible.

Measurements of buoyant forces on a submerged body can be used to estimate the percentage of fat in the body, since the density of fat is of the order of $0.9 \ \mathrm{g \cdot cm^{-3}}$, while the densities of most other body tissues are of the order of $1.0 \ \mathrm{g \cdot cm^{-3}}$, i.e., nearly the same as that of water.

*12–6

FORCES AGAINST A DAM

Suppose water stands at a depth H behind the vertical upstream face of a dam (Fig. 12–10). It exerts a certain resultant horizontal force on the dam, tending to slide it along its foundation, and a certain torque, tending to overturn the dam about the point O. We wish to find the total horizontal force. The *width* of the dam is L.

Figure 12–10b is a view of the upstream face of the dam. The pressure at an elevation y is

$$p = \rho g(H - y).$$

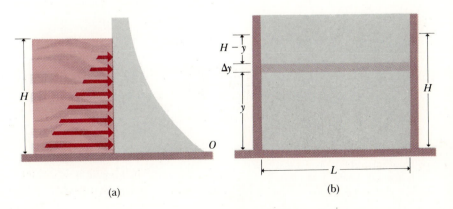

12–10   Forces on a dam.

(a)   (b)

Atmospheric pressure can be omitted, since it also acts upstream against the other face of the dam. The force ΔF against the shaded strip is

$$\Delta F = p \, \Delta A = \rho g (H - y) L \, \Delta y.$$

To find the total force we must add up the forces on all the strips from top to bottom of the dam face. This can be done using integral calculus, but we may also obtain the total force by multiplying the *average* pressure by the total area. Since the pressure increases uniformly with depth, the average pressure is the pressure at half-depth, namely

$$p_{av} = \frac{\rho g H}{2}.$$

Then the total force is

$$F_{tot} = p_{av} A = \left(\frac{\rho g H}{2} \right)(LH) = \tfrac{1}{2}\rho g H^2 L. \tag{12-7}$$

The total force is proportional to the *square* of the depth, a result which is obviously of central importance in the design of dams.

12–7

SURFACE TENSION

A liquid flowing slowly from the tip of a medicine dropper emerges not as a continuous stream but as a succession of drops. A sewing needle, if placed carefully on a water surface, makes a small depression in the surface and rests there without sinking, even though its density may be as much as 10 times that of water. Some insects can walk on the surface of water, their feet making indentations in the surface but not penetrating it. When a clean glass tube of small bore is dipped into water, the water rises in the tube, but if the tube is dipped in mercury, the mercury is depressed. All these phenomena, and many others of a similar nature, are associated with the existence of a *boundary surface* between a liquid and some other substance.

In these phenomena the surface behaves as though it is under *tension;* for any line on the surface, the portions of surface on the two sides of the line exert pulls on it. The situation of Fig. 12–11 demonstrates this effect; a wire ring has a loop of thread attached to it as shown. When the

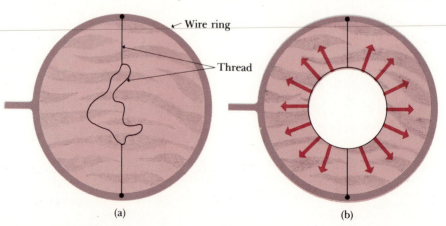

12–11   A wire ring with a flexible loop of thread, dipped in a soap solution, (a) before and (b) after puncturing the surface films inside the loop.

(a)   (b)

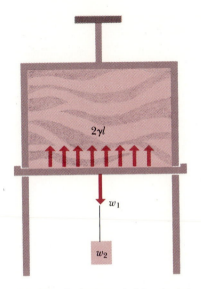

12–12 The horizontal slide wire is in equilibrium under the action of the upward surface force $2\gamma l$ and downward pull $w_1 + w_2$.

ring and thread are dipped in a soap solution and removed, a thin film of liquid is formed in which the thread "floats" freely, as shown in part (a). If the film inside the loop of thread is punctured, the thread springs out into a circular shape as in part (b), as if the surfaces of the liquid were pulling radially outward on it, as shown by the arrows. Indeed, the surface *is* pulling on the thread. The same forces were acting before the film was punctured, but since there was film on *both* sides of the thread the net force exerted by the film on every portion of the thread was zero.

Another simple apparatus for demonstrating surface tension is shown in Fig. 12–12. A piece of wire is bent into the shape of a U and a second piece of wire is used as a slider. When the apparatus is dipped in a soap solution and removed, the slider (if its weight w_1 is not too great) is quickly pulled up to the top of the U. It may be held in equilibrium by adding a second weight w_2. Surprisingly, the same total force $F = w_1 + w_2$ will hold the slider at rest in *any* position, regardless of the area of the liquid film, provided the film remains at constant temperature. That is, the force does not increase as the surface is stretched farther. This is very different from the elastic behavior of a sheet of rubber, for which the force would be greater as the sheet was stretched.

Although a soap film like that in Fig. 12–12 is very thin, its thickness is still enormous compared with the size of a molecule. Hence it can be considered as made up chiefly of bulk liquid, bounded by two surface layers a few molecules thick. When the crossbar in Fig. 12–12 is pulled down and the area of the film is increased, molecules formerly in the main body of the liquid move into the surface layers. That is, these layers are not "stretched" as a rubber sheet would be; more surface is created by molecules moving from the bulk liquid.

Let l be the length of the wire slider. Since the film has two surfaces, the total length along which the surface force acts on the slider is $2l$. The *surface tension* γ in the film is defined as *the ratio of the surface force to the length* (perpendicular to the force) *along which the force acts*. Hence, in this case,

$$\gamma = \frac{F}{2l}. \tag{12–8}$$

The SI unit of surface tension is the newton per meter ($1\ \mathrm{N \cdot m^{-1}}$). This unit is not in common use; the usual unit is the cgs unit, the dyne per centimeter ($1\ \mathrm{dyn \cdot cm^{-1}}$). The conversion factor is

$$1\ \mathrm{N \cdot m^{-1}} = 1000\ \mathrm{dyn \cdot cm^{-1}}.$$

Another example is shown in Fig. 12–13. A circular wire of circumference l is lifted out from the body of a liquid. The additional force F needed to balance the surface forces $2\gamma l$ due to the two surface films on each side is measured either by the stretch of a delicate spring or by the twist of a torsion wire. The surface tension is then given by

$$\gamma = \frac{F}{2l}.$$

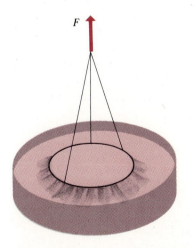

12–13 Lifting a circular wire of length l out of a liquid requires an additional force F to balance the surface forces $2\gamma l$. This method is commonly used to measure surface tension.

Some typical values of surface tension are shown in Table 12–2. Figure 12–14 shows a similar situation, with the net force of surface tension upward rather than downward.

12–14 A paper clip "floating" on water. The clip is supported not by buoyant forces (Section 12–5) but by surface tension of the water. (Nancy Rodgers, Exploratorium.)

TABLE 12–2 EXPERIMENTAL VALUES OF SURFACE TENSION		
Liquid in contact with air	T, °C	Surface tension, dyn·cm^{-1}
Benzene	20	28.9
Carbon tetrachloride	20	26.8
Ethyl alcohol	20	22.3
Glycerin	20	63.1
Mercury	20	465
Olive oil	20	32.0
Soap solution	20	25.0
Water	0	75.6
Water	20	72.8
Water	60	66.2
Water	100	58.9
Oxygen	−193	15.7
Neon	−247	5.15
Helium	−269	0.12

The surface tension of a liquid surface in contact with its own vapor or with air is found to depend *only* on the nature of the liquid and on the temperature. Surface tension usually decreases as temperature increases; Table 12–2 illustrates this behavior for water.

A surface under tension tends to contract until it has the minimum area consistent with any fixed boundaries that may be present and with pressure differences that may be present on opposite sides of the surface. For example, a drop of liquid under no external forces, or in free fall in vacuum, is always spherical in shape because the sphere has smaller surface area for a given volume than has any other geometric shape. Figure 12–15 is a beautiful example of formation of spherical droplets in a very complex phenomenon, the impact of a drop on a rigid surface. This was taken by Dr. Edgerton of M.I.T., one of the pioneers in the development of high-speed photographic technique.

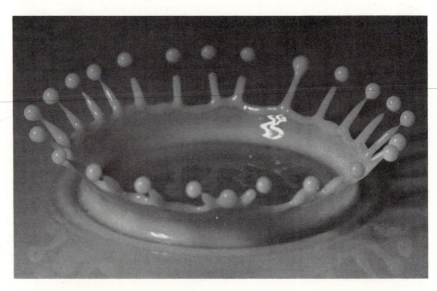

12–15 A drop of milk splashes on a hard surface. (Dr. Harold Edgerton, M.I.T., Cambridge, Massachusetts.)

12–8

PRESSURE DIFFERENCE ACROSS A SURFACE FILM

A soap bubble consists of two spherical surface films very close together, with a thin layer of liquid between. Surface tension causes the films to tend to contract, but as the bubble contracts it compresses the inside air, increasing the interior pressure to a point that prevents further contraction. We may derive a relation between this pressure and the radius.

We assume first that there is no external pressure. Then we consider the equilibrium state of one half the soap bubble, as shown in Fig. 12–16. At the surface where this half joins the upper half, two forces act, the upward force of surface tension and the downward force due to the pressure of air in the upper half. If the radius of the spherical surface is R, the circumference of the circle along which the surface tension acts is $2\pi R$. The total force of surface tension for each surface (inner and outer) is $\gamma(2\pi R)$, so the total force of surface tension is $(2\gamma)(2\pi R)$. The force due to air pressure is the pressure p times the area πR^2 of the circle over which it acts. For the resultant of these forces to be zero, we must have

$$(2\gamma)(2\pi R) = p(\pi R^2),$$

or

$$p = \frac{4\gamma}{R}. \qquad (12\text{–}9)$$

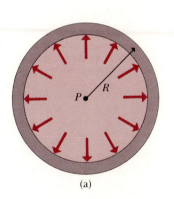

Now the pressure outside the bubble is in general *not* zero. If instead it has the value p_a, then Eq. (12–9) gives the *difference* between inside and outside pressures, and we obtain

$$p - p_a = \frac{4\gamma}{R} \qquad \text{(soap bubble)}. \qquad (12\text{–}10)$$

If the surface tension is constant, the pressure difference is greater for small bubbles than for large ones. If two bubbles are blown at opposite ends of a pipe, the smaller of the two will force air into the larger; the smaller one will get still smaller, and the larger will increase in size.

For a liquid drop, which has only *one* surface film, the difference between pressure of the liquid and that of the outside air is half that for a soap bubble:

$$p - p_a = 2\gamma/R \qquad \text{(liquid drop)}. \qquad (12\text{–}11)$$

EXAMPLE   Calculate the excess pressure inside a drop of mercury of diameter 4 mm at 20°C.

SOLUTION

$$
\begin{aligned}
p - p_a &= \frac{2\gamma}{R} \\
&= \frac{(2)(465 \times 10^{-3}\ \text{N} \cdot \text{m}^{-1})}{0.002\ \text{m}} \\
&= 465\ \text{N} \cdot \text{m}^{-2} \\
&\cong 0.005\ \text{atm}.
\end{aligned}
$$

◀

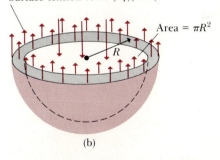

Surface tension force $(2\gamma)(2\pi R)$

Area $= \pi R^2$

12–16   (a) Cross section of soap bubble, showing the two surfaces, the thin layer of liquid, and inside air. (b) Equilibrium of half a soap bubble. The force exerted by the other half is $(2\gamma)(2\pi R)$, and the force exerted by the air inside the bubble is the pressure p times the area πR^2.

*12–9

CONTACT ANGLE AND CAPILLARITY

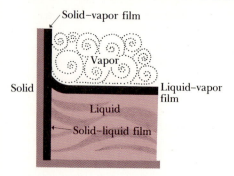

12–17 Surface films exist at the solid–vapor boundary as well as at the liquid–vapor boundary.

The preceding sections have been concerned with surface films lying in the boundary between a liquid and a gas. Surface films also exist between a solid wall and a liquid, and between a solid and a vapor. The three boundaries and their accompanying films are shown schematically in Fig. 12–17. The films are only a few molecules thick. Associated with each film is an appropriate surface tension. Thus,

γ_{SL} = surface tension of the solid-liquid film,

γ_{SV} = surface tension of the solid-vapor film,

γ_{LV} = surface tension of the liquid-vapor film.

The symbol γ without subscripts, defined and used in the preceding sections, now appears as γ_{LV}.

The curvature of the surface of a liquid near a solid wall depends upon the difference between γ_{SV} and γ_{SL}. Consider a portion of a glass wall in contact with methylene iodide, as shown in Fig. 12–18a. When γ_{SV} is greater than γ_{SL}, the line along which the three films meet is pulled upward, as in Figs. 12–17 and 12–18a. Angle θ in Fig. 12–18 is called the *contact angle*, and we see that when γ_{SV} is greater than γ_{SL}, the angle of contact is between zero and 90°. Figure 12–18b shows a case where γ_{SV} is *less* than γ_{SL}, the line of intersection of the films is pulled *downward*, and the contact angle is between 90° and 180°. In Fig. 12–18c, γ_{SV} and γ_{SL} are very nearly equal, and the contact angle is 90°.

Impurities in a liquid may alter the contact angle considerably. Wetting agents or detergents change the contact angle from a large value, greater than 90°, to a value much *smaller* than 90°. Conversely, water-proofing agents applied to cloth cause the contact angle of water in contact with the cloth to be larger than 90°. The effect of a detergent on a drop of water resting on a block of paraffin is shown in Fig. 12–19.

An important phenomenon resulting from surface tension is the elevation of a liquid in an open tube of small cross section. The term *capillarity*, used to describe effects of this sort, originates from the description of such tubes as "capillary" or "hairlike." In the case of a liquid that wets

12–18 The surface of a liquid near a solid wall is curved if the solid–vapor surface tension γ_{SV} differs from the solid–liquid surface tension γ_{SL}.

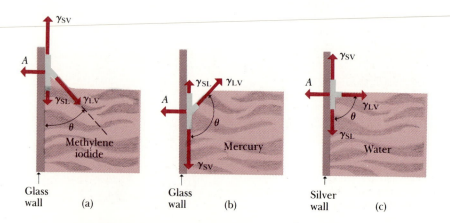

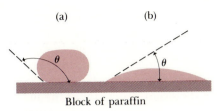

(a)   (b)

Block of paraffin

12–19 Effect of decreasing contact angle by a wetting agent.

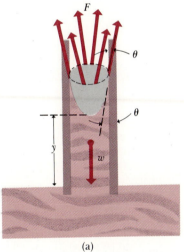

(a)

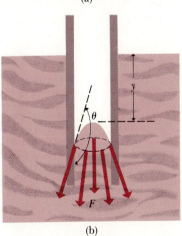

(b)

12–20 Surface tension forces on a liquid in a capillary tube. The liquid rises if $\theta < 90°$ and is depressed if $\theta > 90°$.

the tube, the contact angle is less than 90° and the liquid rises until an equilibrium height y is reached, as shown in Fig. 12–20a. The curved liquid surface in the tube is called a *meniscus*.

If the tube is a cylinder of radius r, the liquid makes contact with the tube along a line of length $2\pi r$. When we isolate the cylinder of liquid of height y and radius r, along with its liquid-vapor film, the total upward force is

$$F = 2\pi r \gamma_{LV} \cos \theta.$$

The downward force is the weight w of the cylinder; this is equal to the weight-density ρg times the volume, which is approximately $\pi r^2 y$, neglecting the small volume of the meniscus itself. Thus

$$w = \rho g \pi r^2 y.$$

Since the cylinder is in equilibrium,

$$\rho g \pi r^2 y = 2\pi r \gamma_{LV} \cos \theta,$$

or

$$y = \frac{2\gamma_{LV} \cos \theta}{\rho g r}. \tag{12–12}$$

The same equation holds for capillary depression, shown in Fig. 12–20b. Capillarity accounts for the rise of ink in blotting paper, the rise of lighter fluid in the wick of a cigarette lighter, and many other common phenomena. Figure 12–21 shows examples of capillarity.

Capillarity is very important in a variety of life processes. A familiar example is the rising of water (actually a dilute aqueous solution) from the roots of a plant to its foliage, due partly to capillarity and osmotic pressure developed in the roots. In the higher animals, including man, blood is pumped through the arteries and veins, but capillarity is still important in the smallest blood vessels, which indeed are called capillaries.

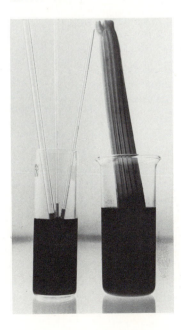

12–21 Water rises in glass tubes and celery stalk as a result of capillary action. The elevation of the liquid level is greater in small-diameter tubes than in larger ones. (Nancy Rogers, Exploratorium.)

A related phenomenon is that of *negative pressure*. Although ordinarily the stress in a liquid is of a compressive nature, there are circumstances in which liquids can sustain actual *tensile* stresses. Imagine a cylindrical tube, closed at one end and having a tight-fitting piston at the other. Suppose the tube is completely filled with a liquid that wets both the inner surface of the tube and the piston face; that is, the molecules of liquid adhere to the surfaces. If we pull the piston out, we expect that the liquid will pull away from the piston, leaving an empty space, but if the surfaces are very clean and the liquid very pure, an actual tensile stress, accompanied by an increase in volume, can be observed. It is as though we are stretching the liquid, and it is prevented from pulling away from the walls of the container by adhesive forces.

With water, tensile stresses as large as 300 atm have been observed. In the laboratory this situation is a highly unstable one; a liquid under tension tends to break up into many small droplets. In tall trees, however, negative pressures are a regular occurrence. Negative pressure is thought to be an important mechanism for transport of water and nutrients from the roots to the leaves in the small xylem tubes (diameter of the order of 0.1 mm) in the growing layers of the tree.

QUESTIONS

12–1 Why are very narrow heels found on some women's shoes more likely to make dents in a soft wood floor than wider heels, even though both may support the same weight?

12–2 A rubber hose is attached to a funnel, and the free end is bent around to point upward. When water is poured into the funnel, it rises in the hose to the same level as in the funnel, despite the fact that the funnel has a lot more water in it than the hose. Why?

12–3 If you drive a car into a lake 10 m deep, do you have a better chance of escaping from the car if you try to open a door or if you open a window and crawl out? Why?

12–4 Is the pressure inside a siphon tube greater or less than atmospheric pressure? Does this suggest a limitation on the maximum height over which a siphon can be used?

12–5 How can one instrument function as both a barometer and an altimeter?

12–6 A pan of water sits on a kitchen scale. You stick your finger into the water, without touching the pan. Does the scale reading change? How?

12–7 In describing the size of a large ship, one uses such expressions as "it displaces 20,000 tons." What does this mean? Can the weight of the ship be obtained from this information?

12–8 A popular, though expensive, sport is hot-air ballooning, in which a large nylon balloon is filled with air heated by a gas burner at the bottom. Why must the air be heated? How does the balloonist control his ascent and descent?

12–9 A submarine is 50 m below the surface of the ocean, with its weight exactly balanced by the buoyant force. If it descends to 100 m, does the buoyant force increase, decrease, or remain the same?

12–10 A rigid lighter-than-air balloon filled with helium cannot continue to rise indefinitely. Why not? What determines the maximum height it can attain?

12–11 Does the buoyant force on an object immersed in water change if the object and the container of water are placed in an elevator that accelerates upward?

12–12 Some people can float in fresh water only by taking a deep breath; when they exhale they sink. Why?

12–13 If a person can barely float in sea water, will he or she float more easily in fresh water, or be unable to float at all? Why?

12–14 The purity of gold can be tested by weighing it in air and in water. How? Why is it not feasible to make a fake gold brick by gold-plating some cheaper material?

12–15 Explain how the admonition "go soak your head" can be taken literally to measure the weight of your head without removing it from your body.

12–16 A possible new technology made feasible by the "zero-gravity" environment of an orbiting space station is the making of metal foam. Why is it easier to make metal foam under zero-g conditions than on earth?

12–17 Hot water running into the sink doesn't splash as much as cold water, and it sounds different. Why?

12–18 Why can't a skin diver obtain an air supply at any desired depth by breathing through a "snorkel," a tube connected to his face mask and having its upper end above the water surface?

12–19 Suppose the door of a room makes an airtight but frictionless fit in its frame. Do you think you could open the door if the air pressure on one side were standard atmospheric pressure and that on the other side differed from standard by 1%?

PROBLEMS

12–1 A rectangular block of an unidentified material has dimensions $5 \times 15 \times 30$ cm, and a mass of 2.0 kg.

a) What is the density?

b) Will the material float in water?

12–2 A lead cube has a total mass of 50 kg. What is the length of a side?

12–3 The piston of a hydraulic automobile lift is 30 cm in diameter. What pressure, in pascals, is required to lift a car of mass 1500 kg? Also express this pressure in atmospheres.

12–4 The expansion tank of a household hot-water heating system is open to the atmosphere and is 10 m above a pressure gauge attached to the furnace. What is the gauge pressure at the furnace, in pascals? In atmospheres?

12–5 The liquid in the open-tube manometer in Fig. 12–4a is mercury, and $y_1 = 3$ cm, $y_2 = 8$ cm. Atmospheric pressure is 970 millibars.

a) What is the absolute pressure at the bottom of the U-tube?

b) What is the absolute pressure in the open tube, at a depth of 5 cm below the free surface?

c) What is the absolute pressure of the gas in the tank?

d) What is the gauge pressure of the gas, in "cm of mercury"?

e) What is the gauge pressure in "cm of water"?

12–6 What gauge pressure must a pump produce in order to pump water from the bottom of Grand Canyon (730 m elevation) to Indian Gardens (1370 m)? Express your results in pascals and in atmospheres.

12–7 A diving bell is to be designed to withstand the pressure of water at a depth of 2000 ft.

a) If sea water weighs 64 lb per cubic foot, what is the pressure at this depth?

b) What is the total force on a circular glass window 6 inches in diameter?

12–8 A common method for testing gold for purity is to measure its density by weighing it in air and then in water, as discussed in Section 12–5. In an era of rising gold prices, a swindler proposed to make a fake gold ingot by using a hollow slab of iridium (density 22.5 g·cm^{-3}) and plating it with a thin layer of gold (density 19.3 g·cm^{-3}).

a) To make a fake ingot of total mass 0.5 kg, what should be the total volume?

b) What is the volume of the interior air space?

c) Is iridium really any cheaper than gold?

12–9 A brass statue has a weight in air of 125 N.

a) What is its volume?

b) What is its apparent weight when immersed in water?

12–10 A certain body weighs 12 N in air and 10 N when immersed in water. Find its total volume and its density.

12–11 According to advertising claims, a certain small car will float in water.

a) If the mass of the car is 1000 kg and its interior volume 4.0 m^3, what fraction of the car is immersed when it floats? The buoyancy of steel and other materials may be neglected.

b) As water gradually leaks in and displaces the air in the car, what fraction of the interior volume is water when the car sinks?

12–12

a) A small test tube, partially filled with water, is inverted in a large jar of water and floats, as shown in Fig. 12–22. The lower end of the test tube is open and the top of the large jar is covered by a tightly fitting rubber membrane. When the membrane is pressed down the test tube sinks, and when the membrane is released it rises again. Explain. (A hollow glass figure in human form is often used instead of the test tube and is called a "Cartesian diver.")

Figure 12–22

b) A torpedoed ship sinks below the surface of the ocean. If the depth is sufficiently great, is it possible for the ship to remain suspended in equilibrium at some point above the ocean bottom?

12–13 When an iceberg floats in sea water, what fraction of its volume is submerged? (The ice, being of glacial origin, is fresh water.)

12–14 A slab of ice floats on a fresh-water lake. What minimum volume must the slab have in order for an 80-kg man to be able to stand on it without getting his feet wet?

12–15 A tube 1 cm^2 in cross section is attached to the top of a vessel 1 cm high and of cross section 100 cm^2. Water is poured into the system, filling it to a depth of 100 cm above the bottom of the vessel, as in Fig. 12–23.

a) What is the force exerted by the water against the bottom of the vessel?

b) What is the weight of the water in the system?

c) Explain why (a) and (b) are not equal.

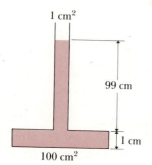

1 cm^2

99 cm

1 cm

Figure 12–23   100 cm^2

12–16 A piece of gold–aluminum alloy weighs 10 lb. When suspended from a spring balance and submerged in water, the balance reads 8 lb. What is the weight of gold in the alloy if the specific gravity of gold is 19.3 and the specific gravity of aluminum is 2.5?

12–17 What is the area of the smallest block of ice 0.5 m thick that will just support a man of mass 100 kg? The density of the ice is 0.917 g·cm^{-3}, and it is floating in fresh water.

12–18 A cubical block of wood 10 cm on a side floats at the interface between oil and water as in Fig. 12–24, with

its lower surface 2 cm below the interface. The density of the oil is 0.6 g·cm^{-3}.

a) What is the mass of the block?

b) What is the gauge pressure at the lower face of the block?

12–19 The densities of air, helium, and hydrogen (at standard conditions) are, respectively, 1.29 kg·m^{-3}, 0.178 kg·m^{-3}, and 0.0899 kg·m^{-3}. What is the volume in cubic meters displaced by a hydrogen-filled dirigible that has a total "lift" of 10,000 kg? What would be the "lift" if helium were used instead of hydrogen?

12–20 A piece of wood is 0.5 m long, 0.2 m wide, and 0.02 m thick. Its specific gravity is 0.6. What volume of lead must be fastened underneath to sink the wood in calm water so that its top is just even with the water level?

12–21 A cubical block of wood 10 cm on a side and of density 0.5 g·cm^{-3} floats in a jar of water. Oil of density 0.8 g·cm^{-3} is poured on the water until the top of the oil layer is 4 cm below the top of the block.

a) How deep is the oil layer?

b) What is the gauge pressure at the lower face of the block?

12–22 A cubical block of steel (density = 7.8 g·cm^{-3}) floats on mercury (density = 13.6 g·cm^{-3}).

a) What fraction of the block is above the mercury surface?

b) If water is poured on the mercury surface, how deep must the water layer be so that the water surface just rises to the top of the steel block?

12–23 Block A in Fig. 12–25 hangs by a cord from spring balance D and is submerged in a liquid C contained in beaker B. The mass of the beaker is 1 kg; the mass of the liquid is 1.5 kg. Balance D reads 2.5 kg and balance E reads 7.5 kg. The volume of block A is 0.003 m^3.

a) What is the mass per unit volume of the liquid?

b) What will each balance read if block A is pulled up out of the liquid?

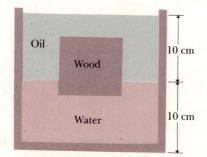

Oil

Wood

10 cm

Water

10 cm

Figure 12–24

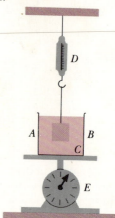

D

A   B

C

E

Figure 12–25

12–24 A hollow sphere of inner radius 9 cm and outer radius 10 cm floats half submerged in a liquid of specific gravity 0.8.

a) Calculate the density of the material of which the sphere is made.

b) What would be the density of a liquid in which the hollow sphere would just float completely submerged?

12–25 Two spherical bodies having the same diameter are released simultaneously from the same height. If the mass of one is 10 times that of the other and if the air resistance on each is the same, show that the heavier body will arrive at the ground first.

12–26 A hydrometer consists of a spherical bulb and a cylindrical stem of cross section 0.4 cm^2. The total volume of bulb and stem is 13.2 cm^3. When immersed in water the hydrometer floats with 8 cm of the stem above the water surface. In alcohol, 1 cm of the stem is above the surface. Find the density of the alcohol.

12–27 The following is quoted from a letter. How would you reply?

It is the practice of carpenters hereabouts, when laying out and leveling up the foundations of relatively long buildings, to use a garden hose filled with water, into the ends of the hose being thrust glass tubes 10 to 12 inches long.

The theory is that the water, seeking a common level, will be of the same height in both the tubes and thus effect a level. Now the question rises as to what happens if a bubble of air is left in the hose. Our greybeards contend the air will not affect the reading from one end to the other. Others say that it will cause important inaccuracies.

Can you give a relatively simple answer to this question, together with an explanation? I include a rough sketch [Fig. 12–26] of the situation that caused the dispute.

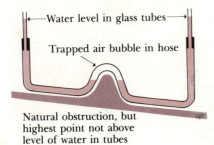

Figure 12–26

Water level in glass tubes

Trapped air bubble in hose

Natural obstruction, but highest point not above level of water in tubes

12–28 A swimming pool measures 25 × 8 × 3 m deep. Compute the force exerted by the water against either end and against the bottom.

12–29 The upper edge of a vertical gate in a dam lies along the water surface. The gate is 2 m wide and is hinged along the bottom edge, which is 3 m below the water surface. What is the torque about the hinge?

12–30 The upper edge of a gate in a dam runs along the water surface. The gate is 2 m high and 3 m wide and is hinged along a horizontal line through its center. Calculate the torque about the hinge.

12–31 Figure 12–27 is a cross-sectional view of a masonry dam whose length perpendicular to the diagram is 30 m. The depth of water behind the dam is 10 m. The masonry of which the dam is constructed has a density of 3000 kg·m^{-3}.

a) Find the dimensions x and $2x$, if the weight of the dam is to be 10 times as great as the horizontal force exerted on it by the water.

b) Is the dam then stable with respect to its overturning about the edge through point O?

c) How does the size of the reservoir behind the dam affect the answers above?

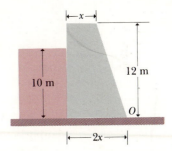

Figure 12–27

12–32 A U-tube of length l (see Fig. 12–28) contains a liquid. What is the difference in height between the liquid columns in the vertical arms,

a) if the tube has an acceleration a toward the right, and

b) if the tube is mounted on a horizontal turntable rotating with an angular velocity ω, with one of the vertical arms on the axis of rotation?

c) Explain why the difference in height does not depend on the density of the liquid, or on the cross-sectional area of the tube. Would it be the same if the vertical tubes did not have equal cross sections? Would it be the same if the horizontal portion were tapered from one end to the other?

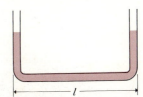

Figure 12–28

12–33 Compare the tension of a soap bubble with that of a rubber balloon in the following respects:

a) Has each a surface tension?

b) Does the surface tension depend on area?

c) Is Hooke's law applicable?

12–34 Water can rise to a height y in a certain capillary. Suppose that this tube is immersed in water so that only a length $y/2$ is above the surface. Will you have a fountain or not? Explain your reasoning.

12–35 A capillary tube is dipped in water with its lower end 10 cm below the water surface. Water rises in the tube to a height of 4 cm above that of the surrounding liquid, and the angle of contact is zero. What gauge pressure is required to blow a hemispherical bubble at the lower end of the tube?

12–36 A glass tube of inside diameter 1 mm is dipped vertically into a container of mercury, with its lower end 1 cm below the mercury surface.

a) What must be the gauge pressure of air in the tube to blow a hemispherical bubble at its lower end?

b) To what height will mercury rise in the tube if the air pressure in the tube is 3000 Pa below atmospheric? The angle of contact between mercury and glass is 140°.

12–37 Find the excess pressure inside a water drop of radius 2 mm, and inside one of radius 0.01 mm, typical of water droplets in fog. Assume $T = 20°C$.

12–38 What radius must a water drop have for the difference between inside and outside pressures to be 0.1 atm?

12–39 On a day when the atmospheric pressure is 950 millibars,

a) what would be the height of the mercury column in a barometric tube of inside diameter 2 mm?

b) What would be the height in the absence of any surface-tension effects?

c) What is the minimum diameter a barometric tube may have in order that the correction for capillary depression shall be less than 0.01 cm of mercury?

12–40

a) Derive the expression for the height of capillary rise in the space between two parallel plates dipping in a liquid.

b) Two glass plates, parallel to each other and separated by 0.5 mm, are dipped in water. To what height will the water rise between them? Assume zero angle of contact.

12–41 Find the gauge pressure, in dynes per square centimeter, in a soap bubble 5 cm in diameter. The surface tension is 25 dyn·cm^{-1} = 25 × 10^{-3} N·m^{-1}.

12–42 When mercury is poured onto a flat glass surface, which it does not wet, it spreads out into a pool of uniform thickness regardless of the size of the pool. Find the thickness of the pool.

12–43 Suppose the xylem tubes in the actively growing outer layer of a tree are uniform cylinders, and that the rising of sap is due entirely to capillarity, with a contact angle of 45° and surface tension of 0.05 N·m^{-1}. What is the maximum radius of the tubes for a tree 20 m tall? Assume the density of sap to be the same as that of water.

12–44 A tree 50 m tall has xylem tubes (sap-carrying pipes) in the form of uniform cylinders of radius 2 × 10^{-4} mm. If the surface tension is 0.05 N·m^{-1} and the contact angle 45°, what minimum pressure must be developed in the roots in order for the sap to reach the top of the tree, if the pressure at the top is one atmosphere? (See Problem 12–43.)

13

FLUID DYNAMICS

Fluid dynamics is the study of fluids in motion. It is one of the most complex branches of mechanics, as illustrated by such familiar examples of fluid flow as a river in flood or a swirling cloud of cigarette smoke. While each drop of water or each smoke particle is governed by Newton's laws of motion, the resulting equations can be exceedingly complex. Fortunately, many situations of practical importance can be represented by idealized models that are simple enough to permit detailed analysis.

We begin this chapter with a discussion of fluids that are incompressible and have no internal friction or viscosity. Such a fluid is called an *ideal fluid*. The assumption of incompressibility is usually a good approximation for liquids. A gas can also be treated as incompressible provided the flow is such that pressure differences are not too great. Internal friction in a fluid gives rise to shear stresses when two adjacent layers of fluid move relative to each other, or when the fluid flows inside a tube or around an obstacle. In some cases these shear forces can be neglected in comparison with gravitational forces and forces arising from pressure differences.

13–1

DESCRIPTION OF FLOW

The path followed by an element of a moving fluid is called a *flow line*. In general, the velocity of the element changes in both magnitude and direction along its line of flow. If every element passing through a given point follows the same flow line as that of preceding elements, the flow is said to be *steady* or *stationary*. When any given flow is first started, it passes through a nonsteady state, but in many instances the flow becomes steady after a certain period of time has elapsed. In steady flow, the velocity at each point of space remains constant in time, although the velocity of a particular particle of the fluid may change as it moves from one point to another.

13–1   A flow tube bounded by streamlines.

A *streamline* is defined as a curve whose tangent, at any point, is in the direction of the fluid velocity at that point. In steady flow, the stream-lines coincide with the flow lines.

If we construct all of the streamlines passing through the periphery of an element of area, such as the area A in Fig. 13–1, these lines enclose a tube called a *flow tube*. From the definition of a streamline, no fluid can cross the side walls of a tube of flow; in steady flow there can be no mixing of the fluids in different flow tubes.

Figure 13–2 illustrates the nature of the flow around a number of obstacles, and in a channel of varying cross section. The photographs were made using an apparatus designed by Pohl, in which alternate streams of clear and colored water flow between two closely spaced glass plates. The obstacles and the channel walls are opaque flat plates that fit between the glass plates. Each obstacle is completely surrounded by a flow tube. The tube splits into two portions at a so-called stagnation point on the upstream side of the obstacle. These portions rejoin at a

(a)

(b)

(c)

(d)

13–2   (a), (b), (c): Streamline flow around obstacles of various shapes. (d) Flow in a channel of varying cross-sectional area.

second stagnation point on the downstream side. The velocity at the stagnation points is zero. The cross sections of all flow tubes decrease at a constriction and increase again when the channel widens.

The patterns of Fig. 13–2 are typical of *streamline* or *laminar* flow, in which adjacent layers of fluid slide smoothly past each other. At sufficiently high flow rates, or when boundary surfaces cause abrupt changes in velocity, the flow becomes irregular and much more complex and is called *turbulent* flow. In turbulent flow there is, strictly speaking, no steady-state pattern, since the flow pattern continuously changes.

13–2

THE EQUATION OF CONTINUITY

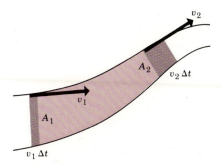

13–3 Flow into and out of a portion of a tube of flow.

Let us consider any stationary, closed surface in a moving fluid; in general, fluid flows into the volume enclosed by the surface at some points and flows out at other points. The *equation of continuity* is a mathematical statement of conservation of mass; the *net* rate of flow of mass *inward* across any closed surface is equal to the rate of increase of the mass within the surface.

For an incompressible fluid in steady flow, the equation takes the following form. Figure 13–3 represents a portion of a flow tube, between two fixed cross sections of areas A_1 and A_2. Let v_1 and v_2 be the speeds at these sections. There is no flow across the side wall of the tube because at every point on the wall the velocity is tangent to the wall. The volume of fluid that flows into the tube across A_1 in a time interval Δt is that contained in the short cylindrical element of base A_1 and height $v_1 \Delta t$, that is, $A_1 v_1 \Delta t$. If the density of the fluid is ρ, the *mass* flowing in is $\rho A_1 v_1 \Delta t$. Similarly, the mass that flows out across A_2 in the same time is $\rho A_2 v_2 \Delta t$. The volume between A_1 and A_2 is constant, and since the flow is steady, the mass flowing out equals that flowing in. Hence

$$\rho A_1 v_1 \, \Delta t = \rho A_2 v_2 \, \Delta t,$$

or

$$A_1 v_1 = A_2 v_2, \tag{13–1}$$

and the product Av is constant along any given tube of flow. It follows that when the cross section of a flow tube decreases, as in the constriction in Fig. 13–2d, the velocity increases. This can readily be shown by introducing small particles into the fluid and observing their motion.

13–3

BERNOULLI'S EQUATION

When an incompressible fluid flows along a horizontal flow tube of varying cross section, its velocity must change, as Eq. (13–1) shows. A force is required to produce this acceleration, and for this force to be caused by the fluid surrounding a particular element of fluid, the pressure must be different in different regions. If the pressure were the same everywhere, the net force on any fluid element would be zero. Thus when the cross section of a flow tube varies, the pressure *must* vary along the tube, even when there is no difference in elevation. If the elevation also changes,

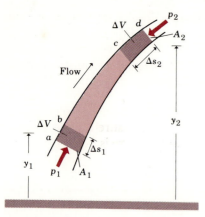

13–4 The net work done on the shaded element equals the increase in its kinetic and potential energy.

there is an additional pressure difference. Bernoulli's equation is a general expression that relates the pressure difference between two points in a flow tube to both velocity changes and elevation changes. Daniel Bernoulli developed this relationship in 1738.

To derive the Bernoulli equation, we apply the work–energy theorem to the fluid in a section of a flow tube. In Fig. 13–4, we consider the fluid that at some initial time lies between the two cross sections a and c. In a time interval Δt, the fluid that was initially at a moves to b, a distance $\Delta s_1 = v_1 \, \Delta t$, where v_1 is the speed at this end of the tube. In the same interval the fluid initially at c moves to d, a distance $\Delta s_2 = v_2 \, \Delta t$. The cross-sectional areas at the two ends are A_1 and A_2, as shown. Because of the continuity relation, the volume of fluid ΔV passing any cross section in time Δt is $\Delta V = A_1 \, \Delta s_1 = A_2 \, \Delta s_2$.

We can compute the *work* done on this fluid during Δt. The force on the cross section at a is $p_1 A_1$, and that at c is $p_2 A_2$, where p_1 and p_2 are the pressures at the two ends. The net *work* done on the element during this displacement is therefore

$$W = p_1 A_1 \, \Delta s_1 - p_2 A_2 \, \Delta s_2$$
$$= (p_1 - p_2) \, \Delta V. \tag{13–2}$$

The negative sign in the second term arises because the force at c is opposite in direction to the displacement.

We now equate this work to the total change in energy, kinetic and potential, of the element. The kinetic energy of the fluid between sections b and c does not change. At the beginning of Δt, the fluid between a and b has volume $A_1 \, \Delta s_1$, mass $\rho A_1 \, \Delta s_1$, and kinetic energy $\frac{1}{2}\rho(A_1 \, \Delta s_1)v_1^2$. Similarly, at the end of Δt, the fluid between c and d has kinetic energy $\frac{1}{2}\rho(A_2 \, \Delta s_2)v_2^2$. Thus the net change in kinetic energy is

$$\Delta K = \tfrac{1}{2}\rho \, \Delta V(v_2^2 - v_1^2). \tag{13–3}$$

The change in potential energy is obtained similarly. The potential energy of the mass entering at a in time Δt is $\Delta m \, gy_1 = \rho \, \Delta V \, gy_1$, and that of the mass leaving at c is $\Delta m \, gy_2 = \rho \Delta V \, gy_2$.

The net change in potential energy is

$$\Delta U = \rho \, \Delta Vg(y_2 - y_1). \tag{13–4}$$

Combining Eqs. (13–2), (13–3), and (13–4) in the work–energy theorem $W = \Delta K + \Delta U$, we obtain

$$(p_1 - p_2) \, \Delta V = \tfrac{1}{2}\rho \, \Delta V(v_2^2 - v_1^2) + \rho \, \Delta V \, g(y_2 - y_1)$$

or

$$p_1 - p_2 = \tfrac{1}{2}\rho(v_2^2 - v_1^2) + \rho g(y_2 - y_1). \tag{13–5}$$

In this form, Bernoulli's equation represents the equality of the work per unit volume of fluid ($p_1 - p_2$) to the sum of the changes in kinetic and potential energies per unit volume that occur during the flow. Or we may interpret Eq. (13–5) in terms of pressures. The second term on the right is the pressure difference arising from the weight of the fluid and the difference in elevation of the two ends of the fluid element. The first term on the right is the additional pressure difference associated with the change of velocity of the fluid. Equation (13–5) can also be written

$$p_1 + \rho g y_1 + \tfrac{1}{2}\rho v_1^2 = p_2 + \rho g y_2 + \tfrac{1}{2}\rho v_2^2 \qquad (13\text{–}6)$$

and since the subscripts 1 and 2 refer to *any* two points along the tube of flow, Bernoulli's equation may also be written

$$p + \rho g y + \tfrac{1}{2}\rho v^2 = \text{constant}. \qquad (13\text{–}7)$$

Note carefully that p is the *absolute* (not gauge) pressure, and that a consistent set of units must be used. In SI units, pressure is expressed in Pa, density in $kg \cdot m^{-3}$, and velocity in $m \cdot s^{-1}$.

EXAMPLE Water enters a house through a pipe 2.0 cm in inside diameter, at an absolute pressure of 4×10^5 Pa (about 4 atm). The pipe leading to the second-floor bathroom 5 m above is 1.0 cm in diameter. When the flow velocity at the inlet pipe is $4 \ m \cdot s^{-1}$, find the flow velocity and pressure in the bathroom.

Solution The flow velocity is obtained from the continuity equation:

$$v_2 = \frac{A_1}{A_2} v_1 = \frac{\pi(1.0 \text{ cm})^2}{\pi(0.5 \text{ cm})^2}(4 \ m \cdot s^{-1}) = 16 \ m \cdot s^{-1}.$$

The pressure is now obtained from Bernoulli's equation:

$$\begin{aligned}
p_2 &= p_1 - \tfrac{1}{2}\rho(v_2^2 - v_1^2) - \rho g(y_2 - y_1) \\
&= 4 \times 10^5 \text{ Pa} - \tfrac{1}{2}(1.0 \times 10^3 \ kg \cdot m^{-3})(256 \ m^2 \cdot s^{-2} - 16 \ m^2 \cdot s^{-2}) \\
&\quad - (1.0 \times 10^3 \ kg \cdot m^{-3})(9.8 \ m \cdot s^{-2})(5 \text{ m}) \\
&= 2.3 \times 10^5 \text{ Pa} \simeq 2.3 \text{ atm}.
\end{aligned}$$

We note that when the water is turned off, the second term on the right vanishes and the pressure rises to 3.5×10^5 Pa. ◀

13–4

APPLICATIONS OF BERNOULLI'S EQUATION

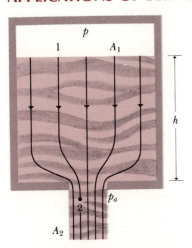

13–5 Flow of a liquid out of an orifice.

1. The equations of hydrostatics are special cases of Bernoulli's equation, when the velocity is everywhere zero. Thus, when v_1 and v_2 are zero, Eq. (13–5) reduces to

$$p_1 - p_2 = \rho g(y_2 - y_1),$$

which is the same as Eq. (12–3).

2. Speed of efflux. Torricelli's theorem. Figure 13–5 represents a tank of cross-sectional area A_1, filled to a depth h with a liquid of density ρ. The space above the top of the liquid contains air at pressure p, and the liquid flows out of an orifice of area A_2. Let us consider the entire volume of moving fluid as a single tube of flow, and let v_1 and v_2 be the speeds at points 1 and 2. The quantity v_2 is called the *speed of efflux*. The pressure at point 2 is atmospheric, p_a. Applying Bernoulli's equation to points 1 and 2, and taking the bottom of the tank as our reference level, we get

$$p + \tfrac{1}{2}\rho v_1^2 + \rho g h = p_a + \tfrac{1}{2}\rho v_2^2,$$

or

$$v_2{}^2 = v_1{}^2 + 2\frac{p - p_a}{\rho} + 2gh. \tag{13–8}$$

From the equation of continuity,

$$v_2 = \frac{A_1}{A_2}v_1. \tag{13–9}$$

Because of the converging of the streamlines as they approach the orifice, the cross section of the stream continues to diminish for a short distance outside the tank. It is the area of smallest cross section, known as the *vena contracta*, that should be used in Eq. (13–9). For a sharp-edged circular opening, the area of the *vena contracta* is about 65 percent as great as the area of the orifice.

Now let us consider some special cases. Suppose the tank is open to the atmosphere, so that

$$p = p_a \qquad \text{and} \qquad p - p_a = 0.$$

Suppose also that $A_1 \gg A_2$. Then $v_1{}^2$ is very much less than $v_2{}^2$ and can be neglected, and from Eq. (13–8),

$$v_2 = \sqrt{2gh}. \tag{13–10}$$

That is, *the speed of efflux is the same as the speed a body would acquire in falling freely through a height h.* This is *Torricelli's theorem.* It is not restricted to an opening in the bottom of a vessel, but applies also to a hole in the side walls at a depth h below the surface.

Now suppose again that the ratio of areas is such that $v_1{}^2$ is negligible and that the pressure p (in a closed vessel) is so large that the term $2gh$ in Eq. (13–8) can be neglected, compared with $2(p - p_a)/\rho$. The speed of efflux is then

$$v_2 = \sqrt{2(p - p_a)/\rho}. \tag{13–11}$$

The density ρ is that of the fluid escaping from the orifice. If the vessel is partly filled with a liquid, as in Fig. 13–5, ρ is the density of the liquid. On the other hand, if the vessel contains only a gas, ρ is the density of the gas. The efflux speed of a gas may be very great, even for small pressures, since its density is small. However, if the pressure is too great, our idealized model is no longer adequate. Compressibility of the gas must be considered, and if the speed is too great, the motion may become turbulent. Bernoulli's equation can no longer be applied to the motion under these conditions.

A flow of fluid out of an orifice in a vessel gives rise to a *thrust* or *reaction force* on the remainder of the system. The mechanics of the problem are the same as those involved in rocket propulsion. The thrust can be computed as follows, provided conditions are such that Bernoulli's equation is applicable. If A is the area of the orifice, ρ the density of the escaping fluid, and v the speed of efflux, the mass of fluid flowing out in time Δt is $\rho A v \, \Delta t$, and its momentum (mass × velocity) is $\rho A v^2 \, \Delta t$. Since we are neglecting the relatively small velocity of the fluid in the container, we can say that the escaping fluid started from rest and acquired

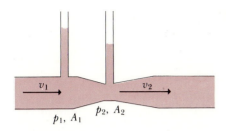

13–6 The Venturi tube.

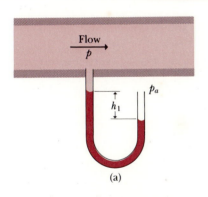

(a)

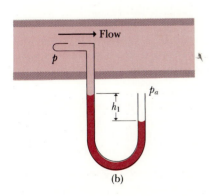

(b)

13–7 Pressure gauges for measuring the static pressure p in a fluid flowing in an enclosed channel.

the momentum above in time Δt. Its *rate of change* of momentum was therefore $\rho A v^2$, and from Newton's second law this equals the force acting on it. By Newton's third law, an equal and opposite reaction force acts on the remainder of the system. Taking the expression for v^2 from Eq. (13–11), we can write the reaction force as

$$F = \rho A v^2 = \rho A \frac{2(p - p_a)}{\rho},$$

or

$$F = 2A(p - p_a). \tag{13–12}$$

Thus, while the *speed* of efflux is inversely proportional to the density, the *thrust* is independent of the density and depends only on the area of the orifice and the gauge pressure $p - p_a$.

3. The *Venturi tube*, illustrated in Fig. 13–6, consists of a constriction or throat inserted in a pipeline and having properly designed tapers at inlet and outlet to avoid turbulence. Bernoulli's equation, applied to the wide and to the constricted portions of the pipe, becomes

$$p_1 + \tfrac{1}{2}\rho v_1^2 = p_2 + \tfrac{1}{2}\rho v_2^2.$$

From the equation of continuity, the speed v_2 is greater than the speed v_1, and hence the pressure p_2 in the throat is *less* than the pressure p_1. Thus a net force to the right acts to accelerate the fluid as it enters the throat, and a net force to the left slows it as it leaves. The pressures p_1 and p_2 can be measured by attaching vertical side tubes as shown in the diagram. From a knowledge of these pressures and of the cross-sectional areas A_1 and A_2, the velocities and the mass rate of flow can be computed. When used for this purpose, the device is called a *Venturi meter*.

The reduced pressure at a constriction finds a number of technical applications. A mixture of air and partly vaporized gasoline is drawn into the manifold of an internal combustion engine by the low pressure produced in a Venturi throat in the carburetor. The *aspirator pump* is a Venturi throat through which water is forced. Air is drawn into the low-pressure water rushing through the constricted portion.

4. *Measurement of pressure in a moving fluid.* The pressure p in a fluid flowing in an enclosed channel can be measured with an open-tube manometer, as shown in Fig. 13–7. In (a), one arm of the manometer is connected to an opening in the channel wall. In (b), a *probe* is inserted in the stream. The probe should be small enough so that the flow is not appreciably disturbed and should be shaped so as to avoid turbulence. The difference h_1 in height of the liquid in the arms of the manometer is proportional to the difference between atmospheric pressure p_a and the fluid pressure p. That is,

$$p_a - p = \rho_m g h_1,$$
$$p = p_a - \rho_m g h_1, \tag{13–13}$$

where ρ_m is the density of the manometer liquid.

The *Pitot tube*, shown in Fig. 13–8, is a probe with an opening at its upstream end. A stagnation point forms at the opening, where the pressure is p_2 and the speed is zero. Applying Bernoulli's equation to the

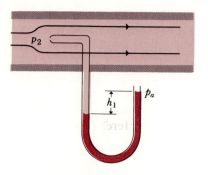

13–8 The Pitot tube.

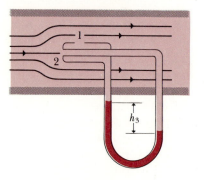

13–9 The Prandtl tube.

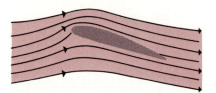

13–10 Flow lines around an airfoil.

stagnation point, and to a point at a large distance from the probe where the pressure is p and the speed is v, we find

$$p_2 = p + \tfrac{1}{2}\rho v^2 \qquad (13\text{–}14)$$

(ρ is the density of the flowing fluid). The pressure p_2 at the stagnation point is therefore the sum of the pressure p and the quantity $\tfrac{1}{2}\rho v^2$.

The quantity $\tfrac{1}{2}\rho v^2$ is sometimes called the *dynamic pressure* and p the *static pressure*. These terms are somewhat misleading, inasmuch as p is always the true pressure (force per unit area) in both static and dynamic situations.

The instrument shown in Fig. 13–9 combines in a single device the functions of the instruments in Figs. 13–7 and 13–8. This device is called a *Prandtl tube;* the term Pitot tube is also used. The pressure at opening 1 corresponds to the pressure p measured in Fig. 13–7, and that at opening 2, a stagnation point, is the quantity $p_2 = p + \tfrac{1}{2}\rho v^2$ measured in Fig. 13–8. The manometric height h_3 is proportional to the difference of these, or to $\tfrac{1}{2}\rho v^2$. Hence

$$\tfrac{1}{2}\rho v^2 = \rho_{\mathrm{m}} g h_3. \qquad (13\text{–}15)$$

This instrument is self-contained, and its reading does not depend on atmospheric pressure. If held at rest, it can be used to measure the velocity of a stream of fluid flowing past it. If mounted on an aircraft, it indicates the velocity of the aircraft relative to the surrounding air and is known as an *airspeed indicator.*

5. Lift on an aircraft wing. Figure 13–10 shows flow lines around a section of an aircraft wing or an airfoil. The orientation of the wing relative to the flow direction causes the flow lines to crowd together above the wing, corresponding to increased flow velocity in this region, much as in the throat of a Venturi. Hence the region above the wing is one of increased velocity and reduced pressure, while below the wing the pressure remains nearly atmospheric. Because the upward force on the underside of the wing is greater than the downward force on the top side, there is a net upward force or *lift.*

This phenomenon can also be understood qualitatively on the basis of Newton's laws. In order for the fluid to exert a net upward force on the wing, the wing must exert a *downward* reaction force on the fluid, deflecting the stream downward. This effect is shown in Fig. 13–10; the fluid suffers a net change in the vertical component of momentum as it passes the wing, corresponding to the downward force the wing exerts on it. The force *on* the wing is thus *upward,* in agreement with the above discussion.

As the angle of the wing relative to the flow increases, turbulent flow occurs in a larger and larger region above the wing, and the pressure drop is no longer as great as that predicted by Bernoulli's principle. The lift on the wing decreases, and in extreme cases the airplane stalls.

6. The curved flight of a spinning ball. This effect is somewhat more subtle than the lift on an airplane wing, but the same basic principles are involved. Figure 13–11a represents a stationary ball in a blast of air moving from right to left. The motion of the air stream around and past the ball is the same as though the ball were moving through still air from

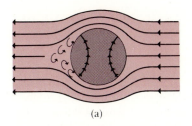

(a)

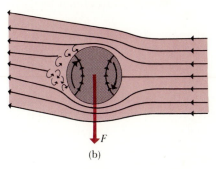

$\downarrow F$

(b)

13–11 (a) Air flow past a stationary ball, showing symmetric region of turbulence behind ball. (b) Air flow past a spinning ball, showing asymmetric region of turbulence and deflection of airstream. The net force on the ball is in the direction shown. Motion of the air from right to left relative to the ball corresponds to motion of a ball through still air from left to right.

left to right. Because of the large velocities ordinarily involved, there is a region of turbulent flow behind the ball, as shown.

When the ball is spinning, as in Fig. 13–11b, the viscosity of air causes layers of air near the ball's surface to be pulled around in the direction of spin. The velocity of air relative to the ball surface is greater at the top than at the bottom. The region of turbulence becomes asymmetric, with turbulence occurring farther forward on the top side than on the bottom. This asymmetry causes a pressure difference; the average pressure at the top becomes greater than that at the bottom. The corresponding net downward force deflects the ball as shown. In a baseball curve pitch, the actual deflection is sideways, and in that case Fig. 13–11 shows a *top* view of the situation.

Corresponding to the downward force on the ball in Fig. 13–11b, the ball must exert an equal and opposite reaction force *upward* on the air. This deflects the air stream upward near the ball, as the figure shows, just as the lift force on an airplane wing corresponds to a downward deflection of the air stream.

A similar effect occurs with golf balls, which always have "backspin" from impact with the slanted club face. The resulting pressure difference between top and bottom of the ball causes a "lift" force that keeps the ball in the air considerably longer than would be possible without spin. A well-hit drive appears from the tee to "float" or even curve *upward* during the initial portion of its flight, and this is a real effect, not an illusion. The dimples on the ball play an essential role; an undimpled ball has a much shorter trajectory than a dimpled one given the same initial velocity and spin. Some manufacturers even claim that the *shape* of the dimples is significant; one feels that polygonal dimples are superior to round ones! Figure 13–12 shows the backspin of a golf ball just after being struck by the club.

13–12 Stroboscopic photograph of a golf ball being struck by a club. The picture was taken at 1000 flashes per second; the ball rotates once in four pictures, corresponding to an angular velocity of 250 revolutions per second. (Dr. Harold Edgerton, M.I.T., Cambridge, Massachusetts.)

13–5

VISCOSITY

Viscosity is internal friction of a fluid. Because of viscosity, a force must be exerted to cause one layer of a fluid to slide past another, or to cause one surface to slide past another if there is a layer of fluid between the surfaces. Both liquids and gases exhibit viscosity, although liquids are much more viscous than gases. The situation is similar to that of shear stress and strain in a solid.

The simplest example of flow of a viscous fluid is flow between two parallel plates, as shown in Fig. 13–13. The bottom plate is stationary, while the top plate moves with constant speed v. It is found that the fluid in contact with each surface has the same speed as that surface; thus at the top surface the fluid has speed v, while the fluid adjacent to the bottom surface is at rest. The speeds of intermediate layers of fluid increase uniformly from one surface to the other, as shown by the arrows.

Flow of this type is called *laminar*. (A lamina is a thin sheet.) The layers of liquid slide over one another, and a portion of the liquid, which at some instant has the shape *abcd*, will a moment later take the shape *abc'd'* and will become more and more distorted as the motion continues. That is, the liquid is in a state of continually increasing shear strain.

In order to maintain the motion, a constant force must be exerted to the right on the upper, moving plate, and hence indirectly on the upper liquid surface. This force tends to drag the fluid, and the lower plate as well, to the right. Therefore an equal force must be exerted toward the left on the lower plate to hold it stationary. These forces are lettered F in Fig. 13–13. If A is the area of the fluid over which these forces are applied (i.e., the surface area of the plates), the ratio F/A is the shear *stress* exerted on the fluid.

When a shear stress is applied to a *solid*, the effect of the stress is to produce a certain displacement of the solid, such as *dd'*. The shear strain is defined as the ratio of this displacement to the transverse dimension l, and within the elastic limit the shear stress is proportional to the shear strain. With a *fluid*, on the other hand, the shear strain increases without limit so long as the stress is applied, and the stress is found by experiment to depend not on the shear strain, but on its *rate of change*. The strain in Fig. 13–13 (at the instant when the volume of fluid has the shape *abc'd'*) is *dd'/ad*, or *dd'/l*. Since l is constant, the *rate of change* of strain equals $1/l$ times the rate of change of *dd'*. But the rate of change of *dd'* is simply the *speed* of point *d'*, that is, the speed v of the moving wall.

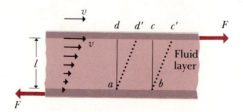

13–13   Laminar flow of a viscous fluid.

Hence

$$\text{Rate of change of shear strain} = \frac{v}{l}.$$

The rate of change of shear strain is also referred to simply as the *strain rate*.

The *coefficient of viscosity* of the fluid, or simply its *viscosity* η, is defined as the ratio of the shear stress, F/A, to the rate of change of shear strain:

$$\eta = \frac{\text{shear stress}}{\text{rate of change of shear strain}} = \frac{F/A}{v/l},$$

or

$$F = \eta A \frac{v}{l}. \tag{13–16}$$

For a liquid that flows readily, such as water or kerosene, the shear stress is relatively small for a given strain rate, and the viscosity also is relatively small. For a liquid such as molasses or glycerine, a greater shear stress is necessary for the same strain rate, and the viscosity is correspondingly greater. Viscosities of gases at ordinary pressures and temperatures are very much *smaller* than those of common liquids. Viscosities of all fluids are markedly dependent on temperature, increasing for gases and decreasing for liquids as the temperature is increased, hence the expression "as slow as molasses in January." An important consideration in the design of oils for engine lubrication is to *reduce* the temperature variation of viscosity as much as possible.

From Eq. (13–16), the unit of viscosity is that of force times distance, divided by area times velocity. In SI units it is

$$1\,\text{N}\cdot\text{m}\cdot\text{m}^{-2}(\text{m}\cdot\text{s}^{-1})^{-1} = 1\,\text{N}\cdot\text{s}\cdot\text{m}^{-2}.$$

The corresponding cgs unit, $1\,\text{dyn}\cdot\text{s}\cdot\text{cm}^{-2}$, is the only viscosity unit in common use and is called 1 *poise* in honor of the French scientist Poiseuille. Thus,

$$1\,\text{poise} = 1\,\text{dyn}\cdot\text{s}\cdot\text{cm}^{-2} = 10^{-1}\,\text{N}\cdot\text{s}\cdot\text{m}^{-2}.$$

Small viscosities are expressed in *centipoises* ($1\,\text{cp} = 10^{-2}$ poise) or *micropoises* ($1\,\mu\text{p} = 10^{-6}$ poise). A few typical values are given in Table 13–1.

TABLE 13–1 TYPICAL VALUES OF VISCOSITY

Temperature, C°	Viscosity of castor oil, poise	Viscosity of water, centipoise	Viscosity of air, micropoise
0	53	1.792	171
20	9.86	1.005	181
40	2.31	0.656	190
60	0.80	0.469	200
80	0.30	0.357	209
100	0.17	0.284	218

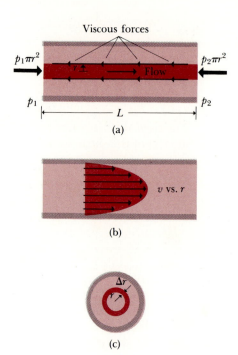

Viscous forces

$p_1\pi r^2$ $p_2\pi r^2$

r Flow

p_1 p_2

L

(a)

v vs. r

(b)

Δr

(c)

13–14 (a) Forces on a cylindrical element of a viscous fluid. (b) Velocity distribution for viscous flow.

Not all fluids behave according to the direct proportionality of force and velocity predicted by Eq. (13–16). An interesting exception is blood, for which velocity increases more rapidly than force. Thus doubling the force in Fig. 13–14 produces *more* than a twofold increase in velocity. This behavior may be understood on the basis that, on a microscopic scale, blood is not a homogeneous fluid but rather a suspension of solid particles in a liquid. The suspended particles have characteristic shapes; for example, red cells are roughly disk-shaped. At small velocities their orientations are random, but as velocity increases they tend to become oriented so as to facilitate flow. The fluids that provide lubrication in human joints exhibit similar behavior.

Fluids for which Eq. (13–16) holds are called *newtonian fluids;* we see that this description is an idealized model, which not all fluids obey. In general, fluids that are suspensions or dispersions are often nonnewtonian in their viscous behavior. Nevertheless, Eq. (13–16) provides a useful model to describe approximately the properties of many pure substances.

13–6

POISEUILLE'S LAW

When a viscous fluid flows in a tube, the flow velocity is different at different points of a cross section. The outermost layer of fluid clings to the walls of the tube, and its velocity is zero. The tube walls exert a backward drag on this layer, which in turn drags backward on the next layer beyond it, and so on. If the velocity is not too great, the flow is *laminar*, with a velocity that is greatest at the center of the tube and decreases to zero at the walls. The flow is like that of a number of telescoping tubes sliding relative to one another, the central tube advancing most rapidly and the outer tube remaining at rest.

Let us consider the variation of velocity with radius for a cylindrical pipe of inner radius R. We consider the flow of a cylindrical element of fluid coaxial with the pipe, of radius r and length L, as shown in Fig. 13–14a. The force on the left end is $p_1\pi r^2$, and that on the right end $p_2\pi r^2$, as shown. The net force is thus

$$F = (p_1 - p_2)\pi r^2.$$

Since the element does not accelerate, this force must just balance the viscous retarding force at the surface of this element. This force is given by Eq. (13–16), but since the velocity does not vary uniformly with distance from the center, we must replace v/l in this expression with $\Delta v/\Delta r$, where Δv is the small change of velocity when we go from distance r to $r + \Delta r$ from the axis. The area over which the viscous force acts is the surface area of the cylinder of radius r; its circumference is $2\pi r$ and its length L, so the area is $A = 2\pi r L$. Thus the viscous force is

$$F = \eta 2\pi r L \frac{\Delta v}{\Delta r}.$$

Equating this to the net force due to pressure on the ends and rearranging, we find that

$$\frac{\Delta v}{\Delta r} = -\frac{(p_1 - p_2)r}{2\eta L}.$$

This shows that the velocity changes more and more rapidly with r as we go from the center ($r = 0$) to the pipe wall ($r = R$). Finding how the velocity itself varies with r requires methods of the calculus; the correct expression is

$$v = \frac{p_1 - p_2}{4\eta L}(R^2 - r^2). \tag{13–17}$$

The velocity decreases from a maximum value $(p_1 - p_2)R^2/4\eta L$ at the center to zero at the wall. Thus we see that the maximum velocity is proportional to the *square* of the pipe radius and is also proportional to the pressure change per unit length $(p_1 - p_2)/L$, called the *pressure gradient*. The curve in Fig. 13–14b is a graph of Eq. (13–17) with the v-axis horizontal and the r-axis vertical.

Equation (13–17) may be used to find the *total* rate of flow of fluid through the pipe. The velocity at each point is proportional to the pressure gradient $(p_1 - p_2)/L$, so the total flow rate must also be proportional to this quantity. The maximum velocity at the center of the pipe is proportional to R^2, so we expect the *average* velocity also to be proportional to R^2. Furthermore, the total flow rate is equal to the total area πR^2 multiplied by the average velocity, and so we expect the total flow rate to be proportional to R^4. Derivation of the complete formula again requires methods of the calculus, but the form of the result is exactly as predicted. The total volume of flow per unit time, denoted by Q, is given by

$$Q = \frac{\pi}{8}\frac{R^4}{\eta}\frac{p_1 - p_2}{L}. \tag{13–18}$$

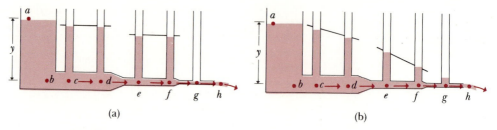

13-15 Pressures along a horizontal tube in which is flowing (a) an ideal fluid, (b) a viscous fluid.

This relation was first derived by Poiseuille* and is called *Poiseuille's law*. The volume rate of flow is inversely proportional to viscosity, as might be expected. It is proportional to the pressure gradient along the pipe, and it varies as the fourth power of the radius. For example, if the radius is halved, the flow rate is reduced by a factor of 16. This relation is familiar to physicians in connection with the selection of needles for hypodermic syringes. Needle size is much more important than thumb pressure in determining the flow rate from the needle; doubling the needle diameter has the same effect as increasing the thumb force sixteenfold. Similarly, blood flow in arteries and veins can be controlled over a wide range by relatively small changes in diameter. This is an important temperature control mechanism in warm-blooded animals. Similarly, the pain of angina pectoris is caused by narrowing of the coronary arteries, resulting in inadequate blood supply to the heart muscle tissue. Nitroglycerin and other medications relieve this condition by dilating these arteries slightly. Because of Poiseuille's law, a slight dilation can result in a substantially increased blood flow.

The difference between the flow of an ideal nonviscous fluid and one having viscosity is shown in Fig. 13–15, where fluid is flowing along a horizontal tube of varying cross section. The height of the fluid in the small vertical tubes is proportional to the gauge pressure.

The fluid in part (a) has no viscosity. The pressure at b is very nearly the static pressure $\rho g y$, since the velocity is small in the large tank. The pressure at c is less than at b because the fluid must accelerate between these points. The pressures at c and d are equal, however, since the velocity and elevation at these points are the same. There is a further pressure drop between d and e, and between f and g. The pressure at g is atmospheric, and the gauge pressure at this point is zero.

Part (b) of the diagram shows the effect of viscosity. Again, the pressure at b is nearly the static pressure $\rho g y$. There is a pressure drop from b to c, due now in part to viscous effects, and also a further drop from c to d. The pressure gradient in this part of the tube is represented by the slope of the black line. The drop from d to e results in part from acceleration and in part from viscosity. The pressure gradient between e and f is greater than between c and d because of the smaller radius in this portion. Finally, pressure at g is somewhat above atmospheric, since there is now a pressure gradient between this point and the end of the tube.

*Approximate pronunciation: Pwah-zoy'.

*13–7

STOKES' LAW

When an ideal fluid of zero viscosity flows past a sphere, or when a sphere moves through a stationary fluid, the streamlines form a perfectly symmetrical pattern around the sphere, as shown in Fig. 13–2a. The pressure at any point on the upstream hemispherical surface is exactly the same as that at the corresponding point on the downstream face, and the resultant force on the sphere is zero. If the fluid has viscosity, however, there will be a viscous drag on the sphere. (A viscous drag is experienced by a body of *any* shape, but only for a sphere is the drag readily calculable.)

We shall not attempt to derive the expression for the viscous force directly from the laws of flow of a viscous fluid. The only quantities on which the force can depend are the viscosity η of the fluid, the radius r of the sphere, and its velocity v relative to the fluid. A complete analysis shows that the force F is given by

$$F = 6\pi\eta r v. \tag{13–19}$$

This equation was first deduced by Sir George Stokes in 1845 and is called *Stokes' law*. We have already used it in Section 4–5, Eq. (4–10), to study the motion of a body falling in a viscous fluid. At that point it was necessary to know only that the viscous force on a body moving in a fluid is proportional to its velocity.

A sphere falling in a viscous fluid reaches a *terminal velocity* v_T at which the viscous retarding force plus the buoyant force equals the weight of the sphere. Let ρ be the density of the sphere and ρ' the density of the fluid. The weight of the sphere is then $(4/3)\pi r^3 \rho g$, and the buoyant force is $(4/3)\pi r^3 \rho' g$; when the terminal velocity is reached, the total force is zero, and

$$\tfrac{4}{3}\pi r^3 \rho' g + 6\pi\eta r v_T = \tfrac{4}{3}\pi r^3 \rho g,$$

or

$$v_T = \frac{2}{9}\frac{r^2 g}{\eta}(\rho - \rho'). \tag{13–20}$$

When the terminal velocity of a sphere of known radius and density is measured, the viscosity of the fluid in which it is falling can be found from the equation above. Conversely, if the viscosity is known, the radius of the sphere can be determined by measuring the terminal velocity. This method was used by Millikan to determine the radius of very small electrically charged oil drops (used to measure the electrical charge of an individual electron) by observing free fall in air.

Even for nonspherical bodies, a relation of the form of Eq. (13–19) holds, with a different numerical coefficient. Biologists call the terminal velocity the *sedimentation velocity* or *sedimentation rate*. This principle can be used to measure the radii of macromolecules, usually using a centrifuge to increase the sedimentation rate, as discussed in Section 5–9. Also, the mass of a macromolecule is proportional to r^3 (i.e., to the volume), while the resisting force is proportional to r. Thus differential sedimentation provides a means of separating molecules of different molecular masses.

REYNOLDS NUMBER

When the velocity of a fluid flowing in a tube exceeds a certain critical value (which depends on the properties of the fluid and the diameter of the tube), the nature of the flow becomes extremely complicated. Within a very thin layer adjacent to the tube walls, called the *boundary layer*, the flow is still laminar. The flow velocity in the boundary layer is zero at the tube walls and increases uniformly throughout the layer. The properties of the boundary layer are of the greatest importance in determining the resistance to flow and the transfer of heat to or from the moving fluid.

Beyond the boundary layer, the motion is highly irregular. Random local circular currents called *vortices* develop within the fluid, with a large increase in the resistance to flow. Flow of this sort is called *turbulent*.

Experiment indicates that a combination of four factors determines whether the flow of a fluid through a tube or pipe is laminar or turbulent. This combination is known as the *Reynolds number*, N_R, and is defined as

$$N_R = \frac{\rho v D}{\eta}, \tag{13–21}$$

where ρ is the density of the fluid, v the average forward velocity, η the viscosity, and D the diameter of the tube. (The average velocity is defined as the uniform velocity over the entire cross section of the tube, which would result in the same volume rate of flow.) The Reynolds number, $\rho v D/\eta$, is a *dimensionless* quantity and has the same numerical value in any consistent system of units. For example, for water at 20°C flowing in a tube of diameter 1 cm with an average velocity of 10 cm·s^{-1}, the Reynolds number is

$$N_R = \frac{\rho v D}{\eta} = \frac{(1 \text{ g·cm}^{-3})(10 \text{ cm·s}^{-1})(1 \text{ cm})}{0.01 \text{ dyn·s·cm}^{-2}} = 1000.$$

Had the four quantities been expressed originally in mks units, the same value of 1000 would have been obtained.

A variety of experiments have shown that when the Reynolds number is less than about 2000 the flow is laminar, whereas above about 3000 the flow is turbulent. In the transition region between 2000 and 3000 the flow is unstable and may change from one type to the other. Thus for water at 20°C flowing in a tube 1 cm in diameter, the flow is laminar when

$$\frac{\rho v D}{\eta} \leq 2000,$$

or when

$$v \leq \frac{(2000)(0.01 \text{ dyn·s·cm}^{-2})}{(1 \text{ g·cm}^{-3})(1 \text{ cm})} = 20 \text{ cm·s}^{-1}.$$

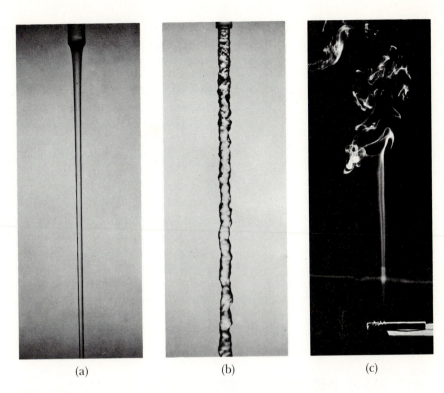

13–16   (a) Laminar flow. (b) Turbulent flow. (c) First laminar, then turbulent.

(a)   (b)   (c)

Above about $30 \text{ cm} \cdot \text{s}^{-1}$ the flow is turbulent. If air at the same temperature were flowing at $30 \text{ cm} \cdot \text{s}^{-1}$ in the same tube, the Reynolds number would be

$$N_\text{R} = \frac{(0.0013 \text{ g} \cdot \text{cm}^{-3})(30 \text{ cm} \cdot \text{s}^{-1})(1 \text{ cm})}{181 \times 10^{-6} \text{ dyn} \cdot \text{s} \cdot \text{cm}^{-2}} = 215.$$

Since this is much less than 3000, the flow would be laminar and would not become turbulent unless the velocity were as great as $420 \text{ cm} \cdot \text{s}^{-1}$.

The distinction between laminar and turbulent flow is shown in the photographs of Fig. 13–16. In (a) and (b) the fluid is water and in (c) air and smoke particles. Figure 13–17 shows the onset of turbulence in the flow of a liquid past a cylindrical obstacle.

13–17   An example of the onset of turbulence produced by a cylindrical obstacle. Upstream the flow is laminar, but downstream flow shows the formation of eddys or vortices. The alternating series of vortices is called a Karman vortex street. The pattern is made visible by injecting smoke filaments into the air upstream from the obstacle. (Courtesy Department of Aeronautics, Imperial College of Science and Technology.)

Blood flow in human arteries is not normally turbulent but can become so if a constriction increases the flow velocity enough. In the usual method of measuring blood pressure, the cuff placed around the arm is pumped up to a pressure greater than the maximum (systolic) pressure. This cuts off the arterial blood flow completely. Then the pressure is gradually decreased until the artery opens for a moment at the time of peak pressure. The noise caused by the resulting turbulent flow can be heard as a tapping sound in the artery at the elbow. The pressure in the cuff is read when this sound is first heard.

The Reynolds number of a system forms the basis for the study of the behavior of real systems through the use of small scale models. A common example is the wind tunnel, in which one measures the aerodynamic forces on a scale model of an aircraft wing. The forces on a full-size wing are then deduced from these measurements.

Two systems are said to be *dynamically similar* if the Reynolds number, $\rho v D/\eta$, is the same for both. The letter D may refer, in general, to any dimension of a system, such as the span or chord of an aircraft wing. Thus the flow of a fluid of given density ρ and viscosity η, about a half-scale model, is dynamically similar to that around the full-size object if the velocity v is twice as great.

QUESTIONS

13–1 Is the continuity relation, Eq. (13–1), valid for compressible fluids? If not, is there a similar relation that *is* valid?

13–2 If the velocity at each point in space in steady-state fluid flow is constant, how can a fluid particle accelerate?

13–3 Whenever possible, airplanes take off and land heading into the wind. Why?

13–4 Does the "lift" of an airplane wing depend on altitude?

13–5 How does a baseball pitcher give the ball the spin that makes it curve? Can he make it curve in either direction? Does it matter whether he is right-handed or left-handed? What is a spitball? Why is it illegal?

13–6 When a car on a highway is passed by a large truck, the car is sometimes pulled *toward* the truck. What does Bernoulli's relation have to say about this?

13–7 In a store-window vacuum cleaner display, a table-tennis ball is suspended in midair in a jet of air blown from the outlet hose of a tank-type vacuum cleaner. The ball bounces around a little but always returns to the center of the jet, even if it is tilted. How does this behavior illustrate the Bernoulli relation?

13–8 Why do jet airplanes usually fly at altitudes of about 30,000 feet, even though it takes a lot of fuel to climb to that altitude?

13–9 What causes the sharp hammering sound sometimes heard from a water pipe when an open faucet is suddenly turned off?

13–10 When a smooth-flowing stream of water comes out of a faucet, it narrows as it falls, and if it falls far enough it eventually breaks up into drops. Why does it narrow? Why does it break up?

13–11 A tornado consists of a rapidly whirling air vortex. Why is the pressure always much lower in the center than at the outside? How does this condition account for the destructive power of a tornado?

13–12 When paddling a canoe, one can attain a certain critical speed with relatively little effort, and then a much greater effort is required to make the canoe go even a little faster. Why?

13–13 Why does air escaping from the open end of a small pipe make a lot more noise than air escaping at the same volume rate from a large pipe?

13–14 Hot water running into a sink makes less noise and splashes less than cold water. Part of the difference is caused by varying surface tension, but part is due to something else. What?

13–15 When water is running in the shower, the shower curtain seems to be pulled in toward the falling water, rather than being blown outward. Why?

PROBLEMS

13–1 A circular hole 2 cm in diameter is cut in the side of a large standpipe, 10 m below the water level in the standpipe. Find (a) the velocity of efflux, and (b) the volume discharged per unit time. Neglect the contraction of the streamlines after emerging from the hole.

13–2 Water stands at a depth H in a large open tank whose side walls are vertical (Fig. 13–18). A hole is made in one of the walls at a depth h below the water surface.

  a) At what distance R from the foot of the wall does the emerging stream of water strike the floor?

  b) At what height above the bottom of the tank could a second hole be cut so that the stream emerging from it would have the same range?

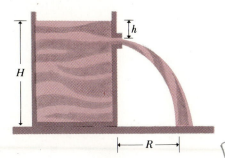

Figure 13–18

13–3 A cylindrical vessel, open at the top, is 20 cm high and 10 cm in diameter. A circular hole whose cross-sectional area is 1 cm^2 is cut in the center of the bottom of the vessel. Water flows into the vessel from a tube above it at the rate of 140 cm$^3 \cdot$s^{-1}. How high will the water in the vessel rise?

13–4 At a certain point in a horizontal pipeline the velocity is 2 m$\cdot$s^{-1} and the gauge pressure is 1.0×10^4 Pa above atmospheric. Find the gauge pressure at a second point in the line if the cross-sectional area at the second point is one-half that at the first. The liquid in the pipe is water.

13–5 Water in an enclosed tank is subjected to a gauge pressure of 2×10^4 Pa, applied by compressed air introduced into the top of the tank. There is a small hole in the side of the tank 5 m below the level of the water. Calculate the speed with which water escapes from this hole.

13–6 What gauge pressure is required in the city mains in order that a stream from a fire hose connected to the mains may reach a vertical height of 20 m?

13–7 A tank of large area is filled with water to a depth of 0.3 m. A hole of 5 cm^2 cross section in the bottom allows water to drain out in a continuous stream.

  a) What is the rate at which water flows out of the tank, in m$^3 \cdot$s^{-1}?

  b) At what distance below the bottom of the tank is the cross-sectional area of the stream equal to one half the area of the hole?

13–8 A sealed tank containing sea water to a height of 2 m also contains air above the water at a gauge pressure of 40 atm. Water flows out from a hole at the bottom. The cross-sectional area of the hole is 10 cm^2.

  a) Calculate the efflux velocity of the water.

  b) Calculate the reaction force on the tank exerted by the water in the emergent stream.

13–9 A pipeline 0.2 m in diameter, flowing full of water, has a constriction of diameter 0.1 m. If the velocity in the 0.2-m portion is 2 m$\cdot$s^{-1}, find (a) the velocity in the constriction, and (b) the discharge rate in cubic meters per second.

13–10 A horizontal pipe of 0.2 m^2 cross section tapers to a cross section of 0.1 m^2. If water is flowing with a velocity of 50 m$\cdot$min^{-1} in the large pipe where a pressure gauge reads 0.8×10^5 Pa, what is the gauge pressure in the adjoining part of the small pipe? The barometer reads 76 cm of mercury.

13–11 At a certain point in a pipeline the velocity is 1 m$\cdot$s^{-1} and the gauge pressure is 3×10^5 Pa. Find the gauge pressure at a second point in the line 20 m lower than the first, if the cross section at the second point is one-half that at the first. The liquid in the pipe is water.

13–12 Water stands at a depth of 1 m in an enclosed tank whose side walls are vertical. The space above the water surface contains air at a gauge pressure of 8×10^5 Pa. The tank rests on a platform 2 m above the floor. A hole of cross-sectional area 1 cm^2 is made in one of the side walls just above the bottom of the tank.

  a) Where does the stream of water from the hole strike the floor?

  b) What is the vertical force exerted on the floor by the stream?

  c) What is the horizontal force exerted on the tank? Assume the water level and the pressure in the tank to remain constant, and neglect any affect of viscosity.

13–13 Water flows steadily in a pipeline of constant cross section leading out of an elevated tank. At a point 2 m below the water level in the tank the gauge pressure in the flowing stream is 10^4 Pa.

  a) What is the velocity of the water at this point?

  b) If the pipe rises to a point 3 m above the level of the water in the tank, what are the velocity and the pressure at the latter point?

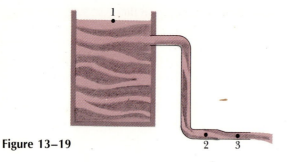

Figure 13–19

13–14 Water flows steadily from a reservoir, as in Fig. 13–19. The elevation of point 1 is 10 m; of points 2 and 3 it is 1 m. The cross section at point 2 is 0.04 m² and at point 3 it is 0.02 m². The area of the reservoir is very large compared with the cross sections of the pipe.

a) Compute the gauge pressure at point 2.

b) Compute the discharge rate in cubic meters per second.

13–15 Water flows through a horizontal pipe of cross-sectional area 10 cm². At one section the cross-sectional area is 5 cm². The pressure difference between the two sections is 300 Pa. How many cubic meters of water will flow out of the pipe in 1 min?

13–16 Two very large open tanks, A and F (Fig. 13–20), both contain the same liquid. A horizontal pipe BCD, having a constriction at C, leads out of the bottom of tank A, and a vertical pipe E opens into the constriction at C and dips into the liquid in tank F. Assume streamline flow and no viscosity. If the cross section at C is one-half that at D, and if D is at distance h_1 below the level of the liquid in A, to what height h_2 will liquid rise in pipe E? Express your answer in terms of h_1. Neglect changes in atmospheric pressure with elevation.

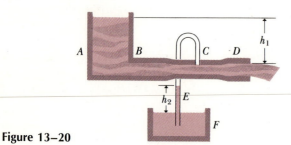

Figure 13–20

13–17 At a certain point in a horizontal pipeline the gauge pressure is 0.5×10^5 Pa. At another point the gauge pressure is 0.3×10^5 Pa. If the areas of the pipe at these two points are 20 cm² and 10 cm², respectively, compute the number of cubic meters of water that flow across any cross section of the pipe per minute.

13–18 Water flowing in a horizontal pipe discharges at the rate of 0.004 m³·s⁻¹. At a point in the pipe where the cross section is 0.001 m², the absolute pressure is 1.2×10^5 Pa. What must be the cross section of a constriction in the pipe such that the pressure there is reduced to 1.0×10^5 Pa?

13–19 The pressure difference between the main pipeline and the throat of a Venturi meter is 10^5 Pa. The areas of the pipe and the constriction are 0.1 m² and 0.05 m², respectively. How many cubic meters per second are flowing through the pipe? The liquid in the pipe is water.

13–20 The section of pipe shown in Fig. 13–21 has a cross section of 40 cm² at the wider portions and 10 cm² at the constriction. The discharge from the pipe is 3000 cm³·s⁻¹.

a) Find the velocities at the wide and the narrow portions.

b) Find the pressure difference between these portions.

c) Find the difference in height between the mercury columns in the U-tube.

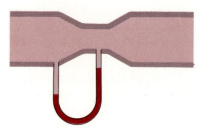

Figure 13–21

13–21 Water is used as the manometric liquid in a Prandtl tube mounted in an aircraft to measure airspeed. If the maximum difference in height between the liquid columns is 0.1 m, what is the maximum airspeed that can be measured? The density of air is 1.3 kg·m⁻³.

13–22 In Problem 13–21, suppose the manometric liquid is mercury. What is the maximum airspeed that can be measured?

13–23 In a wind-tunnel experiment the pressure on the top surface of an airplane wing was 0.90×10^5 Pa, and the pressure on the bottom surface 0.91×10^5 Pa. If the area of each surface is 40 m², what is the net lift force on the wing?

13–24 An airplane of mass 6000 kg has a wing area of 60 m². If the pressure on the lower wing surface is 0.60×10^5 Pa during level flight at an elevation of 4000 m, what is the pressure on the upper wing surface?

13–25 Water at 20°C flows through a pipe of radius 1.0 cm. If the flow velocity at the center is 10 cm·s⁻¹, find the pressure drop along a 2-m section of pipe due to viscosity.

13–26 An aluminum ball of radius 1.0 cm falls through water at 20°C.

a) What is the terminal velocity, assuming laminar flow and including buoyancy?

b) What is the Reynolds number? Is the flow really laminar?

13–27 A viscous liquid flows through a tube with laminar flow, as in Fig. 13–14b. Prove that the volume rate of flow is the same as if the velocity were uniform at all points of a cross section and equal to half the velocity at the axis.

13–28

a) With what terminal velocity will an air bubble 1 mm in diameter rise in a liquid of viscosity 150 cp and density 0.90 $g \cdot cm^{-3}$?

b) What is the terminal velocity of the same bubble in water?

13–29

a) With what velocity is a steel ball 1 mm in radius falling in a tank of glycerine at an instant when its acceleration is one half that of a freely falling body?

b) What is the terminal velocity of the ball? The densities of steel and of glycerine are 8.5 $g \cdot cm^{-3}$ and 1.32 $g \cdot cm^{-3}$, respectively, and the viscosity of glycerine is 8.3 poise.

13–30 Assume that air is streaming horizontally past an aircraft wing such that the velocity is 40 $m \cdot s^{-1}$ over the top surface and 30 $m \cdot s^{-1}$ past the bottom surface. If the wing has a mass of 300 kg and an area of 5 m^2, what is the net force on the wing? The density of air is 1.3 $kg \cdot m^{-3}$.

13–31 Modern airplane design calls for a "lift" of about 1000 N per square meter of wing area. Assume that air flows past the wing of an aircraft with streamline flow. If the velocity of flow past the lower wing surface is 100 $m \cdot s^{-1}$, what is the required velocity over the upper surface to give a "lift" of 1000 $N \cdot m^{-2}$? The density of air is 1.3 $kg \cdot m^{-3}$.

13–32 Water at 20°C flows with a speed of 50 $cm \cdot s^{-1}$ through a pipe of diameter 3 mm.

a) What is the Reynolds number?

b) What is the nature of the flow?

13–33 Water at 20°C is pumped through a horizontal smooth pipe 15 cm in diameter and discharges into the air. If the pump maintains a flow velocity of 30 $cm \cdot s^{-1}$,

a) What is the nature of the flow?

b) What is the discharge rate in liters per second?

13–34 The tank at the left in Fig. 13–15a has a very large cross section and is open to the atmosphere. The depth $y = 40$ cm. The cross sections of the horizontal tubes leading out of the tank are respectively 1 cm^2, 0.5 cm^2, and 0.2 cm^2. The liquid is ideal, having zero viscosity.

a) What is the volume rate of flow out of the tank?

b) What is the velocity in each portion of the horizontal tube?

c) What are the heights of the liquid in the vertical side tubes?

Suppose that the liquid in Fig. 13–15b has a viscosity of 0.5 poise, a density of 0.8 $g \cdot cm^{-3}$, and that the depth of liquid in the large tank is such that the volume rate of flow is the same as in part (a) above. The distance between the side tubes at c and d, and between those at e and f, is 20 cm. The cross sections of the horizontal tubes are the same in both diagrams.

d) What is the difference in level between the tops of the liquid columns in tubes c and d?

e) In tubes e and f?

f) What is the flow velocity on the axis of each part of the horizontal tube?

13–35

a) Is it reasonable to assume that the flow in the second part of Problem 13–34 is laminar?

b) Would the flow be laminar if the liquid were water?

13–36 Oil having a viscosity of 300 centipoise and a density of 0.90 $g \cdot cm^{-3}$ is to be pumped from one large open tank to another through 1 km of smooth steel pipe 15 cm in diameter. The line discharges into the air at a point 30 m above the level of the oil in the supply tank.

a) What gauge pressure, in atmosphere, must the pump exert in order to maintain a flow of 0.05 $m^3 \cdot s^{-1}$?

b) What is the power consumed by the pump?

13–37 In the Reynolds number example following Eq. (13–21), convert the given quantities to SI units; compute the Reynolds number and verify that it has the same value as when the original cgs units are used.

TEMPERATURE AND EXPANSION

With this chapter we begin a study of a class of phenomena called *thermal* or *heat* phenomena. For this study we need to introduce a new fundamental physical quantity, *temperature*, in addition to the three fundamental quantities (mass, length, and time) used in our study of mechanical systems in preceding chapters. We discuss the concept of thermal equilibrium and various temperature-measuring instruments, that is, *thermometers*, along with the various temperature scales in common use. Finally, we consider the expansion and contraction of materials caused by temperature changes.

14–1

TEMPERATURE AND THERMAL EQUILIBRIUM

The concept of temperature springs from the qualitative ideas of "hot" and "cold" based on our sense of touch. A body that feels hot to the touch is said to have a higher temperature than the same body when it feels cold. More generally, many properties of matter depend on temperature. The length of a metal rod increases when the rod becomes hotter. The pressure of gas in a container increases when it becomes hotter; a steam boiler may explode if it becomes too hot. The electrical properties of materials change with temperature; most metals are poorer electrical conductors when hot than when cold. When a material is extremely hot it glows "red-hot" or "white-hot," with a color that depends on its temperature.

The temperature of a body is also directly related to the motions of the molecules in the material. We shall consider this relationship in detail in Chapter 20, but it is also important to understand that temperature can (and should) be defined independently of any consideration of molecular motion.

To define temperature quantitatively, we must devise a scheme to assign numbers to various degrees of hot or cold. To do this, we can make use of some measurable property of a system that varies with its

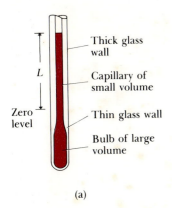

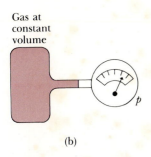

14–1 (a) A system whose state is specified by the value of the length L. (b) A system whose state is given by the value of the pressure p.

hotness or coldness. A simple example is a liquid such as mercury or ethanol in a bulb attached to a thin tube, as in Fig. 14–1a. The significant property of this system is the length L of the liquid column. When the system becomes hotter, the liquid rises in the tube and the value of L increases. Another simple system is a quantity of gas in a constant-volume container, as shown in Fig. 14–1b. The pressure p, measured by the gauge, increases or decreases as the gas becomes hotter or colder. A third example is the electrical resistance R of a conducting wire, which also varies with hotness or coldness.

In each of these examples, a significant quantity that changes with the state of hotness or coldness of the system, such as the length L, the pressure p, or the resistance R, can be used as a quantitative description of the temperature of the system. Thus each device can be used as a *thermometer*.

We have not yet described how such a thermometer can be used to measure the temperature of any other body, because in each case the quantity measured indicates the temperature *of the thermometer*. What good is a thermometer that can measure only its own temperature?

A little thought about how thermometers are actually used shows the way around this obstacle and also leads us to another important concept, that of *thermal equilibrium*. Suppose we want to measure the temperature of a cup of hot coffee with a mercury thermometer initially at room temperature. We immerse the thermometer in the coffee; the thermometer becomes hotter, and the coffee cools off a little. After we observe that no further change is taking place as a result of this interaction, we read the thermometer. This equilibrium state, in which no further change in the states of the system occurs as a result of their interaction with each other, is called a state of *thermal equilibrium*. But how do we know that at thermal equilibrium the temperature of the thermometer is the same as that of the coffee? For this some additional experiments are required.

First we consider the interaction of two systems; let system A be the tube-and-liquid system of Fig. 14–1a and system B be the container of gas and pressure gauge of Fig. 14–1b. When the two systems are brought into contact, in general the values of L and p change as the systems move toward thermal equilibrium. Initially one system was hotter than the other, and each system changes the state of the other.

If, however, the systems are separated by an insulating material such as wood, plastic foam, or fiber glass, they influence each other more slowly. We can think of an idealized insulating material that permits no interactions at all between the two systems. Such an ideal insulator prevents the attainment of thermal equilibrium if the two systems are not intially in thermal equilibrium. This, incidentally, is exactly what the insulating material in a camping cooler does. We don't want the cold beer inside to warm up and attain thermal equilibrium with the hot summer air outside, and the insulation prevents this, although not forever.

Now we consider three systems, A, B, and C, as shown in Fig. 14–2. We need not describe in detail what they are, but we suppose that initially they are *not* in thermal equilibrium. Perhaps one is very hot, one very cold, and the third lukewarm. We surround the three systems with an insulating box to prevent interaction with anything except each other.

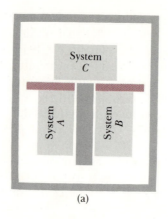

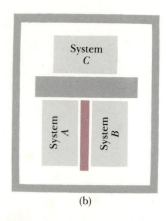

(a)   (b)

14–2   The zeroth law of thermodynamics. (a) If A and B are each in thermal equilibrium with C, then (b) A and B are in thermal equilibrium with each other. Thick gray layers represent insulating walls; thinner colored layers represent conducting walls.

We also separate systems A and B by an insulating wall, the gray slab in Fig. 14–2a, but we let system C interact with both A and B, as indicated symbolically by a thin colored slab, a thermal *conductor*. We wait until no further change is occurring as a result of this interaction, i.e., until thermal equilibrium is attained. We then remove C from A and B by inserting an insulating wall, as in Fig. 14–2b, and we also remove the insulating wall between A and B, replacing it by a conducting wall. What happens?

Experiment shows that *nothing* happens; the states of A and B do not change! Thus the experiment shows that if C is initially in thermal equilibrium with both A and B, then when A and B are placed in contact they are also in thermal equilibrium with each other! This outcome may have seemed obvious to the reader, but that is because of intuition based on everyday experience; this conclusion must be verified experimentally.

These experimental facts may be stated concisely as follows: *Two systems in thermal equilibrium with a third are in thermal equilibrium with each other.* This principle is called the *zeroth law of thermodynamics*. This odd-sounding name refers to the fact that the basic principles of thermodynamics are called the first, second, and third laws of thermodynamics. The principle just discussed is fundamental to all of these, so it deserves to be called the zeroth law.

But what is it about any two systems that determines whether or not they are in thermal equilibrium? It is their *temperature*. Two systems in thermal equilibrium must have the same temperature. As we have seen, the temperature of a system may be represented by a number. Establishing a temperature scale, such as the Fahrenheit or Celsius scale, will be discussed in Section 14–3; this is simply a specific scheme for assigning numbers to temperatures. Once this is done, the condition for thermal equilibrium of two systems is that they have the same temperature. When the temperatures of two systems are different, they *cannot* be in thermal equilibrium.

14–2

THERMOMETERS

One form of thermometer commonly used in everyday life is the tube-and-liquid type shown in Fig. 14–1a. The liquid is usually mercury or colored ethanol. If the liquid level at the freezing point of water is la-

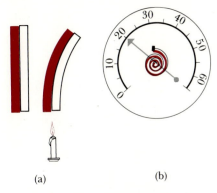

(a)   (b)

14-3 (a) A bimetallic strip bends when heated because one metal expands more than the other. (b) A bimetallic strip, usually in the form of a spiral, may be used as a thermometer.

beled zero and that at the boiling point 100, and if the distance between the two points is divided into 100 equal intervals, the result is the familiar Celsius scale, to be discussed in detail in the next section.

Another type of thermometer in common use contains a metallic strip consisting of two dissimilar metals welded together, as shown in Fig. 14-3a. Because one metal expands more than the other with temperature increases, the strip bends with temperature changes. In practical thermometers this strip is usually made in the form of a spiral, with the outer end anchored to the thermometer case and the inner end attached to a pointer, as shown in Fig. 14-3b. The pointer rotates in response to temperature changes, and a circular scale is provided.

Various other systems are used for thermometers in situations having special requirements such as great precision, speed of attaining equilibrium with the body being measured, or extreme temperature ranges.

One type of electrical thermometer, the *resistance thermometer*, consists of a coil of fine wire, often enclosed in a thin-walled tube for protection. Copper wires lead from the thermometer unit to a resistance-measuring device. Since resistance may be measured with great precision, the resistance thermometer is one of the most precise instruments for the measurement of temperature. In the region of extremely low temperatures, a small carbon cylinder or a small piece of germanium crystal is used instead of a coil of wire.

A different type of electrical thermometer uses a *thermocouple*, a junction of two dissimilar metals that gives rise to an electromotive force or voltage, which depends on the temperature difference between it and a second junction kept at a standard reference temperature.

To measure very high temperatures an *optical pyrometer* can be used. Its principle of operation is to measure the brightness of light emitted by a red-hot or white-hot substance. As shown in Fig. 14-4, it consists of a telescope T with a small electric light bulb L mounted in the tube. The bulb is connected to a variable power supply P, and the brightness of the bulb is varied until it just matches that of the substance being observed.

14-4 An optical pyrometer. The telescope T is pointed toward the hot body B, and the power supply P is adjusted until the brightness of the filament in the lamp L is the same as that of the hot body. The filter F reduces the overall brightness level. The power supply can be calibrated to read temperature directly.

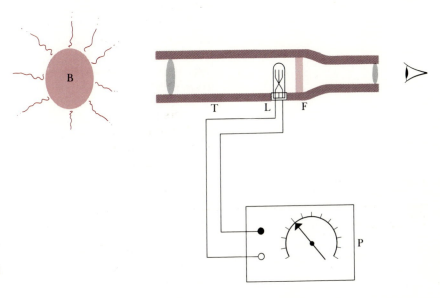

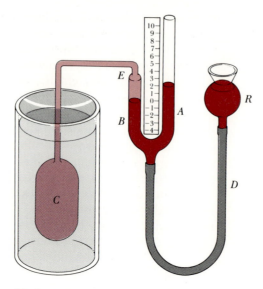

14–5   The constant-volume gas thermometer.

From previous calibration at known temperatures, the power-supply adjustment may be marked to read temperature directly. No part of the instrument contacts the hot substance, so the optical pyrometer can be used at temperatures that would destroy most other thermometers.

Among all temperature-dependent properties, or *thermometric properties*, as they are often called, the pressure of a gas whose volume is kept constant is outstanding in its sensitivity, accuracy of measurement, and reproducibility. The constant-volume gas thermometer is illustrated schematically in Fig. 14–5. The gas, usually helium, is contained in bulb *C* and the pressure exerted by it can be measured by the open-tube mercury manometer. As the temperature of the gas increases, the gas expands, forcing the mercury down in tube *B* and up in tube *A*. Tubes *A* and *B* are connected through a rubber tube *D* to a mercury reservoir *R*. By raising *R*, the mercury level in *B* may be brought back to a reference mark *E*. The gas is thus kept at constant volume. The pressure is determined by the difference in mercury levels in tubes *A* and *B*.

Gas thermometers are used mainly in bureaus of standards and in research laboratories. They are usually large, bulky, and slow in coming to thermal equilibrium.

14–3

THE CELSIUS AND FAHRENHEIT SCALES*

The Celsius temperature scale (formerly called the *centigrade* scale in English-speaking countries) is defined so that the freezing temperature of pure water at normal atmospheric pressure is zero or 0°C (read "zero degrees Celsius") and the boiling temperature of pure water at normal atmospheric pressure is one hundred or 100°C (read "one hundred de-

---

*Named after Anders Celsius (1701–1744) and Gabriel Fahrenheit (1686–1736).

grees Celsius"). Interpolation between or extrapolation beyond these reference temperatures is done using one of the thermometers described in Section 14–2. A temperature corresponding to a condition colder than freezing water is a negative number. The Celsius scale is universally used, both in everyday life and in science and industry, everywhere in the world except for a few English-speaking countries, and all of these except the United States are in the process of converting to the Celsius system.

The scale used in everyday life in the United States is the Fahrenheit scale. It places the freezing temperature of water at 32°F (thirty-two degrees Fahrenheit) and the boiling temperature at 212°F, both at normal atmospheric pressure. Thus there are $212 - 32$, or 180, degrees between freezing and boiling, compared to 100 on the Celsius scale, and one Fahrenheit degree represents only 100/180 or 5/9 as great a temperature change as one Celsius degree.

To convert from Celsius to Fahrenheit, we note that a Celsius temperature T_C is the number of Celsius degrees above freezing; to obtain the number of Fahrenheit degrees above freezing we must multiply this by 9/5. But freezing on the Fahrenehit scale is at 32°F, so to obtain the actual Fahrenheit temperature we must first multiply the Celsius value by 9/5 and then add 32°. Symbolically,

$$T_F = \tfrac{9}{5}T_C + 32°. \tag{14–1}$$

To convert Fahrenheit to Celsius, we solve this equation for T_C, obtaining

$$T_C = \tfrac{5}{9}(T_F - 32°). \tag{14–2}$$

That is, we first substract 32° to obtain the number of Fahrenheit degrees above freezing, and then multiply by $\tfrac{5}{9}$ to obtain the number of Celsius degrees above freezing, i.e., the Celsius temperature.

It is useful to distinguish in our terminology between an actual temperature on a certain scale and a temperature *interval*, representing a difference in the temperatures of two bodies or a change of temperature of a body. It is customary to represent a temperature *interval* of 10° as 10 C° (ten Celsius degrees) and an actual temperature of 20° as 20°C (twenty degrees Celsius). Thus a beaker of water heated from 20°C to 30°C undergoes a temperature change of 10 C°. This usage will be followed throughout this book.

A fundamental problem in defining temperature scales is that when two thermometers, such as a liquid-in-tube system and a resistance thermometer, are calibrated so that they agree to 0°C and 100°C, they do not necessarily agree at intermediate temperatures. Different thermometric properties such as length and resistance don't necessarily change with temperature in precisely the same way. Thus a temperature scale defined in this way always depends somewhat on the specific properties of the material used in the thermometer.

The closest approach to consistency is obtained using a constant-volume gas thermometer with the lowest practical pressures. Very close agreement is found between low-pressure gas thermometers using various gases, and such thermometers are used for high-precision standards. But we are still at the mercy of properties of specific materials. One might well wish for a temperature scale that is completely independent of properties of specific materials. It is in fact possible to define such a scale; we shall return to this fundamental problem in Chapter 19 after the necessary thermodynamic principles have been developed.

The bimetallic strip shown in Fig. 14–3a is also used in most thermostats for controlling heating systems. One end of the bimetallic strip is attached to an electrical contact set, which makes or breaks contact to turn the furnace on or off so as to maintain a constant temperature. This temperature can be adjusted easily by varying the position of the strip. This is a rudimentary example of a *feedback control system;* a physical quantity is measured, and if it does not have the desired value, a system is activated to correct the situation.

A much more sophisticated example of feedback temperature control is found in warm-blooded animals, where body temperature is held constant to within a few tenths of a Celsius degree. The thermometric property that senses blood temperature is provided by a chemical equilibrium condition in a part of the brain called the hypothalamus. This activates appropriate temperature-controlling mechanisms. The most important mechanisms are dilation or contraction of blood vessels near the skin surface (to increase or decrease loss of body heat by conduction) and activation or deactivation of sweat glands to increase or decrease evaporative cooling.

14–4

THE KELVIN AND RANKINE SCALES*

Calibration of a constant-volume gas thermometer might well make use of a graph such as that in Fig. 14–6, showing the relation between gas pressure and temperature. Such a graph suggests that there is a hypothetical temperature at which the pressure would become zero. Surprisingly, this temperature turns out to be the same for a wide variety of gases, namely −273.15°C. It is not possible actually to observe this zero-pressure condition, for several reasons. One is that most gases liquefy and solidify at very low temperatures, and the proportionality of pressure to temperature no longer holds.

This situation invites the speculation that this extrapolated zero-pressure situation may provide a more fundamental way to define the zero-point on a temperature scale than the freezing-point of water. This is in fact the basis of the Kelvin temperature scale. The degrees are the same size as on the Celsius scale, but the zero is shifted so that 0 K = −273.15°C and 273.15 K = 0°C. Thus ordinary room temperature, 20°C, becomes 20 + 273.15, or about 293 K.

In the current SI standard nomenclature, "degree" is not used with the Kelvin scale; the temperature mentioned above is read "293 kelvins," not "degrees Kelvin." Also, Kelvin is capitalized when it refers to the temperature scale, but the *unit* of temperature is the *kelvin,* not capitalized but abbreviated K. (The usage "degrees Kelvin," although officially obsolete, is still fairly common.)

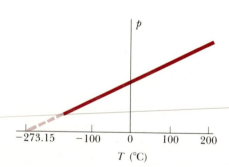

14–6 Graph of pressure versus temperature for a constant-volume gas thermometer. A straight-line extrapolation predicts that the pressure would become zero at −273.15°C if that temperature could be reached and the proportionality of pressure to temperature held exactly.

---

*Named after Baron Kelvin (1824–1907) and William Rankine (1820–1872).

The Kelvin scale has been defined with reference to the Celsius scale, which has two fixed points, the normal freezing and boiling temperatures of water. However, it may also be defined with reference to the gas thermometer (at very low pressure) by means of a single reference temperature. We define the ratio of any two temperatures T_1 and T_2 on the Kelvin scale as the ratio of the corresponding gas-thermometer pressures p_1 and p_2:

$$\frac{T_2}{T_1} = \frac{p_2}{p_1}. \tag{14–3}$$

To complete the definition, we need only to specify the Kelvin temperature of a single reproducible state. For reasons of precision and reproducibility, the state chosen is not a freezing or boiling point but the *triple point* of water, the unique condition under which solid water, liquid, and vapor can all coexist together. This occurs at a temperture of 0.01°C and a pressure of 610 Pa (about 0.006 atm). In modern thermometry the standard fixed point is the triple-point temperature of water, to which is assigned the value

$$273.16 \text{ K}.$$

A cell used to establish the triple-point temperature is shown in Fig. 14–7. Thus, with reference to Eq. (14–3) if p_3 is the pressure in a gas thermometer at the triple-point and p is the pressure at some other temperature T, then T is given on the Kelvin scale by

$$T = T_3 \frac{p}{p_3}, \tag{14–4}$$

with T_3 *defined* to be 273.16 K.

EXAMPLE   Suppose a constant-volume gas thermometer has a pressure of 1.50×10^4 Pa at the triple point of water and a pressure of 1.95×10^4 Pa at some unknown temperature T. What is T?

Solution   From Eq. (14–4),

$$T = (273 \text{ K})\frac{1.95 \times 10^4 \text{ Pa}}{1.50 \times 10^4 \text{ Pa}} = 355 \text{ K}$$

$$= 82°\text{C}. \qquad \blacktriangleleft$$

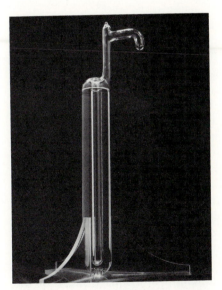

14–7 A cell used to establish the triple-point temperature of water. The cylindrical pyrex container is nearly filled with water of the highest possible purity and then permanently sealed. Partial freezing of the sealed water leads to the triple-point solid-liquid-vapor equilibrium condition and establishes the triple-point temperature of 0.0100°C or 273.1600 K. Calibrations with a single triple-point cell are precise to within 0.0001 K, and triple-point cells agree with each other within 0.0002 K. (Courtesy National Bureau of Standards.)

The Rankine temperature scale has the same relation to the Fahrenheit as Kelvin to Celsius. The zero point on the Kelvin scale is −273.15°C, which on the Fahrenheit scale is $(9/5)(-273.15°) + 32° = -459.7°\text{F}$. Thus we define the Rankine scale so that $0°\text{R} = -459.7°\text{F}$ and $0°\text{F} = 459.7°\text{R}$. The size of a degree is the same on the Rankine and Fahrenheit scales and

$$T_{\text{R}} = T_{\text{F}} + 459.7°. \tag{14–5}$$

The relationships of the four temperature scales discussed above are shown graphically in Fig. 14–8.

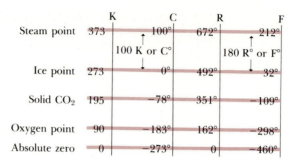

14-8 Relations among Kelvin, Celsius, Rankine, and Fahrenheit temperature scales. Temperatures have been rounded off to the nearest degree.

The Kelvin and Rankine scales are often called *absolute* temperature scales, and their zero point is called *absolute zero*. To define completely what is meant by absolute zero, we must use thermodynamic principles that are developed in the next several chapters. We shall return in Chapter 19 to the concept of absolute zero.

14-5

THERMAL EXPANSION

Most solid materials expand when heated. Suppose a rod of material has a length L_0 at some initial temperature T_0, and that, when the temperature increases by an amount ΔT, the length increases by ΔL. Experiment shows that if ΔT is not too large, ΔL is *directly proportional* to ΔT. Of course, ΔL is also proportional to L_0; if two rods of the same material have the same temperature change, but one is initially twice as long as the other, then the *change* in its length is also twice as great. Introducing a proportionality constant α (which is different for different materials), we may summarize this relation as follows:

$$\Delta L = \alpha L_0 \, \Delta T. \tag{14-6}$$

The constant α, which characterizes the thermal expansion properties of a particular material, is called the *temperature coefficient of linear expansion* or, more briefly, the *coefficient of linear expansion*.

For materials having no preferential directions, every linear dimension changes according to Eq. (14-6). Thus L could equally well represent the thickness of the rod, the side length of a square sheet, or the diameter of a *hole* in the material. There are some exceptional cases. Wood, for example, has different expansion properties along the grain and across the grain; and single crystals of material can have different properties along different crystal-axis directions. We exclude these exceptional cases from the present discussion.

It must be emphasized that the direct proportionality expressed by Eq. (14-6) is not exact, but that it is *approximately* correct for sufficiently small temperature changes. For any temperature one can define a coefficient of thermal expansion by the equation

$$\alpha = \frac{1}{L}\frac{\Delta L}{\Delta T}. \tag{14-7}$$

TABLE 14–1	COEFFICIENT OF LINEAR EXPANSION
Material	α $(C°)^{-1}$
Aluminum	2.4×10^{-5}
Brass	2.0×10^{-5}
Copper	1.7×10^{-5}
Glass	$0.4–0.9 \times 10^{-5}$
Steel	1.2×10^{-5}
Invar	0.09×10^{-5}
Quartz (fused)	0.04×10^{-5}

In this case, α for a given material will be found to vary somewhat with the initial temperature T_0 and the size of the temperature interval. Because Eq. (14–6) is at best an approximation, we shall ignore this variation. Average values of α for several materials are listed in Table 14–1.

EXAMPLE A carpenter uses a steel measuring tape 5 m long calibrated at a temperature of 20°C. What is its length on a hot summer day when the temperature is 35°C?

Solution From Eq (14–6),

$$\Delta L = (5 \text{ m})(1.2 \times 10^{-5}(C°)^{-1})(35°C - 20°C)$$
$$= 0.9 \times 10^{-3} \text{ m} = 0.9 \text{ mm}.$$

Thus the length at 35°C is 5.0009 m.   ◄

Increasing temperature usually causes increases in *volume*, for both solid and liquid materials; experiment shows that if the temperature change ΔT is not too great, the increase in volume ΔV is approximately *proportional* to the temperature change. It is also proportional to the initial volume V_0, just as with linear expressions. The relationship can be expressed as follows:

$$\Delta V = \beta V_0 \, \Delta T. \qquad (14–8)$$

The constant β, which characterizes the volume expansion properties of a particular material, is called the *temperature coefficient of volume expansion*, or the *coefficient of volume expansion*.

Like the coefficient of linear expansion, β varies somewhat with temperature, and Eq. (14–8) should be regarded as an approximate relation, valid for sufficiently small temperature changes. For many substances β decreases as the temperature is lowered, approaching zero as the Kelvin temperature approaches zero. Also, metals with high melting temperatures usually have small volume expansion coefficients.

Some values of β in the neighborhood of room temperature are listed in Table 14–2. Note that the values for liquids are much larger than those for solids.

If there is a hole in a solid body, the volume of the hole increases when the body expands, just as if the hole were a solid of the same

TABLE 14–2   COEFFICIENT OF VOLUME EXPANSION

Solids	β, $(C°)^{-1}$	Liquids	β, $(C°)^{-1}$
Aluminum	7.2×10^{-5}	Ethanol	75×10^{-5}
Brass	6.0×10^{-5}	Carbon disulfide	115×10^{-5}
Copper	5.1×10^{-5}	Glycerin	49×10^{-5}
Glass	$1.2–2.7 \times 10^{-5}$	Mercury	18×10^{-5}
Steel	3.6×10^{-5}		
Invar	0.27×10^{-5}		
Quartz (fused)	0.12×10^{-5}		

material as the body. This remains true even if the hole becomes so large that the surrounding body is reduced to a thin shell. Thus the volume enclosed by a thin-walled glass flask or thermometer bulb increases just as would a solid body of glass of the same size.

EXAMPLE   A glass flask of volume 200 cm^3 is just filled with mercury at 20°C. How much mercury will overflow when the temperature of the system is raised to 100°C? The coefficient of volume expansion of the glass is $1.2 \times 10^{-5}(C°)^{-1}$.

Solution   The increase in the volume of the flask is

$$\Delta V = (1.2 \times 10^{-5}(C°)^{-1})(200 \text{ cm}^3)(100° - 20°) = 0.192 \text{ cm}^3.$$

The increase in the volume of the mercury is

$$\Delta V = (18 \times 10^{-5}(C°)^{-1})(200 \text{ cm}^3)(100° - 20°) = 2.88 \text{ cm}^3.$$

The volume of mercury overflowing is therefore

$$2.88 \text{ cm}^3 - 0.19 \text{ cm}^3 = 2.69 \text{ cm}^3. \qquad \blacktriangleleft$$

Water, in the temperature range from 0°C to 4°C, *decreases* in volume with increasing temperature, contrary to the behavior of most substances. That is, between 0°C and 4°C the coefficient of expansion of water is *negative*. Above 4°C, water expands when heated. Since the volume of a given mass of water is smaller at 4°C than at any other temperature, the density of water is maximum at 4°C. Water also expands upon freezing, unlike most materials.

This anomalous behavior has an important effect on plant and animal life in lakes. When a lake cools, the cooled water at the surface flows to the bottom because of its greater density. But when the temperature reaches 4°C, this flow ceases and the water near the surface remains colder (and less dense) than that at the bottom. As the surface freezes, the ice floats because it is less dense than water. The water at the bottom remains at 4°C until nearly the entire body is frozen. If water behaved like most substances, contracting continuously on cooling and freezing, lakes would freeze from the bottom up, circulation due to density differences would continuously carry warmer water to the surface for efficient cooling, and lakes would freeze solid much more easily, thus destroying all plant and animal life that can withstand cold water but not freezing.

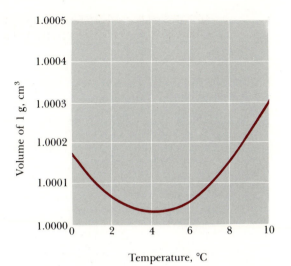

14–9 Volume of 1 gram of water, in the temperature range from 0°C to 10°C. If Eq. (14–8) were obeyed, the curve would be a straight line.

The anomalous expansion of water in the temperature range 0°C to 10°C is shown in Fig. 14–9. Table 14–3 covers a wider range of temperatures.

The volume expansion coefficient for a solid material is related to the linear coefficient. To obtain the relation, we consider a solid body in the form of a rectangular block with dimensions L_1, L_2, and L_3. Then the volume is

$$V_0 = L_1 L_2 L_3.$$

According to Eq. (14–6), when the temperature increases by ΔT, each linear dimension changes, and the new volume is

$$V_0 + \Delta V = L_1 L_2 L_3 (1 + \alpha \, \Delta T)^3 = V_0 (1 + \alpha \, \Delta T)^3$$
$$= L_1 L_2 L_3 [1 + 3\alpha \, \Delta T + 3\alpha^2 (\Delta T)^2 + \alpha^3 (\Delta T)^3].$$

If ΔT is small, the terms containing $(\Delta T)^2$ and $(\Delta T)^3$ are very small and may be neglected. Dropping these and subtracting $V_0 = L_1 L_2 L_3$ from both sides, we find

$$\Delta V = (3\alpha) V_0 \, \Delta T.$$

Comparing this with Eq. (14–8), we obtain

$$\beta = 3\alpha. \tag{14–9}$$

TABLE 14–3 VOLUME AND DENSITY OF WATER		
t, °C	Volume of 1 g, cm^3	Density, g·cm^{-3}
0	1.0002	0.9998
4	1.0000	1.0000
10	1.0003	0.9997
20	1.0018	0.9982
50	1.0121	0.9881
75	1.0258	0.9749
100	1.0434	0.9584

*14–6

THERMAL STRESSES

If the ends of a rod are rigidly held so as to prevent expansion or contraction and the temperature of the rod is changed, tensile or compressive stresses, called *thermal stresses,* are set up in the rod. These stresses may become large enough to stress the rod beyond its elastic limit or even beyond its breaking strength. Hence in the design of any structure that is subject to changes in temperature, provision must be made for expansion. In a long steam pipe this is accomplished by the insertion of expansion joints or a section of pipe in the form of a U. In bridges, one end may be rigidly fastened to its abutment while the other rests on rollers.

It is a simple matter to compute the thermal stress set up in a rod that is not free to expand or contract. Suppose that a rod of length L_0 and cross-sectional area A has its ends rigidly fastened while the temperature is reduced by an amount ΔT, causing a tensile stress.

The fractional change in length if the rod were free to contract would be

$$\frac{\Delta L}{L_0} = \alpha \, \Delta T. \tag{14–10}$$

Both ΔL and ΔT are negative. Since the rod is *not* free to contract, the tension must increase by a sufficient amount to produce an equal and opposite fractional change in length. But from the definition of Young's modulus, Eq. (10–7),

$$Y = \frac{F/A}{\Delta L/L_0}, \qquad \frac{\Delta L}{L_0} = \frac{F}{AY}. \tag{14–11}$$

The tensile force F is determined by the requirement that the *total* fractional change in length, thermal expansion plus elastic strain, must be zero:

$$\alpha \, \Delta T + \frac{F}{AY} = 0,$$

$$F = -AY\alpha \, \Delta T. \tag{14–12}$$

Since ΔT represents a decrease in temperature, it is negative, so F is positive. The tensile *stress* in the rod is

$$\frac{F}{A} = -Y\alpha \, \Delta T. \tag{14–13}$$

If, instead, ΔT represents an *increase* in temperature, then F and F/A become negative, corresponding to *compressive* force and stress, respectively.

Thermal stresses can also be induced by nonuniform temperature. Even if a solid body at uniform temperature has no internal stresses, stress may be induced by nonuniform expansion due to temperature differences. The breaking of a thick glass container when hot water is poured into it is a familiar phenomenon. Heat-resistant glasses such as

Pyrex have exceptionally low expansion coefficients and usually also have high strength to permit thin-wall construction to minimize temperature differences.

Similar phenomena occur with volume expansion. If a bottle is completely filled with water, tightly capped, and then heated, it will break because the thermal expansion coefficient for water is greater than that for glass. If a material is enclosed in a very rigid container so that its volume cannot change, then a rise in temperature ΔT is accompanied by an increase in pressure Δp. An analysis similar to that leading to Eq. (14–13) shows that the pressure increase is given by

$$\Delta p = B\beta \, \Delta T, \tag{14–14}$$

where B is the bulk modulus for the material.

QUESTIONS

14–1 Does it make sense to say that one body is twice as hot as another?

14–2 A student claimed that thermometers are useless because a thermometer always registers *its own* temperature. How would you respond?

14–3 What other properties of matter, in addition to those mentioned in the text, might be used as thermometric properties? How could they be used to make a thermometer?

14–4 A thermometer is laid out in direct sunlight. Does it measure the temperature of the air, or of the sun, or what?

14–5 Thermometers sometimes contain a red or blue liquid, which is often ethanol. What advantages and disadvantages does this have compared with mercury?

14–6 Could a thermometer similar to that shown in Fig. 14–1a be made using water as the liquid? What difficulties would this thermometer present?

14–7 What is the temperature of vacuum?

14–8 Is there any particular reason for constructing a temperature scale with higher numbers corresponding to hotter bodies, rather than the reverse?

14–9 If a brass pin is a little too large to fit in a hole in a steel block, should you heat the pin and cool the block, or the reverse?

14–10 When a block with a hole in it is heated, why doesn't the material around the hole expand into the hole and make it smaller?

14–11 Many automobile engines have cast-iron cylinders and aluminum pistons. What kinds of problems could occur if the engine gets too hot?

14–12 When a hot-water faucet is turned on, the flow often decreases gradually before it settles down. This is an annoyance in the shower; why does it happen?

14–13 Two bodies made of the same material have the same external dimensions and appearance, but one is solid and the other is hollow. When they are heated, is the overall volume expansion the same or different?

14–14 A thermostat for controlling household heating systems often contains a bimetallic element consisting of two strips of different metals, welded together face to face. When the temperature changes, this composite strip bends one way or the other. Why? What determines the direction of bend?

14–15 Why is it sometimes possible to loosen caps on screw-top bottles by dipping the cap briefly in hot water?

14–16 The rate at which a pendulum clock runs depends on the length of the pendulum. Would a pendulum clock gain time in hot weather and lose in cold, or the reverse? Could one design a pendulum, perhaps using two different metals, that would *not* change length with temperature?

PROBLEMS

14–1 The ratio of the pressures of a gas at the melting point of lead and at the triple point of water, when the gas is kept at constant volume, is found to be 2.19816. What is the Kelvin temperature of the melting point of lead?

14–2

a) If you feel sick in France and are told you have a temperature of 40°C, should you be concerned?

b) What is normal body temperature on the Celsius scale?

c) The normal boiling point of liquid oxygen is −182.97°C. What is this temperature on the Kelvin and Rankine scales?

d) At what temperature do the Fahrenheit and Celsius scales coincide?

14–3 When the United States finally converts officially to metric units, the Celsius temperature scale will replace the Fahrenheit scale for everyday use. As a familiarization exercise, find the Celsius temperatures corresponding to

a) a cool room (68°F),

b) a hot summer day (95°F), and

c) a cold winter day (5°F).

14–4 In a rather primitive experiment with a constant-volume gas thermometer, the pressure at the triple point of water was found to be 4.0×10^4 Pa, and the pressure at the normal boiling point 5.4×10^4 Pa. According to these data, what is the temperature of absolute zero on the Celsius scale?

14–5 A gas thermometer of the type shown in Fig. 14–5 registered a pressure corresponding to 5.0 cm of mercury when in contact with water at the triple point. What pressure will it read when in contact with water at the normal boiling point?

14–6 The electrical resistance of some metals varies with temperature (as measured by a gas thermometer) approximately according to $R = R_0[1 + b(T − T_0)]$, where R_0 is the resistance at temperature T_0. For a certain metal, the constant b is found to be 0.004 K^{-1}.

a) If the resistance at 0°C is 100 ohms, what is the resistance at 20°C?

b) At what temperature is the resistance 200 ohms?

14–7 A building with a steel framework is 200 m tall when the temperature is 0°C. How tall is the building on a hot summer day when the temperature is 30°C?

14–8 A steel bridge is built in the summer when the temperature is 25°C. At the time of construction its length is 80.0 m. What is the length of the bridge on a cold winter day when the temperature is −10°C?

14–9 A glass bottle has a capacity of exactly 500 ml at room temperature (20°C). What is its capacity at the freezing temperature of water if its coefficient of linear expansion is 1.2×10^{-5} (C°)$^{-1}$?

14–10 An underground tank with a capacity of 500 l is filled with ethanol that has an initial temperature of 25°C. After the ethanol has cooled off to the temperature of the tank and ground, which is 10°C, how much airspace will there be above the ethanol in the tank?

14–11 Use the data of Table 14–3 to find the average coefficient of volume expansion of water in the temperature range

a) 0° to 4°C;

b) 0° to 10°C;

c) 10° to 20°C;

d) 75° to 100°C.

14–12 What is the average coefficient of volume expansion for water in the range

a) 0° to 4°C?

b) 0° to 10°C?

14–13 The pendulum shaft of a clock is of aluminum. What is the fractional change in length of the shaft when it is cooled from 25°C to 10°C?

14–14 A surveyor's 100-ft steel tape is correct at a temperature of 20°C. The distance between two points, as measured by this tape on a day when the temperature is 35°C, is 86.57 ft. What is the true distance between the points?

14–15 A glass flask whose volume is exactly 1000 cm^3 at 0°C is filled level full of mercury at this temperature. When flask and mercury are heated to 100°C, 15.2 cm^3 of mercury overflow. If the coefficient of volume expansion of mercury is 0.000182 per Celsius degree, compute the coefficient of linear expansion of the glass.

14–16 At a temperature of 20°C, the volume of a certain glass flask, up to a reference mark on the stem of the flask, is exactly 100 cm^3. A chemist filled the flask to this point with a liquid whose coefficient of volume expansion is 120×10^{-5} (C°)$^{-1}$, with both flask and liquid at 20°C. The coefficient of linear expansion of the glass is 8×10^{-6} (C°)$^{-1}$. The cross section of the stem is 1 mm^2 and can be considered constant. How far will the liquid rise or fall in the stem when the temperature is raised to 40°C?

14–17 To ensure a tight fit, the aluminum rivets used in airplane construction are made slightly larger than the rivet holes and cooled by "dry ice" (solid CO_2) before being driven. If the diameter of a hole is 0.5000 cm, what should be the diameter of a rivet at 20°C if its diameter is to equal that of the hole when the rivet is cooled to −78°C, the temperature of dry ice? Assume the expansion coefficient to remain constant at the value given in Table 14–1.

14–18 A steel ring of 3.000 cm inside diameter at 20°C is to be heated and slipped over a brass shaft measuring 3.002 cm in diameter at 20°C.

a) To what temperature should the ring be heated?

b) If the ring and shaft together are cooled by some means such as liquid air, at what temperature will the ring just slip off the shaft?

14–19 A metal rod 30.0 cm long expands by 0.075 cm when its temperature is raised from 0°C to 100°C. A rod of a different metal and of the same length expands by 0.045 cm for the same rise in temperature. A third rod, also 30.0 cm long, is made up of pieces of each of the above metals placed end to end, and expands 0.065 cm between 0°C and 100°C. Find the length of each portion of the composite bar.

14–20 A machinist bores a hole 2.000 cm in diameter in a brass plate at a temperature of 20°C. What is the diameter of the hole when the temperature of the plate is increased to 200°C? Assume the expansion coefficient to remain constant.

14–21 Suppose that a steel hoop could be constructed around the earth's equator, just fitting it at a temperature of 20°C. What would be the thickness of space between the hoop and the earth if the temperature of the hoop were increased by 1 C°?

14–22 A clock whose pendulum makes one vibration in 2 s is correct at 25°C. The pendulum shaft is of steel and its mass may be neglected compared with that of the bob.

a) What is the fractional change in length of the shaft when it is cooled to 15°C?

b) How many seconds per day will the clock gain or lose at 15°C?

14–23 A clock with a brass pendulum shaft keeps correct time at a certain temperature.

a) How closely must the temperature be controlled if the clock is not to gain or lose more than 1 second a day? Does the answer depend on the period of the pendulum?

b) Will an increase of temperature cause the clock to gain or lose?

14–24 The length of a steel bridge is 2000 ft.

a) If it were a continuous span, fixed at one end and free to move at the other, about what would be the range of motion of the free end between a cold winter day (−20°F) and a hot summer day (100°F)?

b) If both ends were rigidly fixed on the summer day, what would be the stress on the winter day?

14–25 The cross section of a steel rod is 10 cm². What is the least force that will prevent it from contracting while cooling from 520°C to 20°C?

14–26

a) A steel wire that is 3 m long at 20°C is found to increase in length by 1.5 cm when heated to 520°C. Compute its average coefficient of linear expansion.

b) Find the stress in the wire if it is stretched taut at 520°C and cooled to 20°C without being allowed to contract.

14–27 A steel rod of length 40 cm and a copper rod of length 36 cm, both of the same diameter, are placed end to end between two rigid supports, with no initial stress in the rods. The temperature of the rods is now raised by 50°C. What is the stress in each rod?

14–28 A heavy brass bar has projections at its ends, as in Fig. 14–10. Two fine steel wires fastened between the projections are just taut (zero tension) when the whole system is at 0°C. What is the tensile stress in the steel wires when the temperature of the system is raised to 300°C? Make any simplifying assumptions you think are justified, but state what they are.

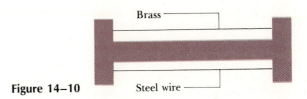

Figure 14–10   Steel wire

14–29 Steel railroad rails 60 ft long are laid on a winter day when the temperature is 20°F.

a) How much space must be left between rails if they are to just touch on a summer day when the temperature is 110°F?

b) If the rails were originally laid in contact, what would be the stress in them on a summer day when the temperature is 110°F?

14–30 Prove that if a body under hydrostatic pressure is raised in temperature but not allowed to expand, the increase in pressure is

$$\Delta p = B\beta \, \Delta T,$$

where the bulk modulus B and the average coefficient of volume expansion β are both assumed positive and constant.

14–31

a) A block of steel at an initial pressure of 1 atm and a temperature of 20°C is kept at constant volume. If the temperature is raised to 32°C, what is the final pressure?

b) If the block is maintained at constant volume by rigid walls that can withstand a maximum pressure of 1200 atm, what is the highest temperature to which the system may be raised? Assume B and β to remain practically constant at the values 1.5×10^{11} Pa and $5.0 \times 10^{-5}(\text{C}°)^{-1}$, respectively.

14–32 What hydrostatic pressure is necessary to prevent a copper block from expanding when its temperature is increased from 20°C to 30°C?

14–33 Table 14–3 lists the volume of 1 g of water, at atmospheric pressure. A steel container is filled with water at 10°C and atmospheric pressure, and the system is heated to 75°C. What is then the pressure in the container? Assume the container to be sufficiently rigid so that its volume is not affected by the increased pressure.

14–34 A liquid is enclosed in a metal cylinder provided with a piston of the same metal. The system is originally at atmospheric pressure and at a temperature of 80°C. The piston is forced down until the pressure on the liquid is increased by 100 atm, and it is then clamped in this position. Find the new temperature at which the pressure of the liquid is again 1 atm. Assume that the cylinder is sufficiently strong so that its volume is not altered by changes in pressure, but only by changes in temperature.

Compressibility of liquid $\kappa = 50 \times 10^{-6}\,\text{atm}^{-1}$.

Coefficient of volume expansion of liquid $\beta = 5.3 \times 10^{-4}(\text{C}°)^{-1}$.

Coefficient of linear expansion of metal $\alpha = 10 \times 10^{-6}(\text{C}°)^{-1}$.

CHAPTER 15

QUANTITY
OF HEAT

In the preceding chapter we discussed the concept of temperature in connection with thermal equilibrium. When two bodies *not* initially in thermal equilibrium are placed in contact, their temperatures change until they reach thermal equilibrium. A study of the interaction that takes place between the bodies during their approach to thermal equilibrium leads to the concept of *heat,* the subject of this chapter. We define quantity of heat and introduce units to measure quantity of heat. Then we study the quantities of heat involved in temperature changes and changes of phase of materials.

15–1

HEAT TRANSFER

Suppose that system A, at a higher temperature than system B, is put in contact with B. When thermal equilibrium has been reached, A is found to have undergone a temperature decrease and B a temperature increase. It is therefore natural to assume that A loses something and that this "something" flows into B. While the temperature changes are taking place, it is customary to refer to a *heat flow* or a *heat transfer* from A to B.

The process of heat transfer was formerly thought to be a flow of an invisible weightless fluid called *caloric.* But during the eighteenth and nineteenth centuries the relationship of heat to mechanical energy and work gradually emerged. Count Rumford (1753–1814) studied the heat evolved during the process of boring cannons, and Sir James Prescott Joule (1818–1889) established a correlation between the heating of water by a paddle wheel and the work needed to turn the wheel. These and many other similar experimental studies established that heat flow is really *energy transfer*, and the concept of caloric was eventually discarded.

When an energy transfer takes place solely because of a temperature difference, it is called a heat flow. For example, water is heated and converted to

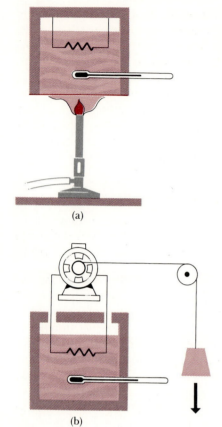

(a)

(b)

15–1 The same temperature change of the same system may be accomplished by either (a) a heat flow or (b) the performance of work.

steam in a steam boiler by contact with a metal pipe or container kept at a high temperature by a hot flame from burning coal or gas. The steam has a greater ability to do *work* (by pushing against a turbine blade or the piston in a steam engine) than it had in the form of cold water, and therefore the water must have received *energy* in a process involving a heat flow from the hot flame (through the container) to the cooler water.

Of course, energy transfer can also occur without heat flow. In an air compressor, a moving piston pushes against a mass of air, doing work on the air as it compresses it to a smaller volume. In this compressed state the gas is capable of doing more work than before and hence has acquired energy.

Finally, if in the air compressor the air and the piston are at different temperatures, heat flow between the piston and the air can occur. Thus we have an example of a process involving two kinds of energy transfer simultaneously, flow of *heat* and performance of *work*. Such processes are of great importance in many practical devices such as internal-combustion engines. We shall study them in detail in Chapters 18 and 19.

In many cases a particular change of state of a quantity of material can be brought about in many different ways. In Fig. 15–1a, the temperature of a quantity of water is heated by a gas flame; in Fig. 15–1b the same change of state is accomplished by a falling weight, which turns a generator and powers an electric heating coil in the water. If we consider the generator as part of our system, we can regard the change as caused by heat flow; if not part of the system, by the performance of mechanical work. Thus in describing various energy-transfer processes, we must be careful always to specify what is and is not included in the system under discussion.

This example also shows that there must be some sort of equivalence between heat and mechanical work, since the same change of state of a system can be produced by either heat flow or work. A detailed study of this equivalence leads to the *first law of thermodynamics*, to be studied in detail in Chapter 18.

15–2

QUANTITY OF HEAT

We shall use the term *heat* only in reference to *flow* or *transfer*. Thus a heat flow is an energy transfer brought about solely by a temperature difference. The amount of energy transferred may be described as a quantity of heat, but we must be careful *not* to take the view that a certain body *contains* a certain quantity of heat. Such a statement would have no meaning. This is a somewhat subtle point, and a thorough discussion of it leads to the concept of *internal energy*, to be introduced in Chapter 18.

A unit of quantity of heat can be defined with reference to any particular heat-flow process with any particular material. The *calorie* was originally defined in the eighteenth century as the amount of heat required to raise the temperature of one gram of water by one Celsius degree (one kelvin). It was found later that this is ambiguous because more heat is required to raise the temperature from, say 90° to 91° than from 20° to 21°. In one of several refinements of the definition, the calorie was defined to be the amount of heat required to heat one gram of water from 14.5°C to 15.5°C. This leads to the "15-degree calorie." Several alternate and slightly different definitions are unfortunately also in current use, so the ambiguity persists even today.

A corresponding unit defined in terms of Fahrenheit degrees and British units is the *British thermal unit*, or Btu. By definition, 1 Btu is the quantity of heat required to raise the temperature of one pound of water from 63°F to 64°F. A third unit in common use, especially in measuring food energy, is the kilocalorie (kcal). The relations among these three units are

$$1 \text{ Btu} = 252 \text{ cal} = 0.252 \text{ kcal}.$$

It should also be noted that the unit of quantity of heat is fundamentally a unit of energy; thus there must be a relation between the above units and the familiar mechanical energy units such as the joule. It has been found experimentally that

$$1 \text{ cal} = 4.186 \text{ joules} = 4.186 \text{ J},$$

$$1 \text{ kcal} = 4186 \text{ J},$$

$$1 \text{ Btu} = 778 \text{ ft-lb}.$$

Figure 15–2 shows an example of this equivalence.

The International Committee on Weights and Measures no longer recognizes the calorie as a fundamental unit, but recommends instead that the joule be used for quantity of heat as well as for all other forms of energy. Although the calorie is convenient in problems involving water, it is awkward when both heat and other forms of energy are involved. There seems little doubt that ultimately the joule will become the universal unit of energy; many of the examples and problems of this and the following chapters will anticipate this usage. The relation between heat and mechanical energy will be explored in detail in Chapters 18 through 20.

15–2 Sugar cubes obtained in a restaurant in Germany. A rough translation is "This package has 22 Calories (i.e., kilocalories), equal to 92 kilojoules" and "Get into the swing with sugar." By law, foods marketed in Germany must show the energy content in joules; the equivalent in calories is optional.

15–3

HEAT CAPACITY

Let us use the symbol Q or ΔQ for quantity of heat. The quantity of heat ΔQ required to increase the temperature of a mass m of a certain material by an amount ΔT is found to be approximately proportional to ΔT. It is also proportional to m; twice as much heat is needed to heat two cups of water to boiling to make tea than to heat only one cup. The quantity of heat needed also depends on the nature of the material; to raise the temperature of one gram of water 1 C° requires over five times as much heat as for the same mass of aluminum and the same temperature increase.

Thus the relationship among all these quantities can be expressed as

$$\Delta Q = mc\,\Delta T, \qquad (15\text{--}1)$$

where c is a constant, different for different materials. The constant c for any material is called the *specific heat capacity* for that material. It is not strictly constant, but depends somewhat on the initial temperature; in solving problems, we shall often ignore this small variation.

Alternatively, we can rearrange Eq. (15–1) and define the specific heat capacity of a material as

$$c = \frac{\Delta Q}{m\,\Delta T}. \qquad (15\text{--}2)$$

The specific heat capacity of water can be taken to be

$$4.19\ \text{J}\cdot\text{g}^{-1}\cdot(\text{C}°)^{-1}, \qquad 4190\ \text{J}\cdot\text{kg}^{-1}\cdot(\text{C}°)^{-1},$$

$$1\ \text{cal}\cdot\text{g}^{-1}\cdot(\text{C}°)^{-1}, \qquad \text{or} \qquad 1\ \text{Btu}\cdot\text{lb}^{-1}\cdot(\text{F}°)^{-1}$$

for most practical purposes.

EXAMPLE   During a bout with the flu, an 80-kg man ran a fever of 2 C° above normal, that is, a body temperature of 39°C or 102.2°F. To raise his temperature by that amount, his body had to generate extra heat by means of chemical reactions in his cells. Assuming the human body is mostly water, how much heat was needed? Express the result in joules and calories.

Solution   From Eq. (15–1),

$$\Delta Q = (80\ \text{kg})(4190\ \text{J}\cdot\text{kg}^{-1}\cdot(\text{C}°)^{-1})(2\ \text{C}°) = 6.70 \times 10^5\ \text{J}$$
$$= (80{,}000\ \text{g})(1\ \text{cal}\cdot\text{g}^{-1}\cdot(\text{C}°)^{-1})(2\ \text{C}°) = 1.60 \times 10^5\ \text{cal}.$$

We also note that this corresponds to 160 kcal, i.e., 160 food-value calories.   ◀

It is often convenient to use the *mole* rather than the *mass* of material to describe the amount of substance. By definition, one mole (1 mol) of any substance is a quantity of matter such that its mass in grams is numerically equal to the *molecular mass M* (often called *molecular weight*).* To calculate the number of moles n, we divide the mass in grams by the molecular mass; thus $n = m/M$. Replacing the mass m in Eq. (15–1) by the product nM, we get

$$Mc = \frac{\Delta Q}{n\,\Delta T}.$$

The product Mc is called the *molar heat capacity* and is represented by the symbol C. Hence, by definition

$$C = Mc = \frac{\Delta Q}{n\,\Delta T}, \tag{15–3}$$

or

$$\Delta Q = nC\,\Delta T. \tag{15–4}$$

The molar heat capacity of water is approximately

$$75.3\ \text{J} \cdot \text{mol}^{-1} \cdot (\text{C}°)^{-1} \quad \text{or} \quad 18\ \text{cal} \cdot \text{mol}^{-1} \cdot (\text{C}°)^{-1}.$$

The quantity defined by Eq. (15–2) is sometimes called simply *specific heat*, and the molar heat capacity defined by Eq. (15–3) is often called the *molar specific heat*. However, the terms *specific heat capacity* and *molar heat capacity* will be used throughout this book. Representative values of specific and molar heat capacity are given in Table 15–1.

TABLE 15–1	MEAN SPECIFIC AND MOLAR HEAT CAPACITIES OF METALS. (CONSTANT PRESSURE, TEMPERATURE RANGE 0°C TO 100°C)			
	Specific (c)		**M,**	**Molar (C)**
Metal	$\text{J} \cdot \text{kg}^{-1} \cdot (\text{C}°)^{-1}$	$\text{cal} \cdot \text{g}^{-1} \cdot (\text{C}°)^{-1}$	$\text{g} \cdot \text{mol}^{-1}$	$\text{J} \cdot \text{mol}^{-1} \cdot (\text{C}°)^{-1}$
Aluminum	910	0.217	27.0	24.6
Beryllium	1970	0.471	9.01	17.7
Copper	390	0.093	63.5	24.8
Iron	470	0.112	55.9	26.3
Lead	130	0.031	207	26.9
Mercury	138	0.033	201	27.7
Silver	234	0.056	108	25.3

From Eq. (15–1), the total quantity of heat Q that must be supplied to a body of mass m to change its temperature from T_1 to T_2 is

$$Q = mc(T_2 - T_1). \tag{15–5}$$

If T_2 is less than T_1, Q is negative, indicating transfer of heat *out of* the body rather than *into* it.

Strictly speaking, the specific heat capacity to be used in Eq. (15–5) should be the *mean* specific heat capacity over the interval $(T_2 - T_1)$, since the specific heat capacity of a material varies somewhat with temperature. When the temperature interval is not too great, this variation

---

*Although the term molecular *weight* is in common use, the quantity measures the *mass*, not the *weight* of a molecule. Thus the term *molecular mass* is preferable. This term is coming into more common use and will be used throughout this book.

may often be ignored. At very low temperatures, specific heat capacities always decrease, and the variation of heat capacity with temperature provides useful information concerning the microscopic structure of materials.

The specific or molar heat capacity of a substance is not the only physical property involving quantity of heat. Heat conductivity, heat of fusion, heat of vaporization, and heat of combustion are a few other examples of *thermal properties* of matter. The field of physics and physical chemistry concerned with the measurement of thermal properties is called *calorimetry*.

15-4

THE MEASUREMENT OF HEAT CAPACITY

To measure a heat capacity we need to add a measured quantity of heat to a measured quantity of a material and observe the resulting temperature change. For maximum precision the thermal measurements are usually made electrically. The heat input is provided by passing a current through a resistance wire in contact with (often wound around) the material and measuring the input of electrical energy. The thermometer is usually a small resistance thermometer or thermocouple embedded in the sample and is chosen for its rapidity of response and its sensitivity.

Readings are taken before the switch in the heater circuit is closed; these are shown on the left of Fig. 15–3. The fact that the temperature rises before energy is supplied to the heater indicates that the surroundings are at a temperature higher than that of the sample. The current is maintained in the heater for a time interval Δt, during which we do not measure the temperature. After the switch in the heater circuit is opened, temperature measurements are resumed, giving rise to the right-hand part of Fig. 15–3, where the positive slope indicates that the surroundings are still at a higher temperature than the sample.

By measuring the voltage across the heater coil, the current through it, and the time interval Δt, one can determine the electrical energy input. This energy is converted completely to heat transferred to the sample, so ΔQ is determined. Then Eq. (15–2) or (15–3) may be used to determine the desired heat capacity, using the value of ΔT from the graph. In careful experiments ΔT may be as small as 0.01 K, and in some work at very low temperatures ΔT may be as small as 10^{-6} K.

Precise measurements of heat capacities require great experimental skill, partly because of the difficulty of avoiding (and compensating for) unwanted heat transfer between the sample and its surroundings. Still, such measurements are very important. The temperature variation of specific heats provides the most direct approach to the understanding of molecular energies in matter. Low-temperature thermal properties in particular are of great interest in contemporary physics.

Figure 15–4 shows the variation of the specific heat capacity of water with temperature. It may be seen that the quantity of heat necessary to raise the temperature of 1 g of water from 14.5°C to 15.5°C is

$$1 \ 15° \ \text{cal} = 4.186 \ \text{J}.$$

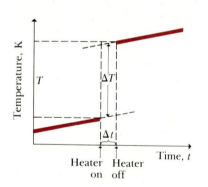

15–3 Temperature–time graph of the data taken during a heat-capacity measurement.

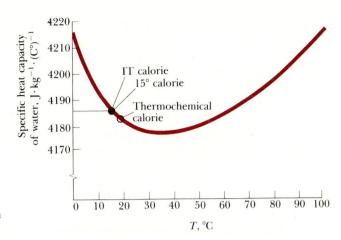

15–4 Specific heat capacity of water as a function of temperature.

Two other calories are frequently used. The *international table calorie* (IT cal) is *defined* to be precisely

$$1 \text{ IT cal} = \frac{1 \text{ W} \cdot \text{hr}}{860} = \frac{3600 \text{ J}}{860} = 4.186 \text{ J},$$

and is almost identical with the 15° calorie. The *thermochemical calorie*, however, is equal to 4.1840 J and may be seen from Fig. 15–4 to correspond to about a 17° calorie. In the absence of any additional information, the word *calorie* should be taken to mean 4.186 J. This variety of definitions of the calorie is an additional argument in favor of *eliminating* the calorie completely and using the joule as the fundamental unit of quantity of heat as well as of all other forms of energy.

Engineers sometimes use the British thermal unit (Btu). The following conversions are useful:

$$1 \text{ Btu} = 778 \text{ ft} \cdot \text{lb}$$
$$= 252.0 \text{ cal} = 1055 \text{ J}.$$

There are as many different definitions of the Btu as there are of the calorie, so these conversion factors are only approximate; the various Btu's differ by as much as 0.5 percent.

15–5

EXPERIMENTAL VALUES OF HEAT CAPACITIES

The amount of heat transferred to or from a system during a temperature change depends on the conditions imposed on the system during the transfer. The most common conditions are for the system to be kept at *constant pressure* or at *constant volume*. The two corresponding specific (or molar) heat capacities are denoted by the respective symbols c_P (or C_P) and c_V (or C_V). In the electrical method of measuring heat capacities, the sample under investigation is usually under constant pressure. As a matter of fact, it would be almost impossible to make a precise determination of c_V because there is no effective way of keeping the volume of a system constant and of correcting for the heat transferred to the containing walls. The two heat capacities differ because if the system is permitted to expand during heating, there is additional energy exchange through performance of mechanical work by the system on its surroundings. If the volume is held constant there is no such work.

A few average values of heat capacities of metals are listed in Table 15–1. The specific heat capacities are all less than that of water, and they decrease with increasing molecular mass. The last column is of particular interest; it shows that the molar heat capacities for all metals except the very lightest are approximately the same (within about 5 percent) and are equal to about $25 \text{ J} \cdot \text{mol}^{-1} \cdot (\text{C}°)^{-1}$. This is called the law of Dulong and Petit, after its discoverers, who first noticed it in 1819. Although only an approximate rule, it contains the germ of a very important idea.

By definition of the mole, the number of molecules in one mole is the same for all substances. It follows that about the same amount of heat is required *per molecule* to raise the temperature of each of these metals by a given amount, even though the *mass* of a molecule of lead (for example) is nearly 10 times as great as that of a molecule of aluminum. To put it in a different way, the heat required for a given temperature increase depends only on *how many* molecules the sample contains, and not on the mass of an individual molecule.

Thus there is a direct relation between this thermal property of matter and its molecular structure. We shall explore this relation in greater detail in Chapter 20. More generally, many different properties of matter can be understood on the basis of molecular structure, and much of contemporary physics is concerned with developing such relationships. This field has enormous practical importance because once these relationships are understood it becomes possible actually to *design* materials having specific mechanical, thermal, electrical, optical, or other properties that may be desired for specific practical applications.

The fact that heat capacities vary with temperature provides additional insight into the microscopic or molecular structure of matter. Figure 15–5 shows the temperature variation of molar heat capacity of several elements and compounds. In all cases the curves approach the Dulong and Petit value of $25 \text{ J} \cdot \text{mol}^{-1} \cdot \text{K}^{-1}$ at sufficiently high temperatures.

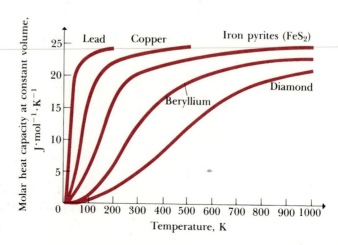

15–5 Temperature variation of C_V of solids. The values used for these curves are computed from measured values of C_P by subtracting corrections for volume changes.

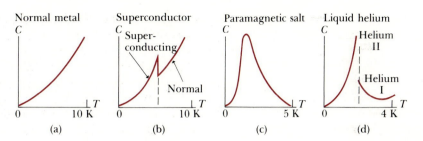

15–6   Different ways in which the molar heat capacity varies with the temperature at low temperatures, indicating widely different atomic processes.

Each curve in Fig. 15–5 has the same shape; it starts at zero, rises rapidly at first, then bends over and approaches the value $25 \, \text{J} \cdot \text{mol}^{-1} \cdot \text{K}^{-1}$. It was shown by Debye that the behavior of the molar head capacity of *nonmetals* over the entire temperature range can be accounted for by the *vibrational motion of the molecules in the crystal lattice*. There are, however, many substances whose temperature variation of heat capacity is more complex. The low-temperature behavior of four "abnormal" types of substance is depicted in Fig. 15–6. There are many more different types of heat-capacity curves; each of them indicates some special molecular or atomic or ionic property of the material.

15–6

CHANGE OF PHASE

The term *phase* as used here refers to the fact that matter exists either as a solid, liquid, or gas. Thus the chemical substance H_2O exists in the *solid phase* as ice, in the *liquid phase* as water, and in the *gaseous phase* as steam. Provided they do not decompose at high temperatures, all substances can exist in any of the three phases under the proper conditions of temperature and pressure. Transitions from one phase to another are usually accompanied by the absorption or liberation of heat and by a change in volume, even when the transition occurs at constant temperature. Such transitions are called *phase changes* or *phase transitions*.

When heat is added to ice at 0°C and normal atmospheric pressure, the temperature of the ice *does not* increase; instead, some of it melts to form liquid water. If the heat is added slowly enough so that thermal equilibrium is maintained between the ice and liquid water, then the temperature remains at 0°C as the heat is added, until all the ice is melted. Thus the effect of adding heat to this system is not to raise its temperature but to change its *phase* from solid to liquid.

Experiment shows that to change 1 kg of ice at 0°C to 1 kg of liquid water at 0°C requires the addition of 3.34×10^5 J of heat. This quantity of heat is called the *heat of fusion* of water. The term *latent heat of fusion* is sometimes used; this is somewhat redundant and will not be used here, but the symbol L will be used for quantities of heat associated with phase transitions. Thus the heat of fusion L_F of water is

$$L_F = 3.34 \times 10^5 \, \text{J} \cdot \text{kg}^{-1}.$$

In other units, the value is

$$L_F = 79.7 \, \text{cal} \cdot \text{g}^{-1}$$
$$= 143 \, \text{Btu} \cdot \text{lb}^{-1}.$$

More generally, to melt a mass m of material requires the addition of a quantity of heat Q given by

$$Q = mL_F. \qquad (15\text{--}6)$$

We have given the value of L_F for water; it is of course different for different materials, and it also varies with pressure.

The process described above is *reversible;* the conversion of liquid water to ice at 0°C requires the *removal* of heat; the amount is again given by Eq. (15–6) but in this case it is considered negative, since it is removed rather than added. We also note that at a given pressure it is possible for liquid water and ice to coexist only at one very specific temperature, which we call the melting temperature. This coexistence of two phases is called *phase equilibrium.*

This entire discussion can be repeated for boiling, a phase transition between liquid and gaseous phases. The corresponding heat of the phase transition is called the *heat of vaporization* L_V; both it and the boiling temperature of a material depend on pressure. Water boils at a lower temperature in Denver than in Pittsburgh because the average atmospheric pressure is less due to the higher elevation. The heat of vaporization is somewhat greater at this lower temperature. At normal atmospheric pressure the heat of vaporization of water is

$$L_V = 2.26 \times 10^6 \text{ J} \cdot \text{kg}^{-1}$$
$$= 539 \text{ cal} \cdot \text{g}^{-1}$$
$$= 970 \text{ Btu} \cdot \text{lb}^{-1}.$$

It is interesting to note that over five times as much heat is required to boil a quantity of water at 100°C as to raise its temperature from 0° to 100°C.

Heats of fusion and vaporization are given for several materials in Table 15–2. Also given are normal melting and boiling temperatures, that is, the melting and boiling temperatures at normal atmospheric pressure.

TABLE 15–2 HEATS OF FUSION AND VAPORIZATION

Substance	Normal melting point		Heat of fusion, L_F, J·kg^{-1}	Normal boiling point		Heat of vaporization, L_V, J·kg^{-1}
	K	°C		K	°C	
Helium	3.5	−269.65	5.23×10^3	4.216	−268.93	20.9×10^3
Hydrogen	13.84	−259.31	58.6×10^3	20.26	−252.89	452×10^3
Nitrogen	63.18	−209.97	25.5×10^3	77.34	−195.81	201×10^3
Oxygen	54.36	−218.79	13.8×10^3	90.18	−182.97	213×10^3
Ethyl alcohol	159	−114	104.2×10^3	351	78	854×10^3
Mercury	234	−39	11.8×10^3	630	357	272×10^3
Water	273.15	0.00	334×10^3	373.15	100.00	2256×10^3
Sulfur	392	119	38.1×10^3	717.75	444.60	326×10^3
Lead	600.5	327.3	24.5×10^3	2023	1750	871×10^3
Antimony	903.65	630.50	165×10^3	1713	1440	561×10^3
Silver	1233.95	960.80	88.3×10^3	2466	2193	2336×10^3
Gold	1336.15	1063.00	64.5×10^3	2933	2660	1578×10^3
Copper	1356	1083	134×10^3	1460	1187	5069×10^3

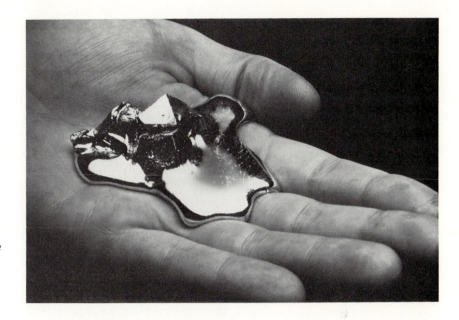

15–7 The metal gallium is one of the few *elements* that melts in the vicinity of room temperature; its melting temperature is 29.8°C, its heat of fusion is 8.04 × 10^4 J·kg^{-1} or 19.2 cal·g^{-1}. A crystal of gallium is shown melting in a person's hand. (Photo by Chip Clark.)

Very few *elements* have melting temperatures in the vicinity of ordinary room temperatures; one of the few is the metal gallium, as shown in Fig. 15–7.

As an illustration of the above discussion, suppose we take crushed ice from a freezer at −25°C, place it in a container with a thermometer, and surround it with a heating coil that supplies heat at a uniform rate. We insulate the system from its surroundings so no other heat enters it, and we observe the rise in temperature with time.

The result is shown in Fig. 15–8, a graph of temperature as a function of time. The temperature of the ice is observed to increase steadily, as shown by the portion of the graph from *a* to *b*, until the temperature has risen to 0°C. In this temperature range the specific heat capacity of ice is approximately

$$2302 \text{ J} \cdot \text{kg}^{-1} \cdot (\text{C}°)^{-1} \quad \text{or} \quad 0.55 \text{ cal} \cdot \text{g}^{-1} \cdot (\text{C}°)^{-1}.$$

As soon as this temperature is reached, some liquid water is observed in the container; the ice begins to *melt*, a *change of phase*, from the solid phase to the liquid phase. The thermometer, however, shows no *increase*

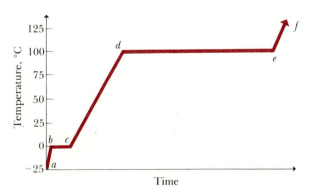

15–8 The temperature remains constant during each change of phase, provided the pressure remains constant.

in temperature, and though heat is being supplied at the same rate as before, the temperature remains at 0°C until all the ice is melted (point *c*).

As soon as the last of the ice has melted, the temperature begins to rise again at a uniform rate (from *c* to *d*) although this rate is *slower* than that from *a* to *b* because the specific heat capacity of water is greater than that of ice. When a temperature of 100°C is reached (point *d*), bubbles of steam (gaseous water or water vapor) start to escape from the liquid surface; the water begins to *boil.* The temperature remains constant at 100°C (at constant atmospheric pressure) until all the water has boiled away. Another change of phase has therefore taken place, from the liquid phase to the gaseous phase.

If all the water vapor had been trapped and not allowed to diffuse away (a very large container would be needed), the heating process could be continued as from *e* to *f*. The gas would now be called "superheated steam."

An essential point in this discussion is that when heat is added slowly (in order to maintain thermal equilibrium) to a substance that can exist in different phases, *either* the temperature rises *or* some of the subtance undergoes a phase change, but *never* both at the same time. Once the temperature for a phase change (e.g., the melting or boiling temperature) has been reached, no further temperature change occurs until *all* the substance has undergone the phase change.

Although we have used water as an example, the same type of curve as in Fig. 15–8 is obtained for many other substances. Some, of course, decompose before reaching a melting or boiling point, and others, such as glass or pitch, do not change phase at a definite temperature but become gradually softer as their temperature is raised. Crystalline substances, such as ice, or a metal, melt at a definite temperature. Glass and tar behave like supercooled liquids of very high viscosity.

When heat is removed from a gas, its temperature falls; at the same temperature at which it boiled, it returns to the liquid phase, or *condenses.* In so doing it gives up to its surroundings the same quantity of heat that was required to vaporize it. The heat so given up, per unit mass, is called the *heat of condensation* and is equal to the heat of vaporization. Similarly, a liquid returns to the solid phase, or freezes, when cooled to the temperature at which it melted, and gives up heat called *heat of solidification* exactly equal to the heat of fusion. Thus the melting point and the freezing point are at the same temperature, and the boiling point and condensation point are at the same temperature.

Under some conditions a material can be cooled below the normal phase-change temperature without a phase change occurring. The resulting state is unstable and is described as *supercooled.* Very pure water can be cooled several degrees below the normal freezing point under ideal conditions; when a small ice crystal is dropped in or the water is agitated, it crystallizes very quickly. Similarly, supercooled water vapor condenses quickly into fog droplets when a disturbance, such as dust particles or ionizing radiation, is introduced. This phenomenon is used in the *cloud chamber,* where charged particles such as protons and electrons induce condensation of supercooled vapor along their path, thus making the path visible.

Similarly, temperatures of liquids can sometimes be raised above their normal boiling temperature, and the liquids are then said to be superheated. Again, any small disturbance such as agitation or the passage of a charged particle through the material causes local boiling with bubble formation. This phenomenon is used in the bubble chamber, to be discussed in Chapter 44, to show the tracks of high-energy particles.

Steam heating systems use a boiling–condensing process to transfer heat from the furnace to the radiators. Each kilogram of water that is turned to steam in the furnace absorbs 2.26×10^6 J (the heat of vaporization of water) from the furnace, and gives up this same amount when it condenses in the radiators. (This figure is correct if the steam pressure is 1 atm. It is slightly smaller at higher pressures.) Thus the steam-heating system does not need to circulate as much water as a hot-water heating system. If water leaves a hot-water furnace at 60°C and returns at 40°C, dropping 20 C°, about 27 kg of water must circulate to carry the same heat as is carried in the form of heat of vaporization by 1 kg of steam.

The temperature-control mechanisms of many warm-blooded animals operate on a similar principle. When the hypothalamus detects a slight rise in blood temperature, the sweat glands are activated. As sweat (chiefly water) evaporates from the surface of the body, it removes heat from the body as heat of vaporization. The heat is brought to the surface of the skin by the blood in the vascular system. This plays the same role as the pipes connecting the furnace to the radiators, with the heart playing the role of the circulating pump in a forced-circulating hot-water or steam system.

The same principle is used in evaporative cooling systems. Hot dry air is pushed through wet filters. Some of the water evaporates, taking its heat of vaporization from the air, which thus becomes cooler, by as much as 20 C° in hot dry climates. Evaporative cooling is also used to condense and recirculate "used" steam in coal-fired or nuclear-powered electric power plants.

Under proper conditions of temperature and pressure, a substance can change directly from the solid to the gaseous phase without passing through the liquid phase. The transfer from solid to vapor is called *sublimation*, and the solid is said to *sublime*. "Dry ice" (solid carbon dioxide) sublimes at atmospheric pressure. Liquid carbon dioxide cannot exist at a pressure lower than about 5×10^5 Pa (about 5 atm). Heat is absorbed in the process of sublimation, and liberated in the reverse process. The quantity of heat per unit mass is called the *heat of sublimation*.

Definite quantities of heat are involved in chemical reactions. Perhaps the most familiar are those associated with combustion; complete combustion of one gram of gasoline produces about 46,000 J or about 11,000 cal, and the *heat of combustion* is said to be

$$46,000 \text{ J} \cdot \text{g}^{-1} \quad \text{or} \quad 46 \times 10^6 \text{ J} \cdot \text{kg}^{-1}.$$

Energy values of foods are defined similarly; the unit of food energy, although called a calorie, is really a *kilocalorie*, equal to 1000 cal or 4186 J. When we say that a gram of peanut butter "contains" 12 calories, we mean that, when it reacts with oxygen, with the help of enzymes, to convert the carbon and hydrogen completely to CO_2 and H_2O, the total energy liberated as heat is 12,000 cal or 50,200 J. Not all of this energy is directly useful for mechanical work; the matter of *efficiency* of utilization of energy will be discussed in detail in Chapter 19.

15–7

EXAMPLES

The basic principle in calculations involving quantity of heat is that when heat flow occurs between two bodies in thermal contact, the amount of heat lost by one body must equal that gained by the other. To avoid confusion about algebraic signs when implementing this principle, the student should use Eqs. (15–1) and (15–6) consistently for each body, noting that each Q is positive when heat enters a body and negative when it leaves. Then the algebraic sum of all the Q's must be zero. The following examples illustrate this principle in the context of phenomena discussed in this chapter.

EXAMPLE 1   An ethnic restaurant down by the docks serves coffee in copper mugs. A waiter fills a cup of mass 0.1 kg, initially at 20°C, with 0.2 kg of coffee initially at 70°C. What is the final temperature after the coffee and cup attain thermal equilibrium?

Solution   Let the final temperature be T. The (negative) heat added to the coffee (which is assumed to have a specific heat capacity equal to that of water) is

$$Q = (0.2 \text{ kg})(4186 \text{ J} \cdot \text{kg}^{-1} \cdot (\text{C}°)^{-1})(T - 70°\text{C}).$$

Similarly, the heat gained by the copper cup is

$$Q = (0.1 \text{ kg})(390 \text{ J} \cdot \text{kg}^{-1} \cdot (\text{C}°)^{-1})(T - 20°\text{C}).$$

Equating the sum of these two quantities of heat to zero yields an algebraic equation for T; solution of this equation gives $T = 67.8°\text{C}$. The final temperature is much closer to the initial temperature of the coffee than that of the cup because of the much larger specific heat capacity of water.   ◀

EXAMPLE 2   A physics student wants to cool 0.25 kg of Omni-cola (mostly water), which is initially at 20°C, by adding ice initially at −20°C. How much ice should she add so the final temperature will be 0°C with all the ice melted, if the heat capacity of the container may be neglected?

Solution   The (negative) heat added to the water is

$$Q = (0.25 \text{ kg})(4186 \text{ J} \cdot \text{kg}^{-1} \cdot (\text{C}°)^{-1})(0°\text{C} - 20°\text{C})$$
$$= -20{,}930 \text{ J}.$$

The specific heat capacity of ice is $2302 \text{ J} \cdot \text{kg}^{-1} \cdot (\text{C}°)^{-1}$. Let the mass of ice be m; then the heat needed to heat it from −20°C to 0°C is

$$Q = m(2302 \text{ J} \cdot \text{kg}^{-1} \cdot (\text{C}°)^{-1})(0°\text{C} - (-20°\text{C}))$$
$$= m(46{,}040 \text{ J} \cdot \text{kg}^{-1}).$$

The additional heat needed to melt the ice is the latent heat of fusion times the mass:

$$Q = m(334,000 \, \text{J} \cdot \text{kg}^{-1}).$$

The sum of these three quantities must equal zero:

$$-20,930 \, \text{J} + m(46,040 \, \text{J} \cdot \text{kg}^{-1}) + m(334,000 \, \text{J} \cdot \text{kg}^{-1}) = 0,$$

from which $m = 0.055$ kg $= 55$ g. This is roughly equivalent to two medium-size ice cubes. ◄

EXAMPLE 3 A certain gasoline camping lantern emits as much light as a 25-watt electric light bulb. Assuming that the efficiency of conversion of heat into light is the same for the lantern and the bulb (which is not actually correct), how much gasoline does the lantern burn in 10 hours?

Solution The rate of energy conversion in the lantern is 25 W, so in 10 hours (36,000 s), the total energy needed is

$$(25 \, \text{J} \cdot \text{s}^{-1})(36,000 \, \text{s}) = 0.9 \times 10^6 \, \text{J}.$$

As mentioned in Section 15–6, combustion of one gram of gasoline produces 46,000 J, so the mass of gasoline required is

$$\frac{0.9 \times 10^6 \, \text{J}}{46,000 \, \text{J} \cdot \text{g}^{-1}} = 19.6 \, \text{g}.$$

Actual lanterns are much less efficient than this, and typically require of the order of 300 to 400 g of gasoline (roughly one pint) to operate for ten hours. ◄

QUESTIONS

15–1 Suppose you have a Thermos bottle half full of cold coffee. Can you warm it up to drinking temperature by shaking it? Is this possible *in principle?* Is it feasible in practice? Are you adding heat to the coffee?

15–2 When the oil in an automatic transmission is churned up by the turbine blades, it becomes hot, and usually an oil-cooling system is required. Is the engine adding heat to the oil?

15–3 A student asserted that a suitable unit for specific heat capacity 1 $\text{m}^2 \cdot \text{s}^{-2} \cdot (\text{C}°)^{-1}$. Is this correct?

15–4 The specific heat capacity of water has the same numerical value when expressed in $\text{cal} \cdot \text{g}^{-1} \cdot (\text{C}°)^{-1}$ as when expressed in $\text{Btu} \cdot \text{lb}^{-1} \cdot (\text{F}°)^{-1}$. Is this a coincidence? Does the same relation hold for heat capacities of other materials?

15–5 A student claimed that when two bodies not initially in thermal equilibrium are placed in contact, the temperature rise of the cooler body must always equal the temperature drop of the warmer. Do you agree? Is there a principle of conservation of temperature, or something like that?

15–6 In choosing a fluid to circulate inside a gasoline engine to cool it (such as water or antifreeze), should you choose a material with a large or small specific heat capacity? Why? What other considerations are important?

15–7 Any heat of a phase transition has a numerical value that is $\frac{5}{9}$ as great when expressed in $\text{cal} \cdot \text{g}^{-1}$ as when expressed in $\text{Btu} \cdot \text{lb}^{-1}$. Why is the conversion so simple?

15–8 Why do you think the heat of vaporization for water is so much larger than the heat of fusion?

15–9 Some household air conditioners used in dry climates cool air by blowing it through a water-soaked filter, evaporating some of the water. How does this work? Would such a system work well in a high-humidity climate?

15–10 Why does food cook faster in a pressure cooker than in boiling water?

15–11 How does the human body maintain a temperature of 37.0°C (98.6°F) in the desert where the temperature is 50°C (122°F)?

15–12 Desert travelers sometimes keep water in a canvas bag. Some water seeps through the bag and evaporates. How does this cool the water inside?

15–13 When water is placed in ice-cube trays in a freezer, why does the water not freeze all at once when the temperature has reached 0°C? In fact, it freezes first in a layer adjacent to the sides of the tray. Why?

15–14 When an automobile engine overheats and the radiator water begins to boil, the car can still be driven some distance before catastrophic engine damage occurs. Why? What determines the onset of really disastrous overheating?

15–15 Why do automobile manufacturers recommend that antifreeze (typically a 50% solution of ethylene glycol in water) be kept in the engine in summer as well as in winter?

15–16 When you step out of the shower you feel cold, but as soon as you are dry you feel warmer, even though the room temperature is the same. Why?

15–17 Suppose the heat of fusion of ice were only $10 \text{ J} \cdot \text{g}^{-1}$ instead of $334 \text{ J} \cdot \text{g}^{-1}$. Would this change the way we make iced tea? Martinis? Lemonade?

15–18 Why is the climate of regions adjacent to large bodies of water usually more moderate than that of regions far from large bodies of water?

PROBLEMS

15–1 A student performs a combustion experiment by burning a mixture of fuel and oxygen in a constant-volume "bomb" surrounded by a water bath. During the experiment the temperature of the water is observed to rise. Considering the mixture of fuel and oxygen as the system,

a) has heat been transferred?

b) Has work been done?

15–2 A liquid is irregularly stirred in a well-insulated container and thereby undergoes a rise in temperature. Considering the liquid as the system,

a) has heat been transferred?

b) Has work been done?

15–3 A taxi driver drives an automobile of mass 1500 kg at a speed of $5 \text{ m} \cdot \text{s}^{-1}$. How many calories of heat are transferred in the brake mechanism when he stops at a stop sign?

15–4 A copper vessel of mass 200 g contains 400 g of water. The water is heated by a friction device used to test brake lining material. The device dissipates mechanical energy supplied by a motor, and it is observed that the temperature of the system rises at the rate of $3 \text{ C}° \cdot \text{min}^{-1}$. Neglect heat losses to surroundings. What power in watts is dissipated in the water?

15–5 How long could a 2000-hp motor be operated on the heat energy liberated by 1 mi^3 of ocean water when the temperature of the water is lowered by $1 \text{ C}°$ if all this heat were converted to mechanical energy? Why do we not utilize this tremendous reservoir of energy?

15–6

a) A home owner in a cold climate has a coal-burning furnace that burns 10 tons of coal during a winter. The heat of combustion of the coal is $11,000 \text{ Btu} \cdot \text{lb}^{-1}$, and 15% of this heat is wasted as hot smoke up the chimney. How many Btu were actually used to heat the house?

b) The home owner proposes to install a solar heating system, heating large tanks of water by solar radiation during the summer and using the stored energy for heating during the winter. Find the required dimensions of the storage tank, assuming it to be a cube, to store a quantity of energy equal to that computed in part (a). Assume that the water is raised to 120°F in the summer and cooled to 80°F in the winter.

15–7 An artificial satellite, constructed of aluminum, encircles the earth at a speed of $9000 \text{ m} \cdot \text{s}^{-1}$.

a) Find the ratio of its kinetic energy to the energy required to raise its temperature by $600 \text{ C}°$. (The melting point of aluminum is 660°C.) Assume a constant specific heat capacity of $0.20 \text{ cal} \cdot \text{g}^{-1} \cdot (\text{C}°)^{-1}$.

b) Discuss the bearing of your answer on the problem of the reentry of a satellite into the earth's atmosphere.

15–8

a) How much heat is required to raise the temperature of 0.20 kg of water from 20° to 30°C?

b) If this amount of heat is added to an equal mass of mercury initially at 20°C, what is the final temperature?

c) If this amount of heat is added to an equal volume of mercury initially at 20°C, what is the final temperature?

15–9 A student uses a 100-watt electric immersion heater to heat 0.2 kg of water from 20° to 100°C to make tea.

a) How much heat must be added to the water?

b) How much time is required?

15–10 An ice-cube tray contains 0.8 kg of water at 20°C. How much heat must be removed to cool the water to 0°C and freeze it?

15–11 An engineer is working on an innovative engine design. One of the moving parts, designed to operate at

150°C, contains 1.2 kg of aluminum and 0.5 kg of iron. How much heat is required to raise its temperature from 20° to 150°C?

15–12 A copper teakettle of mass 2.00 kg, containing 4.0 kg of water, is placed on a stove. How much heat must be added to raise the temperature from 20° to 80°C?

15–13 In a physics lab experiment, a student immersed one hundred copper pennies (mass 3.0 g each) in boiling water. After they reached thermal equilibrium, she fished them out and dropped them into 0.2 kg of water at 20°C. What was the final temperature?

15–14 A calorimeter contains 100 g of water at 0°C. A 1000-g copper cylinder and a 1000-g lead cylinder, both at 100°C, are placed in the calorimeter. Find the final temperature if there is no loss of heat to the surroundings.

15–15

a) Compare the heat capacities (heat capacity is the total heat per unit temperature change) of equal *masses* of water, copper, and lead.

b) Compare the heat capacities of equal *volumes* of water, copper, and lead.

15–16 An aluminum can of mass 500 g contains 117.5 g of water at a temperature of 20°C. A 200-g block of iron at 75°C is dropped into the can. Find the final temperature, assuming no heat loss to the surroundings.

15–17 A metalworker hoists a casting weighing 100 lb from an annealing furnace where its temperature was 900°F and plunges it into a tank containing 800 lb of oil at a temperature of 80°F. The final temperature is 100°F, and the specific heat capacity of the oil is $0.5 \text{ Btu} \cdot \text{lb}^{-1} \cdot (\text{F}°)^{-1}$. What was the specific heat capacity of the casting? Neglect the heat capacity of the tank itself and any heat losses.

15–18 A marksman fires a lead bullet at $350 \text{ m} \cdot \text{s}^{-1}$ into a wood block target, where it comes to rest. What would be the rise in temperature of the bullet if there were no heat loss to the surroundings? Does the bullet melt?

15–19 A copper calorimeter (mass 300 g) contains 500 g of water at a temperature of 15°C. A 560-g block of copper, at a temperature of 100°C, is dropped into the calorimeter and the temperature is observed to increase to 22.5°C. Neglect heat losses to the surroundings. Find the specific heat capacity of copper.

15–20 A laboratory technician dropped a 50-g specimen of an unknown material, at a temperature of 100°C, into a calorimeter containing 200 g of water initially at 20°C. The calorimeter is of copper and its mass is 100 g. The final temperature of the calorimeter is 22°C. Compute the specific heat capacity of the specimen.

15–21 An engineer is developing an electric heater to provide a continuous supply of hot water. One trial design is shown in Fig. 15–9. Water is flowing at the rate of

$300 \text{ g} \cdot \text{min}^{-1}$, the inlet thermometer registers 15°C, the voltmeter reads 120 V and the ammeter 10 A (corresponding to a power input of $(120 \text{ V})(10 \text{ A}) = 1200 \text{ W}$).

a) When a steady state is finally reached, what is the reading of the outlet thermometer?

b) Why is it unnecessary to take into account the heat capacity (mc or nC) of the apparatus itself?

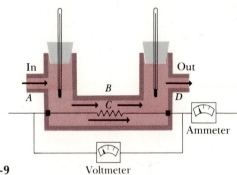

Figure 15–9

15–22 An automobile engine whose output is 40 hp uses 4.5 gallons of gasoline per hour. The heat of combustion is 12×10^7 joules per gallon. What is the efficiency of the engine? That is, what fraction of the heat of combustion is converted to mechanical work?

15–23 The electric power input to a certain electric motor is 0.5 kW, and the mechanical power output is 0.54 hp.

a) What is the efficiency of the motor? (That is, what fraction of the electrical power is converted to mechanical power?)

b) How many calories are developed in the motor in one hour of operation, assuming all electrical energy not converted to mechanical energy is converted to heat?

15–24 Evaporation of perspiration is an important mechanism for temperature control in warm-blooded animals. What mass of water must evaporate from the surface of an 80-kg human body to cool it 1 C°? The specific heat capacity is approximately $1 \text{ cal} \cdot \text{g}^{-1} \cdot (\text{C}°)^{-1}$, and the latent heat of vaporization of water at body temperature (37°C) is $577 \text{ cal} \cdot \text{g}^{-1}$.

15–25 If a person's food-energy intake is 2400 calories per day, and if all this energy eventually is given off as heat, what is the average rate of heat output, in watts? How many people in a room, giving off heat at this rate, would be required to have the same heat output as a 1500-W electric heater?

15–26 A piece of ice at 0°C falls from rest into a lake at 0°C, and one half of one percent of the ice melts. Compute the minimum height from which the ice falls.

15–27 What must be the initial velocity of a lead bullet at a temperature of 25°C, so that the heat developed when it is brought to rest shall be just sufficient to melt it?

15–28 An engineer has designed a continuous-flow calorimeter to measure the heat of combustion of a gaseous fuel. A sketch of his design is shown in Fig. 15–10. Water is supplied at the rate of 12.5 lb·min^{-1} and natural gas at 0.020 ft^3·min^{-1}. In the steady state, the inlet and outlet thermometers register 60°F and 76°F, respectively. What is the heat of combustion of natural gas in Btu·ft^{-3}? Why should the gas flow be made as small as possible?

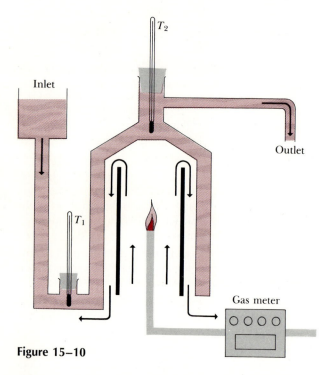

Figure 15–10

15–29 A technician measures the specific heat capacity of an unidentified liquid by immersing an electrical resistor in it. Electrical energy is dissipated for 100 s at a constant rate of 50 W. The mass of the liquid is 530 g, and its temperature increases from 17.64°C to 20.77°C. Find the mean specific heat capacity of the liquid in this temperature range.

15–30 How much heat is required to convert 1 g of ice at −10°C to steam at 100°C?

15–31 A beaker of very small mass contains 500 g of water at a temperature of 80°C. How many grams of ice at a temperature of −20°C must be dropped in the water so that the final temperature of the system will be 50°C?

15–32 An open vessel contains 500 g of ice at −20°C. The mass of the container can be neglected. Heat is supplied to the vessel at the constant rate of 1000 cal·min^{-1}

for 100 min. Plot a curve showing the elapsed time as abscissa and the temperature as ordinate.

15–33 A copper calorimeter of mass 100 g contains 150 g of water and 8 g of ice in thermal equilibrium at atmospheric pressure. 100 g of lead at a temperature of 200°C are dropped into the calorimeter. Find the final temperature if no heat is lost to the surroundings.

15–34 A thirsty farmer cools a one-liter bottle of a soft drink (mostly water) by pouring it into a large copper mug of mass 278 g and adding 500 g of ice initially at −16°C. If soft drink and mug are initially at 20°C, what is the final temperature of the system, assuming no heat losses?

15–35 A tube leads from a flask in which water is boiling under atmospheric pressure to a calorimeter. The mass of the calorimeter is 150 g, its specific heat capacity is 0.1 cal·g^{-1}·(C°)$^{-1}$, and it contains originally 340 g of water at 15°C. Steam is allowed to condense in the calorimeter until its temperature increases to 71°C, after which the total mass of calorimeter and contents is found to be 525 g. Compute the heat of condensation of steam from these data.

15–36 An aluminum canteen whose mass is 500 g contains 750 g of water and 100 g of ice. The canteen is accidentally dropped from an aircraft to the ground. After landing, the temperature of the canteen is found to be 25°C. Assuming that no energy is given to the ground in the impact, what was the velocity of the canteen just before it landed? State any additional assumptions needed.

15–37 A calorimeter contains 500 g of water and 300 g of ice, all at a temperature of 0°C. A metallurgist takes a block of metal of mass 1000 g from a furnace where its temperature was 240°C and drops it quickly into the calorimeter. As a result, all the ice is just melted. What would the final temperature of the system have been if the mass of the block had been twice as great? Neglect heat loss from and the heat capacity of the calorimeter.

15–38 An ice cube whose mass is 50 g is taken from a refrigerator where its temperature was −10°C and dropped into a glass of water at 0°C. If no heat is gained or lost from outside, how much water will freeze onto the cube?

15–39 A copper calorimeter can ($mc = 30$ cal·(C°)$^{-1}$) contains 50 g of ice. The system is initially at 0°C. 12 g of steam at 100°C and 1 atm pressure are admitted into the calorimeter. What is the final temperature of the calorimeter and its contents?

15–40 A vessel whose walls are thermally insulated contains 2100 g of water and 200 g of ice, all at a temperature of 0°C. The outlet of a tube leading from a boiler, in which water is boiling at atmospheric pressure, is inserted into the water. How many grams of steam must condense to raise the temperature of the system to 20°C? Neglect the heat capacity of the container.

15–41 A 2-kg iron block is taken from a furnace where its temperature was 650°C and placed on a large block of ice at 0°C. Assuming that all the heat given up by the iron is used to melt the ice, how much ice is melted?

15–42 In a household hot-water heating system, water is delivered to the radiators at 140°F and leaves at 100°F. The system is to be replaced by a steam system in which steam at atmospheric pressure condenses in the radiators, the condensed steam leaving the radiators at 180°F. How many pounds of steam will supply the same heat as was supplied by 1 lb of hot water in the first installation?

15–43 A "solar house" has storage facilities for 4 million Btu. Compare the space requirements for this storage on the assumption

a) that the heat is stored in water heated from a minimum temperature of 80°F to a maximum at 120°F, and

b) that the heat is stored in Glauber salt ($Na_2SO_4 \cdot 10H_2O$) heated in the same temperature range.

Properties of Glauber Salt	
Specific heat capacity	
Solid	0.46 Btu·lb^{-1}·(F°)$^{-1}$
Liquid	0.68 Btu·lb^{-1}·(F°)$^{-1}$
Specific gravity	1.6
Melting point	90°F
Heat of fusion	104 Btu·lb^{-1}

15–44 The nominal food-energy value of butter is about 6 kcal·g^{-1}, where 1 kcal = 1000 cal = 4186 J. If all this energy could be converted completely to mechanical energy, how much butter would be required to power an 80-kg mountaineer on his journey from Lupine Meadows (elevation 2070 m) to the summit of Grand Teton (4196 m)?

15–45 The capacity of air conditioners is typically expressed in Btu·hr^{-1} or "tons," the latter being the number of tons of ice that can be frozen from water at 0°C in 24 hr by the unit.

a) What is the rating in Btu/hr of a one-ton air conditioner?

b) Express the capacity of a one-ton air conditioner in watts.

15–46 A 6000 Btu·hr^{-1} air conditioner typically consumes about 800 W of electrical power. What is the ratio of the rate of removal of heat to the rate of consumption of electrical energy? Express the two rates in the same units to obtain a dimensionless ratio.

15–47 Suppose 1 liter of gasoline will propel a car a distance of 10 km. The density of gasoline is about 0.7 g·cm^{-3}, and its heat of combustion is about 4.6×10^4 J·g^{-1}.

a) If the engine is 25% efficient, that is, if $\frac{1}{4}$ of the heat of combustion is converted into useful mechanical work, what total work does the engine do during the 10-km displacement?

b) If this work is assumed to be done in opposing a constant resisting force F, find the magnitude of F.

16

HEAT
TRANSFER

Conduction of heat was mentioned in Chapter 15 in connection with the transfer of heat between two bodies initially at different temperatures and their approach to thermal equilibrium. We defined quantity of heat, but it was not necessary to consider in detail the *rate* at which heat is transferred from one body to another. In some problems, however, it is important to know how quickly or slowly heat is transferred, and what quantities determine the rate of heat transfer. In this chapter we shall study in detail the three mechanisms of heat transfer: conduction, convection, and radiation.

16–1

CONDUCTION

If one end of a metal rod is placed in a flame while the other is held in the hand, the end being held becomes hotter and hotter, although it is not itself in direct contact with the flame. Heat is said to reach the cooler end of the rod by *conduction* through the material of the rod. The molecules at the hot end of the rod increase the energy of their vibration as the temperature increases. Then, as they collide with their more slowly moving neighbors farther from the flame, some of their energy of motion is shared with these neighbors, and they in turn pass it along to those still farther from the flame. Hence energy of thermal motion is passed along from one molecule to the next, while each individual molecule remains at its original position.

Most metals are good conductors of electricity and also good conductors of heat. The ability of a metal to conduct an electric current is due to the fact that in the material there are electrons that have become detached from their parent molecules. These "free" electrons also play a part in the conduction of heat, and the reason metals are such good heat conductors is that the free electrons provide an effective mechanism for heat transfer from the hotter to the cooler portions of the metal.

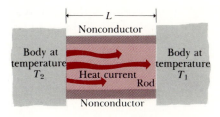

16–1 Steady-state heat flow in a uniform rod.

Conduction of heat can take place in a body only when different parts of the body are at different temperatures, and the direction of heat flow is always from points of higher to points of lower temperature. Figure 16–1 represents a rod of material of cross-sectional area A and length L. Let the left end of the rod be kept at a temperature T_2, and the right end at a lower temperature T_1. The direction of the heat current is then from left to right through the rod. The sides of the rod are covered by an insulating material, so no heat transfer occurs at the sides. This is of course an idealized model; all real materials, even the best thermal insulators, conduct heat to some extent.

After the ends of the rod have been kept at the temperatures T_1 and T_2 for a sufficiently long time, the rod reaches a condition in which the temperature at each point is no longer changing with time and in which heat flows from the hot end to the cold at a constant rate. This condition is called *steady-state heat flow.* If the temperature is then measured at various points along the rod, it is found to decrease uniformly with distance from the hot end.

Experiment shows that the rate of flow of heat through the rod in the steady state is proportional to the area A, proportional to the temperature difference $(T_2 - T_1)$, and inversely proportional to the length L. These proportions may be converted to an equation by introducing a constant k whose numerical value depends on the material of the rod. The quantity k is called the *thermal conductivity* of the material,

$$H = \frac{kA(T_2 - T_1)}{L}, \tag{16–1}$$

where H is the quantity of heat flowing through the rod per unit time, also called the *heat current.* The units of H are energy per unit time, or power. Thus in SI units the unit of H is $1\ \text{J} \cdot \text{s}^{-1}$ or $1\ \text{W}$.

Equation (16–1) may also be used to compute the rate of heat flow through a slab, or *any* homogeneous body having a uniform cross section perpendicular to the direction of flow, provided the flow has attained steady-state conditions and the ends are kept at constant temperatures.

When the cross section is not uniform, or when steady-state conditions are not present, the temperature does not necessarily change uniformly along the direction of heat flow. Equation (16–1) may still be applied to a thin layer of material perpendicular to the direction of flow. If x is the coordinate measured along the flow path, Δx the thickness of the layer, and A the cross-sectional area perpendicular to the path, then Eq. (16–1) may be rewritten

$$H = -kA\frac{\Delta T}{\Delta x}, \tag{16–2}$$

where ΔT is the temperature change across the distance Δx. The negative sign is included because, if the temperature *increases* in the direction of increasing x (Δx and ΔT both positive), the direction of heat flow is the direction of *decreasing* x, and conversely. The temperature change per unit length $\Delta T/\Delta x$ is called the *temperature gradient;* Eq. (16–1) refers to the special case where the temperature gradient is constant, and equal to $(T_2 - T_1)/L$.

The SI unit of rate of heat flow or heat current is the joule per second; other units such as the calorie per second or Btu per second are also sometimes used. The units of k are obtained by solving Eq. (16–2) for k:

$$k = -\frac{H}{A}\frac{\Delta x}{\Delta T}.$$

This shows that in SI units, the unit of k is

$$\frac{(1\ \text{J}\cdot\text{s}^{-1})(1\ \text{m})}{(1\ \text{m})^2(1\ \text{C}°)} = 1\ \text{J}\cdot(\text{s}\cdot\text{m}\cdot\text{C}°)^{-1}.$$

Values of thermal conductivity are sometimes tabulated using cgs units, with the calorie as the energy unit. The unit of k then is $1\ \text{cal}\cdot(\text{s}\cdot\text{cm}\cdot\text{C}°)^{-1}$. The conversion is

$$1\ \text{cal}\cdot(\text{s}\cdot\text{cm}\cdot\text{C}°)^{-1} = 419\ \text{J}\cdot(\text{s}\cdot\text{m}\cdot\text{C}°)^{-1}.$$

For most materials, the thermal conductivity increases slightly with increasing temperature, but the variation is small and can often be neglected. Some numerical values of k, at temperatures near room temperature, are given in Table 16–1. The properties of materials used

TABLE 16–1   THERMAL CONDUCTIVITIES (k)

	$\text{J}\cdot\text{s}^{-1}\cdot\text{m}^{-1}\cdot(\text{C}°)^{-1}$	$\text{cal}\cdot\text{s}^{-1}\cdot\text{cm}^{-1}\cdot(\text{C}°)^{-1}$
Metals		
Aluminum	205	0.49
Brass	109	0.26
Copper	385	0.92
Lead	34.7	0.083
Mercury	8.3	0.020
Silver	406	0.97
Steel	50.2	0.12
Various solids		
(Representative values)		
Insulating brick	0.15	0.00035
Red brick	0.6	0.0015
Concrete	0.8	0.002
Cork	0.04	0.0001
Felt	0.04	0.0001
Fiberglass	0.04	0.0001
Glass	0.8	0.002
Ice	1.6	0.004
Rock wool	0.04	0.0001
Styrofoam	0.01	0.00002
Wood	0.12–0.04	0.0003–0.0001
Gases		
Air	0.024	0.000057
Argon	0.016	0.000039
Helium	0.14	0.00034
Hydrogen	0.14	0.00033
Oxygen	0.023	0.000056

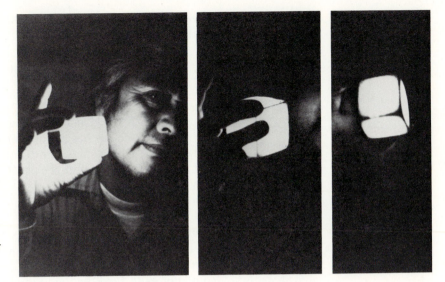

16–2 This protective tile, developed for use on the space shuttle Columbia, has extraordinary thermal properties. The extremely small thermal conductivity and small heat capacity of the material make it possible to hold the tile by its edges, even though it is hot enough to emit the light for these photographs. (Courtesy Lockheed Corp.)

commercially as heat insulators are sometimes expressed in a mixed system in which the unit of heat current is 1 Btu·hr^{-1}, the unit of area is 1 ft^2, and the unit of temperature gradient is one Fahrenheit degree per *inch* (1 F°·in^{-1})! What better argument could there be for adoption of the metric system?

Equation (16–1) shows that the larger the thermal conductivity k, the larger the heat current, if other factors are equal. A material for which k is large is therefore a good heat conductor, while if k is small, the material is a poor conductor or a good insulator. In ordinary circumstances a "perfect heat conductor" ($k = \infty$) or a "perfect heat insulator" ($k = 0$) does not exist. However, Table 16–1 shows that the metals as a group have much greater thermal conductivities than the nonmetals, and that those of gases are extremely small. Figure 16–2 shows a "space-age" ceramic material with very unusual thermal properties.

EXAMPLE 1 A styrofoam box used to keep beer cold at a picnic has total wall area (including the lid) of 0.8 m^2 and wall thickness 2.0 cm. The thermal conductivity of styrofoam is about 0.01 J·m^{-1}·s^{-1}·(C°)$^{-1}$. It is filled with ice and beer at 0°C. What is the rate of heat flow into the box if the outside temperature is 30°C? How much ice melts in one day?

Solution We assume that the total heat flow is approximately the same as it would be through a flat slab of area 0.8 m^2 and thickness 2.0 cm = 0.02 m. The rate of heat flow, from Eq. (16–1), is

$$H = (0.01 \text{ J·m}^{-1}\text{·s}^{-1}\text{·(C°)}^{-1})(0.8 \text{ m}^2) \frac{30 \text{ C°}}{0.02 \text{ m}} = 12 \text{ J·s}^{-1}.$$

There are 86,400 s in one day, so the total heat flow in one day is

$$(12 \text{ J})(86,400 \text{ s}) = 1.04 \times 10^6 \text{ J}.$$

The latent heat of fusion of ice is $334 \text{ J} \cdot \text{g}^{-1}$, so the quantity of ice melted by this quantity of heat is

$$\frac{1.04 \times 10^6 \text{ J}}{334 \text{ J} \cdot \text{g}^{-1}} = 3110 \text{ g} \qquad \text{or about 3.1 kg.} \qquad \blacktriangleleft$$

EXAMPLE 2   A steel bar 10 cm long is welded end to end to a copper bar 20 cm long. Each bar has a square cross section, 2 cm on a side. The free end of the steel bar is placed in contact with steam at 100°C, and the free end of the copper bar with ice at 0°C. Find the temperature at the junction of the two bars and the total rate of heat flow, when steady-state conditions have been reached.

Solution   The key to the solution is the fact that the rates of heat flow in the two bars must be equal; otherwise some sections would have more heat flowing in than out, or the reverse, and steady-state conditions could not exist. Let T be the unknown junction temperature; we use Eq. (16–1) for each bar and equate the two expressions:

$$\frac{k_s A(100°\text{C} - T)}{L_s} = \frac{k_c A(T - 0°\text{C})}{L_c}.$$

The areas A are equal and may be divided out. Substituting numerical values, we find

$$\frac{(50.2 \text{ J} \cdot \text{s}^{-1} \cdot \text{m}^{-1} \cdot (\text{C}°)^{-1})(100°\text{C} - T)}{0.1 \text{ m}} = \frac{(385 \text{ J} \cdot \text{s}^{-1} \cdot \text{m}^{-1} \cdot (\text{C}°)^{-1})(T - 0°\text{C})}{0.2 \text{ m}}.$$

Rearranging and solving for T, we find

$$T = 20.7°\text{C}.$$

Even though the steel bar is shorter, the temperature drop across it is much greater than across the copper bar, because steel is a much poorer conductor.

   The total heat current is now obtained by substituting this value for T back into either of the above expressions:

$$H = \frac{(50.2 \text{ J} \cdot \text{s}^{-1} \cdot \text{m}^{-1} \cdot (\text{C}°)^{-1})(0.02 \text{ m})^2 (100°\text{C} - 20.7°\text{C})}{0.1 \text{ m}}$$

$$= 15.9 \text{ J} \cdot \text{s}^{-1} = 15.9 \text{ W},$$

or

$$H = \frac{(385 \text{ J} \cdot \text{s}^{-1} \cdot \text{m}^{-1} \cdot (\text{C}°)^{-1})(0.02 \text{ m})^2 (20.7°\text{C} - 0°\text{C})}{0.2 \text{ m}}$$

$$= 15.9 \text{ J} \cdot \text{s}^{-1} = 15.9 \text{ W}. \qquad \blacktriangleleft$$

EXAMPLE 3   In Example 2 above, suppose the two bars are separated; one end of each bar is placed in contact with steam at 100°C, and the other end of each with ice at 0°C. What is the *total* rate of heat flow in the two bars?

Solution In this case the bars are in parallel rather than in series. The total heat flow is the sum of the flows in the two bars, and for each bar $T_2 - T_1 = 100°C - 0°C = 100\ C°$. Thus

$$H = \frac{(50.2\ \text{J}\cdot\text{s}^{-1}\cdot\text{m}^{-1}\cdot(\text{C°})^{-1})(0.02\ \text{m})^2(100\ \text{C°})}{0.1\ \text{m}}$$

$$+ \frac{(385\ \text{J}\cdot\text{s}^{-1}\cdot\text{m}^{-1}\cdot(\text{C°})^{-1})(0.02\ \text{m})^2(100\ \text{C°})}{0.2\ \text{m}}$$

$$= 20.1\ \text{J}\cdot\text{s}^{-1} + 77.0\ \text{J}\cdot\text{s}^{-1} = 97.1\ \text{J}\cdot\text{s}^{-1} = 97.1\ \text{W}.$$

The heat flow in the copper bar is much greater than in the steel bar, even though it is longer, because its thermal conductivity is much larger. The total heat flow is much larger than in Example 2 because the full 100 C° temperature difference appears across each bar. ◀

Before steady-state conditions are reached, the temperature at each point may change with time. For example, suppose the rod in Fig. 16–1 is initially at the temperature T_1 throughout, and then at time $t = 0$ the ends are placed in contact with the bodies at temperatures T_1 and T_2. The resulting temperature distributions at various times are shown in Fig. 16–3. In these graphs, $x = 0$ represents the left end of the rod, and $x = L$ the right end. At time t_1 the temperature changes much more rapidly with x near the left end than near the right (i.e., the temperature gradient is greater), corresponding to the fact that the heat flow is much greater near that end to provide a net flow of heat into the middle region and raise its temperature. With increasing time the differences become smaller and smaller, until after a very long time ($t = \infty$) the steady-state uniform temperature distribution is attained. The final steady state is independent of the initial conditions at $t = 0$; it depends only on the temperatures at the ends.

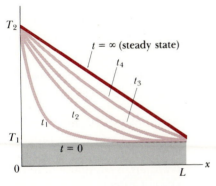

16–3 Transient and steady-state temperature distribution along a rod initially at temperature T_1. The transient distributions were computed for $t_2 = 5t_1$, $t_3 = 10t_1$, $t_4 = 20t_1$.

In the steady state the heat current in the rod is the same in all cross sections; otherwise there would be a net flow of heat *into* certain regions and *out* of others and their temperatures would change, contradicting the assumption of a steady state. Thus steady-state heat flow is somewhat analogous to the flow of an incompressible fluid. In later chapters we will see another analog, that of electric current. In each case there is a basic conservation law; for the incompressible fluid it is conservation of mass. For heat flow it is conservation of energy, and for electric current it is conservation of electric charge.

16–2

CONVECTION

In heat transfer by *conduction*, microscopic molecular vibrations and electron motion are responsible for the energy transfer, but there is no overall bulk motion of the material. Convection, by contrast, is transfer of heat by actual motion of material from one region of space to another. The hot-air furnace, the hot-water heating system, and the flow of blood in the body are examples. If the material is forced to move by a blower or pump, the process is called *forced convection;* if the material flows due to

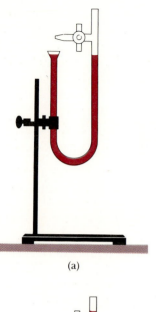

(a)

(b)

16–4 Convection is brought about by differences in density.

322   HEAT TRANSFER

differences in density (caused by thermal expansion), the process is called *natural* or *free convection*. To understand the latter, consider a U-tube, as illustrated in Fig. 16–4.

In (a), the water is at the same temperature in both arms of the U and hence stands at the same level in each. In (b), the right side of the U has been heated. The water in this side expands and therefore has smaller density; thus a longer column is needed to balance the pressure produced by the cold water in the left column. When the stopcock is opened, water flows from the top of the warmer column into the colder column. This increases the pressure at the bottom of the U produced by the cold column, and decreases the pressure at this point due to the hot column. Hence at the bottom of the U, water is forced from the cold to the hot side. If heat is continually applied to the hot side and removed from the cold side, the circulation continues indefinitely. The net result is a continuous transfer of heat from the hot to the cold side of the column. In the common household hot-water heating system, the "cold" side corresponds to the radiators and the "hot" side to the furnace.

The anomalous expansion of water, mentioned in Section 14-5, has an important effect on the way in which lakes and ponds freeze in winter. Suppose a pond is at a temperature of 20°C throughout, and suppose the air temperature at its surface falls to −10°C. The water at the surface becomes cooled; it contracts, becomes more dense than the warmer water below it, and sinks in this less dense water, its place being taken by water at 20°C. The sinking of the cooled water causes a mixing process, which continues until all of the water has been cooled to 4°C. Now, however, when the surface water cools to 3°C, it *expands*, is *less* dense than the water below it, and hence floats on the surface. Convection and mixing then cease, and the remainder of the water can lose heat only by *conduction*. Since water is a poor heat conductor, cooling takes place very slowly after 4°C is reached, with the result that the pond freezes first at its surface. Then, since the density of ice is even smaller than that of water at 0°C, the ice floats on the water below it, and further freezing can result only from heat flow upward by conduction.

There is no simple equation for convective heat transfer as there is for conduction. The heat lost or gained by a surface at one temperature in contact with a fluid at another temperature depends on many factors, such as the shape and orientation of the surface, the mechanical and thermal properties of the fluid, and the nature of the fluid flow, laminar or turbulent.

For practical calculations the usual procedure is to define a convection coefficient h by means of the equation

$$H = hA\, \Delta T, \qquad (16\text{–}3)$$

where H is the heat current due to convection, A is the surface area, and ΔT is the temperature difference between the surface and the main body of the fluid. Values of h are determined experimentally; it is found not to be constant but to depend on ΔT. The results are published as tables or graphs from which a physicist or engineer may obtain convection coefficients appropriate for specific situations.

A situation of practical importance is that of natural convection from a wall or a pipe that is at a constant temperature and is surrounded by air that is cooler than the wall or pipe by an amount ΔT. The convection coefficients applicable in this situation are given in Table 16–2. If the air is *hotter* than a horizontal plate, the first two values of h should be interchanged.

TABLE 16–2 COEFFICIENTS OF NATURAL CONVECTION IN AIR AT ATMOSPHERIC PRESSURE

Equipment	Convection coefficient h, $\mathrm{J \cdot s^{-1} \cdot m^{-2} \cdot (C°)^{-1}}$
Horizontal plate, facing upward	$2.49\,(\Delta T)^{1/4}$
Horizontal plate, facing downward	$1.31\,(\Delta T)^{1/4}$
Vertical plate	$1.77\,(\Delta T)^{1/4}$
Horizontal or vertical pipe $\left(\dfrac{\text{diameter}}{D} \right)$	$1.32\left(\dfrac{\Delta T}{D} \right)^{1/4}$

EXAMPLE The air in a room is at a temperature of 25°C, and the outside air is at -15°C. How much heat is transferred per unit area of a glass windowpane of thermal conductivity $0.80\ \mathrm{J \cdot m^{-1} \cdot s^{-1} \cdot (C°)^{-1}}$ and of thickness 2 mm?

Solution To assume that the inner surface of the glass is at 25°C and the outer surface is at -15°C is erroneous, as anyone can verify by touching the inner surface of a glass windowpane on a cold day. One must expect a much *smaller* temperature difference across the windowpane, so that in the steady state the rates of transfer of heat (1) by convection in the room, (2) by conduction through the glass, and (3) by convection in the outside air, are all equal.

As a first approximation in the solution of this problem, let us assume that the window is at a uniform temperature T. If $T = 5$°C, then the temperature difference between the inside air and the glass is the same as that between the glass and the outside air, or 20°C. Hence the convection coefficient in both cases is

$$h = 1.77(20)^{1/4}\ \mathrm{J \cdot s^{-1} \cdot m^{-2} \cdot (C°)^{-1}}$$
$$= 3.74\ \mathrm{J \cdot s^{-1} \cdot m^{-2} \cdot (C°)^{-1}},$$

and, from Eq. (16–8), the heat transferred per unit area is

$$\frac{H}{A} = (3.74\ \mathrm{J \cdot s^{-1} \cdot m^{-2} \cdot (C°)^{-1}})(20\ \mathrm{C°})$$

$$= 74.9\ \mathrm{J \cdot s^{-1} \cdot m^{-2}}.$$

The glass, however, is *not* at uniform temperature; there must be a temperature difference ΔT across the glass sufficient to provide heat conduction at the rate of $74.9\ \mathrm{J \cdot s^{-1} \cdot m^{-2}}$. Using the conduction equation, Eq.

(16–1), we obtain

$$\Delta T = \frac{LH}{kA} = \frac{(0.002 \text{ m})(74.9 \text{ J}\cdot\text{s}^{-1}\cdot\text{m}^{-2})}{0.80 \text{ J}\cdot\text{m}^{-1}\cdot\text{s}^{-1}\cdot(\text{C}°)^{-1}} = 0.18 \text{ C}°.$$

Thus the inner surface is at about 5.09°C and the outer at 4.91°C. ◄

Heat transfer in the human body involves a combination of mechanisms, which together maintain a remarkably constant and uniform temperature despite large changes in environmental conditions. As mentioned above, the chief *internal* mechanism is forced convection, with the heart serving as the pump and the blood as the circulating fluid. Heat transfer with the surroundings involves conduction, convection, and radiation, in proportions depending on circumstances. Total heat loss from the body is on the order of 2000 to 5000 kcal per day, depending on the amount of activity. A dry nude body in still air loses about half its heat by radiation, but under conditions of vigorous activity and copious perspiration, the dominant mechanism is evaporative cooling. Radiation will be discussed in the following sections.

16–3

RADIATION

When one's hand is placed in direct contact with the surface of a hot-water or steam radiator, heat reaches the hand by *conduction* through the radiator walls. If the hand is held above the radiator but not in contact with it, heat reaches the hand by way of the upward-moving *convection* currents of warm air. If the hand is held at one side of the radiator it still becomes warm, even though conduction through the air is negligible and the hand is not in the path of the convection currents. Energy now reaches the hand by *radiation*.

The term radiation refers to the continuous emission of energy from the surface of all bodies. This energy is called *radiant energy* and is in the form of electromagnetic waves. These waves travel with the speed of light and are transmitted through vacuum as well as through air. (Better, in fact, since they are absorbed by air to some extent.) When they fall on a body that is not transparent to them, such as the surface of one's hand or the walls of the room, they are absorbed, resulting in a transfer of heat to the absorbing material.

The radiant energy emitted by a surface depends on the nature of the surface and on its temperature. At low temperatures the rate of radiation is small and the radiant energy is chiefly of relatively long wavelength. As the temperature is increased, the rate of radiation increases very rapidly, in proportion to the *fourth power* of the absolute temperature. For example, a copper block at a temperature of 100°C (373 K) radiates about 0.03 J·s^{-1} or 0.03 W from each square centimeter of its surface. At a temperature of 500°C (773 K), it radiates about 0.54 W from each square centimeter, and at 1000°C (1273 K), it radiates about 4 W per square centimeter. This rate is 130 times as great as that at a temperature of 100°C.

At any temperature, the radiant energy emitted is a mixture of waves of different wavelengths. Wavelengths of light in the visible spectrum range from 0.4×10^{-6} m (violet) to 0.7×10^{-6} m (red). At a temperature of 300°C, practically all of the radiant energy emitted by a body is carried by waves of *longer* wavelength than these. Such waves are called *infrared,* meaning "beyond the red." As the temperature rises, the wavelengths shift to shorter values. At 800°C a body emits enough visible radiant energy to be self-luminous and appears "red hot." By far the larger part of the energy emitted, however, is still carried by infrared waves. At 3000°C, which is about the temperature of an incandescent lamp filament, the radiant energy contains enough of the shorter wavelengths so that the body appears "white hot."

Radiation is the mechanism by which the sun's energy reaches the earth. Near the earth, the rate of energy transfer due to solar radiation is about 1400 W for each square meter of surface area. Not all of this energy reaches the earth's surface; even on a clear day, about one fourth of this energy is absorbed by the earth's atmosphere.

16–4

STEFAN–BOLTZMANN LAW

Experiment shows that the rate of radiation of energy from a surface is proportional to the surface area A and to the fourth power of the *absolute* temperature T. It also depends on the nature of the surface, described by a dimensionless number e, which is between 0 and 1. Thus the relation may be expressed in terms of the heat current H as

$$H = Ae\sigma T^4, \tag{16–4}$$

where σ is a fundamental physical constant called the Stefan–Boltzmann constant. This relation was deduced by Josef Stefan (1835–1893) on the basis of experimental measurements made by John Tyndall (1820–1893) and was later derived from theoretical considerations by Ludwig Boltzmann (1844–1906).

In Eq. (16–4), H has units of power (energy per unit time). Thus in SI units σ has the units $W \cdot m^{-2} \cdot K^{-4}$. The numerical value is found experimentally to be

$$\sigma = 5.6699 \times 10^{-8} \, W \cdot m^{-2} \cdot K^{-4}.$$

The number e characterizing the emitting properties of a particular surface is called the *emissivity.* It is generally larger for dark, rough surfaces than for light, smooth ones. The emissivity of a smooth copper surface is about 0.3.

EXAMPLE A thin square steel plate, 10 cm on a side, is heated in a blacksmith's forge to a temperature of 800°C. If the emissivity is unity, what is the total rate of radiation of energy?

Solution The total surface area, including both sides, is $2(0.1 \, \text{m})^2 = 0.02 \, \text{m}^2$. The temperature used in Eq. (16–4) must be *absolute* temperature; 800°C = 1073 K. Then Eq. (16–4) gives

$$H = (0.02\ \text{m}^2)(1)(5.67 \times 10^{-8}\ \text{W} \cdot \text{m}^{-2} \cdot \text{K}^{-4})(1073\ \text{K})^4 = 1503\ \text{W}.$$

If the plate were heated by an electric heater instead, an electric power input of 1503 W would be required to maintain it at constant temperature.   ◀

If the surfaces of all bodies continuously emit radiant energy according to Eq. (16–4), then why do they not eventually radiate away *all* their energy and cool down to a temperature of absolute zero? The answer is that they *would* do so if energy were not supplied to them in some way. In the case of an electric heater element or the filament of an electric lamp, energy is supplied electrically to make up for the energy radiated. As soon as this energy supply is cut off, these bodies do, in fact, cool down very quickly to the temperature of their surroundings. The reason they do not cool further is that the surroundings are *also* radiating. Some of this radiated energy is intercepted and absorbed.

The rate at which a body *radiates* energy is determined by the temperature of the *body*, but the rate at which it *absorbs* energy by radiation depends on the temperature of its *surroundings*. When a body is hotter than its surroundings, the rate of emission exceeds the rate of absorption; there is a net loss of energy, and the body cools down, unless its temperature is maintained by some other means. When a body is colder than its surroundings, the rate of absorption is greater than the rate of emission, and its temperature rises. At thermal equilibrium the two rates are equal.

Thus when a body at temperature T is surrounded by walls also at temperature T, maintenance of thermal equilibrium requires that the body's rate of *absorption* of radiant energy from the walls must be equal to the rate of radiation from its surface, which is

$$H = Ae\sigma T^4.$$

Hence, for such a body at a temperature T_1, surrounded by walls at a temperature T_2, (as in Fig. 16–5), the *net* rate of loss (or gain) of energy per unit area by radiation is

$$H_{\text{net}} = Ae\sigma T_1^4 - Ae\sigma T_2^4 = Ae\sigma(T_1^4 - T_2^4). \tag{16–5}$$

Infrared emission from a body can be studied using a camera equipped with infrared-sensitive film, or better with a special device similar in principle to a television camera, sensitive to infrared radiation. The resulting picture is called a *thermograph;* since emission depends on temperature, thermography permits detailed study of temperature distributions. Some present-day instrumentation is sensitive to differences as small as 0.1 C°.

Thermography has a variety of important medical applications. Local temperature variations in the body are associated with various tumors, such as breast cancers, and growths as small as 1 cm in diameter can be detected. Vascular disorders leading to local temperature anomalies can be studied, and many other examples could be cited.

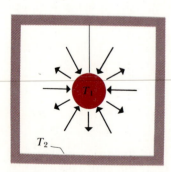

16–5 In thermal equilibrium, the rate of emission of radiant energy equals the rate of absorption. Hence a good absorber is a good emitter.

16–5

THE IDEAL RADIATOR

In Fig. 16–5, a body with initial temperature T_1 is suspended inside an enclosure whose walls are kept at temperature T_2. It is found that eventually the body attains the temperature T_2, irrespective of its emissivity or other characteristics. That is, the body reaches *thermal equilibrium* with its surroundings. When in thermal equilibrium, the body must *emit* radiant energy at the same rate at which it *absorbs* it. But the rate of emission depends on the emissivity; thus a body that is a good emitter (e close to unity) must also be a good absorber, and a poor emitter is a poor absorber. Furthermore, since the radiation striking the body must be either absorbed or reflected, a *good* absorber is a *poor* reflector, and a *poor* absorber is a *good* reflector. Finally we conclude that a *good reflector is a poor emitter*.

This is the reason for silvering the walls of vacuum ("Thermos") bottles. A vacuum bottle is constructed with double glass walls; the spaces between the walls is evacuated so that heat flow by conduction and convection is practically eliminated. To reduce the radiant emission to as low a value as possible, the walls are covered with a coating of silver, which is highly reflecting and hence is a very poor emitter.

Since a good absorber is a good emitter, the *best* emitter is that surface that is the best absorber. But no surface can absorb more than *all* of the radiant energy that strikes it. Any surface that *does* absorb all of the incident energy is the best emitting surface possible. Such a surface would reflect no radiant energy, and hence would appear black in color (provided its temperature is not so high that it is self-luminous). It is called an *ideally black surface*. A body having such a surface is called an ideal blackbody, an ideal radiator, or simply a *blackbody*.

No actual surface is ideally black; the closest approach is lampblack, which reflects only about 1%. Blackbody conditions can be closely approximated, however, by a small opening in the walls of a closed container. Radiant energy entering the opening is in part absorbed by the interior walls. Of the part reflected, only a very little escapes through the opening, the remainder being eventually absorbed by the walls. Hence the *opening* behaves like an ideal absorber.

Conversely, the radiant energy emitted by the walls (or by any body within the enclosure) and escaping through the opening, is, if the walls are of uniform temperature, of the same nature as that emitted by an ideal radiator. This fact is of importance when using an optical pyrometer, described in Section 14–2. The readings of such an instrument are correct only when it is sighted on a blackbody. If used to measure the temperature of a red-hot ingot of iron in the open, its readings will be too low, since iron is a poorer emitter than a blackbody. If, however, the pyrometer is sighted on the iron while still in the furnace, where it is surrounded by walls at the same temperature, "blackbody conditions" are fulfilled and the reading is correct. The failure of the iron to emit as effectively as a blackbody is then just compensated for by the radiant energy that it reflects.

The emissivity e of an ideally black surface is equal to unity. For all real surfaces it is less than unity.

EXAMPLE   Assuming the total surface of the human body is 1.2 m² and the surface temperature is 30°C = 303 K, find the total rate of radiation of energy from the body. If the surroundings are at a temperature of 20°C, what is the net rate of heat loss from the body by radiation?

Solution   Surprisingly, for infrared radiation the body is a very good approximation to an ideal blackbody, irrespective of skin pigmentation. The rate of energy transfer per unit area is given by Eq. (16–4). Taking $e = 1$, we find

$$H = (1.2 \text{ m}^2)(1)(5.67 \times 10^{-8} \text{ W} \cdot \text{m}^{-2} \cdot \text{K}^{-4})(303 \text{ K})^4 = 574 \text{ W}.$$

This loss is partially balanced by *absorption* of radiation, which depends on the temperature of the surroundings. The *net* rate of radiative energy transfer is given by Eq. (16–5):

$$H = (1.2 \text{ m}^2)(1)(5.67 \times 10^{-8} \text{ W} \cdot \text{m}^{-2} \cdot \text{K}^{-4})[(303 \text{ K})^4 - (293 \text{ K})^4]$$
$$= 72 \text{ W}.$$

Heat transfer by radiation is more important than one might have guessed. A premature baby in an incubator can be dangerously cooled by radiation, if the walls of the incubator happen to be cold, even when the air in the incubator is warm. Some incubators regulate the air temperature by measuring the baby's skin temperature.   ◀

16–6

SOLAR ENERGY AND RESOURCE CONSERVATION

The principles of heat transfer discussed in this chapter have many very practical applications in a civilization such as ours with growing energy consumption and dwindling energy resources. A substantial fraction of all the energy used in the United States is used for heating and cooling of homes and other buildings, to maintain comfortable temperature and humidity inside when the outside temperature is much hotter or much colder.

For space heating the objective is to prevent as much heat flow as possible from inside to outside, while using heating units to replace the inevitable loss of heat. Walls insulated with material of low thermal conductivity, storm windows, and multiple-layer glass windows all help to reduce heat loss. It has been estimated that if all buildings used such materials, the total energy used for space heating would be reduced by at least one third.

Air conditioning in summer poses the reverse problem; heat flows from outside to inside, and energy must be expended in refrigerating units to remove it. Again, appropriate insulation can decrease this energy cost considerably.

Direct use of solar energy is a promising development in energy technology. In a typical household heating system, large black plates facing the sun are backed with pipes through which water circulates. The black surface absorbs most of the sun's radiation; the heat is transferred by conduction to the water, and then by convection to radiators inside the house. Heat loss by convection of air near the solar collecting panels is reduced by covering them with glass, with a thin air space. An insulated heat reservoir provides for storage of collected energy for use at night and on cloudy days. Many larger-scale solar-energy conversion systems are also currently under study; a few examples are discussed in Section 19–10.

From an environmental standpoint, solar energy has multiple advantages over the use of fossil fuels (burning of coal or oil) or nuclear power. Fossil fuels are being used up; solar power continues indefinitely. Obtaining fossil fuels may involve strip mining, with its associated destruction of landscape and eliminating of other useful functions of land, such as farming and timber. Offshore oil drilling is a source of pollution of ocean water. Air pollution from combustion products is a familiar problem, as is acid rain, directly attributable in many cases to coal smoke. The long-range effects of excess atmospheric carbon dioxide on our climate are unknown, but some scientists believe they may be serious or even catastrophic. With nuclear power there are radiation hazards and the problem of disposal of radioactive waste material. Solar power avoids all these problems.

Intelligent consideration of the effects of solar radiation in the design of buildings can also reduce energy consumption in heating and cooling. An example is the use of moveable shades on the outsides of buildings, permitting the sun to enter windows in winter but keeping it out in summer. Architects have unfortunately tended in the past to ignore such solutions, but awareness of the need to design buildings with energy cost in mind is gradually growing.

QUESTIONS

16–1 Why does a marble floor feel cooler to the feet than a carpet at the same temperature?

16–2 A beaker of boiling water can be picked up with bare fingers without burning if it is grasped only at the thin turned-out rim at the top. Why?

16–3 Old-time kitchen lore suggests that things cook better (evenly and without burning) in heavy cast-iron pots. What desirable characteristics would such pots have?

16–4 If you have wet hands and pick up a piece of metal that is below freezing, you may stick to it. This doesn't happen with wood. Why not?

16–5 A cold block of metal feels colder than a block of wood at the same temperature, but a hot block of metal feels hotter than a block of wood at the same temperature. Is there any temperature at which they feel equally hot or cold?

16–6 A person pours a cup of hot coffee, intending to drink it five minutes later. To keep it as hot as possible, should he put cream in it now or wait until just before he drinks it?

16–7 In late afternoon when the sun is about to set, swarms of insects such as mosquitoes are sometimes seen in vertical "plumes" above a tree or above the hood of a car with a hot engine. Why?

16–8 Mountaineers caught in a storm sometimes survive by digging a cave in snow or ice. How in the world can you keep warm in an ice cave?

16–9 Old-time pioneers sometimes kept themselves warm on cold winter nights by heating bricks in the fire, wrapping them in thick layers of cloth, and taking them to bed. Discuss the roles of conduction, convection, and radiation in this technique.

16–10 It is well known that a potato bakes faster if a large nail is stuck through it. Why? Is aluminum better than steel? There is also a gadget on the market to hasten roasting of meat, consisting of a hollow metal tube containing a wick and some water; this is claimed to be much better than a solid metal rod. How does this work?

16–11 The temperature in outer space, far from any solid body, is believed to be about 3 K. If you leave a spaceship and go for a space walk, do you get cold very quickly?

16–12 Aluminum foil used for food cooking and storage sometimes has one shiny surface and one dull surface. When food is wrapped for baking, should the shiny side be in or out? Which side should be out when it is to be frozen?

16–13 Some cooks claim that the bottom crust of a pie gets browner if the pie pan is glass than if it is metal. Why should this be?

16–14 Glider pilots in the midwest know that thermal updrafts are likely to occur above freshly plowed fields. Why?

16–15 We're lucky the earth isn't in thermal equilibrium with the sun. But why isn't it?

16–16 On a chilly fall morning, the grass is covered with frost but the concrete sidewalk isn't. Why?

16–17 Why can a microwave oven cook massive objects such as potatoes and beef roasts much faster than a conventional oven? What disadvantages are associated with the reasons for this greater speed?

16–18 Some folks claim that ice cubes freeze faster if the trays are filled with hot water because hot water cools off faster than cold water. What do you think?

PROBLEMS

16–1 A slab of a thermal insulator is 100 cm^2 in cross section and 2 cm thick. Its thermal conductivity is 0.1 J·s^{-1}·m^{-1}·(C°)$^{-1}$. If the temperature difference between opposite faces is 100 C°, how much heat flows through the slab in one day?

16–2 Suppose that the rod in Fig. 16–1 is of copper, of length 10 cm and of cross-sectional area 1 cm^2. Let $T_2 = 100°C$ and $T_1 = 0°C$.

 a) What is the final steady-state temperature gradient along the rod?

 b) What is the heat current in the rod, in the final steady state?

 c) What is the final steady-state temperature at a point in the rod 2 cm from its left end?

16–3 A long rod, insulated to prevent heat losses, has one end immersed in boiling water (at atmospheric pressure) and the other end in a water–ice mixture. The rod consists of 100 cm of copper (one end in steam) and a length, L_2, of steel (one end in ice). Both rods are of cross-sectional area 5 cm^2. The temperature of the copper–iron junction is 60°C, after a steady state has been set up.

 a) How much heat per second flows from the steam bath to the ice-water mixture?

 b) How long is L_2?

16–4 The oven in a kitchen range has total wall area 1.5 m^2 and is insulated with a layer of fiberglass 5 cm thick. The inside is to be kept at a temperature of 200°C, and the outside is at room temperature, 20°C.

 a) What is the heat current through the insulation, assuming it may be treated as a flat slab of area 1.5 m^2?

 b) What electric-power input to the heating element is required to maintain this temperature?

16–5 One experimental method of measuring the thermal conductivity of an insulating material is to construct a box of the material and measure the power input to an electric heater, inside the box, which maintains the interior at a measured temperature above the outside surface. In one experiment an engineer finds that a power input of 120 W is required to keep the interior surface of the box 120 F° above the temperature of the outer surface. The total area of the box is 25 ft^2 and the wall thickness is 1.5 in. Find the thermal conductivity of the material in the commercial system of units.

16–6 A container of wall area 5000 cm^2 and thickness 2 cm is filled with water in which there is a stirrer. The outer surface of the walls is kept at a constant temperature of 0°C. The thermal conductivity of the walls is 0.000478 cal·s^{-1}·cm^{-1}·(C°)$^{-1}$, and the effect of edges and corners can be neglected. The power required to run the stirrer at an angular velocity of 1800 rpm is found to be 100 W. What will be the final steady-state temperature of the water in the container? Assume that the stirrer keeps the entire mass of water at a uniform temperature.

16–7 A boiler with a steel bottom 1.5 cm thick rests on a hot stove. The area of the bottom of the boiler is 1500 cm^2. The water inside the boiler is at 100°C, and 750 g are evaporated every 5 min. Find the temperature of the lower surface of the boiler, which is in contact with the stove.

16–8 An engineer designs a camping icebox, having wall area of 2 m^2 and thickness 5 cm, using insulating material having a thermal conductivity of 0.05 J·s^{-1}·m^{-1}·(C°)$^{-1}$. The outside temperature is 20°C, and the inside of the box is to be maintained at 5°C by ice. The melted ice drains from the box at a temperature of 15°C. If ice costs 10 cents per kilogram, what will it cost to run the icebox for 1 hr?

16–9 A carpenter builds an outer house wall with a layer of wood 3 cm thick on the outside, and a layer of styrofoam insulation 3 cm thick inside. If the wood has $k = 0.04$ J·s^{-1}·m^{-1}·(C°)$^{-1}$, the interior temperature is 20°C, and the exterior temperature is $-10°C$,

a) what is the temperature at the plane where the wood meets the styrofoam?

b) What is the rate of heat flow, per square meter, through this wall?

16–10 A layer of ice 10 cm thick has formed on a lake, where the temperature is $-10°C$. The water under the slab is at $0°C$. At what rate does the thickness of the slab increase if the heat of fusion of the water freezing on the underside of the slab is conducted through the slab?

16–11 Three rods of equal length and cross section are placed end to end, one copper, one steel, one aluminum, in that order. The free end of the copper rod is in contact with ice water at $0°C$, and the free end of the aluminum rod is in contact with steam at $100°C$. Assume no heat transfer occurs at the sides. Find the temperature at each junction.

16–12 A carpenter builds a solid wood door with dimensions $2.0 \text{ m} \times 0.8 \text{ m} \times 4 \text{ cm}$. Its thermal conductivity is

$$k = 0.04 \text{ J} \cdot \text{s}^{-1} \cdot \text{m}^{-1} \cdot (\text{C}°)^{-1}.$$

The inside air temperature is $20°$ and the outside air temperature is $-10°C$.

a) What is the rate of heat flow through the door, assuming the surface temperatures are those of the surrounding air?

b) By what factor is the heat flow increased if a window 0.5 m square is inserted, assuming the glass is 0.5 cm thick and the surface temperatures are again those of the surrounding air?

16–13 Rods of copper, brass, and steel are welded together to form a Y-shaped figure. The cross-sectional area of each rod is 2 cm^2. The end of the copper rod is maintained at $100°C$ and the ends of the brass and steel rods at $0°C$. Assume there is no heat loss from the surfaces of the rods. The lengths of the rods are: copper, 46 cm; brass, 13 cm; steel, 12 cm.

a) What is the temperature of the junction point?

b) What is the heat current in the copper rod?

16–14 A compound bar 2 meters long is constructed of a solid steel core 1 cm in diameter surrounded by a copper casing whose outside diameter is 2 cm. The outer surface of the bar is thermally insulated and one end is maintained at $100°C$, the other at $0°C$.

a) Find the total heat current in the bar.

b) What fraction of this total is carried by each material?

16–15 An electrician installs an electric transformer in a cylindrical tank 60 cm in diameter and 1 m high, with flat top and bottom. If the tank transfers heat to the air only by natural convection, and the electrical losses are to be dissipated at the rate of 1 kW, how many degrees will the tank surface rise above room temperature?

16–16

a) What would be the difference in height between the columns in the U-tube in Fig. 16–4 if the liquid is water and the left arm is 1 m high at $4°C$ while the other is at $75°C$?

b) What is the difference between the pressures at the foot of two columns of water each 10 m high if the temperature of one is $4°C$ and that of the other is $75°C$?

16–17 A flat wall is maintained at constant temperature of $100°C$, and the air on both sides is at atmospheric pressure and at $20°C$. How much heat is lost by natural convection from 1 m^2 of wall (both sides) in 1 hr if

a) the wall is vertical, and

b) the wall is horizontal?

16–18 A vertical steam pipe of outside diameter 7.5 cm and height 4 m has its outer surface at the constant temperature of $95°C$. The surrounding air is at atmospheric pressure and at $20°C$. How much heat is delivered to the air by natural convection in 1 hr?

16–19 What is the rate of energy radiation per unit area of a blackbody at a temperature of

a) 300 K, and

b) 3000 K?

16–20 The emissivity of tungsten is approximately 0.35. A tungsten sphere 1 cm in radius is suspended within a large evacuated enclosure whose walls are at 300 K. What power input is required to maintain the sphere at a temperature of 3000 K if heat conduction along the supports is neglected?

16–21 A small blackened solid copper sphere of radius 2 cm is placed in an evacuated enclosure whose walls are kept at $100°C$. At what rate must energy be supplied to the sphere to keep its temperature constant at $127°C$?

16–22 A physicist uses a cylindrical metal can 10 cm high and 5 cm in diameter to store liquid helium at 4 K, at which temperature its heat of vaporization is $20 \text{ J} \cdot \text{g}^{-1}$. Completely surrounding the helium can are walls maintained at the temperature of liquid nitrogen, 80 K, and the intervening space is evacuated. How much helium is lost per hour? Assume the radiant emittance of the helium can to be 0.2 that of a blackbody at 4 K.

16–23 A solid cylindrical copper rod 10 cm long has one end maintained at a temperature of 20.00 K. The other end is blackened and exposed to thermal radiation from a body at 300 K, and the sides are insulated. When equilibrium is reached, what is the temperature of the blackened end? [*Hint.* Since copper is a very good conductor of heat at low temperature, $K = 4 \text{ cal} \cdot \text{s}^{-1} \cdot \text{cm}^{-1} \cdot (\text{C}°)^{-1}$, the temperature of the blackened end is only slightly greater than 20 K.]

16–24 The operating temperature of a tungsten filament in a long-life incandescent lamp is 2450 K and its emissivity is 0.30. Find the surface area of the filament of a 25-watt lamp.

16–25 A certain incandescent lightbulb used for photography operates at 3000 K. The total surface area of the filament is 0.05 cm^2, and the emissivity is 0.3. What electric power must be supplied to the filament?

16–26 The rate at which radiant energy reaches the surface of the earth from the sun is about 1.4 kW·m^{-2}. The distance from earth to sun is about 1.5×10^{11} m, and the radius of the sun is about 0.7×10^9 m.

  a) What is the rate of radiation of energy, per unit area, from the sun's surface?

  b) If the sun radiates as an ideal blackbody, what is the temperature of its surface?

16–27 In the example in Section 16–5, what is the net rate of heat loss by radiation if the temperature of the surroundings is 0°C?

16–28 In the example of Section 16–5, suppose the total rate of heat loss from the body by all mechanisms is 120 W. What should be the temperature of the surroundings in order for this loss to occur entirely by radiation?

16–29 A well-insulated house of moderate size in temperature climates may require a maximum heat input rate of the order of 20 kW. If this heat is to be supplied by a solar collector with an average (night and day) energy input of 300 W·m^{-2} and a collection efficiency of 60 percent, what area of solar collector is required?

16–30 In Problem 16–29, suppose a heat reservoir is required that can store enough energy for a week of cloudy days, by means of a large tank of water that is heated to 80°C by solar energy and cooled to 40°C as it circulates through the house. What volume of water is required?

16–31 A solar water heater for domestic hot water supply uses solar collecting panels with collection efficiency of 50 percent in a location where the average solar energy input is 200 W·m^{-2}. If the water comes into the house at 15°C and is to be heated to 65°C, how much water can be heated per hour if the collector area is 20 m^2?

16–32 The average rate of energy consumption of all forms in the United States is of the order of 3×10^{11} W. Suppose this energy were to be supplied by solar collectors in the desert. Assume an average solar energy input of 500 W·m^{-2} for twelve hours each day, and a collection efficiency of 50 percent. What total collector area would be required? Express your result in square kilometers and square miles.

CHAPTER 17

THERMAL PROPERTIES OF MATTER

In previous chapters we have discussed several properties of materials, such as elasticity, thermal expansion, specific heat capacity, and thermal conductivity. In these discussions we have made use of several physical quantities that describe the quantity and condition of the material. Among these are total mass m, pressure p, volume V, and temperature T. For example, elasticity is described in terms of the relationship of pressure to volume, thermal expansion in terms of the relationship of volume to temperature. In this chapter we consider a more general formulation of these relationships, using the concept of the *equation of state* of a material. We also examine in greater detail the conditions that determine the *phase* of a material (e.g., solid, liquid, or gas) and the circumstances under which *phase transitions* occur.

17–1

EQUATIONS OF STATE

Quantities such as pressure, volume, temperature, and amount of substance are called *state variables* or *state coordinates*, and together they describe the *state* of the material. Some physical phenomena require the use of additional state variables such as magnetization or electric polarization, but those named above will be sufficient for our present discussion.

The state variables are interrelated; ordinarily it is not possible to change one without also causing a change in one or more of the others. For example, when a gas such as steam is heated in a closed container of constant volume, the pressure *must* increase as the temperature increases. Indeed, if the temperature is high enough, the pressure may cause the container, such as a boiler, to explode.

In some cases the relationship among p, V, T, and m is simple enough that it may be expressed in the form of an equation, which we call the *equation of state*. In other cases the relationship is so complex that it can be described only by the use of graphs or tables of numerical values. Even then, the relation among the variables still exists, and we call it an equation of state, even when it is too complex to be expressed in equation form. In such cases, it is often useful to know how some *one* of the quantities changes when some other is varied, the rest being kept constant. Thus the (isothermal) compressibility k describes the change in volume when the pressure is changed, for a constant mass at a constant temperature. Similarly, the coefficient of volume expansion β gives the change in volume when the temperature is changed, for a constant mass at a constant pressure.

At present we shall consider only *equilibrium states;* the system must be in mechanical and thermal equilibrium. When expansion or compression occurs, there is mass in motion; this requires acceleration and nonuniform pressure. After the system has returned to mechanical equilibrium, the pressure is again the same throughout the system. Similarly, when there is heat transfer within the system, different regions have different temperatures. Only when thermal equilibrium is once again attained is the temperature the same throughout.

17–2

THE IDEAL GAS

Gases at low pressures have particularly simple equations of state. We may study the behavior of a gas by means of a cylinder with a movable piston, equipped with a pressure gauge and a thermometer. We may vary the pressure, volume, and temperature, and pump any desired mass of any gas into the cylinder, in order to explore the relationships between pressure, volume, temperature, and quantity of substance. It is often convenient to measure the amount of gas in terms of the number of *moles n* rather than the mass m. Since the molecular mass M is the mass per mole, the total mass m is given by

$$m = nM.$$

That is, the total mass m of material is the mass of one mole M, multiplied by the number of moles n.

From measurements of the pressure, volume, temperature, and number of moles of various gases, a number of conclusions emerge. First, the volume V is proportional to the number of moles n. If we double the number of moles, keeping pressure and temperature constant, the volume doubles. Second, the volume varies *inversely* with pressure p; if we double the pressure, holding the temperature T and quantity of material n constant, the gas is compressed to one half of its initial volume. Thus

$$pV = \text{constant} \qquad (T \text{ constant}), \qquad (17–1)$$

or, if p_1 and V_1 are the pressure and volume in the initial state and p_2 and V_2 those of the final state, then

$$p_1V_1 = p_2V_2 \qquad (T \text{ constant}). \qquad (17–2)$$

This relation is called Boyle's law, after Robert Boyle (1627–1691), a contemporary of Newton.

Third, the pressure is proportional to the absolute temperature; if we double the absolute temperature, keeping the volume and quantity of material constant, the pressure doubles. Thus

$$p = (\text{constant}) \, T \qquad (V \text{ constant}), \qquad (17\text{--}3)$$

or, if p_1 and T_1 are the pressure and absolute temperature in the initial state and p_2 and T_2 those of the final state, then

$$\frac{p_1}{T_1} = \frac{p_2}{T_2}. \qquad (17\text{--}4)$$

This is called Charles's law, after Jacque Charles (1746–1823).

These three relationships can be combined neatly into a single equation of state:

$$pV = nRT. \qquad (17\text{--}5)$$

The constant of proportionality R might be expected to have different values for different gases, but instead it turns out to have the same value for *all* gases, at least at sufficiently high temperature and low pressure. This quantity is called the *universal gas constant*. The numerical value of R depends, of course, on the units in which p, V, n, and T are expressed. The adjective "universal" means that in any one system of units R has the same value for *all* gases. In SI (mks) units, where the unit of p is 1 Pa or $1 \, \text{N} \cdot \text{m}^{-2}$ and the unit of V is $1 \, \text{m}^3$, the numerical value of R is found to be

$$R = 8.314 \, (\text{N} \cdot \text{m}^{-2}) \cdot \text{m}^3 \cdot \text{mol}^{-1} \cdot \text{K}^{-1} = 8.314 \, \text{J} \cdot \text{mol}^{-1} \cdot \text{K}^{-1}.$$

In the cgs system, where the unit of p is $1 \, \text{dyn} \cdot \text{cm}^{-2}$ and the unit of V is $1 \, \text{cm}^3$,

$$R = 8.314 \times 10^7 \, (\text{dyn} \cdot \text{cm}^{-2}) \cdot \text{cm}^3 \cdot \text{mol}^{-1} \cdot \text{K}^{-1}$$
$$= 8.314 \times 10^7 \, \text{erg} \cdot \text{mol}^{-1} \cdot \text{K}^{-1}.$$

We note that the units of pressure times volume are the same as units of energy, so, in *all* systems of units, R has units of energy per mole, per unit of absolute temperature. In terms of calories,

$$R = 1.99 \, \text{cal} \cdot \text{mol}^{-1} \cdot \text{K}^{-1}.$$

In chemical calculations volumes are commonly expressed in liters (l), pressures in atmospheres, and temperatures in kelvins. In this system,

$$R = 0.08207 \, \text{l} \cdot \text{atm} \cdot \text{mol}^{-1} \cdot \text{K}^{-1}.$$

We now define an *ideal gas* as one for which Eq. (17–5) holds precisely for *all* pressures and temperatures. As the term suggests, the ideal gas is an idealized model, which represents the behavior of gases very well in some circumstances, less well in others. Generally, gas behavior approximates the ideal-gas model most closely at very low pressures, when the gas molecules are far apart. However, the deviations are not very great at moderate pressures and at temperatures not too near those at which the gas liquefies.

For a *fixed mass* (or fixed number of moles) of an ideal gas, the product nR is constant and hence pV/T is constant also. Thus if the subscripts

1 and 2 refer to two states of the same mass of a gas, but at different pressures, volumes, and temperatures,

$$\frac{p_1 V_1}{T_1} = \frac{p_2 V_2}{T_2} = \text{constant}. \tag{17–6}$$

If the temperatures T_1 and T_2 are the same, then

$$p_1 V_1 = p_2 V_2 = \text{constant}, \tag{17–7}$$

which is again Boyle's law.

EXAMPLE 1   The condition called *standard temperature and pressure* (STP) for a gas is defined to be a temperature of 0°C = 273 K and a pressure of 1 atm = 1.013×10^5 Pa. Find the volume of one mole of any gas at STP.

Solution   From Eq. (17–5), using R in $J \cdot mol^{-1} \cdot K^{-1}$,

$$V = \frac{nRT}{p} = \frac{(1 \text{ mol})(8.314 \text{ J} \cdot mol^{-1} \cdot K^{-1})(273 \text{ K})}{1.01 \times 10^5 \text{ Pa}}$$

$$= 0.0224 \text{ m}^3$$
$$= 22{,}400 \text{ cm}^3$$
$$= 22.4 \text{ l}.$$

Alternatively, using R in $l \cdot atm \cdot mol^{-1} \cdot K^{-1}$,

$$V = \frac{nRT}{p} = \frac{(1 \text{ mol})(0.0821 \text{ l} \cdot atm \cdot mol^{-1} \cdot K^{-1})(273 \text{ K})}{1.00 \text{ atm}}$$

$$= 22.4 \text{ l}. \qquad\blacktriangleleft$$

EXAMPLE 2   A tank attached to an air compressor contains 20 l of air at a temperature of 30°C and gauge pressure 4.0×10^5 Pa (about 60 lb·in^{-2}). What is the mass of air, and what volume would it occupy at normal atmospheric pressure and 0°C?

Solution   Air is a mixture of gases, consisting of about 78% nitrogen and 21% oxygen, with small percentages of other gases. The *average* molecular mass, which we may use in the ideal-gas equation of state, is $M = 28.8 \text{ g} \cdot mol^{-1}$. Gauge pressure is the amount in excess of atmospheric pressure (1.01×10^5 Pa), so the total pressure is

$$p = 4.0 \times 10^5 \text{ Pa} + 1.01 \times 10^5 \text{ Pa} = 5.01 \times 10^5 \text{ Pa}.$$

The temperatures must be expressed in kelvins: 30°C = 303 K. We express the volume in cubic meters: 20 l = 20×10^{-3} m^3. To find the mass of gas, we first use Eq. (17–1) to find the number of moles:

$$n = \frac{pV}{RT} = \frac{(5.01 \times 10^5 \text{ Pa})(20 \times 10^{-3} \text{ m}^3)}{(8.314 \text{ J} \cdot mol^{-1} \cdot K^{-1})(303 \text{ K})} = 3.98 \text{ mol};$$

$$m = nM = (3.98 \text{ mol})(28.8 \text{ g} \cdot mol^{-1}) = 115 \text{ g} = 0.115 \text{ kg}.$$

The volume at 1 atm and 0°C (273 K) is

$$V = \frac{nRT}{p} = \frac{(3.98 \text{ mol})(8.314 \text{ J} \cdot \text{mol}^{-1} \cdot \text{K}^{-1})(273 \text{ K})}{1.01 \times 10^5 \text{ Pa}}$$

$$= 89.4 \times 10^{-3} \text{ m}^3 = 89.4 \text{ l}.$$ ◀

EXAMPLE 3 The volume of an oxygen tank is 50 l. As oxygen is withdrawn from the tank, the reading of a pressure gauge drops from 300 lb·in^{-2} to 100 lb·in^{-2}, and the temperature of the gas remaining in the tank drops from 30°C to 10°C.

a) How many kilograms of oxygen were in the tank originally?

b) How many kilograms were withdrawn?

c) What volume would be occupied by the oxygen withdrawn from the tank at a pressure of 1 atm and a temperature of 20°C?

Solution

a) Let us express pressures in atmospheres, volumes in liters, and temperatures in kelvins. Then $R = 0.082 \text{ l} \cdot \text{atm} \cdot \text{mol}^{-1} \cdot \text{K}^{-1}$. The initial and final *gauge* pressures (i.e., the amounts by which the pressures are greater than normal atmospheric pressure), are

$$300 \text{ lb} \cdot \text{in}^{-2} = \frac{300}{14.7} = 20.4 \text{ atm},$$

$$100 \text{ lb} \cdot \text{in}^{-2} = 6.8 \text{ atm}.$$

The corresponding *absolute* or *total* pressures are 21.4 atm and 7.8 atm. Initially,

$$n_1 = \frac{p_1 V}{R T_1} = \frac{(21.4 \text{ atm})(50 \text{ l})}{(0.0821 \text{ l} \cdot \text{atm} \cdot \text{mol}^{-1} \cdot \text{K}^{-1})(303 \text{ K})} = 43.0 \text{ mol}.$$

The original mass was therefore

$$m_1 = (43.0 \text{ mol})(32 \text{ g} \cdot \text{mol}^{-1}) = 1377 \text{ g} = 1.377 \text{ kg}.$$

b) The number of moles remaining in the tank is

$$n_2 = \frac{p_2 V}{R T_2} = \frac{(7.8 \text{ atm})(50 \text{ l})}{(0.0821 \text{ l} \cdot \text{atm} \cdot \text{mol}^{-1} \cdot \text{K}^{-1})(283 \text{ K})} = 16.8 \text{ mol};$$

$$m_2 = (16.8 \text{ mol})(32 \text{ g} \cdot \text{mol}^{-1}) = 5.37 \text{ g} = 0.537 \text{ kg}.$$

The number of kilograms withdrawn is

$$m_1 - m_2 = 1.377 \text{ kg} - 0.537 \text{ kg} = 0.839 \text{ kg}.$$

c) The number of moles withdrawn is $43.0 - 16.8 = 26.2$ mol. The volume occupied would be

$$V = \frac{nRT}{p} = \frac{(26.2 \text{ mol})(0.0821 \text{ l} \cdot \text{atm} \cdot \text{mol}^{-1} \cdot \text{K}^{-1})(293 \text{ K})}{1 \text{ atm}}$$

$$= 630 \text{ l}.$$ ◀

Often we must deal with mixtures of ideal gases; the concept of *partial pressure* is then useful. In a mixture of gases, the partial pressure of each gas is the pressure that gas would exert by itself if it occupied the entire volume and the other gases were not present. It turns out that the actual total pressure of the mixture is then the *sum* of the partial pressures of the components, provided each component behaves as an ideal gas and no chemical reactions occur when the gases are mixed. Thus in air at normal atmospheric pressure, the partial pressure of nitrogen is about 0.8 atm or 0.8×10^5 Pa, and the partial pressure of oxygen is about 0.2 atm or 0.2×10^5 Pa.

The ability of the human body to absorb oxygen from the atmosphere depends critically on the partial pressure of oxygen. This ability begins to drop sharply when the partial pressure of oxygen drops to about 0.13×10^5 Pa (100 mm Hg), corresponding to an elevation above sea level of about 4700 m or 15,000 ft. At partial pressures less than 0.11×10^5 Pa (85 mm Hg), oxygen absorption is not sufficient to maintain life.

17–3

pVT-SURFACE FOR AN IDEAL GAS

Since the equation of state for a fixed mass of a substance is a relation among the three variables p, V, and T, it defines a *surface* in a rectangular coordinate system in which p, V, and T are plotted along the three axes. Figure 17–1 shows the *pVT*-surface of an ideal gas. The dark color lines on the surface show the relation between p and V when T is constant

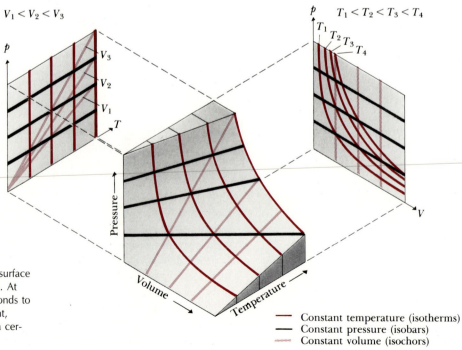

17–1 Projections of ideal-gas *pVT*-surface on the *pT*-plane and on the *pV*-plane. At the left, each light color line corresponds to a certain constant volume; at the right, each dark color line corresponds to a certain constant temperature.

Constant temperature (isotherms)
Constant pressure (isobars)
Constant volume (isochors)

(Boyle's law), the black lines show the relation between V and T when p is constant, and the light color lines show the relation between p and T when V is constant. When viewed perpendicular to the pT-plane, the surface appears as at the left, and its appearance when viewed perpendicular to the pV-plane is shown at the right.

Every possible equilibrium state of a given quantity of gas corresponds to a point on the surface, and every point of the surface corresponds to a possible equilibrium state. The gas cannot exist in a state that is not on the surface. For example, if the volume and temperature are given, thus locating a point in the base plane of Fig. 17–1, the pressure is then determined by the nature of the gas, and it can have only the value represented by the height of the surface above this point.

In any process in which the gas passes through a succession of equilibrium states, the point representing its state moves along a curve lying on the pVT-surface. Such a process must be carried out very slowly to give the temperature and pressure time to remain uniform at all points of the gas.

17–4

pV-DIAGRAMS

As noted in Section 17–3, the equation of state of any material can be represented by a pVT-surface. For ideal gases this surface is fairly simple, but for materials with more complex behavior this representation becomes awkward and impractical. It is more generally useful to draw ordinary two-dimensional graphs; one of the most useful of these is a set of graphs of pressure as a function of volume for each of several constant temperatures. Such a diagram is called a pV-diagram. Each curve, representing a particular constant temperature, is called an *isotherm*, or a pV-*isotherm*.

A pV-diagram for an ideal gas is shown in Fig. 17–2. Each curve is an isotherm showing the relation of pressure to volume at a certain constant temperature. This may be thought of as a graphical representation of Boyle's law, Eq. (17–1), and it is also a graphical representation of the ideal-gas equation of state. Comparison of this diagram to Fig. 17–1 shows that the curves are cross sections of the ideal-gas pVT-surface, made by planes perpendicular to the T axis in the pVT-space.

Figure 17–3 shows a pV-diagram for a material that *does not* obey the ideal-gas equation. At a sufficiently high temperature, say T_4, the behavior is approximately that of an ideal gas, but at lower temperatures the behavior is rather different. At temperatures below some critical temperature T_c, the isotherms develop flat portions in which, as the material is compressed at constant temperature, the pressure does not increase at all. As the volume decreases and the end of the flat portion of an isotherm is reached, the pressure increases very rapidly with further decrease of volume.

Visual inspection of the material shows that in the flat portion of the isotherm the material is *condensing* from vapor to liquid phase; this flat portion represents a condition of phase equilibrium. As compression

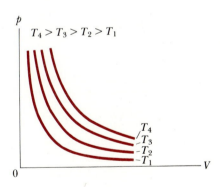

17–2 *pV*-diagram for an ideal gas. Each curve represents the relation between pressure and volume at the indicated constant temperature. This is a graphical representation of Boyle's law; for each curve the product *pV* is constant; the value of that constant is different for different temperatures, increasing with increasing temperature. These curves are called *pV-isotherms* for the substance.

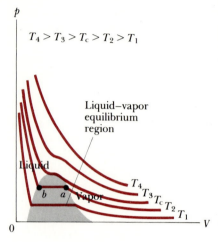

$T_4 > T_3 > T_c > T_2 > T_1$

Liquid–vapor
equilibrium
region

Liquid

T_4
T_3
T_c
T_2
T_1

b   a   Vapor

0   V

17–3 *pV*-diagram for a non-ideal gas, showing *pV*-isotherms for temperatures above and below the critical temperature T_c. The liquid–vapor equilibrium region is shown as a shaded area. At still lower temperature the material might undergo phase transitions from liquid to solid or from gas to solid; these are not shown on this diagram.

occurs, more and more material goes from vapor to liquid, but the pressure does not change. During this process, heat must be removed in order to keep the temperature constant; this is the heat of fusion, discussed in Section 15–6.

Thus when a real (non-ideal) gas is compressed at a constant temperature T_2, it remains in the gaseous phase until point *a* in the diagram is reached. At point *a* it begins to liquefy, and with further decrease in volume more and more material liquefies, with both pressure and temperature remaining constant. At point *b* all the material is in the liquid state; any further compression is accompanied by a rapid rise of pressure, because in general liquids are much less compressible than gases. At a lower constant temperature T_1 similar behavior occurs, but the onset of condensation occurs at lower pressure and greater volume than at the constant temperature T_2.

The shaded area is the region of phase equilibrium for this material. At temperatures greater than T_c, no phase transition occurs as the material is compressed. The temperature T_c is called the *critical temperature* for this material.

We will have many more opportunities to use *pV*-diagrams in the next two chapters in connection with energy calculations. As we shall see, the *area* under a *pV*-curve, whether or not it is an isotherm, represents the *work* done by or on the system during a volume change. This work in turn is directly related to heat transfer and changes in the *internal energy* of the system, which will be defined in Chapter 18.

17–5

PHASE DIAGRAMS

In the discussion of phases of matter in Section 15–6 it was observed that each phase is stable only in certain ranges of temperature and pressure. A transition from one phase to another occurs ordinarily under conditions of phase equilibrium between the two phases, and for a given temperature this occurs only at one specific pressure.

The phases that occur for various combinations of temperature and pressure can be represented conveniently in a *pT*-diagram called a *pT-phase diagram*. Such a diagram shows which phase is present for each pair of values of *p* and *T*.

A typical phase diagram is shown in Fig. 17–4. At each point on the diagram only a single phase can exist, except for points on the lines, where two phases can coexist in *phase equilibrium*. Points on the lines therefore represent conditions under which phase changes occur.

If the temperature of a substance is increased at constant pressure, it passes through a sequence of states represented by points on the broken horizontal line labeled (a). The melting and boiling temperatures at this pressure are the temperatures at which the broken line crosses the fusion and vaporization curves, respectively. Similarly, if material is compressed by increasing the pressure while we hold the temperature constant, as represented by the vertical broken line labeled (b), the material passes from vapor to liquid and then to solid at the points where the broken line crosses the vaporization curve and fusion curve, respectively.

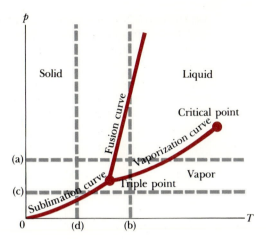

17–4 A pT-phase diagram, showing regions of temperature and pressure at which the various phases exist and where phase changes occur.

When the pressure is sufficiently low, a material may be transformed by constant-pressure heating from solid directly to vapor, as represented by line (c). This process, called *sublimation*, occurs where the line crosses the sublimation curve. At atmospheric pressure solid carbon dioxide undergoes sublimation; there is no liquid phase at this pressure. Similarly, when a substance with initial temperature lower than the triple point is compressed by increasing its pressure, it goes directly from the vapor to the solid phase, as shown by line (d).

Figure 17–4 is characteristic of materials that expand on melting. There are a few materials that *contract* on melting; the most familiar examples are water and the metal antimony. For such a substance the fusion curve always slopes upward to the left, rather than to the right. The melting temperature decreases when the pressure increases, so solid material can be melted by an increase of pressure. Ice-skating is possible because the pressure of the skate blades against the ice (which may be as much as several hundred atmospheres) melts a thin film of water and lubricates the surfaces.

The data needed to plot the fusion curve in Fig. 17–4 for a particular material can be obtained directly from a set of pV-isotherms for that material, as shown in Fig. 17–3. The pressure at the flat portion of each isotherm is the pressure needed for phase equilibrium at the corresponding temperature. Thus a series of pairs of values of p and T can be obtained; a graph of these is the fusion curve for the substance.

17–6

TRIPLE POINT AND CRITICAL POINT

The intersection point of the equilibrium curves in Fig. 17–4 is called the *triple point*. This point is an end view of the *triple line* on the pVT-surface. For any substance there is only one pressure and temperature at which all three phases can coexist, namely, the triple-point pressure and temperature. Triple-point data for a few substances are given in Table 17–1. The role of the triple-point temperature of water in defining a temperature scale has been mentioned in Section 14–4.

TABLE 17–1 TRIPLE-POINT DATA		
Substance	Temperature, K	Pressure, 10^5 Pa
Hydrogen	13.84	0.0704
Deuterium	18.63	0.171
Neon	24.57	0.432
Nitrogen	63.18	0.125
Oxygen	54.36	0.00152
Ammonia	195.40	0.0607
Carbon dioxide	216.55	5.17
Sulfur dioxide	197.68	0.00167
Water	273.16	0.00610

Figure 17–3 shows that the liquid and gas (or vapor) phases can exist together only when the temperature and pressure are less than those at the point lying at the top of the tongue-shaped area labeled "liquid–vapor equilibrium region." This point also corresponds to the endpoint at the top of the vaporization curve in Fig. 17–4. It is called the *critical point,* and the corresponding values of p and T are called the critical pressure and temperature. A gas at a temperature *above* the critical temperature, such as T_4 in Fig. 17–3, does not separate into two phases when compressed isothermally, but its properties change gradually and continuously from those we ordinarily associate with a gas (low density, large compressibility) to those of a liquid (high density, small compressibility). Table 17–2 lists critical constants for a few substances. The very low critical temperatures of hydrogen and helium make it evident why these gases defied attempts to liquefy them for many years.

Many substances can exist in more than one modification of the solid phase. Transitions from one modification to another occur at definite

TABLE 17–2 CRITICAL CONSTANTS				
Substance	Critical temperature, K	Critical pressure, Pa	Critical volume, $m^3 \cdot mol^{-1}$	Critical density, $kg \cdot m^{-3}$
Helium (4)	5.3	2.29×10^5	57.8×10^{-6}	69.3
Helium (3)	3.34	1.17×10^5	72.6×10^{-6}	41.3
Hydrogen (normal)	33.3	13.0×10^5	65.0×10^{-6}	31.0
Deuterium (normal)	38.4	16.6×10^5	60.3×10^{-6}	66.3
Nitrogen	126.2	33.9×10^5	90.1×10^{-6}	311
Oxygen	154.8	50.8×10^5	78×10^{-6}	410
Ammonia	405.5	112.8×10^5	72.5×10^{-6}	235
Freon 12	384.7	40.1×10^5	218×10^{-6}	555
Carbon dioxide	304.2	73.9×10^5	94.0×10^{-6}	468
Sulfur dioxide	430.7	78.8×10^5	122×10^{-6}	524
Water	647.4	221.2×10^5	56×10^{-6}	320
Carbon disulfide	552	79.0×10^5	170×10^{-6}	440

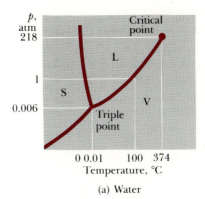

(a) Water

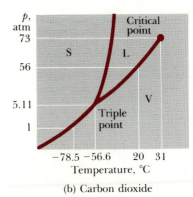

(b) Carbon dioxide

17–5 Pressure–temperature diagrams (not to a uniform scale).

temperatures and pressures, just as with phase changes from liquid to solid, etc. Water is one such substance; at least eight types of ice have been observed at very high pressures.

As numerical examples, consider the pT-diagrams of water and carbon dioxide, in Fig. 17–5. In (a), a horizontal line at a pressure of 1 atm intersects the freezing-point curve at 0°C and the boiling-point curve at 100°C. The boiling point increases with increasing pressure up to the critical temperature of 374°C. Solid, liquid, and vapor can remain in equilibrium only at the triple point, where the vapor pressure is 0.006 atm and the temperature is 0.01°C.

For carbon dioxide, the triple-point temperature is -56.6°C and the corresponding pressure is 5.11 atm. Hence at atmospheric pressure CO_2 can exist only as a solid or vapor. When heat is supplied to solid CO_2, open to the atmosphere, it changes directly to a vapor (as shown in Fig. 17–6) without passing through the liquid phase, whence the name "dry ice." Liquid CO_2 can exist only at a pressure greater than 5.11 atm. The steel tanks in which CO_2 is commonly stored contain liquid and vapor (both saturated). The pressure in these tanks is the vapor pressure of CO_2 at the temperature of the tank. If the latter is 20°C the vapor pressure is 56 atm or about 830 lb·in^{-2}.

The phase diagram for the common isotope of helium (mass number 4) is shown in Fig. 17–7. Two remarkable properties may be seen. First, helium has no triple point at which solid, liquid, and gas coexist in phase equilibrium. Second, there are *two* liquid phases; this is a property unique to helium. Liquid II is the *superfluid* phase; it has several unusual

17–6 At normal atmospheric pressure solid carbon dioxide passes directly to the vapor phase without passing through a liquid phase; the process is called sublimation. In the photograph, CO_2 evaporates from the solid surface; being more dense than air, it flows down along the surface and across the plate. The visible vapor is not the CO_2 itself, but a fog trail caused by water vapor condensing out of the surrounding air as it is chilled by the very cold CO_2 vapor. Liquid CO_2 exists only at pressures greater than 5.11 atm. (Photo by Barry L. Runk/Grant Heilman Photography.)

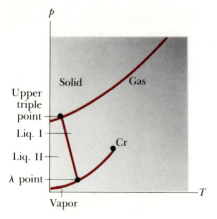

17–7 *pT*-phase diagram for helium.

properties, including essentially zero viscosity, very large thermal conductivity, and the ability to form a surface film that creeps up the sides of a container.

The phase transition between Liquid I and Liquid II is unusual; there is no heat of phase change for this transition. The two phases cannot coexist side by side; the material must be entirely in one phase or entirely in the other, and the corresponding line on the *pT*-diagram shows conditions of temperature and pressure at which the transition can occur. There are two points that would ordinarily be called triple points, but phase equilibrium with three phases cannot occur at either!

Finally, at any pressure less than about 25 atm, helium does not solidify as the temperature is lowered, but remains liquid all the way to absolute zero, or at least to the lowest temperature presently attainable, of the order of 10^{-6} K. To obtain solid helium the pressure must be raised above 25 atm, and then at sufficiently low temperature the helium atoms form a crystal lattice.

17–7

VAPOR PRESSURE

The pressure of the vapor phase of any substance in equilibrium with the liquid or solid phase of the same substance at any temperature is called the *vapor pressure* of the substance at that temperature. For any given substance, it depends only on temperature, not on volume or amount of substance. For example, suppose we have a container with liquid water and water vapor, kept at constant temperature. We try to increase the pressure by decreasing the volume, but if the temperature is constant, then what happens instead is that some of the vapor condenses, and the pressure of vapor does not change. In this case heat of vaporization must be removed to keep the temperature constant.

The vapor pressure of a substance at a particular temperature is independent of other gases that may be present. For example, liquid water may be in equilibrium with air containing water vapor. At 60°C, the vapor pressure of water is about 0.2×10^5 Pa, or about 0.2 atm. Thus if liquid water and water vapor are in equilibrium and the total pressure is 1.0 atm, then the partial pressure of water vapor is 0.2 atm and the partial pressure of air is about 0.8 atm. The latter is of course the sum of partial pressures of the various gases in air.

Table 17–3 shows the vapor pressure of water at various temperatures.

The vapor pressure of a substance always increases rapidly with temperature. The boiling temperature at any given total atmospheric pressure is the temperature at which vapor pressure becomes equal to atmospheric pressure. When water boils in a teakettle, bubbles of steam form on the bottom (where the heat is being added) and rise to the surface. If the vapor pressure inside the bubbles is not as great as atmospheric pressure, then the atmospheric pressure collapses the bubbles, or rather prevents them from forming at all.

Because atmospheric pressure decreases at higher elevations, water boils at lower temperatures at higher elevations. For example, Denver is

	TABLE 17–3 VAPOR PRESSURE OF WATER		
T, °C	Vapor pressure, Pa	Vapor pressure, $lb \cdot in^{-2}$	T, °F
0	0.00610×10^5	0.0886	32
5	0.00868×10^5	0.126	41
10	0.0119×10^5	0.173	50
15	0.0169×10^5	0.245	59
20	0.0233×10^5	0.339	68
40	0.0734×10^5	1.07	104
60	0.199×10^5	2.89	140
80	0.473×10^5	6.87	176
100	$\mathbf{1.01 \times 10^5}$	**14.7**	**212**
120	1.99×10^5	28.8	248
140	3.61×10^5	52.4	284
160	6.17×10^5	89.6	320
180	10.0×10^5	145	356
200	15.5×10^5	225	392
220	23.2×10^5	336	428

about 1600 m above sea level; atmospheric pressure is about 0.82 atm, and the boiling temperature of water is about 94°C. At the summit of Mt. Everest, 8882 m above sea level, atmospheric pressure is about 0.33 atm, and the boiling temperature is about 72°C. Any food-cooking process that involves boiling water takes longer at higher elevations because of the reduced temperatures. Mountaineering expeditions sometimes carry light-weight pressure cookers to permit cooking at pressures of up to 2 atm or so. The extra weight is offset by the saving of fuel.

17–8

EFFECT OF DISSOLVED SUBSTANCES ON FREEZING AND BOILING POINTS

The freezing point of a liquid is lowered when some other substance is dissolved in the liquid. For example, the freezing point of a saturated solution of common salt in water is about −20°C. To understand why a mixture of ice and salt may be used as a freezing mixture, we recall that the freezing point is the only temperature at which the liquid and solid phases can exist in equilibrium. When a saturated salt solution is cooled, it freezes at −20°C, and crystals of ice (pure H_2O) separate from the solution. In other words, ice crystals and a salt solution can exist in equilibrium only at −20°C, just as ice crystals and pure water can exist together only at 0°C.

When ice at 0°C is mixed with a salt solution at 20°C, some of the ice melts, abstracting its heat of fusion from the solution until the temperature falls to 0°C. But ice and salt solution cannot remain in equilibrium at 0°C, so the ice continues to melt. Heat is now supplied by both the ice and the solution, and both cool down until the equilibrium temperature of −20°C is reached. If the melted ice dilutes the salt solution appreciably, the equilibrium temperature rises; but this can be prevented by supplying excess salt so the solution remains saturated. If no heat is supplied from outside, the mixture remains unchanged at −20°C. If the mixture is brought into contact with a warmer body, say an ice cream mixture at 20°C, heat flows from the ice cream mixture to the cold salt solution, melting more of the ice but producing no rise in temperature as long as any ice remains. The flow of heat from the ice cream lowers its temperature to its freezing point (below 0°C, since it is itself a solution). Further loss of heat to the ice–salt mixture causes the ice cream to freeze.

The boiling point of a liquid is also affected by dissolved substances but may be either increased or decreased. Thus the boiling point of a water–alcohol solution is *lower* than that of pure water, while the boiling point of a water–salt solution is *higher* than that of pure water.

17–9

HUMIDITY

Humidity is the pressure of water vapor in the atmosphere. The mass of water vapor per unit volume of atmosphere is called the *absolute humidity*. Water-vapor content can also be described in other ways. As discussed in Section 17–2, the total pressure exerted by the atmosphere is the sum of the partial pressures of the component gases. The partial pressure of water vapor in the atmosphere is ordinarily of the order of 0.002 to 0.05×10^5 Pa, or about 0.002 to 0.05 atm.

The partial pressure of water vapor at any given air temperature can never exceed the vapor pressure of water at that particular temperature. Thus, from Table 17–3, the partial pressure at 10°C cannot exceed 0.0119×10^5 Pa, and at 15°C it cannot exceed 0.0169×10^5 Pa. If the concentration of water vapor, or the absolute humidity, is such that the partial pressure *equals* the vapor pressure, the vapor is *saturated*. If the partial pressure is *less* than the vapor pressure, the vapor is *unsaturated*. The ratio of the partial pressure to the vapor pressure at the same temperature is called the *relative humidity* and is usually expressed as a percentage:

$$\text{Relative humidity (\%)} = (100\%) \frac{\text{partial pressure of water vapor}}{\text{vapor pressure at same temperature}}.$$

The relative humidity is 100% if the vapor is saturated and zero if no water vapor at all is present.

EXAMPLE   The partial pressure of water vapor in the atmosphere is 0.010×10^{-5} Pa and the temperature is 20°C. Find the relative humidity.

Solution   From Table 17–3, the vapor pressure at 20°C is 0.0233×10^5 Pa. Hence,

$$\text{Relative humidity} = \frac{0.010}{0.0233}\,(100\%) = 43\%. \qquad \blacktriangleleft$$

Since the water vapor in the atmosphere is saturated when its partial pressure equals the vapor pressure at the air temperature, saturation can be brought about either by increasing the water vapor content or by lowering the temperature. For example, let the partial pressure of water vapor be 0.01×10^5 Pa when the air temperature is 20°C, as in the preceding example. Saturation or 100% relative humidity could be attained either by adding enough more water vapor (keeping the temperature constant) to increase the partial pressure to 0.0233×10^5 Pa, *or by lowering the temperature* to 7°C, at which, by interpolation from Table 17–3, the vapor pressure is 0.01×10^5 Pa.

If the temperature were to be lowered *below* 7°C, the vapor pressure would be less than 0.01×10^5 Pa. The partial pressure would then be higher than the vapor pressure and enough vapor would condense to reduce the partial pressure to the vapor pressure at the lower temperature. This process brings about the formation of clouds, fog, and rain. At night, when the earth's surface becomes cooled by radiation, the condensed moisture is called *dew.* If the partial pressure is so low that the temperature must fall below 0°C before saturation occurs, the vapor condenses into ice crystals in the form of frost or snow.

The temperature at which the water vapor in a given sample of air becomes saturated is called the *dew point.* One method for measuring the temperature of the dew point is to cool a metal container having a bright polished surface, and to observe its temperature when the surface becomes clouded with condensed moisture. Suppose the dew point is observed to be 10°C, when the air temperature is 20°C. We then know that the water vapor in the air is saturated at 10°C; hence its partial pressure is 0.0119×10^5 Pa, equal to the vapor pressure at 10°C. The pressure necessary for saturation at 20°C is 0.0233×10^5 Pa. The relative humidity is therefore

$$\frac{0.0119}{0.0233} \, (100\%) = 51\%.$$

A simpler but less accurate method of determining relative humidity employs a *wet-and-dry-bulb thermometer.* Two thermometers are placed side by side, the bulb of one being kept moist by a wick dipping in water. The lower the relative humidity, the more rapid will be the evaporation from the wet bulb, and the lower will be its temperature below that of the dry bulb. The relative humidity corresponding to any pair of wet- and dry-bulb temperatures is read from tables. In conditions of very low humidity the wet-bulb temperature may be as much as 15 to 20C° below that of the dry bulb. The principle is of course the same as that of evaporative coolers, mentioned in Section 15–6.

The rate of evaporation of water from a water–air surface depends on the relative humidity. Evaporation is most rapid when the vapor pressure is low, but when the relative humidity is high evaporation is slower. At 100% relative humidity, *no* further evaporation can take place. The added discomfort that high relative humidity brings to a hot summer day is a familiar effect; the reason is that one of the body's important temperature-control mechanisms, cooling by evaporation of sweat from the skin, is inhibited by high relative humidity.

On the other hand, evaporative cooling of the human body makes it possible for humans to maintain normal body temperature (37°C or 98.6°F) in hot dry desert climates where the air temperature may reach

55°C (about 130°F). Evaporation of perspiration keeps the temperature of the skin as much as 20 C° cooler than that of the surrounding air. Adequate water intake is essential under these conditions; a normal person may perspire as much as 8 to 10 liters per day, and unless this lost water is replaced regularly, dehydration, heat stroke, and death result. Old-time desert rats (such as the author) state that in the desert any canteen that holds less than a gallon is to be regarded as a toy!

17–10

THE BUBBLE CHAMBER

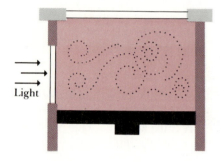

17–8   The bubble chamber.

The bubble chamber is an instrument for studying the interactions of fundamental particles, especially in high-energy collisions, by making their paths visible. As shown in Fig. 17–8, it consists in principle of an enclosure having glass windows and a movable piston. The enclosure contains a liquefied gas, usually hydrogen but occasionally propane, at pressure and temperature just below the boiling point. When the pressure is suddenly decreased slightly by pulling down the piston, the boiling temperature drops below the actual temperature. However, the liquid does not necessarily begin to boil immediately; it may persist in the liquid phase for some time in an unstable *superheated* state.

Bubbles of vapor then tend to form first around ionized molecules or other inhomogeneities in the liquid. Hence when a high-energy charged particle passes through, leaving a string of ions along its path, a string of bubbles forms on the ions. This trail is visible and can be recorded by stereoscopic photography; analysis of this photographic record then determines the precise path and speed of the particle.

A historical forerunner of the bubble chamber was the *cloud chamber*, in which a supercooled vapor is used instead of a superheated liquid. A charged particle passing through leaves a trail of ions, and droplets of liquid condense around these ions, marking the particle's path. In present-day research the cloud chamber has been supplanted by the bubble chamber and other more recent inventions such as the spark chamber, but in the early days of high-energy physics it was of great importance.

QUESTIONS

17–1 If the ideal-gas model were perfectly valid at all temperatures, what would be the volume of a gas when the temperature is absolute zero?

17–2 In the ideal-gas equation of state, could Celsius temperature be used instead of Kelvin if an appropriate numerical value of the gas constant R is used?

17–3 Comment on this statement:
    Equal masses of two different gases, placed in containers of equal volume at equal temperature, must exert equal pressures.

17–4 When a car is driven some distance, the air pressure in the tires increases. Why? Should you let some air out to reduce the pressure?

17–5 Section 17–1 states that pressure, volume, and temperature cannot change individually without one affecting the others. Yet when a liquid evaporates its volume changes, even though its pressure and temperature are constant. Is this inconsistent?

17–6 At low pressure or high temperature, gas behavior deviates from the ideal-gas model. One reason is the finite size of gas molecules. From these deviations, how could one estimate what fraction of the total volume in a gas is occupied by the molecules themselves?

17–7 Radiator caps in automobile engines will release coolant when the pressure reaches a certain value, typically 15 psi or so. Why is it desirable to have the coolant under pressure? Why not just seal the system completely?

17–8 People speak of being angry enough to make their blood boil. Could your blood boil if you go to a high enough altitude? Is the pressure of your blood the same as atmospheric pressure? Does this matter?

17–9 Can it be too cold for good ice skating? What about skiing? Does a thin layer of ice under the skis melt?

17–10 Why does the air cool off more on a clear night than a cloudy night? Desert areas often have exceptionally large day–night temperature variations. Why?

17–11 On a chilly evening you can "see your breath." Can you really? What are you really seeing? Does this phenomenon depend on the temperature of the air, the humidity, or both? How is it related to the dew point?

17–12 Why do frozen water pipes burst? Would a mercury thermometer break if the temperature goes below the freezing temperature of mercury?

17–13 Why does moisture condense on the insides of windows on cold days? If it is cold enough outside, frost will form, but only when the outside temperature is far below freezing. Why is this so? Why are the bottoms frostier than the tops? How do storm windows change the situation?

17–14 "It's not the heat; it's the humidity!" Why are you more uncomfortable on a hot day when the humidity is high than when it is low?

17–15 Why does a shower room get steamy or foggy when the hot water runs for a long time?

17–16 The critical volumes of several substances are listed in Table 17–2, and with a few exceptions they are all in the range of 50 to $100 \times 10^{-6} \ m^3 \cdot mol^{-1}$. Why is there not more variation?

17–17 Why do flames from a fire always go up rather than down?

17–18 Unwrapped food placed in a freezer experiences dehydration, known to housewives as "freezer burn." Why? The same process, carried out intentionally, is called "freeze-drying." For freeze-drying the food is usually frozen first and then placed in a partial vacuum. Why? What advantages might freeze-drying have, compared to ordinary drying?

17–19 An old or burned-out incandescent lightbulb usually has a dark gray area on part of the inside of the bulb. What is this? Why is it not a uniform deposit all over the inside of the bulb?

17–20 Why do glasses containing cold liquids sweat in warm weather? Glasses containing alcoholic drinks sometimes form frost on the outside. How can this happen?

17–21 Why does a person perspire more when working hard than when resting?

17–22 Hiking down a steep trail ought to be much easier than hiking up, yet people perspire a lot when descending a steep trail. Why?

17–23 In a hot desert climate a person can survive a lot longer without food than without water. Why?

PROBLEMS

17–1 A 2-liter tank contains air at 1 atm and 20°C. The tank is sealed and heated until the pressure is 3 atm.

a) What is the temperature now, in degrees Celsius? Assume that the volume of the tank is constant.

b) If the temperature is kept at the value found in (a) and the gas is permitted to expand, what is the volume when the pressure again becomes 1 atm?

17–2 A 20-liter tank contains 0.20 kg of helium at 27°C.

a) What is the number of moles of helium?

b) What is the pressure in pascals? In atmospheres?

17–3 A room with dimensions 3 m × 4 m × 5 m is filled with pure oxygen at 20°C and 1 atm.

a) What is the number of moles of oxygen?

b) What is the mass of oxygen, in kilograms?

17–4 What is the density of air, in kilograms per cubic meter, at 1 atm and 20°C? The average molecular mass of air is $28.8 \ g \cdot mol^{-1}$.

17–5 What is the mass of one cubic meter of nitrogen at 1 atm and 20°C?

17–6 What is the ratio of the density of oxygen to that of water at 1 atm and 0°C?

17–7 A tank contains $0.5 \ m^3$ of nitrogen at 27°C and 1.5×10^5 Pa (absolute pressure). What is the pressure if the volume is increased to $5.0 \ m^3$ and the temperature is increased to 327°C?

17–8 A welder using a tank with a volume of $0.10 \ m^3$ fills it with oxygen at a gauge pressure of 4.0×10^5 Pa and temperature of 47°C. Later it is found that, because of a leak, the gauge pressure has dropped to 3.0×10^5 Pa and the temperature has decreased to 27°C. Find

a) The initial mass of oxygen, and

b) the mass that has leaked out.

17–9 A helium storage tank has a capacity of $0.05 \ m^3$. If the gas pressure is 100 atm at 27°C, determine

a) the number of moles of helium, and

b) the mass of helium.

17–10 A hot-air balloon makes use of the fact that hot air at atmospheric pressure is less dense than cooler air at the same pressure; the buoyant force is the difference between the weight of hot air and that of an equal volume of the cooler surrounding air. If the volume of the balloon is 500 m³ and the surrounding air is at 0°C, what must be the temperature of the air in the balloon in order for it to lift a total mass of 200 kg?

17–11 A flask of volume 2 l, provided with a stopcock, contains oxygen at 300 K and atmospheric pressure. A chemist raises the temperature of the system to a temperature of 400 K, with the stopcock open to the atmosphere. She then closes the stopcock and cools the flask to its original temperature.

a) What is the final pressure of the oxygen in the flask?

b) How many grams of oxygen remain in the flask?

17–12 A balloon whose volume is 500 m³ is to be filled with hydrogen at atmospheric pressure.

a) If the hydrogen is stored in cylinders of volume 0.05 m³ at an absolute pressure of 15×10^5 Pa, how many cylinders does the balloonist need?

b) What is the total weight that can be supported by the balloon, in air at standard conditions?

c) What weight could be supported if the balloon were filled with helium instead of hydrogen?

17–13 Derive from the equation of state of an ideal gas an equation for the density of an ideal gas in terms of pressure, temperature, and appropriate constants.

17–14 At the beginning of the compression stroke, a cylinder of a diesel engine contains 800 cm³ of air at atmospheric pressure and a temperature of 27°C. At the end of the stroke, the air has been compressed to a volume of 50 cm³ and the gauge pressure has increased to 40×10^5 Pa. Compute the temperature.

17–15 A diver observes a bubble of air rising from the bottom of a lake, where the pressure is 3 atm, to the surface, where the pressure is 1 atm. The temperature at the bottom of the lake is 7°C and the temperature at the surface is 27°C. What is the ratio of the volume of the bubble as it reaches the surface to its volume at the bottom?

17–16 A liter of helium under a pressure of 2 atm and at a temperature of 27°C is heated until both pressure and volume are doubled.

a) What is the final temperature?

b) How many grams of helium are there?

17–17 A laboratory technician places 1 g of oxygen in a flask at an absolute pressure of 10 atm and at a temperature of 47°C. At a later time he finds that because of a leak the pressure has dropped to ⅝ of its original value and the temperature has decreased to 27°C.

a) What is the volume of the flask?

b) How many grams of oxygen leaked out between the two observations?

17–18 The submarine *Squalus* sank at a point where the depth of water was 240 ft. The temperature at the surface is 27°C and at the bottom it is 7°C. The density of sea water may be taken as 2 slugs·ft⁻³.

a) If a diving bell in the form of a circular cylinder 8 ft high, open at the bottom and closed at the top, is lowered to this depth, to what height will the water rise within it when it reaches the bottom?

b) At what gauge pressure must compressed air be supplied to the bell while on the bottom to expel all the water from it?

17–19 A bicyclist uses a tire pump whose cylinder is initially full of air at an absolute pressure of 15 lb·in⁻². The length of stroke of the pump is 18 in. At what part of the stroke does air begin to enter a tire in which the gauge pressure is 40 lb·in⁻²? Assume the compression to be isothermal.

17–20 A vertical cylindrical tank 1 m high has its top end closed by a tightly fitting frictionless piston of negligible weight. The air inside the cylinder is at an absolute pressure of 1 atm. The piston is depressed by pouring mercury on it slowly. How far will the piston descend before mercury spills over the top of the cylinder? The temperature of the air is maintained constant.

17–21 A glassblower makes a barometer using a tube 90 cm long and of cross section 1.5 cm². Mercury stands in this tube to a height of 75 cm. The room temperature is 27°C. He then introduces a small amount of nitrogen into the evacuated space above the mercury, and the column drops to a height of 70 cm. How many grams of nitrogen were introduced?

17–22 A large tank of water has a hose connected to it, as shown in Fig. 17–9. The tank is sealed at the top and has compressed air between the water surface and the top. When the water height h_2 is 3 m, the gauge pressure p_1 is 1.0×10^5 N·m⁻². Assume that the air above the water surface expands isothermally.

a) What is the velocity of flow out of the hose when $h_2 = 3$ m?

b) What is the velocity of flow when h_2 has decreased to 2 m? Neglect friction.

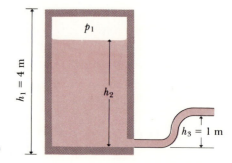

Figure 17–9

17–23 Construct two graphs for a real substance, one showing pressure as a function of volume, and the other showing pressure as a function of temperature. Show on each graph the region in which the substance exists as

a) a gas or vapor,

b) a liquid, and

c) a solid.

Show also the triple point and the critical point.

17–24 A physics student introduces a small amount of liquid into a glass tube, pumps out all the air, and seals the tube. Describe the behavior of the meniscus when the temperature of the system is raised

a) if the volume of the tube is much greater than the critical volume,

b) if the volume of the tube is much less than the critical volume, and

c) if the volume of the tube is only slightly different from the critical volume.

17–25 A physicist places a piece of ice at 0°C alongside a beaker of water at 0°C in a glass vessel, from which all air has been removed. If the ice, water, and vessel are all maintained at a temperature of 0°C by a suitable thermostat, describe the final equilibrium state inside the vessel.

17–26 Compare the density of carbon dioxide at the critical point to the density of solid carbon dioxide ("dry ice") at 1 atm, which is about $1.56 \text{ g} \cdot \text{cm}^{-3}$.

17–27 Determine the density of water at the critical point, and express it as a multiple of the density at 0°C and 1 atm.

17–28 Find the ratio of the density of oxygen to that of water

a) at 0°C and 1 atm, and

b) at their respective critical points.

17–29

a) What is the relative humidity on a day when the temperature is 20°C and the dew point is 5°C?

b) What is the partial pressure of water vapor in the atmosphere?

c) What is the absolute humidity, in grams per cubic meter?

17–30 The temperature in a room is 40°C. A meteorologist cools a can gradually by adding cold water. At 10°C the surface of the can clouds over. What is the relative humidity in the room?

17–31 A pan of water is placed in a sealed room of volume 60 m^3 at a temperature of 27°C and initial relative humidity of 60 percent.

a) How many grams of water will evaporate?

b) What is the absolute humidity in $\text{kg} \cdot \text{m}^{-3}$ after equilibrium has been reached?

c) If the temperature of the room is then increased 1 C°, how many more grams of water will evaporate?

17–32

a) What is the dew-point temperature on a day when the air temperature is 20°C and the relative humidity is 60 percent?

b) What is the absolute humidity, expressed in grams per cubic meter?

17–33 An engineer was hired to design an air-conditioning system to increase the relative humidity of 0.5 m^3 of air per second from 30 percent to 65 percent. The air temperature is 20°C. How many kilograms of water are needed by the system per hour?

17–34 The volume of a closed hospital room, kept at a constant temperature of 20°C, is 60 m^3. The relative humidity in the room is initially 10%. If a nurse brings a pan of water into the room, how many grams of water evaporate?

18

THE FIRST LAW OF THERMODYNAMICS

Thermodynamics is the study of energy relationships involving heat, mechanical work, and other aspects of energy and energy transfer. These relationships are of vital importance in the study of energy-conversion devices such as engines, batteries, and refrigerators, and also in the study of energy relations in living organisms. To state these relationships precisely, we introduce the concept of a *thermodynamic system*, and we discuss *heat* and *work* as two means of transferring energy into or out of such a system. Emerging from this discussion and from a generalized view of the principle of conservation of energy is the concept of *internal energy* of a system.

18–1

ENERGY, HEAT, AND WORK

We have already studied the transfer of energy through mechanical work (Chapter 6) and through heat transfer (Chapters 15 and 16), and now we are ready to combine and generalize these principles. Our analysis will usually refer to some *system*, which will usually be a specified quantity of material, perhaps in the form of a device or an organism. A *thermodynamic system* is a system that can interact with its surroundings in at least two ways, one of which must be heat transfer. A familiar example of a thermodynamic system is a quantity of a gas confined in a cylinder with a piston. Energy can be added to the system by conduction of heat, and the system can also do *work*, since the gas exerts a force on the piston and can move it through a displacement.

We have already encountered the concept of a system. In Chapters 2, 4, and 8 we made extensive use of free-body diagrams to help identify the forces acting on a particular mechanical system so as to include only these in our analysis. Similarly, in Chapter 7 we studied the principle of conservation of momentum for an isolated mechanical system. In both examples it is essential to define clearly at the outset precisely what is and is not included in the system. The same thing is true of thermodynamic systems; it is essential in solving problems to identify the system under discussion and to describe unambiguously the energy transfers in and out of that system.

As already mentioned, thermodynamics has its roots in intensely practical problems. For example, a steam engine or steam turbine makes use of the heat of combustion of coal or other fuel to perform mechanical work to drive an electric generator, pull a train, or perform some other useful function. The gasoline engine in an automobile and the jet engines in an airplane have similar functions. Muscle tissue in living organisms metabolizes chemical energy in food and performs mechanical work on the surroundings of the organism.

In all these problems we shall describe the energy relations in terms of the quantity of heat Q added *to* the system and the work W done *by* the system. Both Q and W may be positive or negative. A positive value of Q represents heat flow *into* the system, with a corresponding input of energy to it; negative Q represents heat flow *out of* the system. A positive value of W represents work done *by* the system against its surroundings, such as an expanding gas, and hence corresponds to energy leaving the system. Negative W, such as compression of a gas in which work is done *on it* by its surroundings, represents energy entering the system. These conventions will be clarified with many examples in this chapter and the next.

Heat can be understood on the basis of microscopic mechanical energy, i.e., the kinetic and potential energies of individual molecules in a material; and it is also possible to develop the principles of thermodynamics from a microscopic viewpoint. In the present chapter that development is deliberately avoided, in order to emphasize that the central principles and concepts of thermodynamics can be treated in a wholly *macroscopic* way, without reference to microscopic models. Indeed, part of the great power and generality of thermodynamics springs from the fact that it is *not* dependent on details of the structure of matter. In Chapter 20 we return to microscopic considerations and examine their relation to the principles of thermodynamics.

18–2

WORK IN VOLUME CHANGES

We have mentioned a gas in a cylinder with a movable piston as a simple example of a thermodynamic system. In the next several sections we consider several kinds of processes for such a system. First we consider the *work* done by the system during a volume change.

When a gas expands, it pushes out on its boundary surfaces as they move outward; hence an expanding gas always does positive work. The same thing is true of any solid or fluid material confined under pressure. Figure 18–1 shows a solid or fluid in a cylinder with a movable piston. Suppose that the cylinder has a cross-sectional area A and that the pressure exerted by the system at the piston face is p. The force F exerted by the system is therefore $F = pA$. If the piston moves out a small distance Δx, the work ΔW of this force is equal to

$$\Delta W = F\,\Delta x = pA\,\Delta x.$$

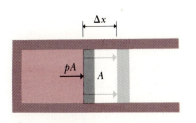

18–1   Force exerted *by* a system during a small expansion.

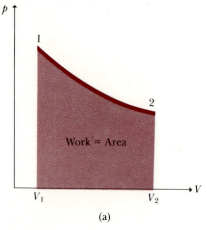

(a)

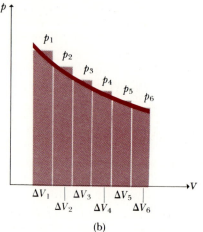

(b)

18–2 The work done when a substance changes its volume and pressure is the area under the curve on a *pV*-diagram.

But

$$A \, \Delta x = \Delta V,$$

where ΔV is the change of volume of the system. Therefore

$$\Delta W = p \, \Delta V. \qquad (18\text{–}1)$$

If the pressure remains constant while the volume changes a finite amount, say from V_1 to V_2, then the work W is

$$W = p(V_2 - V_1) \qquad \text{(constant pressure only)}. \qquad (18\text{–}2)$$

If, on the other hand, the pressure decreases as the volume increases (as would be the case, for example, with an ideal gas at constant temperature), then we may imagine that the whole series of changes consists of a small volume change ΔV_1 while the pressure is p_1, then another small volume change ΔV_2 while the pressure is p_2, and so on. The work will then be

$$W = p_1 \, \Delta V_1 + p_2 \, \Delta V_2 + \cdots.$$

Suppose that pressure is plotted along the vertical axis and volume along the horizontal axis, just as in Section 17–4. The changes of pressure and volume will then be indicated by a smooth curve such as that shown in Fig. 18–2a. If the total volume change $V_2 - V_1$ is divided into a number of small steps ΔV_1, ΔV_2, etc., as shown in Fig. 18–2b, then the smooth curve may be approximated closely by a jagged curve where the pressure is p_1 only during the volume change ΔV_1, and is p_2 only during ΔV_2, and so on. The total work $p_1 \, \Delta V_1 + p_2 \, \Delta V_2 + \cdots$ is thus seen to be the sum of the *areas* of all the rectangles that make up the total area under the jagged curve. If now the number of steps is increased indefinitely, the area under such a jagged curve approaches the area under the smooth curve, and we have the result that

$$W = \text{Area under a curve on a } pV\text{-diagram.} \qquad (18\text{–}3)$$

Note that this method of finding the work of a varying pressure is the same as that for finding the work of a varying force, as explained in Section 6–2.

According to the convention stated above, the work is *positive* when a system *expands*. Thus, if a system expands from 1 to 2 in Fig. 18–2a, the area is regarded as positive. A *compression* from 2 to 1 would give rise to a *negative* area; when a system is compressed, its volume decreases and it does *negative* work on its surroundings.

On the *pV*-diagram in Fig. 18–3 an initial state 1 (characterized by pressure p_1 and volume V_1) and a final state 2 (characterized by pressure p_2 and volume V_2) are represented by the two points 1 and 2. To progress from the initial state to the final state, the system must go through a series of intermediate states, and there are infinitely many possibilities. For example, the pressure might be kept constant at p_1 while the system expands to volume V_2 (point 3 on the diagram), and then the pressure might be reduced to p_2 with the volume remaining constant at V_2 (to point 2 on the diagram). The work during this process is the area under the line $1 \rightarrow 3$; no work is done during the constant-volume process $3 \rightarrow 2$. Or the system might traverse the path $1 \rightarrow 4 \rightarrow 2$, in which case

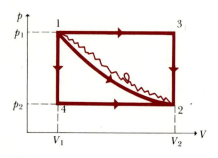

18–3 Work depends on the path.

the work is the area under the line $2 \rightarrow 4$. The wiggly line and the smooth curve from 1 to 2 are two other possibilities, and the work is different for each one. We conclude that *the work depends not only on the initial and final states but also on the intermediate states, i.e., on the path.*

18–3

HEAT IN VOLUME CHANGES

As we have seen in Chapter 15, heat is energy transferred into or out of a system as a result of a temperature difference between the system and its surroundings. The performance of work, discussed in the preceding section, and the transfer of heat are means of energy transfer, that is, processes by which the energy of a system may increase or decrease.

We have seen in Section 18–2 that in a change of state of a thermodynamic system the work done by the system depends not only on the initial and final states but also on the *path* taken, that is, the series of intermediate states. This is also true of the *heat* added to the system in such a process. Here is an example to illustrate this point.

In Fig. 18–4a, a quantity of gas is contained in a cylinder with a piston, with initial volume of two liters, and maintained at a temperature of 300 K by an electric stove. We then let the gas expand slowly; heat flows from the stove to the gas, maintaining the temperature of the gas at the constant value 300 K. Suppose that the gas expands in this slow, controlled, isothermal manner until its volume becomes, say, seven liters. A finite amount of heat is absorbed by the gas during this process.

Figure 18–4b shows a vessel surrounded by insulating walls and divided into two compartments (the lower one of volume two liters, the upper, five liters) by a thin, breakable partition. In the lower compartment the same gas has the same initial volume and temperature as in (a). Suppose the partition is broken and the gas in (b) undergoes a rapid, uncontrolled expansion into the vacuum above, where it encounters no movable piston and therefore exerts no external force. No work is done in this expansion, and also no heat passes through the insulating walls. The final volume is seven liters as in (a). The rapid, uncontrolled expansion of a gas into a vacuum is called a *free expansion* and will be discussed later in this chapter. Experiments have shown that in a free expansion of an ideal gas there is no temperature change; therefore the final state of

18–4 (a) Slow, controlled, isothermal expansion of a gas from an initial state 1 to a final state 2. (b) Rapid, uncontrolled expansion of the same gas starting at the same state 1 and ending at the same state 2.

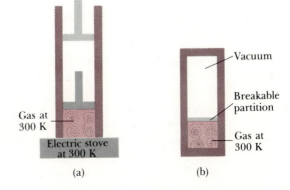

Gas at 300 K

Electric stove at 300 K

(a)

Vacuum

Breakable partition

Gas at 300 K

(b)

the gas is the same as the final state in (a). But the states traversed by the gas (the pressures and volumes) while proceeding from state 1 to state 2 are entirely different, so that (a) and (b) represent *two different paths* connecting the *same states* 1 and 2. During path (a) heat is transferred and work is done; during path (b) *no* heat is transferred and no work is done. *Heat,* like work, *depends not only on the initial and final states but also on the path.*

Thus it would *not* be correct to say that a body *contains* a certain amount of heat. For suppose we assigned an arbitrary value to "the heat in a body" in some standard reference state. The "heat in the body" in some other state would then equal the "heat" in the reference state plus the heat added when the body is carried to the second state. But the heat added depends entirely on the path by which we go from one state to the other, and since there are an infinite number of paths that might be followed, there are an infinite number of values that might equally well be assigned to the "heat in the body" in the second state. Thus there is no consistent way to define "heat in a body," and this is therefore not a useful concept. It does make sense, however, to speak of the amount of *internal energy* in a body, and this important concept is our next topic.

18–4

INTERNAL ENERGY AND THE FIRST LAW

The concept of internal energy of a thermodynamic system can be viewed in various ways, some simple, some subtle. To begin with the simple, we note that matter is made of atoms and molecules; these are in turn made up of particles having kinetic and potential energies. We tentatively define the *internal energy* of a system as the sum of all the kinetic and potential energies of its constituent particles. We shall use the symbol U for internal energy. During a change of state of the system, the internal energy may change from an initial value U_1 to a final value U_2; in that case we denote the change in internal energy in the usual way as $\Delta U = U_2 - U_1$.

We also know the heat transfer is really energy transfer. Thus when a quantity of heat Q is added to a system, the result should be to increase the internal energy by an amount equal to Q; that is, $\Delta U = Q$. Also, when the system does work W by expanding against its surroundings, this represents energy leaving the system. When W is positive, ΔU is negative, and conversely, so $\Delta U = -W$. Finally, if both heat transfer and work occur, the total change in internal energy is

$$U_2 - U_1 = \Delta U = Q - W. \qquad (18\text{–}4)$$

This equation may be rearranged to the form

$$Q = \Delta U + W, \qquad (18\text{–}5)$$

which can be read as follows: When heat Q is added to a system, some of the corresponding energy remains within the system, increasing its internal energy by an amount ΔU; the remainder leaves the system again as the system does work W against its surroundings. Of course, as explained above, W and Q are algebraic quantities that may be either positive or negative, and correspondingly we expect ΔU to be positive for some processes and negative for others.

Equation (18–4) is the *first law of thermodynamics*. We see that in the context of the above discussion it simply represents conservation of energy, generalized to include energy transfer through heat as well as mechanical work. It certainly seems eminently reasonable that conservation of energy, which we encountered in Chapter 6 in a purely mechanical context, should have this more general validity; nevertheless, it is worth a somewhat closer look.

First of all, the above definition of internal energy in terms of microscopic kinetic and potential energies is not entirely convincing. Actually to calculate this total energy for any real system would be hopelessly complicated. Thus our definition is not an *operational* definition because it does not describe how to obtain the defined quantity from physical quantities that can be measured directly. Hence it is necessary to backtrack a little in defining internal energy, in order to show clearly the empirical and experimental basis of the first law of thermodynamics.

Starting over, we *define* the internal energy change ΔU in any change of a system by means of Eq. (18–4). This *is* an operational definition, because Q and W can be measured. It does not define U itself, only ΔU. This is not a serious shortcoming; we can *define* the internal energy of a system to have a specified value in a certain standard state, and then use Eq. (18–4) to define the internal energy in any other state. This procedure is analogous to the definition of potential energy in Chapter 6, where we arbitrarily defined the potential energy of a mechanical system to be zero at a certain position.

This definition trades one difficulty for another. If we define ΔU via Eq. (18–4), then we must ask the following question: If the system goes from some initial state to some final state, by two different paths, is ΔU necessarily the same for the two paths? We have already seen that in such a case Q and W are in general both path-dependent; if ΔU, which is just $Q - W$, is also path-dependent, then ΔU is ambiguous, and the concept of internal energy in a system is subject to the same criticism as the erroneous concept of quantity of heat in a system, as discussed at the end of Section 18–3.

At this point the only recourse is to experiment. We study the properties of a variety of materials; in particular we measure Q and W for various changes of state and various paths, in order to learn whether ΔU is or is not path-dependent. The results of many such investigations are clear and unambiguous; within experimental error, ΔU is found to be path-independent. The change in internal energy in any thermodynamic process is found to depend only on the initial and final states and *not* on the path leading from one to the other.

This, then, is the ultimate justification for believing that in any state a thermodynamic system has a unique internal energy that depends only on the state the system is in. An equivalent statement is that the internal energy U of a system is a function of the state variables p, V, and T (or actually of any two of these, since the three variables are related by the equation of state).

We now return to the first law of thermodynamics. It is perfectly correct to say that the first law, given by either Eq. (18–4) or (18–5), is a statement of conservation of energy for thermodynamic processes. But we must also recognize that part of the content of the first law is the experimental verification that internal energy is a property of the state of a system, and that in changes of state it is path-independent.

All this may seem a little abstract to a reader who is satisfied to regard internal energy as microscopic mechanical energy. There is nothing wrong with that view. But in the interest of precise operational definitions, internal energy, like heat, can and must be defined in a way that is independent of the details of microscopic structure of the material.

If a system is carried through a process that eventually returns it to its initial state (a *cyclic* process), then

$$U_2 = U_1 \quad \text{and} \quad Q = W.$$

Thus, although net work W may be done by the system in the process, energy has not been created, since an equal amount of energy must have flowed into the system as heat Q.

An *isolated* system is one that does no external work and into which there is no flow of heat. Then, for any process taking place in such a system,

$$W = Q = 0$$

and

$$U_2 - U_1 = 0 \quad \text{and} \quad \Delta U = 0.$$

That is, *the internal energy of an isolated system remains constant.* This is the most general statement of the *principle of conservation of energy.* The internal energy of an *isolated* system cannot be changed by any process (mechanical, electrical, chemical, nuclear, or biological) taking place *within* the system. The energy of a system can be changed only by a flow of heat across its boundary, or by the performance of work. If either of these takes place, the system is no longer isolated. The increase in energy of the system is then equal to the energy flowing in as heat, minus the energy flowing out as work.

EXAMPLE   A thermodynamic process is shown in the *pV*-diagram of Fig. 18–5. In process *ab*, 600 J of heat are added, and in process *bd* 200 J of

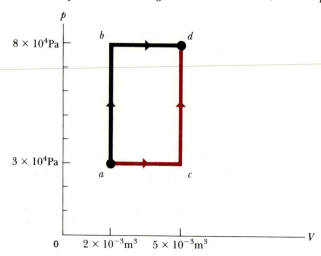

18–5

heat are added. Find (a) the internal energy change in process ab; (b) the internal energy change in process abd; and (c) the total heat added in process acd.

Solution

a) There is no volume change in process ab, so $W = 0$ and $\Delta U = Q = 600$ J.

b) Process bd occurs at constant pressure, so the work done by the system during this expansion is

$$W = p(V_2 - V_1) = (8 \times 10^4 \text{ Pa})(5 \times 10^{-3} \text{ m}^3 - 2 \times 10^{-3} \text{ m}^3)$$
$$= 240 \text{ J}.$$

Thus the total work for abd is $W = 240$ J and the total heat is $Q = 800$ J. Equation (18–4) gives

$$\Delta U = Q - W = 800 \text{ J} - 240 \text{ J} = 560 \text{ J}.$$

c) Because ΔU is independent of path, the internal energy change is the same for path acd as for abd, that is, 560 J. The total work for the path acd is

$$W = (3 \times 10^4 \text{ Pa})(5 \times 10^{-3} \text{ m}^3 - 2 \times 10^{-3} \text{ m}^3) = 90 \text{ J}.$$

Equation (18–4) or (18–5) then gives

$$Q = \Delta U + W = 560 \text{ J} + 90 \text{ J} = 650 \text{ J}.$$

We see that although ΔU is the same for abd and acd, W and Q are quite different for the two processes. ◀

In the next several sections, several general classes of thermodynamic processes are introduced, and the heat, work, and internal energy considerations appropriate for each are discussed.

18–5

ADIABATIC PROCESS

A process in which no heat enters or leaves a system is called an *adiabatic process*. For every adiabatic process, $Q = 0$. This prevention of heat flow may be accomplished either by surrounding the system with a thick layer of heat-insulating material (such as cork, firebrick, or styrofoam), or by performing the process quickly. Heat flow requires finite time, so any process performed quickly enough will be practically adiabatic. Applying the first law to an adiabatic process, we obtain

$$U_2 - U_1 = \Delta U = -W \quad \text{(adiabatic process)}. \quad (18\text{–}6)$$

Thus, the change in the internal energy of a system, in an adiabatic process, is equal *in magnitude* to the work done by the system. If the work W is negative, as when a system is compressed, then $-W$ is positive, U_2 is greater than U_1, and the internal energy of the system increases. If W is positive, as when a system expands, the internal energy of the system decreases. An increase of internal energy is usually (but not always) accompanied by a rise in temperature, and a decrease in internal energy by a temperature drop.

The compression of the mixture of gasoline vapor and air that takes place during the compression stroke of a gasoline engine is an example of an approximately adiabatic process involving a temperature rise. The expansion of the combustion products during the power stroke of the engine is an approximately adiabatic process involving a temperature decrease. Adiabatic processes, therefore, play a very important role in mechanical engineering.

18–6

ISOCHORIC PROCESS

When a substance undergoes a process in which the volume remains unchanged ($\Delta V = 0$), the process is called *isochoric*. The rise of pressure and temperature produced by a flow of heat into a substance contained in a rigid container of fixed volume is an example of an isochoric process. When the volume does not change, no work is done; $W = 0$, and therefore, from the first law,

$$U_2 - U_1 = \Delta U = Q \qquad \text{(isochoric process)}. \qquad (18\text{–}7)$$

All the added heat remains in the system as an increase in the internal energy. The very sudden increase of temperature and pressure accompanying the explosion of gasoline vapor and air in a gasoline engine may be treated mathematically as though it were an isochoric addition of heat.

18–7

ISOTHERMAL PROCESS

An isothermal process is one that takes place at constant temperature. In an isothermal process the system must remain in thermal equilibrium, and this means that the pressure and volume changes must take place slowly enough to maintain thermal equilibrium. In general, *none* of the quantities Q, W, or ΔU is zero.

There are a few special cases in which the internal energy of a system depends only on its temperature, not on pressure or volume. The most familiar system having this property is an ideal gas. For such a system, if the temperature is constant, the internal energy is also; in this case, $\Delta U = 0$ and $Q = W$. This *does not* hold for other systems, since in general the internal energy may depend on pressure as well as temperature, and may vary even when T is constant.

18–8

ISOBARIC PROCESS

A process taking place at constant pressure is called an *isobaric process*. When water enters a steam boiler and is heated to its boiling point and vaporized, and then the steam is superheated, all these processes take place isobarically. Such processes play an important role in mechanical engineering and also in chemistry.

As discussed in Section 18–2,

$$W = p(V_2 - V_1) \qquad \text{(isobaric process)}. \qquad (18\text{--}8)$$

A simple example is the vaporization of a mass m of liquid at constant pressure and temperature. If V_L is the volume of liquid and V_V is the volume of vapor, the work done by the system in expanding from V_L to V_V at constant pressure p is

$$W = p(V_V - V_L).$$

The heat absorbed per unit mass is the heat of vaporization L_V. Hence,

$$Q = mL_V.$$

From the first law,

$$U_V - U_L = mL_V - p(V_V - V_L). \qquad (18\text{--}9)$$

EXAMPLE   One gram of water (1 cm^3) becomes 1671 cm^3 of steam when boiled at a pressure of 1 atm. The heat of vaporization at this pressure is 2256 J·g^{-1}. Compute the external work and the increase in internal energy.

Solution

$$\begin{aligned} W &= p(V_V - V_L) \\ &= (1.013 \times 10^5 \text{ Pa})(1671 \times 10^{-6} \text{ m}^3 - 1 \times 10^{-6} \text{ m}^3) \\ &= 169 \text{ J}. \end{aligned}$$

From Eq. (18–9),

$$\begin{aligned} U_V - U_L &= mL_V - W = 2256 \text{ J} - 169 \text{ J} \\ &= 2087 \text{ J}. \end{aligned}$$

That is, when one gram of water vaporizes, 2256 J of heat are added. Of this added energy, 2087 J remains in the system as an increase in internal energy, and the remaining 169 J leaves the system again as it does work against the surroundings while expanding from liquid to vapor.   ◄

18–9

INTERNAL ENERGY OF AN IDEAL GAS

It was mentioned in Section 18–7 that for an ideal gas the internal energy U depends only on temperature, not on pressure or volume. To explore this special property in more detail, we imagine a thermally insulated container having rigid walls, divided into two compartments by a partition. Suppose that there is a gas in one compartment and that the other contains only vacuum. If the partition is removed, the gas undergoes what is known as a *free expansion*, in which no work is done and no heat is transferred. From the first law, since both Q and W are zero, it follows that *the internal energy remains unchanged during a free expansion*.

Whether or not the *temperature* of a gas changes during a free expansion is an important question, for it provides information concerning the dependence of internal energy on other quantities. If the temperature should change while the internal energy stays the same, one would have to conclude that the internal energy depends on *both* the temperature and the volume, or both the temperature and the pressure, but certainly not on the temperature alone. If, on the other hand, T remains unchanged during a free expansion in which we know U remains unchanged, then the only conclusion that is admissible is that *U is a function of T only*, not of p or V.

Experiment has shown that the internal energy of a real gas does depend to some extent on the pressure or volume as well as on the temperature. At sufficiently low pressures and high temperatures, however, the behavior approaches more and more closely that of an ideal gas, and in this limit the internal energy is found to depend *only* on temperature. In a free expansion an ideal gas undergoes no temperature change.

Thus the assumption that U depends only on T becomes a part of the ideal-gas model, in addition to the ideal-gas equation of state. This property will be used several times in the following sections.

18–10

HEAT CAPACITIES OF AN IDEAL GAS

The temperature of a substance may be changed under a variety of conditions. The volume may be kept constant, or the pressure may be kept constant, or both may be allowed to vary in some definite way. In general the amount of heat per mole needed to cause unit rise in temperature is different for each type of process. In other words, a substance has *many different* molar heat capacities, depending on the conditions under which it is heated or cooled.

The basis of this variation is the first law of thermodynamics. In a constant-volume temperature change, no work is done, and the change in internal energy equals the heat added. In a constant-pressure temperature change, on the other hand, the volume must increase—otherwise the pressure could not remain constant—and as the material expands it does work. Thus the heat input for the constant-pressure process must be *greater* for a given temperature change than for the constant-volume process because, in the former, additional energy must be supplied to account for the work done in the expansion. If the volume were to *decrease* during heating, the heat input would be *less* than in the constant-volume case.

This distinction was ignored in the discussion of heat capacities in Chapter 15. Indeed, for solid and liquid materials it is not usually of great practical importance. The coefficients of volume expansion of such materials are quite small, and ordinarily the work done against the surroundings is so small compared to the total heat added that it may be ignored. The distinction *is* of course important in theoretical studies of heat capacities.

For gases the situation is quite different, and the dependence of heat capacity on the details of the process is quite significant. For air, for example, the heat capacity under constant-pressure conditions is 40 percent greater than under constant-volume conditions. Fortunately, for ideal gases we can derive a simple relation between two heat capacities of special importance.

From a practical standpoint, the two conditions of greatest interest are raising the temperature of a gas at *constant volume,* as in a container whose volume may be considered constant, and at *constant pressure,* where the gas expands just enough so the pressure remains constant. The two corresponding molar heat capacities are called the molar heat capacity at constant volume, denoted as C_v, and the molar heat capacity at constant pressure, denoted as C_p. On the basis of the above discussion, we expect C_p to be greater than C_v. We shall now derive the relation between the two.

We consider the constant-volume process first. We place n moles of the gas at temperature T in a rigid container and bring it in contact with a body at a slightly higher temperature $T + \Delta T$. There is a flow of heat Q into the gas, and by definition of the molar heat capacity at constant volume, C_v,

$$Q = nC_v \, \Delta T. \qquad (18\text{--}10)$$

The pressure of the gas increases during this process, but no *work* is done, since the volume is constant. Hence, $W = 0$, and from the first law, Eq. (18–4),

$$\Delta U = Q - W = Q,$$

we also have

$$\Delta U = nC_v \, \Delta T. \qquad (18\text{--}11)$$

Now we consider a constant-pressure process. We enclose the gas in a cylinder with a piston that moves so as to maintain constant pressure. Again the system is brought in contact with a body at a temperature $T + \Delta T$. As heat flows into the gas it expands at constant pressure and does work. By definition of the molar heat capacity at constant pressure, C_p, the heat Q flowing into the gas is

$$Q = nC_p \, \Delta T. \qquad (18\text{--}12)$$

The work W done by the gas is

$$W = p \, \Delta V.$$

The work W may also be expressed in terms of the temperature change by use of the ideal-gas equation of state. Let the initial volume and temperature be V_1 and T_1 and the final values V_2 and T_2. (The pressure p and number of moles n are the same in the initial and final states.) Thus the ideal-gas equation gives

$$pV_1 = nRT_1, \qquad pV_2 = nRT_2.$$

Subtracting the first equation from the second, we find

$$p(V_2 - V_1) = nR(T_2 - T_1)$$

or

$$W = p \, \Delta V = nR \, \Delta T. \qquad (18\text{--}13)$$

Then from the first law of thermodynamics,

$$Q = \Delta U + W,$$

or, using Eqs. (18–12) and (18–13),

$$nC_p\,\Delta T = \Delta U + nR\,\Delta T. \qquad (18\text{--}14)$$

Now, here comes the crux of the calculation. The internal energy change ΔU is again given by Eq. (18–11), that is, $\Delta U = nC_v\,\Delta T$, *even though now the volume is not constant!* Why is this so? We recall the discussion of Section 18–9; one of the special properties of an ideal gas is that its internal energy depends *only* on temperature. Thus the *change* in internal energy in any process must be determined only by the temperature change. That is, if Eq. (18–11) is valid for an ideal gas for one particular kind of process, it must be valid for an ideal gas for *every* kind of process.

To complete our derivation, we replace ΔU in Eq. (18–14) by $nC_v\,\Delta T$ and divide each term by the common factor $n\,\Delta T$ to obtain

$$nC_p\,\Delta T = nC_v\,\Delta T + nR\,\Delta T,$$

and

$$C_p = C_v + R. \qquad (18\text{--}15)$$

As predicted, the molar heat capacity of an ideal gas at constant pressure is *greater* than that at constant volume; the difference is the universal gas constant R. Of course, R must be expressed in the same units as C_p and C_v, such as $\mathrm{J\cdot mol^{-1}\cdot K^{-1}}$.

Although Eq. (18–15) was derived for an ideal gas, it is very nearly true for many real gases at moderate pressures, where their behavior is very nearly that of an ideal gas. Measured values of C_p and C_v are given in Table 18–1 for some real gases at low pressures; the difference in most cases is approximately $8.31\ \mathrm{J\cdot mol^{-1}\cdot K^{-1}}$.

TABLE 18–1	MOLAR HEAT CAPACITIES OF GASES AT LOW PRESSURE				
Type of gas	**Gas**	$C_p,$ $\mathrm{J\cdot mol^{-1}\cdot K^{-1}}$	$C_v,$ $\mathrm{J\cdot mol^{-1}\cdot K^{-1}}$	$C_p - C_v$	$\gamma = \dfrac{C_p}{C_v}$
Monatomic	He	20.78	12.47	8.31	1.67
	A	20.78	12.47	8.31	1.67
Diatomic	H_2	28.74	20.42	8.32	1.41
	N_2	29.07	20.76	8.31	1.40
	O_2	29.41	21.10	8.31	1.40
	CO	29.16	20.85	8.31	1.40
Polyatomic	CO_2	36.94	28.46	8.48	1.30
	SO_2	40.37	31.39	8.98	1.29
	H_2S	34.60	25.95	8.65	1.33

In the last column of Table 18–1 are listed the values of the *ratio* C_p/C_v, denoted by the Greek letter γ (gamma).

$$\gamma = \frac{C_p}{C_v}. \qquad (18\text{--}16)$$

It is seen that γ is 1.67 for monatomic gases and is about 1.40 for diatomic gases. There is no simple regularity for polyatomic gases. Table 18–1 also shows that the molar heat capacity of a gas seems to be related to its molecular structure; all the *monatomic* gases have approximately equal values of C_v, and all the *diatomic* gases have about equal values. This is not an accident; in Chapter 20 we shall return to a more detailed theoretical study of heat capacities of gases.

Solids and liquids also expand on increase of temperature, if free to do so, and hence perform work. The coefficients of volume expansion of solids and liquids are, however, so much smaller than those of gases that the work is small. Usually for solids and liquids C_p is greater than C_v, but the difference is small and is not expressible as simply as that for a gas. Because of the large stresses set up when solids or liquids are heated and *not* allowed to expand, most heating processes involving them take place at constant pressure, and hence C_p (rather than C_v) is the quantity usually measured for a solid or liquid.

Finally, we emphasize once more that for an ideal gas the internal energy change in *any* process is given by Eq. (18–11),

$$\Delta U = nC_v \, \Delta T,$$

whether the volume is constant or not. This relation holds for other substances *only* when the volume is constant.

The above discussion may be represented graphically, using isotherms on a pV-diagram. Figure 18–6 shows two isotherms for an ideal gas, one representing possible states of the system at temperature T, the other at $T + \Delta T$. Since the internal energy of an ideal gas depends only on the temperature, it is constant if the temperature is constant and the isotherms are also curves of *constant internal energy.* The internal energy therefore has a constant value U at every point on the isotherm at temperature T, and a constant value $U + \Delta U$ at every point on the isotherm at $T + \Delta T$. It follows that the *change* in internal energy, ΔU, is the same in all processes in which the gas is taken from *any* point on one isotherm to *any* point on the other. Thus ΔU is the same for all the processes ab, ac, ad, and ef, in Fig. 18–6.

In particular, ab represents constant-volume heating from temperature T to $T + \Delta T$, and ad represents constant-pressure heating through the same temperature interval. No work is done in process ab; the work in process ad is the area under the curve.

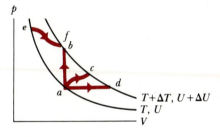

18–6 The change in internal energy of an ideal gas is the same in all processes between two given temperatures.

18–11

ADIABATIC PROCESS FOR AN IDEAL GAS

An adiabatic process, defined in Section 18–5, is a process in which there is no heat transfer between a system and its surroundings. A completely adiabatic process is an idealization, but a process may be approximately adiabatic if the system is well insulated or if the process takes place so rapidly that there is not time for appreciable heat flow to occur. The compression stroke in a gasoline or diesel engine is thus approximately adiabatic.

An adiabatic process for an ideal gas is shown on the pV-diagram of Fig. 18–7. As the gas expands from volume V_a to V_b its temperature drops because of the decrease of internal energy. If the point representing the initial state lies on an isotherm at temperature $T + \Delta T$, then the point for the final state is on a different isotherm at a lower temperature T. Thus an adiabatic curve at any point is always steeper than the iso-

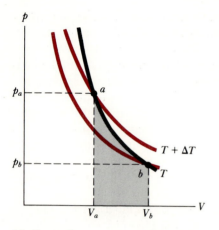

18–7 pV-diagram of an adiabatic process for an ideal gas. As the gas expands from V_a to V_b, its temperature drops from $T + \Delta T$ to T, corresponding to the decrease in internal energy due to the work done by the gas (indicated by shaded area). For an ideal gas, when an isotherm and an adiabatic pass through the same point, the adiabatic is always steeper.

therm passing through the same point. For an adiabatic *compression* from V_b to V_a, the situation is reversed and the temperature rises.

The air in the output pipe of an air compressor, as used in gasoline stations and paint-spraying equipment, is always warmer than that entering the compressor, a result of the approximately adiabatic compression. When air is compressed in the cylinders of a Diesel engine, during the compression stroke, it becomes so hot that when fuel is injected into the cylinders during the power stroke it ignites spontaneously.

It is useful to have quantitative relations between volume and temperature, or volume and pressure, for an adiabatic process. We consider first a *small* change of state in which the temperature changes by ΔT and the volume by ΔV. Equation (18–11) gives the internal energy change for *any* process for an ideal gas, adiabatic or not, so we have $\Delta U = nC_v \Delta T$. Also, the work done by the gas during the process is given by $W = p \Delta V$. (In general p is not constant, but for a small ΔV its variation is negligible.) Then, since $\Delta U = -W$ for an adiabatic process, we have

$$nC_v \Delta T = -p \Delta V. \qquad (18\text{–}17)$$

To obtain a relation containing only the variables T and V, we may eliminate p using the ideal-gas equation in the form $p = nRT/V$. Substituting this in Eq. (18–17) and rearranging, we obtain

$$\frac{\Delta T}{T} = -\frac{R}{C_v} \frac{\Delta V}{V},$$

or, since $R/C_v = (C_p - C_v)/C_v = C_p/C_v - 1 = \gamma - 1$,

$$\frac{\Delta T}{T} = -(\gamma - 1)\frac{\Delta V}{V}.$$

Transforming this into a relation involving T and V alone and not the increments ΔT and ΔV requires methods of the integral calculus. We quote the result without derivation:

$$TV^{\gamma-1} = \text{constant.} \qquad (18\text{–}18)$$

Thus for an initial state (T_1, V_1) and a final state (T_2, V_2),

$$T_1 V_1^{\gamma-1} = T_2 V_2^{\gamma-1}. \qquad (18\text{–}19)$$

In application of Eqs. (18–18) and (18–19) it is important to remember that the T's must be *absolute* temperatures (Kelvin or Rankine).

Equation (18–18) may also be converted into a relation between pressure and volume by eliminating T, using the ideal-gas equation in the form $T = pV/nR$. Since n and R are constant, we obtain

$$\left(\frac{pV}{nR}\right)V^{\gamma-1} = \text{constant,}$$

or, since n and R are constants,

$$pV^{\gamma} = \text{constant.} \qquad (18\text{–}20)$$

Thus for an initial state (p_1, V_1) and a final state (p_2, V_2),

$$p_1 V_1^{\gamma} = p_2 V_2^{\gamma}. \qquad (18\text{–}21)$$

EXAMPLE 1   The compression ratio of a certain diesel engine is 15. This means that air in the cylinders is compressed to 1/15 of its initial volume. If the initial pressure is 1.0×10^5 Pa and the initial temperature is 27°C, find the final pressure and temperature after compression. Air is mostly a mixture of oxygen and nitrogen, and $\gamma = 1.40$.

Solution   From Eq. (18–19),

$$T_2 = T_1\left(\frac{V_1}{V_2}\right)^{\gamma - 1} = (300 \text{ K})(15)^{0.40} = 886 \text{ K} = 613°C.$$

From Eq. (18–21),

$$p_2 = p_1\left(\frac{V_1}{V_2}\right)^{\gamma} = (1.0 \times 10^5 \text{ Pa})(15)^{1.40}$$

$$= 44.3 \times 10^5 \text{ Pa} \simeq 44 \text{ atm.}$$

If the compression had been isothermal, the final pressure would have been 15 atm, but since the temperature also increases, the final pressure is much greater. The high temperature attained during compression causes the fuel to ignite spontaneously without the need for spark plugs, when it is injected into the cylinders at the end of the compression stroke.   ◄

Because $W = -\Delta U$ for an adiabatic process, it is easy to calculate the *work* done by an ideal gas during an adiabatic process. If the initial and final temperatures are known, we have simply

$$W = nC_v(T_1 - T_2); \tag{18–22}$$

or, if the pressures or volumes are known, we may use $pV = nRT$ to obtain

$$W = \frac{C_v}{R}(p_1 V_1 - p_2 V_2) = \frac{1}{\gamma - 1}(p_1 V_1 - p_2 V_2). \tag{18–23}$$

We note that if the process is an expansion, the temperature drops, T_1 is greater than T_2, $p_1 V_1$ is greater than $p_2 V_2$, and the work is positive, as we should expect. If it is a compression, the work is negative.

EXAMPLE 2   In the above example, how much work does the gas do during the compression if the initial volume of the cylinder is $1.0 \text{ l} = 1.0 \times 10^{-3} \text{ m}^3$?

Solution   We may determine the number of moles and then use Eq. (18–22), or we may use Eq. (18–23). In the first case we have

$$n = \frac{p_1 V_1}{RT_1} = \frac{(1.0 \times 10^5 \text{ Pa})(1.0 \times 10^{-3} \text{ m}^3)}{(8.314 \text{ J} \cdot \text{mol}^{-1} \cdot \text{K}^{-1})(300 \text{ K})}$$

$$= 0.040 \text{ mol,}$$

$$W = (0.040 \text{ mol})(20.8 \text{ J} \cdot \text{mol} \cdot \text{K}^{-1})(300 \text{ K} - 886 \text{ K})$$

$$= -488 \text{ J.}$$

In the second,

$$W = \frac{1}{1.40 - 1}\Big[(1.0 \times 10^5 \text{ Pa})(1.0 \times 10^{-3} \text{ m}^3)$$

$$- (44.3 \times 10^5 \text{ Pa})\left(\frac{1.0}{15} \times 10^{-3} \text{ m}^3\right)\Big]$$

$$= -488 \text{ J}.$$

The work is negative because the gas is compressed.   ◄

In the above analysis we have used the ideal-gas equation of state, which holds only when the state of the gas changes slowly enough so that at each step the pressure and temperature are *uniform* throughout the gas. Thus the validity of our results is limited to situations where the process is rapid enough to prevent appreciable heat exchange with the surroundings, yet slow enough so the system does not depart very much from thermal and mechanical equilibrium.

QUESTIONS

18–1 It isn't correct to say that a body contains a certain amount of heat, yet a body can transfer heat to another body. How can a body give away something it doesn't have in the first place?

18–2 Discuss the application of the first law of thermodynamics to a mountaineer who eats food, gets warm and sweats a lot during the climb, and does a lot of mechanical work raising himself to the summit. What about the descent? One also gets warm during the descent. Is the source of this energy the same as during the ascent?

18–3 How can you cool a room (i.e., take heat out of it) by adding energy to it in the form of electric energy supplied to an air conditioner?

18–4 If you are told the initial and final states of a system and the associated change in internal energy, can you determine whether the internal energy change was due to work or to heat transfer?

18–5 Household refrigerators always have arrays or coils of tubing on the outside, usually at the back or bottom. When the refrigerator is running, the tubing becomes quite hot. Where does the heat come from?

18–6 There are a few materials that contract when heated, such as water between 0°C and 4°C. Would you expect C_p for such a material to be greater or less than C_v?

18–7 When ice melts (decreasing its volume), is the internal energy change greater or less than the heat added?

18–8 When one drives a car in cool, foggy weather, ice sometimes forms in the throat of the carburetor, even though the outside air temperature is above freezing. Why?

18–9 On a warm summer day a large cylinder of compressed gas (propane or butane) was used to supply several large gas burners at a cookout. After a while, frost formed on the outside of the tank. Why?

18–10 Air escaping from an air hose at a gas station always feels cold. Why?

18–11 The prevailing winds blow across the central valley of California and up the western slopes of the Sierra Nevada mountains. They cool as they reach the slopes, and the precipitation there is much greater than in the valley. But what makes them cool?

18–12 Applying the same considerations as in Question 18–11, can you explain why Death Valley, on the opposite side of the Sierra from the central valley, is so hot and dry?

18–13 In the situation of Question 18–11, during certain seasons the wind blows in the opposite direction. Although the mountains are cool, the wind in the valley (called the "Santa Ana," after the notorious Mexican general) is always very hot. What heats it? A similar phenomenon in the Alps is called the "Foehn"; by local legend it is blamed for irrational behavior in man and beast, and has even been used as a defense in murder trials.

18–14 When a gas expands adiabatically, it does work on its surroundings. But if there is no heat input to the gas, where does the energy come from?

PROBLEMS

18–1 A student performs a combustion experiment by burning a mixture of fuel and oxygen in a constant-volume "bomb" surrounded by a water bath. During the experiment the temperature of the water is observed to rise. Regard the mixture of fuel and oxygen as the system.

a) Has heat been transferred?

b) Has work been done?

c) What is the sign of ΔU?

18–2 A liquid is irregularly stirred in a well-insulated container and thereby undergoes a rise in temperature. Regard the liquid as the system.

a) Has heat been transferred?

b) Has work been done?

c) What is the sign of ΔU?

18–3 An electrical engineer studying the properties of a resistor immerses it in running water while it is carrying an electric current. Consider the resistor as the system under consideration.

a) Is there a flow of heat into the resistor?

b) Is there a flow of heat into the water?

c) Is work done?

d) Assuming the state of the resistor to remain unchanged, apply the first law to this process.

18–4 In a certain chemical process, a lab technician supplies 500 cal of heat to a system, and at the same time 100 J of work are done on the system. What is the increase in the internal energy of the system?

18–5 A piece of ice falls from a cliff into a lake at 0°C, and one half of one percent of the ice melts. Compute the minimum height from which the ice falls.

18–6 A marksman fires a lead bullet into a target. What must be the initial velocity of the bullet at a temperature of 25°C, so that the heat developed when it is brought to rest shall be just sufficient to melt it?

18–7 In a certain process, 200 Btu are supplied to a system and at the same time the system expands against a constant external pressure of 100 lb·in^{-2}. The internal energy of the system is the same at the beginning and at the end of the process. Find the increase in volume of the system.

18–8 An inventor claims to have developed an engine that takes in 100,000 Btu from its fuel supply, wastes 25,000 Btu in the exhaust, and delivers 25 kWh of mechanical work. Do you advise investing money to put this engine on the market?

18–9 A vessel with rigid walls and covered with styrofoam is divided into two parts by an insulating partition. One part contains a gas at temperature T and pressure p. The other part contains a gas at temperature T' and pressure p'. The partition is removed. What conclusion may be drawn by applying the first law of thermodynamics?

18–10 A combustion engineer places a mixture of hydrogen and oxygen in a rigid insulating container and explodes it with a spark. The temperature and the pressure both increase considerably. Neglecting the small amount of energy provided by the spark itself, what conclusion may be drawn by applying the first law of thermodynamics?

18–11 When water boils under a pressure of 2 atm, the heat of vaporization is 2.20×10^6 J·kg^{-1} and the boiling point is 120°C. At this pressure, one kilogram of water has a volume of 10^{-3} m^3, and one kilogram of steam a volume of 0.824 m^3.

a) Compute the work done when one kilogram of steam is formed at this temperature.

b) Compute the increase in internal energy.

18–12 A gas in a cylinder expands from a volume of 0.4 m^3 to 0.6 m^3. Heat is added just rapidly enough to keep the pressure constant at 2.0×10^5 Pa during the expansion. The total heat added is 1.2×10^5 J.

a) Find the work done by the gas.

b) Find the change in internal energy of the gas.

c) Does it matter whether or not the gas is ideal?

18–13 Nitrogen gas in an expandable container is raised from 0°C to 50°C, with the pressure held constant at 4.0×10^5 Pa. The total heat added is 3.0×10^4 J.

a) Find the number of moles of gas.

b) Find the change in internal energy of the gas.

c) Find the work done by the gas.

d) How much heat would be needed to cause the same temperature change if the volume were constant?

18–14 A gas in a cylinder is cooled and compressed at a constant pressure of 2.0×10^5 Pa, from 1.2 m^3 to 0.8 m^3. A quantity of heat of magnitude 2.4×10^5 is removed from the gas.

a) Find the work done by the gas.

b) Find the change in internal energy of the gas.

c) Does it matter whether the gas is or is not ideal?

18–15 An ideal gas expands slowly to twice its original volume, doing 500 J of work in the process. Find the heat added and the change in internal energy if the process is

a) isothermal, and

b) adiabatic.

18–16 When a system is taken from state a to state b, in Fig. 18–8, along the path acb, 80 J of heat flow into the system, and 30 J of work are done.

a) How much heat flows into the system along path adb if the work is 10 J?

b) When the system is returned from b to a along the curved path, the work is 20 J. Does the system absorb or liberate heat, and how much?

c) If $U_a = 0$ and $U_d = 40$ J, find the heat absorbed in the processes ad and db.

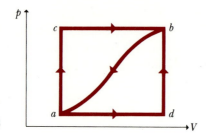

Figure 18–8

18–17 A chemical engineer studying the properties of glycerin uses a steel cylinder of cross-sectional area 0.01 m^2 that contains 1×10^{-2} m^3 of glycerin. The cylinder is equipped with a tightly fitting piston that supports a load of 3×10^4 N. He increases the temperature of the system from 20°C to 70°C. Neglect the expansion of the steel cylinder. Find

a) the increase in volume of the glycerin,

b) the mechanical work of the 3×10^4 N force,

c) the amount of heat added to the glycerin [specific heat of glycerin = 0.58 cal·g^{-1}·(C°)$^{-1}$], and

d) the change in internal energy of the glycerin.

18–18

a) The change in internal energy, ΔU, of a system consisting of n moles of a pure substance, in an infinitesimal process at constant volume, is equal to $nC_v \Delta T$. Explain why the internal energy change in a process at constant pressure is *not* equal to $nC_p \Delta T$.

b) Explain why the change in internal energy of an ideal gas, in *any* infinitesimal process, is given by $nC_v \Delta T$.

18–19 A cylinder contains 1 mole of oxygen gas at a temperature of 27°C. The cylinder is provided with a frictionless piston, which maintains a constant pressure of 1 atm on the gas. The gas is heated until its temperature increases to 127°C.

a) Draw a diagram representing the process in the pV-plane.

b) How much work is done by the gas in this process?

c) On what is this work done?

d) What is the change in internal energy of the gas?

e) How much heat was supplied to the gas?

f) How much work would have been done if the pressure had been 0.5 atm?

18–20 A cylinder with a piston contains 0.5 mol of oxygen at 2×10^5 Pa and 300 K. The gas first expands at constant pressure to twice its original volume; it is then compressed isothermally back to its original volume, and finally it is cooled at constant volume to its original pressure.

a) Show the series of processes on a pV-diagram.

b) Compute the temperature during the isothermal compression.

c) Compute the maximum pressure.

18–21 Use the conditions and processes of Problem 18–20 to compute

a) the work done, the heat added, and the internal energy change during the initial expansion;

b) the work done, the heat added, and the internal energy change during the final cooling;

c) the internal energy change during the isothermal compression.

18–22

a) Compare the quantity of heat required to raise the temperature of 1 g of hydrogen through 1 C°, at constant pressure, with that required to raise the temperature of 1 g of water by the same amount.

b) Of the substances listed in Table 18–1, which has the largest *specific* heat capacity?

18–23 Ten liters of air at atmospheric pressure are compressed isothermally to a volume of 2 l and are then allowed to expand adiabatically to a volume of 10 l. Show the process in a pV-diagram.

18–24 An ideal gas is contained in a cylinder closed with a movable piston. The initial pressure is 1 atm and the initial volume is 1 l. The gas is heated at constant pressure until the volume is doubled, then heated at constant volume until the pressure is doubled, and finally expanded adiabatically until the temperature drops to its initial value. Show the process in a pV-diagram.

18–25 An ideal gas initially at 10 atm and 300 K is permitted to expand adiabatically until its volume doubles. Find the final pressure and temperature if the gas is

a) monatomic,

b) diatomic.

18–26 A gasoline engine takes in air at 20°C and 1 atm, and compresses it adiabatically to $\frac{1}{10}$ the original volume. Find the final temperature and pressure.

18–27 During an adiabatic expansion, the temperature of 0.1 mol of oxygen drops from 30°C to 10°C. How much work does the gas do? How much heat is added?

18–28 A certain ideal gas has $\gamma = 1.33$. Determine the molar heat capacities at constant volume and at constant pressure.

18–29 An engineer is designing an engine that runs on compressed air. Air enters the engine at a pressure of 2×10^6 Pa and leaves at a pressure of 2×10^5 Pa. What must be the temperature of the compressed air in order that there may be no possibility of frost forming in the exhaust ports of the engine? Assume the expansion to be adiabatic. (*Note.* Frost frequently forms in the exhaust ports of an air-driven engine. This happens when the moist air is cooled below 0°C by the expansion that takes place in the engine.)

18–30 Initially at a temperature of 140°F, 10 ft^3 of air expands at a constant gauge pressure of 20 lb·in^{-2} to a volume of 50 ft^3, and then expands further adiabatically to a final volume of 80 ft^3 and a final gauge pressure of 3 lb·in^{-2}. Sketch the process in a pV-diagram and compute the work done by the air.

18–31 A pump compressing air from atmospheric pressure into a very large tank at 60 lb·in^{-2} gauge pressure has a cylinder 10 in. long.

a) At what position in the stroke will air begin to enter the tank? Assume the compression to be adiabatic.

b) If the air is taken into the pump at 27°C, what is the temperature of the compressed air?

18–32 Two moles of helium are initially at a temperature of 27°C and occupy a volume of 20 l. The helium is first expanded at constant pressure until the volume has doubled, and then adiabatically until the temperature returns to its initial value.

a) Draw a diagram of the process in the pV-plane.

b) What is the total heat supplied in the process?

c) What is the total change in internal energy of the helium?

d) What is the total work done by the helium?

e) What is the final volume?

19

THE SECOND LAW OF THERMODYNAMICS

Thus far our discussion of thermodynamics has centered on the first law, expressing conservation of energy in thermodynamic processes. But there is another whole family of questions that the first law cannot answer, having to do with the *directions* of thermodynamic processes. A study of the inherently one-way nature of such processes as heat flow from hot to cold and the conversion of work into heat leads to the *second law of thermodynamics*. This law places limits on the efficiency an engine can have, and on the performance of a refrigerator. We can also use the second law to define a temperature scale that is independent of the properties of any specific material. Finally, we re-state the second law in terms of the concept of *entropy*, a quantitative measure of the degree of disorder or randomness of a system.

19–1

DIRECTIONS OF THERMODYNAMIC PROCESSES

Here are some examples of directions of thermodynamic processes. Heat always flows from a hot body to a cooler body, never the reverse. If it were to flow from cooler to hotter, the first law would still be obeyed, but that does not occur in nature. Suppose all the air in a box were to rush to one side, leaving vacuum in the other side, the reverse of the free expansion described in Section 18–3. This does not happen, either, although the first law does not forbid it. It is easy to convert mechanical energy completely into heat; we do it every time we use a car's brakes to stop it. It is not so easy to convert heat into mechanical energy, and no one has ever built a machine that converts heat *completely* into mechanical energy. The world is full of would-be inventors who would like to cool the air a little, extracting heat from it, and convert that heat to mechanical energy to propel an airplane. No one has ever succeeded with such a scheme.

What these examples have in common is a preferred *direction*. In each case a process proceeds spontaneously in one direction but not in the other. This suggests that there must be some physical law that determines what the preferred direction is for a given process. As we shall see in the following sections, that law is the *second law of thermodynamics*.

Closely related to the concept of directionality is another concept, *reversibility*. A process is said to be *reversible* if it involves a system that is always so close to being in thermodynamic equilibrium within itself and with its surroundings that any change of state that takes place can be reversed (i.e., made to go the other way) by making only an infinitesimal change in the conditions of the system. For example, heat flow between two bodies whose temperatures differ only infinitesimally can be reversed by making only a very small change in one temperature or the other. A gas expanding slowly and adiabatically can be compressed slowly by an infinitesimal increase in pressure.

Reversible processes are thus equilibrium processes. By contrast, the processes cited above, such as heat flow with finite temperature difference, free expansion of a gas, and conversion of work to heat by friction, are all irreversible processes; no small change in conditions could make any of them go the other way. They are also all non-equilibrium processes.

Finally, we will find that there is a relation between the direction of a process and the *disorder* or *randomness* of the resulting state. Imagine a tedious sorting job, such as alphabetizing a thousand book titles written on file cards. Throw the alphabetized stack of cards into the air; do they come down in alphabetical order? No; the tendency is for them to come down in a random or disordered state. In the free expansion example, the air is more disordered after it has expanded into the entire box than when it was confined in one side, because the molecules are scattered over more space.

Similarly, macroscopic kinetic energy is energy associated with organized, coordinated motions of many molecules. Energy associated with heat is random, disordered molecular motion, as we shall see in Chapter 20. Hence, conversion of mechanical energy into heat involves an increase of randomness or disorder.

In the following section we shall explore the essence of the second law of thermodynamics by considering two specific classes of devices, namely *heat engines*, which are partly successful in converting heat into work, and *refrigerators*, which are partly successful in transporting heat from cooler to hotter bodies.

19–2

HEAT ENGINES

The essence of a technological society is its ability to use sources of energy other than muscle power. Sometimes, as with water power, mechanical energy is directly available. Most of our energy, however, comes from the burning of fossil fuels (coal, oil, and gas) and from nuclear reactions; both of these supply energy that is transferred as *heat*. Some heat can be used directly for heating buildings, for cooking, and for chemical and metallurgical processing, but to operate a machine or propel a vehicle we need *mechanical* energy.

Thus a problem of the utmost practical importance is that of taking heat from a source and converting as much of it as possible into mechanical energy or work. This is exactly what happens in gasoline engines in automobiles, in jet engines in airplanes, in steam turbines in electric power plants, and in many other places. Closely related processes occur in the animal kingdom, where food energy is "burned" (i.e., carbohydrates combine with oxygen to yield water, carbon dioxide, and energy) and partly converted to mechanical energy as the animal's muscles do work on their surroundings.

Any device for transforming heat into work or mechanical energy is called a *heat engine*. Typically a certain quantity of matter in the engine undergoes addition and subtraction of heat, expansion and compression, and sometimes change of phase. This matter is called the *working substance* of the engine. In internal-combustion engines it is a mixture of air and burnt fuel; in a steam turbine it is water.

For the sake of simplicity, we will discuss an engine in which the working substance undergoes a *cyclic* process, that is, a sequence of processes that eventually leaves the substance in the same state in which it started. In a steam turbine the water actually is recycled and used over and over. Internal-combustion engines do not use the same air over and over, but they can still be analyzed in terms of cyclic processes that approximate the actual operation.

All the heat engines mentioned absorb heat from a source at a relatively high temperature, perform some mechanical work, and discard some heat at a lower temperature. As far as the engine is concerned, the discarded heat is wasted. In internal-combustion engines it is in the hot exhaust gases; in a steam turbine it is heat that must be taken from the used steam to condense and recycle it.

When a system is carried through a cyclic process, its initial and final internal energies are equal; from the first law, for any number of complete cycles,

$$U_2 - U_1 = 0 = Q - W,$$
$$Q = W.$$

That is, the net heat flowing into the engine in a cyclic process equals the net work done by the engine.

In the analysis of heat engines it is useful to think of two bodies with which the working substance of the engine can interact. One of these, called the *hot reservoir*, can give the working substance large amounts of heat without appreciably changing its temperature. The other, called the *cold reservoir*, can absorb large amounts of discarded heat from the engine. Thus in a steam-turbine system the flames and hot gases in the boiler are the hot reservoir, and the cold water or air used to condense and cool the used steam are the cold reservoir.

Heat transferred from the hot and cold reservoirs will be denoted by Q_H and Q_C, respectively. Each Q is considered positive when heat is transferred *from* a reservoir *to* the working substance, negative when the reverse. Thus in a heat engine Q_H is positive but Q_C is negative, representing heat leaving the working substance.

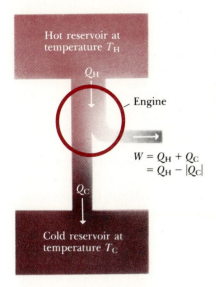

Hot reservoir at
temperature T_H

Q_H

Engine

$W = Q_H + Q_C$
$= Q_H - |Q_C|$

Q_C

Cold reservoir at
temperature T_C

19–1 Schematic flow diagram of a heat engine.

The energy transformations in a heat engine are conveniently represented schematically by the *flow diagram* of Fig. 19–1. The engine itself is represented by the circle. The heat Q_H supplied to the engine by the hot reservoir is proportional to the cross section of the incoming "pipeline" at the top of the diagram. The cross section of the outgoing pipeline at the bottom is proportional to the magnitude $|Q_C|$ of the heat that is rejected as heat in the exhaust. The branch line to the right represents that portion of the heat supplied that the engine converts to mechanical work, W.

As mentioned above, we shall be concerned in this discussion only with engines that repeat the same cycle over and over. In this case Q_H and Q_C will represent the quantities of heat absorbed and rejected by the engine *during one cycle*. The *net* heat absorbed per cycle is

$$Q = Q_H + Q_C = Q_H - |Q_C|, \qquad (19\text{–}1)$$

where Q_C is a negative number. The useful output of the engine is the net work W done by the working substance, and from the first law,

$$W = Q = Q_H + Q_C. \qquad (19\text{–}2)$$

Ideally we would like to convert *all* the heat Q_H into work; in that case we would have $Q_H = W$ and $Q_C = 0$. Experience shows that this is impossible; there is always some heat wasted, and Q_C is never zero. We define the *thermal efficiency* of an engine, denoted by e, as the fraction of Q_H that *is* converted to work; that is, as the quotient W/Q_H:

$$e = \frac{W}{Q_H}. \qquad (19\text{–}3)$$

Using Eq. (19–2) and keeping in mind the signs of Q_H and Q_C, we can write the following equivalent expressions:

$$e = \frac{W}{Q_H} = \frac{Q_H + Q_C}{Q_H} = 1 + \frac{Q_C}{Q_H} = 1 - \left| \frac{Q_C}{Q_H} \right|. \qquad (19\text{–}4)$$

EXAMPLE A gasoline engine takes in 2500 J of heat and delivers 500 J of mechanical work per cycle. The heat is obtained by burning gasoline with a heat of combustion of $5.0 \times 10^4 \text{ J} \cdot \text{g}^{-1}$.

a) What is the thermal efficiency?

b) How much heat is discarded in each cycle?

c) How much gasoline is burned in each cycle?

d) If the engine goes through 100 cycles per second, what is its power output in watts? In horsepower?

e) How much gasoline is burned per second? per hour?

Solution We have $Q_H = 2500$ J and $W = 500$ J.

a) From Eq. (19–3), the thermal efficiency is

$$e = \frac{W}{Q_H} = \frac{500 \text{ J}}{2500 \text{ J}} = 0.20 = 20\%.$$

b) From Eq. (19–2),

$$500 \text{ J} = 2500 \text{ J} + Q_C, \qquad Q_C = -2000 \text{ J}.$$

That is, 2000 J of heat leave the engine during each cycle.

c) If m is the mass of gasoline burned, then

$$2500 \text{ J} = m \ (5.0 \times 10^4 \text{ J} \cdot \text{g}^{-1}),$$

$$m = 0.05 \text{ g}.$$

d) The rate of doing work P is the work per cycle multiplied by the number of cycles per second:

$$P = (500 \text{ J})(100 \text{ s}^{-1}) = 50,000 \text{ W} = 50 \text{ kW}$$
$$= (50,000 \text{ W})(1 \text{ hp}/746 \text{ W}) = 67 \text{ hp}.$$

e) The mass of gasoline burned per second is the mass per cycle multiplied by the number of cycles per second:

$$(0.05 \text{ g})(100 \text{ s}^{-1}) = 5 \text{ g} \cdot \text{s}^{-1}.$$

The mass burned per hour is

$$(5 \text{ g} \cdot \text{s}^{-1})(3600 \text{ s}/1 \text{ hr}) = 18,000 \text{ g} \cdot \text{hr}^{-1} = 18 \text{ kg} \cdot \text{hr}^{-1}.$$

The density of gasoline is about $0.70 \text{ g} \cdot \text{cm}^{-3}$, so this is about $25,700 \text{ cm}^3$, 25.7 l, or 6.8 gallons of gasoline per hour.   ◀

19–3

INTERNAL-COMBUSTION ENGINES

As an example of a heat engine and a calculation of thermal efficiency, let us consider the common gasoline engine, as found in automobiles and many other types of machinery. The sequence of processes in the cycle is shown in Fig. 19–2. First a mixture of air and gasoline vapor flows into a cylinder through an open intake valve while the piston descends, increasing the volume of the cylinder from a minimum of V (when the piston is all the way up) to a maximum of RV (when it is all the way down). The ratio R is called the *compression ratio*, and for present-day automobile engines it is typically about 8. At the end of this *intake stroke*

19–2 Cycle of a four-stroke internal-combustion engine. (a) Intake stroke: piston moves down, causing a partial vacuum in cylinder; gasoline and air are mixed in carburetor and flow through open intake valve into cylinder. (b) Compression stroke: intake valve closes and mixture is compressed as piston moves up. (c) Ignition: spark plug ignites mixture. (d) Power stroke: hot burned mixture pushes piston down, doing work. (e) Exhaust stroke: exhaust valve opens and piston moves up, pushing burned mixture out of cylinder. Engine is now ready for next intake stroke, and the cycle repeats.

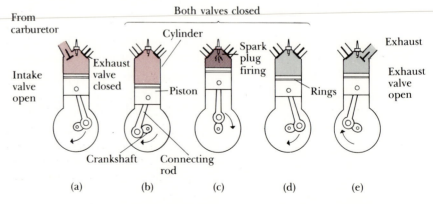

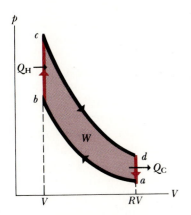

19–3 *pV*-diagram of the Otto cycle, an idealized model of the thermodynamic processes in a gasoline engine.

the intake valve closes and the mixture is compressed to volume V during the *compression stroke*. The mixture is then ignited by the spark plug, and the heated gas expands back to volume RV, pushing on the piston and doing work; this is the *power stroke*. Finally, the exhaust valve opens and the combustion products are pushed out (during the *exhaust stroke*) to prepare the cylinder for the next intake stroke.

Figure 19–3 is a pV-diagram showing an idealized model of the corresponding thermodynamic processes. At a the gasoline–air mixture has entered the cylinder. It is compressed adiabatically along line ab and is then ignited. Heat Q_H is added to the system by the burning gasoline, and the power stroke is the adiabatic expansion cd. The gas is cooled to the temperature of the outside air (da); during this process heat Q_C is exchanged. In practice, of course, this same air does not enter the engine again but, since an equivalent amount does enter, we may consider the process to be cyclic.

It is not difficult to calculate the efficiency of this idealized cycle. Processes bc and da are constant-volume, so the heats Q_H and Q_C are related simply to the temperatures:

$$Q_H = nC_v(T_c - T_b),$$
$$Q_C = nC_v(T_a - T_d).$$

The thermal efficiency is given by Eq. (19–4); inserting the above expressions and cancelling out the common factors nC_v, we obtain

$$e = \frac{T_c - T_b + T_a - T_d}{T_c - T_b}. \tag{19–5}$$

This may be further simplified by using the temperature–volume relation for adiabatic processes, Eq. (18–19). For the two adiabatic processes ab and cd, we find

$$T_a(RV)^{\gamma-1} = T_bV^{\gamma-1},$$
$$T_d(RV)^{\gamma-1} = T_cV^{\gamma-1}.$$

We divide out the common factor $V^{\gamma-1}$ and substitute the resulting expressions for T_b and T_c back into Eq. (19–5). The result is

$$e = \frac{T_dR^{\gamma-1} - T_aR^{\gamma-1} + T_a - T_d}{T_dR^{\gamma-1} - T_aR^{\gamma-1}} = \frac{(T_d - T_a)(R^{\gamma-1} - 1)}{(T_d - T_a)R^{\gamma-1}}.$$

Finally, we divide out the common factor $(T_d - T_a)$, obtaining the simple result

$$e = 1 - \frac{1}{R^{\gamma-1}}. \tag{19–6}$$

The thermal efficiency given by Eq. (19–6) is always less than unity, even for this idealized model. Using $R = 8$ and $\gamma = 1.4$ (the value for air), we find $e = 0.56$, or 56 percent. The efficiency can be increased by increasing R. However, this also increases the temperature at the end of the adiabatic compression of the gas–air mixture. If it is too high, the mixture explodes spontaneously and prematurely, instead of burning evenly after ignition by the spark plug. The mechanical strength and wear of engine parts also pose problems; so the maximum practical compression ratio for ordinary gasoline is about 10. Higher ratios can be used with more exotic fuels.

The cycle just described, called the *Otto cycle*, is of course a highly idealized model. It assumes that the mixture behaves as an ideal gas; it neglects friction, turbulence, loss of heat to cylinder walls, and many other effects that combine to reduce the efficiency of a real engine. Another source of inefficiency is incomplete combustion. A mixture of gasoline vapor with just enough air for complete combustion of the hydrocarbons to H_2O and CO_2 does not ignite readily. Reliable ignition requires a mixture "richer" in gasoline, but this leads to CO and unburned hydrocarbons in the exhaust. The heat obtained from the gasoline is then less than the total heat of combustion; the difference is wasted, and the exhaust contributes to air pollution. One attack on this problem is the stratified-charge engine, in which the concentration of gasoline vapor near the spark plug is greater than in the remainder of the combustion chamber. Efficiencies of real gasoline engines are typically around 20 percent.

The operation of the diesel engine is similar to that of the gasoline engine; the principal difference is that there is no fuel in the cylinder during compression. At the beginning of the power stroke fuel is injected into the cylinder just rapidly enough to keep the pressure approximately constant during the first part of the power stroke. The fuel ignites spontaneously because of the high temperature developed during the adiabatic compression; no spark plugs are needed.

The idealized Diesel cycle is shown in Fig. 19–4. Starting at point a, air is compressed adiabatically to point b, heated at constant pressure to point c, expanded adiabatically to point d, and cooled at constant volume to point a.

Since there is no fuel in the cylinder of a diesel engine during the compression stroke, pre-ignition cannot occur, and the compression ratio R may be much higher than for a gasoline engine. Values of 15 to 20 are typical; with these values and $\gamma = 1.4$, the efficiency of the idealized Diesel cycle is about 0.65 to 0.70. As with the Otto cycle, the efficiency of any actual engine is substantially less than this.

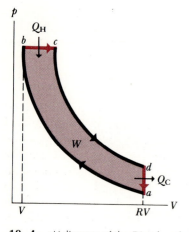

19–4 *pV*-diagram of the Diesel cycle.

19–4

THE REFRIGERATOR

A refrigerator may be considered as a heat engine operating in reverse. A heat engine takes heat from a hot place and gives off heat to a colder place. A refrigerator does the opposite; it takes heat from a cold place (the inside of the refrigerator, containing the food or other material to be cooled) and gives it off to a warmer place (usually the air in the room where the refrigerator is located). A heat engine has a net *output* of mechanical work; the refrigerator requires a net *input* of mechanical work. Thus with the symbols used in Section 19–1, Q_C is positive for a refrigerator, but both W and Q_H are negative.

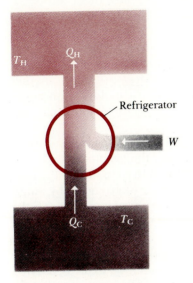

A flow diagram for a refrigerator is shown in Fig. 19–5. From the first law for a cyclic process,

$$Q_H + Q_C - W = 0, \quad \text{or} \quad -Q_H = Q_C - W,$$

or, since both Q_H and W are negative,

$$|Q_H| = Q_C + |W|. \tag{19–7}$$

Thus, as the diagram shows, the heat Q_H leaving the working substance of the engine and given to the hot reservoir is always greater than the heat Q_C taken from the cold reservoir.

From an economic point of view, the best refrigeration cycle is one that removes the greatest amount of heat Q_C from the refrigerator, for the least expenditure of mechanical work W. We therefore define the *performance coefficient* (rather than the efficiency) of a refrigerator as the ratio $K = -Q_C/W$. Since $W = Q_H + Q_C$,

$$\text{Performance coefficient} = K = -\frac{Q_C}{W} = -\frac{Q_C}{Q_H + Q_C}. \tag{19–8}$$

19–5 Schematic flow diagram of a refrigerator.

As always, we assume that Q_H, Q_C, and W are all measured in the same energy units; K is then a dimensionless number.

The principles of the common refrigeration cycle are illustrated schematically in Fig. 19–6. The fluid "circuit" contains a refrigerant fluid (the working substance), typically CCl_2F_2 or another member of the "Freon" family. The left side is at low temperature and low pressure, the right at high temperature and high pressure; ordinarily both sides contain liquid and vapor in phase equilibrium. The compressor takes in fluid, compresses it adiabatically, and delivers it to the condenser coil at high pressure. The temperature is higher than that of the air surrounding the condenser, so the refrigerant gives off heat (Q_H) and partially

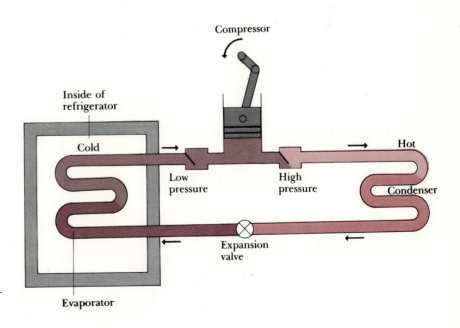

19–6 Principle of the mechanical refrigeration cycle.

condenses to liquid. When the expansion valve is opened, fluid is permitted to expand adiabatically into the evaporator, cooling considerably as it
does so, so that the fluid in the evaporator coils is colder than its surroundings. It absorbs heat (Q_C) from its surroundings, cooling them and
partially vaporizing. It then enters the compressor to begin another
cycle. The compressor, usually driven by an electric motor, requires en-
ergy input and does work $|W|$ on the working substance during each
cycle.

An air conditioner operates on exactly the same principle. The refrigerator box is a room or an entire building; the evaporator coils are
inside, the condenser outside, and fans are used to circulate air through
these. In large installations the condenser coils are often cooled by water.

For air conditioners the figure of greatest practical importance is the
rate of heat removal (i.e., the heat current H) from the region being
cooled. If heat Q_C is removed in time t, then $H = Q_C/t$. Similarly, the
work W is usually expressed in terms of the *power* input $P = W/t$ to the
compressor. The performance coefficient can then be expressed as

$$K = \frac{Q_C}{W} = \frac{Ht}{Pt} = \frac{H}{P}.$$

Typical room air conditioners used in homes have heat removal rates H
of 5000 to 10,000 Btu·h^{-1}, or about 1500 to 3000 W, and require electric power input of about 600 to 1500 W. Typical performance coefficients are roughly 2 to 3, with somewhat larger values for larger-capacity
units.

Unfortunately, K is usually expressed commercially in mixed units,
with H in Btu per hour and P in watts. In these units H/P is called the
energy efficiency rating (EER); for room air conditioners it typically has a
value of 7 to 10. The units, customarily omitted, are Btu·h^{-1}·W^{-1}.

A variation on this theme is the *heat pump*, used to heat buildings by
cooling the outside air. It is like a refrigerator turned inside out; the
evaporator coils are outside, where they take heat from cold air, and the
condenser coils are inside, where they give off heat to the warmer air.
With proper design the heat Q_H delivered to the inside per cycle can be
considerably greater than the work W required to get it there. Hence for
the heat pump the appropriate definition of the performance coefficient, which we shall call K', is

$$K' = \frac{|Q_H|}{|W|} = \frac{Q_H}{W}. \qquad (19\text{--}9)$$

We challenge the reader to show that this is related to Eq. (19–8) by

$$K' = K + 1. \qquad (19\text{--}10)$$

The difference is that with the heat pump the work W is delivered to the
hot side as additional heat and therefore contributes to the goal of heating the building.

If *no* work were needed to operate a refrigerator, the performance
coefficient (heat extracted divided by work done) would be infinite. Performance coefficients of actual refrigerators are typically in the range
from 2 to 6. Experience shows that work is *always* needed to transfer heat

from a colder to a hotter body. Heat flows spontaneously from a hotter to a colder body, and to reverse this flow requires the addition of work from the outside. No one has ever succeeded in making a refrigerator that transports heat from a cold body to a hotter body without the addition of work. Such a mythical device is called a *workless refrigerator*, and experience shows that it is impossible to build such a device.

19–5

THE SECOND LAW OF THERMODYNAMICS

None of the engines described above has a thermal efficiency of 100%. That is, none of them absorbs heat and converts it completely into work. If there were such an engine, it would be consistent with the first law of thermodynamics, provided the work output equals the heat input. However, no one has ever succeeded in making such an engine. Indeed, a very large body of experimental evidence suggests forcefully that it is impossible to do so; there is no experimental evidence to the contrary.

The impossibility of converting heat completely into work forms the basis of one form of the *second law of thermodynamics*, as follows:

> *It is impossible for any system to undergo a process in which it absorbs heat from a reservoir at a single temperature and converts it completely into mechanical work, while ending in the same state in which it began.*

That is, it is impossible in principle for any heat engine to have a thermal efficiency of 100%.

The basis of the second law of thermodynamics lies in the difference between the nature of internal energy and that of macroscopic mechanical energy. The former is the energy of *random* molecular motion, while the latter represents *ordered* or *organized molecular* motion. In a moving body, the molecules have random motion, but superimposed on this is an organized, coordinated motion in the direction of the velocity of the body. The kinetic energy associated with this organized, coordinated motion is what we call the kinetic energy of the moving body. The kinetic and potential energies associated with the random motion constitute the internal energy.

When a moving body makes an inelastic collision or comes to rest as a result of friction, the organized part of the motion becomes converted to random motion. Since we cannot control the motions of individual molecules, it is impossible to reconvert this random motion completely to organized motion. We can, however, convert a part of it, and this is what a heat engine does.

If the second law were *not* true, it would be possible to power an automobile by extracting heat from the air, or to run a power plant by extracting heat from the surrounding air. Neither of these "impossibilities" violates the *first* law of thermodynamics. The second law, therefore, is not a deduction from the first but stands by itself as a separate law of nature, referring to an aspect of nature different from that considered in the first law. The first law denies the possibility of creating or destroying energy; the second limits the ways energy may be used and converted.

The analysis of refrigerators forms the basis for an alternative statement of the second law of thermodynamics. Heat flows spontaneously from hotter to colder bodies, never the reverse. A refrigerator does take heat from a colder to a hotter body, but its operation depends on input of mechanical energy or work. Generalizing this observation, we state:

It is impossible for any process to have as its sole result the transfer of heat from a cooler to a hotter body.

This statement may seem not to be very closely related to the previous statement about the impossibility of converting heat completely into mechanical work. In fact, though, the two statements are completely equivalent. For example, if we could build a workless refrigerator, violating the second or "refrigerator" statement of the second law, we could use it in conjunction with a heat engine, pumping the heat rejected by the engine back to the hot reservoir to be reused, thereby obtaining a composite machine (Fig. 19–7a) that violates the first or "engine" statement of the second law.

Or if we could make an engine with 100% thermal efficiency, in violation of the first statement, we could run it using heat from the hot reservoir, and use the work output to drive a refrigerator (Fig. 19–7b) that pumps heat from the cold reservoir to the hot. This would violate the "refrigerator" statement. Thus any device that violates one form of the second law can also be used to violate the other form; we conclude that if violations of the first form are impossible, so are violations of the second.

Heat flow across a finite temperature gradient is an *irreversible* process, in the sense that no minor change in conditions can make it proceed in the reverse direction. Similarly, conversion of mechanical energy to heat, as in friction or turbulent fluid flow, is an irreversible process. The second law of thermodynamics is an expression of the inherent one-way aspect of these processes and the resulting limitations on heat engines and refrigerators.

All natural, spontaneous processes may be studied in the light of the second law, and in all such cases, this one-sidedness is found. Heat always flows spontaneously from a hotter to a colder body; gases always seep

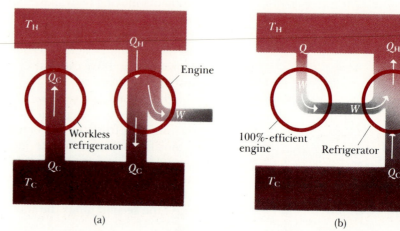

19–7   (a) A workless refrigerator (left), if it existed, could be used in combination with an ordinary heat engine (right) to form a composite device that functions as an engine with 100-percent efficiency, converting heat Q_H completely to work. (b) An engine with 100-percent efficiency (left), if it existed, could be used in combination with an ordinary refrigerator (right) to form a workless refrigerator, transferring heat Q_C from the cold reservoir to the hot with no net input of work. Thus if either of these is impossible, the other must be also.

(a)   (b)

through an opening spontaneously from a region of high pressure to a region of low pressure; gases and liquids left by themselves always tend to mix, not to unmix. Salt dissolves in water but a salt solution does not separate by itself into pure salt and pure water. Rocks weather and crumble; iron rusts; people grow old. These are all examples of *irreversible* processes that take place naturally in only one direction and, by their one-sidedness, express the second law of thermodynamics. An irreversible process is always a *non-equilibrium* process. Irreversible heat flow accompanies departures from thermal equilibrium, a free expansion of a gas involves states that are not in mechanical equilibrium, and so on. In each case the process tends to move the system toward an equilibrium state.

19–6

THE CARNOT CYCLE

According to the second law, no heat engine can have 100% efficiency. But what is the *maximum* possible efficiency of an engine, given two heat reservoirs at temperatures T_H and T_C? This question was answered in 1824 by the French engineer Sadi Carnot, who developed a hypothetical, idealized heat engine that has the maximum possible efficiency consistent with the second law. His engine is called the *Carnot engine*, and its cycle is called the *Carnot cycle*.

To understand the rationale of the Carnot cycle we return to the matter of reversibility, discussed at the end of Section 19–4. *Heat flow* through a finite temperature drop and *conversion of work* into heat are irreversible processes. In maximizing the efficiency of a heat engine (which converts work into heat), we should therefore *avoid* all irreversible processes. During heat transfer there must be no finite temperature difference; when the engine takes heat from the hot reservoir at T_H, the engine itself must be at T_H; otherwise irreversible heat flow would occur. Similarly, when heat is rejected to the cold reservoir at T_C, the engine itself must be at T_C. That is, every process involving heat transfer must be *isothermal*. Conversely, there must be *no* heat transfer in any process where the temperature of the engine changes, for it could not be reversible. In short, every process in our idealized cycle must be either isothermal or adiabatic. In addition, not only thermal but mechanical equilibrium must be maintained at all times, so that each process is completely reversible.

The Carnot cycle consists of two isothermal and two adiabatic processes. A Carnot cycle using an ideal gas as the working substance is shown on a pV-diagram in Fig. 19–8. It consists of the following steps:

1. The gas expands isothermally at temperature T_H, absorbing heat Q_H (*ab*).

2. It expands adiabatically until its temperature drops to T_C (*bc*).

3. It is compressed isothermally at T_C, rejecting heat Q_C (*cd*).

4. It is compressed adiabatically back to its initial state (*da*).

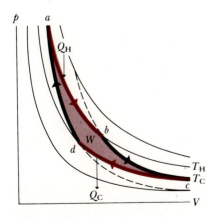

19–8 The Carnot cycle. Color lines are isothermals; black lines are adiabatics.

The heat and work in each of these steps can be calculated in terms of the volumes and temperatures, and from these the thermal efficiency of the engine can be obtained. Surprisingly, it turns out to depend only on the ratio T_C/T_H of the absolute temperatures of the heat reservoirs. Specifically, it is found that

$$\frac{Q_C}{Q_H} = -\frac{T_C}{T_H}, \tag{19-11}$$

and

$$e = 1 - \frac{T_C}{T_H}. \tag{19-12}$$

The efficiency is large when the temperature difference of the two heat reservoirs is large, and it is very small when the temperatures are nearly equal. The efficiency can never be exactly unity, unless $T_C = 0$, and as we shall see this too is impossible.

EXAMPLE 1   A Carnot engine takes 2000 J of heat from a reservoir at 500 K, does some work, and discards some heat to a reservoir at 350 K. How much work is done, how much heat is discarded, and what is the efficiency?

Solution   From Eq. (19-11),

$$Q_C = -Q_H\frac{T_C}{T_H} = -(2000 \text{ J})\frac{350 \text{ K}}{500 \text{ K}} = -1400 \text{ J}.$$

Then from the first law,

$$W = Q_H + Q_C = 2000 \text{ J} + (-1400 \text{ J})$$
$$= 600 \text{ J}.$$

From Eq. (19-12), the efficiency is

$$e = 1 - \frac{350 \text{ K}}{500 \text{ K}} = 0.30 = 30\%.$$

Alternatively, from the basic definition of efficiency,

$$e = \frac{W}{Q_H} = \frac{600 \text{ J}}{2000 \text{ J}} = 0.30 = 30\%. \qquad \blacktriangleleft$$

Because each step in the Carnot cycle is reversible, the entire cycle may be reversed, converting the engine into a refrigerator. The performance coefficient of the Carnot refrigerator is obtained by combining Eqs. (19-8) and (19-11), first rewriting Eq. (19-8) as

$$K = -\left(\frac{Q_H + Q_C}{Q_C}\right)^{-1}.$$

We find

$$K = \frac{T_C}{T_H - T_C}. \tag{19-13}$$

When the temperature difference is small, K is much larger than unity; in that case a lot of heat can be "pumped" from the lower to the higher temperature with only a little expenditure of work. But the greater the temperature difference, the smaller is K, and the more work is required to transfer a given quantity of heat.

EXAMPLE 2 If the cycle described in Example 1 is run backward as a refrigerator, the performance coefficient is given by Eq. (19–8):

$$K = -\frac{Q_C}{W} = -\frac{-1400 \text{ J}}{600 \text{ J}} = 2.33.$$

Because the cycle is a Carnot cycle, we may also use Eq. (19–13):

$$K = \frac{T_C}{T_H - T_C} = \frac{350 \text{ K}}{500 \text{ K} - 350 \text{ K}} = 2.33.$$

For a Carnot cycle, e and K depend only on the temperatures, and it is not necessary to calculate Q and W. For cycles containing irreversible processes, however, more detailed calculations are necessary. ◀

It is easy to prove that no engine can be more efficient than a Carnot engine operating between the same two temperatures. The key to the proof is the above observation that since each step in the Carnot cycle is reversible, the *entire cycle* may be reversed. Run backward, the engine becomes a refrigerator. If any engine is more efficient than a Carnot engine, its work output may be used to drive a Carnot refrigerator and pump the rejected heat back to the hot reservoir, thus violating the engine statement of the second law. Thus still another equivalent statement of the second law is

No engine operating between two given temperatures can be more efficient than a Carnot engine operating between the same two temperatures.

It also follows directly that

All Carnot engines operating between the same two temperatures have the same efficiency, irrespective of the nature of the working substance.

Equation (19–12) points the way to the conditions that a real engine, such as a steam turbine, must fulfill to approach as closely as possible the maximum attainable efficiency. These conditions are that the intake temperature T_H must be made as high as possible and the exhaust temperature T_C as low as possible.

The exhaust temperature cannot be lower than the lowest temperature available for cooling the exhaust. This is usually the temperature of the air, or perhaps of river water if this is available at the plant. The only recourse then is to raise the boiler temperature T_H. Since the vapor pressure of all liquids increases rapidly with increasing temperature, a limit is set by the mechanical strength of the boiler. At 500°C, the vapor pressure of water is about 240×10^5 Pa, 235 atm, or 3450 lb·in^{-2}, and this is about the maximum practical pressure in large present-day steam boilers.

The unavoidable exhaust heat loss in electric power plants creates a serious environmental problem. When a lake or river is used for cooling, the temperature of the body of water may be raised several degrees.

Such a temperature change has a severely disruptive effect on the overall ecological balance, inasmuch as relatively small temperature changes can have significant effects on metabolic rates in plants and animals. Since *thermal pollution*, as this effect is called, is an inevitable consequence of the second law of thermodynamics, careful planning is essential to minimize the ecological impact of new power plants.

*19–7

THE KELVIN TEMPERATURE SCALE

The Carnot cycle can be used to define a temperature scale that is completely independent of the properties of any particular material. As we have seen, the efficiency of a Carnot engine operating between reservoirs at two given temperatures *is independent of the nature of the working substance and is a function only of the temperatures*. If we consider a number of Carnot engines using different working substances and absorbing and rejecting heat to the same two reservoirs, the thermal efficiency is the same for all:

$$e = \frac{Q_H + Q_C}{Q_H} = 1 + \frac{Q_C}{Q_H}$$

$$= \text{constant.}$$

Hence, the ratio Q_H/Q_C is a constant for all of the Carnot engines. Kelvin proposed that the ratio of the temperatures of the reservoirs be *defined* as equal to this constant ratio of the absolute magnitudes of the quantities of heat absorbed and rejected or, since Q_C is a negative quantity, as equal to the negative of the ratio Q_H/Q_C. Thus,

$$\frac{T_H}{T_C} = \frac{|Q_H|}{|Q_C|} = -\frac{Q_H}{Q_C}. \tag{19–14}$$

Equation (19–14) appears identical to Eq. (19–11), but there is a subtle and crucial difference. The temperatures in Eq. (19–11) are those based on an ideal-gas thermometer, as defined in Section (14–3), while Eq. (19–14) defines a temperature scale, based on the Carnot cycle and the second law of thermodynamics, that is completely independent of the behavior of any particular substance. Thus the Kelvin temperature scale is truly *absolute*. To complete the definition of the Kelvin scale we proceed, as in Chapter 14, to assign the arbitrary value of 273.16 K to the temperature of the triple point of water. When a substance is taken around a Carnot cycle, the ratio of the heats absorbed and rejected, $|Q_H|/|Q_C|$, is equal to the ratio of the temperatures of the reservoirs *as expressed on the gas scale*, defined in Chapter 14. Since, in both scales, the triple point of water is chosen to be 273.16 K, it follows that *the Kelvin and the ideal gas scales are identical*.

The zero point on the Kelvin scale is called *absolute zero*. Whether absolute zero may be achieved experimentally is a question of some interest and importance. To achieve temperatures below 4.2 K, at which ordinary helium (mass number 4) liquefies, it is necessary to lower the vapor pressure by pumping away the vapor as fast as possible. The lowest temperature that has ever been reached in this way is 0.7 K, and this required larger pumps and larger pumping tubes than are usually employed in low-temperature laboratories. With the aid of the light isotope of helium (mass number 3), which liquefies at 3.2 K, vigorous pumping yields a temperature of about 0.3 K. Still lower temperatures, as low as 10^{-6} K, may be achieved magnetically, but it becomes apparent that the closer one approaches absolute zero, the more difficult it is to go further. It is generally accepted as a law of nature that, although absolute zero may be approached as closely as we please, it is impossible actually to reach the zero of temperature. This is known as the *"unattainability statement of the third law of thermodynamics."*

Absolute zero can also be interpreted on a molecular level, although this must be done with some discretion. Because of quantum effects, it is *not* correct to say that at $T = 0$ all molecular motion ceases. Rather, at absolute zero the system has its *minimum* possible total energy (kinetic plus potential), although this minimum amount is in general not zero.

*19–8

ENTROPY

The second law of thermodynamics, as stated above, is rather different in form from other familiar physical laws; it is not a quantitative relationship but rather a statement of impossibility. The second law can also be stated in quantitative form, using the concept of *entropy*, which we now introduce.

We have seen in this chapter several examples of the tendency in nature for processes to move in the direction of increasing disorder. Irreversible heat flow increases disorder because initially the molecules are sorted into hotter and cooler regions, and this sorting is lost when the system comes to thermal equilibrium. Adding heat to a body increases its disorder because it increases the randomness of molecular motion. A free expansion of a gas increases its disorder because after the expansion the molecules have greater randomness of position than before.

Entropy provides a quantitative measure of disorder. We first consider the entropy change of a substance undergoing a reversible addition of heat. We use the symbol S for the entropy of the system, and ΔS for the change in entropy during any process. Adding heat increases molecular motion and thus disorder, but the effect is greater if the substance is cold at the beginning than if it is already hot. It turns out that a suitable definition of entropy change in such a process is

$$\Delta S = \frac{Q}{T}, \tag{19–15}$$

where as usual Q is the heat added and T the *absolute* temperature. This definition holds only for reversible, equilibrium processes.

Because entropy is a measure of the disorder of a system in any specific state, we expect it to depend only on the state of the system, and not on the details of any process that occurred on the way to that state. This expectation has been verified by many experimental observations; the entropy of a system depends only on its state. It also follows that when a system proceeds from an initial state with entropy S_1 to a final state with entropy S_2, the change in entropy $\Delta S = S_2 - S_1$ is independent of the path leading from the initial to final state. We recall that internal energy, introduced in Chapter 18, also has this property, although entropy and internal energy are very different quantities.

The fact that entropy is a function only of the state of a system also shows how to compute entropy changes in non-equilibrium processes, where Eq. (19–15) is not applicable. We simply invent a path connecting the given initial and final state that does consist entirely of reversible, equilibrium processes and compute the total entropy change for that path. It is not the actual path, but the entropy change must be the same.

As with internal energy, the above discussion does not define entropy but only the change in entropy in any given process. To complete the definition we may arbitrarily assign a value to the entropy of a system in a specified reference state and then calculate the entropy of any other state with reference to this. The unit of entropy is $1 \text{ J} \cdot \text{K}^{-1}$, $1 \text{ cal} \cdot \text{K}^{-1}$, $1 \text{ Btu} \cdot (\text{R}°)^{-1}$, etc.

EXAMPLE 1 One kilogram of ice at 0°C is melted and converted to water at 0°C. Compute its change in entropy.

Solution The temperature remains constant at 273 K, and

$$S_2 - S_1 = \frac{Q}{T}.$$

But Q is simply the total heat that must be supplied to melt the ice, or 334×10^3 J. Hence

$$S_2 - S_1 = \frac{334 \times 10^3 \text{ J}}{273 \text{ K}} = 1223 \text{ J} \cdot \text{K}^{-1},$$

and the increase in entropy of the system is $1223 \text{ J} \cdot \text{K}^{-1}$. In any *isothermal* reversible process, the entropy change equals the heat added divided by the absolute temperature. ◄

EXAMPLE 2 A gas is allowed to expand adiabatically and reversibly. What is its change in entropy?

Solution In an adiabatic process no heat is allowed to enter or leave the system. Hence $Q = 0$ and there is *no* change in entropy. It follows that every *reversible* adiabatic process is one of constant entropy and may be described as *isentropic*. ◄

EXAMPLE 3 A thermally insulated box is divided by a partition into two compartments, each having volume V. Initially one compartment contains n moles of an ideal gas at temperature T, and the other is evacuated. The partition is then broken, and the gas expands to fill both compartments. What is its entropy change?

Solution For this process, $Q = 0$, $W = 0$, $\Delta U = 0$, and therefore (since it is an ideal gas) $\Delta T = 0$. One might think that the entropy change is zero because there is no heat exchange. But Eq. (19–15) is valid only for *reversible* processes; this free expansion is *not* reversible, and there *is* an entropy change. To calculate it we can use the fact that the entropy change depends only on the initial and final states and not on the intermediate process. Thus we can devise a *reversible* process having the same endpoints, use Eq. (19–15) to calculate its entropy change, and thus obtain the entropy change in the original process. The appropriate reversible process in this case is an isothermal expansion from V to $2V$ at temperature T. The gas does work during this expansion, so heat must be supplied in order to keep the internal energy constant; the total heat equals the total work. The pressure varies during the expansion, so calculating the work requires methods of integral calculus. The result turns out to be simply

$$W = Q = nRT \ln 2.$$

Thus the entropy change is

$$\Delta S = \frac{Q}{T} = nR \ln 2,$$

and this is also the entropy change for the free expansion. For one mole,

$$\Delta S = (1 \text{ mol})(8.314 \text{ J} \cdot \text{mol}^{-1} \cdot \text{K}^{-1})(0.693) = 5.76 \text{ J} \cdot \text{K}^{-1}. \quad \blacktriangleleft$$

EXAMPLE 4 For the Carnot engine in Section 19–6, Example 1, find the total change of entropy in the engine during one cycle.

Solution During the isothermal expansion at 500 K the engine takes in 2000 J, and its entropy change is

$$\Delta S = \frac{Q}{T} = \frac{2000 \text{ J}}{500 \text{ K}} = 4.0 \text{ J} \cdot \text{K}^{-1}.$$

During the isothermal compression at 350 K the engine gives off 1400 J of heat, and its entropy change is

$$\Delta S = \frac{-1400 \text{ J}}{350 \text{ K}} = -4.0 \text{ J} \cdot \text{K}^{-1}.$$

Thus the total entropy change is $4.0 \text{ J} \cdot \text{K}^{-1} - 4.0 \text{ J} \cdot \text{K}^{-1} = 0$. This is to be expected, of course, since the final state is the same as the initial state. The total entropy change of the two heat reservoirs is also zero. This cycle contains no irreversible processes, and the total entropy change is zero. $\quad \blacktriangleleft$

*19–9

ENTROPY AND THE SECOND LAW

One of the features that distinguishes entropy from such concepts as energy, momentum, and angular momentum is that *there is no principle of conservation of entropy*. In fact, the reverse is true. Entropy *can* be created at will and there is an increase in entropy in every natural process, if all systems taking part in the process are considered.

Consider 1 kg of water initially at 0°C placed in thermal contact with 1 kg of water initially at 100°C. The first 1000 cal transferred cools the hot water to 99°C and warms the cold water from 0°C to 1°C. The net change of entropy is approximately

$$\Delta S = -\frac{1000 \text{ cal}}{373 \text{ K}} + \frac{1000 \text{ cal}}{273 \text{ K}} = 0.98 \text{ cal} \cdot \text{K}^{-1}.$$

Further increases in entropy occur as the system approaches thermal equilibrium. It can be shown, in fact, that the *total* increase in entropy is $24 \text{ cal} \cdot \text{K}^{-1}$. Thus an irreversible heat flow is always accompanied by an increase in entropy of the system. The same end state would have been achieved by simply mixing the two quantities of water. This too is an irreversible process, and since entropy depends only on the state of the system, the total entropy change would be the same.

These examples of the mixing of substances at different temperatures, or the flow of heat from a higher to a lower temperature, are characteristic of *all* natural (i.e., irreversible) processes. When all the entropy changes in the process are included, the increases in entropy are always greater than the decreases. In the special case of a reversible process, the increases and decreases are equal. Hence we can formulate the general principle, which is considered a part of the second law of thermodynamics, that *when all systems taking part in a process are included, the entropy either remains constant or increases.* In other words, *no process is possible in which the entropy decreases,* when all systems taking part in the process are included.

What is the significance of the increase of entropy that accompanies every natural process? The answer, or one answer, is that it represents the extent to which the universe becomes more disordered or random in that process. Consider again the example of the mixing of hot and cold water. We *might* have used the hot and cold water as the high- and low-temperature reservoirs of a heat engine, and in the course of removing heat from the hot water and giving heat to the cold water we could have obtained some mechanical work. But once the hot and cold water have been mixed and have come to a uniform temperature, this opportunity of converting heat to mechanical work is lost and, moreover, it is lost irretrievably. The lukewarm water will never *unmix* itself and separate into a hotter and colder portion.* Of course, there is no decrease in *energy* when the hot and cold water are mixed, and what has been "lost" in the mixing process is not *energy,* but *opportunity*—the opportunity to convert a portion of the heat flowing out of the hot water to mechanical work. Hence, when entropy increases, energy becomes more unavailable, and we say that the universe has become more random or "run down" to that extent. This is the true significance of the term "irreversible."

The tendency of all natural processes such as heat flow, mixing, and diffusion, is to bring about a uniformity of temperature, pressure, composition, etc., at all points. One may visualize a distant future in which, as a consequence of these processes, the entire universe has attained a state of absolute uniformity throughout. When and if such a state is reached,

---

*The branch of physics called "statistical mechanics" would modify this statement to read, "It is highly improbable that the water will separate spontaneously into a hotter and a colder portion, but it is not impossible."

although there would have been no change in the energy of the universe, all physical, chemical, and presumably biological processes would have to cease. This fate toward which we appear headed has been described as the "heat death" of the universe!

19–10

ENERGY CONVERSION

(a)

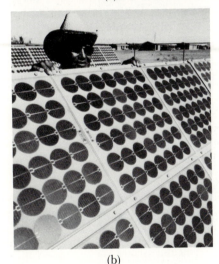

(b)

19–9 (a) A solar energy installation in the Mojave Desert, near Barstow, California. An array of mirrors concentrates the sun's energy on a boiler (central tower) to generate steam for turbines. The mirrors move continuously, controlled by a computer, to track the sun's motion; the inset shows a computer display used to help technicians focus the array. (b) An array of photovoltaic cells for direct conversion of sun power to electric power. Such an array supplies the electrical needs of the Headquarters and Visitors Center of Natural Bridges National Monument, Utah. (Photos by Dan McCoy/Rainbow.)

We have seen in this chapter that the laws of thermodynamics place very general limitations on conversion of energy from one form to another. In this day of increasing energy demand and diminishing resources, these matters are of the utmost practical importance. We conclude this chapter with a brief discussion of a few energy-conversion systems, present and proposed.

Most (over 80 percent) of the electric power generated in the United States is obtained from coal-fired steam-turbine generating plants. Modern boilers can transfer about 80 to 90 percent of the heat of combustion of coal into steam. The theoretical thermodynamic efficiency of the turbine, given by Eq. (19–12), is usually limited to about 0.55, and the actual efficiency is typically 90 percent of the theoretical value, or about 0.50. The efficiency of large electrical generators in converting mechanical power to electrical is very large, typically 99 percent. Thus the overall thermal efficiency of such a plant is roughly (0.85)(0.50)(0.99), or about 40 percent.

In 1970, a generator with a capacity of about 1 gigawatt ($= 1000$ MW $= 10^9$ W) went into operation at the Tennessee Valley Authority's Paradise power plant. The steam is heated to 1003°F, at a pressure of 3650 lb·in^{-2}. The plant consumes 10,500 tons of coal per day, and the overall thermal efficiency is 39.3 percent.

Nuclear power plants have the same theoretical efficiency limit as coal-fired plants. Because at present it is not practical to run nuclear reactors at as high temperatures as coal boilers, the theoretical thermal efficiency is usually lower. The overall thermal efficiency of a nuclear plant is typically 30 percent.

In both coal-fired and nuclear plants, the energy not converted to electrical energy is wasted and must be disposed of. A common practice is to locate such a plant near a lake or river and use the water for disposal of excess heat; this can raise the water temperature several degrees, often with serious ecological consequences.

Solar energy is an inviting possibility. The power in the sun's radiation is about 1.4 kW per square meter, above the earth's atmosphere. A maximum of about 1.0 kW·m^{-2} reaches the surface of the earth on a clear day, and the time average over a 24-hour period is about 0.2 kW·m^{-2}. This radiation could be collected, and focused with mirrors and used to generate steam for a heat engine, as in Fig. 19–9a. A more exotic scheme is to use large banks of photocells for direct conversion to electricity. Such a process is not a heat engine in the usual sense and is not limited by the Carnot efficiency. There are other fundamental limitations on photocell efficiency, but 50% seems attainable in multilayer *p-n* junction photocells. The energy of wind, which is fundamentally

19–10 An array of windmills to collect and convert wind energy. The propellor-like blades turn electric generators, converting the kinetic energy of moving air into electrical energy. (Photo by Kurt Rogers, San Francisco Examiner.)

solar in origin, can be gathered and converted by "forests" of windmills, as shown in Fig. 19–10.

An indirect scheme for collection and conversion of solar energy would use the temperature gradient in the ocean. In the Caribbean, for example, the water temperature near the surface is about 25°C, while at a depth of a few hundred meters it may be 10°C. While the second law of thermodynamics forbids taking heat from the ocean and converting it completely into work, there is nothing to forbid running a heat engine between these two temperatures. The thermodynamic efficiency would be very low, but with such a vast reservoir of energy available this would not be a serious problem.

This is only a small sample of present-day activity in energy-conversion research. Many other processes are under discussion or development, and the principles of thermodynamics outlined in this chapter are of central importance in all of them.

QUESTIONS

19–1 Suppose you want to increase the efficiency of a heat engine. Would it be better to increase T_H or to decrease T_C by an equal amount?

19–2 If an energy-conversion process involves two steps, each with its own efficiency, such as heat to work and work to electrical energy, is the efficiency of the composite process equal to the product of the two efficiencies, or the sum, or the difference, or what?

19–3 In some climates it is practical to heat a house using a heat pump, which acts as an air-conditioner in reverse, cooling the outside air and heating the inside air. Can the heat delivered to the house ever exceed the electric energy input to the pump?

19–4 What irreversible processes occur in a gasoline engine?

19–5 A housewife tries to cool her kitchen on a hot day by leaving the refrigerator door open. What happens?

Would the result be different if an old-fashioned ice box were used?

19–6 Is it a violation of the second law to convert mechanical energy completely into heat?

19–7 A growing plant creates a highly complex and organized structure out of simple materials, such as air, water, and trace minerals. Does this violate the second law of thermodynamics? What is the plant's ultimate source of energy?

19–8 An electric motor has its shaft coupled to that of a generator. The motor drives the generator, and the current from the generator is used to run the motor. The excess current is used to power a television set. What is wrong with this scheme?

19–9 Name some reversible and some irreversible processes in purely mechanical systems, such as blocks sliding on planes, springs, pulleys, strings, and all that stuff.

19–10 Why must a room air conditioner be placed in a window? Why can't it just be set on the floor and plugged in?

19–11 Discuss the following examples of increasing disorder or randomness: mixing of hot and cold water; free expansion of a gas; irreversible heat flow; development of heat through mechanical friction. Are entropy increases involved in all these?

19–12 When the sun shines on a glass-roofed greenhouse, the temperature becomes higher inside than outside. Does this violate the second law?

19–13 When a wet cloth is hung up in a hot wind in the desert, it is cooled by evaporation to a temperature that may be 20°C or so below that of the air. Discuss this process in the light of the second law.

19–14 Are the earth and the sun in thermal equilibrium? Are there entropy changes associated with the transmission of energy from the sun to the earth? Does radiation differ from other modes of heat transfer with respect to entropy changes?

19–15 Discuss the entropy changes involved in the preparation and consumption of a hot-fudge sundae.

PROBLEMS

19–1 A large diesel engine takes in 8000 J of heat and delivers 3000 J of work per cycle. The heat is obtained by burning diesel fuel with a heat of combustion of $5.0 \times 10^4 \text{ J} \cdot \text{g}^{-1}$.

a) What is the thermal efficiency?

b) How much heat is discarded in each cycle?

c) What mass of fuel is burned in each cycle?

d) If the engine goes through 50 cycles per second, what is its power output in watts? In horsepower?

19–2 A gasoline engine has a power output of 20 kW (about 27 hp). Its thermal efficiency is 20%.

a) How much heat must be supplied to the engine per second?

b) How much heat is discarded by the engine per second?

19–3 A coal-fired steam-turbine power plant has a mechanical power output of 500 MW and a thermal efficiency of 40%.

a) At what rate must heat be supplied by burning coal?

b) If the heat of combustion of coal is $2.5 \times 10^4 \text{ J} \cdot \text{g}^{-1}$, what mass of coal is burned per second? Per day?

c) At what rate is heat discarded by the system?

d) If the discarded heat is given to water in a river, and its temperature rises by 5 C°, what volume of water is needed per second?

19–4 A nuclear power plant has a mechanical power output (used to drive an electric generator) of 200 MW. Its rate of heat input from the nuclear reactor is 800 MW.

a) What is the thermal efficiency of the system?

b) At what rate is heat discarded by the system?

19–5 A heat engine carries 0.1 mole of an ideal gas around the cycle shown in the pV-diagram of Fig. 19–11. Process 1–2 is at constant volume, process 2–3 is adiabatic, and process 3–1 is at a constant pressure of 1 atm. The value of γ for this gas is $\frac{5}{3}$.

a) Find the pressure and volume at points 1, 2, and 3.

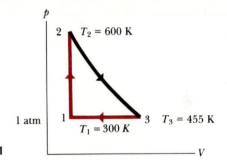

Figure 19–11

b) Find the net work done by the gas in the cycle.

c) If the heat removed in process 3–1 is considered wasted, what is the thermal efficiency of this engine?

19–6 A cylinder contains oxygen at a pressure of 2 atm. The volume is 3 l and the temperature is 300 K. The oxygen is carried through the following processes:

1. Heated at constant pressure to 500 K.
2. Cooled at constant volume to 250 K.
3. Cooled at constant pressure to 150 K.
4. Heated at constant volume to 300 K.

a) Show the four processes above in a pV-diagram, giving the numerical values of p and V at the end of each process.

b) Calculate the net work done by the oxygen.

c) Find the net heat flowing into the oxygen.

d) What is the efficiency of this device as a heat engine?

19–7 What is the thermal efficiency of an engine that operates by taking an ideal gas through the following cycle? Let $C_v = 12 \text{ J} \cdot \text{mol}^{-1} \cdot \text{K}^{-1}$.

1. Start with n moles at P_0, V_0, T_0.
2. Change to $2P_0, V_0$, at constant volume.
3. Change to $2P_0, 2V_0$, at constant pressure.
4. Change to $P_0, 2V_0$, at constant volume.
5. Change to P_0, V_0, at constant pressure.

19-8 A Carnot engine whose low-temperature reservoir is at 280 K has an efficiency of 40%. An engineer is assigned the problem of increasing this to 50%.

a) By how many degrees must the temperature of the high-temperature reservoir be increased if the temperature of the low-temperature reservoir remains constant?

b) By how many degrees must the temperature of the low-temperature reservoir be decreased if that of the high-temperature reservoir remains constant?

19-9 A Carnot engine whose high-temperature reservoir is at 400 K takes in 100 cal of heat at this temperature in each cycle, and gives up 80 cal to the low-temperature reservoir.

a) What is the temperature of the latter reservoir?

b) What is the thermal efficiency of the cycle?

19-10 An ice-making machine operates in a Carnot cycle; it takes heat from water at 0°C and rejects heat to a room at 27°C. Suppose that 50 kg of water at 0°C are converted to ice at 0°C.

a) How much heat is rejected to the room?

b) How much energy must be supplied to the machine?

19-11 A Carnot engine is operated between two heat reservoirs at temperatures of 400 K and 300 K.

a) If the engine receives 1200 cal from the reservoir at 400 K in each cycle, how many calories does it reject to the reservoir at 300 K?

b) If the engine is operated in reverse, as a refrigerator, and receives 1200 cal from the reservoir at 300 K, how many calories does it deliver to the reservoir at 400 K?

c) How many calories would be produced if the mechanical work required to operate the refrigerator in part (b) were converted directly to heat?

19-12

a) Draw a graph of a Carnot cycle, plotting Kelvin temperature vertically and entropy horizontally (a temperature–entropy or TS diagram).

b) Show that the area under any curve in a temperature–entropy diagram represents the heat absorbed by the system.

c) Derive from your diagram the expression for the thermal efficiency of a Carnot cycle.

19-13 A sophomore with nothing better to do adds heat to 0.5 kg of ice at 0°C until it is all melted.

a) What is the change in entropy of the water?

b) If the source of heat is a very massive body at a temperature of 20°C, what is the change in entropy of this body?

c) What is the total change in entropy of the water and the heat source?

19-14 A Carnot engine operates between two heat reservoirs at temperatures T_H and T_C. An inventor proposes to increase the efficiency by running one engine between T_H and an intermediate temperature T', and a second engine between T' and T_C using the heat expelled by the first engine. Compute the efficiency of this composite system, and compare it to that of the original engine.

19-15 A physics student performing a heat-conduction experiment immerses one end of a copper rod in boiling water at 100°C, the other end in an ice–water mixture at 0°C. The sides of the rod are insulated. During a certain time interval, 0.5 kg of ice melts. Find

a) the entropy change of the boiling water,

b) the entropy change of the ice–water mixture,

c) the entropy change of the copper rod,

d) the total entropy change of the entire system.

19-16 An ice-cube maker freezes water at the rate of 5 g every second, starting with water at the freezing point. Heat is given off to the room at 30°C. The system uses an ideal Carnot refrigerator.

a) What electric power input is required?

b) At what rate is heat given off to the room?

19-17 Show that the efficiency e of a Carnot engine and the performance coefficient K of a Carnot refrigerator are related by $K = (1 - e)/e$.

19-18 A solar power plant is to be built with a power output capacity of 1000 MW. Find what land area the solar energy collectors must occupy if they are

a) photocells with 90-percent efficiency, and

b) mirrors that generate steam for a turbine–generator unit with overall efficiency of 30 percent.

19-19 An engine is to be built to extract power from the temperature gradient of the ocean. If the surface and deep-water temperatures are 25°C and 10°C, respectively, what is the maximum theoretical efficiency of such an engine?

19-20 A coal-fired steam-turbine power plant that produces 1000 MW of electrical power is located beside a river. The overall thermal efficiency of the plant is 40 percent.

a) What is the total thermal power input to the plant?

b) At what rate is waste heat discharged from the plant?

c) If the waste heat is delivered to the river, and if the temperature rise must be no greater than 5°C, how much water must be available per second?

d) In part (c), if the river is 100 m wide and 5 m deep, what must be the minimum flow velocity of the water?

20

MOLECULAR PROPERTIES OF MATTER

Several properties of matter in bulk have been discussed in previous chapters; these have included elasticity, density, surface tension, equations of state, heat capacities, phase changes, internal energy, entropy, and others. Although we have referred descriptively to the molecular structure of matter in relation to these properties, our detailed analysis has concentrated entirely on macroscopic or bulk properties. In this chapter we shall study a few examples of how a molecular or microscopic picture can enable us actually to *predict* some properties of matter. We begin with a general discussion of the molecular structure of matter. Then we develop the kinetic theory of gases, which permits understanding of some basic properties of gases, such as the equation of state and heat capacities, on the basis of a molecular model.

20–1

MOLECULAR THEORY OF MATTER

An abundance of physical and chemical evidence has shown conclusively that matter in all phases is made up of particles called molecules. For any substance that is a specific chemical compound, the molecules are all identical. The smallest molecules are of the order of 10^{-10} m in size, the largest at least 10,000 times this large.

In liquids and solids molecules are held together by intermolecular forces that are *electrical* in nature, arising from interactions of the electrically charged fundamental particles that make up the molecules. *Gravitational* forces between molecules are so weak compared with electrical forces that they are completely negligible.

Intermolecular forces do not follow a simple inverse-square law. When molecules are far apart, as in a gas, the intermolecular force is very small and attractive. As a gas is compressed and its molecules are brought closer together the force increases. In liquids, relatively large pressures are required to compress the substance appreciably, and we

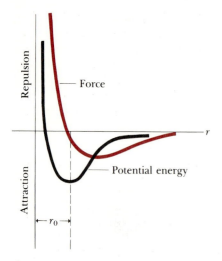

20–1 The force between two molecules (color curve) changes from an attraction when the separation is large, to a repulsion when the separation is small. The potential energy (black curve) is minimum at r_0, where the force is zero.

20–2 Crystal of necrosis virus protein. The actual size of the entire crystal is about two thousandths of a millimeter (0.002 mm). (Courtesy of Ralph W. G. Wyckoff. Reprinted with the permission of Educational Services, Inc., from *Physics*, D. C. Heath and Co., Boston, 1960.)

conclude that at separations between molecules that are only slightly *less* than their normal spacing the force becomes *repulsive* and relatively large.

Thus the force must vary with the distance r between molecules somewhat as shown in Fig. 20–1. At large distances the force is small and attractive. As the molecules are brought closer together, the force of attraction becomes larger, passes through a maximum, and then decreases to zero at an equilibrium separation r_0. When the distance is less than r_0 the force becomes repulsive and increases quite rapidly. The figure also shows the potential energy as a function of r. It has a *minimum* at r_0, where the force is zero. Such a potential-energy function is often called a *potential well*.

In view of the attractive intermolecular forces, why do not all molecules eventually coalesce into matter in the liquid or solid phase? The answer is that the molecules are always in *motion*, and have kinetic energy associated with this motion. Generally speaking, the kinetic energy increases with temperature. At very low temperatures the average kinetic energy of a molecule may be much *less* than the maximum magnitude of potential energy, the "depth" of the potential well in Fig. 20–1. The molecules then condense into a liquid or solid phase with average intermolecular spacing of about r_0. But at higher temperatures the average kinetic energy becomes larger than the depth of the potential well; molecules can then escape the intermolecular force and become free to move independently, as in the gaseous phase of matter.

In *solids* the molecules execute vibratory motion about more or less fixed centers; if the potential well is approximately parabolic in shape near its minimum, the motion is approximately simple harmonic. The amplitudes of the vibratory motions are relatively small, and the fixed centers comprise a space lattice, which gives rise to the repeated spatial patterns and symmetry of *crystals*. A photograph of the individual, very large molecules of *necrosis virus protein* is shown in Fig. 20–2. It was taken with an electron microscope at a magnification of about 80,000. The molecules look like neatly stacked oranges. Each molecule is about 1.2×10^{-8} m in diameter.

In *liquids* the intermolecular distances are usually only a trifle greater than those of a solid. The molecules execute vibratory motion of greater energy about centers that are free to move but remain at approximately the same distances from one another. Liquids show a certain regularity of structure only in the immediate neighborhood of a few molecules. This is called *short-range order*, in contrast with the *long-range order* of a solid crystal.

The molecules of a gas have greater kinetic energy than those of liquids and solids; on the average, they are far away from one another, where only very small attractive forces exist. A molecule of a gas therefore moves with linear motion until a collision takes place either with another molecule or with a wall. In molecular terms, an *ideal gas* is a gas whose molecules exert *no* forces of attraction on each other. The mathematical analysis of a collection of such idealized molecules in random linear motion between collisions is called the *kinetic theory of gases*. The analysis of an ideal gas given in this chapter is a simple example of kinetic theory.

Most common substances exist in the solid phase at low temperatures. When the temperature is raised beyond a definite value, the liquid phase results, and when the temperature of the liquid is raised further, the substance exists in the gaseous phase; i.e., from a large-scale or *macroscopic* point of view, the transition from solid to liquid to gas is in the direction of increasing temperature. From a *molecular* point of view, this transition is in the direction of increasing molecular kinetic energy. Evidently temperature and molecular kinetic energy are related.

20–2

AVOGADRO'S NUMBER

In 1811 John Dalton, often called the father of modern chemistry, first suggested that the molecules of a given chemical substance are all alike. This hypothesis has since been confirmed by an overwhelming variety of evidence; it explains why, in chemical reactions, elements and compounds combine with each other in definite proportions by mass. It also leads directly to the concepts of *molecular mass* and *number of moles*. Specifically, one mole of any pure chemical element or compound contains a definite number of molecules, the same number for all elements and compounds. The official SI definition of the mole is as follows:

The mole is the amount of substance that contains as many elementary entities as there are atoms in 0.012 kilograms of carbon 12.

In our discussion, the "elementary entities" referred to are atoms or molecules.

The number of atoms or molecules in a mole is called *Avogadro's number*, denoted by N_A. (Avogadro was a contemporary of Dalton.) The most precise measurements of N_A have been obtained by using x-rays to measure the distance between layers of molecules in a crystal. The numerical value of N_A is now known with an uncertainty of less than six parts per million; to four significant figures, it is

$$N_A = 6.022 \times 10^{23} \text{ molecules} \cdot \text{mol}^{-1}.$$

The molecular mass M of a compound is the mass of one mole; the mass m of a single molecule is thus given by

$$M = N_A m. \tag{20–1}$$

Once Avogadro's number has been determined, it can be used to compute the mass of a molecule. The mass of 1 mol of atomic hydrogen (i.e., the atomic mass) is 1.008 g. Since by definition the number of molecules in 1 mol is Avogadro's number, it follows that the mass of a single atom of hydrogen is

20–3 How much is one mole? The photograph shows one mole each of sucrose (ordinary sugar, rear pile), iodine (metallic-looking chips), water, mercury, iron (cube), acetylsalicylic acid (aspirin tablets), and sodium chloride (ordinary salt, front pile). (Photo by Chip Clark.)

$$m_H = \frac{1.008 \text{ g} \cdot \text{mol}^{-1}}{6.022 \times 10^{23} \text{ molecules} \cdot \text{mol}^{-1}} = 1.674 \times 10^{-24} \text{ g} \cdot \text{molecule}^{-1}.$$

For an oxygen molecule of molecular mass 32 g·mol⁻¹,

$$m_{O_2} = \frac{32 \text{ g} \cdot \text{mol}^{-1}}{6.022 \times 10^{23} \text{ molecules} \cdot \text{mol}^{-1}} = 53.12 \times 10^{-24} \text{ g} \cdot \text{molecule}^{-1}.$$

Figure 20–3 shows one mole each of several familiar materials.

20–3

PROPERTIES OF MATTER

The usual goal of any molecular theory of matter is to understand the *macroscopic* properties of matter in bulk in terms of the properties, behavior, and interactions of the molecules of which it is composed. Such theories are of tremendous practical importance; once this understanding has been gained, it becomes possible actually to *design* materials with specific desired properties. Thus this kind of analysis has led to the development of high-strength steels, glasses with special optical properties for use in optical instruments, semiconductor materials for solid-state electronic devices, and countless other materials essential to contemporary technology.

In this chapter we shall consider a few simple examples of molecular theories of matter. For example, we can represent a monatomic gas with a model consisting of a large number of particles described completely by their mass and velocities. We can then derive the observed equation of state of a gas at low pressure and can show that C_v for the gas should equal $12.5 \ \mathrm{J \cdot mol^{-1} \cdot K^{-1}}$ (see Table 18–1). If we add the hypothesis that the "particles" are not simply points but have a finite size, then the general features of the viscosity, thermal conductivity, and coefficient of diffusion of a gas can be understood, as well as the fact that polyatomic gases have larger values of C_v than monatomic gases. By assuming that there are forces between the particles, we can refine the equation of state to bring it into better agreement with that of a real gas, and we can begin to understand the phenomena of liquefaction and solidification at low temperatures.

The *electrical* and *magnetic* properties of matter, and the emission and absorption of light by matter, call for a molecular model that is itself an aggregate of subatomic particles. Some of these are electrically charged, and the forces between molecules originate in these electric charges. This development of molecular theory will take us into the area of atomic and molecular structure in Chapter 43. In the next section we return to the starting point and see what properties of a gas at low pressure can be explained by the simplest possible molecular model, an aggregate of particles having mass and velocity.

20–4

KINETIC THEORY OF AN IDEAL GAS

We are now ready to develop in detail the relation between the kinetic–molecular model of an ideal gas and the ideal-gas equation of state, discussed in Section 17–2. We consider a container of volume V, containing N identical molecules each of mass m. The molecules are in constant motion; each molecule collides from time to time with a wall of the container. During such collisions the molecules exert forces on the walls, and this is the origin of the macroscopic *pressure* the gas exerts on the container walls. We assume that each collision of a molecule with a wall of

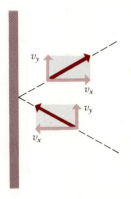

20–4 Elastic collision of a molecule with container wall. The component v_y parallel to the wall does not change; the component v_x perpendicular to the wall reverses direction. The speed v does not change.

the container is perfectly elastic, as shown in Fig. 20–4. In each collision the component of velocity parallel to the wall is unchanged and the component perpendicular to the wall is reversed.

Our program will be to determine for a given wall area A the number of collisions per unit time, the associated momentum change, and the force needed for the momentum change. Then we can obtain an expression for the pressure, which is force per unit area. We let v_x be the *magnitude* of the x-component of velocity of a molecule. At first we shall assume that all molecules have the same v_x. This assumption, though unrealistic, helps to clarify basic ideas; and we shall show soon that it is not really necessary.

In each collision the change in x-component of momentum is $2mv_x$. To find the number of collisions in a time interval Δt, with a given wall area A, we note that, in order to experience a collision during Δt, a molecule must be within a distance $v_x \Delta t$ at the beginning of Δt, as shown in Fig. 20–5, and must be headed toward the wall. Thus to collide with A during Δt, a molecule must, at the beginning of Δt, be within a cylinder of base area A and length $v_x \Delta t$. The volume of such a cylinder is $Av_x \Delta t$.

Assuming the number of molecules per unit volume (N/V) is uniform, the *number* of molecules in this cylinder is $(N/V)(Av_x \Delta t)$. But, on the average, half of these molecules are moving *away from* the wall. Thus the number of collisions with A during Δt is

$$\frac{1}{2}\left(\frac{N}{V}\right)(Av_x \Delta t). \qquad (20–2)$$

The total momentum change ΔP_x due to all these collisions is $2mv_x$ times the *number* of collisions:

$$\Delta P_x = \frac{1}{2}\left(\frac{N}{V}\right)(Av_x \Delta t)(2mv_x) = \frac{NAmv_x^2 \Delta t}{V}, \qquad (20–3)$$

and the *rate* of change of momentum is

$$\frac{\Delta P_x}{\Delta t} = \frac{NAmv_x^2}{V}. \qquad (20–4)$$

According to Newton's second law, this equals the average force exerted by the wall area A on the molecules; from the *third* law, this is the negative of the force exerted *on* the wall *by* the molecules. Finally, pressure p is force per unit area, and we obtain

$$p = \frac{F}{A} = \frac{Nmv_x^2}{V}. \qquad (20–5)$$

Now in fact v_x is *not* the same for all molecules. But we could have sorted them into groups having the same v_x within each group, and added up the resulting contributions to the pressure. The net effect of this is simply to replace v_x^2 in Eq. (20–5) by the *average* value of v_x^2, which we denote by $\overline{v_x^2}$. Furthermore, $\overline{v_x^2}$ is related simply to the *speeds* of the molecules. The speed v (magnitude of velocity) of any molecule is related to the velocity components v_x, v_y, and v_z, by

$$v^2 = v_x^2 + v_y^2 + v_z^2.$$

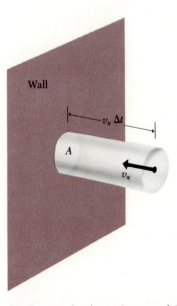

20–5 A molecule moving toward the wall with speed v_x collides with the area A during the time interval Δt only if it is within a distance $v_x \Delta t$ of the wall at the beginning of the interval. All such molecules are contained within a volume $Av_x \Delta t$.

This may be averaged over all molecules:

$$\overline{v^2} = \overline{v_x^2} + \overline{v_y^2} + \overline{v_z^2}.$$

But since the x-, y-, and z-directions are all equivalent,

$$\overline{v_x^2} = \overline{v_y^2} = \overline{v_z^2}.$$

Hence

$$\overline{v_x^2} = \tfrac{1}{3}\overline{v^2},$$

and Eq. (20–5) becomes

$$pV = \tfrac{1}{3}Nm\overline{v^2} = \tfrac{2}{3}(\tfrac{1}{2}Nm\overline{v^2}). \tag{20–6}$$

But $\tfrac{1}{2}m\overline{v^2}$ is the average kinetic energy of a single molecule, and the product of this and the total number of molecules N equals the total random kinetic energy, or internal energy U. Hence the product pV equals two thirds of the internal energy:

$$pV = \tfrac{2}{3}U. \tag{20–7}$$

By experiment, at low pressures, the equation of state of a gas is

$$pV = nRT.$$

The theoretical and experimental laws will therefore be in complete agreement if we set

$$U = \tfrac{3}{2}nRT. \tag{20–8}$$

The average kinetic energy of a single molecule is then

$$\frac{U}{N} = \tfrac{1}{2}m\overline{v^2} = \frac{3nRT}{2N}.$$

But the number of moles, n, equals the total number of molecules, N, divided by Avogadro's number, N_A, the number of molecules per mole:

$$n = \frac{N}{N_A}, \qquad \frac{n}{N} = \frac{1}{N_A}.$$

Hence

$$\tfrac{1}{2}m\overline{v^2} = \frac{3}{2}\frac{R}{N_A}T. \tag{20–9}$$

The ratio R/N_A occurs frequently in molecular theory. It is called the *Boltzmann constant, k:*

$$k = \frac{R}{N_A} = \frac{8.31\ \text{J}\cdot\text{mol}^{-1}\cdot\text{K}^{-1}}{6.02 \times 10^{23}\ \text{molecules}\cdot\text{mol}^{-1}}$$
$$= 1.38 \times 10^{-23}\ \text{J}\cdot\text{molecule}^{-1}\cdot\text{K}^{-1}.$$

Since R and N_A are universal constants, the same is true of k. Then

$$\tfrac{1}{2}m\overline{v^2} = \tfrac{3}{2}kT. \tag{20–10}$$

Thus the average kinetic energy per molecule depends only on the temperature, not on the pressure, volume, or molecular species. An equivalent statement can also be obtained from Eq. (20–9) by using the relation $M = N_A m$:

$$N_A(\tfrac{1}{2}m\overline{v^2}) = \tfrac{1}{2}M\overline{v^2} = \tfrac{3}{2}RT. \qquad (20\text{–}11)$$

That is, the kinetic energy of a *mole* of molecules depends only on T.

From Eqs. (20–10) and (20–11) we can obtain expressions for the square root of $\overline{v^2}$, called the *root-mean-square speed* v_{rms}:

$$v_{rms} = \sqrt{\overline{v^2}} = \sqrt{\frac{3kT}{m}} = \sqrt{\frac{3RT}{M}}. \qquad (20\text{–}12)$$

Finally, it is sometimes convenient to rewrite the ideal-gas equation on a molecular basis. Since $N = N_A n$ and $R = N_A k$, an alternative form of the gas equation is

$$pV = NkT. \qquad (20\text{–}13)$$

Thus k may be regarded as a gas constant on a "per molecule" basis instead of the usual "per mole" basis for R.

EXAMPLE 1   What is the average kinetic energy of a molecule of a gas at a temperature of 300 K?

Solution

$$\tfrac{1}{2}m\overline{v^2} = \tfrac{3}{2}kT = (\tfrac{3}{2})(1.38 \times 10^{-23}\ \text{J} \cdot \text{K}^{-1})(300\ \text{K})$$
$$= 6.21 \times 10^{-21}\ \text{J}. \qquad \blacktriangleleft$$

EXAMPLE 2   What is the total random kinetic energy of the molecules in 1 mole of a gas at a temperature of 300 K?

Solution

$$U = \tfrac{3}{2}nRT = (\tfrac{3}{2})(1\ \text{mol})(8.314\ \text{J} \cdot \text{mol}^{-1} \cdot \text{K}^{-1})(300\ \text{K})$$
$$= 3741\ \text{J} = 894\ \text{cal}. \qquad \blacktriangleleft$$

EXAMPLE 3   What is the root-mean-square speed of a hydrogen molecule at 300 K?

Solution   The mass of a hydrogen molecule (see Section 20–2) is

$$m_{H_2} = (2)(1.674 \times 10^{-27}\ \text{kg}) = 3.347 \times 10^{-27}\ \text{kg}.$$

Hence, from Eq. (20–12),

$$v_{rms} = \sqrt{\frac{3kT}{m}} = \sqrt{\frac{3(1.38 \times 10^{-23}\ \text{J} \cdot \text{K}^{-1})(300\ \text{K})}{3.347 \times 10^{-27}\ \text{kg}}}$$
$$= 1927\ \text{m} \cdot \text{s}^{-1}.$$

Alternatively,

$$v_{rms} = \sqrt{\frac{3RT}{M}} = \sqrt{\frac{3(8.314\ \text{J} \cdot \text{mol}^{-1} \cdot \text{K}^{-1})(300\ \text{K})}{2(1.008 \times 10^{-3}\ \text{kg} \cdot \text{mol}^{-1})}}$$
$$= 1927\ \text{m} \cdot \text{s}^{-1}.$$

We note that in applying Eq. (20–12) using the value of R in SI units, M must be expressed in *kilograms* per mole, not grams per mole. Thus in this example $M = 2.016 \times 10^{-3}$ kg·mol^{-1}, not 2.016 g·mol^{-1}. ◄

EXAMPLE 4 Five gas molecules chosen at random are found to have speeds of 500, 600, 700, 800, and 900 m·s^{-1}. Find the rms speed; is it the same as the *average* speed?

Solution There are five molecules; the average value of v^2 is

$$v^2 = \frac{(500 \text{ m·s}^{-1})^2 + (600 \text{ m·s}^{-1})^2 + (700 \text{ m·s}^{-1})^2 + (800 \text{ m·s}^{-1})^2 + (900 \text{ m·s}^{-1})^2}{5}$$

$$= 510{,}000 \text{ m}^2 \cdot \text{s}^{-2},$$

and v_{rms} is the square root of this:

$$v_{\text{rms}} = 714 \text{ m·s}^{-1}.$$

The *average* speed v is given by

$$v = \frac{500 \text{ m·s}^{-1} + 600 \text{ m·s}^{-1} + 700 \text{ m·s}^{-1} + 800 \text{ m·s}^{-1} + 900 \text{ m·s}^{-1}}{5}$$

$$= 700 \text{ m·s}^{-1}.$$

Clearly, v_{rms} and v are not, in general, the same. ◄

EXAMPLE 5 Find the number of molecules in one cubic meter of air at atmospheric pressure and 0°C.

Solution From Eq. (20–13),

$$N = \frac{pV}{kT} = \frac{(1.013 \times 10^5 \text{ Pa})(1 \text{ m}^3)}{(1.38 \times 10^{-23} \text{ J·K}^{-1})(273 \text{ K})} = 2.69 \times 10^{25}. \quad ◄$$

When a gas expands against a moving piston it does work. This work is accompanied by a decrease in the random kinetic energy of the gas molecules. Conversely, when work is done in compressing a gas, the random kinetic energy of its molecules increases. But if the collisions with the walls are perfectly elastic, as we have assumed, how can a molecule gain or lose energy in a collision with a piston? To understand this, we must consider the collision of a molecule with a moving wall.

When a molecule makes a collision with a *stationary* wall, it exerts a momentary force on the wall but does no work, since the wall does not move. But if the wall is in motion, work *is* done in a collision. Thus, if the wall in Fig. 20–4 is moving to the left, work is done on it by the molecules that strike it, and their velocities (and kinetic energies) after colliding are smaller than they were before a collision. The collision is still completely elastic, since the work done on the moving piston is just equal to the decrease in the kinetic energy of the molecule. Similarly, if the piston is moving toward the right, the kinetic energy of a colliding molecule is *increased,* the increase in kinetic energy being equal to the work done on the molecule.

The assumption that individual molecules undergo *elastic* collisions with the container wall is not strictly correct. More detailed investigation has shown that in most cases molecules actually adhere to the wall for a short time, and then leave again with speeds characteristic of the temperature *of the wall*. However, the gas and the wall are ordinarily in thermal equilibrium, and the validity of our conclusions is not altered by this discovery.

20–5

MOLAR HEAT CAPACITY OF A GAS

When heat flows into a system in a process at constant volume, no work is done and all of the energy inflow goes into an increase in the internal energy U of the system. From the molecular viewpoint, the internal energy of a system is the sum of the kinetic and potential energies of its molecules. If this sum can be computed as a function of temperature, then, from its rate of change with temperature, we can derive a theoretical expression for the molar heat capacity.

The simplest system is a monatomic ideal gas, for which the molecular energy is wholly kinetic and is given by Eq. (20–8):

$$\text{Random kinetic energy} = U = \tfrac{3}{2}nRT.$$

If the temperature increases by ΔT, the random kinetic energy increases by

$$\Delta U = \tfrac{3}{2}nR\,\Delta T.$$

In a process in which the temperature increases by ΔT at constant volume, the energy flowing into a system is, by definition of C_v,

$$\Delta Q = nC_v\,\Delta T = \Delta U.$$

Hence,

$$C_v = \tfrac{3}{2}R. \tag{20–14}$$

In SI units,

$$C_v = \tfrac{3}{2}(8.314\ \text{J}\cdot\text{mol}^{-1}\cdot\text{K}^{-1}) = 12.47\ \text{J}\cdot\text{mol}^{-1}\cdot\text{K}^{-1}.$$

The experimental values of C_v for monatomic gases in Table 18–1 are, in fact, almost exactly equal to $\tfrac{3}{2}R$. This agreement is a striking confirmation of the basic correctness of the kinetic model of a gas and did much to establish this theory at a time when many scientists still refused to accept it.

Since $C_p = C_v + R$, it follows that the theoretical ratio of molar heat capacities for a monatomic ideal gas is

$$\frac{C_p}{C_v} = \gamma = \frac{\tfrac{3}{2}R + R}{\tfrac{3}{2}R} = \frac{5}{3} = 1.67.$$

This also is in good agreement with the experimental values in Table 18–1.

Polyatomic gases present a more complicated problem. We might expect that a molecule composed of two or more atoms could have additional energy associated with *vibratory* motion of the atoms relative to each other and of *rotation* of the molecule about its center of mass. Also, since every atom consists of a positive nucleus and one or more electrons, there may be additional energy associated with position and motion of these electric charges.

The *temperature* of a gas depends only on the average random *translational* kinetic energy of its molecules. When heat flows into a *monatomic* gas, at constant volume, all of this energy goes into an increase in random *translational* molecular kinetic energy, as evidenced by the agreement between the measured values of C_v and the values computed from the increase in translational kinetic energy. But when heat flows into a *diatomic* or *polyatomic* gas, part of the energy may go toward increasing rotational and vibrational motion. Hence for a given temperature change a greater total amount of energy is required, and polyatomic gases have larger molar heat capacities than monatomic gases. The experimental values in Table 18–1 show this effect.

To make the discussion somewhat more quantitative, we introduce the principle of *equipartition of energy*. According to this principle, each velocity component has, on the average, an associated kinetic energy $\frac{1}{2}kT$. The number of velocity components needed to describe the motion of a molecule completely is called the number of *degrees of freedom*. For a monatomic gas it is three; for a diatomic molecule there are two possible axes of rotation, perpendicular to each other and to the molecule's axis. (Rotations about the molecule's own axis are not considered because, in ordinary collisions, there is no way for this rotational motion to change.) Thus if five degrees of freedom are assigned to a diatomic molecule, the average total kinetic energy per molecule is $5kT/2$ instead of $3kT/2$. The total internal energy of n moles is $U = 5nRT/2$, and the molar heat capacity (at constant volume) is

$$C_v = \frac{5}{2}R. \tag{20–15}$$

In SI units,

$$C_v = (5/2)(8.314 \text{ J} \cdot \text{mol}^{-1} \cdot \text{K}^{-1}) = 20.78 \text{ J} \cdot \text{mol}^{-1} \cdot \text{K}^{-1}.$$

Reference to Table 18–1 shows that this is in approximate agreement with measured values for diatomic gases. The corresponding value of γ is

$$\gamma = \frac{C_p}{C_v} = \frac{\frac{5}{2}R + R}{\frac{5}{2}R} = \frac{7}{5} = 1.40,$$

which is again in reasonable agreement with experimental values.

Additional contributions to heat capacities of gases can arise from *vibrational* motion. Molecules are never perfectly rigid; the molecular bonds can stretch and bend, permitting internal vibrations to occur. There are additional degrees of freedom and energies associated with vibrational motion. For most diatomic gases, however, vibrational motion does *not* contribute appreciably to heat capacity, for reasons that require quantum mechanics for their understanding. Briefly, the energy of a vibrational motion can change only in finite steps. If the energy change of the first step is much larger than the energy possessed by most molecules, then nearly all the molecules will remain in the minimum-energy state of motion. In this case changing the temperature does not change their average vibrational energy appreciably, and the vibrational degrees of freedom are said to be "frozen out." But in more complex molecules the gaps between permitted energy levels are sometimes much smaller, and then vibration *does* contribute to heat capacity. In Table 18–1, the large values of C_v for some polyatomic molecules show the contributions of vibrational energy. In addition, a molecule with three or more atoms not in a straight line has three, not two, rotational degrees of freedom.

*20–6

DISTRIBUTION OF MOLECULAR SPEEDS

As mentioned above, the molecules in a gas do not all have the same speed. Direct measurements of the distribution of molecular speeds have been made by a number of methods. Figure 20–6 is a diagram of the apparatus used by Zartman and Ko in 1930–1934, a modification of a technique developed by Stern in 1920. Metallic silver is melted and evaporated in the oven O. A beam of silver atoms escapes through a small opening in the oven and passes through the slits S_1 and S_2 into an evacuated region. The cylinder C can be rotated at approximately 6000 rpm about the axis A. If the cylinder is at rest, the molecular beam enters the cylinder through a slit S_3 and strikes a curved glass plate G. The molecules stick to the glass plate, and the number arriving at any portion can be determined by removing the plate and measuring with a recording microphotometer the darkening that has resulted.

Now suppose the cylinder is rotated. Molecules can enter it only during the short time intervals during which the slit S_3 crosses the molecular beam. If the rotation is clockwise, as indicated, the glass plate moves toward the right while the molecules cross the diameter of the cylinder. They therefore strike the plate to the left of the point of impact when the cylinder is at rest, and the more slowly they travel, the farther to the left is their point of impact. The blackening of the plate is therefore a measure of the "velocity spectrum" of the molecular beam.

Several other techniques have been devised in more recent years for measuring distributions of molecular speeds with greater precision. Results of these experiments can be compared with a predicted speed distribution derived from a more detailed kinetic theory of gases and called the *Maxwell–Boltzmann distribution*. This theoretical distribution is shown in Fig. 20–7 for several different temperatures. At any temperature the curve shows a peak corresponding to the most probable speed; very small and very large values of v occur infrequently. As the temperature increases, the peak shifts to higher and higher speeds, corresponding to the increase in average molecular kinetic energy with temperature. Measured molecular speed distributions are generally in excellent agreement with the theoretical prediction of the Maxwell–Boltzmann distribution.

Rates of chemical reactions are often strongly temperature-dependent, and the Maxwell–Boltzmann distribution contains the reason for this dependence. When two reacting molecules collide, the reaction can occur only when there is a certain amount of interpenetration of the molecules; this requires a minimum energy, called the activation energy,

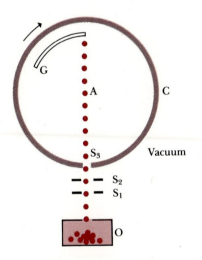

20–6 Apparatus used by Zartman and Ko in studying distribution of velocities.

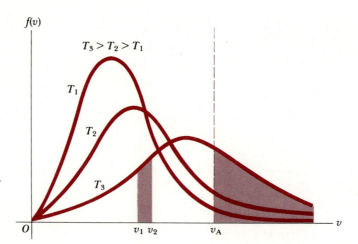

20–7 Maxwell–Boltzmann distribution curves for various temperatures. As the temperature increases, the curve becomes flatter, and its maximum shifts to higher temperature. At temperature T_3, the number of molecules having speeds in the range v_1 to v_2, and the number having speeds greater than v_A, are shown by the shaded areas under the T_3 curve.

and correspondingly a certain minimum speed v_A. As Fig. 20–7 shows, the number of molecules whose speeds exceed v_A, represented by the area under the curve to the right of v_A, increases rapidly with temperature. Thus we expect the rate of any reaction depending on an activation energy to increase with temperature. Similarly, many plant growth processes have strongly temperature-dependent rates.

Correspondingly, chemical *equilibrium* situations involve two competing reactions proceeding in opposite directions, and so such equilibria are often temperature-dependent. Warm-blooded animals depend on many different chemical equilibrium conditions, and so the range of body temperature with which a warm-blooded animal can survive is often very narrow.

20–7

CRYSTALS

Many materials can exist in a variety of solid forms; a familiar example is the element *carbon*. The black soot deposited on a kettle by a smoky campfire is nearly pure carbon. The "lead" in a pencil is not lead at all, but chiefly a different form of carbon called graphite. Diamond is a third form of solid carbon.

The remarkably dissimilar mechanical, thermal, electrical, and optical properties of these three forms of carbon can be understood in terms of the arrangement of the carbon atoms in the solid structure. In soot the atoms have no regular arrangement and are said to form an *amorphous* solid. Graphite and diamond are both crystals, but the orderly arrangement of atoms is different in the two crystal lattices, as shown in Fig. 20–8. The study of the spatial arrangements of atoms (or molecules or ions) in the various types of crystal lattices is a fascinating area of physics. We shall be concerned here with only a few basic ideas that are needed to understand some of the mechanical and thermal properties of single crystals or solids composed of an aggregate of crystals, that is, *polycrystalline solids*.

(a)

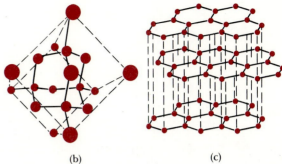

(b)   (c)

20–8 (a) Two different crystalline forms of carbon: graphite and diamond. (Courtesy of General Electric.) (b) Crystal structure of diamond; each atom is located at the center of a regular tetrahedron, with four equidistant nearest-neighbor atoms at the four corners. (c) Crystal structure of graphite, showing two-dimensional hexagonal arrays with strong bonds between atoms in each plane and much weaker bonds between planes. The diamonds shown are in fact synthetic, made by subjecting graphite to extreme temperature and pressure; they are much more costly than natural diamonds of comparable size (about one carat).

The particles comprising a crystal lattice are *closely packed*, as seen in Fig. 20–2 and as suggested by the model in Fig. 20–9, where the sodium and chlorine ions of a sodium chloride crystal are represented as spheres touching one another. Actually, each ion consists of a nucleus surrounded by electrons in motion, so that the boundary of an ion must be thought of as the average positions of the outermost electrons. It follows that the outer electrons of one particle in a crystal lattice come close to and even at times interpenetrate with the outer electrons of a neighbor-

20–9 Close packing of the sodium ions (black spheres) and chlorine ions (white spheres) in a sodium chloride crystal. (Courtesy of Alan Holden, reprinted with permission of Educational Services, Inc., from *Crystals and Crystal Growing*. Doubleday and Co., New York, 1960.)

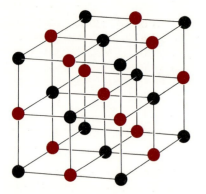

20–10 Symbolic representation of a sodium chloride crystal, with exaggerated distances between ions. Sodium ions are shown as black spheres, chlorine ions as colored spheres.

20–11 The forces between neighboring particles in a crystal may be visualized by imagining every particle to be connected to its neighbors by springs. In the case of a cubic crystal, all springs are assumed to have the same spring constant. Anisotropy is connected with differing spring constants in different directions.

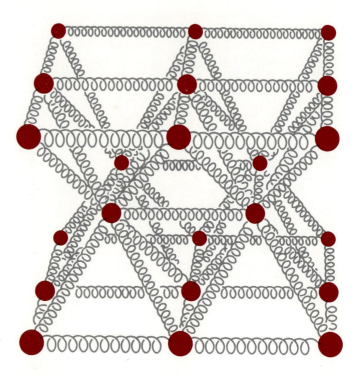

ing particle. In drawings it is customary to exaggerate the distances between neighboring particles, as in the *face-centered cubic* lattice structure shown in Fig. 20–10.

If the chemical composition and density of a very small volume element of a substance are measured at many different places, and are found to be the same at all points, the substance is said to be *homogeneous*. If the physical properties determining the transport of heat, electricity, light, etc., are *the same in all directions,* the substance is said to be *isotropic*. Cubic crystals are both homogeneous and isotropic, whereas all other crystals, although homogeneous, are not isotropic—they are *anisotropic*. Anisotropic crystals have different properties for different directions. Such a crystal may have a larger Young's modulus for stretching in one direction than for a perpendicular direction, it may conduct electricity better in some directions than others, and so on.

The *anisotropy* of noncubic crystals can be understood in terms of the forces between neighboring particles in a lattice. A convenient way of visualizing these forces is shown in Fig. 20–11, an idealized model that represents a crystal lattice as a large number of spheres, each connected to its neighbors by springs. The force constants of parallel springs are imagined equal, but the force constant of each spring pointing in one direction may be different from that of springs pointing in different directions. If each of the particles represented by spheres in Fig. 20–11 is imagined to be vibrating about its equilibrium position, one obtains a rough picture of the dynamic character of a crystal lattice.

The elastic properties of crystals can be partially understood from a diagram such as that in Fig. 20–11. Young's modulus is simply a measure of the stiffness of the springs in the direction in which the crystal is pulled or pushed. The shear modulus depends on the springs that are stretched and those that are compressed when the crystal is twisted. One of the first tests of the correctness of a lattice structure representing a particular material is to compare a calculated value of an elastic modulus with a measured value.

20–8

HEAT CAPACITY OF A CRYSTAL

In Section 20–5 we derived some simple results concerning heat capacities of ideal gases from a kinetic–molecular model. Similar considerations can be applied to crystalline solids. Let us consider a crystal consisting of N identical atoms. Each atom is bound to an equilibrium position by forces that may be pictured as springs, in the same spirit as Fig. 20–11. Thus each atom can vibrate about its equilibrium position; according to the equipartition principle introduced in Section 20–5, each atom should have an average kinetic energy of $\frac{1}{2}kT$ for each of its three degrees of freedom (corresponding to three components of velocity). In addition each atom has on the average some *potential* energy associated with the elastic deformation. Now, for a simple harmonic oscillator (discussed in Chapter 11), it is not hard to show that the average value of the kinetic energy is *equal* to the average potential energy. In our crystal lattice model, each atom is essentially a three-dimensional harmonic oscillator; and it can be shown that the equality of average kinetic and potential energies also holds here, provided the "spring" forces are proportional to the displacement from the equilibrium position.

Thus we expect each atom to have an average kinetic energy $3kT/2$ and an average potential energy $3kT/2$, or an average total energy $3kT$. Thus the total energy of the whole crystal, which may be expressed in either molecular or molar terms, is

$$U = 3NkT = 3nRT. \tag{20–16}$$

From this we conclude that the molar heat capacity of a crystal of N identical atoms should be

$$C_v = 3R. \tag{20–17}$$

In SI units,

$$C_v = (3)(8.314 \text{ J} \cdot \text{mol}^{-1} \cdot \text{K}^{-1}) = 24.9 \text{ J} \cdot \text{mol}^{-1} \cdot \text{K}^{-1}.$$

But this is just the law of Dulong and Petit, which we encountered in Section 15–5; there it was regarded as an *empirical* rule, but now we have *derived* it from kinetic theory. The agreement is only approximate, to be sure, but, considering the very simple nature of the model, it is quite significant.

At low temperatures the heat capacities of most solids *decrease* with decreasing temperature. As with vibrational energy of molecules, this phenomenon requires quantum-mechanical concepts for its understanding. The vibrating atoms in the crystal lattice can gain or lose energy only in certain finite increments. At very low temperatures the quantity kT is much *smaller* than the smallest energy increment the vibrators can accommodate. As a result, at low T most of the systems remain in their lowest energy states because the energy required to make transitions to the next higher energy levels is not available. Thus, at low temperatures the average vibrational energy per vibrator is *less* than kT, and correspondingly the heat capacity per molecule is *less* than k. But in the other limit, where kT is *large* compared to the minimum energy increment, the classical equipartition theorem holds, and the total heat capacity is $3k$ per molecule, or $3R$ per mole, as the Dulong and Petit relation predicts. This behavior is illustrated by Fig. 15–5. Quantitative understanding of the temperature variation of specific heats was one of the triumphs of quantum mechanics during the early years of its development, early in the twentieth century.

20–9

DIFFUSION AND OSMOSIS

When a drop of ink or food coloring is added to a container of water, the color eventually permeates all the water uniformly, even if it is not stirred or agitated. Similarly, perfume sprayed into the air at one end of a room can be detected after a time at the opposite end, even when there is no circulation or mixing of the air. These are two familiar examples of *diffusion*, the migration of one kind of molecules through a material consisting of another kind, caused entirely by the random motions of the molecules. In each example, minority molecules are introduced into a specific region of a material, and they move on the average from regions of greater concentration to those of smaller concentration.

Kinetic theory provides the key to understanding diffusion. In Fig. 20–12 the concentration of perfume molecules (represented by colored dots) is greater in the left side of the box than in the right. In the center is an imaginary surface (not a real physical barrier) dividing the box into two halves. In any specific time interval more molecules cross over this surface from left to right than from right to left because the concentration is greater on the left. Thus there is a net drift from left to right; this apparent flow stops only when the concentration becomes uniform and the left-to-right and right-to-left drifts become equal.

More detailed analysis shows that the rate of diffusion across any given area A is directly proportional to the change in concentration per unit distance. Concentration, denoted by C, is usually measured in moles per unit volume; thus the SI unit of concentration is $1 \text{ mol} \cdot \text{m}^{-3}$. The rate of diffusion of matter, denoted by Q, is measured in amount of substance per unit time; hence in SI units Q is in $\text{mol} \cdot \text{s}^{-1}$. Specifically, it is found that the rate of diffusion Q across any area A between two re-

20–12 The concentration of minority molecules, shown by large colored dots, is greater on the left side, and more molecules from the left pass through the imaginary surface in the middle than from the right. Thus there is a net diffusion of molecules from left to right.

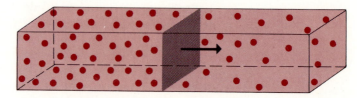

gions a distance L apart where the concentrations are C_1 and C_2 is given by

$$Q = \frac{DA(C_2 - C_1)}{L}, \qquad (20-18)$$

where D is a proportionality constant, different for different combinations of materials, called the *diffusion constant*. We challenge the reader to verify that for consistency of units, D must have units $m^2 \cdot s^{-1}$.

Equation (20–18) is called the *diffusion equation* or *Fick's law*; the physiologist Adolf Fick (1829–1901) discovered this law experimentally, and it has since been derived from kinetic-theory considerations. Its similarity in form to the heat-flow equation, Eq. (16–1), is striking; heat flow can be considered as a kind of diffusion phenomenon, and both relations are obtained from kinetic-theory considerations in similar ways. The quantity $(C_2 - C_1)/L$, the change in concentration per unit length, is called the *concentration gradient*; it is analogous to the temperature gradient in heat flow. The reader may verify that the SI unit of concentration gradient is $1 \, mol \cdot m^{-4}$.

EXAMPLE 1 A hydrogen storage tank is fitted with a horizontal pipe 2.0 m long and 5 cm^2 in cross-sectional area. There is a shut-off valve close to the tank, and the other end of the pipe is open to the atmosphere. The valve is slightly leaky, and the concentration of hydrogen in the pipe near the valve is $0.5 \, mol \cdot m^{-3}$. The diffusion constant for hydrogen in air at one atmosphere and 273 K is $D = 6.3 \times 10^{-5} \, m^2 \cdot s^{-1}$. If the concentration of hydrogen in the outside atmosphere is negligible, what is the rate of diffusion of hydrogen out of the pipe?

Solution From Eq. (20–18), the diffusion rate is

$$Q = \frac{(6.3 \times 10^{-5} \, m^2 \cdot s^{-1})(5 \times 10^{-4} \, m^2)(0.5 \, mol \cdot m^{-3} - 0)}{2.0 \, m}$$

$$= 7.9 \times 10^{-9} \, mol \cdot s^{-1}.$$

The number of *molecules* per second is this number multiplied by Avogadro's number, or about 4.7×10^{15} molecules $\cdot s^{-1}$. We note that the hydrogen concentrations are much smaller than the concentration of air molecules; at STP one mole of an ideal gas occupies a volume of 22.4 l or $0.0224 \, m^3$, so the concentration is $(1/0.0224) \, mol \cdot m^{-3} = 44.6 \, mol \cdot m^{-3}$. ◄

The diffusion constant D depends on the substances involved, and also on temperature. It is generally much smaller for diffusion in liquids and solids than in gases because in condensed matter the diffusing molecules collide with other molecules much more frequently and travel, on the average, much shorter distances between collisions. Diffusion constants usually increase with increasing temperature, because average speeds of random molecular motion increase with temperature.

Diffusion plays an essential role in many life processes. The exchange of oxygen from air to blood in the lungs, and from blood to cells in muscle tissue, are diffusion processes. Diffusion is also the mechanism of exchange of oxygen and carbon dioxide with the atmosphere that is necessary for plant life.

Diffusion phenomena in which fluid molecules pass through a semi-permeable membrane are called *osmosis;* these too are of central importance in many biological processes. The structure or chemical make-up of a membrane may be such that small or special molecules can pass through nearly unimpeded while larger or chemically different ones are blocked. This can result in pressure differences across the membrane as well as other interesting phenomena.

A simple example of osmosis is shown in Fig. 20–13. The left side of the U-tube contains a sugar solution, the right side pure water. The two sides are separated by a semipermeable membrane, permeable to the relatively small water molecules but impermeable to the much larger sugar molecules. Because on the left side the sugar molecules occupy some of the space that would otherwise contain water molecules, the concentration of water molecules is less on the left side than on the right. Thus there is a net diffusion of water from right to left through the membrane. The sugar molecules on the left would like to diffuse into the right side, but the membrane stops them. Thus the fluid level rises in the left side of the tube and drops in the right side.

The total pressure on each side of the membrane can be represented as the sum of partial pressures due to water molecules and to sugar molecules. The diffusion of water from right to left continues until the partial pressures of water are the same on the two sides of the membrane. But then the *total* pressure on the left is greater than on the right because it includes also the pressure due to sugar molecules. It turns out that when equilibrium has been reached the pressure difference is approximately proportional to the concentration C of the dissolved substance. The equilibrium pressure difference is called the *osmotic pressure,* denoted by Π. Specifically, it is found that

$$\Pi = CRT, \tag{20–18}$$

where R is the ideal gas constant and T the absolute temperature. Since C is the number of moles n of solute per unit volume V of solution (i.e., $C = n/V$), this can also be written as

$$\Pi = \frac{nRT}{V}. \tag{20–19}$$

20–13 The concentration of water molecules on the right side of the membrane is greater than on the left, and water molecules diffuse from right to left. The large sugar molecules (color dots) are blocked by the membrane. (b) Equilibrium is reached when the diffusion of water molecules is the same on the two sides of the membrane. In this case the difference of hydrostatic pressure $\rho g h$ is equal to the osmotic pressure Π.

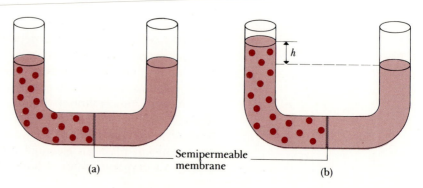

Semipermeable membrane

(a)　(b)

This somewhat surprising result states that the partial pressure of the dissolved substance is the same as though it were an ideal gas! Although the large molecules collide continually with water molecules, they rarely encounter each other. In this respect they act like noninteracting particles, the basic assumption of an ideal gas.

EXAMPLE 2   The left side of the U-tube in Fig. 20–13 contains a sugar solution prepared by dissolving 15.0 g of table sugar in 1 l of water. At equilibrium, at $T = 20°C$, what is the difference in height of fluid on the two sides of the U-tube? Table sugar is sucrose, $C_{12}H_{22}O_{11}$; its molecular mass is $M = 342$ g·mol^{-1}.

Solution   We first find the concentration of the solution. The number of moles of sugar is $(15.0$ g$)/(342$ g·mol$^{-1}) = 0.0439$ mol. This is dissolved in 1.0×10^{-3} m^3 of water, so the concentration is

$$C = \frac{0.0439 \text{ mol}}{1.0 \times 10^{-3} \text{ m}^3} = 43.9 \text{ mol·m}^{-3}.$$

The osmotic pressure is then given by Eq. (20–18):

$$\Pi = CRT = (43.9 \text{ mol·m}^{-3})(8.314 \text{ J·mol}^{-1}\text{·K}^{-1})(293 \text{ K})$$
$$= 1.07 \times 10^5 \text{ Pa}.$$

The difference in height h of the two columns is then given by $\Pi = \rho g h$, where ρ is the density of the liquid and g the acceleration due to gravity:

$$h = \frac{\Pi}{\rho g} = \frac{1.07 \times 10^5 \text{ Pa}}{(1.0 \times 10^3 \text{ kg·m}^{-3})(9.8 \text{ m·s}^{-2})} = 10.9 \text{ m}.$$

(We have ignored the small difference in density of pure water and the sugar solution.) This is a surprisingly great height difference for a rather modest concentration of sugar.   ◀

Osmotic pressure in the roots of maple trees is believed to be responsible for the rising of sap in the trees in spring; pure water diffuses through the semipermeable membranes of the roots, pushing the dilute sugar solution inside the roots up into the tree.

The direction of diffusion in Fig. 20–13a can be reversed by applying a pressure to the surface of the liquid on the left side that exceeds that on the right by more than the osmotic pressure. Pure water then diffuses through the membrane from left to right, leaving the solute molecules behind. This process, called *reverse osmosis*, has been explored as a process for desalinizing sea water. The required pressure turns out to be about 25 atm, and the principal practical difficulty is finding efficient yet durable membranes that can withstand this pressure differential.

Osmosis is of central importance in many biological processes. Cell walls are semipermeable membranes, and the diffusion of water in or out of a cell is determined by the concentrations of dissolved substances on the two sides of the cell wall. When fluids are given to a patient intravenously, the concentrations of impermeable dissolved substances such as salt and glucose (and their osmotic pressures) are often adjusted to balance those of the tissue cells. Such a solution is said to be *isotonic*.

Osmotic pressure in plant cells is important in maintaining the rigidity of the plant's structure, just as an inflated balloon has a definite shape but a deflated one collapses. When insufficient water is available, the plant wilts and droops, or, in the case of cacti and other succulents, takes on a shriveled and emaciated appearance. Addition of water, either through the roots or by diffusion through the semipermeable membranes forming the plant surfaces, restores the osmotic pressure needed for normal firmness of structure.

Finally, we remark that the concept of a semipermeable membrane that blocks large molecules completely is an idealization. Real membranes do permit some passage of large molecules, although more slowly than smaller molecules. A more precise term to describe the behavior of real membranes is *differentially permeable membrane*.

QUESTIONS

20–1 Which has more atoms, a kilogram of hydrogen or a kilogram of lead? Which has more mass?

20–2 Chlorine is a mixture of two isotopes, one having molecular mass $35 \ \mathrm{g \cdot mol^{-1}}$, the other $37 \ \mathrm{g \cdot mol^{-1}}$. Which molecules move faster, on the average?

20–3 The proportion of various gases in the earth's atmosphere changes somewhat with altitude. Would you expect the proportion of oxygen at high altitude to be greater or less than at sea level?

20–4 *Comment on this statement:* When two gases are mixed, if they are to be in thermal equilibrium they must have the same average molecular speed.

20–5 In deriving the ideal-gas equation from the kinetic–molecular model, we ignored potential energy due to the earth's gravity. Is this omission justified?

20–6 In deriving the ideal-gas equation, we assumed the number of molecules to be very large, so that we could compute the average force due to many collisions; however the ideal-gas equation holds accurately only at low pressures, where the molecules are few and far between. Is this inconsistent?

20–7 Some elements of solid crystalline form have molar heat capacities *larger* than $3R$. What effects could account for this?

20–8 Considering molecular speeds, is v_{rms} equal to the *most probable* speed, as indicated by the peak on one of the curves in Fig. 20–7?

20–9 A gas storage tank has a small leak. The pressure in the tank drops more quickly if the gas is hydrogen or helium than if it is oxygen. Why?

20–10 Consider two specimens of gas at the same temperature, both having the same total mass but having different molecular masses. Which has the greater internal energy? Does your answer depend on the molecular structure of the gases?

20–11 A process called *gaseous diffusion* is sometimes used to separate isotopes of uranium, i.e., atoms of the element having different masses, such as ^{235}U and ^{238}U. Can you speculate on how this might work?

20–12 Dried fruit such as prunes or raisins can be rehydrated by soaking in water. Why doesn't this wash the sugar and other nutrients out of the interior of the fruit?

20–13 When ripe grapes or cherries are immersed in water, they swell up and the skins split. Why?

20–14 When food is preserved by freeze-drying, it is first quick-frozen, then placed in a vacuum chamber and irradiated with infrared radiation. What is the purpose of the vacuum? The radiation?

PROBLEMS

Molecular data

N_A = Avogadro's number
   = 6.02×10^{23} molecules·mol^{-1}.
Mass of a hydrogen atom = 1.67×10^{-27} kg.
Mass of a nitrogen molecule = $28 \times 1.66 \times 10^{-27}$ kg.
Mass of an oxygen molecule = $32 \times 1.66 \times 10^{-27}$ kg.

20–1 Consider an ideal gas at 0°C and at 1 atm pressure. Imagine each molecule to be, on the average, at the center of a small cube.

a) What is the length of an edge of this small cube?

b) How does this distance compare with the diameter of a molecule?

20–2 A mole of liquid water occupies a volume of 18 cm^3. Imagine each molecule to be, on the average, at the center of a small cube.

a) What is the length of an edge of this small cube?

b) How does this distance compare with the diameter of a molecule?

20–3 What is the length of the side of a cube, in a gas at standard conditions, that contains a number of molecules equal to the population of the United States (about 200 million)?

20–4 The lowest pressures readily attainable in the laboratory are of the order of 10^{-13} atm. At this pressure and ordinary temperature (say $T = 300$ K), how many molecules are present in a volume of 1 cm^3?

20–5 Experiment shows that the size of an oxygen molecule is of the order of 2×10^{-10} m. Make a rough estimate of the pressure at which the finite volume of the molecules should cause noticeable deviations from ideal-gas behavior at ordinary temperatures ($T = 300$ K).

20–6 How many moles are there in a glass of water (0.2 kg)? How many molecules?

20–7 The density of crystalline sodium chloride is 2.16 g·cm^{-3}, and its crystal structure is shown in Fig. 20–10. Find the spacing of adjacent atoms in the crystal lattice. The atomic mass of sodium is 23.0 g·mol^{-1} and that of chlorine is 35.5 g·mol^{-1}.

20–8 The density of liquid oxygen at 1 atm and $-185°$C is about 1.14 g·cm^{-3}. From this, estimate the volume and size of a single oxygen molecule.

20–9

a) At what temperature is the rms speed of hydrogen molecules equal to the speed of the first earth satellite (about 18,000 mi·hr^{-1})?

b) At what temperature is the rms speed equal to the escape speed from the gravitational field of the earth?

20–10 At what temperature is the rms speed of oxygen molecules equal to the rms speed of hydrogen molecules at 0°C?

20–11 Isotopes of uranium are sometimes separated by gaseous diffusion, using the fact that the rms speeds of their molecules in vapor are slightly different, and hence the vapors diffuse at slightly different rates. Assuming the atomic masses for ^{235}U and ^{238}U are 235 g·mol^{-1} and 238 g·mol^{-1}, respectively, what is the ratio of the rms speed of the ^{235}U atoms in the vapor to that of the ^{238}U atoms, assuming the temperature is uniform?

20–12 A flask contains a mixture of mercury vapor, neon, and helium. Compare

a) the average kinetic energies of the three types of atoms, and

b) the root-mean-square speeds.

20–13 The speed of propagation of a sound wave in air at 27°C is about 350 m·s^{-1}. Compare this with

a) v_{rms} for nitrogen molecules;

b) the root-mean-square value of v_x at this temperature.

(If sound propagation were an isothermal process, which it ordinarily is not, the speed of sound would be equal to $(v_x)_{\mathrm{rms}}$. See Section 21–5.)

20–14

a) What is the average translational kinetic energy of a molecule of oxygen at a temperature of 300 K?

b) What is the average value of the square of its speed?

c) What is the root-mean-square speed?

d) What is the momentum of an oxygen molecule traveling at this speed?

e) Suppose a molecule traveling at this speed bounces back and forth between opposite sides of a cubical vessel 10 cm on a side. What is the average force it exerts on the walls of the container? (Assume that the molecule's velocity is perpendicular to the two sides that it strikes.)

f) What is the average force per unit area?

g) How many molecules traveling at this speed are necessary to produce an average pressure of 1 atm?

h) Compare with the number of oxygen molecules actually contained in a vessel of this size, at 300 K and atmospheric pressure.

i) Your answer for (h) should be three times as large as the answer for (g). Where does this discrepancy arise?

20–15 The surface of the sun has a temperature of about 6000 K and consists largely of hydrogen.

a) Find the rms speed and average kinetic energy of a hydrogen molecule.

b) The escape velocity for a particle to leave the gravitational influence of the sun is given by $(2GM/R)^{1/2}$, where M is the sun's mass, R its radius, and G the gravitational constant (Section 6–5, Example 6). Can appreciable quantities of hydrogen escape from the sun's gravitational field?

20–16 What is the total random kinetic energy of the molecules in 1 mole of helium at a temperature of

a) 300 K?

b) 301 K?

c) Compare the difference between these with the change in internal energy of 1 mole of helium when its temperature is increased by 1 K, as computed from the relation $\Delta U = nC_v \Delta T$.

20–17 Smoke particles in the air typically have masses of the order of 10^{-16} kg. The Brownian motion of these particles resulting from collisions with air molecules can be observed with a microscope.

a) Find the root-mean-square speed of Brownian motion for such a particle in air at 300 K.

b) Would the speed be different if the particle were in hydrogen gas at the same temperature? Explain.

20–18 It can be shown that a projectile of mass m can "escape" from the earth's gravitational field if it is launched vertically upward with a kinetic energy greater than mgR, where g is the acceleration due to gravity at the earth's surface, and R is the earth's radius.

a) At what temperature would an average oxygen molecule have this much energy? An average hydrogen molecule?

b) Can you explain on this basis why the earth's atmosphere contains very little hydrogen?

20–19 Compute the increase in gravitational potential energy of an oxygen molecule for an increase in elevation of 1 m near the earth's surface. At what temperature is this equal to the average kinetic energy of oxygen molecules?

20–20

a) For what mass of molecule or particle is v_{rms} equal to $1 \text{ m} \cdot \text{s}^{-1}$ at a temperature of 300 K?

b) If the particle is an ice crystal, how many molecules does it contain?

c) Estimate the size of the particle. Would it be visible to the naked eye? Through a microscope?

20–21 The oven in Fig. 20–6 contains bismuth at a temperature of 840 K; the drum is 10 cm in diameter and rotates at 6000 rev·min^{-1}. Find the displacement on the glass plate G, measured from a point directly opposite the slit, of the points of impact of the molecules Bi and Bi$_2$. Assume that all molecules of each species travel with the rms speed appropriate to that species.

20–22 Suppose the two atoms in an oxygen molecule are represented as two point masses 2×10^{-10} m apart.

a) Calculate the moment of inertia of a molecule about an axis through the midpoint, perpendicular to the line joining the atoms.

b) If the rotational kinetic energy of the molecule is $\frac{1}{2}kT$, what is the root-mean-square angular velocity of rotation at $T = 300$ K?

20–23 Compute the specific heat capacity at constant volume of hydrogen gas, and compare it with the specific heat capacity of water.

20–24 Calculate the molar heat capacity of water vapor, assuming the triatomic molecule has three translational and three rotational degrees of freedom and that vibrational motion does not contribute. The actual specific heat capacity of water vapor at low pressures is about $0.48 \text{ cal} \cdot \text{g}^{-1} \cdot \text{K}^{-1}$. Compare this with your calculation, and comment on the actual role of vibrational motion.

20–25 For each of the triatomic gases in Table 18–1, compute the value of C_v on the assumption that there is no vibrational energy. Compare with the measured values in the table, and compute the fraction of the total heat capacity that is due to vibration for each of the three gases.

20–26 What is the concentration of

a) pure water at 20°C and 1 atm, in mol·m^{-3}; and

b) air at 20°C and 1 atm, in mol·m^{-3}?

20–27 A horizontal glass tube is filled with water, and a lump of sucrose is inserted at one end. The tube is 0.6 m long and 2.0 cm in diameter. The concentration of sucrose near the lump is found to be 300 mol·m^{-3}. The diffusion constant for sucrose in water is 0.52×10^{-9} m^2·s^{-1}. What is the initial diffusion rate of sucrose along the tube?

20–28 The diffusion constant for water vapor in air at 20°C is 0.24×10^{-4} m^2·s^{-1}. A horizontal glass tube 0.5 m long and 2.0 cm^2 in cross-sectional area is stoppered at one end with a wet sponge; the other end is open to dry air. The temperature is 20°C.

a) If the air in the tube near the sponge is saturated with water vapor, what is the concentration of water vapor there, in mol·m^{-3}?

b) At what rate does water vapor diffuse out the open end of the tube?

c) If the sponge initially contains 4 g of water, how much time is required for it to become completely dry?

d) How does the diffusion rate change if the air outside the tube is not perfectly dry but has 75% relative humidity?

20–29 What minimum concentration of sucrose in maple sap would be needed for the osmotic pressure in the roots to push the sap to a height of 10 m at a temperature of 10°C?

20–30 A ripe cherry is immersed in water at 20°C. The osmotic pressure is found to be 5 atm. If the interior of the cherry is principally a solution of glucose in water, what is the concentration of the solution? (Glucose, $C_6H_{12}O_6$, has a molecular mass of $M = 180$ g·mol^{-1}.)

20–31 Sea water contains about 30 kg of salt per cubic meter. Salt is sodium chloride, NaCl, molecular mass $M = 58.4$ g·mol^{-1}. The molecules are fully dissociated into Na$^+$ and Cl$^-$ ions, so the total ion concentration is twice the number of moles of NaCl per cubic meter.

a) What is the total concentration of ions, in mol·m^{-3}?

b) What is the osmotic pressure across a semipermeable membrane with fresh water on one side and sea water on the other, at a temperature of 20°C?

c) What minimum pressure would be needed for reverse osmosis, i.e., diffusion of water from the salty side to the pure side, leaving the salt ions behind?

CHAPTER 21

MECHANICAL WAVES

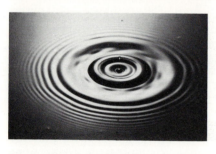

21–1 A small ball dropped vertically into water produces a wave pattern that moves radially outward from the source of the wave. The wave crests and troughs are concentric circles. (Photo by Fundamental Photographs, New York.)

A wave is any disturbance from an equilibrium condition that travels or *propagates* with time from one region of space to another. Examples of wave phenomena are everywhere around us; sound, light, ocean waves, radio and television transmission, and earthquakes are all wave phenomena, and many other examples might be cited. Wave phenomena are found in all branches of physical and biological science, and indeed the concept of waves is one of the most important unifying threads running through the entire fabric of the natural sciences. In this chapter and the next two we consider a particular class of waves called *mechanical waves*, which always travel within some material substance called the *medium* for the wave. Some waves are periodic, in the sense that the particles of the medium undergo periodic motions during wave propagation. We can develop relations between the speed of propagation of a wave and the mechanical properties of the medium. Finally, we study the properties of a type of periodic wave of special importance, a *sinusoidal* wave.

21–1

MECHANICAL WAVE PHENOMENA

In this chapter we consider a particular class of waves called *mechanical waves*. There are various types of mechanical waves; each type has associated with it some material substance called the *medium* for that type. As the wave travels through the medium, the particles that make up the medium undergo displacements of various kinds, depending on the nature of the wave.

There are other types of waves that are *not* mechanical in nature and have no associated medium. Light is an example; many of the observed properties of light are best explained by a wave theory. Light waves are of the same fundamental nature as radio waves, infrared and ultraviolet waves, x-rays, and gamma rays. One of the outstanding developments of twentieth-century physics has been the discovery that fundamental particles have wave properties. A beam of electrons, for example, is reflected

(a) Transverse displacement

(b) Longitudinal displacement

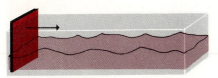

(c) Longitudinal and transverse displacement

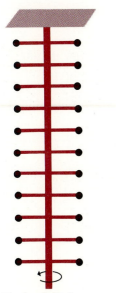

(d) Torsional displacement

21–2 Propagation of disturbances.

by a crystal in much the same way as is a beam of x-rays. We shall study some of these nonmechanical waves in later chapters.

Here are a few examples of mechanical waves. In Fig. 21–2a the medium is a long spring or "Slinky," or even just a wire or cord under tension. If we give the left end a small sideways shake or wriggle, the wriggle travels down the length of the spring, with successive sections of spring going through the same sideways motion that was given at the end, but at successively later times. Because the displacements of the medium are perpendicular (transverse) to the direction of travel of the wave along the medium, this is called a *transverse* wave.

In Fig. 21–2b the medium is a liquid or a gas in a tube with a rigid wall at the right end and a movable piston at the left end. If the piston is given a back-and-forth motion, again a displacement travels down the length of the medium. This time the motions of the particles of the medium are back and forth along the same direction as the travel of the wave, and this type of wave is called a *longitudinal* wave.

In Fig. 21–2c the medium is water in a trough such as an irrigation ditch or a canal. When the flat board at the left end is moved back and forth, a disturbance travels down the length of the trough; careful observation shows that in this case the displacements of the water have both longitudinal and transverse components.

In Fig. 21–2d the medium is a series of dumbbells attached to a steel rod or strip. If the bottom one is given a twisting or *torsional* displacement (by rotating it in a horizontal plane) a *torsion wave* travels upward along the length of the medium.

These examples have several things in common. First, the medium itself does not travel through space; its individual particles undergo back-and-forth motions with respect to their equilibrium positions. What does travel is the wave disturbance. Second, to set any of these systems in motion requires energy input in the form of mechanical work, and the wave motion has the effect of transporting this energy from one region of space to another. Third, the wave motion can be used to transmit information, a function that is familiar with sound waves. Thus waves transport energy from one region to another, but not matter. Finally, observation shows that in each case the disturbance travels with a definite speed through the medium. This speed is called the *wave speed*, and it is determined in each case by the mechanical properties of the medium. Wave speed will be denoted by c.

21–2

PERIODIC WAVES

To introduce basic concepts in as simple a context as possible, we concentrate first on one specific kind of mechanical wave, namely, transverse waves on a stretched string or rope. One end of a long flexible rope is tied to a stationary object and the other end is held, stretching the rope tight. Then this end is given some transverse motion; if it is given a single "flip" or "wiggle," then the result is a single *wave pulse,* which travels down the length of the string.

A more interesting situation arises when the free end of the rope is given a repetitive or *periodic* motion. Suppose, in particular, that the end

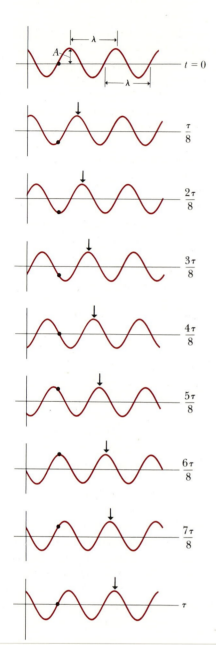

21–3 A sinusoidal transverse wave traveling toward the right. The shape of the string is shown at intervals of one eighth of a period. The vertical scale is exaggerated.

21–4 A sinusoidal longitudinal wave traveling toward the right, shown at intervals of one eighth of a period.

is moved back and forth with *simple harmonic motion* of amplitude A, frequency f, and period τ, where as usual $f = 1/\tau$. A *continuous succession* of transverse sinusoidal waves then advances along the string. The shape of a portion of the string near the end, at intervals of $\frac{1}{8}$ of a period, is shown in Fig. 21–3 for a total time of one period. The waveform advances steadily toward the right, as indicated by the short arrow pointing to one particular wave crest, while any one point on the string (see the black dot) oscillates back and forth about its equilibrium position with simple harmonic motion. It is important to distinguish between the motion of the *waveform*, which moves with constant speed c *along* the string, and the motion of *a particle of the string*, which is simple harmonic and *transverse* to the string.

The distance between two successive maxima (or between any two successive points in the same phase) is the *wavelength* of the wave and is denoted by λ. Since the waveform, traveling with constant speed c, advances a distance of one wavelength in a time interval of one period, it follows that $c = \lambda/\tau$, or since $f = 1/\tau$,

$$c = f\lambda. \tag{21–1}$$

That is, *the speed of propagation equals the product of frequency and wavelength.*

To understand the mechanics of a *longitudinal* wave, we may consider a long tube filled with a fluid and provided with a plunger at the left end, as shown in Fig. 21–4. Suppose the plunger is moved with simple harmonic motion along a line parallel to the direction of the tube. During a part of each oscillation, a region whose pressure is greater than the equilibrium pressure is formed. Such a region is called a *condensation* and is represented by a darkly shaded area. Following the production of a condensation, a region is formed in which the pressure is lower than the equilibrium value. This is called a *rarefaction* and is represented by a lightly shaded area. The condensations and rarefactions move to the right with constant speed c, as indicated by successive positions of the

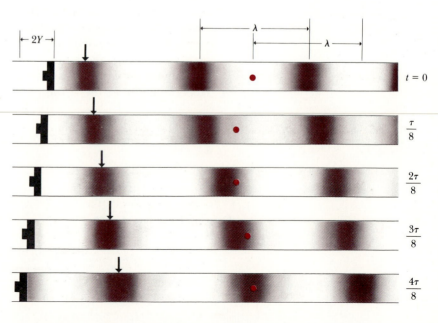

small vertical arrow. The speed of longitudinal waves in air (sound) at 20°C is 344 m·s⁻¹ or 1130 ft·s⁻¹. The motion of a single particle of the medium, shown by a colored dot, is simple harmonic, parallel to the direction of propagation.

The wavelength is the distance between two successive condensations or two successive rarefactions, and the same fundamental equation, $c = f\lambda$, holds in this, as in all types of periodic or repetitive waves.

EXAMPLE What is the wavelength of a sound wave having a frequency of 262 Hz (the approximate frequency of the note "middle C" on the piano)?

Solution At 20°C the speed of sound in air is 344 m·s⁻¹, as mentioned above. From Eq. (21–1),

$$\lambda = \frac{c}{f} = \frac{344 \text{ m·s}^{-1}}{262 \text{ s}^{-1}} = 1.31 \text{ m}.$$

The "high C" sung by coloratura sopranos is two octaves above middle C. The corresponding frequency is four times as large, $f = 4(262 \text{ Hz}) = 1048$ Hz, and the wavelength is one fourth as large, $\lambda = (1.31 \text{ m})/4 = 0.328$ m. ◀

21–3

SPEED OF A TRANSVERSE WAVE

The speed of propagation c of a transverse wave on a string is determined by two mechanical quantities—the *mass* per unit length of the string and the *tension*. We shall now develop that relationship by analyzing a particularly simple type of wave motion.

Consider a perfectly flexible string, as shown in Fig. 21–5, under a tension S and with linear density (mass per unit length) μ. In Fig. 21–5a the string is at rest. At time $t = 0$, a constant transverse force F is applied at the left end of the string. It might be expected that the end would move with constant acceleration; this would certainly occur if the force were applied to a point mass. But here the effect of the force is to set successively more and more mass in motion. Force is the rate of change of momentum mv; in this case mv changes because m changes, not v. Hence the end of the string moves upward with constant *velocity v*.

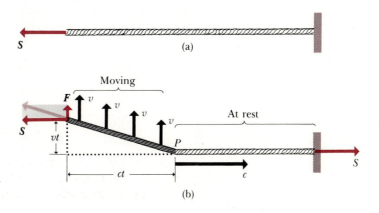

21–5 Propagation of a transverse disturbance in a string.

Figure 21–5b shows the shape of the string after a time t has elapsed. All points of the string at the left of the point P are moving with speed v, whereas all points at the right of P are still at rest. The boundary point P between the moving and the stationary portions is traveling to the right with the *speed of propagation* c. The left end of the string has moved up a distance vt and the boundary point P has advanced a distance ct.

The tension at the left end of the string is the vector sum of the forces S and F. Because there is no motion along the length of the string, there is no unbalanced horizontal force, so S, the magnitude of the horizontal component, does not change when the string is displaced. As a result of the increased tension, the string clearly stretches somewhat. It can be shown that, for small displacements, the amount of stretch is approximately proportional to the increase in tension, as would be expected from Hooke's Law.

The speed of propagation c can be calculated by applying the impulse–momentum relation to the portion of the string that is in motion at time t, that is, the darkly shaded portion in Fig. 21–5. We set the transverse impulse (transverse force × time) equal to the change of transverse momentum of the moving portion (mass × transverse velocity). The *impulse* of the transverse force F in time t is Ft. By similar triangles,

$$\frac{F}{S} = \frac{vt}{ct}, \qquad F = S\frac{v}{c}.$$

Hence,

$$\text{Transverse impulse} = S\frac{v}{c}t.$$

The mass of the moving portion is the product of the mass per unit length μ and the length ct. (In its displaced position, the section of string is stretched somewhat, making its mass per unit length somewhat *less* than μ and its length somewhat *greater* than ct. But the mass of the section is still μct, the same as in the undisplaced position.) Hence,

$$\text{Transverse momentum} = \mu ctv.$$

We note that the momentum increases with time *not* because mass is moving faster, as was usually the case in Chapter 7, but because *more mass* is brought into motion. Nevertheless, the impulse of the force F is still equal to the total change in momentum of the system. Applying this relation, we obtain

$$S\frac{v}{c}t = \mu ctv,$$

and therefore

$$c = \sqrt{\frac{S}{\mu}} \qquad \text{(transverse wave).} \qquad (21–2)$$

Thus the speed of propagation of a transverse pulse in a string depends only on the tension (a force) and the mass per unit length. Although the above calculation of wave speed considered only a very special kind of pulse, a little thought shows that *any* shape of wave disturbance can be considered as a series of pulses with different rates of transverse displacement. Thus, although derived for a special case, Eq. (21–2) is valid for *any* transverse wave motion on a string, including, in particular, the sinusoidal and other periodic waves discussed in Section 21–2.

EXAMPLE   One end of a rubber tube is fastened to a fixed support. The other end passes over a pulley at a distance of 8 m from the fixed end and carries a load of 2 kg. The mass of the tube, between the fixed end and the pulley, is 600 g.

a) What is the speed of a transverse wave in the tube?

b) If a point on the string is given a transverse simple harmonic motion with a frequency of 20 Hz, what is the wavelength of the wave?

Solution

a) The tension in the tube equals the weight of the 2-kg load, or

$$S = (2 \text{ kg})(9.8 \text{ m} \cdot \text{s}^{-2}) = 19.6 \text{ N}.$$

The mass of the tube, per unit length, is

$$\mu = \frac{m}{L} = \frac{0.60 \text{ kg}}{8 \text{ m}} = 0.075 \text{ kg} \cdot \text{m}^{-1}.$$

The speed of propagation is therefore

$$c = \sqrt{\frac{S}{\mu}} = \sqrt{\frac{19.6 \text{ N}}{0.075 \text{ kg} \cdot \text{m}^{-1}}} = 16.2 \text{ m} \cdot \text{s}^{-1}.$$

b) From Eq. (21–1),

$$\lambda = \frac{c}{f} = \frac{16.2 \text{ m} \cdot \text{s}^{-1}}{20 \text{ s}^{-1}} = 0.81 \text{ m}.$$

A transverse wave traveling along a string carries energy from one region of the string to another. Suppose we look at a particular point on the string; if the wave is traveling from left to right, then the portion of string to the left exerts a transverse force on the portion to the right as the point undergoes its vertical displacements. Thus the left portion does work on the right portion and transfers energy to it. A detailed calculation of the rate of doing work (power) shows that for a sinusoidal wave of amplitude A, frequency f, and angular frequency $\omega = 2\pi f$, the power P (or rate of energy transmission) is

$$P = \tfrac{1}{2}SA^2\omega^2, \qquad (21\text{–}3)$$

where S is the string tension. It is interesting to note that the rate of energy transfer is proportional to the *square* of the amplitude, and also that it is proportional to the square of the frequency.

Another property exhibited by transverse waves is *polarization*. In producing a transverse wave on a string, we have a choice of moving the end up and down or sideways; in either case it is perpendicular to the length of the string. In the first case the motion of the entire string is confined to a vertical plane, in the second to a horizontal plane. In either case the wave is said to be *linearly polarized* because the individual particles move back and forth in straight lines perpendicular to the string.

The motion may also be more complex, containing both vertical and horizontal components. Combining two sinusoidal motions of equal amplitude, with a quarter-cycle phase difference, results in a wave in which each particle moves in a circular path perpendicular to the string; such a wave is said to be *circularly polarized.*

It is easy to imagine making a device to separate the various components of motion. We cut a thin slot in a flat board and thread the string through it. When the plane of the board is oriented perpendicular to the string, then any transverse motion parallel to the slot passes through unimpeded, while any motion perpendicular to the slot is blocked. Such a device may be called a *polarizing filter.* Analogous optical devices for polarized light are of considerable practical importance; they form the basis of some kinds of sunglasses and also of polarizing filters used in photography.

21–4

SPEED OF A LONGITUDINAL PULSE

The propagation speed of longitudinal as well as transverse waves is determined by the mechanical properties of the medium. Figure 21–6 shows a fluid (liquid or gas) of density ρ in a tube of cross-sectional area A and under a pressure p. In Fig. 21–6a the fluid is at rest. At time $t = 0$, the piston at the left end of the tube is set in motion toward the right with a speed v. Figure 21–6b shows the fluid after a time t has elapsed. All portions of the fluid at the left of point P are moving with speed v, whereas all portions at the right of P are still at rest. The boundary between the moving and the stationary portions travels to the right with the speed of propagation c. At time t the piston has moved a distance vt and the boundary has advanced a distance ct. As in the case of a transverse disturbance in a string, the speed of propagation can be computed from the impulse–momentum theorem.

The quantity of fluid set in motion in time t is the amount that originally occupied a volume of length ct and of cross-sectional area A. The mass of this fluid is therefore ρctA and the longitudinal momentum it has acquired is

$$\text{Longitudinal momentum} = \rho ctAv.$$

We next compute the increase of pressure, Δp, in the moving fluid. The original volume of the moving fluid, Act, has been decreased by an amount Avt. From the definition of bulk modulus B (see Chapter 10),

$$B = \frac{\text{Change in pressure}}{\text{Fractional change in volume}} = \frac{\Delta p}{Avt/Act}.$$

Therefore,

$$\Delta p = B\frac{v}{c}.$$

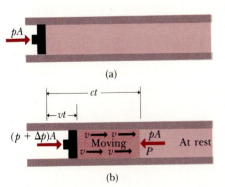

21–6 Propagation of a longitudinal disturbance in a fluid confined in a tube.

The pressure in the moving fluid is therefore $p + \Delta p$, and the force exerted on it by the piston is $(p + \Delta p)A$. The net force on the moving fluid (see Fig. 21–6b) is ΔpA, and the longitudinal impulse is

$$\text{Longitudinal impulse} = \Delta pAt = B\frac{v}{c}At.$$

Applying the impulse–momentum theorem, we find

$$B\frac{v}{c}At = \rho ctAv,$$

and therefore

$$c = \sqrt{\frac{B}{\rho}} \qquad \text{(longitudinal wave)}. \qquad (21\text{–}4)$$

The speed of propagation of a longitudinal pulse in a fluid therefore depends only on the bulk modulus and the density of the medium.

When a longitudinal wave propagates in a solid bar, the situation is somewhat different from that of a fluid confined in a tube of constant cross section, since the bar expands slightly sidewise when it is compressed longitudinally. It can be shown, by the same type of reasoning as that just given, that the velocity of a longitudinal pulse in the bar is given by

$$c = \sqrt{\frac{Y}{\rho}} \qquad \text{(longitudinal wave)}, \qquad (21\text{–}5)$$

where Y is Young's modulus, defined in Chapter 10.

As with the calculation for the transverse wave on a string, Eqs. (21–4) and (21–5) are valid for all wave motions, not just the special case discussed here. In particular, they are valid for sinusoidal and other periodic waves, to be discussed in more detail in Section 21–7.

Since the *shape* of a fluid does not change when a longitudinal wave passes through it, it is not as easy to visualize the relation between particle motion and wave motion as it is for transverse waves in a string. Figure 21–7 may help in correlating these motions. To use this figure, cut a slit about $\frac{1}{16}$ in. wide and 3 in. long in a 3-in. × 5-in. card (or fasten two cards edge-to-edge with a $\frac{1}{16}$-in. gap), place the card over the figure with the slit at the top of the diagram, and move the card downward with constant velocity. The portions of the sine curves that are visible through the slit correspond to a row of particles along which a longitudinal, sinusoidal wave is traveling. Each particle undergoes simple harmonic motion about its equilibrium position, with a phase that increases continuously along the slit, while the regions of maximum compression and expansion move from left to right with constant speed. Moving the card upward simulates a wave traveling from right to left.

EXAMPLE 1   Determine the speed of sound waves in water, and find the wavelength of a wave having a frequency of 262 Hz.

Solution   We use Eq. (21–4) to find the wave speed. From Table 10–2, we find that the compressibility of water, which is the reciprocal of the

21–7 Diagram for illustrating longitudinal traveling waves.

bulk modulus, is $k = 49 \times 10^{-11}$ Pa^{-1}. Thus $B = (1/49) \times 10^{11}$ Pa. The density of water is $\rho = 1.00 \times 10^3$ kg·m^{-3}. We obtain

$$c = \sqrt{\frac{B}{\rho}} = \sqrt{\frac{(1/45.8) \times 10^{11} \text{ Pa}}{1.00 \times 10^3 \text{ kg·m}^{-3}}} = 1478 \text{ m·s}^{-1}.$$

The wavelength is given by

$$\lambda = \frac{c}{f} = \frac{1478 \text{ m·s}^{-1}}{262 \text{ s}^{-1}} = 5.64 \text{ m}.$$

This is over four times the speed of sound in air at ordinary temperatures. A wave of this frequency in air has a wavelength of 1.31 m, as obtained in the example of Section 21–2. ◀

Dolphins emit high-frequency sound waves (typically 100,000 Hz) and use the echoes for guidance and in hunting for prey. The corresponding wavelength is 1.43 cm. With this "sonar" system they can sense objects of about the size of the wavelength, but not much smaller; the high frequency is needed to be able to sense small objects.

EXAMPLE 2 What is the speed of longitudinal sound waves in a steel rod?

Solution We use Eq. (21–5). From Table 10–1, $Y = 2.0 \times 10^{11}$ Pa, and from Table 12–1, $\rho = 7.8$ g·cm^{-3} = 7.8×10^3 kg·m^{-3}. From these,

$$c = \sqrt{\frac{Y}{\rho}} = \sqrt{\frac{2.0 \times 10^{11} \text{ Pa}}{7.8 \times 10^3 \text{ kg·m}^{-3}}} = 5064 \text{ m·s}^{-1}.$$

◀

21–5

ADIABATIC CHARACTER OF A LONGITUDINAL WAVE

It is a familiar fact that compression of a fluid causes a rise in its temperature unless heat is withdrawn in some way. Conversely, an expansion is accompanied by a temperature decrease unless heat is added. As a longitudinal wave advances through a fluid, the regions that are compressed at any instant are slightly warmer than those that are expanded. The condition is present, therefore, for the conduction of heat from a condensation to a rarefaction. The quantity of heat conducted per unit time and per unit area depends on the thermal conductivity of the fluid and upon the distance between a condensation and its adjacent rarefaction (half a wavelength). For ordinary frequencies, say from 20 vibrations per second to 20,000 vibrations per second (the frequency range in which the human ear is sensitive) and for even the best known heat conductors, the wavelength is too large and the thermal conductivity too small for an appreciable amount of heat to flow. The compressions and rarefactions are therefore *adiabatic* rather than isothermal.

The change in volume produced by a given change of pressure depends upon whether the compression (or expansion) is adiabatic or isothermal. Thus there are two bulk moduli, the *adiabatic* bulk modulus B_{ad}

and the *isothermal* bulk modulus. The expression for the speed of a longitudinal wave should therefore be written

$$c = \sqrt{\frac{B_{ad}}{\rho}}. \tag{21-6}$$

In the case of an ideal gas, the relation between pressure p and volume V during an adiabatic process is given (as discussed in Section 18–11) by

$$pV^\gamma = \text{constant}, \tag{21-7}$$

where γ is the ratio of the heat capacity at constant pressure to the heat capacity at constant volume. The bulk modulus is defined in general (Section 10–4) as

$$B = -\frac{\Delta p}{\Delta V/V}$$

$$= -V\frac{\Delta p}{\Delta V}.$$

It can be shown, with the aid of calculus, that for an ideal gas the adiabatic bulk modulus is given by

$$B_{ad} = \gamma p, \tag{21-8}$$

while the isothermal bulk modulus B is given simply by

$$B = p. \tag{21-9}$$

In each case the bulk modulus, characterizing the material's resistance to being compressed, is proportional to the pressure, but the adiabatic modulus is *larger* than the isothermal by a factor $\gamma = C_p/C_v$ the ratio of the molar heat capacities at constant pressure and constant volume.

Combining Eqs. (21–6) and (21–8), we obtain

$$c = \sqrt{\frac{\gamma p}{\rho}} \quad \text{(ideal gas).} \tag{21-10}$$

But, for an ideal gas,

$$\frac{p}{\rho} = \frac{RT}{M},$$

where R is the universal gas constant, M the molecular mass, and T the Kelvin temperature. Therefore

$$c = \sqrt{\frac{\gamma RT}{M}} \quad \text{(ideal gas),} \tag{21-11}$$

and since, for a given gas, γ, R, and M are constants, we see that the speed of propagation is proportional to the square root of the Kelvin temperature.

Let us use Eq. (21–11) to compute the speed of longitudinal waves in air. The mean molecular mass of air is

$$28.8 \text{ g}\cdot\text{mol}^{-1} = 28.8 \times 10^{-3} \text{ kg}\cdot\text{mol}^{-1}.$$

Also, $\gamma = 1.40$ and $R = 8.314 \, \text{J} \cdot \text{mol}^{-1} \cdot \text{K}^{-1}$. At $T = 300 \, \text{K}$ we obtain

$$c = \sqrt{\frac{(1.40)(8.314 \, \text{J} \cdot \text{mol}^{-1} \cdot \text{K}^{-1})(300 \, \text{K})}{28.8 \times 10^{-3} \, \text{kg} \cdot \text{mol}^{-1}}}$$

$$= 348 \, \text{m} \cdot \text{s}^{-1} = 1142 \, \text{ft} \cdot \text{s}^{-1} = 779 \, \text{mi} \cdot \text{hr}^{-1}.$$

This agrees with the measured speed at this temperature to within 0.3 percent.

Longitudinal waves in air give rise to the sensation of *sound*. The ear is sensitive to a range of sound frequencies from about 20 Hz to about 20,000 Hz. From the relation $c = f\lambda$, the corresponding wavelength range is from about 17 m, corresponding to a 20-Hz note, to about 1.7 cm, corresponding to 20,000 Hz.

Like the dolphin, the bat uses high-frequency sound waves for navigation; a typical frequency is 100,000 Hz, and the corresponding wavelength is about 3.5 mm, small enough to permit detection of flying insects useful as food.

The *molecular* nature of a gas has been ignored in the preceding discussion, and a gas has been treated as though it were a continuous medium. Actually, we know that a gas is composed of molecules in random motion, separated by distances that are large compared with their diameters. The vibrations that constitute a wave in a gas are superposed on the random thermal motion. At atmospheric pressure, the average distance traveled between collisions is about 10^{-5} cm, while the displacement amplitude of a faint sound may be only ten thousandths of this amount. An element of gas in which a sound wave is traveling can be compared to a swarm of gnats, where the swarm as a whole can be seen to oscillate slightly while individual insects move about through the swarm, apparently at random.

*21–6

SURFACE WAVES IN LIQUIDS

Probably the most familiar waves are those observed on the surface of a body of water. The equilibrium state is a flat horizontal surface, and disturbances are caused by objects dropped into the water, motion of objects such as fish in the water, and wind blowing across the surface.

Though familiar, water waves are rather complex in the details of their behavior. First, there are two different kinds of restoring forces acting on a disturbed surface. Gravity causes differences in pressure if different points on the surface are at different heights and, hence, tends to restore the surface to a horizontal plane. In addition, surface tension effects arise from curvature of the surface, tending to flatten it. It turns out that with water waves surface tension plays a significant role only for rather short wavelengths, shorter than a few centimeters. Hence surface tension is important in laboratory ripple tanks but unimportant in most ocean waves.

Second, the motions of individual particles of water are neither completely longitudinal nor completely transverse but, instead, have both

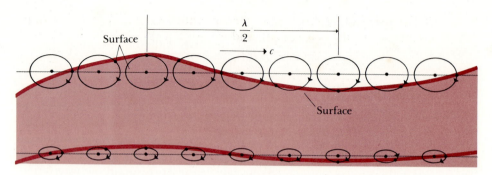

21–8 Particle motion and wave shape for sinusoidal waves on the surface of a liquid.

longitudinal and transverse components. For sinusoidal waves, these components have a quarter-cycle phase difference, and the resulting motion of an individual water particle is *elliptical*. The motion is shown in Fig. 21–8; at greater and greater depths below the surface the ellipses become flatter and the motion more nearly horizontal (longitudinal). When the water is very deep (i.e., depth much greater than the wavelength of the wave) the ellipses are nearly circles, and they become smaller with increasing depth.

Third, the speed of propagation of waves on liquid surfaces depends on the wavelength. This is an effect we have not encountered up to now; transverse waves on a string and longitudinal waves in a pipe containing a fluid or in a solid bar travel with a speed that does not depend on wavelength. It turns out that for water whose depth h is much *greater* than the wavelength λ of the wave, the wave speed c is given by

$$c = \sqrt{\frac{g\lambda}{2\pi}}, \qquad (21\text{–}12)$$

and in the opposite extreme, when the depth h is much *less* than the wavelength (as would be true for ocean waves approaching a beach), the wave speed is given approximately by

$$c = \sqrt{gh}, \qquad (21\text{–}13)$$

where g is as usual the acceleration due to gravity. Thus in this limit c is approximately independent of wavelength, but it depends on the depth of the water. Ocean waves always slow down as they approach the shore. The expression for c in the intermediate cases where h is neither very large nor very small compared to λ, and where surface tension has to be included, is more complex and will not be given here.

EXAMPLE   For an ocean depth of 5 km, find the speed of waves having wavelength 10 m, and of seismic ocean waves (caused by earthquakes) of wavelength 100 km.

Solution   In the first case, h is much greater than the wavelength, and we use Eq. (21–12), obtaining

$$c = \sqrt{\frac{(9.8 \text{ m} \cdot \text{s}^{-2})(10 \text{ m})}{2\pi}} = 3.9 \text{ m} \cdot \text{s}^{-1} = 8.8 \text{ mi} \cdot \text{hr}^{-1}.$$

In the second case, we have the opposite extreme, where h is much smaller than the wavelength, and we use Eq. (21–13):

$$c = \sqrt{(9.8 \text{ m} \cdot \text{s}^{-2})(5 \times 10^3 \text{ m})} = 221 \text{ m} \cdot \text{s}^{-1} = 495 \text{ mi} \cdot \text{hr}^{-1}.$$

Thus seismic ocean waves travel with surprisingly high speed! ◀

Any wave for which the wave speed depends on wavelength is said to be *dispersive*. This term originated in optics; a prism separates or *disperses* a beam of white light into a spectrum, with different colors (wavelengths) emerging in different directions, because the speed of light in glass is different for different wavelengths. Glass is a dispersive medium for wave propagation, and so is the surface of a liquid.

Surface waves can also occur on the surfaces of elastic solids; in this case they are called *Rayleigh waves*. Witnesses to major earthquakes have sometimes described a rolling motion of the surface of the earth; this is probably due to Rayleigh waves. These waves are also taking an increasingly important role in mechanical devices for electronic signal processing. It is possible, for example, to make *mechanical* band-pass filters for the megahertz frequency range that are simpler and more compact than their electrical analogs.

We have not by any means exhausted the list of kinds of mechanical waves. Elastic waves can travel through the interior of solids. There are longitudinal waves, where the restoring forces are compressive, governed by the bulk modulus for the material, and transverse waves, where the restoring forces are shear stresses determined by the shear modulus. Both types, called p and s (or *pressure* and *shear*) waves, respectively, are important in seismology, for earthquake analysis, seismic mineral prospecting, and more general studies of the earth's structure. The two types of waves travel at different speeds and are reflected differently at boundary surfaces between layers of material.

*21–7

MATHEMATICAL REPRESENTATION OF A TRAVELING WAVE

It is often useful to have a detailed description of the motion of a wave medium during wave propagation. We introduce the concept of a *wave function*, a function that describes the position of an arbitrary particle of the medium at any time. From this function, once it is known, one can determine the velocity and acceleration of any particle and much other useful information about the wave. In this discussion we will consider only sinusoidal waves, in which each particle of the medium undergoes simple harmonic motion about its equilibrium position.

Suppose that a wave of any sort travels from left to right in a medium. Let us compare the motion of any one particle of the medium with that of a second particle to the right of the first. We find that the second particle moves in the same manner as the first, but after a lapse of time that is proportional to the distance of the second particle from the first. Hence if one end of a stretched string oscillates with simple harmonic motion, all other points oscillate with simple harmonic motion of the same amplitude and frequency. The *phase* of the motion, however, is different for different points.

Let the displacement of a particle at the origin ($x = 0$) be given by

$$y = A \sin \omega t = A \sin 2\pi f t. \qquad (21\text{--}14)$$

The time required for the wave disturbance to travel with wave speed c from $x = 0$ to some point x to the right of the origin is given by x/c. The motion of point x at time t is the same as the motion of point $x = 0$ at the earlier time $t - x/c$. Thus the displacement of point x at time t is obtained simply by replacing t in Eq. (21–14) by $(t - (x/c))$, and we find

$$y(x,\, t) = A \sin \omega\left(t - \frac{x}{c}\right) = A \sin 2\pi f\left(t - \frac{x}{c}\right). \qquad (21\text{--}15)$$

The notation $y(x, t)$ is a reminder that the displacement y is a function of two variables x and t, corresponding to the fact that the displacement of a point depends on both the location of the point and the time. The function $y(x, t)$ gives the position of any point on the string at any time, and is therefore a complete description of the motion. Any such function is called a *wave function.*

Equation (21–15) can be rewritten in several alternative forms, conveying the same information in different ways. In terms of the period τ and wavelength λ, we find, using Eq. (21–1),

$$y(x,\, t) = A \sin 2\pi\left(\frac{t}{\tau} - \frac{x}{\lambda}\right). \qquad (21\text{--}16)$$

Another convenient form is obtained by defining a quantity k, called the *propagation constant* or the *wave number:*

$$k = \frac{2\pi}{\lambda}. \qquad (21\text{--}17)$$

In terms of k and ω, the wavelength-frequency relation $c = \lambda f$ becomes

$$\omega = ck, \qquad (21\text{--}18)$$

and Eq. (21–16) can be written

$$y(x,\, t) = A \sin(\omega t - kx). \qquad (21\text{--}19)$$

Which of these various forms is used is a matter of convenience in a specific problem; fundamentally they all say the same thing. It should be mentioned in passing that some authors define the wave number as $1/\lambda$ rather than $2\pi/\lambda$. In that case, our k is still called the propagation constant.

At any given *time t,* Eq. (21–16) or (21–19) gives the displacement y of a particle from its equilibrium position, as a function of the *coordinate x* of the particle. If the wave is a transverse wave in a string, the equation represents the *shape* of the string at that instant, as if we had taken a photograph of the string. Thus at time $t = 0$,

$$y = A \sin(-kx) = -A \sin kx = -A \sin 2\pi \frac{x}{\lambda}.$$

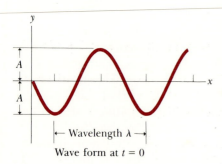

y

A

A

← Wavelength λ →

Wave form at $t = 0$

21–9

This curve is plotted in Fig. 21–9.

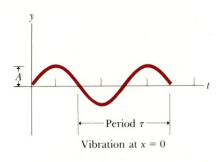

Period τ

Vibration at $x = 0$

21–10

At any given *coordinate x*, Eq. (21–16) or (21–19) gives the displacement y of the particle at that coordinate, as a function of *time*. Thus at the position $x = 0$,

$$y = A \sin \omega t = A \sin 2\pi \frac{t}{\tau}.$$

This curve is plotted in Fig. 21–10.

The above formulas may be used to represent a wave traveling in the negative x-direction by making a simple modification. In this case the displacement of point x at time t is the same as the motion of point $x = 0$ at the *later* time $(t + (x/c))$, and in Eq. (21–14) we must replace t by $(t + (x/c))$. Thus, for a wave traveling in the negative x-direction,

$$y = A \sin 2\pi f\left(t + \frac{x}{c}\right) = A \sin 2\pi\left(\frac{t}{\tau} + \frac{x}{\lambda}\right)$$

$$= A \sin(\omega t + kx).$$

$$(21\text{–}20)$$

EXAMPLE A certain string has a linear mass density of $0.25 \text{ kg} \cdot \text{m}^{-1}$ and is stretched with a tension of 25 N. One end is given a sinusoidal motion with frequency 5 Hz and amplitude 0.01 m. At time $t = 0$ the end has zero displacement and is moving in the $+y$-direction.

a) Find the wave speed, amplitude, angular frequency, period, wavelength, and wave number.

b) Write a wave function describing the wave.

c) Find the position of the point at $x = 0.5$ m at time $t = 0.1$ s.

Solution

a) From Eq. (21–2), the wave speed is

$$c = \sqrt{\frac{S}{\mu}} = \sqrt{\frac{25 \text{ N}}{0.25 \text{ kg} \cdot \text{m}^{-1}}} = 10 \text{ m} \cdot \text{s}^{-1}.$$

The amplitude A is just the amplitude of the motion of the endpoint, $A = 0.01$ m. The angular frequency is

$$\omega = 2\pi f = 2\pi(5 \text{ s}^{-1}) = 31.4 \text{ s}^{-1}.$$

The period is $\tau = 1/f = 0.2$ s. The wavelength is obtained from Eq. (21–1):

$$\lambda = \frac{c}{f} = \frac{10 \text{ m} \cdot \text{s}^{-1}}{5 \text{ s}^{-1}} = 2 \text{ m}.$$

The wave number is obtained from Eq. (21–17) or Eq. (21–18):

$$k = \frac{2\pi}{\lambda} = \frac{2\pi}{2 \text{ m}} = 3.14 \text{ m}^{-1}, \quad \text{or} \quad k = \frac{\omega}{c} = \frac{31.4 \text{ s}^{-1}}{10 \text{ m} \cdot \text{s}^{-1}} = 3.14 \text{ m}^{-1}.$$

b) The wave function, using the form of Eq. (21–16), is

$$y = (0.01 \text{ m}) \sin 2\pi\left(\frac{t}{0.2 \text{ s}} - \frac{x}{2 \text{ m}}\right)$$

$$= (0.01 \text{ m}) \sin[(31.4 \text{ s}^{-1})t - (3.14 \text{ m}^{-1})x].$$

Alternatively, using the form of Eq. (21–19),

$$y = (0.01 \text{ m}) \sin[(31.4 \text{ s}^{-1})t - (3.14 \text{ m}^{-1})x].$$

c) The displacement of the point $x = 0.5$ m at time $t = 0.1$ s is given by substituting these values into either of the above wave equations:

$$y = (0.01 \text{ m}) \sin 2\pi\left(\frac{0.1 \text{ s}}{0.2 \text{ s}} - \frac{0.5 \text{ m}}{2 \text{ m}}\right)$$

$$= (0.01 \text{ m}) \sin 2\pi(0.25) = 0.01 \text{ m}. \qquad \blacktriangleleft$$

QUESTIONS

21–1 The term *wave* has a variety of meanings in ordinary language. Think of several, and discuss their relation to the precise physical sense in which the term is used in this chapter.

21–2 What kinds of energy are associated with waves on a stretched string? How could such energy be detected experimentally?

21–3 Are torsional waves (Fig. 21–2d) longitudinal or transverse?

21–4 Is it possible to have a longitudinal wave on a stretched string? A transverse wave on a steel rod?

21–5 The speed of sound waves in air depends on temperature, but the speed of light waves does not. Why?

21–6 For the wave motions discussed in this chapter, does the speed of propagation depend on the amplitude?

21–7 Children make toy telephones by sticking each end of a long string through a hole in the bottom of a paper cup and knotting it so it won't pull out. When the string is pulled taut, sound can be transmitted from one cup to the other. How does this work? Why is the transmitted sound louder than the sound traveling through air for the same distance?

21–8 An echo is sound reflected from a distant object such as a wall or a cliff. Explain how you can determine how far away the object is by timing the echo.

21–9 Why do you see lightning before you hear the thunder? A familiar rule of thumb is to start counting slowly, once per second, when you see the lightning; when you hear thunder, divide the number you have reached by five, to obtain your distance (in miles) from the lightning. Why does this work? Or does it?

21–10 When ocean waves approach a beach, the crests are always nearly parallel to the shore, despite the fact that they must come from various directions. Why?

21–11 The amplitudes of ocean waves increase as they approach the shore, and often the crests bend over and "break." Why?

21–12 When a rock is thrown into a pond and the resulting ripples spread in ever-widening cirles, the amplitude decreases with increasing distance from the center. Why?

21–13 In speaker systems designed for high-fidelity music reproduction, the "tweeters" that reproduce the high frequencies are always much smaller than the "woofers" used for low frequencies. Why?

21–14 When sound travels from air into water, does the frequency of the wave change? The wavelength? The speed?

21–15 Which of the quantities describing a sinusoidal wave is most closely related to musical pitch? To loudness?

21–16 Musical notes produced by different instruments (such as a flute and an oboe) may have the same pitch and loudness and yet sound different. What is the difference, in physical terms?

PROBLEMS

21–1 The speed of sound in air at 20°C is 344 m·s^{-1}.

a) What is the wavelength of a sound wave of frequency 32 Hz, the lowest pedal note on medium-size pipe organs?

b) What is the frequency of a wave having a wavelength of 4 ft, corresponding approximately to the D above middle C on the piano?

21–2 The speed of radio waves in vacuum (equal to the speed of light) is 3.00×10^8 m·s^{-1}. Find the wavelength

for

a) an AM radio station with frequency 1000 kHz, and

b) an FM radio station with frequency 100 MHz.

21–3 The speed of sound in water is about 1480 m·s^{-1}. Find the frequency of a sound wave such that its wavelength in water is the same as the wavelength in air at 20°C of a sound wave of frequency 1000 Hz, for which $c = 344$ m·s^{-1}.

21–4 Middle C on the piano corresponds to a frequency of 262 Hz = 262 s^{-1}.

a) Find the wavelength of the corresponding sound wave in air at 20°C.

b) What is the angular frequency?

c) What is the wavelength of the wave in water at 20°C, if the wave speed is 1480 m·s^{-1}?

21–5 A steel wire 6 m long has a mass of 60 g and is stretched with a tension of 1000 N. What is the speed of propagation of a transverse wave in the wire?

21–6 One end of a horizontal string is attached to a prong of an electrically driven tuning fork whose frequency of vibration is 240 Hz. The other end passes over a pulley and supports a mass of 5 kg. The linear mass density of the string is 0.02 kg·m^{-1}.

a) What is the speed of a transverse wave in the string?

b) What is the wavelength?

21–7 One end of a stretched rope is given a periodic transverse motion with a frequency of 10 Hz. The rope is 50 m long, has total mass 0.5 kg, and is stretched with a tension of 400 N.

a) Find the wave speed and the wavelength.

b) If the tension is doubled, how must the frequency be changed to maintain the same wavelength?

21–8 One end of a rubber tube 20 m long, of total mass 1 kg, is fastened to a fixed support. A cord attached to the other end passes over a pulley and supports a body of mass 10 kg. The tube is struck a transverse blow at one end. Find the time required for the pulse to reach the other end.

21–9 A metal wire has these properties:

Coefficient of linear expansion = 1.5 × 10^{-5}(C°)$^{-1}$.

Young's modulus = 2.0 × 10^{11} Pa.

Density = 9.0 × 10^3 kg·m^{-3}.

At each end are rigid supports. If the tension is zero at 20°C, what will be the speed of a transverse wave at 8°C?

21–10 A metal wire, of density 5 × 10^3 kg·m^{-3}, with a Young's modulus equal to 2.0 × 10^{11} Pa, is stretched between rigid supports. At one temperature the speed of a transverse wave is found to be 200 m·s^{-1}. When the temperature is raised 100 C°, the speed decreases to 160 m·s^{-1}. What is the coefficient of linear expansion?

21–11 Provided the amplitude is sufficiently great, the human ear can respond to longitudinal waves over a range of frequencies from about 20 Hz to about 20,000 Hz. Compute the wavelengths corresponding to these frequencies

a) for waves in air, and

b) for waves in water.

(See Problem 21–13.) Assume that in air c = 345 m·s^{-1}.

21–12 What must be the stress (F/A) in a stretched wire of a material whose Young's modulus is Y, in order that the speed of longitudinal waves shall equal 10 times the speed of transverse waves?

21–13 The speed of longitudinal waves in water is approximately 1450 m·s^{-1} at 20°C. Compute the adiabatic compressibility (1/B_{ad}) of water and compare with the isothermal compressibility in Table 10–2.

21–14 A longitudinal wave propagates in a steel bar having density 7.0 g·cm^{-3} and Young's modulus 2 × 10^{11} Pa.

a) What is the wave speed?

b) By what factor is this speed greater than the speed of sound in air at 20°C?

21–15 At a temperature of 27°C, what is the speed of longitudinal waves in

a) argon, and

b) hydrogen?

c) Compare with the speed in air at the same temperature.

21–16 What is the difference between the speeds of longitudinal waves in air at −3°C and at 57°C?

21–17 The sound waves from a loudspeaker spread out nearly uniformly in all directions when their wavelength is large compared with the diameter of the speaker. When the wavelength is small compared with the diameter of the speaker, much of the sound energy is concentrated in the forward direction. For a speaker of diameter 25 cm, compute the frequency for which the wavelength of the sound waves, in air, is

a) 10 times the diameter of the speaker,

b) equal to the diameter of the speaker, and

c) $\frac{1}{10}$ the diameter of the speaker.

21–18 A steel pipe 100 m long is struck at one end. A person at the other end hears two sounds as a result of two longitudinal waves, one in the pipe and the other in the air. What is the time interval between the two sounds? Take Young's modulus of steel to be 2 × 10^{11} Pa, and the speed of sound in air to be 345 m·s^{-1}.

21–19 What is the ratio of the speed of sound in a diatomic gas to the rms speed of gas molecules at the same temperature?

21–20

a) If the propagation of sound waves in gases were characterized by isothermal rather than adiabatic expansions and compressions, show that the speed of sound would be given by (RT/M)$^{1/2}$.

b) What would be the speed of sound in air at 20°C in this case?

c) Under what circumstances might the wave propagation be expected to be isothermal?

21–21 Find the speed of surface waves on water, of wavelength 10 m, if the body of water is

a) a lake 300 m deep, and

b) a pond 0.5 m deep.

21–22 A canoeist on a shallow pond notices that long-wavelength waves travel with a speed of 2.8 m·s^{-1}. What is the depth of the pond?

21–23 Surface waves on a very deep lake were observed to travel with a speed of 4.0 m·s^{-1}.

a) What is the wavelength of these waves?

b) A wave with this wavelength might also travel with the same speed on the surface of a very shallow lake. How deep would such a lake be?

21–24 Suppose the next group of moon explorers builds a swimming pool on the moon, where $g = 1.67$ m·s^{-2}. Due to lack of water, the pool is only 0.8 m deep.

a) For waves whose wavelength is several meters, what is the wave speed?

b) By what factor does this differ from the value that would be observed on earth?

21–25 The equation of a transverse traveling wave on a string is

$$y = 2 \cos[\pi(0.5x - 200t)],$$

where x and y are in cm and t is in seconds.

a) Find the amplitude, wavelength, frequency, period, and speed of propagation.

b) Sketch the shape of the string at the following values of t: 0, 0.0025 s, and 0.005 s.

c) If the mass per unit length of the string is 5 g·cm^{-1}, find the tension.

21–26 The equation of a certain traveling transverse wave is

$$y = 2 \sin 2\pi\left(\frac{t}{0.01} - \frac{x}{30}\right),$$

where x and y are in centimeters and t is in seconds. What are

a) the amplitude,

b) the wavelength,

c) the frequency, and

d) the speed of propagation of the wave?

21–27 Show that Eq. (21–16) may be written

$$y = -A \sin\frac{2\pi}{\lambda}(x - ct).$$

21–28 A traveling transverse wave on a string is represented by the equation in Problem 21–27. Let $A = 8$ cm, $\lambda = 16$ cm, and $c = 2$ cm·s^{-1}.

a) At time $t = 0$, compute the transverse displacement y at 2-cm intervals of x (that is, at $x = 0$, $x = 2$ cm, $x = 4$ cm, etc.) from $x = 0$ to $x = 32$ cm. Show the results in a graph. This is the shape of the string at time $t = 0$.

b) Repeat the calculations, for the same values of x, at times $t = 1$ s, $t = 2$ s, $t = 3$ s, and $t = 4$ s. Show on the same graph the shape of the string at these instants. In what direction is the wave traveling?

21–29 A transverse wave of amplitude 10 cm and wavelength 200 cm travels from left to right along a long horizontal stretched string with a speed of 100 cm·s^{-1}. Take the origin at the left end of the undisturbed string. At time $t = 0$, the left end of the string is at the origin and is moving downward.

a) What is the frequency of the wave?

b) What is the angular frequency?

c) What is the propagation constant?

d) What is the equation of the wave?

e) What is the equation of motion of the left end of the string?

f) What is the equation of motion of a particle 150 cm to the right of the origin?

g) What is the (absolute) maximum transverse velocity of any particle of the string?

h) Find the transverse displacement and the transverse velocity of a particle 150 cm to the right of the origin, at time $t = 3.25$ s.

i) Make a sketch of the shape of the string, for a length of 400 cm, at time $t = 3.25$ s.

CHAPTER 22

VIBRATING BODIES

In Chapter 21 we studied the propagation of mechanical waves in *unbounded* media; we were not concerned with the possibility that a wave may arrive at an end or boundary of the elastic medium in which it propagates. But there are many wave phenomena in which such boundaries do play a significant role. A familiar example is the echo that results when a sound wave is reflected from a rigid wall. Such reflections lead to overlap or *superposition* of two or more waves in the same region of the medium. For a wave medium with boundaries, sinusoidal wave motion is possible only when the frequency has one of a series of special "allowed" values, determined by the dimensions and mechanical properties of the medium. These special motions are called *normal modes;* many familiar phenomena such as the pitch of a musical instrument and various mechanical vibrations are associated with normal modes.

22–1

BOUNDARY CONDITIONS FOR A STRING

As a simple example of reflections and the role of boundaries, we return to transverse waves on a stretched string. Let us consider what happens when a wave pulse or a continuous succession of waves (such as a sinusoidal wave) arrives at the end of the string. If the end is fastened to a rigid support, it must remain at rest. The arriving wave exerts a force on the support; the reaction to this force "kicks back" on the string and sets up a *reflected* pulse or wave traveling in the reverse direction.

At the opposite extreme from a rigidly fixed end is one that is perfectly free to move in the direction transverse to the length of the string. For example, the string might be tied to a light ring, which slides on a smooth rod perpendicular to the length of the string; the ring and rod maintain the tension of the string but exert no transverse force. At a free end the arriving pulse or wave train causes the end to "overshoot," and again a reflected wave is set up. The conditions imposed on the motion

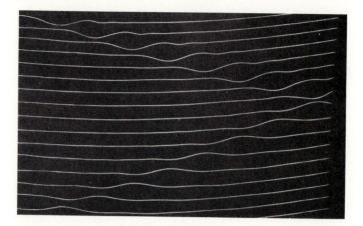

22–1 A pulse starts in the upper left corner, travels to the right, and is reflected from the fixed end of the string at the right.

of the end of the string, such as attachment to a rigid support or the complete absence of transverse force, are called *boundary conditions.*

The multiflash photograph of Fig. 22–1 shows the reflection of a pulse at a fixed end of a string. (The camera was tipped vertically while the photographs were taken so that successive images lie one under the other. The "string" is a rubber tube, and it sags somewhat.) The pulse is reflected with both its displacement and its direction of propagation reversed. When reflection takes place at a *free* end, the direction of propagation is reversed but the direction of the displacement is unchanged.

It is helpful to think of the process of reflection in the following way. Imagine the string to be extended indefinitely beyond its actual end. The actual pulse can be considered to continue on into the imaginary portion as though the support were not there, while at the same time a "virtual" pulse, which has been traveling in the imaginary portion, moves out into the real string and forms the reflected pulse. The nature of the reflected pulse depends on whether the end is fixed or free. The two cases are shown in Fig. 22–2.

The displacement at a point where the actual and virtual pulses cross each other is the *algebraic sum* of the displacements in the individual

22–2 Description of reflection of a pulse (a) at a fixed end of a string, and (b) at a free end, in terms of an imaginary "virtual" pulse.

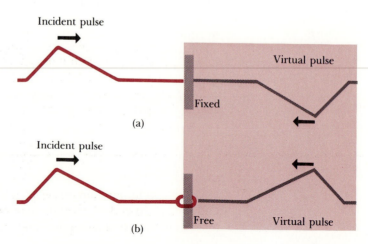

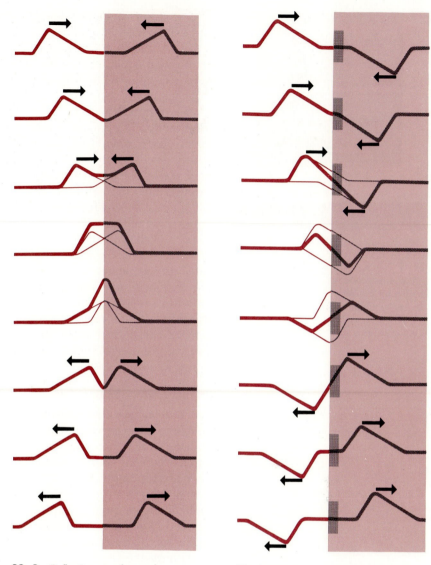

22–3   Reflection at a free end.   **22–4**   Reflection at a fixed end.

pulses. Figures 22–3 and 22–4 show the shape of the region near the end of the string for both types of reflected pulses, Fig. 22–3 for a free end and Fig. 22–4 for a fixed end. In the latter case, the incident and reflected pulses must combine in such a way that the displacement of the end of the string is always zero.

22–2

SUPERPOSITION AND STANDING WAVES

When a continuous succession of waves, such as a sinusoidal wave, arrives at a fixed end of a string, a corresponding continuous succession of *reflected* waves originates at this end and travels in the opposite direction.

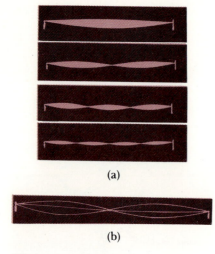

22–5 (a) Standing waves in a stretched string (time exposure). (b) Multiflash photograph of a standing wave, with nodes at the center and at the ends.

The resulting motion of the string is determined by an extremely important principle known as the *principle of superposition,* which states that the actual displacement of any point on the string, at any time, is obtained by adding the displacement that point would have if only the first wave were present, and the displacement it would have with only the second wave.

The principle of superposition is of central importance in all types of wave motion; it applies not only to waves on a string, but also to sound waves, electromagnetic waves (such as light), and all other wave phenomena in which the wave equation is linear. The general term *interference* is used to describe phenomena that result from two or more waves passing through the same region at the same time.

When a sinusoidal wave is reflected by a fixed point on a string, the appearance of the resulting motion gives no evidence that there are two waves traveling in opposite directions. If the frequency is sufficiently great so that the eye cannot follow the motion, the string appears subdivided into a number of segments, as in the time exposure photograph of Fig. 22–5a. A multiflash photograph of the same string, in Fig. 22–5b, indicates a few of the instantaneous shapes of the string. At every instant its shape is a sine curve, but whereas in a traveling wave the amplitude remains constant while the wave progresses, here the waveform remains fixed in position (longitudinally) while the amplitude fluctuates. Certain points known as the *nodes* remain always at rest. Midway between these points, at the *loops* or *antinodes,* the fluctuations are a maximum. The vibration as a whole is called a *standing wave.*

To understand the formation of a standing wave, consider the separate graphs of the waveform at four instants one tenth of a period apart, shown in Fig. 22–6. The gray curves represent a wave traveling to the right. The light color curves represent a wave of the same propagation speed, wavelength, and amplitude traveling to the *left.* The dark color curves represent the resultant waveform, obtained by applying the principle of superposition, that is, by adding displacements. At those places marked N at the bottom of Fig. 22–6, the resultant displacements are *always* zero. These are the **nodes.** Midway between the nodes, the vibrations have the *largest* amplitude. These are the **antinodes.** It is evident from the figure that

$$\left.\begin{array}{c} \text{Distance between adjacent nodes} \\ \text{or} \\ \text{Distance between adjacent antinodes} \end{array}\right\} = \frac{\lambda}{2}.$$

The wave function for the standing wave of Fig. 22–6 may be obtained by adding the displacements of two waves of equal amplitude, period, and wavelength, traveling in opposite directions. Thus, if

$$y_1 = A \sin(\omega t - kx) \qquad \text{(positive } x\text{-direction)},$$

$$y_2 = -A \sin(\omega t + kx) \qquad \text{(negative } x\text{-direction)},$$

then

$$y_1 + y_2 = A[\sin(\omega t - kx) - \sin(\omega t + kx)].$$

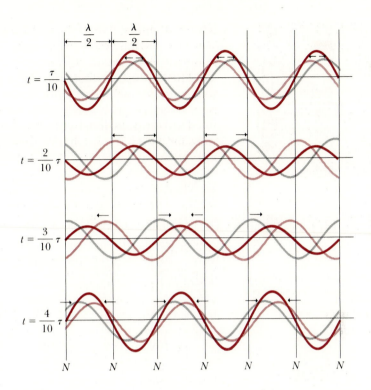

22–6 The formation of a standing wave. A wave traveling to the right (gray lines) combines with a wave traveling to the left (light color lines) to form a standing wave (dark color line).

Introducing the expressions for the sine of the sum and difference of two angles and combining terms, we obtain

$$y_1 + y_2 = -[2A \cos \omega t] \sin kx. \tag{22–1}$$

The shape of the string at each instant is, therefore, a sine curve whose amplitude (the expression in brackets) varies with time. The appearance is not that of a traveling wave shape, but of a sinusoidal displacement in one position which grows larger and smaller with time. Each point in the string still undergoes simple harmonic motion, but instead of the progressively increasing phase differences between motions of adjacent points, all points move *in phase* (or 180° out of phase).

The positions of the nodes can be obtained from Eq. (22–1). Wherever $\sin kx = 0$, the displacement is always zero. This occurs when $kx = 0$, π, 2π, 3π, . . . , or

$$\begin{aligned} x &= 0, \quad \pi/k, \quad 2\pi/k, \quad 3\pi/k, \quad . . . \\ &= 0, \quad \lambda/2, \quad \lambda, \quad 3\lambda/2, \quad . . . . \end{aligned} \tag{22–2}$$

22–3

VIBRATION OF A STRING FIXED AT BOTH ENDS

We have mentioned the reflection or echo of a sound wave from a rigid wall, and the analogous reflection of a transverse wave on a string from a rigidly held end. Suppose now we consider repeated reflection of sound waves between two parallel walls. If a sharp sound such as a hand-clap is originated midway between the walls, the result is a regular procession of echoes resulting from repeated reflection between the walls. In room acoustics this phenomenon is called "flutter echo" and is the bane of acoustical engineers.

The analogous situation with transverse waves on a string is a string with some definite length L, rigidly held at *both* ends. If a sinusoidal wave is produced on such a string, the wave is reflected and re-reflected; and since the string is fixed at both ends, both ends must be nodes. Since adjacent nodes are one-half wavelength apart, the length of the string may be

$$\frac{\lambda}{2}, \quad \frac{2\lambda}{2}, \quad \frac{3\lambda}{2},$$

or, in general, *any* integer number of half-wavelengths. Or (to put it differently) if one considers a particular string of length L, standing waves may be set up in the string by vibrations of a number of different frequencies, namely, those that give rise to waves of wavelengths

$$\lambda = 2L, \quad \frac{2L}{2}, \quad \frac{2L}{3}, \quad \cdots = \frac{2L}{n} \qquad (n = 1, 2, 3, \ldots). \quad (22\text{–}3)$$

The wave speed c is the same for all frequencies. From the relation $f = c/\lambda$, the possible frequencies are

$$f = \frac{c}{2L}, \quad \frac{2c}{2L}, \quad \frac{3c}{2L}, \quad \cdots = \frac{nc}{2L} \qquad (n = 1, 2, 3, \ldots). \quad (22\text{–}4)$$

The lowest frequency, $c/2L$, is called the *fundamental* frequency f_1 and the others are *overtones*. The frequencies of the latter are therefore $2f_1$, $3f_1$, $4f_1$, and so on. Overtones whose frequencies are integral multiples of the fundamental are said to form a *harmonic series*. The fundamental is the *first harmonic*. The frequency $2f_1$ is the *first overtone* or the *second harmonic*, the frequency $3f_1$ is the *second overtone* or the *third harmonic*, and so on.

The above results may also be obtained directly from Eq. (22–1). The boundary conditions require that $y_1 + y_2 = 0$ at the ends of the string, that is at $x = 0$ and $x = L$. Since the sine of zero is zero, the first condition is satisfied automatically. The second requires that $\sin kL = 0$, and this is true only when k has certain special values. The sine of an angle is zero only when the angle is zero or an integer multiple of π (180°). Thus we must have

$$kL = n\pi \qquad (n = 1, 2, 3, \ldots). \qquad (22\text{–}5)$$

We do not include the possibility $n = 0$ because that gives $k = 0$, i.e., a wave with zero displacement *everywhere* (a possible case, to be sure, but not a very interesting one!).

Replacing k above by $2\pi/\lambda$, we obtain

$$\frac{2\pi L}{\lambda} = n\pi \qquad \text{or} \qquad \lambda = \frac{2L}{n}, \qquad (22\text{–}6)$$

in agreement with Eq. (22–3).

Each of the frequencies given by Eq. (22–4) corresponds to a possible *normal mode* of motion, that is, a motion in which each particle of the string moves sinusoidally, all with the same frequency. As the above analysis shows, there is an infinite number of normal modes, each with its characteristic frequency. This situation is in striking contrast with the simpler harmonic oscillator system, a single mass and a spring. The harmonic oscillator has only *one* normal mode and one characteristic frequency, while the vibrating string has an infinite number.

When a body suspended from a spring is pulled down and released, only one frequency of vibration will ensue. If a string is initially distorted so that its shape is the same as *any one* of the possible harmonics, it will vibrate, when released, at the frequency of that particular harmonic. But when a piano string is struck or a guitar string is plucked, not only the fundamental but many of the overtones are present in the resulting vibration. This motion is therefore a combination or *superposition* of normal modes. Several frequencies and motions are present simultaneously, and the displacement of any point on the string is the sum (or superposition) of displacements associated with the individual modes.

The fundamental frequency of the vibrating string is $f_1 = c/2L$, where, from Eq. (22–12), $c = \sqrt{S/\mu}$. It follows that

$$f_1 = \frac{1}{2L} \sqrt{\frac{S}{\mu}}. \tag{22–7}$$

Stringed instruments afford many examples of the implications of this equation. For example, all such instruments are "tuned" by varying the tension S, an increase of tension increasing the frequency or pitch, and vice versa. The inverse dependence of frequency on length L is illustrated by the long strings of the bass section of the piano or the bass viol compared with the shorter strings of the piano treble or the violin. Pressing the strings against the fingerboard of a violin or guitar with the fingers to change the length of the vibrating portion of the string is the usual means of varying pitch when playing these instruments. One reason for winding the bass strings of a piano with wire is to increase the mass per unit length μ, so as to obtain the desired low frequency without resorting to a string that is inconveniently long.

22–4

LONGITUDINAL STANDING WAVES

Longitudinal waves traveling along a tube of finite length are reflected at the ends of the tube in much the same way that transverse waves in a string are reflected at its ends. Interference between the waves traveling in opposite directions again gives rise to standing waves.

When reflection takes place at a closed end, the displacement of the particles there must always be zero. Hence a closed end is a *displacement node*. If the end of the tube is open, the nature of the reflection is more complex and depends on whether the tube is wide or narrow compared with the wavelength. If the tube is narrow compared with the wavelength, which is the case in most musical instruments, the reflection is such that the open end is a *displacement antinode*, for reasons discussed below. Thus the longitudinal waves in a column of fluid are reflected at the closed and open ends of a tube in the same way that transverse waves in a string are reflected at fixed and free ends, respectively.

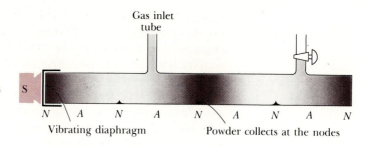

Gas inlet
tube

S

N | A   N   A   N | A   N   A   N

Vibrating diaphragm   Powder collects at the nodes

22–7 Kundt's tube for determining the velocity of sound in a gas. The shading represents the density of the gas molecules at an instant when the pressure at the displacement nodes is a maximum or a minimum.

Standing longitudinal waves in a column of gas may be demonstrated conveniently with the aid of the apparatus shown in Fig. 22–7, known as Kundt's tube. A glass tube a meter or so long is closed at one end with glass and at the other with a flexible diaphragm. The gas to be studied is admitted to the tube at a known temperature and at atmospheric pressure. A source of longitudinal waves, S, whose frequency may be varied, causes vibration of the flexible diaphragm. A small amount of light powder or cork dust is sprinkled uniformly along the tube.

When a frequency is found at which the air column is in resonance, the amplitude of the standing waves becomes large enough for the gas particles to sweep the cork dust along the tube, at all points where the gas is in motion. The powder therefore collects at the displacement nodes, where the gas remains at rest. Sometimes a wire, running along the axis of the tube, is maintained at a dull red heat by an electric current. The nodes show themselves as hot points, compared with the antinodes, which are cooled by moving gas.

In a standing wave, the distance between two adjacent nodes is one-half (not one!) wavelength; the wavelength λ is obtained by measuring the distance between alternate clumps of powder. When the frequency f is known, the velocity c is then

$$c = f\lambda.$$

At a displacement node, the pressure variations above and below the average are *maximum*, whereas at a displacement antinode, pressure does not vary. This may be understood easily when it is realized that two small masses of gas on opposite sides of a displacement node vibrate in *opposite phase*. When they approach each other, the pressure at the node is maximum, and when they recede from each other, the pressure at the node is minimum. Two small masses of gas, however, on opposite sides of a displacement *antinode* vibrate *in phase* and, hence, cause no pressure variations at the antinode.

We may speak of *pressure nodes*, the points where the pressure does not vary, and *pressure antinodes*, the points where its variation is maximum. With this language, a pressure node always is a displacement antinode, and a pressure antinode is a displacement node. An open end of a tube or a pipe is a pressure node because such an end is open to the atmosphere and is thus at constant pressure. But for this reason also an open end is always a displacement *antinode*, as stated above.

Longitudinal standing waves may be visualized with the help of Fig. 22–8, which is analogous to Fig. 21–6 for longitudinal traveling waves. Again use a 3-in. × 5-in. card with a slit 1/16 in. wide and 3 in. long, or

22–8 Diagram for illustrating longitudinal standing waves.

two cards fastened edge-to-edge with a 1/16-in. gap. Place the card over the diagram with the slit horizontal and move it vertically with constant velocity. The portions of the sine curves that appear in the slit correspond to the oscillations of the particles in a longitudinal standing wave. Each particle moves with simple harmonic motion about its equilibrium positions. The particles at the nodes do not move, and the nodes are regions of maximum compression and expansion. Midway between nodes are the antinodes, regions of maximum displacement but zero compression and expansion.

22–5

VIBRATIONS OF ORGAN PIPES

If one end of a pipe is open and a stream of air is directed against an edge, vibrations are set up and the tube resonates at its natural frequencies. As in the case of a plucked string, the fundamental and overtones exist at the same time. In the case of a pipe open at both ends, the fundamental frequency f_1 corresponds to an antinode at each end and a node in the middle, as shown at the top of Fig. 22–9. Succeeding diagrams of Fig. 22–9 show two of the overtones, which are seen to be the second and third harmonics. *In an open pipe the fundamental frequency is c/2L and all harmonics are present.*

The properties of a *stopped* pipe (open at one end, closed at the other) are shown in the diagrams of Fig. 22–10. The fundamental frequency is seen to be $c/4L$, which is one half that of an open pipe of the same length. In the language of music, the *pitch* of a closed pipe is one octave lower (a factor of two in frequency) than that of an open pipe of equal length. From the remaining diagrams of Fig. 22–10, it may be seen that the second, fourth, etc., harmonics are missing. Hence, *in a stopped pipe, the fundamental frequency is c/4L and only the odd harmonics are present.*

In an organ pipe several modes are almost always present at once. The extent to which modes higher than the fundamental are present depends on the cross section of the pipe, the ratio of length to width, whether the pipe is straight or tapered, and to a slight extent on the

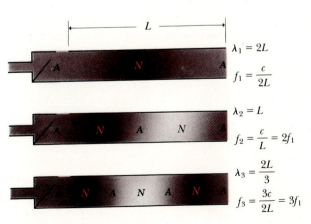

$\lambda_1 = 2L$

$f_1 = \dfrac{c}{2L}$

$\lambda_2 = L$

$f_2 = \dfrac{c}{L} = 2f_1$

$\lambda_3 = \dfrac{2L}{3}$

$f_3 = \dfrac{3c}{2L} = 3f_1$

22–9 Modes of vibration of an open organ pipe.

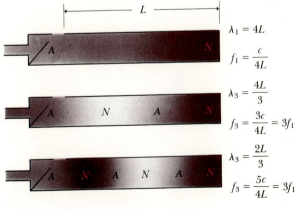

$\lambda_1 = 4L$

$f_1 = \dfrac{c}{4L}$

$\lambda_3 = \dfrac{4L}{3}$

$f_3 = \dfrac{3c}{4L} = 3f_1$

$\lambda_3 = \dfrac{2L}{3}$

$f_3 = \dfrac{5c}{4L} = 3f_1$

22–10 Modes of vibration of a stopped organ pipe.

material. The harmonic content of the tone is an important factor in determining the "tone quality" or timbre. A very thin pipe produces a tone rich in higher harmonics, which the ear perceives as thin and "stringy," while a fatter pipe produces principally the fundamental tone, perceived as a softer, more flutelike tone.

*22–6

VIBRATIONS OF RODS AND PLATES

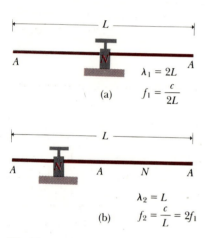

$\lambda_1 = 2L$

(a) $f_1 = \dfrac{c}{2L}$

$\lambda_2 = L$

(b) $f_2 = \dfrac{c}{L} = 2f_1$

22–11   Modes of vibration of a rod.

A rod may be set in longitudinal vibration by clamping it at some point and applying a time-varying force of appropriate frequency. In Fig. 22–11a the rod is clamped in the middle; consequently, a standing wave is set up with a node in the middle and antinodes at each end, exactly the same as the fundamental mode of an open organ pipe. The fundamental frequency of the rod is then $c/2L$, where c is the velocity of a longitudinal wave in the rod. Since the speed of a longitudinal wave in a solid is much greater than that in air, a rod has a higher fundamental frequency than an open organ pipe of the same length.

When the rod is clamped at a point $\frac{1}{4}$ of its length from one end, as shown in Fig. 22–11b, the second harmonic may be produced, and so on.

When a stretched flexible membrane, such as a drumhead, is struck a blow, a two-dimensional pulse travels outward from the struck point and is reflected and re-reflected at the boundary of the membrane. If some point of the membrane is forced to vibrate periodically, continuous waves travel along the membrane. Just as with the stretched string, standing waves can be set up in the membrane; each of these waves has a certain natural frequency. The lowest frequency is the fundamental and the others are overtones. In general, when the membrane is vibrating, a number of overtones are present.

The nodes of a vibrating membrane are *lines* (nodal lines) rather than points. The boundary of the membrane is evidently one such line. Some of the other possible nodal lines of a circular membrane are indicated by arrows in Fig. 22–12. The natural frequency of each mode is

22–12   Modes of vibration of a membrane, showing nodal lines. The frequency of each mode is given in terms of the fundamental frequency, f_1. (Adapted from *Vibration and Sound*, by Philip M. Morse, second edition, McGraw-Hill Book Company, Inc., 1948. By permission of the publishers.)

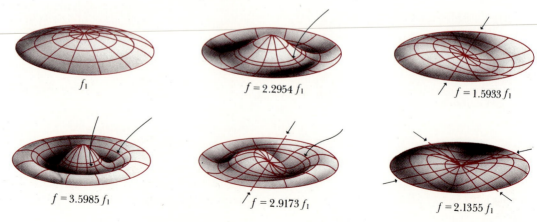

f_1

$f = 2.2954 f_1$

$f = 1.5933 f_1$

$f = 3.5985 f_1$

$f = 2.9173 f_1$

$f = 2.1355 f_1$

given in terms of the fundamental f_1. It will be noted that the frequencies of the overtones are *not* integer multiples of f_1. That is, they are not harmonics.

When a drumhead is struck, the resulting motion is not simply one of the normal-mode patterns of Fig. 22–12 but a combination or superposition of many such patterns, each with its own associated frequency. The mix of normal modes that results from a particular blow depends on how and where the drumhead is struck. Part of the art of playing tympani is in knowing how and where to strike the drumhead to produce the clearest tone with the most definite pitch sensation.

The restoring force in a vibrating flexible membrane arises from the tension with which it is stretched. A metal plate, if sufficiently thick, vibrates in a similar way, the restoring force being produced by bending stresses in the plate. The study of vibrations of membranes and plates is of importance in connection with the design of loudspeaker diaphragms, the diaphragms of telephone receivers and microphones, and many other acoustical devices.

22–7

INTERFERENCE OF LONGITUDINAL WAVES

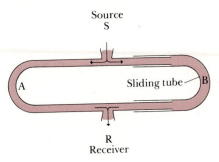

22–13 Apparatus for demonstrating interference of longitudinal waves.

The phenomenon of *interference* between two longitudinal waves in air may be demonstrated with the aid of the apparatus shown in Fig. 22–13. A wave emitted by a source S is sent into a tube, where it divides into two waves, one following the constant path SAR, the other the path SBR, which may be varied by sliding the tube B to the right. Suppose the frequency of the source is 350 Hz. Then the wavelength is $\lambda = c/f = 1.0$ m. If both paths are of equal length, the two waves arrive at R at the same time and the vibrations set up by both waves are *in phase*. The resulting vibration has an amplitude equal to the sum of the two individual amplitudes and the phenomenon of *reinforcement* or *constructive interference* may be detected either with the ear at R or with the aid of a microphone, amplifier, and loudspeaker.

Now suppose the tube B is moved out a distance of 0.25 m, thereby making the path SBR 0.5 m longer than the path SAR. The right-hand wave travels a distance $\lambda/2$ greater than the left-hand wave and the vibration set up at R by the right-hand wave is therefore in opposite phase to that set up by the left-hand wave. The consequent *destructive interference* or *cancellation* is shown by a marked *reduction* in sound at R.

If the tube B is now pulled out another 0.25 m, so that the *path difference*, SBR minus SAR, is 1.0 m (1 wavelength), the two vibrations at R again reinforce each other. Thus

$$\left\{\begin{array}{l}\text{Reinforcement takes place}\\\text{when the path difference}\end{array}\right\} = 0,\ \lambda,\ 2\lambda,\ \text{etc.}$$

$$\left\{\begin{array}{l}\text{Cancellation takes place}\\\text{when the path difference}\end{array}\right\} = \frac{\lambda}{2},\ \frac{3\lambda}{2},\ \frac{5\lambda}{2},\ \text{etc.}$$

A similar phenomenon can be demonstrated with two loudspeakers driven by the same amplifier, as shown in Fig. 22–14. Suppose the speakers both emit a pure sinusoidal sound wave of constant frequency.

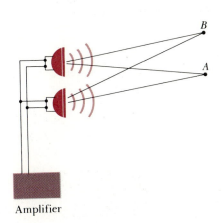

22–14

When a microphone is placed at point A in the figure, equidistant from the speakers, it receives a strong acoustic signal, corresponding to the fact that two waves emitted from the speakers in phase also arrive at point A in phase and add to each other. But at point B the signal is much *weaker* than when only one speaker is present, corresponding to the fact that the wave from one speaker travels a half-wavelength farther than that from the other; the two arrive a half-cycle out of phase and cancel each other out almost completely. An experiment closely analogous to this one, but using light waves, provided the first conclusive evidence of the wave nature of light. This is discussed in detail in Chapter 39.

22–8

RESONANCE

In previous sections we have discussed several examples of mechanical systems having normal modes of oscillation. In each mode, every particle of the system oscillates with simple harmonic motion of the same frequency as the frequency of this mode. The systems discussed have an infinite series of normal modes, but the basic concept is closely related to motion of a simple harmonic oscillator, which has only a single normal mode.

Now suppose a periodically varying force is applied to a system having normal modes. The system vibrates with a frequency equal to that of the *force;* this motion is called a *forced oscillation.* In general the amplitude of this motion is relatively small, but if the frequency of the force is close to one of the normal-mode frequencies, the amplitude can become quite large.

If the frequency of the force were precisely *equal* to a normal-mode frequency, and if there were no friction or other energy-dissipating mechanism, then the force would continue to add energy to the system, and the amplitude would increase indefinitely. In any real system there is always some dissipation of energy, but the "response" of the system (that is, the amplitude of the forced oscillation) is greatest when the force frequency is equal to one of the normal-mode frequencies. This behavior is called *resonance.*

A common example of mechanical resonance is provided by pushing a swing. The swing is a pendulum with a single natural frequency depending on its length. If a series of regularly spaced pushes is given to the swing, with a frequency equal to that of the swing, the motion may be made quite large. If the frequency of the pushes differs from the natural frequency of the swing, or if the pushes occur at irregular intervals, the swing will hardly execute any vibration at all.

Unlike a simple pendulum, which has only one natural frequency, a stretched string (and other systems discussed in this chapter) has a large number of natural frequencies. Suppose that one end of a stretched string is fixed while the other is moved back and forth in a transverse direction. The amplitude at the driven end is fixed by the driving mechanism. Standing waves are set up in the string, whatever the value of the frequency f. If the frequency is *not* equal to one of the natural frequencies of the string, the amplitude at the antinodes will be fairly small.

However, if the frequency is equal to *any one* of the natural frequencies, the string is in resonance and the amplitude at the antinodes will be very much *larger* than that at the driven end. In other words, although the driven end is not a node, it lies much closer to a node than to an antinode when the string is in resonance. In Fig. 22–5a, the right end of the string was held fixed and the left end was forced to oscillate vertically with small amplitude. The photographs show the standing waves of relatively large amplitude that resulted when the frequency of oscillation of the left end was equal to the fundamental frequency or to one of the first three overtones.

A steel bridge or, for that matter, any elastic structure, is capable of vibrating with certain natural frequencies. If the regular footsteps of a marching band were to have a frequency equal to one of the natural frequencies of a bridge that the band is crossing, a vibration of dangerously large amplitude might result. Therefore, in crossing a bridge, the group always "breaks step."

Tuning a radio or television receiver is an example of *electrical* resonance. By turning a dial, one adjusts the natural frequency of an alternating current in the receiving circuit to be equal to the frequency of the waves broadcast by the desired station. *Optical* resonance may also take place between atoms in a gas at low pressure and light waves from a lamp containing the same atoms. Thus, light from a sodium lamp may cause the sodium atoms in a glass bulb to glow with characteristic yellow sodium light.

The phenomenon of resonance may be demonstrated with the aid of the longitudinal waves set up in air by a vibrating plate or tuning fork. If two identical tuning forks are placed some distance apart and one is struck, the other will be heard when the first is suddenly damped. Should a small piece of wax or modeling clay be put on one of the forks, the frequency of that fork will be altered enough to destroy the resonance.

A similar phenomenon can be demonstrated with a piano. With the damper pedal depressed so the dampers are lifted and the strings free to vibrate, one sings a steady tone into the piano. When the singing stops the piano seems to continue to sing the same note. The sound waves from the voice excite vibrations in those strings with natural frequencies close to those (fundamental or harmonics) of the note sung initially.

QUESTIONS

22–1 When you inhale helium, your voice becomes high and squeaky. Why? (Don't try this with hydrogen; you might explode like the Hindenburg.) What happens when you inhale carbon dioxide?

22–2 A musical interval of an octave corresponds to a factor of two in frequency. By what factor must the tension in a piano or violin string be increased to raise its pitch one octave?

22–3 The pitch (or frequency) of an organ pipe changes with temperature. Does it increase or decrease with increasing temperature? Why?

22–4 By touching a string lightly at its center while bowing, a violinist can produce a note exactly one octave above the note to which the string is tuned, i.e., a note with exactly twice the frequency. Why is this possible?

22–5 In most modern wind instruments the pitch is changed by changing the length of the vibrating air column using keys or valves. The bugle, however, has no valves or keys, yet it can play many notes. How? Are there restrictions on what notes it can play?

22–6 Kettledrums (tympani) have a pitch, determined by the frequencies of several of the normal modes. How can a kettledrum be tuned?

22–7 Energy can be transferred along a string by wave motion. However, in a standing wave no energy can ever be transferred past a node. Why?

22–8 Can a standing wave be produced on a string by superposing two waves with the same frequency but different amplitudes, traveling in opposite directions?

22–9 In discussing standing longitudinal waves in an organ pipe, we spoke of pressure nodes and antinodes. What physical quantity is analogous to pressure for standing transverse waves on a stretched string?

22–10 A glass of water sits on a kitchen counter just above a dishwasher that is running. The surface of the water has a set of concentric stationary circular ripples. What is happening?

22–11 Consider the peculiar sound of a large bell. Do you think the overtones are harmonic?

22–12 What is the difference between music and noise?

22–13 Is the mass per unit length the same for all strings on a piano? On a guitar? Why?

22–14 When a heavy truck drives up a steep hill, the windows in adjacent houses sometimes vibrate. Why?

22–15 A popular children's pastime is lining up a row of dominoes, standing on edge, so that when the first one is pushed over the whole row falls, one after another. Is this a wave motion? In what respects is it similar to waves on a stretched string? In what respects is it different?

22–16 A series of traffic lights along a street is timed so that a car going at the right speed can hit a succession of green lights and not have to stop. Considering several cars on the same street, does this situation have wavelike properties?

22–17 Some opera singers are reputed to be able to break a glass by singing the appropriate note. What physical phenomenon could account for this?

22–18 A pipe organ in a church is tuned when the temperature is 20°C. On a cold winter day when the temperature in the room is 10°C the organ still sounds "in tune," despite the phenomenon mentioned in Question 22–3. Why?

PROBLEMS

22–1 A piano tuner stretches a steel piano "string" 50 cm long, of mass 5 g, with a tension of 400 N.

a) What is the frequency of its fundamental mode of vibration?

b) What is the number of the highest overtone that could be heard by a person who is capable of hearing frequencies up to 10,000 Hz?

22–2 A steel wire of length $L = 100$ cm and density $\rho = 8$ g·cm^{-3} is stretched tightly between two rigid supports. Vibrating in its fundamental mode, the frequency is $f = 200$ Hz.

a) What is the speed of transverse waves on this wire?

b) What is the longitudinal stress (force per unit area) in the wire?

c) If the maximum acceleration at the midpoint of the wire is 80,000 cm·s^{-2}, what is the amplitude of vibration at the midpoint?

22–3 A physics student observes that a stretched string vibrates with a frequency of 30 Hz in its fundamental mode when the supports are 60 cm apart. The amplitude at the antinode is 3 cm. The string has a mass of 30 g.

a) What is the speed of propagation of a transverse wave in the string?

b) Compute the tension in the string.

22–4 A cellist tunes the A-string of her instrument to a fundamental frequency of 220 Hz. The vibrating portion of the string is 68 cm long and has a mass of 1.29 g. With what tension must it be stretched?

22–5 A standing wave of frequency 1100 Hz in a column of methane at 20°C produces nodes that are 20 cm apart. What is the ratio of the heat capacity at constant pressure to that at constant volume (γ)?

22–6 An aluminum sculpture is hung from a steel wire. The fundamental frequency for transverse standing waves on the wire is 300 Hz. The object is then immersed in water so that one half of its volume is submerged. What is the new fundamental frequency?

22–7 A student in a physics lab sets up standing waves in a Kundt's tube by using the longitudinal vibration of an iron rod 1 m long, clamped at the center. If the frequency of the iron rod is 2480 Hz and the powder heaps within the tube are 6.9 cm apart, what is the speed of the waves

a) in the iron rod, and

b) in the gas?

22–8 The speed of a longitudinal wave in a mixture of helium and neon at 300 K was found to be 758 m·s^{-1}. What is the composition of the mixture?

22–9 The atomic mass of iodine is 127 g·mol^{-1}. A standing wave in iodine vapor at 400 K produces nodes that are 6.77 cm apart when the frequency is 1000 Hz. Is iodine vapor monatomic or diatomic?

22–10 A copper rod 1 m long, clamped at the $\frac{1}{4}$ point, is set in longitudinal vibration and is used to produce standing waves in a Kundt's tube containing air at 300 K. Heaps of cork dust within the tube are found to be 4.95 cm apart. What is the speed of the longitudinal waves in copper?

22–11 Find the fundamental frequency and the first four overtones of a 6-in. pipe

a) if the pipe is open at both ends, and

b) if the pipe is closed at one end.

c) How many overtones may be heard by a person having normal hearing for each of the above cases? Assume $c = 1130 \text{ ft} \cdot \text{s}^{-1}$.

22–12 A long tube contains air at a pressure of 1 atm and temperature 77°C. The tube is open at one end and closed at the other by a movable piston. A tuning fork near the open end is vibrating with a frequency of 500 Hz. Resonance is produced when the piston is at distances 18.0, 55.5, and 93.0 cm from the open end.

a) From these measurements, what is the speed of sound in air at 77°C?

b) From the above result, what is the ratio of the specific heat capacities (γ) for air?

22–13 An organ pipe A of length 2 ft, closed at one end, is "overblown" to make it vibrate in the second mode (first overtone). Another organ pipe B of length 1.35 ft, open at both ends, is vibrating in its fundamental mode. Take the speed of sound in air as $1120 \text{ ft} \cdot \text{s}^{-1}$. Neglect end corrections.

a) What is the frequency of the tone from A?

b) What is the frequency of the tone from B?

22–14 A plate cut from a quartz crystal is often used to control the frequency of an oscillating electrical circuit. Longitudinal standing waves are set up in the plate with displacement antinodes at opposite faces. The fundamental frequency of vibration is given by the equation

$$f_1 = \frac{2.87 \times 10^5}{s},$$

where f_1 is in hertz and s is the thickness of the plate in centimeters. The density of quartz is $2.66 \text{ g} \cdot \text{cm}^{-3}$.

a) Find the speed of longitudinal waves in the plate.

b) Compute Young's modulus for the quartz plate.

c) Compute the thickness of plate required for a frequency of 1200 kHz. (1 kHz = 1000 Hz.)

22–15 In Problem 22–4, what percent increase in tension is needed to increase the frequency from 220 Hz to 233 Hz, corresponding to a rise in pitch from A to A-sharp?

22–16 The longest pipe found in most medium-size pipe organs is 16 feet long. What is the frequency of the corresponding note if the pipe is

a) open at both ends, and

b) open at one end, closed at the other?

Assume $T = 293$ K.

22–17 The frequency of middle C is 262 Hz.

a) If an organ pipe is open at both ends, what length must it have to produce this note at 20°C?

b) At what temperature will the frequency be 6 percent higher, corresponding to a rise in pitch from C to C-sharp?

22–18 A certain organ pipe produces a frequency of 440 Hz in air. If the pipe is filled with helium at the same temperature, what frequency does it produce?

23

ACOUSTIC PHENOMENA

In this chapter we are concerned primarily with longitudinal waves in air that, when striking the ear, give rise to the sensation of *sound*. The human ear is sensitive to waves in the frequency range from about 20 to 20,000 Hz, although the term *sound wave* is sometimes applied also to similar waves with frequencies outside the range of human audibility. Properties of sound waves include frequency, intensity, and harmonic content, perceived by the ear as pitch, loudness, and timbre, respectively. We study the frequency relations involved in musical intervals and scales, and the interaction of two tones with slightly different frequencies. When the source and observer are in relative motion, frequency shifts occur. Finally, we discuss various sources of sound and practical applications of acoustic phenomena.

23–1

SOUND WAVES

The simplest sound waves are *sinusoidal* waves with definite frequency, amplitude, and wavelength. When such a wave arrives at the ear, it causes a vibration of the air particles at the eardrum with a definite frequency and amplitude. This vibration may also be described in terms of the variation of *air pressure* at the same point. The air pressure rises above atmospheric pressure and then drops below atmospheric pressure, with simple harmonic variation of the same frequency as that of an air particle's motion.

The simplest periodic sound wave in an elastic medium is described by a wave function of the form introduced in Section 21–7:

$$y = A \sin (\omega t - kx), \qquad (23\text{–}1)$$

where y is the displacement from equilibrium of a point in the medium, and the amplitude A is as usual the *maximum* displacement from equilibrium.

From a practical standpoint it is nearly always easier to measure the *pressure* variations in a sound wave than the displacements. For sinusoidal waves the pressure variations are also described by sinusoidal functions, and it can be shown that, for the wave described by Eq. (23–1), the pressure variation at any point is given by

$$p = BkA \cos(\omega t - kx), \qquad (23\text{–}2)$$

where B is the bulk modulus for the material and k is the wave number ($k = 2\pi/\lambda$) defined in Section 21–7. This expression shows that the quantity BkA represents the maximum pressure variation. This maximum is called the *pressure amplitude* and is denoted by p_{max}. Thus

$$p_{max} = BkA. \qquad (23\text{–}3)$$

We note that the pressure amplitude is directly proportional to the displacement amplitude A, as might be expected, and that it also depends on the wavelength. Waves of shorter wavelength (larger k) have greater pressure variations for a given amplitude because the maxima and minima are squeezed closer together.

It is also useful to develop a relation between the pressure amplitude p_{max} and the maximum *velocity* of a particle in the medium. This maximum value, called the *velocity amplitude* (not to be confused with the wave speed c), is denoted by V and is related to A just as for any sinusoidal motion: $V = A\omega$. (See, for example, Section 11–3.) Expressing Eq. (23–3) in terms of V, and using the relation $\omega = ck$ from Eq. (21–17), we obtain

$$p_{max} = BkA = \frac{BkV}{\omega} = \frac{BV}{c}. \qquad (23\text{–}4)$$

Thus the relation between pressure amplitude and velocity amplitude does not depend on frequency. Recalling that the propagation speed c is given by Eq. (21–4),

$$c = \sqrt{B/\rho},$$

we may rewrite Eq. (23–4) as follows:

$$p_{max} = \sqrt{B\rho}\, V. \qquad (23\text{–}5)$$

The quantity $(B\rho)^{1/2}$ is called the *mechanical impedance* of the system, in reference to an analogy with electric circuits in which pressure is analogous to voltage, velocity to current, and mechanical impedance to resistance or impedance. In terms of this analogy, Eq. (23–5) is the mechanical analog of Ohm's law, discussed in Section 28–3.

EXAMPLE Measurements of sound waves show that the maximum pressure variations in the loudest sounds that the ear can tolerate without pain are of the order of 30 Pa (above and below atmospheric pressure, which is about 100,000 Pa). Find the corresponding maximum displacement, if the frequency is 1000 Hz and $c = 350 \text{ m}\cdot\text{s}^{-1}$.

Solution   We have $\omega = (2\pi)(1000 \text{ Hz}) = 6283 \text{ s}^{-1}$ and

$$k = \frac{\omega}{c} = \frac{6283 \text{ s}^{-1}}{350 \text{ m}\cdot\text{s}^{-1}} = 18.0 \text{ m}^{-1}.$$

The adiabatic bulk modulus for air is

$$B = \gamma p = (1.4)(1.01 \times 10^5 \text{ Pa}) = 1.42 \times 10^5 \text{ Pa}.$$

From Eq. (23–3) we find

$$A = \frac{p_{max}}{Bk} = \frac{(30 \text{ Pa})}{(1.42 \times 10^5 \text{ Pa})(18.0 \text{ m}^{-1})}$$
$$= 1.17 \times 10^{-5}\text{m} = 0.0117 \text{ mm}.$$

Thus the displacement amplitude of even the loudest sound is *extremely* small. The maximum pressure variation in the *faintest* audible sound of frequency 1000 Hz is only about 3×10^{-5} Pa. The corresponding displacement amplitude is about 10^{-9} cm. For comparison, the wavelength of yellow light is 6×10^{-5} cm, and the diameter of a molecule is about 10^{-8} cm. The ear is an extremely sensitive organ!   ◄

23–2

INTENSITY

An essential aspect of wave propagation of all sorts is transfer of *energy*. A familiar example is the energy supply of the earth, which reaches us from the sun via electromagnetic waves. The intensity I of a traveling wave is defined as *the time average rate at which energy is transported by the wave, per unit area*, across a surface perpendicular to the direction of propagation. More briefly, the intensity is the average *power* transported per unit area.

We have seen that the power developed by a force equals the product of force and velocity. Hence the power per unit area in a sound wave equals the product of the excess pressure (force per unit area) and the *particle* velocity. The *maximum* power per unit area is the product of the maximum pressure variation p_{max} and the maximum velocity V. From Eq. (23–3) and $V = A\omega$,

$$p_{max}V = (BkA)(A\omega) = \omega BkA^2.$$

The *average* value of the product pv is, of course, less than this, since each quantity is a sinusoidal function. It turns out that the average value is precisely one half the maximum value. This simple result is related to the fact that the average value of $\sin^2 x$ or $\cos^2 x$ over a complete cycle is $\frac{1}{2}$. Thus the intensity, or average power per unit area, is given by

$$I = \tfrac{1}{2}\omega BkA^2. \tag{23–6}$$

It is usually more convenient to express I in terms of the pressure amplitude p_{max}. Using Eq. (23–5), we find

$$I = \tfrac{1}{2}p_{max}V = \frac{p_{max}^2}{2\sqrt{\rho B}}. \tag{23–7}$$

By use of the wave speed relation $c = (B/\rho)^{1/2}$, this may also be written in the alternative forms

$$I = \frac{p_{max}^2 c}{2B} = \frac{p_{max}^2}{2\rho c}. \tag{23–8}$$

Derivation of these expressions is left as an exercise.

The intensity of a sound wave of the largest amplitude tolerable to the human ear (about $p_{max} = 30$ Pa) is

$$I = \frac{(30 \text{ Pa})^2}{2(1.22 \text{ kg} \cdot \text{m}^{-3})(346 \text{ m} \cdot \text{s}^{-1})}$$

$$= 1.07 \text{ J} \cdot \text{s}^{-1} \cdot \text{m}^{-2} = 1.07 \text{ W} \cdot \text{m}^{-2}$$

$$= 1.07 \times 10^{-4} \text{ W} \cdot \text{cm}^{-2}.$$

The unit 1 W $\cdot$ cm^{-2} is a mixed unit, neither cgs nor mks. We mention it here because it is unfortunately in general use among acousticians.

The pressure amplitude of the *faintest* sound wave that can be heard is about 3×10^{-5} Pa, and the corresponding intensity is about 10^{-12} W $\cdot$ m^{-2} or 10^{-16} W $\cdot$ cm^{-2}.

The *total* power carried across a surface by a sound wave equals the product of the intensity at the surface and the surface area, if the intensity over the surface is uniform. The average total power developed as sound waves by a person speaking in an ordinary conversational tone is about 10^{-5} W, while a loud shout corresponds to about 3×10^{-2} W. Since the population of the city of New York is about eight million persons, the acoustical power developed, if all were to speak at the same time, would be about 80 W, or enough to operate a medium-size electric light. On the other hand, the power required to fill a large auditorium with loud sound is considerable. Suppose the intensity over the surface of a hemisphere 20 m in radius is 1 W $\cdot$ m^{-2}. The area of the surface is about 2500 m^2. Hence the acoustic power output of a speaker at the center of the sphere would have to be

$$(1 \text{ W} \cdot \text{m}^{-2})(2500 \text{ m}^2) = 2500 \text{ W},$$

or 2.5 kW. The electrical power input to the speaker would need to be considerably larger, since the efficiency of such devices is not very high.

Because of the extremely large range of intensities over which the ear is sensitive, a *logarithmic* rather than an arithmetic intensity scale is convenient. Accordingly, the *intensity level* β of a sound wave is defined by the equation

$$\beta = 10 \log \frac{I}{I_0}, \tag{23–9}$$

where I_0 is an arbitrary reference intensity, taken as 10^{-12} W $\cdot$ m^{-2}, which corresponds roughly to the faintest sound that can be heard. Intensity levels are expressed in *decibels*, abbreviated dB.*

---

*Originally, a scale of intensity levels in *bels* was defined by the relation

$$\text{Intensity level} = \log (I/I_0).$$

This unit proved inconveniently large, and hence the *decibel*, one tenth of a bel, has come into general use. The unit was named in honor of Alexander Graham Bell.

If the intensity of a sound wave equals I_0 or 10^{-12} W·m^{-2}, its intensity level is 0 dB. The maximum intensity that the ear can tolerate, about 1 W·m^{-2}, corresponds to an intensity level of 120 dB. Table 23–1 gives the intensity levels in decibels of a number of familiar noises. It is taken from a survey made by the New York City Noise Abatement Commission.

TABLE 23–1	NOISE LEVELS DUE TO VARIOUS SOURCES (REPRESENTATIVE VALUES)	
Source or description of noise	Noise level, dB	Intensity, W·m^{-2}
Threshold of pain	120	1
Riveter	95	3.2×10^3
Elevated train	90	10^{-3}
Busy street traffic	70	10^{-5}
Ordinary conversation	65	3.2×10^{-6}
Quiet automobile	50	10^{-7}
Quiet radio in home	40	10^{-8}
Average whisper	20	10^{-10}
Rustle of leaves	10	10^{-11}
Threshold of hearing	0	10^{-12}

Within the range of audibility, the sensitivity of the ear varies with frequency. The *threshold of audibility* at any frequency is the minimum intensity of sound at that frequency that can be detected. For a young adult with normal hearing, the threshold of audibility at 1000 Hz is about 0 dB; at 200 and 15,000 Hz it is about 20 dB; and at 50 and 18,000 Hz it is about 50 dB. Thus the ear's sensitivity drops off at the low and high ends of the frequency scale. Frequencies above 20,000 Hz (20 kHz) are not audible to humans at *any* intensity, and such frequencies are referred to as *ultrasonic*.

23–3

THE EAR AND HEARING

Figure 23–1 shows a section of the right ear. The scale of the inner ear has been exaggerated to show details. Sound waves traveling down the ear canal strike the eardrum. A linkage of three small bones, the hammer, anvil, and stirrup, transmits the vibrations to the oval window. The oval window in turn transmits them to the *cochlea*, shaped something like a snail shell. The terminals of the auditory nerve, of which there are about 30,000 in each ear, are distributed along the *basilar membrane* that divides the cochlea into two canals. The vibration at the oval window induces a wave that propagates down one canal of the fluid-filled cochlea and back in the other, terminating finally at the round window (not shown). The 30,000 nerve terminals, called *hair cells*, actually occupy an area only about 30 millimeters long and $\frac{1}{3}$ of a millimeter wide.

Individual hair cells do not correspond to specific frequencies; instead, each frequency produces a characteristic pattern of excitation of the entire basilar membrane. Figure 23–2a shows the spiral shape of the

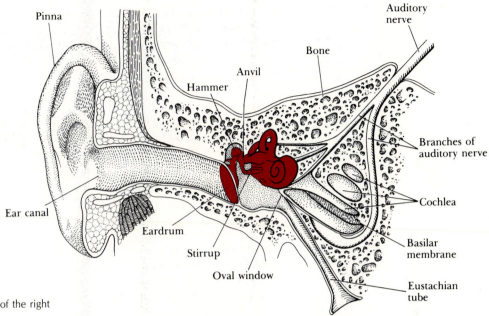

Pinna

Hammer

Anvil

Bone

Auditory nerve

Branches of auditory nerve

Cochlea

Ear canal

Eardrum

Stirrup

Oval window

Basilar membrane

Eustachian tube

23–1   Diagrammatic section of the right ear.

cochlea, and Fig. 23–2b shows some of the hair cells on the basilar membrane.

The range of frequencies and intensities to which the ear is sensitive is conveniently represented by a diagram such as that of Fig. 23–3, which is a graph of the *auditory area* of a person of good hearing. The height of the lower colored curve at any frequency represents the intensity level of

(a)

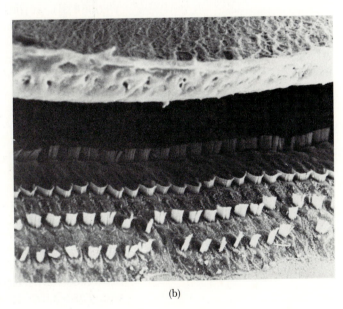

(b)

23–2   (a) Spiral form of a guinea pig cochlea, which is nearly identical to that of the human ear. (b) Some of the hair cells, or nerve terminals, on the basilar membrane, as shown by an electron microscope. The magnification is roughly 2000×. (Photos by Molly Webster, © 1982 DISCOVER Magazine, Time Inc.)

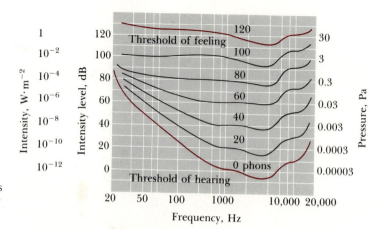

23–3 Thresholds of hearing and feeling (color curves) and contours of constant loudness (black curves). All sinusoidal tones on a contour sound equally loud. (Courtesy of Dr. Harvey Fletcher.)

the faintest pure (sinusoidal) tone of that frequency that can be heard. It will be seen from the diagram that the ear is most sensitive to frequencies between 2000 and 3000 Hz where the *threshold of hearing,* as it is called, is about −5 dB. The upper colored curve represents the intensity level of the loudest pure tone that can be tolerated. At intensities above this curve, which is called the *threshold of feeling or pain,* the sensation changes from one of hearing to discomfort or even pain. The height of the upper curve is approximately constant at a level of about 120 dB for all frequencies. Every pure tone that can be heard may be represented by a point lying somewhere in the area between these two curves. For a loud tone of intensity level 80 dB, the range of audibility is from 20 to 20,000 Hz, but at a level of 20 dB it is only from about 200 to about 15,000 Hz. At 1000 Hz the range of intensity level is from 0 dB to 120 dB, but at 100 Hz it is only from 30 dB to 120 dB.

The term *loudness* refers to a listener's subjective perception of a sound sensation. Loudness increases with intensity, but because of the varying sensitivity of the ear it also depends on frequency. Experiments with human subjects are necessary to establish the relation of loudness to intensity and frequency. The results of many such experiments are summarized in the black curves of Fig. 23–3. Each curve is drawn through points representing pure (sinusoidal) tones of equal perceived loudness. For example, a 100-Hz tone at 62 dB, a 1000-Hz tone at 40 dB, and a 10,000-Hz tone at 52 dB all sound equally loud. These tones are said to have a *loudness level* of 40 phon, meaning that they all sound as loud as a 40-dB tone at the reference frequency of 1000 Hz. Similarly, the other curves show loudness levels of 0, 20, 60, 80, 100, and 120 phons.

The standard unit of perceived loudness is the *sone;* by definition, one sone is the loudness experienced by a listener with normal hearing when a 1000-Hz tone at 40 dB is presented to both ears. As the above discussion shows, a 100-Hz tone at 62 dB and a 10,000-Hz tone at 52 dB would also have a loudness of 1 sone. Then a 2-sone tone is defined to be a tone sounding twice as loud as a 1-sone tone. The corresponding loudness level (in phons) must again be determined by experiment with human subjects who are asked to judge "twice as loud." The resulting correlation between loudness level (in phons) and loudness (in sones) is

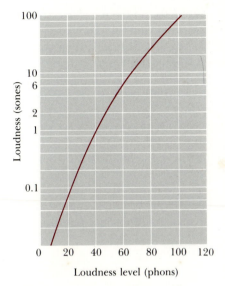

23–4   Graph showing the relation between loudness (in sones) and loudness level (in phons). The loudness level of any tone is equal to the sound intensity level (in dB) of a 1000-Hz tone having the same perceived loudness as that tone.

given in Fig. 23–4. For example, a loudness of 6 sones corresponds to a loudness level of 60 phon. Reference to Fig. 23–3 shows that this might be a 1000-Hz tone at 60 dB, a 100-Hz tone at 72 dB, a 10,000-Hz tone at 75 dB, and so on. All these have the same loudness level and the same loudness, and the average listener would judge any of them to be six times as loud as a 1000-Hz tone at 40 dB.

Sounds that are not pure tones do not have a single frequency and hence cannot be represented by a single point on the diagram of Fig. 23–3. A sound such as that from a musical instrument, consisting of a mixture of a relatively few frequencies (the fundamental and overtones), can be represented by a set of points, each point giving the intensity and frequency of one particular overtone. A sound such as a street noise cannot be considered as made up of a fundamental and overtones, but it can still be represented on the diagram. The sound is picked up by a microphone and sent through an electrical network that selects a narrow range of frequencies and measures the average intensity within this range. By repeating the process at a large number of frequencies throughout the audible range, we obtain a series of points that can be plotted. A continuous curve drawn through them is called the *spectrogram* of the sound.

A typical spectrogram of street noise is shown in Fig. 23–5. The units on the vertical axis are intensity per unit frequency interval or pressure per unit frequency interval. Each point on the curve is obtained by measuring the intensity within a narrow frequency interval, dividing the result by the width of the interval (in Hz), and, for the dB and pressure scales, converting the resulting number to equivalent dB or p_{max}, using Eq. (23–9) or (23–8), respectively.

The term "spectrogram" is borrowed from optics. The process just described is entirely analogous to the optical one of dispersing a beam of light waves into a spectrum by means of a prism and measuring the intensity at a number of wavelengths throughout the spectrum. The light emitted by a gas in an electrical discharge is a mixture of waves of a number of definite frequencies and corresponds to the sound emitted by a musical instrument. Most light beams, however, are a mixture of all frequencies and are therefore the optical analog of noise.

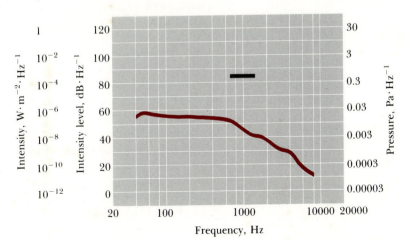

23–5   Spectrogram of street noise. (Courtesy of Dr. Harvey Fletcher.)

The *total* intensity level of a noise can be found from its spectrogram by an integration process. There are also instruments known as noise meters that measure this level directly. The level of the street noise in Fig. 23–5 is about 85 dB and is shown by the short heavy line.

How we sense the *direction* of a sound source is an interesting study. We sometimes turn our heads to locate a sound source; the ear nearer the source hears a louder sound because the head casts a "sound shadow" for the ear farther from the source. However, this effect disappears when the wavelength of the sound becomes appreciably larger than the distance between the ears (around 15 cm) because then the sound wave bends around the head appreciably. We also perceive directions by means of the relative *phase* of the two waves arriving at the two ears. This effect is most noticeable when the wavelength is comparable to the distance between the ears; for 15 cm spacing, the corresponding frequency is about 2000 Hz.

Experiments have shown that for frequencies up to about 1000 Hz the phase-difference effect is most important; for frequencies above 4000 Hz the loudness effect is dominant. In between, direction finding is less accurate, indicating that there is rather little overlap between the two processes. We invite the reader to consider the relevance of these experiments for stereo sound reproduction.

23–4

QUALITY AND PITCH

A string that has been plucked or a plate that has been struck, if allowed to vibrate freely, will vibrate with many frequencies at the same time. It is a rare occurrence for a body to vibrate with only one frequency. A carefully made tuning fork struck lightly on a rubber block may vibrate with only one frequency, but in the case of musical instruments, the fundamental and many harmonics are usually present at the same time. The impulses that are sent from the ear to the brain give rise to one net effect that is characteristic of the instrument. Suppose, for example, that the sound spectrum of a tone consisted of a fundamental of 200 Hz and harmonics 2, 3, 4, and 5, all of different intensity, whereas the sound spectrum of another tone consisted of exactly the same frequencies but with a different intensity distribution. The two tones would sound different; they are said to differ in *quality* or *timbre*.

Adjectives used to describe the quality of musical tones are purely subjective in character, such as reedy, golden, round, mellow, tinny. The quality of a sound is determined in part by the number of overtones present and their respective intensity-versus-time curves. Typical sound spectra of several musical instruments are shown in Fig. 23–6.

Another important factor in determining tone quality is the behavior at the beginning and end of a tone. A piano tone begins percussively with a thump and then dies away gradually; a harpsichord tone, in addition to having different harmonic content, begins much more quickly and incisively with a click, the higher harmonics beginning before the lower ones. The cessation of tone when the key is released is also much prompter with the harpsichord than the piano. Similar effects are present in other musical instruments; with wind and string instruments the

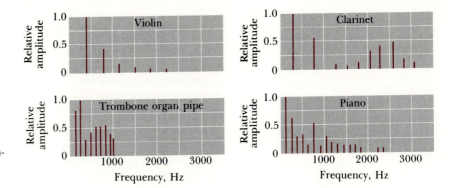

23–6 Sound spectra of some musical instruments. (Courtesy of Dr. Harvey Fletcher.)

player has considerable control over the attack and decay of the tone, and these characteristics help define the unique characteristics of each instrument.

The term *pitch* refers to the attribute of a sound sensation that enables one to classify a note as "high" or "low." Like loudness, it is a subjective quantity and cannot be measured with instruments. Pitch is closely related to the objective quantity *frequency*, but there is not an exact one-to-one correspondence. For a pure tone of constant intensity, the pitch becomes higher as the frequency is increased, but the pitch of a pure tone of constant frequency becomes lower as the intensity level is raised.

Many of the notes played on musical instruments are rich in harmonics, some of which may be more prominent than the fundamental. Presented with an array of frequencies constituting a harmonic series, the ear assigns a characteristic pitch to the combination, this pitch being that associated with the fundamental frequency of the series. So definite is this pitch sensation that it is possible to eliminate the fundamental frequency entirely, by means of filters, without changing the pitch! The ear supplies the fundamental, provided the correct harmonics are present. It is this rather surprising property of the ear that enables a small loudspeaker that does not radiate low frequencies well to give nevertheless the impression of good bass response. Because the speaker is a fairly efficient radiator for the frequencies of the harmonics, the listener believes he is actually hearing the low frequencies, when instead he is hearing only multiples of these frequencies and his ear is supplying the fundamental.

The sound spectra in Fig. 23–6 should not be taken too seriously, for several reasons. First, the spectrum is never exactly the same for higher notes as for lower notes, within the register of the instrument. Second, the spectrum often changes considerably with loudness. A French horn played very softly produces mostly fundamental, with rather little harmonic content. As the tone grows louder it also becomes much richer in harmonics. This effect is inherent in the instrument, and the player has only limited control over it. Finally, as already mentioned, the sound spectra of percussion instruments vary with time as the various harmonics die away at different rates. For all these reasons, quantitative description of the tone colors of musical instruments is an extremely complex problem.

*23–5

MUSICAL INTERVALS AND SCALES

When certain musical tones are produced in succession or together, even an untrained listener recognizes a relationship among them. Such relations are described in musical language by words such as octave, major third, minor third. The listener recognizes something basic in these combinations of tones; and experimental measurement of the fundamental frequencies of the separate tones discloses that their relationships, one to the other, may be approximated by simple whole-number ratios. For example, the fundamental frequencies of middle C of the piano and the C an octave above have a ratio of 1 to 2. Another basic set of tones is obtained by playing C, E, G. These constitute what is known as a *major triad,* and the frequencies are found to be approximately proportional to 4, 5, and 6.

Starting at middle C of the piano, and playing only the white keys toward the right, it is possible to find, within only nine notes, three major triads. These are shown in Table 23–2; they enable one to calculate the frequencies of all the notes, once one of them has been chosen arbitrarily. The frequency of the A above middle C is usually chosen to be 440 Hz. The frequencies of all the notes, shown in row 5 of Table 23–2, are those that correspond to the *just diatonic scale* of the key of C. The

TABLE 23–2 FREQUENCY RELATIONS IN THE JUST DIATONIC SCALE AND IN THE EQUALLY TEMPERED SCALE

Frequency relations	Do Middle' C	Re D	Mi E	Fa F	Sol G	La A	Ti B	Do' C'	Re' D'
Octave, key of C	1							2	
Major triad, key of C	4		5		6			(8)	
Major triad, key of F				4		5		6	
Major triad, key of G		(3)			4		5		6
Just scale, key of C	264	297	330	352	396	440	495	528	594
Intervals in just scale		9/8 whole	10/9 whole	16/15 half	9/8 whole	10/9 whole	9/8 whole	16/15 half	9/8 whole
Equally tempered scale suitable for all keys	261.6	293.7	329.6	349.2	392.0	440	493.9	523.3	587.4
Intervals in equally tempered scale		$\sqrt[6]{2}$	$\sqrt[6]{2}$	$\sqrt[12]{2}$	$\sqrt[6]{2}$	$\sqrt[6]{2}$	$\sqrt[6]{2}$	$\sqrt[12]{2}$	$\sqrt[6]{2}$

frequency ratios of adjacent notes are seen to be either 9/8, 10/9, or 16/15. The interval between two tones whose frequencies bear the ratio 9/8 or 10/9 is called a *whole tone*, whereas the interval between two notes of frequency ratio 16/15 is a *half tone*.

If a major diatonic scale were constructed starting at D instead of at C, four new notes would be needed for a perfect diatonic scale. If all possible musical keys were to be provided for, many additional notes would be needed for each octave. To avoid this tremendous complication, what is known as the *equally tempered scale* has been devised. In this scheme there are 12 half-tone intervals in every range of an octave, adjacent notes a half tone apart bearing the constant ratio of the twelfth root of 2, i.e., 1.05946.

Simple as this scheme is, it results in no one scale being exactly diatonic. Since the ratios of the diatonic scale were originally selected to suit the preferences of the ear (being made up of three sets of major triads, each of which constitutes a harmonious combination), this means that an instrument tuned to the equally tempered scale is not quite so pleasant to the ear. Thus a piano whose white keys are tuned to the "just" scale sounds better (i.e., more "in tune") than one tuned to equal temperament when played in the key of C, but in some other keys it sounds worse. Hence an instrument to be used for compositions in all keys is usually tuned to equal temperament.

In the Baroque period, however, keys with more than three sharps or flats were used relatively infrequently, and various compromise temperaments were devised by Kirnberger, Werckmeister, and others, to favor the commonly used keys. Organs and harpsichords intended primarily for music of this period are sometimes tuned to one of these temperaments. There is evidence that J. S. Bach favored these rather than equal temperament. His composition "The Well-Tempered Clavier" contains preludes and fugues in all the major and minor keys, but it is important not to misconstrue *well-tempered* as *equally tempered!*

23–6

BEATS

Standing waves in an air column have been cited as one example of *interference*. They arise when two waves of the same amplitude and frequency are traveling through the same region in opposite directions. We now wish to consider another type of interference that results when two waves of equal amplitude but slightly different frequency travel through the same region. Such a condition exists when two tuning forks of slightly different frequency are sounded simultaneously or when two piano strings struck by the same key are slightly "out of tune."

Let us consider a particular point in space through which the two waves pass simultaneously. The individual displacements are plotted as functions of time in Fig. 23–7a. If the total length of the time axis represents about one second, the frequencies are 16 Hz and 18 Hz. Applying the principle of superposition, we add the two displacements at each instant of time to find the total displacement at that time, obtaining the graph of Fig. 23–7b. At certain times the two waves are in phase; their maxima coincide and their amplitudes add. But as time goes on they

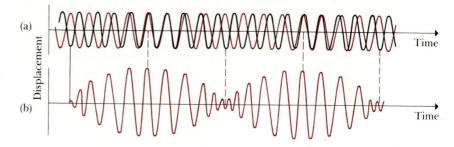

(a)

Displacement

Time

(b)

Time

23–7 Beats are fluctuations in amplitude produced by two sound waves of slightly different frequency. (a) Individual waves. (b) Pattern formed by superposition of the two waves.

become more and more out of phase because of their slightly different frequencies. Eventually a maximum of one wave coincides with a maximum in the opposite direction for the other, they cancel each other, and the total amplitude is zero.

Thus the amplitude of the composite wave varies from a maximum value to zero and back, as shown in Fig. 23–7b. The appearance is that of a single sinusoidal wave with a varying amplitude; in this example, the amplitude goes through two maxima and two minima in one second, and thus the frequency of this amplitude variation is 2 Hz. The amplitude variation causes variations of loudness that are called *beats*, and the frequency with which the amplitude varies is called the *beat frequency*, which in this example is the *difference* of the two frequencies. If the beat frequency is a few hertz, it is perceived as a waver or pulsation in the tone.

We can prove that the beat frequency is *always* the difference of the two frequencies f_1 and f_2. Suppose f_1 is larger than f_2; the corresponding periods are τ_1 and τ_2, with $\tau_1 < \tau_2$. If the two waves start out in phase at time $t = 0$, they will again be in phase at a time T such that the first wave has gone through exactly one more cycle than the second. Let n be the number of cycles of the first wave in time T; then the number of cycles of the second in the same time is $(n - 1)$, and we have the relations

$$T = n\tau_1 = (n - 1)\tau_2.$$

We solve for n and substitute the result back into the first equation, obtaining

$$T = \frac{\tau_1\tau_2}{\tau_2 - \tau_1}.$$

Now T is just the *period* of the beat, and its reciprocal is the beat *frequency*, $f_{\text{beat}} = 1/T$. Thus we find

$$f_{\text{beat}} = \frac{\tau_2 - \tau_1}{\tau_1\tau_2} = \frac{1}{\tau_1} - \frac{1}{\tau_2},$$

and finally

$$f_{\text{beat}} = f_1 - f_2. \tag{23–10}$$

Beats between two tones can be detected by the ear up to a beat frequency of 6 Hz or 7 Hz. Two piano strings or two organ pipes differing in frequency by two or three hertz sound wavery and "out of tune," although some organ stops contain two sets of pipes deliberately tuned to beat frequencies of about 1 Hz to 2 Hz for a gently undulating effect.

Listening for beats is an important technique in tuning all musical instruments.

At higher frequencies, individual beats can no longer be distinguished and the sensation merges into one of *consonance* or *dissonance* depending on the frequency ratio of the tones. A beat frequency, even though it lies within the frequency range of the ear, is not necessarily interpreted by the ear as a tone of that frequency. Nevertheless, a tone can be heard of frequency equal to the frequency difference between two others sounded simultaneously. Such a tone is called a *difference* tone.

*23–7

THE DOPPLER EFFECT

When a source of sound, or a listener, or both, are in motion relative to the air, the pitch of the sound, as heard by the listener, is in general not the same as when source and listener are at rest. The most common example is the sudden drop in pitch of the sound from an automobile horn as one meets and passes a car proceeding in the opposite direction. This phenomenon is called the *Doppler effect*.

Let v_L and v_S represent the velocities of a listener and a source, relative to the air. We shall consider only the special case in which the velocities lie along the line joining listener and source. Since these velocities may be in the same or opposite directions, and the listener may be either ahead of or behind the source, we need a sign convention. We shall take the positive directions of v_L and v_S as that *from* the position of the listener, toward the position of the source. The speed of propagation of sound waves, c, will always be considered positive.

We consider first a listener L moving with velocity v_L toward a stationary source S, as in Fig. 23–8. The source emits a wave with frequency

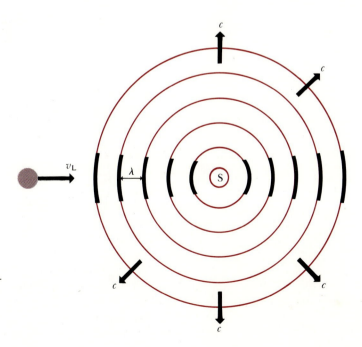

23–8 A listener moving toward a stationary source hears a frequency higher than the source frequency because the relative velocity of listener and wave is greater than c.

f_S and wavelength $\lambda = c/f_S$. The figure shows several wave crests, separated by equal distances λ. The waves approaching the moving listener have a speed of propagation *relative to him* of $(c + v_L)$. Thus the frequency f_L with which the listener encounters wave crests, i.e., the frequency he *hears*, is

$$f_L = \frac{c + v_L}{\lambda} = \frac{c + v_L}{c/f_S},$$
(23–11)

or

$$f_L = f_S\left(\frac{c + v_L}{c}\right) = f_S\left(1 + \frac{v_L}{c}\right).$$
(23–12)

Thus an observer moving toward a source hears a larger frequency and higher pitch than a stationary observer. Similarly, a listener moving away from the source ($v_L < 0$) hears a lower pitch.

Now suppose the source is also moving, with velocity v_S, as in Fig. 23–9. The wave speed relative to the air is still c; the speed of propagation of a wave is not altered by the motion of the source but is a property of the wave medium alone. But the wavelength is no longer given by c/f_S. The time for emission of one cycle of the wave is the period $\tau = 1/f_S$. During this time the wave travels a distance $c\tau = c/f_S$, and the source moves a distance $v_S\tau = v_S/f_S$. The wavelength is the distance between successive wave crests, and this is determined by the *relative* displacement of source and wave. In the region to the right of the source, the wavelength is

$$\lambda = \frac{c}{f_S} - \frac{v_S}{f_S} = \frac{c - v_S}{f_S},$$
(23–13)

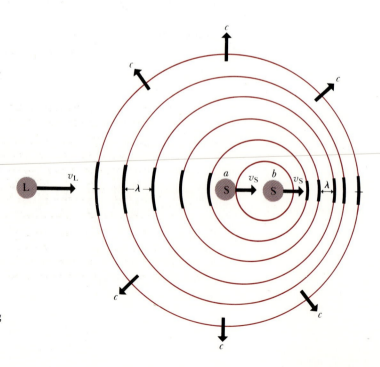

23–9 Wave surfaces emitted by a moving source are crowded together in front of the source and stretched out behind it.

and in the region to the left it is

$$\lambda = \frac{c + v_S}{f_S}. \qquad (23\text{–}14)$$

The waves are, respectively, compressed and stretched out by the motion of the source.

The frequency measured by the listener is now given by substituting Eq. (23–14) into the first form of Eq. (23–11):

$$f_L = \frac{c + v_L}{\lambda} = \frac{c + v_L}{(c + v_S)/f_S},$$

or

$$\frac{f_L}{c + v_L} = \frac{f_S}{c + v_S}, \qquad (23\text{–}15)$$

which expresses the frequency f_L heard by the listener in terms of the frequency f_S of the source.

This general relation includes all possibilities for collinear motion of source and listener relative to the medium. If one happens to be at rest in the medium, the corresponding velocity is zero, and of course when both are at rest or have the same velocity relative to the medium, then $f_L = f_S$. Whenever source or listener is moving in the opposite direction to that which we have designated as positive, the corresponding velocity to be used in Eq. (23–15) is negative. The examples illustrate these sign conventions.

EXAMPLES Let $f_S = 300$ Hz, and $c = 300$ m·s^{-1}. The wavelength of the waves emitted by a stationary source is then $c/f_S = 1.00$ m.

a) What are the wavelengths ahead of and behind the moving source in Fig. 23–9 if its velocity is 30 m·s^{-1}?

In front of the source,

$$\lambda = \frac{c - v_S}{f_S} = \frac{300 \text{ m·s}^{-1} - 30 \text{ m·s}^{-1}}{300 \text{ Hz}} = 0.90 \text{ m.}$$

Behind the source,

$$\lambda = \frac{c + v_S}{f_S} = \frac{300 \text{ m·s}^{-1} + 30 \text{ m·s}^{-1}}{300 \text{ Hz}} = 1.10 \text{ m.}$$

b) If the listener L in Fig. 23–9 is at rest and the source is moving away from L at 30 m·s^{-1}, what is the frequency as heard by the listener?

Since

$$v_L = 0 \quad \text{and} \quad v_S = 30 \text{ m·s}^{-1},$$

we have

$$f_L = f_S \frac{c}{c + v_S}$$

$$= 300 \text{ Hz} \left(\frac{300 \text{ m·s}^{-1}}{300 \text{ m·s}^{-1} + 30 \text{ m·s}^{-1}} \right) = 273 \text{ Hz.}$$

c) If the source in Fig. 23–9 is at rest and the listener is moving toward the left at 30 m·s^{-1}, what is the frequency as heard by the listener?

The positive direction (from listener to source) is still from left to right, so

$$v_L = -30 \text{ m·s}^{-1}, \qquad v_S = 0,$$

$$f_L = f_S \frac{c + v_L}{c}$$

$$= 300 \text{ Hz} \left(\frac{300 \text{ m·s}^{-1} - 30 \text{ m·s}^{-1}}{300 \text{ m·s}^{-1}} \right) = 270 \text{ Hz}.$$

Thus, while the frequency f_L heard by the listener is less than the frequency f_S both when the source moves away from the listener and when the listener moves away from the source, the decrease in frequency is not the same for the same speed of recession.   ◄

In the preceding equations, the velocities v_L, v_S, and c are all *relative to the air,* or more generally, to the medium in which the waves are traveling. The Doppler effect exists also for electromagnetic waves in empty space, such as light waves or radio waves. In this case, there is no "medium" relative to which a velocity can be defined, and we can speak only of the *relative* velocity v *of source and receiver.*

To derive the expression for the Doppler frequency shift for light requires the use of relativistic kinematic relations. These will be derived in Chapter 40, but meanwhile we quote the result without derivation. The wave speed c is the speed of light and is the same for both source and listener. In the frame of reference in which the listener is at rest, the source is moving away from the listener with velocity v. (If, instead, the source is *approaching* the listener, v is negative.) The source frequency is again f_S. The frequency f_L measured by the listener (i.e., the frequency of arrival of the waves at L) is then given by

$$f_L = \left(\sqrt{\frac{c - v}{c + v}} \right) f_S. \tag{23–16}$$

When v is positive, the source moves *away* from the listener and f_L is always *less* than f_S; when v is negative, the source moves *toward* the listener and f_L is *greater* than f_S. Thus, the qualitative effect is the same as for sound, although the quantitative relationship is different.

The Doppler effect provides a convenient means of tracking a satellite that is emitting a radio signal of constant frequency f_S. The frequency f_L of the signal received on the earth decreases as the satellite is passing, since the velocity component *toward* the earth decreases from position 1 to position 2 in Fig. 23–10, and then points *away* from the earth from 2 to 3. If the received signal is combined with a constant signal generated in the receiver to give rise to *beats,* then the beat frequency may be such as to produce an audible note whose pitch decreases as the satellite passes overhead.

A similar technique is used by law-enforcement officers to measure automobile speeds. An electromagnetic wave is emitted by a source at the side of the road, typically attached to a police car. The wave is reflected

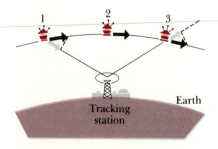

23–10 Change of velocity component along the line of sight of a satellite passing a tracking station.

from a moving car, which thus acts as a moving source; the reflected wave is Doppler-shifted in frequency. Measurement of the frequency shift using beats, as with satellite tracking, permits simple measurement of the speed.

The Doppler effect for *light* is important in astronomy. Analysis of the spectra of light from distant stars shows shifts in wavelength compared to spectra from the same elements on earth. These can be interpreted as Doppler shifts due to motion of the stars. The shift is nearly always toward the longer wavelength or red end of the spectrum, and is therefore called the *red shift*. Such observations have provided practically all the evidence for the "exploding universe" cosmological theories, which represent the universe as having evolved from a great explosion several billion years ago in a relatively small region of space.

23–8

SOURCES OF SOUND

Although this chapter has been concerned primarily with the nature and characteristics of sound, the *production* of sound is also an important area of study, posing interesting physical problems. To produce sound we must produce back-and-forth motion of air molecules; this can be accomplished by means of an acoustic vibration in an enclosed space, or by the vibration of a solid body that sets the adjacent air molecules into motion.

The acoustic-vibration category includes organ pipes, discussed in Section 22–5, and the flute and recorder families of instruments. None of these has any moving parts, in the usual sense; sound is produced by conversion of a steady stream of air into vibrating motion of air within the device, some of which is transmitted through openings to the outside. The pitch of such an instrument is determined by the frequencies of the normal modes, as discussed in Section 22–5. Each pipe in a pipe organ normally produces only a single note, but flutes and recorders have series of holes that are opened and closed by the player's fingers to provide a variety of pitches.

A less musical sound, the exhaust noise from an internal-combustion engine, is caused by the sudden release of hot gases under high pressure. The release is caused by mechanical motion of a valve, but the actual sound wave is generated by the motion of the hot gases out of the engine and through the exhaust system.

Instruments in the string and percussion families, by contrast, make use of solid bodies whose vibrations are transmitted to the surrounding air. In string instruments, including the violin family, the guitar, banjo, piano, and several others, a string or several strings are set into vibration by bowing, plucking, or striking with hammers. They vibrate with their normal-mode frequencies, as discussed in Section 22–3. The thinness of the strings makes them ineffective in transmitting their vibration to the surrounding air, and so some means is provided to couple this vibration to a solid body having larger surface area, to provide more effective coupling to the air. In the violin family the string motion is transmitted through the bridge to the body of the instrument, which vibrates with

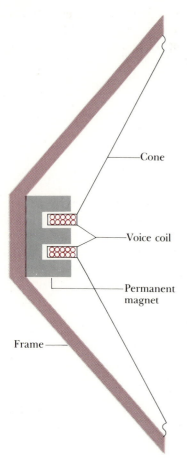

the same frequency as the string. The sounding board on a piano or harpsichord serves the same function. A piano without a sounding board has a very tinny, weak, and unsatisfying sound.

The wind instruments of the orchestra combine acoustic vibrations in an air column with a vibrating reed that vibrates in response to air blown through or past it. In the clarinet, oboe, reed organ pipes, and similar instruments, the reed is part of the instrument; in brass instruments, such as the trumpet, trombone, and tuba, the role of the reed is played by the player's lips in the mouthpiece. In these instruments the interaction between the reed vibration and that of the air column is a complex process; the pitch of the tone is usually determined primarily by the size and shape of the air column, although the relationships are complex and require rather sophisticated analysis.

The human voice is similar; the vocal cords play the role of the reed. In this case the pitch is determined entirely by the mechanical properties of the vocal cords (including the varying tension placed on them). The changing shape of the vocal tract (including the throat, mouth, and nose) plays an important role in changing the timbre of the voice to produce the various vowel sounds.

In sound reproduction or amplification systems, the source of sound is the familiar loudspeaker. The structure of a common type of loudspeaker is shown in Fig. 23–11. The stiff paper cone is made flexible at its outer edge by folds. Attached to its center is a coil of wire (the voice coil) that moves within the field of a permanent magnet. Fluctuating current in the voice coil causes the magnetic field to exert corresponding fluctuating forces on the coil, and thus the mechanical vibration of the speaker cone reproduces the electrical vibration of the current in the voice coil.

Microphones have exactly the opposite function, the conversion of sound waves to electrical waves. In principle the speaker in Fig. 23–11 could also be used as a microphone, but actual microphones can be made much more compact. In a sound-amplification system sound is converted to electrical vibrations, amplified electronically, and then converted back to sound by means of loudspeakers.

In recording and reproducing sound there are at least two additional steps. In the familiar phonograph record the amplified electrical signal from the microphone is fed to a device similar in principle to a speaker but equipped with a stylus that scratches a wavy spiral line in a rotating disk. The signal is thus stored mechanically; in playback it is traced by another stylus attached to a device that plays the role of a microphone, converting the mechanical vibrations once again to electrical for amplification and final conversion to sound in the speaker. In magnetic tape recording the record is replaced by a moving tape that can be permanently magnetized; the signal is then stored in the form of a series of regions of greater and lesser magnetization.

Interference effects play an important role in the behavior of loudspeakers. Each element of area of the speaker cone acts as a source of sound, and the sounds from these individual sources are at different distances from the listener; hence they do not all arrive in phase, and there are interference effects. If the wavelength of the emitted sound is long compared to the dimensions of the cone, interference effects are

23–11 Cross section of a loudspeaker. The folds at the outer edge of the paper cone permit the voice coil to move freely in and out of the magnet. When a fluctuating current passes through the voice coil, the magnetic forces exerted by the field of the permanent magnet cause the coil, and with it the cone, to vibrate back and forth from left to right. For low-frequency sound reproduction, the cone is typically 20 cm to 30 cm in diameter. For high-frequency reproduction the cone is often replaced by a rigid dome a few centimeters in diameter, with a flexible mounting at its edge. High-quality speaker systems often use several speakers for optimum reproduction of the entire audible frequency range.

negligible, but at shorter wavelengths and higher frequencies they become significant. The most familiar effect is that the higher frequencies tend to be focused in a narrow cone along the speaker axis. It can be shown that for a cone in the form of a flat disk of diameter D, sound of wavelength λ is concentrated along the axis, with about 85 percent of the sound energy concentrated in directions within an angle α from the axis, given by

$$\sin \alpha = 1.22 \frac{\lambda}{D}. \tag{23-17}$$

When broad dispersion of high-frequency sound is required, very small speakers are used; the cones are often convex or are fitted with cellular horns to help disperse the sound. For low frequencies, on the other hand, large cones are required to provide enough surface area for effective coupling to the air.

In general, when the wavelength of the wave is small compared to the size of the loudspeaker, or of an aperture through which waves are emerging, the spreading of the waves is small, while if the wavelength is large, the waves spread out in all directions.

When there is an obstacle in the path of a wave, the resultant effect at the far side of the obstacle is due to those portions of the advancing wave that are *not* obstructed. In very general terms, if the wavelength is relatively small, the spreading is small and the obstacle casts a sharp "shadow." The larger the wavelength, the greater the spreading or bending of the waves. Thus one can hear around the corner of a wall, but the effect is greater for waves of long wavelength (or low frequency) than for waves of short wavelength (or high frequency). The general term for the phenomena described above, in which one is concerned with the resultant effect of a large number of waves from different parts of a source, is *diffraction*.

23–9

APPLICATIONS OF ACOUSTIC PHENOMENA

The relevance of acoustic principles is not by any means limited to human hearing. It is well known that bats depend primarily on sound rather than sight for guidance during flight. They emit short pulses of ultrasonic sound of variable frequency ranging roughly from 30 kHz to 150 kHz, and the returning echos give information about the location and size of obstacles and potential prey such as flying insects. Direction sensing is believed to involve the same mechanisms as discussed at the end of Section 23–3.

Dolphins use an analogous system for underwater navigation; again the frequencies are ultrasonic, of the order of 100 kHz. With such a system the animal can sense objects of about the size of the wavelength of the sound, but not much smaller; hence the high frequency is needed in order to sense small objects. An example is discussed in Section 21–4, Example 1. The corresponding manufactured systems, used for submarine navigation, depth measurements, location of fish or wrecked ships, and so on, are known as *sonar*. By use of the Doppler effect, sonar systems can measure motion as well as position of submerged objects.

Analysis of elastic waves in the earth provides important information about its structure. The interior of the earth may be pictured crudely as made of concentric spherical shells, with mechanical properties such as density and elastic moduli that are different in different shells. Waves produced by explosions or earthquakes are reflected and refracted at the interfaces between these shells, and analysis of these waves permits partial determination of the dimensions and properties of the shells. Local anomalies such as oil deposits can also be detected by study of anomalous wave-propagation behavior.

Some of the most interesting recent applications of acoustics lie in the field of medicine, where sound waves are used for both diagnosis and therapy. In diagnosis, ultrasonic frequencies are usually used because their short wavelengths permit study of smaller-scale phenomena than with audible sound. Reflection of such waves from regions in the interior of the body can be used to detect a wide variety of anomalous conditions such as tumors, and to study various phenomena such as heart-valve action. Ultrasound is more sensitive than x-rays in distinguishing various kinds of tissues; it is believed to be less hazardous than x-rays, although possible hazards of ultrasound have not yet been thoroughly explored. At much higher power levels, ultrasound appears to have promise as a selective destroyer of pathological tissue, and may find usefulness in the treatment of certain cancers as well as arthritis and related diseases. Pulverizing of gallstones and kidney stones is yet another possibility.

In medical and other applications of ultrasonics, a variety of electronic instrumentation is used. The sound is always produced by first generating an electrical wave, and using this to drive a *transducer* that converts electrical to mechanical (i.e., sound) waves. Thus the *transducer* is similar in action to a loudspeaker except that the frequency ranges may be different; frequencies as high as several megahertz (i.e., several million cycles per second) are routinely used. The detecting instruments include transducers, functioning in the role of microphones, as well as amplifiers and devices for displaying reflected signals. A common technique is to send out pulses and observe the transmitted and reflected pulse "blips" on an oscilloscope, which permits a direct measurement of the time interval between transmission and reflection. In recent years, techniques have been developed in which transducers move over or *scan* the region of interest and a computer-reconstructed image is produced. This is a rapidly developing and very promising area of research in medical instrumentation. An example of such techniques is shown in Fig. 23–12.

The applications of acoustic principles to environmental problems are obvious. It is now recognized that noise is an important aspect of environmental degradation. The design of quiet mass-transit vehicles, for example, involves detailed study of sound generation and propagation in the motors, wheels, and supporting structures of vehicles. Excessive noise levels can lead to permanent hearing impairment; recent studies have shown that many young "rock" musicians have suffered hearing losses typical of persons 65 years of age, and even prolonged listening to high-level rock music (90 dB to 100 dB) can lead to permanent damage. Stereo headsets used at high volume levels pose similar threats to hearing.

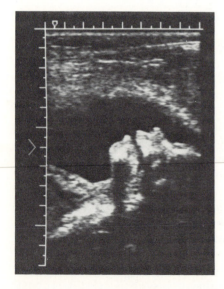

23–12 Sonogram of a human fetus in the womb showing the facial profile of a 19-week fetus. (Courtesy of Acuson Computed Sonography.)

QUESTIONS

23–1 Ultrasonic cleaners use ultrasonic waves of high intensity in water or a solvent to clean dirt from dishes, engine parts, and so on. How do they work? What advantages and disadvantages does this process have, compared with other cleaning methods?

23–2 Lane dividers on highways sometimes have regularly spaced ridges or ripples. When the tires of a moving car roll along such a divider, a musical note is produced. Why? Could this phenomenon be used to measure the car's speed? How?

23–3 Two tuning forks have identical frequencies, but one is stationary while the other is mounted on a rotating record turntable. What does a listener hear?

23–4 The organist in a cathedral plays a loud chord and then releases it. The sound persists for a few seconds and gradually dies away. Why does it persist? What happens to the energy when it dies away?

23–5 Why do foghorns always have very low pitches?

23–6 Some stereo amplifiers intended for home use have a maximum power output of 600 W or more. What would happen if you had 600 W of actual sound power in a moderate-size room?

23–7 Why does your voice sound different over the telephone than in person?

23–8 Why does a guitar have a sharper, more metallic sound when played with a hard pick than when plucked with the bare fingertips?

23–9 When a record is being played with the volume control turned down all the way, a faint sound can be heard coming directly from the stylus. It seems to be all treble and no bass. How is the sound produced, and why are the frequencies so unbalanced?

23–10 The tone quality of a violin is different when the bow is near the bridge (the ends of the strings) than when it is nearer the centers of the strings. Why?

23–11 An engineer who likes to express everything in technical terms remarked that the price of gasoline rose by 2 dB in 1979. What do you think he meant?

23–12 When you are shouting at someone a fair distance away, it is easier for him to hear you if the wind is blowing from you to him, than if it is in the opposite direction. What is the physical basis for this difference?

23–13 Two notes an octave apart on a piano have a frequency ratio of $2:1$. When one is slightly out of tune, beats are heard. How are they produced?

23–14 A large church has part of the organ in front and part in back. When both divisions are playing at once and a person walks rapidly down the center aisle, the two divisions sound out of tune. Why?

23–15 Can you think of circumstances in which a Doppler effect would be observed for surface waves in water? For elastic waves propagating in a body of water?

23–16 How does a person perceive the direction from which a sound comes? Is our directional perception more acute for some frequency ranges than for others? Why?

23–17 A sound source and a listener are both at rest on the earth, but a strong wind is blowing. Is there a Doppler effect?

PROBLEMS

Unless otherwise indicated, assume the speed of sound in air to be $c = 345 \text{ m} \cdot \text{s}^{-1}$.

23–1

a) If the pressure amplitude in a sound wave is tripled, by what factor is the intensity of the wave increased?

b) By what factor must the pressure amplitude of a sound wave be increased in order to increase the intensity by a factor of 16?

23–2

a) Two sound waves of the same frequency, one in air and one in water, are equal in intensity. What is the ratio of the pressure amplitude of the wave in water to that of the wave in air?

b) If the pressure amplitudes of the waves are equal, what is the ratio of their intensities?

c) What is the difference between their intensity levels? (The speed of sound in water may be taken as $1490 \text{ m} \cdot \text{s}^{-1}$.)

23–3

a) Relative to the arbitrary reference intensity of $10^{-12} \text{ W} \cdot \text{m}^{-2}$, what is the intensity level in decibels of a sound wave whose intensity is $10^{-6} \text{ W} \cdot \text{m}^{-2}$?

b) What is the intensity level of a sound wave in air whose pressure amplitude is 0.2 Pa?

23–4

a) Show that if β_1 and β_2 are the intensity levels in dB of sounds of intensities I_1 and I_2, respectively, the difference in intensity levels of the sounds is

$$\beta_2 - \beta_1 = 10 \log \frac{I_2}{I_1}.$$

b) Show that if $(p_{max})_1$ and $(p_{max})_2$ are the pressure amplitudes of two sound waves, the difference in intensity levels of the waves, in dB, is

$$\beta_2 - \beta_1 = 20 \log \frac{(p_{max})_1}{(p_{max})_2}$$

23–5

a) Show that if the reference level of intensity is $I_0 = 10^{-12}$ W·m^{-2}, the intensity level (in dB) of a sound of intensity I (in W·m^{-2}) is

$$\beta = 120 + 10 \log I.$$

b) Is this relation valid if I is expressed in W·cm^{-2} instead of W·m^{-2}? If not, how should it be modified?

23–6 The intensity due to a number of independent sound sources is the sum of the individual intensities. How many decibels greater is the intensity level when all five quintuplets cry simultaneously than when a single one cries? How many more crying babies would be required to produce a further increase in the intensity level of an equal number of decibels?

23–7 A window whose area is 1 m^2 opens on a street where the street noises result in an intensity level, at the window, of 60 dB. How much "acoustic power" enters the window via the sound waves?

23–8 A very noisy chain saw operated by a tree surgeon emits a total sound power of 10 W, uniformly in all directions. At what distance from the source is the sound level

a) 100 dB, and

b) 60 dB?

23–9 A certain sound source radiates uniformly in all directions in air. At a distance of 5 m the sound level is 80 dB. The frequency is 440 Hz.

a) What is the displacement amplitude at this distance?

b) What is the pressure amplitude?

c) At what distance is the sound level 60 dB?

23–10 Derive Eq. (23–8) from the preceding equations.

23–11 Derive an expression for the intensity of a sinusoidal sound wave in terms of the velocity amplitude V and relevant constants such as c, B, and so on.

23–12

a) What are the upper and lower limits of intensity level of a person whose auditory area is represented by the graph of Fig. 23–3?

b) What are the highest and lowest frequencies that a person can hear when the intensity level is 40 dB?

23–13 Approximately what is the sound intensity level (dB) of a 100-Hz tone that sounds as loud as a 1000-Hz tone at sound intensity level 40 dB?

23–14 A 10,000-Hz tone at 52 dB sounds as loud as a 1000-Hz tone at what intensity level?

23–15 Two loudspeakers, A and B, radiate sound uniformly in all directions. The output of acoustic power from A is 8×10^{-4} W, and from B it is 13.5×10^{-4} W. Both loudspeakers are vibrating in phase at a frequency of 173 Hz.

a) Determine the difference in phase of the two signals at a point C along the line joining A and B, 3 m from B and 4 m from A.

b) Determine the intensity at C from speaker A if speaker B is turned off, and the intensity at C from speaker B if speaker A is turned off.

c) With both speakers on, what are the intensity and intensity level at C?

23–16 What should be the diameter of a sound source in the form of a circular piston set in a wall, if the central lobe of the diffraction pattern is to have a half-angle of 45°, for a frequency of 10,000 Hz?

23–17 The sound source of a sonar system operates at a frequency of 50,000 Hz. Approximate the source by a circular disk set in the hull of a ship. The velocity of sound in water can be taken as 1450 m·s^{-1}.

a) What is the wavelength of the waves emitted by the source?

b) What must be the diameter of the source if the half-angular divergence of the main beam is not to be more than 10°?

c) What is the difference in frequency between the directly radiated waves and the waves reflected from a whale traveling directly away from the ship at 25 km·hr^{-1}?

23–18 Two identical piano strings when stretched with the same tension have a fundamental frequency of 400 Hz. By what fractional amount must the tension in one string be increased in order that 4 beats·s^{-1} shall occur when both strings vibrate simultaneously?

23–19 In Table 23–2, the interval from C to G is called a *fifth*. Compare the frequency ratio for this interval in the just scale to that in the equally tempered scale. Which is larger, and by what percent?

23–20 In Table 23–2, compare the frequency ratio for the interval C to G with the frequency ratio for the interval D to A, in the just scale.

a) By what percent do the two ratios differ?

b) Show that the two ratios are equal in the equally tempered scale.

23–21 The frequency ratio of a half-tone interval on the equally tempered scale is 1.059. Find the speed of an automobile passing a listener at rest in still air, if the pitch of the car's horn drops a half tone between the times when the car is coming directly toward him and when it is moving directly away from him.

23–22

a) Refer to Fig. 23–9 and the examples in Section 23–7. Suppose that a wind of speed $15 \text{ m} \cdot \text{s}^{-1}$ is blowing in the same direction as that in which the source is moving. Find the wavelengths ahead of and behind the source.

b) Find the frequency heard by a listener at rest when the source is moving away from him.

23–23 A railroad train is traveling at $30 \text{ m} \cdot \text{s}^{-1}$ in still air. The frequency of the note emitted by the locomotive whistle is 500 Hz. What is the wavelength of the sound waves

a) in front of, and

b) behind the locomotive?

What would be the frequency of the sound heard by a stationary listener

c) in front of, and

d) behind the locomotive?

What frequency would be heard by a passenger on another train traveling at $15 \text{ m} \cdot \text{s}^{-1}$ and

e) approaching the first,

f) receding from the first?

g) How would each of the preceding answers (a) through (f) be altered if a wind of speed $10 \text{ m} \cdot \text{s}^{-1}$ were blowing in the same direction as that in which the locomotive was traveling?

23–24 A plane sound wave of frequency f_0 and wavelength λ_0 travels horizontally toward the right. It strikes and is reflected from a large, rigid, vertical plane surface, perpendicular to the direction of propagation of the wave and moving toward the left with a speed v.

a) How many positive wave crests strike the surface in a time interval t?

b) At the end of this time interval, how far to the left of the surface is the wave that was reflected at the beginning of the time interval?

c) What is the wavelength of the reflected waves, in terms of λ_0?

d) What is the frequency, in terms of f_0?

e) A listener is at rest at the left of the moving surface. Describe the sensation of sound that he hears as a result of the combined effect of the incident and reflected wave trains.

23–25 Two whistles, A and B, each have a frequency of 500 Hz. A is stationary and B is moving toward the right

(away from A) at a speed of $50 \text{ m} \cdot \text{s}^{-1}$. An observer is between the two whistles, moving toward the right with a speed of $25 \text{ m} \cdot \text{s}^{-1}$. Take the speed of sound in air as $350 \text{ m} \cdot \text{s}^{-1}$.

a) What is the frequency from A as heard by the observer?

b) What is the frequency B as heard by the observer?

c) What is the beat frequency heard by the observer?

23–26 A man stands at rest in front of a large smooth wall. Directly in front of him, between him and the wall, he holds a vibrating tuning fork of frequency 400 Hz. He now moves the fork toward the wall with a speed of $1 \text{ m} \cdot \text{s}^{-1}$. How many beats per second will he hear between the sound waves reaching him directly from the fork, and those reaching him after being reflected from the wall?

23–27 A source of sound waves, S, emitting waves of frequency 1000 Hz, is traveling toward the right in still air with a speed of $30 \text{ m} \cdot \text{s}^{-1}$. At the right of the source is a large, smooth, reflecting surface toward the left with a speed of $120 \text{ m} \cdot \text{s}^{-1}$.

a) How far does an emitted wave travel in 0.001 s?

b) What is the wavelength of the emitted waves in front of (i.e., at the right of) the source?

c) How many waves strike the reflecting surface in 0.001 s?

d) What is the speed of the reflected waves?

e) What is the wavelength of the reflected waves?

23–28

a) Show that Eq. (23–21) can be written

$$f_L = f_S \left(1 - \frac{v}{c}\right)^{1/2} \left(1 + \frac{v}{c}\right)^{-1/2}.$$

b) Use the binomial theorem to show that if $v \ll c$, this is approximately equal to

$$f_L = f_S \left(1 - \frac{v}{c}\right).$$

c) An earth satellite emits a radio signal of frequency 10^8 Hz. An observer on the ground detects beats between the received signal and a local signal also of frequency 10^8 Hz. At a particular moment, the beat frequency is 2400 Hz. What is the component of the satellite's velocity directed toward the earth at this moment?

COULOMB'S LAW

Interactions between electrically charged bodies form one of the fundamental classes of interactions found in nature. In this chapter we study several basic phenomena involving interactions of electric charges at rest, that is, *electrostatic* phenomena. The basic law of force for interaction of electric charges at rest is called *Coulomb's law;* we study a variety of applications of this law.

24–1

ELECTRIC CHARGES

It was known to the ancient Greeks as long ago as 600 B.C. that amber, rubbed with wool, acquires the property of attracting light objects. In describing this property today, we say that the amber is *electrified*, or possesses an *electric charge*, or is *electrically charged*. These terms are derived from the Greek word *elektron*, meaning amber. It is possible to impart an electric charge to any solid material by rubbing it with any other material. Thus a person becomes electrified by scuffing his shoes across a nylon carpet, a comb is electrified in passing through dry hair, an electric charge is developed on a sheet of paper moving through a printing press, and so on.

Plastic rods and fur are often used in demonstrations. Suppose a plastic rod is electrified by rubbing it with fur and is then touched to two small, light balls of cork or pith, suspended by thin silk or nylon threads. The balls are found to be *repelled* by the plastic rod, and they also repel each other.

A similar experiment performed with a glass rod that has been rubbed with silk gives rise to the same result; pith balls electrified by contact with such a glass rod are repelled not only by the rod but by each other. On the other hand, when a pith ball that has been in contact with electrified plastic is placed near one that has been in contact with electrified glass, the pith balls *attract* each other. Thus there must be *two kinds* of electric charge—that possessed by plastic after being rubbed with fur, called a *negative* charge, and that possessed by glass after being rubbed with silk, called a *positive* charge. (This designation of positive and negative was first proposed by Benjamin Franklin.) The experiments on pith balls described above lead to the fundamental results that (1) *like charges repel*, (2) *unlike charges attract*.

These repulsive or attractive forces of electrical origin are distinct from *gravitational* attraction, and usually so much larger that the gravitational interaction between two electrically charged bodies may be neglected. Of course, a charged body may be acted on by the gravitational field of a large body such as the earth, and in such cases the electrical and gravitational forces may be of comparable magnitude.

Two bodies may also interact by means of magnetic interaction; the most familiar example is attraction of iron objects to a permanent magnet. Although magnetic forces were once believed to be a fundamentally different type of interaction, it is now known that magnetic interactions are really the interactions between charged particles in motion. Indeed, an electromagnet shows magnetic interactions when an electric current passes through its coils. Magnetic interactions will be discussed in later chapters; for the present we concentrate on *charges at rest*, that is, on *electrostatics*.

Suppose a plastic rod is rubbed with fur and then touched to a suspended pith ball. Both the rod and the pith ball are then negatively charged. If the *fur* is now brought near the pith ball, the ball is *attracted*, indicating that the fur is *positively* charged. It follows that when plastic is rubbed with fur, opposite charges appear on the two materials. This is found to happen whenever any substance is rubbed with any other substance. Thus, glass becomes positive, while the silk with which the glass was rubbed becomes negative. This suggests strongly that electric charges are not generated or created, but that the process of acquiring an electric charge consists of *transferring* something from one body to another, so that one body has an excess and the other a deficiency of that something. It was not until the end of the nineteenth century that this "something" was found to consist of negatively charged particles, known today as *electrons*.

24–2

ATOMIC STRUCTURE

The interactions responsible for the structure of atoms and molecules, and hence of all matter, are primarily electrical interactions between electrically charged particles. The fundamental building blocks are three kinds of particles, the negatively charged *electron*, the positively charged *proton*, and the neutral *neutron*. The negative charge of the electron is of the same magnitude as the positive charge of the proton. No charges of smaller magnitude have ever been observed, although one model of fundamental particles includes entities called *quarks*, having charges of $\pm\frac{1}{3}$ and $\pm\frac{2}{3}$ of the electron charge. The charge of a proton or an electron is the ultimate, natural unit of charge. Measurements of the charge of the electron are discussed in Section 26–6.

The particles are arranged in the same general way in all atoms. The protons and neutrons always form a closely packed group called the *nucleus,* which has a net positive charge due to the protons. The diameter of the nucleus, if we think of it as roughly spherical, is of the order of 10^{-14} m. Outside the nucleus, at relatively large distances from it, are the electrons; the number of electrons is equal to the number of protons within the nucleus. If the atom is undisturbed, and no electrons are removed from the space around the nucleus, the atom as a whole is electrically *neutral.* That is, *the algebraic sum of the positive charges of the nucleus and the negative charges of the electrons is zero,* just as equal positive and negative numbers sum to zero. If one or more electrons are removed, the remaining positively charged structure is called a *positive ion.* A *negative* ion is an atom that has *gained* one or more extra electrons. The process of losing or gaining electrons is called *ionization.*

In the atomic model proposed by the Danish physicist Niels Bohr in 1913, the electrons were pictured as whirling about the nucleus in circular or elliptical orbits. More recent research has shown that the electrons are more accurately represented as spread-out distributions of electric charge, governed by the principles of quantum mechanics, which will be discussed in Chapter 42. Nevertheless, the Bohr model is still useful for visualizing the structure of an atom. The diameters of the electron charge distributions, which the Bohr model pictures as orbits, determine the overall size of the atom as a whole, and these are of the order of 10^{-10} m, or about ten thousand times as great as the diameter of the nucleus. A Bohr atom is analogous to a solar system in miniature, with electrical forces taking the place of gravitational forces. The massive, positively charged central nucleus corresponds to the sun, while the electrons, moving around the nucleus under the electrical force of its attraction, correspond to the planets moving around the sun under the influence of its gravitational attraction.

The masses of the proton and neutron are nearly equal, and the mass of the proton is about 1836 times that of the electron. Nearly all the mass of an atom, therefore, is concentrated in its nucleus. Since one mole of monatomic hydrogen consists of 6.022×10^{23} particles (Avogadro's number) and its mass is 1.008 g, the mass of a single hydrogen atom is

$$\frac{1.008 \text{ g}}{6.022 \times 10^{23}} = 1.674 \times 10^{-24} \text{ g} = 1.674 \times 10^{-27} \text{ kg}.$$

The nucleus of a hydrogen atom is a single proton, and around it there is a single electron. Hence, of the total mass of the hydrogen atom, $\frac{1}{1837}$ part is the mass of the electron and the remainder is the mass of a proton. To four significant figures,

$$\text{Mass of electron} = 9.110 \times 10^{-31} \text{ kg},$$
$$\text{Mass of proton} = 1.673 \times 10^{-27} \text{ kg},$$
$$\text{Mass of neutron} = 1.675 \times 10^{-27} \text{ kg}.$$

After hydrogen, the atom with the next simplest structure is that of helium. Its nucleus consists of two protons and two neutrons, and it has two electrons outside the nucleus. When these two electrons are removed, the doubly charged helium ion (which is the helium nucleus itself) is often called an *alpha particle,* or α-particle. The next element, lithium, has three protons in its nucleus and has thus a nuclear charge of three units. In the un-ionized state the lithium atom has three extranuclear electrons. Each element has a different number of nuclear protons and therefore a different positive nuclear charge. In the table of elements listed at the end of this book, known as the *periodic table,* each element occupies a box with which is associated a number, called the *atomic number.*

> *The atomic number is the number of nuclear protons; in the un-ionized state, this is equal to the number of extranuclear electrons.*

Every material body contains a tremendous number of charged particles, positively charged protons in the nuclei of its atoms and negatively charged electrons outside the nuclei. When the total number of protons equals the total number of electrons, the body as a whole is electrically neutral.

To give a body an excess negative charge, we may either *add* a number of *negative* charges to a neutral body, or *remove* a number of *positive* charges from the body. Similarly, either an *addition* of *positive* charge or a *removal* of *negative* charge results in an excess positive charge. In most instances, it is negative charges (electrons) that are added or removed, and a "positively charged body" is one that has lost some of its normal complement of electrons.

The "charge" of a body refers to its *excess* charge only. The excess charge is always a very small fraction of the total positive or negative charge in the body.

Implicit in the above statements is the *principle of conservation of charge.* This principle states that the algebraic sum of all the electric charges in any closed system is constant. Charge can be transferred from one body to another, but it cannot be created or destroyed. Conservation of charge is believed to be a *universal* conservation law; there is no experimental evidence for any violation of this principle.

24–3

CONDUCTORS AND INSULATORS

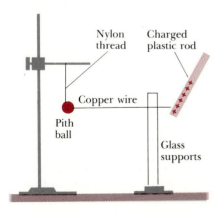

24–1 Copper is a conductor of electricity.

Some materials permit electric charge to move from one region of the material to another, while other materials do not. Suppose we touch one end of a copper wire to an electrified plastic rod and the other end to an initially uncharged pith ball, as in Fig. 24–1. The pith ball is found (by its subsequent interaction with other charged pith balls) to become charged. Thus charge has moved along the wire from the rod to the ball. The copper wire is called a *conductor* of electricity. If the experiment is repeated using a rubber band or nylon thread in place of the wire, *no* transfer of charge is observed; and these materials are called *insulators.* The motion of charge through a material substance will be studied in more detail in Chapter 28, but for the present it is sufficient to state that most substances fall into one or the other of the two classes above. Conductors permit the passage of charge through them, whereas insulators do not.

Metals in general are good conductors, while most nonmetals are insulators. The positive valency of metals and the fact that they form positive ions in solution indicate that the atoms of a metal can easily lose one or more of their outer electrons. Within a metallic conductor such as a copper wire, a few outer electrons become detached from each atom and can move freely throughout the metal in much the same way that the molecules of a gas can move through the spaces between grains of sand in a sand-filled container. In fact, these free electrons are often referred to as an "electron gas." The positive nuclei and the remainder of the electrons remain fixed in position. Within an insulator, on the other hand, there are no (or at most very few) free electrons.

24–4

CHARGING BY INDUCTION

When a pith ball is charged by contact with a plastic rod that has been rubbed with fur, some of the extra electrons on the plastic are transferred to the ball, leaving the plastic with a smaller negative charge. There is, however, another way to use the plastic rod to charge other bodies, in which the plastic may impart a charge of *opposite* sign and lose none of its own charge. This process, called *charging by induction,* is illustrated in Fig. 24–2.

In Fig. 24–2a, two neutral metal spheres are in contact, both supported on insulating stands. When a negatively charged rod is brought near one of the spheres but without touching it, as in (b), the free electrons in the metal spheres are repelled and drift slightly away from the rod, toward the right. Since the electrons cannot escape from the spheres, an excess negative charge accumulates at the right surface of the right sphere. This leaves a deficiency of negative charge, or an excess positive charge, at the left surface of the left sphere. These excess charges are called *induced* charges.

Not *all* of the free electrons in the spheres are driven to the surface of the right sphere. As soon as any induced charges develop, they also exert forces on the free electrons within the spheres. In this case this force is toward the left (a repulsion by the negative induced charge and an attraction by the positive induced charge). Within an extremely short time the system reaches an equilibrium state in which, at every point in the interior of the spheres, the force on an electron toward the right, exerted by the charged rod, is just balanced by a force toward the left exerted by the induced charges.

The induced charges remain on the surfaces of the spheres as long as the rod is held nearby. When the rod is removed, the electron cloud in the spheres moves to the left and the original neutral condition is restored.

Suppose that the spheres are separated slightly, as shown in (c), while the plastic rod is nearby. If the rod is now removed, as in (d), we are left with two oppositely charged metal spheres whose charges attract each other. When the two spheres are separated by a great distance, as in (e), each of the two charges becomes uniformly distributed over its sphere. It should be noticed that the negatively charged rod has lost none of its charge in the steps from (a) to (e).

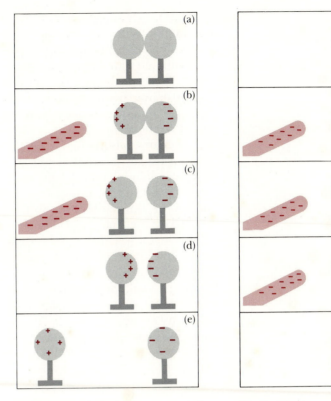

24–2 Two metal spheres are oppositely charged by induction.

24–3 Charging a single metal sphere by induction.

The steps from (a) to (e) in Fig. 24–3 should be self-explanatory. In this figure, a single metal sphere (on an insulating stand) is charged by induction. The symbol lettered "ground" in part (c) simply means that the sphere is connected to the earth (a conductor). The earth thus takes the place of the second sphere in Fig. 24–2. In step (c), electrons are repelled to ground either through a conducting wire, or along the moist skin of a person who touches the sphere with his finger. The earth thus acquires a negative charge equal to the induced positive charge remaining on the sphere.

The process taking place in Figs. 24–2 and 24–3 could be explained equally well if the mobile charges in the spheres were *positive* or, in fact, if *both* positive and negative charges were mobile. Although we now know that in a metallic conductor it is actually the *negative* charges that move, it is often convenient to describe a process *as if* the positive charges moved.

24–5

COULOMB'S LAW

The electrical interaction between two charged particles is described in terms of the *forces* they exert on each other. The first quantitative investigation of the behavior of this force was carried out by Augustin de Coulomb (1736–1806) in 1784, using for force measurements a torsion balance of the type used 13 years later by Cavendish to study (much weaker) gravitational forces (Section 4–3). Coulomb studied the force of attraction or repulsion between two "point charges," that is, charged bodies whose dimensions are small compared with the distance between them.

Coulomb found that the force grows weaker with increasing separation between the bodies. When the distance doubles, the force decreases to $\frac{1}{4}$ of its initial value. Thus it varies inversely with the square of the distance. If r is the distance between the particles, then the force is proportional to $1/r^2$.

The force also depends on the quantity of charge on each body, which is usually denoted by q or Q. In Coulomb's time, no *unit* of charge had been defined, nor had any method been developed for comparing a given charge with a unit. Despite this, Coulomb devised an ingenious method of showing how the force exerted on or by a charged body depended on the amount of charge. He reasoned that if a charged spherical conductor were brought in contact with a second identical conductor, originally *uncharged*, the charge on the first would, by symmetry, be shared equally between the conductors. He thus had a method for obtaining one half, one quarter, and so on, of any given charge. The results of his experiments were consistent with the conclusion that the force between two point charges q and q' is proportional to each charge and hence proportional to the *product* of the charges. The complete expression for the magnitude of the force between two point charges is therefore

$$F = k\frac{|qq'|}{r^2}, \tag{24–1}$$

where k is a proportionality constant. The numerical value of k depends on the units in which F, q, q', and r are expressed. Equation (24–1) is the mathematical statement of what is known today as *Coulomb's law:*

The force of attraction or repulsion between two point charges is directly proportional to the product of the charges and inversely proportional to the square of the distance between them.

The *direction* of the force on each particle is always along the line joining the two particles, pulling each particle toward the other in the case of attractive forces on unlike charges, and pushing them apart in the case of repulsive forces on like charges.

The form of Eq. (24–1) is the same as that of the law of gravitation, discussed in Section 4–3, but electrical and gravitational interactions are two distinct classes of phenomena. The proportionality of the force to $1/r^2$ has been verified with great precision. There is no reason to suspect, for example, that the electrical force might vary as $1/r^{2.0001}$.

The charges q and q' are *algebraic* quantities, corresponding to the existence of two kinds of charge (positive and negative), but Eq. (24–1) gives the magnitude of the interaction force in all cases. When the charges are of like sign the forces are repulsive; when they are unlike, the forces are attractive. In either case, the forces obey Newton's third law; the force that q exerts on q' is the negative of the force q' exerts on q. The "absolute value" bars in Eq. (24–1) are needed because F, the magnitude of a vector quantity, is by definition always positive, while the product qq' is negative whenever the two charges have opposite signs.

When two or more charges exert forces simultaneously on a given charge, the total force experienced by that charge is found to be the *vector sum* of the forces that the various charges would exert individually. This important property, called the *principle of superposition*, permits the application of Coulomb's law to arrays of charge of any degree of complexity, although the computational problems can be very great. Example 3 below illustrates the application of the superposition principle.

If there is matter in the space between the charges, the *net* force acting on each is altered because charges are induced in the molecules of the intervening material. This effect will be described later. As a practical matter, the law can be used as stated for point charges in air, since even at atmospheric pressure the effect of the air is to alter the force from its value in vacuum by only about one part in two thousand.

In the chapters of this book dealing with electrical phenomena, we shall use the SI (mks) system of units exclusively. The SI electrical units include all the familiar electrical units such as the volt, the ampere, the ohm, and the watt. The cgs system is also used, more so in scientific work than in commerce and industry, but there is *no* British system of electrical units. This is one of many reasons for abandoning the British system and adopting metric units universally, and there seems little doubt that SI units will eventually receive worldwide adoption.

To the three basic SI units (the meter, kilogram, and second) we now add a fourth, the unit of electric charge. This unit is called one *coulomb* (1 C). The electrical constant k in Eq. (24–1) is, in this system,

$$k = 8.98755 \times 10^9 \ \text{N} \cdot \text{m}^2 \cdot \text{C}^{-2}$$
$$\approx 9.0 \times 10^9 \ \text{N} \cdot \text{m}^2 \cdot \text{C}^{-2}.$$

Later, in connection with the study of electromagnetic radiation, we shall show that k is closely related to the speed of light in vacuum,

$$c = 2.998 \times 10^8 \ \text{m} \cdot \text{s}^{-1}.$$

Specifically,

$$k = 10^{-7} c^2.$$

The relationship is not accidental, but results from the definition of the unit of current, which in turn is related to the interaction of electric and magnetic fields, to be studied later.

In the cgs system of electrical units (not used in this book) the constant k is defined to be unity, without units. This defines a unit of electric charge called the *statcoulomb* or the *esu* (electrostatic unit). The conversion factor is

$$1 \ \text{C} = 2.998 \times 10^9 \ \text{esu}.$$

In SI units the constant k in Eq. (24–1) is usually written not as k but as $1/4\pi\epsilon_0$, where ϵ_0 is another constant. This appears to complicate matters, but it actually simplifies many formulas to be encountered later. Thus Coulomb's law is usually written as

$$F = \frac{1}{4\pi\epsilon_0} \frac{|qq'|}{r^2}, \tag{24–2}$$

with

$$\frac{1}{4\pi\epsilon_0} = 8.987552 \times 10^9 \text{ N}\cdot\text{m}^2\cdot\text{C}^{-2}$$

and

$$\epsilon_0 = 8.854188 \times 10^{-12} \text{ C}^2\cdot\text{N}^{-1}\cdot\text{m}^{-2}.$$

In examples and problems we shall often use the approximate value

$$\frac{1}{4\pi\epsilon_0} = 9.0 \times 10^9 \text{ N}\cdot\text{m}^2\cdot\text{C}^{-2},$$

which is within about 0.1% of the correct value.

The "natural" unit of charge is the magnitude of charge of an electron or a proton. This quantity is denoted by e; the most precise measurements to date yield the value

$$e = 1.602192 \times 10^{-19} \text{ C} \approx 1.60 \times 10^{-19} \text{ C}.$$

One coulomb therefore represents the negative of the total charge carried by about 6×10^{18} electrons. For comparison, the population of the earth is estimated to be about 5×10^9 persons, while a cube of copper 1 cm on a side contains about 2.4×10^{24} electrons.

In electrostatics problems, charges as large as one coulomb are unusual; a more typical range of magnitude is 10^{-9} to 10^{-6} C. The microcoulomb ($1 \ \mu\text{C} = 10^{-6}$ C) is often used as a practical unit of charge.

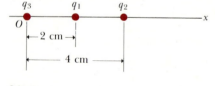

24–4

EXAMPLE 1 Two charges are located on the positive x-axis of a coordinate system, as shown in Fig. 24–4. Charge $q_1 = 2 \times 10^{-9}$ C is 2 cm from the origin, and charge $q_2 = -3 \times 10^{-9}$ C is 4 cm from the origin. What is the total force exerted by these two charges on a charge $q_3 = 5 \times 10^{-9}$ C located at the origin?

Solution The total force on q_3 is the vector sum of the forces due to q_1 and q_2 individually. Converting distances to meters, we use Eq. (24–2) to find the magnitude F_1 of the force on q_3 due to q_1:

$$F_1 = \frac{(9.0 \times 10^9 \text{ N}\cdot\text{m}^2\cdot\text{C}^{-2})(2 \times 10^{-9} \text{ C})(5 \times 10^{-9} \text{ C})}{(0.02 \text{ m})^2} = 2.25 \times 10^{-4} \text{ N}.$$

This force has a negative x-component because q_3 is repelled (i.e., pushed in the $-x$-direction) by q_1, which has the same sign. Similarly, the force due to q_2 is found to have magnitude

$$F_2 = \frac{(9.0 \times 10^9 \text{ N}\cdot\text{m}^2\cdot\text{C}^{-2})(3 \times 10^{-9} \text{ C})(5 \times 10^{-9} \text{ C})}{(0.04 \text{ m})^2} = 0.84 \times 10^{-4} \text{ N}.$$

This force has a positive x-component because q_3 is attracted (i.e., pulled in the positive x-direction) by the opposite charge q_2. The sum of the x-components is

$$\sum F_x = -2.25 \times 10^{-4} \text{ N} + 0.84 \times 10^{-4} \text{ N} = -1.41 \times 10^{-4} \text{ N}.$$

There are no y- or z-components. Thus the total force on q_3 is directed to the left, with magnitude 1.41×10^{-4} N. ◄

EXAMPLE 2 An α-particle is a nucleus of doubly ionized helium. It has a mass m of 6.68×10^{-27} kg and a charge q of $+2e$ or 3.2×10^{-19} C. Compare the force of the electrostatic repulsion between two α-particles with the force of gravitational attraction between them.

Solution The electrostatic force F_e is

$$F_e = \frac{1}{4\pi\epsilon_0} \frac{q^2}{r^2},$$

and the gravitational force F_g is

$$F_g = G\frac{m^2}{r^2}.$$

The ratio of the electrostatic to the gravitational force is

$$\frac{F_e}{F_g} = \frac{1}{4\pi\epsilon_0 G} \frac{q^2}{m^2} = \frac{(9.0 \times 10^9 \text{ N} \cdot \text{m}^2 \cdot \text{C}^{-2})}{(6.67 \times 10^{-11} \text{ N} \cdot \text{m}^2 \cdot \text{kg}^{-2})} \frac{(3.2 \times 10^{-19} \text{ C})^2}{(6.68 \times 10^{-27} \text{ kg})^2}$$
$$= 3.1 \times 10^{35}.$$

Thus the gravitational force is negligible compared to the electrostatic force. This is always the case for interactions of atomic and subatomic particles, while for objects the size of the earth the *electrical* interactions are usually much smaller than the gravitational. ◄

EXAMPLE 3 In Fig. 24–5, two equal positive charges $q = 2.0 \times 10^{-6}$ C interact with a third charge $Q = 4.0 \times 10^{-6}$ C. Find the magnitude and direction of the total (resultant) force on Q.

Solution The key word is *total;* we must compute the force each charge exerts on Q, and then obtain the *vector sum* of the forces. This is most easily accomplished using components. The figure shows the force on Q due to the upper charge q. From Coulomb's law,

$$F = (9.0 \times 10^9 \text{ N} \cdot \text{C}^{-2} \cdot \text{m}^{-2})\frac{(4.0 \times 10^{-6} \text{ C})(2.0 \times 10^{-6} \text{ C})}{(0.5 \text{ m})^2} = 0.29 \text{ N}.$$

The components of this force are given by

$$F_x = F \cos\theta = (0.29 \text{ N})\left(\frac{0.4 \text{ m}}{0.5 \text{ m}}\right) = 0.23 \text{ N},$$

$$F_y = -F \sin\theta = -(0.29 \text{ N})\left(\frac{0.3 \text{ m}}{0.5 \text{ m}}\right) = -0.17 \text{ N}.$$

The lower charge q exerts a force of the same magnitude, but in a different direction. From symmetry we see that its x-component is the same as that due to the upper charge, but its y-component is opposite. Hence,

$$\sum F_x = 2(0.23 \text{ N}) = 0.46 \text{ N},$$
$$\sum F_y = 0.$$

The total force on Q is horizontal, with magnitude 0.46 N. How would this solution differ if the lower charge were *negative*? ◄

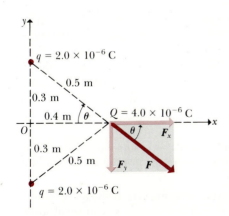

24–5 *F* is the force on Q due to the upper charge q.

24–6

ELECTRICAL INTERACTIONS

Because matter is made up of charged particles, it is not surprising that electrical interactions play a central and dominant role in all aspects of the structure of matter. The forces that hold atoms together in a molecule or in a solid crystal lattice, the adhesive force of glue, the forces associated with surface tension—all these are basically electrical in nature, arising from the electrical forces between the charged particles making up the interacting atoms. A complete description of the detailed behavior of these forces requires new *mechanical* principles and the introduction of quantum-mechanical concepts, to be discussed in Chapter 42. Nevertheless, Coulomb's law and the additional effects resulting from relative motion of the charges still describe the basic electrical interactions involved.

Electrical interactions alone are *not* sufficient to understand the structure of the *nuclei* of atoms, however. A nucleus is made up of protons, which repel each other, and neutrons, which have no electrical charge. For nuclei to be stable there must be additional forces, attractive in nature, to hold their constituent parts together despite the electrical repulsion. This new kind of interaction, not seen outside the nucleus, is called the *nuclear force;* many phenomena associated with the stability or instability of nuclei pivot around the competition between the repulsive electrical forces and the attractive nuclear forces. These matters will be discussed in greater detail in Chapter 44.

QUESTIONS

24–1 Plastic food wraps can be used to cover a container by simply stretching the material across the top and pressing the overhanging material against the sides. What makes it stick? Does it stick to itself with equal tenacity? Why? Does it matter whether the container is metallic or not?

24–2 Bits of paper are attracted to an electrified comb or rod, even though they have no net charge. How is this possible?

24–3 How do we know that the magnitudes of electron and proton charge are *exactly* equal? With what precision is this really known?

24–4 When you walk across a nylon rug and then touch a large metallic object, you may get a spark and a shock. Why does this tend to happen more in the winter than the summer? Why do you not get a spark when you touch a *small* metal object?

24–5 The free electrons in a metal have mass and therefore weight and are gravitationally attracted toward the earth. Why, then, do they not all settle to the bottom of the conductor, as sediment settles to the bottom of a river?

24–6 Simple electrostatics experiments, such as picking up bits of paper with an electrified comb, never work as well on rainy days as on dry days. Why?

24–7 High-speed printing presses sometimes use gas flames to reduce electric charge build-up on the paper passing through the press. Why does this help? (An added benefit is rapid drying of the ink.)

24–8 What similarities do electrical forces have to gravitational forces? What are the most significant differences?

24–9 Given two identical metal objects mounted on insulating stands, describe a procedure for placing charges of equal magnitude and opposite sign on the two objects.

24–10 How do we know that protons have positive charge and electrons negative charge, rather than the reverse?

24–11 Gasoline transport trucks sometimes have chains that hang down and drag on the ground at the rear end. What are these for?

24–12 When a nylon sleeping bag is dragged across a rubberized cloth air mattress in a dark tent, small sparks are sometimes seen. What causes them?

24–13 Atomic nuclei are made of protons and neutrons. This fact by itself shows that there must be another kind of interaction in addition to the electrical forces. Explain.

24–14 When transparent plastic tape is pulled off a roll and one tries to position it precisely on a piece of paper, it often jumps over and sticks where it isn't wanted. Why does it do this?

24–15 When a thunderstorm is approaching, sailors at sea sometimes observe a phenomenon called "St. Elmo's fire," a bluish flickering light at the tips of masts and along wet rigging. What causes this?

PROBLEMS

24–1 Two equal point charges of $+1.0 \times 10^{-6}$ C are placed 0.1 m apart. What is the magnitude of the force each exerts on the other? What are the directions of the forces?

24–2 A negative charge -0.50×10^{-6} C exerts a repulsive force of magnitude 0.20 N on an unknown charge 0.20 m away. What is the unknown charge (magnitude and sign)?

24–3 At what distance would the repulsive force between two electrons have a magnitude of one newton? Between two protons?

24–4 Two small plastic balls are given positive electrical charges. When they are 5 cm apart the repulsive forces between them have magnitude 0.10 N. What is the charge on each ball

a) if the two charges are equal, and

b) if one ball has twice the charge of the other?

24–5 How many excess electrons must be placed on each of two small spheres spaced 3 cm apart if the force of repulsion between the spheres is to be 10^{-19} N?

24–6 Each of two small spheres is positively charged, the combined charge totaling 4×10^{-8} C. What is the charge on each sphere if they are repelled with a force of 27×10^{-5} N when placed 0.1 m apart?

24–7 6.02×10^{23} atoms of monatomic hydrogen have a mass of one gram. How far would the electron of a hydrogen atom have to be removed from the nucleus for the force of attraction to equal the weight of the atom?

24–8 What is the total positive charge, in coulombs, of all the protons in 1 mol of hydrogen atoms?

24–9 If all the positive charges in a mole of hydrogen atoms were lumped into a single charge, and all the negative charges into a single charge, what force would the two lumped charges exert on each other at a distance of

a) 1 m,

b) 10^7 m (comparable to the diameter of the earth)?

24–10 An alpha particle consists of two protons and two neutrons bound together. What is the repulsive force between two alpha particles at a distance of 10^{-15} m, comparable to the sizes of nuclei?

24–11 Two copper spheres, each having mass 1 kg, are separated by 1 m.

a) How many electrons does each sphere contain?

b) How many electrons would have to be removed from one sphere and added to the other to cause an attractive force of 10^4 N (roughly one ton)?

c) What fraction of all the electrons on a sphere does this represent?

24–12 Point charges of 2×10^{-9} C are situated at each of three corners of a square whose side is 0.20 m. What would be the magnitude and direction of the resultant force on a point charge of -1×10^{-9} C if it were placed

a) at the center of the square, and

b) at the vacant corner of the square?

24–13 Two charges of $+10^{-9}$ C each are 8 cm apart in air. Find the magnitude and direction of the force exerted by these charges on a third charge of $+5 \times 10^{-11}$ C that is 5 cm distant from each of the first two charges.

24–14 Two positive point charges, each of magnitude q, are located on the y-axis at points $y = +a$ and $y = -a$. A third positive charge of the same magnitude is located at some point on the x-axis.

a) What is the force exerted on the third charge when it is at the origin?

b) What is the magnitude and direction of the force on the third charge when its coordinate is x?

c) Sketch a graph of the force on the third charge as a function of x, for values of x between $+4a$ and $-4a$. Plot forces to the right upward, forces to the left downward.

24–15 A negative point charge of magnitude q is located on the y-axis at the point $y = +a$, and a positive charge of the same magnitude is located at $y = -a$. A third positive charge of the same magnitude is located at some point on the x-axis.

a) What is the magnitude and direction of the force exerted on the third charge when it is at the origin?

b) What is the force on the third charge when its coordinate is x?

c) Sketch a graph of the force on the third charge as a function of x, for values of x between $+4a$ and $-4a$.

24–16 Two small balls, each of mass 10 g, are attached to silk threads 1 m long and hung from a common point. When the balls are given equal quantities of negative charge, each thread makes an angle of 4° with the vertical.

a) Draw a diagram showing all of the forces on each ball.

b) Find the magnitude of the charge on each ball.

24–17 A certain metal sphere of volume 1 cm^3 has a mass of 7.5 g and contains 8.2×10^{22} free electrons.

a) How many electrons must be removed from each of two such spheres so that the electrostatic force of repulsion between them just balances the force of gravitational attraction? Assume the distance between the spheres is great enough so that the charges on them can be treated as point charges.

b) Express the number of electrons removed as a fraction of the total number of free electrons.

24–18 In the Bohr model of atomic hydrogen, an electron of mass 9.11×10^{-31} kg revolves about a proton in a circular orbit of radius 5.29×10^{-11} m. The proton has a positive charge equal in magnitude to the negative charge on the electron and its mass is 1.67×10^{-27} kg.

a) What is the radial acceleration of the electron?

b) What is its velocity?

c) What is its angular velocity?

24–19 One gram of monatomic hydrogen contains 6.02×10^{23} atoms, each consisting of an electron with charge -1.60×10^{-19} C and a proton with charge $+1.60 \times 10^{-19}$ C.

a) Suppose all these electrons could be located at the north pole of the earth and all the protons at the south pole. What would be the total force of attraction exerted on each group of charges by the other? The diameter of the earth is 12,800 km.

b) What would be the magnitude and direction of the force exerted by the charges in part (a) on a third positive charge, equal in magnitude to the total charge at one of the poles and located at a point on the equator? Draw a diagram.

24–20 The dimensions of atomic nuclei are of the order of 10^{-14} m. Suppose that two α-particles are separated by this distance.

a) What is the force exerted on each α-particle by the other?

b) What is the acceleration of each? (See Example 2 in Section 24–5 for numerical data.)

24–21 The pair of equal and opposite charges in Problem 24–15 is called an *electric dipole*.

a) Show that when the x-coordinate of the third charge in Problem 24–15 is large compared with the distance a, the force on it is inversely proportional to the *cube* of its distance from the midpoint of the dipole.

b) Show that if the third charge is located on the y-axis, at a y-coordinate large compared with the distance a, the force on it is also inversely proportional to the cube of its distance from the midpoint of the dipole.

24–22 A small ball having a positive charge q_1 hangs by an insulating thread. A second ball with a negative charge $q_2 = -q_1$ is kept at a horizontal distance a to the right of the first. (The distance a is large compared with the diameter of the ball.)

a) Show in a diagram all of the forces on the hanging ball in its final equilibrium position.

b) You are given a third ball having a positive charge $q_3 = 2q_1$. Find at least two points at which this ball can be placed so that the first ball will hang vertically.

24–23 Two point charges are located in the xy-plane, as follows: A charge 2.0×10^{-9} C is at the point ($x = 0$, $y = 4$ cm), and a charge -3.0×10^{-9} C is at the point ($x = 3$ cm, $y = 4$ cm).

a) If a third charge of 4.0×10^{-9} C is placed at the origin, find the x- and y-components of the total force on this third charge.

b) Find the magnitude and direction of the total force on the charge at the origin in (a).

24–24 A charge -3×10^{-9} C is placed at the origin of an xy-coordinate system, and a charge 2×10^{-9} C is placed on the positive y-axis, at $y = 4$ cm. If a third charge 4×10^{-9} C is now placed at the point ($x = 3$ cm, $y = 4$ cm), find the components of the total force exerted on this charge by the other two. Also find the magnitude and direction of this force.

25

THE ELECTRIC FIELD

The electrical interaction between charged particles can be reformulated using the concept of *electric field*. An electric charge is thought of as creating an electric field in the surrounding space; that field in turn exerts a force on any other charge in that space. In this chapter the electric fields caused by various arrangements of charge are discussed. When a charge distribution has a high degree of symmetry, determining the resulting electric field can be simplified by means of a principle called *Gauss's law*.

25–1

THE ELECTRIC FIELD

To introduce the concept of electric field, we consider the mutual repulsion of two positively charged bodies A and B, as shown in Fig. 25–1a. In particular, the force on B is labeled $\boldsymbol{F}$ in the figure. This is an "action-at-a-distance" force; it can act across empty space and does not need any matter in the intervening space to transmit the force.

Now let us think of body A as having the effect of modifying some of the properties of the space in its vicinity. We remove body B and label its former position as point P. The charged body A is said to produce or cause an *electric field* at point P (and at all other points in its vicinity). Then when body B is placed at point P and experiences the force $\boldsymbol{F}$, we take the point of view that the force is exerted on B *by the field*, rather than directly by A. Since B would experience a force at any point in space around A, the electric field exists at all points in the region around A. (One could equally well consider that body B sets up an electric field, and that the force on body A is exerted by the field due to B.)

The experimental test for the existence of an electric field at any point is simply to place a charged body, which will be called a *test charge*, at the point. If a force (of electrical origin) is exerted on the test charge, then an electric field exists at the point.

An electric field is said to exist at a point if a force of electrical origin is exerted on a charged body placed at the point.

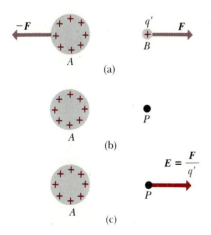

25–1 The space around a charged body is an electric field.

Since force is a vector quantity, the electric field is a *vector quantity* whose properties are determined when both the magnitude and the direction of an electric force are specified. We define the *electric field* E at a point as the quotient obtained when the force F, acting on a positive test charge, is divided by the magnitude q' of the test charge. Thus,

$$E = \frac{F}{q'},\tag{25-1}$$

and the direction of E is the direction of F. It follows that

$$F = q'E.$$

The force on a *negative* charge, such as an electron has a direction *opposite* to that of the direction of the electric field.

The electric field is sometimes called *electric intensity* or *electric field intensity*. In SI (mks) units, where the unit of force is 1 N and the unit of charge is 1 C, the unit of electric field magnitude is 1 newton per coulomb ($1\ \text{N}\cdot\text{C}^{-1}$). Electric field may also be expressed in other units, to be defined later.

The force experienced by the test charge q' varies from point to point, and so the electric field is also different at different points. Thus in general E is not a single vector quantity, but an infinite set of vector quantities, one associated with each point in space. It is an example of a *vector field*. Another example of a vector field is the description of motion of a flowing fluid. In general, different points in the fluid have different velocities, so the velocity is a vector field. If a rectangular coordinate system is used, then in principle each component of E can be expressed as a *function* of the coordinates (x, y, z) of a point in space. Vector fields are an important part of the mathematical language used in many areas of physics, particularly in electricity and magnetism.

One difficulty with our definition of electric field is that in Fig. 25–1 the force exerted by the test charge q' may change the charge distribution A, especially if the body is a conductor on which charge is free to move, so that the electric field around A when q' is present is not the same as when it is absent. However, when q' is very small, the redistribution of charge on body A is also very small; thus the difficulty can be avoided by refining the definition of electric field to be *the limiting value of the force per unit charge on a test charge q' at the point, as the charge q' approaches zero:*

$$E = \lim_{q' \to 0} \frac{F}{q'}.$$

If an electric field exists within a *conductor*, a force is exerted on every charge in the conductor. The motion of the free charges brought about by this force is called a *current*. Conversely, if there is *no* current in a conductor, and hence no motion of its free charges, *the electric field in the conductor must be zero*.

In most instances, the magnitude and direction of an electric field vary from point to point. If the magnitude and direction are constant throughout a certain region, the field is said to be *uniform* in this region.

EXAMPLE 1   What is the electric field 30 cm from a charge $q = 4 \times 10^{-9}$ C?

Solution   From Coulomb's law, the *force* on a test charge q' 30 cm from q has magnitude

$$F = \frac{(9 \times 10^9 \text{ N} \cdot \text{m}^2 \cdot \text{C}^{-2})(4 \times 10^{-9} \text{ C})(q')}{(0.3 \text{ m})^2}$$

$$= (400 \text{ N} \cdot \text{C}^{-1})(q').$$

Then from Eq. (25–1), the magnitude of $\boldsymbol{E}$ is

$$E = \frac{F}{q'} = 400 \text{ N} \cdot \text{C}^{-1}.$$

The *direction* of $\boldsymbol{E}$ at this point is along the line joining q and q', away from q.   ◄

EXAMPLE 2   When the terminals of a 100-V battery are connected to two large parallel plates 1 cm apart, the field in the region between the plates is very nearly uniform and the electric field magnitude E is 10^4 N·C^{-1}. Suppose the direction of $\boldsymbol{E}$ is vertically upward. Compute the force on an electron in this field and compare with the weight of the electron.

Solution

$$\text{Electron charge } e = 1.60 \times 10^{-19} \text{ C,}$$

$$\text{Electron mass } m = 9.1 \times 10^{-31} \text{ kg.}$$

$$F_{\text{elec}} = eE = (1.60 \times 10^{-19} \text{ C})(10^4 \text{ N} \cdot \text{C}^{-1})$$

$$= 1.60 \times 10^{-15} \text{ N;}$$

$$F_{\text{grav}} = mg = (9.1 \times 10^{-31} \text{ kg})(9.8 \text{ m} \cdot \text{s}^{-2})$$

$$= 8.9 \times 10^{-30} \text{ N.}$$

The ratio of the electrical to the gravitational force is therefore

$$\frac{1.60 \times 10^{-15} \text{ N}}{8.9 \times 10^{-30} \text{ N}} = 1.8 \times 10^{14}.$$

The gravitational force is negligibly small compared to the electrical force.   ◄

EXAMPLE 3   If released from rest, what speed will the electron of Example 2 acquire while traveling 1 cm? What will then be its kinetic energy? How long a time is required?

Solution   The force is constant, so the electron moves with a constant acceleration of

$$a = \frac{F}{m} = \frac{eE}{m} = \frac{1.60 \times 10^{-15} \text{ N}}{9.1 \times 10^{-31} \text{ kg}}$$

$$= 1.8 \times 10^{15} \text{ m} \cdot \text{s}^{-2}.$$

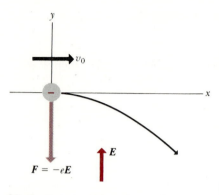

25–2 Trajectory of an electron in an electric field.

Its speed after traveling 1 cm, or 10^{-2} m, is

$$v = \sqrt{2ax} = 6.0 \times 10^6 \text{ m} \cdot \text{s}^{-1}.$$

Its kinetic energy is

$$\tfrac{1}{2}mv^2 = 1.6 \times 10^{-17} \text{ J}.$$

The time required is

$$t = \frac{v}{a} = 3.3 \times 10^{-9} \text{ s}. \qquad \blacktriangleleft$$

EXAMPLE 4   If the electron of Example 2 is projected into the field with an initial horizontal velocity v_0, as in Fig. 25–2, find the equation of its trajectory.

Solution   The direction of the field is upward in Fig. 25–2, so the force on the electron is downward. The initial velocity is along the positive x-axis. The x-acceleration is zero, the y-acceleration is $-(eE/m)$. Hence, after a time t,

$$x = v_0 t,$$

$$y = -\frac{1}{2}\left(\frac{eE}{m}\right)t^2.$$

Elimination of t gives

$$y = -\frac{1}{2}\left(\frac{eE}{2mv_0{}^2}\right)x^2,$$

which is the equation of a parabola. The motion is the same as that of a body projected horizontally in the earth's gravitational field. The deflection of electrons by an electric field is used to control the direction of an electron stream in many electronic devices, such as the cathode-ray oscilloscope, discussed in Section 26–8. $\qquad \blacktriangleleft$

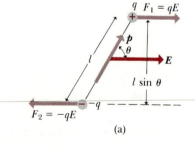

(a)

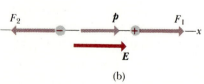

(b)

25–3   (a) The torque on the dipole is $\Gamma = pE \sin\theta$. (b) The dipole is in equilibrium in a uniform field when p and E are parallel.

EXAMPLE 5   Figure 25–3 represents two point charges, $+q$ and $-q$, of equal magnitude but of opposite sign, separated by a distance l. Such a pair of charges is called an *electric dipole*. The dipole is in a *uniform* electric field E, whose direction makes an angle θ with the line joining the two charges, called the *dipole axis*. A force F_1, of magnitude qE, in the direction of the field, is exerted on the positive charge, and a force F_2, of the same magnitude but in the opposite direction, is exerted on the negative charge. The resultant *force* on the dipole is zero, but since the two forces do not have the same line of action, they constitute a *couple* (see Section 8–4). The moment of the couple is

$$\Gamma = (qE)(l \sin\theta),$$

since $l \sin\theta$ is the perpendicular distance between the action lines of the forces.

The product ql of the charge q and the distance l is called the *electric dipole moment*, and is represented by p:

$$p = ql.$$

The torque exerted by the couple is therefore

$$\Gamma = pE \sin \theta. \qquad (25\text{–}2)$$

The *vector dipole moment* of the dipole, $\boldsymbol{p}$, is defined as a vector of magnitude p lying along the dipole axis and pointing from the negative toward the positive charge.

The effect of the torque is to tend to rotate the dipole to a position in which the dipole moment $\boldsymbol{p}$ is parallel to the electric vector $\boldsymbol{E}$, as in Fig. 25–3b. If the field is uniform, the dipole is in equilibrium in this position. ◀

25–2

CALCULATION OF ELECTRIC FIELD

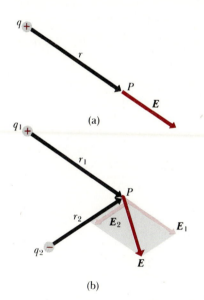

25–4 (a) The electric field $\boldsymbol{E}$ is in the same direction as the vector $\boldsymbol{r}$ when q is positive. (b) The resultant electric field at point P is the vector sum of $\boldsymbol{E}_1$ and $\boldsymbol{E}_2$.

The preceding section has described an experimental way to measure the electric field at a point. The method consists of placing a small test charge at the point, *measuring* the force on it, and taking the ratio of the force to the charge. The electric field at a point may also be *computed* from Coulomb's law if the magnitudes and positions of all charges contributing to the field are known. Thus, to find the magnitude of the electric field at a point P, at a distance r from a point charge q, we imagine a test charge q' to be placed at P. The force on the test charge, by Coulomb's law, has magnitude

$$F = \frac{1}{4\pi\epsilon_0} \frac{qq'}{r^2},$$

and hence the electric field at P has magnitude

$$E = \frac{F}{q'}$$
$$= \frac{1}{4\pi\epsilon_0} \frac{q}{r^2}.$$

The direction of the field is away from the charge q if the latter is positive, toward q if it is negative.

If a number of point charges q_1, q_2, etc., are at distances r_1, r_2, etc., from a given point P, as in Fig. 25–4b, each exerts a force on a test charge q' placed at the point, and the resultant force on the test charge is the *vector sum* of these forces. The resultant electric field is the *vector sum* of the individual electric fields, and

$$\boldsymbol{E} = \boldsymbol{E}_1 + \boldsymbol{E}_2 + \cdots \qquad (25\text{–}3)$$

Because each term to be summed is a vector, the sum is a vector sum. The fact that the total field is the sum of the separate fields that would be caused by the individual charges is a direct result of the *principle of superposition*, discussed in Section 24–5.

EXAMPLE 1 Point charges q_1 and q_2 of $+12 \times 10^{-9}$ C and -12×10^{-9} C, respectively, are placed 0.1 m apart, as in Fig. 25–5. Compute the electric fields due to these charges at points a, b, and c.

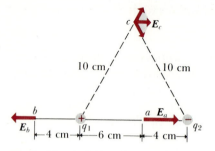

25–5 Electric field at three points, *a*, *b*, and *c*, in the field set up by charges q_1 and q_2.

Solution At point *a*, the vector due to the positive charge q_1 is directed toward the right, and its magnitude is

$$E_1 = (9.0 \times 10^9 \text{ N}\cdot\text{m}^2\cdot\text{C}^{-2})\frac{(12 \times 10^{-9} \text{ C})}{(0.06 \text{ m})^2}$$

$$= 3.00 \times 10^4 \text{ N}\cdot\text{C}^{-1}.$$

The vector due to the negative charge q_2 is also directed toward the right; its magnitude is

$$E_2 = (9.0 \times 10^9 \text{ N}\cdot\text{m}^2\cdot\text{C}^{-2})\frac{(12 \times 10^{-9} \text{ C})}{(0.04 \text{ m})^2}$$

$$= 6.75 \times 10^4 \text{ N}\cdot\text{C}^{-1}.$$

Hence, at point *a*,

$$E_a = (3.00 + 6.75) \times 10^4 \text{ N}\cdot\text{C}^{-1}$$

$$= 9.75 \times 10^4 \text{ N}\cdot\text{C}^{-1}, \qquad \text{toward the right.}$$

At point *b*, the vector due to q_1 is directed toward the left, with magnitude

$$E_1 = (9.0 \times 10^9 \text{ N}\cdot\text{m}^2\cdot\text{C}^{-2})\frac{(12 \times 10^{-9} \text{ C})}{(0.04 \text{ m})^2}$$

$$= 6.75 \times 10^4 \text{ N}\cdot\text{C}^{-1}.$$

The vector due to q_2 is directed toward the right, with magnitude

$$E_2 = (9.0 \times 10^9 \text{ N}\cdot\text{m}^2\cdot\text{C}^{-2})\frac{(12 \times 10^{-9} \text{ C})}{(0.14 \text{ m})^2}$$

$$= 0.55 \times 10^4 \text{ N}\cdot\text{C}^{-1}.$$

Hence, at point *b*

$$E_b = (6.75 - 0.55) \times 10^4 \text{ N}\cdot\text{C}^{-1}$$

$$= 6.20 \times 10^4 \text{ N}\cdot\text{C}^{-1}, \qquad \text{toward the left.}$$

At point *C*, the magnitude of each vector is

$$E = (9.0 \times 10^9 \text{ N}\cdot\text{m}^2\cdot\text{C}^{-2})\frac{(12 \times 10^{-9} \text{ C})}{(0.1 \text{ m})^2}$$

$$= 1.08 \times 10^4 \text{ N}\cdot\text{C}^{-1}.$$

The directions of these vectors are shown in the figure; their resultant is easily seen to be

$$E_c = 1.08 \times 10^4 \text{ N}\cdot\text{C}^{-1}, \qquad \text{toward the right.} \qquad \blacktriangleleft$$

In practical situations, electric fields are often set up by charges distributed over the surfaces of conductors of finite size, rather than by point charges. The electric field must then be calculated by imagining the charge distribution to be subdivided into many small elements of charge Δq, at varying distances from the point *P* at which the field is to be calculated, and adding the separate contributions of these elements to the total field. In most cases this requires use of the methods of the

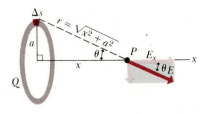

25–6 Electric field due to ring of charge.

integral calculus, but in some examples it is relatively easy to carry out the summation.

EXAMPLE 2   A ring-shaped conductor of radius a carries a total charge Q. Find the electric field at point P, a distance x from the center, along the line perpendicular to the plane of the ring, through its center.

Solution   The situation is shown in Fig. 25–6. We consider first a small segment Δs of the ring; we call the charge on this segment ΔQ. At the point P, this element of charge produces an electric field of magnitude

$$E = \frac{1}{4\pi\epsilon_0}\frac{\Delta Q}{r^2} = \frac{1}{4\pi\epsilon_0}\frac{\Delta Q}{x^2+a^2}.$$

The component of this field along the x-axis is given by

$$E_x = E\cos\theta = \frac{1}{4\pi\epsilon_0}\frac{\Delta Q}{x^2+a^2}\frac{x}{\sqrt{x^2+a^2}} = \frac{1}{4\pi\epsilon_0}\frac{\Delta Qx}{(x^2+a^2)^{3/2}}. \quad (25\text{–}4)$$

Now x is the same for every element of charge around the ring, so to find the *total* x-component of electric field from all of the charge elements ΔQ we simply replace ΔQ in the above expression by the *total* charge Q. The result is

$$E_x = \frac{1}{4\pi\epsilon_0}\frac{Qx}{(x^2+a^2)^{3/2}}. \quad (25\text{–}5)$$

The y-component of the total field at point P is zero, because the y-components of field produced by two segments at opposite sides of the ring cancel each other out, and similarly the contributions of all such pairs add to zero.

Equation (25–5) shows that, at the center of the ring ($x = 0$), the total field is zero, as might be expected; charges on opposite sides pull in opposite directions, and their fields cancel. When x is much larger than a, Eq. (25–5) becomes approximately equal to $Q/4\pi\epsilon_0 x^2$, corresponding to the fact that at distances much greater than the dimensions of the ring it appears as a point charge.

25–3

FIELD LINES

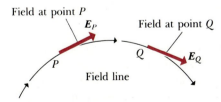

25–7   The direction of the electric field at any point is tangent to the field line through that point.

The concept of field lines was introduced by Michael Faraday (1791–1867) as an aid in visualizing electric (and magnetic) fields. A *field line* (in an electric field) is *an imaginary line drawn in such a way that its direction at any point* (i.e., the direction of its tangent) *is the same as the direction of the field at that point.* (See Fig. 25–7.) Since, in general, the direction of a field varies from point to point, field lines are usually curves. Faraday called these lines "lines of force," but the term "field line" is preferable.

Figure 25–8 shows some of the field lines in two planes containing (a) a single positive charge; (b) two equal charges, one positive and one negative (an electric dipole); and (c) two equal positive charges. The direction of the resultant field at every point in each diagram is along the

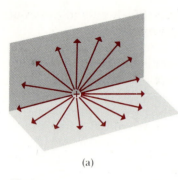

(a)

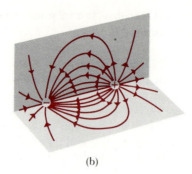

(b)

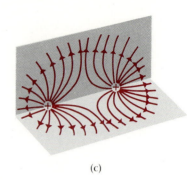

(c)

25–8   The mapping of an electric field with the aid of field lines.

tangent to the field line passing through the point. Arrowheads on the field lines indicate the direction in which the tangent is to be drawn.

No field lines begin or end in the space surrounding a charge. Every field line in an *electrostatic* field is a continuous line terminated by a positive charge at one end and a negative charge at the other.* While sometimes for convenience we speak of an "isolated" charge and draw its field as in Fig. 25–8a, this simply means that the charges on which the lines end are at large distances from the charge under consideration. For example, if the charged body in Fig. 25–8a is a small sphere suspended by a thread from the laboratory ceiling, the negative charges on which its field lines end would be found on the walls, floor, or ceiling, as well as on other objects in the laboratory.

At any one point, the electric field can have but one direction. Hence, only one field line can pass through each point of the field. In other words, *field lines never intersect.*

If a field line were to be drawn through every point of an electric field, all of space and the entire surface of a diagram would be filled with lines, and no individual line could be distinguished. By limiting the number of field lines, we can use them to indicate the *magnitude* of a field as well as its *direction.* This is accomplished by spacing the lines in such a way that *the number per unit area crossing a surface at right angles to the direction of the field is at every point proportional to the electric field.* In a region where the field is large, such as that between the positive and negative charges of Fig. 25–8b, the field lines are closely spaced, whereas in a region where the field is small, such as that between the two positive charges of Fig. 25–8c, the lines are widely separated. In a *uniform* field, the field lines are straight, parallel, and uniformly spaced.

25–4

FIELDS OF VARIOUS CHARGE DISTRIBUTIONS

There are several charge distributions for which the electric fields can be described in terms of simple formulas. Deriving these relationships often requires the use of calculus or of Gauss's law, to be discussed in Section 25–5. In this section we state and discuss several of these results without formal proof.

*We shall see in Chapter 32 that a *changing magnetic field* sets up an electric field whose lines *do not* terminate on electric charges, but close on themselves.

1. Point charge. From the discussion in Section 25–1, it follows directly that the field due to a point charge q has magnitude E given by

$$E = \frac{1}{4\pi\epsilon_0}\frac{q}{r^2}, \qquad (25\text{–}6)$$

where r is the distance from the charge. The direction of the field at each point is along the line from the charge to the point (opposite if q is negative).

2. Uniformly charged spherical shell. Suppose charge is distributed uniformly over the surface of a sphere of radius R, with total charge q. It can be shown that the field at any point *inside* the sphere is zero; such a point is surrounded by charge, and it turns out that the fields from charges on opposite sides of the point cancel out exactly. At points *outside* the sphere the field is the same as though all the charge were concentrated at the center.

$$E = 0 \qquad\qquad \text{when } r < R,$$
$$\qquad\qquad\qquad\qquad\qquad\qquad\qquad (25\text{–}7)$$
$$E = \frac{1}{4\pi\epsilon_0}\frac{q}{r^2} \qquad \text{when } r \ge R.$$

3. Long straight line charge. Suppose charge is distributed uniformly along a very long straight line. If a certain section of the line, of length l, has charge q, we define the charge per unit length (or linear charge density) λ as $\lambda = q/l$. The magnitude of the field at a distance r from the line is then given by

$$E = \frac{1}{2\pi\epsilon_0}\frac{\lambda}{r}. \qquad (25\text{–}8)$$

The direction of the field at each point is radially outward away from the wire (or radially inward if λ is negative). The field magnitude drops off with distance as $1/r$ rather than as $1/r^2$ as for a point charge.

4. Cylindrical shell of charge. Charge is distributed uniformly over the surface of a cylindrical shell or pipe of radius R. The total charge per unit length of cylinder is λ, as in (3). The result is similar to that for the spherical shell; at points inside the cylinder the field is zero, and at points outside it is the same as though the charge were concentrated along the axis of the cylinder. Thus

$$E = 0 \qquad\qquad \text{when } r < R,$$
$$\qquad\qquad\qquad\qquad\qquad\qquad\qquad (25\text{–}9)$$
$$E = \frac{1}{2\pi\epsilon_0}\frac{\lambda}{r} \qquad \text{when } r \ge R.$$

5. Charge distributed over a plane. Charge is distributed uniformly over a very large flat sheet. If an area A of the sheet has total charge q, we define the charge per unit area (or surface charge density), σ as $\sigma = q/A$. The field magnitude near this plane, on either side, is given by

$$E = \frac{\sigma}{2\epsilon_0}. \qquad (25\text{–}10)$$

Of course, an infinitely large plane does not exist in nature, but Eq. (25–10) is valid for points not too near the edge of the plane and at

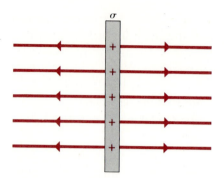

25–9 Cross section of part of a large plane with uniform surface charge density σ. The electric field is uniform on each side.

25–10 Cross section of parts of two large parallel planes with uniform surface charge densities σ and $-\sigma$. The field is uniform between the planes, except near the edges, and is zero outside.

distances from it that are small compared to the plane's dimensions. For such points the electric field is *uniform*, that is, it has the same magnitude and direction for all points in the region. It has opposite directions on the two sides of the plane; Fig. 25–9 shows field lines for this situation.

6. *Field between two large parallel charged planes.*   In the situation shown in Fig. 25–10, two large parallel planes have surface charge densities σ and $-\sigma$, that is, equal magnitude but opposite sign. The fields due to the two surface charges add in the region between the planes but cancel in the space outside this region. Hence the field between the planes has magnitude

$$E = \frac{\sigma}{\epsilon_0}. \qquad (25\text{–}11)$$

The field lines are shown in Fig. 25–10; the field between the planes (at points not too near the edges) is uniform in magnitude and direction.

*25–5

GAUSS'S LAW

Karl Friedrich Gauss (1777–1855) was one of the greatest mathematical geniuses of all time. Many diverse areas of mathematics, from number theory and geometry to the theory of differential equations, bear the mark of his influence, and he made many equally significant contributions to theoretical physics.

   The content of Gauss's law is suggested by consideration of field lines, discussed in Section 25–3. The field of an isolated positive point charge q is represented by lines radiating out in all directions. Suppose we imagine this charge as surrounded by a spherical surface of radius R, with the charge at its center. The area of this imaginary surface is $4\pi R^2$, so if the total number of field lines emanating from q is N, then the number of lines *per unit surface area* on the spherical surface is $N/4\pi R^2$. We imagine a second sphere concentric with the first, but with radius $2R$.

Its area is $4\pi(2R)^2 = 16\pi R^2$, and the number of lines per unit area on this sphere is $N/16\pi R^2$, one fourth that of the first sphere. This corresponds to the fact that, at distance $2R$, the field has only one fourth the magnitude it has at distance R, and verifies our qualitative statement in Section 25–3 that the number of lines per unit area is proportional to the magnitude of the field.

The fact that the *total* number of lines at distance $2R$ is the same as at R can be expressed another way. The field is inversely proportional to R^2, but the *area* of the sphere is proportional to R^2, so the *product* of the two is independent of R. For a sphere of arbitrary radius r, the magnitude of E on the surface is

$$E = \frac{1}{4\pi\epsilon_0}\frac{q}{r^2},$$

the surface area is

$$A = 4\pi r^2,$$

and the product of the two is

$$EA = \frac{q}{\epsilon_0}. \qquad (25\text{--}12)$$

This is independent of r and depends *only* on the charge q.

Although Eq. (25–12) has been derived only for spherical surfaces, it can be generalized to *any* closed surface surrounding an electric charge. Without deriving this generalization in detail we can state the basic ideas as follows: We imagine any surface surrounding a charge as divided into small elements of area ΔA. For each element we determine the component of electric field $E_\perp$ perpendicular to the area. Then the product $E_\perp \Delta A$ is proportional to the number of field lines passing through the area, and the *sum* of all such products for the entire surface is proportional to the *total* number of field lines emanating from the charge. The final result may be stated as

$$\sum E_\perp \Delta A = \frac{q}{\epsilon_0}. \qquad (25\text{--}13)$$

Furthermore, this result is also valid if there are many charges inside the volume, because the total $\boldsymbol{E}$ field at any point is simply the vector sum of the $\boldsymbol{E}$ fields of the individual charges. Thus if Q represents the *total* charge enclosed by the surface, the general statement of Gauss's law is

$$\sum E_\perp \Delta A = \frac{Q}{\epsilon_0}. \qquad (25\text{--}14)$$

The quantity $\Sigma E_\perp \Delta A$ is also called the *electric flux* through the surface, denoted by Ψ:

$$\sum E_\perp \Delta A = \Psi. \qquad (25\text{--}15)$$

Evaluating the sum in Eq. (25–14) may appear to be an impossible task, but there are some cases where the symmetry of the situation may be exploited to accomplish this calculation quite simply. The following general observations are useful:

1. If E is perpendicular to a surface at every point and has the same magnitude at every point, then for that surface the sum in Eq. (25–14) is simply EA.

2. If E is parallel to a surface at every point, then $E_\perp = 0$ and the sum is zero.

3. The surface surrounding the charges need not be a real physical surface. In many applications of Gauss's law it is useful to consider an imaginary or purely geometric surface that may be in empty space, embedded in a solid body, or partly in space and partly within a body.

The following examples will illustrate applications of Gauss's law and the usefulness of the points listed above.

EXAMPLE 1 *Coulomb's law.* We have considered Coulomb's law as the fundamental equation of electrostatics and have derived Gauss's law from it. An alternative procedure is to consider Gauss's law as a fundamental experimental relation. Coulomb's law can then be derived from Gauss's law, by using this law to obtain the expression for the electric field E due to a point charge.

Consider the electric field of a single positive point charge q. By *symmetry*, the field is everywhere radial (there is no reason why it should deviate to one side of a radial direction rather than to another) and its magnitude is the same at all points at the same distance r from the charge (any point at this distance is like any other). Hence, if we select as a gaussian surface a spherical surface of radius r, $E_\perp = E = constant$ at all points of the surface. Then

$$\sum E_\perp \, \Delta A = EA = 4\pi r^2 E.$$

From Gauss's law

$$4\pi r^2 E = \frac{q}{\epsilon_0} \quad \text{and} \quad E = \frac{1}{4\pi\epsilon_0} \frac{q}{r^2}.$$

The force on a point charge q' at a distance r from the charge q is then

$$F = q'E = \frac{1}{4\pi\epsilon_0} \frac{qq'}{r^2},$$

which is Coulomb's law.   ◄

EXAMPLE 2 *Uniformly charged spherical shell.* A total charge Q is distributed uniformly over the surface of a sphere of radius R. We apply Gauss's law to an imaginary spherical surface of radius r, concentric with the sphere of charge. Then by symmetry the electric field is uniform over this imaginary sphere and is everywhere perpendicular to it. If r is less than R, the imaginary sphere is inside the charge distribution, and the total charge it encloses is zero. Then from Gauss's law, EA is zero and E is zero. When r is greater than R, the imaginary sphere is outside the charge, and the total charge enclosed is Q. The surface area is $A = 4\pi r^2$, so from Gauss's law,

$$EA = E(4\pi r^2) = \frac{Q}{\epsilon_0}$$

and

$$E = \frac{1}{4\pi\epsilon_0}\frac{Q}{r^2}.$$

Thus the field is zero at all points inside the spherical charge distribution, and at points outside of it the field is the same as it would be if the total charge Q were concentrated at the center of the sphere. These results thus confirm Eqs. (25–7). ◀

EXAMPLE 3 *Long straight line charge.* Charge is distributed uniformly along a very long straight line such as a thin wire; the charge per unit length (linear charge density) is λ. If the wire is very long and we are not too near either end, then, by symmetry, the field lines outside the wire are *radial* and lie in planes perpendicular to the wire. Also, the field has the same magnitude at all points at the same radial distance from the wire. This suggests that we use as a gaussian surface a *cylinder* of arbitrary radius r and arbitrary length l, with its ends pependicular to the wire, as in Fig. 25–11. If λ is the charge *per unit length* on the wire, the charge within the gaussian surface is λl. Since E is at right angles to the wire, the component of E normal to the end faces is zero. Thus the end faces make no contribution to the sum in Gauss's law. At all points of the curved surface $E_\perp = E =$ constant, and since the area of this surface is $2\pi rl$, we have

$$EA = (E)(2\pi rl) = \frac{\lambda l}{\epsilon_0},$$

and

$$E = \frac{1}{2\pi\epsilon_0}\frac{\lambda}{r},$$

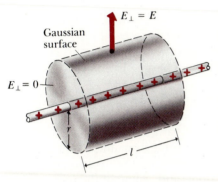

25–11 Cylindrical gaussian surface for calculating the electric field due to a long charged wire.

in agreement with Eq. (25–8).

It should be noted that although the *entire* charge on the wire contributes to the field E, only that portion of the total charge lying within the gaussian surface is used when we apply Gauss's law. This feature of the law is puzzling at first; it appears as though we had somehow obtained the right answer by ignoring a part of the charge, and that the field of a *short* wire of length l would be the same as that of a very long wire. The existence of the entire charge on the wire *is*, however, taken into account when we consider the *symmetry* of the problem. Suppose the wire had been a short one, of length l. Then we could *not* conclude by symmetry that the field at one end of the cylinder, say, would equal that at the center, or that the lines of force would everywhere be perpendicular to the wire. So the entire charge on the wire actually *is* taken into account, but in an indirect way.

It is left as a problem (1) to show that the field outside a long charged cylinder is the same as though the charge on the cylinder were concen-

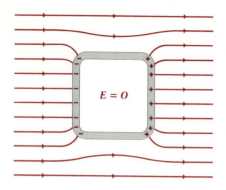

25–12 A conducting box in a uniform electric field. The field pushes electrons toward the left, leaving a net negative charge on the left side and a net positive charge on the right. The total electric field at every point inside the box is zero; the shapes of the exterior field lines near the box are somewhat changed.

trated in a line along its axis, and (2) to calculate the electric field in the interspace between a charged cylinder and a coaxial hollow cylinder that surrounds it. ◄

As the above examples show, Gauss's law may be used to derive relationships for electric fields caused by various charge distributions. In addition, it is useful in working out various general principles concerning charge distributions and fields. Here are a few examples.

First, when there are no charges in motion, the electric field inside any conducting material is always zero. If there were a field in a conductor, the mobile charges in the conductor would move until an equilibrium situation was established in which there would be no further motion and hence no electric field.

Second, Gauss's law may be used to show that if there is never any electric field inside a conductor, when the charges are at rest, then there cannot be any net electric charge in the interior of a conductor. Hence in an electrostatic situation any excess electric charge placed on a conductor must be entirely on the surface.

Finally, if a volume of space contains no charge and is surrounded by a conducting surface, the electric field everywhere within this surface must be zero. This principle forms the basis for *electrostatic shielding*.

Suppose we have a very sensitive electronic instrument that must be protected from stray electric fields. We surround the instrument with a conducting box, or we line the walls, floor, and ceiling of the room with a conducting material such as sheet copper. The external electric field redistributes the free electrons in the conductor, leaving a net positive charge in some regions and a net negative charge in others, as shown in Fig. 25–12. This charge distribution causes an additional electric field, and the total field at a point inside the box is the vector sum of this new field and the external field. Gauss's law can be used to show that this sum is zero; the two fields exactly cancel at all points within the box, so that the total field at every point inside the box is zero. The charge distribution on the conductor also alters the shapes of the external field lines near the box, as the figure shows.

QUESTIONS

25–1 It was shown in the text that the electric field inside a spherical hole in a conductor is zero. Is this also true for a cubical hole? Can the same argument be used?

25–2 Coulomb's law and Newton's law of gravitation have the same *form*. Can Gauss's law be applied to gravitational fields as well as electric fields? Is so, what modifications are needed?

25–3 By considering how the field lines must look near a conducting surface, can you see why the charge density and electric field magnitude at the surface of an irregularly shaped solid conductor must be greatest in regions where the surface curves most sharply, and least in flat regions?

25–4 The electric field and the velocity field in a moving fluid are two examples of vector fields. Think of several other examples. There are also *scalar* fields, which associate a single number with each point in space. Temperature is an example; think of several others.

25–5 Consider the electric field caused by two point charges separated by some distance. Suppose there is a point where the field is zero; what does this tell you about the *signs* of the charges?

25–6 Does an electric charge experience a force due to the field that the charge itself produces?

25–7 A particle having electric charge and mass moves in an electric field. If it starts from rest, does it always move along the field line that passes through its starting point? Explain.

25–8 If the exponent 2 in the r^2 of Coulomb's law were 3 instead, would Gauss's law still be valid?

25–9 A student claimed that an appropriate unit for electric field magnitude is $1\,\mathrm{J\cdot C^{-1}\cdot m^{-1}}$. Is this correct?

25–10 A certain region of space bounded by an imaginary closed surface contains no charge. Is the electric field always zero everywhere on the surface? If not, under what circumstances is it zero on the surface?

25–11 A student claimed that the electric field produced by a dipole is represented by field lines that cross each other. Is this correct? Is there any simple rule governing field lines for a superposition of fields due to point charges, if the field lines due to the separate charges are known? Do field lines *ever* cross?

25–12 Nineteenth-century physicists liked to give everything mechanical attributes. Faraday and his contemporaries thought of field lines as elastic strings that repelled each other and arranged themselves in equilibrium under the action of their elastic tension and mutual repulsion. Try this picture on several examples and decide whether it

makes any sense. (These mechanical properties are of course now known to be completely fictitious.)

25–13 Are Coulomb's law and Gauss's law *completely* equivalent? Are there any situations in electrostatics where one is valid and the other isn't?

25–14 The text states that, in an electrostatic field, every field line must start on a positive charge and terminate on a negative charge. But suppose the field is that of a single positive point charge. Then what?

25–15 Is the total (net) electric charge in the universe positive, negative, or zero?

25–16 A lightning rod is a pointed copper rod mounted on top of a building and welded to a heavy copper cable running down into the ground. Lightning rods are used in prairie country to protect houses and barns from lightning; the lightning current runs through the copper rather than the barn. Why? Why should the end of the rod be pointed? (The answer to Question 25–3 may be helpful.)

25–17 When a high-voltage power line falls on your car, you are safe as long as you stay in the car, but when you step out you may be electrocuted. Why? What about a car as a safe place in a thunderstorm?

PROBLEMS

25–1 Find the magnitude and direction of the electric field at a point 0.5 m directly above a particle having an electric charge of $+2.0 \times 10^{-6}$ C.

25–2 The electric field caused by a certain point charge has a magnitude $5.0 \times 10^3\,\mathrm{N\cdot C^{-1}}$ at a distance of 0.10 m from the charge. What is the magnitude of the charge?

25–3 At what distance from a particle of charge 5.0×10^{-9} C does the electric field of that charge have a magnitude of $2.00\,\mathrm{N\cdot C^{-1}}$?

25–4 Two particles having charges $q_1 = 1.00 \times 10^{-9}$ C and $q_2 = 2.00 \times 10^{-9}$ C are separated by a distance of 2.00 m. At what point is the total electric field due to the two charges equal to zero?

25–5 A small object carrying a charge of -5×10^{-9} C experiences a downward force of 20×10^{-9} N when placed at a certain point in an electric field.

a) What is the electric field at the point?

b) What would be the magnitude and direction of the force acting on an electron placed at the point?

25–6 What must be the charge on a particle of mass 2 g for it to remain stationary in the laboratory when placed in a downward-directed electric field of $500\,\mathrm{N\cdot C^{-1}}$?

25–7 A uniform electric field exists in the region between two oppositely charged plane parallel plates. An electron is released from rest at the surface of the negatively

charged plate and strikes the surface of the opposite plane, 2 cm distant from the first, in a time interval of 1.5×10^{-8} s.

a) Find the electric field.

b) Find the velocity of the electron when it strikes the second plate.

25–8 An electron is projected with an initial velocity $v_0 = 10^7\,\mathrm{m\cdot s^{-1}}$ into the uniform field between the parallel plates in Fig. 25–13. The direction of the field is vertically downward, and the field is zero except in the space between the plates. The electron enters the field at a point midway between the plates. If the electron just misses the upper plate as it emerges from the field, find the magnitude of the electric field.

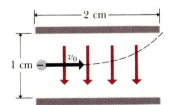

Figure 25–13

25–9 An electron is projected into a uniform electric field of $5000\,\mathrm{N\cdot C^{-1}}$. The direction of the field is vertically upward. The initial velocity of the electron is 10^7 m·s^{-1}, at an angle of 30° above the horizontal.

a) Find the maximum distance the electron rises vertically above its initial elevation.

b) After what horizontal distance does the electron return to its original elevation?

c) Sketch the trajectory of the electron.

25–10 In a rectangular coordinate system a charge of 25×10^{-9} C is placed at the origin of coordinates, and a charge of -25×10^{-9} C is placed at the point $x = 6$ m, $y = 0$. What is the electric field at

a) $x = 3$ m, $y = 0$?

b) $x = 3$ m, $y = 4$ m?

25–11 A charge of 16×10^{-9} C is fixed at the origin of coordinates, a second charge of unknown magnitude is at $x = 3$ m, $y = 0$, and a third charge of 12×10^{-9} C is at $x = 6$ m, $y = 0$. What is the magnitude of the unknown charge if the resultant field at $x = 8$ m, $y = 0$ is 20.25 N·C^{-1} directed to the right?

25–12 In a rectangular coordinate system, two positive point charges of 10^{-8} C each are fixed at the points $x = +0.1$ m, $y = 0$, and $x = -0.1$ m, $y = 0$. Find the magnitude and direction of the electric field at the following points:

a) the origin;

b) $x = 0.2$ m, $y = 0$;

c) $x = 0.1$ m, $y = 0.15$ m;

d) $x = 0$, $y = 0.1$ m.

25–13 Same as Problem 25–12, except that the first point charge is positive and the other negative.

25–14

a) What is the electric field of a gold nucleus, at a distance of 10^{-12} cm from the nucleus?

b) What is the electric field of a proton, at a distance of 5.28×10^{-9} cm from the proton?

25–15 A small sphere whose mass is 0.1 g carries a charge of 3×10^{-10} C and is attached to one end of a silk fiber 5 cm long. The other end of the fiber is attached to a large vertical conducting plate, which has a surface charge of 25×10^{-6} C·m^{-2} on each side. Find the angle the fiber makes with the vertical.

25–16 How many excess electrons must be added to an isolated spherical conductor 10 cm in diameter to produce a field of intensity 1300 N·C^{-1} just outside the surface?

25–17 What is the magnitude of an electric field in which the force on an electron is equal in magnitude to the weight of the electron?

25–18 A wire is bent into a ring of radius R and given a charge q.

a) What is the magnitude of the electric field at the center of the ring?

b) Derive the expression for the electric field at a point on a line perpendicular to the plane of the ring and passing through its center, at a distance r from the center of the ring. What is the direction of the **E**-vector at points on this line?

c) Sketch a graph of the magnitude of **E** as a function of r, from $r = 0$ to $r = 2R$.

25–19 The electric field in the region between a pair of oppositely charged plane parallel plates, each 100 cm^2 in area, is 10^4 N·C^{-1}. What is the charge on each plate? Neglect edge effects.

25–20 A small conducting sphere of radius a, mounted on an insulating handle and having a positive charge q, is inserted through a hole in the walls of a hollow conducting sphere of inner radius b and outer radius c. The hollow sphere is supported on an insulating stand and is initially uncharged, and the small sphere is placed at the center of the hollow sphere. Neglect any effect of the hole.

a) Show that the electric field at a point in the region between the spheres, at a distance r from the center, is equal to

$$E = \frac{1}{4\pi\epsilon_0} \frac{q}{r^2}.$$

b) What is the electric field at a point outside the hollow sphere?

c) Sketch a graph of the magnitude of **E** as a function of r, from $r = 0$ to $r = 2c$.

d) Represent the charge on the small sphere by four + signs. Sketch the field lines of the system, within a spherical volume of radius $2c$.

e) The small sphere is moved to a point near the inner wall of the hollow sphere. Sketch the field lines.

25–21 Prove that the electric field outside an infinitely long cylindrical conductor with a uniform surface charge is the same as if all the charge were on the axis.

25–22 A long coaxial cable consists of an inner cylindrical conductor of radius a and an outer coaxial cylinder of inner radius b and outer radius c. The outer cylinder is mounted on insulating supports and has no net charge. The inner cylinder has a uniform positive charge λ per unit length. Calculate the electric field

a) at any point between the cylinders, and

b) at any external point.

c) Sketch a graph of the magnitude of **E** as a function of the distance r from the axis of the cable, from $r = 0$ to $r = 2c$.

d) Find the charge per unit length on the inner surface of the outer cylinder, and on the outer surface.

25–23 Suppose that positive charge is uniformly distributed throughout a spherical volume of radius R, the charge per unit volume being ρ.

a) Use Gauss's law to prove that the magnitude of the electric field inside the volume, at a distance r from the center, is

$$E = \frac{\rho r}{3\epsilon_0}.$$

b) What is the electric field at a point outside the spherical volume at a distance r from the center? Express your answer in terms of the total charge q within the spherical volume.

c) Compare the answers to (a) and (b) when $r = R$.

d) Sketch a graph of the magnitude of $\boldsymbol{E}$ as a function of r, from $r = 0$ to $r = 3R$.

25–24 Suppose that positive charge is uniformly distributed throughout a very long cylindrical volume of radius R, the charge per unit volume being ρ.

a) Derive the expression for the electric field inside the volume at a distance r from the axis of the cylinder, in terms of the charge density ρ.

b) What in the electric field at a point outside the volume, in terms of the charge per unit length λ in the cylinder?

c) Compare the answers to (a) and (b) when $r = R$.

d) Sketch a graph of the magnitude of $\boldsymbol{E}$ as a function of r, from $r = 0$ to $r = 3R$.

25–25 A conducting spherical shell of inner radius a and outer radius b has a positive point charge Q located at its center. The total charge on the shell is zero, and it is insulated from its surroundings.

a) Derive expressions for the electric field magnitude in terms of the distance r from the center, for the regions $r < a$, $a < r < b$, and $r > b$.

b) What is the surface charge density on the inner surface of the conducting shell?

c) Draw a sketch showing electric field lines and the location of all charges.

d) Draw a graph of E as a function of r.

25–26 The earth has net electric charge that causes a field at points near its surface of the order of $100 \ \mathrm{N \cdot C^{-1}}$.

a) If the earth is regarded as a conducting sphere of radius $6.38 \times 10^6 \ \mathrm{m}$, what is the magnitude of its charge?

b) How much charge would a 70-kg human have to acquire to overcome his or her weight by repulsion by the earth's charge? Is electrostatic repulsion a feasible means of flight?

26

POTENTIAL

In Chapter 6 the concepts of *work* and *energy* were introduced, and we saw that they provide a very important method of analysis for many mechanical problems. Energy plays an equally fundamental role in electricity and magnetism; in this chapter we shall apply work and energy considerations to the electric field. When a charged particle moves in an electric field, the field does *work* on the particle. We shall see that the work can always be expressed in terms of a potential energy, which in turn is associated with a new concept called *electrical potential* or simply *potential*. Several examples of calculations of potential and their applications to the motions of charged particles will be discussed.

26–1

ELECTRICAL POTENTIAL ENERGY

When a charged particle moves in a region of space where there is an electric field, the field exerts a force on the particle, and thus does *work* on it. Figure 26–1 shows a simple example; a pair of charged parallel metal plates sets up a uniform electric field of magnitude E in the region between them, and the resulting force on a test charge q' has magnitude $F = q'E$. When the charge moves from point a to point b, the work done by this force on the test charge is

$$W_{a \to b} = q'Ed. \tag{26–1}$$

This work can be represented by a potential-energy function, just as for gravitational potential energy, discussed in Section 6–5. If we take the potential energy to be zero at point b, then at point a it has the value $q'Ed$, and we may say

$$W_{a \to b} = U_a - U_b. \tag{26–2}$$

More generally, the potential energy at any point a distance y above the bottom plate is given by

$$U(y) = q'Ey, \tag{26–3}$$

26–1 A test charge q' moving from a to b experiences a force of magnitude $q'E$; the work done by this force is $W_{a \to b} = q'Ed$, and is independent of the particle's path.

and when the test charge moves from height y_1 to height y_2, the work done by the field is given by

$$W_{1 \to 2} = U(y_1) - U(y_2) = q'Ey_1 - q'Ey_2. \qquad (26\text{--}4)$$

When y_1 is greater than y_2, U decreases and the field does positive work; when y_1 is less than y_2, U increases and the field does negative work. This whole situation is directly analogous to the motion of a particle in a uniform gravitational field, with its associated work and potential energy.

An important characteristic of the work done by electric-field forces is that in any displacement of a charge from point a to point b, the work is independent of the path. In Fig. 26–1, the work depends only on the change in the coordinate y. This property is characteristic of *conservative force fields;* such fields are discussed in Section 6–7, and the reader would do well to review that discussion. Although thus far this property has been discussed only for the uniform field, it can be shown that *every* electric field caused by electric charges at rest is conservative, and hence can be associated with a potential-energy function.

We next consider the work done on a test charge q' by the electric field due to a single point charge q, as shown in Fig. 26–2. Such a field is clearly *not* uniform, and somewhat more elaborate calculations are required. We consider first the work done on the charge q' as it passes through the points shown, at distances r_1, r_2, and r_3 from q.

The magnitude of the force on q' is given as usual by $F = qq'/4\pi\epsilon_0 r^2$. To calculate the work done by this force we cannot simply multiply force by distance, because the force is not constant but depends on r. The work can be obtained very easily by use of methods of integral calculus, but even without calculus it can be computed fairly simply. Consider the first displacement $\Delta s = r_1 - r_a$. The force varies from $qq'/4\pi\epsilon_0 r_a^2$ to $qq'/4\pi\epsilon_0 r_1^2$, so an approximate average force for the interval is $qq'/4\pi\epsilon_0 r_a r_1$. Thus the work done in this small displacement is approximately

$$W_{a \to 1} = F_{av}\,\Delta s = \frac{qq'}{4\pi\epsilon_0 r_a r_1}(r_1 - r_a).$$

This can be rearranged to give

$$W_{a \to 1} = \frac{qq'}{4\pi\epsilon_0}\left(\frac{1}{r_a} - \frac{1}{r_1}\right). \qquad (26\text{--}5)$$

Similarly, the work for each of the other intervals can be written

$$W_{1 \to 2} = \frac{qq'}{4\pi\epsilon_0}\left(\frac{1}{r_1} - \frac{1}{r_2}\right),$$

$$W_{2 \to 3} = \frac{qq'}{4\pi\epsilon_0}\left(\frac{1}{r_2} - \frac{1}{r_3}\right),$$

$$W_{3 \to b} = \frac{qq'}{4\pi\epsilon_0}\left(\frac{1}{r_3} - \frac{1}{r_b}\right).$$

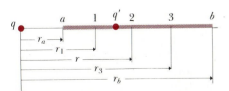

26–2 Charge q' moves along a straight line extending radially from charge q. As it moves from a to b, the distance varies from r_a to r_b.

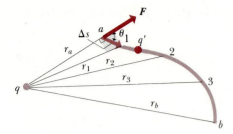

26–3 Charge q' moves along a curved path from a to b. The work done on it by the field of charge q is found to depend only on the initial and final distances, r_a and r_b.

The *total* work done on q' is then given approximately by the sum of these contributions. We note that all the terms containing r_1, r_2, and r_3 cancel out, leaving

$$W_{a \to b} = \frac{qq'}{4\pi\epsilon_0} \left(\frac{1}{r_a} - \frac{1}{r_b} \right). \tag{26–6}$$

It might seem that this result is only approximate because of the use of an average force in each interval. But we could just as well divide the path into many more intervals, so that the force would change very little in each interval, and the same cancellation would occur. In the limit, when we take a very large number of intervals, the error disappears completely; thus Eq. (26–6) is in fact *exactly* correct.

Thus the work for this particular path depends only on the end-points. We have not yet proved that the work is the same for *all possible* paths from a to b. To do this, we consider a more general displacement, in which a and b do *not* lie on the same radial line, as in Fig. 26–3. To find the work in the displacement from point a to point 1 we must now use the more general definition of work: $W = F \Delta s \cos \theta$. But as the figure shows, $\Delta s \cos \theta$ is equal to $r_1 - r_a$. Thus Eq. (26–5) still gives the work in this interval correctly even though the displacement is no longer along a radial line. Furthermore, in the calculation of the total work, the same cancellation of the intermediate distance occurs, and Eq. (26–6) gives the work for this more general displacement.

This result shows that the work done on q' by the E field depends only on the initial and final distances r_a and r_b and not on the path connecting these points. The work is the same for *any* path between these points, or between any pair of points at distances r_a and r_b from the charge q. Also, if the test charge is returned from b to a along any path, the work done by the E field is the negative of that done in the displacement from a to b. Thus, the total work done during a loop displacement returning to the starting point is always zero.

Comparing Eqs. (26–2) and (26–6), we see that the term $qq'/4\pi\epsilon_0 r_a$ is the potential energy U_a when q' is at point a, at distance r_a from q, and $qq'/4\pi\epsilon_0 r_b$ the potential energy U_b when it is at point b, at distance r_b from q. Thus the potential energy U of the test charge q' at *any* distance r from charge q is given by

$$U = \frac{1}{4\pi\epsilon_0} \frac{qq'}{r}. \tag{26–7}$$

If the field in which charge q' moves is not that of a single point charge but is due to a more general charge distribution, we may always divide the charge distribution into small elements and treat each element as a point charge. Then, since the total field is the vector sum of the fields due to the individual elements, and since the total work on q' is the sum of the contributions from the individual charge elements, we may conclude that *every* electric field due to a static charge distribution is a conservative force field. Furthermore, the potential energy of a test charge q' at point a in Fig. 26–4, due to a collection of charges q_1, q_2, q_3,

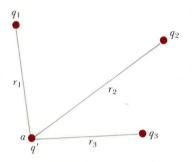

26–4 Potential energy of a charge q' at point a depends on charges q_1, q_2, and q_3, and on their distances r_1, r_2, r_3 from point a.

etc., at distances r_1, r_2, r_3, etc., from the test charge q', is given by

$$U = \frac{q'}{4\pi\epsilon_0}\left(\frac{q_1}{r_1} + \frac{q_2}{r_2} + \frac{q_3}{r_3} + \cdots\right),$$

$$U = \frac{q'}{4\pi\epsilon_0}\sum \frac{q_i}{r_i}. \qquad (26\text{–}8)$$

At a second point b, the potential energy is given by the same expression except that r_1, r_2, . . . now represent the distances from the respective charges to point b. The work of the electric force in moving the test charge from a to b along any path is equal to the difference $U_a - U_b$ between its potential energies at a and at b.

In the discussion of potential energy in Chapter 6, it was observed that it is always possible to add an arbitrary constant C to a potential-energy function U, since in Eq. (26–2) adding the same constant C to both U_a and U_b does not change the physically significant quantity, namely the difference $U_a - U_b$. Thus we are always free to choose the constant C so that U is zero at some convenient reference position. In Chapter 6, the potential energy of a body in a uniform gravitational field could be taken to be zero at some convenient reference level, often the surface of the earth. When the body is above this reference level, its potential energy is positive; when below, negative.

In Eqs. (26–7) and (26–8), the reference position for electrical potential energy, the position at which $U = 0$, has been chosen implicitly to be that at which all the distances r_1, r_2, . . . are *infinite*. That is, the potential energy of the test charge is zero when it is very far removed from all the charges setting up the field. This is the most convenient reference level for most electrostatic problems. When dealing with electrical *circuits*, other reference levels are more convenient, which simply means that a constant term is added to the potential energy. This is not significant, since it is only *differences* in potential energy that are of practical significance or even measurable.

From the above definitions, *the potential energy of a test charge at any point in an electric field is equal to the work done by the electric force when the test charge is brought from the point in question to a zero reference level, often taken at infinity.*

26–2

POTENTIAL

Instead of dealing directly with the potential energy U of a charged particle, it is useful to introduce the more general concept of *potential energy per unit charge*. This quantity is called the *potential*, and *the potential at any point of an electrostatic field is defined as the potential energy per unit charge at the point.* Potential is represented by the letter V:

$$V = \frac{U}{q'}, \qquad \text{or} \qquad U = q'V. \qquad (26\text{–}9)$$

Potential energy and charge are both *scalars*, so potential is a scalar quantity. Its basic SI unit is 1 *joule per coulomb* $(1\ \text{J}\cdot\text{C}^{-1})$. For brevity, a potential of $1 \text{ J} \cdot \text{C}^{-1}$ is called 1 *volt* (1 V). The unit is named in honor of the Italian scientist Alessandro Volta (1745–1827). Commonly used multiples of the volt are the kilovolt (1 kV = 10^3 V), the megavolt (1 MV = 10^6 volts), and the gigavolt (1 GV = 10^9 V). Submultiples are the millivolt (1 mV = 10^{-3} V) and the microvolt (1 μV = 10^{-6} V).

To put Eq. (26–2) on a "work per unit charge" basis, we divide both sides by q', obtaining

$$\frac{W_{a \to b}}{q'} = \frac{U_a}{q'} - \frac{U_b}{q'} = V_a - V_b, \qquad (26\text{–}10)$$

where $V_a = U_a/q'$ is the potential energy per unit charge at point a, and similarly for V_b. V_a and V_b are called the *potential at point a* and *potential at point b*, respectively. From Eq. (26–8), the potential V at a point due to an arbitrary collection of point charges is given by

$$V = \frac{U}{q'} = \frac{1}{4\pi\epsilon_0} \sum \frac{q_i}{r_i}. \qquad (26\text{–}11)$$

The difference $V_a - V_b$ is called the *potential of a with respect to b*, and is often abbreviated V_{ab}. It should be noted that *potential, like electric field, is independent of the test charge q' used to define it*.

A *voltmeter* is an instrument that measures the potential *difference* between the points to which its terminals are connected. The principle of the common type of moving-coil voltmeter will be described later. There are also much more sensitive potential-measuring devices that make use of electronic amplification, and devices that can measure a potential difference of 1 μV are used routinely.

EXAMPLE 1   A particle having a charge $q = 3 \times 10^{-9}$ C moves from point a to point b along a straight line, a total distance $d = 0.5$ m. The electric field is uniform along this line, in the direction from a to b, with magnitude $E = 200 \text{ N} \cdot \text{C}^{-1}$. Determine the force on q, the work done on it by the field, and the potential difference $V_a - V_b$.

Solution   The force is in the same direction as the electric field, and its magnitude is given by

$$F = qE = (3 \times 10^{-9} \text{ C})(200 \text{ N} \cdot \text{C}^{-1}) = 600 \times 10^{-9} \text{ N}.$$

The work done by this force is

$$W = Fd = (600 \times 10^{-9} \text{ N})(0.5 \text{ m}) = 300 \times 10^{-9} \text{ J}.$$

The potential difference is the work per unit charge, which is

$$V_a - V_b = \frac{W}{q} = \frac{300 \times 10^{-9} \text{ J}}{3 \times 10^{-9} \text{ C}}$$
$$= 100 \text{ J} \cdot \text{C}^{-1} = 100 \text{ V}.$$

Alternatively, since E is force per unit charge, the work per unit charge is obtained by multiplying E by the distance d:

$$V_a - V_b = Ed = (200 \text{ N} \cdot \text{C}^{-1})(0.5 \text{ m})$$
$$= 100 \text{ J} \cdot \text{C}^{-1} = 100 \text{ V}. \qquad \blacktriangleleft$$

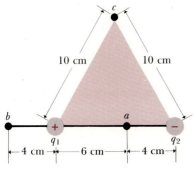

26–5

EXAMPLE 2   Point charges of $+12 \times 10^{-9}$ C and -12×10^{-9} C are placed 10 cm apart, as in Fig. 26–5. Compute the potentials at points a, b, and c.

Solution   We must evaluate the *algebraic* sum $(1/4\pi\epsilon_0)\Sigma(q_i/r_i)$ at each point. At point a, the potential due to the positive charge is

$$(9.0 \times 10^9 \text{ N} \cdot \text{m}^2 \cdot \text{C}^{-2}) \frac{12 \times 10^{-9} \text{ C}}{0.06 \text{ m}} = 1800 \text{ N} \cdot \text{m} \cdot \text{C}^{-1}$$

$$= 1800 \text{ J} \cdot \text{C}^{-1} = 1800 \text{ V},$$

and the potential due to the negative charge is

$$(9.0 \times 10^9 \text{ N} \cdot \text{m}^2 \cdot \text{C}^{-2}) \frac{-12 \times 10^{-9} \text{ C}}{0.04 \text{ m}} = -2700 \text{ J} \cdot \text{C}^{-1} = -2700 \text{ V}.$$

Hence

$$V_a = 1800 \text{ V} - 2700 \text{ V} = -900 \text{ V} = -900 \text{ J} \cdot \text{C}^{-1}.$$

At point b, the potential due to the positive charge is $+2700$ V and that due to the negative charge is -770 V. Hence,

$$V_b = 2700 \text{ V} - 770 \text{ V} = 1930 \text{ V} = 1930 \text{ J} \cdot \text{C}^{-1}.$$

At point c the potential is

$$V_c = 1080 \text{ V} - 1080 \text{ V} = 0. \qquad \blacktriangleleft$$

EXAMPLE 3   Compute the potential energy of a point charge of $+4 \times 10^{-9}$ C if placed at points a, b, and c in Fig. 26–5.

Solution   First,

$$U = qV.$$

Hence at point a,

$$U = qV_a = (4 \times 10^{-9} \text{ C})(-900 \text{ J} \cdot \text{C}^{-1}) = -36 \times 10^{-7} \text{ J}.$$

At point b,

$$U = qV_b = (4 \times 10^{-9} \text{ C})(1930 \text{ J} \cdot \text{C}^{-1}) = 77 \times 10^{-7} \text{ J}.$$

At point c,

$$U = qV_c = 0.$$

(All relative to a point at infinity.) $\blacktriangleleft$

EXAMPLE 4   In Fig. 26–6, a particle having mass $m = 5$ g and charge $q' = 2 \times 10^{-9}$ C starts from rest at point a and moves in a straight line to point b, under the influence of the electric fields of the two charges. What is its speed v at point b?

Solution   Conservation of energy gives

$$K_a + U_a = K_b + U_b.$$

3×10^{-9} C a b -3×10^{-9} C

←1 cm→|←1 cm→|←1 cm→

26–6

For this situation, $K_a = 0$ and $K_b = \frac{1}{2}mv^2$. The potential energies are given in terms of the potentials by Eq. (26–9): $U_a = q'V_a$ and $U_b = q'V_b$. Thus the energy equation becomes

$$0 + q'V_a = \tfrac{1}{2}mv^2 + q'V_b.$$

When solved for v, this yields

$$v = \sqrt{\frac{2q'(V_a - V_b)}{m}}.$$

We obtain the potentials just as we did in the preceding examples:

$$V_a = (9.0 \times 10^9 \ \text{N} \cdot \text{m}^2 \cdot \text{C}^{-2}) \left(\frac{3 \times 10^{-9} \ \text{C}}{0.01 \ \text{m}} + \frac{-3 \times 10^{-9} \ \text{C}}{0.02 \ \text{m}} \right) = 1350 \ \text{V},$$

$$V_b = (9.0 \times 10^9 \ \text{N} \cdot \text{m}^2 \cdot \text{C}^{-2}) \left(\frac{3 \times 10^{-9} \ \text{C}}{0.02 \ \text{m}} + \frac{-3 \times 10^{-9} \ \text{C}}{0.01 \ \text{m}} \right) = -1350 \ \text{V}.$$

Finally,

$$v = \sqrt{\frac{2(2 \times 10^{-9} \ \text{C})(2700 \ \text{V})}{5 \times 10^{-3} \ \text{kg}}} = 4.65 \times 10^{-2} \ \text{m} \cdot \text{s}^{-1} = 4.65 \ \text{cm} \cdot \text{s}^{-1}.$$

Consistency of units may be checked by noting that $1 \ \text{V} = 1 \ \text{J} \cdot \text{C}^{-1}$, and thus the numerator under the radical has units of J or $\text{kg} \cdot \text{m}^2 \cdot \text{s}^{-2}$. ◀

26–3

CALCULATION OF POTENTIAL DIFFERENCES

In principle the potential difference between any two points can be calculated by using Eq. (26–11) to find the potential at each point, but in many cases, especially where the electric field is easily obtained, it is easier to use Eq. (26–10), calculating from the known electric field the work done on a test charge during a displacement from one point to the other. In some problems a combination of these two approaches is useful. The following problems illustrate these remarks.

EXAMPLE 1 *Charged spherical conductor.* We consider first a solid conducting sphere of radius R, with total charge q. From Gauss's law, as discussed in Chapter 25, we conclude that, at all points *outside* the sphere, the field is the same as that of a point charge q at the center of the sphere. *Inside* the sphere the field is zero everywhere; otherwise charge would move within the sphere.

As mentioned above, it is convenient to take the reference level of potential (the point where $V = 0$) at a very large distance from all charges. Since the field at any point r greater than R is the same as for a point charge, the work on a test charge moving from a finite value of r to infinity is also the same as for a point charge, so the potential V at a radial distance r, relative to a point at infinity, is

$$V = \frac{1}{4\pi\epsilon_0} \frac{q}{r}. \tag{26–12}$$

The potential is positive if q is positive, negative if q is negative.

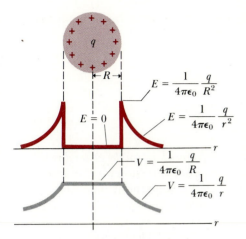

26–7 Electric field magnitude E and potential V at points inside and outside a charged spherical conductor.

Equation (26–12) applies to the field of a charged spherical conductor only when r is greater than or equal to the radius R of the sphere. The potential *at* the surface is

$$V = \frac{1}{4\pi\epsilon_0}\frac{q}{R}. \qquad (26\text{–}13)$$

Inside the sphere the field is zero everywhere, and no work is done on a test charge displaced from any point to any other in this region. Thus the potential is the same at all points inside the sphere, and is equal to its value $q/4\pi\epsilon_0 R$ at the surface. The field and potential are shown as functions of r in Fig. 26–7.

The electric field $\boldsymbol{E}$ at the surface has magnitude

$$E = \frac{1}{4\pi\epsilon_0}\frac{q}{R^2}. \qquad (26\text{–}14)$$

The maximum potential to which a conductor in air can be raised is limited by the fact that air molecules become ionized, and hence the air becomes a conductor, at an electric field of about $0.8 \times 10^6 \text{ N} \cdot \text{C}^{-1}$. Comparing Eqs. (26–13) and (26–14), we note that at the surface the field and potential are related by $V = ER$. Thus if E_m represents the upper limit of electric field, known as the *dielectric strength*, the maximum potential to which a spherical conductor can be raised is

$$V_m = RE_m.$$

For a sphere 1 cm in radius, in air,

$$V_m = (10^{-2} \text{ m})(0.8 \times 10^6 \text{ N} \cdot \text{C}^{-1}) = 8000 \text{ V},$$

and no amount of "charging" could raise the potential of a sphere of this size, in air, higher than about 8000 V. It is this fact that necessitates the use of large spherical terminals on high-voltage machines. If we make $R = 2$ m, then

$$V_m = (2 \text{ m})(0.8 \times 10^6 \text{ N} \cdot \text{C}^{-1}) = 1.6 \times 10^6 \text{ V} = 1.6 \text{ MV}.$$

At the other extreme is the effect produced by a surface of very *small* radius of curvature, such as a sharp point. Since the maximum potential is proportional to the radius, even relatively small potentials applied to sharp points in air will produce sufficiently high fields just outside the point to result in ionization of the surrounding air.   ◀

EXAMPLE 2 *Parallel plates.* The electric field between two parallel plates was discussed at the beginning of Section 26–1 and illustrated in Fig. 26–1. To obtain the potential difference between points y and b, also called the potential of y with respect to b, we note that the force on a test charge q' is $q'E$, the work done on the charge by the field during a displacement from y to b is

$$W_{y \to b} = q'Ey. \qquad (26\text{–}15)$$

Thus the potential difference, or work per unit charge, is

$$V_y - V_b = Ey. \qquad (26\text{–}16)$$

The potential thus decreases linearly with y as we move from the upper to the lower plate. At point a, where $y = d$ and $V_y = V_a$,

$$V_a - V_b = Ed,$$

and

$$E = \frac{V_a - V_b}{d} = \frac{V_{ab}}{d}. \qquad (26\text{--}17)$$

That is, *the electric field equals the potential difference between the plates divided by the distance between them.* This is a more useful expression for E than Eq. (25–11) is, because the potential difference V_{ab} can readily be measured with a voltmeter, while there are no instruments that read surface density of charge directly.

Equation (26–17) also shows that the unit of electric field can be expressed as 1 *volt per meter* (1 V $\cdot$ m^{-1}), as well as 1 N $\cdot$ C^{-1}. In practice, the volt per meter is most commonly used as the unit of E. ◄

EXAMPLE 3 A simple type of vacuum tube known as a *diode* consists essentially of two electrodes within a highly evacuated enclosure. One electrode, the cathode, is maintained at a high temperature and emits electrons from its surface. A potential difference of a few hundred volts is maintained between the cathode and the other electrode, known as the *anode* or *plate*, with the anode at the higher potential. Suppose that, in a certain diode, the plate potential is 250 V above that of the cathode, and an electron is emitted from the cathode with no initial velocity. What is its velocity when it reaches the anode?

Solution Let V_c and V_a represent the cathode and anode potentials, respectively, and let the charge on the electron be $-e$. Since the charge is negative, the electron goes from the cathode, where the potential is lower, to the anode, where the potential is higher. Hence, the *decrease* in electrical potential energy is $-e(V_c - V_a)$ or eV_{ac}, and the electron *gains* a corresponding amount of kinetic energy during its travel from cathode to anode. Letting v_c and v_a be the speeds at anode and cathode, respectively, we have

$$eV_{ac} = \tfrac{1}{2}mv_a{}^2 - \tfrac{1}{2}mv_c{}^2.$$

Since $v_c = 0$,

$$v_a = \sqrt{2eV_{ac}/m}$$

$$= \sqrt{\frac{2(1.6 \times 10^{-19}\,\text{C})(250\,\text{J}\cdot\text{C}^{-1})}{9.1 \times 10^{-31}\,\text{kg}}}$$

$$= 9.4 \times 10^6 (\text{J}\cdot\text{kg}^{-1})^{1/2} = 9.4 \times 10^6\ \text{m}\cdot\text{s}^{-1}.$$

Note that the shape or separation of the electrodes need not be known. The final velocity depends only on the difference of potential between cathode and anode. However, the *time* of transit from cathode to anode depends on the geometry of the tube. ◄

26–4

EQUIPOTENTIAL SURFACES

The potential distribution in an electric field may be represented graphically by *equipotential surfaces*. An equipotential surface is one on which the potential has the same value at all points on the surface. An equipotential surface may be constructed through every point of an electric field, but it is usual to show only a few of the equipotentials in a diagram.

Since the potential energy of a charged body is the same at all points of a given equipotential surface, it follows that no (electrical) work is needed to move a charged body over such a surface. Hence, the equipotential surface through any point must be at right angles to the direction of the field at that point. If this were not so, the field would have a component lying in the surface and work would have to be done against electrical forces to move a charge in the direction of this component. The field lines and the equipotential surfaces thus form a mutually perpendicular network. In general, the field lines of a field are curves and the equipotentials are curved surfaces. For the special case of a uniform field, where the field lines are straight and parallel, the equipotentials are parallel planes perpendicular to the field lines.

Figure 26–8 shows several arrangements of charges. The field lines are represented by colored lines, and cross sections of the equipotential

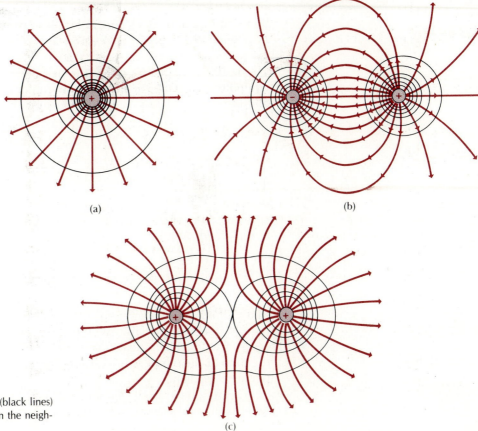

(a)   (b)

(c)

26–8 Equipotential surfaces (black lines) and field lines (colored lines) in the neighborhood of point charges.

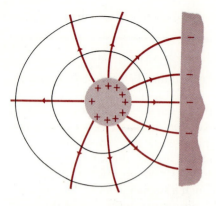

26–9 Field lines always meet charged conducting surfaces at right angles.

surfaces as black lines. The actual field is of course three-dimensional. At each crossing of an equipotential and a field line, the two are perpendicular.

When charges at rest reside on the surface of a conductor, the electric field just outside the conductor must be perpendicular to the surface at every point. That is, at the surface there can never be a component of **E** *parallel* to the surface. If there were such a component, it would do work on a charge moving close to and parallel to the surface. But then the charge could go through the surface and return to its starting point by a path close to the surface but inside the conductor. No work would be done in the return trip because **E** is zero everywhere inside the conductor. Thus when it returned to the starting point the total work would not be zero; this contradicts the general principle that an electrostatic field is always conservative. We must conclude that the existence of a parallel component of **E** at the surface is an impossibility.

It also follows that when a test charge is moved from point to point on the surface of a conductor, the electric field does no work on the charge; hence when all charges are at rest on the surface *a conducting surface is always an equipotential surface.* These principles are shown in Fig. 26–9.

26–5

POTENTIAL GRADIENT

Suppose that a given electric field has been mapped by its network of field lines and equipotential surfaces, with the (electrical) spacing between the equipotentials equal to some constant difference ΔV such as 1 V or 100 V. Let Δs represent the perpendicular distance between two equipotentials. Since the potential difference is the work per unit charge, and the electric field is the force per unit charge,

$$\Delta V = E \, \Delta s, \qquad \text{or} \qquad \Delta s = \frac{\Delta V}{E}.$$

That is, the greater the electric field E, the smaller the perpendicular distance Δs between the equipotentials. The equipotentials are therefore crowded close together in a strong field and are more widely separated in a weak field. Another way of saying the same thing is to write

$$E = \frac{\Delta V}{\Delta s}, \tag{26–18}$$

which expresses the magnitude of the electric field in terms of the difference of potential per unit distance in a direction perpendicular to the equipotentials. The ratio $\Delta V/\Delta s$ is also called the *potential gradient*, just as the ratio of the temperature difference to the distance is called the temperature gradient.

Clearly, the units of potential gradient (volt per meter) must be the same as those of electric field (newtons per coulomb). This may be verified by noting that potential is potential energy per unit charge and that $1 \text{ V} = 1 \text{ J} \cdot \text{C}^{-1}$. Then

$$1 \text{ V} \cdot \text{m}^{-1} = 1 \text{ J} \cdot \text{C}^{-1} \cdot \text{m}^{-1} = 1 \text{ N} \cdot \text{m} \cdot \text{C}^{-1} \cdot \text{m}^{-1}$$
$$= 1 \text{ N} \cdot \text{C}^{-1}.$$

26–6

THE MILLIKAN OIL-DROP EXPERIMENT

We can now describe one of the classical physical experiments of all time. In a brilliant series of investigations carried out at the University of Chicago in the period 1909–1913, Robert Andrews Millikan not only demonstrated conclusively the discrete nature of electric charge but actually measured the charge of an individual electron.

Millikan's apparatus is shown schematically in Fig. 26–10a. Two parallel horizontal metal plates, A and B, are insulated from each other and separated by a few millimeters. Oil is sprayed in fine droplets from an atomizer above the upper plate, and a few of the droplets are allowed to fall through a small hole in this plate. A beam of light is directed horizontally between the plates, and a telescope is set up with its axis at right angles to the light beam. The oil drops, illuminated by the light beam and viewed through the telescope, appear like tiny bright stars, falling slowly with a terminal velocity determined by their weight and by the viscous air-resistance force opposing their motion.

It is found that some of the oil droplets are electrically charged, presumably because of frictional effects. Charges can also be given the drops if the air in the apparatus is ionized by x-rays or a bit of radioactive material. Some of the electrons or ions then collide with the drops and stick to them. The drops are usually negatively charged, but occasionally one with a positive charge is found.

The simplest method, in principle, for measuring the charge on a drop is as follows. Suppose a drop has a negative charge and the plates are maintained at a potential difference such that a downward electric field of intensity E ($=V_{AB}/l$) is set up between them. The forces on the drop are then its weight mg and the upward force of qE. By adjusting the field E, qE can be made just equal to mg, so that the drop remains at rest, as indicated in Fig. 26–10b. Under these circumstances,

$$q = \frac{mg}{E}.$$

The mass of the drop equals the product of its density ρ and its volume, $4\pi r^3/3$, and $E = V_{AB}/l$, so

$$q = \frac{4\pi}{3} \frac{\rho r^3 g l}{V_{AB}}. \tag{26–19}$$

All the quantities on the right are readily measurable with the exception of the drop radius r, which is of the order of 10^{-5} cm and is much too small to be measured directly. It can be calculated, however, by cutting off the electric field and measuring the terminal velocity v_T of the

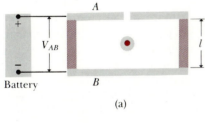

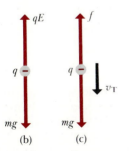

26–10 (a) Schematic diagram of Millikan apparatus. (b) Forces on a drop at rest. (c) Forces on a drop falling with its terminal velocity v_T.

drop as it falls through a known distance d defined by reference lines in the ocular of the telescope.

The terminal velocity is that at which the weight mg is just balanced by the viscous force f. The viscous force on a sphere of radius r, moving with a velocity v through a fluid of viscosity η, is given by Stokes' law, discussed in Section 13–7:

$$f = 6\pi\eta rv.$$

If Stokes' law applies, and the drop is falling with its terminal velocity v_T,

$$mv = f,$$

$$\frac{4}{3}\pi r^3 \rho g = 6\pi\eta r v_T,$$

and

$$r = 3\sqrt{\eta v_T / 2\rho g}.$$

When this expression for r is inserted in Eq. (26–19), we have

$$q = 18\pi \frac{l}{V_{AB}} \sqrt{\frac{\eta^3 v_T{}^3}{2\rho g}}, \tag{26–20}$$

which expresses the charge q in terms of measurable quantities.

In actual practice this procedure is modified somewhat. To correct for the buoyant force of the air through which the drop falls, the density ρ of the oil should be replaced by $(\rho - \rho_g)$, where ρ_g is the density of air. A correction to Stokes' law is also required, because air is not a continuous fluid but a collection of molecules separated by distances that are of the same order of magnitude as the dimensions of the drops.

Millikan and his co-workers measured the charges of thousands of drops and found that, within the limits of their experimental error, every drop had a charge equal to some small integer multiple of a basic charge e. That is, drops were observed with charges of e, $2e$, $3e$, etc., but never with such values as $0.76e$ or $2.49e$. The evidence is conclusive that electric charge is not something that can be divided indefinitely, but that it exists in nature only in units of magnitude e. When a drop is observed with charge e, we conclude it has acquired one extra electron; if its charge is $2e$, it has two extra electrons, and so on.

As stated in Section 24–5, the best experimental value of the charge e is

$$e = 1.602192 \times 10^{-19} \text{ C} \approx 1.60 \times 10^{-19} \text{ C}.$$

The quark theory of fundamental particle structure, discussed in Section 44–12, incorporates particles called *quarks* having fractional charges $\pm e/3$ and $\pm 2e/3$. There are conflicting views among physicists as to whether properties (such as charge) of an individual quark should be directly observable. In 1979 the American physicist William Fairbank reported an experiment in which fractionally charged particles were observed. His results have not been duplicated in other laboratories, and their significance is still uncertain.

26–7

THE ELECTRONVOLT

The change in potential energy of a particle having a charge q, when it moves from a point where the potential is V_a to a point where the potential is V_b, is

$$\Delta U = q(V_a - V_b) = qV_{ab}.$$

In particular, if the charge q equals the electronic charge $e = 1.60 \times 10^{-19}$ C, and the potential difference $V_{ab} = 1$ V, the change in energy is

$$\Delta U = (1.60 \times 10^{-19} \text{ C})(1 \text{ V}) = 1.60 \times 10^{-19} \text{ J}.$$

This quantity of energy is called 1 *electronvolt* (1 eV):

$$1 \text{ eV} = 1.60 \times 10^{-19} \text{ J}.$$

Other commonly used units are

$$1 \text{ keV} = 10^3 \text{ eV},$$

$$1 \text{ MeV} = 10^6 \text{ eV},$$

$$1 \text{ GeV} = 10^9 \text{ eV}.$$

The electronvolt is a convenient energy unit when one is dealing with the motions of electrons and ions in electric fields, because the charge in potential energy between two points on the path of a particle having a charge e, when expressed in electronvolts, is *numerically* equal to the potential difference between the points, in volts. If the charge is some multiple of e, say Ne, the change in potential energy in electronvolts is, numerically, N times the potential difference in volts. For example, if a particle having a charge $2e$ moves between two points for which the potential difference is 1000 V, the change in its potential energy is

$$\Delta U = qV_{ab} = (2)(1.6 \times 10^{-19} \text{ C})(10^3 \text{ V})$$
$$= 3.2 \times 10^{-6} \text{ J}$$
$$= 2000 \text{ eV}.$$

Although the electronvolt was defined above in terms of *potential* energy, energy of *any* form, such as the kinetic energy of a moving particle, can be expressed in terms of the electronvolt. Thus one may speak of a "one-million-volt electron," meaning an electron having a kinetic energy of one million electronvolts (1 MeV).

One of the principles of the special theory of relativity, to be developed in Chapter 40, is that the mass m of a particle is equivalent to a quantity of energy mc^2, where c is the speed of light. The rest mass of an electron is 9.108×10^{-31} kg, and the energy equivalent to this is

$$E_0 = mc^2 = (9.108 \times 10^{-31} \text{ kg})(3 \times 10^8 \text{ m} \cdot \text{s}^{-1})^2$$
$$= 82 \times 10^{-15} \text{ J}.$$

Since $1 \text{ eV} = 1.60 \times 10^{-19}$ J, this is equivalent to

$$E_0 = 511{,}000 \text{ eV} = 0.511 \text{ MeV}.$$

When the kinetic energy of a particle becomes of comparable magnitude to its rest energy, Newton's laws of motion are not strictly valid and must be replaced by the more general relations of relativistic mechanics, to be developed in detail in Chapter 40. For example, an electron accelerated through a potential difference of 500 kV acquires a kinetic energy approximately *equal* to its rest energy, and a correct analysis of the motion of the particle requires the use of relativistic mechanics.

*26–8

THE CATHODE-RAY TUBE

Figure 26–11 is a schematic diagram of the elements of a cathode-ray tube. Such tubes are found in oscilloscopes and in computer terminal CRT displays, and the principle of the TV picture tube is very similar. A cathode-ray tube uses an electron beam that, before its basic nature was well understood, was called a *cathode-ray* beam. Cathode rays are now known to be electrons.

The interior of the tube is highly evacuated. The *cathode* at the left end is raised to a high temperature by the *heater,* and electrons evaporate from its surface. The *accelerating anode,* which has a small hole at its center, is maintained at a high positive potential V_1 relative to the cathode, so that there is an electric field, directed from right to left between the accelerating anode and cathode. This field is confined to the cathode–anode region, and electrons passing through the hole in the anode travel with constant horizontal velocity from the anode to the *fluorescent screen.*

The function of the *control grid* is to regulate the number of electrons that reach the anode (and hence the brightness of the spot on the screen). The *focusing anode* ensures that electrons leaving the cathode in slightly different directions are focused down to a narrow beam and all arrive at the same spot on the screen. These two electrodes need not be considered in the following analysis. The complete assembly of cathode, control grid, focusing anode, and accelerating electrode is referred to as an *electron gun.*

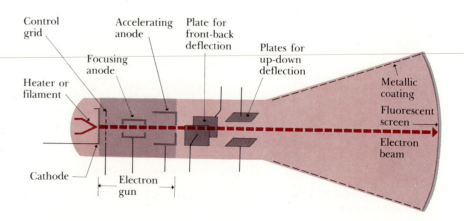

26–11 Basic elements of a cathode-ray tube.

The accelerated electrons pass between two pairs of *deflecting plates*. An electric field between the first pair of plates deflects them to the front or back, and a field between the second pair deflects them up or down. In the absence of such fields the electrons travel in a straight line from the hole in the accelerating anode to the *fluorescent screen* and produce a bright spot on the screen where they strike it.

We first calculate the speed v imparted to the electrons by the electron gun, just as in Example 4 of Section 26-2. We assume the electrons leave the cathode with zero initial speed; in fact, the electrons do have some small initial speed when they evaporate from the cathode, but it is very small compared to the final speed and can be neglected. As in the example, we find

$$v = \sqrt{\frac{2eV_1}{m}}. \tag{26-21}$$

As a numerical example, if $V_1 = 2000$ V,

$$v = \sqrt{\frac{2(1.6 \times 10^{-19}\,\text{C})(2 \times 10^3\,\text{V})}{9.11 \times 10^{-31}\,\text{kg}}}$$

$$= 2.65 \times 10^7\ \text{m} \cdot \text{s}^{-1}.$$

We note that the kinetic energy of an electron at the anode depends only on the *potential difference* between anode and cathode, and not at all on the details of the fields within the electron gun or on the shape of the electron trajectory within the gun.

If there is no electric field between the plates for front–back deflection, the electrons enter the region between the other plates with a speed equal to v and represented by v_x in Fig. 26–12. If there is a potential difference V_2 between the plates, and the upper plate is positive, a downward electric field of magnitude $E = V_2/l$ is set up between the plates. A constant upward force eE then acts on the electrons, and their upward acceleration is

$$a_y = \frac{eE}{m} = \frac{eV_2}{ml}. \tag{26-22}$$

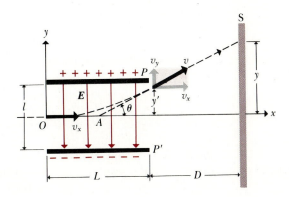

26–12 Electrostatic deflection of electron beam in cathode-ray tube.

The *horizontal* component of velocity v_x remains constant, so the time required for the electrons to travel the length L of the plates is

$$t = \frac{L}{v_x}. \tag{26-23}$$

In this time, they acquire an upward velocity component given by

$$v_y = a_y t, \tag{26-24}$$

and are displaced upward by an amount

$$y' = \tfrac{1}{2} a_y t^2.$$

On emerging from the deflecting field, their velocity v makes an angle θ with the x-axis, where

$$\tan \theta = \frac{v_y}{v_x};$$

and from this point on they travel in a straight line to the screen. It is not difficult to show that this straight line, if projected backward, intersects the x-axis at a point A that is midway between the ends of the plates. Then if y is the vertical coordinate of the point of impact with screen S,

$$\tan \theta = \frac{y}{D + (L/2)}.$$

When this is combined with Eqs. (26-22), (26-23), and (26-24), we obtain

$$y = \left[\frac{L}{2l} \left(D + \frac{L}{2} \right) \right] \frac{V_2}{V_1}. \tag{26-25}$$

The term in brackets is a purely geometrical factor. If the accelerating voltage V_1 is held constant, *the deflection y is proportional to the deflecting voltage V_2.*

If a field is also set up between the *horizontal* deflecting plates, the beam is also deflected in the horizontal direction, perpendicular to the plane of Fig. 26–12. The coordinates of the luminous spot on the screen are then proportional, respectively, to the horizontal and vertical deflecting voltages. This is the principle of the *cathode-ray oscilloscope.* If the horizontal deflection voltage sweeps the beam from left to right at a uniform rate, the beam traces out a graph of the vertical voltage as a function of time. Oscilloscopes are extremely useful laboratory instruments in many areas of pure and applied science.

The picture tube in a television set is similar, but the beam is deflected by magnetic fields (to be discussed in later chapters) rather than electric fields. The electron beam traces out the area of the picture 30 times per second in an array of 525 horizontal lines, as the intensity of the beam is varied to make bright and dark areas on the screen. Computer terminal displays operate on the same principle, using a magnetically deflected electron beam to trace out images on a fluorescent screen. The accelerating voltage in TV picture tubes (V_1 in the above discussion) is typically of the order of 20 kV.

QUESTIONS

26-1 Are there cases in electrostatics where a conducting surface is *not* an equipotential surface? If so, give an example.

26-2 If the electric potential at a single point is known, can the electric field at that point be determined?

26-3 If two points are at the same potential, is the electric field necessarily zero everywhere between them?

26-4 If the electric field is zero throughout a certain region of space, is the potential also zero in this region? If not, what *can* be said about the potential?

26-5 A student said: "Since electrical potential is always proportional to potential energy, why bother with the concept of potential at all?" How would you respond?

26-6 Is potential gradient a scalar or vector quantity?

26-7 A conducting sphere is to be charged by bringing in positive charge, a little at a time, until the total charge is Q. The total work required for this process is alleged to be proportional to Q^2. Is this correct? Why or why not?

26-8 The potential (relative to a point at infinity) midway between two charges of equal magnitude and opposite sign is zero. Can you think of a way to bring a test charge from infinity to this midpoint in such a way that no work is done in any part of the displacement?

26-9 A high-voltage dc power line falls on a car, so the entire metal body of the car is at a potential of 10,000 V with respect to the ground. What happens to the occupants

a) when they are sitting in the car, and

b) when they step out of the car?

26-10 In electronics it is customary to define the potential of ground (thinking of the earth as a large conductor) as zero. Is this consistent with the fact that the earth has a net electric charge that is not zero? (Cf. Prob. 25-26.)

26-11 A positive point charge is placed near a very large conducting plane. A professor of physics asserted that the field caused by this configuration is the same as would be obtained by removing the plane and placing a negative point charge of equal magnitude in the mirror-image position behind the initial position of the plane. Is this correct?

26-12 It is easy to produce a potential of several thousand volts on your body by scuffing your shoes across a nylon carpet, yet contact with a power line of comparable voltage would probably be fatal. What's the difference?

PROBLEMS

26-1 Two large parallel metal sheets carrying equal and opposite electric charges are separated by a distance of 0.05 m. The electric field between them is approximately uniform and has magnitude 600 $N \cdot C^{-1}$. What is the potential difference between the plates? Which plate is at higher potential?

26-2 Two large parallel metal plates carry opposite charges. They are separated by 0.1 m, and the potential difference between them is 500 V.

a) What is the magnitude of the electric field, if it is uniform, in the region between the plates?

b) Compute the work done by this field on a charge of 2.0×10^{-9} C as it moves from the higher-potential plate to the lower.

c) Compare the result of (b) to the change of potential energy of the same charge, computed from the electrical potential.

26-3 A total electric charge of 4×10^{-9} C is distributed uniformly over the surface of a sphere of radius 0.20 m. If the potential is zero at a point at infinity, what is the value of the potential

a) at a point on the surface of the sphere, and

b) at a point inside the sphere, 0.1 m from the center?

26-4 A particle of charge $+3 \times 10^{-9}$ C is in a uniform electric field directed to the left. It is released from rest and moves a distance of 5 cm, after which its kinetic energy is found to be $+4.5 \times 10^{-5}$ J.

a) What work was done by the electrical force?

b) What is the magnitude of the electric field?

c) What is the potential of the starting point with respect to the endpoint?

26-5 In Problem 26-4, suppose that another force in addition to the electrical force acts on the particle, so that when it is released from rest it moves to the right. After it has moved 5 cm, the additional force has done 9×10^{-5} J of work and the particle has 4.5×10^{-5} J of kinetic energy.

a) What work was done by the electrical force?

b) What is the magnitude of the electric field?

c) What is the potential of the starting point with respect to the endpoint?

26-6 A charge of 2.5×10^{-8} C is placed in an upwardly directed uniform electric field having magnitude 5×10^{4} $N \cdot C^{-1}$. What is the work of the electrical force when the charge is moved

a) 45 cm to the right?

b) 80 cm downward?

c) 260 cm at an angle of 45° upward from the horizontal?

26–7

a) Show that $1 \text{ N} \cdot \text{C}^{-1} = 1 \text{ V} \cdot \text{m}^{-1}$.

b) A potential difference of 2000 V is established between parallel plates in air. If the air becomes electrically conducting when the electric intensity exceeds $3 \times 10^6 \text{ N} \cdot \text{C}^{-1}$, what is the minimum separation of the plates?

26–8 A small sphere of mass 0.2 g hangs by a thread between two parallel vertical plates 5 cm apart. The charge on the sphere is $6 \times 10^{-9} \text{ C}$. What potential difference between the plates will cause the thread to assume an angle of 30° with the vertical?

26–9 Two point charges $q_1 = +40 \times 10^{-9} \text{ C}$ and $q_2 = -30 \times 10^{-9} \text{ C}$ are 10 cm apart. Point A is midway between them, point B is 8 cm from q_1 and 6 cm from q_2. Find

a) the potential at point A,

b) the potential at point B,

c) the work done by the field of these charges when a charge of $25 \times 10^{-9} \text{ C}$ moves from point B to point A.

26–10 Three equal point charges of $3 \times 10^{-7} \text{ C}$ are placed at the corners of an equilateral triangle whose side is one meter. What is the potential energy of the system? Take as zero potential energy the energy of the three charges when they are infinitely far apart.

26–11 The potential at a certain distance from a point charge is 600 V, and the electric field is $200 \text{ N} \cdot \text{C}^{-1}$.

a) What is the distance to the point charge?

b) What is the magnitude of the charge?

26–12 Two point charges whose magnitudes are $+20 \times 10^{-9} \text{ C}$ and $-12 \times 10^{-9} \text{ C}$ are separated by a distance of 5 cm. An electron is released from rest between the two charges, 1 cm from the negative charge, and moves along the line connecting the two charges. What is its velocity when it is 1 cm from the positive charge?

26–13 Two positive point charges, each of magnitude q, are fixed on the y-axis at the points $y = +a$ and $y = -a$.

a) Draw a diagram showing the positions of the charges.

b) What is the potential V_0 at the origin?

c) Show that the potential at any point on the x-axis is

$$V = \frac{1}{4\pi\epsilon_0} \frac{2q}{\sqrt{a^2 + x^2}}.$$

d) Sketch a graph of the potential on the x-axis as a function of x over the range from $x = +4a$ to $x = -4a$.

e) At what value of x is the potential one-half that at the origin?

26–14 Consider the same distribution of charges as in Problem 26–13.

a) Sketch a graph of the potential on the y-axis as a function of y, over the range from $y = +4a$ to $y = -4a$.

b) Discuss the physical meaning of the graph at the points $+a$ and $-a$.

c) At what point or points on the y-axis is the potential equal to that at the origin?

d) At what points on the y-axis is the potential equal to half its value at the origin?

26–15 Consider the same charge distribution as in Problem 26–13.

a) Suppose a positively charged particle of charge q' and mass m is placed precisely at the origin and released from rest. What happens?

b) What will happen if the charge in part (a) is displaced slightly in the direction of the y-axis?

c) What will happen if it is displaced slightly in the direction of the x-axis?

26–16 Again consider the charge distribution in Problem 26–13. Suppose a positively charged particle of charge q' and mass m is displaced slightly from the origin in the direction of the x-axis.

a) What is its speed at infinity?

b) Sketch a graph of the velocity of the particle as a function of x.

c) If the particle is projected toward the left along the x-axis from a point at a large distance to the right of the origin, with a velocity half that acquired in part (a), at what distance from the origin will it come to rest?

d) If a negatively charged particle were released from rest on the x-axis, at a very large distance to the left of the origin, what would be its velocity as it passed the origin?

26–17 A positive charge $+q$ is located at the point $x = -a$, $y = -a$, and an equal negative charge $-q$ is located at the point $x = +a$, $y = -a$.

a) Draw a diagram showing the positions of the charges.

b) What is the potential at the origin?

c) What is the expression for the potential at a point on the x-axis, as a function of x?

d) Sketch a graph of the potential as a function of x, in the range from $x = +4a$ to $x = -4a$. Plot positive potentials upward, negative potentials downward.

26–18 A potential difference of 1600 V is established between two parallel plates 4 cm apart. An electron is released from the negative plate at the same instant that a proton is released from the positive plate.

a) How far from the positive plate will they pass each other?

b) How do their velocities compare when they strike the opposite plates?

c) How do their energies compare when they strike the opposite plates?

26–19 Consider the same charge distribution as in Problem 26–13.

a) Constuct a graph of the potential energy of a positive point charge on the x-axis, as a function of x.

b) Construct a graph of the potential energy of a negative point charge on the axis, as a function of x.

c) What is the potential gradient at the origin, in the direction of the x-axis?

26–20 In the Bohr model of the hydrogen atom, a single electron revolves around a single proton in a circle of radius R.

a) By equating the electrical force to the electron mass times its acceleration, derive an expression for the electron's speed.

b) Obtain an expression for the electron's kinetic energy, and show that its magnitude is just half that of the electrical potential energy.

c) Obtain an expression for the total energy, and evaluate it using $R = 0.528 \times 10^{-10}$ m.

26–21 A vacuum diode consists of a cylindrical cathode of radius 0.05 cm, mounted coaxially within a cylindrical anode 0.45 cm in radius. The potential of the anode is 300 V above that of the cathode. An electron leaves the surface of the cathode with zero initial speed. Find its speed when it strikes the anode.

26–22 A vacuum triode may be idealized as follows. A plane surface (the cathode) emits electrons with negligible initial velocities. Parallel to the cathode and 3 mm away from it is an open grid of fine wire at a potential of 18 V above the cathode. A second plane surface (the anode) is 12 mm beyond the grid and is at a potential of 15 V above the cathode. Assume that the plane of the grid is an equipotential surface, and that the potential gradients between cathode and grid, and between grid and anode, are uniform. Assume also that the structure of the grid is sufficiently open for electrons to pass through it freely.

a) Draw a diagram of potential vs. distance, along a line from cathode to anode.

b) With what velocity will electrons strike the anode?

26–23 A positively charged ring of radius R is placed with its plane perpendicular to the x-axis and with its center at the origin.

a) Derive an expression for the potential V at points on the x-axis, as a function of x. Construct a graph of this function.

b) Construct, in the same diagram, a graph of the magnitude of the electric field E.

c) How is the second graph related geometrically to the first?

26–24 Refer to Problem 24–23.

a) Calculate the potential at the origin, and at the point (3 cm, 0) due to the first two point charges.

b) Calculate the work the electric field would do on the third charge if it moved from the origin to the point (3 cm, 0).

26–25 Refer to Problem 24–24.

a) Calculate the potential at the point (3 cm, 0) and at the point (3 cm, 4 cm) due to the first two charges.

b) If the third charge moves from the point (3 m, 0) to the point (3 cm, 4 cm), calculate the work done on it by the field of the first two charges. Comment on the *sign* of this work; is your result reasonable?

26–26 In an apparatus for measuring the electronic charge e by Millikan's method, an electric intensity of 6.34×10^4 V·m^{-1} is required to maintain a certain charged oil drop at rest. If the plates are 1.5 cm apart, what potential difference between them is required?

26–27 An oil droplet of mass 3×10^{-11} g and of radius 2×10^{-4} cm carries 10 excess electrons. What is its terminal velocity

a) when falling in a region in which there is no electric field?

b) When falling in an electric field of magnitude 3×10^5 N·C^{-1} directed downward?

The viscosity of air is 180×10^{-7} N·s·m^{-2}. (Neglect the buoyant force of the air.)

26–28 A charged oil drop, in a Millikan oil-drop apparatus, is observed to fall through a distance of 1 mm in a time of 27.4 s, in the absence of any external field. The same drop can be held stationary in a field of 2.37×10^4 N·C^{-1}. How many excess electrons has the drop acquired? The viscosity of air is 180×10^{-7} N·s·m^{-2}. The density of oil is 824 kg·m^{-3}, and the density of air is 1.29 kg·m^{-3}.

26–29 Find the energy equivalent of the rest mass of the proton; express your result in MeV.

26–30 Find the potential energy of the interaction of two protons at a distance of 10^{-15} m, typical of the dimensions of atomic nuclei. Express your result in MeV.

26–31 An alpha particle with kinetic energy 10 MeV makes a head-on collision with a gold nucleus at rest. What is the distance of closest approach of the two particles?

26–32

a) Prove that when a particle of constant mass and charge is accelerated from rest in an electric field, its final velocity is proportional to the square root of the potential difference through which it is accelerated.

b) Find the magnitude of the proportionality constant if the particle is an electron, the velocity is in meters per second, and the potential difference is in volts.

c) What is the final velocity of an electron accelerated through a potential difference of 1136 V if it has an initial velocity of 10^7 m·s^{-1}?

26–33

a) What is the maximum potential difference through which an electron can be accelerated if its kinetic energy is not to exceed 1% of the rest energy?

b) What is the speed of such an electron, expressed as a fraction of the speed of light, c?

c) Make the same calculations for a *proton*.

26–34 The electric field in the region between the deflecting plates of a certain cathode-ray oscilloscope is 30,000 N·C^{-1}.

a) What is the force on an electron in this region?

b) What is the acceleration of an electron when acted on by this force?

26–35 In Fig. 26–13, an electron is projected along the axis midway between the plates of a cathode-ray tube with an initial velocity of 2×10^7 m·s^{-1}. The uniform electric field between the plates has an intensity of 20,000 N·C^{-1} and is upward.

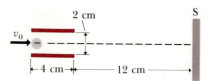

Figure 26–13

a) How far below the axis has the electron moved when it reaches the end of the plates?

b) At what angle with the axis is it moving as it leaves the plates?

c) How far below the axis will it strike the fluorescent screen S?

26–36 An electron with kinetic energy 100 MeV collides head-on with a gold nucleus at rest. Assuming that the gold nucleus can be treated as a uniform distribution of charge through a sphere of radius 7×10^{-15} m and that the electron can penetrate into the nucleus, what is its kinetic energy when it reaches the center of the nucleus?

26–37 Some cell walls in the human body have a double layer of surface charge, with a layer of negative charge inside and a layer of positive charge of equal magnitude on the outside. If the surface charge densities are $\pm 0.5 \times 10^{-3}$ C·m^{-2} and the cell wall is 5×10^{-9} m thick, find

a) the electric field magnitude in the wall, between the two charge layers;

b) the potential difference between inside and outside the cell. Which is at higher potential?

CAPACITANCE
AND
DIELECTRICS

A capacitor is a device consisting of two conductors separated by vacuum or an insulating material. When charges of equal magnitude and opposite sign are placed on the conductors, an electric field is established in the region between them, and there is a potential difference between the conductors. For a given capacitor the ratio of charge to potential difference is a constant, called the *capacitance*. Placing the charges on the conductors requires an input of energy, and this energy is stored in the capacitor. When the space between conductors contains an insulating material rather than vacuum, the capacitance is increased. This change can be understood on the basis of redistribution of charge within the material. Capacitors have many practical uses, including electronic circuits, power distribution and conversion systems, and others.

27–1

CAPACITORS

Any two conductors separated by an insulator are said to form a *capacitor*. In most cases of practical interest the conductors usually have charges of equal magnitude and opposite sign, so that the *net* charge on the capacitor as a whole is zero. The electric field in the region between the conductors is then proportional to the magnitude of this charge, and it follows that the *potential difference* V_{ab} between the conductors is also proportional to the charge magnitude Q.

The *capacitance* C of a capacitor is defined as the ratio of the magnitude of the charge Q on *either* conductor to the magnitude of the potential difference V_{ab} between the conductors:

$$C = \frac{Q}{V_{ab}}. \tag{27-1}$$

It follows from its definition that the unit of capacitance is one *coulomb per volt* (1 C · V^{-1}). A capacitance of one coulomb per volt is called one *farad* (1 F) in honor of Michael Faraday. A capacitor is represented by the symbol

When one speaks of a capacitor as having charge Q, what is really meant is that the conductor at higher potential has a charge Q, and the conductor at lower potential a charge $-Q$ (assuming Q is a positive quantity). This interpretation should be kept in mind in the following discussion and examples.

Capacitors find many applications in electrical circuits. Capacitors are used for tuning radio circuits and for "smoothing" the rectified current delivered by a power supply. A capacitor is used to eliminate sparking when a circuit containing inductance is suddenly opened. The ignition systems of some internal-combustion engines contain capacitors to eliminate sparking of the "points" when they open and close. The efficiency of alternating-current power transmission can often be increased by the use of large capacitors.

27–2

THE PARALLEL-PLATE CAPACITOR

The most common type of capacitor consists in principle of two conducting plates parallel to each other and separated by a distance that is small compared with the linear dimensions of the plates (see Fig. 27–1). Practically the entire field of such a capacitor is localized in the region between the plates, as shown. There is a slight "fringing" of the field at the edges, but if the plates are sufficiently close, the fringing may be neglected. The field between the plates is then uniform, and the charges on the plates are uniformly distributed over their opposing surfaces. This arrangement is known as a *parallel-plate capacitor*.

The electric field magnitude between a pair of closely spaced parallel plates in vacuum, as discussed in Section 25–4, is

$$E = \frac{\sigma}{\epsilon_0} = \frac{Q}{\epsilon_0 A},$$

where σ is the magnitude of surface charge density on either plate, A is the area of each plate, and Q is the magnitude of total charge on each plate. Since the electric field (potential gradient) between the plates is uniform, the potential difference between the plates is

$$V_{ab} = Ed = \frac{1}{\epsilon_0} \frac{Qd}{A},$$

where d is the separation of the plates. Hence the capacitance of a parallel-plate capacitor in vacuum is

$$C = \frac{Q}{V_{ab}} = \epsilon_0 \frac{A}{d}. \qquad (27\text{–}2)$$

Since ϵ_0, A, and d are constants for a given capacitor, the capacitance is a constant independent of the charge on the capacitor, and is directly

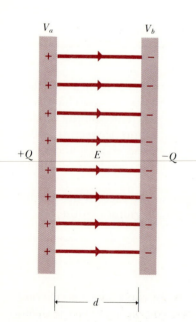

27–1   Parallel-plate capacitor.

proportional to the area of the plates and inversely proportional to their separation. If SI units are used, A is expressed in square meters and d in meters. The capacitance C is then in farads.

As an example, let us compute the area of the plates of a 1-F parallel-plate capacitor if the separation of the plates is 1 mm and the plates are in vacuum:

$$C = \epsilon_0 \frac{A}{d},$$

$$A = \frac{Cd}{\epsilon_0} = \frac{(1 \text{ F})(10^{-3} \text{ m})}{8.85 \times 10^{-12} \text{ C}^2 \cdot \text{N}^{-1} \cdot \text{m}^{-2}}$$

$$= 1.13 \times 10^8 \text{ m}^2.$$

This corresponds to a square 10,600 m, or 34,800 ft, or about $6\frac{1}{2}$ miles on a side!

Since the farad is such a large unit of capacitance, units of more convenient size are the *microfarad* (1 μF = 10^{-6} F), and the *picofarad* (1 pF = 10^{-12} F). For example, a common radio set contains in its power supply several capacitors whose capacitances are of the order of ten or more microfarads, while the capacitances of the tuning capacitors are of the order of ten to one hundred picofarads.

EXAMPLE The plates of a parallel-plate capacitor are 5 mm apart and 2 m^2 in area. The plates are in vacuum. A potential difference of 10,000 V is applied across the capacitor. Compute (a) the capacitance, (b) the charge on each plate, and (c) the electric intensity in the space between them.

Solution

a) From Eq. (27–2),

$$C = \epsilon_0 \frac{A}{d} = \frac{(8.85 \times 10^{-12} \text{ C}^2 \cdot \text{N}^{-1} \cdot \text{m}^{-2})(2 \text{ m}^2)}{5 \times 10^{-3} \text{ m}}$$

$$= 3.54 \times 10^{-9} \text{ C}^2 \cdot \text{N}^{-1} \cdot \text{m}^{-1}.$$

But

$$1 \text{ C}^2 \cdot \text{N}^{-1} \cdot \text{m}^{-1} = 1 \text{ C}^2 \cdot \text{J}^{-1} = 1 \text{ C} \left(\frac{\text{J}}{\text{C}}\right)^{-1}$$

$$= 1 \text{ C} \cdot \text{V}^{-1} = 1 \text{ F},$$

so

$$C = 3.54 \times 10^{-9} \text{ F} = 0.00354 \ \mu\text{F}.$$

b) The charge on the capacitor is

$$Q = CV_{ab} = (3.54 \times 10^{-9} \text{ C} \cdot \text{V}^{-1})(10^4 \text{ V})$$

$$= 3.54 \times 10^{-5} \text{ C} = 35.4 \ \mu\text{C}.$$

That is, the plate at higher potential has charge $+35.4 \ \mu$C, and the other plate has charge $-35.4 \ \mu$C.

c) The electric field is

$$E = \frac{\sigma}{\epsilon_0} = \frac{Q}{\epsilon_0 A} = \frac{3.54 \times 10^{-5}\ \text{C}}{(8.85 \times 10^{-12}\ \text{C}^2 \cdot \text{N}^{-1} \cdot \text{m}^{-2})(2\ \text{m}^2)}$$
$$= 20 \times 10^5\ \text{N} \cdot \text{C}^{-1};$$

or, since the electric field equals the potential gradient,

$$E = \frac{V_{ab}}{d} = \frac{10^4\ \text{V}}{5 \times 10^{-3}\ \text{m}} = 20 \times 10^5\ \text{V} \cdot \text{m}^{-1}.$$

The newton per coulomb and the volt per meter are equivalent units. ◀

27–3

CAPACITORS IN SERIES AND PARALLEL

In Fig. 27–2a, two capacitors are connected in *series* between points a and b, maintained at a constant potential difference V_{ab}. The capacitors are both initially uncharged. In this connection, both capacitors always have the same charge Q. One might ask whether the lower plate of C_1 and the upper plate of C_2 might have charges different from those on the remaining two plates, but in this case the net charge on each capacitor would not be zero, and the resulting electric field in the conductor connecting the two capacitors would cause a current to flow until the total charge on each capacitor is zero. Hence, *in a series connection the magnitude of charge on all plates is the same.*

Referring again to Fig. 27–2a, we have

$$V_{ac} \equiv V_1 = \frac{Q}{C_1}, \qquad V_{cb} \equiv V_2 = \frac{Q}{C_2},$$

$$V_{ab} \equiv V = V_1 + V_2 = Q\left(\frac{1}{C_1} + \frac{1}{C_2}\right),$$

and

$$\frac{V}{Q} = \frac{1}{C_1} + \frac{1}{C_2}. \tag{27–3}$$

The *equivalent* capacitance C of the series combination is defined as the capacitance of a *single* capacitor for which the charge Q is the same as for the combination, when the potential difference V is the same. For such a capacitor, shown in Fig. 27–2b,

$$Q = CV, \qquad \frac{V}{Q} = \frac{1}{C}. \tag{27–4}$$

Hence, from Eqs. (27–3) and (27–4),

$$\frac{1}{C} = \frac{1}{C_1} + \frac{1}{C_2}.$$

Similarly, for any number of capacitors in series,

$$\frac{1}{C} = \frac{1}{C_1} + \frac{1}{C_2} + \frac{1}{C_3} + \cdots. \tag{27–5}$$

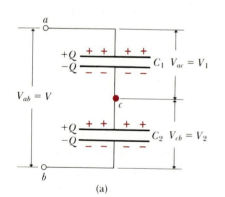

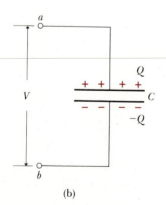

27–2 (a) Two capacitors in series, and (b) their equivalent.

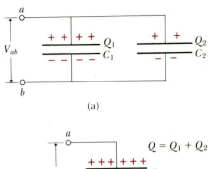

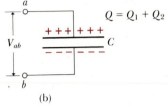

(a)

(b)

27–3 (a) Two capacitors in parallel, and (b) their equivalent.

The reciprocal of the equivalent capacitance equals the sum of the reciprocals of the individual capacitances.

In Fig. 27–3a, two capacitors are connected in *parallel* between points a and b. In this case, the upper plates of the two capacitors are connected together to form an equipotential surface, and the lower plates another. The potential difference is the same for both capacitors and is equal to $V_{ab} = V$. The charges Q_1 and Q_2, not necessarily equal, are

$$Q_1 = C_1 V, \qquad Q_2 = C_2 V.$$

The *total* charge Q supplied by the source was

$$Q = Q_1 + Q_2 = V(C_1 + C_2),$$

and

$$\frac{Q}{V} = C_1 + C_2. \tag{27–6}$$

The *equivalent* capacitance C of the parallel combination is defined as that of a single capacitor, shown in Fig. 27–3b, for which the total charge is the same as in part (a). For this capacitor,

$$\frac{Q}{V} = C,$$

and hence

$$C = C_1 + C_2.$$

In the same way, for any number of capacitors in parallel

$$C = C_1 + C_2 + C_3 + \cdots . \tag{27–7}$$

The equivalent capacitance equals the *sum* of the individual capacitances.

EXAMPLE In Figs. 27–2 and 27–3, let $C_1 = 6\ \mu\text{F}$, $C_2 = 3\ \mu\text{F}$, $V_{ab} = 18$ V.

The equivalent capacitance of the series combination in Fig. 27–2a is given by

$$\frac{1}{C} = \frac{1}{6\ \mu\text{F}} + \frac{1}{3\ \mu\text{F}}, \qquad C = 2\ \mu\text{F}.$$

The charge Q is

$$Q = CV = (2\ \mu\text{F})(18\ \text{V}) = 36\ \mu\text{C}.$$

The potential differences across the capacitors are

$$V_{ac} \equiv V_1 = \frac{Q}{C_1} = \frac{36\ \mu\text{C}}{6\ \mu\text{F}} = 6\ \text{V}, \qquad V_{cb} \equiv V_2 = \frac{Q}{C_2} = \frac{36\ \mu\text{C}}{3\ \mu\text{F}} = 12\ \text{V}.$$

The *larger* potential difference appears across the *smaller* capacitor.

The equivalent capacitance of the parallel combination in Fig. 27–3a is

$$C = C_1 + C_2 = 9\ \mu\text{F}.$$

The charges Q_1 and Q_2 are

$$Q_1 = C_1 V = (6\,\mu\text{F})(18\,\text{V}) = 108\,\mu\text{C},$$

$$Q_2 = C_2 V = (3\,\mu\text{F})(18\,\text{V}) = 54\,\mu\text{C}.$$

◄

27–4

ENERGY OF A CHARGED CAPACITOR

The process of charging a capacitor consists of transferring charge from the plate at lower potential to the plate at higher potential. The charging process therefore requires doing work and adding energy to the system. Imagine the charging process to be carried out by starting with both plates completely uncharged, and then repeatedly removing small positive charges from one plate and transferring them to the other plate. The final charge Q and the final potential difference V are related by

$$Q = CV.$$

Since the potential difference increases proportionally with the charge, the *average* potential difference V_{av} during the charging process is just one half the maximum value, or

$$V_{\text{av}} = \frac{Q}{2C}.$$

This is the *average* work per unit charge, so the *total* work W required is just V_{av} multiplied by the total charge, or

$$W = V_{\text{av}}Q = \frac{Q^2}{2C}.$$

Using the relation $Q = CV$, we may also write

$$W = \frac{Q^2}{2C} = \frac{1}{2}CV^2 = \frac{1}{2}QV. \qquad (27\text{–}8)$$

The work is expressed in joules when Q is in coulombs and V is in volts.

A charged capacitor is the electrical analog of a stretched spring, whose elastic potential energy equals $\frac{1}{2}kx^2$. The charge Q is analogous to the elongation x, and the *reciprocal* of the capacitance, $1/C$, is analogous to the force constant k. The energy supplied to a capacitor in the charging process is stored by the capacitor and is released when the capacitor discharges.

It is often useful to consider the stored energy to be localized in the *electric field* between the capacitor plates. The capacitance of a parallel-plate capacitor in vacuum is

$$C = \epsilon_0 \frac{A}{d}.$$

The electric field fills the space between the plates, of volume Ad, and its magnitude E is given by

$$E = \frac{V}{d}.$$

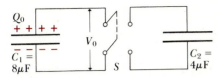

27-4 When the switch S is closed, the charged capacitor C_1 is connected to an uncharged capacitor C_2.

The energy per unit volume, or the *energy density*, denoted by u, is

$$u = \text{Energy density} = \frac{\frac{1}{2}CV^2}{Ad}.$$

Making use of the preceding equations, we can express this as

$$u = \tfrac{1}{2}\epsilon_0 E^2. \tag{27-9}$$

Although this relation has been derived with reference to capacitors, it describes the energy stored in *any* electric field, whether associated with a capacitor or not.

EXAMPLE In Fig. 27-4 capacitor C_1 is initially charged by connecting it to a source of potential difference V_0 (not shown in the figure). Let $C_1 = 8 \ \mu\text{F}$ and $V_0 = 120$ V. The charge Q_0 is

$$Q_0 = C_1 V_0 = 960 \ \mu\text{C},$$

and the energy of the capacitor is

$$\tfrac{1}{2}Q_0 V_0 = \tfrac{1}{2}(960 \times 10^{-6} \ \text{C})(120 \ \text{V}) = 0.0576 \ \text{J}.$$

After the switch S is closed, the positive charge Q_0 is distributed over the upper plates of both capacitors, and the negative charge $-Q_0$ is distributed over the lower plates of both. Let Q_1 and Q_2 represent the magnitudes of the final charges on the respective capacitors. Then

$$Q_1 + Q_2 = Q_0.$$

When the motion of charges has ceased, both upper plates are at the same potential and both lower plates are at the same potential (different from that of the upper plates). The final potential difference between the plates, V, is therefore the same for both capacitors, and

$$Q_1 = C_1 V, \qquad Q_2 = C_2 V.$$

When this is combined with the preceding equation, we find that

$$V = \frac{Q_0}{C_1 + C_2} = \frac{960 \ \mu\text{C}}{12 \ \mu\text{F}} = 80 \ \text{V},$$

$$Q_1 = 640 \ \mu\text{C}, \qquad Q_2 = 320 \ \mu\text{C}.$$

The final energy of the system is

$$\tfrac{1}{2}Q_0 V = \tfrac{1}{2}(960 \times 10^{-6} \ \text{C})(80 \ \text{V}) = 0.0384 \ \text{J}.$$

This is less than the original energy of 0.0576 J; the difference is converted to energy of some other form. If the resistance of the connecting wires is large, most of the energy is converted to heat. If the resistance is small, most of the energy is radiated in the form of electromagnetic waves.

This process is exactly analogous to an inelastic collision of a moving car with a stationary car. In the electrical case, the charge $Q = CV$ is conserved. In the mechanical case, the momentum $p = mv$ is conserved. The electrical energy $\tfrac{1}{2}CV^2$ is *not* conserved, and the mechanical energy $\tfrac{1}{2}mv^2$ is *not* conserved. ◀

*27–5

EFFECT OF A DIELECTRIC

Most capacitors have a solid, nonconducting material or *dielectric* between their plates. A common type of capacitor incorporates strips of metal foil, forming the plates, separated by strips of wax-impregnated paper or plastic sheet such as Mylar, which serves as the dielectric. A sandwich of these materials is rolled up, forming a compact unit that can provide a capacitance of several microfarads in a relatively small volume.

Electrolytic capacitors have as their dielectric an extremely thin layer of nonconducting oxide between a metal plate and a conducting solution. Because of the thinness of the dielectric, electrolytic capacitors of relatively small dimensions may have a capacitance of the order of 100 to 1000 μF. (From Eq. (27–2), the capacitance is inversely proportional to the distance d between plates.)

The function of a solid dielectric between the plates of a capacitor is threefold. First, it solves the mechanical problem of maintaining two large metal sheets at an extremely small separation but without actual contact.

Second, any dielectric material, when subjected to a sufficiently large electric field, experiences *dielectric breakdown,* a partial ionization, which permits conduction through a material that is supposed to insulate. Many insulating materials can tolerate stronger electric fields without breakdown than can air.

Third, the capacitance of a capacitor of given dimensions is *larger* when there is dielectric material between the plates than when the plates are separated only by air or vacuum. This effect can be demonstrated with the aid of a sensitive electrometer, a device that can measure the potential difference between two conductors without permitting any charge to flow from one to the other. In Fig. 27–5a, a capacitor is charged, with magnitude of charge Q on each plate and potential difference V_0. When a sheet of dielectric, such as glass, paraffin, or polystyrene, is inserted between the plates, the potential difference is found to *decrease* to a smaller value V. When the dielectric is removed, the potential difference returns to its original value, showing that the original charges on the plates have not been affected by insertion of the dielectric.

The original capacitance of the capacitor, C_0, was

$$C_0 = \frac{Q}{V_0}.$$

Since Q does not change and V is observed to be less than V_0, it follows that C is *greater* than C_0. The ratio of C to C_0 is called the *dielectric constant* of the material, K:

$$K = \frac{C}{C_0}. \tag{27–10}$$

Since C is always greater than C_0, the dielectric constants of all dielectrics are greater than unity. Some representative values of K are given in Table 27–1. For vacuum, of course, $K = 1$ by definition, and K for air is

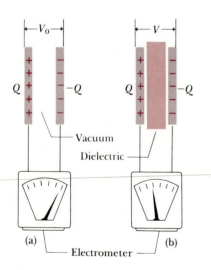

(a) (b)

27–5 Effect of a dielectric between the plates of a parallel-plate capacitor. The electrometer measures potential difference. (a) With a given charge, the potential difference is V_0. (b) With the same charge, the potential difference V is smaller than V_0.

TABLE 27–1	DIELECTRIC CONSTANT K AT 20°C			
Material	K		**Material**	K
Vacuum	1		Strontium titanate	310
Glass	5–10		Titanium dioxide (rutile)	173($\perp$),
Mica	3–6			86($\parallel$)
Mylar	3.1		Water	80.4
Neoprene	6.70		Glycerin	42.5
Plexiglas	3.40		Liquid ammonia	
Polyethylene	2.25		($-78°C$)	25
Polyvinyl chloride	3.18		Benzene	2.284
Teflon	2.1		Air (100 atm)	1.00059
Germanium	16		Air (100 atm)	1.0548

so nearly equal to 1 that for most purposes an air capacitor is equivalent to one in vacuum; the original measurement of V_0 in Fig. 27–5a could have been made with the plates in air instead of in vacuum.

With vacuum (or air) between the plates, the electric field E_0 in the region between the plates of a parallel-plate capacitor is

$$E_0 = \frac{V_0}{d} = \frac{\sigma}{\epsilon_0}.$$

The observed reduction in potential difference, when a dielectric is inserted between the plates, implies a reduction in the electric field, which in turn implies a reduction in the charge per unit area. Since no charge has leaked off the plates, such a reduction could be caused only by induced charges of opposite sign appearing on the two surfaces of the *dielectric*. That is, the dielectric surface adjacent to the positive plate must have an *induced negative charge* and that adjacent to the negative plate an *induced positive charge of equal magnitude*, as shown in Fig. 27–6. These induced surface charges are a result of redistribution of charge within the dielectric material, a phenomenon called *polarization*.

If σ_i is the magnitude of induced charge per unit area on the surfaces of the dielectric, then the *net* surface charge on each side that contributes to the electric field within the dielectric has magnitude $(\sigma - \sigma_i)$, and the electric field in the dielectric is

$$E = \frac{V}{d} = \frac{\sigma - \sigma_i}{\epsilon_0}. \tag{27–11}$$

But

$$K = \frac{C}{C_0} = \frac{Q/V}{Q/V_0} = \frac{V_0}{V} = \frac{E_0}{E} = \frac{\sigma}{\sigma - \sigma_i}; \tag{27–12}$$

and therefore

$$\sigma - \sigma_i = \frac{\sigma}{K}. \tag{27–13}$$

Substituting Eq. (27–13) into Eq. (27–11), we obtain

$$E = \frac{\sigma}{K\epsilon_0}. \tag{27–14}$$

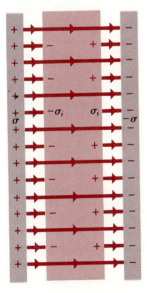

27–6 Induced charges on the faces of a dielectric in an external field.

The product $K\epsilon_0$ is called the *permittivity* of the dielectric and is represented by ϵ:

$$\epsilon = K\epsilon_0. \tag{27-15}$$

The electric field within the dielectric may therefore be written

$$E = \frac{\sigma}{\epsilon}. \tag{27-16}$$

Also,

$$C = KC_0 = K\epsilon_0\frac{A}{d},$$

and the capacitance of a parallel-plate capacitor with a dielectric between its plates is therefore

$$C = \epsilon\frac{A}{d}. \tag{27-17}$$

In empty space, where $K = 1$, $\epsilon = \epsilon_0$, and therefore ϵ_0 may be described as the "permittivity of empty space" or the "permittivity of vacuum." Since K is a pure number, the units of ϵ and ϵ_0 are evidently the same, $C^2 \cdot N^{-1} \cdot m^{-2}$.

EXAMPLE The parallel plates in Fig. 27–6 have an area of 2000 cm² or 2×10^{-1} m², and are 1 cm or 10^{-2} m apart. The original potential difference between them, V_0, is 3000 V, and it decreases to 1000 V when a sheet of dielectric is inserted between the plates. Compute (a) the original capacitance C_0, (b) the charge Q on each plate, (c) the capacitance C after insertion of the dielectric, (d) the dielectric constant K of the dielectric, (e) the permittivity ϵ of the dielectric, (f) the induced charge Q_i on each face of the dielectric, (g) the original electric field E_0 between the plates, and (h) the electric field E after insertion of the dielectric.

Solution

a) $C_0 = \epsilon_0\dfrac{A}{d} = (8.85 \times 10^{-12}\ C^2 \cdot N^{-1} \cdot m^{-2})\dfrac{2 \times 10^{-1}\ m^2}{10^{-2}\ m}$

$= 17.7 \times 10^{-11}\ F = 177\ pF.$

b) $Q = C_0V_0 = (17.7 \times 10^{-11}\ F)(3 \times 10^3\ V) = 53.1 \times 10^{-8}\ C.$

c) $C = \dfrac{Q}{V} = \dfrac{53.1 \times 10^{-8}\ C}{10^3\ V} = 53.1 \times 10^{-11}\ F = 531\ pF.$

d) $K = \dfrac{C}{C_0} = \dfrac{53.1 \times 10^{-11}\ F}{17.7 \times 10^{-11}\ F} = 3.$

The dielectric constant could also be found from Eq. (27–12),

$$K = \frac{V_0}{V} = \frac{3000\ V}{1000\ V} = 3.$$

e) $\epsilon = K\epsilon_0 = (3)(8.85 \times 10^{-12}\ C^2 \cdot N^{-1} \cdot m^{-2})$

$= 26.6 \times 10^{-12}\ C^2 \cdot N^{-1} \cdot m^{-2}.$

f) $Q_i = A\sigma_i$, $\qquad Q = A\sigma$,

$$\sigma - \sigma_i = \frac{\sigma}{K}, \qquad \sigma_i = \sigma\left(1 - \frac{1}{K}\right),$$

$$Q_i = Q\left(1 - \frac{1}{K}\right) = (53.1 \times 10^{-8} \text{ C})\left(1 - \frac{1}{3}\right)$$

$$= 35.4 \times 10^{-8} \text{ C}.$$

g) $E_0 = \dfrac{V_0}{d} = \dfrac{3000 \text{ V}}{10^{-2} \text{ m}} = 3 + 10^5 \text{ V}\cdot\text{m}^{-1}.$

h) $E = \dfrac{V}{d} = \dfrac{1000 \text{ V}}{10^{-2} \text{ m}} = 1 \times 10^5 \text{ V}\cdot\text{m}^{-1};$

or

$$E = \frac{\sigma}{\epsilon} = \frac{Q}{A\epsilon} = \frac{53.1 \times 10^{-8} \text{ C}}{(2 \times 10^{-1} \text{ m}^2)(26.6 \times 10^{-12} \text{ C}^2 \cdot \text{N}^{-1} \cdot \text{m}^{-2})}$$

$$= 1 \times 10^5 \text{ V}\cdot\text{m}^{-1};$$

or

$$E = \frac{\sigma - \sigma_i}{\epsilon_0} = \frac{Q - Q_i}{A\epsilon_0}$$

$$= \frac{(53.1 - 35.4) \times 10^{-8} \text{ C}}{(2 \times 10^{-1} \text{ m}^2)(8.85 \times 10^{-12} \text{ C}^2 \cdot \text{N}^{-1} \cdot \text{m}^{-2})}$$

$$= 1 \times 10^5 \text{ V}\cdot\text{m}^{-1};$$

or, from Eq. (27–12),

$$E = \frac{E_0}{K} = \frac{3 \times 10^5 \text{ V}\cdot\text{m}^{-1}}{3} = 1 \times 10^5 \text{ V}\cdot\text{m}^{-1}. \qquad \blacktriangleleft$$

Any dielectric material, when subjected to a sufficiently strong electric field, becomes a conductor, a phenomenon known as *dielectric breakdown*. The onset of conduction, associated with cumulative ionization of molecules of the material, is often quite sudden and may be characterized by spark or arc discharges. When a capacitor is subjected to excessive voltage, an arc may be formed through a layer of dielectric, burning or melting a hole in it, permitting the two metal foils to come in contact, creating a short circuit, and rendering the device permanently useless as a capacitor.

The maximum electric field a material can withstand without the occurrence of breakdown is called the *dielectric strength*. Because dielectric breakdown is affected significantly by impurities in the material, small irregularities in the metal electrodes, and other factors difficult to control, only approximate figures can be given for dielectric strengths. The dielectric strength of dry air is about $0.8 \times 10^6 \text{ V}\cdot\text{m}^{-1}$. Typical values for plastic and ceramic materials commonly used to insulate capacitors and current-carrying wires are of the order of $10^7 \text{ V}\cdot\text{m}^{-1}$. For example, a layer of such a material, 10^{-4} m in thickness, could withstand a maximum voltage of 1000 V, for $1000 \text{ V}/10^{-4} \text{ m} = 10^7 \text{ V}\cdot\text{m}^{-1}$.

*27–6

MOLECULAR MODEL OF INDUCED CHARGES

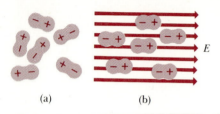

27–7 Behavior of polar molecules (a) in the absence and (b) in the presence of an electric field.

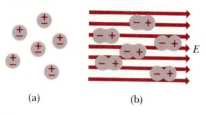

27–8 Behavior of nonpolar molecules (a) in the absence and (b) in the presence of an electric field.

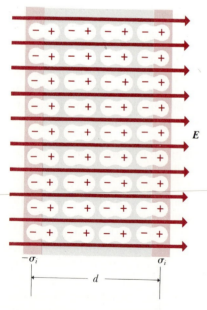

27–9 Polarization of a dielectric in an electric field gives rise to thin layers of bound charges on the surfaces.

We now discuss briefly how surface charges on a dielectric, described in the preceding section, can come about. If the material were a *conductor* the answer would be simple; conductors contain charge that is free to move, and in the presence of an electric field some of the charge redistributes itself on the surface so that there is no electric field inside the conductor. But dielectrics have no charges that are free to move, so how can there be a surface charge?

To understand this we have to look at rearrangement of charge at the *molecular* level. First, some molecules have equal amounts of positive and negative charge but a lopsided distribution, with excess positive charge concentrated on one side and negative on the other. Such a molecule has an electric dipole moment (cf. Section 25–1, Example 5) and is called a *polar molecule*. Placed in an electric field, such a molecule tends to orient itself as in Fig. 27–7, as a result of the electric-field forces and resulting torque. Even if the molecule is not polar, the electric field causes a redistribution of charge within the molecule, as shown in Fig. 27–8. In either case, the result is a layer of charge on each surface of the dielectric material, as shown in Fig. 27–9, and these layers are the surface charges described in Section 27–5. The charges are not free to move indefinitely because each is bound to a molecule; they are in fact called *bound charges*. In the *interior* of the material the net charge per unit volume remains zero. The material in this state is said to be *polarized*.

The four parts of Fig. 27–10 illustrate the behavior of a sheet of dielectric when inserted in the field between a pair of oppositely charged capacitor plates. Part (a) shows the original field. Part (b) is the situation

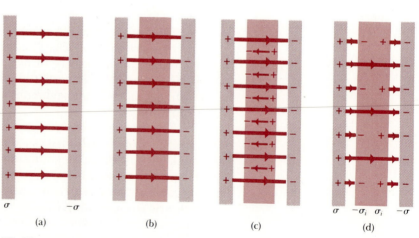

27–10 (a) Electric field between two charged plates. (b) Introduction of a dielectric. (c) Induced surface charges and their field. (d) Resultant field when a dielectric is between charged plates.

after the dielectric has been inserted but before any rearrangement of charges has occurred. Part (c) shows by dotted lines the additional field set up in the dielectric by its induced surface charges. This field is *opposite* to the original field but not enough to cancel it completely, since the charges in the dielectric are not free to move indefinitely. The field in the dielectric is therefore decreased in magnitude.

The resultant field is shown in Fig. 27–10d. Some of the field lines leaving the positive plate penetrate the dielectric; others terminate on the induced charges on the faces of the dielectric.

The charges induced on the surface of a dielectric in an external field afford an explanation of the attraction of an *uncharged* object such as a pith ball or bit of paper by a charged rod of rubber or glass. Figure 27–11 shows an uncharged dielectric sphere B in the radial field of a positive charge A. The induced positive charges on B experience a force toward the right, while the force on the negative charges is toward the left. Since the negative charges are closer to A and therefore in a stronger field than are the positive, the force toward the left exceeds that toward the right, and B, although its net charge is zero, experiences a resultant force toward A. The sign of A's charge does not affect the conclusion, as may readily be seen. Furthermore, the effect is not limited to dielectrics; a conducting sphere would be similarly attracted.

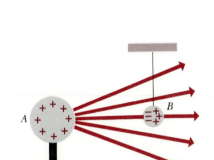

27–11 An uncharged dielectric sphere B in the radial field of a positive charge A.

QUESTIONS

27–1 A student claimed that two capacitors connected in parallel always have an effective capacitance *greater* than that of either capacitor, but that when they are in series the effective capacitance is always *less* than that of either capacitor. Confirm or refute these claims.

27–2 Could one define a capacitance for a single conductor? What would be a reasonable definition?

27–3 Suppose the two plates of a capacitor have different areas. When the capacitor is charged by connecting it to a battery, do the charges on the two plates have equal magnitude, or may they be different?

27–4 Can you think of a situation in which the two plates of a capacitor *do not* have equal magnitudes of charge?

27–5 A capacitor is charged by connecting it to a battery, and is then disconnected from the charging agency. The plates are then pulled apart a little. How does the electric field change? The potential difference? The total energy?

27–6 According to the text, the energy in a charged capacitor can be considered to be located in the field between the plates. But suppose there is vacuum between the plates; can there be energy in vacuum?

27–7 The charged plates of a capacitor attract each other. To pull the plates farther apart therefore requires work by some external force. What becomes of the energy added by this work?

27–8 A solid slab of metal is placed between the plates of a capacitor without touching either plate. Does the capacitance increase, decrease, or remain the same?

27–9 The two plates of a capacitor are given charges $\pm Q$, and then they are immersed in a tank of oil. Does the electric field between them increase, decrease, or remain the same? How might this field be measured?

27–10 Is dielectric strength the same thing as dielectric constant?

27–11 Liquid dielectrics having polar molecules (such as water) always have dielectric constants that decrease with increasing temperature. Why?

27–12 A capacitor made of aluminum foil strips separated by Mylar film was subjected to excessive voltage, and the resulting dielectric breakdown melted holes in the Mylar. After this the capacitance was found to be about the same as before, but the breakdown voltage was much less. Why?

27–13 Two capacitors have equal capacitance, but one has a higher maximum voltage rating than the other. Which one is likely to be bulkier? Why?

27–14 A capacitor is made by rolling a sandwich of aluminum foil and Mylar, as described in Section 27–5. A student claimed that the capacitance when the sandwich is rolled up is twice the value when it is flat. Discuss this allegation.

PROBLEMS

27–1 An air capacitor is made from two flat parallel plates 0.5 mm apart. The magnitude of charge on each plate is 0.01 μC when the potential difference is 200 V.

a) What is the capacitance?

b) What is the area of each plate?

c) What maximum voltage can be applied without dielectric breakdown?

d) When the charge is 0.01 μC, what total energy is stored?

27–2 A parallel-plate air capacitor has a capacitance of 0.001 μF.

a) What potential difference is required for a charge of 0.5 μC on each plate?

b) In (a), what is the total stored energy?

c) If the plates are 1.0 mm apart, what is the area of each plate?

d) What potential difference is required for dielectric breakdown?

27–3 An air capacitor consisting of two closely spaced parallel plates has a capacitance of 1000 pF. The charge on each plate is 1 μC.

a) What is the potential difference between the plates?

b) If the charge is kept constant, what will be the potential difference between the plates if the separation is doubled?

c) How much work is required to double the separation?

27–4 A capacitor has a capacitance of 8.85 μF. How much charge must be removed to decrease the potential difference between its plates by 50 V?

27–5 A parallel-plate air capacitor has a capacitance of 500 pF and a charge of magnitude 0.2 μC on each plate. The plates are 0.2 mm apart.

a) What is the potential difference between plates?

b) What is the area of each plate?

c) What is the electric-field magnitude between plates?

d) What is the surface-charge density on each plate?

27–6 A 20-μF capacitor is charged to a potential difference of 1000 V. The terminals of the charged capacitor are then connected to those of an uncharged 5-μF capacitor. Compute

a) the original charge of the system,

b) the final potential difference across each capacitor,

c) the final energy of the system, and

d) the decrease in energy when the capacitors are connected.

27–7 In Fig. 27–12, each capacitance C_3 is 3 μF and each capacitance C_2 is 2 μF.

a) Compute the equivalent capacitance of the network between points a and b.

b) Compute the charge on each of the capacitors nearest a and b when $V_{ab} = 900$ V.

c) With 900 V across a and b, compute V_{cd}.

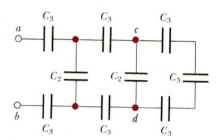

Figure 27–12

27–8 A number of 0.5-μF capacitors are available. The voltage across each is not to exceed 400 V. A capacitor of capacitance 0.5 μF is needed; it is to be connected across a potential difference of 600 V.

a) Show in a diagram how an equivalent capacitor having the desired properties can be obtained.

b) No dielectric is a perfect insulator, of infinite resistance. Suppose that the dielectric in one of the capacitors in your diagram is a moderately good conductor. What will happen?

27–9 A 1-μF capacitor and a 2-μF capacitor are connected in series across a 1200-V supply line.

a) Find the charge on each capacitor and the voltage across each.

b) The charged capacitors are disconnected from the line and from each other, and reconnected with terminals of like sign together. Find the final charge on each and the voltage across each.

27–10 A 1-μF capacitor and a 2-μF capacitor are connected in parallel across a 1200-V supply line.

a) Find the charge on each capacitor and the voltage across each.

b) The charged capacitors are then disconnected from the line and from each other, and reconnected with terminals of unlike sign together. Find the final charge on each and the voltage across each.

27–11 In Fig. 27–3a, let $C_1 = 6 \mu$F, $C_2 = 3 \mu$F, and $V_{ab} = 18$ V. Suppose that the charged capacitors are disconnected from the source and from each other, and reconnected with plates of *opposite* sign connected together. By how much does the energy of the system decrease?

27–12 Three capacitors having capacitances of 8, 8, and 4 μF are connected in series across a 12-V line.

a) What is the charge on the 4-μF capacitor?

b) What is the total energy of all three capacitors?

c) The capacitors are disconnected from the line and reconnected in parallel with the positively charged plates connected together. What is the voltage across the parallel combination?

d) What is the energy of the combination?

27–13 A 500-μF capacitor is charged to 120 V. How many joules of heat are produced on discharging the capacitor if all of the energy goes into heating the wire?

27–14 A parallel-plate capacitor is to be constructed using, as a dielectric, rubber having a dielectric constant of 3 and a dielectric strength of 2×10^5 V·cm^{-1}. The capacitor is to have a capacitance of 0.15 μF and must be able to withstand a maximum potential difference of 6000 V. What is the minimum area the plates of the capacitor may have?

27–15 The paper dielectric in a paper and aluminum-foil capacitor is 0.005 cm thick. Its dielectric coefficient is 2.5 and its dielectric strength is 50×10^6 V·m^{-1}.

a) What area of paper, and of foil, is required for a 0.1-μF capacitor?

b) If the electric field in the paper is not to exceed one half the dielectric strength, what is the maximum potential difference that can be applied across the capacitor?

27–16 A Mylar–aluminum-foil capacitor is to be designed for a capacitance of 0.1 μF and a voltage rating of 500 V. It is to be made from strips of aluminum foil and Mylar about 4 cm wide, and the foil has a thickness of 0.02 mm.

a) What thickness Mylar film is required?

b) What total area of Mylar and aluminum is required?

c) What is the diameter of the rolled-up capacitor?

27–17 A parallel-plate air capacitor is made using two plates 0.2 m square, spaced 1 cm apart. It is connected to a 50-V battery.

a) What is the capacitance?

b) What is the charge on each plate?

c) What is the electric field between the plates?

d) What is the energy stored in the capacitor?

e) If the battery is disconnected and then the plates are pulled apart to a separation of 2 cm, what are the answers to parts (a), (b), (c), and (d)?

27–18 In Problem 27–17, suppose the battery remains connected while the plates are pulled apart. What are the answers to parts (a), (b), (c), and (d)?

27–19 A parallel-plate capacitor with plate area A and separation x is charged to a charge of magnitude q on each plate.

a) What is the total energy stored in the capacitor?

b) The plates are now pulled apart an additional distance Δx; now what is the total energy?

c) If F is the force with which the plates attract each other, then the difference in the two energies above must equal the work $W = F\Delta x$ done in pulling the plates apart. Hence show that $F = q^2/2\epsilon_0 A$.

27–20 A parallel-plate capacitor has the space between the plates filled with a slab of dielectric with constant K_1 and one with constant K_2, each of thickness $d/2$, where d is the plate separation. Show that the capacitance is

$$C = \frac{2\epsilon_0 A}{d}\left(\frac{K_1 K_2}{K_1 + K_2}\right).$$

27–21 Three square metal plates A, B, and C, each 10 cm on a side and 3 mm thick, are arranged as in Fig. 27–13. The plates are separated by sheets of paper 0.5 mm thick and of dielectric constant 5. The outer plates are connected together and connected to point b. The inner plate is connected to point a.

a) Copy the diagram, and show by $+$ and $-$ signs the charge distribution on the plates when point a is maintained at a positive potential relative to point b.

b) What is the capacitance between points a and b?

Figure 27–13

27–22 Two parallel plates have equal and opposite charges. When the space between the plates is evacuated, the electric field is 2×10^5 V·m^{-1}. When the space is filled with dielectric, the electric field is 1.2×10^5 V·m^{-1}.

a) What is the charge density on the surface of the dielectric?

b) What is its dielectric constant?

27–23 Two oppositely charged conducting plates, having numerically equal quantities of charge per unit area, are separated by a dielectric 5 mm thick, of dielectric constant 3. The resultant electric field in the dielectric is 10^6 V·m^{-1}. Compute

a) the charge per unit area on the conducting plate, and

b) the charge per unit area on the surfaces of the dielectric.

27–24 A capacitor consists of two parallel plates of area 25 cm^2 separated by a distance of 0.2 cm. The material between the plates has a dielectric constant of 5. The plates of the capacitor are connected to a 300-V battery.

a) What is the capacitance of the capacitor?

b) What is the charge on either plate?

c) What is the energy in the charged capacitor?

d) What is the energy density in the dielectric?

27–25 Two parallel plates of $100 \, \text{cm}^2$ area are given equal and opposite charges of 10^{-7} C. The space between the plates is filled with a dielectric material, and the electric field within the dielectric is $3.3 \times 10^5 \, \text{V} \cdot \text{m}^{-1}$.

a) What is the dielectric constant of the dielectric?

b) What is the total induced charge on either face of the dielectric?

27–26 An air capacitor is made using two flat plates of area A separated by a distance d. Then a metal slab having thickness a (less than d) and the same shape and size as the plates is inserted between them, parallel to the plates and not touching either plate.

a) What is the capacitance of this arrangement?

b) Express the capacitance as a multiple of the capacitance when the metal slab is not present.

27–27 The capacitors in Fig. 27–14 are initially uncharged, and are connected as in the diagram with switch S open.

a) What is the potential difference V_{ab}?

b) What is the potential of point b after switch S is closed?

c) How much charge flowed through the switch when it was closed?

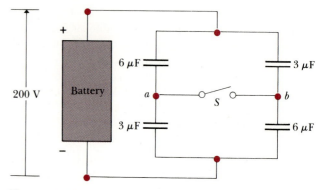

Figure 27–14

28

CURRENT, RESISTANCE, AND ELECTROMOTIVE FORCE

An electric current is a flow of charge within a conductor. A continuous closed path for an electric current is called an electric circuit; current is maintained in a circuit by electromotive force, which can be supplied by a battery, a generator, a photocell, or other devices. Electric charge moves more readily in some conductors than others; this variation is described by a property called resistivity, and the corresponding property of electric-circuit devices is called resistance. Both resistivity and resistance vary significantly with the temperature of the material. The concepts of current, resistance, and electromotive force provide the essential language for the analysis of many electric circuits of practical importance.

28–1

CURRENT

Any motion of charge from one region of a conductor to another is a *current*. When an isolated conductor is placed in an electrostatic field, the charges in the conductor rearrange themselves so as to make the entire interior of the conductor a field-free region and its surface an equipotential surface. The motion of charges during this rearrangement is a temporary or *transient* current. To maintain a *continuous* current we must maintain a steady force on the mobile charge in the conductor, either with an electrostatic field or by other means to be described later. For the present we assume that there is an electric field E within the conductor, so that a particle of charge q experiences a force $F = qE$.

The motion of a charged particle in a conductor is very different from the smoothly accelerated trajectory that results when a constant field is applied to a charge in vacuum. In a conductor, a charge bumps along randomly, accelerating until it collides with a stationary particle, gives up some of its kinetic energy, again accelerates until it bumps into something else, and so on. Thus there is much back-and-forth motion, with a gradual *drift* in the direction of the electric-field force. The inelastic collisions with the stationary charges transfer energy to them; this increases their vibrational energy and hence the temperature of the conductor.

It is useful to think of the motion of charge in a conductor, across or through some imaginary area, which might be, for example, a cross section of a wire used as a conductor. The current across an area is defined quantitatively as *the net charge flowing across the area per unit time*. Thus if a net charge ΔQ flows across an area in a time Δt, the current I across the area is

$$I = \frac{\Delta Q}{\Delta t}. \tag{28-1}$$

Current is a *scalar* quantity.

The SI unit of current, *one coulomb per second*, is called *one ampere* (1 A), in honor of the French scientist André Marie Ampère (1775–1836). Small currents are more conveniently expressed in *milliamperes* (1 mA = 10^{-3} A) or in *microamperes* (1 μA = 10^{-6} A).

The current across an area can be expressed in terms of the drift velocity v of the moving charges. Consider a portion of a conductor of cross-sectional area A within which there is an electric field E from left to right. We suppose first that the conductor contains free *positively* charged particles; these move in the same direction as the field. A few positive particles are shown in Fig. 28–1. Suppose there are n such particles per unit volume, all moving with a drift velocity v. In a time Δt each advances a distance $v \Delta t$. Hence, all of the particles within the shaded cylinder of length $v \Delta t$, and only those particles, will flow across the end of the cylinder during time Δt. The volume of the cylinder is $Av \Delta t$, the number of particles within it is $nAv \Delta t$, and if each has a charge q, the charge ΔQ flowing across the end of the cylinder in time Δt is

$$\Delta Q = nqvA \, \Delta t,$$

The current carried by the positively charged particles is therefore

$$I = \frac{\Delta Q}{\Delta t} = nqvA. \tag{28-2}$$

If the moving charges are negative rather than positive, the electric-field force is opposite to E, and so the drift velocity is right to left, opposite to the direction shown in Fig. 28–1. Positive particles crossing from left to right *increase* the *positive* charge at the right of the section, while negative particles crossing from right to left *decrease* the *negative* charge at the right of the section. But a *decrease* of *negative* charge is equivalent to an *increase* of *positive* charge, so the motion of *both* kinds of charge has the same effect, namely, to increase the positive charge at the right of the section. In either case, particles flowing out through an end of the cylindrical section are continuously replaced by particles flowing *in* through the opposite end.

In general, a conductor may contain several different kinds of charged particles having charges $q_1, q_2, \ldots$, densities $n_1, n_2, \ldots$, and drift velocities $v_1, v_2, \ldots$; the total current is then

$$I = A(n_1 q_1 v_1 + n_2 q_2 v_2 + \cdots). \tag{28-3}$$

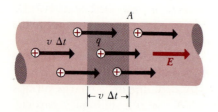

28–1 All of the particles, and only those particles, within the shaded cylinder will cross its base in time Δt.

In metals the moving charges are always (negative) electrons, while an ionized gas has moving electrons and positively charged ions. In a semiconductor material such as germanium or silicon, conduction is partly by electrons and partly by motion of *vacancies,* also known as *holes;* these are sites of missing electrons and act like positive charges. Conduction by motion of holes in semiconductors is discussed in Section 43–8.

The current *per unit cross-sectional area* is called the *current density J;* for each kind of charged particle $J = I/A = nqv$, and in general

$$J = \frac{I}{A} = n_1 q_1 v_1 + n_2 q_2 v_2 + \cdots . \tag{28–4}$$

We can also define a vector current density $\boldsymbol{J}$ that includes the directions of the drift velocities:

$$\boldsymbol{J} = n_1 q_1 \boldsymbol{v}_1 + n_2 q_2 \boldsymbol{v}_2 + \cdots . \tag{28–5}$$

The direction of the drift velocity $\boldsymbol{v}$ of a positive charge is the same as that of the electric field $\boldsymbol{E}$, and the direction of the velocity of a negative charge is opposite to $\boldsymbol{E}$. But since the charge q is negative, each of the vectors $nq\boldsymbol{v}$ is in the same direction as $\boldsymbol{E}$, and hence the *vector current density J always has the same direction as the field E*. Thus even in a metallic conductor, where the moving charges are negative electrons only and move in the *opposite* direction to $\boldsymbol{E}$, the *vector* current density $\boldsymbol{J}$ is in the *same* direction as $\boldsymbol{E}$.

Thus the effect of a current is the same, whether it consists of positive charges moving in the direction of $\boldsymbol{E}$, of negative charges moving in the opposite direction, or a combination of the two. In describing circuit behavior it is customary to describe currents as though they consisted entirely of positive charge flow, even in cases where the actual current is known to be due to electrons. This convention will be followed consistently in the following sections. When the current consists of a flow of positive charges, the direction of the current is the same as that of the motion of the charges; when it is a flow of negative charges, the direction of the current is *opposite* to that of the motion of the charges. These statements are consistent with Eq. (28–5); when q is negative, $\boldsymbol{J}$ and $\boldsymbol{v}$ have opposite directions. In Chapter 30, when we study the effect of a *magnetic* field on a moving charge, we shall consider a phenomenon, the Hall effect, in which the sign of the moving charges *is* important.

When there is a steady current in a closed loop (a "complete circuit"), the total charge in every portion of a conductor remains constant. Hence, if we consider a portion between two fixed cross sections, the rate of flow of charge *out* of the portion at one end equals the rate of flow of charge *into* the portion at the other end. In other words, the current is the same at any two cross sections and hence is *the same at all cross sections*. Current is *not* something that squirts out of the positive terminal of a battery and gets all used up by the time it reaches the negative terminal! These statements are direct consequences of the principle of conservation of charge, introduced in Section 24–2.

EXAMPLE A copper conductor of square cross section 1 mm on a side carries a constant current of 20 A. The density of free electrons is 8 ×

10^{28} electrons per cubic meter. Find the current density and the drift velocity.

Solution The current density in the wire is

$$J = \frac{I}{A} = 20 \times 10^6 \,\mathrm{A \cdot m^{-2}}.$$

From Eq. (28–4),

$$v = \frac{J}{nq} = \frac{(20 \times 10^6 \,\mathrm{A \cdot m^{-2}})}{(8 \times 10^{28} \,\mathrm{m^{-3}})(1.6 \times 10^{-19} \,\mathrm{C})}$$
$$= 1.6 \times 10^{-3} \,\mathrm{m \cdot s^{-1}},$$

or about $1.6 \,\mathrm{mm \cdot s^{-1}}$. At this speed, an electron would require 625 s or about 10 min to travel the length of a wire 1 m long. Thus the drift velocity is *very* small compared with the velocity of propagation of a current pulse along a wire, about $3 \times 10^8 \,\mathrm{m \cdot s^{-1}}$. ◄

28–2

RESISTIVITY

The current density J in a conductor depends on the electric field E, and on the nature of the conductor. In general the dependence of J on E can be quite complex but, for some materials, especially the metals, it can be represented quite well by a direct proportionality. For such materials the ratio of E to J is constant; we define the *resistivity* ρ of a particular material as the ratio of electric field to current density:

$$\rho = \frac{E}{J}. \tag{28–6}$$

That is, the resistivity is the *electric field per unit current density*. The greater the resistivity, the greater the field needed to establish a given current density, or the smaller the current density for a given field. Representative values are given in Table 28–1. The unit $\Omega \cdot \mathrm{m}$ (ohm·meter) will be explained in the following section. A "perfect" conductor would have zero resistivity and a "perfect" insulator an infinite resistivity. Metals and alloys have the lowest resistivities and are the best conductors. The resistivities of insulators exceed those of the metals by a factor of the order of 10^{22}.

Comparison with Table 16–1 shows that *thermal* insulators have thermal resistivities (the reciprocals of their thermal conductivities) that differ from those of good thermal conductors by factors of only about 10^3. By the use of electrical insulators, electric currents can be confined to well-defined paths in good electrical conductors, while it is impossible to confine heat currents to a comparable extent. It is also interesting to note that the metals, as a class, are also the best *thermal* conductors. The free electrons in a metal that carry charge in electrical conduction also play an important role in the conduction of heat; hence, a correlation can be expected between electrical and thermal conductivity. It is a familiar fact that good electrical conductors, such as the metals, are also good conduc-

TABLE 28–1 RESISTIVITIES AT ROOM TEMPERATURE				
Substance		ρ, $\Omega \cdot$m	**Substance**	ρ, $\Omega \cdot$m
Conductors			Semiconductors	
Metals	Silver	1.47×10^{-8}	Pure Carbon	3.5×10^{-5}
	Copper	1.72×10^{-8}	Germanium	0.60
	Gold	2.44×10^{-8}	Silicon	2300
	Aluminum	2.63×10^{-8}	Insulators	
	Tungsten	5.51×10^{-8}	Amber	5×10^{14}
	Steel	$20 \ \times 10^{-8}$	Glass	10^{10}–10^{14}
	Lead	$22 \ \times 10^{-8}$	Lucite	$> 10^{13}$
	Mercury	$95 \ \times 10^{-8}$	Mica	10^{11}–10^{15}
Alloys	Manganin	$44 \ \times 10^{-8}$	Quartz (fused)	75×10^{16}
	Constantan	$49 \ \times 10^{-8}$	Sulfur	10^{15}
	Nichrome	$100 \ \times 10^{-8}$	Teflon	$> 10^{13}$
			Wood	10^{8}–10^{11}

tors of heat, while poor electrical conductors, such as ceramic and plastic materials, are also poor thermal conductors.

The *semiconductors* form a class intermediate between the metals and the insulators. They are of importance not primarily because of their resistivities, but because of the way in which these are affected by temperature and by small amounts of impurities.

The discovery that J is proportional to E for a metallic conductor at constant temperatures was made by G. S. Ohm (1789–1854) and is called *Ohm's law*. A material obeying Ohm's law is called an *ohmic* conductor or a *linear* conductor. If Ohm's law is *not* obeyed, the conductor is called *nonlinear*. Thus Ohm's law, like the ideal gas equation, Hooke's law, and many other relations describing the properties of materials, is an *idealized model* that describes the behavior of certain materials reasonably well but is by no means a general property of all matter.

The resistivity of all *metallic* conductors increases with increasing temperature, as shown in Fig. 28–2a. Over a temperature range that is not too great, the resistivity of a metal can be represented approximately by the equation

$$\rho_T = \rho_0[1 + \alpha(T - T_0)], \qquad (28-7)$$

where ρ_0 is the resistivity at a reference temperature T_0 (often taken as 0°C or 20°C) and ρ_T the resistivity at temperature T°C. The factor α is called the *temperature coefficient of resistivity*. Some representative values are given in Table 28–2. The resistivity of carbon (a nonmetal) *decreases* with increasing temperature, and its temperature coefficient of resistivity is negative. The resistivity of the alloy manganin is practically independent of temperature.

A number of materials have been found to exhibit the property of *superconductivity*. As the temperature is decreased, the resistivity at first decreases regularly, like that of any metal. At the *critical* temperature, usually in the range 0.1 K to 20 K, a phase transition occurs, and the resistivity suddenly drops to zero, as shown in Fig. 28–2b. A current once established in a superconducting ring will continue indefinitely, without the presence of any driving field.

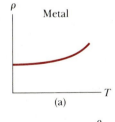
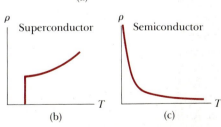

28–2 Variation of resistivity with temperature for three conductors: (a) an ordinary metal, (b) a superconducting metal, alloy, or compound, and (c) a semiconductor.

TABLE 28–2	TEMPERATURE COEFFICIENTS OF RESISTIVITY (APPROXIMATE VALUES NEAR ROOM TEMPERATURE)		
Material	α, °C^{-1}	**Material**	α, °C^{-1}
Aluminum	0.0039	Lead	0.0043
Brass	0.0020	Manganin (Cu 84,	0.000000
Carbon	− 0.0005	Mn 12, Ni 4)	
Constantan (Cu 60, Ni 40)	+ 0.000002	Mercury	0.00088
Copper (Commercial	0.00393	Nichrome	0.0004
annealed)		Silver	0.0038
Iron	0.0050	Tungsten	0.0045

The resistivity of a *semiconductor* decreases rapidly with increasing temperature, as shown in Fig. 28–2c. A tiny bead of semiconducting material, called a *thermistor,* serves as a sensitive thermometer.

28–3

RESISTANCE

The current density J, at a point within a conductor where the electric field is E, is given by Eq. (28–6):

$$E = \rho J.$$

It is often difficult to measure E and J directly, and it is useful to put this relation in a form involving readily measured quantities such as total current and potential difference. To do this we consider a conductor with uniform cross-sectional area A and length l, as shown in Fig. 28–3. Assuming a constant current density over a cross section, and a uniform electric field along the length of the conductor, the total current I is given by

$$I = JA,$$

and the potential difference V between the ends is

$$V = El. \tag{28–8}$$

Solving these equations for J and E, respectively, and substituting the results in Eq. (28–6), we obtain

$$\frac{V}{l} = \frac{\rho I}{A}. \tag{28–9}$$

Thus the total current is proportional to the potential difference.

The quantity $\rho l/A$ for a particular specimen of material is called its *resistance R:*

$$R = \frac{\rho l}{A}. \tag{28–10}$$

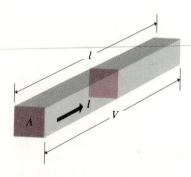

28–3 A conductor of uniform cross section. The current density is uniform over any cross section, and the electric field is constant along the length.

Equation (28–9) then becomes

$$V = IR. \tag{28–11}$$

This relation is often referred to as *Ohm's law;* in this form it refers to a specific piece of material, not to a general property of the material as with Eq. (28–6).

Equation (28–10) shows that the resistance of a wire or other conductor of uniform cross section is directly proportional to its length and inversely proportional to its cross-sectional area. It is of course also proportional to the resistivity of the material of which the conductor is made.

The SI unit of resistance is one *volt per ampere* ($1 \ \text{V} \cdot \text{A}^{-1}$). A resistance of $1 \ \text{V} \cdot \text{A}^{-1}$ is called 1 *ohm* ($1 \ \Omega$). The unit of resistivity is therefore one *ohm · meter* ($1 \ \Omega \cdot \text{m}$).

Large resistances are conveniently expressed in *kilohms* ($1 \ \text{k}\Omega = 10^3 \ \Omega$) or *megohms* ($1 \ \text{M}\Omega = 10^6 \ \Omega$), and small resistances in microhms ($1 \ \mu\Omega = 10^{-6} \ \Omega$). Resistivities are also expressed in a variety of hybrid units, most common of which is the *ohm · centimeter* ($1 \ \Omega \cdot \text{cm} = 10^{-2} \ \Omega \cdot \text{m}$).

Because the resistance of any specimen of material is proportional to its resistivity, which varies with temperature, resistance also varies with temperature. For temperature ranges that are not too great, this variation may be represented approximately as a linear relation analogous to Eq. (28–7):

$$R_T = R_0[1 + \alpha(T - T_0)]. \qquad (28\text{–}12)$$

Here R_T is the resistance at temperature T, and R_0 is the resistance at the temperature T_0, often taken to be 20°C or 0°C. Within the limits of validity of Eq. (28–12), the *change* in resistance resulting from a temperature change $T - T_0$ is given by $\alpha(T - T_0)$, where α is the temperature coefficient of resistivity, given for several common materials in Table 28–2.

EXAMPLE 1   For the example at the end of Section 28–1, find the electric field, and the potential difference between two points 100 m apart.

Solution   From Eq. (28–6), the electric field is given by

$$E = \rho J = (1.72 \times 10^{-8} \ \Omega \cdot \text{m})(20 \times 10^6 \ \text{A} \cdot \text{m}^{-2})$$
$$= 0.344 \ \text{V} \cdot \text{m}^{-1}.$$

The potential difference is given by

$$V = El = (0.344 \ \text{V} \cdot \text{m}^{-1})(100 \ \text{m}) = 34.4 \ \text{V}.$$

Thus the resistance of a piece of this wire 100 m in length is

$$R = \frac{V}{I} = \frac{34.4 \ \text{V}}{20 \ \text{A}} = 1.72 \ \Omega.$$

This result can also be obtained directly from Eq. (28–10):

$$R = \frac{\rho l}{A} = \frac{(1.72 \times 10^{-8} \ \Omega \cdot \text{m})(100 \ \text{m})}{(1 \times 10^{-3} \ \text{m})^2} = 1.72 \ \Omega. \qquad \blacktriangleleft$$

EXAMPLE 2   In the previous example, suppose the resistance is $1.72 \ \Omega$ at a temperature of 20°C. Find the resistance at 0°C and at 100°C.

Solution   We use Eq. (28–12). In this instance $T_0 = 20°C$ and $R_0 = 1.72\ \Omega$. From Table 28–2, the temperature coefficient of resistivity of copper is $\alpha = 0.00393(°C)^{-1}$. Thus at $T = 0°C$,

$$R = 1.72\ \Omega[1 + (0.00393°C^{-1})(0°C - 20°C)]$$
$$= 1.58\ \Omega,$$

and at $T = 100°C$,

$$R = 1.72\ \Omega[1 + (0.00393°C^{-1})(100°C - 20°C)]$$
$$= 2.26\ \Omega.$$ ◀

28–4

ELECTROMOTIVE FORCE AND CIRCUITS

In order for a steady current to exist in a conducting path, that path must form a closed loop or *complete circuit*. Otherwise charge would accumulate at the ends of the conductor, the resulting electric field would change with time, and the current could not be constant.

However, such a path cannot consist solely of resistance. Current in a resistor requires an electric field and an associated potential. The field always does *positive* work on the charge, which moves always in the direction of *decreasing* potential. But after a complete trip around the loop, a charge returns to its starting point, and the potential then must be the same as when it left that point. This cannot be so if its travel around the loop involves only *decreases* in potential.

Thus there must be some portion of the loop where a charge travels "uphill," from lower to higher potential, despite the electrostatic force trying to push it from higher to lower potential. The influence that makes charge move from lower to higher potential is called *electromotive force*. Every complete circuit in which there is a steady current must have some device that provides electromotive force. This term is usually abbreviated emf (pronounced "ee-em-eff").

Batteries, generators, photovoltaic cells, and thermocouples are examples of sources of emf. Any such device has the ability to convert energy of some form (mechanical, chemical, heat, and so on) into electrical energy and transfer it into the circuit into which it is connected.

Figure 28–4 is a schematic diagram of a source of emf, such as a battery or generator. Such a device can maintain a potential difference between conductors a and b, called the *terminals* of the device. Terminal a, marked +, is maintained at *higher* potential than terminal b, marked −. Associated with this potential difference is an electric field E in the region around the terminals, both inside and outside the source. The electric field inside the device is directed from a to b, as shown. A charge q within the source experiences a force $F_e = qE$ caused by this field. But the source is itself a conductor, and if this were the *only* force on the free charges, then positive charge would move from a toward b and negative charge from b toward a. The excess charges on the conductors would decrease, and the potential difference would eventually decrease to zero.

But this is *not* the way batteries and generators actually act; they maintain a potential difference even when there is a steady current

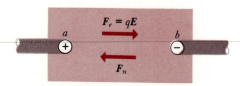

28–4   Schematic diagram of a source of emf in an "open-circuit" situation. The electric-field force $F_e = qE$ and the nonelectrostatic force F_n on a charge q are shown. The work done by F_n on a charge q moving from b to a is equal to $q\mathcal{E}$, where $\mathcal{E}$ is the electromotive force. In the open-circuit situation, F_e and F_n have equal magnitude.

through them from b to a. Thus there must be some *additional* force on the charges in the source, tending to push positive charges from lower to higher potential, *opposite* to the direction of the electric-field force. The origin of this additional force depends on the nature of the source. In a generator it results from the action of a *magnetic* field on moving charges. In a battery it is associated with varying electrolyte concentrations arising from chemical reactions. In an electrostatic machine such as a Van de Graaff or Wimshurst generator, it is a mechanical force applied by a moving belt or wheel.

When there is no external current path, that is, under "open-circuit" conditions, as in Fig. 28–4, the charges are in equilibrium and F_e and F_n must cancel each other at every point. The potential V_{ab} of point a with respect to point b is defined, as always, as the work per unit charge performed by the electrostatic force F_e on a charge moving from a to b. Similarly, the emf of the source, denoted by $\mathcal{E}$, is the work per unit charge performed by the nonelectrostatic force F_n during the "uphill" displacement from b to a. For a source on open circuit, the potential difference V_{ab} is equal to the electromotive force $\mathcal{E}$:

$$V_{ab} = \mathcal{E} \qquad \text{(source on open circuit).} \qquad (28\text{–}13)$$

The SI unit of emf is the same as that of potential or potential difference, namely $1 \, \text{J} \cdot \text{C}^{-1}$ or $1 \, \text{V}$. There is a subtle but important distinction between emf and potential difference; emf refers to work done by the nonelectrostatic forces within a source, while potential difference is associated with electrostatic fields caused by distributions of electric charge. The nonelectrostatic force and the associated emf may often be taken to be constant, independent of current, for a given source, while V_{ab} may depend on current. In the following discussion the emf of a source will usually be assumed to be constant.

Now suppose that the terminals of a source are connected by a wire, as shown schematically in Fig. 28–5, forming a *complete circuit*. The driving force on the free charges *in the wire* is due solely to the electrostatic field E_e set up by the charged terminals a and b of the source. This field sets up a current *in the wire* from a toward b. If the wire has resistance R, then from Eq. (28–11) the current I in the circuit is determined by

$$V_{ab} = IR. \qquad (28\text{–}14)$$

If current could travel through the source without impediment (that is, if the source had no *internal* resistance), charge entering the external circuit through terminal a would be replaced immediately by charge flow through the source. In this case the internal electrostatic field in the source would not change under complete-circuit conditions, and the terminal potential difference V_{ab} would still be equal to $\mathcal{E}$. Since V_{ab} is also related to the current and resistance in the external circuit by Eq. (28–14), we would then have

$$\mathcal{E} = V_{ab} = IR. \qquad (28\text{–}15)$$

This relation determines the current in the circuit, once $\mathcal{E}$ and R are specified.

We say "if" in the above paragraph because every real source has some *internal* resistance, which we may denote as r. Under closed-circuit

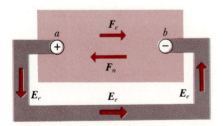

28–5 Schematic diagram of a source with a complete circuit. The vectors F_n and F_e represent the directions of the corresponding forces. The current is everywhere the same and is in the direction of the *resultant* field, that is, from a to b in the external circuit and from b to a within the source. $V_{ab} = IR = \mathcal{E} - Ir$.

conditions the total force $\boldsymbol{F}_e + \boldsymbol{F}_n$ on a charge inside the source cannot be precisely zero because some net field is required in order to push the charge through the internal resistance r. Thus $\boldsymbol{F}_e$ must have somewhat smaller magnitude than $\boldsymbol{F}_n$, and correspondingly V_{ab} is less than $\mathcal{E}$; the difference is equal to the work per unit charge done by the total force, which is simply Ir. Thus the terminal potential difference under closed-circuit conditions is given by

$$V_{ab} = \mathcal{E} - Ir. \qquad (28\text{--}16)$$

The current in the external circuit is still determined by Eq. (28–14); combining it with Eq. (28–16), we find

$$\mathcal{E} - Ir = IR,$$

or

$$I = \frac{\mathcal{E}}{R + r}. \qquad (28\text{--}17)$$

That is, the current equals the source emf divided by the *total* circuit resistance, external plus internal.

If the terminals of a source are connected by a conductor of zero (or negligible) resistance, the source is said to be *short circuited*. (This would be an *extremely* dangerous procedure to carry out with the storage battery of your car, or with the terminals of the power line! Don't try it!) Then $R = 0$, and from the circuit equation the *short-circuit current* I_s is

$$I_s = \frac{\mathcal{E}}{r}. \qquad (28\text{--}18)$$

The terminal voltage is then zero:

$$V_{ab} = \mathcal{E} - \left(\frac{\mathcal{E}}{r}\right) r = 0. \qquad (28\text{--}19)$$

The *electrostatic* field within the source is zero, and the driving force on the charges within it is due to the *non*electrostatic field only.

This discussion of sources and electromotive force has been somewhat abstract, yet the basic ideas are quite simple. We have shown that a source with its associated emf can be represented in terms of two components, an emf $\mathcal{E}$, which supplies a constant potential difference independent of current, in series with an internal resistance r. Thus a source is completely described by its emf $\mathcal{E}$ and its internal resistance r. These properties may be found (at least in principle) from measurements of the open-circuit terminal voltage, which equals $\mathcal{E}$, and the short-circuit current, which enables r to be calculated from Eq. (28–18).

The diagrams in the preceding sections have been more or less pictorial, in order to show the electric fields within sources and conductors. We now introduce the usual symbols used in electric-circuit diagrams. In principle every conductor (except a superconductor) has resistance, but in practice devices that are made to have certain resistances are usually connected with wires or other conductors whose resistance is negligibly small. A resistor is represented by the symbol

$$\mathrm{-\!\!\!-\!\!\!-\!\!\backslash\!\!\backslash\!\!\backslash\!\!\!-\!\!\!-\!\!\!-}$$

Conductors having negligible resistance are shown by straight lines.

A variable resistor is called a *rheostat* or, particularly in electronics, a *potentiometer*. A common type consists of a resistor with a sliding contact that can be moved along its length and is represented by the symbol

Connections are made to either end of the resistor and to the sliding contact. The symbol

is also used for a variable resistor.

A source is represented by the symbol

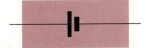

The longer vertical line always corresponds to the + terminal. We shall modify this in the following examples to

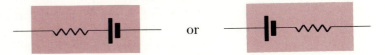

or

so as to show explicitly that a source has an internal resistance. The order of the two parts of this symbol is immaterial.

EXAMPLE 1 Consider a source whose emf $\mathcal{E}$ is constant and equal to 12 V, and whose internal resistance r is 2 Ω. (The internal resistance of a commercial 12-V lead storage battery is only a few thousandths of an ohm.) Figure 28–6 represents the source with a voltmeter V connected between its terminals a and b. A voltmeter reads the potential difference between its terminals. If it is of the conventional type, the voltmeter provides a conducting path between the terminals and so there is a current in the source (and through the voltmeter). We shall assume, however, that the resistance of the voltmeter is so large (essentially infinite) that it draws no appreciable current. The source is then on *open circuit*, corresponding to the source in Fig. 28–4, and the voltmeter reading V_{ab} equals the emf $\mathcal{E}$ of the source, or 12 V. ◄

EXAMPLE 2 In Fig. 28–7, an ammeter A and a resistor of resistance $R = 4$ Ω have been connected to the terminals of the source to form a complete circuit. The total resistance of the circuit is the sum of the resistance R, the internal resistance r, and the resistance of the ammeter. The ammeter resistance, however, can be made very small, and we shall assume it so small (essentially zero) that it can be neglected. The ammeter (whatever its resistance) reads the current I through it. The circuit corresponds to that in Fig. 28–5.

The wires connecting the resistor to the source and the ammeter, shown by the straight lines, have zero resistance and hence there is no

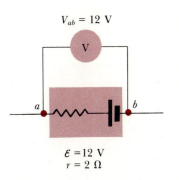

$V_{ab} = 12$ V

a b

$\mathcal{E} = 12$ V
$r = 2$ Ω

28–6 A source on open circuit.

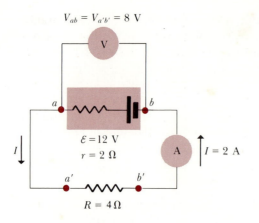

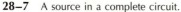

28–7   A source in a complete circuit.

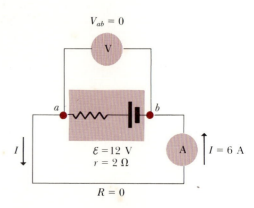

28–8   A source on short circuit.

potential difference between their ends. Thus, points a and a' are at the same potential and are electrically equivalent, as are points b and b'. The potential differences V_{ab} and $V_{a'b'}$ are therefore equal. In the future, we shall use the same symbol to represent all points in a circuit that are connected by resistanceless conductors and are at the same potential.

The current I in the resistor (and hence at all points of the circuit) could be found from the relation $I = V_{ab}/R$, if the potential difference V_{ab} were known. However, V_{ab} is the terminal voltage of the source, equal to $\mathcal{E} - Ir$, and since this depends on I it is unknown at the start. We can, however, calculate the current from the circuit equation:

$$I = \frac{\mathcal{E}}{R + r} = \frac{12 \text{ V}}{4\,\Omega + 2\,\Omega} = 2 \text{ A.}$$

The potential difference V_{ab} can now be found by considering a and b either as the terminals of the resistor or as those of the source. If we consider them as the terminals of the resistor,

$$V_{a'b'} = IR = (2 \text{ A})(4\,\Omega) = 8 \text{ V.}$$

If we consider them as the terminals of the source,

$$V_{ab} = \mathcal{E} - Ir = 12 \text{ V} - (2 \text{ A})(2\,\Omega) = 8 \text{ V.}$$

The voltmeter therefore reads 8 V and the ammeter reads 2 A. ◄

EXAMPLE 3   In Fig. 28–8, the source is short circuited. The current is

$$I = \frac{\mathcal{E}}{r} = \frac{12 \text{ V}}{2\,\Omega} = 6 \text{ A.}$$

The terminal voltage is

$$V_{ab} = \mathcal{E} - Ir = 12 \text{ V} - (6 \text{ A})(2\,\Omega) = 0.$$

The ammeter reads 6 A and the voltmeter reads zero.   ◄

The rule represented by Eq. (28–17), for finding the current in a circuit, can be recast in a form that is particularly useful in more complex circuits having several seats of emf or several branches. The electrostatic field is a *conservative* force field. Suppose we go around a loop, measuring potential differences across successive circuit elements. When we return to the starting point we must find that the *algebraic sum* of these differences is zero; otherwise we could not say that the potential at this point has a definite value.

Thus, in Fig. 28–7, if we start at point *b* and travel counterclockwise (the same direction as *I*) we find a *rise* in potential (a *positive* change) due to the battery emf, a *drop* (a *negative* change) due to the battery's internal resistance, and an additional drop due to the 4-Ω resistor. The algebraic sum of these must be zero. Thus

$$12 \text{ V} - I(2 \text{ Ω}) - I(4 \text{ Ω}) = 0, \qquad I = 2 \text{ A},$$

in agreement with the result obtained in Example 2.

To allow for the possibility of several seats of emf in the loop, we generalize this procedure as follows: *The algebraic sum of the potential differences around a complete circuit, including those corresponding to the emf's of the sources and those due to the IR products, must equal zero.* This is called *Kirchhoff's loop rule.* In applying it, we need some sign conventions. We first assume a direction for the current, and mark it on the diagram. Then, starting at any point in the circuit, we go around the circuit in the direction of the assumed current, adding emf's and *IR* products as we come to them. When a source is traversed in the direction from (−) to (+) (the direction of E_n in that source), the emf is considered *positive,* and when from (+) to (−), negative. The *IR* products are all negative because the direction of current is always that of *decreasing* potential. Of course, we could also traverse the loop in the direction *opposite* to that of the assumed current. In that case all the emf's have opposite sign, and all the *IR* products are *positive* because in going in the opposite direction to the current we are going "uphill" from lower to higher potential.

For the circuit in Fig. 28–7, if we start at point *b* and go clockwise, the resulting equation is

$$I(4 \text{ Ω}) + I(2 \text{ Ω}) - 12 \text{ V} = 0,$$

which is the same as the previous equation except for an overall factor of (−1) that does not change the value of *I*. But if we assume that *I* is clockwise, then starting at *b* and going counterclockwise around the circuit yields the equation

$$12 \text{ V} + I(2 \text{ Ω}) + I(4 \text{ Ω}) = 0,$$
$$I = -2 \text{ A}.$$

In this case the negative sign on the result shows that our initial assumption about the current direction was wrong; the actual direction is counterclockwise.

This same bookkeeping system may be used to determine the potential difference between any two points *a* and *b* in a circuit. To find $V_{ab} = V_a - V_b$ (the potential at *a* with respect to *b*), we start at *b* and add the potential changes encountered in going from *b* to *a*. Again an emf is considered positive when we go from − to +, negative otherwise. An *IR* product is positive when we go "uphill," against the current direction, negative when "downhill," in the same direction as the current.

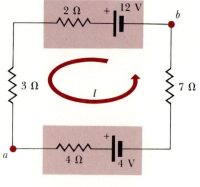

28–9

EXAMPLE 4   The circuit shown in Fig. 28–9 contains two batteries, each having an emf and an internal resistance, and two resistors. Find the current in the circuit and the potential difference V_{ab}.

Solution   We assume a direction for the current, as shown; then, starting at a and going counterclockwise, we add potential increases and decreases and equate the sum to zero. The resulting equation is

$$-I(4\ \Omega) - 4\ V - I(7\ \Omega) + 12\ V - I(2\ \Omega) - I(3\ \Omega) = 0.$$

Collecting terms containing I and solving for I, we find

$$8\ V = I(16\ \Omega) \quad \text{and} \quad I = 0.5\ A.$$

The result for I is positive, showing that our assumed current direction is correct. The reader may verify that if one assumes the opposite direction the result is $I = -0.5$ A, indicating that the actual current is opposite to this assumption.

To find V_{ab}, the potential at a with respect to b, we start at b and go toward a, adding potential changes. There are two possible paths from b to a; taking the lower one first, we find

$$V_{ab} = (0.5\ A)(7\ \Omega) + 4\ V + (0.5\ A)(4\ \Omega) = 9.5\ V.$$

Point a is at 9.5 V higher potential than b. All the terms in this sum are positive because each represents an *increase* in potential as we go from b toward a. If instead we use the upper path, the resulting equation is

$$V_{ab} = 12\ V - (0.5\ A)(2\ \Omega) - (0.5\ A)(3\ \Omega) = 9.5\ V.$$

Here the IR terms are negative because our path goes in the direction of the current, with potential decreases through the resistors. The result is the same as for the other path, as it must be in order for the total potential change around the complete loop to be zero. In each case potential rises are taken as positive, and drops as negative.   ◀

In the foregoing analysis, emf's have been treated on the same basis as potential differences in circuit elements. This procedure might well be questioned; an emf is, strictly speaking, *not* a potential difference, inasmuch as it represents work done (per unit charge) by a force of *non*electrostatic origin. But as we have seen the (electrostatic) potential difference between terminals of a source can always be expressed in terms of its emf and the drop across its internal resistance. Thus the emf terms in the above analysis really *do* represent true potential differences.

28–5

CURRENT–VOLTAGE RELATIONS

The current through a device such as a resistor depends on the potential difference between its terminals. For a device obeying Ohm's law, the current is directly proportional to voltage, as shown in Eq. (28–11). But as mentioned in Section 28–2, there are many devices for which this simple model is not an adequate description. Current may depend on voltage in a more complicated way, and the current resulting from a given potential difference may depend on the polarity of the potential

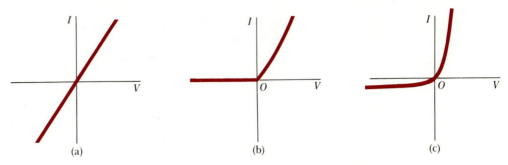

28–10 Current–voltage relations for (a) a resistor obeying Ohm's law; (b) a vacuum diode; (c) a semiconductor diode.

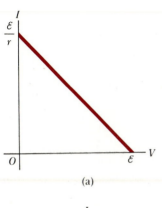

(a)

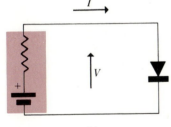

(b)

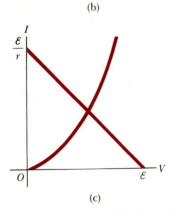

(c)

28–11 (a) Current–voltage relation for a source with emf $\mathcal{E}$ and internal resistance r; (b) a circuit containing a source and a nonlinear element; (c) simultaneous solution of I–V equations for this circuit.

difference. This is the case with *diodes*, devices constructed deliberately to conduct much better in one direction than the other.

It is convenient to represent the current–voltage relation as a graph, and Fig. 28–10 shows several examples. Part (a) shows the behavior of a resistor that obeys Ohm's law, for which the graph is a straight line. Part (b) shows the relation for a vacuum diode. For positive potentials of anode with respect to cathode, I is approximately proportional to $V^{3/2}$, while for negative potentials the current is several orders of magnitude smaller and for most purposes may be assumed to be zero. Germanium diode behavior (c) is somewhat different but still strongly asymmetric, acting as a one-way valve in a circuit. Diodes are used to convert alternating current to direct and to perform a wide variety of logic functions in computer circuitry. The microscopic basis of diode behavior will be explored in some detail in later chapters.

An additional consideration is that for nearly all materials the current–voltage relation is temperature-dependent. Thus at low temperatures the curve in Fig. 28–10c rises more steeply for positive V than at higher temperatures, and at successively higher temperatures the asymmetry in the curve becomes less and less pronounced.

The current–voltage relation for a source may also be represented graphically. For a source represented by Eq. (28–16), that is,

$$V = \mathcal{E} - Ir,$$

the graph appears as in Fig. 28–11a. The intercept on the V-axis, corresponding to the open-circuit condition ($I = 0$), is at $V = \mathcal{E}$, and the intercept on the I-axis, corresponding to a short-circuit situation ($V = 0$), is at $I = \mathcal{E}/r$.

This line may be used to find the current in a circuit containing a nonlinear device, as in Fig. 28–11b. Its current–voltage relation is shown in Fig. 28–11c, and Eq. (28–16) is also plotted on this graph. Each curve represents a current–voltage relation that must be satisfied, so the intersection represents the only possible values of V and I. This amounts to a graphical solution of two simultaneous equations for V and I, one of which is nonlinear.

Finally, we remark that Eq. (28–16) is not always an adequate representation of the behavior of a source. What we have described as an internal resistance may actually be a more complex voltage–current relation. Nevertheless, the concept of internal resistance frequently provides an adequate description of batteries, generators, and other energy converters. The difference between a fresh flashlight battery and an old one is not in the emf, which decreases only slightly with use, but principally in the internal resistance, which may increase from a few ohms when fresh to as much as 1000 Ω or more after long use. Similarly, the current a car battery can deliver to the starter motor on a cold morning is less than when the battery is warm, not because the emf is appreciably less but because the internal resistance is temperature-dependent, decreasing with increasing temperature. Residents of northern Wisconsin have been known to soak their car batteries in warm water to provide greater starting power on very cold mornings!

28–6

WORK AND POWER IN ELECTRICAL CIRCUITS

We are now ready to consider in detail the energy relations in electric circuits. The rectangle in Fig. 28–12 represents a portion of a circuit having current I and potential difference $V_a - V_b = V_{ab}$ between the two conductors leading to and from this point of the circuit. The detailed nature of this circuit element does not matter. As charge passes through, the electric field does work on it. In a time interval Δt, an amount of charge $\Delta Q = I\,\Delta t$ passes through, and the work ΔW done by the electric field is given by the product of the potential difference (work per unit charge) and the quantity of charge:

$$\Delta W = V_{ab}\,\Delta Q = V_{ab}I\,\Delta t.$$

By means of this work the electric field transfers energy into this portion of the circuit.

Rate of transfer of energy is *power*, denoted by P; dividing the above relation by Δt, we obtain the rate at which energy enters this part of the circuit:

$$P = \frac{\Delta W}{\Delta t} = V_{ab}I. \tag{28–20}$$

It may happen that the potential at b is higher than that at a; in this case V_{ab} is negative. The charge then *gains* potential energy (at the expense of some other form of energy), and there is a corresponding transfer of electrical energy *out of* this portion of the circuit.

Equation (28–20) is the general expression for the magnitude of the electrical power input to (or the power output from) any portion of an electrical circuit. The unit of V_{ab} is one volt, or one joule per coulomb, and the unit of I is one ampere or one coulomb per second. The SI unit

28–12 The power input P to the portion of the circuit between a and b is $P = V_{ab}I$.

of power is therefore

$$(1 \text{ J} \cdot \text{C}^{-1})(1 \text{ C} \cdot \text{s}^{-1}) = 1 \text{ J} \cdot \text{s}^{-1} = 1 \text{ W} = 1 \text{ watt}.$$

We now consider some special cases.

1. Pure resistance. If the portion of the circuit in Fig. 28–12 is a pure resistance, the potential difference is given by $V_{ab} = IR$ and

$$P = V_{ab}I = I^2R = \frac{V^2}{R}. \tag{28–21}$$

The potential at a is necessarily higher than at b and there is a power *input* to the resistor. The circulating charges give up energy to the atoms of the resistor when they collide with them, and the temperature of the resistor increases unless there is a flow of heat out of it. We say that energy is *dissipated* in the resistor at a rate I^2R.

Because of this heat, every resistor has a maximum power rating, the maximum power that can be dissipated without overheating the device. When this rating is exceeded the resistance may change unpredictably; in more extreme cases the resistor may melt or even explode. In practical applications, the power rating of a resistor is just as important a characteristic as its resistance value.

2. Power output of a source. The upper rectangle in Fig. 28–13 represents a source having emf $\mathcal{E}$ and internal resistance r, connected by ideal (resistanceless) conductors to an external circuit represented by the lower rectangle; the precise nature of the external circuit does not matter. We assume only that there is a current I in the circuit in the direction shown, from a to b in the external circuit and from b to a within the source. The letters a and b can be considered to represent either the terminals of the source or those of the external circuit, and $V_a > V_b$. The external circuit then corresponds to the rectangle in Fig. 28–12, and the power input to it is

$$P = V_{ab}I.$$

If a and b are considered as the terminals of the source, then as we have shown,

$$V_{ab} = \mathcal{E} - Ir$$

and hence,

$$P = V_{ab}I = \mathcal{E}I - I^2r. \tag{28–22}$$

The terms $\mathcal{E}I$ and I^2r have the following significance. The emf $\mathcal{E}$ has been defined as the work per unit charge performed on the charges by the nonelectrostatic field $\mathbf{E}_n$ as the charges are pushed "uphill" from b to a in the source. Hence, if a charge ΔQ flows in time Δt, this field does work $\Delta W = \mathcal{E}\Delta Q$, and its *rate* of doing work, or power, is

$$\frac{\Delta W}{\Delta t} = \mathcal{E}\frac{\Delta Q}{\Delta t} = \mathcal{E}I. \tag{28–23}$$

Hence, the product $\mathcal{E}I$ is the rate at which work is done on the circulating charges by the agency that maintains the nonelectrostatic field.

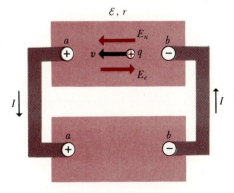

28–13 The rate of conversion of nonelectrical to electrical energy in the source equals $\mathcal{E}I$. The rate of energy dissipation in the source is I^2r. The difference $\mathcal{E}I - I^2r$ is the power output of the source.

The term I^2r is the rate at which energy is *dissipated* in the internal resistance of the source, and the difference $\mathcal{E}I - I^2r$ is the rate at which energy is delivered by the source to the remainder of the circuit. In other words, the power P in Eq. (28–22) represents the net *power output* of the source, or the power delivered to the remainder of the circuit.

3. Power input to a source. Suppose that the lower rectangle in Fig. 28–13 is itself a source of emf *larger* than that of the upper source and with its emf opposite to that of the upper source. The current I in the circuit is then *opposite* to that shown in Fig. 28–13. That is, the lower source pushes current backward through the upper source. The power output of the lower source is

$$P = V_{ab}I.$$

Considering a and b as the terminals of the upper source, we have

$$V_{ab} = \mathcal{E} + Ir,$$

and

$$P = V_{ab}I = \mathcal{E}I + I^2r. \tag{28–24}$$

The terms $\mathcal{E}I$ and I^2r have the following interpretation. The charges in the upper source now move from left to right, in a direction opposite the field $\mathbf{E}_n$, and the work done by the nonelectrostatic force $\mathbf{F}_n$ is *negative*. That is, work is done *on* the agent maintaining the nonelectrostatic field. The product $\mathcal{E}I$ equals the rate at which work is done on this agent, and the term I^2r again equals the rate of dissipation of energy in the internal resistance of the source. The sum $\mathcal{E}I + I^2r$ is therefore the total *power input* to the upper source.

Thus the upper source has become a sink, absorbing electrical energy and converting it to nonelectrical energy. This is exactly what happens when a rechargeable battery (a storage battery) is connected to a charger. The charger supplies electrical energy to the battery; part of it is stored in the battery as chemical energy, to be reconverted later, and the remainder is dissipated as heat in the battery's internal resistance.

EXAMPLE 1   The rate of energy conversion in the source in Fig. 28–7 is

$$\mathcal{E}I + (12 \text{ V})(2 \text{ A}) = 24 \text{ W}.$$

The rate of dissipation of energy in the source is

$$I^2r = (2 \text{ A})^2(2 \text{ }\Omega) = 8 \text{ W}.$$

The power *output* of the source is the difference between these, or 16 W.

The power output is also given by

$$IV_{ab} = (2 \text{ A})(8 \text{ V}) = 16 \text{ W}.$$

The power input to the resistor is

$$V_{a'b'}I = (2 \text{ A})(8 \text{ V}) = 16 \text{ W}.$$

This equals the rate of dissipation of energy in the resistor:

$$I^2R = (2 \text{ A})^2(4 \text{ }\Omega) = 16 \text{ W}.$$ ◄

EXAMPLE 2 The rate of energy conversion in the source in Fig. 28–8 is

$$\mathcal{E}I = (12 \text{ V})(6 \text{ A}) = 72 \text{ W}.$$

The rate of dissipation of energy in the source is

$$I^2r = (6 \text{ A})^2(2 \text{ }\Omega) = 72 \text{ W}.$$

The power *output* of the source (also given by $V_{ab}I$) equals zero. *All of the energy converted is dissipated within the source.* ◄

EXAMPLE 3 The rate of energy conversion in the upper source in Fig. 28–9 is

$$\mathcal{E}I = (12 \text{ V})(0.5 \text{ A}) = 6 \text{ W}.$$

The rate of dissipation of energy in this source is

$$I^2r = (0.5 \text{ A})^2(2 \text{ }\Omega) = 0.5 \text{ W}.$$

The power output of the upper source is 6 W − 0.5 W = 5.5 W.
The rates of dissipation of energy in the 3-Ω and 7-Ω resistors are, respectively,

$$(0.5 \text{ A})^2(3 \text{ }\Omega) = 0.75 \text{ W},$$
$$(0.5 \text{ A})^2(7 \text{ }\Omega) = 1.75 \text{ W}.$$

The rate of energy conversion in the lower source is

$$\mathcal{E}I = (4 \text{ V})(0.5 \text{ A}) = 2 \text{ W},$$

and the rate of energy dissipation in this source is

$$I^2r = (0.5 \text{ A})^2(4 \text{ }\Omega) = 1 \text{ W}.$$

The power *input* to this source is 2 W + 1 W = 3 W. Thus, of the 5.5-W output of the upper source, 2.5 W is dissipated in the two resistors, and the remaining 3 W is partly converted, partly dissipated in the lower source. Conversion of energy in the sources is often partly *reversible*, as in a storage battery where electrical energy is converted to chemical for later retrieval as electrical energy. Energy dissipated in resistors is converted *irreversibly* to heat. ◄

28–7

PHYSIOLOGICAL EFFECTS OF CURRENTS

Electrical potential differences and currents play a vital role in the nervous systems of animals. Conduction of nerve impulses is basically an electrical process, although the mechanism of conduction is much more complex than in simple materials such as metals. A nerve fiber or *axon*, along which an electrical impulse can travel, includes a cylindrical membrane with one conducting fluid (electrolyte) inside and another outside. By mechanisms similar to those in batteries, a potential difference of the order of 0.1 V is maintained between these fluids.

When a pulse is initiated, the membrane temporarily becomes more permeable to the ions in the fluids, leading to a local drop in potential. As the pulse passes, with a typical speed of the order of 30 m·s^{-1}, the membrane recovers and the potential returns to its initial value. There are several aspects of this process that are not yet well understood.

The basically electrical nature of nerve-impulse conduction is responsible for the great sensitivity of the body to externally supplied electrical currents. Currents through the body as small as 0.1 A, much too small to produce significant heating, are fatal because they interfere with nerve processes essential for vital functions such as heartbeat. The *resistance* of the human body is highly variable; body fluids are usually quite good conductors because of their substantial ion concentrations, but the conductivity of skin is relatively low. The resistance between two electrodes grasped by dry hands is of the order 5 kΩ to 10 kΩ. For $R = 10$ kΩ, a current of 0.1 A requires a potential difference $V = IR = (0.1$ A$)(10$ kΩ$) = 1000$ V.

Even much smaller currents can be very dangerous. A current of 0.01 A causes strong, convulsive muscle action and considerable pain, and with 0.02 A the person typically is unable to release a conductor inflicting the shock. Currents of this magnitude and even as small as 0.001 A can cause ventricular fibrillation, a disorganized twitching of heart muscles that occurs instead of regular beating, and that unfortunately pumps very little blood. Surprisingly, very large currents (over 0.1 A) are somewhat *less* likely to cause fatal fibrillation because the heart muscle is "clamped" in one position and is more likely to resume normal beating when the current is removed. Severe burns are, of course, more likely with large currents.

The moral of this rather morbid story, if there is one, is that under certain conditions voltages as small as 10 V can be dangerous, and should not be regarded with anything but respect and caution.

On the positive side, rapidly alternating currents can have beneficial effects. Alternating currents with frequencies the order of 10^6 Hz do not interfere appreciably with nerve processes and can be used for therapeutic heating for arthritic conditions, sinusitis, and a variety of other disorders. If one electrode is made very small, the resulting concentrated heating can be used for local destruction of tissue, such as tumors, or even for cutting tissue in certain surgical procedures.

Study of particular nerve impulses is also an important *diagnostic* tool in medicine. The most familiar examples are electrocardiography (EKG) and electroencephalography (EEG). Electrocardiograms, obtained by attaching electrodes to the chest and back and recording the regularly varying potential differences, are used to study heart function. Similarly, electrodes attached to the scalp permit study of potentials in the brain, and the resulting patterns can be helpful in diagnosing epilepsy, brain tumors, and other disorders.

*28–8

THEORY OF METALLIC CONDUCTION

Additional insight into the phenomenon of conduction can be gained by examining the microscopic mechanisms of conductivity. Here we consider only a crude and primitive model that treats the electrons as classi-

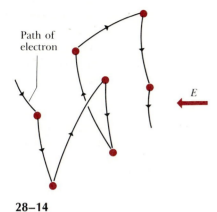

Path of
electron

28–14

cal particles and ignores their inherently quantum-mechanical behavior in solids. Thus this model is not correct conceptually; yet it is useful in helping to develop an intuitive idea of the microscopic basis of conduction.

In the simplest microscopic model of metallic conduction, each atom in the crystal lattice is assumed to give up one or more of its outer electrons. These electrons are then free to move through the crystal lattice, colliding at intervals with the stationary positive ions. Their motion is like that of the molecules of gas in a container and they are often referred to as an "electron gas." In the absence of an electric field, the electrons move in straight lines between collisions, but if there is an electric field, the paths are slightly curved, as in Fig. 28–14, which represents schematically a few free paths of an electron in an electric field directed from right to left. At each collision, the electron is assumed to lose any energy it may have acquired from the field and to make a fresh start. The energy given up in these collisions increases the thermal energy of vibration of the positive ions.

A force $F = eE$ is exerted on each electron by the field, and produces an acceleration a in the direction of the force given by

$$a = \frac{F}{m} = \frac{eE}{m},$$

where m is the electron mass. Let u represent the average *random* speed of an electron, and λ the mean free path (the average distance traveled between collisions). The average time t between collisions, called the *mean free time,* is

$$t = \frac{\lambda}{u}.$$

In this time, the electron acquires a final velocity component v_f in the direction of the force, given by

$$v_f = at = \frac{eE}{m}\frac{\lambda}{u}.$$

Its *average* velocity v in the direction of the force, which is superposed on its random velocity and which we interpret as the *drift* velocity, is one half the final velocity, so

$$v = \frac{1}{2}v_f = \frac{1}{2}\frac{e\lambda}{mu}E.$$

The drift velocity is therefore proportional to the electric field E.

The current density is

$$J = nev = \frac{ne^2\lambda}{2mu}E,$$

and the resistivity is

$$\rho = \frac{E}{J} = \frac{2mu}{ne^2\lambda}. \qquad (28\text{–}25)$$

This is the theoretical expression for the resistivity, and it is in *qualitative* agreement with experiment. At a given temperature, the quantities m, u, n, e, and λ are constant. The resistivity is then constant and Ohm's law is obeyed. When the temperature is increased, the random speed u increases and the theory predicts that the resistivity of a metal increases with increasing temperature.

In a semiconductor, the number of charge carriers per unit volume, n, increases rapidly with increasing temperature. The increase in n far outweighs any increase in u, and the resistivity decreases. At low temperatures, n is very small and the resistivity becomes so large that the material can be considered an insulator.

The modern theory of superconductivity predicts that, in effect, at temperatures below the critical temperature the electrons move freely throughout the lattice. The mean free path λ then becomes very large and the resistivity very small.

QUESTIONS

28–1 A rule of thumb used to determine the internal resistance of a source is that it is the open-circuit voltage divided by the short-circuit current. Is this correct?

28–2 The energy that can be extracted from a storage battery is always less than the energy that goes into it while it is being charged. Why?

28–3 In circuit analysis one often assumes that a wire connecting two circuit elements has no potential difference between its ends; yet there must be an electric field within the wire to make the charges move, and so there must be a potential difference. How do you resolve this discrepancy?

28–4 Long-distance electric-power transmission lines always operate at very high voltage, sometimes as much as 750 kV. What are the advantages of such high voltages? The disadvantages?

28–5 Ordinary household electric lines usually operate at 110 V. Why is this a desirable voltage, rather than a value considerably larger or smaller? What about cars, which usually have 12-V electrical systems?

28–6 What is the difference between an emf and a potential difference?

28–7 Electric power for household and commercial use always uses *alternating current*, which reverses direction 120 times each second. A student claimed that the power conveyed by such a current would have to average out to zero, since it is going one way half the time and the other way the other half. What is your response?

28–8 As discussed in the text, the drift velocity of electrons in a good conductor is very slow. Why then does the light come on so quickly when the switch is turned on?

28–9 A fuse is a device designed to break a circuit, usually by melting, when the current exceeds a certain value. What characteristics should the material of the fuse have?

28–10 What considerations determine the maximum current-carrying capacity of household wiring?

28–11 The text states that good thermal conductors are also good electrical conductors. If so, why don't the cords used to connect toasters, irons, and similar heat-producing appliances get hot by conduction of heat from the heating element?

28–12 Eight flashlight batteries in series have an emf of about 12 V, about the same as that of a car battery. Could they be used to start a car with a dead battery?

28–13 High-voltage power supplies are sometimes designed intentionally to have rather large internal resistance, as a safety precaution. Why is such a power supply with a large internal resistance safer than one with the same voltage but lower internal resistance?

28–14 How would you expect the resistivity of a good insulator such as glass or polystyrene to vary with temperature? Why?

28–15 A would-be inventor proposed to increase the power supplied by a battery to a light bulb by using thick wire near the battery and thinner wire near the bulb. In the thin wire, the electrons from the thick wire would become more densely packed, more electrons per second would reach the bulb, and the energy received by the bulb would be greater than that emitted by the battery. What do you think of this scheme?

PROBLEMS

28–1 A silver wire 1 mm in diameter transfers a charge of 90 C in 1 hr and 15 min. Silver contains 5.8×10^{28} free electrons per m³.

a) What is the current in the wire?

b) What is the drift velocity of the electrons in the wire?

28–2 When a sufficiently high potential difference is applied between two electrodes in a gas, the gas ionizes, electrons moving toward the positive electrode and positive ions toward the negative electrode.

a) What is the current in a hydrogen discharge if, in each second, 4×10^{18} electrons and 1.5×10^{18} protons move in opposite directions past a cross section of the tube?

b) What is the direction of the current?

28–3 A vacuum diode can be approximated by a plane cathode and a plane anode, parallel to each other and 5 mm apart. The area of both cathode and anode is 2 cm². In the region between cathode and anode the current is carried solely by electrons. If the electron current is 50 mA, and the electrons strike the anode surface with a speed of 1.2×10^7 m·s⁻¹, find the number of electrons per cubic millimeter in the space just outside the surface of the anode.

28–4 In the Bohr model of the hydrogen atom the electron makes about 6×10^{15} rev·s⁻¹ around the nucleus. What is the average current at a point on the orbit of the electron?

28–5 Refer to the example at the end of Section 28–1. Assume there are 10^{29} free electrons per cubic meter in the wire, and that the current in the wire is 10 A.

a) What is the current density in the wire?

b) What is the electric field?

c) How long a time is required for an electron to travel the length of the wire?

28–6 A wire 100 m long and 2 mm in diameter has a resistivity of 4.8×10^{-8} Ω·m.

a) What is the resistance of the wire?

b) A second wire of the same material has the same weight as the 100-m length, but twice its diameter. What is its resistance?

28–7 A certain electrical conductor has a square cross section, 2.0 mm on a side, and is 12 m long. The resistance between its ends is 0.072 Ω.

a) What is the resistivity of the material?

b) If the electric field magnitude in the conductor is 0.12 V·m⁻¹, what is the total current?

c) If the material has 8.0×10^{28} free electrons per cubic meter, find the average drift velocity under conditions of part (b).

28–8 In household wiring a copper wire commonly known as "12-gauge" is often used. Its diameter is 2.05 mm. Find the resistance of a 50-m length of this wire.

28–9 What length of copper wire 1.0 mm in diameter would have a resistance of 1.00 Ω?

28–10 The following measurements of current and potential difference were made on a resistor constructed of Nichrome wire:

I, A	V_{ab}, V
0.5	2.18
1.0	4.36
2.0	8.72
4.0	17.44

a) Make a graph of V_{ab} as a function of I.

b) Does the Nichrome obey Ohm's law?

c) What is the resistance of the resistor, in ohms?

28–11 The following measurements were made on a Thyrite resistor:

I, A	V_{ab}, V
0.5	4.76
1.0	5.81
2.0	7.05
4.0	8.56

a) Make a graph of V_{ab} as a function of I. Does Thyrite have constant resistance?

b) Construct a graph of the resistance R as a function of I.

c) Find the dynamic resistance at 2.0 A, and compare with the resistance values found in part (b).

28–12 An aluminum bar 2.5 m long has a rectangular cross section 1 cm by 5 cm.

a) What is its resistance?

b) What would be the length of an iron wire 15 mm in diameter having the same resistance?

28–13 A solid cube of silver has a mass of 84.0 g. What is its resistance between opposite faces?

28–14 The two parallel plates of a capacitor have equal and opposite charges Q. The dielectric has a dielectric constant K and a resistivity ρ. Show that the "leakage" current carried by the dielectric is given by the relationship $i = Q/K\epsilon_0\rho$.

28–15

a) What is the resistance of a Nichrome wire at 0°C, if its resistance is 100.00 Ω at 12°C?

b) What is the resistance of a carbon rod at 30°C, if its resistance is 0.0150 Ω at 0°C?

28–16 A copper wire is initially at 0°C. How much must its temperature rise for the resistance to increase by 10 percent?

28–17 A carbon resistor is to be used as a thermometer. On a winter day when the temperature is 0°C, its resistance is 217.3 Ω. What is the temperature on a hot summer day when the resistance is 214.2 Ω?

28–18 The resistance of a coil of copper wire is 200 Ω at 20°C. What is its resistance at 50°C?

28–19 A toaster using a Nichrome heating element operates on 120 volts. When it is switched on at 0°C, it carries an initial current of 1.5 A. A few seconds later the current reaches the steady value of 1.33 A. What is the final temperature of the element? The average value of the temperature coefficient of Nichrome over the temperature range is 0.00045 (C°)$^{-1}$.

28–20 A certain resistor has a resistance of 150.4 Ω at 20°C and a resistance of 162.4 Ω at 40°C. What is its temperature coefficient of resistivity?

28–21 A resistance thermometer using a platinum wire is used to measure the temperature of a liquid. The resistance is 2.42 Ω at 0°C, and when immersed in the liquid it is 2.98 Ω. The temperature coefficient of resistivity of platinum is 0.0038 (C°)$^{-1}$. What is the temperature of the liquid?

28–22 A piece of wire has a resistance R. It is cut into three pieces of equal length, and the pieces are twisted together in parallel. What is the resistance of the resulting wire?

28–23 What diameter must an aluminum wire have if it is to have the same resistance as an equal length of copper wire of diameter 2.0 mm?

28–24 When switch S is open, the voltmeter V, connected across the terminals of the dry cell in Fig. 28–15, reads

1.52 V. When the switch is closed, the voltmeter reading drops to 1.37 V and the ammeter A reads 1.5 A. Find the emf and internal resistance of the cell. Neglect meter corrections.

28–25 The potential difference across the terminals of a battery is 8.5 V when there is a current of 3 A in the battery from the negative to the positive terminal. When the current is 2 A in the reverse direction, the potential difference becomes 11 V.

a) What is the internal resistance of the battery?

b) What is the emf of the battery?

28–26

a) What is the potential difference V_{ad} in the circuit of Fig. 28–16?

b) What is the terminal voltage of the 4-V battery?

c) A battery of emf 17 V and internal resistance 1 Ω is inserted in the circuit at d, with its positive terminal connected to the positive terminal of the 8-V battery. What is now the difference of potential V_{bc} between the terminals of the 4-V battery?

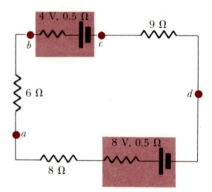

Figure 28–16

28–27 A closed circuit consists of a 12-V battery, a 3.7-Ω resistor, and a switch. The internal resistance of the battery is 0.3 Ω. The switch is opened. What would a high-resistance voltmeter read when placed

a) across the terminals of the battery,

b) across the resistor,

c) across the switch?

Repeat (a), (b), and (c) for the case when the switch is closed.

28–28 The internal resistance of a dry cell increases gradually with age, even though the cell is not used. The emf, however, remains fairly constant at about 1.5 V. Dry cells may be tested for age at the time of purchase by connecting an ammeter directly across the terminals of the cell and reading the current. The resistance of the ammeter is so small that the cell is practically short circuited.

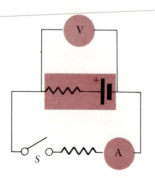

Figure 28–15

a) The short-circuit current of a fresh No. 6 dry cell (1.5-V emf) is about 30 A. Approximately what is the internal resistance?

b) What is the internal resistance if the short-circuit current is only 10 A?

c) The short-circuit current of a 6-V storage battery may be as great as 1000 A. What is its internal resistance?

28–29 The open-circuit terminal voltage of a source is 10 V and its short-circuit current is 4.0 A.

a) What will be the current when the source is connected to a linear resistor of resistance 2 Ω?

b) What will be the current in the Thyrite resistor of Problem 28–11 when connected across the terminals of this source?

c) What is the terminal voltage at this current?

28–30 A "660-W" electric heater is designed to operate from 120-V lines.

a) What is its resistance?

b) What current does it draw?

c) What is the rate of dissipation of energy, in cal·s^{-1}?

d) If the line voltage drops to 110 V, what power does the heater take, in watts? (Assume the resistance constant. Actually, it will change because of the change in temperature.)

28–31 A resistor develops heat at the rate of 360 W when the potential difference across its ends is 180 V. What is its resistance?

28–32 A motor operating on 120 V draws a current of 2 A. If heat is developed in the motor at the rate of 40 J·s^{-1}, what is its efficiency?

28–33

a) Express the rate of dissipation of energy in a resistor in terms of (1) potential difference and current, (2) resistance and current, (3) potential difference and resistance.

b) Energy is dissipated in a resistor at the rate of 40 W when the potential difference between its terminals is 60 V. What is its resistance?

28–34 A storage battery whose emf is 12 V and whose internal resistance is 0.1 Ω is to be charged from a 112-V dc supply. [*Caution:* Don't try to charge a battery directly from a power line, either dc or the more usual ac in household wiring. Real 12-V car batteries have internal resistance of only a few milliohms, and a serious explosion is possible.]

a) Should the + or the − terminal of the battery be connected to the + side of the line?

b) What will be the charging current if the battery is connected directly across the line?

c) Compute the resistance of the series resistor required to limit the current to 10 A.

With this resistor in the circuit, compute

d) the potential difference between the terminals of the battery,

e) the power taken from the line,

f) the power dissipated in the series resistor, and

g) the *useful* power input to the battery.

h) If electrical energy costs 3 cents per kWh, what is the cost of operating the circuit for 2 hr when the current is 10 A?

28–35 In the circuit in Fig. 28–17, find

a) the rate of conversion of internal energy to electrical energy within the battery,

b) the rate of dissipation of energy in the battery,

c) the rate of dissipation of energy in the external resistor.

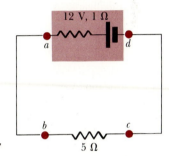

Figure 28–17

28–36 A source whose emf is $\mathcal{E}$ and whose internal resistance is r is connected to an external circuit.

a) Show that the power output of the source is maximum when the current in the circuit is one-half the short-circuit current of the source.

b) If the external circuit consists of a resistance R, show that the power output is maximum when $R = r$, and that the maximum power is $\mathcal{E}^2/4r$.

28–37 A typical small flashlight contains two batteries, each having an emf of 1.5 V, connected in series with a bulb with resistance 15 Ω.

a) If the internal resistance of the batteries is negligible, what is the power delivered to the bulb?

b) If the batteries last for 5 hours, what is the total energy delivered to the bulb?

c) The resistance of real batteries increases as they run down. If the initial internal resistance is negligible, what is the internal resistance of each battery when the power to the bulb has decreased to half its initial value?

28-38 The capacity of a storage battery, such as those used in automobile electrical systems, is rated in amper-hours (A·hr). A 50-A·hr battery can supply a current of 50 A for one hour, or 25 A for two hours, and so on.

a) What total energy is stored in a 12-V, 50-A·hr battery if its internal resistance is negligible?

b) What quantity of gasoline has a total heat of combustion equal to the energy obtained in (a)? (See Section 15-6, near end.)

c) If a windmill-powered generator has an average electrical power output of 300 W, how much time would be required for it to charge the battery fully?

28-39 A certain 12-V storage battery has a capacity of 60 A·hr. (See Problem 28-38.) Its internal resistance is 0.2 Ω. The battery is charged by passing a 15-A current through it for 4 hr.

a) What is the terminal voltage during charging?

b) What total energy is supplied to the battery during charging?

c) What energy is dissipated in the internal resistance during charging?

The battery is now completely discharged through a resistor, again with a constant current of 15 A.

d) What is the external circuit resistance?

e) What total energy is supplied to the external resistor?

f) What total energy is dissipated in the internal resistance?

g) Why are the answers to (b) and (e) not equal?

28-40 Repeat Problem 28-39 with charge and discharge currents of 30 A. What differences in performance do you see?

28-41 A person with body resistance between hands (as described in Section 28-7) of 10 kΩ accidentally grasps the terminals of a 20-kV power supply.

a) If the internal resistance of the power supply is 1000 Ω, what is the current through the person's body? Is this likely to be lethal?

b) What is the power dissipated in the body?

c) If the power supply is to be made safer by increasing its internal resistance, what should the internal resistance be in order for the maximum current in the above situation to be 0.001 A or less?

28-42 The average bulk resistivity of the human body (apart from surface resistance of the skin) is about 5 Ω·m. The conducting path between the hands can be represented approximately as a cylinder 1.6 m long and 0.1 m in diameter. The skin resistance can be made negligible by soaking the hands in salt water (or sea water).

a) What is the resistance between the hands if skin resistance is negligible?

b) What potential difference between the hands is needed for a lethal shock current of 100 mA?

c) With the current in (b), what power is dissipated in the body?

d) Does the result of (b) increase your respect for electrical shock hazards?

29

DIRECT-CURRENT CIRCUITS

Many practical electric circuits contain several sources, resistors, and other circuit elements such as capacitors and motors interconnected in a *network*. In this chapter we study general methods for analyzing networks, including finding unknown voltages, currents, and properties of circuit elements. When several resistors are connected in series or in parallel, they can always be represented as a single equivalent resistor. For more general networks we need two rules called *Kirchhoff's rules*. One is basically the principle of conservation of charge applied to a junction; the other is based on the sum of potential differences around a closed loop. Also discussed are instruments for measuring various electrical quantities. Finally, we discuss the concept of *displacement current* in the charging of capacitors.

29–1

RESISTORS IN SERIES AND IN PARALLEL

Figure 29–1 shows four different ways in which three resistors having resistances R_1, R_2, and R_3 might be connected between points a and b. In (a), the resistors provide only a single path between the points and are said to be connected in *series* between these points. Any number of circuit elements such as resistors, cells, and motors are in series with one another between two points if connected as in (a) so as to provide only a single current path between the points. The *current* is the same in each element.

The resistors in Fig. 29–1b are said to be in *parallel* between points a and b. Each resistor provides an alternative path between the points, and any number of circuit elements similarly connected are in parallel with one another. The *potential difference* is the same across each element.

In Fig. 29–1c, resistors R_2 and R_3 are in parallel with each other, and this combination is in series with the resistor R_1. In Fig. 29–1d, R_2 and R_3 are in series, and this combination is in parallel with R_1.

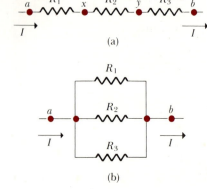

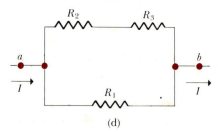

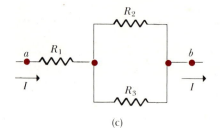

29–1 Four different ways of connecting three resistors.

It is always possible to find a single resistor that could replace a combination of resistors and leave unaltered the potential difference between the terminals of the combination and the current in the rest of the circuit. The resistance of this single resistor is called the *equivalent* resistance of the combination. If any one of the networks in Fig. 29–1 were replaced by its equivalent resistance R, we could write

$$V_{ab} = IR \quad \text{or} \quad R = \frac{V_{ab}}{I},$$

where V_{ab} is the potential difference between the terminals a and b of the network and I is the current at point a or b. To compute an equivalent resistance we assume a potential difference V_{ab} across the actual network, compute the corresponding current I, and take the ratio of one to the other.

If the resistors are in *series*, as in Fig. 29–1a, the current I must be the same in all. Hence

$$V_{ax} = IR_1, \qquad V_{xy} = IR_2, \qquad V_{yb} = IR_3,$$

and

$$V_{ab} = V_{ax} + V_{xy} + V_{yb} = I(R_1 + R_2 + R_3),$$

$$\frac{V_{ab}}{I} = R_1 + R_2 + R_3.$$

But V_{ab}/I is, by definition, the equivalent resistance R. Therefore

$$R = R_1 + R_2 + R_3. \tag{29–1}$$

The equivalent resistance of *any number* of resistors in series equals the sum of their individual resistances.

If the resistors are in *parallel*, as in Fig. 29–1b, the potential difference between the terminals of each must be the same and equal to V_{ab}. If the currents in each are denoted by I_1, I_2, and I_3, respectively,

$$I_1 = \frac{V_{ab}}{R_1}, \qquad I_2 = \frac{V_{ab}}{R_2}, \qquad I_3 = \frac{V_{ab}}{R_3}.$$

Charge is delivered to point a by the current I and removed from a by the currents I_1, I_2, and I_3. Since charge is not accumulating at a, it follows that

$$I = I_1 + I_2 + I_3 = V_{ab}\left(\frac{1}{R_1} + \frac{1}{R_2} + \frac{1}{R_3}\right),$$

or

$$\frac{I}{V_{ab}} = \frac{1}{R_1} + \frac{1}{R_2} + \frac{1}{R_3}.$$

But

$$\frac{I}{V_{ab}} = \frac{1}{R},$$

so

$$\frac{1}{R} = \frac{1}{R_1} + \frac{1}{R_2} + \frac{1}{R_3}. \tag{29–2}$$

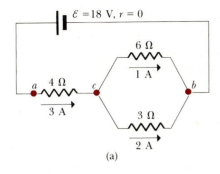

(a)

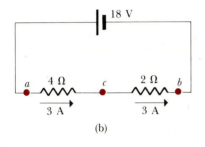

(b)

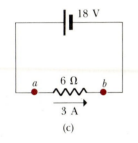

(c)

29–2

For *any number* of resistors in parallel, the *reciprocal* of the equivalent resistance equals the *sum of the reciprocals* of their individual resistances.

For the special case of *two* resistors in parallel,

$$\frac{1}{R} = \frac{1}{R_1} + \frac{1}{R_2} = \frac{R_2 + R_1}{R_1 R_2}$$

and

$$R = \frac{R_1 R_2}{R_1 + R_2}.$$

Also, since $V_{ab} = I_1 R_1 = I_2 R_2$,

$$\frac{I_1}{I_2} = \frac{R_2}{R_1}, \qquad (29\text{--}3)$$

and the currents carried by two resistors in parallel are *inversely proportional* to their resistances.

The equivalent resistances of the networks in Figs. 29–1c and 29–1d could be found by the same general method, but it is simpler to consider them as combinations of series and parallel arrangements. Thus, in (c) the combination of R_2 and R_3 in parallel is first replaced by its equivalent resistance, which then forms a simple series combination with R_1. In (d), the combination of R_2 and R_3 in series forms a simple parallel combination with R_1.

EXAMPLE Compute the equivalent resistance of the network in Fig. 29–2, and find the current in each resistor.

Solution Successive stages in the reduction to a single equivalent resistance are shown in parts (b) and (c). The 6-Ω and the 3-Ω resistors in part (a) are equivalent to the single 2-Ω resistor in part (b), and the series combination of this with the 4-Ω resistor results in the single equivalent 6-Ω resistor in part (c).

In the simple series circuit of part (c), the current is 3 A, and hence the current in the 4-Ω and 2-Ω resistors in part (b) is 3 A also. The potential difference V_{cb} is therefore 6 V, and since it must be 6 V in part (a) as well, the currents in the 6-Ω and 3-Ω resistors in part (a) are 1 A and 2 A, respectively.   ◀

29–2
KIRCHHOFF'S RULES

Not all networks can be reduced to simple series–parallel combinations. An example is a resistance network with a cross connection, as in Fig. 29–3a. A circuit like that in Fig. 29–3b, which contains sources in parallel paths, is another example. No new *principles* are required to compute the currents in these networks, but there are a number of techniques that enable such problems to be handled systematically. We shall describe only one of these, first developed by Gustav Robert Kirchhoff (1824–1887).

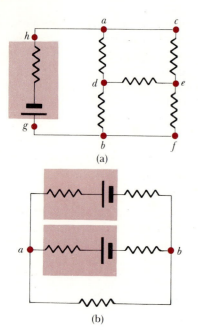

(a)

(b)

29–3 Two networks that cannot be reduced to simple series–parallel combinations of resistors.

We first define two terms. A *branch point* in a network is a point where three or more conductors are joined. A *loop* is any closed conducting path. In Fig. 29–3a, for example, points a, d, e, and b are branch points but c and f are not. In Fig. 29–3b there are only two branch points, a and b.

Some possible loops in Fig. 29–3a are the closed paths *aceda, defbd, hadbgh,* and *hadefbgh.*

Kirchhoff's rules consist of the following two statements:

Point rule   *The algebraic sum of the currents **toward** any branch point is zero:*

$$\sum I = 0. \tag{29–4}$$

Loop rule   *The algebraic sum of the potential differences in any loop,* including those associated with emf's and those of resistive elements, *must equal zero.*

The point rule is an application of the principle of conservation of electric charge. Since no charge can accumulate at a branch point, the total current entering the point must equal the total leaving, or (considering those entering as positive and those leaving as negative) the algebraic sum of currents into a point must be zero.

The loop rule, which we have already encountered in Section 28–4, is an expression of an *energy* relationship; as a charge traverses a loop and returns to its starting point, the sum of the *rises* in potential associated with emf's in the loop must equal the sum of *drops* in potential associated with resistors (or, in more general problems, other circuit elements).

These basic rules are sufficient for the solution of a wide variety of network problems. Usually in such problems some of the emf's, currents, and resistances are known and others are unknown. The number of equations obtained from Kirchhoff's rules must always be equal to the number of unknowns, to permit simultaneous solution of the equations. The principle difficulty is not in understanding the basic ideas but in keeping track of algebraic signs! The following procedures should be followed carefully.

First, all quantities, known and unknown, should be labeled carefully, including an assumed sense of direction for each unknown current and emf. Often one does not know in advance the *actual* direction of an unknown current or emf, but this does not matter. The solution is carried out using the assumed directions, and if the actual direction of a particular quantity is opposite to the assumed direction, the value of the quantity will emerge from the analysis with a negative sign. Hence Kirchhoff's rules, correctly used, give the directions as well as the magnitudes of unknown currents and emf's. This point will be illustrated in the examples to follow.

Usually in labeling currents it is advantageous to use the point rule immediately to express the currents in terms of as few quantities as possible. For example, Fig. 29–4a shows a circuit correctly labeled, and Fig. 29–4b shows the same circuit, relabeled by applying the point rule to point a to eliminate I_3.

The following guidelines will help with the problem of signs:

1. Choose any closed loop in the network, and designate a direction (clockwise or counterclockwise) to traverse the loop in applying the loop rule.

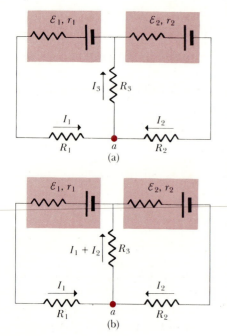

(a)

(b)

29–4 Application of the point rule to point a reduces the number of unknown currents from three to two.

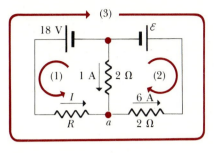

29–5

2. Go around the loop in the designated direction, adding emf's and potential differences. An emf is counted as positive when it is traversed from (−) to (+) (in the direction of the non-electrostatic field E_n in the source) and negative when from (+) to (−). An IR product is counted as negative if the resistor is traversed in the *same* direction as the assumed current, positive if in the opposite direction.

3. Equate the sum of step (2) to zero.

4. If necessary, choose another loop to obtain a different relation between the unknowns, and continue until there are as many equations as unknowns, or until every circuit element has been included in at least one of the chosen loops.

EXAMPLE 1 In the circuit shown in Fig. 29–5, find the unknown current I, resistance R, and emf $\mathcal{E}$.

Solution Application of the point rule to point a yields the relation

$$I + 1 \text{ A} - 6 \text{ A} = 0,$$
$$I = 5 \text{ A}.$$

To determine R we apply the loop rule to the loop labeled (1), obtaining

$$18 \text{ V} - (5 \text{ A})R + (1 \text{ A})(2 \text{ }\Omega) = 0,$$
$$R = 4 \text{ }\Omega.$$

The term for resistance R is negative because our loop traverses that element in the same direction as the current and hence finds a potential *drop*, while the term for the 2-Ω resistor is positive because in traversing it in the direction opposite to the current we find a potential *rise*. If we had chosen to traverse loop (1) in the opposite direction, every term would have had the opposite sign, and the result for R would have been the same.

To determine $\mathcal{E}$, we apply the loop rule to loop (2);

$$\mathcal{E} + (6 \text{ A})(2 \text{ }\Omega) + (1 \text{ A})(2 \text{ }\Omega) = 0,$$
$$\mathcal{E} = -14 \text{ V}.$$

This shows that the actual polarity of this emf is opposite to that assumed, and that the positive terminal of this source is really on the left side. Alternatively, one could use loop (3), obtaining the equation

$$\mathcal{E} + (6 \text{ A})(2 \text{ }\Omega) + (5 \text{ A})(4 \text{ }\Omega) - 18 \text{ V} = 0,$$

from which again $\mathcal{E} = -14$ V. ◄

EXAMPLE 2 In Fig. 29–6, find the current in each resistor, and find the equivalent resistance.

Solution As pointed out at the beginning of this section, it is not possible to represent this network in terms of series and parallel combinations. There are five different currents to determine, but by applying the point rule to junctions a and b we represent them in terms of three

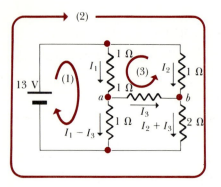

29–6

unknown currents, as indicated in the figure. The current in the battery is of course $(I_1 + I_2)$.

We now apply the loop rule to the three loops shown, obtaining the following three equations:

$$13 \text{ V} - I_1(1 \ \Omega) - (I_1 - I_3)(1 \ \Omega) = 0; \tag{1}$$

$$-I_2(1 \ \Omega) - (I_2 + I_3)(2 \ \Omega) + 13 \text{ V} = 0; \tag{2}$$

$$-I_1(1 \ \Omega) - I_3(1 \ \Omega) + I_2(1 \ \Omega) = 0. \tag{3}$$

This is a set of three simultaneous equations for the three unknown currents. They may be solved by various methods; one straightforward procedure is to solve the third for I_2, obtaining $I_2 = I_1 + I_3$, and then substitute this expression into the first two equations to eliminate I_2. When this is done one obtains the two equations

$$13 \text{ V} = I_1(2 \ \Omega) - I_3(1 \ \Omega), \tag{1'}$$

$$13 \text{ V} = I_1(3 \ \Omega) + I_3(5 \ \Omega). \tag{2'}$$

Now I_3 may be eliminated by multiplying the first of these by 5 and adding the two to obtain

$$78 \text{ V} = I_1(13 \ \Omega),$$

from which $I_1 = 6$ A. This result may be substituted back into (1') to obtain $I_3 = -1$ A, and finally from (3) we find $I_2 = 5$ A. We note that the direction of I_3 is opposite that of the initial assumption.

The total current through the network is $I_1 + I_2 = 11$ A, and the potential drop across it is equal to the battery emf, namely, 13 V. Thus the equivalent resistance of the network is

$$R = \frac{13 \text{ V}}{11 \text{ A}} = 1.18 \ \Omega. \qquad \blacktriangleleft$$

Once the currents in a circuit are determined, the potential difference between any two points a and b can be found by the procedure discussed in Section 28–4, just preceding Example 4. To find $V_{ab} = V_a - V_b$ (the potential at a with respect to b) we start at b and add the potential changes encountered as we go from b to a. An emf is considered positive when we go from $(-)$ to $(+)$, negative otherwise. An IR term is positive when we go "uphill," against the current direction, negative when in the same direction as the current.

EXAMPLE 3 In the circuit of Example 2 (Fig. 29–6) find the potential difference V_{ab}.

Solution Starting at point b, we follow a path to point a, adding potential rises and drops as we go. The simplest path is through the center 1-Ω resistor. We have found $I_3 = -1$ A, showing that the actual current direction in this branch is from right to left. Thus as we go from b to a there is a drop of potential of magnitude $IR = (1 \text{ A})(1 \ \Omega) = 1$ V, and $V_{ab} = -1$ V. Alternatively, we may go around the lower loop. We then have

$$I_2 + I_3 = 5 \text{ A} + (-1 \text{ A}) = 4 \text{ A},$$

$$I_1 - I_3 = 6 \text{ A} - (-1 \text{ A}) = 7 \text{ A},$$

and

$$V_{ab} = -(4 \text{ A})(2 \text{ }\Omega) + (7 \text{ A})(1 \text{ }\Omega) = -1 \text{ V}.$$

The reader may try other paths from b to a to verify that they also give this result. ◄

29–3

AMMETERS AND VOLTMETERS

The most common instruments for measuring potential difference ("voltage") or current use a device called a *d'Arsonval galvanometer*. A pivoted coil of fine wire is placed in the magnetic field of a permanent magnet, as shown in Fig. 29–7. When there is a current in the coil, the magnetic field exerts on the coil a *torque* that is proportional to the current. (This magnetic interaction is discussed in detail in Chapter 30.) This torque is opposed by a spring, similar to the hairspring on the balance wheel of a watch, which exerts a restoring torque proportional to the angular displacement.

Thus the angular deflection of the indicator needle attached to the pivoted coil is directly proportional to the coil current, and the device can be calibrated to measure current. The maximum deflection for which the meter is designed, typically 90° to 120°, is called *full-scale deflection*. The current required to produce full-scale deflection (typically of the order of 10 μA to 10 mA) and the resistance of the coil (typically of the order of 10 to 1000 Ω) are the essential electrical characteristics of the meter.

The meter deflection is proportional to the *current* in the coil, but if the coil obeys Ohm's law, the current is proportional to the *potential difference* between the terminals of the coil. Thus the deflection is also proportional to this potential difference. For example, consider a meter whose

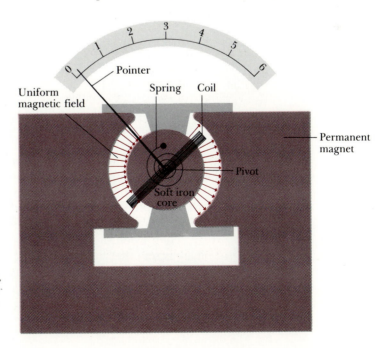

29–7 A d'Arsonval meter movement, showing pivoted coil with attached pointer, permanent magnet supplying uniform magnetic field, and spring to provide restoring torque, which opposes magnetic-field torque.

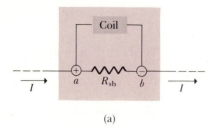

(a)

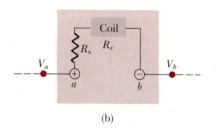

(b)

29–8 (a) Internal connections of an ammeter. (b) Internal connections of a voltmeter.

coil has a resistance of 20 Ω, and which deflects full scale with a current of 1 mA in its coil. The corresponding potential difference is

$$V_{ab} = IR = (10^{-3} \text{ A})(20 \text{ }\Omega) = 0.020 \text{ V} = 20 \text{ mV}.$$

A current-measuring device is usually called an *ammeter* (or milliammeter, microammeter, etc., depending on the range). Such a device always measures the current passing through it. A meter such as the one just described can be adapted to measure currents larger than its full-scale reading by connecting a resistor in parallel with it, as in Fig. 29–8a, so that some of the current bypasses the meter. The parallel resistor is called a *shunt resistor* or simply a *shunt*, symbol R_{sh}.

For example, suppose we need an ammeter with a range of 0 A to 10 A, based on the 1-mA meter described above. We must choose a shunt so that the total current I through both meter and shunt is 10 A when the current through the meter itself is 1 mA = 0.001 A. Thus the current in the shunt is 9.999 A at full-scale deflection. The potential difference across the shunt is the same as that across the meter, namely 0.020 V, so the shunt resistance must be (from Ohm's law)

$$R_{sh} = \frac{0.020 \text{ V}}{9.999 \text{ A}} = 0.00200 \text{ }\Omega.$$

The equivalent resistance R of the instrument is

$$\frac{1}{R} = \frac{1}{R_c} + \frac{1}{R_{sh}},$$

and

$$R = 0.00200 \text{ }\Omega.$$

Thus we have a low-resistance instrument with the desired range of 0 to 10 A. Of course, if the current I is *less* than 10 A, the coil current (and the deflection) is correspondingly less, also.

An ideal ammeter would have zero resistance, so that including it in a branch of a circuit would not affect the current in that branch. Real ammeters always have some finite resistance, but it is always desirable for an ammeter to have the smallest practical resistance.

This same 1-mA meter may also be used to measure potential difference or *voltage;* any voltage-measuring device is usually called a *voltmeter* (or millivoltmeter, etc., depending on range). A voltmeter always measures the potential difference between two points, and its terminals must be connected to these points. Our 1-mA meter may be used as a voltmeter, but the maximum voltage it can measure is 20 mV. The range may be extended by connecting a resistor R_s in series with the meter, as in Fig. 29–8b, so that only some fraction of the total potential difference appears across the meter itself, and the remainder across R_s.

For example, suppose we need a voltmeter with a maximum range of 10 V. Then when the voltage across the meter is 20 mV = 0.020 V, the voltage across the series resistor R_s must be 10 V − 0.020 V, or 9.98 V. The current through the meter at full-scale deflection is still 1 mA or 0.001 A, so from Ohm's law the value of R_s must be

$$R_s = \frac{9.98 \text{ V}}{0.001 \text{ A}} = 9980 \text{ }\Omega.$$

The equivalent resistance of the device is then

$$R = R_c + R_s = 10,000 \ \Omega.$$

An ideal voltmeter would have infinite resistance, so that connecting it between two points in a circuit would not alter any of the currents. Real voltmeters always have finite resistance; to be useful a voltmeter must have large enough resistance so that connecting it in a circuit does not change the other currents appreciably.

A voltmeter and an ammeter can be used together to measure *resistance* and *power*. The resistance of a resistor equals the potential difference V_{ab} between its terminals, divided by the current I:

$$R = \frac{V_{ab}}{I},$$

and the power input to any portion of a circuit equals the product of the potential difference across this portion and the current:

$$P = V_{ab}I.$$

The most straightforward method of measuring R or P is therefore to measure V_{ab} and I simultaneously.

In Fig. 29–9a, ammeter A reads correctly the current I in the resistor R. Voltmeter V, however, reads the *sum* of the potential difference V_{ab} across the resistor and the potential difference V_{bc} across the ammeter.

If we transfer the voltmeter terminal from c to b, as in Fig. 29–9b, the voltmeter reads correctly the potential difference V_{ab} but the ammeter now reads the *sum* of the current I in the resistor and the current I_V in the voltmeter. Thus, whichever connection is used, we must correct the reading of one instrument or the other to obtain the true values of V_{ab} or I (unless, of course, the corrections are small enough to be neglected).

EXAMPLE 1 The circuit of Fig. 29–9a is to be used to measure an unknown resistor R. The meter resistances are $R_V = 10,000 \ \Omega$ and $R_A = 2.0 \ \Omega$. If the voltmeter reads 12.0 V and the ammeter reads 0.10 A, what is the true resistance?

Solution If the meters were ideal (i.e., $R_V = \infty$ and $R_A = 0$), the resistance would be simply $R = V/I = (12.0 \text{ V})/(0.10 \text{ A}) = 120 \ \Omega$. But the voltmeter reading includes the potential V_{bc} across the ammeter as well as that (V_{ab}) across the resistor. We have $V_{bc} = IR_A = (0.10 \text{ A})(2.0 \ \Omega) = 0.2 \text{ V}$, so the actual potential drop V_{ab} across the resistor is 12.0 V − 0.2 V = 11.8 V, and the resistance is

$$R = \frac{V_{ab}}{I} = \frac{11.8 \text{ V}}{0.10 \text{ A}} = 118 \ \Omega.$$ ◀

EXAMPLE 2 Suppose the same meters are connected instead as shown in Fig. 29–9b, and the above readings are obtained. What is the true resistance?

Solution In this case the voltmeter measures the potential across the resistor correctly; the difficulty is that the ammeter measures the voltme-

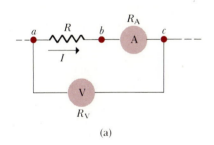

(a)

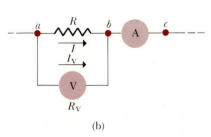

(b)

29–9 Ammeter–voltmeter method for measuring resistance or power.

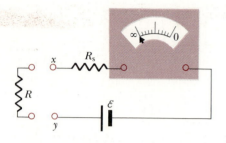

29–10 Ohmmeter circuit. The backward scale on the meter is calibrated to read resistance directly.

ter current I_V as well as the current I in the resistor. We have $I_V = V/R_V = (12.0 \text{ V})/(10,000 \text{ }\Omega) = 1.2$ mA; so the actual current I in the resistor is $I = 0.10 \text{ A} - 0.0012 \text{ A} = 0.0988$ A. Thus the resistance is

$$R = \frac{V_{ab}}{I} = \frac{12.0 \text{ V}}{0.0988 \text{ A}} = 121.5 \text{ }\Omega. \quad \blacktriangleleft$$

An alternate method for measuring resistance is to use a d'Arsonval meter in an arrangement called an *ohmmeter*. It consists of a meter, a resistor, and a source (often a flashlight cell) connected in series, as in Fig. 29–10. The resistance R to be measured is connected between terminals x and y.

The series resistance R_s is chosen so that when terminals x and y are short-circuited (that is, when $R = 0$), the meter deflects full scale. When the circuit between x and y is open (that is, when $R = \infty$), the galvanometer shows no deflection. For a value of R between zero and infinity, the galvanometer deflects to some intermediate point depending on the value of R, and hence the meter scale can be calibrated to read the resistance R. Larger currents correspond to smaller resistances, so this scale reads backward compared to the current scale.

*29–4

THE *R–C* SERIES CIRCUIT

In the circuits considered thus far, we have assumed that the emf's and resistances are constant, so that all potentials and currents are constant, independent of time. Figure 29–11 shows a simple example of a circuit in which the current and voltages are *not* constant. The capacitor is initially uncharged; at some time $t = 0$ we close the switch, completing the circuit and permitting current around the loop to begin charging the capacitor.

Because there is no initial charge on the capacitor, the initial potential difference across it is zero. The entire battery voltage appears across the resistor, causing an initial current $I = V/R$. As the capacitor charges, its voltage increases, and the potential difference across the resistor decreases, corresponding to a decrease in current. After a long time the capacitor has become fully charged, the entire battery voltage V appears across the capacitor, there is no potential difference across the resistor, and the current becomes zero.

Let q represent the charge on the capacitor and i the charging current at some instant after the switch is closed. (As is customary in circuit analysis, we use small letters for quantities that vary with time, and capital letters for constant quantities.) The instantaneous potential differences v_{ac} and v_{cb} are

$$v_{ac} = iR, \qquad v_{cb} = \frac{q}{C}. \tag{29–5}$$

Therefore,

$$V_{ab} = V = v_{ac} + v_{cb} = iR + \frac{q}{C}, \tag{29–6}$$

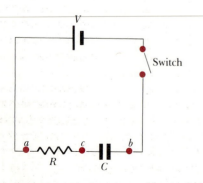

29–11

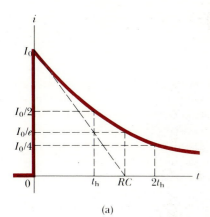

(a)

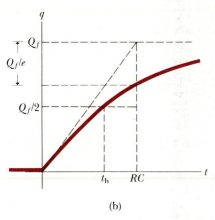

(b)

29–12

where V = constant. The current i is then

$$i = \frac{V}{R} - \frac{q}{RC}. \tag{29–7}$$

At the instant the switch is closed, $q = 0$; the *initial current* is $I_0 = V/R$, which would be the steady current if the capacitor were not present.

As the charge q increases, the term q/RC becomes larger, and the current decreases and eventually becomes zero. When $i = 0$,

$$\frac{V}{R} = \frac{q}{RC}, \qquad q = CV = Q_f, \tag{29–8}$$

where Q_f is the final charge on the capacitor.

The behavior of the current and capacitor charge as functions of time are shown in Fig. 29–12. At the instant the switch is closed ($t = 0$), the current jumps from zero to its initial value $I_0 = V/R$, and then gradually decreases to zero. The capacitor charge starts at zero and gradually approaches the final value $Q_f = CV$. Obtaining actual expressions describing the variation of i and q with time requires methods of calculus, and the expressions contain exponential functions. We quote the results below, but those who are unfamiliar with exponential and logarithmic functions may omit the following discussion.

The current and charge are given as functions of time by

$$i = I_0 e^{-t/RC}, \tag{29–9}$$

$$q = Q_f(1 - e^{-t/RC}). \tag{29–10}$$

Figure 29–12a is a graph of Eq. (29–9), and Fig. 29–12b is a graph of Eq. (29–10). At a time $t = RC$, the current has decreased to $1/e$ (about 0.368) of its initial value and the charge has increased to *within* $1/e$ of its final value. The product RC is called the *time constant*, or the *relaxation time*, of the circuit, denoted by τ:

$$\tau = RC. \tag{29–11}$$

The *half-life* of the circuit, t_h, is the time for the current to decrease to half its initial value, or for the capacitor to acquire half its final charge. Setting $i = I_0/2$ in Eq. (29–9), we find

$$t_h = RC \ln 2 = \tau \ln 2 = 0.693\, RC.$$

The half-life depends only on the time constant RC, and not on the initial current. Thus if the current decreases from I_0 to $I_0/2$ in a time t_h, as shown in Fig. 29–12a, it decreases to half this value, or to $I_0/4$, in another half-life, and so on.

EXAMPLE   A resistor of resistance $R = 10$ MΩ is connected in series with a capacitor of capacitance $1\,\mu$F. The time constant is

$$\tau = RC = (10 \times 10^6\ \Omega)(10^{-6}\ \text{F}) = 10\ \text{s},$$

and the half-life is

$$t_h = (10\ \text{s})(\ln 2) = 6.9\ \text{s}.$$

On the other hand, if $R = 10\ \Omega$, the time constant is only 10×10^{-6} s, or $10\ \mu$s.   ◀

*29–5

DISPLACEMENT CURRENT

For a *conducting* circuit, in the steady state, the total current *into* any given portion must be equal to the current *out of* that portion. This statement forms the basis of Kirchhoff's point rule, discussed in Section 29–2. However, this rule is *not* obeyed for a capacitor that is being charged. In Fig. 29–13, there is a conduction current *into* the left plate but no conduction current *out of* this plate; similarly, there is conduction current out of the right plate, but none into it.

James Clerk Maxwell (1831–1879) showed that it is possible to generalize the definition of current so that one can still say that the current out of each plate is equal to the current into it. As the capacitor charges, the conduction current increases the charge on each plate, and this in turn increases the electric field between the plates. The *rate* of increase of field is proportional to the conduction current; Maxwell's scheme was to associate an equivalent current density with this rate of increase of field.

Assuming, for simplicity, that the capacitor consists of two parallel plates with uniform charge density σ, the field E between the plates is given by

$$E = \frac{\sigma}{\epsilon_0} = \frac{Q}{\epsilon_0 A}. \tag{29–12}$$

In a time interval Δt, the charge Q increases by $\Delta Q = I\,\Delta t$; the corresponding change in E is

$$\Delta E = \frac{\Delta Q}{\epsilon_0 A} = \frac{I\,\Delta t}{\epsilon_0 A},$$

and the *rate* of change of E is

$$\frac{\Delta E}{\Delta t} = \frac{I}{\epsilon_0 A}. \tag{29–13}$$

We now define an *equivalent current density* J_D between the plates as

$$J_D = \epsilon_0 \frac{\Delta E}{\Delta t}. \tag{29–14}$$

Then the total equivalent current between the plates, which we may call I_D, is

$$I_D = J_D A = \epsilon_0 \left(\frac{\Delta E}{\Delta t}\right) A = I. \tag{29–15}$$

This equivalent current may also be expressed in terms of electric flux, as defined in Section 25–5. In this case the electric flux is simply $\Psi = EA$, so the change in electric flux $\Delta\Psi$ during the time interval Δt is $(\Delta E)A$, and Eq. (29–15) may be written as

$$I_D = \epsilon_0 \frac{\Delta\Psi}{\Delta t}. \tag{29–16}$$

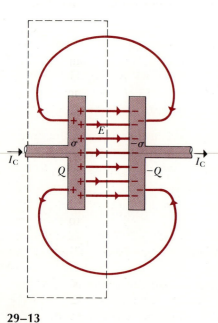

29–13

Thus, if we include the "equivalent current" as well as the conduction current, the current *into* the region bounded by the broken line in Fig. 29–13 equals the current *out* of this region. The subscript D is chosen because Maxwell originally called this equivalent current the *displacement current*.

If the space between capacitor plates contains a dielectric, Eq. (29–12) must be replaced by the more general relation derived in Section 27–5:

$$E = \frac{\sigma}{\epsilon} = \frac{Q}{\epsilon A} = \frac{Q}{K\epsilon_0 A}. \tag{29-17}$$

The corresponding modification of the definition of displacement current density is

$$J_D = \epsilon \left(\frac{\Delta E}{\Delta t} \right) = K\epsilon_0 \left(\frac{\Delta E}{\Delta t} \right). \tag{29-18}$$

Maxwell's generalized view of current and current density may appear to be merely an artifice introduced to preserve Kirchhoff's current rule even in cases where charge is accumulating in a certain region of space, such as a capacitor plate. However, it is much more than this. In a later chapter we shall study the role of current as a source of *magnetic* field, and we shall see that a *displacement* current sets up a magnetic field in exactly the same way as an ordinary *conduction* current. Thus displacement current, far from being an artifice, is a fundamental fact of nature, and Maxwell's discovery of it was the bold step of an extraordinary genius. The concept of displacement current provided the necessary basis for the understanding of electromagnetic waves in the last third of the nineteenth century.

29–6

POWER DISTRIBUTION SYSTEMS

We conclude this chapter with a brief discussion of practical household and automotive electric-power distribution systems. Automobiles use direct-current systems, while nearly all household, commercial, and industrial systems use alternating current (ac). Most of the same basic wiring concepts apply to both. Alternating-current circuits are discussed in greater detail in Chapter 34.

The various lamps, motors, and other appliances to be operated are always connected in *parallel* to the power source, the wires from the power company in the case of houses, the battery and alternator for a car. The basic idea of house wiring is shown in Fig. 29–14. One side of the "line," as the pair of conductors is called, is always connected to "ground." For houses this is an actual electrode driven into the earth (usually a good conductor) and also connected to the household water pipes. Electricians speak of the "hot" side and the "ground" side of the line.

The voltage–current–power relations are the same as those discussed in Section 28–6. Household voltage is nominally 120 V in the United States and Canada, often 240 V in Europe. These are actually the

29–14 Schematic diagram of house wiring system. Only two branch circuits are shown; an actual system might have four to thirty branch circuits. The conventional symbol for ground is shown. Various lamps and appliances may be plugged into the outlets. The grounding conductors, which normally carry no current, are not shown.

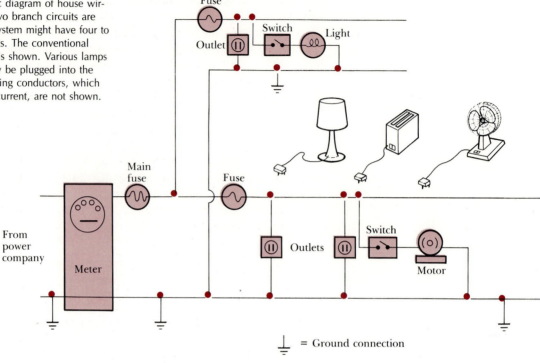

root-mean-square voltages, as discussed in Section 34–4. The current in a 100-W light bulb, for example, can be determined from Eq. (28–21):

$$I = \frac{P}{V} = \frac{100 \text{ W}}{120 \text{ V}} = 0.83 \text{ A}.$$

The resistance of this bulb at operating temperature is given by

$$R = \frac{V}{I} = \frac{120 \text{ V}}{0.83 \text{ A}} = 144 \ \Omega,$$

or

$$R = \frac{V^2}{P} = \frac{(120 \text{ V})^2}{100 \text{ W}} = 144 \ \Omega.$$

Similarly, a 1500-W waffle iron draws a current of (1500 W)/(120 V) = 12.5 A and has a resistance at operating temperature of 9.6 Ω.

The maximum current available from an individual circuit is determined by the resistance of the wires; the power loss in the wires from I^2R heat causes them to become hot, and in extreme cases this heat can cause a fire or melt the wires. Ordinary lighting and outlet wiring in houses is usually 12-gauge wire, which has a diameter of 2.05 mm and can carry a maximum current of 20 A without overheating. Larger sizes such as 8-gauge (3.26 mm) or 6-gauge (4.11 mm) are used for high-current appliances such as ranges and clothes dryers, and 2-gauge (6.54 mm) or larger for the main entrance lines.

Protection against overloading and overheating of circuits is provided by fuses or circuit breakers. A *fuse* contains a link of a lead-tin alloy with a very low melting temperature; the link melts and breaks the circuit when its rated current is exceeded. A *circuit breaker* is an electromechanical device that performs the same function, using an electromagnet or a bimetallic strip to "trip" the breaker and interrupt the circuit when the current exceeds a specified value. Circuit breakers have the advantage that they can be reset after they are tripped; a blown fuse must be replaced, but fuses are somewhat more reliable in operation than circuit breakers.

The reader may have had the experience of blowing a fuse by plugging too many high-current appliances into the same outlet or into two outlets fed by the same circuit. *Do not* replace the fuse with one of larger rating; to do so risks overheating wires and starting a fire. The only safe solution is to distribute the appliances among several circuits. Modern kitchens often have three or four separate 20-A circuits.

Accidental contact between the hot and ground sides of the line is called a *short circuit;* such a situation, which can be caused by faulty insulation or a variety of kinds of mechanical malfunction, provides a very low-resistance current path, permitting a very large current that would quickly melt the wires and ignite their insulation if the current were not interrupted by a fuse or circuit breaker. The opposite situation, a broken wire that interrupts the current path, is called an *open circuit*. An open circuit can be hazardous because of the sparking that can occur at the point of intermittent contact.

In approved wiring practice a fuse or breaker is placed *only* in the hot side of the line, never in the ground side. Otherwise, if a short circuit should develop because of faulty insulation or other malfunction, the ground-side fuse could blow, leaving the hot side alive and posing a shock hazard to a person who touched the live conductor and a grounded object such as a water pipe. For similar reasons, the wall switch for a light fixture is always in the hot side of the line, never the ground side.

Further protection against shock hazard is provided by a third conductor included in all present-day wiring. This conductor (corresponding to the long round prong of the three-prong connector plug on an appliance or power tool) normally carries no current, but it connects the metal case or frame of the device to ground. If a conductor on the hot side of the line accidentally contacts the frame or case, the grounding conductor provides a current path and the fuse blows. Without the ground wire, the frame could become "live," i.e., at a potential 120 V above ground; then touching it and a water pipe (or even a damp basement floor) at the same time would offer a shock hazard. In some special situations, especially outlets located outdoors or near a sink or other water pipes, a special kind of circuit breaker called a *ground-fault interrupter* is used. This device senses the current in the grounding conductor (which is normally zero) and trips when it exceeds some very small value, typically 10 mA.

All the above discussion can be applied directly to automobile wiring. The voltage is 12 V; the power is supplied by the battery and by the

29–15 Simplified schematic diagram of a 120–240-V house wiring system. Only one circuit on each side is shown; in actual systems there would be several 120-V circuits on each side of the neutral. Grounding wires are not shown.

alternator, which charges the battery when the engine is running. The ground side of the circuits is connected to the body and frame of the vehicle. There is no separate grounding conductor, and indeed the frame itself sometimes provides the ground side of the circuit. The fuse or circuit breaker arrangement is the same in principle as in household wiring. Because of the lower voltage, more current is required for the same power; a 100-W headlight bulb requires a current of (100 W)/(12 V) = 8.3 A.

To help prevent wiring errors, household wiring uses a standardized color code in which the hot side of a line has black or red insulation, the ground side has white insulation, and the grounding conductor is bare or has green insulation. In electronic devices and equipment, on the other hand, the ground side of the line is always black! Beware!

Most household wiring systems actually use a slight elaboration of the system described above. The power company provides *three* conductors. One is grounded or "neutral"; the other two are both at 120 V with respect to the neutral, but with opposite polarity, giving them a voltage between them of 240 V. Thus 120-V lamps and appliances can be connected between neutral and either hot conductor, and high-power devices requiring 240 V are connected between the two hot lines, as shown in Fig. 29–15. Ranges and dryers usually are designed for 240-V power input.

Although we have spoken of *power* in the above discussion, what we buy from the power company is *energy*. Power is energy transferred per unit time, so energy is power multiplied by time. The usual unit of energy sold by the power company is the kilowatt-hour (1 kWh):

$$1 \text{ kWh} = (10^3 \text{ W})(3600 \text{ s}) = 3.6 \times 10^6 \text{ W} \cdot \text{s} = 3.6 \times 10^6 \text{ J}.$$

One kilowatt-hour typically costs 5 to 10 cents, depending on location and quantity of energy purchased. To operate a 1500-W (1.5-kW) waffle iron for one hour requires 1.5 kWh of energy and costs 7.5 to 15 cents (not including the cost of flour and eggs in the waffle batter). The cost of operating any lamp or appliance for a specified time can be calculated in the same way if the power rating is known.

QUESTIONS

29–1 Can the potential difference between terminals of a battery ever be opposite in direction to the emf?

29–2 Why do the lights on a car become dimmer when the starter is operated?

29–3 What determines the maximum current that can be carried safely by household wiring? (Typical limits are 15 A for 14-gauge wire, 20 A for 12-gauge, and so on.)

29–4 Lights in a house often dim momentarily when a motor, such as a washing machine or a power saw, is turned on. Why does this happen?

29–5 Compare the formulas for resistors in series and parallel with those for capacitors in series and parallel. What similarities and differences do you see? Sometimes in circuit analysis one uses the quantity *conductance*, denoted as g and defined as the reciprocal of resistance: $g = 1/R$. What is the corresponding comparison for conductance and capacitance?

29–6 Is it possible to connect resistors together in a way that cannot be reduced to some combination of series and parallel combinations? If so, give examples; if not, state why not.

29–7 In a two-cell flashlight, the batteries are usually connected in series. Why not connect them in parallel?

29–8 Some Christmas-tree lights have the property that, when one bulb burns out, all the lights go out, while with others only the burned-out bulb goes out. Discuss this difference in terms of series and parallel circuits.

29–9 What possible advantage could there be in connecting several identical batteries in parallel?

29–10 Two 110-V lightbulbs, one 25-W and one 200-W, were connected in series across a 220-V line. It seemed like a good idea at the time, but one bulb burned out almost instantaneously. Which one burned out, and why?

29–11 When the direction of current in a battery reverses, does the direction of its emf also reverse?

29–12 Under what conditions would the terminal voltage of a battery be zero?

29–13 What sort of meter should be used to test the condition of a dry cell (such as a flashlight battery) having constant emf but internal resistance that increases with age and use?

29–14 For very large resistances, it is easy to construct RC circuits having time constants of several seconds or minutes. How might this fact be used to measure very large resistances, too large to measure by more conventional means?

PROBLEMS

29–1 Three resistors having resistances of 1 Ω, 2 Ω, and 3 Ω are connected in series to a 12-V battery having negligible internal resistance.

a) Find the equivalent resistance of the combination.

b) Find the current in each resistor.

c) Find the total current through the battery.

d) Find the voltage across each resistor.

e) Find the power dissipated in each resistor.

29–2 In Problem 29–1, the same three resistors are connected in parallel to the same battery. Answer the same five questions for this situation.

29–3 Two lamps, marked "60 W, 120 V" and "40 W, 120 V," are connected in series across a 120-V line. What power is consumed in each lamp? Assume that the resistance of the filaments does not vary with current.

29–4 Three equal resistors are connected in series. When a certain potential difference is applied across the combination, the total power dissipated is 10 W. What power would be dissipated if the three resistors were connected in parallel across the same potential difference?

29–5

a) The power rating of a 10,000-Ω resistor is 2 W. (The power rating is the maximum power the resistor can safely dissipate without too great a rise in temperature.) What is the maximum allowable potential difference across the terminals of the resistor?

b) A 20,000-Ω resistor is to be connected across a potential difference of 300 V. What power rating is required?

c) It is desired to connect an equivalent resistance of 1000 Ω across a potential difference of 200 V. A number of 10-W, 1000-Ω resistors are available. How should they be connected?

29–6 A 60-Ω resistor and a 90-Ω resistor are connected in parallel and the combination is connected across a 120-V dc line.

a) What is the resistance of the parallel combination?

b) What is the total current through the parallel combination?

c) What is the current through each resistor?

29–7 A 1000-Ω 2-W resistor is needed, but only several 1000-Ω 1-W resistors are available.

a) How can the required resistance and power rating be obtained by a combination of the available units?

b) What power is then dissipated in each resistor?

29–8 A 25-W, 120-V lightbulb and a 100-W, 120-V lightbulb are connected in series across a 240-V line. Assume that the resistance of each bulb does not vary with current.

a) Find the current through the bulbs.

b) Find the power dissipated in each bulb.

c) One bulb burns out very quickly. Which one? Why?

29–9

a) Calculate the equivalent resistance of the circuit of Fig. 29–16 between x and y.

b) What is the potential difference between x and a if the current in the 8-Ω resistor is 0.5 A?

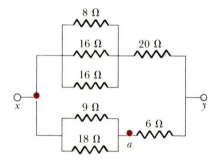

Figure 29–16

29–10 Each of three resistors in Fig. 29–17 has a resistance of 2 Ω and can dissipate a maximum of 18 W without becoming excessively heated. What is the maximum power the circuit can dissipate?

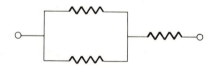

Figure 29–17

29–11 An electric heating element was designed for a power rating of 550 W when connected to a 110-V line. If the line voltage is 120 V, what power is consumed?

29–12 Prove that the resistance of the infinite network shown in Fig. 29–18 is equal to $(1 + \sqrt{3})r$.

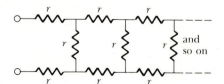

Figure 29–18

29–13

Note: Figure 29–19 employs a convention often used in circuit diagrams. The battery (or other power supply) is not shown explicitly. It is understood that the point at the top, labeled "36 V," is connected to the positive terminal of a 36-V battery having negligible internal resistance, and the "ground" symbol at the bottom to its negative terminal. The circuit is completed through the battery, even though it is not shown on the diagram.

a) In Fig. 29–19a, what is the potential difference V_{ab} when switch S is open?

b) What is the current through switch S, when it is closed?

c) In Fig. 29–19b, what is the potential difference V_{ab} when switch S is open?

d) What is the current through switch S when it is closed?

What is the equivalent resistance in Fig. 29–19b,

e) when switch S is open?

f) when switch S is closed?

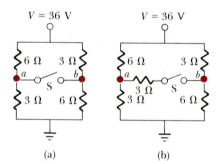

Figure 29–19   (a)   (b)

29–14 (See note with Problem 29–13.)

a) What is the potential of point a with respect to point b in Fig. 29–20 when switch S is open?

b) Which point, a or b, is at the higher potential?

c) What is the final potential of point b when switch S is closed?

d) How much charge flows through switch S when it is closed?

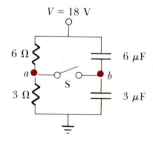

Figure 29–20

29–15 (See note with Problem 29–13.)

a) What is the potential of point a with respect to point b in Fig. 29–21 when switch S is open?

b) Which point, a or b, is at the higher potential?

c) What is the final potential of point b when switch S is closed?

d) How much does the charge on each capacitor change when S is closed?

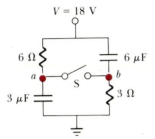

Figure 29–21

29–16 Calculate the three currents indicated in the circuit diagram of Fig. 29–22.

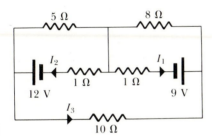

Figure 29–22

29–17 Find the emf's $\mathcal{E}_1$ and $\mathcal{E}_2$ in the circuit of Fig. 29–23, and the potential of point a with respect to point b.

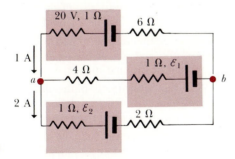

Figure 29–23

29–18

a) Find the potential of point a with respect to point b in Fig. 29–24.

b) If a and b are connected, find the current in the 12-V cell.

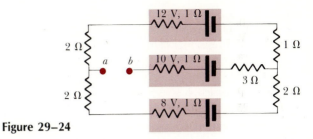

Figure 29–24

29–19 Find the current in each branch of the circuit shown in Fig. 29–25.

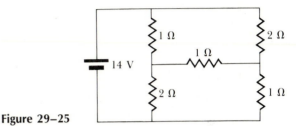

Figure 29–25

29–20 In the circuit shown in Fig. 29–26, find

a) the current in resistor R,

b) the resistance R,

c) the unknown emf $\mathcal{E}$.

d) If the circuit is broken at point x, what is the current in the 28-V battery?

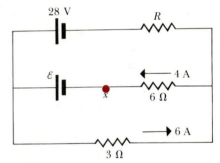

Figure 29–26

29–21 In the circuit shown in Fig. 29–27, find

a) the current in each branch,

b) the potential difference V_{ab}.

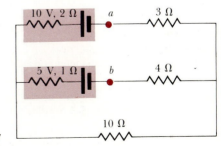

Figure 29–27

29–22 Suppose a resistor R lies along each edge of a cube (12 resistors in all) with connections at the corners. Find the equivalent resistance between two diagonally opposite corners of the cube.

29–23 A 600-Ω resistor and a 400-Ω resistor are connected in series across a 90-V line. A voltmeter across the 600-Ω resistor reads 45 V.

a) Find the voltmeter resistance.

b) Find the reading of the same voltmeter if connected across the 400-Ω resistor.

29–24 Point a in Fig. 29–28 is maintained at a constant potential of 300 V above ground. (See note with Problem 29–13.)

a) What is the reading of a voltmeter of the proper range, and of resistance $3 \times 10^4 \, \Omega$, when connected between point b and ground?

b) What would be the reading of a voltmeter of resistance $3 \times 10^6 \, \Omega$?

c) Of a voltmeter of infinite resistance?

Figure 29–28

29–25 Two 150-V voltmeters, one of resistance 15,000 Ω and the other of resistance 150,000 Ω, are connected in series across a 120-V dc line. Find the reading of each voltmeter.

29–26 A 100-V battery has an internal resistance of 5 Ω.

a) What is the reading of a voltmeter having a resistance of 500 Ω when placed across the terminals of the battery?

b) What maximum value may the ratio r/R_V have if the error in the reading of the emf of a battery is not to exceed 5 percent?

29–27 The resistance of a meter coil is 50 Ω and the current required for full-scale deflection is 500 μA.

a) Show in a diagram how to convert the galvanometer to an ammeter reading 5 A full-scale, and compute the shunt resistance.

b) Show how to convert the galvanometer to a voltmeter reading 150 V full-scale, and compute the series resistance.

29–28 The resistance of the moving coil of the meter G in Fig. 29–29 is 25 Ω; the meter deflects full-scale with a current of 0.010 A. Find the magnitudes of the resistances R_1, R_2, and R_3 required to convert the galvanometer to a multirange ammeter deflecting full-scale with currents of 10 A, 1 A, and 0.1 A.

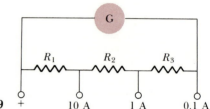

Figure 29–29

29–29 Figure 29–30 shows the internal wiring of a "three-scale" voltmeter whose binding posts are marked +, 3 V, 15 V, 150 V. The resistance of the moving coil, R_G, is 15 Ω, and a current of 1 mA in the coil causes it to deflect full-scale. Find the resistances R_1, R_2, R_3, and the overall resistance of the meter on each of its ranges.

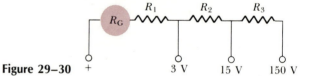

Figure 29–30

29–30 A certain meter has a resistance of 200 Ω and deflects full-scale with a current of 1 mA in its coil. It is desired to replace this with a second meter whose resistance is 50 Ω and that deflects full-scale with a current of 50 μA in its coil. Devise a circuit incorporating the second meter such that

a) the equivalent resistance of the circuit equals the resistance of the first meter, and

b) the second meter will deflect full-scale when the current into and out of the circuit equals the full-scale current of the first meter.

29–31 Suppose the meter of the ohmmeter in Fig. 29–10 has a resistance of 50 Ω and deflects full-scale with a current of 1 mA in its coil. The emf $\mathcal{E} = 1.5$ V.

a) What should be the value of the series resistance R_S?

b) What values of R correspond to meter deflections of $\frac{1}{4}$, $\frac{1}{2}$, and $\frac{3}{4}$ full-scale?

c) Does the ohmmeter have a linear scale?

29–32 In the ohmmeter in Fig. 29–31, M is a 1-mA meter having a resistance of 100 Ω. The battery B has an emf of 3 V and negligible internal resistance. R is so chosen that when the terminals a and b are shorted ($R_x = 0$), the meter reads full scale. When a and b are open ($R_x = \infty$), the meter reads zero.

a) What should be the value of the resistor R?

b) What current would indicate a resistance R_x of 600 Ω?

c) What resistances correspond to meter deflections of $\frac{1}{4}$, $\frac{1}{2}$, and $\frac{3}{4}$ full-scale?

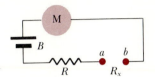

Figure 29–31

29–33 In Fig. 29–32, a resistor of resistance 75 Ω is connected between points a and b. The resistance of the galvanometer G is 90 Ω. What should be the resistance between b and the sliding contact c, if the galvanometer current I_G is to be $\frac{1}{3}$ of the current I?

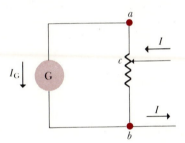

Figure 29–32

29–34 The circuit shown in Fig. 29–33 is called a *Wheatstone bridge;* it is used to determine the value of an unknown resistor X by comparison with three precisely known resistors M, N, and P. With switches K_1 and K_2 closed, these resistors are varied until the current in the galvanometer G is zero; the bridge is then said to be *balanced.* Show that under this condition the unknown resistance is given by $X = MP/N$. (This method permits very high precision in comparing resistors.)

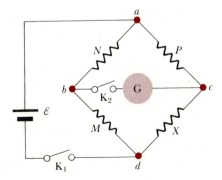

Figure 29–33

29–35 In Fig. 29–33, the galvanometer G shows zero deflection when $M = 1000\ \Omega$, $N = 10.00\ \Omega$, and $P = 27.49\ \Omega$. What is the unknown resistance X?

29–36 The circuit shown in Fig. 29–34 is called a *potentiometer.* It permits measurements of potential difference without drawing current from the circuit being measured, and hence acts as an infinite-resistance voltmeter. The resistor between a and b is a uniform wire of length l, with a sliding contact c at a distance x from b. An unknown potential difference V_x is measured by sliding the contact until the galvanometer G reads zero.

a) Show that under this condition the unknown potential difference is given by $V_x = (x/l)\mathcal{E}$.

b) Why is the internal resistance of the galvanometer not important?

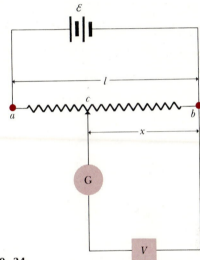

Figure 29–34

29–37 In Fig. 29–34, suppose $\mathcal{E} = 12.00$ V and $l = 1.000$ m. The galvanometer G reads zero when $x = 0.793$ m.

a) What is the potential difference V?

b) Suppose V is the emf of a battery; can its internal resistance be determined by this method?

29–38 A 0.05-μF capacitor is charged to a potential of 200 V and is then permitted to discharge through a 10-MΩ resistor. How much time is required for the charge to decrease to

a) $1/e$, and

b) $1/e^2$

of its initial value?

29–39 A capacitor is charged to a potential of 10 V and is then connected to a voltmeter having an internal resistance of 1.0 MΩ. After a time of 5 s, the voltmeter reads 5 V. What is the capacitance?

29–40 A 10-μF capacitor is connected through a 1-MΩ resistor to a constant potential difference of 100 V.

a) Compute the charge on the capacitor at the following times after the connections are made: 0, 5 s, 10 s, 20 s, 100 s.

b) Compute the charging current at the same instants.

c) How long a time would be required for the capacitor to acquire its final charge if the charging current remained constant at its initial value?

d) Find the time required for the charge to increase from zero to 5×10^{-4} C.

e) Construct graphs of the results of parts (a) and (b) for a time interval of 20 s.

29–41 Two capacitors are charged in series by a 12-V battery (Fig. 29–35).

a) What is the time constant of the charging circuit?

b) After being closed for the length of time determined in (a), the switch S is opened. What is the voltage across the 6-μF capacitor?

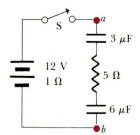

Figure 29–35

29–42 A capacitor of capacitance C is charged by connecting it through a resistance R to the terminals of a battery of emf $\mathcal{E}$ and of negligible internal resistance.

a) How much energy is supplied by the battery in the charging process?

b) What fraction of this energy appears as heat in the resistor?

29–43 Suppose that the parallel plates in Fig. 29–13 have an area of 2 m^2 and are separated by a sheet of dielectric 1 mm thick, of dielectric constant 3. (Neglect edge effects.) At a certain instant, the potential difference between the plates is 100 V and the current I_C equals 2 mA.

a) What is the charge Q on each plate?

b) What is the rate of change of charge on the plates?

c) What is the displacement current in the dielectric?

29–44 A capacitor is made from two cylindrical rods of radius R, placed end to end with a gap of width d between them, where d is much smaller than R. There is a current i in each rod, and the charge q on each end surface changes at a rate given by $\Delta q/\Delta t = i$.

a) Find the conduction current density in each rod.

b) Find the electric field between the surfaces, in terms of q.

c) Find the displacement current density in the gap, and show that it is equal to the conduction current density in the rods.

29–45 A parallel-plate capacitor is made from two plates, each having area A, spaced a distance d apart. The space between plates is filled with a material having dielectric constant K. This material is not a perfect insulator but has resistivity ρ. The capacitor is initially charged with charge of magnitude Q_0 on each plate, which gradually discharges by conduction through the dielectric. Show that at any instant the displacement current density in the dielectric is equal in magnitude to the conduction current density but opposite in direction, so that the *total* current density is zero at every instant.

29–46 How many 100-W light bulbs can be connected to a 20-A, 120-V circuit without tripping the circuit breaker?

29–47 A 1500-W toaster, a 1200-W electric frypan, and a 100-W lamp are plugged into the same outlet in a 20-A, 120-V circuit.

a) What current is drawn by each device?

b) Will this combination blow a fuse?

29–48 An electric dryer is rated at 5.0 kW when connected to a 240-V line.

a) What is the current in the dryer? Is 12-gauge wire large enough to supply this current?

b) What is the resistance of the dryer's heating element?

c) At 10 cents per kWh, how much does it cost to operate the dryer for one hour?

29–49 A 100-W driveway light is left on night and day for a year.

a) What total energy is required? Express your result in joules and in kilowatt-hours.

b) What does this energy cost if the power company's rate is 10 cents per kWh?

30

THE MAGNETIC FIELD AND MAGNETIC FORCES

The most familiar aspects of magnetism are those associated with permanent magnets, which attract unmagnetized iron objects and can also attract or repel other magnets. The fundamental nature of magnetism, however, is to be found in interactions of electric charges in motion. A magnetic field is established by a permanent magnet, by an electric current in a conductor, or by moving charges. This field in turn exerts forces on moving charges and current-carrying conductors. In this chapter we study these magnetic forces, and in Chapter 31 we examine the ways in which magnetic fields are produced by moving charges.

30–1

MAGNETISM

Magnetic phenomena were first observed at least 2000 years ago, in fragments of magnetized iron ore found near the ancient city of Magnesia. It was found that when an iron rod was brought in contact with a natural magnet, the rod also became a magnet. Such a rod, when suspended by a string from its center, tended to line itself up in a north-south direction, like a compass needle; magnets have been used for navigation at least since the eleventh century.

In 1819 the Danish scientist Hans Christian Oersted discovered that a compass needle was deflected by a current-carrying wire. A few years later Michael Faraday and Joseph Henry discovered that moving a magnet near a conducting loop can cause a current in the loop, and that a changing current in one conducting loop can cause a current in another separate loop. These observations were the first evidence of the relationship of magnetism to moving charges.

Before the relation of magnetic interactions to moving charges was understood, the interactions of bar magnets and compass needles was

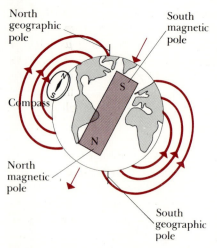

30–1 A sketch of the earth's magnetic field. A compass placed at any point in this field would point in the direction of the field line at that point. Representing the earth's field as that of a tilted bar magnet is only a crude approximation of the actual fairly complex field configuration.

described in terms of *magnetic poles*. The end of a compass needle that points north is called a *north pole* or *N-pole*, and the other end is a south or *S*-pole. Two opposite poles attract each other, and two like poles repel. The concept of magnetic poles is of limited usefulness and is somewhat misleading because a single isolated magnetic pole has never been found; poles always appear in pairs. If a bar magnet is broken in two, each broken end becomes a pole. The existence of an isolated magnetic pole or *magnetic monopole* would have sweeping implications for theoretical physics; extensive experimental searches for magnetic monopoles have been carried out, so far without success.

A compass needle points north because the earth is a magnet; its north geographic pole is a magnetic *south* pole! The earth's magnetic axis is not quite parallel to its geographic axis (the axis of rotation), so a compass reading deviates somewhat from geographic north; this deviation is called *magnetic declination*. The earth's magnetic field can be represented in terms of magnetic-field lines, to be discussed in Section 30–3. A sketch of the earth's magnetic field is shown in Fig. 30–1. The direction of the field at any point can be defined as the direction of the force the field exerts on a magnetic north pole, although there are more fundamental ways to define magnetic field, as we shall see in Section 30–2.

30–2

THE MAGNETIC FIELD

In describing the interaction of two charges *at rest*, we found it useful to introduce the concept of *electric field*, and to describe the interaction in two stages:

1. One charge sets up or creates an electric field E in the space surrounding it;

2. The electric field E exerts a force $F = qE$ on a charge q placed in the field.

We shall follow the same pattern in describing the interactions of charges in motion:

1. A *moving* charge or a current sets up or creates a *magnetic field* in the space surrounding it;

2. The magnetic field exerts a *force* on a *moving* charge or a current in the field.

Like electric field, magnetic field is a *vector field*, that is, a vector quantity associated with each point in space. We shall use the symbol B for magnetic field.

In this chapter we consider the *second* aspect of the interaction: Given the presence of a magnetic field, what force does it exert on a moving charge or a current? Then in Chapter 31 we return to the problem of how magnetic fields are *created* by moving charges and currents.

Some aspects of the magnetic force on a moving charge are analogous to corresponding properties of the electric-field force. Both have magnitude proportional to the charge. If a 1-μC charge and a 2-μC charge move through a given magnetic field with the same velocity, the force on the 2-μC charge is twice as great as that on the 1-μC charge. Also, both are proportional to the magnitude or "strength" of the magnetic field; if a given charge moves with the same velocity in two magnetic fields, one having twice the magnitude (and the same direction) as the other, it experiences twice as great a force in the larger field as in the smaller.

The dependence of the magnetic force on the particle's *velocity* is quite different from the electric-field case. The *electric* force on a charge *does not* depend on velocity; it is the same whether the charge is moving or not. The *magnetic* force, conversely, is found to have a magnitude that increases with speed. Furthermore, the *direction* of the force depends on the directions of the magnetic field B and the velocity v in an interesting way. The force F *does not* have the same direction as B, but instead is always *perpendicular* to both B and v. The magnitude F of the force is found to be proportional to the component of v perpendicular to the field; when that component is zero, that is, when v and B are parallel or antiparallel, there is *no* force!

These characteristics of the magnetic force can be summarized, with reference to Fig. 30–2, as follows: The direction of F is perpendicular to the plane containing v and B; its magnitude is given by

$$F = qv_{\perp}B = qvB \sin \phi, \qquad (30\text{--}1)$$

where q is the charge and ϕ is the angle between the vectors v and B, as shown in the figure.

This prescription does not specify the direction of F completely; there are always two opposite directions both perpendicular to the plane of v and B. To complete the description we use a "right-hand" rule. We imagine turning v into B, through the smaller of the two possible angles; the direction of F is the direction in which a right-hand-thread screw would advance if turned the same way. Alternatively, one can wrap the fingers of the right hand around the line perpendicular to the plane of v and B so that they curl around with the sense of rotation; the thumb then points in the direction of F. This rule is similar to the one used in connection with vector representation of angular quantities in Section 9–13.

Spatial relations for magnetic forces can be troublesome; the following review may help. First, always draw the vectors v and B from a common point, that is, with their tails together. Second, visualize and if possible draw the *plane* in which these two vectors lie. The force vector always lies along a line perpendicular to this plane. To determine its direction along this line, imagine rotating v into B, as in Fig. 30–2. The direction of F is the direction of advance of a right-hand-thread screw rotated in this direction, or the direction of the extended thumb of the right hand if the fingers are curled around the perpendicular line with this direction of rotation.

Finally, we note that Eq. (30–1) can be interpreted in a different but equivalent way. Recalling that ϕ is the angle between the direction of vectors v and B, we may interpret ($B \sin \phi$) as the component of B perpendicular to v, that is, $B_{\perp}$. With this notation, the force expression

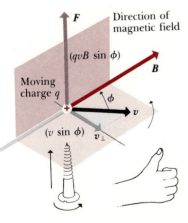

30–2 The magnetic force F acting on a charge q moving with velocity v is perpendicular to both the magnetic field B and to v.

becomes

$$F = qvB_\perp. \tag{30–2}$$

Although equivalent to Eq. (30–1), this is sometimes more convenient, especially in problems involving currents rather than individual particles.

The *units* of B can be deduced from Eq. (30–1); they must be the same as the units of F/qv. Therefore the SI unit of B is one newton second per coulomb meter, or $1 \text{ N}\cdot\text{s}\cdot\text{C}^{-1}\cdot\text{m}^{-1}$, or, since one ampere is one coulomb per second, $1 \text{ N}\cdot\text{A}^{-1}\cdot\text{m}^{-1}$. This unit is called *one tesla* (1 T). The cgs unit of B, the *gauss* ($1 \text{ G} = 10^{-4} \text{ T}$) is also in common use. To summarize,

$$1 \text{ T} = 1 \text{ N}\cdot\text{A}^{-1}\cdot\text{m}^{-1} = 10^4 \text{ G}.$$

In this discussion we have assumed that q is a *positive* charge. If q is negative, the direction of $\boldsymbol{F}$ is opposite to that given by the right-hand rule shown in Fig. 30–2. Thus if two charges of equal magnitude and opposite sign move in the same $\boldsymbol{B}$ field with the same velocity, the forces on the two charges have equal magnitude and opposite direction.

An unknown *electric* field can be "explored" by measuring the magnitude and direction of the force on a test charge. To "explore" an unknown *magnetic* field, we must measure the magnitude and direction of the force on a *moving* test charge. The cathode-ray tube, discussed in Section 26–8, is a convenient device for studying the behavior of moving charges in a magnetic field. At one end of this tube an electron gun shoots out a narrow electron beam at a speed that can be controlled and calculated. At the other end is a fluorescent screen that emits light at the small spot where the electron beam strikes it. In the absence of any deflecting force, the beam strikes the center of the screen.

In a magnetic field, the electron beam is in general deflected. By rotating the cathode-ray tube, however, one finds that there are always two directions of the beam for which *no* deflection takes place. *One of the directions of motion of a charge on which a magnetic field exerts **no** force is the direction of the **B**-vector*. Thus, in Fig. 30–3, the electron beam is unde-

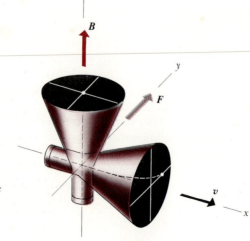

30–3 The electron beam of the cathode-ray tube is undeflected when the beam is parallel to the z-axis. The **B**-vector then points either up or down. When the tube axis is parallel to the x-axis, the beam is deflected in the positive y-direction. Then the **B**-vector points upward, and the force **F** on the electrons points along the positive y-axis, opposite to the rule of Fig. 30–2 because q is negative.

flected when its direction is parallel to the z-axis and hence the **B**-vector must point either up or down.

We now turn the tube by 90°, as shown in Fig. 30–3. The beam is now found to be deflected in a direction showing a force perpendicular to the plane of **B** and **v**. This observation thus corroborates our previous discussion. Additional experiments in which the angle between **B** and **v** is between zero and 90° confirm Eq. (30–1) and the accompanying discussion. We note that the electron has a negative charge, and the force in Fig. 30–3 is opposite in direction to the force on a positive charge.

30–3

MAGNETIC FIELD LINES; MAGNETIC FLUX

A magnetic field can be represented by lines, such that the direction of a line through a given point is the same as that of the magnetic field vector **B** at that point. We shall call these *magnetic field lines;* they are sometimes called magnetic lines of force, but this term is unfortunate because, unlike electric field lines, they *do not* point in the direction of the force on a charge.

In a uniform magnetic field, where the **B** vector has the same magnitude and direction at every point in a region, the field lines are straight and parallel. If the poles of an electromagnet are large, flat, and close together, there is a region between them where the magnetic field is approximately uniform.

Magnetic-field lines produced by a few examples of magnetic-field sources are shown in Figs. 30–4 and 30–5.

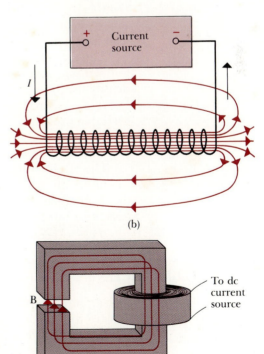

30–4 Magnetic-field lines produced by (a) a bar-shaped permanent magnet, (b) a coil of wire wound on a cylindrical form, i.e., a solenoid, and (c) a laboratory electromagnet with an iron core. In (c) the magnetic field is nearly uniform in the gap in the core. In all three cases the field lines are continuous curves closing on themselves; there are no endpoints.

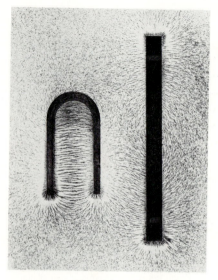

30–5 Magnetic fields can be visualized by use of iron filings, which orient themselves parallel to the field direction. In this photograph the concentration of the fields near the poles is shown. (Photo by Fundamental Photographs, New York.)

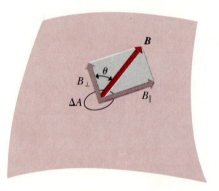

30–6 The magnetic flux through an area element ΔA is defined to be $\Phi = B_\perp \, \Delta A$.

The *magnetic flux* across a surface is defined in analogy to the electric-field flux used with Gauss's law (Section 25–5). Any surface can be divided into elements of area, as in Fig. 30–6. For an element ΔA, we obtain the components of B normal and tangent to the surface, as shown. From the figure, $B_\perp = B \cos \theta$. The magnetic flux Φ through this area is defined as

$$\Phi = B_\perp \, \Delta A = B \cos \theta \, \Delta A. \qquad (30\text{–}3)$$

The total magnetic flux through the surface is the sum of the contributions from the individual area elements. In the special case where B is uniform over a plane surface with total area A,

$$\Phi = BA \cos \theta. \qquad (30\text{–}4)$$

If B happens to be perpendicular to the surface, $\cos \theta = 1$, and this expression reduces to $\Phi = BA$. The chief usefulness of the concept of magnetic flux is in the study of electromagnetic induction, to be considered in Chapter 32.

The SI unit of magnetic field B is one tesla = one newton per ampere meter; hence the unit of magnetic flux is *one newton meter per ampere* $(1 \; \text{N} \cdot \text{m} \cdot \text{A}^{-1})$. In honor of Wilhelm Weber (1804–1890), $1 \; \text{N} \cdot \text{m} \cdot \text{A}^{-1}$ is called one *weber* (1 Wb).

If the element of area ΔA in Eq. (30–3) is at right angles to the field lines, $B_\perp = B$, and hence

$$B = \frac{\Delta \Phi}{\Delta A}.$$

That is, the magnetic field equals the *flux per unit area* across an area at right angles to the magnetic field. Since the unit of flux is 1 weber, the unit of field, 1 tesla, is equal to one *weber per square meter* $(1 \; \text{Wb} \cdot \text{m}^{-2})$. The magnetic field B is sometimes called *flux density*.

The total flux across a surface can then be pictured as proportional to the number of field lines crossing the surface, and the field (the flux density) as the number of lines *per unit area*.

In the cgs system, the unit of magnetic flux is one *maxwell* and the corresponding unit of flux density, one maxwell per square centimeter, is called one *gauss* (1 G). (Instruments for measuring flux density are often referred to as *gaussmeters*.)

The largest values of magnetic field that can be produced in the laboratory are of the order of 30 T = 30 $\text{Wb} \cdot \text{m}^{-2}$ or 300,000 G, while the magnetic field of the earth is only a few hundred-thousandths of a tesla, or a few tenths of a gauss.

30–4

MOTION OF CHARGED PARTICLES IN MAGNETIC FIELDS

In Fig. 30–7, suppose a particle with positive charge q is at point O, moving with velocity v in a uniform magnetic field directed into the plane of the figure. An upward force F, of magnitude qvB, is exerted on the particle at this point. Since the force is at right angles to the velocity, it does not change the *magnitude* of this velocity but changes its *direction*. At points such as P and Q the directions of force and velocity will have

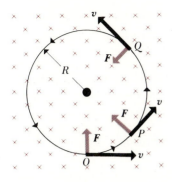

30–7 The orbit of a charged particle in a uniform magnetic field is a circle when the initial velocity is perpendicular to the field. The crosses represent a uniform magnetic field directed *away* from the reader.

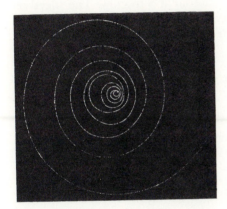

30–8 A track in a liquid-hydrogen bubble chamber made by an electron with initial kinetic energy of about 27 MeV. The magnetic field strength is 1.8 T. As the particle loses energy the radius of curvature decreases; the maximum radius is about 5 cm. (Courtesy of Lawrence Berkeley Laboratory, University of California.)

30–9 Electron–positron pair production in a liquid-hydrogen bubble chamber. A high-energy gamma ray coming in from above scatters on an atomic electron. The paths of the recoil electron and of the electron–positron pair can be seen; the directions of curvature in the magnetic field show the signs of the charges. (Courtesy of Lawrence Berkeley Laboratory, University of California.)

changed as shown; the magnitude of the force is constant, since the magnitudes of q, v, and B are constant. The particle therefore moves under the influence of a force whose *magnitude* is constant but whose *direction* is always at right angles to the velocity of the particle. The orbit of the particle is therefore a *circle* described with constant tangential speed v. Since the centripetal acceleration is v^2/R, as shown in Section 5–5, we have, from Newton's second law,

$$F = qvB = m\left(\frac{v^2}{R}\right),$$

where m is the mass of the particle. The radius of the circular orbit is

$$R = \frac{mv}{Bq}. \qquad (30\text{--}5)$$

If the direction of the initial velocity is *not* perpendicular to the field, the velocity component parallel to the field remains constant and the particle moves in a helix. In this case, v in Eq. (30–5) is the component of velocity perpendicular to the field.

Note that the radius is proportional to the *momentum* of the particle, mv. Note also that the *work* of the magnetic force acting on a charged particle is always *zero*, because the force is always at *right angles* to the motion. The only effect of a magnetic force is to change the *direction* of motion, never to increase or decrease the *magnitude* of the velocity. Thus, *motion of a charged particle under the action of a magnetic field alone is always motion with constant speed.*

Figure 30–8 is a photograph of a track made in a liquid-hydrogen bubble chamber by a high-energy electron moving in a magnetic field perpendicular to the plane of the paper. As the particle loses energy (and speed) the radius of curvature decreases according to Eq. (30–5). Figure 30–9 shows electron–positron pair production, also in a bubble chamber. Similar experiments using a cloud chamber provided the first experimental evidence in 1932 for the existence of the *positron*, or positive electron.

30–5

THOMSON'S MEASUREMENT OF *e/m*

The charge-to-mass ratio of an electron, *e/m*, was first measured by Sir J. J. Thomson in 1897 at the Cavendish Laboratory in Cambridge, England. The discovery that this ratio is constant provided the best experimental evidence available at that time of the *existence* of electrons, particles with definite mass and charge. Thomson's term for these particles was "cathode corpuscles."

Thomson's apparatus (Fig. 30–10) is very similar in principle to the cathode-ray tube discussed in Section 26–8. It consists of a highly evacuated glass tube into which several metal electrodes are sealed. Electrons from the hot cathode C are formed into a beam by the anodes A, A', and the beam passes into the region between the two plates P and P'. After passing between the plates, the electrons strike the end of the tube, where they cause fluorescent material at S to glow. The speed of the electrons is related to the accelerating potential V; we simply equate the kinetic energy $\frac{1}{2}mv^2$ to the loss of potential energy eV (where e is the magnitude of the electron charge):

$$\tfrac{1}{2}mv^2 = eV \qquad \text{or} \qquad v = \sqrt{\frac{2eV}{m}}. \qquad (30\text{–}6)$$

If now a potential difference is established between the two deflecting plates P and P', as shown, the resulting downward electric field deflects the negatively charged electrons *upward*. Alternatively, we may impose a *magnetic* field directed into the plane of the figure, shown by the ×'s; this field results in a *downward* deflection of the beam; the reader should verify this direction. Finally, if *both* **E** and **B** fields are applied simultaneously, it is possible to adjust their relative magnitudes so that the two forces cancel and the beam is undeflected. The condition that must be satisfied, obtained by equating the two force magnitudes, is

$$eE = evB, \qquad \text{or} \qquad v = \frac{E}{B}. \qquad (30\text{–}7)$$

Finally, we may combine this with Eq. (30–6) to eliminate v and obtain an expression for the charge-to-mass ratio *e/m* in terms of the other

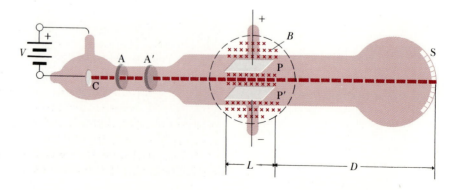

30–10 Thomson's apparatus for measuring the ratio *e/m* for cathode rays.

quantities:

$$\frac{E}{B} = \sqrt{\frac{2eV}{m}},$$

$$\frac{e}{m} = \frac{E^2}{2B^2V}.$$

$$(30\text{–}8)$$

All the quantities on the right side can be measured, so e/m can be determined. We note that it is *not* possible to measure e or m separately by this method, but only their ratio.

Thomson measured e/m for his "cathode corpuscles" and found a unique value for this quantity that was independent of the cathode material and the residual gas in the tube. This independence indicated that cathode corpuscles are a common constituent of all matter. The modern accepted value of e/m is $(1.758803 \pm 0.000003) \times 10^{11}$ C·kg^{-1}. Thus Thomson is credited with discovery of the first subatomic particle, the electron. He also found that the speed of the electrons in the beam was about one tenth the speed of light, much larger than any previously measured material particle speed.

Fifteen years after Thomson's experiments, Millikan succeeded in measuring the charge of the electron with his famous oil-drop experiment, described in Section 26–6. The magnitude of the electron charge e is 1.602×10^{-19} C. Thus the *mass* of the electron can be obtained:

$$m = \frac{1.602 \times 10^{-19} \text{ C}}{1.759 \times 10^{11} \text{ C·kg}^{-1}} = 9.109 \times 10^{-31} \text{ kg}.$$

*30–6

ISOTOPES AND MASS SPECTROSCOPY

Thomson devised a method similar to the above e/m measurement, for measuring the charge–mass ratio for positive ions. An added difficulty was that in Thomson's day it was difficult to produce a beam of ions all having the same speed. Because the e/m electron experiment depends on the particles having a common speed (in order for the electric and magnetic field forces to balance), this method is not directly applicable for a beam of particles having various velocities. Thomson's method was to make the electric and magnetic fields *parallel,* so that the deflections due to these fields are in perpendicular directions. The net deflection can then never be zero, but it turns out that the relation between the x- and y-deflections for a beam permits determination of the charge–mass ratio of the particles.

Thomson assumed that each positive ion had a charge equal in magnitude to that of the electron because each was an atom that had lost one electron. He could then identify particular values of q/m with particular ions. Positive particles move more slowly than electrons and have lower values of q/m because they are much more massive. The *largest* q/m for positive particles is that for the *lightest* element, hydrogen. From the value of q/m it was found that the mass of the *hydrogen ion* or *proton* is 1836.13 ± 0.01 times the mass of an electron. Electrons contribute only a small fraction of the mass of material objects.

The most striking result of these experiments was that certain chemically pure gases had *more than one* value of q/m. Most notable was the case of neon, which has atomic mass $20.2 \text{ g} \cdot \text{mol}^{-1}$. Thomson obtained *two* values of q/m, corresponding to 20 and $22 \text{ g} \cdot \text{mol}^{-1}$, and after trying and discarding various explanations he concluded that there must be two kinds of neon atoms with different masses.

Soon afterward Aston, a student of Thomson, succeeded in actually separating these two atomic species. Aston permitted the gas to diffuse repeatedly through a porous plug between two containers. At a given temperature T, each atom has an average kinetic energy $3kT/2$, as discussed in Section 20–4. Thus the more massive atoms have, on the average, somewhat smaller speeds; because of this speed difference, the gas emerging from the plug had a slightly greater concentration of the less massive atoms than the gas entering the plug. Thus Aston demonstrated directly the existence of two species of neon atoms with different masses.

Since these experiments in the early years of the twentieth century, many other elements have been shown to have several kinds of atoms, identical in their chemical behavior but differing in mass. Such forms of an element are called *isotopes*. We shall see in a later chapter that the mass differences are due to differing numbers of neutrons in the *nuclei* of the atoms.

A detailed search for the isotopes of all the elements required precise experimental technique. Aston built the first of many instruments called *mass spectrometers* in 1919. His instrument had a precision of one part in 10,000, and he found that many elements have isotopes. We shall describe here a variation built by Bainbridge. The Bainbridge mass spectrometer (Fig. 30–11) has a source of ions (not shown) situated above S_1. The ions under study pass through slits S_1 and S_2 and move down into the electric field between the two plates P and P'. In the region of the electric field there is also a magnetic field $\boldsymbol{B}$, perpendicular to the paper. Thus the ions enter a region of crossed electric and magnetic fields like those used by Thomson to measure the velocity of electrons in his determination of e/m. Only those ions whose speed is equal to E/B pass undeviated through this region; ions with other speeds are stopped by the slit S_3. Thus all ions that emerge from S_3 have the same velocity. The region of crossed fields is called a *velocity selector*.

Below S_3 the ions enter a region where there is another magnetic field $\boldsymbol{B'}$, perpendicular to the page, but no electric field. Here the ions move in circular paths of radius R. From Eq. (30–5), we find that

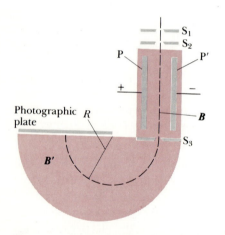

30–11 Bainbridge's mass spectrograph, utilizing a velocity selector.

$$m = \frac{qB'R}{v}. \tag{30–9}$$

Assuming equal charges on each ion, the mass of each ion is proportional to the radius of its path. Ions of different isotopes converge at different points on the photographic plate. The relative abundance of the isotopes is measured from the densities of the photographic images they produce.

In present-day mass spectrometers the photographic plate is replaced by a more sophisticated particle detector. Figure 30–12 shows a modern mass spectrometer and a typical isotope analysis. For each isotope the number given represents the *mass number*, equal to the total

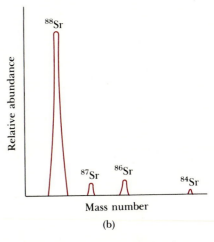

(a)

(b)

30–12 (a) A modern solid-source mass spectrometer. The specimen is placed in the cylindrical chamber in the center; the ion beam is deflected by the large magnet toward the collector and detector (right front). The instrument is controlled and the data processed by the equipment in front of the operator. (Courtesy of Institute of Geological Studies.) (b) A typical isotope analysis, showing several isotopes of strontium.

number of protons and neutrons in the nucleus of the atom; different isotopes of an element have the same number of protons but differing numbers of neutrons in the nucleus.

Masses of atoms are often expressed in *atomic mass units*. By definition, one atomic mass unit (1 u) is $\frac{1}{12}$ the mass of one atom of the most abundant isotope of carbon, ^{12}C. Since the mass of an atom in grams is equal to its atomic mass divided by Avogadro's number, it follows that

$$1 \text{ u} = \frac{(1/12)(12 \text{ g} \cdot \text{mol}^{-1})}{6.022 \times 10^{23} \text{ mol}^{-1}}$$

$$= 1.661 \times 10^{-24} \text{ g} = 1.661 \times 10^{-27} \text{ kg}.$$

30–7

FORCE ON A CONDUCTOR

When a current-carrying conductor lies in a magnetic field, magnetic forces are exerted on the moving charges within the conductor. These forces are transmitted to the material of the conductor, and the conductor as a whole experiences a force distributed along its length. The electric motor and the moving-coil galvanometer both depend for their operation on the magnetic forces on conductors carrying currents.

Figure 30–13 represents a portion of a conducting wire of length l and cross-sectional area A, in which the current density J is from left to right. The wire is in a magnetic field $\boldsymbol{B}$, perpendicular to the plane of the diagram, and directed *into* the plane. For generality we assume there are moving charges of both signs. In metals the moving charges are always negatively charged electrons, but in semiconductors charges of both signs may be present. A positive charge q_1 within the wire, moving with its drift velocity v_1, is acted on by an upward force $\boldsymbol{F}_1$ of magnitude $q_1 v_1 B$.

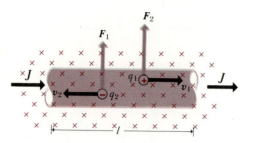

30–13 Forces on the moving charges in a current-carrying conductor. The forces on both positive and negative charges are in the same direction.

The drift velocity v_2 of a negative-charge carrier q_2 is opposite to v_1 and the force F_2 on it has magnitude $q_2 v_2 B$. Because q_1 and q_2 have opposite signs and v_1 and v_2 opposite directions, F_2 has the *same* direction as F_1, as shown.

The *total* force on all the moving charges in a length l of conductor can be expressed in terms of the current, using the considerations of Section 28–1. Let n_1 and n_2 represent the numbers of positive and negative charge carriers, respectively, per unit volume. The numbers of carriers in the portion are then $n_1 Al$ and $n_2 Al$; the total force F on all carriers (and hence the total force on the wire) has magnitude

$$F = (n_1 Al)(q_1 v_1 B) + (n_2 Al)(q_2 v_2 B)$$
$$= (n_1 q_1 v_1 + n_2 q_2 v_2) AlB.$$

But $n_1 q_1 v_1 + n_2 q_2 v_2$ equals the current density J, and the product JA equals the current I, so finally

$$F = IlB. \tag{30–10}$$

If the B field is not perpendicular to the wire but makes an angle ϕ with it, the situation is just like that discussed in Section 30–2 for a single charge. The component of B parallel to the wire (and thus to the drift velocities of the charges) exerts no force; the component perpendicular to the wire is given by $B_\perp = B \sin \phi$ so, in general,

$$F = IlB_\perp = IlB \sin \phi. \tag{30–11}$$

To find the direction of the force on a current-carrying conductor placed in a magnetic field, we may use the same right-hand-screw rule that was used in the case of a moving positive charge. Rotate a right-hand screw from the direction of I toward B; the direction of advance will be the direction of F. The situation is the same as that shown in Fig. 30–2, but with v replaced by the direction of I.

EXAMPLE   A straight horizontal wire carries a current of 50 A from west to east, in a region where the magnetic field is toward the northeast (i.e., 45° north of east) with magnitude 1.2 T. Find the magnitude and direction of the force on a 1-m section of wire.

Solution   The angle ϕ between the directions of current and field is 45°. From Eq. (30–11) we obtain

$$F = (50 \text{ A})(1 \text{ m})(1.2 \text{ T})(\sin 45°) = 42.4 \text{ N}.$$

Consistency of units may be checked by observing, as noted in Section 30–2, that $1 \text{ T} = 1 \text{ N} \cdot \text{A}^{-1} \cdot \text{m}^{-1}$. The direction of the force is perpendicular to the plane of the current and the field, both of which lie in the horizontal plane. Thus the force must be along the vertical; the right-hand rule shows that it is vertically upward. ◀

30–8

FORCE AND TORQUE ON A CURRENT LOOP

Any current-carrying conductor can be represented as a number of straight line segments; curved portions can be represented by a large number of very small segments. Thus Eq. (30–11) can be used to calculate the total magnetic-field force on *any* conductor. As an example we shall consider a rectangular current loop; as we shall see, the total *force* is zero, but there is a *torque* that tends to turn the loop to a particular orientation with respect to the magnetic field.

Figure 30–14 shows a rectangular loop of wire with sides of lengths a and b. The normal to the plane of the loop makes an angle α with the direction of a uniform magnetic field, and the loop carries a current I. (Provision must be made for leading the current into and out of the loop, or for inserting a seat of emf. This is omitted from the diagram, for simplicity.)

The force $\boldsymbol{F}$ on the right side, of length a, is in the direction of the *x*-axis, toward the right, as shown. In this side $\boldsymbol{B}$ is perpendicular to the current direction, and the total force on this side (the sum of the forces distributed along the side) has magnitude

$$F = IaB.$$

A force of the same magnitude but in the opposite direction acts on the opposite side, as shown in the figure.

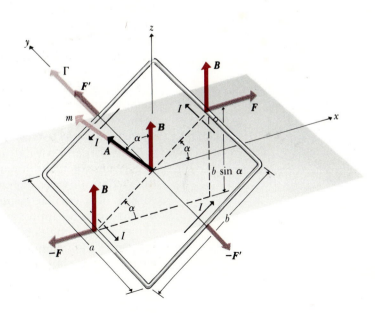

30–14 Forces on the sides of a current-carrying loop in a magnetic field. The resultant of the set of forces is a couple of moment $\Gamma = IAB \sin \alpha$.

The forces on the sides of length b, represented by the vectors $\boldsymbol{F'}$, have magnitude

$$IbB \sin (90° - \alpha), \quad \text{or} \quad IbB \cos \alpha.$$

The lines of action of both lie along the y-axis.

The resultant *force* on the loop is zero, since the forces on opposite sides are equal and opposite. The resultant *torque*, however, is not zero. Each of the forces on the sides of length a has a moment arm $\frac{1}{2}b \sin \alpha$ with respect to the center of the loop, and the total torque is

$$\Gamma = (IBa)(b \sin \alpha). \tag{30-12}$$

The torque is maximum when $\alpha = 90°$ (when the plane of the coil is parallel to the field), and it is zero when $\alpha = 0$ and the plane of the coil is perpendicular to the field. What is the position of stable equilibrium?

Since ab is the area A of the coil, Eq. (30–12) may also be written

$$\Gamma = IBA \sin \alpha. \tag{30-13}$$

The product IA is called the *magnetic moment m* of the loop. It is analogous to the dipole moment of an electric dipole, introduced in Section 25–1. It may be regarded as a vector quantity having a direction perpendicular to the plane of the loop, with a sense determined by the right-hand rule, as shown in Fig. 30–14. Thus the torque on a magnetic moment corresponding to a current loop can be expressed simply as

$$\Gamma = mB \sin \alpha. \tag{30-14}$$

Because of the directional relations indicated, the effect of the torque Γ is to tend to rotate the loop toward its equilibrium position, in which it lies in the xy-plane with its vector magnetic moment $\boldsymbol{m}$ in the same direction as the field $\boldsymbol{B}$.

When a magnetic dipole changes its orientation in a magnetic field, the field does *work* on it, and there is a corresponding potential energy. As the above discussion suggests, the potential energy is least when $\boldsymbol{m}$ and $\boldsymbol{B}$ are parallel and greatest when they are antiparallel. It can be shown that the potential energy U of this interaction is given by

$$U = -mB \cos \alpha. \tag{30-15}$$

Although Eq. (30–13) has been derived for a rectangular current loop, it can be shown that it is valid for a loop of any shape at all. The loop may be circular, polygonal, or anything else; the torque on *any* current loop in a magnetic field is given by Eqs. (30–13) or (30–14), and Eq. (30–15) also is valid for any shape of loop.

An arrangement of particular interest is the *solenoid*, which is a helical winding of wire, such as that obtained by winding wire around the surface of a circular cylinder. If the windings are closely spaced, the solenoid can be approximated by a number of circular loops lying in planes at right angles to its long axis. The total torque on a solenoid in a magnetic field is simply the sum of the torques on the individual turns. Hence, for a solenoid of N turns in a uniform field B,

$$\Gamma = NIAB \sin \alpha, \tag{30-16}$$

where α is the angle between the axis of the solenoid and the direction of the field. The torque is maximum when the magnetic field is parallel to the planes of the individual turns or perpendicular to the long axis of the solenoid. The effect of this torque, if the solenoid is free to turn, is to rotate it into a position in which each turn is perpendicular to the field and the axis of the solenoid is parallel to the field.

Although little has been said thus far regarding permanent magnets, the behavior of the solenoid as described above closely resembles that of a bar magnet or compass needle, in that both the solenoid and the magnet will, if free to turn, set themselves with their axes parallel to a magnetic field. The behavior of a bar magnet or compass is sometimes described by ascribing the torque on it to magnetic forces exerted on "poles" at its ends. We see, however, that no such interpretation is demanded in the case of the solenoid. In fact, the whirling electrons in a bar of magnetized iron play the same role as the current in the windings of a solenoid, and the observed torque arises from the same cause in both instances. We shall return to this matter later.

EXAMPLE 1   A circular coil of wire 0.05 m in radius, having 30 turns, lies in a horizontal plane, as shown in Fig. 30–15. It carries a current of 5 A, in a counterclockwise sense when viewed from above. The coil is in a magnetic field directed toward the right, with magnitude 1.2 T. Find the magnetic moment and the torque on the coil.

Solution   The area of the coil is

$$A = \pi r^2 = \pi(0.05 \text{ m})^2 = 7.85 \times 10^{-3} \text{ m}^2.$$

The magnetic moment of each turn of the coil is

$$m = IA = (5 \text{ A})(7.85 \times 10^{-3} \text{ m}^2) = 3.93 \times 10^{-2} \text{ A} \cdot \text{m}^2,$$

and the total magnetic moment of all 30 turns is

$$m = (30)(3.93 \times 10^{-2} \text{ A} \cdot \text{m}^2) = 1.18 \text{ A} \cdot \text{m}^2.$$

The angle α between the direction of $\boldsymbol{B}$ and the normal to the plane of the coil is 90°, and from Eq. (30–13) the torque on each turn of the coil is

$$\Gamma = IBA \sin \alpha = (5 \text{ A})(1.2 \text{ T})(7.85 \times 10^{-3} \text{ m}^2)(\sin 90°) = 0.047 \text{ N} \cdot \text{m},$$

and the total torque on the coil is

$$\Gamma = (30)(0.047 \text{ N} \cdot \text{m}) = 1.41 \text{ N} \cdot \text{m}.$$

Alternatively, from Eq. (30–14),

$$\Gamma = mB \sin \alpha = (1.18 \text{ A} \cdot \text{m}^2)(1.2 \text{ T})(\sin 90°) = 1.41 \text{ N} \cdot \text{m}.$$

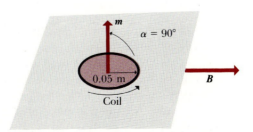

30–15

The direction of the torque is such as to tend to push the right side of the coil down and the left side up and to rotate it into a position where the normal to its plane is parallel to **B**.   ◄

EXAMPLE 2   If the coil in Example 1 rotates from its initial position to a position where its magnetic moment is parallel to **B**, what is the change in potential energy?

Solution   From Eq. (30–15), the initial potential energy U_1 is

$$U_1 = -(1.18 \text{ A} \cdot \text{m}^2)(1.2 \text{ T})(\cos 90°) = 0,$$

and the final potential energy U_2 is

$$U_2 = -(1.18 \text{ A} \cdot \text{m}^2)(1.2 \text{ T})(\cos 0°) = -1.41 \text{ J}.$$

Thus the change in potential energy is -1.41 J.   ◄

EXAMPLE 3   What vertical forces applied to the left and right edges of the coil of Fig. 30–15 would be required to hold it in equilibrium in its initial position?

Solution   An upward force of magnitude F at the right side and a downward force of equal magnitude on the left side would have a total torque (with respect to the coil's center) of

$$\Gamma = (2)(0.05 \text{ m})F.$$

This must be equal to the magnitude of the magnetic-field torque of 1.41 N·m, and we find that the required forces have magnitude 14.1 N.   ◄

*30–9

THE DIRECT-CURRENT MOTOR

The direct-current motor is illustrated schematically in Fig. 30–16. The armature or rotor, A, is a cylinder of soft steel mounted on a shaft so that it can rotate about its axis. Embedded in longitudinal slots in the surface of the rotor are a number of copper conductors C. Current is led into and out of these conductors through graphite brushes making contact with a cylinder of the shaft called the *commutator* (not shown in Fig. 30–16). The commutator is an automatic switching arrangement, which maintains the currents in the conductors in the directions shown in the figure, whatever the position of the rotor. The current in the field coils F and F' sets up a magnetic field in the motor frame M and in the gap between the pole pieces P and P' and the rotor. Some of the magnetic field lines are shown as broken lines. With the directions of field and rotor currents shown, the side thrust on each conductor in the rotor is such as to produce a *counterclockwise* torque on the rotor.

If the rotor and the field windings are connected in series, we have a *series* motor; if they are connected in parallel, we have a *shunt* motor. In some motors the field windings are in two parts, one in series with the rotor and the other in parallel with it; the motor is then *compound*.

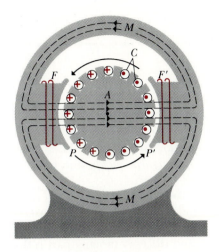

30–16 Schematic diagram of a dc motor. The armature or rotor *A* rotates on a shaft through its center, perpendicular to the plane of the figure. The conductors on the rotor are shown in cross section; those with dots at their centers carry current out of the plane, those with crosses, into the plane.

A motor converts electrical energy to mechanical energy or work, and thus requires electrical energy input. If the potential difference between terminals is V_{ab} and the current is I, then the power input is $P = V_{ab}I$. Even if the motor coils have no resistance, there must be a potential difference if P is to be different from zero. This potential difference results principally from magnetic forces exerted on the charges in the conductors of the rotor as they rotate through the magnetic field. These forces are an example of nonelectrostatic forces, and give rise to a corresponding equivalent nonelectrostatic field (concepts introduced in Section 28–4). In the present context the resulting electromotive force $\mathcal{E}$ is called an *induced* emf or sometimes a *back* emf, referring to the fact that its sense is opposite to that of the current. Induced emf's resulting from motion of conductors in magnetic fields will be considered in greater detail in Chapter 32.

The situation is analogous to that of a battery being charged, as discussed in Section 28–4. In that case, electrical energy is converted to chemical rather than mechanical energy. But in either case, if the device has internal resistance r, then V_{ab} is greater than $\mathcal{E}$, the difference being the drop Ir across the internal resistance of the device. Thus, for either a motor or a battery being charged,

$$V_{ab} = \mathcal{E} + Ir. \tag{30–17}$$

For a battery, $\mathcal{E}$ is often constant or nearly so; for a motor, $\mathcal{E}$ is *not* constant but depends on the speed of rotation of the rotor.

EXAMPLE A dc motor with its rotor and field coils connected in series has an internal resistance of 2.0 Ω. When running at full load on a 120-V line, it draws a current of 4.0 A.

a) When is the emf in the rotor?

$$V_{ab} = \mathcal{E} + Ir,$$
$$120 \text{ V} = \mathcal{E} + (2.0 \ \Omega)(4.0 \text{ A}),$$
$$\mathcal{E} = 112 \text{ V}.$$

b) What is the power delivered to the motor?

$$P = IV_{ab} = (4.0 \text{ A})(120 \text{ V}) = 480 \text{ W}.$$

c) What is the rate of dissipation of energy in the motor?

$$P = I^2r = (4.0 \text{ A})^2(2.0 \ \Omega) = 32 \text{ W}.$$

d) What is the mechanical power developed?

Mechanical power = total power − rate of dissipation of energy
$$= 480 \text{ W} - 32 \text{ W}$$
$$= 448 \text{ W}.$$

The mechanical power may also be calculated from the relation

Mechanical power = $\mathcal{E}I$ = (112 V)(4.0 A)
$$= 448 \text{ W}.$$ ◀

*30–10

THE ELECTROMAGNETIC PUMP

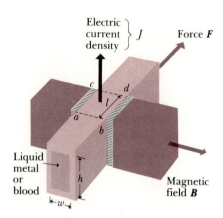

30–17   An electromagnetic pump.

Magnetic forces acting on conducting fluids provide a convenient means of pumping these fluids. An example is found in nuclear-reactor design; in some types of reactors, heat is transferred from the reactor core to the point where it is to be utilized, by a circulating flow of liquid metal. (Sodium, lithium, bismuth, and sodium–potassium alloys have been used.) The flow is maintained by an *electromagnetic pump*, one form of which is shown schematically in Fig. 30–17.

A current is sent transversely through the liquid metal, in a direction at right angles to a transverse magnetic field. The resulting side thrust on the current-carrying metal drives it along the pipe in which it is contained. The system is completely sealed, and the only moving part is the metal itself.

Magnetic pumps are also finding a variety of applications in medical technology, particularly for pumping blood. Ordinary mechanical pumps with moving parts can damage blood cells. Blood contains ions, making it an electrical conductor, and magnetic pumping is possible. The pump is again completely sealed, reducing danger of contamination, and there are no moving parts except the blood itself. Magnetic pumps are now in use in some heart-lung machines and artificial kidney machines.

*30–11

THE HALL EFFECT

The reality of the forces on the moving charges in a conductor in a magnetic field is strikingly demonstrated by the *Hall effect*. The conductor in Fig. 30–18 is in the form of a flat strip. The current carriers within it are driven toward the upper edge of the strip by the magnetic force qvB exerted on them. Here v is the drift velocity of the moving charges in the material.

If the charge carriers are electrons, as in Fig. 30–18a, an excess negative charge accumulates at the upper edge of the strip, leaving an excess positive charge at its lower edge, until the resulting transverse electrostatic field E_e causes a force of magnitude qE_e that is equal and opposite to the magnetic field force of magnitude qvB. Associated with this electric field is a potential difference between opposite edges of the strip, which can be measured with a potentiometer and which is called the *Hall voltage* or the *Hall emf*. Experiment shows that for metals the upper edge of the strip in Fig. 30–18a *does* become negatively charged, showing that the charge carriers in a metal are indeed negative electrons.

If however the charge carriers are *positive*, as in Fig. 30–18b, then *positive* charge accumulates at the upper edge, and the potential difference is *opposite* to that resulting from the deflection of negative charges. Soon after the discovery of the Hall effect, in 1879, it was observed that some materials, notably the *semiconductors*, exhibit a Hall emf opposite to that of the metals, *as if* their charge carriers were *positively* charged. We now know that these materials conduct by a process known as *hole conduc-*

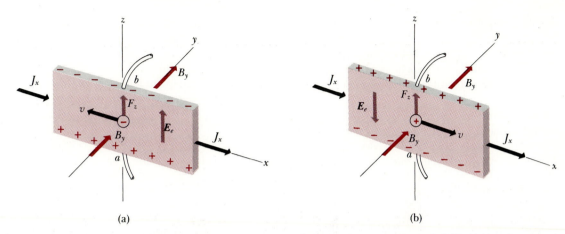

(a)   (b)

30–18 Forces on charge carriers in a conductor in a magnetic field. (a) Negative current carriers (electrons) are pushed toward top of slab, leading to charge distribution as shown. Point *a* is at higher potential than point *b*. (b) Positive current carriers; polarity of potential difference is opposite to that of (a).

tion. There are sites within the material that would normally be occupied by an electron but are actually empty, and a *missing negative* charge is equivalent to a *positive* charge. When an electron moves in one direction to fill a hole, it leaves another hole behind it, and the result is that the *hole* (equivalent to a positive charge) migrates in the direction *opposite* to that of the electron.

In terms of the coordinate axes in Fig. 30–18a, the electrostatic field E_e is in the z-direction, and we write it as E_z. The magnetic field is in the y-direction, and we write it as B_y. The magnetic-field force (in the $-z$-direction) is qvB_y. The current density J_x is in the x-direction.

In the final steady state, when the forces qE_z and qvB_y are equal,

$$E_z = vB_y.$$

The current density J_x is

$$J_x = nqv.$$

When v is eliminated, we have

$$nq = \frac{J_x B_y}{E_z}. \tag{30–18}$$

Thus, from measurements of J_x, B_y, and E_z, one can compute the product nq. In both metals and semiconductors, q is equal in magnitude to the electron charge, so the Hall effect permits a direct measurement of n, the density of current-carrying charges in the material.

QUESTIONS

30–1 Does the earth's magnetic field have a significant effect on the electron beam in a TV picture tube?

30–2 If an electron beam in a cathode-ray tube travels in a straight line, can you be sure there is no magnetic field present?

30–3 If the magnetic-field force does no work on a charged particle, how can it have any effect on the particle's motion? Are there other examples of forces that do no work but have a significant effect on a particle's motion?

30–4 A permanent magnet can be used to pick up a string of nails, tacks, or paper clips, even though these are not magnets by themselves. How can this be?

30–5 How could a compass be used for a *quantitative* determination of magnitude and direction of magnetic field at a point?

30–6 The direction in which a compass points (magnetic north) is not in general exactly the same as the direction toward the north pole (true north). The difference is called *magnetic declination;* it varies from point to point on the earth and also varies with time. What are some possible explanations for magnetic declination?

30–7 Can a charged particle move through a magnetic field without experiencing any force? How?

30–8 Could the electron beam in an oscilloscope tube (cathode-ray tube) be used as a compass? How? What advantages and disadvantages would it have, compared with a conventional compass?

30–9 Does a magnetic field exert forces on the electrons within atoms? What observable effect might such interaction have on the behavior of the atom?

30–10 How might a loop of wire carrying a current be used as a compass? Could such a compass distinguish between north and south?

30–11 How could the direction of a magnetic field be determined by making only *qualitative* observations of the magnetic force on a straight wire carrying a current?

30–12 Do the currents in the electrical system of a car have a significant effect on a compass placed in the car?

30–13 A student claimed that, if lightning strikes a metal flagpole, the force exerted by the earth's magnetic field on the current in the pole can be large enough to bend it. Typical lightning currents are of the order of 10^4 to 10^5 A; is the student's opinion justified?

30–14 A student tried to make an electromagnetic compass by suspending a coil of wire from a thread (with the plane of the coil vertical) and passing a current through it. He expected the coil to align itself perpendicular to the horizontal component of the earth's magnetic field; but instead, the coil went into what appeared to be angular simple harmonic motion (cf. Section 11–5), turning back and forth past the expected direction. What was happening? Was the motion truly simple harmonic?

30–15 When the polarity of the voltage applied to a dc motor is reversed, the direction of rotation *does not* reverse. Why not? How *could* the direction of rotation be reversed?

30–16 If an emf is produced in a dc motor, would it be possible to use the motor somehow as a *generator* or *source*, taking power out of it instead of putting power into it? How might this be done?

30–17 In an electromagnetic pump used to pump liquid sodium, as in Fig. 30–17, it was found that there was a difference in pressure between the top and bottom of the pipe passing through the pump. Gravity is one reason for this difference, but there is also an electromagnetic effect. What is it?

30–18 It was noted in Section 30–4 that a magnetic force on a moving charge does no work on the charge. Yet work is required to push a fluid through a pipe, and so a magnetic pump must do work on the pumped fluid. How can this apparent inconsistency be resolved?

30–19 Hall-effect voltages are much *larger* for relatively poor conductors such as germanium than for good conductors such as copper, for comparable currents, fields, and dimensions. Why?

30–20 Is it possible that, in a Hall-effect experiment, *no* transverse potential difference will be observed? Under what circumstances might this happen?

PROBLEMS

30–1 Each of the lettered circles at the corners of the cube in Fig. 30–19 represents a positive charge q moving with a velocity of magnitude v in the direction indicated. The region in the figure is a uniform magnetic field B, parallel to the x-axis and directed toward the right. Copy the figure, find the magnitude and direction of the force on each charge, and show the force in your diagram.

30–2 The magnetic field B in a certain region is 2 T and its direction is that of the positive x-axis in Fig. 30–20.

a) What is the magnetic flux across the surface *abcd* in the figure?

b) What is the magnetic flux across the surface *befc*?

c) What is the magnetic flux across the surface *aefd*?

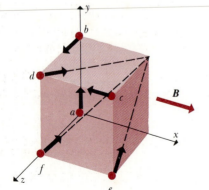

Figure 30–19

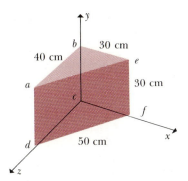

Figure 30–20

30–3 A particle having a mass of 0.5 g carries a charge of 2.5×10^{-8} C. The particle is given an initial horizontal velocity of 6×10^4 m·s^{-1}. What are the magnitude and direction of the minimum magnetic field that will keep the particle moving in a horizontal direction?

30–4 A deuteron, an isotope of hydrogen whose mass is very nearly 2 u, travels in a circular path of radius 40 cm in a magnetic field of magnitude 1.5 T.

  a) Find the speed of the deuteron.

  b) Find the time required for it to make one half a revolution.

  c) Through what potential difference would the deuteron have to be accelerated to acquire this velocity?

30–5 An electron at point A in Fig. 30–21 has a speed v_0 of 10^7 m·s^{-1}. Find

  a) the magnitude and direction of the magnetic field that will cause the electron to follow the semicircular path from A to B, and

  b) the time required for the electron to move from A to B.

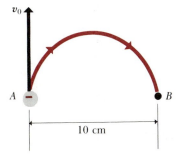

Figure 30–21

30–6 In Problem 30–5, suppose the particle is a proton rather than an electron. Answer the same questions as in that problem.

30–7 In a magnetic field directed vertically upward, a particle initially moving north is deflected toward the west. What is the sign of the charge on the particle?

30–8 In a TV picture tube, an electron in the beam is accelerated by a potential difference of 20,000 V. Then it passes through a region of transverse magnetic field

where it moves in a circular arc with radius 12 cm. What is the magnitude of the field?

30–9 Estimate the effect of the earth's magnetic field on the electron beam in a TV picture tube. Suppose the accelerating voltage is 20,000 V; calculate the approximate deflection of the beam over a distance of 0.4 m from the electron gun to the screen, under the action of a transverse field of magnitude 0.5×10^{-4} T (comparable to the magnitude of the earth's field), assuming there are no other deflecting fields. Is this deflection significant?

30–10 A particle carries a charge of 4×10^{-9} C. When it moves with a velocity v_1 of 3×10^4 m·s^{-1} at 45° above the x-axis in the xy-plane, a uniform magnetic field exerts a force F_1 along the z-axis. When the particle moves with a velocity v_2 of 2×10^4 m·s^{-1} along the z-axis, there is a force F_2 of 4×10^{-5} N exerted on it along the x-axis. What are the magnitude and direction of the magnetic field? (See Fig. 30–22.)

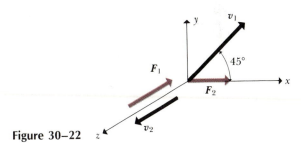

Figure 30–22

30–11 An electron moves in a circular path of radius 1.2 cm perpendicular to a uniform magnetic field. The speed of the electron is 10^6 m·s^{-1}. What is the total magnetic flux encircled by the orbit?

30–12 An electron and an alpha particle (a doubly ionized helium atom) both move in circular paths in a magnetic field with the same tangential speed. Compare the number of revolutions they make per second. The mass of the alpha particle is 6.68×10^{-27} kg.

30–13 In the Bainbridge mass spectrometer (Fig. 30–11), suppose the magnetic field B in the velocity selector is 1.0 T, and ions having a speed of 0.4×10^7 m·s^{-1} pass through undeflected.

  a) What should be the electric field between the plates P and P'?

  b) If the separation of the plates is 0.5 cm, what is the potential difference between plates?

30–14

  a) What is the velocity of a beam of electrons when the simultaneous influence of an electric field of 34×10^4 V·m^{-1} and a magnetic field of 2×10^{-3} T, both normal to the beam and to each other, produces no deflection of the electrons?

b) Show in a diagram the relative orientation of the vectors v, E, and B.

c) What is the radius of the electron orbit when the electric field is removed?

30–15 A singly charged Li^7 ion has a mass of 1.16×10^{-23} g. It is accelerated through a potential difference of 500 V and then enters a magnetic field of 0.4 T, moving perpendicular to the field. What is the radius of its path in the magnetic field?

30–16 Suppose the electric field between the plates P and P′ in Fig. 30–11 is 150 V·cm^{-1}, and the magnetic field is 0.5 T. If the source contains the three isotopes of magnesium, $^{24}_{12}Mg$, $^{25}_{12}Mg$, and $^{26}_{12}Mg$, and the ions are singly charged, find the distance between the lines formed by the three isotopes on the photographic plate. Assume the atomic masses of the isotopes equal to their mass numbers.

30–17 The electric field between the plates of the velocity selector in a Bainbridge mass spectrograph is 1200 V·cm^{-1}, and the magnetic field in both regions is 0.6 T. A stream of singly charged neon atoms moves in a circular path of 7.28-cm radius in the magnetic field. Determine the mass number of the neon isotope.

30–18 A particle of mass m and charge $+q$ starts form rest at the origin in Fig. 30–23. There is a uniform electric field E in the positive y-direction and a uniform magnetic field B directed toward the reader. It is shown in more advanced books that the path is a *cycloid* whose radius of curvature at the top points is twice the y-coordinate at that level.

a) Explain why the path has this general shape and why it is repetitive.

b) Prove that the speed at any point is equal to $\sqrt{2qEy/m}$. (*Hint.* Use energy conservation.)

c) Applying Newton's second law at the top point, prove that the speed at this point is $2E/B$.

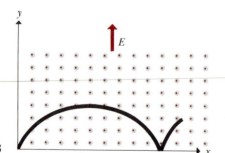

Figure 30–23

30–19 The cube in Fig. 30–24, 0.5 m on a side, is in a uniform magnetic field of 0.6 T, parallel to the x-axis. The wire *abcdef* carries a current of 4 A in the direction indicated. Determine the magnitude and direction of the force acting on the segments *ab*, *bc*, *cd*, *de*, and *ef*.

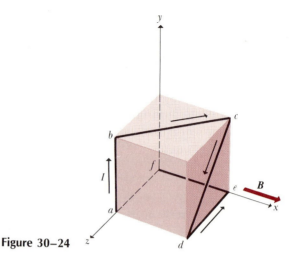

Figure 30–24

30–20 The plane of a rectangular loop of wire 5 cm × 8 cm is parallel to a magnetic field of magnitude 0.15 T.

a) If the loop carries a current of 10 A, what torque acts on it?

b) What is the magnetic moment of the loop?

c) What is the maximum torque that can be obtained with the same total length of wire carrying the same current in this magnetic field?

30–21 An electromagnet produces a magnetic field of 1.2 T in a cylindrical region of radius 5 cm between its poles. A wire carrying a current of 20 A passes through this region, intersecting the axis of the cylinder and perpendicular to it. What force is exerted on the wire?

30–22 A wire 0.5 m long lies along the y-axis and carries a current of 10 A in the $+y$ direction. The magnetic field is uniform and has components $B_x = 0.3$ T, $B_y = -1.2$ T, and $B_z = 0.5$ T.

a) Find the components of force on the wire.

b) What is the magnitude of the total force on the wire?

30–23 A horizontal rod 0.2 m long is mounted on a balance and carries a current. In the vicinity of the wire is a uniform horizontal magnetic field of magnitude 0.05 T, perpendicular to the wire. The force on the rod is measured by the balance and is found to be 0.24 N. What is the current?

30–24 In Problem 30–23, suppose the magnetic field is horizontal but makes an angle of 30° with the rod. What is the current in the rod?

30–25 The rectangular loop in Fig. 30–25 is pivoted about the y-axis and carries a current of 10 A in the direction indicated.

a) If the loop is in a uniform magnetic field of magnitude 0.2 T, parallel to the x-axis, find the force on

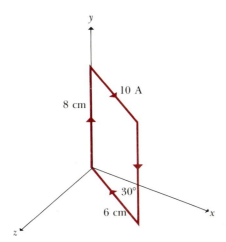

Figure 30–25

each side of the loop and the torque required to hold the loop in the position shown.

b) Same as (a) except that the field is parallel to the z-axis.

c) What torque would be required if the loop were pivoted about an axis through its center, parallel to the y-axis?

30–26 The rectangular loop of wire in Fig. 30–26 has a mass of 0.1 g per centimeter of length and is pivoted about side ab as a frictionless axis. The current in the wire is 10 A in the direction shown.

a) Find the magnitude and sense of the magnetic field, parallel to the y-axis, that will cause the loop to swing up until its plane makes an angle of 30° with the yz-plane.

b) Discuss the case where the field is parallel to the x-axis.

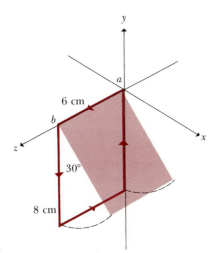

Figure 30–26

30–27 What is the maximum torque on a coil 5 × 12 cm, of 600 turns, when carrying a current of 10^{-5} A in a uniform field of magnitude 0.1 T?

30–28 The coil of a pivoted-coil galvanometer has 50 turns and encloses an area of 6 cm². The magnetic field in the region in which the coil swings is 0.01 T and is radial. The torsional constant of the hairsprings is

$$k' = 1.0 \times 10^{-8} \text{ N} \cdot \text{m} \cdot (\text{degree})^{-1}.$$

Find the angular deflection of the coil for a current of 1 mA.

30–29 A circular coil of wire 8 cm in diameter has 12 turns and carries a current of 5 A. The coil is in a region where the magnetic field is 0.6 T.

a) What is the maximum torque on the coil?

b) In what position would the torque be one half as great as in (a)?

30–30 In a shunt-wound dc motor (Fig. 30–27), the resistance R_f of the field coils is 150 Ω and the resistance R_r of the rotor is 2 Ω. When a difference of potential of 120 V is applied to the brushes, and the motor is running at full speed delivering mechanical power, the current supplied to it is 4.5 A.

a) What is the current in the field coils?

b) What is the current in the rotor?

c) What is the emf developed by the motor?

d) How much mechanical power is developed by this motor?

30–31 A shunt-wound dc motor (Fig. 30–27) operates from a 120-V dc power line. The resistance of the field windings, R_f, is 240 Ω. The resistance of the rotor, R_r, is 3 Ω. When the motor is running the rotor develops an emf $\mathcal{E}$. The motor draws a current of 4.5 A from the line. Compute

a) the field current,

b) the rotor current,

c) the emf $\mathcal{E}$,

d) the rate of development of heat in the field windings,

e) the rate of development of heat in the rotor,

f) the power input to the motor, and

g) the efficiency of the motor, if friction and windage losses amount to 50 W.

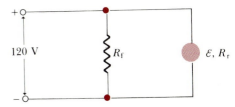

Figure 30–27

30–32 A horizontal tube of rectangular cross section (height h, width w) is placed at right angles to a uniform magnetic field B, so that a length l is in the field. (See Fig. 30–17.) The tube is filled with liquid sodium, and an electric current of density J is maintained in the third mutually perpendicular direction.

a) Show that the difference of pressure between a point in the liquid on a vertical plane through ab (Fig. 30–17) and a point in the liquid on another vertical plane through cd, under conditions in which the liquid is prevented from flowing, is

$$\Delta p = JlB.$$

b) What current density would be needed to provide a pressure difference of 1 atm between these two points if $B = 1$ T and $l = 0.1$ m?

30–33 Figure 30–28 shows a portion of a silver ribbon with $z_1 = 2$ cm and $y_1 = 1$ mm, carrying a current of 200 A in the positive x-direction. The ribbon lies in a uniform magnetic field, in the y-direction, of magnitude 1.5 T. If

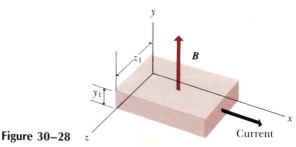

Figure 30–28

there are 7.4×10^{28} free electrons per m^3, find

a) the drift velocity of the electrons in the x-direction,

b) the magnitude and direction of the electric field in the z-direction due to the Hall effect, and

c) the Hall emf.

30–34 Let Fig. 30–28 represent a strip of copper of the same dimensions as those of the silver ribbon of the preceding problem. When the magnetic field is 5 T and the current is 100 A, the Hall emf is found to be 45.4 μV. What is the density of free electrons in the copper?

31

SOURCES
OF MAGNETIC
FIELD

In Chapter 30 we discussed one aspect of the magnetic interaction of moving charges; we took the existence of the magnetic field as a given fact and studied the *forces* exerted by the magnetic field on moving charges and on currents in conductors. We now return to the other aspect of this interaction, the principles that describe how magnetic fields are *created* by moving charges and by currents. In most practical situations the moving charges are usually those associated with electric current in a conductor, and these situations form the principal topic of this chapter. We first study the magnetic fields around a few particular shapes of conductors and the magnetic forces between currents in parallel current-carrying conductors. More general methods for computing magnetic fields of more complex current distributions are discussed briefly, and then we examine the role of *displacement current*, introduced in Section 29–5, as a source of magnetic field. In this discussion, diagrams showing magnetic-field lines, as described in Section 30–3, are useful in visualizing magnetic fields.

31–1

MAGNETIC FIELD OF A LONG STRAIGHT CONDUCTOR

The simplest current configuration to describe is that of a very long, straight conductor carrying a steady current *I*. The magnetic field caused by such a conductor has the general shape shown in Fig. 31–1. The magnetic field lines are *circles* centered on the conductor and lying in planes perpendicular to it. As the figure shows, the magnetic field ***B*** at any point is perpendicular to the direction of the conductor and is also perpendicular to a radial line from the center of the circle to the point.

The *magnitude* of the magnetic field, as shown by the experiments of Biot and Savart, is inversely proportional to the distance *r* from the wire, and is directly proportional to the current. This relationship is expressed

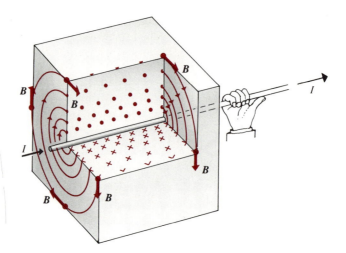

31–1 Magnetic field around a long straight conductor. The field lines are circles, with directions determined by the right-hand rule.

by the equation

$$B = \frac{\mu_0 I}{2\pi r} \qquad \text{(long straight wire).} \qquad (31\text{--}1)$$

In this equation μ_0 is a constant that depends on the system of units used. Its units are those of magnetic field times distance, divided by current. Thus, in SI units, the units of μ_0 are $(\text{T}\cdot\text{m}\cdot\text{A}^{-1})$. The numerical value, which is determined by the definition of the unit of current, is *exactly* $4\pi \times 10^{-7}$. Thus

$$\mu_0 = 4\pi \times 10^{-7}\,\text{T}\cdot\text{m}\cdot\text{A}^{-1}.$$

From Eq. (30–11), $1\,\text{T} = 1\,\text{N}\cdot\text{A}^{-1}\cdot\text{m}^{-1}$; thus an alternative expression is $\mu_0 = 4\pi \times 10^{-7}\,\text{N}\cdot\text{A}^{-2}$.

We recall from Section 24–5 that the constant ϵ_0 in Coulomb's law has the value

$$\epsilon_0 = 8.854185 \times 10^{-12}\,\text{C}^2\cdot\text{N}^{-1}\cdot\text{m}^{-2}.$$

The product $\mu_0\epsilon_0$ is also of interest:

$$\begin{aligned}
\mu_0\epsilon_0 &= (4\pi \times 10^{-7}\,\text{N}\cdot\text{A}^{-2})(8.854185 \times 10^{-12}\,\text{C}^2\cdot\text{N}^{-1}\cdot\text{m}^{-2}) \\
&= 1.1126497 \times 10^{-17}\,\text{C}^2\cdot\text{A}^{-2}\cdot\text{m}^{-2},
\end{aligned}$$

or, since $1\,\text{A} = 1\,\text{C}\cdot\text{s}^{-1}$,

$$\mu_0\epsilon_0 = 1.1126497 \times 10^{-17}\,\text{s}^2\cdot\text{m}^{-2}.$$

Taking reciprocals of both sides, we find

$$\frac{1}{\mu_0\epsilon_0} = 8.987554 \times 10^{16}\,\text{m}^2\cdot\text{s}^{-2} = (2.997925 \times 10^8\,\text{m}\cdot\text{s}^{-1})^2.$$

Thus $1/\mu_0\epsilon_0$ is equal to the square of the measured value of the speed of light $c = 2.997925 \times 10^8\,\text{m}\cdot\text{s}^{-1}$:

$$\frac{1}{\mu_0\epsilon_0} = c^2. \qquad (31\text{--}2)$$

This result cannot be mere coincidence; it shows that there must be a close relationship between electricity, magnetism, and light. We shall return to this relationship in Chapter 35.

EXAMPLE   A long straight conductor carries a current of 100 A. At what distance from the conductor is the magnetic field caused by the current equal in magnitude to the earth's magnetic field in Pittsburgh (about 0.5×10^{-4} T)?

Solution   We use Eq. (31–1); everything except r is known, so we solve for r and insert the appropriate numbers:

$$r = \frac{\mu_0 I}{2\pi B} = \frac{(4\pi \times 10^{-7}\ \text{T}\cdot\text{m}\cdot\text{A}^{-1})(100\ \text{A})}{2\pi(0.5 \times 10^{-4}\ \text{T})} = 0.4\ \text{m}.$$

At smaller distances the field becomes stronger; when $r = 0.2$ m, $B = 1.0 \times 10^{-4}$ T, and so on.   ◀

The shape of the magnetic-field lines in this situation is completely different from that of the electric-field lines in the analogous electrical situation. Electric-field lines radiate outward from the charges that are their sources (or inward for negative charges). By contrast, magnetic-field lines *encircle* the current that acts as their source. Electric-field lines begin and end at charges, while magnetic-field lines *never* have endpoints! This property can be interpreted as saying that there is no such thing as a single "magnetic charge" to act as a source of **B**. The sources of **B** are moving *electric* charges.

Finally, we note that the direction of the **B** lines around a straight conductor is given by a right-hand rule: Grasp the conductor, with the thumb extended in the direction of the current; the fingers then curl around the conductor in the direction of the **B** lines. This rule is illustrated in Fig. 31–1.

31–2

FORCE BETWEEN PARALLEL CONDUCTORS; THE AMPERE

Figure 31–2 shows portions of two long straight parallel conductors separated by a distance r and carrying currents I and I', respectively, in the same direction. Since each conductor lies in the magnetic field set up by the other, each experiences a force. The diagram shows some of the field lines set up by the current in the *lower* conductor.

From Eq. (31–1), the magnitude of the **B**-vector at the upper conductor is

$$B = \frac{\mu_0 I}{2\pi r}.$$

From Eq. (30–10), the force on a length l of the upper conductor is

$$F = IlB = \frac{\mu_0 lII'}{2\pi r},$$

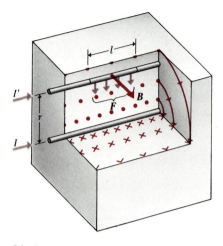

31–2 Parallel conductors carrying currents in the same direction attract each other.

and the force *per unit length* is therefore

$$\frac{F}{l} = \frac{\mu_0 I I'}{2\pi r}.$$ (31–3)

The right-hand rule shows that the direction of the force on the upper conductor is downward. There is an equal and opposite force per unit length on the lower conductor, as may be seen by considering the field set up by the upper conductor. So, the conductors *attract* one another.

If the direction of either current is reversed, the forces reverse also. Parallel conductors carrying currents in opposite directions repel one another.

The fact that two straight parallel conductors exert forces of attraction or repulsion on one another is made the basis of the SI definition of the ampere. The ampere is defined as follows:

One ampere is that unvarying current which, if present in each of two parallel conductors of infinite length and one meter apart in empty space, causes each conductor to experience a force of exactly 2×10^{-7} newtons per meter of length.

It follows from this and the preceding equation that, by definition, the constant μ_0 is *exactly* $4\pi \times 10^{-7}$ N·A^{-2}, as stated in Section 31–1.

From the definition above, current can be measured in principle with a meter stick and a spring balance. For the practical standardization of the ampere, coils of wire are used instead of straight wires, and their separation is made only a few centimeters. The complete instrument, which is capable of measuring currents with a high degree of precision, is called a *current balance.*

Having defined the ampere, we can now define the coulomb as *the quantity of charge that in one second crosses a section of a circuit in which there is a constant current of one ampere.*

Mutual forces of attraction exist not only between *wires* carrying currents in the same direction but also between each of the various regions of a current-carrying conductor. If the conductor is a liquid or an ionized gas (a plasma), these forces result in a constriction of the conductor as if its surface were acted on by an external, inward, pressure force. The constriction of the conductor is called the *pinch effect;* the high temperature produced by the pinch effect in a plasma is used to bring about nuclear fusion.

31–3

MAGNETIC FIELD OF A CIRCULAR LOOP

In many devices in which a current is used to establish a magnetic field, as in an electromagnet or a transformer, the wire carrying the current is wound into a *coil,* often consisting of many circular loops. We therefore consider next the magnetic field set up by a single circular loop of wire carrying a current.

Figure 31–3 shows a circular conducting loop of wire of radius R, carrying a current I that is led into and out of the loop through two long straight wires side by side. The currents in the straight wires are in opposite directions and annul each other's magnetic effects.

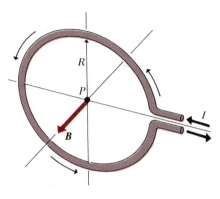

31–3 Field at the center of a circular loop.

Biot and Savart found experimentally that the magnetic field at the center of such a loop has the direction shown in the figure and that its magnitude is directly proportional to the current and inversely proportional to the radius of the loop. Specifically, it is found that

$$B = \frac{\mu_0 I}{2R} \qquad \text{(center of circular loop).} \qquad (31\text{–}4)$$

If, instead of a single loop as in Fig. 31–3, one has a coil of N closely spaced loops, all having the same radius, each loop contributes equally to the field, and Eq. (31–4) becomes

$$B = \frac{\mu_0 N I}{2R} \qquad \text{(center of } N \text{ circular loops).} \qquad (31\text{–}5)$$

Some of the magnetic-field lines surrounding a circular loop and lying in planes through the axis are shown in Fig. 31–4. Again we see that the field lines encircle the conductor, and that their directions are given by the right-hand rule, as for the long straight conductor. The field lines for the circular loop are *not* circles, but they are closed curves that link the conductor.

By methods to be described later, one can derive an expression for the magnetic field at any point on the axis of the loop, such as point P, a distance x from the center, in Fig. 31–4. As the figure shows, the *direction* of **B** at all points on the axis is along the axis; that is, in fact, required by the axial symmetry of the problem. The field magnitude turns out to be given by the expression

$$B = \frac{\mu_0 I R^2}{2(x^2 + R^2)^{3/2}}. \qquad (31\text{–}6)$$

The corresponding result for N circular loops is obtained by inserting the factor N into Eq. (31–6), in the same way we obtained Eq. (31–5) from Eq. (31–4).

We note that when $x = 0$, Eq. (31–6) reduces to Eq. (31–4), as of course it must. For points on the axis very far away from the center, that

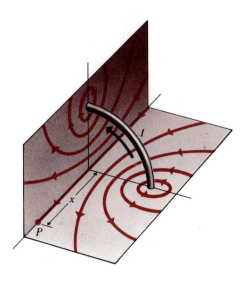

31–4 Field lines surrounding a circular loop.

is, x much greater than R, the term R^2 in the denominator of Eq. (31–6) becomes negligible compared to x^2, and the expression becomes approximately equal to $\mu_0 I R^2 / 2x^3$. Thus at large distances the field along the axis is approximately proportional to $1/x^3$.

It was pointed out in Chapter 30 that a loop of wire of *any* shape, carrying a current, is acted on by a torque when placed in an external magnetic field. Stable equilibrium occurs when the plane of the loop is perpendicular to the external field. In this position, the **B** field produced by the loop itself is in the same direction inside the loop as that of the external field. In other words, a loop, if free to turn, will orient itself so that the flux passing through the area enclosed by it has its *maximum* possible value. This is found to be true in all instances and is a useful general principle. For example, if a current is sent through an irregular loop of flexible wire in an external magnetic field, the loop will assume a circular form with its plane perpendicular to the field and with its own flux adding to that of the field.

31–4

MAGNETIC FIELD OF A SOLENOID

A solenoid is constructed by winding wire in a helical coil around the surface of a cylindrical form, usually of circular cross section. The turns of the winding are ordinarily closely spaced and may consist of one or more layers. For simplicity, we have represented a solenoid in Fig. 31–5 by a relatively small number of circular turns, each carrying a current I. The resultant field at any point is the vector sum of the **B**-vectors due to the individual turns. The diagram shows the field lines in the xy- and yz-planes. Exact calculations show that for a long, closely wound solenoid, half of the lines passing through a cross section at the center emerge from the ends and half "leak out" through the windings between center and end.

If the length of the solenoid is large compared with its cross-sectional diameter, the *internal* field near its center is very nearly uniform and

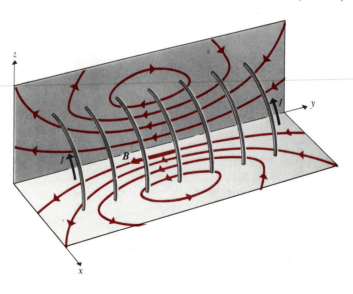

31–5 Magnetic-field lines surrounding a solenoid.

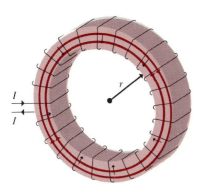

31–6 A toroidal solenoid. The field is very nearly zero at all points except those within the space enclosed by the windings.

parallel to the axis, and the *external* field near the center is very small. The magnetic-field magnitude B at the center depends on the number of turns *per unit length* of the solenoid, which we shall call n. That is, if there are N turns distributed uniformly over a total length L, then $n = N/L$. It is found that B at the center of the solenoid is given by

$$B = \mu_0 nI \qquad \text{(solenoid).} \qquad (31\text{–}7)$$

At the center, the field is very nearly uniform over the cross section of the cylinder.

A variation is the *toroidal* (doughnut-shaped) *solenoid,* shown in Fig. 31–6. This shape has the interesting property that the magnetic field is confined entirely to the space enclosed by the windings; there is no field at all outside this region. If there are N turns in all, then the field magnitude at a point at a distance r from the center of the torus (*not* the center of its cross section) is given by

$$B = \frac{\mu_0 NI}{2\pi r} \qquad \text{(toroidal solenoid).} \qquad (31\text{–}8)$$

The magnetic field is *not* uniform over a cross section of the core. However, if the radial thickness of the core is small compared with the toroid radius r, the field varies only slightly across a section. In that case, considering that $2\pi r$ is the *circumferential* length of the toroid and that $N/2\pi r$ is the number of turns per unit length n, the field may be written

$$B = \mu_0 nI,$$

just as at the center of a long *straight* solenoid.

31–5

MAGNETIC-FIELD CALCULATIONS

In the preceding sections we have quoted equations for the magnetic fields caused by several shapes of conductors. Deriving these relations usually involves the use of integral calculus and vector algebra, and such derivations are outside the scope of this text. We can however give some hint of the two general methods that are used. The first uses a principle called the law of Biot and Savart, which gives the magnetic field caused by a short segment of conductor. The other uses Ampère's law, a principle that relates the field at various points around a closed curve to the current enclosed or linked by that curve.

Biot and Savart found that the magnetic field due to any shape of conductor could be found by imagining it to be divided into short segments, calculating the field due to each segment, and adding all these to obtain the total field. The rule for finding the field caused by a single segment is shown in Fig. 31–7. A segment having length Δl, carrying current I, sets up a magnetic field $\Delta \boldsymbol{B}$, as shown. At any point P, the direction of $\Delta \boldsymbol{B}$ is perpendicular to the plane formed by the axis of Δl and the line from Δl to P.

The field lines, to which the vectors $\Delta \boldsymbol{B}$ are tangent, are *circles* lying in planes perpendicular to the axis of the element. The direction of these lines is *clockwise* when viewed along the direction of the current in Δl. The direction may also be described by a right-hand rule: grasp the

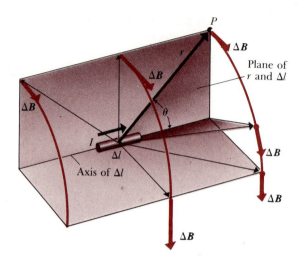

31–7 Magnetic-field vectors at various points due to a current segment. The field lines are circles with centers along the line of Δl.

element in the right hand with the extended thumb pointing in the direction of the current. The fingers then encircle the element in the direction of the field lines.

The *magnitude* of $\Delta \mathbf{B}$ is given by

$$\Delta B = \frac{\mu_0}{4\pi} \frac{I \, \Delta l \sin \theta}{r^2}, \qquad (31\text{–}9)$$

where r is the distance between Δl and the point P, and θ is the angle between r and Δl. This relation is called the *law of Biot and Savart*.

It follows from Eq. (31–9) that the field ΔB due to a current segment is zero at all points on the axis of the segment, since $\sin \theta = 0$ at all such points. At a given distance r from the segment, the magnetic field is greatest in a plane passing through the segment perpendicular to its axis, since for points in this plane $\theta = 90°$ and $\sin \theta = 1$.

The expression for the *total* field at any point of space, due to current I in a complete circuit, is obtained by forming the vector sum of all the $\Delta \mathbf{B}$'s due to the various segments of the conductor. In most cases this is a fairly elaborate mathematical problem, but there are a few cases where it is fairly simple.

EXAMPLE 1   Find the magnetic field at the center of a circular conducting loop of radius R, carrying current I.

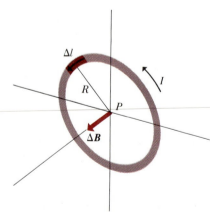

31–8 Magnetic field $\Delta \mathbf{B}$ caused by a segment Δl of a circular conducting loop. The segment, the radial line, and the direction of $\Delta \mathbf{B}$ are all mutually perpendicular.

Solution   We imagine the loop as divided into a large number of segments having lengths Δl_1, Δl_2, and so on. A typical segment of length Δl is shown in Fig. 31–8; all segments are at the same distance R from the point P at the center, and each makes a right angle with the line joining it to P. The vectors $\Delta \mathbf{B}_1$, $\Delta \mathbf{B}_2$, etc., due to each segment, are all in the same direction perpendicular to the plane of the loop. Thus in Eq. (31–9), $\sin \theta = 1$ and $r = R$ for every segment, and the total $\mathbf{B}$ field magnitude is given by

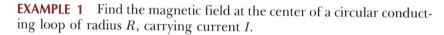

$$B = \frac{\mu_0 I}{4\pi R^2} (\Delta l_1 + \Delta l_2 + \cdots).$$

But $\Delta l_1 + \Delta l_2 + \cdots$ is equal to the circumference of the loop, $2\pi R$. Therefore,

$$B = \frac{\mu_0 I}{2R},$$

in agreement with Eq. (31–4). ◀

Ampère's law provides an alternative basis for magnetic-field calculations. To introduce the basic idea, we consider again the field caused by a long straight conductor; it was stated in Section 31–1 that the field magnitude at distance r from the conductor is given by

$$B = \frac{\mu_0 I}{2\pi r},$$

and that the magnetic field lines are circles centered on the conductor. The circumference of such a circle is $2\pi r$, and we note that for any r the product of B and the circumference is

$$B(2\pi r) = \left(\frac{\mu_0 I}{2\pi r}\right)(2\pi r) = \mu_0 I.$$

That is, this product is independent of r and dependent only on the current in the conductor.

This is a special case of Ampère's law. The general statement is as follows: We construct any imaginary closed curve encircling one or more conductors. The conductors then pass through the area enclosed by the curve. We divide this curve into segments, calling a typical segment Δs. At each segment, we take the component of **B** parallel to the segment; we call this component $B_\parallel$. We take all the products $B_\parallel \Delta s$ and add them as we go completely around the closed curve. The result of this sum is always equal to μ_0 times the total current in all the conductors. Symbolically,

$$\sum B_\parallel \Delta s = \mu_0 I. \tag{31–10}$$

In computing the total current, we need a sign rule. Looking at the surface, we proceed counterclockwise around the boundary in evaluating the sum in Eq. (31–10). Any current moving toward us through the surface is counted as positive, and any away from us as negative.

If we reverse the above discussion and take Ampère's law as a starting point, we can then *derive* the expression for the field of a long straight wire by taking as our imaginary curve a circle of radius r.

EXAMPLE 2 Find the magnetic-field magnitude inside a toroidal solenoid, at a distance r from the center, if the solenoid has N turns carrying a current I.

Solution We take as our imaginary curve a circle of radius r inside the volume enclosed by the solenoid. According to the discussion of Section 31–4, the **B** field is everywhere tangent to such a circle, and it has the same magnitude at every point on the circle. Thus $B_\parallel = B$, and the sum in Eq. (31–10) becomes simply $B(2\pi r)$. The total current enclosed by the curve is NI, so Eq. (31–10) becomes

$$B(2\pi r) = \mu_0 NI,$$

or

$$B = \frac{\mu_0 NI}{2\pi r},$$

in agreement with Eq. (31–8).

◀

We note that in this example, as well as with the long straight wire, the symmetry of the situation makes it easy to evaluate the sum in Ampère's law. When there is no symmetry Ampère's law is much less useful, and then field calculations are usually based on the law of Biot and Savart.

*31–6

MAGNETIC MATERIALS

In the preceding discussion of magnetic fields caused by currents, we have assumed that the space surrounding the conductors contains only vacuum. If matter is present in the surrounding space, the magnetic field is changed. The atoms of which all matter is made contain electrons in motion, and these electrons form microscopic current loops capable of producing magnetic fields of their own. In many materials these currents are randomly oriented so as to cause no net magnetic field. But in the presence of an externally caused field the loops can become oriented preferentially with the field, so that their magnetic fields *add* to the external field. The material is said to become *magnetized*.

A material having this behavior is said to be *paramagnetic*. The result is that the magnetic field at any point in such a material is greater by a constant factor K_m than it would be if the material were not present. The value of K_m is different for different materials; it is called the *relative permeability* of the material. Values of K_m for common paramagnetic materials are typically 1.0001 to 1.0020, as shown in Table 31–1.

Thus all the equations in this chapter for magnetic fields caused by currents can be adapted to the situation where the conductor is embedded in a paramagnetic material by simply replacing μ_0 everywhere by $K_m\mu_0$. This product is usually denoted as μ; it is called the *permeability* of the material:

$$\mu = K_m\mu_0. \tag{31–11}$$

The amount by which the relative permeability differs from unity is called the *magnetic susceptibility*, symbol χ_m:

$$\chi_m = K_m - 1. \tag{31–12}$$

In some materials, the total field due to the electrons in each atom adds to zero when there is no external field, and these materials have *no* net atomic current loops. But even in these materials magnetic effects are present because an external field alters the electron motions to cause *induced* current loops. In this case the additional field caused by the induced current loops is always *opposite* in direction to the external field.

TABLE 31–1	MAGNETIC SUSCEPTIBILITIES OF PARAMAGNETIC AND DIAMAGNETIC MATERIALS, AT $T = 20°C$
Materials	$\chi_m = K_m - 1$
Paramagnetic	
Iron ammonium alum	66×10^{-5}
Uranium	40×10^{-5}
Platinum	26×10^{-5}
Aluminum	2.2×10^{-5}
Sodium	0.72×10^{-5}
Oxygen gas	0.19×10^{-5}
Diamagnetic	
Bismuth	-16.6×10^{-5}
Mercury	-2.9×10^{-5}
Silver	-2.6×10^{-5}
Carbon (diamond)	-2.1×10^{-5}
Lead	-1.8×10^{-5}
Rock salt	-1.4×10^{-5}
Copper	-1.0

Such materials are said to be *diamagnetic;* they always have relative permeabilities slightly less than unity, typically of the order of 0.99990 to 0.99999.

There is a third class of materials, called *ferromagnetic* materials, in which the atomic current loops tend to line up paria[parallel] to each other even when there is no external field. This cooperative phenomenon leads to a relative permeability that is much larger than unity, typically of the order of 1,000 to 10,000. Iron, cobalt, and nickel, and many alloys containing these elements are ferromagnetic.

Ferromagnetic materials differ in their behavior from paramagnetic and diamagnetic materials in two other important ways. First, the additional field caused by the microscopic current loops is not directly proportional to the external field except at relatively small fields. As the external field increases, a point is reached where nearly all the microscopic current loops have their axes parallel to the external field. This condition is called *saturation magnetization;* after it is reached, a further increase in the external field causes no increase in magnetization or in the additional field caused by the material.

Second, some ferromagnetic materials retain their magnetization even when there is no externally caused field at all. These materials can thus be permanent magnets. Many kinds of steel and many alloys such as Alnico are commonly used for permanent magnets. The magnetic field in such a material, when it is magnetized to near saturation, is typically of the order of 1 T.

Ferromagnetic materials are widely used in electromagnets, transformer cores, and motors and generators, where it is desirable to have as large a magnetic field as possible for a given current. In these applications it is usually desirable for the material *not* to have permanent magnetization. Soft iron is often used; it has high permeability without appreciable permanent magnetization.

*31–7

MAGNETIC FIELD AND DISPLACEMENT CURRENT

Throughout this chapter we have discussed the magnetic fields caused by conduction currents, that is, currents corresponding to motion of electric charge in conducting materials or in space. But in Section 29–5 we encountered another kind of current, called *displacement current*, that is associated with a changing electric field in a region of space, rather than with a transport of charge through that region. In the context of the present chapter, it is natural to ask whether displacement current acts as a source of magnetic field in the same way as does conduction current.

The answer turns out to be affirmative. Ordinarily, displacement currents are not confined to well-defined paths as the conduction currents in wires are, but it has been found that the same basic principles determine the magnetic fields caused by *both* conduction and displacement currents. In particular, the law of Biot and Savart and Ampère's law, Eqs. (31–9) and (31–10), respectively, can be generalized to include displacement current I_D as well as conduction current I_C. The generalized form of the law of Biot and Savart is

$$\Delta B = \frac{\mu_0}{4\pi} \frac{(I_C + I_D) \, \Delta l \sin \theta}{r^2}, \tag{31–13}$$

and the generalized Ampère's law is

$$\sum B_{\parallel} \, \Delta s = \mu_0 (I_C + I_D). \tag{31–14}$$

Note that in the Biot law, the symbols I_C and I_D refer to the currents in an element Δl of a *conductor;* in Ampère's law they refer to the *total* currents across an area bounded by the imaginary path around which $\sum B_{\parallel} \, \Delta s$ is computed, and Δs is an element of that path.

EXAMPLE   Figure 31–9 shows a cross section of a parallel-plate capacitor consisting of two circular conducting plates of radius R separated by a narrow gap. The capacitor is being charged by the conduction currents I_C as shown; the *E* field in the gap is increasing, so there is a displacement current in the gap. Find the magnetic field in the gap caused by this displacement current.

Solution   If the gap is small compared to R, the *E* field caused by the charge distributions on the plates is confined almost entirely to the region between plates and is nearly uniform, as shown. Let us consider the *B* field at some point in the midplane, shown in the figure as a thin line. There is no *conduction* current across the midplane; the current across this plane is a *displacement* current only. Also, since every *E*-field line crosses the plane at some point, the *total* displacement current across the plane is the *displacement* current *out of* the left plate; from the definition of displacement current in Section 29–5, this must equal the *conduction* current *into* the plate. Furthermore, by symmetry, the *B* field lines are circles with centers on the axis. Thus, at any point on a circle perpendicular to the axis and passing through points such as a and b, the *B*-vector has the same magnitude and is tangent to the circle. At point a, the

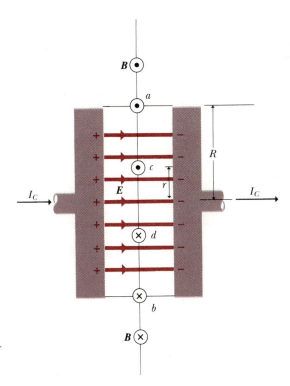

31–9 A capacitor being charged. The changing electric field corresponds to a displacement current, and this causes a *magnetic* field in the gap between the plates.

B-vector points toward the reader and at point *b* it points away from the reader, as indicated by the symbols • and ✕.

The displacement current *density* J_D is given by Eq. (29–14):

$$ J_D = \epsilon_0 \frac{\Delta E}{\Delta t}. $$

At any instant E is uniform over the region between plates, so J_D is also uniform; the *total* displacement current I_D between plates is just J_D times the area πR^2 of the plates. As remarked above, this must also equal the total *conduction* current I_C. Thus

$$ I_D = I_C = \pi R^2 J_D, $$

$$ J_D = \frac{I_C}{\pi R^2}. $$

(31–15)

Now we may find the magnetic field at a point in the region between plates, at a distance r from the axis. We apply Ampère's law to a circle of radius r passing through the point; such a circle passes through points c and d in Fig. 31–9. The total current enclosed by the circle is J_D times its area, or $(I_C/\pi R^2)(\pi r^2)$. The sum $\Sigma B_\parallel \Delta s$ in Ampère's law is just B times the circumference $2\pi r$ of the circle, and Ampère's law becomes

$$ \sum B_\parallel \, \Delta s = 2\pi r B = \mu_0 \left(\frac{r^2}{R^2} \right) I_C, $$

or

$$ B = \frac{\mu_0}{2\pi} \left(\frac{r}{R^2} \right) I_C. $$

(31–16)

This result shows that in the region between the plates **B** is zero at the axis, increasing linearly with distance from the axis. A similar calculation shows that *outside* the region between the plates, **B** is the same as though the wire were continuous and the plates not present at all.   ◄

The fact that *displacement* current acts as a source of magnetic field plays an essential role in the understanding of electromagnetic *waves*. A changing *electric* field in a region of space induces a *magnetic* field in neighboring regions, even when no conduction current and no matter are present. This relationship, first proposed by Maxwell in 1865, provides the key to a theoretical understanding of electromagnetic radiation and of light as a particular example of this radiation. We return to this topic in Chapter 35.

QUESTIONS

31–1 Streams of charged particles emitted from the sun during unusual sunspot activity create a disturbance in the earth's magnetic field. How does this happen?

31–2 A topic of current interest in physics research is the search (thus far unsuccessful) for an isolated magnetic pole or magnetic *monopole*. If such an entity were found, how could it be recognized? What would its properties be?

31–3 What are the relative advantages and disadvantages of Ampère's Law and the Biot–Savart Law for practical calculations of magnetic fields?

31–4 A student proposed to obtain an isolated magnetic pole by taking a bar magnet (N pole at one end, S at the other) and breaking it in half. Will this work?

31–5 Pairs of conductors carrying current into and out of the power-supply components of electronic equipment are sometimes twisted together to reduce magnetic-field effects. Why does this help?

31–6 The text discusses the magnetic field of an infinitely long straight conductor carrying a current. Of course, there's no such thing as an infinitely long *anything*. How do you decide whether a particular wire is long enough to be considered infinite?

31–7 Suppose one has three long parallel wires, arranged so that in cross section they are at the corners of an equilateral triangle. Is there any way to arrange the currents so that all three wires attract each other? So that all three wires repel each other?

31–8 Two parallel conductors carrying current in the same direction attract each other. If they are permitted to move toward each other, the forces of attraction do work. Where does the energy come from? Does this contradict the assertion in Chapter 30 that magnetic forces on moving charges do no work?

31–9 Considering the magnetic field of a circular loop of wire, would you expect the field to be greatest at the cen-

ter, or is it greater at some points in the plane of the loop but off-center?

31–10 Two concentric circular loops of wire, of different diameter, carry currents in the same direction. Describe the nature of the forces exerted on the inner loop.

31–11 A current was sent through a helical coil spring. The spring appeared to contract, as though it had been compressed. Why?

31–12 Using the fact that magnetic-field lines never have a beginning or end, explain why it is reasonable for the field of a toroidal solenoid to be confined entirely to its interior, while a straight solenoid *must* have some field outside it.

31–13 Can one have a displacement current as well as a conduction current within a conductor?

31–14 A character in a popular comic strip has at various times proposed the possibility of "harnessing the earth's magnetic field" as a nearly inexhaustible source of energy. Comment on this concept.

31–15 Why should the permeability of a paramagnetic material be expected to decrease with increasing temperature?

31–16 In the discussion of magnetic forces on current loops in Section 30–8, it was found that no net force is exerted on a complete loop in a uniform magnetic field, only a torque. Yet magnetized materials, which contain atomic current loops, certainly *do* experience net forces in magnetic fields. How is this discrepancy resolved?

31–17 What features of atomic structure determine whether an element is diamagnetic or paramagnetic?

31–18 The magnetic susceptibility of paramagnetic materials is quite strongly temperature-dependent, while that of diamagnetic materials is nearly independent of temperature. Why the difference?

PROBLEMS

31-1 Two hikers are reading a compass under an overhead power transmission line that is 5 m above the ground and carries a current of 400 A from south to north.

a) Find the magnetic field (magnitude and direction) at a point on the ground directly under the conductor.

b) One hiker suggests they walk on another 50 m to avoid inaccurate compass readings caused by the current. Considering that the magnitude of the earth's field is of the order of 0.5×10^{-4} T, is the current really a problem?

31-2 A long straight conductor passes vertically through the center of a laboratory; the direction of the current is upward. The magnetic field at a point 0.20 m from the wire is found to have magnitude 5.0×10^{-4} T. What is the current in the conductor?

31-3 A magnetic field of magnitude 5.0×10^{-4} T is to be produced at a distance of 5 cm from a long straight wire.

a) What current is required to produce this field?

b) With the current found in (a), what is the magnitude of the field at a distance of 10 cm from the wire? At 20 cm?

31-4 A long straight telephone cable contains six wires, each carrying a current of 0.5 A. The distances between wires can be neglected.

a) If the currents in all six wires are in the same direction, what is the magnitude of the magnetic field 10 cm from the cable?

b) If four wires carry currents in one direction and the other two in the opposite direction, what is the field magnitude 10 cm from the cable?

31-5 Figure 31-10 is an end view of two parallel wires perpendicular to the xy-plane, each carrying a current I but in opposite directions.

a) Copy the diagram, and show by vectors the **B** field of each wire, and the resultant **B** field at point P.

b) Derive the expression for the magnitude of **B** at any point on the x-axis, in terms of the coordinate x of the point.

c) Construct a graph of the magnitude of **B** at any point on the x-axis.

d) At what value of x is B a maximum?

31-6 Same as Problem 31-5, except that the current in both wires is *away* from the reader.

31-7 Two long, straight, horizontal parallel wires, one above the other, are separated by a distance $2a$. If the wires carry equal currents in opposite directions, what is the field magnitude in the plane of the wires at a point

a) midway between them, and

b) at a distance a above the upper wire?

If the wires carry equal currents in the same direction, what is the field magnitude in the plane of the wires at a point

c) midway between them, and

d) at a distance a above the upper wire?

31-8 Two long, straight, parallel wires are 1.0 m apart, as in Fig. 31-11. The upper wire carries a current I_1 of 6 A into the plane of the paper.

a) What must be the magnitude and direction of the current I_2 for the resultant field at point P to be zero?

b) What is then the resultant field at Q?

c) At S?

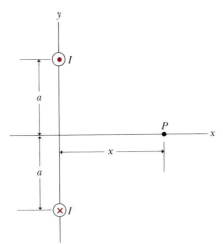

Figure 31-10

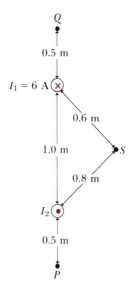

Figure 31-11

31–9 In Fig. 31–10 suppose a third long straight wire, parallel to the other two, passes through point P, and that each wire carries a current $I = 20$ A. Let $a = 30$ cm and $x = 40$ cm. Find the magnitude and direction of the force per unit length on the third wire,

a) if the current in it is away from the reader, and

b) if the current is *toward* the reader.

31–10 A long straight wire carries a current of 1.5 A. An electron travels with a speed of 5×10^6 cm·s^{-1} parallel to the wire, 10 cm from it, and in the same direction as the current. What force does the magnetic field of the current exert on the moving electron?

31–11 A long horizontal wire AB rests on the surface of a table. (See Fig. 31–12.) Another wire CD vertically above the first is 100 cm long and is free to slide up and down on the two vertical metal guides C and D. The two wires are connected through the sliding contacts and carry a current of 50 A. The mass per unit length of the wire CD is 0.05 g·cm^{-1}. To what equilibrium height h will the wire CD rise, assuming the magnetic force on it to be due wholly to the current in the wire AB?

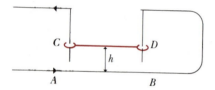

Figure 31–12

31–12 Two long parallel wires are hung by cords of 4 cm length from a common axis, as shown in Fig. 31–13. The wires have a mass per unit length of 50 g·m^{-1} and carry the same current in opposite directions. What is the current if the cords hang at an angle of 6° with the vertical?

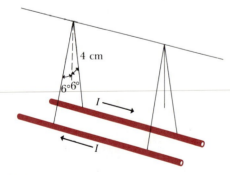

Figure 31–13

31–13 The long straight wire AB in Fig. 31–14 carries a current of 20 A. The rectangular loop whose long edges are parallel to the wire carries a current of 10 A. Find the magnitude and direction of the resultant force exerted on the loop by the magnetic field of the wire.

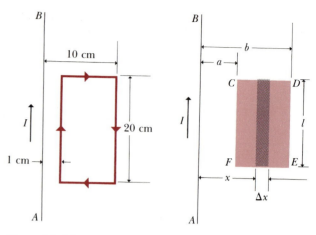

Figure 31–14   **Figure 31–15**

31–14 The long straight wire AB in Fig. 31–15 carries a constant current I.

a) What is the magnetic field at the shaded area at a perpendicular distance x from the wire?

b) What is the magnetic flux $\Delta\Phi$ through the shaded area?

31–15 Refer to Fig. 31–4. Sketch a graph of the magnitude of the B field on the axis of the coil, from $x = -3R$ to $x = +3R$.

31–16 Figure 31–16 is a sectional view of two circular coils of radius a, each wound with N turns of wire carrying a current I, circulating in the same direction in both coils. The coils are separated by a distance a equal to their radii.

a) Derive the expression for the magnetic field at point P, midway between the coils.

b) Calculate the magnitude of $\boldsymbol{B}$ if $N = 100$ turns, $I = 5$ A, and $a = 30$ cm.

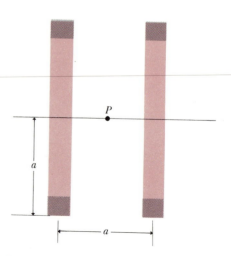

Figure 31–16

31–17 Considering the magnetic field along the axis of a circular loop of radius R, at what distance from the center of the loop is the field $\frac{1}{10}$ of its value at the center?

31–18 A circular coil of radius 5 cm has 200 turns and carries a current of 0.2 A. What is the magnetic field

a) at the center of the coil?

b) at a point on the axis of the coil, 10 cm from its center?

31–19 A closely wound coil has a diameter of 40 cm and carries a current of 2.5 A. How many turns does it have if the magnetic field at the center of the coil is 1.26×10^{-4} T?

31–20 A wire of circular cross section and radius R carries a current I, uniformly distributed over its cross-sectional area.

a) In terms of I, R, and r_1, what is the current through a circular area of radius r_1, inside the wire?

b) Use Ampère's law to find B inside the wire, at a distance r_1 from the axis.

c) What is B outside the wire, at a distance r_2 from the axis?

d) What would the field be at this distance if the current were concentrated in a very fine wire along the axis?

e) Sketch a graph of the magnitude of $\boldsymbol{B}$ as a function of r, from $r = 0$ to $r = 2R$.

31–21 A coaxial cable consists of a solid conductor of radius R_1, supported by insulating disks on the axis of a tube of inner radius R_2 and outer radius R_3. If the central conductor and the tube carry equal currents in opposite directions, find the magnetic field

a) at points outside the axial conductor but inside the tube, and

b) at points outside the tube.

31–22 A solenoid of length 20 cm and radius 2 cm is closely wound with 200 turns of wire. The current in the winding is 5 A. Compute the magnetic field at a point near the center of the solenoid.

31–23 A wooden ring whose mean diameter is 10 cm is wound with a closely spaced toroidal winding of 500 turns. Compute the field at a point on the mean circumference of the ring when the current in the windings is 0.3 A.

31–24 A conductor is made in the form of a hollow cylinder with inner and outer radii a and b, respectively. It carries a current I, uniformly distributed over the cross section. Derive expressions for the magnetic field in the regions $r < a$, $a < r < b$, and $r > b$.

31–25 A solenoid is to be designed to produce a magnetic field of 0.1 T at its center. The radius is to be 5 cm and the length 50 cm, and the available wire can carry a maximum current of 10 A.

a) How many turns per unit length should the solenoid have?

b) What total length of wire is required?

31–26 A long straight wire, carrying a current of 200 A, runs through a cubical wooden box, entering and leaving through holes in the centers of opposite faces, as in Fig. 31–17. The length of each side of the box is 20 cm. Consider an element of the wire 1 cm long at the center of the box. Compute the magnitude of the magnetic field ΔB produced by this element at the points lettered a, b, c, d, and e in Fig. 31–17. Points a, c, and d are at the centers of the faces of the cube, point b is at the midpoint of one edge, and point e is at a corner. Copy the figure and show by vectors the directions and relative magnitudes of the field vectors.

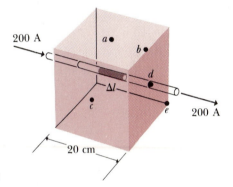

Figure 31–17

31–27 A long straight wire carries a current of 10 A directed along the negative y-axis, as shown in Fig. 31–18. A uniform magnetic field $\boldsymbol{B}_0$ of magnitude 10^{-6} T is directed parallel to the x-axis. What is the resultant magnetic field at the following points?

a) $x = 0$, $z = 2$ m;

b) $x = 2$ m, $z = 0$;

c) $x = 0$, $z = -0.5$ m.

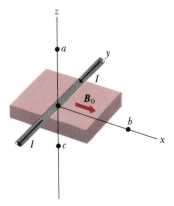

Figure 31–18

31–28 A toroidal solenoid having 500 turns of wire and a mean circumferential length of 50 cm carries a current of 0.3 A. The relative permeability of the core is 600.

a) What is the magnetic field in the core?

b) What part of the magnetic field is due to atomic currents?

31–29 The current in the windings on a toroidal solenoid is 2.0 A. There are 400 turns and the mean circumferential length is 40 cm. The magnetic field is found to be 1.0 T. Calculate the magnetic susceptibility and the relative permeability.

31–30 A toroidal solenoid with 400 turns is wound on a ring having a cross section of 2 cm^2 and a mean circumference of 30 cm. Find the current in the winding that is required to set up a flux density of 0.1 T in the ring,

a) if the ring is of annealed iron ($K_m = 1400$),

b) if the ring is of silicon steel ($K_m = 5200$).

31–31 A long straight copper wire with circular cross section of area 4 mm^2 carries a current of 50 A. The resistivity of the material is $2.0 \times 10^{-8} \, \Omega \cdot m$.

a) What is the electric field in the material?

b) What is the magnetic field 5 cm from the wire?

c) If the current is changing at the rate of 5000 A·s^{-1}, at what rate is the electric field in the material changing?

d) What is the displacement current density in the material in (c)?

e) What is the additional magnetic field 5 cm from the wire, due to the displacement current? Is it significant compared to that due to the conduction current?

31–32 A capacitor has two parallel plates of area A separated by a distance d. The capacitor is given an initial charge Q but, because the damp air between the plates is slightly conductive, the charge slowly leaks through the air. The charge initially changes at a rate $\Delta Q/\Delta t$.

a) In terms of $\Delta Q/\Delta t$, what is the initial rate of change of electric field between the plates?

b) Show that the displacement current density has the same magnitude as the conduction current density, but the opposite direction. Hence show that the magnetic field in the material is exactly zero at all times.

32

INDUCED ELECTROMOTIVE FORCE

We return in this chapter to the subject of electromotive force, introduced in Chapter 28 in connection with the study of current, energy, and power in electric circuits. Our present concern is electromotive force of *magnetic* origin. The present large-scale production, distribution, and use of electrical energy would not be economically feasible if the only seats of emf available were those of *chemical* nature, such as dry cells. The development of electrical engineering, as we now know it, began with Faraday and Henry, who discovered the principles of *magnetically* induced emf's and the methods by which mechanical energy can be converted directly to electrical energy.

32–1

MOTIONAL ELECTROMOTIVE FORCE

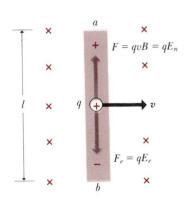

32–1 Conducting rod in uniform magnetic field.

To introduce the concept of magnetically induced electromotive force, we consider a conductor moving in a magnetic field. Figure 32–1 represents a conductor of length l in a uniform magnetic field perpendicular to the plane of the diagram and directed away from the reader. If the conductor is set in motion toward the right with a velocity v, perpendicular both to its own length and to the magnetic field, a charged particle q within it experiences a force F of magnitude qvB directed along the length of the conductor. The direction of the force on a negative charge is from a toward b in Fig. 32–1, while the force on a positive charge is from b toward a. Because this force is of nonelectrostatic origin, we denote it by F_n:

$$F_n = qvB. \qquad (32\text{--}1)$$

The state of affairs within the conductor is the same as though it had been inserted in an *electric* field of magnitude vB, directed from b toward a. Following the discussion of Section 28–4, we define the equivalent

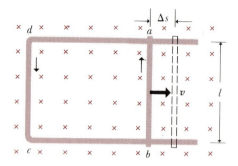

32–2 Current produced by the motion of a conductor in a magnetic field.

nonelectrostatic field E_n as the nonelectrostatic force per unit charge:

$$E_n = vB. \tag{32–2}$$

The free charges in the conductor move in the direction of the force acting on them until the accumulation of excess charges at the ends of the conductor establishes an electrostatic field E_e such that the *resultant* force on every charge within the conductor is zero. The charges are then in equilibrium. The upper end of the wire acquires an excess positive charge, and the lower end an excess negative charge. Thus an electrical potential difference is established between the ends of the conductor. The upper end a is at higher potential than the lower end b, and the potential of a with respect to b is given by

$$V_{ab} = E_e l = vBl. \tag{32–3}$$

Now suppose that the moving conductor slides along a stationary U-shaped conductor, as in Fig. 32–2. There is no magnetic force on the charges within the stationary conductor but, since it lies in the electrostatic field surrounding the moving conductor, a current will be established within it; the direction of this current (defined as usual as the direction of positive-charge motion) is counterclockwise, or from a through d and c to b. As long as the motion of the conductor is maintained there is a continuous current in a counterclockwise direction. The moving conductor corresponds to a seat of electromotive force and is said to have *induced* within it a *motional electromotive force*.

The magnitude of this emf can be found as follows. When a charge q moves from b to a through a distance l, the work of the force F is

$$W = Fl = vBql.$$

The emf $\mathcal{E}$ is the work per unit charge, so

$$\mathcal{E} = \frac{W}{q} = vBl. \tag{32–4}$$

Thus the emf is equal to the open-circuit potential difference V_{ab} given in Eq. (32–3). This emf is the same whether or not a complete circuit exists, while in general V_{ab} will be different under complete-circuit conditions because of resistance in the conductors.

If the velocity of the conductor is not perpendicular to the field but makes an angle ϕ with it, then only the component of velocity $v \sin \phi$ perpendicular to the field appears in the magnetic force expression, and the induced emf is then

$$\mathcal{E} = v \sin \phi \, Bl. \tag{32–5}$$

If v is expressed in $\mathrm{m \cdot s^{-1}}$, B in T, and l in meters, the emf is in joules per coulomb or volts, as the reader should verify. The sense of $\mathcal{E}$ is given by the right-hand-screw rule: rotate a right-hand screw from the direction of v toward B and the direction of advance will be the direction of the emf, as shown in Fig. 32–3.

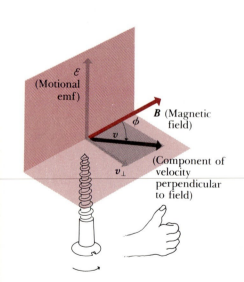

32–3 The vectors $v_\perp$, B, and $\mathcal{E}$ form a mutually perpendicular set. If the vector v is rotated toward the vector B, the direction of $\mathcal{E}$ is that in which a right-hand screw advances.

EXAMPLE 1 Let the length l in Fig. 32–2 be 0.1 m, the velocity v be $0.1 \ \mathrm{m \cdot s^{-1}}$, the resistance of the loop be 0.01 Ω, and let $B = 1$ T.

The emf $\mathcal{E}$ is

$$\mathcal{E} = vBl = 0.01 \text{ V.}$$

The current in the loop is

$$I = \frac{\mathcal{E}}{R} = 1 \text{ A.}$$

Because of this current, there is a force F on the loop, in the opposite direction to its motion, and equal to

$$F = IBl = 0.1 \text{ N.}$$

The power necessary to move the loop against this force is

$$P = Fv = 0.01 \text{ W.}$$

The product $\mathcal{E}I$ is

$$\mathcal{E}I = 0.01 \text{ W.}$$

Thus, the rate of energy conversion, $\mathcal{E}I$, equals the mechanical power input, Fv, to the system.   ◄

EXAMPLE 2   The rectangular loop in Fig. 32–4, of length a and width b, is rotating with uniform angular velocity ω about the y-axis. The entire loop lies in a uniform, constant $\boldsymbol{B}$-field, parallel to the z-axis. We wish to calculate the induced emf in the loop, from Eq. (32–5).

Solution   The velocity v of the sides of the loop of length a is

$$v = \omega\left(\frac{b}{2}\right).$$

The motional emf in each of the two sides of length a is

$$\mathcal{E} = vB \sin \theta \, a = \tfrac{1}{2}\omega B \, ab \sin \theta.$$

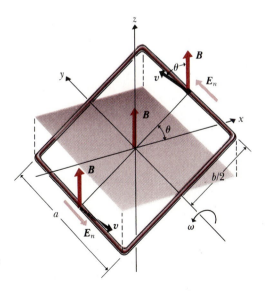

32–4   Rectangular loop rotating with constant angular velocity in a uniform magnetic field.

These two emf's add, so the total emf due to these two sides is

$$\mathcal{E} = \omega B\, ab\, \sin\theta.$$

The magnetic forces on the other two sides of the loop are transverse to these sides and so contribute nothing to the emf. Since ab is the area A of the loop, the total emf may be written

$$\mathcal{E} = \omega BA \sin\theta.$$

If the loop lies in the xy-plane at $t = 0$, then $\theta = \omega t$. Hence,

$$\mathcal{E} = \omega AB \sin \omega t. \tag{32–6}$$

The emf therefore varies *sinusoidally* with time. The *maximum* emf $\mathcal{E}_m$, which occurs when $\sin \omega t = 1$, is

$$\mathcal{E}_m = \omega AB,$$

so we can write Eq. (32–6) as

$$\mathcal{E} = \mathcal{E}_m \sin \omega t. \tag{32–7}$$

The rotating loop is the prototype of the *alternating-current generator,* or *alternator;* we say it develops a *sinusoidal alternating emf.*

The emf is a maximum (in absolute value) when $\theta = 90°$ or $270°$ and the long sides are moving at right angles to the field. The emf is zero when $\theta = 0$ or $180°$ and the long sides are moving parallel to the field. The emf depends on the *area A* of the loop but not on its *shape.* This is most easily verified by use of Faraday's law, which will be introduced in the next section. ◄

The rotating loop in Fig. 32–4 can be utilized as the source in an external circuit by making connections to *slip rings* S, S, which rotate with the loop, as shown in Fig. 32–5a. Stationary brushes bearing against the rings are connected to the output terminals a and b. The instantaneous terminal voltage v_{ab}, on open circuit, equals the instantaneous emf. Figure 32–5b is a graph of v_{ab} as a function of time.

A terminal voltage that always has the same sign, although it fluctuates in *magnitude,* can be obtained by connecting the loop to a split ring or *commutator,* as in Fig. 32–6a. At the position in which the emf reverses, the connections to the external circuit are interchanged. Figure 32–6b is a graph of the terminal voltage, and the device is the prototype of a dc generator.

Commercial dc generators have a large number of coils and commutator segments, and their terminal voltage is not only unidirectional but also practically constant.

EXAMPLE 3   A disk of radius R, shown in Fig. 32–7, lies in the xz-plane and rotates with uniform angular velocity ω about the y-axis. The disk is in a uniform, constant $\boldsymbol{B}$ field parallel to the y-axis. We wish to find the induced emf between the center and the rim of the disk.

Solution   We consider a short portion of a narrow radial segment of the disk, of length Δr. Its velocity is $v = \omega r$, and since $\boldsymbol{v}$ is at right angles to $\boldsymbol{B}$,

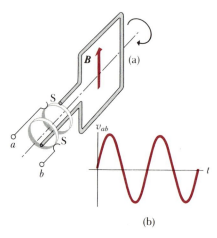

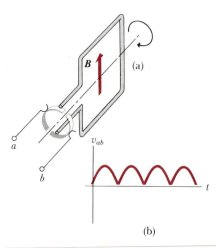

32–5 (a) Prototype of an alternator; (b) voltage as a function of time.

32–6 (a) Prototype of a direct-current generator; (b) voltage as a function of time.

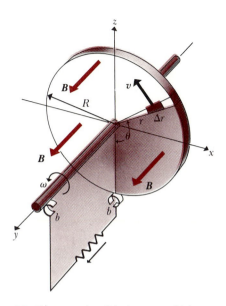

32–7 A Faraday disk dynamo, which supplies a steady emf.

the motional emf due to the segment is

$$\Delta \mathcal{E} = vB \,\Delta r = \omega \, Br \,\Delta r.$$

The *total* emf between center and rim is

$$\mathcal{E} = \sum \omega \, Br \,\Delta r = \omega B \sum r \,\Delta r.$$

The elements Δr near the edge of the disk, where r is nearly equal to R, contribute more to the total emf than do those closer to the axis, where r is smaller. Since these contributions increase proportionally with r, we may replace r in the sum by its *average* value $(R/2)$. Then

$$\sum r \,\Delta r = \sum \left(\frac{R}{2}\right) \Delta r = \left(\frac{R}{2}\right) \sum \Delta r$$

$$= \left(\frac{R}{2}\right) R = \tfrac{1}{2}R^2,$$

and the total emf is

$$\mathcal{E} = \tfrac{1}{2}\omega BR^2. \tag{32–8}$$

The emf between center and rim equals that in any radial segment. The entire disk can therefore be considered a *source* for which the emf between center and rim equals $\omega BR^2/2$. The source can be included in a closed circuit by completing the circuit through sliding contacts of brushes b, b.

The emf in such a disk was studied by Faraday, and the device is called a *Faraday disk dynamo*. ◄

32–2

FARADAY'S LAW

The induced emf in the circuit of Fig. 32–2 may be considered from another viewpoint. When the conductor moves toward the right a distance Δs, the area enclosed by the circuit *abcd* increases by

$$\Delta A = l \,\Delta s,$$

and the change in magnetic *flux* through the circuit is

$$\Delta \Phi = B \,\Delta A = Bl \,\Delta s.$$

When both sides are divided by Δt, we obtain

$$\frac{\Delta \Phi}{\Delta t} = \left(\frac{\Delta s}{\Delta t}\right) Bl = vBl.$$

But the product vBl equals the induced emf $\mathcal{E}$, so the preceding equation states that *the induced emf in the circuit is numerically equal to the rate of change of the magnetic flux through it.*

The right-hand-screw convention of sign is usually used with this equation. If one faces the circuit, an emf is considered positive if it results in a conventional current in a clockwise direction, and $\Delta \Phi / \Delta t$ is considered positive if there is an increase in the flux directed away from the observer. (Then, a decrease in flux away from the observer is negative,

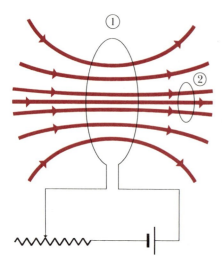

32–8   As the current in circuit 1 is varied, the magnetic flux through circuit 2 changes.

an increase in flux toward the observer is negative, and a decrease in flux toward the observer is positive.) In Fig. 32–2 the current is counterclockwise, so the emf is negative, while the flux is away from the observer and is increasing, so $\Delta\Phi/\Delta t$ is positive. A study of other possibilities shows that $\mathcal{E}$ and $\Delta\Phi/\Delta t$ always have *opposite* signs. Hence, we write the equation

$$\mathcal{E} = -\frac{\Delta\Phi}{\Delta t}. \qquad (32\text{–}9)$$

Equation (32–9) is known as *Faraday's law*. As it stands, it appears to be merely an alternative form of Eqs. (32–4) or (32–5) for the emf in a moving conductor. It turns out, however, that the relation has a much deeper significance than might be expected from its derivation. That is, it is found to apply to circuits through which the flux is caused to vary, even though there is no *motion* of any part of the circuit and hence no emf directly related to a force on a moving charge.

Suppose, for example, that two loops of wire are located as in Fig. 32–8. A current in circuit 1 sets up a magnetic field whose magnitude at all points is proportional to this current. A part of this flux passes through circuit 2, and if the current in circuit 1 is increased or decreased, the flux through circuit 2 will also vary. Circuit 2 is not moving in a magnetic field and hence no "motional" emf is induced in it, but there is a change in the flux through it, and it is found experimentally that an emf appears in circuit 2 of magnitude $\mathcal{E} = \Delta\Phi/\Delta t$. In such a situation, no one portion of circuit 2 can be considered the seat of emf; the *entire circuit* constitutes the seat.

Here is another example. Suppose we set up a magnetic field within the toroidal winding of Fig. 32–9, link the toroid with a conducting ring, and vary the current in the winding. We have shown that the flux lines set up by a steady current in a toroidal solenoid are wholly confined to the space enclosed by the winding; not only is the ring not *moving* in a magnetic field, but if the current were steady it would not even be *in* a magnetic field! However, field lines do pass through the area bounded by the ring, and the flux changes as the current in the windings changes. Equation (32–9) predicts an induced emf in the ring, and we find by experiment that the emf actually exists. In case the reader has not identified the apparatus in Fig. 32–9, it may be pointed out that it is merely a *transformer* with a one-turn secondary, so that the phenomenon we are now discussing is the basis of the operation of every transformer.

To sum up, then, an emf is induced in a circuit whenever the magnetic flux through the circuit varies with time. The flux may be caused to change in various ways: For example, (1) a conductor may move through a stationary magnetic field, as in Figs. 32–2 and 32–4, or (2) the magnetic field through a stationary conducting loop may change with time, as in Figs. 32–8 and 32–9. For case (1), the emf may be computed *either* from

$$\mathcal{E} = vBl$$

or from

$$\mathcal{E} = -\frac{\Delta\phi}{\Delta t}.$$

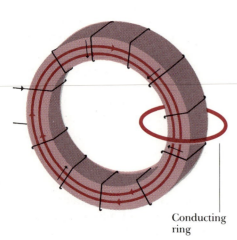

Conducting ring

32–9   An emf is induced in the ring when the flux in the toroid varies.

For case (2), the emf may be computed *only* by

$$\mathcal{E} = -\frac{\Delta\phi}{\Delta t}.$$

If we have a coil of N turns and the flux varies at the same rate through each, the induced emf's in the turns are in series, and the total emf is

$$\mathcal{E} = -N\left(\frac{\Delta\Phi}{\Delta t}\right). \qquad (32–10)$$

EXAMPLE 1   A certain coil of wire consists of 500 circular loops of radius 4 cm. It is placed between the poles of a large electromagnet, where the magnetic field is uniform, perpendicular to the plane of the coil, and increasing at a rate of 0.2 T·s^{-1}. What is the magnitude of the resulting induced emf?

Solution   The flux Φ at any time is given by $\Phi = BA$, and the rate of change of flux by $\Delta\Phi/\Delta t = (\Delta B/\Delta t)A$. In our problem, $A = \pi \times (0.04\text{ m})^2 = 0.00503$ m^2 and

$$\frac{\Delta\Phi}{\Delta t} = (0.2\text{ T·s}^{-1})(0.00503\text{ m}^2)$$

$$= 0.00101\text{ T·m}^2\text{·s}^{-1} = 0.00101\text{ Wb·s}^{-1}.$$

From Eq. (32–10), the magnitude of the induced emf is

$$|\mathcal{E}| = N\frac{\Delta\Phi}{\Delta t} = (500)(0.00101\text{ Wb·s}^{-1}) = 0.503\text{ V}.$$

If the coil is tilted so that a line perpendicular to its plane makes an angle of 30° with $\mathbf{B}$, then only the component $B\cos 30°$ contributes to the flux through the coil. In that case the induced emf has magnitude $\mathcal{E} = (0.503\text{ V})(\cos 30°) = 0.435$ V.   ◀

EXAMPLE 2   We consider again the rotating rectangular loop in Fig. 32–4, discussed in Example 2 of Section 32–1. The emf may also be computed from Faraday's law. The complete quantitative calculation requires use of calculus, but it is easy to see *qualitatively* that the result must agree with the calculation in Section 32–1. For example, when θ is zero or 180° the flux is *maximum*. That is, at these instants Φ is neither increasing nor decreasing, and, correspondingly, $\mathcal{E}$ is instantaneously zero. Similarly, when θ is 90° or 270°, the plane of the coil is parallel to B, the flux is changing at its *maximum* rate, and $\mathcal{E}$ is largest, in agreement with the above results. The emf depends not on the flux through the loop, but on its *rate of change*.   ◀

EXAMPLE 3   We consider again the rotating disk in Fig. 32–7, discussed in Example 3 of Section 32–1. To compute the emf from Faraday's law, Eq. (32–9), we consider the circuit to be the periphery of the shaded areas in Fig. 32–7. The rectangular portion in the yz-plane is stationary. The area of the shaded section in the xz-plane is $\frac{1}{2}R^2\theta$, and the flux through it is

$$\Phi = \tfrac{1}{2}BR^2\theta.$$

As the disk rotates, the shaded area increases. In a time Δt, the angle θ increases by $\Delta\theta = \omega\,\Delta t$. The flux increases by

$$\Delta\Phi = \tfrac{1}{2}BR^2\,\Delta\theta = \tfrac{1}{2}BR^2\omega\,\Delta t,$$

and the induced emf is

$$\mathcal{E} = \frac{\Delta\Phi}{\Delta t} = \tfrac{1}{2}BR^2\omega,$$

in agreement with the previous result.   ◀

EXAMPLE 4   *The search coil.* A useful experimental method of measuring magnetic field strength uses a small, closely wound coil of N turns called a *search coil.* Suppose that such a coil is placed with its plane perpendicular to a magnetic field **B**. If the area enclosed by the coil is A, the flux Φ through it is $\Phi = BA$. Now if the coil is quickly given a quarter-turn about one of its diameters so that its plane becomes parallel to the field, or if it is quickly snatched from its position to another where the field is known to be zero, the flux through it decreases rapidly from BA to zero. During the time that the flux is decreasing, an emf of short duration is induced in the coil and there is a momentary induced current in the external circuit to which the coil is connected.

The current at any instant is

$$i = \frac{\mathcal{E}}{R},$$

where R is the combined resistance of external circuit and search coil, $\mathcal{E}$ is the instantaneous induced emf, and i the instantaneous current.

From Faraday's law, we obtain

$$\mathcal{E} = -N\left(\frac{\Delta\Phi}{\Delta t}\right), \qquad i = -\frac{N}{R}\left(\frac{\Delta\Phi}{\Delta t}\right),$$

and

$$i\,\Delta t = -\frac{N}{R}\,\Delta\Phi = \frac{N}{R}(\Phi - 0).$$

But $i\,\Delta t = q$, so

$$\Phi = \frac{Rq}{N}, \tag{32–11}$$

and

$$B = \frac{\Phi}{A} = \frac{Rq}{NA}. \tag{32–12}$$

It is not difficult to construct an instrument that measures the total charge that passes through it. Thus if the external circuit contains such an instrument, q may be measured directly. Equation (32–12) may then be used to compute B. Strictly speaking, this method gives only the *average* field over the area of the coil. But, if the area is sufficiently small, this approximates closely the field at, say, the center of the coil.   ◀

32–3

INDUCED ELECTRIC FIELDS

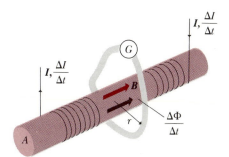

32–10 The windings of a long solenoid carry a current I that is increasing at a rate $\Delta I/\Delta t$. The magnetic flux in the solenoid is increasing at a rate $\Delta\Phi/\Delta t$, and this changing flux passes through a wire loop of arbitrary size and shape. An emf $\mathcal{E}$ is induced in the loop, given by $\mathcal{E} = -\Delta\Phi/\Delta t$.

The examples of induced emf considered thus far have all involved conductors moving in a magnetic field, but we have also pointed out, in introducing Faraday's law, that induced emf can also occur with *stationary* conductors. An example is shown in Fig. 32–10, a solenoid encircled by a conducting loop. A current I in the windings of the solenoid sets up a magnetic field **B** along the solenoid axis, and a magnetic flux $\Phi = BA$ passes through any surface bounded by the loop. Suppose that a small galvanometer G is inserted in the loop. It is found that if the current I is changed (and hence if the flux Φ is changed), the galvanometer indicates an emf $\mathcal{E}$ in the wire *during the time that the flux is changing*, and that this emf is equal to the *time rate of change of flux* through the surface bounded by the wire. That is,

$$\mathcal{E} = -\frac{\Delta\Phi}{\Delta t}. \qquad (32\text{--}13)$$

The reason for the minus sign is the same as in Section 32–2. With the sign convention introduced there we note that in Fig. 32–10, $\Delta\Phi/\Delta t$ is positive if I is increasing; in that case, according to Eq. (32–13), $\mathcal{E}$ must be negative. Since positive $\mathcal{E}$ corresponds to a *clockwise* induced current, we conclude that when I is increasing the actual induced current is *counterclockwise*.

Equation (32–13) is Faraday's law; the essential difference between this and preceding examples is that here it is applied not to a moving conductor but to a stationary one in which there is a changing flux because the magnetic field in the area bounded by the conductor is changing.

EXAMPLE Suppose the long solenoid in Fig. 32–10 is wound with 1000 turns per meter, and the current in its windings is increasing at the rate of $100 \text{ A}\cdot\text{s}^{-1}$. The cross-sectional area of the solenoid is $4 \text{ cm}^2 = 4 \times 10^{-4} \text{ m}^2$.

The flux Φ in the solenoid, through a cross section not too near its ends, is

$$\Phi = BA = \mu_0 nIA.$$

The rate of change of flux is

$$\frac{\Delta\Phi}{\Delta t} = \mu_0 nA \frac{\Delta I}{\Delta t}$$

$$= (4\pi \times 10^{-7} \text{ Wb}\cdot\text{A}^{-1}\cdot\text{m}^{-1})(1000 \text{ turns}\cdot\text{m}^{-1}) \times$$
$$(4 \times 10^{-4} \text{ m}^2)(100 \text{ A}\cdot\text{s}^{-1})$$
$$= 16\pi \times 10^{-6} \text{ Wb}\cdot\text{s}^{-1}.$$

The magnitude of the induced emf is

$$|\mathcal{E}| = \frac{\Delta\Phi}{\Delta t} = 16\pi \times 10^{-6} \text{ V} = 16\pi \ \mu\text{V}. \qquad \blacktriangleleft$$

In situations where an emf is induced by a changing magnetic flux through a circuit rather than by motion of a conductor, there is no magnetic-field force on the mobile charges in the conductor. Instead, we must conclude that the changing flux causes an electric field within the conductors, and that this electric field pushes the charges through the conductor. This is an example of a *nonelectrostatic field* E_n, a concept introduced in Section 28–4 in the discussion of electromotive force.

In summary, Eq. (32–13) is valid for two rather different situations. In the first, an emf is induced by the *magnetic* forces on charges in a conductor moving through a magnetic field, while in the second a time-varying magnetic field induces an *electric* field of nonelectrostatic nature in a stationary conductor and hence induces an emf. The E_n field in the latter case differs from an *electrostatic* field in two significant ways. First, the work done by E_n on a charged particle undergoing a closed-loop displacement (returning to the starting point) is *not* zero, so it is not a *conservative* field. In contrast, an electrostatic field is *always* conservative, as discussed in Section 26–1. Second, the nonelectrostatic field is *not* of a sort that could be produced by static charges.

Thus, a changing magnetic field acts as a source of electric field, but one that differs from an electrostatic field in two ways: It is nonconservative, and it is not related to electric charge in a way that can be described by Coulomb's law or Gauss's law. The reader may also note that this situation is analogous to that of the displacement current discussed in Section 29–5, in which a changing *electric* field acts as a source of *magnetic* field. These relations thus exhibit a kind of symmetry in the behavior of the two fields. We shall return to this relationship in Chapter 35, in connection with the analysis of electromagnetic waves.

32–4

LENZ'S LAW

We return now to the question of determining the sign or direction of an induced emf or current or the direction of the associated nonelectrostatic field. A very useful principle is *Lenz's law*. H. F. E. Lenz (1804–1864) was a German scientist who, without knowledge of the work of Faraday and Henry, duplicated many of their discoveries nearly simultaneously. The law states:

> *The direction of an induced current is such as to oppose the cause producing it.*

The "cause" of the current may be the motion of a conductor in a magnetic field, or it may be the change of flux through a stationary circuit. In the first case, the direction of the induced current in the moving conductor is such that the direction of the side-thrust exerted on the conductor by the magnetic field is opposite in direction to its motion. The motion of the conductor is therefore "opposed."

In the second case, the induced current sets up a magnetic field of its own that within the area bounded by the circuit is (a) *opposite* to the original field if this is *increasing*, but (b) is in the *same* direction as the original field if the latter is *decreasing*. Thus it is the *change in flux* through the circuit (not the flux itself) that is "opposed" by the induced current.

In order to have an induced current, we must have a closed circuit. If a conductor does not form a closed circuit, then we mentally complete the circuit between the ends of the conductor and use Lenz's law to determine the direction of the current. The polarity of the ends of the open-circuit conductor may then be deduced.

Lenz's law is not really a separate principle, but conveys the same information as the sign convention in Faraday's law. Nevertheless, it is a convenient rule to use in many problems involving induced emf's and currents.

EXAMPLE 1 In Fig. 32–2, when the conductor moves to the right, a counterclockwise current is induced in the loop. The force exerted by the field on the moving conductor as a result of this current is to the left, *opposing* the conductor's motion. ◀

EXAMPLE 2 In Fig. 32–4, the direction of the induced current in the loop is the same as that of the nonelectrostatic field E_n. The magnetic field exerts forces on the loop as a result of this current; the force on the right side of the loop (of length a) is in the $+x$-direction, and that on the left side is in the $-x$-direction. The resulting torque is opposite in direction to ω and thus opposes the motion. ◀

EXAMPLE 3 In Fig. 32–10, when the solenoid current is increasing, the induced current in the loop is counterclockwise. The additional field caused by this current, at points inside the loop, is opposite in direction to that of the solenoid; hence the induced current tends to oppose the increase in flux through the loop by causing flux in the opposite sense. ◀

Lenz's law is also directly related to energy conservation. For instance, in Example 1 above, the induced current in the loop dissipates energy at the rate I^2R, and this energy must be supplied by the force that makes the conductor move despite the magnetic force opposing its motion. The work done by this applied force, in fact, must equal the energy dissipated in the circuit resistance. If the induced current were to have the opposite direction, the resulting force on the moving conductor would make it move faster and faster, violating energy conservation.

32–5

EDDY CURRENTS

Thus far we have considered only instances in which the currents resulting from induced emf's were confined to well-defined paths provided by the wires and apparatus of the external circuit. In many pieces of electrical equipment, however, one finds masses of metal moving in a magnetic field or located in a changing magnetic field, with the result that induced currents circulate throughout the volume of the metal. Because of their general circulatory nature, these are referred to as *eddy currents*.

Consider a disk rotating in a magnetic field perpendicular to the plane of the disk but confined to a limited portion of its area, as in Fig. 32–11a. Element Ob is moving across the field and has an emf induced in

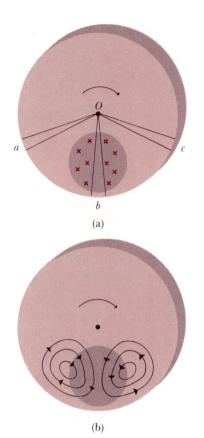

(a)

(b)

32–11   Eddy currents in a rotating disk.

it. Elements Oa and Oc are not in the field but, in common with all other elements located outside the field, provide return conducting paths along which charges displaced along Ob can return from b to O. A general eddy-current circulation is therefore set up in the disk, somewhat as sketched in Fig. 32–11b.

The currents in the neighborhood of radius Ob experience a side thrust that *opposes* the motion of the disk, while the return currents, since they lie outside the field, do not experience such a thrust. The interaction between the eddy currents and the field therefore results in a braking action on the disk. The apparatus finds some technical applications and is known as an "eddy current brake."

As a second example of eddy currents, consider the core of an alternating-current transformer, shown in Fig. 32–12. The alternating current in the primary winding P sets up an alternating flux within the core, and an induced emf develops in the secondary winding S because of the continual change in flux through them. The iron core, however, is also a conductor, and any section such as that at AA can be thought of as a number of closed conducting circuits, one within the other. The flux through each of these circuits is continually changing, so that there is an eddy-current circulation in the entire volume of the core, the lines of flow lying in planes perpendicular to the flux. These eddy currents are very undesirable both because of the energy they dissipate and because of the flux they themselves set up.

In all actual transformers, the eddy currents are greatly reduced by the use of a *laminated* core, that is, one built up of thin sheets or laminae. The electrical resistance between the surfaces of the laminations (due either to a natural coating of oxide or to an insulating varnish) effectively confines the eddy currents to individual laminae. The resulting length of path is greatly increased, with consequent increase in resistance. Hence, although the induced emf is not altered, the currents and their heating effects are minimized.

In small transformers where eddy-current losses must be kept to an absolute minimum, the cores are sometimes made of *ferrites*, which are complex oxides of iron and other metals. These materials are ferromagnetic but have relatively high resistivity.

32–12   Reduction of eddy currents by use of a laminated core.

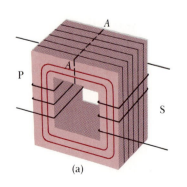

(a)

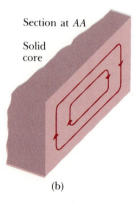

Section at AA

Solid core

(b)

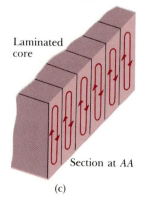

Laminated core

Section at AA

(c)

QUESTIONS

32-1 In most parts of the northern hemisphere the earth's magnetic field has a vertical component directed *into* the earth. An airplane flying east generates an emf between its wingtips. Which wingtip acquires an excess of electrons and which a deficiency?

32-2 A sheet of copper is placed between the poles of an electromagnet, so the magnetic field is perpendicular to the sheet. When it is pulled out, a considerable force is required, and the force required increases with speed. What's happening?

32-3 In Fig. 32-4, if the angular velocity ω of the loop is doubled, the frequency with which the induced current changes direction doubles, and the maximum emf also doubles. Why? Does the torque required to turn the loop change?

32-4 Comparing the conventional dc generator (Fig. 32-6) and the Faraday disk dynamo (Fig. 32-7), what are some advantages and disadvantages of each?

32-5 Some alternating-current generators use a rotating permanent magnet and stationary coils. What advantages does this scheme have? What disadvantages?

32-6 When a conductor moves through a magnetic field, the magnetic forces on the charges in the conductor cause an emf. But if this phenomenon is viewed in a frame of reference moving with the conductor, there is no motion, but there is still an emf. How is this paradox resolved?

32-7 Two circular loops lie adjacent to each other. One is connected to a source that supplies an increasing current; the other is a simple closed ring. Is the induced current in the ring in the same direction as that in the ring connected to the source, or opposite? What if the current in the first ring is decreasing?

32-8 A farmer claimed that the high-voltage transmission lines running parallel to his fence induced dangerously large voltages on the fence. Is this within the realm of possibility?

32-9 Small one-cylinder gasoline engines sometimes use a device called a *magneto* to supply current to the spark plug. A permanent magnet is attached to the flywheel, and there is a stationary coil mounted adjacent to it. What happens when the magnet passes the coil?

32-10 A current-carrying conductor passes through the center of a metal ring, perpendicular to its plane. If the current in the conductor increases, is a current induced in the ring?

32-11 A student asserted that if a permanent magnet is dropped down a vertical copper pipe, it eventually reaches a terminal velocity, even if there is no air resistance. Why should this be? Or should it?

PROBLEMS

32-1 In Fig. 32-2, a rod of length $l = 0.40$ m moves in a magnetic field of magnitude $B = 1.2$ T. The emf induced in the moving rod is found to be 2.40 V.

a) What is the speed of the rod?

b) If the total circuit resistance is 1.2 Ω, what is the induced current?

c) What force (magnitude and direction) does the field exert on the rod as a result of this current?

32-2 In Fig. 32-1 a rod of length $l = 0.25$ m moves with constant speed of 6.0 m·s⁻¹ in the direction shown. The induced emf is found to be 3.0 V.

a) What is the magnitude of the magnetic field?

b) Which point is at higher potential, *a* or *b*?

c) What is the direction of the electrostatic field in the rod? The nonelectrostatic field?

d) What is the magnitude of the nonelectrostatic field?

32-3 In Fig. 32-1, let $l = 1.5$ m, $B = 0.5$ T, $v = 4$ m·s⁻¹.

a) What is the motional emf in the rod?

b) What is the potential difference between its terminals?

c) Which end is at the higher potential?

32-4 The cube in Fig. 32-13, 1 m on a side, is in a uniform magnetic field of 0.2 T, directed along the positive *y*-axis. Wires *A*, *C*, and *D* move in the directions indicated, each with a speed of 0.5 m·s⁻¹.

a) What is the motional emf in each wire?

b) What is the potential difference between the terminals of each?

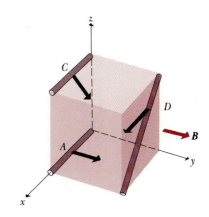

Figure 32-13

32–5 A conducting rod AB in Fig. 32–14 makes contact with the metal rails CA and DB. The apparatus is in a uniform magnetic field 0.5 T, perpendicular to the plane of the diagram.

a) Find the magnitude and direction of the emf induced in the rod when it is moving toward the right with a speed $4 \text{ m} \cdot \text{s}^{-1}$.

b) If the resistance of the circuit $ABCD$ is 0.2 Ω (assumed constant), find the force required to maintain the rod in motion. Neglect friction.

c) Compare the rate at which mechanical work is done by the force (Fv) with the rate of development of heat in the circuit (i^2R).

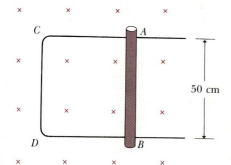

Figure 32–14

32–6 A closely wound rectangular coil of 50 turns has dimensions of 12 cm × 25 cm. The plane of the coil is rotated from a position where it makes an angle of 45° with a magnetic field 2 T to a position perpendicular to the field in time $t = 0.1$ s. What is the average emf induced in the coil?

32–7 A coil of 1000 turns enclosing an area of 20 cm^2 is rotated from a position where its plane is perpendicular to the earth's magnetic field to one where its plane is parallel to the field, in 0.02 s. What average emf is induced if the earth's magnetic field is 6×10^{-5} T?

32–8 A square loop of wire with resistance R is moved at constant velocity v across a uniform magnetic field confined to a square region whose sides are twice the length of those of the square loop. (See Fig. 32–15.)

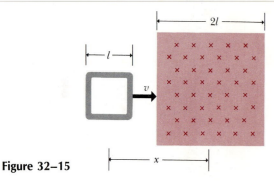

Figure 32–15

a) Sketch a graph of the external force F needed to move the loop at constant velocity, as a function of the distance x, from $x = -2l$ to $x = +2l$.

b) Sketch a graph of the induced current in the loop as a function of x, plotting clockwise currents upward and counterclockwise currents downward.

32–9 The long rectangular loop in Fig. 32–16, of width l, mass m, and resistance R, starts from rest in the position shown, and is acted on by a constant force F. At all points in the colored area there is a uniform magnetic field B, perpendicular to the plane of the diagram.

a) Sketch a graph of the velocity of the loop as a function of time.

b) Find the terminal velocity.

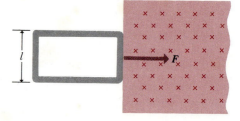

Figure 32–16

32–10 A slender rod 1 m long rotates about an axis through one end and perpendicular to the rod, with an angular velocity of 2 rev·s^{-1}. The plane of rotation of the rod is perpendicular to a uniform magnetic field 0.5 T.

a) What is the induced emf in the rod?

b) What is the potential difference between its terminals?

32–11 A flat square coil of 10 turns has sides of length 12 cm. The coil rotates in a magnetic field 0.025 T.

a) What is the angular velocity of the coil if the maximum emf produced is 20 mV?

b) What is the average emf at this velocity?

32–12 A rectangular coil of wire having 10 turns with dimensions of 20 cm × 30 cm rotates at a constant speed of 600 rpm in a magnetic field 0.10 T. The axis of rotation is perpendicular to the field. Find the maximum emf produced.

32–13 The rectangular loop in Fig. 32–17, of area A and resistance R, rotates at uniform angular velocity ω about the y-axis. The loop in a uniform magnetic field of flux density B in the direction of the x-axis. Sketch the following graphs:

a) the flux Φ through the loop as a function of time (let $t = 0$ in the position shown in Fig. 32–17);

b) the rate of change of flux, $\Delta\Phi/\Delta t$;

c) the induced emf in the loop;

d) the torque Γ needed to keep the loop rotating at constant angular velocity;

e) the induced emf if the angular velocity is doubled. (Neglect the self-inductance of the loop.)

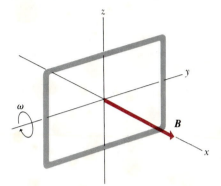

Figure 32–17

32–14 In Problem 32–13 and Fig. 32–17, let $A = 400\ cm^2$, $R = 2\ \Omega$, $\omega = 10\ rad \cdot s^{-1}$, $B = 0.5\ T$. Find

a) the maximum flux through the loop,

b) the maximum induced emf,

c) the maximum torque.

d) Show that the work of the external torque in one revolution is equal to energy dissipated in the loop.

32–15 Suppose the loop in Fig. 32–17 is

a) rotated about the z-axis,

b) rotated about the x-axis,

c) rotated about an edge parallel to the y-axis.

What is the maximum induced emf in each case, if the angular velocity is the same as in Problem 32–14?

32–16 A flexible circular loop 10 cm in diameter lies in a magnetic field 1.2 T, directed into the plane of the diagram in Fig. 32–18. The loop is pulled at the points indicated by the arrows, forming a loop of zero area in 0.2 s.

a) Find the average induced emf in the circuit.

b) What is the direction of the current in R?

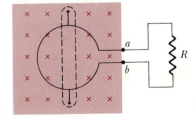

Figure 32–18

32–17 The magnetic field B at all points within the crossed circle of Fig. 32–19 is 0.5 T. It is directed into the plane of the diagram and is decreasing at the rate of $0.1\ T \cdot s^{-1}$.

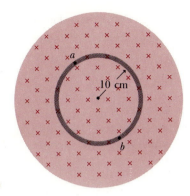

Figure 32–19

a) What is the shape of the field lines of the induced E_n-field in Fig. 32–19, within the crossed circle?

b) What are the magnitude and direction of this field at any point of the circular conducting ring of radius 10 cm, and what is the emf in the ring?

c) What is the current in the ring, if its resistance is 2 Ω?

d) What is the potential difference between points a and b of the ring?

e) How do you reconcile your answers to (c) and (d)?

f) If the ring is cut at some point and the ends separated slightly, what will be the potential difference between the ends?

32–18 A long, straight solenoid of cross-sectional area 6 cm^2 is wound with 10 turns of wire per centimeter, and the windings carry a current of 0.25 A. A secondary winding of 2 turns encircles the solenoid. When the primary circuit is opened, the magnetic field of the solenoid becomes zero in 0.05 s. What is the average induced emf in the secondary?

32–19 A cardboard tube is wound with two windings of insulated wire, as in Fig. 32–20. Terminals a and b of winding A may be connected to a battery through a reversing switch. State whether the induced current in the resistor R is from left to right, or from right to left, in the following circumstances:

a) the current in winding A is from a to b and is increasing;

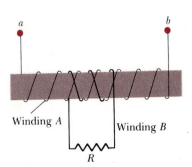

Figure 32–20

b) the current is from b to a and is decreasing;

c) the current is from b to a and is increasing.

32–20 The cross-sectional area of a closely wound search coil having 20 turns is 1.5 cm^2 and its resistance is 4 Ω. The coil is connected through leads of negligible resistance to a charge-measuring instrument having internal resistance 16 Ω. Find the quantity of charge displaced when the coil is pulled quickly out of a region where $B = 1.8$ T to a point where the magnetic field is zero. The plane of the coil, when in the field, makes an angle of 90° with the magnetic field.

32–21 A solenoid 50 cm long and 8 cm in diameter is wound with 500 turns. A closely wound coil of 20 turns of insulated wire surrounds the solenoid at its midpoint, and the terminals of the coil are connected to a charge-measuring instrument. The total circuit resistance is 25 Ω.

a) Find the quantity of charge displaced through the instrument when the current in the solenoid is quickly decreased from 3 A to 1 A.

b) Draw a sketch of the apparatus, showing clearly the directions of windings of the solenoid and coil, and of the current in the solenoid. What is the direction of the current in the coil when the solenoid current is decreased?

32–22 A closely wound search coil has an area of 4 cm^2, 160 turns, and a resistance of 50 Ω. It is connected to a charge-measuring instrument whose resistance is 30 Ω. When the coil is rotated quickly from a position parallel to a uniform magnetic field to one perpendicular to the field, the instrument indicates a charge of 4×10^{-5} C. What is the magnitude of the field?

32–23 A Faraday disk dynamo is to be used to supply current to a large electromagnet that requires 20,000 A at 1.0 V. The disk is to be 0.6 m in radius, and it turns in a magnetic field of 1.2 T, supplied by a smaller electromagnet.

a) How many revolutions per second must the disk turn?

b) What torque is required to turn the disk, assuming that all the mechanical energy is dissipated as heat in the electromagnet?

32–24 A search coil used to measure magnetic fields is to be made with a radius of 2 cm. It is to be designed so that flipping it 180° in a field of 0.1 T causes a total charge of 10^{-4} C to flow in a charge-measuring instrument when the total circuit resistance is 50 Ω. How many turns should the coil have?

32–25 Using Lenz's law, determine the direction of the current in resistor ab of Fig. 32–21 when

a) switch S is opened,

b) coil B is brought closer to coil A,

c) the resistance of R is decreased.

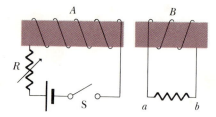

Figure 32–21

33

INDUCTANCE

The phenomena of induction and induced emf, discussed in Chapter 32, have immediate practical applications in a variety of electric-circuit devices, including transformers and inductors. When two coils are adjacent, a changing current in one induces an emf in the other; this is the principle of the transformer, and the coupling between the coils is characterized by the quantity *mutual inductance*. A changing current in a single coil also causes an induced emf in that coil, and the relationship of current to emf is described by the *self-inductance* of the coil. A study of energy relationships in such situations leads to the concept of *energy stored in magnetic fields*. A few simple circuits containing inductors are considered.

33–1

MUTUAL INDUCTANCE

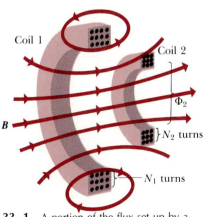

33–1 A portion of the flux set up by a current in coil 1 links with coil 2.

An emf is induced in a stationary circuit whenever the magnetic flux through the circuit varies with time. If the variation in flux is brought about by a varying current in a second circuit, it is convenient to express the induced emf in terms of the varying *current*, rather than in terms of the varying *flux*. We shall use the symbol i to represent the instantaneous value of a varying current.

Figure 33–1 is a sectional view of two closely wound coils of wire. A current i_1 in coil 1 sets up a magnetic field, as indicated by the color lines, and some of these lines pass through coil 2. Let the resulting flux through coil 2 be Φ_2. The magnetic field is proportional to i_1, so Φ_2 is also proportional to i_1. When i_1 changes, Φ_2 changes; this changing flux induces an emf $\mathcal{E}_2$ in coil 2, given by

$$\mathcal{E}_2 = -N_2 \frac{\Delta\Phi_2}{\Delta t}. \tag{33–1}$$

The proportionality of Φ_2 and i_1 could be represented in the form $\Phi_2 =$ (constant) i_1, but it is more convenient to include the number of turns N_2

in the relation. Introducing a proportionality constant M, we write

$$N_2\Phi_2 = Mi_1. \qquad (33\text{--}2)$$

From this,

$$N_2 \frac{\Delta\Phi_2}{\Delta t} = M \frac{\Delta i_1}{\Delta t},$$

and Eq. (33–1) may be rewritten

$$\mathcal{E}_2 = -M \frac{\Delta i_1}{\Delta t}. \qquad (33\text{--}3)$$

The constant M, which depends only on the geometry of the two coils, is called their *mutual inductance*. It is defined by Eq. (33–2), which may be rewritten

$$M = \frac{N_2\Phi_2}{i_1}. \qquad (33\text{--}4)$$

The entire discussion can be repeated for the case where a changing current i_2 in coil 2 causes a changing flux Φ_1 and hence an emf $\mathcal{E}_1$ in coil 1. It might be expected that the constant M would be different in this case, since the two coils are not, in general, symmetric. It turns out, however, that M is always the same in this case as in the case considered above, so a single value of the mutual inductance characterizes completely the induced-emf interaction of two coils.

The SI unit of mutual inductance, from Eq. (33–4), is *one weber per ampere*. An equivalent unit, obtained by reference to Eq. (33–3), is *one volt-second per ampere*. These two equivalent units are called *one henry* (1 H), in honor of Joseph Henry (1797–1878), a pioneer in the development of electromagnetic theory. Thus the unit of mutual inductance is

$$1 \text{ H} = 1 \text{ Wb} \cdot \text{A}^{-1} = 1 \text{ V} \cdot \text{s} \cdot \text{A}^{-1}.$$

EXAMPLE   A long solenoid of length l and cross-sectional area A is closely wound with N_1 turns of wire. A small coil of N_2 turns surrounds it at its center, as in Fig. 33–2. A current i_1 in the solenoid sets up a $\boldsymbol{B}$ field at its center, of magnitude

$$B = \mu_0 n i_1 = \frac{\mu_0 N_1 i_1}{l}.$$

The flux through the central section is equal to BA, and since all of this flux links with the small coil, the mutual inductance is

$$M = \frac{N_2\Phi_2}{i_1} = \frac{N_2}{i_1}\left(\frac{\mu_0 N_1 i_1}{l}\right)A = \frac{\mu_0 A N_1 N_2}{l}.$$

If $l = 0.50$ m, $A = 10$ cm$^2 = 10^{-3}$ m^2, $N_1 = 1000$ turns, $N_2 = 10$ turns,

$$M = \frac{(4\pi \times 10^{-7} \text{ Wb} \cdot \text{A}^{-1} \cdot \text{m}^{-1})(10^{-3} \text{ m}^2)(1000)(10)}{0.5 \text{ m}}$$

$$= 25.1 \times 10^{-6} \text{ Wb} \cdot \text{A}^{-1} = 25.1 \times 10^{-6} \text{ H} = 25.1 \ \mu\text{H}. \qquad \blacktriangleleft$$

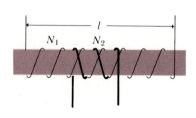

33–2

33–2

SELF-INDUCTANCE

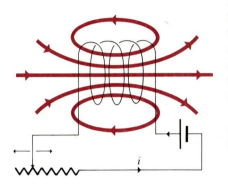

33–3 A flux Φ linking a coil of N turns. When the current in the circuit changes, the flux changes also, and a self-induced emf appears in the circuit.

In the preceding section, the circuit acting as the source of the magnetic field linking a circuit was assumed to be independent of the circuit in which the induced emf appeared. But whenever there is a current in any circuit, this current sets up a magnetic field that links with *the same* circuit and varies when the current varies. Hence any circuit in which there is a varying current has induced in it an emf due to the variation in *its own* magnetic field. Such an emf is called a *self-induced electromotive force*.

Suppose that a circuit has N turns of wire and that a flux Φ passes through each turn, as in Fig. 33–3. In analogy to Eq. (33–4), we define the *self-inductance* L of the circuit:

$$L = \frac{N\Phi}{i}. \tag{33–5}$$

This can be written as

$$N\Phi = Li.$$

If Φ and i change with time, then

$$N\frac{\Delta\Phi}{\Delta t} = L\frac{\Delta i}{\Delta t},$$

and, since the self-induced emf $\mathcal{E}$ is

$$\mathcal{E} = -N\frac{\Delta\Phi}{\Delta t},$$

it follows that

$$\mathcal{E} = -L\frac{\Delta i}{\Delta t}. \tag{33–6}$$

The self-inductance of a circuit is therefore *the self-induced emf per unit rate of change of current.* The SI unit of self-inductance is 1 henry.

A circuit, or part of a circuit, that has inductance is called an *inductor.* An inductor is represented by the symbol

The *direction* of a self-induced (nonelectrostatic) field can be found from Lenz's law. We consider the "cause" of this field, and hence of the emf associated with it, to be the *changing current* in the conductor. If the current is *increasing,* the direction of the induced field is *opposite* to that of the current. If the current is *decreasing,* the induced field is in the *same* direction as the current. Thus it is the *change* in current, not the current itself, that is "opposed" by the induced field.

EXAMPLE 1 An air-core toroidal solenoid of cross-sectional area A and mean circumferential length l is closely wound with N turns of wire. We neglect the variation of B across the cross section, assuming its average value to be very nearly equal to the value at the center of the cross section, as given by Eq. (31–8). Then the flux in the toroid is

$$\Phi = BA = \frac{\mu_0 NiA}{l}.$$

Since all of the flux links with each turn, the self-inductance is

$$L = \frac{N\Phi}{i} = \frac{\mu_0 N^2 A}{l}.$$

Thus, if $N = 100$ turns, $A = 10 \text{ cm}^2 = 10^{-3} \text{ m}^2$, $l = 0.50$ m,

$$L = \frac{(4\pi \times 10^{-7} \text{ Wb} \cdot \text{A}^{-1} \cdot \text{m}^{-1})(100)^2(10^{-3} \text{ m}^2)}{0.50 \text{ m}}$$

$$= 25.1 \times 10^{-6} \text{ H} = 25.1 \ \mu\text{H}. \qquad \blacktriangleleft$$

EXAMPLE 2   If the current in the coil above increases uniformly from zero to 1 A in 0.1 s, find the magnitude and direction of the self-induced emf.

Solution

$$\mathcal{E} = L \frac{\Delta i}{\Delta t} = (25 \times 10^{-6} \text{ H}) \frac{1 \text{ A}}{0.1 \text{ s}}$$

$$= 2.5 \times 10^{-4} \text{ V}.$$

Since the current is increasing, the direction of this emf is opposite to that of the current. For example, if the inductor terminals are a and b and an increasing current is in the direction a to b in the inductor, then the emf in the inductor is in the direction b to a, tending to cause a current *in the external circuit* in the direction a to b. $\qquad \blacktriangleleft$

When Kirchhoff's loop rule (Section 28–4) is used with circuits containing inductors, this emf is treated as though it were a potential difference, with a at higher potential than b. That is, if the defined positive direction of the current i is from a to b in the inductor, then

$$V_{ab} = L \frac{\Delta i}{\Delta t}. \qquad (33\text{--}7)$$

In the above examples the magnetic field has been calculated by assuming that the conductor was surrounded by vacuum. If matter is present, then in calculating the field the constant μ_0, the permeability of vacuum, must be replaced by the permeability of the material, $\mu = K_m\mu_0$ as discussed in Section 31–6. If the material is diamagnetic or paramagnetic this will make very little difference, but if the material is ferromagnetic the difference is of crucial importance. An inductor wound on a soft iron core having $K_m = 5000$ has an inductance approximately 5000 times as great as the same coil with an air core. Iron-core inductors are very widely used in a variety of electronic and electric-power applications.

33–3

ENERGY IN AN INDUCTOR

To establish a current in an inductor requires an input of energy. A changing current in an inductor causes an emf, so to establish the current there must be a potential difference between the terminals of the source and hence a transfer of energy into the circuit containing the inductor. When an inductor carries an instantaneous current i that is changing at the rate $\Delta i/\Delta t$, the induced emf is equal to $\mathcal{E} = L \, \Delta i/\Delta t$, and the power P (rate of energy supply) to the inductor is

$$P = \mathcal{E}i = Li \frac{\Delta i}{\Delta t}.$$

The energy ΔW supplied in time Δt is $P \, \Delta t$, or

$$\Delta W = Li \, \Delta i.$$

To calculate the *total* energy supplied while the current increases from zero to a final value I, we note that the *average* value of Li during the entire increase is $LI/2$. Multiplying this by the total increase I gives the total energy W supplied, and we obtain

$$W = \tfrac{1}{2}LI^2. \tag{33–8}$$

After the current has reached its final steady value, $\Delta i/\Delta t = 0$, and the power input is *zero*. No energy input is required to maintain a steady current in an inductor. The energy that has been supplied to the inductor is used to establish the magnetic field around the inductor, where it is "stored" as a form of potential energy so long as the current is maintained. When the circuit is opened, the magnetic field disappears, and this energy is returned to the circuit. It is this release of energy that maintains the arc often seen when a switch is opened in an inductive circuit.

The energy can be considered as associated with the magnetic field itself, and a relationship can be developed that is analogous to that obtained for electric-field energy in Section 27–4, Eq. (27–9). As in that discussion, we consider only one simple case, the toroidal solenoid. As in the preceding example, we assume that the cross-sectional area A is small enough so the magnetic field may be considered constant over the area, and so the volume in the toroid is approximately equal to the circumferential length $l = 2\pi r$ multiplied by the area A. From the preceding example, the self-inductance of the toroidal solenoid is

$$L = \frac{\mu_0 N^2 A}{l},$$

and the energy stored in the toroid when the current in the windings is I is

$$W = \frac{1}{2}LI^2 = \frac{1}{2}\left(\frac{\mu_0 N^2 A}{l}\right)I^2.$$

We can think of this energy as localized in the volume enclosed by the windings, equal to lA. The energy *per unit volume u* is then

$$u = \frac{W}{lA} = \frac{1}{2}\mu_0\left(\frac{N^2I^2}{l^2}\right).$$

But $N^2I^2/l^2 = B^2/\mu_0^2$, so

$$u = \frac{B^2}{2\mu_0}, \tag{33–9}$$

which is the analog of the expression for the energy per unit volume in the electric field of an air capacitor, $\frac{1}{2}\epsilon_0 E^2$, discussed in Section 27–4.

33–4

THE *R–L* CIRCUIT

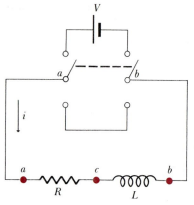

33–4

Every inductor necessarily has some resistance (unless its windings are superconducting). To distinguish between the effects of the resistance R and the self-inductance L, we represent the inductor as in Fig. 33–4, replacing it with an ideal resistanceless inductor in series with a noninductive resistor. The same diagram can also represent a resistor in series with an inductor, in which case R is the *total* resistance of the combination. By means of the switch, the $R–L$ circuit may be connected to a source of constant terminal voltage V, or it may be shorted by the conductor across the lower switch terminals.

Suppose the switch in the diagram is suddenly closed in the "up" position. Because of the self-induced emf, the current does not rise to its final value at the instant the circuit is closed, but grows at a rate that depends on the inductance and resistance of the circuit.

At some instant after the switch is closed, let i represent the current in the circuit and $\Delta i/\Delta t$ its rate of increase. The potential difference across the inductor is

$$V_{cb} = L\frac{\Delta i}{\Delta t},$$

and that across the resistor is

$$V_{ac} = iR.$$

Since $V = V_{ac} + V_{cb}$, it follows that

$$V = L\frac{\Delta i}{\Delta t} + iR. \tag{33–10}$$

The rate of increase of current is therefore

$$\frac{\Delta i}{\Delta t} = \frac{V - iR}{L} = \frac{V}{L} - \frac{R}{L}i. \tag{33–11}$$

At the instant the circuit is first closed, $i = 0$ and the current starts to grow at the rate

$$\left(\frac{\Delta i}{\Delta t}\right)_{initial} = \frac{V}{L}.$$

The greater the self-inductance L, the more slowly does the current start to increase.

As the current increases, the term Ri/L increases also, and hence the *rate* of increase of current becomes smaller and smaller, as Eq. (33–11) shows. When the current reaches its final *steady-state* value I, its rate of increase is zero. Then

$$0 = \frac{V}{L} - \left(\frac{R}{L}\right)I$$

and

$$I = \frac{V}{R}.$$

That is, the *final* current does not depend on the self-inductance and is the same as it would be in a pure resistance R connected to a cell of emf V.

Figure 33–5a illustrates the variation of current with time. The instantaneous current i first rises rapidly, then increases more slowly and approaches asymptotically the final value $I = V/R$. The *time constant* of the circuit is defined as the time τ at which $R\tau/L = 1$, or when

$$\tau = \frac{L}{R}. \tag{33–12}$$

Deriving an expression for the current at any instant as a function of time requires use of calculus, and the result contains an exponential function. We quote the result below, but those who are unfamiliar with exponential functions may omit the following discussion.

The current is given as a function of time by

$$i = \frac{V}{R}(1 - e^{-Rt/L}) = I(1 - e^{-Rt/L}). \tag{33–13}$$

At a time equal to L/R, the current has risen to $(1 - 1/e)I$ or about 0.63 of its final value. For a circuit with a given resistance, this time is longer for larger inductance, and vice versa. Thus although the graph of i vs. t has the same general shape whatever the inductance, the current rises rapidly to its final value if L is small, and slowly if L is large. For example, if $R = 100\ \Omega$ and $L = 10$ H,

$$\frac{L}{R} = \frac{10\ \text{H}}{100\ \Omega} = 0.1\ \text{s},$$

and the current increases to about 63% of its final value in 0.1 s. On the other hand, if $L = 0.01$ H,

$$\frac{L}{R} = \frac{0.01\ \text{H}}{100\ \Omega} = 10^{-4}\ \text{s},$$

and only 10^{-4} s is required for the current to increase to 63% of its final value.

If there is a steady current I in the circuit of Fig. 33–4 and the switch is then quickly thrown to the "down" position, the current decays, as

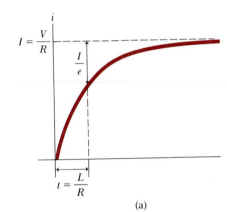

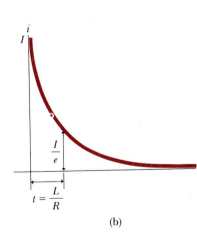

33–5 (a) Growth of current in a circuit containing inductance and resistance. (b) Decay of current in a circuit containing inductance and resistance.

shown in Fig. 33–5b. The equation of the decaying current is

$$i = Ie^{-Rt/L}, \tag{33–14}$$

and the time constant $\tau = L/R$, is the time for the current to decrease to $1/e$ or about 37% of its original value.

The energy necessary to maintain the current during this decay is provided by the energy stored in the magnetic field of the inductor.

33–5

THE L–C CIRCUIT

The behavior of an R–C circuit was discussed in Section 29–4, and that of an R–L circuit in Section 33–4. We now consider the L–C circuit shown in Fig. 33–6, a resistanceless inductor connected between the terminals of a charged capacitor. At the instant connections are made, in Fig. 33–6a, the capacitor starts to discharge through the inductor. At a later instant, represented in Fig. 33–6b, the capacitor has completely discharged and the potential difference between its terminals (and those of the inductor) has decreased to zero. The current in the inductor has meanwhile established a magnetic field in the space around it. This magnetic field now decreases, inducing an emf in the inductor in the same direction as the current. The current therefore persists, although with diminishing magnitude, until the magnetic field has disappeared and the capacitor has been charged in the opposite sense to its initial polarity, as in Fig. 33–6c. The process now repeats itself in the reversed direction, and, in the absence of energy losses, the charges on the capacitor surge back and forth indefinitely. This process is called an *electrical oscillation*.

From the energy standpoint, the oscillations of an electrical circuit consist of a transfer of energy back and forth from the electric field of the capacitor to the magnetic field of the inductor, the *total* energy associated with the circuit remaining constant. This is analogous to the transfer of energy in an oscillating mechanical system from kinetic to potential, and vice versa.

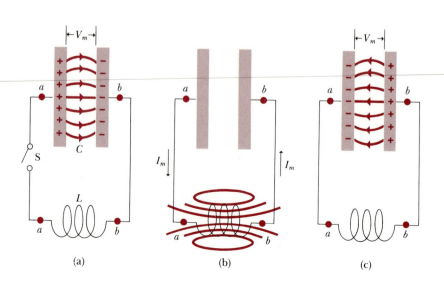

33–6 Energy transfer between electric and magnetic fields in an oscillating L–C circuit.

(a)   (b)   (c)

TABLE 33-1 OSCILLATION OF A MASS ON A SPRING COMPARED WITH THE ELECTRICAL OSCILLATION IN AN L–C CIRCUIT	
Mass on a spring	**Circuit containing inductance and capacitance**
Kinetic energy $= \frac{1}{2}mv^2$	Magnetic energy $= \frac{1}{2}Li^2$
Potential energy $= \frac{1}{2}kx^2$	Electrical energy $= \dfrac{q^2}{2C}$
$\frac{1}{2}mv^2 + \frac{1}{2}kx^2 = \frac{1}{2}kA^2$	$\frac{1}{2}Li^2 + \dfrac{q^2}{2C} = \dfrac{Q^2}{2C}$
$v = \pm\sqrt{k/m}\sqrt{A^2 - x^2}$	$i = \pm\sqrt{1/LC}\sqrt{Q^2 - q^2}$
$v = \dfrac{\Delta x}{\Delta t}$	$i = \dfrac{\Delta q}{\Delta t}$
$x = A\cos\sqrt{k/m}\,t = A\cos\omega t$	$q = Q\cos\sqrt{1/LC}\,t = Q\cos\omega t$
$v = -\omega A\sin\omega t = -v_{\max}\sin\omega t$	$i = -\omega Q\sin\omega t = -I\sin\omega t$

The frequency of the electrical oscillations of a circuit containing inductance and capacitance only (a so-called L–C circuit) may be calculated in exactly the same way as the frequency of oscillation of a body suspended from a spring, discussed in Chapter 11. In the mechanical problem, a body of mass m is attached to a spring of force constant k. Let the body be displaced a distance A from its equilibrium positions and released from rest at time $t = 0$. Then, as shown in the left column of Table 33–1, the kinetic energy of the system at any later time is $\frac{1}{2}mv^2$, and its elastic potential energy is $\frac{1}{2}kx^2$. Because the system is conservative, the sum of these equals the initial energy of the system, $\frac{1}{2}kA^2$. The velocity v at any position x is therefore

$$v = \pm\sqrt{\frac{k}{m}}\sqrt{A^2 - x^2}. \tag{33–15}$$

The velocity equals $\Delta x/\Delta t$, and the coordinate x as a function of t has been shown (Chapter 11) to be

$$x = A\cos\left(\sqrt{\frac{k}{m}}\right)t = A\cos\omega t, \tag{33–16}$$

where the angular frequency ω is

$$\omega = \sqrt{\frac{k}{m}}.$$

We recall also that the ordinary frequency f, the number of cycles per unit time, is given by $f = \omega/2\pi$.

In the electrical problem, also a conservative system, a capacitor of capacitance C is given an initial charge Q and, at time $t = 0$, is connected to the terminals of an inductor of self-inductance L. The magnetic energy of the inductor at any later time corresponds to the kinetic energy of the vibrating body and is given by $\frac{1}{2}Li^2$. The electrical energy of the capacitor corresponds to the elastic potential energy of the spring and is given by $q^2/2C$, where q is the charge on the capacitor. The sum of these equals the initial energy of the system, $Q^2/2C$. That is,

$$\tfrac{1}{2}Li^2 + \frac{q^2}{2C} = \frac{Q^2}{2C}.$$

Solving for i, we find that when the charge on the capacitor is q, the current i is

$$i = \sqrt{\frac{1}{LC}}\sqrt{Q^2 - q^2}. \qquad (33\text{--}17)$$

Comparing this with Eq. (33–15), we see that the current $i = \Delta q/\Delta t$ varies with time in the same way as the velocity $v = \Delta x/\Delta t$ in the mechanical problem. Continuing the analogy, we conclude that q is given as a function of time by

$$q = Q \cos\left(\sqrt{\frac{1}{LC}}\right)t = Q \cos \omega t. \qquad (33\text{--}18)$$

The angular frequency ω of the electrical oscillations is therefore

$$\omega = \sqrt{\frac{1}{LC}}. \qquad (33\text{--}19)$$

This is called the *natural frequency* of the L–C circuit.

   The striking parallel between the mechanical and electrical systems displayed in Table 33–1 is only one of many such examples in physics. So close is the parallel between electrical and mechanical (and acoustical) systems that it has been found possible to solve complicated mechanical and acoustical problems by setting up analogous electrical circuits and measuring the currents and voltages that correspond to the desired mechanical and acoustical "unknowns." This is the basic principle of the *analog computer*.

33–6

THE R–L–C CIRCUIT

In the above discussion we have assumed that the L–C circuit contains no *resistance*. This is an idealization, of course; for every real inductor there is resistance associated with the windings, and there may be resistance in the connecting wires as well. The effect of resistance is to dissipate the electromagnetic energy and convert it to heat; thus, resistance in an electric circuit plays a role analogous to that of friction in a mechanical system.

   Suppose an inductor of self-inductance L and a resistor of resistance R are connected in series across the terminals of a capacitor. If the capacitor is initially charged, it starts to discharge at the instant the connections are made but, because of i^2R losses in the resistor, the energy of the inductor when the capacitor is completely discharged is *less* than the original energy of the capacitor. In the same way, the energy of the capacitor when the magnetic field has collapsed is still smaller, and so on.

   If the resistance R is relatively small, the circuit oscillates, but with *damped harmonic motion*, as illustrated in Fig. 33–7a. As R is increased, the

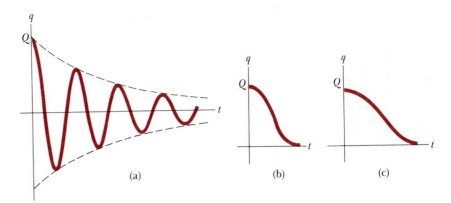

33–7 Graphs of q versus t in an R–L–C circuit. (a) Small damping. (b) Critically damped. (c) Overdamped.

oscillations die out more rapidly. At a sufficiently large value of R, the circuit no longer oscillates and is said to be *critically damped*, as in Fig. 33–7b. For still larger resistances it is *overdamped*, as in Fig. 33–7c.

With the proper electronic circuitry, energy can be fed *into* an R–L–C circuit at the same rate as that at which it is dissipated by i^2R or radiation losses. In effect, a *negative resistance* is inserted in the circuit so that its total resistance is zero. The circuit then oscillates with *sustained* oscillations, as does the idealized circuit with no resistance.

QUESTIONS

33–1 A resistor is to be made by winding a wire around a cylindrical form. In order to make the inductance as small as possible, it is proposed that we wind half the wire in one direction and the other half in the opposite direction. Would this achieve the desired result? Why or why not?

33–2 In Fig. 33–1, if coil 2 is turned 90° so that its axis is vertical, does the mutual inductance increase or decrease?

33–3 The toroidal solenoid is one of the few configurations for which it is easy to calculate self-inductance. What features of the toroidal solenoid give it this simplicity?

33–4 Two identical closely wound circular coils, each having self-inductance L, are placed side by side, close together. If they are connected in series, what is the self-inductance of the combination? What if they are connected in parallel? Can they be connected so that the total inductance is zero?

33–5 If two inductors are separated enough so that practically no flux from either links the coils of the other, show that the equivalent inductance of two inductors in series or parallel is obtained by the same rules used for combining resistance.

33–6 Two closely wound circular coils have the same number of turns, but one has twice the radius of the other. How are the self-inductances of the two coils related?

33–7 One of the great problems in the field of energy resources and utilization is the difficulty of storing electri-

cal energy in large quantities economically. Discuss the possibility of storing large amounts of energy by means of currents in large inductors.

33–8 In what regions in a toroidal solenoid is the energy density greatest? Least?

33–9 Suppose there is a steady current in an inductor. If one attempts to reduce the current to zero instantaneously by opening a switch, a big fat arc appears at the switch contacts. Why? What happens to the induced emf in this situation? Is it physically possible to stop the current instantaneously?

33–10 In the L–R circuit of Fig. 33–4, is the current in the resistor always the same as that in the inductor? How do you know?

33–11 In the R–L circuit of Fig. 33–4, when the switch is in the up position and is then thrown to the down position, the potential V_{ab} changes suddenly and discontinuously, but the current does not. Why can the voltage change suddenly but not the current?

33–12 In the R–L–C circuit, what criteria could be used to decide whether the system is overdamped or underdamped? For example, could one compare the maximum energy stored during one cycle to the energy dissipated during one cycle?

PROBLEMS

33–1 A solenoid of length 10 cm and radius 2 cm is wound uniformly with 1000 turns. A second coil of 50 turns is wound around the solenoid at its center. What is the mutual inductance of the two coils?

33–2 A toroidal solenoid (cf. Section 31–4) has a radius of 10 cm and a cross-sectional area of 5 cm^2, and is wound uniformly with 1000 turns. A second coil with 500 turns is wound uniformly on top of the first. What is the mutual inductance?

33–3 Find the self-inductance of the toroidal solenoid in Problem 33–2 if only the 1000-turn coil is used. How would your answer change if the two coils were connected in series?

33–4 A toroidal solenoid has two coils with n_1 and n_2 turns, respectively; it has radius r and cross-sectional area A.

a) Derive an expression for the self-inductance L_1 when only the first coil is used, and that for L_2 when only the second coil is used.

b) Derive an expression for the mutual inductance of the two coils.

c) Show that $M^2 = L_1L_2$. This result is valid whenever all the flux linked by one coil is also linked by the other.

33–5 Two coils have mutual inductance $M = 0.01$ H. The current i_1 in the first coil increases at a uniform rate of 0.05 A·s^{-1}.

a) What is the induced emf in the second coil? Is it constant?

b) Suppose that the current described is that in the second coil rather than the first; what is the induced emf in the first coil?

33–6 An inductor of inductance 5 H carries a current that decreases at a uniform rate, $\Delta i/\Delta t = -0.02$ A·s^{-1}. Find the self-induced emf; what is its polarity?

33–7 An inductor used in a dc power supply has an inductance of 20 H and a resistance of 200 Ω, and carries a current of 0.1 A.

a) What is the energy stored in the magnetic field?

b) At what rate is energy dissipated in the resistor?

33–8

a) Show that the two expressions for self-inductance, namely,

$$\frac{N\Phi}{i} \quad \text{and} \quad \frac{\mathcal{E}}{\Delta i/\Delta t},$$

have the same units.

b) Show that L/R and RC both have the units of time.

c) Show that 1 Wb·s^{-1} equals 1 V.

33–9 The current in a resistanceless inductor is caused to vary with time as in the graph of Fig. 33–8. Sketch a graph of voltage across the inductor, as a function of time.

Figure 33–8

33–10 The 1000-turn toroidal solenoid described in Problem 33–2 carries a current of 5.0 A.

a) What is the energy density in the magnetic field?

b) What is the total magnetic-field energy? Find using Eq. (33–8) and also by multiplying the energy density from (a) by the volume of the toroid, which is $2\pi rA$; compare the two results.

33–11 A toroidal solenoid has a mean radius of 0.12 m and a cross-sectional area of 20×10^{-4} m^2. It is found that when the current is 20 A, the energy stored is 0.1 J. How many turns does the winding have?

33–12 An inductor of inductance 3 H and resistance 6 Ω is connected to the terminals of a battery of emf 12 V and of negligible internal resistance. Find

a) the initial rate of increase of current in the circuit,

b) the rate of increase of current at the instant when the current is 1 A,

c) the current 0.2 s after the circuit is closed,

d) the final steady-state current.

33–13 The resistance of a 10-H inductor is 200 Ω. The inductor is suddenly connected across a potential difference of 10 V.

a) What is the final steady current in the inductor?

b) What is the initial rate of increase of current?

c) At what rate is the current increasing when its value is one half the final current?

d) At what time after the circuit is closed does the current equal 99% of its final value?

e) Compute the current at the following times after the circuit is closed: 0, 0.025 s, 0.05 s, 0.075 s, 0.10 s. Show the results in a graph.

33–14 Refer to Problem 33–12.

a) What is the power input to the inductor at the instant when the current in it is 0.5 A?

b) What is the rate of dissipation of energy at this instant?

c) What is the rate at which the energy of the magnetic field is increasing?

d) How much energy is stored in the magnetic field when the current has reached its final steady value?

33–15 An inductor having inductance L and resistance R carries a current I. Show that the time constant is equal to *twice the ratio* of the energy stored in the magnetic field to the rate of dissipation of energy in the resistance.

33–16 Show that the quantity $(L/C)^{1/2}$ has units of resistance (ohms).

33–17 The maximum capacitance of a variable air capacitor is 35 pF.

a) What should be the self-inductance of a coil to be connected to this capacitor if the natural frequency of the L–C circuit is to be 550 kHz, corresponding to one end of the broadcast band?

b) The frequency at the other end of the broadcast band is 1550 kHz. What must be the minimum capacitance of the capacitor if the natural frequency is to be adjustable over the range of the broadcast band?

33–18 An inductor having $L = 40\ \mu\text{H}$ is to be combined with a capacitor to make an L–C circuit with natural frequency 2×10^6 Hz. What value of capacitance should be used?

33–19 An inductor is made with two coils wound close together on a form, so that all the flux linking one coil also links the other. The number of turns is the same in each. If the inductance of one coil is L, what is the inductance when the two coils are connected

a) in series?

b) in parallel?

c) If an L–C circuit using this inductor has natural frequency ω using one coil, what is the natural frequency when the two coils are used in series?

33–20 An L–C circuit in an AM radio tuner uses an inductor with inductance 0.1 mH and a variable capacitor. If the natural frequency of the circuit is to be adjustable over the range 0.5 to 1.5 MHz, corresponding to the AM broadcast band, what range of capacitance must the variable capacitor cover?

ALTERNATING CURRENTS

In Chapters 28 and 29 we have studied the behavior of *direct-current* (dc) circuits, in which all the currents, voltages, and emfs are *constant*, that is, not varying with time. However, many electric circuits of practical importance use *alternating current* (ac), in which the voltages and currents vary with time, often in a *sinusoidal* manner. Alternating currents are of utmost importance in technology and industry. Transmission of power over long distance is much easier and more economical with alternating than with direct currents. Circuits used in modern communication equipment, including radio and television, make extensive use of alternating currents. Many life processes involve alternating voltages and currents. The beating of the heart induces alternating currents in the surrounding tissues; the detection and study of these currents, called electrocardiography, provide valuable information concerning the health or pathology of the heart. Electroencephalograms provide analogous information regarding brain function. In this chapter we study the properties of circuits in which there are sinusoidally varying voltages and currents. Many of the same principles used in Chapters 28 and 29 are applicable here as well, and there are some new ideas related to the circuit behavior of inductors and capacitors.

34–1

ac SOURCES AND PHASORS

We have already seen several sources of alternating emf or voltage. A coil of wire, rotating with constant angular velocity in a magnetic field, develops a sinusoidal alternating emf, as explained in Section 32–1. This simple device is the prototype of the commercial alternating current generator, or *alternator*. An *L–C* circuit, as discussed in Section 33–5, oscillates sinusoidally, and with the proper circuitry provides an alternating potential difference between its terminals having a frequency, depending on the purpose for which it is designed, that may range from a few hertz to many millions of hertz.

We consider first a number of circuits connected to an alternator or oscillator that maintains between its terminals a sinusoidal alternating potential difference

$$v = V \cos \omega t. \tag{34–1}$$

In this expression V is the maximum potential difference or the *voltage amplitude*, v is the *instantaneous* potential difference, and ω the *angular frequency*, equal to 2π times the frequency f. For brevity, the alternator or oscillator will be referred to as an *ac source*. The circuit-diagram symbol for an ac source is $\bigodot$.

Analysis of alternating-current circuits is facilitated by use of vector diagrams similar to those used in the study of harmonic motion (Section 11–5). In such diagrams, the instantaneous value of a quantity that varies sinusoidally with time is represented by the *projection* onto a horizontal axis of a vector of length corresponding to the amplitude of the quantity and rotating counterclockwise with angular velocity ω. In the context of ac-circuit analysis, these rotating vectors are often called *phasors*, and diagrams containing them are called *phasor diagrams*.

34–2

CIRCUITS CONTAINING RESISTANCE, INDUCTANCE, OR CAPACITANCE

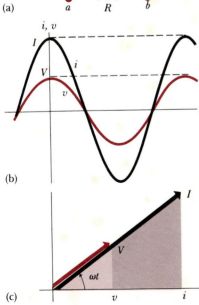

(a)

(b)

(c)

34–1 (a) Resistance R connected across an ac source. (b) Graphs of instantaneous voltage and current. (c) Phasor diagram; current and voltage in phase.

Let a resistor of resistance R be connected between the terminals of an ac source, as in Fig. 34–1a. The instantaneous potential of point a with respect to point b is $v_{ab} = V \cos \omega t$, and the instantaneous current in the resistor is

$$i = \frac{v_{ab}}{R} = \frac{V}{R} \cos \omega t.$$

The maximum current I, or the *current amplitude*, is evidently

$$I = \frac{V}{R}, \tag{34–2}$$

and we can therefore write

$$i = I \cos \omega t. \tag{34–3}$$

The current and voltage are both proportional to $\cos \omega t$, so the current is *in phase* with the voltage. The current and voltage amplitudes, from Eq. (34–2), are related in the same way as in a dc circuit.

Figure 34–1b shows graphs of i and v as functions of time. The fact that the curve representing the current has the greater amplitude in the diagram is of no significance, because the choice of vertical scales for i and v is arbitrary. The corresponding phasor diagram is given in Fig. 34–1c. Because i and v are *in phase* and have the same frequency, the current and voltage phasors rotate together.

Next, suppose that a capacitor of capacitance C is connected across the source, as in Fig. 34–2a. The instantaneous charge q on the capacitor is

$$q = C v_{ab} = CV \cos \omega t. \tag{34–4}$$

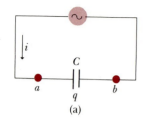

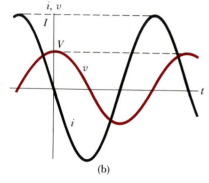

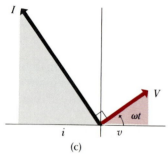

34–2 (a) Capacitor C connected across an ac source. (b) Graphs of instantaneous voltage and current. (c) Phasor diagram; current *leads* voltage by 90°.

In this case the instantaneous current is equal to *the rate of change* of the capacitor charge and therefore to the rate of change of voltage:

$$i = \frac{\Delta q}{\Delta t} = C\left(\frac{\Delta v}{\Delta t}\right). \tag{34–5}$$

Referring to Fig. 34–2b, we see that the current is greatest at times when the v curve is rising or falling most steeply, and is zero at times when the v curve "levels off," that is, when it reaches its maximum and minimum values. Thus the current must vary with time as shown by the i curve in Fig. 34–2b. This curve has the same shape as the graph of $-\sin \omega t$, and in fact it can be shown by use of calculus that the current is given by

$$i = -\omega CV \sin \omega t. \tag{34–6}$$

The negative sign is required if the current is to have the proper direction; the reader may check this by comparing the curves in Fig. 34–2b.

Thus the voltage and current are "out of step" or *out of phase* by a quarter-cycle, with the current a quarter-cycle ahead. The peaks of current occur a quarter-cycle *before* the corresponding voltage peaks. This result is also represented by the vector diagram of Fig. 34–2c, which shows the current vector ahead of the voltage vector by a quarter-cycle, or 90°. This *phase difference* between current and voltage can also be obtained by rewriting Eq. (34–6), using the trigonometric identity $\cos (A + 90°) = -\sin A$:

$$i = \omega CV \cos (\omega t + 90°). \tag{34–7}$$

This shows that the expression for i can be viewed as a cosine function with a "head start" of 90° compared with that for the voltage. We say that the current *leads* the voltage by 90° or that the voltage *lags* the current by 90°.

Equations (34–6) and (34–7) show that the maximum current I is given by

$$I = \omega CV. \tag{34–8}$$

This expression can be put in the same *form* as that for the maximum current in a resistor ($I = V/R$) if we write Eq. (34–8) as

$$I = \frac{V}{1/\omega C},$$

and define a quantity X_C, called the *capacitive reactance* of the capacitor, as

$$X_C = \frac{1}{\omega C}. \tag{34–9}$$

Then

$$I = \frac{V}{X_C}. \tag{34–10}$$

It will be seen from Eq. (34–10) that the unit of capacitive reactance is one *volt per ampere* (1 V·A^{-1}), or one *ohm* (1 Ω).

The reactance of a capacitor is inversely proportional both to the capacitance C and to the angular frequency ω; the greater the capacitance, and the higher the frequency, the *smaller* is the reactance X_C.

EXAMPLE 1 At an angular frequency of 1000 rad·s⁻¹, the reactance of a 1-μF capacitor is

$$X_C = \frac{1}{\omega C} = \frac{1}{(10^3 \text{ rad·s}^{-1})(10^{-6} \text{ F})} = 1000 \ \Omega.$$

At frequency of 10,000 rad·s⁻¹, the reactance of the same capacitor is only 100 Ω, and at a frequency of 100 rad·s⁻¹ it is 10,000 Ω. ◀

Finally, suppose a pure inductor having a self-inductance L and zero resistance is connected to an ac source as in Fig. 34–3. Since the potential difference v between the terminals of an inductor equals $L\,\Delta i/\Delta t$, the instantaneous voltage is proportional to the *rate of change* of the instantaneous current. The points of maximum voltage thus must correspond on the graphs to points of maximum steepness of the current curve, and the points of zero voltage correspond to the points where the current curve levels off at its maximum and minimum values. These relations are shown in Fig. 34–3b, which also shows that the general shape of the i curve is that of a *sine* function. It can be shown (again using calculus) that in this case i is given by

$$i = \frac{V}{\omega L} \sin \omega t. \tag{34–11}$$

Again, the voltage and current are a quarter-cycle out of phase, but this time the current *lags* the voltage (or the voltage *leads* the current) by 90°. This may also be seen by rewriting Eq. (34–11) using the trigonometric identity $\cos(A - 90°) = \sin A$:

$$i = \frac{V}{\omega L} \cos(\omega t - 90°). \tag{34–12}$$

This result and the vector diagram of Fig. 34–3c show that the current can be viewed as a cosine function with a "late start" of 90°.

The maximum current is

$$I = \frac{V}{\omega L}, \tag{34–13}$$

and

$$i = I \sin \omega t. \tag{34–14}$$

The *inductive reactance* X_L of an inductor is defined as

$$X_L = \omega L, \tag{34–15}$$

and Eq. (34–13) can be written in the same form as that for a resistor:

$$I = \frac{V}{X_L}. \tag{34–16}$$

The unit of inductive reactance is also 1 *ohm*.

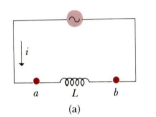

(a)

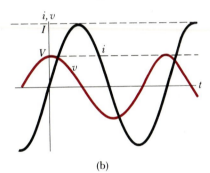

(b)

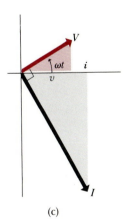

(c)

34–3 (a) Inductance L connected across an ac source. (b) Graphs of instantaneous voltage and current. (c) Vector diagram; current lags voltage by 90°.

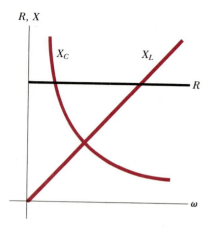

34–4 Graphs of R, X_L, and X_C as functions of frequency.

The reactance of an inductor is directly proportional both to its inductance L and to the angular frequency ω; the greater the inductance, and the higher the frequency, the *larger* is the reactance.

EXAMPLE 2 At an angular frequency of 1000 rad·s⁻¹, the reactance of a 1-H inductor is

$$X_L = \omega L = (10^3 \text{ rad·s}^{-1})(1 \text{ H}) = 1000 \text{ }\Omega.$$

At a frequency of 10,000 rad·s⁻¹ the reactance of the same inductor is 10,000 Ω, while at a frequency of 100 rad·s⁻¹ it is only 100 Ω. ◄

The graphs in Fig. 34–4 summarize the variations with frequency of the resistance of a resistor, and of the reactances of an inductor and of a capacitor. As the frequency increases, the reactance of the inductor approaches infinity, and that of the capacitor approaches zero. As the frequency decreases, the inductive reactance approaches zero and the capacitive reactance approaches infinity. The limiting case of zero frequency corresponds to a dc circuit.

34–3

THE *R–L–C* SERIES CIRCUIT

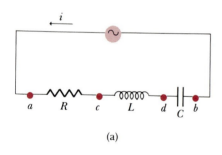

(a)

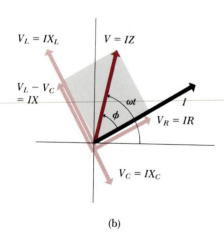

(b)

34–5 (a) A series *R–L–C* circuit. (b) Phasor diagram.

In many instances, ac circuits include resistance, inductive reactance, and capacitive reactance. A simple series circuit is shown in Fig. 34–5a. Analysis of this and similar circuits is facilitated by use of a phasor diagram, which includes the voltage and current vectors for the various components. In this instance the instantaneous *total* voltage across all three components is equal to the source voltage at that instant, and its phasor is the *vector sum* of the phasors for the individual voltages. The complete phasor diagram for this circuit is shown in Fig. 34–5b. This may appear complex, so we shall explain it step by step.

The instantaneous current i has the same value at all points of the circuit. Thus, a *single phasor I*, of length proportional to the current amplitude, suffices to represent the current in each circuit element.

Let us use the symbols v_R, v_L, and v_C for the instantaneous voltages across R, L, and C, and V_R, V_L, and V_C for their maximum values. The instantaneous and maximum voltages across the source will be represented by v and V. Then $v = v_{ab}$, $v_R = v_{ac}$, $v_L = v_{cd}$, and $v_C = v_{db}$.

We have shown that the potential difference between the terminals of a resistor is *in phase* with the current in the resistor, and that its maximum value V_R is

$$V_R = IR.$$

Thus the phasor V_R in Fig. 34–5b, in phase with the current vector, represents the voltage across the resistor. Its projection on the horizontal axis, at any instant, gives the instantaneous potential difference v_R.

The voltage across an inductor *leads* the current by 90°. The voltage amplitude is

$$V_L = IX_L.$$

The phasor V_L in Fig. 34–5b represents the voltage across the inductor, and its projection at any instant onto the horizontal axis equals v_L.

The voltage in a capacitor *lags* the current by 90°. The voltage amplitude is

$$V_C = IX_C.$$

The phasor V_C in Fig. 34–5b represents the voltage across the capacitor, and its projection at any instant onto the horizontal axis equals v_C.

The instantaneous potential difference v between terminals a and b equals at every instant the (algebraic) sum of the potential differences v_R, v_L, and v_C. That is, it equals the sum of the projections of the phasors V_R, V_L, and V_C. But the *projection* of the *vector sum* of these phasors is equal to the *sum* of their *projections*, so this vector sum V must be the phasor that represents the source voltage. To form the vector sum, we first subtract the phasor V_C from the phasor V_L (since these always lie in the same straight line), giving the phasor $V_L - V_C$. Since this is at right angles to the phasor V_R, the magnitude of the phasor V is

$$
\begin{aligned}
V &= \sqrt{V_R{}^2 + (V_L - V_C)^2} \\
&= \sqrt{(IR)^2 + (IX_L - IX_C)^2} \\
&= I\sqrt{R^2 + (X_L - X_C)^2}.
\end{aligned}
$$

The quantity $X_L - X_C$ is called the reactance of the circuit, denoted by X:

$$X = X_L - X_C. \tag{34–17}$$

Finally, we define the *impedance Z* of the circuit as

$$Z = \sqrt{R^2 + (X_L - X_C)^2} = \sqrt{R^2 + X^2}, \tag{34–18}$$

so we can write

$$V = IZ \qquad \text{or} \qquad I = \frac{V}{Z}. \tag{34–19}$$

Thus, the *form* of the equation relating current and voltage *amplitudes* is the same as that for a dc circuit, the impedance Z playing the same role as the resistance R of the dc circuit. Note, however, that the impedance is actually a function of R, L, and C, as well as of the frequency ω. The complete expression for Z, for a series circuit, is

$$
\begin{aligned}
Z &= \sqrt{R^2 + X^2} \\
&= \sqrt{R^2 + (X_L - X_C)^2} \\
&= \sqrt{R^2 + [\omega L - (1/\omega C)]^2}.
\end{aligned}
\tag{34–20}
$$

The unit of impedance, from Eq. (34–19), is evidently one *volt per ampere* $(1 \text{ V} \cdot \text{A}^{-1})$ or one *ohm*.

The expressions for the impedance Z of (a) an $R–L$ series circuit, (b) an $R–C$ series circuit, and (c) an $L–C$ series circuit can be obtained from Eq. (34–20) by letting (a) $X_C = 0$, (b) $X_L = 0$, and (c) $R = 0$.

Equation (34–20) gives the impedance Z only for a *series* $R–L–C$ circuit. But whatever the nature of an $R–L–C$ network, its impedance can be *defined* by Eq. (34–19) as the ratio of the voltage amplitude to the current amplitude.

The angle ϕ, in Fig. 34–5b, is the phase angle of the source voltage V with respect to the current I. It should be evident from the diagram that

$$\tan \phi = \frac{V_L - V_C}{V_R} = \frac{I(X_L - X_C)}{IR} = \frac{X}{R}. \tag{34–21}$$

Hence, if the source voltage is represented by a cosine function,

$$v = V \cos \omega t,$$

the line current lags by an angle ϕ between 0 and 90°, and its equation is

$$i = I \cos (\omega t - \phi).$$

Figure 34–5b has been constructed for a circuit in which $X_L > X_C$. If $X_L < X_C$, vector V lies on the opposite side of the current vector I and the current *leads* the voltage. In this case, $X = X_L - X_C$ is a *negative* quantity, $\tan \phi$ is negative, and ϕ is a negative angle between 0 and −90°.

To summarize, we can say that the *instantaneous* potential differences in an ac series circuit add *algebraically*, just as in a dc circuit, while the voltage *amplitudes* add *vectorially*.

EXAMPLE   In the series circuit of Fig. 34–5, let $R = 300\ \Omega$, $L = 0.9$ H, $C = 2.0\ \mu$F, and $\omega = 1000$ rad$\cdot$s^{-1}. Then

$$X_L = \omega L = 900\ \Omega, \qquad X_C = \frac{1}{\omega C} = 500\ \Omega.$$

The reactance X of the circuit is

$$X = X_L - X_C = 400\ \Omega,$$

and the impedance Z is

$$Z = \sqrt{R^2 + X^2} = 500\ \Omega.$$

If the circuit is connected across an ac source of voltage amplitude 50 V, the current amplitude is

$$I = \frac{V}{Z} = 0.10\ \text{A}.$$

The lag angle ϕ is

$$\phi = \tan^{-1} \frac{X}{R} = 53°.$$

The voltage amplitude across the resistor is

$$V_R = IR = 30\ \text{V}.$$

The voltage amplitudes across the inductor and capacitor are, respectively,

$$V_L = IX_L = 90\ \text{V}, \qquad V_C = IX_C = 50\ \text{V}. \qquad \blacktriangleleft$$

The above analysis describes the *steady-state* condition of a circuit, the condition that prevails after the circuit has been connected to the source for a long time. When the source is first connected, there may be additional voltages and currents, called *transients*, whose nature depends on the time in the cycle when the circuit is initially completed. A detailed analysis of transients is beyond our scope. In any event, they always die out after a sufficiently long time and therefore do not affect the steady-state behavior of the circuit.

34–4

AVERAGE AND ROOT-MEAN-SQUARE VALUES; ac INSTRUMENTS

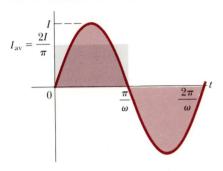

34–6   The average value of a sinusoidal current over a half-cycle is $2I/\pi$. The average over a complete cycle is zero.

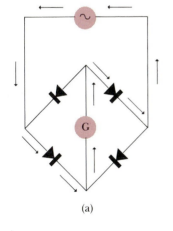

(a)

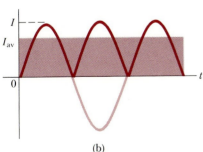

(b)

34–7   (a) A full-wave rectifier. (b) Graph of a full-wave rectified current and its average value.

The *instantaneous* potential difference between two points of an ac circuit can be measured by connecting a calibrated oscilloscope between the points, and the instantaneous current by connecting an oscilloscope across a resistor in the circuit. The usual moving-coil galvanometer, however, has too large a moment of inertia to follow the instantaneous values of an alternating current. It averages out the fluctuating torque on its coil, and its deflection is proportional to the *average* current.

Let us consider a current given by

$$i = I \sin \omega t.$$

It can be shown that the *average value* of $\sin \omega t$, for the *half-cycle* from $t = 0$ to $t = \pi/\omega$, is $2/\pi$; thus

$$I_{\text{av}} = \left(\frac{2}{\pi}\right) I. \tag{34–22}$$

That is, the average current is $2/\pi$ (about $\frac{2}{3}$) times the maximum current, and the *area* under the rectangle in Fig. 34–6 equals the area under one loop of the sine curve.

During the second half-cycle, the direction reverses, and the average current for a *complete cycle* (or any number of complete cycles) is zero. The *positive* area of the loop between 0 and π/ω is equal to the *negative* area of the loop between π/ω and $2\pi/\omega$. Hence, if a sinusoidal current is sent through a moving-coil galvanometer, the meter reads zero!

Such a meter can be used in an ac circuit, however, if it is connected in the *full-wave rectifier circuit* shown in Fig. 34–7a. As explained in Section 28–5, an ideal rectifier offers a constant finite resistance to current in the forward direction and an infinite resistance to current in the opposite direction. When the current is in the direction shown, two of the rectifiers are conducting and two are nonconducting, and the current in the galvanometer is upward. When the current is in the opposite direction, the rectifiers that are nonconducting in Fig. 34–7a carry the current, and the current in the galvanometer is still upward. Thus, if the line current alternates sinusoidally, the current in the galvanometer has the waveform shown as a color curve in Fig. 34–7b. Although pulsating, it is always in the same direction and its average value is *not* zero. Then, when provided with the necessary series resistance or shunt, as for a dc voltmeter or ammeter, the galvanometer can serve as an ac voltmeter or ammeter.

The average value of the rectified current, in any number of complete cycles, is the same as the average current in the first half-cycle in Fig. 34–6, or $2/\pi$ times the maximum current I. Hence, if the meter deflects full-scale with a steady current I_0 through it, it will also deflect full-scale when the *average value* of the rectified current, $2I/\pi$, is equal to

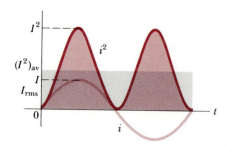

34–8 The average value of the square of a sinusoidally varying current, over any number of half-cycles, is $I^2/2$. The root-mean-square value is $I/\sqrt{2}$.

I_0. The current amplitude I, when the meter deflects full-scale, is then

$$I = \frac{\pi I_0}{2}.$$

For example if $I_0 = 1$ A, $I = 1.57$ A.

Most ac meters are calibrated to read not the maximum value of the current or voltage, but the *root-mean-square* value, that is, the *square root* of the *average* value of the *square* of the current or voltage, abbreviated as the rms value. Since i^2 is always positive, the average of i^2 is never zero even when the average of i is zero.

Figure 34–8 shows graphs of a sinusoidally varying current and of its square. If $i = I \sin \omega t$, then

$$i^2 = I^2 \sin^2 \omega t = I^2 \left[\frac{1}{2}(1 - \cos 2\omega t) \right]$$

$$= \frac{1}{2}I^2 - \frac{1}{2}I^2 \cos 2\omega t.$$

The average value of i^2, or the *mean square current*, is equal to the constant term $\frac{1}{2}I^2$, since the average value of $\cos 2\omega t$, over any number of complete cycles, is zero:

$$(I^2)_{av} = \frac{I^2}{2}.$$

The root-mean-square current is the square root of this, or

$$I_{rms} = \sqrt{(I^2)_{av}} = \frac{I}{\sqrt{2}}. \tag{34–23}$$

Similarly, the root-mean-square value of a sinusoidal voltage is

$$V_{rms} = \frac{V}{\sqrt{2}}.$$

Voltages and currents in power distribution systems are always referred to in terms of their rms values. Thus, when we speak of our household power supply as "115-volt ac," this means that the rms voltage is 115 V. The voltage amplitude is

$$V = \sqrt{2}\, V_{rms} = 163 \text{ V}.$$

All the voltage-current relations developed in preceding sections have been stated in terms of amplitudes (maximum values) of voltage and current, but they remain valid when rms values are used; in each case the difference is simply a multiplicative factor of $1/\sqrt{2}$.

34–5

POWER IN ac CIRCUITS

The *instantaneous* power input p to an ac circuit is

$$p = vi,$$

where v is the instantaneous source potential difference and i is the instantaneous current. We consider first some special cases.

If the circuit consists of a pure resistance R, as in Fig. 34–1, i and v are *in phase*. The graph representing p is obtained by multiplying together at every instant the ordinates of the graphs of v and i in Fig. 34–1b, and it is shown by the full curve in Fig. 34–9a. (The product vi is positive when v and i are both positive or both negative.) That is, energy is supplied *to* the resistor at all instants, whatever the direction of the current, although the *rate* at which it is supplied is not constant.

The power curve is symmetrical about a value equal to one half its maximum ordinate VI, so the *average* power P is

$$P = \frac{1}{2}VI. \tag{34–24}$$

The average power can also be written

$$P = \frac{V}{\sqrt{2}}\left(\frac{I}{\sqrt{2}}\right) = V_{\text{rms}}\,I_{\text{rms}}. \tag{34–25}$$

Furthermore, since $V_{\text{rms}} = I_{\text{rms}}R$, we have

$$P = I_{\text{rms}}{}^{2}R. \tag{34–26}$$

Note that Eqs. (34–25) and (34–26) have the same *form* as those for a dc circuit.

Suppose next that the circuit consists of a capacitor, as in Fig. 34–2. The current and voltage are then 90° out of phase. When the curves of v and i are multiplied together (the product vi is *negative* when v and i have *opposite* signs), we get the power curve in Fig. 34–9b, which is symmetrical about the horizontal axis. The average power is therefore zero.

To see why this is so, we recall that positive power means that energy is supplied *to* a device and that negative power means that energy is supplied *by* a device. The process we are considering is merely the charge

34–9 Graphs of voltage, current, and power as functions of time, for various circuits. (a) Instantaneous power input to a resistor. The average power is $\frac{1}{2}VI$. (b) Instantaneous power input to a capacitor. The average power is zero. (c) Instantaneous power input to a pure inductor. The average power is zero. (d) Instantaneous power input to an arbitrary ac circuit. The average power is $\frac{1}{2}VI\cos\phi = V_{\text{rms}}I_{\text{rms}}\cos\phi$.

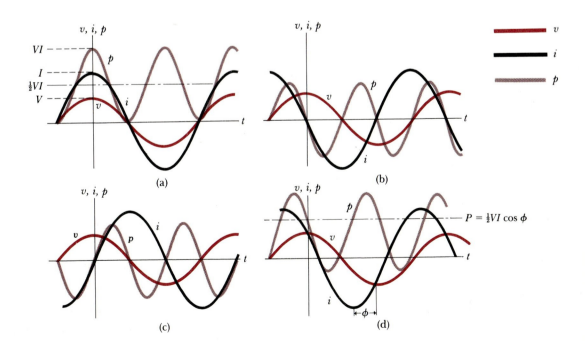

and discharge of a capacitor. During the intervals when p is positive, energy is supplied to charge the capacitor, and when p is negative the capacitor is discharging and returning energy to the source.

Figure 34–9c is the power curve for a pure inductor. As with a capacitor, the current and voltage are out of phase by 90°, and the average power is zero. Energy is supplied to establish a magnetic field around the inductor and is returned to the source when the field collapses.

In the most general case, current and voltage differ in phase by an angle ϕ and

$$p = [V \sin \omega t][I \sin (\omega t - \phi)]. \tag{34–27}$$

The instantaneous power curve has the form shown in Fig. 34–9d. The area under the positive loops is greater than that under the negative loops and the net average power is positive.

The preceding analyses have shown that when v and i are *in phase*, the average power equals $\frac{1}{2} VI$; and when v and i are 90° *out of phase*, the average power is zero. Hence, in the general case, when v and i differ by an angle ϕ, the average power equals $\frac{1}{2} V$, multiplied by $I \cos \phi$, the component of I that is *in phase* with V. That is,

$$P = \frac{1}{2} VI \cos \phi = V_{\text{rms}} I_{\text{rms}} \cos \phi. \tag{34–28}$$

This is the general expression for the power input to *any* ac circuit. The factor $\cos \phi$ is called the *power factor* of the circuit. For a pure resistance, $\phi = 0$, $\cos \phi = 1$, and $P = V_{\text{rms}} I_{\text{rms}}$. For a capacitor or inductor, $\phi = 90°$, $\cos \phi = 0$, and $P = 0$.

A low power factor (large angle of lag or lead) is usually undesirable in power circuits because, for a given potential difference, a large current is needed to supply a given amount of power with correspondingly large heat losses in the transmission lines. Since many types of ac machinery draw a lagging current, this situation is likely to arise. It can be corrected by connecting a capacitor in parallel with the load. The leading current drawn by the capacitor compensates for the lagging current in the other branch of the circuit. The capacitor itself takes no net power from the line.

34–6

SERIES RESONANCE

The impedance of an R–L–C series circuit depends on the frequency, since the inductive reactance is directly proportional to frequency, and the capacitive reactance inversely proportional. This dependence is illustrated in Fig. 34–10a, where a logarithmic frequency scale has been used because of the wide range of frequencies covered. Note that there is one particular frequency at which X_L and X_C are numerically equal. At this frequency, $X = X_L - X_C$ is zero. Hence the impedance Z, equal to $\sqrt{R^2 + X^2}$, is *minimum* at this frequency and is equal to the resistance R.

If an ac source of constant voltage amplitude but variable frequency is connected across the circuit, the current amplitude I varies with frequency, as shown in Fig. 34–10b, and is *maximum* at the frequency at

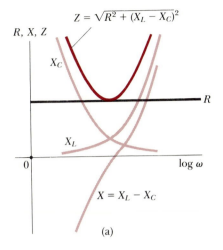

$$Z = \sqrt{R^2 + (X_L - X_C)^2}$$

R, X, Z

X_C

R

X_L

0

log ω

$X = X_L - X_C$

(a)

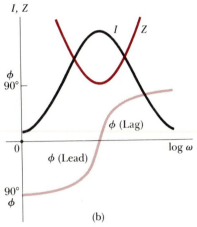

I, Z

I   Z

ϕ
90°

ϕ (Lag)

0

log ω

ϕ (Lead)

90°
ϕ

(b)

34–10 (a) Reactance, resistance, and impedance as functions of frequency (logarithmic frequency scale). (b) Impedance, current, and phase angle as functions of frequency (logarithmic frequency scale).

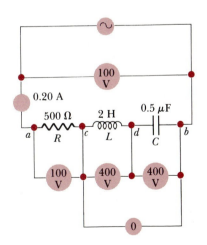

100 V

0.20 A

500 Ω   2 H   0.5 μF

a   R   c   L   d   C   b

100 V   400 V   400 V

0

34–11 Series resonant circuit.

which the impedance Z is *minimum*. The same diagram also shows the phase difference ϕ as a function of frequency. At low frequencies, where capacitive reactance X_C predominates, the current *leads* the voltage. At the frequency where $X = 0$, the current and voltage are *in phase*, and at high frequencies, where X_L predominates, the current *lags* the voltage.

The behavior of the current in an ac series circuit, as the source frequency is varied, is exactly analogous to the response of a spring–mass system having a viscous damping force as the frequency of the driving force is varied. The frequency ω_0 at which the current is maximum is called the *resonant frequency* and is easily computed from the fact that, at this frequency, $X_L = X_C$:

$$X_L = X_C, \qquad \omega_0 L = \frac{1}{\omega_0 C}, \qquad \omega_0 = \frac{1}{\sqrt{LC}}. \qquad (34\text{--}29)$$

Note that this is equal to the natural frequency of oscillation of an L–C circuit, as derived in Sec. 33–5.

If the inductance L or the capacitance C of a circuit can be varied, the resonant frequency can be varied also. This is the procedure by which a radio or television receiving set may be "tuned" to receive the signal from a desired station.

EXAMPLE The series circuit in Fig. 34–11 is connected to the terminals of an ac source whose frequency is variable but whose rms terminal voltage is constant and equal to 100 V. The resonant frequency is

$$\omega_0 = \frac{1}{\sqrt{LC}} = \frac{1}{\sqrt{(2 \text{ H})(0.5 \times 10^{-6} \text{ F})}} = 1000 \text{ rad} \cdot \text{s}^{-1}.$$

At this frequency,

$$X_L = (1000 \text{ rad} \cdot \text{s}^{-1})(2 \text{ H}) = 2000 \text{ Ω},$$

$$X_C = \frac{1}{(1000 \text{ rad} \cdot \text{s}^{-1})(0.5 \times 10^{-6} \text{ F})} = 2000 \text{ Ω},$$

$$X = X_L - X_C = 0.$$

The impedance Z is then equal to the resistance R, and the rms current is

$$I = \frac{V}{Z} = \frac{V}{R} = 0.20 \text{ A}.$$

The rms potential difference across the resistor is

$$V_R = IR = 100 \text{ V}.$$

The rms potential differences across the inductor and capacitor are, respectively,

$$V_L = IX_L = (0.20 \text{ A})(2000 \text{ Ω}) = 400 \text{ V},$$
$$V_C = IX_C = (0.20 \text{ A})(2000 \text{ Ω}) = 400 \text{ V}.$$

The rms potential difference across the inductor–capacitor combination (V_{cb}) is

$$V = IX = I(X_L - X_C) = 0.$$

This is true because the instantaneous potential differences across the inductor and the capacitor have equal amplitudes but are 180° out of phase. Although the rms value of each is substantial, their *resultant* at each instant is zero.   ◄

34–7

PARALLEL CIRCUITS

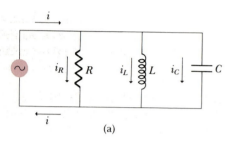

(a)

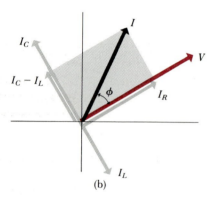

(b)

34–12

Circuit elements connected in parallel across an ac source can be analyzed by the same procedure as for elements in series. The phasor diagram for the circuit in Fig. 34–12a is given in Fig. 34–12b. In this case, the instantaneous potential difference across each element is the same, and a *single* phasor V represents the common terminal voltage. The phasor I_R, of amplitude V/R and in phase with V, represents the current in the resistor. Phasor I_L, of amplitude V/X_L and lagging V by 90°, represents the current in the inductor, and phasor I_C, of amplitude V/X_C and leading V by 90°, represents the current in the capacitor.

The instantaneous current i, by Kirchhoff's point rule, equals the (algebraic) sum of the instantaneous currents i_R, i_L, and i_C and is represented by the phasor I, the vector sum of phasors I_R, I_L, and I_C. Angle ϕ is the phase angle of current with respect to source voltage.

From Fig. 34–12,

$$I = \sqrt{I_R^2 + (I_C - I_L)^2} = \sqrt{\left(\frac{V}{R}\right)^2 + \left(\omega C V - \frac{V}{\omega L}\right)^2}$$

$$= V\sqrt{\frac{1}{R^2} + \left(\omega C - \frac{1}{\omega L}\right)^2}. \tag{34–30}$$

The maximum current I is frequency-dependent, as expected. It is minimum when the second factor in the radical is zero; this occurs when the two reactances have equal magnitudes, at the resonant frequency ω_0 given by Eq. (34–29).

Thus at resonance the total current in the parallel R–L–C circuit is *minimum*, in contrast to the R–L–C *series* circuit, which has *maximum* current at resonance. This can be understood by noting that, in the parallel circuit, the currents in L and C are *always* exactly a half-cycle out of phase; when they also have equal *magnitudes,* they cancel each other completely, and the total current is simply that through R. Indeed, when $\omega C = 1/\omega L$, Eq. (34–30) becomes simply $I = V/R$. This does *not* mean that there is *no* current in L or C at resonance, but only that the two currents cancel. If R is large, the equivalent impedance of the circuit near resonance is much *larger* than the individual reactances X_L and X_C.

*34–8

THE TRANSFORMER

For reasons of efficiency, it is desirable to transmit electrical power at high voltages and small currents, with consequent reduction of I^2R heating in the transmission line. On the other hand, considerations of safety and of insulation of moving parts require relatively low voltages in generating equipment and in motors and household appliances. One of the most useful features of ac circuits is the ease and efficiency with which voltages (and currents) may be changed from one value to another by *transformers*.

In principle, the transformer consists of two coils electrically insulated from each other and wound on the same iron core. An alternating current in one winding sets up an alternating flux in the core, and the induced electric field produced by this varying flux (see Section 33–1) induces an emf in the other winding. Energy is thus transferred from one winding to another via the core flux and its associated induced electric field. The winding to which power is supplied is called the *primary;* that from which power is delivered is called the *secondary*. The circuit symbol for an iron core transformer is

The power output of a transformer is necessarily less than the power input because of unavoidable losses. These losses consist of I^2R losses in the primary and secondary windings, and hysteresis and eddy-current losses in the core. Hysteresis losses are minimized by the use of iron having a narrow hysteresis loop; and eddy currents are minimized by laminating the core. In spite of these losses, transformer efficiencies are usually well over 90%, and in large installations may reach 99%.

For simplicity, we shall consider only an idealized transformer in which there are *no* losses and in which all of the flux is confined to the iron core, so that the same flux links both primary and secondary. The transformer is shown schematically in Fig. 34–13. A primary winding of N_1 turns and a secondary winding of N_2 turns both encircle the core in the same sense. An ac source voltage amplitude V_1 is connected to the primary and, to begin with, we assume the secondary to be open, so that there is no secondary current. The primary winding then functions merely as an inductor.

The primary current, which is small, lags the primary voltage by 90° and is called the *magnetizing* current. The power input to the transformer is zero. The core flux is in phase with the primary current. Since the same flux links both primary and secondary, the induced emf *per turn* is the same in each. The ratio of primary to secondary induced emf is therefore equal to the ratio of primary to secondary turns, or

$$\frac{\mathcal{E}_2}{\mathcal{E}_1} = \frac{N_2}{N_1}.$$

Since the windings are assumed to have zero resistance, the induced emf's $\mathcal{E}_1$ and $\mathcal{E}_2$ are numerically equal to the corresponding terminal voltages V_1 and V_2, and

$$\frac{V_2}{V_1} = \frac{N_2}{N_1}. \tag{34–31}$$

Hence by properly choosing the turn ratio N_2/N_1, we may obtain any desired secondary voltage from a given primary voltage. If $V_2 > V_1$, we have a *step-up* transformer; if $V_2 < V_1$, a *step-down* transformer.

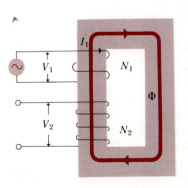

34–13 Schematic diagram of a transformer with secondary open.

Consider next the effect of closing the secondary circuit. The secondary current I_2 and its phase angle ϕ_2 will, of course, depend on the nature of the secondary circuit. As soon as the latter is closed, some power must be delivered by the secondary (except when $\phi_2 = 90°$); and, from energy considerations, an equal amount of power must be supplied to the primary. The process by which the transformer is enabled to draw the requisite amount of power is as follows. When the secondary circuit is open, the core flux is produced by the primary current only. But when the secondary circuit is closed, *both* primary and secondary currents set up a flux in the core. The secondary current, by Lenz's law, tends to weaken the core flux and therefore to decrease the back-emf in the primary. But (in the absence of losses) the back-emf in the primary must equal the primary terminal voltage, which is assumed to be fixed. The primary current therefore increases until the core flux is restored to its original no-load magnitude.

If the secondary circuit is completed by a resistance R, $I_2 = V_2/R$. From energy considerations, the power delivered to the primary equals that taken out of the secondary (neglecting losses), so

$$V_1 I_1 = V_2 I_2. \tag{34–32}$$

Combining this with the above expression for I_2 and Eq. (34–31) to eliminate V_2 and I_2, we find

$$I_1 = \frac{V_1}{(N_1/N_2)^2 R}. \tag{34–33}$$

Thus, when the secondary circuit is completed through a resistance R, the result is the same as if the *source* had been connected directly to a resistance equal to R multiplied by the reciprocal of the *square* of the turns ratio. In other words, the transformer "transforms" not only voltages and currents but resistances (more generally, impedances) as well. It can be shown that maximum power is supplied by a source to a resistor when its resistance equals the internal resistance of the source. The same principle applies in ac circuits, with resistance replaced by impedance. When a high-impedance ac source must be connected to a low-impedance circuit, as when an audio amplifier is connected to a loudspeaker, the impedance of the source can be *matched* to that of the circuit by the insertion of a transformer having the correct turns ratio.

QUESTIONS

34–1 Some electric-power systems formerly used 25-Hz alternating current instead of the 60-Hz that is now standard. The lights flickered noticeably. Why is this not a problem with 60-Hz?

34–2 Power-distribution systems in airplanes sometimes use 400-Hz ac. What advantages and disadvantages does this have compared to the standard 60-Hz?

34–3 Fluorescent lights often use an inductor, called a "ballast," to limit the current through the tubes. Why is it better to use an inductor than a resistor for this purpose?

34–4 At high frequencies a capacitor becomes a short circuit. Discuss.

34–5 At high frequencies an inductor becomes an open circuit. Discuss.

34–6 Household electric power in most of western Europe is 220 volts, rather than the 110-V which is standard in the United States and Canada. What advantages and disadvantages does each system have?

34–7 The current in an ac power line changes direction 120 times per second, and its average value is zero. So how is it possible for power to be transmitted in such a system?

34–8 Electric-power connecting cords, such as lamp cords, always have two conductors that carry equal currents in opposite directions. How might one determine, using measurements only at the midpoints along the lengths of the wires, the direction of power transmission in the cord?

34–9 Are the equations for the average and rms values of current, Eqs. (34–22) and (34–23), correct when the variation with time is not sinusoidal? Explain.

34–10 Electric power companies like to have their power factors (cf. Section 34–5) as close to unity as possible. Why?

34–11 Some electrical appliances operate equally well on ac or dc, while others work only on ac or only on dc. Give examples of each, and explain the differences.

34–12 When a series-resonant circuit is connected across a 110-V ac line, the voltage rating of the capacitor may be exceeded, even if it is rated at 200 or 400 V. How can this be?

34–13 In a parallel-resonant circuit connected across a 110-V line, it is possible for the maximum current rating in the inductor to be exceeded, even if the total current through the circuit is very small. How can this be?

34–14 Can a transformer be used with dc? What happens if a transformer designed for 110-V ac is connected to a 110-V dc line?

34–15 During the last quarter of the nineteenth century there was a great and acrimonious controversy concerning whether ac or dc should be used for power transmission. Edison favored dc, George Westinghouse ac. What arguments might each proponent have used to promote his scheme?

PROBLEMS

$$\text{Angular frequency} = 2\pi \ (\text{frequency}),$$
$$\omega(\text{rad} \cdot \text{s}^{-1}) = 2\pi f(\text{Hz}).$$

34–1

a) At what frequency would a 5-H inductor have a reactance of 4000 Ω?

b) At what frequency would a 5-μF capacitor have the same reactance?

34–2 What is the reactance of a 0.015-μF capacitor at

a) 1 Hz?

b) 5 kHz?

c) 2 MHz?

34–3

a) What is the reactance of a 1-H inductor at a frequency of 60 Hz?

b) What is the inductance of an inductor whose reactance is 1 Ω at 60 Hz?

c) What is the reactance of a 1-μF capacitor at a frequency of 60 Hz?

d) What is the capacitance of a capacitor whose reactance is 1 ohm at 60 Hz?

34–4

a) Compute the reactance of a 10-H inductor at frequencies of 60 Hz and 600 Hz.

b) Compute the reactance of a 10-μF capacitor at the same frequencies.

c) At what frequency is the reactance of a 10-H inductor equal to that of a 10-μF capacitor?

34–5 A 1-μF capacitor is connected across an ac source whose voltage amplitude is kept constant at 50 V, but whose frequency can be varied. Find the current amplitude when the angular frequency is

a) 100 rad $\cdot$ s^{-1},

b) 1000 rad $\cdot$ s^{-1},

c) 10,000 rad $\cdot$ s^{-1}.

d) Construct a log-log plot of current amplitude versus frequency.

34–6 The voltage amplitude of an ac source is 50 V and its angular frequency is 1000 rad $\cdot$ s^{-1}. Find the current amplitude if the capacitance of a capacitor connected across the source is

a) 0.01 μF,

b) 1.0 μF,

c) 100 μF.

d) Construct a log-log plot of current amplitude versus capacitance.

34–7 An inductor of self-inductance 10 H and of negligible resistance is connected across the source in Problem 34–5. Find the current amplitude when the angular frequency is

a) 100 rad $\cdot$ s^{-1},

b) 1000 rad $\cdot$ s^{-1},

c) 10,000 rad $\cdot$ s^{-1}.

d) Construct a log-log plot of current amplitude versus frequency.

34–8 Find the current amplitude if the self-inductance of a resistanceless inductor connected across the source of Problem 34–6 is

a) 0.01 H,

b) 1.0 H,

c) 100 H.

d) Construct a log-log plot of current amplitude versus self-inductance.

34–9 The expression for the impedance Z of an R–L series circuit can be obtained from Eq. (34–20) by setting $X_C = 0$, which corresponds to $C = \infty$. Explain.

34–10 In an R–L–C series circuit, the source has a constant voltage amplitude of 50 V and a frequency of 1000 rad·s^{-1}. $R = 300\ \Omega$, $L = 0.9$ H, and $C = 2.0\ \mu$F. Suppose a series circuit contains only the resistor and the inductor in series.

a) What is the impedance of the circuit?

b) What is the current amplitude?

c) What are the voltage amplitudes across the resistor and across the inductor?

d) What is the phase angle ϕ? Does the current lag or lead?

e) Construct the phasor diagram.

34–11 Same as Problem 34–10, except that the circuit consists of the resistor and the capacitor in series.

34–12 Same as Problem 34–10, except that the circuit consists of the inductor and capacitor in series.

34–13

a) Compute the impedance of an R–L–C series circuit at an angular frequency of 500 rad·s^{-1}, using the numbers in Problem 34–10.

b) Describe how the current amplitude varies as the frequency of the source is slowly reduced from 1000 rad·s^{-1} to 500 rad·s^{-1}.

c) What is the phase angle when $\omega = 500$ rad·s^{-1}? Construct the phasor diagram when $\omega = 500$ rad·s^{-1}.

34–14 Five infinite impedance voltmeters, calibrated to read rms values, are connected as shown in Fig. 34–14. What does each voltmeter read? Use the numbers given in Problem 34–10.

34–15 What is the reading of each voltmeter in Fig. 34–14 if the angular frequency $\omega = 500$ rad·s^{-1}? Use the values of V, R, L, and C in Problem 34–10, and assume that the voltmeters measure rms values.

34–16 A 400-Ω resistor is in series with a 0.1-H inductor and a 0.5-μF capacitor. Compute the impedance of the circuit and draw the vector impedance diagram

a) at a frequency of 500 Hz,

b) at a frequency of 1000 Hz.

Compute, in each case, the phase angle between line current and line voltage, and state whether the current lags or leads.

34–17

a) Construct a graph of the current amplitude in the circuit of Fig. 34–5 as the angular frequency of the source is increased from 500 rad·s^{-1} to 1000 rad·s^{-1}. Use the values of V, R, L, and C in Problem 34–10.

b) At what frequency is the circuit in resonance?

c) What is the power factor at resonance?

d) What is the reading of each voltmeter in Fig. 34–14 when the frequency equals the resonant frequency?

e) What would be the resonant frequency if the resistance were reduced to 100 Ω?

f) What would then be the rms current at resonance?

34–18

a) At what frequency is the voltage amplitude across the inductor in an R–L–C series circuit a maximum?

b) At what frequency is the voltage amplitude across the capacitor a maximum?

34–19 In an R–L–C series circuit, $R = 250\ \Omega$, $L = 0.5$ H, and $C = 0.02\ \mu$F.

a) What is the resonant frequency of the circuit?

b) The capacitor can withstand a peak voltage of 350 V. What maximum effective terminal voltage may the generator have at resonant frequency?

34–20 A resistor of 500 Ω and a capacitor of 2 μF are connected in parallel to an ac generator supplying a constant voltage amplitude of 282 V and having an angular frequency of 377 rad·s^{-1}. Find

a) the current amplitude in the resistor,

b) the current amplitude in the capacitor,

c) the phase angle, and

d) the line current amplitude.

34–21 A coil has a resistance of 20 Ω. At a frequency of 100 Hz, the voltage across the coil leads the current in it by 30°. Determine the inductance of the coil.

34–22 An inductor having a reactance of 25 Ω gives off heat at the rate of 2.39 cal·s^{-1} when it carries a current of 0.5 A (rms). What is the impedance of the inductor?

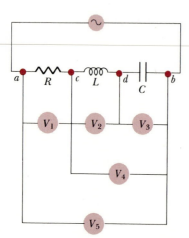

Figure 34–14

34–23 The circuit of Problem 34–16 carries an rms current of 0.25 A with frequency 100 Hz.

a) What power is consumed in the circuit?

b) In the resistor?

c) In the capacitor?

d) In the inductor?

e) What is the power factor of the circuit?

34–24 A series circuit has a resistance of 75 Ω and an impedance of 150 Ω. What power is consumed in the circuit when a voltage of 120 V (rms) is impressed across it?

34–25 A circuit draws 330 W from a 110-V, 60-Hz ac line. The power factor is 0.6 and the current lags the voltage.

a) Find the capacitance of the series capacitor that will result in a power factor of unity.

b) What power will then be drawn from the supply line?

34–26 A series circuit has an impedance of 50 Ω and a power factor of 0.6 at 60 Hz, the voltage lagging the current.

a) Should an inductor or a capacitor be placed in series with the circuit to raise its power factor?

b) What size element will raise the power factor to unity?

34–27 The internal resistance of an ac source is 10,000 Ω.

a) What should be the turns ratio of a transformer to match the source to a load of resistance 10 Ω?

b) If the voltage amplitude of the source is 100 V, what is the voltage amplitude of the secondary on open circuit?

34–28 A 100-Ω resistor, a 0.1-μF capacitor, and a 0.1-H inductor are connected in parallel to a voltage source with amplitude 100 V.

a) What is the resonant frequency? The resonant angular frequency?

b) What is the maximum total current through the parallel combination at the resonant frequency?

c) What is the maximum current in the resistor at resonance?

d) What is the maximum current in the inductor at resonance?

e) What is the maximum energy stored in the inductor at resonance? In the capacitor?

34–29 The same three components as in Problem 34–28 are connected in *series* to a voltage source with amplitude 100 V.

a) What is the resonant frequency? The resonant angular frequency?

b) What is the maximum current in the resistor at resonance?

c) What is the maximum voltage across the capacitor at resonance?

d) What is the maximum energy stored in the capacitor at resonance?

34–30 A transformer connected to a 120-V ac line is to supply 12 V to a low-voltage lighting system for a model-railroad village. The total equivalent resistance of the system is 2 Ω.

a) What should be the turns ratio of the transformer?

b) What current must the secondary supply?

c) What power is delivered to the load?

d) What resistance connected directly across the 120-V line would draw the same power as the transformer? Show that this is equal to 2 Ω times the square of the turns ratio.

34–31 A step-up transformer connected to a 120-V ac line is to supply 18,000 V for a neon sign. To reduce shock hazard, a fuse is to be inserted in the primary circuit; the fuse is to blow when the current in the secondary circuit exceeds 10 mA.

a) What is the turns ratio of the transformer?

b) What power must be supplied to the transformer when the secondary current is 10 mA?

c) What current rating should the fuse in the primary circuit have?

35

ELECTROMAGNETIC WAVES

An electromagnetic wave consists of time-varying electric and magnetic fields that travel or propagate through space with a definite speed. The theoretical basis for electromagnetic waves, discovered by Maxwell in 1864, rests on the ability of a time-varying electric field and its associated displacement current to act as a source of magnetic field, and that of a time-varying magnetic field to act as a source of electric field, according to Faraday's law. Electromagnetic waves carry energy and momentum and show the property of polarization. The simplest electromagnetic waves are sinusoidal waves, in which electric and magnetic fields vary sinusoidally with time.

35–1

INTRODUCTION

Our discussion of electric and magnetic fields in the last several chapters can be classified in two general categories. The first includes fields that do not vary with time. The electrostatic field of a distribution of charges at rest and the magnetic field of a steady current in a conductor are examples of fields that vary from point to point in space but do not vary with time at any individual point. For such situations we found it possible to treat the electric and magnetic fields independently, without worrying unduly about interactions between the two fields.

The second category includes situations in which the fields *do* vary with time, and in all such cases we have found that it is *not* possible to treat the fields independently. Faraday's law tells us that a time-varying magnetic field acts as a source of electric field. This field is manifested in the induced emf's in inductances and transformers. Similarly, in developing the general formulation of Ampère's law (Section 31–7), which is valid for charging capacitors and similar situations as well as for ordinary conductors, we found it necessary to regard a changing electric field as a source of magnetic field.

Thus, when *either* field is changing with time, a field of the other kind is induced in adjacent regions of space, and we are led naturally to consider the possibility of an electromagnetic disturbance, consisting of time-varying electric and magnetic fields, that can propagate through space from one region to another even when there is no matter in the intervening region. Such a disturbance, if it exists, will have the properties of a *wave*, and an appropriate descriptive term is *electromagnetic wave*. Such waves do exist, of course, and it is a familiar fact that radio and television transmission, light, x-rays, and many other phenomena are examples of electromagnetic radiation. In the following pages we shall show how the existence of such waves is related to the principles of electromagnetism studied thus far and shall examine their properties.

As so often happens in the development of science, the theoretical understanding of electromagnetic waves originally took a considerably more devious path than the one outlined above. In the early days of electromagnetic theory (the early nineteenth century), two different units of electric charge were used, one for electrostatics, the other for magnetic phenomena involving currents. The particular system of units in common use at the time had the property that these two units of charge had different physical dimensions; their ratio turned out to have units of velocity. This in itself is not so astounding, but experimental measurements revealed that the ratio had a numerical value precisely equal to the speed of light, $3.00 \times 10^8 \ \text{m} \cdot \text{s}^{-1}$. At the time, physicists regarded this as an extraordinary coincidence and had no idea how to explain it. It was the search for an explanation of this result that led Maxwell, in 1864, to prove by theoretical reasoning that an electrical disturbance should propagate in free space with a speed equal to that of light, and hence to postulate that light waves were *electromagnetic* waves.

It remained for Heinrich Hertz, in 1887, actually to *produce* electromagnetic waves with the aid of oscillating circuits and to receive and detect these waves with other circuits tuned to the same frequency. Hertz then produced stationary electromagnetic waves and measured the distance between adjacent nodes, to measure the wavelength. Knowing the frequency of his resonators, he then found the velocity of the waves from the fundamental wave equation $c = f\lambda$ and verified Maxwell's theoretical value directly.

The possible use of electromagnetic waves for purposes of long-distance communication does not seem to have occurred to Hertz. It remained for the enthusiasm and energy of Marconi and others to make radio communication a familiar household phenomenon.

The unit of frequency, one cycle per second, is named one *hertz* (1 Hz) in honor of Hertz.

35–2

SPEED OF AN ELECTROMAGNETIC WAVE

In developing the relation of electromagnetic wave propagation to familiar electromagnetic principles, we begin with a particularly simple and somewhat artificial example of an electromagnetic wave. We shall postulate a particular configuration of fields and then test whether it is consis-

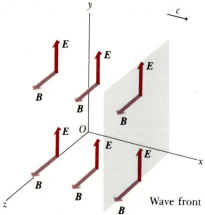

35–1 An electromagnetic wave front. The *E* and *B* fields are uniform over the region to the left of the plane, but are zero everywhere to the right of it. The plane representing the wave front moves to the right with speed *c*.

tent with the principles mentioned above, Faraday's law and Ampère's law including displacement current.

Using an *xyz*-coordinate system, as shown in Fig. 35–1, we imagine that all space is divided into two regions by a plane perpendicular to the *x*-axis (parallel to the *yz*-plane). At all points to the right of this plane there are no electric or magnetic fields; at all points to the left there is a uniform electric field *E* in the +*y*-direction and a uniform magnetic field *B* in the +*z*-direction, as shown. Furthermore, we suppose that the boundary surface, which may also be called the *wave front*, moves to the right with a constant speed *c*, as yet unknown. Thus the *E* and *B* fields travel to the right into previously field-free regions, with a definite speed. The situation, in short, does describe a rudimentary electromagnetic wave, provided it is consistent with the laws of electromagnetism.

We shall not worry about how such a field configuration can be produced. It can be shown that, in principle, an infinitely large sheet of charge in the *yz*-plane, which is initially at rest and suddenly starts moving with constant velocity in the +*y*-direction, gives such fields, but of course there is no practical way to realize an infinitely large charge sheet. At any rate, we now apply Faraday's law to a rectangle in the *xy*-plane, as in Fig. 35–2, located so that at some instant the wave front has progressed partway through the rectangle, as shown. In a time Δt, the boundary surface moves a distance $c\,\Delta t$ to the right, sweeping out an area $ac\,\Delta t$ of the rectangle, and in this time the magnetic flux increases by $B(ac\,\Delta t)$.

Thus the rate of change of magnetic flux is given by

$$\frac{\Delta \Phi}{\Delta t} = Bac. \tag{35–1}$$

According to Faraday's law, this must equal $\Sigma E_{\parallel}\,\Delta l$, where the sum is taken around the boundary of the area. The field at the right end is zero, and the top and bottom sides do not contribute because there the component of *E* along Δl is zero. Thus the sum becomes simply *Ea*, and Faraday's law gives

$$E = cB. \tag{35–2}$$

Thus the postulated wave is consistent with Faraday's law only if *E*, *B*, and *c* are related as in Eq. (35–2).

Next we consider a rectangle in the *xz*-plane, as shown in Fig. 35–3. We wish to apply Ampère's law to this rectangle; the general form of this law, including displacement current, was stated in Chapter 31, Eq. (31–14):

$$\sum B_{\parallel}\,\Delta s = \mu_0 (I_C + I_D). \tag{35–3}$$

In the present situation there is no conduction current, so $I_C = 0$. Previously we have defined the displacement current I_D in terms of a displacement current *density*, as in Eq. (29–17):

$$J_D = \epsilon_0 \frac{\Delta E}{\Delta t}. \tag{35–4}$$

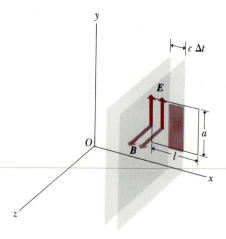

35–2 In time Δt the wave front moves to the right a distance $c\,\Delta t$. The magnetic flux through the rectangle in the *xy*-plane increases by an amount $\Delta \Phi$ equal to the flux through the shaded rectangle of area $ac\,\Delta t$, that is, $\Delta \Phi = Bac\,\Delta t$.

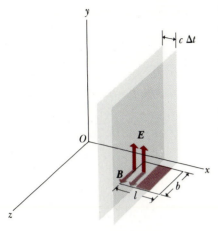

35–3 In time Δt the electric flux through the rectangle in the xz-plane increases by an amount equal to E times the area ($bc\ \Delta t$) of the shaded rectangle; that is, $\Delta\Psi = Ebc\ \Delta t$. Thus $\Delta\Psi/\Delta t = Ebc$.

Then if the E field lines pass through an area A normal to the direction of E, we have

$$I_D = J_D A = \epsilon_0 \frac{\Delta E}{\Delta t} A. \qquad (35\text{–}5)$$

Now it is useful to rephrase this relation in a language analogous to that used for Faraday's law. In Chapter 25 we defined *electric flux*, analogous to magnetic flux but using the *electric* field rather than the magnetic field. If an area A lies perpendicular to an electric field E, the electric flux through the area, denoted by Ψ, is defined, as in Eq. (25–15), as

$$\Psi = EA.$$

In terms of electric flux, Eq. (35–5) can be rewritten as in Eq. (29–16):

$$I_D = \epsilon_0 \frac{\Delta\Psi}{\Delta t}, \qquad (35\text{–}6)$$

and Ampère's law with displacement current but no conduction current is

$$\sum B_\| \Delta s = \mu_0 \epsilon_0 \frac{\Delta\Psi}{\Delta t}. \qquad (35\text{–}7)$$

This is the form we shall apply to the rectangle in Fig. 35–3. The change in electric flux $\Delta\Psi$ in time Δt is the area $bc\ \Delta t$ swept out by the wave front, multiplied by E. In evaluating the sum $\sum B_\| \Delta s$, we note that B is zero on the right end and $B_\| = 0$ on the front and back sides. Thus only the left end contributes, and we find $\sum B_\| \Delta s = Bb$. Combining these results with Eq. (35–7) and dividing through by b, we obtain

$$B = \mu_0 \epsilon_0 c E. \qquad (35\text{–}8)$$

Thus Ampère's law is obeyed only if B, c, and E are related as in Eq. (35–8).

Since *both* Ampère's law and Faraday's law must be obeyed simultaneously, Eqs. (35–2) and (35–8) must both be satisfied. This can occur only when $\mu_0\epsilon_0 c = 1/c$, or

$$c = \frac{1}{\sqrt{\epsilon_0\mu_0}}. \qquad (35\text{–}9)$$

Inserting the numerical values of these quantities, we find

$$c = \frac{1}{\sqrt{(8.85 \times 10^{-12})(4\pi \times 10^{-7})}}$$
$$= 3.00 \times 10^8 \text{ m·s}^{-1}.$$

The postulated field configuration *is* consistent with the laws of electrodynamics, provided the wave front moves with the speed given above, which of course is recognized as the speed of light.

We have chosen a particularly simple wave for study in order to avoid undue mathematical complexity, but this special case nevertheless illustrates several important features of *all* electromagnetic waves:

1. The wave is *transverse;* both *E* and *B* are perpendicular to the direction of propagation of the wave, and to each other.
2. There is a definite ratio between *E* and *B*.
3. The waves travel in vacuum with a definite and unchanging speed.

It is not difficult to generalize the above discussion to a more realistic situation. Suppose we have several wave fronts in the form of parallel planes perpendicular to the axis and all moving to the right with speed c. Suppose that within a single region between two planes, the *E* and *B* fields are the same at all points in the region, but that they differ from region to region. An extension of the above development shows that such a situation is also consistent with Ampère's and Faraday's laws, provided the wave fronts all move with the speed c given by Eq. (35–9). From this picture it is only a short additional step to a wave picture in which the *E* and *B* fields at any instant vary smoothly, rather than in steps, as we move along the x-axis and the entire field pattern moves to the right with speed c. In Section 35–5 we consider waves in which the dependence of *E* and *B* on position and time is *sinusoidal;* first, however, we consider the *energy* associated with an electromagnetic wave.

*35–3

ENERGY IN ELECTROMAGNETIC WAVES

Analysis of energy needed to charge a capacitor and to establish a current in an inductor has led us (Sections 27–4 and 33–3) to associate an energy density (energy per unit volume) with electric and magnetic fields. Specifically, Eqs. (27–9) and (33–9) show that the total energy density u in a region of space where *E* and *B* fields are present is given by

$$u = \frac{1}{2}\epsilon_0 E^2 + \frac{1}{2\mu_0}B^2. \tag{35–10}$$

This and subsequent expressions become somewhat more symmetrical when expressed in terms of a field *H* defined as $H = B/\mu_0$ and called "magnetic intensity." In terms of *E* and *H*, the energy density is

$$u = \frac{1}{2}\epsilon_0 E^2 + \frac{1}{2}\mu_0 H^2. \tag{35–11}$$

Now we have found above that *E* and *B* in an electromagnetic wave are not independent but are related by

$$B = \frac{E}{c} = \sqrt{\epsilon_0\mu_0}E \quad \text{or} \quad H = \sqrt{\frac{\epsilon_0}{\mu_0}}E. \tag{35–12}$$

Thus the energy density may also be expressed

$$u = \frac{1}{2}\epsilon_0 E^2 + \frac{1}{2}\mu_0\left(\sqrt{\frac{\epsilon_0}{\mu_0}}E\right)^2 = \frac{1}{2}\epsilon_0 E^2 + \frac{1}{2}\epsilon_0 E^2 = \epsilon_0 E^2, \tag{35–13}$$

which shows that in a wave, the energy density associated with the **E** field is equal to that of the **H** field.

Equation (35–12) may also be written

$$\frac{E}{H} = \sqrt{\frac{\mu_0}{\epsilon_0}}. \tag{35-14}$$

The unit of E is $1 \text{ V} \cdot \text{m}^{-1}$ and that of H is $1 \text{ A} \cdot \text{m}^{-1}$, so the unit of the ratio E/H is $1 \text{ V} \cdot \text{A}^{-1}$ or 1Ω. When numerical values of ϵ_0 and μ_0 are inserted, we find

$$\frac{E}{H} = 377 \ \Omega. \tag{35-15}$$

For reasons that we cannot discuss in detail, this ratio is called the *impedance of free space*.

Because the **E** and **H** fields in the simple wave considered above advance with time into regions where originally there were no fields, it is clear that the wave transports energy from one region to another. This energy transfer is conveniently characterized by considering the energy transferred *per unit time, per unit cross-sectional area* for an area perpendicular to the direction of wave travel. This quantity will be denoted by S. This is analogous to the concept of current density, the charge per unit time transferred across unit area perpendicular to the direction of flow.

To see how the energy flow is related to the fields, we consider a stationary plane perpendicular to the x-axis, which at a certain time coincides with the wave front. In a time Δt after this time, the wave front moves a distance $c \ \Delta t$ to the right. Considering an area A on the stationary plane, we note that the energy in the space to the right of this area must have passed through it to reach its new location. The volume ΔV of the relevant region is the base area A times the length $c \ \Delta t$, and the total energy ΔU in this region is the energy density u times this volume:

$$\Delta U = \epsilon_0 E^2 A c \ \Delta t. \tag{35-16}$$

Since this much energy passed through area A in time Δt, the energy flow S per unit time, per unit area is

$$S = \frac{\Delta U}{A \ \Delta t} = \epsilon_0 c E^2.$$

Using Eqs. (35–9) and (35–12), we obtain the alternative forms

$$S = \frac{\epsilon_0}{\sqrt{\epsilon_0 \mu_0}} E^2 = \sqrt{\frac{\epsilon_0}{\mu_0}} E^2 = EH. \tag{35-17}$$

The quantity S is called the *intensity* of the radiation. The unit of S is energy per unit time, per unit area. The SI unit of S is $1 \text{ J} \cdot \text{s}^{-1} \cdot \text{m}^{-2}$ or $1 \text{ W} \cdot \text{m}^{-2}$. In more advanced treatments, it is customary to define a vector quantity **S** called the *Poynting vector*, with magnitude given by Eq. (35–17) and directed along the direction of propagation of the wave. The magnitude EH gives the flow of energy across a cross section perpendicular to the propagation direction, per unit area and per unit time.

EXAMPLE   In the wave discussed above, suppose

$$E = 100 \text{ V} \cdot \text{m}^{-1} = 100 \text{ N} \cdot \text{C}^{-1}.$$

Find the values of B and H, the energy density, and the energy flow S.

Solution   From Eq. (35–2),

$$B = \frac{E}{c} = \frac{100 \text{ V} \cdot \text{m}^{-1}}{3.0 \times 10^8 \text{ m} \cdot \text{s}^{-1}} = 3.33 \times 10^{-7} \text{ T};$$

$$H = \frac{B}{\mu_0} = \frac{3.33 \times 10^{-7} \text{ T}}{4\pi \times 10^{-7} \text{ Wb} \cdot \text{A}^{-1} \cdot \text{m}^{-1}} = 0.265 \text{ A} \cdot \text{m}^{-1}.$$

From Eq. (35–13),

$$u = \epsilon_0 E^2 = (8.85 \times 10^{-12} \text{ C}^2 \cdot \text{N}^{-1} \cdot \text{m}^{-2})(100 \text{ N} \cdot \text{C}^{-1})^2$$
$$= 8.85 \times 10^{-8} \text{ N} \cdot \text{m}^{-2}$$
$$= 8.85 \times 10^{-8} \text{ J} \cdot \text{m}^{-3};$$

$$S = EH = (100 \text{ V} \cdot \text{m}^{-1})(0.265 \text{ A} \cdot \text{m}^{-1})$$
$$= 26.5 \text{ W} \cdot \text{m}^{-2}.$$

◀

The idea that energy can travel through empty space without the aid of any matter in motion may seem strange, yet this is the very mechanism by which energy reaches us from the sun. The conclusion that electromagnetic waves transport energy is as inescapable as the conclusion that energy is required to establish electric and magnetic fields. It is also possible to show that electromagnetic waves carry *momentum*, with a corresponding momentum density of magnitude

$$\frac{EH}{c^2} = \frac{S}{c^2}. \tag{35–18}$$

This momentum is a property of the field alone and is not associated with moving mass. There is also a corresponding momentum flow rate; just as the energy density u corresponds to S, the rate of energy flow per unit area, the momentum density given by Eq. (35–18) corresponds to the momentum flow rate

$$\frac{1}{A} \frac{\Delta p}{\Delta t} = \frac{S}{c} = \frac{EH}{c}, \tag{35–19}$$

which represents the momentum transferred per unit surface area, per unit time.

This momentum is responsible for the phenomenon of *radiation pressure;* when an electromagnetic wave is absorbed by a surface perpendicular to the propagation direction, the time rate of change of momentum is equal to the force on the surface. Thus the force per unit area, or pressure, is equal to S/c. If the wave is totally reflected, the momentum change is twice as great, and the pressure is $2S/c$. For example, the value of S for direct sunlight is about $1.4 \text{ kW} \cdot \text{m}^{-2}$, and the corresponding pressure on a completely absorbing surface is

$$p = \frac{1.4 \times 10^3 \text{ W} \cdot \text{m}^{-2}}{3.0 \times 10^8 \text{ m} \cdot \text{s}^{-1}} = 4.7 \times 10^{-6} \text{ Pa}$$
$$= 4.7 \times 10^{-6} \text{ N} \cdot \text{m}^{-2}.$$

Radiation pressure is important in the structure of stars. Gravitational attractions tend to shrink a star, but this tendency is balanced by radiation pressure in maintaining the size of a star through most stages of its evolution.

*35–4

ELECTROMAGNETIC WAVES IN MATTER

The above analysis can be extended to include electromagnetic waves in dielectrics. The wave speed is not the same as in vacuum, and we denote it by w instead of c. Faraday's law is unaltered, but Eq. (35–2) is replaced by $E = wB$. In Ampère's law, the displacement current density is given not by $\epsilon_0 \, \Delta E/\Delta t$ but by $\epsilon \, \Delta E/\Delta t = K\epsilon_0 \, \Delta E/\Delta t$. In addition, the constant μ_0 in Ampère's law must be replaced by $\mu = K_m\mu_0$. Thus, Eq. (35–8) is replaced by

$$B = \mu\epsilon wE,$$

and the wave speed is given by

$$w = \frac{1}{\sqrt{\epsilon\mu}} = \frac{1}{\sqrt{KK_m}} \frac{1}{\sqrt{\epsilon_0\mu_0}}. \tag{35–20}$$

For many dielectrics the relative permeability K_m is very nearly equal to unity; in such cases we can say

$$w \approx \frac{1}{\sqrt{K}} \frac{1}{\sqrt{\epsilon_0\mu_0}} = \frac{c}{\sqrt{K}}.$$

Because K is always greater than unity, the speed of electromagnetic waves in a dielectric is always *less* than the speed in vacuum, by a factor of $1/\sqrt{K}$. The ratio of the speed in vacuum to the speed in a material is known in optics as the *index of refraction n* of the material; we see that this is given by

$$\frac{c}{w} = n = \sqrt{KK_m} \simeq \sqrt{K}. \tag{35–21}$$

Corresponding modifications are required in the expressions for the energy density and the Poynting vector; these are developed in detail in more advanced texts on electromagnetic theory.

The waves described above cannot propagate in a *conducting* material because the E field leads to currents that provide a mechanism for dissipating the energy of the wave. For an ideal conductor with zero resistivity, E must be zero everywhere inside the material. When an electromagnetic wave strikes such a material, it is totally reflected. Real conductors with finite resistivity permit some penetration of the wave into the material, with partial reflection. A polished metal surface is usually a good *reflector* of electromagnetic waves, but metals are not *transparent* to radiation.

35–5

SINUSOIDAL WAVES

Sinusoidal electromagnetic waves are closely analogous to sinusoidal transverse mechanical waves on a stretched string, as discussed in Chapter 21. At any point in space, the E and H fields are sinusoidal functions of time, and at any instant of time, the spatial variation of the fields is also sinusoidal.

The simplest sinusoidal electromagnetic waves share with the wave of Section 35–2 the property that any instant the fields are uniform over any plane perpendicular to the direction of propagation. The entire pattern travels to the right with speed c. Since E and H are at right angles to the direction of propagation, the wave is *transverse*.

The frequency f, the wavelength λ, and the speed of propagation c are related by the equation applicable to any sort of periodic wave motion, namely, $c = f\lambda$. If the frequency f is the power-line frequency of 60 Hz, the wavelength is

$$\lambda = \frac{c}{f} = \frac{3 \times 10^8 \text{ m} \cdot \text{s}^{-1}}{60 \text{ Hz}} = 5 \times 10^6 \text{ m} = 5000 \text{ km},$$

which is of the order of the earth's radius! Hence at this frequency even a distance of many miles includes only a small fraction of a wavelength. On the other hand, if the frequency is 10^8 Hz (100 MHz), typical of commercial FM radio stations, the wavelength is

$$\lambda = \frac{3 \times 10^8 \text{ m} \cdot \text{s}^{-1}}{10^8 \text{ Hz}} = 3 \text{ m},$$

and a moderate distance can include a number of complete waves.

Figure 35–4 represents a sinusoidal electromagnetic wave traveling in the +x-direction. The E and H vectors are shown only for a few points

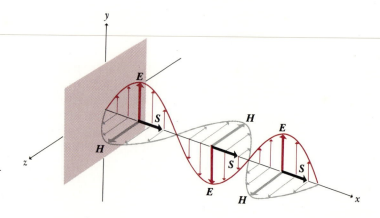

35–4 *E-, H-*, and *S*-vectors in a sinusoidal electromagnetic wave traveling in the x-direction.

on the x-axis; if a plane is constructed perpendicular to the x-axis at a particular point, the fields have the same values at all points in that plane. The values are of course different on different planes. In those planes in which the **E** vector is in the $+y$-direction, **H** is in the $+z$-direction; where **E** is in the $-y$-direction, **H** is in the $-z$-direction. In both cases the direction of the Poynting vector, given by Eq. (35–17), is along the $+x$-direction.

We recall from Section 21–7, Eq. (21–19) that one form of the equation of a transverse wave traveling to the right along a stretched string is $y = A \sin (\omega t - kx)$, where y is the transverse displacement from its equilibrium position, at the time t, of a point of the string whose coordinate is x. The quantity A is the maximum displacement or the *amplitude* of the wave, ω is its *angular frequency*, equal to 2π times the frequency f, and k is the *propagation constant*, equal to $2\pi/\lambda$, where λ is the wavelength.

Let E and H represent the instantaneous values, and E_{max} and H_{max} the maximum values, or *amplitudes*, of the electric and magnetic fields in Fig. 35–4. The equations of the traveling electromagnetic wave are then

$$E = E_{max} \sin (\omega t - kx), \qquad H = H_{max} \sin (\omega t - kx). \quad (35–22)$$

The sine curves in Fig. 35–4 represent instantaneous values of E and H, as functions of x at the time $t = 0$. The wave travels to the right with speed c.

The instantaneous value S of the Poynting vector is

$$S = EH = E_{max}H_{max} \sin^2 (\omega t - kx)$$
$$= \tfrac{1}{2}E_{max}H_{max}[1 - \cos 2(\omega t - kx)].$$

The time average value of $\cos 2(\omega t - kx)$ is zero, so the average value S_{av} of the Poynting vector is

$$S_{av} = \tfrac{1}{2}E_{max}H_{max}. \quad (35–23)$$

This is the *average power* transmitted per unit area, and, as noted in Section 35–3, it is called the *intensity* of the radiation.

Figure 35–5 represents schematically the electric and magnetic fields of a wave traveling in the *negative x*-direction. At points where **E** is in the positive y-direction, **H** is in the *negative z*-direction, and where **E** is in the

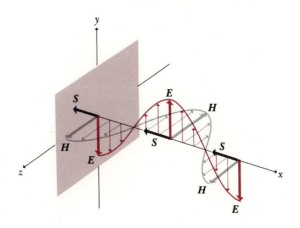

35–5 Electric and magnetic fields of a sinusoidal wave traveling in the negative x-direction.

negative y-direction, H is in the *positive* z-direction. The Poynting vector is in the negative x-direction at all points. (Compare with Fig. 35–4, which represents a wave traveling in the *positive* x-direction.) The equations of the wave are

$$E = -E_{max} \sin (\omega t + kx), \qquad H = H_{max} \sin (\omega t + kx). \quad (35\text{–}24)$$

*35–6

STANDING WAVES

Electromagnetic waves can be reflected, and the superposition of an incident wave and a reflected wave can form a *standing wave* analogous to standing waves on a stretched string, discussed in Section 22–2. A conducting surface can serve as a reflector for electromagnetic waves.

Suppose a sheet of an ideal conductor, having zero resistivity, is placed in the yz-plane of Fig. 35–5, and that the wave shown, traveling in the negative x-direction, is incident on it. The essential characteristic of an ideal conductor is that there can never be any electric field in it; any attempt to establish a field is immediately canceled by rearrangement of the mobile charges in the conductor. Thus E must always be zero everywhere in this plane, and the E field of the incident wave induces sinusoidal currents in the conductor so that E is zero inside it.

These induced currents then produce a reflected wave, traveling out from the plane, to the right. From the superposition principle, the total E field at any point to the right of the plane is the vector sum of the E fields of the incident and reflected waves; the same is true for the total H field.

Suppose the incident wave is described by Eqs. (35–24) and the reflected wave by Eqs. (35–22). Then the *total* fields at any point are given by

$$E = E_{max}[-\sin (\omega t + kx) + \sin (\omega t - kx)],$$
$$H = H_{max}[\sin (\omega t + kx) + \sin (\omega t - kx)].$$

These expressions may be expanded and simplified, using the identity

$$\sin (A \pm B) = \sin A \cos B \pm \cos A \sin B.$$

The results are

$$E = -2E_{max} \cos \omega t \sin kx,$$
$$H = 2H_{max} \sin \omega t \cos kx. \qquad (35\text{–}25)$$

The first of these is analogous to Eq. (22–1) for the stretched string. We see that at $x = 0$, E is *always* zero; this is required by the nature of the ideal conductor, which plays the same role as a fixed point at the end of the string. Furthermore, E is zero at all times in those planes for which $\sin kx = 0$, that is, $kx = 0, \pi, 2\pi, \ldots$, or

$$x = 0, \quad \frac{\lambda}{2}, \quad \lambda, \quad \frac{3\lambda}{2}, \quad \ldots.$$

These are called *nodal planes* of the E field.

The magnetic field is zero at all times in those planes for which $\cos kx = 0$, or at which

$$x = \frac{\lambda}{4}, \quad 3\frac{\lambda}{4}, \quad 5\frac{\lambda}{4}, \quad \text{etc.}$$

These are the nodal planes of the **H** field. The magnetic field is *not* zero at the conducting surface ($x = 0$), and there is no reason it should be. The nodal planes of one field are midway between those of the other, and the nodal planes of either field are separated by one-half wavelength. Figure 35–6 shows a standing-wave pattern at one instant of time.

The electric field is a *cosine* function of t and the magnetic field is a *sine* function of t. The fields are therefore 90° out of phase. At times when $\cos \omega t = 0$, the electric field is zero *everywhere* and the magnetic field is maximum. When $\sin \omega t = 0$, the magnetic field is zero *everywhere* and the electric field is maximum.

Pursuing the stretched-string analogy, we may now insert a second conducting plane, parallel to the first and a distance L from it, along the $+x$-axis. This is analogous to the stretched string held at the points $x = 0$ and $x = L$. A standing wave can exist only when L is an integer multiple of $\lambda/2$. Thus the possible wavelengths are

$$\lambda_n = \frac{2L}{n}, \qquad n = 1, 2, 3, \ldots, \tag{35–26}$$

and the corresponding frequencies are

$$f_n = \frac{c}{\lambda_n} = n\frac{c}{2L}, \qquad n = 1, 2, 3, \ldots. \tag{35–27}$$

Thus there is a set of *normal modes*, each with a characteristic frequency, wave shape, and node pattern. Measurement of the node positions makes it possible to measure the wavelength. If the frequency is known, the wave speed can be determined. This technique was in fact used by Hertz in his pioneering investigations of electromagnetic waves.

Conducting surfaces are not the only reflectors of electromagnetic waves; reflections also occur at an interface between two insulating mate-

35–6 *E*- and *H*-vectors in a standing wave. The pattern does not move along the x-axis, but the *E*- and *H*-vectors grow and diminish with time at each point. At each point *E* is maximum when *H* is minimum, and conversely.

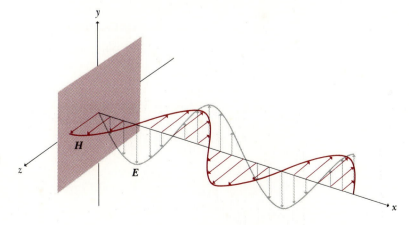

rials having different dielectric or magnetic properties. The mechanical analog is a junction of two strings with equal tension but different linear mass density. In general, a wave incident on such a boundary surface is partly transmitted into the second material and partly reflected back into the first. The partial transmission and reflection of light at a glass surface is a familiar phenomenon.

35–7

THE ELECTROMAGNETIC SPECTRUM

Electromagnetic waves cover an extremely broad spectrum of wavelength and frequency, as indicated in Fig. 35–7. Radio and TV transmission, visible light, infrared and ultraviolet radiation, and gamma rays all form parts of the electromagnetic spectrum. It is well established that light consists of electromagnetic waves; it is a small part of a very broad class of electromagnetic radiations, all having the general characteristics described in preceding sections, including the common speed of propagation $c = 3.00 \times 10^8$ m·s^{-1} but differing in frequency f and wavelength λ. The general wave relation $c = \lambda f$ holds for each.

The wavelengths of visible light (i.e., of electromagnetic waves that are perceived by the sense of sight) can be measured by methods to be discussed in Chapter 39 and are found to lie in the range 4 to 7×10^{-7}m (400 to 700 nm). The corresponding range of frequencies is about 7.5 to 4.3×10^{14} Hz.

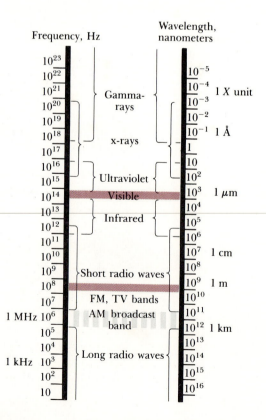

35–7   A chart of the electromagnetic spectrum.

Because of the very small magnitudes of light wavelengths, it is convenient to measure them in small units of length. Three such units are commonly used: the micrometer (1 μm), the nanometer (1 nm) (both accented on the *first* syllable), and the angstrom (1 Å):

$$1 \ \mu m = 10^{-6} \ m = 10^{-4} \ cm,$$

$$1 \ nm = 10^{-9} \ m = 10^{-7} \ cm,$$

$$1 \ Å = 10^{-10} \ m = 10^{-8} \ cm.$$

In older literature the micrometer is sometimes called the *micron,* and the nanometer is sometimes called the *millimicron;* these terms are now obsolete. Most workers in the fields of optical instrument design, color, and physiological optics express wavelengths in *nanometers.* For example, the wavelength of the yellow light from a sodium-vapor lamp is 589 nm, but many spectroscopists would identify this same wavelength as 5890 Å.

Different parts of the visible spectrum evoke the sensations of different colors. Wavelengths for colors in the visible spectrum are (very approximately) as follows:

400 to 450 nm	Violet
450 to 500 nm	Blue
500 to 550 nm	Green
550 to 600 nm	Yellow
600 to 650 nm	Orange
650 to 700 nm	Red

By the use of special sources or special filters, it is possible to limit the wavelength spread to a small band, say from 1 to 10 nm. Such light is called roughly *monochromatic light,* meaning light of a single color. Light consisting of only one wavelength is an idealization that is useful in theoretical calculations but represents an experimental impossibility. When the expression "monochromatic light of wavelength 550 nm" is used in theoretical discussions, it refers to one wavelength, but in descriptions of laboratory experiments it means a small band of wavelengths *around* 550 nm. One distinguishing characteristic of light from a *laser* is that it is much more nearly monochromatic than light obtainable in any other way.

*35–8

RADIATION FROM AN ANTENNA

The waves discussed above are called *plane waves,* referring to the fact that at any instant the fields are uniform over a plane perpendicular to the direction of propagation of the wave. Although these waves are the simplest to describe and analyze, they are by no means the simplest to produce experimentally. Any charge or current distribution that oscillates sinusoidally with time produces sinusoidal electromagnetic waves. The simplest example of such a source is an oscillating dipole, a pair of electric charges of equal magnitude and opposite sign, with the charge magnitude varying sinusoidally with time. Such an oscillating dipole can be constructed in a number of ways, the details of which need not concern us.

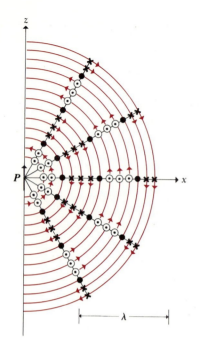

The radiation from an oscillating dipole is *not* a plane wave but travels out in all directions from the source. At points far from the source the *E* and *H* fields are perpendicular to the direction from the source and to each other; in this sense the wave is still transverse. The value of *S* drops off as the square of the distance from the source and also depends on the direction from the source; the radiation is most intense in directions perpendicular to the dipole axis, and $S = 0$ in directions parallel to the axis. The radiation pattern from a dipole source is shown schematically in Fig. 35–8.

Electromagnetic waves can be *reflected* by conducting surfaces; when the surface is large compared to the wavelength of the radiation, the reflection behaves like reflection of light rays from a mirror, as discussed in Chapter 37. Large parabolic mirrors several meters in diameter are used for both transmitting and receiving antennas for microwave communication signals; typical wavelengths are a few centimeters. The transmitting reflector produces a wave that radiates in a narrow, well-defined beam; the receiving reflector gathers wave energy over its whole area and reflects it to the focus of the parabola, where a detecting device is placed. Figure 35–9 shows an antenna installation using a large parabolic reflector.

35–8 Cross section in the *xz*-plane of radiation from an oscillating electric dipole, showing electric-field vectors at one instant of time. At the points with circles the **B**-field comes out of the plane of the figure; at the points with crosses, it is into the plane.

35–9 A microwave receiving antenna 64 m in diameter at Goldstone Tracking Station, California, the prime station of a worldwide network of stations used by NASA to monitor unmanned interplanetary spacecraft.

QUESTIONS

35–1 In Ampère's law, is it possible to have both a conduction current and a displacement current at the same time? Is it possible for the effects of the two kinds of current to cancel each other exactly, so that there is *no* magnetic field produced?

35–2 By measuring the electric and magnetic fields at a point in space where there is an electromagnetic wave, can one determine the direction from which the wave came?

35–3 Sometimes neon signs located near a powerful radio station are seen to glow faintly at night, even though they aren't turned on. What is happening?

35–4 Light can be *polarized;* is this a property of all electromagnetic waves, or is it unique to light? What about polarization of sound waves? What fundamental distinction in wave properties is involved?

35–5 How does a microwave oven work? Why does it heat materials that are conductors of electricity, including most foods, but not insulators, such as glass or ceramic dishes? Why do the directions forbid putting anything metallic in the oven?

35–6 Electromagnetic waves can travel through vacuum, where there is no matter. We usually think of vacuum as empty space, but is it *really* empty if electric and magnetic fields are present? What *is* vacuum, anyway?

35–7 Give several examples of electromagnetic waves that are encountered in everyday life. How are they all alike? How do they differ?

35–8 We are surrounded by electromagnetic waves emitted by many radio and television stations. How is a radio or television receiver able to select a single station among all this mishmash of waves? What happens inside a radio receiver when the dial is turned to change stations?

35–9 The metal conducting rods on a television antenna are always in a horizontal plane. Would they work as well if they were vertical?

35–10 If a light beam carries momentum, should a person holding a flashlight feel a recoil analogous to the recoil

of a rifle when it is fired? Why is this recoil not actually observed?

35–11 The nineteenth-century inventor Nikolai Tesla proposed to transmit large quantities of electrical energy across space using electromagnetic waves instead of conventional transmission lines. What advantages would this scheme have? What disadvantages?

35–12 Does an electromagnetic *standing wave* have energy? Momentum? What distinction can be drawn between a standing wave and a propagating wave on this basis?

35–13 If light is an electromagnetic wave, what is its frequency? Is this a proper question to ask?

35–14 When an electromagnetic wave is reflected from a moving reflector, the frequency of the reflected wave is different from that of the initial wave. Explain physically how this can happen. (Some radar systems used for highway speed control operate on this principle.)

35–15 The ionosphere is a layer of ionized air 100 km or so above the earth's surface. It acts as a reflector of radio waves of frequency less than about 30 MHz, but not of higher frequency. How does this reflection occur? Why does it work better for lower frequencies than for higher?

PROBLEMS

35–1 A certain radio station broadcasts at a frequency of 1020 kHz. At a point some distance from the transmitter, the maximum magnetic field of the electromagnetic wave it emits is found to be 1.6×10^{-11} T.

  a) What is the speed of propagation of the wave?

  b) What is the wavelength?

  c) What is the maximum electric field?

35–2 A certain plane electromagnetic wave emitted by a microwave antenna has a wavelength of 3.0 cm and a maximum magnitude of electric field 2.0×10^{-4} V·m^{-1}.

  a) What is the frequency of the wave?

  b) What is the maximum magnetic field?

  c) What is the intensity (power per unit area) if the *average* power is half the power computed from the *maximum* values of E and B?

35–3 An electromagnetic wave propagates in a ferrite material having $K = 10$ and $K_m = 1000$. Find

  a) the speed of propagation;

  b) the wavelength of a wave having a frequency of 100 MHz.

35–4 The maximum electric field in the vicinity of a certain radio transmitter is 1.0×10^{-3} V·m^{-1}. What is the

maximum magnitude of the B field? How does this compare in magnitude with the earth's field?

35–5 The energy flow to the earth associated with sunlight is about 1.4 kW·m^{-2}.

  a) Find the maximum values of E and B for a wave of this intensity.

  b) The distance from the earth to the sun is about 1.5×10^{11} m. Find the total power radiated by the sun.

35–6 For a 50,000-W radio station, find the maximum magnitudes of E and B at a distance of 100 km from the antenna, assuming that the antenna radiates equally in all directions (which is probably not actually the case).

35–7 The nineteenth-century inventor Nikolai Tesla proposed to transmit electric power via electromagnetic waves. Suppose power is to be transmitted in a beam of cross-sectional area 100 m^2; what E and B strengths are required to transmit an amount of power comparable to that handled by modern transmission lines (of the order of 500 kV and 1000 A)?

35–8 A bar magnet is mounted on an insulating support, as in the side and end views of Fig. 35–10(a) and (b), and is given a positive electric charge. Copy the diagram, sketch

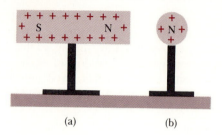

(a)   (b)

Figure 35–10

the lines of the E and H fields around the magnet, and draw the Poynting vector at a number of points.

a) What is the general nature of the energy flow in the electromagnetic field?

b) What can you say about the electromagnetic-field momentum in this case?

35–9 A cylindrical conductor of circular cross section has a radius a and a resistivity ρ and carries a constant current I.

a) What are the magnitude and direction of the E-vector at a point inside the wire, at a distance r from the axis?

b) What are the magnitude and direction of the H-vector at the same point?

c) What are the magnitude and direction of the Poynting vector S at the same point?

d) Compare your answer to (c) with the rate of dissipation of energy within a volume of the conductor of length l and radius r.

35–10 A wire of radius 1 mm whose resistance per unit length is 3×10^{-3} $\Omega \cdot m^{-1}$ carries a current of 25.1 A. At a point very near the surface of the wire, calculate

a) the magnitude of H,

b) the component of E parallel to the wire,

c) the component of S perpendicular to the wire.

35–11 Assume that 10% of the power input to a 100-W lamp is radiated uniformly as light of wavelength 500 nm (1 nm $= 10^{-9}$ m). At a distance of 2 m from the source, the electric and magnetic intensities vary sinusoidally according to the equations $E = E_{max} \sin{(\omega t + \phi)}$ and $H = H_{max} \sin{(\omega t + \phi)}$. Calculate E_{max} and H_{max}.

35–12 A capacitor consists of two circular plates of radius r separated by a distance l. Neglecting fringing, show that while the capacitor is being charged, the rate at which energy flows into the space between the plates is equal to the rate at which the electrostatic energy increases.

35–13 A very long solenoid of n turns per unit length and radius a carries an increasing current i.

a) Calculate the induced electric field at a point inside the solenoid at a distance r from the solenoid axis.

b) Compute the magnitude and direction of the Poynting vector at this point.

35–14 An FM radio station antenna radiates a power of 10 kW at a wavelength of 3 m. Assume for simplicity that the radiated power is confined to, and is uniform over, a hemisphere with the antenna at its center. What are the amplitudes of E and H in the radiation field at a distance of 10 km from the antenna?

35–15 In a TV picture, ghost images are formed when the signal from the transmitter travels directly to the receiver and also indirectly after reflection from a building or other large metallic mass. In a 25-inch set the ghost is about 1 cm to the right of the principal image if the reflected signal arrives 1 μs after the principal signal. In this case, what is the difference in path length for the two signals?

35–16 For a sinusoidal electromagnetic wave in vacuum, such as that described by Eqs. (35–22), show that the average density of energy in the electric field is the same as that in the magnetic field.

35–17 A plane electromagnetic wave has a wavelength of 3.0 cm and an E-field amplitude of 30 V $\cdot$ m^{-1}.

a) What is the frequency?

b) What is the B-field amplitude?

c) What is the intensity?

d) What average force does the radiation pressure exert on a totally absorbing surface of area 0.5 m^2 perpendicular to the direction of propagation?

35–18 If the intensity of direct sunlight is 1.4 kW $\cdot$ m^{-2}, find the radiation pressure (in pascals) on

a) a totally absorbing surface;

b) a totally reflecting surface.

Also express your results in atmospheres.

CHAPTER 36

THE NATURE AND PROPAGATION OF LIGHT

Light is electromagnetic radiation of very short wavelength and is thus a *wave* phenomenon. Many aspects of the propagation of light can be described more simply by a *ray* model; rays travel in straight lines in homogeneous materials but can be reflected and also bent or *refracted* at interfaces between materials. In this chapter we study several basic phenomena associated with reflection and refraction of light. Like all transverse waves, light waves are *polarized;* we examine several aspects of polarization phenomena.

36–1

NATURE OF LIGHT

Until the time of Newton (1642–1727), most scientists (including Newton) thought that light consisted of streams of some sort of particles or *corpuscles* emanating from light sources. But at about this time Huygens and others suggested that light might be a *wave* phenomenon. Indeed, diffraction effects that are now recognized as wave phenomena were observed by Grimaldi as early as 1665, but their significance was not understood then. However, by the early nineteenth century evidence that light is a wave phenomenon grew more persuasive. The experiments of Fresnel, Thomas Young, and others revealed many phenomena that can be understood on the basis of a wave picture but not on the corpuscular model. These phenomena will be discussed in detail in Chapter 39.

The next great step was taken in 1873 by Maxwell, who predicted the existence of electromagnetic waves and calculated their speed of propagation, as discussed in Chapter 35. In 1887 Hertz succeeded in producing short-wavelength electromagnetic waves and showing that they have all the properties of light waves. They could be reflected, refracted, focused by a lens, polarized, and so on. Thus the evidence grew more and more conclusive that light is indeed an electromagnetic-wave phenomenon.

Successful as the wave picture of light is, it is not the whole story. There are several phenomena associated with emission and absorption of light that reveal a particle aspect, in the sense that the energy carried by light waves is packaged in discrete bundles called *photons* or *quanta*. Some of these phenomena will be explored in Chapter 41. These apparently contradictory wave and particle properties have been reconciled only since 1930 with the development of quantum electrodynamics, a comprehensive theory that includes *both* wave and particle properties. *Propagation* of light is best handled by a wave model, but emission and absorption phenomena require a particle approach.

36–2

SOURCES OF LIGHT

The fundamental sources of all electromagnetic radiation are electric charges in motion. All bodies emit electromagnetic radiation as a result of thermal motion of their molecules; this radiation, called *thermal radiation*, is a mixture of different wavelengths. At a temperature of 300°C the most intense of these waves has a wavelength of 5000×10^{-9} m or 5000 nm, which is the *infrared* region. At a temperature of 800°C a body emits enough visible radiant energy to be self-luminous and appears "red hot." By far the larger part of the energy emitted, however, is still carried by infrared waves. At 3000°C, which is about the temperature of an incandescent lamp filament, the radiant energy contains enough of the "visible" wavelengths, between 400 nm and 700 nm, that the body appears "white hot." In modern incandescent lamps, the filament is a coil of fine tungsten wire. An inert gas such as argon is introduced to reduce evaporation of the filament. Incandescent lamps vary in size from one no larger than a grain of wheat to one with a power input of 5000 W, used for illuminating airfields.

One of the brightest sources of light is the *carbon arc*. Two carbon rods, typically 10 cm to 20 cm long and 1 cm in diameter, are connected to a 110-V or 220-V dc source. They are touched together momentarily and then are pulled apart a few millimeters. Intense electron bombardment of the positive rod causes an extremely hot crater to form at its end; this crater, whose temperature is typically 4000°C, is the source of light. Carbon-arc lights are used in most theater motion-picture projectors and in large searchlights and lighthouses.

Some light sources use an arc discharge in a metal vapor, such as mercury or sodium, in a sealed bulb containing two electrodes, which are connected to a power source. Some argon is sometimes added to permit a glow discharge that helps vaporize and ionize the metal. The bluish light of mercury-arc lamps and the bright orange-yellow of sodium-vapor lamps are familiar in highway and other outdoor lighting.

An important variation of the mercury-arc lamp is the *fluorescent* lamp. This consists of a glass tube containing argon and a droplet of mercury. The electrodes are of tungsten. When an electric discharge takes place in the mercury–argon mixture, only a small amount of visible light is emitted by the mercury and argon atoms. There is, however, considerable *ultraviolet* light (light of wavelength shorter than that of visible violet). This ultraviolet light is absorbed in a thin layer of material, called a *phosphor*, which is the white coating on the interior walls of the glass tube. The phosphor has the property of *fluorescence*, which means that it emits visible light when illuminated by light of shorter wavelength. Lamps may be obtained that will fluoresce with any desired color, depending on the nature of the phosphor.

A special light source that has attained prominence in the last twenty years is the *laser*, which can produce a very narrow beam of enormously intense radiation. High-intensity lasers have been used to cut through steel, fuse high-melting-point materials, and bring about many other effects that are important in physics, chemistry, biology, and engineering. An equally significant characteristic of laser light is that it is much more nearly *monochromatic*, or single-frequency, than any other light source. Operation of one type of laser will be discussed in Chapter 41.

36–3

THE SPEED OF LIGHT

The speed of light in empty space is a fundamental constant of nature. Its magnitude is so large (3.00×10^8 m·s^{-1}) that in ancient times light was thought to travel with infinite speed. The first successful observation of the finite speed of light was made in 1676 by the Danish astronomer Olaf Roemer, who measured the period of revolution of one of Jupiter's satellites. The period (about 42 hr) appeared longer when the earth was moving in its orbit *away from* Jupiter than when it was moving *toward* Jupiter. Roemer correctly concluded that the difference was due to the displacement of the earth during one revolution of the satellite. When the earth is moving away from Jupiter, light leaving the satellite at the end of a particular revolution has to travel farther than light leaving at the beginning of the same revolution. In that case the apparent time for one revolution, measured on earth, is a few seconds longer than the actual time. Six months later, when the earth is moving toward Jupiter, the situation is reversed, and the apparent period is a few seconds shorter than the actual value.

The first successful determination of the speed of light from purely *terrestrial* measurements was made by the French scientist Fizeau in 1849. A schematic diagram of his apparatus is given in Fig. 36–1. Lens L_1

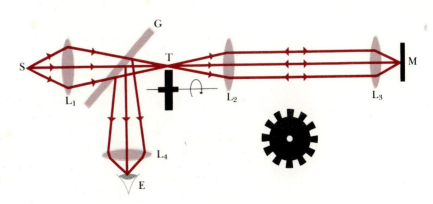

36–1 Fizeau's toothed-wheel method for measuring the velocity of light. S is a light source; L_1, L_2, L_3, and L_4 are lenses. T is the toothed wheel, M is a mirror, and G is a glass plate.

forms an image of the light source S at a point near the rim of a toothed wheel T, which can be rotated at high speed. G is an inclined plate of clear glass. Suppose first that the wheel is stationary and the light passes through one of the openings between the teeth. Lenses L_2 and L_3, which are separated by about 8.6 km, form a second image on the mirror M. The light is reflected from M, retraces its path, and is in part reflected from the glass plate G through the lens L_4 into the eye of an observer at E.

If the wheel T is set in rotation, the light from S is "chopped up" into a succession of wave pulses of limited length. A pulse travels through an opening between two teeth of the wheel, is reflected by the mirror M, and returns to the wheel. At a certain speed of rotation, the wheel will have turned just enough during this time so the return path is blocked by a tooth of the wheel. In this case, *no* reflected light is seen by the observer at E. At twice this angular velocity, the light transmitted through any one opening will return through the next, and an image of S will again be observed. From a knowledge of the angular velocity and radius of the wheel, the distance between openings, and the distance from wheel to mirror, the speed of light may be computed. Fizeau obtained the value 3.15×10^8 m·s^{-1}.

Fizeau's apparatus was modified by Foucault, who replaced the toothed wheel with a rotating mirror. The most precise measurements by the Foucault method were made by the American physicist Albert A. Michelson (1852–1931). His first experiments were performed in 1878; the last, under way at the time of his death, were completed in 1935 by Pease and Pearson.

From analysis of all measurements up to 1983, the most probable value for the speed of light is

$$c = 2.99792458 \times 10^8 \text{ m·s}^{-1}.$$

This number is based on the definition of the meter in terms of the krypton wavelength and the definition of the second in terms of the cesium clock, as described in Section 1–2. The definition of the second is precise to within one part in 10 trillion (10^{13}), whereas the definition of the meter is much less precise, about four parts in a billion (10^9). Thus it has become advantageous to redefine the meter in terms of the unit of time by *defining* the speed of light to have a specific value and then defining the meter in terms of the distance traveled by light in one second.

Accordingly, in November 1983 the General Conference on Weights and Measures *defined* the speed of light in vacuum to be precisely 299,792,458 m·s^{-1} and defined one meter to be the distance traveled by light in a time of 1/299,792,458 s, with the second defined by the cesium clock.

As shown in Chapter 35, the speed of any electromagnetic wave in empty space is given by

$$c = \frac{1}{\sqrt{\mu_0 \epsilon_0}}.$$

In SI units, μ_0 is assigned the value of precisely $4\pi \times 10^{-7}$ N·s^2·C^{-2}.

Thus the preceding equation provides a precise means of evaluating the electrical constant ϵ_0:

$$\epsilon_0 = \frac{1}{\mu_0 c^2} = 8.85418782 \times 10^{-12}\ \mathrm{C^2 \cdot N^{-1} \cdot m^{-2}}.$$

This value has much greater precision than could be obtained from direct measurements of forces between electric charges.

36–4

WAVES, WAVE FRONTS, AND RAYS

It is convenient to represent a wave of any sort by means of *wave fronts. A wave front is defined as the locus of all points at which the phase of vibration of a physical quantity is the same.* A familiar example is a crest of a water wave. In the case of sound waves spreading out in all directions from a point source, any spherical surface concentric with the source is a possible wave front. Some of these spherical surfaces are the loci of points at which the pressure is a *maximum*, others where it is a *minimum*, and so on; but the *phase* of the pressure variations is the same over any spherical surface. It is customary to draw only a few wave fronts, usually those that pass through the maxima and minima of the disturbance. Such wave fronts are separated from one another by one-half wavelength.

If the wave is a light wave, the quantity corresponding to the pressure in a sound wave is the electric or magnetic field. It is usually not necessary to indicate in a diagram either the magnitude or direction of the field; instead we simply show the *shapes* of the wave fronts or their intersections with some reference plane. For example, the electromagnetic waves radiated by a small light source may be represented by *spherical* surfaces concentric with the source or, as in Fig. 36–2a, by the intersections of these surfaces with the plane of the diagram. At a sufficiently great distance from the source, where the radii of the spheres have become very large, the spherical surfaces can be considered planes and we have a *plane* wave, as in Fig. 36–2b.

For some purposes, especially in that branch of optics called *geometrical* optics, it is convenient to represent a light wave by *rays* rather than by wave fronts. Indeed, rays were used to describe light long before its wave nature was firmly established and, in a corpuscular theory of light, rays are merely the paths of the corpuscles. From the wave viewpoint, *a ray is an imaginary line drawn in the direction in which the wave is traveling.* Thus, in Fig. 36–2a the rays are the radii of the spherical wave fronts, and in Fig. 36–2b they are straight lines perpendicular to the wave fronts. In fact, in every case where waves travel in a homogeneous isotropic material, the rays are straight lines normal to the wave fronts. At a boundary surface between two materials, such as the surface between a glass plate and the air outside it, the direction of a ray may change suddenly, but it is a straight line both in the air and in the glass. If the material is *not* homogeneous, for instance, if one is considering the passage of light through the earth's atmosphere, where the density and hence the velocity vary with elevation, the rays are curved but are still normal to the wave fronts.

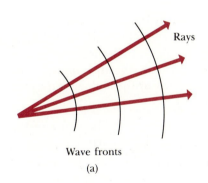

Rays

Wave fronts

(a)

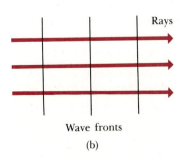

Rays

Wave fronts

(b)

36–2 Wave fronts and rays.

36–5

REFLECTION AND REFRACTION

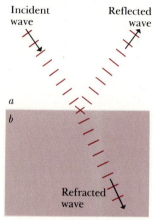

(a)

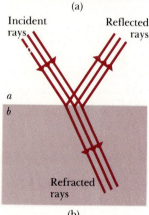

(b)

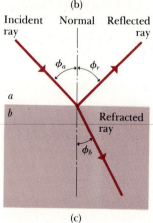

(c)

36–3 (a) A plane wave is in part reflected and in part refracted at the boundary between two media. (b) The waves in (a) are represented by rays. (c) For simplicity, only one example of incident, reflected, and refracted rays is drawn.

Many familiar optical phenomena involve the behavior of a wave that strikes an interface between two optical materials, such as air and glass or water and glass. When the interface is smooth, i.e., when its irregularities are small compared with the wavelength, the wave is in general partly reflected and partly transmitted into the second material, as shown in Fig. 36–3a. For example, when one looks into a store window from the street, one sees light coming from inside by transmission through the glass and also a reflection of the street scene caused by the air–glass interface.

The segments of plane waves shown in Fig. 36–3a can be represented by bundles of rays forming *beams* of light, as in Fig. 36–3b; and for simplicity in discussing the various angles we often consider only one ray in each beam, as in Fig. 36–3c. Representing these waves in terms of rays is the basis of that branch of optics called *geometrical optics*. This chapter and the two following chapters are devoted entirely to geometrical optics and are concerned with optical phenomena that can be understood on the basis of rays, without having to use the wave nature of light explicitly. It must be understood that geometrical optics, like any idealized model, has its limitations, and that there are optical phenomena that require a more detailed model embodying the *wave* properties of light for their understanding. We return to such phenomena in Chapter 39.

The directions of the incident, reflected, and refracted beams of light are specified in terms of the angles they make with the *normal* to the surface at the point of incidence. For this purpose it is sufficient to indicate one ray, as in Fig. 36–3c, although a single ray of light is a geometrical abstraction. A careful experimental study of the directions of the incident, reflected, and refracted beams leads to the following results:

1. *The incident, reflected, and refracted rays, and the normal to the surface, all lie in the same plane.* Thus, if the incident ray is in the plane of the diagram, and the surface of separation is perpendicular to this plane, the reflected and refracted rays are in the plane of the diagram.

2. *The angle of reflection ϕ_r is equal to the angle of incidence ϕ_a for all colors and any pair of substances.* Thus

$$\phi_r = \phi_a. \tag{36–1}$$

The experimental result that $\phi_r = \phi_a$, and that the incident and reflected rays and the normal all lie in the same plane, is known as the *law of reflection.*

3. For monochromatic light and for a given pair of substances, *a* and *b*, on opposite sides of the surface of separation, *the ratio of the sine of the angle ϕ_a (between the ray in substance a and the normal) and the sine of the angle ϕ_b (between the ray in substance b and the normal) is a constant.* Thus

$$\frac{\sin \phi_a}{\sin \phi_b} = \text{constant.} \tag{36–2}$$

This experimental result, together with the fact that the incident and refracted rays and the normal to the surface all lie in the same plane, is known as the *law of refraction*. The discovery of this law is usually credited to Willebrord Snell (1591–1626), although there appears to be some doubt that it was actually original with him. It is called *Snell's law*.

The laws of reflection and refraction relate only to the *directions* of the corresponding rays but say nothing about an equally important question, namely, the *intensities* of the reflected and refracted rays. These depend on the angle of incidence; for the present we simply state that the fraction reflected is smallest at *normal* incidence, where it is a few percent, and that it increases with increasing angle of incidence to almost 100% at grazing incidence or when $\phi_a = 90°$.

When a ray of light is directed from *below* the surface in Fig. 36–3, there are again reflected and refracted rays; these, together with the incident ray and the normal, all lie in the same plane. The same law of reflection applies as when the ray is originally traveling in air, and the same law of refraction, Eq. (36–2). *The passage of a ray of light in going from one medium to another is reversible.* It follows the same path in going from b to a as when going from a to b.

Let us now consider a beam of monochromatic light traveling *in vacuum*, making an angle of incidence ϕ_0 with the normal to the surface of a substance a, and let ϕ_a be the angle of refraction in the substance. The constant in Snell's law is then called the *index of refraction* of substance a and is designated by n_a:

$$\frac{\sin \phi_0}{\sin \phi_a} = n_a. \tag{36–3}$$

The index of refraction (also called *refractive index*) depends not only on the substance but on the wavelength of the light. If no wavelength is stated, the index is often assumed to be that corresponding to the yellow light from a sodium lamp, of wavelength 589 nm. This wavelength is near the middle of the visible spectrum, and sodium lamps are simple, inexpensive, and easy to use.

The index of refraction of most of the common glasses used in optical instruments lies between 1.46 and 1.96. There are very few substances having indices larger than this value, diamond being one, with an index of 2.42, and rutile (crystalline titanium dioxide) another, with an index of 2.7. The values for a number of solids and liquids are given in Table 36–1.

The index of refraction of *air* at standard conditions is about 1.0003; for most purposes the *index of refraction of air can be assumed to be unity*. The index of refraction of a gas increases uniformly as the density of the gas increases.

It follows from Eq. (36–3) that the angle of refraction ϕ_a is always *less* than the angle of incidence ϕ_0 for a ray passing from vacuum into one of the materials listed in Table 36–1, where all indices are seen to be *greater* than unity. In such a case, the ray is bent *toward* the normal. If the light is traveling in the opposite direction, the reverse is true and the ray is bent *away from* the normal.

To relate the index of refraction to the constant in Eq. (36–2), we now consider two parallel-sided plates of substances a and b placed paral-

TABLE 36–1 INDEX OF REFRACTION FOR YELLOW SODIUM LIGHT (λ = 589 NM)

Substance	Index of refraction
Solids	
Ice (H_2O)	1.309
Fluorite (CaF_2)	1.434
Rock salt (NaCl)	1.544
Quartz (SiO_2)	1.544
Zircon ($ZrO_2 \cdot SiO_2$)	1.923
Diamond (C)	2.417
Fabulite ($SrTiO_3$)	2.409
Glasses (typical values)	
Crown	1.52
Light flint	1.58
Medium flint	1.62
Dense flint	1.66
Lanthanum flint	1.80
Liquids at 20°C	
Methyl alcohol (CH_3OH)	1.329
Water (H_2O)	1.333
Ethyl alcohol (C_2H_5OH)	1.36
Carbon tetrachloride (CCl_4)	1.460
Turpentine	1.472
Glycerine	1.473
Benzene	1.501
Carbon disulfide (CS_2)	1.628

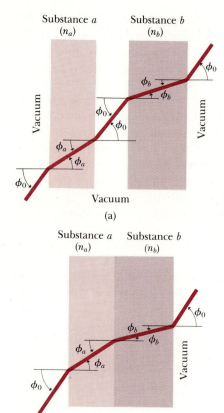

(a)

(b)

36–4 The transmission of light through parallel plates of different substances. The incident and emerging rays are parallel, regardless of direction and regardless of the thickness of the space between adjacent slabs.

lel to each other with an arbitrary space between them, as shown in Fig. 36–4a. Let the medium surrounding both plates be vacuum, although the behavior of light would be practically the same if the plates were surrounded by air. If a ray of monochromatic light starts at the lower left with an angle of incidence ϕ_0, the angle between ray and normal in substance a is ϕ_a, and the light emerges from substance a at an angle ϕ_0 equal to its incident angle. The light ray therefore enters plate b with an angle of incidence ϕ_0, makes an angle ϕ_b in substance b, and emerges again at an angle ϕ_0. Exactly the same path would be traversed if the same light ray were to start at the upper right and enter substance b at an angle ϕ_0. Moreover, *the angles are independent of the thickness of the space between the two plates*, and are the same when the space shrinks to nothing, as in Fig. 36–4b.

Applying Snell's law to the refractions at the surface between vacuum and substance a, and at the surface between vacuum and substance b, we have

$$\frac{\sin \phi_0}{\sin \phi_a} = n_a, \qquad \frac{\sin \phi_0}{\sin \phi_b} = n_b.$$

Dividing the second equation by the first, we obtain

$$\frac{\sin \phi_a}{\sin \phi_b} = \frac{n_b}{n_a}, \tag{36–4}$$

which shows that *the constant in Snell's law for the refraction between substances a and b is the ratio of the indices of refraction*. From Eq. (36–4) we see that the simplest and most symmetrical way of writing Snell's law for any two substances a and b, and for any direction, is

$$n_a \sin \phi_a = n_b \sin \phi_b. \tag{36–5}$$

EXAMPLE In Fig. 36–3, let material a be water and material b a glass with index of refraction 1.50. If the incident ray makes an angle of 60° with the normal, find the directions of the reflected and refracted rays.

Solution The reflected-ray angle with the normal is the same as that of the incident ray, according to Eq. (36–1). Hence $\phi_r = \phi_a = 60°$. To find the direction of the refracted ray we use Eq. (36–5), with $n_a = 1.33$, $n_b = 1.50$, and $\phi_a = 60°$. We find

$$(1.33)(\sin 60°) = (1.50)(\sin \phi_b),$$

$$\phi_b = 50.2°.$$

The second material has larger refractive index than the first, and the refracted ray is bent toward the normal. As Eq. (36–5) shows, this is always the case when the second index is larger than the first. In the opposite case, where the second index is *smaller* than the first, the ray is always bent *away from* the normal. ◀

An additional significance of the index of refraction, to be developed in Section 36–13, is the following. Light always travels more slowly in a material than in vacuum, and the ratio of the two speeds is equal to the index of refraction. Thus the speed v of light in a material having index of refraction n is given by

$$v = \frac{c}{n}, \quad \text{or} \quad n = \frac{c}{v}. \tag{36–6}$$

The wavelength of a light wave in a material is also reduced by a factor of n, while the frequency does not change. The nature of the interaction of light with matter consists of absorption of radiation by charged particles in the material and re-radiation *with the same frequency*. Thus, since $v = \lambda f$, when v is less than in vacuum, λ is also correspondingly reduced. Thus the wavelength λ of light in a material is less than the wavelength λ_0 of the same light in vacuum, by a factor n:

$$\lambda = \frac{\lambda_0}{n}. \tag{36–7}$$

Two final comments about reflection and refraction need to be added. First, reflection occurs at a highly polished surface of an *opaque* material such as a metal. There is then no refracted ray, but the reflected ray still behaves according to Eq. (36–1). Second, if the reflecting surface of either a transparent or an opaque material is rough, with irregularities on a scale comparable to or larger than the wavelength of light, reflection occurs not in a single direction but in all directions; such reflection is called *diffuse* reflection. Conversely, reflections in a single direction from smooth surfaces are called *regular* reflections or *specular* reflections.

36–6

TOTAL INTERNAL REFLECTION

Figure 36–5a shows a number of rays diverging from a point source P in medium a of index n_a and striking the surface of a second medium b of index n_b, where $n_a > n_b$. From Snell's law,

$$\sin \phi_b = \frac{n_a}{n_b} \sin \phi_a.$$

Since n_a/n_b is greater than unity, $\sin \phi_b$ is larger than $\sin \phi_a$. Thus there must be some value of ϕ_a less than 90° for which $\sin \phi_b = 1$ and $\phi_b = 90°$. This is illustrated by ray 3 in the diagram, which emerges just grazing the surface at an angle of refraction of 90°. The angle of incidence for which the refracted ray emerges tangent to the surface is called the *critical angle* and is designated by ϕ_{crit} in the diagram. If the angle of incidence is greater than the critical angle, the sine of the angle of refraction, as computed by Snell's law, would have to be greater than unity. This may be interpreted to mean that beyond the critical angle the ray *does not pass* into the upper medium but is *totally internally reflected* at the boundary surface. Total internal reflection can occur only when a ray is incident on the surface of a medium whose index is *smaller* than that of the medium in which the ray is traveling.

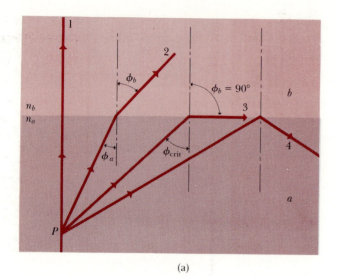

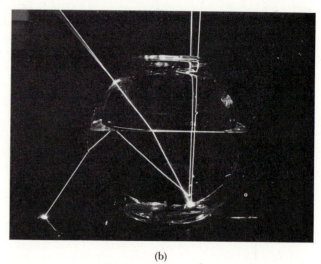

(a)   (b)

36–5   (a) Total internal reflection. The angle of incidence ϕ_a for which the angle of refraction is 90°, is called the critical angle. (b) Rays of laser light enter the water in the
fishbowl from above; they are reflected at the bottom by mirrors tilted at slightly different
angles, and one ray undergoes total internal reflection at the air–water interface. (The
Exploratorium.)

The critical angle for two given substances may be found by setting
$\phi_b = 90°$ or $\sin \phi_b = 1$ in Snell's law. We then have

$$\sin \phi_{\text{crit}} = \frac{n_b}{n_a}. \tag{36–8}$$

The critical angle of a glass–air surface, taking 1.50 as a typical index
of refraction of glass, is

$$\sin \phi_{\text{crit}} = \frac{1}{1.50} = 0.67, \qquad \phi_{\text{crit}} = 42°.$$

The fact that this angle is slightly less than 45° makes it possible to
use a prism of angles 45°–45°–90° as a totally reflecting surface. The
advantages of totally reflecting prisms over metallic surfaces as reflectors
are, first, that the light is *totally* reflected, while no metallic surface reflects 100% of the light incident on it, and second, the reflecting properties are permanent and not affected by tarnishing. Offsetting these is the
fact that there is some loss of light by reflection at the surfaces where
light enters and leaves the prism, although coating the surfaces with
so-called nonreflecting films can reduce this loss considerably.

A 45°–45°–90° prism, used as in Fig. 36–6a, is called a *Porro* prism.
Light enters and leaves at right angles to the hypotenuse and is totally
reflected at each of the shorter faces. The deviation is 180°. Two Porro
prisms are sometimes combined, as in Fig. 36–6b, an arrangement often
found in binoculars.

If a beam of light enters one end of a transparent rod, as in Fig.
36–7, the light is totally reflected internally and is "trapped" within the
rod even if it is curved, provided the curvature is not too great. The rod
is sometimes referred to as a *light pipe*. A bundle of fine fibers will behave

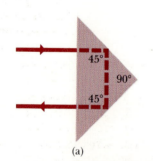

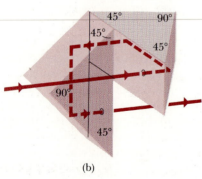

36–6   (a) A Porro prism. (b) A combination of two Porro prisms.

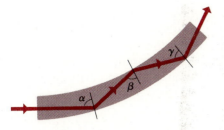

36–7 A light ray "trapped" by internal reflections.

in the same way and has the advantage of being flexible. A bundle may consist of thousands of individual fibers, each of the order of 0.002 mm to 0.01 mm in diameter. If the fibers are assembled in the bundle so that the relative positions of the ends are the same (or mirror images) at both ends, the bundle can transmit an image, as shown in Fig. 36–8.

Fiber-optic devices have found a wide range of applications in medicine in instruments called *endoscopes*, which can be inserted directly into the bronchial tubes, the urinary bladder, the colon, and so on, for direct visual examination. A bundle can be enclosed in a hypodermic needle for study of tissues and blood vessels far beneath the skin.

Fiber optics are now also finding applications in communication systems, where they are used to transmit a modulated laser beam. Because

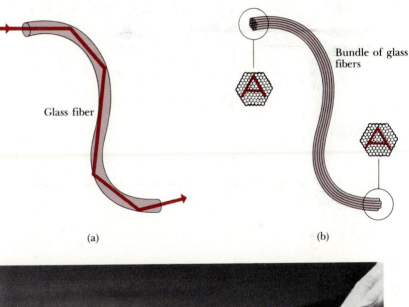

Glass fiber

Bundle of glass fibers

(a)

(b)

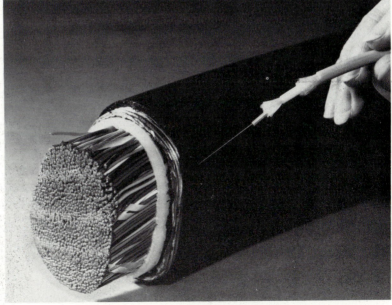

36–8 (a) Total internal reflection in a single fiber. (b) Image transmission by a bundle of fibers. (c) A fiber-optic cable used to transmit a modulated laser beam for communication purposes compared to a larger copper cable that has equal information-transmitting capacity. (Courtesy of Corning Glass Works.)

(c)

the frequency is very much higher than those used in wire or radio communication, an enormous amount of information can be transmitted through one fiber-optic cable. For example, the Carnegie-Mellon University computer system, consisting of several thousand microcomputers networked with mainframe computers, will be linked partly by fiber-optic cables as shown in Fig. 36–8c. In the future, most telephone systems are likely to be connected by fiber optics.

36–7

DISPERSION

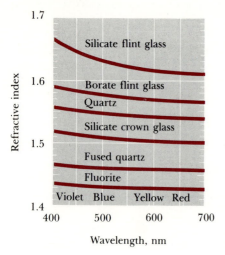

36–9 Variation of index with wavelength.

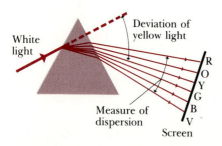

36–10 Dispersion by a prism. The band of colors on the screen is called a spectrum.

Most light beams are a mixture of waves whose wavelengths extend throughout the visible spectrum. While the speed of light waves in a vacuum is the same for all wavelengths, the speed in a material substance is different for different wavelengths. Hence the index of refraction of a substance is a function of wavelength. The dependence of wave speed on wavelength was mentioned in connection with water waves, in Section 21–6. A substance in which the speed of a wave varies with wavelength is said to exhibit *dispersion*. Figure 36–9 is a diagram showing the variation of index of refraction with wavelength for a number of the more common optical materials.

Consider a ray of white light, a mixture of all visible wavelengths, incident on a prism, as in Fig. 36–10. Since the deviation produced by the prism increases with increasing index of refraction, violet light is deviated most and red least, with other colors occupying intermediate positions. On emerging from the prism, the light is spread out into a fan-shaped beam, as shown. The light is said to be *dispersed* into a spectrum. Dispersion depends on the *difference* between the refractive index for violet light and that for red light. From Fig. 36–9 it can be seen that for a substance such as fluorite, whose index for yellow light is small, the difference between the indices for red and violet is also small. On the other hand, in the case of silicate flint glass, both the index for yellow light and the difference between extreme indices are large. In other words, for most transparent materials the greater the deviation, the greater the dispersion.

The brilliance of diamond is due in part to its large dispersion and in part to its unusually large refractive index. In recent years synthetic crystals of titanium dioxide and of strontium titanate, with about eight times the dispersion of diamond, have been produced.

*36–8

ILLUMINATION

We have defined the *intensity* of light and other electromagnetic radiation as power per unit area, measured in watts per square meter. Similarly, the *total* rate of radiation of energy from any of the sources of light discussed in Section 36–2 is called the *radiant power* or *radiant flux*, measured in watts. These quantities are not adequate to measure the visual sensation of *brightness*, however, for two reasons: First, not all the radiation from a source lies in the visible spectrum; an ordinary incandescent light bulb radiates more energy in the infrared than in the visible spectrum. Second, the eye is not equally sensitive to all wavelengths; a bulb emitting 1 watt of yellow light appears brighter than one emitting one watt of blue light.

The quantity analogous to radiant power but compensated to include the above effects is called *luminous flux*, denoted by F. The unit of luminous flux is the *lumen*, abbreviated lm, defined as that quantity of light of any frequency that is equal to a quantity of light of frequency 5.40×10^{14} Hz (wavelength 555 nm) having radiant flux of 1/683 W. As examples, the total light output (luminous flux) of a 40-watt incandescent light bulb is about 500 lm, while that of a 40-watt fluorescent tube is about 2300 lm.

When luminous flux strikes a surface, the surface is said to be *illuminated*. The intensity of illumination, analogous to the intensity of electromagnetic radiation (which is power per unit area) is the *luminous flux per unit area*, called the *illuminance*, denoted by E. The unit of illuminance is the lumen per square meter, also called the *lux:*

$$1 \text{ lux} = 1 \text{ lm} \cdot \text{m}^{-2}.$$

An older unit, the lumen per square foot, or foot-candle, has become obsolete. If luminous flux F falls at normal incidence on an area A, the illuminance E is given by

$$E = \frac{F}{A}. \qquad (36\text{–}9)$$

Most light sources do not radiate equally in all directions; it is useful to have a quantity that describes the intensity of a source in a specific direction, without using any specific distance from the source. We place the source at the center of an imaginary sphere of radius R. A small area A of the sphere subtends a solid angle Ω given by $\Omega = A/R^2$. If the luminous flux passing through this area is F, we define the *luminous intensity I* in the direction of the area as

$$I = \frac{F}{\Omega}. \qquad (36\text{–}10)$$

The unit of luminous intensity is one lumen per steradian, also called one *candela*, abbreviated cd:

$$1 \text{ cd} = 1 \text{ lm} \cdot \text{sr}^{-1}.$$

The term "luminous intensity" is somewhat misleading. The usual usage of *intensity* connotes power per unit area, and the intensity of radiation from a point source decreases as the square of the distance. Luminous intensity, however, is flux per unit *solid angle*, not per unit *area*, and the luminous intensity of a source in a particular direction *does not* decrease with increasing distance.

EXAMPLE   A certain 100-watt lightbulb emits a total luminous flux of 1200 lm, distributed uniformly over a hemisphere. Find the illuminance and the luminous intensity at a distance of 1 m, and at 5 m.

Solution The area of a half-sphere of radius 1 m is

$$(2\pi)(1 \text{ m})^2 = 6.28 \text{ m}^2.$$

The illuminance at 1 m is

$$E = \frac{1200 \text{ lm}}{6.28 \text{ m}^2} = 191 \text{ lm} \cdot \text{m}^{-2} = 191 \text{ lux}.$$

Similarly, the area of a half-sphere of radius 5 m is

$$(2\pi)(5 \text{ m})^2 = 157 \text{ m}^2,$$

and the illuminance at 5 m is

$$E = \frac{1200 \text{ lm}}{157 \text{ m}^2} = 7.64 \text{ lm} \cdot \text{m}^{-2} = 7.64 \text{ lux}.$$

This is smaller by a factor of 5^2 than the illuminance at 1 m and illustrates the inverse-square law for illuminance from a point source.

The solid angle subtended by a hemisphere is 2π sr. The luminous intensity is

$$I = \frac{1200 \text{ lm}}{2\pi \text{ sr}} = 191 \text{ lm} \cdot \text{sr}^{-1} = 191 \text{ cd}.$$

The luminous intensity does not depend on distance. ◀

36–9

POLARIZATION

Polarization phenomena occur only with transverse waves. Our principal concern in this chapter is with electromagnetic waves, especially light, but to introduce basic concepts we first consider a mechanical example, transverse waves on a string, as discussed in Chapter 21. For a string whose equilibrium position is along the x-axis, the displacements may be along the y-direction, as in Fig. 36–11a, in which case the string always lies in the x-y plane. But the displacements might instead be along the z-axis, as in Fig. 36–11b, so that the string lies in the x-z plane.

The wave having only y-displacements in the above discussion is said to be *linearly polarized* in the y-direction, and the one with only z-displacements is linearly polarized in the z-direction. Any transverse wave on a string may be regarded as a superposition of waves polarized in the y- and z-directions, respectively. It is also easy in principle to construct a mechanical *filter* that permits only waves with a certain polarization direction to pass. An example is shown in Fig. 36–11c; the string can slide vertically in the slot without friction, but no horizontal motion is possible. Thus this filter passes waves polarized in the y-direction but blocks those polarized in the z-direction.

This same language can be applied to light and other electromagnetic waves, which also exhibit polarization. An electromagnetic wave consists of fluctuating electric and magnetic fields, perpendicular to each other and to the direction of propagation, as described in detail in Chapter 35. By convention, the direction of polarization is taken to be that of the *electric*-field vector, not the magnetic field, because most mechanisms

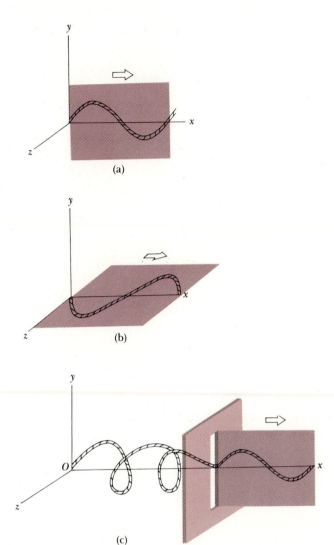

36–11 (a) Transverse wave on a string, polarized in the y-direction. (b) Wave polarized in the z-direction. (c) Barrier with a frictionless vertical slot passes components polarized in the y-direction but blocks those polarized in the z-direction, acting as a polarizing filter.

for detecting electromagnetic waves employ principally the electric-field forces on electrons in materials. The most common manifestations of electromagnetic radiation are due chiefly to the electric-field force, not the magnetic-field force.

Polarizing filters can be made for electromagnetic waves; the details of construction depend on the wavelength. For microwaves having a wavelength of a few centimeters, a grid of closely spaced, parallel conducting wires insulated from each other will pass waves whose *E*-fields are perpendicular to the wires but not those with *E*-fields parallel to the wires. For light, the most common polarizing filter is a material known by the trade name Polaroid, widely used for sunglasses and polarizing filters for camera lenses. This material works on a principle of preferential absorption, passing waves polarized parallel to a certain axis in the material (called the *polarizing axis*) with 80 percent or more transmission, but offering only one percent or less transmission to waves with polarization perpendicular to this axis. The action of such a polarizing filter is shown schematically in Fig. 36–12.

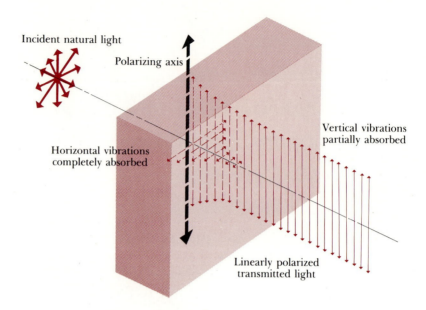

Incident natural light

Polarizing axis

Vertical vibrations
partially absorbed

Horizontal vibrations
completely absorbed

Linearly polarized
transmitted light

36–12 Linearly polarized light transmitted by a polarizing filter.

Waves emitted by a radio transmitter are usually linearly polarized; a vertical rod antenna of the type widely used for CB radios emits waves that in a horizontal plane around the antenna are polarized in the vertical direction (parallel to the antenna). Light from ordinary sources is *not* polarized, for a slightly subtle reason. The "antennas" that radiate light waves are the molecules of which light sources are composed. The electrically charged particles in the molecules acquire energy in some way and radiate this energy as electromagnetic waves of short wavelength. The waves from any one molecule may be linearly polarized, like those from a radio antenna; but since any actual light source contains a tremendous number of molecules, oriented at random, the light emitted is a random mixture of waves linearly polarized in all possible transverse directions.

36–10

POLARIZING FILTERS

An ideal polarizing filter or *polarizer* has the property that it passes 100% of the incident light polarized in the direction of the filter's polarizing axis but blocks completely all light polarized perpendicular to this axis. Such a device is an unattainable idealization, but the concept is useful in clarifying the basic ideas. Some real polarizers approximate this ideal behavior very closely. In Fig. 36–13, unpolarized light is incident on a polarizer whose axis is represented by the broken line. As the polarizer is rotated about an axis parallel to the incident ray, the intensity does not change. The polarizer transmits the components of the incident waves in which the E-vector is parallel to the transmission direction of the polarizer, and by symmetry the components are equal for all azimuths.

The transmitted intensity is found to be exactly one half that of the incident light. This may be understood as follows: The incident light can always be resolved into components polarized parallel to the polarizer

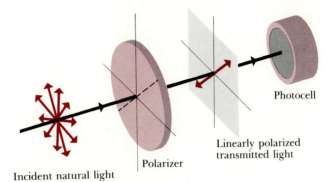

36–13 The intensity of the transmitted linearly polarized light, measured by the photocell, is the same for all orientations of the polarizing filter.

axis and components polarized perpendicular to it. Because the incident light is a random mixture of all states of polarization, these two components are, on the average, equal. Thus (in the ideal polarizer) exactly half of the incident intensity, that corresponding to the component parallel to the polarizer axis, is transmitted.

Suppose now that a second polarizer is inserted between the first polarizer and the photocell, as in Fig. 36–14. Let the transmission direction of the second polarizer, or *analyzer*, be vertical, and let that of the first polarizer make an angle θ with the vertical. The linearly polarized light transmitted by the polarizer may be resolved into two components as shown, one parallel and the other perpendicular to the transmission direction of the analyzer. Only the parallel component, of amplitude $E \cos \theta$, will be transmitted by the analyzer. The transmitted intensity is maximum when $\theta = 0$ and is zero when $\theta = 90°$, or when polarizer and analyzer are *crossed*. At intermediate angles, since the quantity of energy is proportional to the *square* of the amplitude, we have

$$I = I_{max} \cos^2 \theta, \qquad (36\text{–}11)$$

where I_{max} is the maximum amount of light transmitted and I is the amount transmitted at the angle θ. This relation, discovered experimentally by Etienne Louis Malus in 1809, is called *Malus's law*.

The angle θ is the angle between the transmission directions of polarizer and analyzer. If either the analyzer or the polarizer is rotated, the amplitude of the transmitted beam varies with the angle between them according to Eq. (36–11).

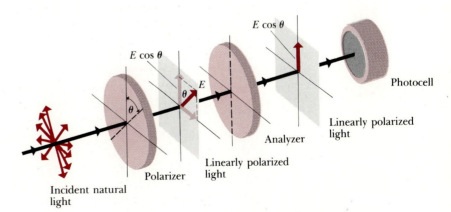

36–14 The analyzer transmits only the component parallel to its transmission direction or polarizing axis.

EXAMPLE   In Fig. 36–14, let the incident unpolarized light have intensity I_0. Find the intensity transmitted by the first polarizer and by the second, if the angle θ is 30°.

Solution   As explained above, the intensity after the first filter is $(I_0/2)$. According to Eq. (36–11), the second filter reduces the intensity by a factor of $\cos^2 30° = \frac{3}{4}$. Thus the intensity transmitted by the second polarizer is $(I_0/2)(\frac{3}{4}) = 3I_0/8$. ◀

If, in Fig. 36–13, the incident light is completely linearly polarized rather than unpolarized, then of course the transmitted intensity varies from zero to some maximum value as the polarizer axis is turned from perpendicular to parallel to the polarization direction of the light. Or the incident light may be a mixture of linearly polarized and unpolarized, with intensities I_{pol} and I_{un}, respectively. The *percent polarization* is then defined as the ratio of the polarized intensity to the total intensity, times 100%:

$$\% \text{ polarization} = \frac{I_{pol}}{I_{pol} + I_{un}} \times 100\%. \qquad (36\text{–}12)$$

In this case, in Fig. 36–13 the transmitted intensity varies from a minimum (but not zero) when the polarizer axis is perpendicular to the polarized component of incident light, to a maximum when it is parallel.

Unpolarized light can be partially polarized by *reflection*. When unpolarized light strikes a reflecting surface between two optical materials, there is found to be a preferential reflection for those waves in which the electric-field vector is perpendicular to the plane of incidence (the plane containing the incident ray and the normal to the surface). At one particular angle of incidence, known as the *polarizing angle* ϕ_p, no light whatever is reflected except that in which the electric vector is perpendicular to the plane of incidence. This case is shown in Fig. 36–15.

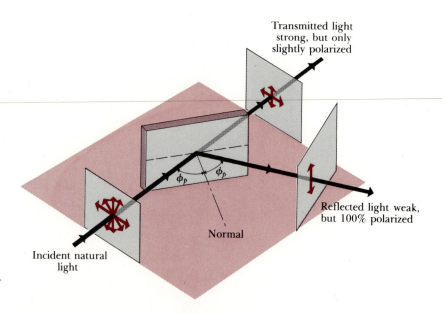

36–15   When light is incident at the polarizing angle, the reflected light is linearly polarized.

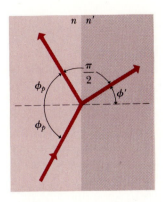

36–16 At the polarizing angle the reflected and transmitted rays are perpendicular to each other.

When light is incident at the polarizing angle, *none* of the component parallel to the plane of incidence is reflected; this component is 100% transmitted in the *refracted* beam. Of the component perpendicular to the plane of incidence, about 15% is reflected if the reflecting surface is glass. The fraction reflected depends on the index of the reflecting material. Hence the *reflected* light is weak and *completely* polarized. The *refracted* light is a mixture of the parallel component, all of which is refracted, and the remaining 85% of the perpendicular component. It is therefore strong, but only *partially* polarized.

At angles of incidence other than the polarizing angle, some of the component parallel to the plane of incidence is reflected, so that, except at the polarizing angle, the reflected light is not completely linearly polarized.

In 1812, Sir David Brewster noticed that when the angle of incidence is equal to ϕ_p, the reflected ray and the refracted ray are perpendicular to each other, as shown in Fig. 36–16. When this is the case, the angle of refraction ϕ' becomes the complement of ϕ_p, so that $\sin \phi' = \cos \phi_p$. Since

$$n \sin \phi_p = n' \sin \phi',$$

we find $n \sin \phi_p = n' \cos \phi_p$, and

$$\tan \phi_p = \frac{n'}{n}, \qquad (36\text{--}13)$$

a relation known as *Brewster's law.*

Some crystalline materials have different refractive indices for different directions of polarization. A common example is calcite; when the crystal is oriented approximately in a beam of unpolarized light, its refractive index (for $\lambda = 589$ nm) is 1.658 for one direction of polarization and 1.486 for the perpendicular direction. Such crystals are said to be *doubly refracting* or *birefringent;* they can be used in arrangements of prisms that separate spatially the two polarized components of an unpolarized beam and reflect one component out of the beam direction. These polarizing filters are more efficient than Polaroid filters and also much more expensive.

Some doubly refracting crystals exhibit *dichroism;* that is, one of the polarized components is *absorbed* much more strongly than the other. Hence, if the crystal is cut of the proper thickness, one component is practically extinguished by absorption, while the other is transmitted in appreciable amount, as indicated in Fig. 36–12. Tourmaline is one example of such a dichroic crystal.

An early form of Polaroid, invented by Edwin H. Land in 1928, consists of a thin layer of tiny needlelike dichroic crystals of herapathite (iodoquinine sulfate), in parallel orientation, embedded in a plastic matrix and enclosed for protection between two transparent plates. A more recent modification, developed by Land in 1938 and known as an H-sheet, is a *molecular* polarizer. It consists of long polymeric molecules of polyvinyl alcohol (PVA) that have been given a preferred direction by stretching and have been stained with an ink containing iodine that causes the sheet to exhibit dichroism. The PVA sheet is laminated to a support sheet of cellulose acetate butyrate.

Polaroid sheet is widely used in sunglasses where, from the standpoint of its polarizing properties, it plays the role of the analyzer in Fig. 36–14. We have seen that when unpolarized light is reflected, there is a preferential reflection for light polarized perpendicular to the plane of incidence. When sunlight is reflected from a horizontal surface, the plane of incidence is vertical. Hence, in the reflected light there is a preponderance of light polarized in the horizontal direction. When such reflection occurs at smooth asphalt road surfaces or the surface of a lake, it causes unwanted "glare," and vision is improved by eliminating it. The transmission direction of the Polaroid sheet in the sunglasses is vertical, so none of the horizontally polarized light is transmitted to the eyes.

Apart from this polarizing feature, these glasses serve the same purpose as any dark glasses, absorbing 50% of the incident light; in an unpolarized beam, half the light can be considered as polarized horizontally and half vertically, and only the vertically polarized light is transmitted. The sensitivity of the eye is independent of the state of polarization of the light.

*36–11

SCATTERING OF LIGHT

The sky is blue. Sunsets are red. Skylight is partially polarized, as can readily be verified by looking at the sky directly overhead through a polarizing filter. It turns out that one phenomenon is responsible for all three of these effects.

In Fig. 36–17, sunlight (unpolarized) comes from the left along the z-axis and passes over an observer looking vertically upward along the y-axis. Molecules of the earth's atmosphere are located at point O. The electric field in the beam of sunlight sets the electric charges in the mole-

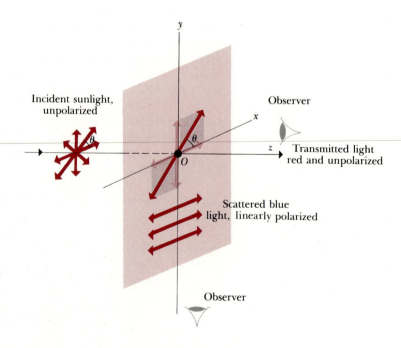

36–17 Scattered light is linearly polarized.

cules in vibration. Since light is a transverse wave, the direction of the electric field in any component of the sunlight lies in the *xy*-plane and the motion of the charges takes place in this plane. There is no field, and hence no vibration, in the direction of the *z*-axis.

A component of the incident light at an angle θ with the *x*-axis sets the electric charges in the molecules vibrating in the same direction, as indicated by the heavy line through point O. We can resolve this vibration into two components, one along the *x*-axis and the other along the *y*-axis. Each component in the incident light produces the equivalent of two molecular "antennas," oscillating with the frequency of the incident light, and lying along the *x*- and *y*-axes.

It has been explained in Section 35–7 that an antenna does not radiate in the direction of its own length. Hence, the antenna along the *y*-axis does not send any light to the observer directly below it. It does, of course, send out light in other directions. The only light reaching the observer comes from the component of vibration along the *x*-axis, and, as is the case with the waves from any antenna, this light is linearly polarized, with the electric field parallel to the antenna. The vectors on the *y*-axis below point O show the direction of polarization of the light reaching the observer.

The process described above is called *scattering*. The energy of the scattered light is removed from the original beam, which becomes weakened in the process. The intensity of the scattered light increases with increasing frequency; blue light is scattered more than red, with the result that the hue of the scattered light is blue.

Polarizers are commonly used in photography. Because skylight is partially polarized, the sky can be darkened in a photograph by appropriate orientation of the polarizer axis. The effect of atmospheric haze can be reduced in exactly the same way, and unwanted reflections can be controlled just as with polarizing sunglasses, discussed in Section 36–10.

Toward evening, when sunlight has to travel a large distance through the earth's atmosphere to reach a point over or nearly over an observer, a large proportion of the blue light in sunlight is removed from it by scattering. White light minus blue light is yellow or red in hue. Thus when sunlight, with the blue component removed, is incident on a cloud, the light reflected from the cloud to the observer has the yellow or red hue so commonly seen at sunset. If the earth had no atmosphere we would receive *no* skylight at the earth's surface, and the sky would appear as black in the daytime as it does at night. To an astronaut in a spaceship or on the moon, the sky appears black, not blue.

*36–12

CIRCULAR AND ELLIPTICAL POLARIZATION

Up to this point we have discussed polarization phenomena in terms of *linearly polarized* light. Light (and all other electromagnetic radiation) may also have *circular* or *elliptical* polarization. To introduce these new concepts, we return again to mechanical waves on a stretched string. In Fig. 36–11, suppose the two linearly polarized waves in parts (a) and (b) have equal amplitude and are *superposed;* then each point in the string has simultaneously y- and z-displacements of equal magnitude, and a little thought shows that the resultant wave lies in a plane oriented at 45° to the x- and z-axes. The amplitude of the resultant wave is larger by a factor of $\sqrt{2}$ than that of either component wave, and the resultant wave is again linearly polarized.

But now suppose one component wave differs in phase by a quarter-cycle from the other. Then the resultant motion of each point corresponds to a superposition of two simple harmonic motions at right angles with a quarter-cycle phase difference. The motion is then no longer confined to a single plane, and it can be shown that each point on the rope moves in a *circle* in a plane parallel to the yz-plane. Successive points on the rope have successive phase differences, and the overall motion of the string then has the appearance of a rotating helix. This particular superposition of two linearly polarized waves is called *circular polarization*. By convention, the wave is said to be *right circularly polarized* when, as in the present instance, the sense of motion of a particle, to an observer looking *backwards* along the direction of propagation, is *clockwise*, and *left circularly polarized* when it appears counterclockwise to that observer. The latter would be the result if the phase difference between y- and z-components were opposite to that in our example.

If the phase difference between the two component waves is something other than a quarter-cycle, or if the two component waves have different amplitudes, then each point on the string traces out not a circle but an *ellipse,* and the resulting wave is said to be *elliptically polarized.*

For electromagnetic waves of radio frequencies, circular or elliptical polarization can be produced by using two antennas at right angles, fed from the same transmitter but with a phase-shifting network that introduces the appropriate phase difference. For light, the phase shift can be introduced by use of a birefringent material. If two waves with perpendicular directions of polarization are in phase as they enter the material, they travel with different speeds because the refractive index is different. Thus in general they are no longer in phase when they emerge from the crystal. If the thickness of the crystal is such as to introduce a quarter-cycle phase difference, then the crystal converts linearly polarized light to circularly polarized. Such a crystal is called a *quarter-wave plate*. We challenge the reader to show that such a plate also converts circularly polarized light to linearly polarized light!

When a polarizer and an analyzer are mounted in the "crossed" position, i.e., with their transmission directions at right angles to each other, no light is transmitted through the combination. But if a doubly refracting crystal is inserted between polarizer and analyzer, the light after passing through the crystal is, in general, elliptically polarized, and some light will be transmitted by the analyzer. Thus the field of view, dark in the absence of the crystal, becomes light when the crystal is inserted.

Some substances, such as glass and various plastics, while not normally doubly refracting become so when subjected to mechanical stress. This is the basis of the science of *photoelasticity.* Stresses in opaque engineering materials such as girders, boiler plates, and gear teeth can be analyzed by constructing a transparent model of the object, usually of a plastic, and examining it between a polarizer and an analyzer in the crossed position. Very complicated stress distributions, such as those around a hole or a gear tooth, that would be practically impossible to analyze mathematically may thus be studied by optical methods. Figure 36–18 is a photograph of a photoelastic model under stress.

Liquids are not normally doubly refracting, but some become so when an electric field is established within them. This phenomenon is known as the *Kerr effect*. The existence of the Kerr effect makes it possible to construct an electrically controlled "light valve." A cell with transparent walls contains the liquid between a pair of parallel plates. The cell is inserted between crossed Polaroid disks. Light is transmitted when an electric field is set up between the plates and is cut off when the field is removed.

When a beam of linearly polarized light is sent through certain types of crystals and certain liquids, the direction of polarization of the emerging linearly polarized light is found to be different from the original direction. This phenomenon is called *rotation of the direction of polarization*, and substances that exhibit the effect are called *optically active*. Those that rotate the direction of polarization to the right, looking along the advancing beam, are called *dextrorotatory* or right-handed; those that rotate it to the left, *levorotatory* or left-handed.

Optical activity may be due to an asymmetry of the molecules of a substance, or it may be a property of a crystal as a whole. For example, solutions of cane sugar are dextrorotatory, indicating that the optical activity is a property of the sugar molecule. The molecules of the sugars dextrose and levulose are mirror images, and their optical activities are opposite. The rotation of the plane of polarization by a sugar solution is used commercially as a method of determining the proportion of cane sugar in a given sample. Crystalline quartz is also optically active, some natural crystals being right-handed and others left-handed. Here the optical activity is a consequence of the crystalline structure, since it disappears when the quartz is melted and allowed to resolidify into a glassy, noncrystalline state called fused quartz.

36–18 (a) Photoelastic stress analysis. (Courtesy of Dr. W. M. Murray, Massachusetts Institute of Technology.) (b) Stress analysis of a model of a cross section of a Gothic cathedral. The masonry construction used for this kind of building had great strength in compression but very little in tension. Inadequate buttressing and high winds sometimes caused tensile stresses in normally compressed structural elements, leading to some spectacular collapses. (Sepp Seitz/Woodfin Camp.)

(a)

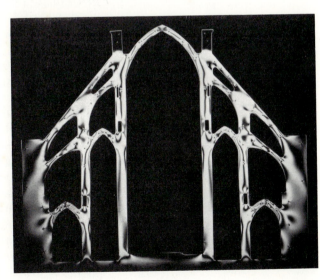

(b)

*36–13

HUYGENS' PRINCIPLE

$r = vt$

36-19 Huygens' principle.

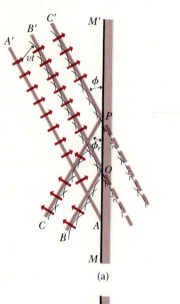

(a)

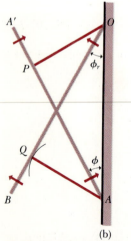

(b)

36-20 (a) Successive positions of a plane wave AA' as it is reflected from a plane surface. (b) A portion of (a).

The principles governing reflection and refraction of light rays, discussed in the preceding section, were discovered experimentally long before the wave nature of light was firmly established. These principles may, however, be derived from *wave* considerations and thus shown to be consistent with the wave nature of light. To establish this connection we use a principle called *Huygens' principle;* as stated originally by Christian Huygens in 1678, this principle is a geometrical method for finding, from the known shape of a wave front at some instant, the shape of the wave front at some later time. Huygens assumed that *every point of a wave front may be considered the source of secondary wavelets that spread out in all directions with a speed equal to the speed of propagation of the waves.* The new wave front is then found by constructing a surface *tangent* to the secondary wavelets or, as it is called, the *envelope* of the wavelets.

Huygens' principle is illustrated in Fig. 36–19. The original wave front, AA', is traveling as indicated by the small arrows. We wish to find the shape of the wave front after a time interval t. Let v represent the speed of propagation. We construct a number of circles (traces of spherical wavelets) of radius $r = vt$, with centers along AA'. The trace of the envelope of these wavelets, which is the new wave front, is the curve BB'. The speed v has been assumed the same at all points and in all directions.

To derive the law of reflection from Huygens' principle, we consider a plane wave approaching a plane reflecting surface. In Fig. 36–20a, the lines AA', BB', and CC' represent successive positions of a wave front approaching the surface MM'. The actual planes are perpendicular to the plane of the figure. Point A on the wave front AA' has just arrived at the reflecting surface. The position of the wave front after a time interval t may be found by applying Huygens' principle. With points on AA' as centers, draw a number of secondary wavelets of radius vt, where v is the speed of propagation of the wave. Those wavelets originating near the upper end of AA' spread out unhindered, and their envelope gives that portion of the new wave surface OB'. The wavelets originating near the lower end of AA', however, strike the reflecting surface. If the latter had not been there, they would have occupied the positions shown by the broken circular arcs. The effect of the reflecting surface is to *reverse the direction* of travel of those wavelets that strike it, so that that part of a wavelet that would have penetrated the surface actually lies to the left of it, as shown by the full lines. The envelope of these reflected wavelets is then that portion of the wave front OB. The trace of the entire wave front at this instant is the broken line BOB'. A similar construction gives the line CPC' for the wave front after another interval t.

The angle ϕ between the incident *wave front* and the *surface* is the same as that between the incident *ray* and the *normal* to the surface and is therefore the angle of incidence. Similarly, ϕ_r is the angle of reflection. To find the relation between these angles, we consider Fig. 36–20b. From O, draw $OP = vt$, perpendicular to AA'. Now OB, by construction, is tangent to a circle of radius vt with center at A. Hence, if AQ is drawn from A to the point of tangency, the triangles APO and OQA are equal

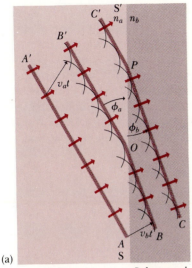

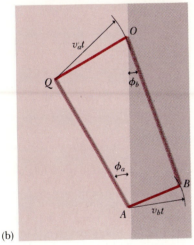

36–21 (a) Successive positions of a plane wave front AA' as it is refracted by a plane surface. (b) A portion of (a). The case $v_b < v_a$ is shown.

(right triangles with the side AO in common and with $AQ = OP$). The angle ϕ therefore equals the angle ϕ_r, and we have the law of reflection.

The law of refraction is obtained by a similar procedure. In Fig. 36–21a, we consider a wave front, represented by line AA', for which point A has just arrived at the boundary surface MM' between two transparent materials a and b, of indices of refraction n_a and n_b. (The *reflected* waves are not shown in the figure; they proceed exactly as in Fig. 36–20.) Let us apply Huygens' principle to find the position of the refracted wave front after a time t.

With points on AA' as centers, we draw a number of secondary wavelets. Those originating near the upper end of AA' travel with speed v_a and, after a time interval t, are spherical surfaces of radius $v_a t$. The wavelet originating at point A, however, is traveling in the second material b with speed v_b and at time t is a spherical surface of radius $v_b t$. The envelope of the wavelets from the original wave front is the plane whose trace is the broken line BOB'. A similar construction leads to the trace CPC' after a second interval t.

The angles ϕ_a and ϕ_b between the surface and the incident and refracted wave fronts are, respectively, the angle of incidence and the angle of refraction. To find the relation between these angles, refer to Fig. 36–21b. Draw $OQ = v_a t$, perpendicular to AQ, and draw $AB = v_b t$, perpendicular to BO. From the right triangle AOQ,

$$\sin \phi_a = \frac{v_a t}{AO},$$

and from the right triangle AOB,

$$\sin \phi_b = \frac{v_b t}{AO}.$$

Hence

$$\frac{\sin \phi_a}{\sin \phi_b} = \frac{v_a}{v_b}. \qquad (36\text{–}14)$$

Since v_a/v_b is a constant, Eq. (36–14) express Snell's law, and we have derived Snell's law from a wave theory.

The most general form of Snell's law is given by Eq. (36–4), namely

$$\frac{\sin \phi_a}{\sin \phi_b} = \frac{n_b}{n_a}.$$

Comparing this with Eq. (36–14), we see that

$$\frac{v_a}{v_b} = \frac{n_b}{n_a},$$

and

$$n_a v_a = n_b v_b.$$

When either material is vacuum, $n = 1$ and the speed is c. Hence

$$n_a = \frac{c}{v_a}, \qquad n_b = \frac{c}{v_b}, \qquad (36\text{–}15)$$

showing that *the index of refraction of any material is the ratio of the speed of light in a vacuum to the speed in the material.* For any material, n is always greater than unity; hence the speed of light in a material is always *less* than in vacuum.

In Fig. 36–21, if t is chosen to be the period T of the wave, the spacing is vT, which is the wavelength λ. The figure shows that when v_b is less than v_a, the wavelength in the second material is smaller than that in the first. When a light wave proceeds from one material to another, where the speed is different, the wavelength changes *but not the frequency.* Since

$$v_a = f\lambda_a \quad \text{and} \quad v_b = f\lambda_b,$$

$$\frac{\lambda_a}{v_a} = \frac{\lambda_b}{v_b} \quad \text{and} \quad \lambda_a \frac{c}{v_a} = \lambda_b \frac{c}{v_b}.$$

Therefore,

$$\lambda_a n_a = \lambda_b n_b.$$

If either material is vacuum, the index is 1 and the wavelength in vacuum is represented by λ_0. Hence

$$\lambda_a = \frac{\lambda_0}{n_a}, \qquad \lambda_b = \frac{\lambda_0}{n_b}, \qquad (36\text{–}16)$$

showing that *the wavelength in any material is the wavelength in a vacuum divided by the index of refraction of the medium.*

QUESTIONS

36–1 During a thunderstorm one always sees the flash of lightning before hearing the accompanying thunder. Discuss this in terms of the various wave speeds. Can this phenomenon be used to determine how far away the storm is?

36–2 When hot air rises around a radiator or from a heating duct, objects behind it appear to shimmer or waver. What is happening?

36–3 Light requires about 8 minutes to travel from the sun to the earth. Is it delayed appreciably by the earth's atmosphere?

36–4 Sometimes when looking at a window one sees two reflected images, slightly displaced from each other. What causes this?

36–5 An object submerged in water appears to be closer to the surface than it actually is. Why? Swimming pools are always deeper than they look; is this the same phenomenon?

36–6 A ray of light in air strikes a glass surface. Is there a range of angles for which total reflection occurs?

36–7 As shown in Table 36–1, diamond has a much larger refractive index than glass. Is there a larger or smaller range of angles for which total internal reflection occurs for diamond, than for glass? Does this have any-

thing to do with the fact that a real diamond has more sparkle than a glass imitation?

36–8 Sunlight or starlight passing through the earth's atmosphere is always bent toward the vertical. Why? Does this mean that a star isn't really where it appears to be?

36–9 The sun or moon usually appears flattened just before it sets. Is this related to refraction in the earth's atmosphere, mentioned in Question 36–8?

36–10 A student claimed that, because of atmospheric refraction (cf. Question 36–8), the sun can be seen after it has set, and that the day is therefore longer than it would be if the earth had no atmosphere. First, what does he mean by saying the sun can be seen after it has set? Second, comment on the validity of his conclusion.

36–11 It has been proposed that automobile windshields and headlights should have polarizing filters, to reduce the glare of oncoming lights during night driving. Would this work? How should the polarizing axes be arranged? What advantages would this scheme have? What disadvantages?

36–12 A salesperson at a bargain counter claims that a certain pair of sunglasses has Polaroid filters; you suspect they are just tinted plastic. How could you find out for sure?

36–13 When unpolarized light is incident on two crossed polarizers, no light is transmitted. A student asserted that if a third polarizer is inserted between the other two, some transmission may occur. Does this make sense? How can adding a third filter *increase* transmission?

36–14 How could you determine the direction of the polarizing axis of a single polarizer?

36–15 In three-dimensional movies, two images are projected on the screen, and the viewers wear special glasses to sort them out. How does this work?

36–16 In Fig. 36–17, since the light scattered out of the incident beam is polarized, why is the transmitted beam not also partially polarized?

36–17 Light from blue sky is strongly polarized because of the nature of the scattering process described in Section 36–11. But light scattered from white clouds is usually *not* polarized. Why not?

36–18 When a sheet of plastic food wrap is placed between two crossed polarizers, no light is transmitted. When the sheet is stretched in one direction, some light passes through. What is happening?

36–19 Television transmission usually uses plane-polarized waves. It has been proposed to use circularly polarized waves to improve reception. Why? (*Hint.* See Problem 36–39.)

36–20 Can sound waves be reflected? Refracted? Give examples. Does Huygens' principle apply to sound waves?

36–21 Why should the wavelength of light change, but not its frequency, in passing from one material to another?

36–22 When light is incident on an interface between two materials, the angle of the refracted ray depends on the wavelength, but the angle of the reflected ray does not. Why should this be?

PROBLEMS

36–1 What is the wavelength in meters, microns, nanometers, and angstrom units of

a) soft x-rays of frequency 2×10^{17} Hz?

b) green light of frequency 5.6×10^{14} Hz?

36–2 The visible spectrum includes a wavelength range from about 400 nm to about 700 nm. Express these wavelengths in inches.

36–3 Fizeau's measurements of the speed of light were continued by Cornu, using Fizeau's apparatus but with the distance between mirrors increased to 22.9 km. One of the toothed wheels used was 40 mm in diameter and had 180 teeth. Find the angular velocity at which it should rotate so that light transmitted through one opening will return through the next.

36–4 Prove that a ray of light reflected from a plane mirror rotates through an angle 2θ when the mirror rotates through an angle θ about an axis perpendicular to the plane of incidence.

36–5 A parallel beam of light is incident on a prism, as shown in Fig. 36–22. Part of the light is reflected from one face and part from another. Show that the angle θ between the two reflected beams is twice the angle A between the two reflecting surfaces.

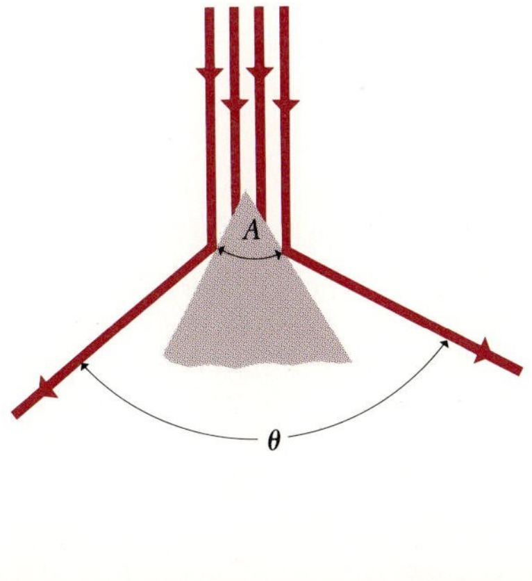

36–6 A parallel beam of light makes an angle of 30° with the surface of a glass plate having a refractive index of 1.50.

a) What is the angle between the refracted beam and the surface of the glass?

b) What should be the angle of incidence ϕ with this plate for the angle of refraction to be $\phi/2$?

36–7 Light strikes a glass plate at an angle of incidence of 60°; part of the beam is reflected and part refracted. It is observed that the reflected and refracted portions make an angle of 90° with each other. What is the index of refraction of the glass?

36–8 A ray of light is incident on a plane surface separating two transparent substances of indices 1.60 and 1.40. The angle of incidence is 30° and the ray originates in the medium of higher index. Compute the angle of refraction.

36–9 A parallel-sided plate of glass having a refractive index of 1.60 is held on the surface of water in a tank. A ray coming from above makes an angle of incidence of 45° with the top surface of the glass.

a) What angle does the ray make with the normal in the water?

b) How does this angle vary with the refractive index of the glass?

36–10 An inside corner of a cube is lined with mirrors. A ray of light is reflected successively from each of three mutually perpendicular mirrors; show that its final direction is always exactly opposite to its initial direction. This

Figure 36–22

principle is used in tail-light lenses and reflecting highway signs.

36–11

a) What is the speed of light of wavelength 500 nm (in vacuum) in glass whose index at this wavelength is 1.50?

b) What is the wavelength of these waves in the glass?

36–12 A glass plate 3 mm thick, of index 1.50, is placed between a point source of light of wavelength 600 nm (in vacuum) and a screen. The distance from source to screen is 3 cm. How many waves are there between source and screen?

36–13 The speed of light of wavelength 656 nm in heavy flint glass is 1.60×10^8 m·s^{-1}. What is the index of refraction of this glass?

36–14 Light of a certain frequency has a wavelength in water of 442 nm. What is the wavelength of this light when it passes into carbon disulfide?

36–15 A 45°–45°–90° prism is immersed in water. What is the minimum index of refraction the prism may have if it is to totally reflect a ray incident normally on one of its shorter faces?

36–16 A point source of light is 20 cm below the surface of a body of water. Find the diameter of the largest circle at the surface through which light can emerge from the water.

36–17 A ray of light is incident at an angle with the normal of 60° on one surface of a glass plate 2 cm thick, of index 1.50. The medium on either side of the plate is air. Find the transverse displacement between the incident and emergent rays.

36–18 The prism of Fig. 36–23 has a refractive index of 1.414, and the angles A are 30°. Two light rays m and n are parallel as they enter the prism. What is the angle between them after they emerge?

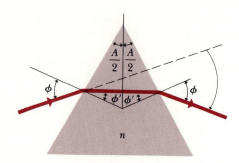

Figure 36–24

directions of the ray) is given by

$$\sin \frac{A + \delta}{2} = n \sin \frac{A}{2}.$$

36–20 Use the result of Problem 36–19 to find the angle of deviation for a ray of light passing symmetrically through a prism having three equal angles ($A = 60°$) and $n = 1.50$.

36–21 A certain glass has a refractive index of 1.50 for red light (700 nm) and 1.52 for violet light (400 nm). If both colors pass through symmetrically, as described in Problem 38–20, and if $A = 60°$, find the difference between the angles of deviation for the two colors, using the result of Problem 36–19.

36–22 The glass vessel shown in Fig. 36–25a contains a large number of small, irregular pieces of glass and a liquid. The dispersion curves of the glass and of the liquid are shown in Fig. 36–25b. Explain the behavior of a parallel beam of white light as it traverses the vessel. (This is known as a *Christiansen filter*.)

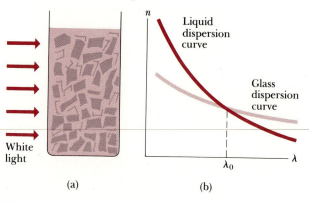

(a)   (b)

Figure 36–25

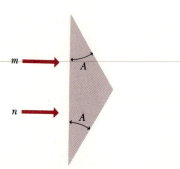

Figure 36–23

36–19 Light passes symmetrically through a prism having apex angle A, as shown in Fig. 36–24. Show that the angle of deviation δ (the angle between the initial and final

36–23 The illuminance of direct sunlight is about 10^5 lux. If a photoflash lamp has an intensity in a certain direction of 5×10^6 cd, at what distance from a surface should it be placed to produce illuminance equal to that of sunlight?

36–24 The *luminous efficiency* of a lamp is defined as the ratio of luminous flux to electric power input. A certain lamp mounted 3 m above a desk top has a luminous efficiency of 20 lm·W^{-1}. What is the power input to the lamp if the illuminance on the desk is equal to that of sunlight, about 10^5 lux? Assume that the lamp radiates uniformly over its lower half-sphere.

36–25 A certain baseball field in the shape of a square 140 m on a side is to be illuminated for night games by six towers supporting banks of 1000-watt incandescent lamps with luminous efficiency (see Problem 36–24) of 30 lm·W^{-1}. The illuminance required on the playing field is 200 lux. Assume that 50% of the luminous flux from the lamps reaches the field.

a) How many lamps are required in each tower?

b) What is the electric power input to each tower?

c) If power for all six towers is supplied by a generator driven by a gasoline engine, what must be the power capacity of the engine?

36–26 Unpolarized light of intensity I_0 is incident on a polarizing filter, and the emerging light strikes a second polarizing filter with its axis at 45° to that of the first. Determine

a) the intensity of the emerging beam, and

b) its state of polarization.

36–27 Three polarizing filters are stacked, with the polarizing axes of the second and third at 45° and 90°, respectively, with that of the first.

a) If unpolarized light of intensity I_0 is incident on the stack, find the intensity and state of polarization of light emerging from each filter.

b) If the second filter is removed, how does the situation change?

36–28 Three polarizing filters are stacked, with the polarizing axes of the second and third at angles θ and 90°, respectively, with that of the first. Unpolarized light of intensity I_0 is incident on the stack.

a) Derive an expression for the intensity of light transmitted through the stack, as a function of I_0 and θ.

b) Show that maximum transmission occurs when $\theta = 45°$.

36–29 It is desired to rotate the direction of polarization of linearly polarized light 90°, using two Polaroid filters. Explain how this can be done, and find the final intensity in terms of the incident intensity.

36–30 A polarizer and an analyzer are oriented so that the maximum amount of light is transmitted. To what fraction of its maximum value is the intensity of the trans-

mitted light reduced when the analyzer is rotated through

a) 30°,

b) 45°,

c) 60°?

36–31 A beam of light is incident on a liquid of 1.40 refractive index. The reflected rays are completely polarized. What is the angle of refraction of the beam?

36–32 The critical angle of light in a certain substance is 45°. What is the polarizing angle?

36–33

a) At what angle above the horizontal must the sun be in order that sunlight reflected from the surface of a calm body of water shall be completely polarized?

b) What is the plane of the **E**-vector in the reflected light?

36–34 A parallel beam of "natural" light is incident at an angle of 58° on a plane glass surface. The reflected beam is completely linearly polarized.

a) What is the angle of refraction of the transmitted beam?

b) What is the refractive index of the glass?

36–35 The refractive index of a certain flint glass is 1.65. For what incident angle is light reflected from the surface of this glass completely polarized if the glass is immersed in

a) air?

b) water?

36–36 A beam of light, after passing through the Polaroid disk P_1 in Fig. 36–26, traverses a cell containing a scattering medium. The cell is observed at right angles through another Polaroid disk P_2. Originally, the disks are oriented until the brightness of the field as seen by the observer is a maximum.

a) Disk P_2 is now rotated through 90°. Is extinction produced?

b) Disk P_1 is now rotated through 90°. Is the field bright or dark?

c) Disk P_2 is then restored to its original position. Is the field bright or dark?

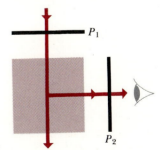

Figure 36–26

36–37 In Fig. 36–27, A and C are Polaroid sheets whose transmission directions are as indicated. B is a sheet of doubly refractive material whose optic axis is vertical. All three sheets are parallel. Unpolarized light enters from the left. Discuss the state of polarization of the light at points 2, 3, and 4.

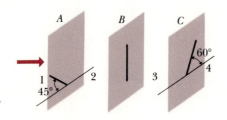

Figure 36–27

36–38 A certain birefringent material has indexes of refraction n_1 and n_2 for the two perpendicular components of linearly polarized light passing through it. The corresponding wavelengths are $\lambda_1 = \lambda_0/n_1$ and $\lambda_2 = \lambda_0/n_2$, where λ_0 is the wavelength in vacuum.

a) For a crystal of thickness d, show that the numbers of wavelengths of each component within the material at any time are d/λ_1 and d/λ_2.

b) If the crystal is to function as a quarter-wave plate, the two expressions in (a) must differ by $\frac{1}{4}$. Thus show that the minimum thickness for a quarter-wave plate is

$$d = \frac{\lambda_0}{4(n_1 - n_2)}.$$

c) Find the minimum thickness of a quarter-wave plate made of calcite, if the indexes of refraction are 1.658 and 1.486 and the wavelength is $\lambda_0 = 589$ nm.

36–39 A beam of right circularly polarized light is reflected at normal incidence from a reflecting surface. Is the reflected beam right or left circularly polarized? Explain.

36–40 What is the state of polarization of the light transmitted by a quarter-wave plate when the electric vector of the incident linearly polarized light makes an angle of 30° with the optic axis?

37

IMAGES FORMED BY A SINGLE SURFACE

In the preceding chapter we discussed the principles governing reflection and refraction of a ray of light at a reflecting surface or an interface between two materials. We now consider the behavior of several rays that diverge from a common point and strike a reflecting or refracting surface. A central concept in this discussion is that of *image*. After reflection or refraction the rays emerge with directions characteristic of having passed through some other common point called the *image point*. This discussion lays the foundation for analysis of many familiar optical instruments, including camera lenses, magnifiers, the human eye, microscopes, and telescopes.

37–1

REFLECTION AT A PLANE SURFACE

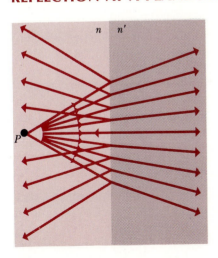

37–1 Reflection and refraction of rays at a plane interface between two transparent materials.

We begin with the situation shown in Fig. 37–1. Rays diverge from point P and are reflected or refracted (or both) at the interface. The direction of each *reflected* ray is given by the law of reflection, and that of each transmitted or *refracted* ray by Snell's law. Here and in the following discussion we denote the two refractive indexes as n and n', without using the more elaborate subscript notation of Chapter 36. We concentrate first on the *reflected* rays. The reflection can occur at an interface between two transparent materials or at a highly polished surface of an opaque material such as a metal, in which case the surface is usually called a *mirror*. We shall often use the term mirror to include all these possibilities.

As mentioned above, the key concept is that of *image*. After reflection the rays appear to have diverged from some common point, which we call the *image point*. In some cases the emerging rays really *do* meet at a common point and then diverge again after passing it; such a point is called a *real* image point. In other cases the rays diverge *as though* they had passed through such a point, which is then called a *virtual* image point. In many situations the image point exists only in an approximate

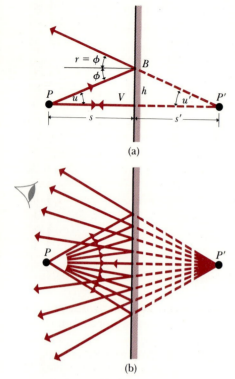

(a)

(b)

37-2 After reflection at a plane surface, all rays originally diverging from the object point P now diverge from the point P', although they do not *originate* at P'. Point P' is called the *virtual image* of point P. The eye sees some of the outgoing rays and perceives them as having come from point P'.

sense, that is, when certain approximations are used in the calculations. This chapter and the next are devoted primarily to a study of the formation and properties of images.

Figure 37–2a shows two rays diverging from a point P at a distance s to the left of a plane mirror. The ray PV, incident normally on the mirror, returns along its original path. The ray PB, making an angle u with PV, strikes the mirror at an angle of incidence $\phi = u$ and is reflected at an angle $r = \phi = u$. When extended backward, the reflected ray intersects the normal to the suface at point P'. The angle u' is equal to r and hence to u.

Figure 37–2b shows several rays diverging from P. The construction of Fig. 37–2a can be repeated for each of them, and we see that the directions of the outgoing rays are *as though* they had originated at point P', which is therefore the *image* of P. The rays do not, of course, actually pass through this point; in fact, if the mirror is opaque there is no light at all on the right side. Thus P' is a *virtual* image. Nevertheless P' is a very real point in the sense that it describes the final directions of all the rays that originally diverged from P.

From the symmetry of the figure, we see that P' lies on a line perpendicular to the mirror passing through P, and that P and P' are equidistant from the mirror, on opposite sides. Thus *for a plane mirror the image of an object point lies on the extension of the normal from the object point to the mirror, and the object and image points are equidistant from the mirror.*

Before proceeding further, we pause here to introduce some conventions that anticipate later situations in which the object and image may be on either side of a reflecting or refracting surface. We adopt the following:

1. *When the object is on the same side of the reflecting surface as the incoming light, the object distance s is positive; otherwise it is negative.*

2. *When the image is on the same side of the reflecting surface as the outgoing light, the image distance s' is positive; otherwise it is negative.*

For a mirror the incoming and outgoing sides are always the same; in Fig. 37–2 they are both the left side. The above rules have been stated in this form so they may be applied also to *refracting* surfaces, for which light comes in one side and goes out the opposite side.

In Fig. 37–2 the object distance s is *positive* because the object point P is on the incoming side, that is, the left side, of the reflecting surface. The image distance s' is *negative* because the image point P' is *not* on the outgoing side of the surface. Thus object and image distances are related simply by

$$s = -s'. \tag{37–1}$$

Next we consider an object of finite size, parallel to the mirror, represented by the arrow PQ in Fig. 37–3. Point P', the image of P, is found as in Fig. 37–2. Two of the rays from Q are shown, and *all* rays from Q diverge from its image Q' after reflection. Other points of the object PQ have image points between P' and Q'. As the figure shows, the object PQ and image $P'Q'$ have the same size and orientation; $y = y'$.

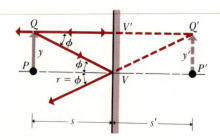

37-3 Construction for determining the height of an image formed by reflection at a plane surface.

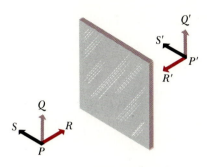

37–4 The image formed by a plane mirror is virtual, erect, and perverted, and is the same size as the object.

In general, if a transverse object is represented by an arrow, its image may point in the same direction as the object, or in the opposite direction. When, as in Fig. 37–3, the directions are the same, the image is called *erect;* if they are opposite, the image is *inverted.*

The three-dimensional virtual image of a three-dimensional object, formed by a plane mirror, is shown in Fig. 37–4. The imges $P'Q'$ and $P'S'$ are parallel to their objects, but $P'R'$ is reversed relative to PR. The image of a three-dimensional object formed by a plane mirror is the same size as the object in both its lateral and transverse dimensions. However, the image and object are not identical in all respects but are related in the same way as are a right hand and a left hand. To verify this, point your thumbs along PR and $P'R'$, your forefingers along PQ and $P'Q'$, and your middle fingers along PS and $P'S'$. When an object and its image are related in this way the image is said to be *perverted.* When the transverse dimensions of object and image are in the same direction, the image is erect. Thus a plane mirror forms an erect but perverted image.

37–2

REFLECTION AT A SPHERICAL MIRROR

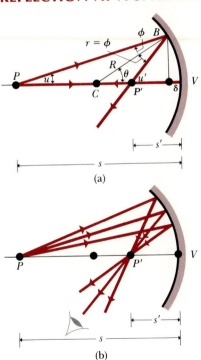

37–5 (a) Construction for finding the position of the image P' of a point object P, formed by a concave spherical mirror. (b) If the angle u is small, *all* rays from P intersect at P'. The eye sees some of the outgoing rays and perceives them as having come from P'.

Next we consider the formation of an image by a *spherical* mirror. Figure 37–5a shows a spherical mirror of radius of curvature R, with its concave side facing the incident light. The center of curvature of the surface is at C. Point P is an object point; for the moment we assume that P is farther from V than is the center of curvature. The ray PV, passing through C, strikes the mirror normally and is reflected back on itself. Point V is called the *vertex* of the mirror, and the line PCV the *optic axis.*

Ray PB, at an angle u with the axis, strikes the mirror at B, where the angle of incidence is ϕ and the angle of reflection is $r = \phi$. The reflected ray intersects the axis at point P'. We shall show shortly that if the angle u is small, *all* rays from P intersect the axis at the *same* point P', as in Fig. 37–5b, no matter what u is. Point P' is therefore the *image* of an object point P. The object distance, measured from the vertex V, is s, and the image distance is s'.

The object point P is on the same side as the incident light, so the object distance s is positive. The image point P' is on the same side as the reflected light, so the image distance s' is also positive.

Unlike the reflected rays in Fig. 37–2, the reflected rays in Fig. 37–5b actually intersect at point P' and then diverge from P' *as if* they had originated at this point. The image P' is called *real,* and a real image corresponds to a *positive* image distance.

Making use of the fact that an exterior angle of a triangle equals the sum of the two opposite interior angles, and considering the triangles PBC and $P'BC$ in Fig. 39–5a, we have

$$\theta = u + \phi, \qquad u' = \theta + \phi.$$

Eliminating ϕ between these equations gives

$$u + u' = 2\theta. \tag{37–2}$$

We may now introduce a sign convention for radii of curvature.

When the center of curvature C is on the same side as the outgoing (re-flected) light, the radius of curvature is positive; otherwise, it is negative.

In Fig. 37–5, R is positive, because the center of curvature C is on the same side of the mirror as the reflected light. This is always the case for reflection from the concave side of a surface; for a convex surface, the center of curvature is on the opposite side from the reflected light, and R is negative.

We may now compute the image position s'. Let h represent the height of point B above the axis, and δ the short distance from V to the foot of this vertical line. Now write expressions for the tangents of u, u', and θ, remembering that s, s', and R are all positive quantities:

$$\tan u = \frac{h}{s - \delta}, \qquad \tan u' = \frac{h}{s' - \delta}, \qquad \tan \theta = \frac{h}{R - \delta}.$$

These *trigonometric* equations cannot be solved as simply as the corre-sponding *algebraic* equations for a plane mirror. However, *if the angle u is small*, the angles u' and θ will be small also. Since the tangent of a small angle is nearly equal to the angle itself (in radians), we can replace $\tan u'$ by u', etc., in the equations above. Also if u is small, the distance δ can be neglected compared with s', s, and R. Hence, *approximately, for small angles,*

$$u = \frac{h}{s}, \qquad u' = \frac{h}{s'}, \qquad \theta = \frac{h}{R}.$$

Substituting in Eq. (37–2) and canceling h, we obtain

$$\frac{1}{s} + \frac{1}{s'} = \frac{2}{R} \tag{37–3}$$

as a general relation among the three quantities s, s', and R. Since the equation above does not contain the angle u, *all* rays from P making sufficiently small angles with the axis will, after reflection, intersect at P'. Such rays, nearly parallel to the axis, are called *paraxial* rays.

It must be understood that Eq. (37–3), as well as many similar rela-tions to be derived later, is the result of a calculation containing approxi-mations and is valid only for paraxial rays. (The term *paraxial approxima-tion* is also commonly used.) As the angle increases, the point P' moves closer to the vertex; a spherical mirror, unlike a plane mirror, does not form precisely a point image of a point object. This property of a spheri-cal mirror is called *spherical aberration*.

If $R = \infty$, the mirror becomes *plane* and Eq. (37–3) reduces to Eq. (37–1), previously derived for this special case.

Now suppose we have an object of finite size, represented by the arrow PQ in Fig. 37–6, perpendicular to the axis PV. The image of P formed by paraxial rays is at P'. Since the object distance for point Q is very nearly equal to that for the point P, the image $P'Q'$ is nearly straight and is perpendicular to the axis. Herein lies another approximation; if

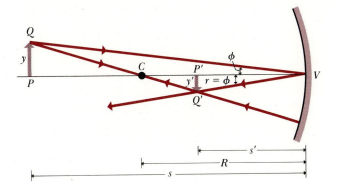

37–6 Construction for determining the position, orientation, and height of an image formed by a concave spherical mirror.

the height PQ is not sufficiently small compared to the object distance s, the image $P'Q'$ is not a straight line but is curved; this is another aberration of spherical surfaces, called *curvature of field*. We shall assume height PQ is small enough that this effect is negligible.

We note that object and image have different sizes, y and y', respectively, and that they have opposite orientation. It is useful to define the *lateral magnification m* as the ratio of image to object size:

$$m = \frac{y'}{y}.$$

Because triangles PVQ and $P'VQ'$ in Fig. 37–6 are *similar*, we also have the relation $y/s = -y'/s'$. The negative sign is needed because object and image are on opposite sides of the optic axis; if y is positive, y' is negative. Hence,

$$m = \frac{y'}{y} = -\frac{s'}{s}. \tag{37–4}$$

A negative value of m indicates that the image is *inverted* relative to the object, as the figure shows. In cases to be considered later, where m may be either positive or negative, a positive value always corresponds to an erect image, a negative value to an inverted one. For a plane mirror, $s = -s'$ and hence $y' = y$, as we have already shown.

Although the ratio of image size to object size is referred to as the *magnification*, the image formed by a mirror or lens may be *smaller* than the object. The magnification is then a small fraction and might more appropriately be called the *reduction*. The image formed by an astronomical telescope mirror, or by a camera lens, is much smaller than the object. It is interesting to note that for three-dimensional objects the ratio of image-to-object distances measured *along* the optic axis is different from the ratio of *lateral* distances (which we have called the lateral magnification). In particular, if m is a small fraction, the three-dimensional image of a three-dimensional object is reduced *longitudinally* much more than it is reduced *transversely*. Figure 37–7 illustrates this effect. Also, the image formed by a spherical mirror, like that of a plane mirror, is always perverted.

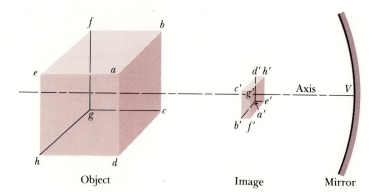

37–7 Schematic diagram of an object and its real, inverted, reduced image formed by a concave mirror.

Object   Image   Mirror

EXAMPLE   (a) What type of mirror is required to form an image, on a wall 3 m from the mirror, of the filament of a headlight lamp 10 cm in front of the mirror? (b) What is the height of the image if the height of the object is 5 mm?

Solution

a)
$$s = 10 \text{ cm}, \qquad s' = 300 \text{ cm},$$

$$\frac{1}{10 \text{ cm}} + \frac{1}{300 \text{ cm}} = \frac{2}{R},$$

$$R = 19.4 \text{ cm}.$$

Since the radius is positive, a concave mirror is required.

b)
$$m = \frac{y'}{y} = -\frac{s'}{s} = -\frac{300 \text{ cm}}{10 \text{ cm}} = -30.$$

The image is therefore inverted (m is negative) and is 30 times the height of the object, or $(30)(5 \text{ mm}) = 150 \text{ mm}$.   ◀

37–8 Construction for finding (a) the position and (b) the magnification of the image formed by a convex mirror.

In Fig. 37–8a, the *convex* side of a spherical mirror faces the incident light so that R is negative. Ray PB is reflected with the angle of reflection r equal to the angle of incidence ϕ, and the reflected ray, projected backward, intersects the axis at P'. As in the case of a concave mirror, *all* rays

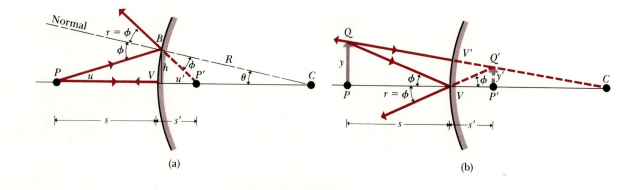

(a)   (b)

from P will, after reflection, diverge from the same point P', provided that the angle u is small; so P' is the image of P. The object distance s is positive, the image distance s' is negative, and the radius of curvature R is negative.

Figure 37–8b shows two rays diverging from the head of the arrow PQ, and the virtual image $P'Q'$ of this arrow. It is left to the reader to show, by the same procedure as that used for a concave mirror, that

$$\frac{1}{s} + \frac{1}{s'} = \frac{2}{R},$$

and the lateral magnification is

$$m = \frac{y'}{y} = -\frac{s'}{s}.$$

These expressions are exactly the same as those for a concave mirror, as they should be when a consistent sign convention is used.

37–3

FOCAL POINT AND FOCAL LENGTH

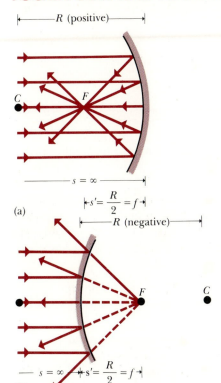

(a)

(b)

37–9 Incident rays parallel to the axis (a) converge to the focus F of a concave mirror, (b) diverge as though coming from the focus F of a convex mirror.

When an object point is a very large distance from a mirror, all rays from that point that strike the mirror are parallel to one another. The object distance is $s = \infty$ and, from Eq. (37–3),

$$\frac{1}{\infty} + \frac{1}{s'} = \frac{2}{R}, \qquad s' = \frac{R}{2}.$$

When R is positive (concave mirror), the situation is as shown in Fig. 37–9a. A beam of incident parallel rays converges after reflection at a point F at a distance $R/2$ from the vertex of the mirror. Point F is called the *focal point* or simply the *focus,* and its distance from the vertex, denoted by f, is called the *focal length.*

When R is negative (convex mirror), as in Fig. 37–9b, the image point is behind the mirror and f is negative. In that case the outgoing rays do not converge at a point but instead diverge as though they had come from the point F behind the mirror. In this case F is called a *virtual focus.*

The entire discussion may be reversed, as shown in Fig. 37–10. If the image distance s' is very large, then the outgoing rays are parallel to the optic axis. In this case the object distance s is given by

$$\frac{1}{s} + \frac{1}{\infty} = \frac{2}{R}, \qquad s = \frac{R}{2}.$$

In Fig. 37–10a, rays coming in toward the mirror pass through the focus and diverge from it, and after reflection they are parallel to the optic axis. In Fig. 37–10b, the incoming rays are converging as though they would meet at the virtual focus F, and they are reflected parallel to the optic axis.

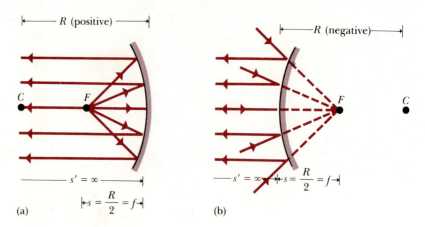

37–10 Rays from a point object at the focus of a spherical mirror are parallel to the axis after reflection. The object in part (b) is virtual.

Thus for both concave and convex mirrors the focal length f is related to the radius of curvature R by

$$f = \frac{R}{2}.$$

(37–5)

For a concave mirror both are positive, and for a convex mirror both are negative.

The relation between object and image distances, Eq. (37–3), for a mirror may now be written as

$$\frac{1}{s} + \frac{1}{s'} = \frac{1}{f}.$$

(37–6)

37–4

GRAPHICAL METHODS

The position and size of the image formed by a mirror may be found by a simple graphical method. This method consists of finding the point of intersection, after reflection from the mirror, of a few particular rays diverging from some point of the object *not* on the mirror axis, such as point Q in Fig. 37–11. Then (neglecting aberrations) *all* rays from this point that strike the mirror will intersect at the same point. Four rays whose paths may readily be traced are shown in Fig. 37–11. These are often called *principal rays*.

1. *A ray parallel to the axis,* after reflection, passes through the focus of a concave mirror or appears to come from the focus of a convex mirror.

2. *A ray from (or proceeding toward) the focus* is reflected parallel to the axis.

3. *A ray along the radius* (extended if necessary) intersects the surface normally and is reflected back along its original path.

4. *A ray to the vertex* is reflected with equal angles with the optic axis.

Once the position of the image point has been found by means of the intersection of any two of these principal rays (1, 2, 3, 4) the paths of all other rays from the same object point may be drawn.

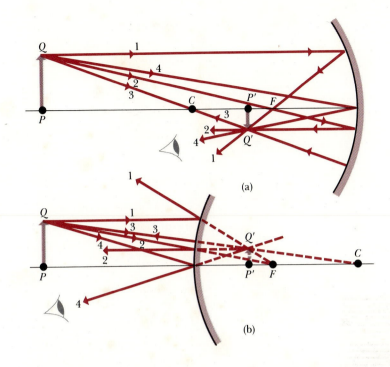

(a)

(b)

37–11 Rays used in the graphical method of locating an image.

EXAMPLE A concave mirror has a radius of curvature of magnitude 20 cm. Find graphically the image of an object in the form of an arrow perpendicular to the axis of the mirror at each of the following object distances: 30 cm, 20 cm, 10 cm, and 5 cm. Check the construction by computing the size and magnification of the image.

Solution The graphical constructions are shown in the four parts of Fig. 37–12. The reader should study each of these diagrams carefully, comparing each numbered ray with the description above. Several points are worth noting. First, in (b) the object and image distances are equal; ray 3 cannot be drawn in this case because a ray from Q through the center of curvature C does not strike the mirror. For the same reason, ray 2 cannot be drawn in (c); in this case the outgoing rays are parallel, corresponding to an infinite image distance. In (d), the outgoing rays have no real intersection point; they must be extended backwards to find the point from

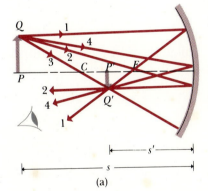

(a)

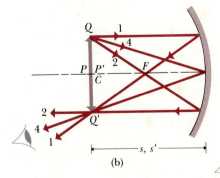

(b)

37–12 Image of an object at various distances from a concave mirror, showing principal-ray construction.

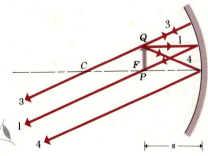

(c)

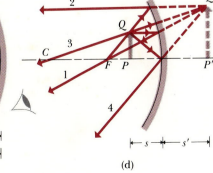

(d)

which they appear to diverge, that is, from the *virtual image point Q'*. The scale in (d) is somewhat distorted.

Measurement of the figures, with appropriate scaling, gives the following approximate image distances: (a) 15 cm, (b) 20 cm, (c) ∞ or $-\infty$, (d) -10 cm. To *compute* these distances, we first note that $f = R/2 = 10$ cm; then we use Eq. (37–6):

a) $\dfrac{1}{30 \text{ cm}} + \dfrac{1}{s'} = \dfrac{1}{10 \text{ cm}}, \qquad s' = 15 \text{ cm},$

b) $\dfrac{1}{20 \text{ cm}} + \dfrac{1}{s'} = \dfrac{1}{10 \text{ cm}}, \qquad s' = 20 \text{ cm},$

c) $\dfrac{1}{10 \text{ cm}} + \dfrac{1}{s'} = \dfrac{1}{10 \text{ cm}}, \qquad s' = \infty \text{ (or } -\infty\text{)},$

d) $\dfrac{1}{5 \text{ cm}} + \dfrac{1}{s'} = \dfrac{1}{10 \text{ cm}}, \qquad s' = -10 \text{ cm}.$

The lateral magnifications measured from the figures are approximately (a) $-1/2$, (b) -1, (c) ∞ or $-\infty$, (d) $+2$. *Computing* the magnifications from Eq. (37–4), we find:

a) $m = -(15 \text{ cm})/(30 \text{ cm}) = -1/2,$

b) $m = -(20 \text{ cm})/(20 \text{ cm}) = -1,$

c) $m = -(\pm\infty \text{ cm})/(10 \text{ cm}) = \mp\infty,$

d) $m = -(-10 \text{ cm})/(5 \text{ cm}) = +2.$

In (a) and (b) the image is inverted; in (c) and (d) it is erect.   ◄

In every problem involving formation of an image by a reflecting surface or a lens, the student should *always* draw a principal-ray diagram. This is useful not only as a graphical check on numerical calculations but also as an aid to understanding the basic concepts, especially the concept of *image*.

37–5

REFRACTION AT A PLANE SURFACE

The method of finding the image of a point object formed by rays *refracted* at a plane or spherical surface is essentially the same as for reflection; the only difference is that Snell's law replaces the law of reflection. We let n represent the refractive index of the material on the "incoming" side of the surface and n' that of the material on the "outgoing" side. The same convention of signs is used as in reflection.

Consider first a plane surface, shown in Fig. 37–13, and assume $n' > n$. A ray from the object point P toward the vertex V is incident normally and passes into the second material without deviation. A ray making an angle u with the axis is incident at B with an angle of incidence $\phi = u$. The angle of refraction, ϕ', is found from Snell's law,

$$n \sin \phi = n' \sin \phi'.$$

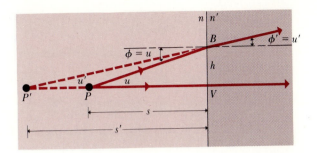

37–13 Construction for finding the position of the image P' of a point object P, formed by refraction at a plane surface.

The two rays both appear to come from the image point P' after refraction. From the triangles PVB and $P'VB$,

$$\tan \phi = \frac{h}{s}, \qquad \tan \phi' = \frac{h}{-s'}. \qquad (37\text{–}7)$$

We must write $-s'$, since the image point is on the side *opposite* to that of the refracted (outgoing) light.

If the angle u is small, the angles ϕ, u', and ϕ' are small also, and therefore, approximately,

$$\tan \phi = \sin \phi, \qquad \tan \phi' = \sin \phi'.$$

Then Snell's law can be written as

$$n \tan \phi = n' \tan \phi',$$

and from Eq. (37–7), after cancelling h, we have

$$\frac{n}{s} = \frac{n'}{s'},$$

or

$$\frac{s'}{s} = -\frac{n'}{n}. \qquad (37\text{–}8)$$

This is an *approximate* relation, *valid for paraxial rays only*. That is, a plane refracting surface does *not* image all rays from a point object at the same image point.

Consider next the image of a finite object, as in Fig. 37–14. The two rays diverging from point Q appear to diverge from its image Q' after refraction, and $P'Q'$ is the image of the object PQ. As the figure shows,

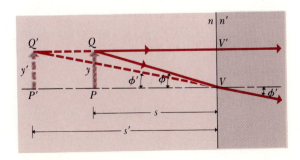

37–14 Construction for determining the height of an image formed by refraction at a plane surface.

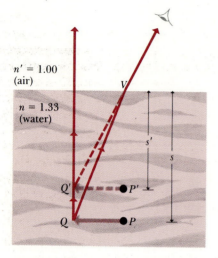

the object and the image are the same size, and so the lateral magnification is unity:

$$m = \frac{y'}{y} = 1. \tag{37-9}$$

The image *distance* is greater than the object distance, but image and object are the same height.

Here is a familiar example of refraction at a plane surface. When one looks vertically downward into the quiet water of a pond or a swimming pool, the apparent depth is less than the actual depth. Figure 37–15 illustrates this situation. Two rays are shown diverging from a point Q at a distance s below the surface. Here, n' (air) is less than n (water) and the ray through V is deviated *away from* the normal. The rays after refraction appear to diverge from Q', and the arrow PQ, to an observer looking vertically downward, appears lifted to the position $P'Q'$. From Eq. (37–8),

$$s' = -\frac{n'}{n}s = -\frac{1.00}{1.33}s = -0.75\,s.$$

37–15 Arrow $P'Q'$ is the image of the underwater object PQ. The angles of the rays with the vertical are exaggerated for clarity.

The apparent depth s' is therefore only three fourths of the actual depth s. The same phenomenon accounts for the apparent sharp bend in an oar when a portion of it extends below a water surface. The submerged portion appears lifted above its actual position.

*37–6

REFRACTION AT A SPHERICAL SURFACE

Finally, we consider refraction at a spherical surface. In Fig. 37–16, P is an object point at a distance s to the left of a spherical surface of radius R, with center of curvature C. The refractive indexes at the left and right of the surface are n and n', respectively. Ray PV, incident normally, passes into the second material without deviation. Ray PB, making an angle u with the axis, is incident at an angle ϕ with the normal and is refracted at an angle ϕ'. These rays intersect at P' at a distance s' to the right of the vertex.

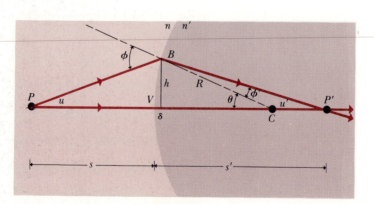

37–16 Construction for finding the position of the image P' of a point object P, formed by refraction at a spherical surface.

We shall show that if the angle u is small, *all* rays from P intersect at the same point P', so P' is the *real image* of P. The object and image distances are both positive. The radius of curvature is positive also, because the center of curvature is on the side of the outgoing or refracted light.

When we considered rays from P that are *reflected* at the surface, as in Fig. 37–8, the radius of curvature was negative. Thus the radius of curvature of a given surface has one sign for reflected light and the opposite sign for refracted light. This apparent inconsistency is resolved by noting that in both cases R is positive when the center of curvature is on the "outgoing" side of the reflecting or refracting surface and negative when it is on the "incoming" side.

From the triangles PBC and $P'BC$, we have

$$\phi = \theta + u, \qquad \theta = u' + \phi'. \tag{37–10}$$

From Snell's law,

$$n \sin \phi = n' \sin \phi'.$$

Also, the tangents of u, u', and θ are

$$\tan u = \frac{h}{s + \delta}, \qquad \tan u' = \frac{h}{s' - \delta'}, \qquad \tan \theta = \frac{h}{R - \delta}.$$

For paraxial rays we may approximate both the sine and tangent of an angle by the angle itself, and neglect the small distance δ. Snell's law then becomes

$$n\phi = n'\phi',$$

and, combining with the first of Eqs. (37–10), we obtain

$$\phi' = \frac{n}{n'}(u + \theta).$$

Substituting this in the second of Eqs. (37–10) gives

$$nu + n'u' = (n' - n)\theta.$$

Using the small-angle approximations and canceling h, we obtain

$$\frac{n}{s} + \frac{n'}{s'} = \frac{n' - n}{R}. \tag{37–11}$$

This equation does not contain the angle u, so the image distance is the same for all paraxial rays from P.

If the surface is plane, $R = \infty$ and this equation reduces to Eq. (37–8), already derived for the special case of a plane surface.

The magnification is found from the construction in Fig. 37–17. From point Q draw two rays, one through the center of curvature C and the other incident at the vertex V. From the triangles PQV and $P'Q'V$,

$$\tan \phi = \frac{y}{s}, \qquad \tan \phi' = \frac{-y'}{s'}.$$

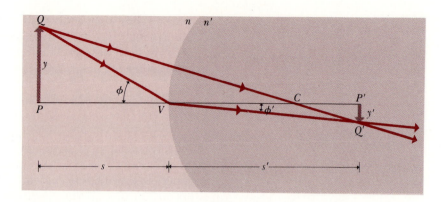

37–17 Construction for determining the height of an image formed by refraction at a spherical surface.

and from Snell's law

$$n \sin \phi = n' \sin \phi'.$$

For small angles,

$$\tan \phi = \sin \phi, \qquad \tan \phi' = \sin \phi',$$

and hence

$$\frac{ny}{s} = -\frac{n'y'}{s'},$$

or

$$m = \frac{y'}{y} = -\frac{ns'}{n's}. \tag{37–12}$$

This is the same relation previously derived for a *plane* surface!

EXAMPLE 1   One end of a cylindrical glass rod (Fig. 37–18) is ground to a hemispherical surface of radius $R = 20$ mm. Find the image distance of a point object on the axis of the rod, 80 mm to the left of the vertex. The rod is in air.

Solution

$$n = 1, \qquad n' = 1.5,$$
$$R = +20 \text{ mm}, \qquad s = +80 \text{ mm}.$$

From Eq. (37–11),

$$\frac{1}{80 \text{ mm}} + \frac{1.5}{s'} = \frac{1.5 - 1}{+20 \text{ mm}},$$
$$s' = +120 \text{ mm}.$$

The image is therefore formed at the right of the vertex (s' is positive) and at a distance of 120 mm from it. Suppose that the object is an arrow 1 mm high, perpendicular to the axis. Then, from Eq. (37–12),

$$m = -\frac{ns'}{n's} = -\frac{(1)(120 \text{ mm})}{(1.5)(80 \text{ mm})} = -1.$$

That is, the image is the same height as the object but is inverted.   ◄

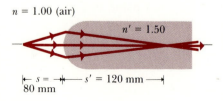

$n = 1.00$ (air)

$n' = 1.50$

$s = 80$ mm $s' = 120$ mm

37–18

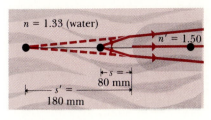

$n = 1.33$ (water)

$n' = 1.50$

$s = 80$ mm

$s' = 180$ mm

37–19

EXAMPLE 2 The same rod is immersed in water of index 1.33; the other quantities have the same values as before. Find the image distance (Fig. 37–19).

Solution

$$\frac{1.33}{80 \text{ mm}} + \frac{1.5}{s'} = \frac{1.5 - 1.33}{+20 \text{ mm}}, \qquad s' = -180 \text{ mm}.$$

The fact that s' is negative means that the rays, after refraction by the surface, are not converging but *appear* to diverge from a point 180 mm to the *left* of the vertex. We have met a similar case before in the refraction of spherical waves by a plane surface and have called the point a *virtual image*. In this example, then, the surface forms a virtual image 180 mm to the left of the vertex. ◄

Equations (37–11) and (37–12) can be applied to both convex and concave refracting surfaces when a consistent sign convention is used, and they apply whether n' is greater or less than n. The reader should construct diagrams like Figs. 37–16 and 37–17, when R is negative and $n' < n$, and use them to derive Eqs. (37–11) and (37–12).

37–7

SUMMARY

The results of this chapter are summarized in Table 37–1. Note that if we let $R = \infty$, the equation for a plane surface follows immediately from the corresponding equation for a spherical surface.

TABLE 37–1

	Plane mirror	Spherical mirror	Plane refracting surface	Spherical refracting surface
Object and image distances	$\dfrac{1}{s} + \dfrac{1}{s'} = 0$	$\dfrac{1}{s} + \dfrac{1}{s'} = \dfrac{2}{R} = \dfrac{1}{f}$	$\dfrac{n}{s} + \dfrac{n'}{s'} = 0$	$\dfrac{n}{s} + \dfrac{n'}{s'} = \dfrac{n' - n}{R}$
Lateral magnification	$m = -\dfrac{s'}{s} = 1$	$m = -\dfrac{s'}{s}$	$m = -\dfrac{ns'}{n's} = 1$	$m = -\dfrac{ns'}{n's}$

Sign conventions:
   s positive when object is on incoming side of surface, negative otherwise.
   s' positive when image is on outgoing side of surface, negative otherwise.
   R positive when center of curvature is on outgoing side of surface, negative otherwise.
   m is positive when image is erect, negative when inverted.

QUESTIONS

37–1 Can a person see a real image by looking backward along the direction from which the rays come? A virtual image? Can you tell by looking whether an image is real or virtual? How *can* the two be distinguished?

37–2 Why does a plane mirror reverse left and right but not top and bottom?

37–3 For a spherical mirror, if $s = f$, then $s' = \infty$, and the lateral magnification m is infinite. Does this make sense? If so, what does it mean?

37–4 According to the discussion of the preceding chapter, light rays are reversible. Are the formulas in Table 37–1 still valid if object and image are interchanged? What does reversibility imply with respect to the *forms* of the various formulas?

37–5 If a spherical mirror is immersed in water, does its focal length change?

37–6 For what range of object positions does a concave spherical mirror form a real image? What about a convex spherical mirror?

37–7 If a piece of photographic film is placed at the location of a real image, the film will record the image. Can this be done with a virtual image? How might one record a virtual image?

37–8 When a room has mirrors on two opposing walls, an infinite series of reflections can be seen. Discuss this phenomenon in terms of images. Why do the distant images appear darker?

37–9 When observing fish in an aquarium filled with water, one can see clearly only when looking nearly perpendicularly to the glass wall; objects viewed at an oblique angle always appear blurred. Why? Do the fish have the same problem when looking at you?

37–10 Can an image formed by one reflecting or refracting surface serve as an object for a second reflection or refraction? Does it matter whether the first image is real or virtual?

37–11 A concave mirror (sometimes surrounded by lights) is often used as an aid for applying cosmetics to the face. Why is such a mirror always concave rather than convex? What considerations determine its radius of curvature?

37–12 A student claimed that one can start a fire on a sunny day by use of the sun's rays and a concave mirror. How is this done? Is the concept of image relevant? Could one do the same thing with a convex mirror?

37–13 A person looks at his reflection in the concave side of a shiny spoon. Is it right side up or inverted? What if he looks in the convex side?

37–14 In the example in Section 37–4, there appears to be an ambiguity for the case $s = 10$ cm, as to whether s' is ∞ or $-\infty$, and as to whether the image is erect or inverted. How is this resolved? Or is it?

37–15 "See yourself as others see you." Can you do this with an ordinary plane mirror? If not, how *can* you do it?

37–16 The shadow formed under a tree by sunlight partially blocked by leaves ordinarily is sprinkled with small circles of light and irregular patterns. During a solar eclipse, however, the shadow is sprinkled instead with a pattern of overlapping crescents that copy the shape of the occluded sun. Why?

PROBLEMS

37–1 What is the size of the smallest vertical plane mirror in which an observer standing erect can see his full-length image?

37–2 The image of a tree just covers the length of a 5-cm plane mirror when the mirror is held 30 cm from the eye. The tree is 100 m from the mirror. What is its height?

37–3 An object is placed between two mirrors arranged at right angles to each other.

a) Locate all of the images of the object.

b) Draw the paths of rays from the object to the eye of an observer.

37–4 An object 1 cm high is 20 cm from the vertex of a concave spherical mirror whose radius of curvature is 50 cm. Compute the position and size of the image. Is it real or virtual? Erect or inverted?

37–5 A concave mirror is to form an image of the filament of a headlight lamp on a screen 4 m from the mirror. The filament is 5 mm high, and the image is to be 40 cm high.

a) What should be the radius of curvature of the mirror?

b) How far in front of the vertex of the mirror should the filament be placed?

37–6 The diameter of the moon is 3480 km and its distance from the earth is 386,000 km. Find the diameter of the image of the moon formed by a spherical concave telescope mirror of focal length 4 m.

37–7 A spherical concave shaving mirror has a radius of curvature of 30 cm.

a) What is the magnification when the face is 10 cm from the vertex of the mirror?

b) Where is the image?

37–8 A concave spherical mirror has a radius of curvature of 10 cm. Make a diagram of the mirror to scale, and show rays incident on it parallel to the axis and at distances of 1, 2, 3, 4, and 5 cm from the axis. Using a protractor, construct the reflected rays and indicate the points at which they cross the axis.

37–9 An object is 16 cm from the center of a silvered spherical glass Christmas tree ornament 8 cm in diameter. What are the position and magnification of its image?

37–10 A concave mirror of radius 5 cm has a radius of curvature of 20 cm.

a) What is its focal length?

b) If the mirror is immersed in water (refractive index 1.33), what is its focal length?

37–11 An object 2 cm high is placed 5 cm away from a concave spherical mirror having radius of curvature of 20 cm.

a) Draw a principal-ray diagram showing formation of the image.

b) Determine the position, size, orientation, and nature of the image.

37–12 Prove that the image formed of a real object by a convex mirror is always virtual, no matter what the object position.

37–13 If light striking a convex mirror does not diverge from an object point but instead is converging toward a point at a (negative) distance s to the right of the mirror, this point is called a *virtual object.*

a) For a convex mirror having radius of curvature 10 cm, for what range of virtual-object positions is a real image formed?

b) What is the orientation of the image?

c) Draw a principal-ray diagram showing formation of such an image.

37–14 A tank whose bottom is a mirror is filled with water to a depth of 20 cm. A small object hangs motionless 8 cm under the surface of the water. What is the apparent depth of its image when viewed at normal incidence?

37–15 A ray of light in air makes an angle of incidence of 45° at the surface of a sheet of ice. The ray is refracted within the ice at an angle of 30°.

a) What is the critical angle for the ice?

b) A speck of dirt is embedded 2 cm below the surface of the ice. What is its apparent depth when viewed at normal incidence?

37–16 A microscope is focused on the upper surface of a glass plate. A second plate is then placed over the first. In order to focus on the bottom surface of the second plate, the microscope must be raised 1 mm. In order to focus on the upper surface it must be raised 2 mm *farther.* Find the index of refraction of the second plate. (This problem illustrates one method of measuring index of refraction.)

37–17 A layer of ether ($n = 1.36$) 2 cm deep floats on water ($n = 1.33$) 4 cm deep. What is the apparent distance from the ether surface to the bottom of the water layer, when viewed at normal incidence?

37–18 The end of a long glass rod 8 cm in diameter has a hemispherical surface 4 cm in radius. The refractive index of the glass is 1.50. Determine each position of the image if an object is placed on the axis of the rod at the following distances from its end:

a) infinitely far,

b) 16 cm,

c) 4 cm.

37–19 The rod of Problem 37–18 is immersed in a liquid. An object 60 cm from the end of the rod and on its axis is imaged at a point 100 cm inside the rod. What is the refractive index of the liquid?

37–20 What should be the index of refraction of a transparent sphere in order that paraxial rays from an infinitely distant object will be brought to a focus at the vertex of the surface opposite the point of incidence?

37–21 The left end of a long glass rod 10 cm in diameter, of index 1.50, is ground and polished to a convex hemispherical surface of radius 5 cm. An object in the form of an arrow 1 mm long, at right angles to the axis of the rod, is located on the axis 20 cm to the left of the vertex of the convex surface. Find the position and magnification of the image of the arrow formed by paraxial rays incident on the convex surface.

37–22 A transparent rod 40 cm long is cut flat at one end and rounded to a hemispherical surface of 12 cm radius at the other end. A small object is embedded within the rod along its axis and halfway between its ends. When viewed from the flat end of the rod the apparent depth of the object is 12.5 cm. What is its apparent depth when viewed from the curved end?

37–23 A solid glass hemisphere having a radius of 10 cm and a refractive index of 1.50 is placed with its flat face downward on a table. A parallel beam of light of circular cross section 1 cm in diameter travels directly downward and enters the hemisphere along its diameter. What is the diameter of the circle of light formed on the table?

37–24 A small tropical fish is at the center of a spherical fish bowl 30 cm in diameter.

a) Find its apparent position and magnification to an observer outside the bowl. The effect of the thin walls of the bowl may be neglected.

b) The owner of the bowl was advised by a friend to keep it out of direct sunlight to avoid blinding the fish, who might swim into the focal point of the parallel rays from the sun. Is the focal point actually within the bowl? If not, where is it?

37–25 In Fig. 37–20a the first focal length f is seen to be the value of s corresponding to $s' = \infty$; in (b) the second focal length f' is the value of s' when $s = \infty$.

a) Prove that $n/n' = f/f'$.

b) Prove that the general relation between object and image distances is

$$\frac{f}{s} + \frac{f'}{s'} = 1.$$

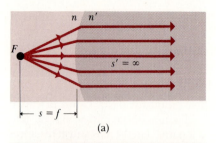

(a)

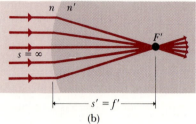

(b)

Figure 37–20

38

LENSES AND OPTICAL INSTRUMENTS

The most familiar and most significant simple optical device is the lens. We introduce the concepts of focus and focal length for a lens and then study the formation of images by lenses. Graphical methods for the analysis of image formation are developed. Lenses are central to the operation of many familiar optical devices, including the human eye, magnifiers, projectors, and cameras. Other devices, such as telescopes and compound microscopes, use combinations of lenses or of lenses and mirrors. Here, as in the preceding chapter, the concept of image provides the key to analyzing and understanding these optical devices.

38-1

THE THIN LENS

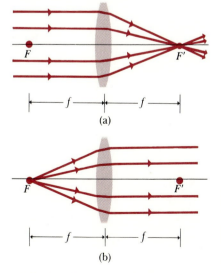

(a)

(b)

38-1 Focal points of a thin lens.

A lens is an optical system including two refracting surfaces. In this chapter we concentrate primarily on lenses with two *spherical* surfaces sufficiently close together so that the distance between them (the thickness of the lens) can be neglected; such a device is therefore called a *thin lens*. The behavior of a lens can be derived in detail by repeated application of the analysis in Section 37–6, for refraction by a single spherical surface. We postpone this derivation until Section 38–4, in order to move immediately into a discussion of the properties of thin lenses.

A lens of the type shown in Fig. 38–1 has the property that a beam of parallel rays converges, after passing through the lens, at a point F', as shown in Fig. 38–1a. Similarly, rays passing through point F emerge from the lens as a beam of parallel rays, as shown in Fig. 38–1b. The points F and F' are called the first and second *focal points* or *foci* (plural form of *focus*), and f is called the *focal length*. The concepts of focus and focal length are directly analogous to the same concepts used for spherical mirrors in Section 37–3. The two focal lengths in Fig. 38–1, both labeled f, are always equal for a thin lens, even when the two sides have different curvatures. This may be surprising, but it's true! The central horizontal line is called the *optic axis,* as with spherical mirrors. The centers of curvature of the two spherical surfaces lie on the optic axis.

The focal length f depends on the refractive index of the lens material (and on that of the surrounding matter if it is not vacuum) and on the radii of curvature of the spherical surfaces. This relationship, called the lensmaker's equation, is derived in Section 38–4.

A thin lens forms an image of an object of finite size. Figure 38–2 shows how the position of the image may be calculated. Let s and s' be the object and image distances, respectively, and let y and y' be the object and image heights. This notation is the same as used in Chapter 37. Ray QA, parallel to the optic axis before refraction, passes through the second focus F' after refraction. Ray QOQ' passes undeflected straight through the center of the lens because at the center the two surfaces are parallel and (we have assumed) very close together.

The angles labeled α in Fig. 38–2 are vertical angles and are equal. Therefore the two right triangles PQO and $P'Q'O'$ are *similar,* and ratios of corresponding sides are equal. Thus

$$\frac{y}{s} = -\frac{y'}{s'}, \quad \text{or} \quad \frac{y'}{y} = -\frac{s'}{s}. \tag{38–1}$$

(The negative sign arises because y' is negative, according to the convention used in Chapter 37.) Also, the two angles labeled β are equal, and the two right triangles OAF' and $P'Q'F'$ are similar, so

$$\frac{y}{f} = -\frac{y'}{s'-f}, \quad \text{or} \quad \frac{y'}{y} = -\frac{s'-f}{f}. \tag{38–2}$$

We now equate Eqs. (38–1) and (38–2), divide by s', and rearrange, to obtain

$$\frac{1}{s} + \frac{1}{s'} = \frac{1}{f}. \tag{38–3}$$

This analysis also gives us immediately the magnification $m = y'/y$ of the system; from Eq. (38–1),

$$m = -\frac{s'}{s}. \tag{38–4}$$

The negative sign tells us that when s and s' are both positive the image is *inverted,* and y and y' have opposite signs.

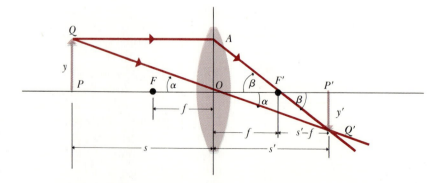

38–2 Construction used to find image position for a thin lens. The ray QAQ' is shown as bent at the midplane of the lens rather than at the two surfaces, to emphasize that the thickness of the lens is assumed to be very small.

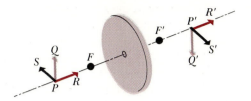

38–3   A lens forms a three-dimensional image of a three-dimensional object.

Equations (38–3) and (38–4) are the basic equations for thin lenses, and it is pleasing to note that their form is exactly the same as the corresponding equations for spherical mirrors, Eqs. (37–4) and (37–6).

The three-dimensional image of a three-dimensional object, formed by a lens, is shown in Fig. 38–3. Since point R is nearer the lens than point P, its image, from Eq. (38–3), is farther from the lens than is point P', and the image $P'R'$ points in the same direction as the object PR. Arrows $P'S'$ and $P'Q'$ are reversed in space, relative to PS and PQ. Although we speak of the image as "inverted," only its transverse dimensions are reversed.

Figure 38–3 should be compared with Fig. 37–4, showing the image formed by a plane mirror. Note that the image formed by a lens, although it is inverted, is *not* perverted. That is, if the object is a left hand, its image is a left hand also. This may be verified by pointing the left thumb along PR, the left forefinger along PQ, and the left middle finger along PS. A rotation of 180° about the thumb as an axis then brings the fingers into coincidence with $P'Q'$ and $P'S'$. In other words, *inversion* of an image is equivalent to a rotation of 180° about the lens axis.

38–2

DIVERGING LENSES

A bundle of parallel rays incident on the lens shown in Figure 38–1 converges to a real image after passing through the lens. The lens is called a *converging lens*. Its focal length is a positive quantity and therefore the lens is also called a *positive lens*.

A bundle of parallel rays incident on the lens in Fig. 38–4 *diverges* after refraction, and the lens is called a *diverging lens*. Its focal length is a negative quantity and therefore the lens is also called a *negative lens*. The focal points of a negative lens are reversed, relative to those of a positive lens. The second focal point, F', of a negative lens is the point from

38–4   Focal points and focal length of a diverging lens.

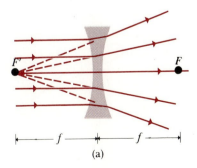

(a)

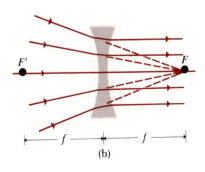

(b)

38–5 (a) Meniscus, plano-convex, and double-convex converging lenses. (b) Meniscus, plano-concave, and double concave diverging lenses. A converging lens is always thicker at its center than at its edges, and the reverse is true for all diverging lenses.

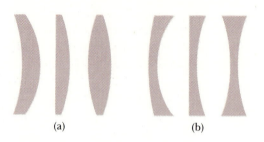

(a)   (b)

which rays, originally parallel to the axis, appear to diverge after refraction, as in Fig. 38–4a. Incident rays converging toward the first focal point *F*, as in Fig. 38–4b, emerge from the lens parallel to its axis. Equations (38–3) and (38–4) apply both to negative and to positive lenses. Various types of lenses, both converging and diverging, are shown in Fig. 38–5.

38–3

GRAPHICAL METHODS

The position and size of the image of an object formed by a thin lens may be found by a simple graphical method. This method consists of finding the point of intersection, after passing through the lens, of a few rays (called *principal rays*) diverging from some chosen point of the object *not* on the lens axis, such as point *Q* in Fig. 38–6. Then (neglecting lens aberrations) all rays from this point that pass through the lens will intersect at the same point. In using the graphical method, the entire deviation of any ray is assumed to take place at a plane through the center of the lens. Three principal rays whose paths may readily be traced are shown in Fig. 38–6.

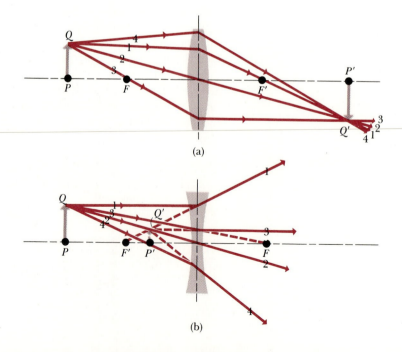

(a)

(b)

38–6 Principal-ray diagram, showing graphical method of locating an image. (a) A converging lens; (b) a diverging lens.

1. *A ray parallel to the axis,* after refraction by the lens, passes through the second focal point of a converging lens, or appears to come from the second focal point of a diverging lens.

2. *A ray through the center of the lens* is not appreciably deviated, since the two lens surfaces through which the central ray passes are very nearly parallel if the lens is thin.

3. *A ray through (or proceeding toward) the first focal point* emerges parallel to the axis.

Once the position of the image point has been found by means of the intersection of any two rays 1, 2, and 3, the paths of all other rays from the same point, such as ray 4 in Fig. 38–6, may be drawn. A few examples of this procedure are given in Fig. 38–7. The unnumbered rays in Fig. 38–7 are not principal rays.

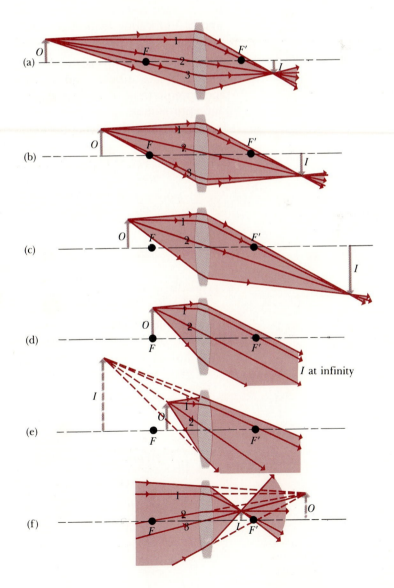

38–7 Formation of an image by a thin lens. The principal rays are labeled.

The reader should study each of these diagrams carefully, comparing each numbered ray with the above description. Several points are worth noting. In (d) the object is at the focus; ray 3 cannot be drawn because it does not pass through the lens. In (e) the object distance is less than the focal length. The outgoing rays are divergent, and the *virtual image* is located by extending the outgoing rays backward. In this case the image distance s' is negative. Part (f) corresponds to a *virtual object*. The incoming rays are not diverging from a real object point but converging as though they would meet at the virtual object point O on the right side. The object distance s is negative in this case. The image is real, and the image distance s' is positive and less than f.

EXAMPLE   A converging lens has a focal length of 20 cm. Find graphically the image location for an object at each of the following distances from the lens: 50 cm, 20 cm, 15 cm, −40 cm. Determine the magnification in each case. Check your results by calculating the image position and magnification from Eqs. (38–3) and (38–4).

Solution   The appropriate principal-ray diagrams are shown in Fig. 38–7, parts (a), (d), (e), and (f). The approximate image distances, from measurements of these diagrams, are 35 cm, ∞, −40 cm, and 15 cm, and the approximate magnifications are −2/3, ∞, +3, and +1/3.
Calculating the image positions from Eq. (38–3), we find

a) $\dfrac{1}{50\text{ cm}} + \dfrac{1}{s'} = \dfrac{1}{20\text{ cm}},\qquad s' = 33.3\text{ cm},$

d) $\dfrac{1}{20\text{ cm}} + \dfrac{1}{s'} = \dfrac{1}{20\text{ cm}},\qquad s' = \infty,$

e) $\dfrac{1}{15\text{ cm}} + \dfrac{1}{s'} = \dfrac{1}{20\text{ cm}},\qquad s' = -60\text{ cm},$

f) $\dfrac{1}{-40\text{ cm}} + \dfrac{1}{s'} = \dfrac{1}{20\text{ cm}},\qquad s' = 13.3\text{ cm}.$

The graphical results are fairly close to these except for (e), where the precision of the diagram is limited by the fact that the rays extended backward have nearly the same direction.
From Eq. (38–4), the magnifications are

a) $m = -\dfrac{33.3\text{ cm}}{30\text{ cm}} = -2/3,$

d) $m = -\dfrac{\infty}{20\text{ cm}} = -\infty,$

e) $m = -\dfrac{-60\text{ cm}}{15\text{ cm}} = +4,$

f) $m = -\dfrac{13.3\text{ cm}}{-40\text{ cm}} = 1/3.$

In every problem involving formation of an image by a lens, the student should *always* draw a principal-ray diagram; this serves not only as a graphical check on numerical calculations but also as an aid to the understanding of basic concepts, especially the concept of *image*.

38–4

IMAGES AS OBJECTS

An image formed by one lens or refracting surface can serve as the object for a second lens or refracting surface. Many optical systems, such as camera lenses, microscopes, and telescopes, use more than one lens. In each case the image formed by any one lens serves as the object for the next lens. Figure 38–8 shows various possibilities. Lens 1 forms a real image at P of a real object at O. This real image serves as a real object for lens 2. The virtual image at Q formed by lens 2 is a real object for lens 3. If lens 4 were not present, lens 3 would form a real image at R. Although this image is never formed, it serves as a *virtual object* for lens 4, which forms a final real image at I.

These considerations can also be applied to images formed by individual refracting surfaces, as discussed in Section 37–6. In particular, the thin-lens equation, Eq. (38–3), can be derived by repeated application of the single-surface equation, Eq. (37–11). For this derivation we consider the situation of Fig. 38–9. The figure shows several rays diverging from point Q of an object PQ. The first surface forms a virtual image of Q at Q'. This virtual image serves as a real object for the second surface of the lens, which forms a real image of Q' at Q''. Distance s_1 is the object distance for the first surface; s_1' is the corresponding image distance. The object distance for the second surface is s_2, equal to the sum of s_1' and the lens thickness t, and s_2' is the image distance for the second surface.

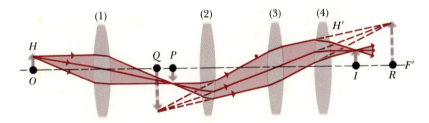

38–8 The object for each lens, after the first, is the image formed by the preceding lens.

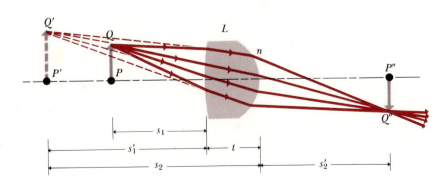

38–9 The image formed by the first surface of a lens serves as the object for the second surface.

If, as is often the case, the lens is so thin that its thickness t is negligible in comparison with the distances s_1, s_1', s_2, s_2', we may assume that s_1' equals $-s_2$ and measure object and image distances from either vertex of the lens. We shall also assume the medium on both sides of the lens to be air, with index of refraction 1.00. For the first refraction, Eq. (37–11) becomes

$$\frac{1}{s_1} + \frac{n}{s_1'} = \frac{n-1}{R_1}.$$

Refraction at the second surface yields the equation

$$\frac{n}{s_2} + \frac{1}{s_2'} = \frac{1-n}{R_2}.$$

Adding these two equations, and remembering that the lens is so thin that $s_2 = -s_1'$, we find

$$\frac{1}{s_1} + \frac{1}{s_2'} = (n-1)\left(\frac{1}{R_1} - \frac{1}{R_2}\right).$$

Since s_1 is the object distance for the thin lens and s_2' is the image distance, the subscripts may be omitted, and we finally obtain

$$\frac{1}{s} + \frac{1}{s'} = (n-1)\left(\frac{1}{R_1} - \frac{1}{R_2}\right). \tag{38–5}$$

The usual sign conventions apply to this equation. Thus, in Fig. 38–10, s, s', and R_1 are positive quantities, but R_2 is negative.

Comparing Eq. (38–5) with our previous form of the thin-lens equation, Eq. (38–3), we see that the focal length is given by

$$\frac{1}{f} = (n-1)\left(\frac{1}{R_1} - \frac{1}{R_2}\right), \tag{38–6}$$

which is known as the *lensmaker's equation*. Thus in this derivation we obtain the thin-lens equation and also find how the focal length is related to the refractive index and the radii of curvature of the surfaces.

The focal length can also be obtained directly from Eq. (38–5) by recalling that the focus is the image of an infinitely distant object; when $s = \infty$, $s' = f$. Inserting these values into Eq. (38–5) yields Eq. (38–6) immediately.

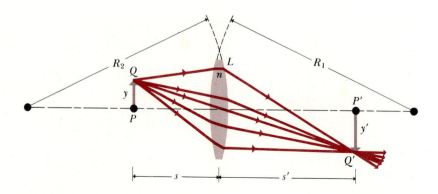

38–10 A thin lens.

EXAMPLE In Fig. 38–10, let the absolute magnitudes of the radii of curvature of the lens surfaces be respectively 20 cm and 5 cm. Since the center of curvature of the first surface is on the side of the outgoing light, $R_1 = +20$ cm, and since that of the second surface is not, $R_2 = -5$ cm. Let $n = 1.50$. Then

$$\frac{1}{f} = (1.50 - 1)\left(\frac{1}{20\text{ cm}} - \frac{1}{-5\text{ cm}}\right), \qquad f = +8\text{ cm.} \qquad \blacktriangleleft$$

This same method of analysis can be extended to systems consisting of several *thick* lenses; an example is shown in Fig. 38–11. The arrow at point O represents a small object at right angles to the axis. A narrow cone of rays diverging from the head of the arrow is traced through the system. Surface 1 forms a real image of the arrow at point P. Distance OV_1 is the object distance for the first surface and distance V_1P is the image distance. Both of these are positive.

The image at P, formed by surface 1, serves as the object for surface 2. The object distance is PV_2 and is positive, since the direction from P to V_2 is the same as that of the incident light. The second surface forms a virtual image at point Q. The image distance is V_2Q and is negative because the direction from V_2 to Q is opposite to that of the refracted light.

The image at Q, formed by surface 2, serves as the object for surface 3. The object distance is QV_3 and is positive. The image at Q, although virtual, constitutes a *real object* so far as surface 3 is concerned. The rays incident on surface 3 are rendered converging and, except for the interposition of surface 4, would converge to a real image at point R. Even though this image is never formed, distance V_3R is the image distance for surface 3 and is positive.

The rays incident on surfaces 1, 2, and 3 have all been diverging, and the object distance has been the distance from the surface to the point from which the rays were actually or apparently diverging. The rays incident on surface 4, however, are *converging* and there is no point at the left of the vertex from which they diverge or appear to diverge. The *image at R, toward which the rays are converging,* is the object for surface 4, and since the direction from R to V_4 is opposite to that of the incident light, the object distance RV_4 is negative. The image at R is called a *virtual object* for surface 4. In general, whenever a *converging* cone of rays is incident on a surface, the point toward which the rays are converging serves as the object, the object distance is negative, and the point is called a virtual object.

Finally, surface 4 forms a real image at I, the image distance being V_4I and positive.

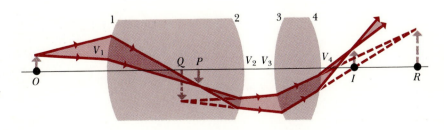

38–11 The object for each surface, after the first, is the image formed by the preceding surface.

*38–5

LENS ABERRATIONS

The relatively simple equations we have derived relating object and image distances, focal lengths, radii of curvature, etc., were based upon the *paraxial approximation;* all rays were assumed to be *paraxial,* that is, to make small angles with the axis. In general, however, a lens must image not only points on its axis, but points that lie off the axis as well. Furthermore, because of the finite size of the lens, the cone of rays that forms the image of any point is of finite size. Nonparaxial rays proceeding from a given object point *do not,* in general, all intersect at precisely the same point after refraction by a lens. Consequently, the image formed by these rays is not a perfectly sharp one. Furthermore, the focal length of a lens depends upon its index of refraction, which varies with wavelength. Therefore, if the light proceeding from an object is not monochromatic, a lens forms a number of colored images, which lie in different positions and are of different sizes, even if formed by paraxial rays.

The departures of an actual image from the predictions of simple theory are called *aberrations.* Those caused by the variation of index with wavelength are the *chromatic aberrations.* The others, which would arise even if the light were monochromatic, are the *monochromatic aberrations.* Lens aberrations are not caused by faulty construction of the lens, such as the failure of its surfaces to conform to a truly spherical shape, but are simply consequences of the laws of refraction at spherical surfaces.

The monochromatic aberrations are all related to the failure of the paraxial-ray approximation for lenses of finite aperture, but it is customary to distinguish various aspects of this difficulty, each with its characteristic effect. *Spherical aberration* is the failure of rays from a point object on the optic axis to converge to a point image; instead, the rays converge to a circle of minimum radius, called *the circle of least confusion,* and then diverge again, as shown in Fig. 38–12. The corresponding effect for points off the axis produces images that are comet-shaped figures rather than circles; this is called *coma.*

Astigmatism is the imaging of a point off the axis as a *line;* in this aberration the rays from a point object converge at some distance from the lens to a line in the place defined by the optic axis and the object point and, at a somewhat different distance from the lens, to a line *perpendicular* to this plane. The circle of least confusion appears between these two positions, at a location that depends on the object point's dis-

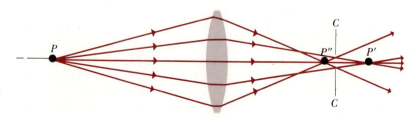

38–12 Spherical aberration. The circle of least confusion is shown by C–C.

tance from the axis as well as its distance from the lens. As a result, object points lying in a plane are, in general, imaged not in a plane but in some curved surface; this effect is called *curvature of field.* Finally, the image of a straight line that does not pass through the axis may be curved; as a result the image of a square with the axis through its center may resemble a barrel (sides bent outward) or a pincushion (sides bent inward). This effect, called *distortion,* is not related to lack of sharpness of the image but results from a change in lateral magnification with distance from the axis.

Chromatic aberrations result directly from the variation of index of refraction with wavelength. Even in the absence of all monochromatic aberration, different wavelengths are imaged at different points, and when an object is illuminated with white light containing a mixture of wavelengths, there is no single point at which a point object is imaged. The magnification of a lens also varies with wavelength; this effect is responsible for the rainbow-fringed images seen with inexpensive binoculars or telescopes.

It is impossible to eliminate these aberrations from a single lens, but in a compound lens of several elements, the aberrations of one element may partially cancel those of another element. Design of such lenses is an extremely complex problem, which has been aided greatly in recent years by the use of high-speed computers. It is still impossible to eliminate all aberrations, but it *is* possible to decide which ones are most troublesome for a particular application and to design accordingly.

38–6

THE EYE

The essential parts of the eye, considered as an optical system, are shown in Fig. 38–13. The eye is very nearly spherical in shape, and about 2.5 cm in diameter. The front portion is somewhat more sharply curved and is covered by a tough, transparent membrane *C,* called the *cornea.* The region behind the cornea contains a liquid *A* called the *aqueous humor.* Next comes the *crystalline lens L,* a capsule containing a fibrous jelly, hard at the center and progressively softer at the outer portions. The crystalline lens is held in place by ligaments that attach it to the ciliary muscle *M.* Behind the lens, the eye is filled with a thin watery jelly *V,* called the *vitreous humor.* The indices of refraction of both the aqueous humor and the vitreous humor are nearly equal to that of water, about 1.336. The crystalline lens, while not homogeneous, has an "average" index of 1.437. This is not very different from the indices of the aqueous and vitreous humors; most of the refraction of light entering the eye occurs at the cornea.

Refraction at the cornea and the surfaces of the lens produces a real image of the object being viewed, on the light-sensitive *retina R,* lining the rear inner surface. The *rods* and *cones* in the retina act like an array of miniature photocells; they sense the image and transmit it via the *optic nerve O* to the brain. Vision is most acute in a small central region called the *fovea centralis Y,* about 0.25 mm in diameter.

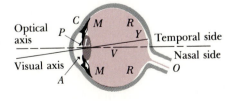

38–13 The eye.

In front of the lens is the *iris,* containing an aperture of variable diameter called the *pupil,* which opens and closes to adapt to changing light intensity. The receptors of the retina also have intensity adaptation mechanisms.

For an object to be seen sharply, the image must be formed exactly at the location of the retina. The lens-to-retina distance, corresponding to s', does not change, but the eye accommodates to different object distances s by changing the focal length of its lens. When the ciliary muscle surrounding the lens contracts, the lens bulges and the radii of curvature of its surfaces decrease, decreasing the focal length. For the normal eye, an object at infinity is sharply focused when the ciliary muscle is relaxed, and maximum tension in this muscle provides the appropriate focal length for an object about 25 cm distance. This process is called *accommodation.*

The extremes of the range over which distinct vision is possible are known as the *far point* and the *near point* of the eye. The far point of a normal eye is at infinity. The position of the near point depends on the amount the curvature of the crystalline lens may be increased. The range of accommodation gradually diminishes with age as the crystalline lens loses its flexibility. For this reason the near point gradually recedes as one grows older. This recession of the near point with age is called *presbyopia.* Following is a table of the approximate average position of the near point at various ages:

Age, years	Near point, cm
10	7
20	10
30	14
40	22
50	40
60	200

*38–7

DEFECTS OF VISION

Several common defects of vision result from an incorrect relation between the parts of the optical system of the eye. A normal eye forms an image on the retina of an object at infinity when the eye is relaxed, as in Fig. 38–14a. In the *myopic* (nearsighted) eye, the eyeball is too long from front to back in comparison with the radius of curvature of the cornea, and rays from an object at infinity are focused in front of the retina. The most distant object for which an image can be formed on the retina is then nearer than infinity. In the *hyperopic* (farsighted) eye, the eyeball is too short and the image of an infinitely distant object would be formed behind the retina. The myopic eye produces too much convergence in a parallel bundle of rays for an image to be formed on the retina; the hyperopic eye, not enough convergence.

Astigmatism refers to a defect in which the surface of the cornea is not spherical but is more sharply curved in one plane than another. Astigma-

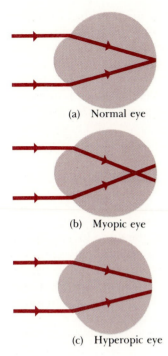

(a) Normal eye

(b) Myopic eye

(c) Hyperopic eye

38–14 Refractive errors for myopic (near-sighted) and hyperopic (farsighted) eye viewing a very distant object.

tism makes it impossible, for example, to focus clearly on the horizontal and vertical bars of a window at the same time.

These defects can be corrected by the use of corrective lenses ("glasses"). The near point of either a presbyopic or a hyperopic eye is farther from the eye than normal. To see clearly an object at normal reading distance (usually assumed to be 25 cm) we must place in front of the eye a lens of such focal length that it forms a virtual image of the object at or beyond the near point. Thus the function of the lens is not to make the object appear larger, but in effect to move the object farther away from the eye to a point where a sharp retinal image can be formed.

EXAMPLE 1 The near point of a certain hyperopic eye is 100 cm in front of the eye. What lens should be used to see clearly an object 25 cm in front of the eye?

Solution We have

$$s = +25 \text{ cm}, \qquad s' = -100 \text{ cm},$$

$$\frac{1}{f} = \frac{1}{s} + \frac{1}{s'} = \frac{1}{+25 \text{ cm}} + \frac{1}{-100 \text{ cm}},$$

$$f = +33 \text{ cm}.$$

That is, a converging lens of focal length 33 cm is required. ◄

The far point of a *myopic* eye is nearer than infinity. To see clearly objects beyond the far point, a lens must be used that will form an image of such objects, not farther from the eye than the far point.

EXAMPLE 2 The far point of a certain myopic eye is 1 m in front of the eye. What lens should be used to see clearly an object at infinity?

Solution Assume the image to be formed at the far point. Then

$$s = \infty, \qquad s' = -100 \text{ cm},$$

$$\frac{1}{f} = \frac{1}{s} + \frac{1}{s'} = \frac{1}{\infty} + \frac{1}{-100 \text{ cm}},$$

$$f = -100 \text{ cm}.$$

A *diverging* lens of focal length 100 cm is required. ◄

Astigmatism is corrected by means of a *cylindrical* lens. The curvature of the cornea in a horizontal plane may have the proper value such that rays from infinity are focused on the retina. In the vertical plane, however, the curvature may not be sufficient to form a sharp retinal image. When a cylindrical lens with axis horizontal is placed before the eye, the rays in a horizontal plane are unaffected, while the additional convergence of the rays in a vertical plane now causes these to be sharply imaged on the retina.

The optometrist describes the converging or diverging effect of lenses in terms, not of the focal length, but of its *reciprocal*. The reciprocal of the focal length of a lens is called its *power*, and if the focal length is in meters the power is in *diopters*. Thus the power of a positive lens whose focal length is 1 m is 1 diopter; if the focal length is 2 m the power is 0.5 diopter, and so on. If the focal length is negative, the power is negative also. Thus in the two examples above, the required powers are +3.0 diopter and −1.0 diopter, respectively.

38–8

THE MAGNIFIER

The apparent size of an object is determined by the size of its retinal image; if the eye is unaided, this depends upon the *angle* subtended by the object at the eye, called its *angular size*. When one wishes to examine a small object in detail one brings it close to the eye, in order that the angle subtended and the retinal image may be as large as possible. Since the eye cannot focus sharply on objects closer than the near point, a particular object subtends the maximum possible angle at an unaided eye when placed at this point. (We assume here that the near point is 25 cm from the eye.)

When a converging lens is placed in front of the eye, it forms a virtual image farther from the eye than the object. Thus the object may be moved closer to the eye, and the angular size of the image may be substantially larger than the angular size of the object at 25 cm without the lens. A lens used in this way is called a *magnifying glass* or simply a *magnifier*. Viewing of the virtual image is most comfortable when it is placed at infinity, and in the following discussion we assume that is done.

The magnifier is illustrated in Fig. 38–15. In (a), the object is at the near point, where it subtends an angle u at the eye. In (b), a magnifier in front of the eye forms an image at infinity, and the angle subtended at the magnifier is u'. The *angular magnification M* (not to be confused with the *lateral magnification m*) is defined as the ratio of the angle u' to the angle u. The value of M may be found as follows.

From Fig. 38–15, u and u' are given (in radians) by

$$u = \frac{y}{25 \text{ cm}} \quad \text{(approximately)},$$

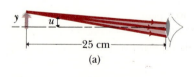

(a)

$$u' = \frac{y}{f} \quad \text{(approximately)}.$$

Hence,

$$M = \frac{u'}{u} = \frac{y/f}{y/25} = \frac{25}{f} \quad (f \text{ in centimeters}). \qquad (38\text{–}7)$$

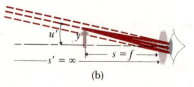

(b)

38–15   A simple magnifier.

While it appears at first that the angular magnification may be made as large as desired by decreasing the focal length f, the aberrations of a simple double convex lens set a limit to M of about 2× or 3×. If these aberrations are corrected, the magnification may be carried as high as 20×.

38–9

THE CAMERA

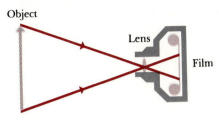

Object

Lens

Film

38–16 Essential elements of a camera.

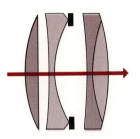

38–17 Zeiss "Tessar" lens design.

The essential elements of a camera are a lens equipped with a shutter, a light-tight enclosure, and a light-sensitive film to record an image. An example is shown in Fig. 38–16. The lens forms a real image in the plane of the film of the object being photographed, just as the lens of the human eye forms a real image on the retina, the eye's "film." The lens may be moved closer to or farther from the film to provide proper image distances for various object distances. All but the most inexpensive lenses have several elements, to permit partial correction of various aberrations. An example is the Zeiss "Tessar" design, shown in Fig. 38–17.

In order for the image to be recorded properly on the film, the total light energy per unit area reaching the film (the "exposure") must fall within certain limits; this is controlled by the shutter and the lens aperture. The shutter controls the time during which light enters the lens, typically adjustable in steps corresponding to factors of about two, from 1 s to $\frac{1}{1000}$ s or thereabouts. The light-gathering capacity of the lens is proportional to its effective area; this may be varied by means of an adjustable aperture or *diaphragm,* which is a nearly circular hole of variable diameter. The aperture size is usually described in terms of its "*f*-number," which is the focal length of the lens divided by the diameter of the aperture. Thus a lens having $f = 50$ mm and an aperture diameter of 25 mm would be said to have an aperture of $f/2$.

Because the light-gathering capacity of the lens is proportional to its area and thus to the *square* of its diameter, changing the diameter by a factor of $\sqrt{2}$ corresponds to a factor of two in exposure. Thus, adjustable apertures usually have scales labeled with successive numbers related by factors of $\sqrt{2}$, such as

$$f/2, \quad f/2.8, \quad f/4, \quad f/5.6, \quad f/8, \quad f/11, \quad f/16,$$

and so on, with the larger numbers representing smaller apertures and exposures, each step corresponding to a factor of two in exposure.

The choice of focal length for a camera lens depends on the film size and the desired angle of view, or *field.* For the popular 35-mm cameras, with image size of 24 × 36 mm, the normal lens is usually about 50 mm in focal length and permits an angle of about 45°. A longer focal-length lens, used with the same film size, provides a smaller angle of view and a larger image of part of the object, compared with a normal lens; this gives the impression that the camera is closer than it really is, and such a lens is called a *telephoto* lens. At the other extreme, a lens of shorter focal length, such as 35 mm or 28 mm, permits a wider angle of view and is called a *wide-angle* lens.

38–10

THE PROJECTOR

A projector for slides or motion pictures operates very much like a camera in reverse. The essential elements are shown in Fig. 38–18. Light from the source (an incandescent lamp bulb or in large motion-picture

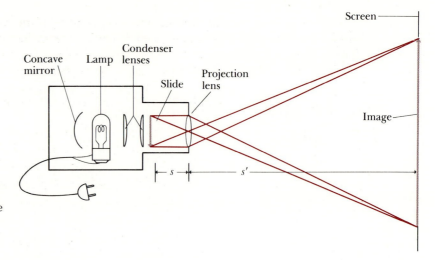

38–18 A slide projector. The concave mirror and condenser lenses gather and direct the light from the lamp so it will enter the projection lens after passing through the slide. The cooling fan, usually needed to take away excessive heat from the lamp, is not shown.

projectors a carbon-arc lamp) shines through the film, and the projection lens forms a real, enlarged, inverted image of the film on the projection screen. Additional lenses called *condenser* lenses are placed between lamp and film to direct the light from source so that most of it enters the projection lens after passing through the film. A concave mirror behind the lamp also helps. The condenser lenses must be large enough to cover the entire area of the film. The image on the screen is always inverted; that's why slides have to be put into a projector upside-down.

The position and size of the image projected on the screen are determined by the position and focal length of the projection lens.

EXAMPLE An ordinary 35-mm color slide has a picture area 24 × 36 mm. What focal-length projection lens would be needed to project an image 1.2 × 1.8 m on a screen 5 m from the lens?

Solution We need a lateral magnification (apart from sign) of 1.2 m/24 mm = 50. Thus, from Eq. (38–4) the ratio s'/s must also be 50. We are given $s' = 5$ m, so $s = 5$ m/50 = 0.1 m. Then from Eq. (38–3),

$$\frac{1}{f} = \frac{1}{0.1 \text{ m}} + \frac{1}{5 \text{ m}}, \qquad f = 0.098 \text{ m} = 98 \text{ mm}.$$

A commercially available lens would probably have $f = 100$ mm; this is a popular focal length for home slide projectors. ◄

38–11

THE COMPOUND MICROSCOPE

When an angular magnification larger than that attainable with a simple magnifier is desired, it is necessary to use a *compound microscope*, usually called merely a *microscope*. The essential elements of a microscope are illustrated in Fig. 38–19. The object O to be examined is placed just

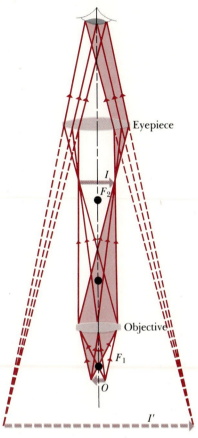

38–19 The microscope.

beyond the first focal point F_1 of the *objective* lens, which forms a real and enlarged image I. This image lies just within the first focal point F_2 of the *eyepiece*, which forms a final virtual image of I at I'. As was stated earlier, the position of I' may be anywhere between the near and far points of the eye. Although both the objective and eyepiece of an actual microscope are highly corrected compound lenses, they are shown as simple thin lenses for simplicity.

Since the objective lens forms an enlarged real image that is viewed by the eyepiece, the overall angular magnification M of the compound microscope is the product of the *lateral* magnification m_1 of the objective and the *angular* magnification M_2 of the eyepiece. The former is given by

$$m_1 = -\frac{s_1'}{s_1},$$

where s_1 and s_1' are the object and image distances for the objective lens. Ordinarily the object is very close to the focus, resulting in an image whose distance from the objective is much larger than its focal length f_1. Thus s_1 is approximately equal to f_1, and $m_1 = -s_1'/f_1$, approximately. The angular magnification of the eyepiece, from Eq. (38–7), is $M_2 = (25 \text{ cm})/f_2$, where f_2 is the focal length of the eyepiece, considered as a simple lens. Hence the overall magnification M of the compound microscope is, apart from a negative sign, which is customarily ignored,

$$M = m_1 M_2 = \frac{(25 \text{ cm})s_1'}{f_1 f_2}, \tag{38–8}$$

where s_1', f_1, and f_2 are measured in centimeters. Microscope manufacturers customarily specify the values of m_1 and M_2 for microscope components rather than the focal lengths of the objective and eyepiece.

38–12

TELESCOPES

The optical system of a refracting telescope is similar to that of a compound microscope. In both instruments, the image formed by an objective is viewed through an eyepiece. The difference is that the telescope is used to examine large objects at large distances and the microscope to examine small objects close at hand.

The *astronomical* telescope is illustrated in Fig. 38–20. The objective lens forms a real, reduced image I of the object, and a virtual image of I

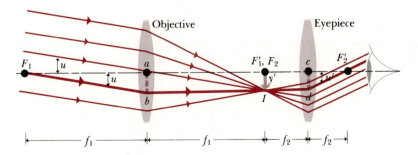

38–20 Telescope; final image at infinity.

38–21   The prism binocular. (Courtesy of Bushnell Division of Bausch & Lomb Optical Co.)

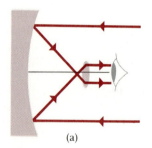

(a)

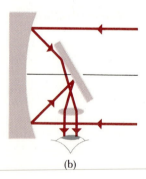

(b)

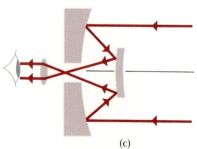

(c)

38–22   The reflecting telescope.

is formed by the eyepiece. As with the microscope, the image I' may be formed anywhere between the near and far points of the eye. In practice, the objects examined by a telescope are at such large distances from the instrument that the image I is formed very nearly at the second focal point of the objective. Furthermore, if the image I' is at infinity, the image I is at the first focal point of the eyepiece. The distance between objective and eyepiece, or the length of the telescope, is therefore the *sum* of the focal lengths of objective and eyepiece, $f_1 + f_2$.

The angular magnification M of a telescope is defined as the ratio of the angle subtended at the eye by the final image I', to the angle subtended at the (unaided) eye by the object. This ratio may be expressed in terms of the focal lengths of objective and eyepiece as follows. In Fig. 38–20, the ray passing through F_1, the first focal point of the objective, and through F_2', the second focal point of the eyepiece, has been emphasized. The object (not shown) subtends an angle u at the objective and would subtend essentially the same angle at the unaided eye. Also, since the observer's eye is placed just to the right of the focal point F_2', the angle subtended at the eye by the final image is very nearly equal to the angle u'. The distances ab and cd are evidently equal to each other and to the height y' of the image I. Since u and u' are small, they may be approximated by their tangents. From the right triangles F_1ab and $F_2'cd$,

$$u = \frac{-y'}{f_1}, \qquad u' = \frac{y'}{f_2}.$$

Hence,

$$M = \frac{u'}{u} = -\frac{y'/f_2}{y'/f_1} = -\frac{f_1}{f_2}. \qquad (38\text{–}9)$$

The angular magnification M of a telescope is therefore equal to the ratio of the focal length of the objective to that of the eyepiece. The negative sign denotes an inverted image.

An inverted image is not a disadvantage if the instrument is to be used for astronomical observations, but it is desirable that a terrestrial telescope form an erect image. This is accomplished in the *prism binocular* by a pair of 45°–45°–90° totally reflecting prisms inserted between objective and eyepiece, as shown in Fig. 38–21. The image is inverted by the four reflections from the inclined faces of the prisms. It is customary to stamp on a flat metal surface of a binocular two numbers separated by a multiplication sign, thus, 7 × 50. The first number is the magnification and the second is the diameter of the objective lenses in millimeters, which determine the brightness of the image.

In the *reflecting telescope* the objective lens is replaced by a concave mirror, as shown in Fig. 38–22. In large telescopes this scheme has many advantages, both theoretical and practical. The mirror is intrinsically free of chromatic aberrations, and spherical aberrations are much easier to correct than with a lens. The material need not be transparent, and the reflector can be made more rigid than a lens, which has to be supported only at its edges. The largest reflecting telescope in the world has a mirror over 5 m in diameter.

38–23 The reflector of the 200-inch Hale telescope at Palomar Observatory, being cleaned in preparation for the application of a new aluminum reflective surface. (Palomar Observatory Photograph.)

Because the image is formed in a region traversed by incoming rays, this image can be observed directly with an ocular only by blocking off part of the incoming beam; this is practical only for the very largest telescopes. Alternative schemes use a mirror to reflect the image out the side or through a hole in the mirror; as shown in Figs. 38–22b and 38–22c. The latter is also used in some long-focal-length telephoto lenses for cameras. In the context of photography such a system is called a *catadioptric lens,* which is a fancy name for an optical system containing both reflecting and refracting elements. The reflector for a large astronomical telescope is shown in Fig. 38–23.

QUESTIONS

38–1 Sometimes a wine glass filled with white wine forms an image of an overhead light on a white tablecloth. Would the same image be formed with an empty glass? With a glass of gin? Gasoline?

38–2 How could one very quickly make an approximate measurement of the focal length of a converging lens? Could the same method be applied if we wished to use a diverging lens?

38–3 If you look closely at a shiny Christmas-tree ball, you can see nearly the entire room. Does the room appear right-side-up or upside-down? Discuss your observations in terms of images.

38–4 A student asserted that any lens with spherical surfaces has a positive focal length if it is thicker at the center than at the edge, and negative if it is thicker at the edge. Do you agree?

38–5 The focal length of a simple lens depends on the color (wavelength) of light passing through it. Why? Is it possible for a lens to have a positive focal length for some colors and negative for others?

38–6 The human eye is often compared to a camera. In what ways is it similar to a camera? In what ways does it differ?

38–7 How could one make a lens for sound waves?

38–8 A student proposed to use a plastic bag full of air, immersed in water, as an underwater lens. Is this possible? If the lens is to be a converging lens, what shape should the air pocket have?

38–9 When a converging lens is immersed in water, does its focal length increase or decrease, compared with the value in air?

38–10 You are marooned on a desert island and want to use your eyeglasses to start a fire. Can this be done if you are nearsighted? If you are farsighted?

38–11 While lost in the mountains a person who was nearsighted in one eye and farsighted in the other made a crude emergency telescope from the two lenses of his eyeglasses. How did he do this?

38–12 In using a magnifying glass, is the magnification greater when the glass is close to the object or when it is close to the eye?

38–13 When a slide projector is turned on without a slide in it, and the focus adjustment is moved far enough in one direction, a gigantic image of the lightbulb filament can be seen on the screen. Explain how this happens.

38–14 A spherical air bubble in water can function as a lens. Is it a converging or diverging lens? How is its focal length related to its radius?

38–15 As discussed in the text, some binoculars use prisms to invert the final image. Why are prisms better than ordinary mirrors for this purpose?

38–16 There have been reports of round fishbowls starting fires by focusing the sun's rays coming in a window. Is this possible?

38–17 How does a person judge distance? Can a person with vision in only one eye judge distance? What is meant by "binocular vision"?

38–18 Zoom lenses are widely used in television cameras and conventional photography. Such a lens has, effectively, a variable focal length; changes in focal length are accomplished by moving some lens elements relative to others. Try to devise a scheme to accomplish this effect.

38–19 Why can't you see clearly when your head is under water? Could you wear glasses so you *could* see under water? Would the lenses be converging or diverging?

PROBLEMS

38–1 A converging lens of focal length 10 cm forms a real image 1 cm high, 12 cm to the right of the lens. Determine the position and size of the object. Is the image erect or inverted? Draw a principal-ray diagram for this situation.

38–2 An object is 10 cm away from a lens that forms an image 15 cm on the opposite side of the lens from the object.

a) What is the focal length of the lens?

b) If the object is 1 cm high, how high is the image?

c) Draw a principal-ray diagram.

38–3 A lens forms an image of an object 20 cm from it. The image is 4 cm from the lens, on the same side as the object.

a) What is the focal length of the lens? Is it converging or diverging?

b) If the object is 2 cm high, how high is the image? Is it erect or inverted?

c) Draw a principal-ray diagram.

38–4 Prove that the image of a real object formed by a diverging lens is *always* virtual.

38–5 An object is placed 18 cm from a screen.

a) At what points between object and screen may a lens of 4 cm focal length be placed to obtain an image on the screen?

b) What is the magnification of the image for these positions of the lens?

38–6 An object is imaged by a lens on a screen placed 12 cm from the lens. When the lens is moved 2 cm farther from the object, the screen must be moved 2 cm closer to the object to refocus it. What is the focal length of the lens?

38–7 A converging lens has a focal length of 10 cm. For object distances of 30 cm, 20 cm, 15 cm, and 5 cm determine

a) image position,

b) magnification,

c) whether the image is real or virtual,

d) whether the image is erect or inverted.

38–8 Sketch the various possible thin lenses obtainable by combining two surfaces whose radii of curvature are, in absolute magnitude, 10 cm and 20 cm. Which are converging and which are diverging? Find the focal length of each lens if made of glass of index 1.50.

38–9 The radii of curvature of the surfaces of a thin lens are +10 cm and +30 cm. The index is 1.50.

a) Compute the position and size of the image of an object in the form of an arrow 1 cm high, perpendicular to the lens axis, 40 cm to the left of the lens.

b) A second similar lens is placed 160 cm to the right of the first. Find the position of the final image.

c) Same as (b), except the second lens is 40 cm to the right of the first.

d) Same as (c), except the second lens is diverging, of focal length −40 cm.

38–10 A diverging meniscus lens of 1.48 refractive index has concave spherical surfaces whose radii are 2.5 and 4 cm. What would be the position of the image if an object were placed 15 cm in front of the lens?

38–11 Two thin lenses, both of 10 cm focal length, the first converging, the second diverging, are placed 5 cm apart. An object is placed 20 cm in front of the first (converging) lens.

a) How far from this lens will the image be formed?

b) Is the image real or virtual?

38–12

a) Prove that when two thin lenses of focal lengths f_1 and f_2 are placed *in contact,* the focal length f of the combination is given by the relation

$$\frac{1}{f} = \frac{1}{f_1} + \frac{1}{f_2}.$$

b) A converging meniscus lens has an index of refraction of 1.50, and the radii of its surfaces are 5 and 10 cm. The concave surface is placed upward and filled with water. What is the focal length of the water–glass combination?

38–13 When an object is placed at the proper distance in front of a converging lens, the image falls on a screen 20 cm from the lens. A diverging lens is now placed halfway between the converging lens and the screen, and it is found that the screen must be moved 20 cm farther away from the lens to obtain a sharp image. What is the focal length of the diverging lens?

38–14 A transparent rod 40 cm long is cut flat at one end and rounded to a hemispherical surface of 12 cm radius at the other end. An object is placed on the axis of the rod, 10 cm from the hemispherical end.

a) What is the position of the final image?

b) What is its magnification? Assume the refractive index to be 1.50.

38–15 Both ends of a glass rod 10 cm in diameter, of index 1.50, are ground and polished to convex hemispherical surfaces of radius 5 cm at the left end and radius 10 cm at the right end. The length of the rod between vertices is 60 cm. An arrow 1 mm long, at right angles to the axis and 20 cm to the left of the first vertex, constitutes the object for the first surface.

a) What constitutes the object for the second surface?

b) What is the object distance for the second surface?

c) Is the object real or virtual?

d) What is the position of the image formed by the second surface?

e) What is the height of the final image?

38–16 The same rod as in Problem 38–15 is now shortened to a distance of 10 cm between its vertices, the curvatures of its ends remaining the same.

a) What is the object distance for the second surface?

b) Is the object real or virtual?

c) What is the position of the image formed by the second surface?

d) Is the image real or virtual, erect or inverted, with respect to the original object?

e) What is the height of the final image?

38–17 A glass rod of refractive index 1.50 is ground and polished at both ends to hemispherical surfaces of 5 cm radius. When an object is placed on the axis of the rod, 20 cm from one end, the final image is formed 40 cm from the opposite end. What is the length of the rod?

38–18 Three thin lenses, each of focal length 20 cm, are aligned on a common axis and adjacent lenses are separated by 30 cm. Find the position of the image of a small object on the axis, 60 cm to the left of the first lens.

38–19 The picture size on ordinary 35-mm camera film is 24×36 mm. Focal lengths of lenses available for 35-mm cameras typically include 28 mm, 35 mm, 50 mm (the "standard" lens), 85 mm, 100 mm, 135 mm, 200 mm, and 300 mm, among others. Which of these lenses should be used to photograph the following objects, assuming the object is to fill most of the picture area?

a) A cathedral 100 m high and 150 m long, at a distance of 150 m.

b) An eagle with a wingspan 2.0 m, at a distance of 15 m.

38–20 During a lunar eclipse, a picture of the moon (diameter 3.48×10^6 m, distance from earth 3.8×10^8 m) is taken with a camera whose lens has focal length 50 mm. What is the diameter of the image on the film?

38–21 The *resolution* of a camera lens can be defined as the maximum number of lines per millimeter in the image that can barely be distinguished as separate lines. A certain lens has a focal length of 50 mm and resolution of 100 lines mm^{-1}. What is the minimum separation of two lines in an object 100 m away if they are to be visible in the image as separate lines?

38–22 Show that when two thin lenses are placed in contact, the *power* of the combination in diopters, as defined in Section 38–7, is the sum of the powers of the separate lenses. Is this relation valid even when one lens has positive power and the other negative?

38–23 An eyepiece consists of two similar positive thin lenses having focal lengths of 6 cm, separated by a distance of 3 cm. Where are the focal points of the eyepiece?

38–24 In a simplified model of the human eye, the aqueous and vitreous humors and the lens all have refractive index 1.40, and all the refraction occurs at the cornea, about 2.50 cm from the retina. What should be the radius of curvature of the cornea in order to focus an infinitely distant object on the retina?

38–25 For the model of the eye described in Problem 38–24, what should be the radius of curvature of the cornea in order to focus an object located 25 cm away from the cornea?

38–26 A certain very nearsighted person cannot focus anything farther than 10 cm from the eye. If the radius of curvature of the cornea is 0.70 cm and the indexes of refraction are as described in Problem 38–24, what is the cornea-to-retina distance? What does this tell you about the shape of the nearsighted eye?

38–27 A person with normal vision cannot focus his or her eyes underwater.

a) Why not?

b) With the simplified model described in Problem 38–24, what corrective lenses would be needed to enable a person underwater to focus an infinitely distant object? (Be careful; the focal length of a lens underwater is not the same as in air! Assume the corrective lens has a refractive index in air of 1.60.)

38–28 When two thin lenses are closely spaced, the power of the combination is the sum of the powers of the individual lenses. Two thin lenses of 25 cm and 40 cm focal lengths are in contact. What is the power of the combination?

38–29 What is the power of the spectacles required

a) by a hyperopic eye whose near point is at 125 cm?

b) by a myopic eye whose far point is at 50 cm?

38–30

a) What spectacles are required for reading purposes by a person whose near point is at 200 cm?

b) The far point of a myopic eye is at 30 cm. What spectacles are required for distant vision?

38–31

a) Where is the near point of an eye for which a spectacle lens of power +2 diopters is prescribed?

b) Where is the far point of an eye for which a spectacle lens of power −0.5 diopter is prescribed for distant vision?

38–32 A thin lens of focal length 10 cm is used as a simple magnifier.

a) What angular magnification is obtainable with the lens?

b) When an object is examined through the lens, how close may it be brought to the eye?

38–33 The focal length of a simple magnifier is 10 cm.

a) How far in front of the magnifier should an object to be examined be placed if the image is formed at the observer's near point, 25 cm in front of his eye?

b) If the object is 1 mm high, what is the height of its image formed by the magnifier? Assume the magnifier to be a thin lens.

38–34 A camera lens is focused on a distant point source of light, the image forming on a screen at a (Fig. 38–24). When the screen is moved backward a distance of 2 cm to b, the circle of light on the screen has a diameter of 4 mm. What is the f/number of the lens?

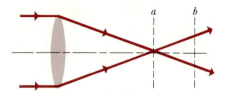

Figure 38–24

38–35 Camera A, having an $f/8$ lens 2.5 cm in diameter, photographs an object using the correct exposure of $\frac{1}{100}$ s. What exposure should camera B use in photographing the same object if it has an $f/4$ lens 5 cm in diameter?

38–36 The focal length of an $f/2.8$ camera lens is 8 cm.

a) What is the diameter of the lens?

b) If the correct exposure of a certain scene is $\frac{1}{200}$ s at $f/2.8$, what would be the correct exposure at $f/5.6$?

38–37 The dimensions of the picture on a 35-mm color slide are 24 mm × 36 mm. It is desired to project an image of the slide, enlarged to 2 m × 3 m, on a screen 10 m from the projection lens.

a) What should be the focal length of the projection lens?

b) Where should the slide be placed?

38–38 The image formed by a microscope objective of focal length 4 mm is 180 mm from its second focal point. The eyepiece has a focal length of 31.25 mm.

a) What is the magnification of the microscope?

b) The unaided eye can distinguish two points as separate if they are about 0.1 mm apart. What is the minimum separation, using this microscope?

38–39 A certain microscope is provided with objectives of focal lengths 16 mm, 4 mm, and 1.9 mm, and with eyepieces of angular magnification 5× and 10×. What is

a) the largest, and

b) the least overall magnification obtainable?

Each objective forms an image 160 mm beyond its second focal point.

38–40 The focal length of the eyepiece of a certain microscope is 2.5 cm. The focal length of the objective is 16 mm. The distance between objective and eyepiece is 22.1 cm. The final image formed by the eyepiece is at infinity. Treat all lenses as thin.

a) What should be the distance from the objective to the object viewed?

b) What is the linear magnification produced by the objective?

c) What is the overall magnification of the microscope?

38–41 A microscope with an objective of focal length 9 mm and an eyepiece of focal length 5 cm is used to project an image on a screen 1 m from the eyepiece. What is the lateral magnification of the image? Let the image distance of the objective be 18 cm.

38–42 The moon subtends an angle at the earth of approximately $\frac{1}{2}°$. What is the diameter of the image of the moon produced by the objective of the Lick Observatory telescope, a refractor having a focal length of 18 m?

38–43 The eyepiece of a telescope has a focal length of 10 cm. The distance between objective and eyepiece is 2.1 m. What is the angular magnification of the telescope?

38–44 A crude telescope is constructed of two spectacle lenses of focal lengths 100 cm and 20 cm, respectively.

a) Find its angular magnification.

b) Find the height of the image formed by the objective of a building 80 m high and distant 2 km.

38–45 Figure 38–25 is a diagram of a *Galilean telescope*, or *opera glass*, with both the object and its final image at infin-

ity. The image I serves as a virtual object for the eyepiece. The final image is virtual and erect. Prove that the angular magnification $M = -f_1/f_2$.

38–46 A Galilean telescope is to be constructed, using the same objective as in Problem 38–40.

a) What type lens should be used as an eyepiece and what focal length should it have, if the telescopes are to have the same magnification?

b) Compare the lengths of the telescopes.

38–47 A reflecting telescope is made using a mirror of radius of curvature 0.50 m and an eyepiece of focal length 1.0 cm. What is the angular magnification? What should be the position of the eyepiece if both the object and the final image are at infinity?

38–48 A certain reflecting telescope has a mirror 10 cm in diameter, with radius of curvature 1.0 m, and an eyepiece of focal length 1.0 cm. If the angular magnification is 48 and the object is at infinity, find the position of the lens and the position and nature of the final image.

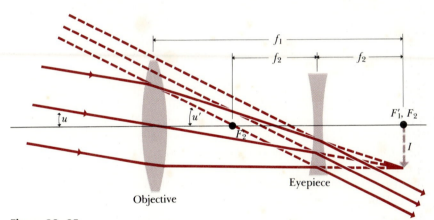

Figure 38–25

INTERFERENCE
AND DIFFRACTION

In analyzing the formation of images by lenses and mirrors, we have represented light as *rays* that travel in straight lines in a homogeneous medium and that are deviated in accordance with simple laws at a reflecting surface or an interface between two optical media. This simple model forms the basis of *geometrical optics,* which, as we have seen, is adequate for understanding a wide variety of phenomena involving lenses and mirrors.

In this chapter we shall discuss the phenomena of *interference* and *diffraction,* for whose understanding the principles of geometrical optics *do not* suffice. Instead, we must return to the more fundamental point of view that light is a *wave motion,* and that the total effect of a number of waves arriving at one point depends on the *phases* of the waves as well as upon their amplitudes. This part of the subject is called *physical optics.* When light passes through apertures or around obstacles, patterns are formed that could not be predicted on the basis of a ray model but instead depend directly on the wave nature of light. In several simple situations we can predict the characteristics of these patterns in detail.

39–1

INTERFERENCE AND COHERENT SOURCES

In our discussions of mechanical waves in Chapter 21 and electromagnetic waves in Chapter 35, we have often considered *sinusoidal* waves having a single frequency and a single wavelength. Such a wave is called a *monochromatic* (single-color) wave. Common sources of light, such as an incandescent lightbulb or a flame, *do not* emit monochromatic light but rather a continuous distribution of wavelengths. A strictly monochromatic light wave is an unattainable idealization.

Monochromatic light can be *approximated* in the laboratory. Continuous-spectrum light can be passed through a filter that blocks all but a narrow range of wavelengths. Gas-discharge lamps, such as the mercury-arc lamp, emit line spectra in which the light consists of a discrete set of colors, each having a narrow band of wavelengths called a *spectrum line*. For example, the bright green line in the mercury spectrum has an average wavelength of 546.1 nm, with a spread of wavelength of the order of ± 0.001 nm, depending on the pressure and temperature of the mercury vapor in the lamp. By far the most nearly monochromatic source available at present is the *laser*, to be discussed in Chapter 41. The familiar helium–neon laser, inexpensive and readily available, emits visible light at 632.8 nm with a line width (wavelength range) of the order of ± 0.000001 nm, or about one part in 10^9. Laser light also has much greater *coherence* (to be discussed later) than ordinary light.

The term *interference* refers to any situation in which two or more waves overlap in space. This term was introduced in Section 22–2 in connection with standing waves on a stretched string, formed by the superposition of two sinusoidal waves traveling in opposite directions. In such cases, the total displacement at any point at any instant of time is governed by the *principle of linear superposition*, introduced in Section 22–2. This principle, the most important in all of physical optics, states that *when two or more waves overlap, the resultant displacement at any point and at any instant may be found by adding the instantaneous displacements that would be produced at the point by the individual waves if each were present alone.* The term "displacement," as used here, is a general one. If one is considering surface ripples on a liquid, the displacement means the actual displacement of the surface above or below its normal level. If the waves are sound waves, the term refers to the excess or deficiency of pressure. If the waves are electromagnetic, the displacement means the magnitude of the electric or magnetic field. When light of extremely high intensity passes through matter, the principle of linear superposition is *not* precisely obeyed, and the resulting phenomena are classified under the heading *nonlinear optics*. (These effects are beyond the scope of this book.)

To introduce the essential ideas of interference, we consider first the problem of two identical sources of monochromatic waves, S_1 and S_2, separated in space by a certain distance. The two sources are permanently *in phase*, so that at every point in space there is a definite and unchanging phase relation between waves from the two sources. They might be, for example, two agitators in a ripple tank, two loudspeakers driven by the same amplifier, two radio antennas powered by the same transmitter, or two small apertures in an opaque screen, illuminated by the same monochromatic light source.

We locate the sources S_1 and S_2 along the y-axis and equidistant from the origin, as shown in Fig. 39–1. Let P_0 be any point on the x-axis. From symmetry, the two distances S_1P_0 and S_2P_0 are equal; waves from the two sources thus require equal times to travel to P_0, and having left S_1 and S_2 in phase, they arrive at P_0 in phase. The total amplitude at P_0 is thus twice the amplitude of each individual wave.

Next, we consider a point P_1, located so that its distance from S_2 is exactly one wavelength greater than its distance from S_1. That is,

$$S_2P_1 - S_1P_1 = \lambda.$$

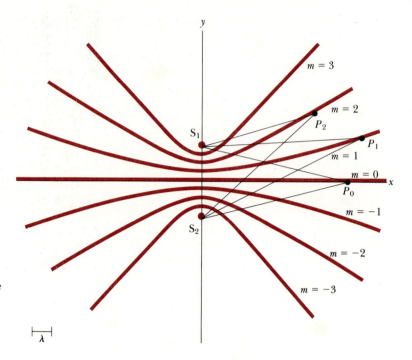

39–1 Curves of maximum intensity in the interference pattern of two monochromatic point sources. In this example the distance between sources is four times the wavelength.

Then any given wave crest from S_1 arrives at P_1 exactly one cycle earlier than the crest emitted at the same time from S_2, and again the two waves arrive in phase. Similarly, waves arrive in phase at all points P_2 for which the path difference is *two* wavelengths ($S_2P_2 - S_1P_2 = 2\lambda$), or indeed for *any* positive or negative integer number of wavelengths.

The addition of amplitudes that results when waves from two or more sources arrive at a point *in phase* is often called *constructive interference* or *reinforcement,* and the above discussion shows that constructive interference occurs whenever the path difference for the two sources is an integer multiple of the wavelength:

$$S_2P - S_1P = m\lambda \qquad (m = 0, \pm 1, \pm 2, \pm 3, \ldots). \qquad (39\text{--}1)$$

In our example, the points satisfying this condition lie on the set of curves shown in Fig. 39–1.

Intermediate between these lines is a set of other lines for which the path difference for the two sources is a *half-integer* number of wavelengths. Waves from the two sources arrive at a point on one of these lines exactly a half-cycle out of phase, and the resultant amplitude is the *difference* of the two individual amplitudes. If the amplitudes are equal, which is approximately the case when the distance from either source to P is much greater than the distance between sources, then the total amplitude at such a point is zero! This condition is called *destructive interference* or *cancellation.* In our example, the condition for destructive interference is

$$S_2P - S_1P = (m + \tfrac{1}{2})\lambda \qquad (m = 0, \pm 1, \pm 2, \pm 3, \ldots). \qquad (39\text{--}2)$$

An example of the interference pattern just described is the familiar ripple-tank pattern shown in Fig. 39–2. The two wave sources are two agitators driven by the same vibrating mechanism. The regions of both

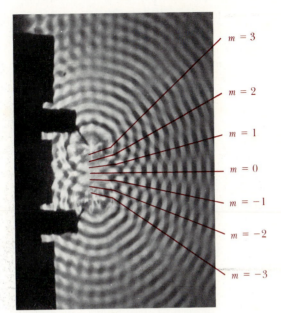

39–2 Photograph of an interference pattern produced by water waves in a shallow ripple tank. The two wave sources are small balls moved up and down by the same vibrating mechanism. As waves move outward from the sources, they overlap and produce an interference pattern. The lines of maximum amplitude in the pattern are shown by the superimposed color lines. (*PSSC Physics,* second edition, 1965; D. C. Heath and Co., with Educational Development Center, Inc., Newton, Mass.)

maximum and zero amplitude are clearly visible. The superimposed color lines, corresponding to those in Fig. 39–1, show the lines of maximum amplitude.

A ripple tank is an inherently two-dimensional situation, but in some cases, such as two loudspeakers or two radio-transmitter antennas, the pattern is three-dimensional; we may think of rotating the color curves of Fig. 39–1 about the horizontal line through the center; the resulting surfaces are the points where maximum constructive interference occurs.

In the above discussion the constant *phase* relationship between the sources is an essential requirement. If the relative phase of the sources changes, the positions of the maxima and minima in the resulting interference pattern also change. When the radiation is light, it is possible for the two sources to have a definite and constant phase relation *only when they both emit light coming from a single primary source;* it is *not* possible with two separate sources. The reason is a fundamental one associated with the mechanism of light emission.

In ordinary light sources, atoms of the material of the source are given excess energy by thermal agitation or impact with accelerated electrons. An atom thus "excited" begins to radiate and continues until it has lost all the energy it can, typically in a time of the order of 10^{-8} s. A source ordinarily contains a very large number of atoms, which radiate in an unsynchronized and random phase relationship. Thus emission from two such sources has a rapidly varying phase relation; the result is a constantly changing interference pattern that with ordinary observations does not reveal a visible interference pattern at all.

However, if the light from a single source is split so that parts of it emerge from two or more regions of space, forming two or more *secondary sources,* any random phase change in the source affects these secondary sources equally and does not change their *relative* phase. Two such sources derived from a single source and having a definite phase relation are said to be *coherent*.

The distinguishing feature of light from a *laser* is that the emission of light from many atoms is *synchronized* in frequency and phase, by mechanisms to be discussed in Chapter 41. As a result, the random phase changes mentioned above occur *much* less frequently. Definite phase relations are preserved over correspondingly much greater lengths in the beam. Accordingly, laser light is said to be much more *coherent* than ordinary light.

39–2

YOUNG'S EXPERIMENT

One of the earliest demonstrations of the fact that light can produce interference effects was performed in 1800 by the English scientist Thomas Young. The experiment was a crucial one at the time, since it added further evidence to the growing belief in the wave nature of light. A corpuscular theory was quite inadequate to account for the effects observed.

Young's apparatus is shown in Fig. 39–3a. Monochromatic light issuing from a narrow slit S_0 is divided into two parts by falling upon a screen in which are cut two other narrow slits S_1 and S_2, very close together. The dimensions in this figure are distorted for clarity. The distance from the source slit S_0 to the screen containing S_1 and S_2 is 20 cm to 100 cm. The distance from the double-slit screen to the final screen is usually from 1 m to 5 m. The slits are 0.1 mm to 0.2 mm wide and the separation of the slits S_1 and S_2 is less than 1 mm. In short, all slit widths and slit separations are fractions of millimeters; all other distances are hundreds or thousands of millimeters.

According to Huygens' principle (Section 36–13), cylindrical wavelets spread out from slit S_0 and reach slits S_1 and S_2 in phase, because they travel equal distances from S_0. A succession of Huygens wavelets diverges from each slit; and the two sets of wavelets leave in phase, and therefore act as *coherent* sources. But they do not necessarily arrive at point P in phase because of the path difference $(r_1 - r_2)$ for the two waves.

39–3 (a) Interference of light waves passing through two slits. (b) Young's experiment.

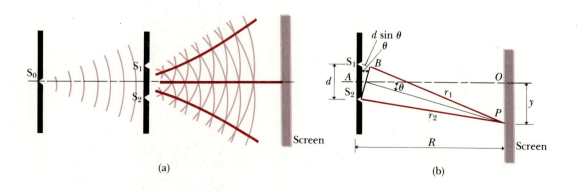

(a)   (b)

When the distance R is very large compared to the distance d between slits, as is usually the case, the path difference can be expressed simply in terms of the angle θ. With P as a center and PS_2 as radius, we draw an arc that intersects line PS_1 at point B. The path difference is then the distance S_1B. When R is much larger than d, the arc S_2B is very nearly a straight line, and it is perpendicular to PS_2, PA, and PS_1. Then triangle BS_1S_2 is a right triangle, similar to POA, and the distance S_1B is equal to $d \sin \theta$. Thus the path difference is given by

$$r_1 - r_2 = d \sin \theta. \qquad (39\text{--}3)$$

According to the principles discussed in the preceding section, complete reinforcement occurs at point P (that is, P lies at the center of a bright fringe) only when the path difference $d \sin \theta$ is some integral number of wavelengths, say $m\lambda$ ($m = 0, 1, 2, 3$, etc.). Thus,

$$d \sin \theta = m\lambda \qquad \text{or} \qquad \sin \theta = \frac{m\lambda}{d}. \qquad (39\text{--}4)$$

Now λ is of the order of 5×10^{-5} cm, while d cannot be made much smaller than about 10^{-2} cm. As a rule, only the first five to ten fringes are bright enough to be seen, so that m is at most, say, 10. Therefore the very largest value of $\sin \theta$ is

$$\sin \theta \text{ (maximum)} = \frac{(10)(5 \times 10^{-5} \text{ cm})}{10^{-2} \text{ cm}} = 0.05,$$

which corresponds to an angle of only 3°. Figure 39–4 shows a typical pattern.

The central bright fringe at point O, or zeroth fringe ($m = 0$), corresponds to zero path difference, or $\sin \theta = 0$. If point P is at the center of the mth fringe, the distance y_m from the zeroth to the mth fringe is, from Fig. 39–3b,

$$y_m = R \tan \theta_m.$$

If, as is often the case in optical interference experiments, y is much smaller than R, then the angle θ_m for all values of m is extremely small. In that case, $\tan \theta_m \approx \sin \theta_m$ and

$$y_m = R \sin \theta_m.$$

Zeroth fringe

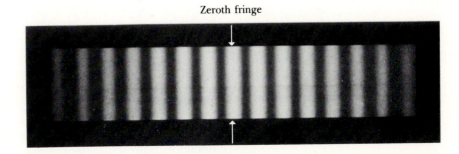

39–4   Interference fringes produced by Young's double-slit interferometer.

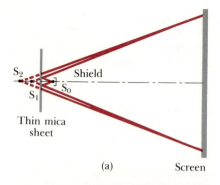

(a)   Screen

(b)

39–5 (a) Pohl's mica-sheet interferometer. (b) Circular interference fringes produced by Pohl's interferometer. The dark rectangle is the shadow of the mercury arc housing.

Therefore,

$$y_m = R\frac{m\lambda}{d},$$

and

$$\lambda = \frac{y_m d}{mR}. \tag{39–5}$$

Hence, by measuring the distance d between the slits, the distance R to the screen, and the distance y_m from the center of the zeroth fringe to the center of the mth fringe on either side, one may compute the wavelength of the light producing the interference pattern. Such an experiment, first performed by Thomas Young in 1800, provided the first direct measurement of wavelengths of light and, of course, also gave very strong confirmation of the wave nature of light.

EXAMPLE   With two slits spaced 0.2 mm apart, and a screen at a distance of 1 m, the third bright fringe is found to be displaced 7.5 mm from the central fringe. Find the wavelength of the light used.

Solution   Let λ be the unknown wavelength. Then

$$\lambda = \frac{y_m d}{mR} = \frac{(0.75\ \text{cm})(0.02\ \text{cm})}{(3)(100\ \text{cm})} = 5 \times 10^{-5}\ \text{cm}$$
$$= 500 \times 10^{-9}\ \text{m} = 500\ \text{nm}. \qquad \blacktriangleleft$$

Circular interference fringes may be produced very easily with the aid of a simple apparatus suggested by Robert Pohl and shown in Fig. 39–5a. A small arc lamp S_0 is placed a few centimeters away from a sheet of mica of thickness about 0.05 mm. Some light is reflected from the first surface, as though it were issuing from the virtual image S_1. An approximately equal amount of light is reflected from the back surface, as though it were coming from the virtual image S_2. The circular interference fringes formed by the light issuing from these two mutually coherent sources may be shown on the entire wall of a room, as shown in Fig. 39–5b.

39–3

INTERFERENCE IN THIN FILMS. NEWTON'S RINGS

The brilliant colors that are often seen when light is reflected from a soap bubble or from a thin layer of oil floating on water are produced by interference effects between the two light waves reflected at opposite surfaces of the thin films of soap solution or of oil. In Fig. 39–6, the line ab is one ray in a beam of monochromatic light incident on the upper surface of a thin film. A part of the incident light is reflected at the first surface, as indicated by ray bc, and a part, represented by bd, is transmitted. At the second surface a part is again reflected, and, of this, a part emerges as represented by ray ef. The rays bc and ef come together at a

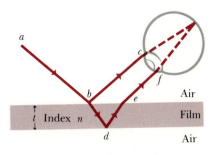

39–6 Interference between rays reflected from the upper and lower surface of a thin film.

39–7 Interference between two light waves reflected from the two sides of an air wedge separating two glass plates. The path difference is 2d.

point on the retina of the eye. Depending on the phase relationship, they may interfere constructively or destructively. Because different colors have different wavelengths, the interference may be constructive for some colors and destructive for others; hence the appearance of colored rings or fringes.

To keep things as simple as possible, let us consider interference of *monochromatic* light reflected from two nearly parallel surfaces. Figure 39–7 shows two plates of glass separated by a wedge of air; we want to consider interference between the two light waves reflected from the surfaces adjacent to the air wedge, as shown. If the observer is at a great distance compared to the other dimensions of the experiment and the rays are nearly perpendicular to the reflecting surfaces, then the path difference between the two waves (corresponding to $r_1 - r_2$ in the preceding section) is just twice the thickness d of the air wedge at each point. At points for which this path difference is an integer number of wavelengths, we expect to see constructive interference and a bright area, and where it is a half-integer number of wavelengths, destructive interference and a dark area. Along the line where the plates are in contact there is *no* path difference and we expect a bright area.

It is easy enough to carry out this experiment; the bright and dark fringes appear as expected, but they are interchanged! Along the line of contact a *dark* fringe, not a bright one, is found. What happened? The inescapable conclusion is that one of the waves has undergone a half-cycle phase shift during its reflection, so that the two reflected waves are a half-cycle out of phase even though they have the same path length.

Further experiments show that a half-cycle phase change occurs whenever the material in which the wave is initially traveling before reflection has a *smaller* refractive index than the second material forming the interface. But when the first material has *greater* refractive index than the second, such as a wave in glass reflected internally at a glass–air interface, there is found to be *no* phase change. Thus, in Fig. 39–6 the waves reflected at points b and e undergo the half-cycle phase shift, while that reflected at d does not. Similarly, in Fig. 39–7 the wave reflected from the upper surface of the lower plate has a half-cycle phase shift, while the other wave has none.

EXAMPLE Suppose the two glass plates in Fig. 39–7 are two microscope slides 10 cm long; at one end they are in contact, while at the other end they are separated by a thin piece of tissue paper 0.02 mm thick. What is the spacing of the resulting interference fringes? Is the fringe adjacent to the line of contact bright or dark? Assume monochromatic light with $\lambda = 500$ nm.

Solution To answer the second question first, the fringe at the line of contact is dark because the wave reflected from the lower surface of the air wedge has undergone a half-cycle phase shift, while that from the upper surface has not. For this reason, the condition for *destructive* interference (a dark fringe) is that the path difference ($2d$) be an integer number of wavelengths:

$$2d = m\lambda \qquad (m = 0, 1, 2, 3, \ldots). \qquad (39\text{–}6)$$

From similar triangles, d is proportional to the distance x from the line of contact:

$$\frac{d}{x} = \frac{h}{l}.$$

Combining this with Eq. (39–6), we find

$$\frac{2xh}{l} = m\lambda,$$

or

$$x = m\frac{l\lambda}{2h} = m\frac{(0.1 \text{ m})(500 \times 10^{-9} \text{ m})}{(2)(0.02 \times 10^{-3} \text{ m})} = m(1.25 \text{ mm}).$$

Thus successive dark fringes, corresponding to successive integer values of m, are spaced 1.25 mm apart.

In this example, if the space between plates contains water ($n = 1.33$) instead of air, the phase changes are the same but the wavelength is $\lambda = \lambda_0/n = 376$ nm, and the fringe spacing is 0.94 mm. But now suppose the top plate is a glass with $n = 1.4$, the wedge is filled with a silicone grease having $n = 1.5$, and the bottom plate has $n = 1.6$. In this case there are half-cycle phase shifts at *both* surfaces bounding the wedge, and the line of contact corresponds to a *bright* fringe, not a dark one. The fringe spacing is again obtained using the wavelength in the wedge, $\lambda = 500$ nm/1.5 = 333 nm, and is found to be 0.83 nm. ◄

If the arrangement in Fig. 39–7 is illuminated first by blue, then by red light, the spacing of the red fringes is greater than that of the blue, as is to be expected from the greater wavelength of the red light. The fringes produced by intermediate wavelengths occupy intermediate positions. If it is illuminated by white light, the color at any point is that due to the mixture of those colors that may be reflected at that point, while the colors for which the thickness is such as to result in destructive interference are absent. Just those colors that are absent in the reflected light, however, are found to predominate in the transmitted light. At any point, the color of the wedge by reflected light is *complementary* to its color by transmitted light!

If the convex surface of a lens is placed in contact with a plane glass plate, as in Fig. 39–8, a thin film of air is formed between the two surfaces. The thickness of this film is very small at the point of contact, gradually increasing as one proceeds outward. The interference bands are circular, concentric with the point of contact. When viewed by reflected light, the center of the pattern is black; there is no phase reversal of the light reflected from the upper surface of the film (which here is of smaller index than that of the medium in which the light is traveling before reflection), but the phase of the wave reflected from the lower surface is reversed. These interference fringes were studied by Newton, and they are called *Newton's rings*. Figure 39–9 is a photograph of Newton's rings formed by the air film between a convex and a plane surface.

The surface of an optical part that is being ground to some desired curvature may be compared with that of another surface, known to be

39–8 Air film between a convex and a plane surface.

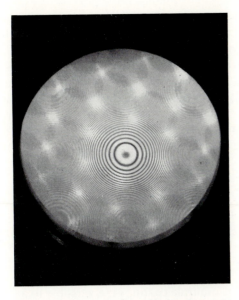

39–9 Newton's rings formed by interference in the air film between a convex and a plane surface. (Courtesy of Bausch & Lomb Optical Co.)

39–10 The surface of a telescope objective under inspection during manufacture. (Courtesy of Bausch & Lomb Optical Co.)

correct, by bringing the two in contact and observing the interference fringes. Figure 39–10 is a photograph made at one stage of the process of manufacturing a telescope objective. The lower, larger-diameter, thicker disk is the master. The smaller upper disk is the objective under test. The "contour lines" are Newton's interference fringes, and each one indicates an additional departure of the specimen from the master of $\frac{1}{2}$ wavelength of light. That is, at 10 lines from the center spot the space between the specimen and master is 5 wavelengths, or about 0.0001 inch. This specimen is very poor; high-quality lenses are routinely ground with a precision of less than a wavelength!

39–4

THIN COATINGS ON GLASS

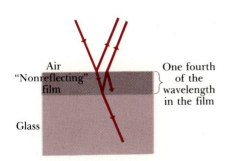

Air
"Nonreflecting" film

One fourth of the wavelength in the film

Glass

39–11 Destructive interference results when the film thickness is one quarter of the wavelength in the film.

The phenomenon of interference is utilized in nonreflective coatings for glass. A thin layer or film of hard transparent material with an index of refraction smaller than that of the glass is deposited on the surface of the glass, as in Fig. 39–11. If the coating has the proper index of refraction, equal quantities of light will be reflected from its outer surface and from the boundary surface between it and the glass. Furthermore, since in both reflections the light is reflected from a medium of greater index than that in which it is traveling, the same phase change occurs in each reflection. It follows that if the film thickness is $\frac{1}{4}$ wavelength *in the film* (normal incidence is assumed), the light reflected from the first surface will be 180° out of phase with that reflected from the second, and complete destructive interference will result.

The thickness can, of course, be $\frac{1}{4}$ wavelength for only one particular wavelength. This is usually chosen in the yellow-green portion of the spectrum (about 550 mm) where the eye is most sensitive. Some reflection then takes place at both longer and shorter wavelengths, and the reflected light has a purple hue. The overall reflection from a lens or prism surface can be reduced in this way from 4% or 5% to a fraction of 1%. The treatment is extremely effective in eliminating stray reflected light and increasing the contrast in an image formed by highly corrected lenses having a large number of air–glass surfaces. A commonly used material is magnesium fluoride, MgF_2, with an index of 1.38. With this coating, the wavelength of green light in the coating is

$$\lambda = \frac{\lambda_0}{n} = \frac{550 \times 10^{-9} \text{ m}}{1.38} = 4 \times 10^{-5} \text{ cm},$$

and the thickness of a "nonreflecting" film of MgF_2 is 10^{-5} cm.

If a material whose index of refraction is *greater* than that of glass is deposited on glass to a thickness of $\frac{1}{4}$ wavelength, then the reflectivity is *increased*. For example, a coating of index 2.5 will allow 38% of the incident energy to be reflected, instead of the usual 4% when there is no coating. With the aid of multiple coatings it is possible to achieve reflectivity for a particular wavelength of almost 100%.

*39–5

THE MICHELSON INTERFEROMETER

Young's double-slit experiment uses light from two very narrow sources. The Michelson interferometer, by contrast, uses light from an extended source. Figure 39–12 is a diagram of its principal features. The figure shows the path of one ray from a point A of an extended but monochromatic source. This ray strikes a glass plate C, the right side of which has a thin coating of silver. Part of the light is reflected from the silvered surface at point P to the mirror M_2 and back through C to the observer's eye. The remainder of the light passes through the silvered surface and

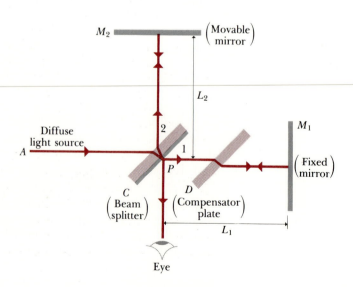

39–12   The Michelson interferometer.

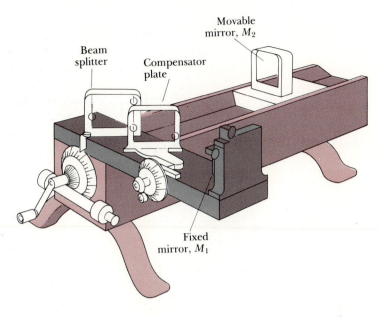

Beam
splitter

Compensator
plate

Movable
mirror, M_2

Fixed
mirror, M_1

39–13   A common type of Michelson interferometer.

the compensator plate D and is reflected from mirror M_1. It then returns through D and is reflected from the silvered surface of C to the observer. The compensator plate D is cut from the same piece of glass as plate C, so that its thickness will not differ from that of C by more than a fraction of a wavelength. Its purpose is to ensure that rays 1 and 2 pass through the same thickness of glass. Plate C is called a *beam splitter*.

The whole apparatus is mounted on a heavy rigid frame and a fine, very accurate screw thread is used to move the mirror M_2. A common commercial model of the interferometer is shown in Fig. 39–13. The source is placed to the left, and the observer is directly in front of the handle that turns the screw.

If the distances L_1 and L_2 in Fig. 39–12 are exactly equal, and the mirrors M_1 and M_2 are exactly at right angles, the virtual image of M_1 formed by reflection at the silvered surface of plate C will coincide with mirror M_2. If L_1 and L_2 are not exactly equal, the image of M_1 will be displaced slightly from M_2; and if the angle between the mirrors is not exactly a right angle, the image of M_1 will make a slight angle with M_2. Then the mirror M_2, and the virtual image of M_1, play the same roles as the two surfaces of a thin film, discussed in Section 39–3, and the same sort of interference fringes result from the light that is reflected from these surfaces.

Suppose that the extended source in Fig. 39–12 is monochromatic with wavelength λ, and that the angle between mirror M_2 and the virtual image of M_1 is such that five or six vertical fringes are present in the field of view. If the mirror M_2 is now moved slowly either backward or forward by a distance $\lambda/2$, the effective film thickness will change by λ and each of the fringes will move either to the right or to the left through a distance equal to the spacing of the fringes. If the fringes are observed through a telescope whose eyepiece is equipped with a crosshair, and m

fringes cross the crosshair when the mirror is moved a distance x, then

$$x = m\frac{\lambda}{2} \quad \text{or} \quad \lambda = \frac{2x}{m}. \quad (39\text{–}7)$$

If m is as large as several thousand, the distance x is sufficiently great so that it can be measured with good precision and hence a precise value of the wavelength λ can be obtained.

Until recently the meter was defined as a length equal to a specified number of wavelengths of the orange-red light of krypton-86. Before this standard could be established, it was necessary to measure as accurately as possible the number of these wavelengths in the *former* standard meter, defined as the distance between two scratches on a bar of platinum-iridium. The measurement was made with a modified Michelson interferometer, many times and under very carefully controlled conditions. The number of wavelengths in a distance equal to the old standard meter was found to be 1,650,763.73 wavelengths. The meter was then defined as *exactly* this number of wavelengths. As mentioned in Section 1–2, this has recently been superseded by a new length standard based on the unit of *time*.

Another application of the Michelson interferometer of considerable historical interest is the Michelson–Morley experiment. To understand the purpose of this experiment, we must recall that before the electromagnetic theory of light and Einstein's special theory of relativity became established, physicists believed that light waves were propagated in a medium called *the ether*.

In 1887, Michelson and Morley utilized the Michelson interferometer in an attempt to detect the relative motion of the earth and the ether. Suppose the interferometer in Fig. 39–12 is moving from left to right relative to the ether. According to nineteenth-century theory, this would lead to changes in the speed of light in the horizontal portions of the path, and corresponding fringe shifts relative to the positions the fringes would have if the instrument were at rest in the ether. Then, when the entire instrument was rotated 90°, the vertical paths would be similarly affected, giving a fringe shift in the opposite direction.

Michelson and Morley expected a fringe shift of about four tenths of a fringe when the instrument was rotated. The shift actually observed was less than a hundredth of a fringe, and within the limits of experimental uncertainty appeared in fact to be zero! Despite its orbital velocity the earth appeared to be at rest relative to the ether. This negative result baffled physicists of the time, and to this day the Michelson–Morley experiment is the most significant "negative-result" experiment ever performed.

Understanding of this result had to wait for Einstein's special theory of relativity, published in 1905. Einstein realized that the velocity of a light wave has the same magnitude c relative to *all* reference frames, whatever their velocity may be relative to other frames. The presumed ether then plays no role, and the very concept of an ether has been given up. The underlying principle of Einstein's theory of special relativity may be stated as follows: *There is no preferred frame of reference for the measurement of the speed of light; the speed is the same for every observer, without regard to the magnitude or direction of his velocity relative to other observers.* This theory, a well-established cornerstone of modern physics, is discussed in detail in Chapter 40.

39–6

FRESNEL DIFFRACTION

According to *geometrical* optics, if an opaque object is placed between a point light source and a screen, as in Fig. 39–14, the edges of the object cast a sharp shadow on the screen. No light reaches the screen at points within the geometrical shadow, while outside the shadow the screen is uniformly illuminated. But as we have seen, geometrical optics is an idealized model of the behavior of light, and there are many situations in which the ray model of light is inadequate. Another important class of such phenomena is grouped under the heading *diffraction*.

The photograph in Fig. 39–15 was made by placing a razor blade halfway between a pinhole illuminated by monochromatic light and a photographic film, so that the film made a record of the shadow cast by the blade. Figure 39–16 is an enlargement of a region near the shadow of an edge of the blade. The boundary of the *geometrical* shadow is indicated by the short arrows. Note that a small amount of light has "bent" around the edge, into the geometrical shadow, which is bordered by alternating bright and dark bands. Note also that the first bright band, just outside the geometrical shadow, is actually *brighter* than in the region of uniform illumination to the extreme left. This simple experimental setup serves to give some idea of the true complexity of what is often considered the most elementary of optical phenomena, the shadow cast by a small source of light.

Diffraction patterns such as that in Fig. 39–15 are not commonly observed in everyday life because most ordinary light sources are not point sources of monochromatic light. If the shadow of a razor blade is cast by a frosted-bulb incandescent lamp, for example, the light from every point of the surface of the lamp forms its own diffraction pattern, but these overlap to such an extent that no individual pattern can be observed.

The term *diffraction* is applied to problems in which one is concerned with *the resultant effect produced by a limited portion of a wave front.* Since in

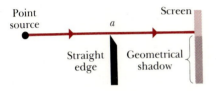

39–14 Geometrical shadow of a straight edge.

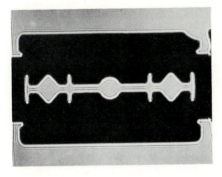

39–15 Shadow of a razor blade.

39–16 Shadow of a straight edge.

most diffraction problems some light is found within the region of geometrical shadow, diffraction is sometimes defined as "the bending of light around an obstacle." It should be emphasized, however, that the process by which diffraction effects are produced is going on continuously in the propagation of *every* wave. Only if a part of the wave is cut off by some obstacle are the effects commonly called "diffraction effects" observed. But every optical instrument makes use of only a limited portion of a wave; a telescope, for example, utilizes only that portion of a wave admitted by the objective lens. Thus diffraction plays a role in practically all optical phenomena.

Figure 39–17 shows a diffraction pattern formed by a steel ball about 3 mm in diameter; we note rings in the pattern both outside and inside the geometrical shadow area, and at the very center of the shadow is a prominent bright spot! The existence of this spot was predicted in 1818 by the French mathematician Poisson. Ironically, Poisson himself was not a believer in the wave theory of light, and he published this *apparently* absurd result as a final ridicule of the wave theory. But almost immediately the bright spot was observed experimentally by Arago. It had in fact been observed much earlier, in 1723 by Maraldi, but its significance was not recognized then.

The essential features observed in diffraction effects can be predicted with the help of Huygens' principle; every point of a wave surface can be considered the source of a secondary wavelet that spreads out in all directions. At every point we must combine the displacements that would be produced by the secondary wavelets, taking into account their amplitudes and relative phases. The mathematical operations are often quite complicated.

In Fig. 39–14 both the point source and the screen are at large but finite distances (say several meters) from the obstacle forming the diffraction pattern, and no lenses are used. This situation is described as *Fresnel diffraction* (after Augustin Jean Fresnel, 1788–1827), and the resulting pattern on the screen is called a *Fresnel diffraction pattern*. If the

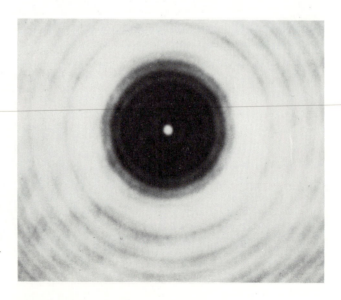

39–17 Fresnel diffraction pattern formed by a steel ball 3 mm in diameter. The Poisson bright spot is seen at the center of the shadow area. (Courtesy Prof. E. Hecht.)

source is far enough away so that the diffraction pattern appears on a screen in the second focal plane of the lens, the phenomenon is called *Fraunhofer diffraction* (after Joseph von Fraunhofer, 1787–1826.) The latter situation is simpler to treat in detail, and we shall consider it first.

39–7

FRAUNHOFER DIFFRACTION FROM A SINGLE SLIT

Suppose a beam of parallel monochromatic light is incident from the left on an opaque plate having a narrow vertical slit. According to geometrical optics, the transmitted beam should have been the same cross section as the slit, and a screen in the path of the beam would be illuminated uniformly over an area of the same size and shape as the slit. What is actually observed is the pattern shown in Fig. 39–18b. The beam spreads out horizontally after passing through the slit, and the diffraction pattern consists of a central bright band, which may be much wider than the slit width, bordered by alternating dark bands and bright bands of decreasing intensity. A diffraction pattern of this nature can readily be observed by looking at a point source such as a distant street light through a narrow slit formed between two fingers in front of the eye. The retina of the eye then corresponds to the screen.

According to Huygens' principle, each element of area in the opening of the slit can be considered as a source of secondary wavelets. In particular, we may imagine elements of area by subdividing the slit into narrow strips parallel to the long edges, perpendicular to the page in Fig. 39–19a. From each strip secondary wavelets spread out, as shown.

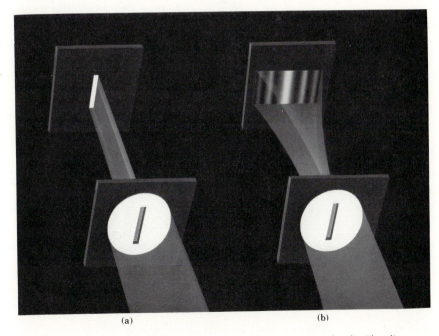

(a)   (b)

39–18 (a) Geometrical "shadow" of a slit. (b) Diffraction pattern of a slit. The slit width has been greatly exaggerated.

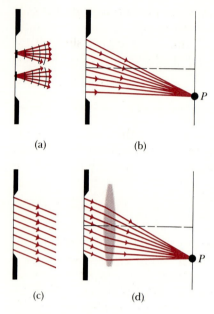

(a)   (b)

(c)   (d)

39–19 Diffraction by a single slit. Relation of Fresnel diffraction to Fraunhofer diffraction by a single slit.

In Fig. 39–19b a screen is placed to the right of the slit; the resultant intensity at a point P is calculated by adding the contributions from the individual wavelets, with due regard for their differing phases and amplitudes. The problem becomes much simpler if the screen is far away, so the rays from the slit to the screen are parallel, as in Fig. 39–19c, or when a lens is placed as in Fig. 39–19d, in which case the lens forms in its focal plane a reduced *image* of the pattern that would be formed on an infinitely distant screen without the lens. It might be questioned whether the lens may introduce additional phase shifts that are different for different parts of the wave front; it can be shown quite generally that the lens *does not* cause any additional phase shifts.

The situation of Fig. 39–19b is Fresnel diffraction; those in Figs. 39–19c and 39–19d are Fraunhofer diffraction. Some aspects of Fraunhofer diffraction from a single slit can be deduced easily. We first consider two narrow strips, one just below the top edge of the slit and one at its center, as in Fig. 39–20. The difference in path length to point P is $(a/2) \sin \theta$, where a is the slit width. Suppose this path difference happens to be equal to $\lambda/2$; then light from these two strips arrives at point P with a half-cycle phase difference, and cancellation occurs. Similarly, light from two strips just below these two will also arrive a half-cycle out of phase; and, in fact, light from *every* strip in the top half cancels out that from a corresponding strip in the bottom half, resulting in complete cancellation, and giving a dark fringe in the interference pattern. Thus, a dark fringe occurs whenever

$$\left(\frac{a}{2}\right) \sin \theta = \frac{\lambda}{2} \qquad \text{or} \qquad \sin \theta = \frac{\lambda}{a}. \qquad (39\text{–}8)$$

We may also divide the screen into quarters, sixths, and so on, and use the above argument to show that a dark fringe occurs whenever $\sin \theta = 2\lambda/a$, $3\lambda/a$, and so on. Thus the condition for a *dark* fringe is

$$\sin \theta = \frac{n\lambda}{a}, \quad \text{where } n = 1, 2, 3, \ldots. \qquad (39\text{–}9)$$

For example, if the slit width is equal to ten wavelengths, dark fringes occur at $\sin \theta = \frac{1}{10}, \frac{2}{10}, \frac{3}{10}, \ldots$. Midway between the dark fringes are bright fringes. We also note that $\sin \theta = 0$ is a *bright* band, since then light

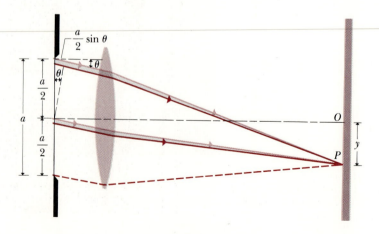

39–20 The wavefront is divided into a large number of narrow strips.

from the entire slit arrives at P in phase. Thus the central bright fringe is twice as wide as the others, as Fig. 39–18 shows.

Since λ is of the order of 5×10^{-5} cm, and a typical slit width is 10^{-2} cm, $\sin \theta_1$ is ordinarily so small that $\sin \theta = \theta$, and Eq. (39–9) becomes

$$\theta = \frac{n\lambda}{a}, \qquad n = 1, 2, 3, \ldots.$$

Also, if the distance from lens to screen is R, and the height of the nth dark band above the center of the pattern is y_n, then $\tan \theta = y_n/R$. Again, if θ is very small we approximate $\tan \theta = \theta$. Combining this with the above relation, we find

$$y_n = \frac{Rn\lambda}{a}. \tag{39–10}$$

This equation has the same form as that obtained for the two-slit pattern, Eq. (39–5), but here it gives the *dark* fringes in a *single-slit* pattern rather than the *bright* fringes in a *double-slit* pattern.

It is possible to calculate the complete intensity distribution in the single-slit Fraunhofer diffraction pattern by adding the contributions from the Huygens wavelets from narrow strips of the slit, but such calculations are beyond our scope. Figure 39–21a shows the result of such a calculation, and Fig. 39–21b is a photograph of an actual single-slit diffraction pattern.

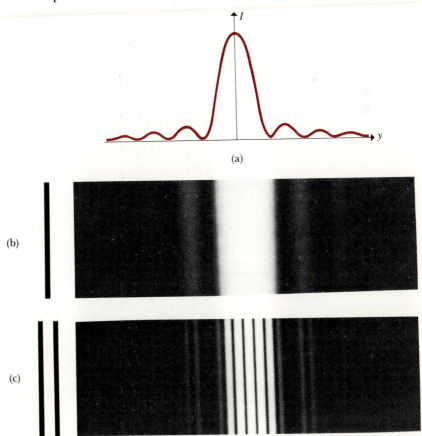

(a)

(b)

(c)

39–21   (a) Intensity distribution. (b) Photograph of the Fraunhofer diffraction pattern of a single slit. (c) Fraunhofer diffraction pattern of a double slit.

The photograph in Fig. 39–21c shows the Fraunhofer diffraction pattern of *two* slits each of the same width a as the one above, but separated by a distance $d = 4a$. Note that the interference fringes due to cooperation between the slits have intensities that follow the diffraction pattern of each separate slit. The interference fringes are modulated by the shape of the curve shown in part (a) of the figure, because of the finite width of the slits S_1 and S_2. In our earlier discussion of the two-slit pattern in Section 39–2, we assumed in effect that the width of each slit is much smaller than the distance between them; in that case the single-slit pattern is not seen.

39–8

THE DIFFRACTION GRATING

Suppose that instead of a single slit, or two slits side by side as in Young's experiment, we have a very large number of parallel slits, all of the same width and spaced at regular intervals. Such an arrangement, known as a *diffraction grating*, was first constructed by Fraunhofer. The earliest gratings were of fine wires, 0.04 mm to 0.6 mm in diameter. Gratings are now made by using a diamond point to rule a large number of equidistant grooves on a glass or metal surface.

Let GG in Fig. 39–22 represent the grating, the slits of which are perpendicular to the plane of the paper. While only five slits are shown in the diagram, an actual grating contains several thousand, with a grating spacing d of the order of 0.002 mm. Let a plane wave be incident normally on the grating from the left. The problem of finding the intensity of the light transmitted by the grating then combines the principles of interference and diffraction. That is, each slit gives rise to a diffracted beam, and these diffracted beams then interfere with one another to produce the final pattern.

Let us consider first the light proceeding from elements at the center of each opening, and traveling in a direction making an angle θ with that of the incident beam, as in Fig. 39–22. A lens at the right of the grating

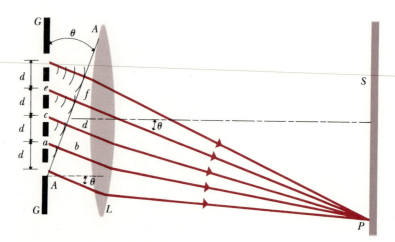

39–22   The plane diffraction grating.

forms in its focal plane a diffraction pattern similar to that which would appear on a screen at infinity.

Suppose the angle θ in Fig. 39–22 is taken so that the distance ab equals λ, the wavelength of the incident light. Then $cd = 2\lambda$, $ef = 3\lambda$, etc. The waves from all of these elements, since they are in phase at the plane of the grating, are also in phase along the plane AA and therefore reach the point P in phase. The same holds true for any set of elements in corresponding positions in the various slits.

If the angle θ is increased slightly, the disturbances from the grating elements no longer arrive at AA in phase with one another, and even an extremely small change in angle results in almost complete destructive interference among them, provided there is a large number of slits in the grating. Hence, the maximum at the angle θ is an extremely *sharp* one, differing from the rather broad maxima and minima that result from interference or diffraction effects with a small number of openings.

As the angle θ is increased still further, a position is eventually reached in which the distance ab in Fig. 39–22 becomes equal to 2λ. Then cd equals 4λ, ef equals 6λ, and so on. The disturbances at AA are again all in phase, the path difference between them now being 2λ, and another maximum results. Evidently still others will appear when $ab = 3\lambda, 4\lambda, \ldots$. Maxima will also be observed at corresponding angles on the opposite side of the grating normal, as well as along the normal itself, since in the latter position the phase difference between disturbances reaching AA is zero.

The angles of deviation for which the maxima occur may readily be found from Fig. 39–23. Consider the right angle Aba. Let d be the distance between successive grating elements, called the *grating spacing*. The necessary condition for a maximum is that $ab = m\lambda$, where $m = 0, 1, 2, 3$, etc. It follows that $d \sin \theta = m\lambda$, or

$$\sin \theta = m \frac{\lambda}{d} \qquad (m = 0, 1, 2, \ldots), \qquad (39\text{–}11)$$

is the necessary condition for a maximum. The angle θ is also the angle by which the rays corresponding to the maxima have been *deviated* from the direction of the incident light.

In practice, the parallel beam incident on the grating is usually produced by a lens having a narrow illuminated slit at its first focal point. Each of the maxima is then a sharp image of the slit, of the same color as that of the light illuminating the slit, assumed thus far to be monochromatic. If the slit is illuminated by light consisting of a mixture of wavelengths, the lens will form a number of images of the slit in different positions; each wavelength in the original light gives rise to a set of slit images deviated by the appropriate angles. If the slit is illuminated with white light, a continuous group of images is formed side by side or, in other words, the white light is dispersed into continuous spectra. In contrast with the single spectrum produced by a prism, a grating forms a number of spectra on either side of the normal. Those that correspond to $m = 1$ in Eq. (39–11) are called *first-order*, those that correspond to $m = 2$ are called *second-order*, and so on. Since for $m = 0$ the deviation is

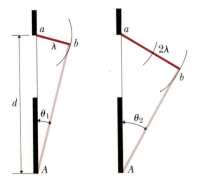

39–23 First-order maximum when $ab = \lambda$, second-order maximum when $ab = 2\lambda$.

zero, all colors combine to produce a white image of the slit in the direction of the incident beam.

For substantial deviation to occur, the grating spacing should be of the same order of magnitude as the wavelength of light. Gratings for use in or near the visible spectrum are ruled with from about 500 to 1500 lines per millimeter.

The diffraction grating is widely used in spectrometry, instead of a prism, as a means of dispersing a light beam into spectra. If the grating spacing is known, then from a measurement of the angle of deviation of any wavelength, the value of this wavelength may be computed. In the case of a prism this is not so; the angles of deviation are not related in any simple way to the wavelengths but depend on the characteristics of the material of which the prism is constructed. Since the index of refraction of optical glass varies more rapidly at the violet end than at the red end of the spectrum, the spectrum formed by a prism is always spread out more at the violet end than it is at the red. Also, while a prism deviates red light the least and violet the most, the reverse is true of a grating.

EXAMPLE 1 The wavelengths of the visible spectrum are approximately 400 nm to 700 nm. Find the angular breadth of the first-order visible spectrum produced by a plane grating having 6000 lines per centimeter, when light is incident normally on the grating.

Solution The grating spacing d is

$$d = \frac{1}{6000 \text{ lines} \cdot \text{cm}^{-1}}$$

$$= 1.67 \times 10^{-6} \text{ m}.$$

The angular deviation of the violet is

$$\sin \theta = \frac{400 \times 10^{-9} \text{ m}}{1.67 \times 10^{-6} \text{ m}} = 0.240,$$

$$\theta = 13.9°.$$

The angular deviation of the red is

$$\sin \theta = \frac{700 \times 10^{-9} \text{ m}}{1.67 \times 10^{-6} \text{ m}} = 0.420,$$

$$\theta = 24.8°.$$

Hence, the first-order visible spectrum includes an angle of

$$24.8° - 13.9° = 10.9°.$$ ◀

EXAMPLE 2 Show that the violet of the third-order spectrum overlaps the red of the second-order spectrum.

Solution The angular deviation of the third-order violet is

$$\sin \theta = \frac{(3)(400 \times 10^{-9} \text{ m})}{d}$$

and of the second-order red it is

$$\sin \theta = \frac{(2)(700 \times 10^{-9} \text{ m})}{d}.$$

Since the first angle is smaller than the second, whatever the grating spacing, the third order will *always* overlap the second. ◀

DIFFRACTION OF X-RAYS BY A CRYSTAL

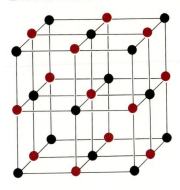

39–24 Model of arrangement of ions in a crystal of NaCl. Black circles, Na; color circles, Cl.

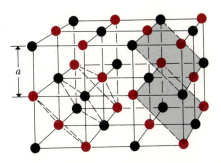

39–25 Cubic crystal lattice, showing two different families of crystal planes. The spacing of the planes on the left is $a/\sqrt{3}$; that of the planes on the right is $a/\sqrt{2}$. There are also three sets of planes parallel to the cube faces, with spacing a.

Although x-rays were discovered by Roentgen in 1895, it was not until 1913 that x-ray wavelengths were measured with any degree of precision. Experiments had indicated that these wavelengths might be of the order of 10^{-8} cm, which is about the same as the interatomic spacing in a solid. It occurred to Laue in 1913 that if the atoms in a crystal were arranged in a regular way, a crystal might serve as a three-dimensional diffraction grating for x-rays. The experiment was performed by Friederich and Knipping and it succeeded, thus verifying in a single stroke both the hypothesis that x-rays *are* waves (or, at any rate, wavelike in some of their properties) and that the atoms in a crystal *are* arranged in a regular manner. Since that time, the phenomenon of x-ray diffraction by a crystal has proved an invaluable tool of the physicist, both as a way to measure x-ray wavelengths and as a method of studying the structure of crystals.

Figure 39–24 is a diagram of a simple type of crystal, that of sodium chloride (NaCl). The black circles represent the sodium, and the color circles the chloride ions. The atoms in the crystal can be thought of as forming planes, as in Fig. 39–25; many different sets of planes are possible.

Figure 39–26 is a photograph made by directing a narrow beam of x-rays at a thin section of a crystal of quartz and allowing the diffracted beams to strike a photographic plate. Each atom in the crystal scatters some of the incident radiation, and interference occurs between the

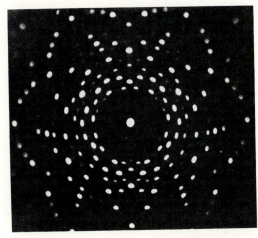

39–26 Laue diffraction pattern formed by directing a beam of x-rays at a thin section of quartz crystal. (Courtesy of Dr. B. E. Warren.)

waves scattered by the various atoms. Just as with the diffraction grating, nearly complete cancellation occurs for all but certain very specific directions for which constructive interference is possible; hence, the spots in Fig. 39–26. The interference pattern is often described in terms of *reflections* from various planes of atoms as in Fig. 39–25, but the basic phenomenon is one of *interference*.

If the crystal lattice spacing is known, the wavelength may be calculated, just as the wavelengths of visible light can be determined from measurements on patterns formed by slits or diffraction gratings. For example, the crystal lattice spacing for sodium chloride can be determined from its density and Avogadro's number, both of which can be determined independently. Conversely, once the x-ray wavelength is known, the structure and lattice spacing of crystals of unknown structure may be explored by x-ray diffraction.

Indeed, crystal x-ray diffraction has been by far the most important experimental tool in the investigation of crystal structure. More recently, x-ray diffraction has played an important role in the investigation of structure of liquids and of organic molecules. But the basic principles, superposition and interference, are no different from those of the other phenomena discussed in this chapter.

*39–10

THE RESOLVING POWER OF OPTICAL INSTRUMENTS

The expressions for the magnification of a telescope or a microscope involve (except for certain numerical factors) only the focal lengths of the lenses making up the optical system of the instrument. It appears, at first glance, as though any desired magnification might be attained by a proper choice of these focal lengths. Beyond a certain point, however, while the image formed by the instrument becomes larger (or subtends a larger angle) it *does not gain in detail*, even though all lens aberrations have been corrected. This limit to the useful magnification is set by the fact that light is a wave motion and the laws of geometrical optics do not hold strictly for a wave surface of limited extent. Physically, the image of a point source is not the intersection of *rays* from the source, but the diffraction pattern of those *waves* from the source that pass through the lens system.

It is an important experimental fact that the light from a point source, diffracted by a circular opening, is focused by a lens not as a geometrical point but as a disk of finite radius surrounded by dark and bright rings. The larger the wave surface admitted (i.e., the larger the lenses or diaphragms in an optical system), the smaller the diffraction pattern of a point source and the closer together may two point sources be before their diffraction disks overlap and become indistinguishable. An optical system is said to be able to *resolve* two point sources if the corresponding diffraction patterns are sufficiently small or sufficiently separated to be distinguished. The numerical measure of the ability of the system to resolve two such points is called its *resolving power* or its *resolution*.

39–27 Diffraction patterns of four "point" sources, with a circular opening in front of the lens. In (a), the opening is so small that the patterns at the right are just resolved, by Rayleigh's criterion. Increasing the aperture decreases the size of the diffraction patterns, as in (b) and (c).

Figure 39–27a is a photograph of four point sources made with the camera lens "stopped down" to an extremely small aperture. The nature of the diffraction patterns is clearly evident, and it is obvious that further magnification of the picture would not aid in resolving the sources. What is necessary is not to make the image *larger* but to make the diffraction patterns *smaller*. Figures 39–27b and 39–27c show how the resolving power of the lens is increased by increasing its aperture. In (b) the diffraction patterns are sufficiently small for all four sources to be distinguished. In (c) the full aperture of the lens was utilized.

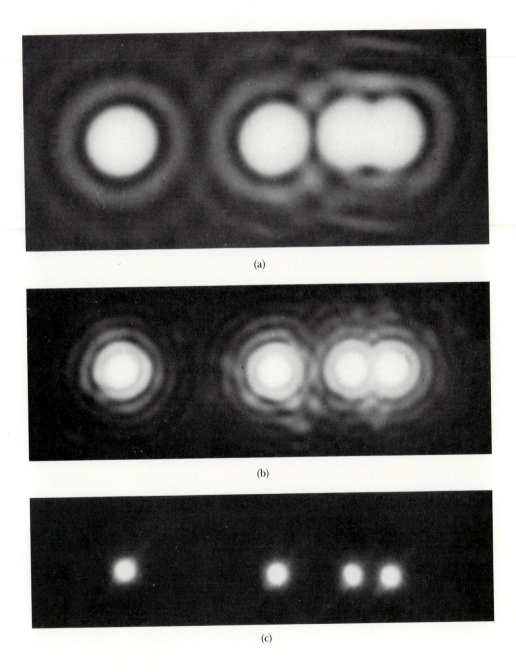

(a)

(b)

(c)

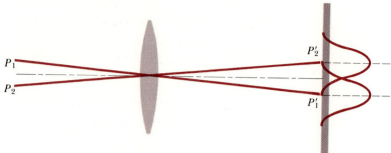

39–28 Two point sources are just resolved if the maximum of the diffraction pattern of one source just coincides with the first minimum of the other (Rayleigh's criterion).

An arbitrary criterion proposed by Lord Rayleigh is that two point sources are just resolvable if the central maximum of the diffraction pattern of one source just coincides with the first *minimum* of the other. Let P_1 and P_2 in Fig. 39–28 be the central rays in the two parallel beams coming from two very distant point sources. Let P_1' and P_2' be the centers of their diffraction patterns, formed by some optical instrument such as a microscope or telescope, which for simplicity is represented in the diagram as a single lens. If the images are just resolved, that is, if according to Rayleigh's criterion the first minimum of one pattern coincides with the center of the other, the separation of the centers of the patterns equals the radius of the central bright disk. For the diffraction pattern from a circular aperture of radius a, Rayleigh showed that the angular size of the central maximum, that is, the angular separation θ between the center of the pattern and the first minimum, is given by

$$\sin \theta = 1.22 \frac{\lambda}{a}. \tag{39–12}$$

Thus the smaller the aperture, the larger the diffraction pattern, as with the single-slit pattern. Equation (39–12) gives the minimum angular separation for two images to be resolved, according to Rayleigh's criterion.

The minimum separation of two points that can just be resolved is called the *limit of resolution* of the instrument. The smaller this distance, the greater is said to be the *resolving power*. The resolving power increases with the size of the aperture, and is inversely proportional to the wavelength of the light used. Light waves are the "tools" with which our optical system is provided; the smaller the tools, the finer the work the system can do!

*39–11

HOLOGRAPHY

Holography is a technique for recording and reproducing an image of an object without the use of lenses or mirrors. Unlike the two-dimensional images recorded by an ordinary photograph or television system, a holographic image is truly three-dimensional. Such an image can be viewed from different directions to reveal different sides, and from various distances to reveal changing perspective.

The basic procedure for making a hologram is very simple in principle. A possible arrangement is shown in Fig. 39–29a. The object to be

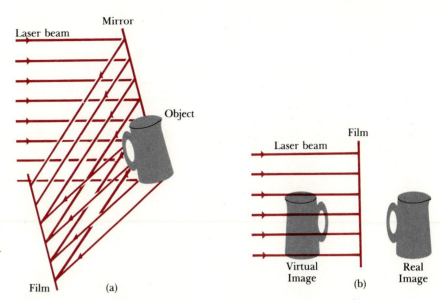

39–29 (a) The hologram is the record on film of the interference pattern formed with light directly from the source and light scattered from the object. (b) Images are formed when light is projected through the hologram.

39–30 (a) Constructive interference of the plane and spherical waves occurs in the plane of the film at every point Q for which the distance d_n from P is greater than the distance d_0 from P to O by an integer number of wavelengths $n\lambda$. For the point shown, $n = 2$. (b) When a plane wave strikes the developed film, the diffracted wave consists of a wave converging to P' and then diverging again, and a diverging wave that appears to originate at P. These waves form the real and virtual images, respectively.

holographed is illuminated by monochromatic light, and a photographic film is located so that it is struck by scattered light from the object and also by direct light from the source. In practice the source must be a laser, for reasons to be discussed later. Interference between the direct and scattered light leads to the formation and recording of a complex interference pattern on the film.

To form the images, one simply projects laser light through the developed film, as shown in Fig. 39–29b. Two images are formed, a virtual image on the side of the film nearer the source, and a real image on the opposite side.

A complete analysis of holography is beyond our scope, but we can gain some insight into the process by examining how a single point is holographed and imaged. We consider the interference pattern formed on a screen by the superposition of an incident plane wave and a spherical wave, as shown in Fig. 39–30a. The spherical wave originates at a point source P a distance d_0 from the screen; P may in fact be a small

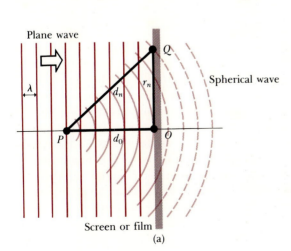

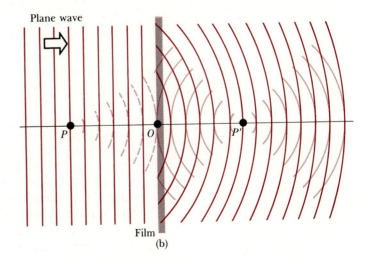

object that scatters part of the incident plane wave. In any event, we assume that the two waves are monochromatic and coherent, and that the phase relation is such that constructive interference occurs at point O on the diagram. Then constructive interference will *also* occur at any point Q on the screen that is farther from P than O is, by an integer number of wavelengths. That is, if $d_n - d_0 = n\lambda$, where n is an integer, constructive interference occurs. The points where this condition is satisfied form circles centered at O, with radii r_n given by

$$d_n - d_0 = \sqrt{d_0{}^2 + r_n{}^2} - d_0 = n\lambda, \quad (n = 1, 2, 3, \ldots). \quad (39\text{--}13)$$

Solving this for $r_n{}^2$, we find

$$r_n{}^2 = \lambda \ (2nd_0 + n^2\lambda).$$

Ordinarily d_0 is very much larger than λ, so we neglect the second term in parentheses, obtaining

$$r_n = \sqrt{2n\lambda d_0}. \qquad (39\text{--}14)$$

Since n may be any integer, the interference pattern consists of a series of concentric bright circular fringes, with the radii of the brightest regions given by Eq. (39–14). Between these bright fringes will be darker fringes. This pattern may be recorded on film by simply placing the film in the position of the screen.

Now the film is developed and a transparent positive print is made, so the bright-fringe areas have the greatest transparency on the film. It is then illuminated with monochromatic plane-wave light of the same wavelength as that used initially. In Fig. 39–30b, consider a point P' at a distance d_0 along the axis from the film; the centers of successive bright fringes differ in their distances from P' by an integer number of wavelengths, and therefore a strong *maximum* in the diffracted wave occurs at P'. That is, light converges to P' and then diverges from it on the opposite side, and P' is therefore a *real image* of point P.

This is not the entire diffracted wave, however; there is also a diverging spherical wave, which would represent a continuation of the wave originally emanating from P if the film had not been present. Thus the *total* diffracted wave is a superposition of a converging spherical wave forming a real image at P', and a diverging spherical wave shaped as though it had originated at P, forming a virtual image at P.

Because of the principle of linear superposition, what is true for the imaging of a single point is also true for the imaging of any number of points. The film records the superposed interference pattern from the various points, and when light is projected through the film the various image points are reproduced simultaneously. Thus the images of an extended object can be recorded and reproduced just as for a single point object.

Several practical problems must be overcome. First, the light used must be *coherent* over the distances large compared to the dimensions of the object and its distance from the film. Ordinary light sources *do not* satisfy this requirement, for reasons discussed in Section 39–1, and laser light is essential. Second, extreme mechanical stability is needed. If there is any relative motion of source, object, or film during exposure, even by as much as a wavelength, the interference pattern on the film is blurred enough to prevent satisfactory image formation. These obstacles are not insurmountable, however, and holography promises to become increasingly important in research, entertainment, and a wide variety of technological applications.

QUESTIONS

39–1 Could an experiment similar to Young's two-slit experiment be performed with sound? How might this be carried out? Does it matter that sound waves are longitudinal and electromagnetic waves transverse?

39–2 At points of constructive interference between waves of equal amplitude, the intensity is four times that of either individual wave. Does this violate energy conservation? If not, why not?

39–3 In using the superposition principle to calculate intensities in interference and diffraction patterns, could one add the intensities of the waves instead of their amplitudes? What's the difference?

39–4 A two-slit interference experiment is set up and the fringes displayed on a screen. Then the whole apparatus is immersed in the nearest swimming pool. How does the fringe pattern change?

39–5 Would the headlights of a distant car form a two-source interference pattern? If so, how might it be observed? If not, why not?

39–6 A student asserted that it is impossible to observe interference fringes in a two-source experiment if the distance between sources is less than half the wavelength of the wave. Do you agree? Explain.

39–7 An amateur scientist proposed to record a two-source interference pattern using only one source, placing it first in position S_1 in Fig. 39–1 and turning it on for a certain time, then placing it at S_2 and turning it on for an equal time. Does this work?

39–8 When a thin oil film spreads out on a puddle of water, the thinnest part of the film looks lightest in the resulting interference pattern. What does this tell you about the relative magnitudes of the refractive indexes of oil and water?

39–9 A glass windowpane with a thin film of water on it reflects less than when it is perfectly dry. Why?

39–10 In high-quality camera lenses, the resolution in the image is determined by diffraction effects. Is the resolution best when the lens is "wide open" or when it is "stopped down" to a smaller aperture? How does this behavior compare with the effect of aperture size on the depth of focus, that is, on the limit of resolution due to imprecise focusing?

39–11 If a two-slit interference experiment were done with white light, what would be seen?

39–12 Why is a diffraction grating better than a two-slit setup for measuring wavelengths of light?

39–13 Would the interference and diffraction effects described in this chapter still be seen if light were a longitudinal wave instead of transverse?

39–14 One sometimes sees rows of evenly spaced radio antenna towers. A student remarked that these act like diffraction gratings. What did he mean? Why would one *want* them to act like a diffraction grating?

39–15 Could x-ray diffraction effects with crystals be observed using visible light instead of x-rays? Why or why not?

39–16 Does a microscope have better resolution with red light or blue light?

39–17 How could an interference experiment, such as one using a Michelson interferometer or fringes caused by a thin air space between glass plates, be used to measure the refractive index of air?

PROBLEMS

39–1 Two slits are spaced 0.3 mm apart and are placed 50 cm from a screen. What is the distance between the second and third dark lines of the interference pattern when the slits are illuminated with light of 600-nm wavelength?

39–2 In Problem 39–1, suppose the entire apparatus is immersed in water. Then what is the distance between the second and third dark lines?

39–3 Young's experiment is performed with sodium light ($\lambda = 589$ nm). Fringes are measured carefully on a screen 100 cm away from the double slit, and the center of the twentieth fringe is found to be 11.78 mm from the center of the zeroth fringe. What is the separation of the two slits?

39–4 An FM radio station has a frequency of 100 MHz and uses two identical antennas mounted at the same elevation, 12 m apart. The resulting radiation pattern has maximum intensity along a horizontal line perpendicular to the line joining the antennas. At what other angles (measured from the line of maximum intensity) is the intensity maximum? At what angles is it zero?

39–5 An AM radio station has a frequency of 1000 kHz; it uses two identical antennas at the same elevation, 150 m apart.

a) In what directions is the intensity maximum, considering points in a horizontal plane?

b) Calling the maximum intensity in (a) I_0, determine in terms of I_0 the intensity in directions making angles of 30°, 45°, 60°, and 90° to the direction of maximum intensity.

39–6 Light from a mercury-arc lamp is passed through a filter that blocks everything except for one spectrum line in the green region of the spectrum. It then falls on two slits separated by 0.6 mm. In the resulting interference pattern on a screen 2.5 mm away, adjacent bright fringes are separated by 2.27 mm. What is the wavelength?

39–7 Light of wavelength 500 nm is incident perpendicularly from air on a film 10^{-4} cm thick and of refractive index 1.375. Part of the light is reflected from the first surface of the film, and part enters the film and is reflected back at the second face.

a) How many waves are contained along the path of this light in the film?

b) What is the phase difference between these waves as they leave the film?

39–8 A glass plate, 0.40 μm thick, is illuminated by a beam of white light normal to the plate. The index of refraction of the glass is 1.50. What wavelengths within the limits of the visible spectrum ($\lambda = 400$ nm to $\lambda = 700$ nm) will be intensified in the reflected beam?

39–9 How far must the mirror of M_2 of the Michelson interferometer be moved in order that 3000 fringes of krypton-86 light ($\lambda = 606$ nm) move across a line in the field of view?

39–10 Two rectangular pieces of plane glass are laid one upon the other on a table. A thin strip of paper is placed between them at one edge so that a very thin wedge of air is formed. The plates are illuminated by a beam of sodium light at normal incidence ($\lambda = 589$ nm). Interference fringes are formed, there being ten per centimeter length of wedge measured normal to the edges in contact. Find the angle of the wedge.

39–11 A sheet of glass 10 cm long is placed in contact with a second sheet, and is held at a small angle with it by a metal strip 0.1 mm thick placed under one end. The glass is illuminated from above with light of 546-nm wavelength. How many interference fringes are observed per cm in the reflected light?

39–12 The radius of curvature of the convex surface of a plano-convex lens is 120 cm. The lens is placed convex side down on a plane glass plate and illuminated from above with red light of wavelength 650 nm. Find the diameter of the third bright ring in the interference pattern.

39–13 What is the thinnest film of 1.40 refractive index in which destructive interference of the violet component (400 nm) of an incident white beam in air can take place by reflection? What is, then, the residual color of the beam?

39–14 The surfaces of a prism of index 1.52 are to be made "nonreflecting" by coating them with a thin layer of transparent material of index 1.30. The thickness of the layer is such that, at a wavelength of 550 nm (in vacuum), light reflected from the first surface is $\frac{1}{2}$ wavelength out of phase with that reflected from the second surface. Find the thickness of the layer.

39–15

a) Is a thin film of quartz suitable as a nonreflecting coating for fabulite? (See Table 36–1.)

b) If so, how thick should the film be?

39–16 Parallel rays of green mercury light of wavelength 546 nm pass through a slit of width 0.437 mm covering a lens of focal length 40 cm. In the focal plane of the lens, what is the distance from the central maximum to the first minimum?

39–17 Light of wavelength 589 nm from a distant source is incident on a slit 1.0 mm wide, and the resulting diffraction pattern is observed on a screen 2.0 m away. What is the distance between the two dark fringes on either side of the central bright fringe?

39–18 In Problem 39–17, suppose the entire apparatus is immersed in water. Then what is the distance between the two dark fringes?

39–19 Plane monochromatic waves of wavelength 600 nm are incident normally on a plane transmission grating having 500 lines·mm^{-1}. Find the angles of deviation in the first, second, and third orders.

39–20 A plane transmission grating is ruled with 4000 lines·cm^{-1}. Compute the angular separation in degrees, in the second-order spectrum, between the α and δ lines of atomic hydrogen, whose wavelengths are, respectively, 656 nm and 410 nm. Assume normal incidence.

39–21

a) What is the wavelength of light that is deviated in the first order through an angle of 20° by a transmission grating having 6000 lines·cm^{-1}?

b) What is the second-order deviation of this wavelength? Assume normal incidence.

39–22 A slit of width a was placed in front of a lens of focal length 80 cm. The slit was illuminated by parallel light of wavelength 600 nm and the diffraction pattern of Fig. 39–21b was formed on a screen in the second focal plane of the lens. If the photograph of Fig. 39–21b represents an enlargement to twice the actual size, what was the slit width?

39–23 What is the longest wavelength that can be observed in the fourth order for a transmission grating having 5000 lines·cm^{-1}? Assume normal incidence.

39–24 In Fig. 39–31, two point sources of light, a and b, at a distance of 50 m from lens L and 6 mm apart, produce images at c that are just resolved by Rayleigh's criterion. The focal length of the lens is 20 cm. What is the diameter of the diffraction circles at c?

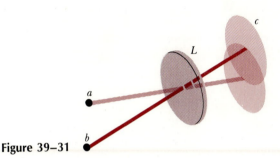

Figure 39–31

39–25 A telescope is used to observe two distant point sources 1 m apart. ($\lambda = 500$ nm.) The objective of the telescope is covered with a slit of width 1 mm. What is the maximum distance in meters at which the two sources may be distinguished?

39–26 A typical lens in a 35-mm camera has a focal length of 50 mm and a diameter of 25 mm ($f/2$). The resolution of such lenses is expressed as the number of lines per millimeter in the image that can be resolved. If the resolution of this lens is limited by diffraction effects, approximately what is its resolution in lines·mm^{-1}?

39–27 If a hologram is made using 600-nm light and then viewed using 500-nm light, how will the images look compared to those observed with 500-nm light?

39–28 A hologram is made using 600-nm light and is then viewed using continuous-spectrum white light from an incandescent bulb. What will be seen?

39–29 Ordinary photographic film reverses black and white, in the sense that the most brightly illuminated areas become blackest upon development (hence the term *negative*). Suppose a hologram negative is viewed directly, without making a positive transparency. How will the resulting images differ from those obtained with the positive hologram?

40

RELATIVISTIC MECHANICS

In preceding chapters we have stressed the importance of inertial frames of reference. Newton's laws of motion are valid only in inertial frames, but they are valid in *all* inertial frames. Any frame moving with constant velocity with respect to an inertial frame is itself an inertial frame, and all such frames are equivalent with respect to expressing the basic principles of mechanics. *The laws of mechanics are the same in every inertial frame of reference.*

Einstein proposed in 1905 that this principle should be extended to include *all* the basic laws of physics. This innocent-sounding proposition has far-reaching and startling consequences, a few of which have already been mentioned. If the principle of conservation of momentum is to be valid in all inertial systems, for example, the definition of momentum for particles moving at speeds comparable to the speed of light must be changed from mv to $mv/(1 - v^2/c^2)^{1/2}$, and kinetic energy must also be redefined. Equally fundamental are the modifications needed in the *kinematic* aspects of motion. The resulting generalizations of the laws of mechanics are part of the *special theory of relativity*, the subject of this chapter. In studying this material the reader must be prepared to confront some ideas that will seem at first sight not to make sense, and to mistrust intuition when dealing with phenomena far removed from everyday experience!

40–1

INVARIANCE OF PHYSICAL LAWS

Einstein's *principle of relativity* states that *the laws of physics are the same in every inertial frame of reference.* A familiar example of this principle in electromagnetism is the induced electromotive force in a coil of wire resulting from motion of a nearby permanent magnet. In a frame of reference where the *coil* is stationary, the moving magnet causes a change of magnetic flux through the coil and hence an induced emf. In a frame where the *magnet* is stationary, the coil is moving through a magnetic field, and this motion causes magnetic-field forces on the mobile charges in the conductor, inducing an emf. According to the principle of relativity, both of these points of view have equal validity and both must predict the same result for the induced emf. As we have seen in Chapter 32, Faraday's law of electromagnetic induction does indeed satisfy this requirement.

Of equal significance is the prediction of the speed of light and other electromagnetic radiation, emerging from the development in Chapter 35. The principle of relativity requires that the speed of propagation of this radiation must be independent of the frame of reference and must be the same for all inertial frames.

Indeed, the speed of light plays a special role in the theory of relativity. Light travels in vacuum with speed c that is independent of the motion of the source. Its numerical value is known very precisely; to six significant figures it is

$$c = 2.99792 \times 10^8 \text{ m} \cdot \text{s}^{-1}.$$

The approximate value $c = 3.00 \times 10^8 \text{ m} \cdot \text{s}^{-1}$ is in error by less than one part in 1000 and is often used when greater precision is not required.

At one time it was thought that light traveled through a hypothetical medium called *the ether*, just as sound waves travel through air. During the late nineteenth and early twentieth centuries, intensive efforts were made to find experimental evidence for the existence of the ether. The Michelson–Morley experiment, described in Section 39–5, was an effort to detect motion of the earth relative to the ether. This and all similar experiments yielded consistently negative results, and the ether concept has been discarded. With this result in mind, let us suppose the speed of light is measured by two observers, one at rest with respect to the source, the other moving away from it. Both are in inertial frames of reference, and according to Einstein's principle of relativity, the laws of physics, and in particular the speed of light, must be the same in both frames.

For example, suppose a light source is located in a spaceship moving with respect to Earth. An observer moving with the spaceship measures the speed of the light and obtains the value c. But after the light has left the source its motion cannot be influenced by the motion of the source, so an observer on Earth measuring the speed of this same light must also obtain the value c, despite the fact that there is relative motion between the two observers. This conclusion may appear not to be consistent with common sense; but it is important to recognize that "common sense" is intuition based on everyday experience, and this does not usually include measurements of the speed of light. We must be prepared to accept results that seem to be in conflict with common sense when they involve realms far removed from everyday observation.

Thus the speed of light (in vacuum) is independent of the motion of the source and is the same in all frames of reference. To explore the consequences of this statement, we consider the newtonian relationship between two inertial frames, labeled S and S' in Fig. 40–1. Let the x-axes of the two frames lie along the same line, but let the origin O' of S' move relative to the origin O of S with constant velocity u along the common x-axis. If the two origins coincide at time $t = 0$, then their separation at a later time t is ut.

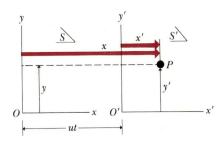

40–1 The position of point P can be described by the coordinates x and y in frame of reference S, or by x' and y' in S'. S' moves relative to S with constant velocity u along the common x–x' axis. The two origins O and O' coincide at time $t = t' = 0$.

A point P may be described by the coordinates (x, y, z) in S or by coordinates (x', y', z') in S'. Reference to the figure shows that these are related by

$$x = x' + ut, \qquad y = y', \qquad z = z'. \tag{40-1}$$

These equations are called the *galilean coordinate transformation*.

If the point P moves in the x-direction, its velocity v relative to S is given by $v = \Delta x/\Delta t$, and its velocity v' relative to S' is $v' = \Delta x'/\Delta t$. Intuitively it is clear that these are related by

$$v = v' + u. \tag{40-2}$$

This relation may also be obtained formally from Eqs. (40–1). Suppose the particle is at a point described by coordinate x_1 or x_1' at time t_1, and at x_2 or x_2' at time t_2. Then $\Delta t = t_2 - t_1$, and, from Eq. (40–1),

$$\Delta x = x_2 - x_1 = (x_2' - x_1') + u(t_2 - t_1) = \Delta x' + u\,\Delta t,$$

$$\frac{\Delta x}{\Delta t} = \frac{\Delta x'}{\Delta t} + u,$$

and

$$v = v' + u,$$

in agreement with Eq. (40–2).

A fundamental problem now appears. Applied to the speed of light, Eq. (40–2) says $c = c' + u$. Einstein's principle of relativity, supported by experimental observations, says $c = c'$. This is a genuine inconsistency, not an illusion, and it demands resolution. If we accept the principle of relativity, we are forced to conclude that Eqs. (40–1) and (40–2), intuitively appealing as they are, cannot be correct, but need to be modified to bring them into harmony with this principle.

The resolution involves modifications of our basic kinematic concepts. The first of these involves an assumption so fundamental that it might seem unnecessary, namely the assumption that the same *time scale* is used in frames S and S'. This may be stated formally by adding to Eqs. (40–1) a fourth equation $t = t'$. Obvious though this assumption may seem, it is not strictly correct when the relative speed u of the two frames of reference is comparable to the speed of light. The difficulty lies in the concept of *simultaneity*, which we examine next.

40–2

RELATIVE NATURE OF SIMULTANEITY

Measuring times and time intervals involves the concept of *simultaneity*. When a person says he awoke at seven o'clock he means that two *events*, his awakening and the arrival of the hour hand of his clock at the number seven, occurred *simultaneously*. The fundamental problem in measuring time intervals is that, in general, two events that appear simultaneous in one frame of reference *do not* appear simultaneous in a second frame that is moving relative to the first, even if both are inertial frames.

The following thought experiment, devised by Einstein, illustrates this point. Consider a long train moving with uniform velocity, as shown in Fig. 40–2a. Two lightning bolts strike the train, one at each end. Each

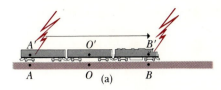

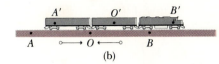

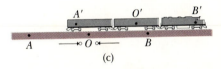

40-2 (a) To the stationary observer at point *O*, two lightning bolts appear to strike simultaneously. (b) The moving observer at point *O'* sees the light from the front of the train first and thinks that the bolt at the front struck first. (c) The two light pulses arrive at *O* simultaneously.

bolt leaves a mark on the train and one on the ground at the same instant. The points on the ground are labeled *A* and *B* in the figure, and the corresponding points on the train *A'* and *B'*. An observer on the ground is located at *O*, midway between *A* and *B*; another observer is at *O'*, moving with the train and midway between *A'* and *B'*. Both observers use the light signals from the lightning to observe the events.

Suppose the two light signals reach the observer at *O* simultaneously; he concludes that the two events took place at *A* and *B* simultaneously. But the observer at *O'* is moving with the train, and the light pulse from *B'* reaches him before the light pulse from *A'* does; he concludes that the event at the front of the train happened *earlier* than that at the rear. Thus the two events appear simultaneous to one observer, but not to the other. *Whether two events at different space points are simultaneous depends on the state of motion of the observer.* It follows that *the time interval between two events at different space points is in general different for two observers in relative motion.*

It might be argued that in this example the lightning bolts really *are* simultaneous, and that if the observer at *O'* could communicate with the distant points without time delay, he would realize this. But the finite speed of information transmission is not the problem. If *O'* is really midway between *A'* and *B'*, then, in his frame of reference, the time for a signal to travel from *A'* to *O'* is the same as from *B'* to *O'*. Two signals arrive simultaneously at *O'* only if they were emitted simultaneously at *A'* and *B'*; in this example they *do not* arrive simultaneously at *O'*, and so *O'* must conclude that the events at *A'* and *B'* were *not* simultaneous.

Furthermore, there is no basis for saying either that *O* is right and *O'* is wrong, or the reverse since, according to the principle of relativity, no inertial frame of reference is preferred over any other in the formulation of physical laws. Each observer is correct *in his own frame of reference,* but simultaneity is not an absolute concept. Whether two events are simultaneous depends on the frame of reference, and the time interval between two events also depends on the frame of reference.

40-3

RELATIVITY OF TIME

To derive a quantitative relation between time intervals in different coordinate systems, we consider another thought experiment. As before, a frame of reference *S'* moves with velocity *u* relative to a frame *S*. An observer in *S'* directs a source of light at a mirror a distance *d* away, as shown in Fig. 40-3, and measures the time interval $\Delta t'$ for light to make the "round trip" to the mirror and back. The total distance is 2*d*, so the time interval is

$$\Delta t' = \frac{2d}{c}. \tag{40-3}$$

As measured in frame *S*, the time for the round trip is a different interval Δt. During this time, the source moves relative to *S* a distance $u\,\Delta t$, and the total round-trip distance is not just 2*d* but is 2*l*, where

$$l = \sqrt{d^2 + \left(\frac{u\,\Delta t}{2}\right)^2}.$$

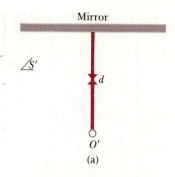

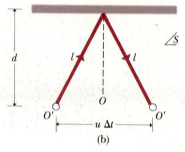

40–3 (a) Light pulse emitted from source at O' and reflected back along the same line, as observed in S'. (b) Path of the same light pulse, as observed in S. The positions of O' at the times of departure and return of the pulse are shown. The speed of the pulse is the same in S as in S', but the path is longer in S.

The speed of light is the same for both observers, so the relation in S analogous to Eq. (40–3) is

$$\Delta t = \frac{2l}{c} = \frac{2}{c}\sqrt{d^2 + \left(\frac{u\,\Delta t}{2}\right)^2}. \qquad (40\text{--}4)$$

To obtain a relation between Δt and $\Delta t'$ that does not contain d, we solve Eq. (40–3) for d and substitute the result into Eq. (40–4), obtaining

$$\Delta t = \frac{2}{c}\sqrt{\left(\frac{c\,\Delta t'}{2}\right)^2 + \left(\frac{u\,\Delta t}{2}\right)^2}.$$

This may now be squared and solved for Δt; the result is

$$\Delta t = \frac{\Delta t'}{\sqrt{1 - u^2/c^2}}. \qquad (40\text{--}5)$$

We may generalize this important result: If a time interval $\Delta t'$ separates two events occurring at the same space point in a frame of reference S' (in this case, the departure and arrival of the light signal at O'), then the time interval Δt between these two events as observed in S is *larger* than $\Delta t'$, and the two are related by Eq. (40–5). Thus when the rate of a clock at rest in S' is measured by an observer in S, the rate measured in S is *slower* than the rate observed in S'. This effect is called *time dilation*.

It is important to note that the observer in S measuring the time interval Δt cannot do so with a single clock. In Fig. 40–3 the points of departure and return of the light pulse are different space points in S, although they are the same point in S'. If S tries to use a single clock, the finite time of communication between two points will becloud the issue. To avoid this, S may use two clocks at the two relevant points. There is no difficulty in synchronizing two clocks in the same frame of reference; one procedure is to send a light pulse simultaneously to two clocks from a point midway between them, with the two operators setting their clocks to a predetermined time when the pulses arrive. In thought experiments it is often helpful to imagine a large number of synchronized clocks distributed conveniently in a single frame of reference. Only when a clock is moving relative to a given frame of reference do ambiguities of synchronization or simultaneity arise.

EXAMPLE A spaceship flies past earth with a speed of $0.99c$ (about 2.97×10^8 m·s^{-1}). A high-intensity signal light (perhaps a pulsed laser) blinks on and off, each pulse lasting 2×10^{-6} s. At a certain instant the ship appears to an earthling observer to be directly overhead at an altitude of 1000 km, and to be traveling perpendicular to the line of sight. What is the duration of each light pulse, as measured by this observer, and how far does the ship travel relative to earth during each pulse?

Solution The observer does not see the pulse at the instant it is emitted, because the light signal requires a time equal to $(1000 \times 10^3$ m$)/(3 \times 10^8$ m·s$^{-1})$, or $(1/300)$ s, to travel from the ship to earth. But if the distance from the spaceship to observer is essentially constant during the emission of a pulse, the time delays at beginning and end of the pulse are equal and the time *interval* is not affected.

Let S be the earth's frame of reference, S' that of the spaceship. Then, in the notation of Eq. (40–5), $\Delta t' = 2 \times 10^{-6}$ s. This interval refers to two events occurring at the same point relative to S', namely, the starting and stopping of the pulse. The corresponding interval in S is given by Eq. (40–5):

$$\Delta t = \frac{\Delta t'}{\sqrt{1 - u^2/c^2}} = \frac{2 \times 10^{-6} \text{ s}}{\sqrt{1 - (0.99)^2}} = 14.2 \times 10^{-6} \text{ s}.$$

Thus the time dilation in S is about a factor of seven. The distance D traveled in S during this interval is

$$D = u\,\Delta t = (0.99)(3 \times 10^8 \text{ m} \cdot \text{s}^{-1})(14.2 \times 10^{-6} \text{ s})$$

$$= 4220 \text{ m} = 4.20 \text{ km}.$$

If the spaceship is traveling directly *toward* the observer, the time interval cannot be measured directly by a single observer because the time delay is not the same at the beginning and end of the pulse. One possible scheme, at least in principle, is to use *two* observers at rest in S, with synchronized clocks, one at the position of the ship when the pulse starts, the other at its position at the end of the pulse. These observers will again measure a time interval in S of 14.2×10^{-6} s. ◀

Time-dilation effects are not observed in everyday life because the speeds of all practical modes of transportation are much less than the speed of light. For example, for a jet airplane flying at 600 mi·hr^{-1} (about 270 m·s^{-1}),

$$\frac{v^2}{c^2} = \left(\frac{270 \text{ m} \cdot \text{s}^{-1}}{3 \times 10^8 \text{ m} \cdot \text{s}^{-1}} \right)^2 = 8.1 \times 10^{-13},$$

and the time-dilation factor in Eq. (40–5) is approximately $1 + (4 \times 10^{-13})$. Thus to observe time dilation in this situation requires a clock with a precision of the order of one part in 10^{13}. But, as noted in Section 1–2, atomic clocks capable of this precision have recently been developed, and in the past few years experiments with such clocks in jet airplanes have verified Eq. (40–5) directly.

From the derivation of Eq. (40–5) and the spaceship example, it can be seen that a time interval between two events occurring *at the same point* in a given frame of reference is a more fundamental quantity than an interval between events at different points. The term *proper time* is used to denote an interval between two events occurring at the same space point. Thus Eq. (40–5) may be used *only* when $\Delta t'$ is a proper time interval in S', in which case Δt is *not* a proper time interval in S. If, instead, Δt is proper in S, then Δt and $\Delta t'$ must be interchanged in Eq. (40–5).

When the relative velocity u of S and S' is very small, the factor $(1 - u^2/c^2)$ is very nearly equal to unity, and Eq. (40–5) approaches the newtonian relation $\Delta t = \Delta t'$ (i.e., the same time scale for all frames of reference). This assumption, therefore, retains its validity in the limit of small relative velocities.

Equation (40–5) suggests an apparent paradox. Consider two identical-twin astronauts; one remains on earth while the other takes off on a high-speed trip through the galaxy. Because of time dilation, the earthbound twin sees the other twin's heartbeat and all other life processes proceeding more slowly than his own. Thus he thinks the roving twin not only ages more slowly but that when he returns to earth he is younger than the one who stayed home.

Now here is the paradox: Cannot the roving twin make exactly the same arguments, since all inertial frames are equivalent, to conclude that the earthbound twin is in fact the younger? Thus each twin would think the other is younger, which is a paradox. But the two twins are *not* identical in all respects; if the earthbound twin remains in an inertial frame at all times, the other must at times have an acceleration with respect to inertial frames in order to turn around and come back. Thus there is a real physical difference between the circumstances of the two twins. Careful analysis shows that the earthbound twin is correct; when the rover returns he is younger than the earthbound one.

40–4

RELATIVITY OF LENGTH

Just as the time interval between two events depends on the frame of reference, the *distance* between two points also depends on the frame of reference. To measure a distance one must, in principle, observe the positions of two points, such as the two ends of a ruler, simultaneously; but what is simultaneous in one frame is not in another.

To develop a relation between lengths in various coordinate systems, we consider another thought experiment. We attach a source of light pulses to one end of a ruler and a mirror to the other, as shown in Fig. 40–4. Let the ruler be at rest in S' and its length in this frame be l'. Then the time $\Delta t'$ required for a light pulse to make the round trip from source to mirror and back is given by

$$\Delta t' = \frac{2l'}{c}. \qquad (40\text{–}6)$$

This is a proper time interval, since departure and return occur at the same point in S'.

In S the ruler is displaced during this travel of the light pulse. Let the length of the ruler in S be l, and let the time of travel from source to mirror, as measured in S, be Δt_1. During this interval the mirror moves a distance $u\,\Delta t_1$, and the total length of path d from source to mirror is not l but

$$d = l + u\,\Delta t_1. \qquad (40\text{–}7)$$

But since the pulse travels with speed c, it is also true that

$$d = c\,\Delta t_1. \qquad (40\text{–}8)$$

Combining Eqs. (40–7) and (40–8) to eliminate d,

$$c\,\Delta t_1 = l + u\,\Delta t_1,$$

or

$$\Delta t_1 = \frac{l}{c - u}. \qquad (40\text{–}9)$$

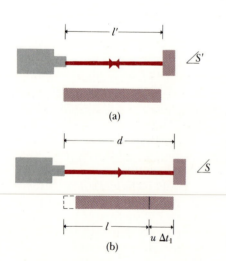

(a)

(b)

40–4   (a) A light pulse is emitted from a source at one end of a ruler, reflected from a mirror at the opposite end, and returned to the source position. (b) Motion of the light pulse as seen by an observer in S. The distance traveled from source to mirror is greater than the length l measured in S, by the amount $u\,\Delta t_1$, as shown.

In the same way it can be shown that the time Δt_2 for the return trip from mirror to source is

$$\Delta t_2 = \frac{l}{c + u}. \tag{40-10}$$

The *total* time $\Delta t = \Delta t_1 + \Delta t_2$ for the round trip, as measured in S, is

$$\Delta t = \frac{l}{c - u} + \frac{l}{c + u} = \frac{2l}{c(1 - u^2/c^2)}. \tag{40-11}$$

It is also known that Δt and $\Delta t'$ are related by Eq. (40-5), since $\Delta t'$ is proper in S'. Thus Eq. (40-6) becomes

$$\Delta t \sqrt{1 - \frac{u^2}{c^2}} = \frac{2l'}{c}. \tag{40-12}$$

Finally, combining this with Eq. (40-11) to eliminate Δt, and simplifying, we obtain

$$l = l' \sqrt{1 - \frac{u^2}{c^2}}. \tag{40-13}$$

Thus the length measured in S, in which the ruler is moving, is *shorter* than in S', where it is at rest. A length measured in the rest frame of the body is called a proper length; thus, l' above is a proper length in S', and the length measured in any other frame is less than l'. This effect is called *contraction of length*.

EXAMPLE In the spaceship example of Section 40-3, what distance does the spaceship travel during emission of a pulse, as measured in its rest frame?

Solution The question is somewhat ambiguous since, of course, in its own frame of reference the ship is at rest. But suppose it leaves markers in space, such as small smoke bombs, at the instants when the pulse starts and stops, and measures the distance between these markers, with the aid of observers behind the ship but moving with it, each with a clock synchronized with that of the ship. The distance d between the markers is a proper length in the earth's frame S. In the spaceship's frame S', the distance d' is contracted by the factor given in Eq. (40-13):

$$d' = d \sqrt{1 - \frac{u^2}{c^2}} = (4220 \text{ m}) \sqrt{1 - (0.99)^2}$$

$$= 595 \text{ m}.$$

(Note that because d, not d', is a proper length, we must reverse the roles of l and l'.) An observer in the spaceship can calculate his speed relative to earth from this data:

$$u = \frac{d'}{\Delta t'} = \frac{594 \text{ m}}{2 \times 10^{-6} \text{ s}} = 2.97 \times 10^8 \text{ m} \cdot \text{s}^{-1},$$

which agrees with the initial data. ◄

When u is very small compared to c, the contraction factor in Eq. (40–13) approaches unity, and in the limit of small speeds we recover the newtonian relation $l = l'$. This and the corresponding result for time dilation show that Eqs. (40–1) retain their validity in the limit of small speeds; only at speeds comparable to c are modifications needed.

Lengths measured perpendicular to the direction of motion are *not* contracted; this may be verified by constructing a thought experiment for measuring in S and S' the length of a ruler oriented perpendicular to the direction of relative motion. The details of this discussion are not essential for our further work and will not be given here.

*40–5

THE LORENTZ TRANSFORMATION

The galilean coordinate transformation given by Eqs. (40–1) is valid only in the limit when u is much smaller than c, but we are now in position to derive a more general transformation not subject to this limitation. The more general relations are called the *Lorentz transformation*. When u is small they reduce to the galilean transformation, but they may also be used when u is comparable to c.

The basic problem is this: When an event occurs at point (x, y, z) at time t, as observed in a frame of reference S, what are the coordinates (x', y', z') and time t' of the event as observed in a second frame S' moving relative to S with constant velocity u along the x-direction?

To derive the transformation equations, we return to Fig. 40–1. As before, we assume that the origins coincide at the initial time $t = t' = 0$. Then in S the distance from O to O' at time t is still ut. The coordinate x' is a proper length in S', so in S it appears contracted by the factor given in Eq. (40–13). Thus the distance x from O to P in S is given not simply by $x = ut + x'$ as in the galilean transformation, but by

$$x = ut + x' \sqrt{1 - \frac{u^2}{c^2}}. \tag{40–14}$$

Solving this equation for x', we obtain

$$x' = \frac{x - ut}{\sqrt{1 - u^2/c^2}}. \tag{40–15}$$

This is half of the Lorentz transformation; the other half is the equation giving t' in terms of x and t. To obtain this we note that the principle of relativity requires that the *form* of the transformation from S to S' must be identical to that from S' to S, the only difference being a change in the sign of the relative velocity u. Thus, from Eq. (40–14), it must be true that

$$x' = -ut' + x \sqrt{1 - \frac{u^2}{c^2}}. \tag{40–16}$$

We may now equate Eqs. (40–15) and (40–16) to eliminate x' from this expression, obtaining the desired relation for t' in terms of x and t.

The algebraic details will be omitted; the result is

$$t' = \frac{t - ux/c^2}{\sqrt{1 - u^2/c^2}}.$$

(40–17)

As remarked previously, lengths perpendicular to the direction of relative motion are not affected by the motion, so $y' = y$ and $z' = z$.

Collecting all the transformation equations, we have

$$x' = \frac{x - ut}{\sqrt{1 - u^2/c^2}},$$

$$y' = y,$$

$$z' = z,$$

(40–18)

$$t' = \frac{t - ux/c^2}{\sqrt{1 - u^2/c^2}}.$$

These are the *Lorentz transformation equations*, the relativistic generalization of the galilean transformation, Eqs. (40–1). When u is much smaller than c, the two transformations become identical.

Next we consider the relativistic generalization of the velocity transformation relation, Eq. (40–2), which, as previously noted, is valid only when u is very small. The relativistic expression can easily be obtained from the Lorentz transformation. Suppose a body observed in S' is at point x_1' at time t_1' and point x_2' at time t_2'. Then its speed v' in S' is given by

$$v' = \frac{x_2' - x_1'}{t_2' - t_1'} = \frac{\Delta x'}{\Delta t'}.$$

(40–19)

To obtain the speed in S we use Eqs. (40–18) to translate this expression into terms of the corresponding positions x_1 and x_2 and times t_1 and t_2 observed in S. We find

$$x_2' - x_1' = \frac{x_2 - x_1 - u(t_2 - t_1)}{\sqrt{1 - u^2/c^2}} = \frac{\Delta x - u\,\Delta t}{\sqrt{1 - u^2/c^2}},$$

$$t_2' - t_1' = \frac{t_2 - t_1 - u(x_2 - x_1)/c^2}{\sqrt{1 - u^2/c^2}} = \frac{\Delta t - u\,\Delta x/c^2}{\sqrt{1 - u^2/c^2}}.$$

Using these results in Eq. (40–19), we find

$$v' = \frac{\Delta x - u\,\Delta t}{\Delta t - u\,\Delta x/c^2} = \frac{(\Delta x/\Delta t) - u}{1 - (u/c^2)(\Delta x/\Delta t)}.$$

Now $\Delta x/\Delta t$ is just the velocity v measured in S, so we finally obtain

$$v' = \frac{v - u}{1 - uv/c^2}.$$

(40–20)

We note that when u and v are much smaller than c, the denominator becomes equal to unity, and we obtain the nonrelativistic result $v' = v - u$. The opposite extreme is the case $v = c$; then we find

$$v' = \frac{c - u}{1 - uc/c^2} = c.$$

That is, anything moving with speed c relative to S also has speed c relative to S', despite the relative motion of the two frames. This result demonstrates the consistency of Eq. (40–20) with our initial assumption that the speed of light is the same in all frames of reference.

Equation (40–20) may also be rearranged to give v in terms of v'. The algebraic details are left as a problem; the result is

$$v = \frac{v' + u}{1 + uv'/c^2}. \tag{40–21}$$

EXAMPLE A spaceship moving away from earth with a speed $0.9c$ fires a missile in the same direction as its motion, with a speed $0.9c$ relative to the spaceship. What is the missile's speed relative to earth?

Solution Let the earth's frame of reference be S, the spaceship's S'. Then $v' = 0.9c$ and $u = 0.9c$. The nonrelativistic velocity addition formula would give a velocity relative to earth of $1.8c$. The correct relativistic result, obtained from Eq. (40–21), is

$$v = \frac{0.9c + 0.9c}{1 + (0.9c)(0.9c)/c^2}$$

$$= 0.994c. \qquad \blacktriangleleft$$

When u is less than c, a body moving with a speed less than c in one frame of reference also has a speed less than c in *every other* frame of reference. This is one reason for thinking that no material body may travel with a speed greater than that of light, relative to any frame of reference. The relativistic generalizations of energy and momentum, to be considered next, give further support to this hypothesis.

40–6

MOMENTUM

We have discussed the fact that Newton's laws of motion are *invariant* under the galilean coordinate transformation, but that to satisfy the principle of relativity this transformation must be replaced by the more general Lorentz transformation. This requires corresponding generalizations in the laws of motion and the definitions of momentum and energy.

The principle of conservation of momentum states that *when two bodies collide, the total momentum is constant*, provided there is no interaction except that of the two bodies with each other. However, when one considers a collision in one coordinate system S, in which momentum is conserved, and then uses the Lorentz transformation to obtain the velocities in a second system S', it is found that if the newtonian definition of momentum ($\boldsymbol{p} = m\boldsymbol{v}$) is used, momentum is *not* conserved in the second system. Thus if the Lorentz transformation is correct, and if we believe in the principle of relativity (i.e., momentum conservation must hold in *all* systems), the only choice remaining is to modify the *definition* of momentum.

Deriving the correct relativistic generalization is beyond our scope, and we simply quote the result:

$$p = \frac{mv}{\sqrt{1 - v^2/c^2}}. \tag{40–22}$$

We note that, as usual when the particle's speed v is much less than c, this reduces to the newtonian expression $p = mv$, but that in general the momentum is greater in magnitude than mv.

In newtonian mechanics the second law of motion can be stated in the form

$$F = \frac{\Delta p}{\Delta t}. \tag{40–23}$$

That is, force equals time rate of change of momentum. Experiment shows that this result is still valid in relativistic mechanics, provided we use the relativistic momentum given by Eq. (40–22). This has the effect that a body under the action of a constant force *does not* experience a constant acceleration; as the particle's speed increases the acceleration for a given force continuously *decreases*. As the speed approaches c, the acceleration approaches zero, no matter how great the force. Thus it is impossible to accelerate a particle from a state of rest to a speed equal to or greater than c, and the speed of light is sometimes referred to as "the ultimate speed."

Equation (40–22) is sometimes interpreted to mean that a rapidly moving particle undergoes an increase in mass. If the mass at zero velocity is denoted by m_0, called the *rest mass*, then the "relativistic mass" m_{rel} is given by

$$m_{rel} = \frac{m_0}{\sqrt{1 - v^2/c^2}}.$$

Indeed, when we consider the motion of a system of particles (such as gas molecules in a moving container), the total mass of the system must be taken to be the sum of the relativistic masses of the particles rather than the sum of their rest masses.

But the concept of relativistic mass also has its pitfalls. It is *not* correct to say that the relativistic generalization of Newton's second law is $F = m_{rel}a$, and it is *not* correct that the relativistic kinetic energy of a particle is $K = \frac{1}{2}m_{rel}v^2$. Thus this concept must be approached with great caution. For the present discussion it is better to regard Eq. (40–22) as a generalized definition of momentum and to retain the meaning of m as an inertial quantity characteristic of each particle and independent of its state of motion.

40–7

WORK AND ENERGY

The work–energy relation developed in Chapter 6 made use of Newton's laws of motion. Since these must be generalized to bring them into accord with the principle of relativity, it is not surprising that the work–energy relation also requires generalization.

Because a constant force on a body no longer causes a constant acceleration (except at very small velocities), even the simplest dynamics problems require the use of calculus. Thus we will not be able to discuss in detail the relativistic generalization of the discussion of Section 6–3, which led to the work–energy theorem stated in Eq. (6–8). Instead, we must be content with quoting the result. Retaining the classical definition of work done by a force,

$$W = \sum_i F_i \, \Delta x_i,$$

we find

$$W = \frac{mc^2}{\sqrt{1 - v_2^2/c^2}} - \frac{mc^2}{\sqrt{1 - v_1^2/c^2}},$$

where v_1 and v_2 are the initial and final velocities of the particle.

This result suggests defining kinetic energy as

$$K = \frac{mc^2}{\sqrt{1 - v^2/c^2}}. \qquad (40\text{--}24)$$

But this expression is not zero when $v = 0$; instead it becomes equal to mc^2. Thus the correct relativistic generalization of kinetic energy K is

$$K = \frac{mc^2}{\sqrt{1 - v^2/c^2}} - mc^2. \qquad (40\text{--}25)$$

This expression, if correct, must reduce to the newtonian expression $K = \frac{1}{2}mv^2$ when v is much smaller than c. It is not obvious that this is the case; to demonstrate that it is so, we can expand the radical using the binomial theorem:

$$\left(1 - \frac{v^2}{c^2}\right)^{-1/2} = 1 + \frac{1}{2}\frac{v^2}{c^2} + \frac{3}{8}\frac{v^4}{c^4} + \frac{5}{16}\frac{v^6}{c^6} + \cdots.$$

Combining this with Eq. (40–25), we find

$$K = mc^2\left(1 + \frac{1}{2}\frac{v^2}{c^2} + \frac{3}{8}\frac{v^4}{c^4} + \cdots\right) - mc^2$$

$$= \frac{1}{2}mv^2 + \frac{3}{8}m\frac{v^4}{c^2} + \cdots. \qquad (40\text{--}26)$$

In each expression the dots stand for omitted terms. When v is much smaller than c, all the terms in the series except the first are negligibly small, and we obtain the classical $\frac{1}{2}mv^2$.

But what is the significance of the term mc^2 that had to be subtracted in Eq. (40–25)? Although Eq. (40–24) does not give the *kinetic energy* of the particle, perhaps it represents some kind of *total* energy, including both the kinetic energy and an additional energy mc^2, which the particle possesses even when it is not moving. Calling this total energy E and using Eq. (40–25), we find

$$E = K + mc^2 = \frac{mc^2}{\sqrt{1 - v^2/c^2}}. \qquad (40\text{--}27)$$

The hypothetical energy mc^2 associated with mass rather than motion may be called the *rest energy* of the particle. This speculation does not prove that the concept of rest energy is meaningful, but it points the way toward further investigation.

There is in fact direct experimental evidence of the existence of rest energy. The simplest example is the decay of the π^0 meson, an unstable particle that "decays"; in the decay process, the particle disappears and electromagnetic radiation appears. When the particle is at rest (and therefore with no kinetic energy) before its decay, the total energy of the radiation produced is found to be exactly equal to mc^2. There are many other examples of fundamental particle transformations in which the total mass of the system changes, and in every case there is a corresponding energy change consistent with the assumption of a rest energy mc^2 associated with a mass m.

Although the principles of conservation of mass and of energy originally developed quite independently, the theory of relativity shows that they are but two special cases of a single broader conservation principle, the *principle of conservation of mass and energy*. There are physical phenomena where neither mass nor energy is separately conserved, but where the changes in these quantities are governed by the more general relation that a change m in the mass of the system must be accompanied by an opposite change mc^2 in its energy.

The term *mass* as used here always means the rest mass of a particle, the inertial mass m measured through Eq. (40–22). For a given particle, m is a constant, independent of the state of motion of the particle. It is sometimes useful to introduce the concept of a variable, velocity-dependent relativistic mass. This concept is not needed in the present analysis, and it is not used in this discussion.

The possibility of conversion of mass into energy is the fundamental principle involved in the generation of power through nuclear reactions, a subject to be discussed in later chapters. When a uranium nucleus undergoes fission in a nuclear reactor, the total mass of the resulting fragments is less than that of the parent nucleus, and the total kinetic energy of the fragments is equal to this mass deficit times c^2. This kinetic energy can be used to produce steam to operate turbines for electric power generators or in a variety of other ways.

The total energy E (kinetic plus rest) of a particle is related simply to its momentum, as shown by combining Eqs. (40–22) and (40–27) to eliminate the particle's velocity. This is most easily accomplished by rewriting these equations in the following forms:

$$\left(\frac{E}{mc^2}\right)^2 = \frac{1}{1 - v^2/c^2}; \qquad \left(\frac{p}{mc}\right)^2 = \frac{v^2/c^2}{1 - v^2/c^2}.$$

Subtracting the second of these from the first and rearranging, we find

$$E^2 = (mc^2)^2 + (pc)^2. \tag{40–28}$$

Again we see that for a particle at rest ($p = 0$), $E = mc^2$. Equation (40–28) also suggests that a particle may have energy and momentum even when it has no rest mass. In such a case, $m = 0$ and

$$E = pc. \tag{40–29}$$

Massless particles, including photons, the quanta of electromagnetic radiation, and others, will be discussed in Chapter 44. The existence of such particles is well established, and they will be discussed in greater detail in later chapters. These particles always travel with the speed of light; they are emitted and absorbed during changes of state of atomic systems, accompanied by corresponding changes in the energy and momentum of these systems.

40–8

RELATIVITY AND NEWTONIAN MECHANICS

The sweeping changes required by the principle of relativity go to the very roots of newtonian mechanics, including the concepts of length and time, the equations of motion, and the conservation principles. Thus it may appear that foundations on which Newton's mechanics are built have been destroyed. While this is true in one sense, it is essential to keep in mind that the newtonian formulation still retains its validity whenever speeds are small compared with the speed of light. In such cases time dilation, length contraction, and the modifications of the laws of motion are so small that they are unobservable. In fact, every one of the principles of newtonian mechanics survives as a special case of the more general relativistic formulation.

Relativity does not *contradict* the older mechanics but *generalizes* it. After all, Newton's laws rest on a very solid base of experimental evidence, and it would be very strange indeed to advance a new theory inconsistent with this evidence. So it always is with the development of physical theory. Whenever a new theory is in partial conflict with an older, established theory, it nevertheless must yield the same predictions as the old in areas where the old theory is supported by experimental evidence. Every new physical theory must pass this test, called the *correspondence principle,* which has come to be regarded as a fundamental procedural rule in all physical theory. There are many problems for which newtonian mechanics is clearly inadequate, including all situations where particle speeds approach that of light or there is direct conversion of mass to energy. But there is still a large area, including nearly all the behavior of macroscopic bodies in mechanical systems, in which newtonian mechanics is still perfectly adequate.

At this point it is legitimate to ask whether the relativistic mechanics just discussed is the final word on this subject or whether *further* generalizations are possible or necessary. For example, inertial frames of reference have occupied a privileged position in all our discussion thus far. Should the principle of relativity be extended to noninertial frames as well?

Here is an example to illustrate some implications of this question. A man decides to go over Niagara Falls while enclosed in a large wooden box. During his free fall over the falls he can, in principle, perform experiments inside the box. An object released inside the box does not fall to the floor because both the box and the object are in free fall with a downward acceleration of $9.8 \text{ m} \cdot \text{s}^{-2}$. But an alternative interpretation, from this man's point of view, is that the force of gravity has suddenly been turned off. Provided he remains in the box and it remains in free fall, he cannot tell whether he is indeed in free fall or whether the force of gravity has vanished. A similar problem appears in a space station in orbit around the earth. Objects in the spaceship appear weightless, but without going outside the ship there is no way to determine whether gravity has disappeared or the spaceship is in an accelerated (i.e., noninertial) frame of reference.

These considerations form the basis of Einstein's *general theory of relativity*. If one cannot distinguish experimentally between a gravitational field and an accelerated reference system, then there can be no real distinction between the two. Pursuing this concept, we may try to represent *any* gravitational field in terms of special characteristics of the coordinate system. This turns out to require even more sweeping revisions of our space–time concepts than did the special theory of relativity, and we find that, in general, the geometric properties of the space are noneuclidean.

The basic ideas of the general theory of relativity are now well established, but some of the details remain speculative in nature. Its chief application is in cosmological investigations of the structure of the universe, the formation and evolution of stars, and related matters. It is not believed to have any relevance for atomic or nuclear phenomena or macroscopic mechanical problems of less than astronomical dimensions.

QUESTIONS

40–1 What do you think would be different in everyday life if the speed of light were $10 \text{ m} \cdot \text{s}^{-1}$ instead of its actual value?

40–2 The average life span in the United States is about 70 years. Does this mean that it is impossible for an average person to travel a distance greater than 70 light-years away from the earth? (A light-year is the distance light travels in a year.)

40–3 What are the fundamental distinctions between an inertial frame of reference and a noninertial frame?

40–4 A physicist claimed that it is impossible to define what is meant by a rigid body in a relativistically correct way. Why?

40–5 Two events occur at the same space point in a particular frame of reference and appear simultaneous in that frame. Is it possible that they may not appear simultaneous in another frame?

40–6 Does the fact that simultaneity is not an absolute concept also destroy the concept of *causality*? If event A is to *cause* event B, A must occur first. Is it possible that in some frames A may appear to cause B, and in others B appears to cause A?

40–7 A social scientist who has done distinguished work in fields far removed from physics has written a book purporting to refute the special theory of relativity. He begins with a premise that might be paraphrased as follows: "Either two events occur at the same time, or they don't; that's just common sense." How would you respond to this in the light of our discussion of the relative nature of simultaneity?

40–8 When an object travels across an observer's field of view at a relativistic speed, it appears not only foreshortened but also slightly rotated, with the side toward the observer shifted in the direction of motion relative to the side away from him. How does this come about?

40–9 According to the twin paradox mentioned in Section 40–3, if one twin stays on earth while the other takes off in a spaceship at relativistic speed and then returns, one will be older than the other. Can you think of a practical experiment, perhaps using two very precise atomic clocks, that would test this conclusion?

40–10 When a monochromatic light source moves toward an observer, its wavelength appears to be shorter than the value measured when the source is at rest. Does this contradict the hypothesis that the speed of light is the same for all observers? What about the apparent frequency of light from a moving source?

40–11 A student asserted that a massive particle must always have a speed less than that of light, while a massless particle must always travel at exactly the speed of light. Is he correct? If so, how do massless particles such as photons and neutrinos acquire this speed? Can't they start from rest and accelerate?

40–12 The theory of relativity sets an upper limit on the speed a particle can have. Are there also limits on its energy and momentum?

40–13 In principle, does a hot gas have more mass than the same gas when it is cold? Explain. In practice would this be a measurable effect?

40–14 Why do you think the development of newtonian mechanics preceded the more refined relativistic mechanics by so many years?

PROBLEMS

40–1 The π^+ meson, an unstable particle, lives, on the average, about 2.6×10^{-8} s (measured in its own frame of reference) before decaying.

a) If such a particle is moving with respect to the laboratory with a speed of $0.8c$, what lifetime is measured in the laboratory?

b) What distance, measured in the laboratory, does the particle move before decaying?

40–2 The μ^+ meson (or positive muon) is an unstable particle with a lifetime of about 2.3×10^{-6} s (measured in the rest frame of the muon).

a) If a muon travels with a speed of $0.99c$ relative to a laboratory, what is the lifetime as measured in the laboratory?

b) What distance, measured in the laboratory, does the particle travel during its lifetime?

40–3 For the two trains discussed in Section 40–2, suppose the two lightning bolts appear simultaneous to an observer on the train. Show that they *do not* appear simultaneous to an observer on the ground. Which appears to come first?

40–4 Solve Eqs. (40–18) to obtain x and t in terms of x' and t', and show that the resulting transformation has the same form as the original one except for a change of sign for u.

40–5 A light pulse is emitted at the origin of a frame of reference S' at time $t' = 0$. Its distance x' from the origin after a time t' is given by $x'^2 = c^2 t'^2$. Use the Lorentz transformation to transform this equation to an equation in x and t, and show that the result is $x^2 = c^2 t^2$; that is, the motion appears exactly the same in the frame of reference S of x and t, as it does in S'.

40–6 Two events observed in a frame of reference S have positions and times given by (x_1, t_1) and (x_2, t_2), respectively. Show that in a frame S' moving just fast enough so the two events occur at the same point in S', the time interval $\Delta t'$ between the two events is given by

$$\Delta t' = \sqrt{(\Delta t)^2 - \left(\frac{\Delta x}{c}\right)^2},$$

where $\Delta x = x_2 - x_1$, and $\Delta t = t_2 - t_1$. Hence show that, if $\Delta x \geq c\,\Delta t$, there is *no* frame S' in which the two events occur at the same point. The interval $\Delta t'$ is sometimes called the *proper time interval* for the events; is this term appropriate?

40–7 For the two events in Problem 40–6, show that if $\Delta x > c\,\Delta t$ there is a frame of reference S' in which the two events occur *simultaneously*. Find the distance between the two events in S'. This distance is sometimes called a *proper length;* is this term appropriate?

40–8 Two events are observed in a frame of reference S to occur at the same space point, the second occurring 2 s after the first. In a second frame S' moving relative to S, the second event is observed to occur 3 s after the first. What is the distance between the positions of the two events as measured in S'?

40–9 Two events are observed in a frame of reference S to occur simultaneously, at points separated by a distance of 1 m. In a second frame S' moving relative to S along the line joining the two points in S, the two events appear to be separated by 2 m. What is the time interval between the events, as measured in S'?

40–10 A particle is said to be in the *extreme relativisitic range* when its kinetic energy is much larger than its rest energy.

a) What is the speed of a particle (expressed as a fraction of c) such that the total energy is ten times the rest energy?

b) For such a particle, what percent error in the energy–momentum relation of Eq. (40–28) results if the term $(mc^2)^2$ is neglected?

40–11 A photon of energy E is emitted by an atom of mass m, which recoils in the opposite direction. Assuming that the atom can be treated nonrelativistically, compute the recoil velocity of the atom. Hence, show that the recoil velocity is much smaller than c whenever E is much smaller than the rest energy mc^2 of the atom.

40–12 Two particles emerge from a high-energy accelerator in opposite directions, each with a speed $0.6c$. What is the relative velocity of the particles?

40–13 At what speed is the momentum of a particle twice as great as the result obtained from the nonrelativistic expression mv?

40–14

a) At what speed does the momentum of a particle differ from the value obtained using the nonrelativistic expression mv by 1 percent?

b) Is the correct relativistic value greater or less than that obtained from the nonrelativistic expression?

40–15 The mass of an electron is 9.11×10^{-31} kg. Comparing the classical definition of momentum with its relativistic generalization, by how much is the classical expression in error if

a) $v = 0.01c$,

b) $v = 0.5c$,

c) $v = 0.9c$?

40–16 A radioactive isotope of cobalt, ^{60}Co, emits an electromagnetic photon (γ ray) of wavelength 0.932×10^{-12} m. The cobalt nucleus contains 27 protons and 33 neutrons, each with a mass of about 1.66×10^{-27} kg.

a) If the nucleus is at rest before emission, what is the speed afterward?

b) Is it necessary to use the relativistic generalization of momentum?

40–17 In Problem 40–16, suppose the cobalt atom is in a metallic crystal containing 0.01 mol of cobalt (about 6.02×10^{21} atoms) and that the entire crystal recoils as a unit, rather than just the single nucleus. Find the recoil velocity. (This recoil of the entire crystal rather than a single nucleus is called the *Mössbauer effect,* in honor of its discoverer, who first observed it in 1958.)

40–18 What is the speed of a particle whose kinetic energy is equal to its rest energy?

40–19 At what speed is the kinetic energy of a particle equal to $10 \, mc^2$?

40–20 How much work must be done to accelerate a particle from rest to a speed $0.1c$? From a speed $0.9c$ to a speed $0.99c$?

40–21 In *positron annihilation,* an electron and a positron (a positively charged electron) collide and disappear, producing electromagnetic radiation. If each particle has a mass of 9.1×10^{-31} kg and they are at rest just before the annihilation, find the total energy of the radiation.

40–22 The total consumption of electrical energy per year in the United States is of the order of 10^{19} joules. If matter could be converted completely into energy, how many kilograms of matter would have to be converted to produce this much energy?

40–23 Compute the kinetic energy of an electron (mass 9.11×10^{-31} kg) using both the nonrelativistic and relativistic expressions, and compare the two results, for speeds of

a) 1.0×10^8 m·s^{-1},

b) 2.0×10^8 m·s^{-1}.

40–24 In a hypothetical nuclear-fusion reactor, two deuterium nuclei combine or "fuse" to form one helium nucleus. The mass of a deuterium nucleus, expressed in atomic mass units (u), is 2.0147 u; that of a helium nucleus is 4.0039 u. (1 u = 1.66×10^{-27} kg.)

a) How much energy is released when 1 kg of deuterium undergoes fusion?

b) The annual consumption of electrical energy in the United States is of the order of 10^{19} J. How much deuterium must react to produce this much energy?

40–25 A nuclear bomb containing 20 kg of plutonium explodes. The rest mass of the products of the explosion is less than the original rest mass by one part in 10^4.

a) How much energy is released in the explosion?

b) If the explosion takes place in 1 μs, what is the average power developed by the bomb?

c) How much water could the released energy lift to a height of 1 km?

40–26 Construct a right triangle in which one of the angles is α, where $\sin \alpha = v/c$. (v is the speed of a particle, c the speed of light.) If the base of the triangle (the side adjacent to α) is the rest energy mc^2, show that

a) the hypotenuse is the total energy, and

b) the side opposite α is c times the relativistic momentum.

c) Describe a simple graphical procedure for finding the kinetic energy K.

40–27 An electron in a certain x-ray tube is accelerated from rest through a potential difference of 180,000 V in going from the cathode to the anode. When it arrives at the anode, what is

a) its kinetic energy in eV,

b) its total energy,

c) its velocity?

d) What is the velocity of the electron, calculated classically?

40–28 A certain electron accelerator accelerates electrons through a potential difference of 6.5×10^9 V, so that their kinetic energy is 6.5×10^9 eV.

a) What is the ratio of the speed v of an electron having this energy to the speed of light, c?

b) What would the speed be if computed from the principles of classical mechanics?

41

PHOTONS, ELECTRONS, AND ATOMS

The past several chapters have been concerned with understanding the propagation of light on the basis of an electromagnetic wave theory. The work of Maxwell, Hertz, and others has established the electromagnetic nature of light. Interference, diffraction, and polarization can be understood on the basis of a *wave* model; when interference effects can be neglected, the further simplification of *ray* optics permits analysis of the behavior of lenses and mirrors. These phenomena together form the subject of *classical optics*.

But there are many other phenomena that show a different aspect of the nature of light, in which it seems to behave as a stream of *particles*. Among these are the photoelectric effect, where electrons are liberated from a surface by absorption of energy from light, line spectra of elements, and the production and scattering of x-rays. All these point to a model in which the energy of light is carried in packages of a definite size, called *quanta* or *photons*. The energy of a single photon is proportional to the frequency of the radiation, and photons also carry momentum. Understanding the role of photons in emission and absorption phenomena requires some radical changes in our view of the nature of radiation and of matter itself. The new theory that emerges is called *quantum mechanics;* it provides the key to understanding many aspects of the structure and behavior of atoms and molecules that cannot be understood on the basis of the older "classical" theories of mechanics and electromagnetism.

41–1

EMISSION AND ABSORPTION OF LIGHT

The electromagnetic wave model of light provides an adequate basis for understanding many aspects of the propagation of light, including interference, diffraction, and polarization phenomena. It is considerably *less* successful with respect to the emission and absorption of light and other electromagnetic radiation. An example is the emission of light from matter. The electromagnetic waves of Hertz were discussed in Section 35–1; they were produced by oscillations with frequencies of the order of 10^8 Hz in a resonant L–C circuit similar to those studied in Chapter 33. Frequencies of visible light are much larger, of the order of 10^{15} Hz, far higher than the highest frequencies attainable with conventional electronic equipment.

In the mid-nineteenth century it was speculated that visible light might be produced by motion of electric charge within individual atoms rather than in macroscopic circuits. In fact, in 1862 Faraday placed a light source in a strong magnetic field in an attempt to determine whether the emitted radiation was changed by the field. He was not able to detect any change, but when his experiments were repeated thirty years later by Zeeman with greatly improved equipment, changes *were* observed.

Particularly puzzling was the existence of *line spectra*. We have seen how a prism or grating spectrograph functions to disperse a beam of light into a *spectrum*. If the light source is an incandescent solid or liquid, the spectrum is *continuous;* that is, light of all wavelengths is present. If, however, the source is a gas through which an electrical discharge is passing, or a flame into which a volatile salt has been introduced, the spectrum is of an entirely different character. Instead of a continuous band of color, only a few colors appear, in the form of isolated parallel lines. (Each "line" is an image of the spectrograph slit, deviated through an angle dependent on the frequency of the light forming the image.) A spectrum of this sort is termed a *line spectrum.*

The wavelengths of the lines are characteristic of the element emitting the light. That is, hydrogen always gives a set of lines in the same position, sodium another set, iron still another, and so on. The line structure of the spectrum extends into both the ultraviolet and infrared regions, where photographic or other means are required for its detection.

The existence of a characteristic spectrum for each element, discovered early in the nineteenth century, suggested that there is a direct relation between the characteristics and internal structure of an atom and its spectrum. Attempts to understand this relation on the basis of newtonian mechanics and classical electricity and magnetism were not successful, however.

There were other mysteries. The *photoelectric effect*, discovered by Hertz in 1887 during his investigations of electromagnetic wave propagation, is the liberation of electrons from the surface of a conductor when light strikes the surface. When light is absorbed by the surface it transfers energy to electrons near the surface, and some of the electrons acquire enough energy to surmount the potential-energy barrier at the surface and escape from the material into space. More detailed investigation revealed some puzzling features that could *not* be understood on the basis of classical optics.

Still another area of unsolved problems centered around the production and scattering of *x-rays*, electromagnetic radiation with wavelengths shorter than those of visible light by a factor of the order of 10^4 and with correspondingly greater frequencies. These rays were produced in high-voltage glow discharge tubes, but the details of this process eluded understanding. Even worse, when these rays collided with matter, the scattered ray sometimes had a longer wavelength than the original ray. This is like directing a beam of blue light at a mirror and having it reflect as red!

All these phenomena, and several others, pointed forcefully to the conclusion that classical optics, successful though it was in explaining ray optics, interference, and polarization, nevertheless had its limitations. Understanding the phenomena cited above would require at least some generalization of the classical theory. In fact, it has required something much more radical than that. All these phenomena are concerned with the *quantum* theory of radiation, which includes the assumption that despite the *wave* nature of electromagnetic radiation, it nevertheless has some properteis akin to those of *particles*. In particular, the *energy* conveyed by an electromagnetic wave is always carried in units whose magnitude is proportional to the frequency of the wave. These units of energy are called *photons* or *quanta*.

Thus, electromagnetic radiation emerges as an entity with a dual nature, having both wave and particle aspects. The remainder of this chapter will be devoted to the applications of this duality to some of the phenomena mentioned above, and to a study of this seemingly (but not actually) inconsistent nature of electromagnetic radiation.

41–2

THE PHOTOELECTRIC EFFECT

The photoelectric effect is the liberation of electrons from the surface of a conductor by light striking the surface. The electrons absorb energy from the incident radiation and are thus able to climb over the potential-energy barrier that normally confines them in the material.

The photoelectric effect was first observed in 1887 by Heinrich Hertz, who noticed that a spark would jump more readily between two spheres when their surfaces were illuminated by the light from another spark. The effect was investigated in detail in the following years by Hallwachs and Lenard.

A modern phototube is shown schematically in Fig. 41–1. A beam of light, indicated by the arrows, falls on a photosensitive surface K called the *cathode*. The battery or other source of potential difference creates an electric field in the direction from A (called the *collector* or *anode*) toward K, and electrons emitted from K are pushed by this field to the anode A. Anode and cathode are enclosed in an evacuated container; the photoelectric current is measured by the galvanometer G.

It is found that with a given material as emitter, no photoelectrons at all are emitted unless the wavelength of the light is *shorter* than some critical value. The corresponding *minimum* frequency is called the *threshold frequency* of the particular surface. The threshold frequency for most metals is in the ultraviolet (corresponding to wavelengths of 200 to 300 nm), but for potassium and cesium oxide it lies in the visible spectrum (400 to 700 nm).

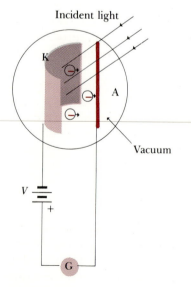

41–1   Schematic diagram of a photocell circuit.

Some of the electrons are emitted with substantial initial speeds; this is shown by the fact that, even with *no* emf in the external circuit, a few electrons reach the collector, causing a small current in the external circuit. Indeed, even when the polarity of the potential difference V is reversed and the associated electric-field force on the electrons is back toward the cathode, some electrons still reach the anode. Only when the reversed potential V is made large enough so that the potential energy eV is greater than the maximum kinetic energy $\frac{1}{2}mv_{max}^2$ with which the electrons leave the cathode does the electron flow stop completely. The critical reversed potential is called the *stopping potential*, denoted by V_0, and it provides a direct measurement of the maximum kinetic energy with which electrons leave the cathode, through the relation

$$\tfrac{1}{2}mv_{max}^2 = eV_0. \tag{41–1}$$

Surprisingly, it turns out that the maximum kinetic energy of a photoelectron does not depend on the *intensity* of the incident light, but does depend on its *wavelength*. When the light intensity increases, the photoelectric current also increases, but only because the *number* of emitted electrons increases, not the energy of an individual electron.

The correct explanation of the photoelectric effect was given by Einstein in 1905. Extending a proposal made two years earlier by Planck, Einstein postulated that a beam of light consisted of small bundles of energy that are now called *light quanta* or *photons*. The energy E of a photon is porportional to its frequency f or is equal to its frequency multiplied by a constant. That is,

$$E = hf, \tag{41–2}$$

where h is a universal constant, called *Planck's constant*, whose value is 6.626×10^{-34} J·s. When a photon collides with an electron at or just within the surface of a metal, it may transfer its energy to the electron. This transfer is an "all-or-none" process, the electron getting all the photon's energy or none at all. The photon then simply drops out of existence. The energy acquired by the electron may enable it to escape from the surface of the metal if it is moving in the right direction.

In leaving the surface of the metal, the electron loses an amount of energy ϕ (the work function of the surface). Some electrons may lose more than this if they start at some distance below the metal surface, but the *maximum* energy with which an electron can emerge is the energy gained from a photon minus the work function. Hence the maximum kinetic energy of the photoelectrons ejected by light of frequency f is

$$\tfrac{1}{2}mv_{max}^2 = hf - \phi. \tag{41–3}$$

Combining this with Eq. (41–1) leads to the relation

$$eV_0 = hf - \phi. \tag{41–4}$$

Thus by measuring the stopping potential V_0 required for each of several values of frequency f for a given cathode material, one can determine both the work function ϕ for the material and the value of the quantity h/e. Thus this experiment provides a direct measurement of the value of Planck's constant in addition to a direct confirmation of Einstein's interpretation of photoelectric emission.

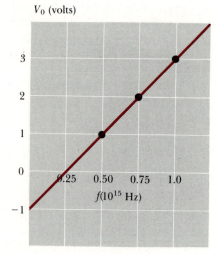

41–2 Stopping potential as a function of frequency. For a different cathode material having a different work function, the line would be displaced up or down but would have the same slope.

EXAMPLE   For a certain cathode material used in a photoelectric-effect experiment, a stopping potential of 3.0 V was required for light of wavelength 300 nm, 2.0 V for 400 nm, and 1.0 V for 600 nm. Determine the work function for this material, and the value of Planck's constant.

Solution   According to Eq. (41–4), a graph of V_0 as a function of f should be a straight line. We rewrite this equation as

$$V_0 = \frac{h}{e}f - \frac{\phi}{e}.$$

In this form we see that the *slope* of the line is h/e and the *intercept* on the vertical axis (corresponding to $f = 0$) is at $-\phi/e$. The frequencies, obtained from $f = c/\lambda$ and $c = 3.00 \times 10^8$ m·s^{-1}, are 0.5, 0.75, and 1.0×10^{15} s^{-1}, respectively. The graph is shown in Fig. 41–2. From it we find

$$-\frac{\phi}{e} = -1.0 \text{ V}, \qquad \phi = 1.0 \text{ eV},$$

and

$$\frac{h}{e} = \frac{1.0 \text{ V}}{0.25 \times 10^{15} \text{ s}^{-1}} = 4 \times 10^{-15} \text{ J·C}^{-1}\text{·s},$$

$$h = (4.0 \times 10^{-15} \text{ J·C}^{-1}\text{·s})(1.6 \times 10^{-19} \text{ C})$$
$$= 6.4 \times 10^{-34} \text{ J·s}.$$

This experimental value differs by about 3.4% from the presently accepted value of 6.626×10^{-34} J·s.   ◀

The particle-like nature of electromagnetic radiation, used in the above analysis of the photoelectric effect, has been established beyond any reasonable doubt. A photon of electromagnetic radiation of frequency f and corresponding wavelength $\lambda = c/f$ has energy E given by

$$E = hf = \frac{hc}{\lambda}. \tag{41–5}$$

Furthermore, according to relativity theory, every particle having energy must also have momentum, even if it has no rest mass. Photons have zero rest mass; according to Eq. (40–29), the momentum $\boldsymbol{p}$ of a photon of energy E has magnitude p given by

$$E = pc. \tag{41–6}$$

Thus the wavelength of a photon and its momentum are related simply by

$$p = \frac{h}{\lambda}. \tag{41–7}$$

These relations will be used frequently in the remainder of this chapter.

Figure 41–3 shows a very direct illustration of the particle aspects of light.

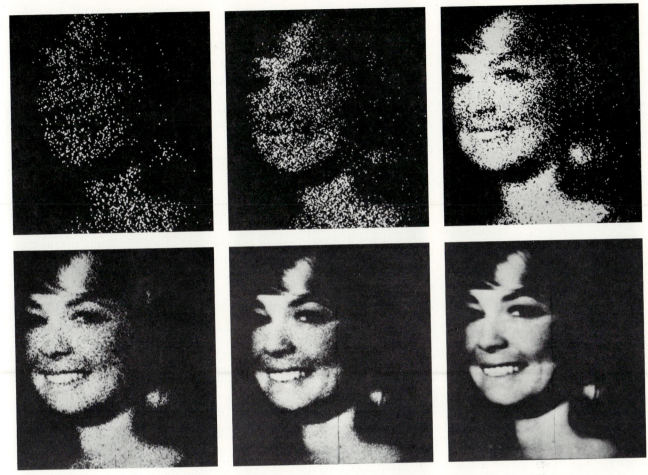

41–3   These photographs are images made by small numbers of photons, using electronic image amplification. In the upper pictures each spot corresponds to one photon; under extremely faint light (few photons) the pattern seems almost random, but it emerges more distinctly as the light level increases (lower pictures). (Used with permission of RCA Corporation.)

41–3

LINE SPECTRA

The quantum hypothesis, used in the preceding section for the analysis of the photoelectric effect, also plays an important role in the understanding of atomic spectra, particularly the line spectra described in Section 41–1. It might be expected that the spectrum frequencies of the light emitted by a particular element would be arranged in some regular way. For instance, a radiating atom might be analogous to a vibrating string, emitting a fundamental frequency and its harmonics. At first sight there does not seem to be any semblance of order or regularity in the lines of a typical spectrum; for many years unsuccessful attempts were made to correlate the observed frequencies with those of a fundamental and its overtones. Finally, in 1885, Johann Jakob Balmer (1825–1898) found a simple formula that gave the frequencies of a group of lines emitted by atomic hydrogen. Since the spectrum of this element is relatively simple, and fairly typical of a number of others, we shall consider it in more detail.

Under the proper conditions of excitation, atomic hydrogen may be made to emit the sequence of lines illustrated in Fig. 41–4. This sequence is called a *series*. There is evidently a certain order in this spectrum, the lines becoming crowded more and more closely together as the limit of the series is approached. The line of longest wavelength or lowest frequency, in the red, is known as H_α; the next, in the blue-green, as H_β; the third as H_γ, and so on. Balmer found that the wavelengths of these lines were given accurately by the simple formula

$$\frac{1}{\lambda} = R\left(\frac{1}{2^2} - \frac{1}{n^2}\right), \tag{41–8}$$

where λ is the wavelength, R is a constant called the Rydberg constant, and n may have the integral values 3, 4, 5, etc. If λ is in meters,

$$R = 1.097 \times 10^7 \text{ m}^{-1}.$$

Letting $n = 3$ in Eq. (41–8), one obtains the wavelength of the H_α-line:

$$\frac{1}{\lambda} = 1.097 \times 10^7 \text{ m}^{-1}\left(\frac{1}{4} - \frac{1}{9}\right)$$

$$= 1.524 \times 10^6 \text{ m}^{-1},$$

whence

$$\lambda = 656.3 \text{ nm}.$$

For $n = 4$, one obtains the wavelength of the H_β-line, etc. For $n = \infty$, one obtains the limit of the series, at $\lambda = 364.6$ nm. This is the *shortest* wavelength in the series.

Other series spectra for hydrogen have since been discovered. These are known, after their discoverers, as the Lyman, Paschen, Brackett, and Pfund series. The formulas for these are

Lyman series:

$$\frac{1}{\lambda} = R\left(\frac{1}{1^2} - \frac{1}{n^2}\right), \qquad n = 2, 3, \ldots,$$

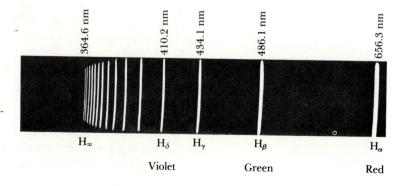

41–4 The Balmer series of atomic hydrogen. (Reproduced by permission from *Atomic Spectra and Atomic Structures* by Gerhard Herzberg. Copyright 1937 by Prentice-Hall, Inc.)

Paschen series:

$$\frac{1}{\lambda} = R\left(\frac{1}{3^2} - \frac{1}{n^2}\right), \qquad n = 4, 5, \ldots,$$

Brackett series:

$$\frac{1}{\lambda} = R\left(\frac{1}{4^2} - \frac{1}{n^2}\right), \qquad n = 5, 6, \ldots,$$

Pfund series:

$$\frac{1}{\lambda} = R\left(\frac{1}{5^2} - \frac{1}{n^2}\right), \qquad n = 6, 7, \ldots.$$

The Lyman series is in the ultraviolet, and the Paschen, Brackett, and Pfund series are in the infrared. The Balmer series evidently fits into the scheme between the Lyman and the Paschen series.

The Balmer formula, Eq. (41–8), may also be written in terms of the frequency of the light, by use of the relation

$$c = f\lambda, \qquad \text{or} \qquad \frac{1}{\lambda} = \frac{f}{c}.$$

Thus, Eq. (41–8) becomes

$$f = Rc\left(\frac{1}{2^2} - \frac{1}{n^2}\right), \tag{41–9}$$

or

$$f = \frac{Rc}{2^2} - \frac{Rc}{n^2}. \tag{41–10}$$

Each of the fractions on the right-hand side of Eq. (41–10) is called a *term*, and the frequency of every line in the series is given by the difference between two terms.

There are only a few elements (hydrogen, singly ionized helium, doubly ionized lithium) whose spectra can be represented by a simple formula of the Balmer type. Nevertheless, it is possible to separate the more complicated spectra of other elements into series, and to express the frequency of each line in the series as the difference of two *terms*. The first term is constant for any one series, while the various values of the second term can be labeled by values of an integer index n analogous to the n appearing in Eq. (41–10). In a few simple cases the numerical values of the terms can be *calculated* from theoretical considerations, as we shall see in Chapter 42. For complex atoms, however, the term values must be determined experimentally by analysis of spectra, often an extremely complex problem.

41–4

ENERGY LEVELS

Every element has a characteristic line spectrum, which must somehow result from the characteristics and structure of the *atoms* of that element. Part of the key to understanding the relation of atomic structure to atomic spectra was supplied in 1913 by the Danish physicist Niels Bohr, who applied to spectra the same concept of light quanta or *photons* that Einstein had used earlier in analysis of the photoelectric effect.

Bohr's hypothesis was as follows: Each atom, as a result of its internal structure (and presumably internal motion), can have a variable amount of *internal energy*. But the energy of an atom cannot change by any arbitrary amount; rather, each atom has a series of discrete *energy levels*, such that an atom can have an amount of internal energy corresponding to any one of these levels, but it cannot have an energy *intermediate* between two levels. All atoms of a given element have the same set of energy levels, but atoms of different elements have different sets. While an atom is in one of these states corresponding to a definite energy, it does not radiate, but an atom can make a transition from one energy level to a lower level by emitting a photon, whose energy is equal to the energy difference between the initial and final states. If E_i is the initial energy of the atom, before such a transition, and E_f its final energy, after the transition, then, since the photon's energy is hf, we have

$$hf = E_i - E_f. \qquad (41\text{--}11)$$

For example, a photon of orange light of wavelength 600 nm has a frequency f given by

$$f = \frac{c}{\lambda} = \frac{3.00 \times 10^8 \text{ m} \cdot \text{s}^{-1}}{600 \times 10^{-9} \text{ m}} = 5.00 \times 10^{14} \text{ Hz}$$
$$= 5.00 \times 10^{14} \text{ s}^{-1}.$$

The corresponding photon energy is

$$E = hf = (6.63 \times 10^{-34} \text{ J} \cdot \text{s})(5.00 \times 10^{14} \text{ s}^{-1})$$
$$= 3.31 \times 10^{-19} \text{ J} = 2.07 \text{ eV}.$$

Thus, this photon must be emitted in a transition between two states of the atom differing in energy by 2.07 eV.

The Bohr hypothesis, if correct, would shed new light on the analysis of spectra on the basis of *terms*, as described in Section 41–3. For example, Eq. (41–10) gives the frequencies of the Balmer series in the hydrogen spectrum. Multiplied by Planck's constant h, this becomes

$$hf = \frac{Rch}{2^2} - \frac{Rch}{n^2}. \qquad (41\text{--}12)$$

If we now compare Eqs. (41–11) and (41–12), identifying $-Rch/n^2$ with the initial energy of the atom E_i and $-Rch/2^2$ with its final energy, E_f, before and after a transition in which a photon of energy $hf = E_i - E_f$ is emitted, then Eq. (41–12) takes on the same form as Eq. (41–11). More generally, if we assume that the possible energy levels for the hydrogen atom are given by

$$E_n = -\frac{Rch}{n^2}, \qquad n = 1, 2, 3, \ldots, \qquad (41\text{--}13)$$

then *all* the series spectra of hydrogen can be understood on the basis of transitions from one energy level to another. For the Lyman series the final state is always $n = 1$, for the Paschen series it is $n = 3$, and so on.

Similarly, complex spectra of other elements, represented by *terms*, are understood on the basis that each term corresponds to an energy level; and a frequency, represented as a difference of two terms, corresponds to a transition between the two corresponding energy levels.

In 1914 a series of experiments by Franck and Hertz provided more direct experimental evidence for energy levels in atoms. In studying the motion of electrons through mercury vapor, under the action of an electric field, they found that a spectrum line at 254 nm was emitted by the vapor when the electron kinetic energy was greater than 4.9 eV but not when it was less. This strongly suggests the existence of an energy level 4.9 eV above the ground state; a mercury atom is excited to this level by collision with an electron and subsequently decays to the ground state by emitting a photon. The energy of the photon, according to Eq. (41–2), should be

$$E = hf = \frac{hc}{\lambda} = \frac{(6.63 \times 10^{-34} \text{ J} \cdot \text{s})(3.00 \times 10^{8} \text{ m} \cdot \text{s}^{-1})}{254 \times 10^{-9} \text{ m}}$$

$$= 7.82 \times 10^{-19} \text{ J} = \frac{7.82 \times 10^{-19} \text{ J}}{1.60 \times 10^{-19} \text{ J} \cdot \text{eV}^{-1}}$$

$$= 4.9 \text{ eV},$$

in excellent agreement with the measured electron energy.

Although the Bohr hypothesis permits partial understanding of line spectra on the basis of energy levels of atoms, it is not yet complete because it provides no basis for *predicting* what the energy levels for any particular kind of atom should be. We shall return to this problem in Chapter 42, where the new mechanical principles required for the understanding of the structure and energy levels of atoms will be developed. These principles constitute the subject of *quantum mechanics*.

41–5

ATOMIC SPECTRA

As we have seen, the key to the understanding of atomic spectra is the concept of atomic *energy levels*. Every spectrum line corresponds to a specific transition between two energy levels of an atom, and the corresponding frequency is given in each case by Eq. (41–11).

Thus the fundamental problem of the spectroscopist is to determine the energy levels of an atom from the measured values of the wavelengths of the spectral lines emitted when the atom proceeds from one energy level to another. In the case of complicated spectra emitted by the heavier atoms, this is a task requiring tremendous ingenuity. Nevertheless, almost all atomic spectra have been analyzed, and the resulting energy levels have been tabulated or plotted with the aid of diagrams similar to the one shown for sodium in Fig. 41–5.

As we shall see, every atom has a lowest energy level, representing the *minimum* energy the atom can have. This lowest energy level is called the *ground state*, and all higher levels are called *excited states*. As we have seen, a spectral line is emitted when an atom proceeds from an excited state to a lower state. The only means discussed so far for raising the atom from the normal state to an excited state has been with the aid of an

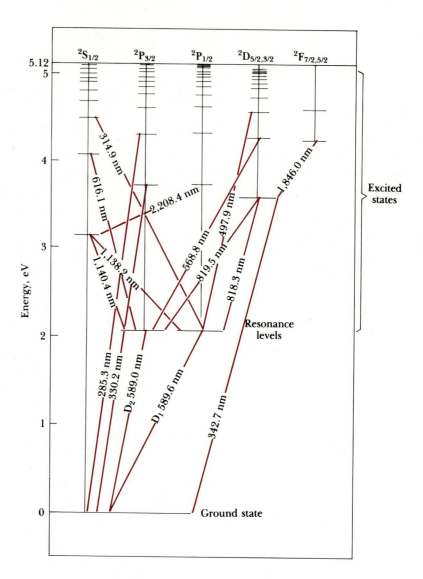

41–5 Energy levels of the sodium atom. Numbers on the lines between levels are wavelengths.

electric discharge. Let us consider now another method, involving absorption of radiant energy.

From Fig. 41–5 it may be seen that a sodium atom emits the characteristic yellow light of wavelengths 589.0 and 589.6 nm (the D_1- and D_2-lines) when it undergoes the transitions from the two levels marked *resonance levels* to the ground state. Suppose a sodium atom in the ground state were to *absorb* a quantum of radiant energy of wavelength 589.0 or 589.6 nm. It would then undergo a transition in the opposite direction and be raised to one of the resonance levels. After a short time, the average value of which is called the *lifetime* of the excited state, the atom returns to the ground state and emits this quantum. For the resonance levels of the sodium atom, the lifetime is $\sim 1.6 \times 10^{-8}$ s.

This emission process is called *resonance radiation* and may be easily demonstrated as follows. A strong beam of the yellow light from a sodium arc is concentrated on a glass bulb that has been highly evacuated and into which a small amount of pure metallic sodium has been dis-

41–6   Absorption spectrum of sodium.

41–7   A portion of the solar spectrum, between 390 nm and 460 nm, showing the Fraunhofer lines corresponding to the absorption spectrum. (Courtesy of Mt. Wilson & Las Campanas Observatories, Carnegie Institute of Washington.)

tilled. If the bulb is warmed to increase the sodium vapor pressure, resonance radiation will take place throughout the whole bulb, which glows with the yellow light characteristic of sodium.

A sodium atom in the ground state may absorb radiant energy of wavelengths other than the yellow resonance lines. All wavelengths corresponding to spectral lines *emitted* when the sodium atom returns to its normal state may also be absorbed. Thus, from Fig. 41–5, wavelengths 330.2 nm, 285.3 nm, etc., may be *absorbed* by a normal sodium atom. If, therefore, the continuous-spectrum light from a carbon arc is sent through an absorption tube containing sodium vapor, and then examined with a spectroscope, there will be a series of dark lines corresponding to the wavelengths absorbed, as shown in Fig. 41–6. This is known as an *absorption spectrum*.

The sun's spectrum is an absorption spectrum. The main body of the sun emits a continuous spectrum, whereas the cooler vapors in the sun's atmosphere emit line spectra corresponding to all the elements present. When the intense light from the main body of the sun passes through the cooler vapors, the lines of these elements are absorbed. The light *emitted* by the cooler vapors is so small compared with the unabsorbed continuous spectrum that the continuous spectrum appears to be crossed by many faint *dark* lines. These were first observed by Fraunhofer and are therefore called *Fraunhofer lines*. They may be observed with any student spectroscope pointed toward any part of the sky. Figure 41–7 shows Fraunhofer lines in a portion of the sun's spectrum.

41–6

THE LASER

If the energy difference between the normal and the first excited state of an atom is E, the atom is capable of absorbing a photon whose frequency f is given by the Planck equation $E = hf$. The *absorption* of a photon by a normal atom A is depicted schematically in Fig. 41–8a. After absorbing the photon, the atom becomes an excited atom A^*. A short time later, *spontaneous emission* takes place and the excited atom becomes normal again by emitting a photon of the same frequency as that which was originally absorbed but in a random direction and with a random phase, as shown in Fig. 41–8b. There is also a third process, first proposed by

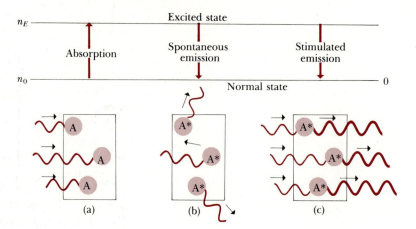

41–8 Three interaction processes between an atom and radiation.

Einstein, called *stimulated emission*, shown schematically in Fig. 41–8c. *Stimulated emission takes place when a photon encounters an excited atom and forces it to emit another photon of the same frequency, in the same direction, and in the same phase.* The two photons go off together as *coherent* radiation.

Consider an absorption cell containing a large number of atoms of the type depicted in Fig. 41–8. In the absence of an external beam of radiation, most of the atoms are in the ground state; there are only a few excited atoms present in the cell. The ratio of the number n_E of excited atoms to the number n_0 of normal atoms is extremely small.

Now suppose a beam of radiation is sent through the cell with frequency f corresponding to the energy difference E. The ratio of the numbers n_E and n_0, that is, the ratio of the *populations* of the energy levels, is increased. Since the population of the normal state was so much larger than that of the excited state, an enormously intense beam of light would be required to increase the population of the excited state to a value comparable to or greater than that of the normal state. Therefore, the rate at which energy is extracted from the beam by absorption of normal atoms far outweighs the rate at which energy is added to the beam by stimulated emission of excited atoms.

If a condition can be created in which n_E is substantially increased compared to the normal equilibrium value, creating a condition known as *population inversion*, the rate of energy radiation by stimulated emission may *exceed* the rate of absorption. The system then acts as a *source* of radiation with photon energy E. Furthermore, since the photons are the result of stimulated emission, they all have the same frequency, phase, polarization, and direction. The resulting radiation is therefore very much more *coherent* than light from ordinary sources, in which the emissions of individual atoms are *not* coordinated.

The necessary population inversion can be achieved in a variety of ways. As an example we consider the helium–neon laser, a simple, inexpensive laser available in many undergraduate laboratories. A mixture of helium and neon, each typically at a pressure of the order of 10^2 Pa (or 10^{-3} atm), is sealed in a glass enclosure provided with two electrodes. When a sufficiently high voltage is applied, a glow discharge occurs.

Collisions between ionized atoms and electrons carrying the discharge current excite atoms to various energy states.

Figure 41–9 shows an energy-level diagram for the system. The notation used to label the various energy levels, such as $1s2s$ or $5s$, will be discussed in Section 43–1, and need not concern us here. Helium atoms excited to the $1s2s$ state cannot return to the ground state by emitting a 20.61-eV photon, as might be expected, because both the states have zero total angular momentum, while a photon must carry away at least one unit ($h/2\pi$) of angular momentum. Such a state, in which radiative decay is impossible, is called a *metastable state*.

The helium atoms *can*, however, lose energy by energy-exchange collisions with neon atoms initially in the ground state. A $1s2s$ helium atom, with its internal energy of 20.61 eV and a little additional kinetic energy, can collide with a neon atom in the ground state, exciting it to the $5s$ excited state at 20.66 eV and leaving the helium atom in the $1s^2$ ground state. Thus, we have the necessary mechanism for a population inversion in neon, with the population in the $5s$ state substantially enhanced. Stimulated emission from this state then results in the emission of highly coherent light at 632.8 nm, as shown on the diagram. In practice the beam is sent back and forth through the gas many times by a pair of parallel mirrors, so as to stimulate emission from as many excited atoms as possible. One of the mirrors is partially transparent, so a portion of the beam emerges as an external beam.

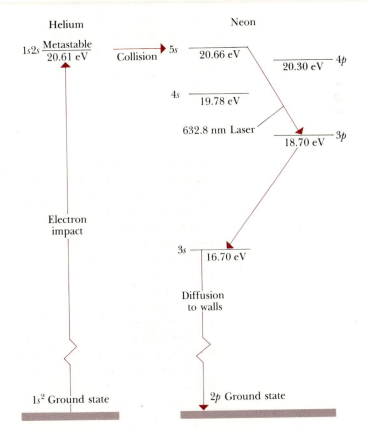

41–9 Energy-level diagram for helium–neon laser.

The net effect of all the processes taking place in a laser tube is a beam of radiation that is (1) very intense, (2) almost perfectly parallel, (3) almost monochromatic, and (4) spatially *coherent* at all points within a given cross section. To understand this fourth characteristic, we recall the simple double-slit interference experiment. A mercury arc placed directly behind the double slit would not give rise to interference fringes because the light issuing from the two slits would come from different points of the arc and would not retain a constant phase relationship. In the use of the usual laboratory arc-lamp sources, it is necessary to use the light from a very small portion of the source to illuminate the double slit. The slightly diverging beam from a laser, however, may be allowed to fall directly on a double slit (or other interferometer) because the light rays from any two points of a cross section are in phase, and are said to exhibit "spatial coherence."

In recent years lasers have found a wide variety of practical applications. The high intensity of a laser beam makes it a convenient drill. A very small hole can be drilled in a diamond for use as a die in drawing very small-diameter wire. The ability of a laser beam to travel long distances without appreciable spreading makes it a very useful tool for surveyors, especially in situations where great precision is required over long distances, as in the case of a long tunnel drilled from both ends.

Lasers are finding increasing application in medical science. A laser can produce a very *narrow* beam with extremely *high intensity*, high enough to vaporize anything in its path. This property is used in the treatment of a detached retina; a short burst of radiation damages a small area of the retina, and the resulting scar tissue "welds" the retina back to the choroid from which it has become detached. Laser beams are also used in surgery; blood vessels cut by the beam tend to seal themselves off, making it easier to control bleeding. The use of laser radiation in the teatment of skin cancer is an active area of research, as shown in Fig. 41–10.

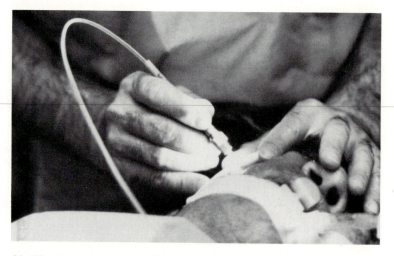

41–10 A laser beam, directed by a fiber-optic conductor, is used to treat a skin malignancy. (Photo by Dan McCoy/Rainbow.)

41–7

X-RAY PRODUCTION AND SCATTERING

X-rays are produced when rapidly moving electrons that have been accelerated through potential differences of the order of 10^3 to 10^6 V are allowed to strike a metal target. They were first observed by Wilhelm K. Röntgen (1845–1923) in 1895 and were originally called *Röntgen rays*.

X-rays are of the same nature as light or any other electromagnetic wave and, like light waves, they are governed by quantum relations in their interaction with matter. Hence, one may speak of x-ray photons or quanta, the energy of such a photon being given by the relation $E = hf$. Wavelengths of x-rays range from approximately 0.001 to 1 nm (10^{-12} to 10^{-9} m). X-ray wavelengths can be measured quite precisely by crystal diffraction techniques, described in Section 39–9.

A common x-ray tube is the Coolidge type, invented in 1913 by W. D. Coolidge of the General Electric laboratories. A Coolidge tube is shown in Fig. 41–11. A heated cathode and an anode are enclosed in a glass tube that has been pumped down to an extremely low pressure. Electrons emitted from the cathode can then travel directly to the anode with only a small probability of a collision on the way, and they reach the anode with a speed corresponding to the full potential difference across the tube. X-radiation is emitted from the anode surface as a consequence of its bombardment by the electron stream.

Two distinct processes are involved in x-ray emission. Some of the electrons are stopped by the target and their kinetic energy is converted directly to x-radiation. Others transfer their energy in whole or in part to the atoms of the target, which retain it temporarily as "energy of excitation" but very quickly emit it as x-radiation. The latter radiation is characteristic of the material of the target, while the former is not.

The atomic energy levels associated with x-ray excitation are rather different in character from those associated with visible spectra. To understand them we need some understanding of the arrangement of electrons in complex atoms, a topic to be discussed in more detail in Chapter 43. For the present we state simply that in a many-electron atom the electrons are always arranged in concentric *shells* at increasing distances from the nucleus. These shells are labeled K, L, M, N, etc., the K shell being closest to the nucleus, the L shell next, and so on. For any given atom in the ground state there is a definite number of electrons in each shell.

41–11 Coolidge-type x-ray tube.

For reasons to be discussed later, each shell has a maximum number of electrons it can accommodate, and we may speak of *filled shells* and *partially filled shells*. The K shell can contain at most two electrons. The next, the L shell, can contain eight. The third, the M shell, has a capacity for 18 electrons, while the N shell may hold 32. The sodium atom, for example, which contains 11 electrons, has two in the K shell, eight in the L shell, and a single electron in the M shell. Molybdenum, with 42 electrons, has two in the K shell, eight in the L shell, 18 in the M shell, 13 in the N shell, and one in the O shell.

The *outer* electrons of an atom are the ones responsible for the optical spectra of the elements. Relatively small amounts of energy suffice to remove these to excited states, and on their return to their normal states, wavelengths in or near the visible region are emitted. The inner electrons, being closer to the nucleus, are more tightly bound, and much more energy is required to displace them from their normal levels. As a result, we would expect a photon of much larger energy, and hence much higher frequency, to be emitted when the atom returns to its normal state after the displacement of an inner electron. This is, in fact, the case, and it is the displacement of the inner electrons that gives rise to the emission of x-rays.

On colliding with the atoms of the anode, some of the electrons accelerated in an x-ray tube, provided they have acquired sufficient energy, will dislodge one of the inner electrons of a target atom, say one of the K electrons. This leaves a vacant space in the K shell, which is immediately filled by an electron from either the L, M, or N shell. The readjustment of the electrons is accompanied by a decrease in the energy of the atom, and an x-ray photon is emitted with energy just equal to this decrease. Since the energy change is perfectly definite for atoms of a given element, the emitted x-rays should have definite frequencies. In other words, the x-ray spectrum should be a *line spectrum*. We can predict further that there should be just three lines in the series, corresponding to the three possibilities that the vacant space may have been filled by an L, M, or N electron.

This is precisely what is observed. Figure 41–12 illustrates the so-called K series of the elements tungsten, molybdenum, and copper. Each series consists of three lines, known as the K_α-, K_β-, and K_γ- lines. The K_α-line is produced by the transition of an L electron to the vacated space in the K shell, the K_β-line by an M electron, and the K_γ-line by an N electron.

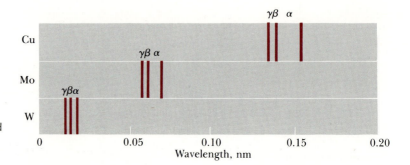

41–12 Wavelengths of the K_α-, K_β-, and K_γ-lines of copper, molybdenum, and tungsten.

In addition to the K series, there are other series known as the L, M, and N series, produced by the ejection of electrons from the L, M, and N shells rather than the K shell. As would be expected, the electrons in these outer shells, being farther away from the nucleus, are not held as firmly as those in the K shell. Consequently, the other series may be excited by more slowly moving electrons, and the photons emitted are of lower energy and longer wavelength.

In addition to the x-ray *line* spectrum there is a background of *continuous* x-radiation from the target of an x-ray tube. This is due to the sudden deceleration of those "cathode rays" (bombarding electrons) that do not happen to eject an atomic electron. The remarkable feature of the continuous spectrum is that while it extends indefinitely toward the *long* wavelength end, it is cut off very sharply at the *short* wavelength end. The quantum theory furnishes a simple explanation of the short-wave limit of the continuous x-ray spectrum.

A bombarding electron may be brought to rest in a single process if the electron happens to collide head on with an atom of the target; or it may make a number of collisions before coming to rest, giving up part of its energy each time. If we assume that the energy lost at each collision is radiated as an x-ray photon, these photons may have any energy up to a certain maximum, namely, that of an electron that gives up all of its energy in a single collision. Hence there is a short-wave limit length to the spectrum. The frequency of this limit is found by setting the energy of the electron equal to the energy of the x-ray photon:

$$hf = \tfrac{1}{2}mv^2. \tag{41–14}$$

This is precisely the same equation as that for the photoelectric effect except for the work-function term, which is negligible here since the energies of the x-ray photons are so large. In fact, the emission of x-rays may be described as an *inverse photoelectric effect*. In photoelectric emission the energy of a photon is transformed into kinetic energy of an electron; here, the kinetic energy of an electron is transformed into that of a photon.

EXAMPLE Compute the potential difference through which an electron must be accelerated in order that the short-wave limit of the continuous x-ray spectrum shall be exactly 0.1 nm.

Solution The frequency corresponding to 0.1 nm (10^{-10} m) is given by

$$f = \frac{c}{\lambda} = \frac{3 \times 10^8 \text{ m} \cdot \text{s}^{-1}}{10^{-10} \text{ m}} = 3 \times 10^{18} \text{ s}^{-1} = 3 \times 10^{18} \text{ Hz}.$$

The energy of the photon is

$$hf = (6.62 \times 10^{-34} \text{ J} \cdot \text{s})(3 \times 10^{18} \text{ s}^{-1}) = 19.9 \times 10^{-16} \text{ J}.$$

This must equal the kinetic energy of the electron, $\tfrac{1}{2}mv^2$, which is also equal to the product of the electronic charge and the accelerating voltage, V:

$$\tfrac{1}{2}mv^2 = eV = 19.9 \times 10^{-16} \text{ J}.$$

Since

$$e = 1.60 \times 10^{-19} \text{ C},$$

$$V = \frac{19.9 \times 10^{-16} \text{ J}}{1.60 \times 10^{-19} \text{ C}} = 12,400 \text{ V}.$$ ◀

A phenomenon called *Compton scattering*, first observed in 1924 by A. H. Compton, provides additional direct confirmation of the quantum nature of electromagnetic radiation. When x-rays impinge on matter, some of the radiation is *scattered*, just as visible light falling on a rough surface undergoes diffuse reflection. Observation shows that some of the scattered radiation has smaller frequency and longer wavelength than the incident radiation, and that the change in wavelength depends on the angle through which the radiation is scattered. Specifically, if the scattered radiation emerges at an angle ϕ with respect to the incident direction, and if λ and λ' are the wavelengths of the incident and scattered radiation, respectively, it is found that

$$\lambda' - \lambda = \frac{h}{mc}(1 - \cos \phi) \tag{41-15}$$

where m is the electron mass.

Compton scattering cannot be understood on the basis of classical electromagnetic theory. On the basis of classical principles, the scattering mechanism is induced motion of electrons in the material, caused by the incident radiation. This motion must have the same frequency as that of the incident wave, and so the scattered wave radiated by the oscillating charges should have the same frequency. There is no way the frequency can be *shifted* by this mechanism.

The quantum theory, by contrast, provides a beautifully simple explanation. We imagine the scattering process as a collision of two *particles*, the incident photon and an electron initially at rest, as in Fig. 41–13. The

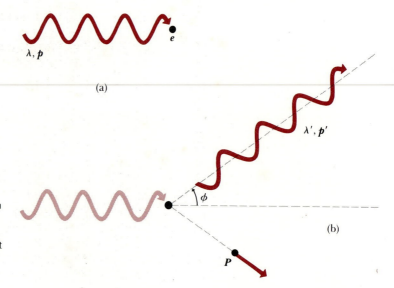

(a)

(b)

41–13 Schematic diagram of Compton scattering showing (a) electron initially at rest with incident photon of wavelength λ and momentum $\boldsymbol{p}$; (b) scattered photon with longer wavelength λ' and momentum $\boldsymbol{p}'$ and recoiling electron with momentum $\boldsymbol{P}$. The direction of the scattered photon makes an angle ϕ with that of the incident photon, and the angle between $\boldsymbol{p}$ and $\boldsymbol{p}'$ is also ϕ.

photon gives up some of its energy and momentum to the electron, which recoils as a result of this impact; and the final photon has less energy, smaller frequency, and longer wavelength than the initial one.

Equation (41–15) can be derived from the principles of conservation of energy and of momentum. We sketch the derivation below; the details of the calculation are left as a problem. Because relativistic energies may be involved, we use relativistic energy and momentum relations for the electron; its initial energy is mc^2, and its final energy E is given by $E^2 = (mc^2)^2 + (Pc)^2$. The energies of the incident and scattered photons are respectively pc and $p'c$. Thus energy conservation yields the relation

$$pc + mc^2 = p'c + E,$$

or

$$(pc - p'c + mc^2)^2 = E^2 = (mc^2)^2 + (Pc)^2. \qquad (41\text{–}16)$$

We may eliminate the electron momentum $\boldsymbol{P}$ from this equation by using momentum conservation:

$$\boldsymbol{p} = \boldsymbol{p}' + \boldsymbol{P},$$

or

$$\boldsymbol{p} - \boldsymbol{p}' = \boldsymbol{P}. \qquad (41\text{–}17)$$

By using the law of cosines with the vector diagram representing the vector subtraction of Eq. (41–17), we obtain the relation

$$P^2 = p^2 + p'^2 - 2pp' \cos \phi. \qquad (41\text{–}18)$$

This expression for P^2 may now be substituted into Eq. (41–16) and the left side multiplied out. A common factor c^2 is divided out; several terms cancel, and when the resulting equation is divided through by (pp'), the result is

$$\frac{mc}{p'} - \frac{mc}{p} = 1 - \cos \phi. \qquad (41\text{–}19)$$

Finally, we substitute $p = h/\lambda$ and $p' = h/\lambda'$ and rearrange again to obtain Eq. (41–15).

X-rays have many practical applications in medicine and industry. Because they can penetrate several centimeters of solid matter, they can be used to visualize the interiors of materials opaque to ordinary light, such as broken bones or defects in welds in structural steel. The object to be visualized is placed between an x-ray source and a large sheet of photographic film; the darkening of the film is proportional to the radiation exposure. A crack or air bubble allows greater transmission and shows as a dark area. Bones appear lighter than the surrounding flesh because they contain greater proportions of elements with high atomic number (and greater absorption) than flesh, where the light elements carbon, hydrogen, and oxygen predominate. This technique is not very effective in discriminating slightly different absorption characteristics, as found with many kinds of tumors.

In the past decade several vastly improved x-ray imaging techniques have been developed. One widely used system is *computerized axial tomography;* the corresponding instrument is called a CAT-scanner. The x-ray source produces a thin fan-shaped beam that is detected on the opposite side of the subject by an array of several hundred detectors in a line.

Each detector measures absorption along a thin line through the subject. The entire apparatus is rotated around the subject in the plane of the beam during a few seconds. The changing reactions of the detectors are recorded digitally; a computer processes this information and reconstructs a picture of density over an entire cross section of the subject. Density differences as small as one percent can be detected with CAT scans, and tumors and other anomalies much too small to be seen with older techniques are readily seen.

X-rays cause damage to living tissues. As x-ray photons are absorbed in tissues, they break molecular bonds and create highly reactive free radicals (such as neutral H and OH), which in turn can disturb the molecular structure of proteins and especially genetic material. Young and rapidly growing cells are particularly susceptible; hence x-rays are useful for selective destruction of cancer cells. Conversely, however, a cell may be damaged by radiation but survive, continue dividing, and produce generations of defective cells; hence x-rays can *cause* cancer. Even when the organism itself shows no apparent damage, radiation exposure can cause changes in the reproductive system that will affect the organism's offspring. The use of x-rays in medical diagnosis has become an area of great concern in recent years; a careful assessment of the balance between risks and benefits of radiation exposure is essential in each individual case.

QUESTIONS

41–1 In analyzing the photoelectric effect, how can we be sure that each electron absorbs only *one* photon?

41–2 In what ways do photons resemble other particles such as electrons? In what ways do they differ? Do they have mass? Electric charge? Can they be accelerated? What mechanical properties do they have?

41–3 Considering a two-slit interference experiment, if the photons are not synchronized with each other (i.e., are not coherent) and if half go through each slit, how can they possibly interfere with each other? Is there any way out of this paradox?

41–4 Can you devise an experiment to measure the work function of a material?

41–5 How might the energy levels of an atom be measured directly, i.e., without recourse to analysis of spectra?

41–6 Would you expect quantum effects to be generally more important at the low-frequency end of the electromagnetic spectrum (radio waves) or at the high-frequency end (x-rays and gamma rays)? Why?

41–7 Most black-and-white photographic film (with the exception of some special-purpose films) is less sensitive at the far red end of the visible spectrum than at the blue end and has almost no sensitivity to infrared. How can these properties be understood on the basis of photons?

41–8 Human skin is relatively insensitive to visible light, but ultraviolet radiation can be quite destructive. Does this have anything to do with photon energies?

41–9 Does the concept of photon energy shed any light (no pun intended) on the question of why x-rays are so much more penetrating than visible light?

41–10 The phosphorescent materials that coat the inside of a fluorescent lamp tube convert ultraviolet radiation (from the mercury-vapor discharge inside the tube) to visible light. Could one also make a phosphor that converts visible light to ultraviolet?

41–11 As a body is heated to very high temperature and becomes self-luminous, the apparent color of the emitted radiation shifts from red to yellow and finally to blue as the temperature increases. Why the color shift?

41–12 Elements in the gaseous state emit line spectra with well-defined wavelengths; but hot solid bodies usually emit a continuous spectrum, that is, a continuous smear of wavelengths. Can you account for this difference?

41–13 Could Compton scattering occur with protons as well as electrons? Suppose, for example, one directed a beam of x-rays at a liquid-hydrogen target. What similarities and differences in behavior would be expected?

PROBLEMS

$$e = 1.602 \times 10^{-19} \text{ C}$$
$$m = 9.109 \times 10^{-31} \text{ kg}$$
$$h = 6.626 \times 10^{-34} \text{ J} \cdot \text{s}$$
$$N_0 = 6.022 \times 10^{23} \text{ atoms} \cdot \text{mol}^{-1}$$

Energy equivalent of 1 u = 931.5 MeV

$$\frac{e}{m} = 1.758 \times 10^{11} \text{ C} \cdot \text{kg}^{-1}$$
$$k = 1.381 \times 10^{-23} \text{ J} \cdot \text{K}^{-1}$$
$$1 \text{ eV} = 1.602 \times 10^{-19} \text{ J}$$
$$1 \text{ u} = 1.661 \times 10^{-27} \text{ kg}$$
$$\epsilon_0 = 8.854 \times 10^{-12} \text{ C}^2 \cdot \text{N}^{-1} \cdot \text{m}^{-2}$$

41–1 A sodium-vapor lamp emits light of wavelength 589 nm. If the total power of the emitted light is 6 W, how many photons are emitted per second?

41–2 A photon of orange light has a wavelength of 600 nm. Find the frequency, momentum, and energy of the photon; express the energy both in joules and in electronvolts.

41–3 A laser used to weld detached retinas emits light of wavelength 633 nm with a power of 0.5 W, in pulses 20 ms in duration.

a) How much energy is in each pulse, in joules? In electronvolts?

b) What is the energy of one photon, in joules? In electronvolts?

c) How many photons are in each pulse?

41–4 A radio station broadcasts at a frequency of 100 MHz, with a total power output of 50 kW.

a) What is the energy of a photon, in joules? In electronvolts?

b) How many photons are emitted per second?

41–5 If the average wavelength emitted by a 100-W light bulb is 600 nm, and 10% of the input power is emitted as visible light, approximately how many photons are emitted per second? At what distance would this correspond to 100 photons per square centimeter, per second, if the light is emitted uniformly in all directions?

41–6 A nucleus in an excited state emits a γ-ray photon of energy 1 MeV.

a) What is the photon frequency?

b) What is the photon wavelength?

c) How does the wavelength compare with typical nuclear radii (of the order of 10^{-15} m)?

41–7 In the photoelectric effect, what is the relation between the threshold frequency f_0 and the work function ϕ?

41–8 The photoelectric threshold wavelength of tungsten is 2.73×10^{-5} cm. Calculate the maximum kinetic energy of the electrons ejected from a tungsten surface by ultraviolet radiation of wavelength 1.80×10^{-5} cm. (Express the answer in electronvolts.)

41–9 A photoelectric surface has a work function of 4.00 eV. What is the maximum speed of the photoelectrons emitted by light of frequency 3×10^{15} Hz?

41–10 When ultraviolet light of wavelength 2.54×10^{-5} cm from a mercury arc falls upon a clean copper surface, the retarding potential necessary to stop the emission of photoelectrons is 0.59 V. What is the photoelectric threshold wavelength for copper?

41–11 When a certain photoelectric surface is illuminated with light of different wavelengths, the stopping potentials in the table below are observed:

Wavelength, nm	Stopping potential, V
366	1.48
405	1.15
436	0.93
492	0.62
546	0.36
579	0.24

Plot the stopping potential as ordinate against the frequency of the light as abscissa. Determine

a) the threshold frequency,

b) the threshold wavelength,

c) the photoelectric work function of the material, and

d) the value of Planck's constant h (assuming the value of e is known).

41–12 The photoelectric work function of potassium is 2.25 eV. If light having a wavelength of 360 nm falls on potassium, find

a) the stopping potential,

b) the kinetic energy in electronvolts of the most energetic electrons ejected, and

c) the speeds of these electrons.

41–13 What will be the change in the stopping potential for photoelectrons emitted from a surface if the wavelength of the incident light is reduced from 400 nm to 360 nm?

41–14 The photoelectric work functions for particular samples of certain metals are as follows: cesium, 2.00 eV; copper, 4.00 eV; potassium, 2.25 eV; and zinc, 3.60 eV.

a) What is the threshold wavelength for each metal?

b) Which of these metals could not emit photoelectrons when irradiated with visible light?

41–15 The light-sensitive compound on most photographic films is silver bromide, AgBr. A film is "exposed" when the light energy absorbed dissociates this molecule into its atoms. (The actual process is more complex, but the quantitative result does not differ greatly.) The dissociation energy of AgBr is 23.9 kcal·mol^{-1}. Find

a) the energy in electronvolts,

b) the frequency of the photon that is just able to dissociate a molecule of silver bromide, and

c) the wavelength of the photon.

d) What is the energy in electronvolts of a quantum of radiation having a frequency of 100 MHz?

e) Explain the fact that light from a firefly can expose a photographic film, whereas the radiation from a TV station transmitting 50,000 W at 100 MHz cannot.

f) Will photographic films stored in a light-tight container be ruined (exposed) by the radio waves constantly passing through them? Explain. What about x-ray baggage scanners at airports?

41–16

a) Show that the energy E (in electronvolts) of a photon of wavelength λ (in nanometers) is given by E (eV) = $(1240/\lambda)$(nm).

b) What is the energy in electronvolts of a photon having a wavelength of 91.2 nm?

c) What is the momentum of the photon?

41–17 If 5% of the energy supplied to an incandescent lightbulb is radiated as visible light, how many visible photons are emitted per second by a 100-W bulb? Assume the wavelength of all the visible light to be 560 nm.

41–18 Find the longest and shortest wavelengths in the Lyman, Balmer, and Paschen series for hydrogen. In what region of the electromagnetic spectrum does each series lie?

41–19 Calculate (a) the frequency, and (b) the wavelength of the H_β-line of the Balmer series for hydrogen. This line is emitted in the transition from $n = 4$ to $n = 2$.

41–20

a) What is the least amount of energy in electronvolts that must be given to a hydrogen atom so that it can emit the H_β-line (see Problem 41–19) in the Balmer series?

b) How many different possibilities of spectral line emission are there for this atom when the electron goes from $n = 4$ to the ground state?

41–21 In Fig. 41–9, compute the energy difference for the $5s$–$3p$ transition; express your result in electronvolts and in joules. Compute the wavelength of a photon having this energy, and compare your result with the observed wavelength of the laser light.

41–22 In the helium–neon laser, what wavelength corresponds to the $3p$–$3s$ transition in neon? Why is this not observed in the beam with the same intensity as the 632.8-nm laser line?

41–23

a) What is the minimum potential difference between the filament and the target of an x-ray tube if the tube is to produce x-rays of wavelength 0.05 nm?

b) What is the shortest wavelength produced in an x-ray tube operated at 2×10^6 V?

41–24 An electron in a certain x-ray tube is accelerated from rest through a potential difference of 180,000 V in going from the cathode to the anode. When it arrives at the anode, what is

a) its kinetic energy in electronvolts,

b) its relativistic speed?

c) What is the speed of the electron, calculated classically?

41–25 An x-ray tube is operating at 150,000 V and 10 mA.

a) If only 1% of the electric power supplied is converted into x-rays, at what rate is the target being heated in calories per second?

b) If the target has a mass of 300 g and a specific heat of 0.035 cal·g^{-1}·°C^{-1}, at what average rate would its temperature rise if there were no thermal losses?

c) What must be the physical properties of a practical target material? What would be some suitable target elements?

41–26 If electrons in a metal had the same energy distribution as molecules in a gas at the same temperature (which is not actually the case), at what temperature would the average electron kinetic energy equal 1 eV, typical of work functions of metals?

41–27 If hydrogen were monatomic, at what temperature would the average translational kinetic energy be equal to the energy required to raise a hydrogen atom from the ground state to the $n = 2$ excited state?

41–28 Complete the derivation of the Compton-scattering formula, Eq. (41–15), following the outline given in Eqs. (41–16) through (41–19).

41–29 Calculate the maximum increase in x-ray wavelength that can occur during Compton scattering.

41–30 X-rays with initial wavelength 0.5×10^{-10} m undergo Compton scattering. For what scattering angle is the wavelength of the scattered x-rays greater than that of the incident x-rays by one percent?

41–31 X-rays are produced in a tube operating at 50 kV. After emerging from the tube, some x-rays strike a target and are Compton-scattered through an angle of 20°.

a) What is the original x-ray wavelength?

b) What is the wavelength of the scattered x-rays?

c) What is the energy of the scattered x-rays (in electron-volts)?

41–32 What is the energy (in electronvolts) of the smallest-energy x-ray photon for which Compton scattering could result in doubling the original wavelength?

41–33 A photon of wavelength 0.12 nm is Compton-scattered through an angle of 180°.

a) What is the wavelength of the scattered photon?

b) How much energy is given to the electron?

c) What is the recoil speed of the electron? Is it necessary to use the relativistic kinetic energy relationship?

42

QUANTUM MECHANICS

We have seen in the preceding chapter that some aspects of emission and absorption of light, including atomic spectra, can be understood on the basis of the photon concept, together with the concept of discrete energy levels in atoms. But a complete theory should also offer some means of *predicting*, on theoretical grounds, the values of these energy levels for any particular atom. Bohr attempted with some success to do this for the hydrogen atom, but his method could not be generalized to atoms having more than one electron. It became clear that more drastic departures from nineteenth-century concepts were needed; in particular, the wave–particle duality that had become firmly established for electromagnetic radiation must be extended to particles as well. That is, in some situations the entities, such as electrons, that we are accustomed to calling *particles* may exhibit *wavelike* behavior.

Developing a new theory based on this dual nature of particles requires fundamental changes in the language used to describe the state of a mechanical system. A particle can no longer be described as a localized point moving in space but, instead, must be regarded as an inherently spread-out entity. As it moves, the spread-out character has some of the properties of a *wave;* for example, particles can undergo *diffraction!* This new theory is called *quantum mechanics;* in it is found the key to the analysis of the structure of atoms and molecules, including their spectra, chemical behavior, and many other properties. Quantum mechanics has the happy effect of restoring unity to our description of both particles and radiation, and wave concepts are central to the entire theory.

42–1

THE BOHR ATOM

At the same time (1915) that Bohr advanced his hypothesis about the relation of spectrum-line frequencies to energy levels of atoms, he also proposed a mechanical model of the simplest atom, hydrogen. By combining some classical mechanics with a postulate that had no basis in previous physics, he was able to calculate the energy levels of hydrogen and obtain agreement with values determined from spectra.

Bohr's was not by any means the first attempt to understand the internal structure of atoms. Starting in 1906, Rutherford and his co-workers had performed experiments on the scattering of alpha particles (helium nuclei emitted from radioactive elements) by thin metallic foils. These experiments, which will be discussed in Chapter 44, showed that each atom contains a massive nucleus whose size (of the order of 10^{-15} m) is very much smaller than the overall size of the atom (of the order of 10^{-10} m). The nucleus is surrounded by a swarm of electrons.

To account for the fact that the electrons remain at relatively large distances from the positively charged nucleus despite the electrostatic attraction the nucleus exerts on the electrons, Rutherford postulated that the electrons *revolve* about the nucleus in orbits, more or less as the planets in the solar system revolve around the sun, but with the electrical attraction providing the necessary centripetal force.

This assumption, however, has an unfortunate consequence. A body moving in a circle is continuously accelerated toward the center of the circle and, according to classical electromagnetic theory, an accelerated electron radiates energy. The total energy of the electrons would therefore gradually decrease, their orbits would become smaller and smaller, and eventually they would spiral into the nucleus and come to rest. Furthermore, according to classical theory, the *frequency* of the electromagnetic waves emitted by a revolving electron is equal to the frequency of revolution. As the electrons radiated energy, their angular velocities would change continuously and they would emit a *continuous* spectrum (a mixture of all frequencies), in contradiction to the *line* spectrum actually observed.

Faced with the dilemma that electromagnetic theory predicted an unstable atom emitting radiant energy of all frequencies, while observation showed stable atoms emitting only a few frequencies, Bohr concluded that, in spite of the success of electromagnetic theory in explaining large-scale phenomena, it could not be applied to processes on an atomic scale. He therefore postulated that an electron in an atom can revolve in certain stable orbits, each having a definite associated energy, without emitting radiation, contrary to the predictions of classical electromagnetic theory. According to Bohr, an atom radiates only when it makes a transition from one of these special orbits to another, emitting (or absorbing) a photon of appropriate energy, given by Eq. (41–11), at the same time.

To determine the radii of the "permitted" orbits, Bohr introduced what must be regarded in hindsight as a brilliant intuitive guess. He noted that the *units* of Planck's constant h, usually written as $\text{J} \cdot \text{s}$, are the same as the units of angular momentum, usually written as $\text{kg} \cdot \text{m}^2 \cdot \text{s}^{-1}$, and he postulated that only those orbits are permitted for which the angular momentum is an integer multiple of $h/2\pi$. We recall from Chapter 9 that the angular momentum of a particle of mass m, moving with tangential speed v in a circle of radius r, is mvr. Hence the above condition may be stated as

$$mvr = n\frac{h}{2\pi},$$

where $n = 1, 2, 3$, etc. For each value of n there is a permitted value of orbit radius, which we denote from now on by r_n, and a corresponding speed v_n. With this notation, the above equation becomes

$$mv_nr_n = n\frac{h}{2\pi}. \tag{42-1}$$

We now incorporate this condition into the analysis of the hydrogen atom. This atom consists of a single electron of charge $-e$, revolving about a single proton of charge $+e$. The proton, being nearly 2000 times as massive as the electron, will be assumed stationary in this discussion. The electrostatic force of attraction between the charges,

$$F = \frac{1}{4\pi\epsilon_0}\frac{e^2}{r_n^2},$$

provides the centripetal force and, from Newton's second law,

$$\frac{1}{4\pi\epsilon_0}\frac{e^2}{r_n^2} = \frac{mv_n^2}{r_n}. \tag{42-2}$$

When Eqs. (42–1) and (42–2) are solved simultaneously for r and v, we obtain

$$r_n = \epsilon_0\frac{n^2h^2}{\pi me^2}, \tag{42-3}$$

$$v_n = \frac{1}{\epsilon_0}\frac{e^2}{2nh}. \tag{42-4}$$

Let

$$\epsilon_0\frac{h^2}{\pi me^2} = r_1. \tag{42-5}$$

Then Eq. (42–3) becomes

$$r_n = n^2r_1,$$

and the permitted, nonradiating orbits have radii r_1, $4r_1$, $9r_1$, etc. The appropriate value of n is called the *quantum number* of the orbit.

The numerical values of the quantities on the left side of Eq. (45–5) are

$$\epsilon_0 = 8.854 \times 10^{-12}\,\text{C}^2\cdot\text{N}^{-1}\cdot\text{m}^{-2},$$

$$h = 6.626 \times 10^{-34}\,\text{J}\cdot\text{s},$$

$$m = 9.109 \times 10^{-31}\,\text{kg},$$

$$e = 1.602 \times 10^{-19}\,\text{C}.$$

Hence r_1, the radius of the first Bohr orbit, is

$$r_1 = \frac{(8.854 \times 10^{-12}\,\text{C}^2\cdot\text{N}^{-1}\cdot\text{m}^{-2})(6.626 \times 10^{-34}\,\text{J}\cdot\text{s})^2}{(3.14)(9.109 \times 10^{-31}\,\text{kg})(1.602 \times 10^{-19}\,\text{C})^2}$$

$$= 0.53 \times 10^{-10}\,\text{m} = 0.53 \times 10^{-8}\,\text{cm}.$$

This is in good agreement with atomic diameters as estimated by other methods, namely, about 10^{-8} cm.

The kinetic energy of the electron in any orbit is

$$K_n = \tfrac{1}{2}mv_n^2 = \frac{1}{\epsilon_0^2}\frac{me^4}{8n^2h^2},$$

and the potential energy is

$$U_n = -\frac{1}{4\pi\epsilon_0}\frac{e^2}{r_n} = -\frac{1}{\epsilon_0^2}\frac{me^4}{4n^2h^2}.$$

The total energy, E_n, is therefore

$$E_n = K_n + U_n = -\frac{1}{\epsilon_0^2}\frac{me^4}{8n^2h^2}. \qquad (42\text{–}6)$$

The total energy has a negative sign because the reference level of potential energy is taken with the electron at an infinite distance from the nucleus. Since we are interested only in energy *differences*, this is not of importance.

The energy of the atom is least when its electron is revolving in the orbit for which $n = 1$, for then E_n has its largest negative value. For $n = 2, 3, \ldots$, the absolute value of E_n is smaller; hence the energy is progressively larger in the outer orbits. The *normal* state of the atom, called the *ground state*, is that of lowest energy, with the electron revolving in the orbit of smallest radius, r_1. As a result of collisions with rapidly moving electrons in an electrical discharge, or for other causes, the atom may temporarily acquire sufficient energy to raise the electron to some outer orbit. The atom is then said to be in an *excited* state. This state is an unstable one, and the electron soon falls back to a state of lower energy, emitting a photon in the process.

Let n_1 be the quantum number of some excited state, and n_2 the quantum number of the lower state to which the electron returns after the emission process. Then E_i, the initial energy, is

$$E_i = -\frac{1}{\epsilon_0^2}\frac{me^4}{8n_1^2h^2},$$

and E_f, the final energy, is

$$E_f = -\frac{1}{\epsilon_0^2}\frac{me^4}{8n_2^2h^2}.$$

The decrease in energy, $E_i - E_f$, which we place equal to the energy hf of the emitted photon, is

$$E_i - E_f = hf = -\frac{1}{\epsilon_0^2}\frac{me^4}{8n_1^2h^2} + \frac{1}{\epsilon_0^2}\frac{me^4}{8n_2^2h^2},$$

or

$$f = \frac{1}{\epsilon_0^2}\frac{me^4}{8h^3}\left(\frac{1}{n_2^2} - \frac{1}{n_1^2}\right). \qquad (42\text{–}7)$$

This equation is of precisely the same form as the Balmer formula, Eq. (41–9), for the frequencies in the hydrogen spectrum if we place

$$\frac{1}{\epsilon_0^2}\frac{me^4}{8h^3} = Rc, \qquad (42\text{–}8)$$

and let $n_2 = 1$ for the Lyman series, $n_2 = 2$ for the Balmer series, etc. The Lyman series is therefore the group of lines emitted by electrons returning from some excited state to the ground state. The Balmer series is the group emitted by electrons returning from some higher state, but stopping in the *second orbit* instead of falling at once to that of lowest energy. That is, an electron returning from the third orbit ($n = 3$) to the second orbit ($n = 2$) emits the H_α-line. One returning from the fourth orbit ($n = 4$) to the second ($n = 2$) emits the H_β-line, etc. These transitions are shown in Fig. 42−1.

Every quantity in Eq. (42−8) may be determined quite independently of the Bohr theory, and apart from this theory we have no reason to expect these quantities to be related in this particular way. The quantities m and e, for instance, are found from experiments on free electrons, h may be found from the photoelectric effect, and R by measurements of wavelengths, while c is the speed of light. However, if we substitute in Eq. (42−8) the values of these quantities, obtained by such diverse means, we find that it *does* hold exactly, within the limits of experimental error, providing direct confirmation of Bohr's theory.

The interaction of atoms with a magnetic field can be analyzed with the Bohr model. It was suggested first by Faraday that shifts in spectrum wavelengths might occur when atoms are placed in a magnetic field. Faraday's spectroscopic techniques were not refined enough to observe such shifts, but Zeeman, using improved instruments, *was* able to detect shifts, an effect now called the *Zeeman effect*.

The interaction is most easily treated using the concept of magnetic moment, introduced in Section 30−8. Let us consider the electron in the first Bohr orbit ($n = 1$). The orbiting charge is equivalent to a current

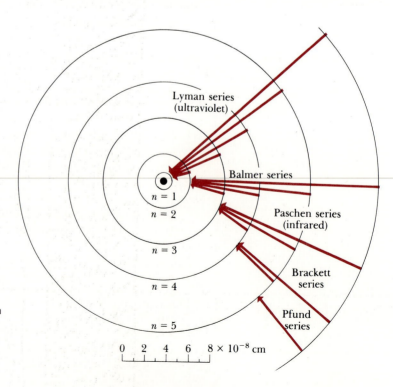

42−1 "Permitted" orbits of an electron in the Bohr model of a hydrogen atom. The transitions responsible for some of the lines of the various series are indicated by arrows.

loop of radius r and area πr^2. The average charge per unit time passing a point of the orbit is the average current I, and this is given by e/τ, where τ is the time for one revolution: $\tau = 2\pi r/v$. The magnetic moment, which we denote by μ to avoid confusion with the electron mass m, is given by

$$\mu = IA = \tfrac{1}{2}evr.$$

But according to Bohr's theory, the angular momentum mvr is equal to $h/2\pi$, or

$$vr = \frac{h}{2\pi m}.$$

Therefore

$$\mu = \frac{h}{4\pi} \cdot \frac{e}{m} = \frac{(6.626 \times 10^{-34}\ \text{J}\cdot\text{s})(1.758 \times 10^{11}\ \text{C}\cdot\text{kg}^{-1})}{12.57}$$

$$= 9.27 \times 10^{-24}\ \text{A}\cdot\text{m}^2.$$

EXAMPLE   Find the interaction potential energy when the hydrogen atom described above is placed in a magnetic field of 2 T.

Solution   According to Eq. (30–15), the interaction energy U when $\alpha = 0$ is

$$U = -\mu B = -(9.27 \times 10^{-24}\ \text{A}\cdot\text{m}^2)(2\ \text{T})$$

$$= -1.85 \times 10^{-23}\ \text{J}$$

$$= -1.16 \times 10^{-4}\ \text{eV}.$$

When $\boldsymbol{\mu}$ and $\boldsymbol{B}$ are antiparallel, the energy is $+1.16 \times 10^{-4}$ eV. We note that these energies are much *smaller* than the energy of the electron. The Zeeman effect is discussed further in Section 42–7.   ◀

Although the Bohr model was successful in predicting the energy levels of the hydrogen atom, it raised as many questions as it answered. It combined elements of classical physics with new postulates inconsistent with classical ideas. It provided no insight into what happens during a transition from one orbit to another, and the stability of certain orbits was bought at the expense of discarding the only picture available at the time of the mechanism of radiation of energy. There was no clear justification (except that it led to the right answer) for restricting the angular momentum to multiples of $h/2\pi$. Furthermore, attempts to extend the model to atoms with two or more electrons were not successful. We shall see in the next section that an even more radical departure from classical concepts was required before the understanding of atomic structure could progress further.

42–2

WAVE NATURE OF PARTICLES

The next advance in understanding atomic structure came in 1923, about 10 years after the Bohr theory. This was a suggestion by de Broglie that since light is dualistic in nature, behaving in some aspects like waves and in others like particles, the same might be true of matter. That is, electrons and protons, which until that time had been thought to be purely corpuscular, might in some circumstances behave like *waves*. Specifically, de Broglie postulated that a free electron of mass m, moving with speed v, should have a wavelength λ related to its momentum $p = mv$ in exactly the same way the wavelength and momentum of a photon are related, as expressed by Eq. (41–7). Thus, according to the de Broglie hypothesis, the wavelength of an electron is given by

$$\lambda = \frac{h}{mv}, \tag{42–9}$$

where h is the same Planck's constant that appears in the frequency–energy relation for photons.

This wave hypothesis, unorthodox though it seemed at the time, almost immediately received very direct experimental confirmation. We have described in Section 39–9 how the layers of atoms in a crystal serve as a diffraction grating for x-rays. An x-ray beam is strongly reflected when it strikes a crystal at such an angle that the waves scattered from the atomic layers combine to reinforce one another. The essential point here is that the existence of these strong reflections is evidence of the *wave* nature of x-rays.

In 1927, Davisson and Germer, working in the Bell Telephone Laboratories, were studying the nature of the surface of a crystal of nickel by directing a beam of *electrons* at the surface and observing the electrons reflected at various angles. It might be expected that even the smoothest surface attainable would still look rough to an electron, and that the electron beam would therefore be diffusely reflected. But Davisson and Germer found that the electrons were reflected in almost the same way that x-rays would be reflected from the same crystal. The wavelengths of the electrons in the beam were computed from their known speed, with the help of Eq. (42–9); and the angles at which strong reflection took place were found to be the same as those at which x-rays of the same wavelength would be reflected. This result gave strong support to de Broglie's hypothesis.

This wave hypothesis clearly requires sweeping revisions of our fundamental concepts regarding the description of matter. What we are accustomed to calling a *particle* actually looks like a particle only if we do not look too closely. In general, a particle has to be regarded as a spread-out entity that is not entirely localized in space; and at least in some cases this spreading out appears as a periodic pattern suggesting wavelike properties. The wave and particle aspects are not inconsistent, but the particle model is an *approximation* of a more general wave picture. We are reminded of the ray picture of geometrical optics, a special case of the more general wave picture of physical optics; indeed, there is a very close analogy between optics and the description of particles.

Within a few years, after 1923, the wave hypothesis of de Broglie was developed by Heisenberg, Schrödinger, and many others, into a complete theory called *wave mechanics* or *quantum mechanics*. In the following sections we shall attempt to outline the main lines of thought in a nonmathematical way, describe some of the experimental evidence of the wave nature of material particles, and show how the quantum numbers that were introduced in such an artificial way by Bohr now enter naturally into the theory of atomic structure.

One of the essential features of quantum mechanics is that material particles are no longer regarded as geometrical points, localized in space, but are intrinsically spread-out entities. The spatial distribution of a *free* electron may have a recurring pattern characteristic of a wave that propagates through space. Electrons *in atoms* are visualized as diffuse clouds surrounding the nucleus. The idea that the electrons in an atom move in definite orbits such as those in Fig. 42–1 has been abandoned. The orbits themselves, however, were never an essential part of Bohr's theory, since the quantities that determine the frequencies of the emitted photons are the *energies* corresponding to the orbits. The new theory still assigns definite energy states to an atom. In the hydrogen atom the energies are the same as those given by Bohr's theory; in more complicated atoms where the Bohr theory did not work, the quantum mechanical picture is in excellent agreement with observation.

We shall illustrate how quantization arises in atomic structure by an analogy with the classical mechanical problem of a vibrating string fixed at its ends. When the string vibrates, the ends must be nodes, but nodes may occur at other points also, and the general requirement is that the length of the string shall equal some *integer* number of half-wavelengths. The point of interest is that the solution of the problem of the vibrating string leads to the appearance of *integer numbers.*

In a similar way, the principles of quantum mechanics lead to a wave equation (Schrödinger's equation) that must be satisfied by an electron in an atom, subject also to certain boundary conditions. Let us think of an electron as a wave extending in a circle around the nucleus. In order that the wave may "come out even," the circumference of this circle must include some *integer number* of wavelengths. The wavelength of a particle of mass m, moving with speed v, is given, according to wave mechanics, by Eq. (42–9), $\lambda = h/mv$. Then if r is the radius and $2\pi r$ the circumference of the circle occupied by the wave, we must have $2\pi r = n\lambda$, where $n = 1, 2, 3$, etc. Since $\lambda = h/mv$, this equation becomes

$$2\pi r = n\frac{h}{mv}, \qquad mvr = n\frac{h}{2\pi}. \tag{42–10}$$

But mvr is the angular momentum of the electron, so we see that the wave-mechanical picture leads naturally to Bohr's postulate that the angular momentum equals some integral multiple of $h/2\pi$.

To be sure, the idea of wrapping a wave around in a circular orbit is a rather vague notion. But the agreement of Eq. (42–10) with Bohr's hypothesis is much too remarkable to be a coincidence; and it strongly suggests that the wave properties of electrons do indeed have something to do with atomic structure.

EXAMPLE Find the speed and kinetic energy of a neutron ($m = 1.675 \times 10^{-27}$ kg) having a de Broglie wavelength of 0.1 nm. Compare the energy with the average kinetic energy of a gas molecule at room temperature ($T = 20°C$).

Solution   From Eq. (42–9),

$$v = \frac{h}{\lambda m} = \frac{6.626 \times 10^{-34}\,\text{J}\cdot\text{s}}{(0.1 \times 10^{-9}\,\text{m})(1.675 \times 10^{-27}\,\text{kg})}$$
$$= 3.96 \times 10^3\,\text{m}\cdot\text{s}^{-1}.$$

$$K = \tfrac{1}{2}mv^2 = \tfrac{1}{2}(1.675 \times 10^{-27}\,\text{kg})(3.96 \times 10^3\,\text{m}\cdot\text{s}^{-1})^2$$
$$= 1.31 \times 10^{-20}\,\text{J} = 0.0828\,\text{eV}.$$

The average translational kinetic energy of a molecule of an ideal gas is given by Eq. (20–10):

$$K = \tfrac{3}{2}kT = (3/2)(1.38 \times 10^{-23}\,\text{J}\cdot\text{K}^{-1})(293\,\text{K})$$
$$= 6.06 \times 10^{-21}\,\text{J} = 0.0378\,\text{eV}.$$

Thus the two energies are of comparable magnitude, and indeed a neutron with this energy range is called a *thermal neutron*. Diffraction of thermal neutrons can be used to study crystal and molecular structure in the same way as x-ray diffraction; neutron diffraction has proved especially useful in the study of large organic molecules. ◀

*42–3

THE ELECTRON MICROSCOPE

An electron beam can be used to form an image of an object in exactly the same way as a light beam. A ray of light is bent by reflection or refraction, and an electron trajectory is bent by an electric or magnetic field. Rays of light diverging from a point on an object can be brought to convergence by a converging lens, and electrons diverging from a small region can be brought to convergence by an electrostatic or magnetic lens. Figure 42–2 shows the behavior of a simple type of electrostatic lens, and Fig. 42–3 shows the analogous optical system. In each case the image can be made larger than the object by appropriate design; hence both devices can act as *magnifiers.*

The analogy between light rays and electrons goes deeper. The ray model of geometrical optics is an approximate representation of the more general wave picture, and geometrical optics is valid whenever

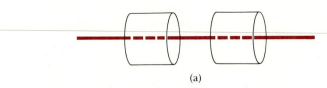

(a)

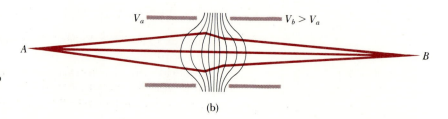

(b)

42–2   (a) An electrostatic lens. The two cylinders are at different electrical potentials V_a and V_b; there is an electric field in the region between them, and the corresponding equipotential lines are shown in black. (b) Cross-sectional view. The electron trajectories are shown in color; electrons diverging from point A are brought to a focus at point B. The behavior of magnetic lenses is similar.

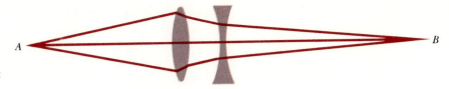

42–3 Optical analog of the electrostatic lens in Fig. 42–2.

interference and diffraction effects can be neglected. Similarly, we have seen in Section 42–2 that the model of an electron as a point particle following a line trajectory is an approximate description of the actual behavior of the electron, valid when effects associated with the wave nature of electrons can be neglected.

Herein lies the value of the electron microscope. The resolution of an optical microscope is limited by diffraction effects, as discussed in Section 39–10. With a wavelength of 500 nm, typical of visible light, no optical microscope can resolve objects smaller than a few hundred nanometers, no matter how carefully its lenses are made. The resolution of an electron microscope is similarly limited by the wavelengths of the electrons, but these may be many thousands of times *smaller* than wavelengths of visible light. Hence the useful magnification of an electron microscope can be thousands of times as great as that of an optical microscope.

It is important to understand that the ability of the electron microscope to form an image *does not* depend on the wave properties of electrons. Their trajectories can be computed by treating them as charged particles under the action of electric- and magnetic-field forces. It is only in the matter of *resolution* that the wave properties become significant.

EXAMPLE An electron beam is formed by a set-up similar to that of the cathode-ray tube, discussed in Section 26–8. If the accelerating voltage is 10 kV (10,000 V), what is the wavelength of the electrons?

Solution The wavelength is determined by Eq. (42–9). To find the speed we use conservation of energy; the kinetic energy $\frac{1}{2}mv^2$ of an electron equals the loss of potential energy eV. Thus

$$\tfrac{1}{2}mv^2 = eV, \qquad v = \frac{\sqrt{2eV}}{m}.$$

Inserting this result in Eq. (42–9), we find

$$\lambda = \frac{h}{m}\sqrt{\frac{m}{2eV}} = \frac{h}{\sqrt{2meV}}$$

$$= \frac{6.63 \times 10^{-34}\,\text{J}\cdot\text{s}}{\sqrt{2(9.11 \times 10^{-31}\,\text{kg})(1.60 \times 10^{-19}\,\text{C})(10^4\,\text{V})}} \qquad (42\text{–}11)$$

$$= 1.23 \times 10^{-11}\,\text{m} = 0.0123\,\text{nm}.$$

This is smaller than typical wavelengths of visible light (around 500 nm) by a factor of about 40,000. ◀

Most electron microscopes use magnetic rather than electrostatic lenses, for practical reasons. A common setup uses three lenses in a "compound-microscope" arrangement, as shown in Fig. 42–4. Electrons are emitted from a hot cathode and accelerated by a potential difference, typically 10 kV to 100 kV. The arrangement is similar to that of the cathode-ray tube, discussed in Section 26–8. They pass through a condenser lens and are formed into a parallel beam before passing through the specimen or object to be viewed. The objective lens then forms an intermediate image of this object, and the projection lens a final real image of that image. These lenses play the roles of the objective and eyepiece lenses, respectively, of a compound optical microscope. The final image is recorded on photograph film or projected on a fluorescent screen for viewing or photographing. The entire apparatus, including the specimen, must of course be enclosed in a vacuum container just as with the cathode-ray tube; otherwise electrons would collide with air molecules, muddling up the image. The specimen to be viewed is very thin, typically 10 nm to 100 nm, so the electrons are not slowed appreciably as they pass through.

It might be thought that with electrons of wavelength 0.01 nm, as in the above example, the resolution would be 0.01 nm or less. In fact, it is seldom better than 0.5 nm, for several reasons. Large-aperture magnetic lenses have aberrations analogous to those of optical lenses, as discussed in Section 38–5. In addition, the focal length of a magnetic lens depends on the current in the coil, which must be controlled precisely, and on the electron speed, which is never a single precise value; the latter effect is the equivalent of chromatic aberration.

A very useful variation is the scanning electron microscope. The electron beam is focused to a very fine line and is swept across the specimen just as the electron beam in a TV picture tube traces out the picture. As it scans the specimen, electrons are knocked off; these are collected by a collecting anode kept at a potential a few hundred volts positive with respect to the specimen. The current in the collecting anode is amplified and used to modulate the electron beam in a cathode-ray tube, which is swept in synchronism with the microscope beam. Thus the cathode-ray tube traces out a greatly magnified image of the specimen. This scheme has the advantages that the beam need not pass through the specimen and that knock-off electron production depends on the angle at which the beam strikes the surface. Thus scanning electron micrographs have a much greater three-dimensional appearance than conventional ones. The resolution is not as great, typically of the order of 10 nm, still a great improvement over the optical microscope. A scanning electron microscope is shown in Fig. 42–5, and a photograph made with such an instrument is shown in Fig. 42–6.

42–4 An electron microscope. The magnetic lenses, coils of wire carrying currents, are shown in cross section. The condensing lens forms a parallel beam of electrons that strikes the object. The objective lens forms an intermediate image that serves as the object for the final image formed by the projection lens. All images are *real*. The final image is projected on photographic film or a fluorescent screen. The magnification of each lens may be of the order of 100×, and the overall magnification of the order of 10,000×. For greater magnification an additional intermediate lens may be used. The color lines are not continuous electron trajectories through the instrument but are drawn from lens to lens to show the formation of images. The angles of the electron paths with the optic axis are greatly exaggerated; in actual instruments these angles are usually less than 0.01 rad or 0.5°. The entire apparatus is enclosed in a vacuum chamber, not shown in the diagram.

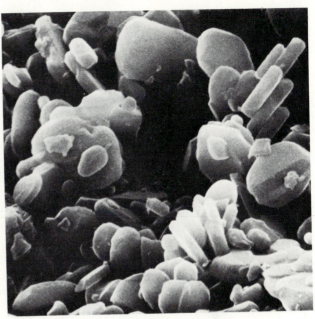

42–5 A scanning electron microscope, showing a vacuum chamber in the background and cathode-ray tube monitors in the foreground. (Courtesy of General Electric.)

42–6 Photograph of powdered aluminum oxide. The overall magnification is about 4500×. (Manfred Kage-Peter Arnold, Inc.)

42–4

PROBABILITY AND UNCERTAINTY

As we have seen, the entities that we are accustomed to calling *particles* can, in some experiments, behave like *waves*. Figure 42–7 shows a *wave packet* or *wave pulse* that has both wave and particle properties. The regular spacing λ_{av} between successive maxima is characteristic of a wave, but there is also a particle-like localization in space. To be sure, the wave pulse is not localized at a single point, but in any experiment that detects only dimensions much larger than Δx, the wave pulse appears to be a localized particle. Although it would be simplistic to say that Fig. 42–7 is a picture of an electron, the figure does suggest that wave and particle properties are not necessarily incompatible.

It must be emphasized that it would *not* be correct to regard the wave of Fig. 42–7 as having only a single wavelength. A sinusoidal wave with a definite wavelength has no beginning and no end; to make a wave *pulse* it is necessary to superpose many sinusoidal waves having various wavelengths. Thus Fig. 42–7 is a wave having such a distribution of wavelengths, of which λ_{av} is an average value. This distribution has additional important implications, which will be explored at the end of this section.

The discovery of the dual wave–particle nature of matter has forced a drastic revision of the language used to describe the behavior of a particle. In classical newtonian mechanics we think of a particle as an idealized geometrical point that, at any instant of time, has a perfectly definite location in space and is moving with a perfectly definite velocity. As we shall see, such a specific description is, in general, not possible; on

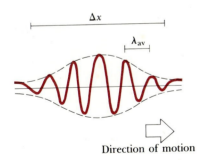

42–7 A wave pulse or packet. There is an average wavelength λ_{av}, the distance between adjacent peaks, but the wave is localized at any instant in a region with length of the order of Δx. The broken lines are called the envelope of the pulse; all the peaks lie on the envelope curves, which approach zero at both ends of the pulse.

Direction of motion

a sufficiently small scale, there are fundamental limitations on the precision with which the position and velocity of a particle can be described; and some aspects of a particle's behavior can be stated only in terms of *probabilities*.

To illustrate the nature of the problem, let us consider again the single-slit diffraction experiment described in Section 39–7. As Fig. 39–22b suggests, most of the intensity in the diffraction pattern is concentrated in the central maximum; the angular size of this maximum is determined by the positions of the first intensity minimum on either side of the central maximum. Using Eq. (39–9), with $n = 1$, we find that the angle θ in Fig. 42–8 between the central peak and the minimum on either side is given by $\sin \theta = \lambda/a$, where a is the slit width. If λ is much smaller than a, then θ is very small, $\sin \theta$ is very nearly equal to θ, and this may be simplified further to

$$\theta = \frac{\lambda}{a}. \tag{42–12}$$

Now we perform the same experiment again, but using a beam of *electrons* instead of a beam of monochromatic light. The apparatus must be evacuated to avoid collisions of electrons with air molecules; and there are other experimental details that need not concern us. The electron beam can be produced with a setup similar in principle to the electron gun in a cathode-ray oscilloscope, described in Section 26–8, which produces a narrow beam of electrons all having the same direction and speed, and therefore also the same wavelength. Such an experiment is shown schematically in Fig. 42–8.

The result of this experiment, as again recorded on photographic film or by means of more sophisticated detectors, is a diffraction pattern identical to that shown in Fig. 39–22b, providing additional direct evidence of the wave nature of electrons. Most of the electrons strike the film in the vicinity of the central maximum, but a few strike farther from the center, near the edges of that maximum and also in the subsidiary maxima on both sides. Thus the wave behavior in this experiment presents no surprises.

Interpreted in terms of *particles*, however, this experiment poses very serious problems. First, although the electrons all have the same initial

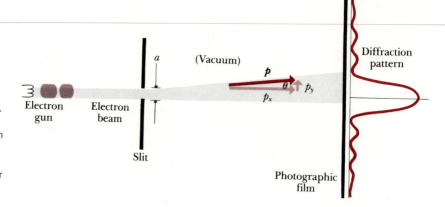

42–8 An electron diffraction experiment. The graph at the right shows the degree of blackening of the film, which in any region is proportional to the number of electrons striking that region. The components of momentum of an electron striking the outer fringe of the central maximum are shown.

state of motion, they do not all follow the same path; and in fact the trajectory of an individual electron cannot be predicted from knowledge of its initial state. The best we can do is to say that most of the electrons go to a certain region, fewer to other regions, and so on; alternatively, we can describe the *probability* for an individual electron to strike each of various areas on the film. This fundamental indeterminacy has no counterpart in newtonian mechanics, where the motion of a particle or a system is always completely predictable if the initial position and motion are known with sufficient precision.

Second, there are fundamental *uncertainties* in both position and momentum of an individual particle, and these two uncertainties are related inseparably. To illustrate this point, we note that, in Fig. 42–8, an electron striking the film at the outer edge of the central maximum, at angle θ, must have a component of momentum p_y in the y-direction, as well as a component p_x in the x-direction, despite the fact that initially the beam was directed along the x-axis. From the geometry of the situation, the two components are related by $p_y/p_x = \tan \theta$; and if θ is small we may approximate $\tan \theta = \theta$, obtaining

$$p_y = p_x \, \theta. \tag{42–13}$$

Neglecting any electrons striking the film outside the central maximum (that is, at angles greater than λ/a), we see that the y-component of momentum may be as large as

$$p_y = p_x \left(\frac{\lambda}{a} \right). \tag{42–14}$$

Hence the *uncertainty* Δp_y in the y-component of momentum is of the order of $p_x \lambda/a$. Thus the narrower the slit width a, the broader is the diffraction pattern and the greater the uncertainty in the y-component of momentum.

Now the electron wavelength λ is related to the momentum $p_x = mv_x$ by the de Broglie relation, Eq. (42–9), which may be rewritten as $\lambda = h/p_x$. Using this result in Eq. (42–14) and simplifying, we find

$$\Delta p_y = p_x \left(\frac{h}{a p_x} \right) = \frac{h}{a},$$

or

$$\Delta p_y \, a = h. \tag{42–15}$$

To interpret this result, we note that the slit width a represents the uncertainty in *position* of an electron as it passes through the slit; we do not know through which particular part of the slit each particle passes. Thus the y-components of *both* position and momentum have uncertainties, and the two uncertainties are related by Eq. (42–15). We can reduce the *momentum* uncertainty only by increasing the slit width, which increases the *position* uncertainty; and conversely, when we decrease the position uncertainty by narrowing the slit, the diffraction pattern broadens and the corresponding momentum uncertainty increases.

All of this may be bitter medicine for a reader steeped in the tradition of nearly three centuries of newtonian mechanics; but the weight of experimental evidence leaves us no alternatives. To those who protest that the lack of a definite position and momentum is contrary to common sense, we reply that what we call common sense is based on familiarity gained through experience, and that our usual experience includes very little contact with the microscopic behavior of particles. Thus we must sometimes be prepared to accept conclusions that seem not to make sense when we are dealing with areas far removed from everyday experience.

In more general discussions of uncertainty relations, it is customary to describe the uncertainty of a quantity in terms of the statistical concept of *standard deviation*, a measure of the spread or dispersion of a set of numbers around their average. If a coordinate x has an uncertainty Δx, defined in this way, and if the corresponding momentum component p_x has an uncertainty Δp_x, then the two uncertainties are found to be related in general by the inequality

$$\Delta x \, \Delta p_x \geq \frac{h}{2\pi}.$$

(42–16)

Equation (42–16) is one form of the *Heisenberg uncertainty principle;* it states that, in general, neither the momentum nor the position of a particle can be known with arbitrarily great precision, but that the two play complementary roles as described above. It might be protested that greater precision could be attained by using more sophisticated particle detectors in various areas of the slit or by other means, but this turns out to be not possible. To detect a particle the detector must *interact* with it, and this interaction unavoidably changes the state of motion of the particle. A more detailed analysis of such hypothetical experiments shows that the uncertainty is a fundamental and intrinsic one that cannot be circumvented even in principle by any experimental technique, no matter how sophisticated.

Additional insight into the uncertainty principle is provided by the wave packet shown in Fig. 42–7. Such a packet can be constructed by superposing several sinusoidal waves (each of which has no beginning or end but extends indefinitely in both directions) with various wavelengths, choosing the wavelengths and the amplitude of each component wave so that constructive interference occurs only in a small region of width Δx, as shown in the figure, and the interference is destructive everywhere else. It turns out that a rather broad wave packet can be obtained by superposing waves with a relatively small range of wavelengths; but to make a narrow packet requires a wider range of wavelengths. But a range of wavelengths also means a corresponding range of values of momentum, because of the de Broglie relation. So again we see that a small uncertainty in position must be accompanied by a large uncertainty in momentum, and conversely.

There is also an uncertainty principle involving *energy*. It turns out that the energy of a system as well as its position and momentum always has uncertainty. The uncertainty ΔE is found to depend on the time interval Δt during which the system remains in the given state. The relation is

$$\Delta E \, \Delta t \geq \frac{h}{2\pi}.$$

(42–17)

Thus a system that remains in a certain state for a very long time can have a very well-defined energy, but if it remains in that state for only a short time, the uncertainty in energy must be correspondingly greater.

EXAMPLE   A sodium atom in one of the "resonance levels" shown in Fig. 41–5 remains in that state for an average time of 1.6×10^{-8} s before making a transition to the ground state by emitting a photon of wavelength 589 nm and energy 2.109 eV. What is the uncertainty in energy of the resonance level?

Solution   From Eq. (42–17),

$$\Delta E = \frac{h}{2\pi \, \Delta t} = \frac{(6.626 \times 10^{-34} \text{ J} \cdot \text{s})}{(2\pi)(1.6 \times 10^{-8} \text{ s})}$$
$$= 6.59 \times 10^{-27} \text{ J} = 4.11 \times 10^{-8} \text{ eV}.$$

The atom remains an indefinitely long time in the ground state, so there is *no* uncertainty there; the uncertainty of the resonance level energy and of the photon energy amounts to about two parts in 10^8. This irreducible uncertainty is called the *natural line width* of this particular spectrum line. Ordinarily the natural line width is much smaller than line broadening from other causes such as collisions among atoms.   ◀

*42–5

WAVE FUNCTIONS

As we have seen, the dual wave–particle nature of electrons and other fundamental particles requires generalization of the kinematic language used to describe the position and motion of a particle. The classical notion of a particle as a point having at each instant a definite position in space (described by three coordinates) and a definite velocity (described by three components) must be replaced by a more general language.

In making the needed generalizations we are guided by the language of classical wave motion. In Chapter 21, in studying transverse waves on a string, we decribed the motion of the string by specifying the position of each point on the string at each instant of time. This was done by means of a *wave function*, introduced in Section 21–7. If y represents the displacement from equilibrium at time t of a point on the string whose equilibrium position is a distance x from the origin, then the function $y = f(x, t)$ or, more briefly, $y(x, t)$ represents the displacement of point x at time t. If we know the wave function for a given motion, we know everything there is to know about the motion; from the function the shape of the string at any time, the slope at each point, the velocity and acceleration of each point, and any other needed information can be obtained.

Similarly, in Chapter 23 we discussed a sound wave propagating in the x-direction. Letting p represent the variation in air pressure from its equilibrium value at any point, we write $p(x, t)$ as the pressure variation at any point x at any time t; again this is a *wave function*. If the wave is three-dimensional, we can describe p at a space point with coordinates (x, y, z) at any time t by means of a wave function $p(x, y, z, t)$, which contains all the space coordinates and time. The same pattern reappears once more in the description of *electromagnetic* waves in Section 35–5, where we use two wave functions to describe the electric and magnetic fields at any point in space, at any time.

Thus it is natural to use a wave function as the central element of our generalized language for describing particles. The symbol usually used for this wave function is Ψ, and it is, in general, a function of all the space coordinates and time. Just as the wave function $y(x, t)$ for mechanical waves on a string provides a complete description of the motion, the wave function $\Psi(x, y, z, t)$ for a particle contains all the information that can be known about the particle.

Two questions immediately arise. First, what is the *meaning* of the wave function Ψ for a particle? Second, how is Ψ determined for any given physical situation? Our answers to both these questions must be qualitative and incomplete. The wave function describes the distribution of the particle in space. It is related to the *probability* of finding the particle in each of various regions; the particle is most likely to be found in regions where Ψ is large, and so on. If the particle has charge, the wave function can be used to find the *charge density* at any point in space. In addition, from Ψ one can calculate the *average* position of the particle, its average velocity, and dynamic quantities such as momentum, energy, and angular momentum. The required techniques are far beyond the scope of this discussion, but they are well established and no longer subject to any reasonable doubt.

The answer to the second question is that the wave function must be one of a set of solutions of a certain differential equation called the *Schrö-dinger equation*, developed by Erwin Schrödinger in 1925. One can set up a Schrödinger equation for any given physical situation, such as the electron in a hydrogen atom; the functions that are solutions of this equation represent various possible physical states of the system. Furthermore, it turns out for some systems that acceptable solutions exist only when some physical quantity such as the energy of the system has certain special values. Thus the solutions of the Schrödinger equation are also associated with *energy levels*. This discovery is of the utmost importance; before the discovery of the Schrödinger equation there was no way to predict energy levels from any fundamental theory, except for the very limited success of the Bohr model for hydrogen.

Soon after its discovery the Schrödinger equation was applied to the problem of the hydrogen atom. The predicted energy levels turned out to be identical to those from the Bohr model, Eq. (42–6), and thus to agree with experimental values from spectrum analysis. The energy levels are labeled with the quantum number n. In addition, the solutions have *quantized* values of angular momentum; that is, only certain discrete values of the magnitude of angular momentum and its components are possible. We recall that angular momentum quantization was put into the Bohr model as an *ad hoc* assumption with no fundamental justification; with the Schrödinger equation it comes out automatically!

Specifically, it is found that the magnitude L of the angular momentum of an electron in the hydrogen atom in a state with energy E_n and quantum number n must be given by

$$L = \sqrt{l\,(l + 1)}\left(\frac{h}{2\pi}\right), \qquad (42\text{–}18)$$

where l is zero or a positive integer no larger than $n - 1$. The *component* of L in a given direction, say the z-component L_z, can have only the set of values

$$L_z = mh/2\pi, \qquad (42\text{–}19)$$

where m can be zero or a positive or negative integer up to but no larger than l.

The quantity $h/2\pi$ appears so often in quantum mechanics that it is given a special symbol, $\hbar$. That is,

$$\hbar = \frac{h}{2\pi} = 1.054 \times 10^{-34}\,\text{J} \cdot \text{s}.$$

In terms of $\hbar$, the two preceding equations become

$$L = \sqrt{l(l + 1)}\,\hbar \qquad (l = 0, 1, 2, \ldots, n - 1), \qquad (42\text{–}20)$$

and

$$L_z = m\hbar \qquad (m = 0, \pm 1, \pm 2, \ldots, \pm l). \qquad (42\text{–}21)$$

We note that the component L_z can never be quite as large as L. For example, when $l = 4$ and $m = 4$, we find

$$L = \sqrt{4(4 + 1)}\,\hbar = 4.47\hbar,$$

$$L_z = 4\hbar.$$

This inequality arises from the uncertainty principle, which makes it impossible to know the *direction* of the angular momentum vector with complete certainty. Thus the component of $\mathbf{L}$ in a given direction can never be quite as large as the magnitude L, except when $l = 0$ and both L and L_z are zero. Unlike the Bohr model, the Schrödinger equation gives values for the magnitude L of angular momentum that are *not* integer multiples of $\hbar$.

Another interesting feature of Eqs. (42–20) and (42–21) is that there are states for which the angular momentum is *zero*. This is a result that has no classical analog; in the Bohr model the electron always moved in an orbit and thus had nonzero angular momentum; but in the new mechanics we find states having zero angular momentum.

The possible wave functions for the hydrogen atom can be labeled according to the values of the three integers n, l, and m, called, respectively, the *principal* quantum number, the *angular momentum* quantum number, and the *magnetic* quantum number. For each energy level E_n there are several distinct states with the same energy but different values of l and m, the only exception being the ground state $n = 1$, for which only $l = 0$, $m = 0$ is possible. Examination of the spatial extent of the wave functions shows that, in each case, the wave function is confined primarily to a region of space around the nucleus having a radius of the same order of magnitude as the corresponding Bohr radius. For increasing values of n the electron is, on the average, farther away from the nucleus and hence has less negative potential energy and becomes less tightly bound.

The new mechanics described above is called *quantum mechanics;* it is much more complex, both conceptually and mathematically, than newtonian mechanics. To a reader who yearns for the comparative simplicity of newtonian mechanics, we can only reply that quantum mechanics enables us to understand physical phenomena and to analyze physical problems for which classical mechanics is completely powerless. Added complexity is the price we pay for this greatly expanded understanding.

42–6

ELECTRON SPIN

One additional concept, that of *electron spin*, is needed for the analysis of more complex atoms, and even for certain details of the spectrum of hydrogen. To illustrate this concept, we consider the motion of the earth around the sun. The earth travels in a nearly circular orbit and at the same time *rotates* on its axis. There is angular momentum associated with each motion, called the *orbital* and *spin* angular momentum, respectively, and the total angular momentum of the system is the sum of the two. If we were to model the earth as a single point, no spin angular momentum would be possible; but a more refined model including an earth of finite size includes the possibility of spin angular momentum.

This discussion can be translated to the language of the Bohr model. Suppose the electron is not a point charge moving in an orbit but a small spinning sphere in orbit. Then again there is additional angular momentum associated with the spin motion. Because the sphere carries an electric charge, the spinning motion leads to current loops and to a magnetic moment, as discussed in Section 30–8. If this magnetic moment really exists, then when the atom is placed in a magnetic field there is an interaction with the field. Associated with the interaction is a potential energy that causes shifts in the energy levels of the atom and thus in the wavelengths of the spectrum lines.

Such shifts *are* indeed observed in precise spectroscopic analysis; this and other experimental evidence have shown conclusively that the electron *does* have angular momentum and magnetic moment that are not related to the orbital motion but are intrinsic to the particle itself. Like orbital angular momentum, spin angular momentum (usually denoted by S) is found to be *quantized*. Denoting the z-component of S by S_z, we find that the only possible values are

$$S_z = \pm\tfrac{1}{2}\hbar. \qquad (42\text{–}22)$$

This relation is reminiscent of Eq. (42–21) for the z-component of orbital angular momentum, but the component is one half of $\hbar$ instead of an *integer* multiple.

In quantum mechanics, where the Bohr orbits are superseded by wave functions, it is not so easy to picture electron spin. If the wave functions are visualized as clouds surrounding the nucleus, then one can imagine many tiny arrows distributed through the cloud, all pointing in the same direction, either all $+z$ or all $-z$. Of course, this picture should not be taken too seriously; there is no hope of actually seeing an atom's structure because it is thousands of times smaller than wavelengths of light and because interactions with light photons would seriously disturb the very structure one is trying to observe.

In any event, the concept of electron spin is well established by a variety of experimental evidence. To label completely the state of the electron in a hydrogen atom, we now need a fourth quantum number s, which specifies the electron spin orientation. If s can take the values $+1$ or -1, then the z-component of spin angular momentum is given by

$$S_z = \tfrac{1}{2}s\hbar. \tag{42–23}$$

The corresponding component of magnetic moment, which we may denote by μ_z, turns out to be related to S_z, in a simple way:

$$\mu_z = \frac{e}{m_e}S_z, \tag{42–24}$$

where e and m_e are the charge and mass of the electron, respectively. The ratio of μ to S, in this case (e/m_e), is called the *gyromagnetic ratio*.

In this connection it should be mentioned that there is also magnetic moment associated with the *orbital* motion of the electron; in this case the gyromagnetic ratio is found to be just *half* as great as for electron spin, $e/2m_e$. Thus the z-component of magnetic moment associated with the orbital angular momentum component L_z is given by

$$\mu_z = \frac{e}{2m_e}L_z = m\frac{e\hbar}{2m_e}. \tag{42–25}$$

Care must be taken to distinguish between the electron mass m_e and the magnetic quantum number m.

*42–7

THE ZEEMAN EFFECT

When an atom is placed in a magnetic field, there is a potential energy associated with the interaction of the magnetic moment with the field. This interaction was mentioned in Secton 42–1 in connection with the Bohr model of the hydrogen atom. The interaction energy is given by Eq. (30–15), with magnetic moment denoted by μ instead of m. That is, $U = -\mu B \cos \alpha$. Suppose B is directed along the z-axis; then $\mu \cos \alpha$ is the component of magnetic moment along the z-direction; $\mu_z = \mu \cos \alpha$. Then, using Eq. (42–25), we find

$$U = -\mu_z B = -m\frac{e\hbar B}{2m_e}. \tag{42–26}$$

The effect of the magnetic field is to shift the energy level by an amount U, given by Eq. (42–26), which depends on the orientation of the angular momentum, that is, on the value of the magnetic quantum number m. A level having a certain value of the quantum number l is thus split into $2l + 1$ levels, with adjacent levels differing in energy by $e\hbar B/2m_e$. Spectrum lines corresponding to transitions involving this level are correspondingly split and appear as a series of closely spaced spectrum lines replacing a single line. This effect is called the *Zeeman effect*, in honor of the Dutch physicist Pieter Zeeman, who first observed it in 1896. The effect had been anticipated by Faraday, who looked for it as early as 1862; the resolution of Faraday's spectrometer was not great enough to reveal the line splitting. The Zeeman effect is a very direct experimental confirmation of the quantization of angular momentum. Zeeman was awarded the Nobel prize in Physics (along with H. A. Lorentz) in 1904 for this work.

EXAMPLE   An atom in a state having $l = 1$ emits a photon with wavelength 600 nm as it decays to its ground state (with $l = 0$). If the atom is placed in a magnetic field of magnitude $B = 1.0$ T, determine the shifts in the energy level and in the wavelength, if the effect of electron spin can be ignored (which is not actually the case).

Solution   The energy of a 600-nm photon, as obtained in the example of Section 41–4, is 3.31×10^{-19} J or 2.07 eV. The ground-state level has $l = 0$ and is not split by the field. The splitting of adjacent levels in the $l = 1$ state is given by Eq. (42–26):

$$U = -m \frac{(1.60 \times 10^{-19} \text{ C})(1.05 \times 10^{-34} \text{ J} \cdot \text{s})(1.0 \text{ T})}{2(9.11 \times 10^{-31} \text{ kg})}$$

$$= -m(9.27 \times 10^{-24} \text{ J}) = -m(5.79 \times 10^{-5} \text{ eV}).$$

When $l = 1$, the possible values of m are -1, 0, and $+1$, and the three resulting levels are split by equal intervals of 5.79×10^{-4} eV. This is a small fraction of the photon energy: $(5.79 \times 10^{-4}$ eV$)/(2.07$ eV $= 2.80 \times 10^{-4})$. Thus we expect the corresponding *wavelength* shifts to be approximately $(2.80 \times 10^{-4})(600 \text{ nm}) = 0.17$ nm, and the original 600 nm is split into a triplet with wavelengths 599.83, 600.00, and 600.17 nm. This splitting is well within the limit of resolution of modern spectrometers.   ◀

Somewhat more subtle is the level-splitting caused by the electron spin magnetic moment even when there is no external field. In the Bohr model, an observer moving with the electron sees the positively charged nucleus moving around him; this moving charge causes a magnetic field at the location of the electron, and the resulting interaction energy with the spin magnetic moment causes a twofold splitting of this level, corresponding to the two possible orientations of electron spin. Although the Bohr model is now known to be inadequate, a similar result can be derived from a more complete quantum-mechanical treatment. The effect is called *spin-orbit coupling;* it is responsible for the two closely spaced "resonance levels" of sodium shown in Fig. 41–5, and for the familiar doublet (589.0, 589.6 nm) in the spectrum of sodium.

The various line splittings resulting from these magnetic interactions are collectively called *fine structure.* There are further, much smaller splittings resulting from the fact that the nucleus of the atom also has a magnetic moment, and that it interacts with the orbital and spin angular momentum of the electrons. These effects are called *hyperfine structure.*

QUESTIONS

42–1 In analyzing the absorption spectrum of hydrogen at room temperature, one finds absorption lines corresponding to wavelengths in the Lyman series but not to those in the Balmer series. Why not?

42–2 A singly ionized helium atom has one of its two electrons removed, and the energy levels of the remaining electron are closely related to those of the hydrogen atom. The nuclear charge for helium is $+2e$ instead of just $+e$;

exactly how are the energy levels related to those of hydrogen? How is the size of the ion in the ground state related to that of the hydrogen atom?

42–3 Consider the line spectrum emitted from a gas discharge tube such as a neon sign or a sodium-vapor or mercury-vapor lamp. It is found that when the pressure of the vapor is increased the spectrum lines spread out, i.e., are less sharp and less monochromatc. Why?

42–4 Suppose a two-slit interference experiment is carried out using an electron beam. Would the same interference pattern result if one slit at a time is uncovered instead of both at once? If not, why not? Doesn't each electron go through one slit or the other? Or does every electron go through both slits? Does the latter possibility make sense?

42–5 Is the wave nature of electrons significant in the function of a television picture tube? For example, do diffraction effects limit the sharpness of the picture?

42–6 A proton and an electron have the same speed. Which has longer wavelength?

42–7 A proton and an electron have the same kinetic energy. Which has longer wavelength?

42–8 Does the uncertainty principle have anything to do with marksmanship? That is, is the accuracy with which a bullet can be aimed at a target limited by the uncertainty principle?

42–9 Is the Bohr model of the hydrogen atom consistent with the uncertainty principle?

42–10 If the energy of a system can have uncertainty, as stated by Eq. (42–17), does this mean that the principle of conservation of energy is no longer valid?

42–11 If quantum mechanics replaces the language of newtonian mechanics, why do we not have to use wave functions to describe the motion of macroscopic objects such as baseballs and cars?

42–12 Why is analysis of the helium atom much more complex than that of the hydrogen atom, either in a Bohr-type model or using the Schrödinger equation?

42–13 Do gravitational forces play a significant role in atomic structure?

PROBLEMS

$$e = 1.602 \times 10^{-19} \text{ C}$$
$$m_e = 9.109 \times 10^{-31} \text{ kg}$$
$$h = 6.626 \times 10^{-34} \text{ J} \cdot \text{s}$$
$$N_0 = 6.022 \times 10^{23} \text{ atoms} \cdot \text{mol}^{-1}$$
Energy equivalent of 1 u = 931.5 MeV
$$\frac{e}{m_e} = 1.758 \times 10^{11} \text{ C} \cdot \text{kg}^{-1}$$
$$k = 1.381 \times 10^{-23} \text{ J} \cdot \text{K}^{-1}$$
$$1 \text{ eV} = 1.602 \times 10^{-19} \text{ J}$$
$$1 \text{ u} = 1.661 \times 10^{-27} \text{ kg}$$
$$\epsilon_0 = 8.854 \times 10^{-12} \text{ C}^2 \cdot \text{N}^{-1} \cdot \text{m}^{-2}$$

42–1 For a hydrogen atom in the ground state, determine, in electron volts,

a) the kinetic energy,
b) the potential energy,
c) the total energy, and
d) the energy required to remove the electron completely.

42–2 Referring to Problem 42–1, what are the energies if the atom is a singly ionized helium atom (i.e., a helium atom with one electron removed)?

42–3 A singly ionized helium ion (a helium atom with one electron removed) behaves very much like a hydrogen atom, except that the nuclear charge is twice as great.

a) How do the energy levels differ in magnitude from those of the hydrogen atom?
b) Which spectral series for He$^+$ have lines in the visible spectrum?

42–4 Refer to the example in Section 42–1. Suppose a hydrogen atom makes a transition from the $n = 3$ state to the $n = 2$ state (the Balmer Hα line at 656.3 nm) while in a magnetic field of magnitude 2 T. If the magnetic moment of the atom is parallel to the field in both the initial and final states,

a) by how much is each energy level shifted from the zero-field value, and
b) by how much is the wavelength of the spectrum line shifted?

42–5 The negative μ-meson (or muon) has a charge equal to that of an electron but a mass about 207 times as great. Consider a hydrogenlike atom consisting of a proton and a muon.

a) What is the ground-state energy?
b) What is the radius of the $n = 1$ Bohr orbit?
c) What is the wavelength of the radiation emitted in the transition from the $n = 2$ state to the $n = 1$ state?

42–6 According to Bohr, the Rydberg constant R is equal to $me^4/8\epsilon_0^2 h^3 c$. Calculate R in m^{-1} and compare with the experimental value.

42–7

a) Show that the frequency of revolution of an electron in its circular orbit in the Bohr model of the hydrogen atom is $f = me^4/4\epsilon_0^2 n^3 h^3$.

b) Show that when n is very large, the frequency of revolution equals the radiated frequency calculated from Eq. (42–7) for a transition from $n + 1$ to n.

(This problem illustrates Bohr's *correspondence principle*, which is often used as a check on quantum calculations. When n is small, quantum physics gives results that are very different from those of classical physics. When n is large, the differences are not significant and the two methods then "correspond.")

42–8 A 10-kg satellite circles the earth once every 2 hr in an orbit having a radius of 8000 km.

a) Assuming that Bohr's angular-momentum postulate applies to satellites just as it does to an electron in the hydrogen atom, find the quantum number of the orbit of the satellite.

b) Show from Bohr's first postulate and Newton's law of gravitation that the radius of an earth-satellite orbit is directly proportional to the square of the quantum number, $r = kn^2$, where k is the constant of proportionality.

c) Using the result from part (b), find the distance between the orbit of the satellite in this problem and its next "allowed" orbit.

d) Comment on the possibility of observing the separation of the two adjacent orbits.

e) Do quantized and classical orbits correspond for this satellite? Which is the "correct" method for calculating the orbits?

42–9 A hydrogen atom initially in the ground state absorbs a photon, which excites it to the $n = 4$ state. Determine the wavelength and frequency of the photon.

42–10 In Problem 42–3,

a) how much energy would be required to remove the remaining electron completely from a helium ion?

b) What wavelength would a photon of this energy have?

c) In what region of the electromagnetic spectrum would it lie?

42–11 Calculate the energy of the longest-wavelength line in the Balmer series for hydrogen, and of the shortest-wavelength line.

42–12 Approximately what range of photon energies (in electron volts) corresponds to the visible spectrum? Approximately what range of wavelengths would electrons in this energy range have?

42–13 For crystal-diffraction experiments, wavelengths of the order of 0.1 nm are often appropriate. Find the energy, in electron volts, for a particle with this wavelength if the particle is

a) a photon, and

b) an electron.

42–14

a) An electron moves with a velocity of 3×10^8 cm·s^{-1}. What is its de Broglie wavelength?

b) A proton moves with the same velocity. Determine its de Broglie wavelength.

42–15 The average kinetic energy of a thermal neutron is $(3/2)kT$. What is the de Broglie wavelength associated with the neutrons in thermal equilibrium with matter at 300 K? (The mass of a neutron is approximately 1 u.)

42–16

a) What is the de Broglie wavelength of an electron that has been accelerated through a potential difference of 200 V?

b) Would this electron exhibit particlelike or wavelength characteristics on meeting an obstacle or opening 1 mm in diameter?

42–17 What is the de Broglie wavelength of an electron accelerated through 20,000 V, a typical voltage for color-television picture tubes?

42–18 Suppose the uncertainty of position of an electron is equal to the radius of the $n = 1$ Bohr orbit, about 0.5×10^{-10} m. Find the uncertainty in its momentum, and compare this with the magnitude of the momentum of the electron in the $n = 1$ Bohr orbit.

42–19 Suppose that the uncertainty in position of a particle is equal to its de Broglie wavelength; show that in this case the uncertainty in its momentum is equal to its momentum.

42–20 In a certain television picture tube the accelerating voltage is 20,000 V and the electron beam passes through an aperture 0.5 mm in diameter to a screen 0.3 m away.

a) What is the uncertainty in position of the point where the electrons strike the screen?

b) Does this uncertainty affect the clarity of the picture significantly?

42–21 If the Bohr model applies to the motion of the earth around the sun as well as to the motion of an electron around a proton, find the quantum number for the earth's motion.

42–22 The radii of atomic nuclei are of the order of 10^{-15} m. Find the uncertainty of speed of an electron if it is confined within a nucleus. On the basis of this result, is it likely that there could be electrons within the nucleus?

42-23 A certain atom has an energy level 2.0 eV above the ground state. When excited to this state it remains on the average 2.0×10^{-6} s before emitting a photon and returning to the ground state.

a) What is the energy of the photon? Its wavelength?

b) What is the uncertainty in energy of the photon? Of its wavelength?

42-24 The π^0 meson is an unstable particle produced in high-energy particle collisions. Its mass is about 264 times that of the electron, and it exists for an average lifetime of 0.8×10^{-16} s before decaying into two gamma-ray photons. Assuming that the mass and energy of the particle are related by the Einstein relation $E = mc^2$, find the uncertainty in the mass of the particle, and express it as a fraction of the mass.

42-25 Suppose an unstable particle produced in a high-energy collision has a mass three times that of the proton and an uncertainty in mass that is 1% of the particle's mass. Assuming mass and energy are related by $E = mc^2$, find the lifetime of the particle.

42-26 A hydrogen atom is placed in a magnetic field of magnitude 1.2 T. Find the interaction energy with the field due to the electron spin.

42-27 Make a chart showing all the possible sets of quantum numbers for the states of the electron in the hydrogen atom when $n = 3$.

42-28 Show that the total number of hydrogen-atom states for a given value of the principal quantum number n is $2n^2$.

43

ATOMS, MOLECULES, AND SOLIDS

The principles of quantum mechanics introduced in the preceding chapter provide the basis for understanding a wide variety of phenomena associated with the structure of atoms, molecules, and solid materials. One additional principle is needed, the *exclusion principle*, which forbids two electrons to occupy the same quantum-mechanical state. With this principle the most important features of the structure and chemical behavior of multi-electron atoms, including the periodic table of the elements, can be derived. The nature of *molecular bonds*, by which two or more atoms combine in a stable structure, can be understood on the basis of the electron configurations of the atoms. Molecular *spectra* arise from transitions among vibrational and rotational energy states of molecules. Interatomic forces are also responsible for the large-scale binding of atoms into solid structures. Several different types of interatomic interactions are possible, and many properties of solid materials can be understood at least qualitatively on the basis of the types of bonding that are present. *Semiconductors*, a particular class of solid materials, are discussed in some detail because of their inherent interest and their great practical importance in present-day technology.

43–1

THE EXCLUSION PRINCIPLE

The hydrogen atom, discussed in Chapter 42, is the simplest of all atoms, containing one electron and one proton. Analysis of atoms with more than one electron increases in complexity very rapidly; each electron interacts not only with the positively charged nucleus but also with all the other electrons. In principle the motion of the electrons is governed by the Schrödinger equation, mentioned in Section 42–5, but the mathematical problem of finding appropriate solutions of this equation is so complex that it has not been accomplished exactly even for the helium atom, with two electrons.

Various approximation schemes can be used; the simplest (and most drastic) is to ignore completely the interactions between electrons, and regard each electron as influenced only by the electric field of the nucleus, considered to be a point charge. A less drastic and more useful approximation is to think of all the electrons together as making up a charge cloud that is, on the average, *spherically symmetric,* and to think of each individual electron as moving in the total electric field due to the nucleus and this averaged-out electron cloud. This is called the *central-field* approximation; it provides a useful starting point for the understanding of atomic structure.

An additional principle is also needed, the *exclusion principle.* To understand the need for this principle, we consider the lowest-energy state or *ground state* of a many-electron atom. The central-field model suggests that each electron has a lowest energy state (roughly corresponding to the $n = 1$ state for the hydrogen atom). It might be expected that, in the ground state of a complex atom, all the electrons should be in this lowest state. If this is the case, then when we examine the behavior of atoms with increasing numbers of electrons, we should find gradual changes in physical and chemical properties of elements as the number of electrons in the atoms increases.

A variety of evidence shows conclusively that this is *not* what happens at all. For example, the elements of fluorine, neon, and sodium have, respectively, 9, 10, and 11 electrons per atom. Fluorine is a halogen and tends strongly to form compounds in which each atom acquires an extra electron. Sodium, an alkali metal, forms compounds in which it *loses* an electron, and neon is an inert gas, forming no compounds at all. This and many other observations show that in the ground state of a complex atom the electrons *cannot* all be in the lowest energy states.

The key to this puzzle, discovered by the Swiss physicist Wolfgang Pauli in 1925, is called the *Pauli exclusion principle.* Briefly, it states that *no two electrons can occupy the same quantum-mechanical state.* Since different states correspond to different spatial distributions, including different distances from the nucleus, this means that in a complex atom there is not room for all the electrons in states near the nucleus; some are forced into states farther away, having higher energies.

To apply the Pauli principle to atomic structure, we first review some results quoted in Section 42–5. The quantum-mechanical state of the electron in the hydrogen atom is identified by the four quantum numbers n, l, m, and s, which determine the energy, angular momentum, and components of orbital and spin angular momentum in a particular direction. It turns out that this scheme can also be used when the electron moves not in the electric field of a point charge, as in the hydrogen atom, but in the electric field of *any spherically symmetric charge distribution,* as in the central-field approximation. One important difference is that the energy of a state is no longer given by Eq. (42–6); in general the energy corresponding to a given set of quantum numbers depends on *both* n and l, usually increasing with increasing l for a given value of n.

We can now make a list of the possible sets of quantum numbers and thus of the possible states of electrons in an atom. Such a list is given in Table 43–1, which also indicates two alternative notations. It is custom-

TABLE 43–1

n	l	m	Spectroscopic notation	Maximum number of electrons	Shell
1	0	0	$1s$	2	K
2	0	0	$2s$	2 $\big\}$ 8	L
2	1	-1			
2	1	0	$2p$	6	
2	1	1			
3	0	0	$3s$	2	M
3	1	-1			
3	1	0	$3p$	6	
3	1	1			
3	2	-2			18
3	2	-1			
3	2	0	$3d$	10	
3	2	1			
3	2	2			
4	0	0	$4s$	2	N
4	1	-1			
4	1	0	$4p$	6 $\big\}$ 32	
4	1	1			
	etc.				

ary to designate the value of l by a letter, according to this scheme:

$$l = 0: \quad s \text{ state}$$
$$l = 1: \quad p \text{ state}$$
$$l = 2: \quad d \text{ state}$$
$$l = 3: \quad f \text{ state}$$
$$l = 4: \quad g \text{ state}$$

The origins of these letters are rooted in the early days of spectroscopy, and need not concern us. A state for which $n = 2$ and $l = 1$ is called a $2p$ state, and so on, as shown in Table 43–1. This table also shows the relation between values of n and the x-ray levels ($K, L, M, \ldots$) described in Section 41–7. The $n = 1$ levels are designated as K, $n = 2$ as L, and so on. Because the average electron distance from the nucleus increases with n, each value of n corresponds roughly to a region of space around the nucleus in the form of a spherical shell. Hence, one speaks of the *L shell* as that region occupied by the electrons in the $n = 2$ states, and so on. States with the same n but different l are said to form *subshells*, such as the $3p$ subshell.

We are now ready for a more precise statement of the exclusion principle: *In any atom, only one electron can occupy any given quantum state.* That is, no two electrons in an atom can have all four quantum numbers the same. Since each quantum state corresponds to a certain distribution of the electron "cloud" in space, the principle says in effect: "Not more than two electrons (with opposite values of the spin quantum number s) can occupy the same region of space." This statement must be interpreted rather broadly, since the clouds and associated wave functions describing electron distributions do not have definite, sharp boundaries; but the exclusion principle limits the degree of overlap of electron wave functions that is permitted. The maximum numbers of electrons in each shell and subshell are shown in Table 43–1.

The exclusion principle plays an essential role in the understanding of the structure of complex atoms. In the next section we shall see how the periodic table of the elements can be understood on the basis of this principle.

43–2

ATOMIC STRUCTURE

The number of electrons in an atom in its normal state is called the *atomic number*, denoted by Z. The nucleus contains Z protons and some number of neutrons. The proton and electron charges have the same magnitude but opposite sign, so in the normal atom the net electric charge is zero. Because the electrons are attracted to the nucleus, we expect the quantum states corresponding to regions near the nucleus to have lowest energy. We may imagine starting with a bare nucleus with Z protons, and adding electrons one by one until the normal complement of Z electrons for a neutral atom is reached. We expect the lowest-energy states, ordinarily those with the smallest values of n and l, to fill first, and we use successively higher states until all electrons are accommodated.

The chemical properties of an atom are determined principally by interactions involving the outermost electrons, so it is of particular interest to find out how these are arranged. For example, when an atom has one electron considerably farther from the nucleus (on the average) than the others, this electron will be rather loosely bound; the atom will tend to lose this electron and form what chemists call an *electrovalent* or *ionic* bond, with valence +1. This behavior is characteristic of the alkali metals lithium, sodium, potassium, and so on.

We now proceed to describe ground-state electron configurations for the first few atoms (in order of increasing Z). For hydrogen the ground state is 1s; the single electron is in the state $n = 1$, $l = 0$, $m = 0$, and $s = \pm 1$. In the helium atom (Z = 2) *both* electrons are in 1s states, with opposite spins; this state is denoted as $1s^2$. For helium, the K shell is completely filled and all others are empty.

Lithium (Z = 3) has three electrons; in the ground state two are in the 1s state, and one in a 2s state. We denote this state as $1s^2 2s$. On the average, the 2s electron is considerably farther from the nucleus than the 1s electrons, as shown schematically in Fig. 43–1. Thus, according to Gauss's law, the *net* charge influencing the 2s electron is +e, rather than +3e as it would be without the 1s electrons present. Thus the 2s electron is loosely bound, as the chemical behavior of lithium suggests. An alkali metal, it forms ionic compounds with a valence of +1, in which each atom loses an electron.

Next is beryllium (Z = 4); its ground-state configuration is $1s^2 2s^2$, with two electrons in the L shell. Beryllium is the first of the *alkaline-earth* elements, forming ionic compounds with a valence of +2.

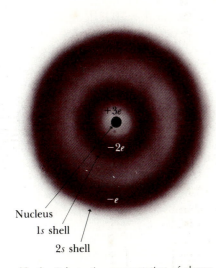

43–1 Schematic representation of charge distribution in lithium atom. The nucleus has a charge 3e; the two 1s electrons are closer to the nucleus than the 2s electron, which moves in a field approximately equal to that of a point charge 3e − 2e, or simply e.

		TABLE 43–2 GROUND-STATE ELECTRON CONFIGURATIONS	
Element	**Symbol**	**Atomic number (Z)**	**Electron configuration**
Hydrogen	H	1	$1s$
Helium	He	2	$1s^2$
Lithium	Li	3	$1s^2 2s$
Beryllium	Be	4	$1s^2 2s^2$
Boron	B	5	$1s^2 2s^2 2p$
Carbon	C	6	$1s^2 2s^2 2p^2$
Nitrogen	N	7	$1s^2 2s^2 2p^3$
Oxygen	O	8	$1s^2 2s^2 2p^4$
Fluorine	F	9	$1s^2 2s^2 2p^5$
Neon	Ne	10	$1s^2 2s^2 2p^6$
Sodium	Na	11	$1s^2 2s^2 2p^6 3s$
Magnesium	Mg	12	$1s^2 2s^2 2p^6 3s^2$
Aluminum	Al	13	$1s^2 2s^2 2p^6 3s^2 3p$
Silicon	Si	14	$1s^2 2s^2 2p^6 3s^2 3p^2$
Phosphorus	P	15	$1s^2 2s^2 2p^6 3s^2 3p^3$
Sulfur	S	16	$1s^2 2s^2 2p^6 3s^2 3p^4$
Chlorine	Cl	17	$1s^2 2s^2 2p^6 3s^2 3p^5$
Argon	Ar	18	$1s^2 2s^2 2p^6 3s^2 3p^6$
Potassium	K	19	$1s^2 2s^2 2p^6 3s^2 3p^6 4s$
Calcium	Ca	20	$1s^2 2s^2 2p^6 3s^2 3p^6 4s^2$

Table 43–2 shows the ground-state electron configurations of the first twenty elements. The L shell can hold a total of eight electrons, as the reader should verify from the rules in Section 43–1; at $Z = 10$ the K and L shells are filled and there are no electrons in the M shell. This is expected to be a particularly stable configuration, with little tendency to gain or lose electrons, and in fact this element is neon, an inert gas with no known compounds. The next element after neon is sodium ($Z = 11$), with filled K and L shells and one electron in the M shell. Thus its "filled-shell-plus-one-electron" structure resembles that of lithium; both are alkali metals. The element *before* neon is fluorine, with $Z = 9$. It has a vacancy in the L shell and might be expected to have an affinity for an electron, forming ionic compounds with a valence of -1. This behavior is characteristic of the *halogens* (fluorine, chlorine, bromine, iodine, astatine), all of which have "filled-shell-minus-one" configurations.

By similar analysis, one can understand all the regularities in chemical behavior exhibited by the periodic table of the elements, on the basis of electron configurations. With the M and N shells there is a slight complication because the $3d$ and $4s$ subshells ($n = 3$, $l = 2$, and $n = 4$, $l = 0$, respectively) overlap in energy. Thus argon ($Z = 18$) has all the $1s$, $2s$, $2p$, $3s$, and $3p$ states filled, but in potassium ($Z = 19$) the additional electron goes into a $4s$ rather than $3d$ level. The next several elements have one or two electrons in the $4s$ states and increasing numbers in the $3d$ states. These elements are all metals with rather similar properties, and form the first *transition series*, starting with scandium ($Z = 21$) and ending with zinc ($Z = 30$), for which the $3d$ levels are filled.

Thus, the similarity of elements in each *group* of the periodic table reflects corresponding similarity in electron configuration. All the inert gases (helium, neon, argon, krypton, xenon, and radon) have filled-shell configurations. All the alkali metals (lithium, sodium, potassium, rubidium, cesium, and francium) have "filled-shell-plus-one" configurations. All the alkaline-earth metals (beryllium, magnesium, calcium, strontium, barium, and radium) have "filled-shell-plus-two" configurations, and all the halogens (fluorine, chlorine, bromine, iodine, and astatine) have "filled-shell-minus-one" structures. And so it goes.

This whole theory can, of course, be refined to account for differences between elements in a group and to account for various aspects of chemical behavior. Indeed, some physicists like to claim that all of chemistry is contained in the Schrödinger equation! This statement may be a bit extreme, but we can see that even this qualitative discussion of atomic structure takes us a considerable distance toward understanding the atomic basis of many chemical phenomena. Next we shall examine in somewhat more detail the nature of the chemical bond.

43–3

DIATOMIC MOLECULES

As indicated in Section 43–2, the study of electron configurations in atoms provides valuable insight into the nature of the chemical bond, that is, the interaction that holds atoms together to form stable structures such as molecules and crystalline solids. There are a number of types of chemical bonds; the simplest to understand is the *ionic* bond, also called the *electrovalent* or *heteropolar* bond. The most familiar example is sodium chloride (NaCl), in which the sodium atom gives its $3s$ electron to the chlorine atom, filling the vacancy in the $3p$ subshell of chlorine. Energy is required to make this transfer if the atoms are far apart, but the result is two ions, one positively charged and one negatively, that attract each other. As they come together, their potential energy decreases, so that the final bound state of Na^+Cl^- has lower total energy than the state in which the two atoms are separated and neutral.

Removing the $3s$ electron from the sodium requires 5.1 eV of energy; this is called the *ionization energy* or *ionization potential*. Chlorine actually has an electron *affinity* of 3.8 eV. That is, the neutral chlorine atom can attract an extra electron that, once it takes its place in the $3p$ level, requires 3.8 eV for its removal. Thus, creating the separate Na^+ and Cl^- ions requires a net expenditure of only $5.1 - 3.8 = 1.3$ eV. The negative potential energy associated with the mutual attraction of the ions is determined by the closeness to which they can approach each other; this in turn is determined by the exclusion principle, which forbids extensive overlap of the electron clouds of the two atoms. The minimum potential energy turns out to be -5.5 eV at a separation of 0.24 nm; at distances less than this, the interaction becomes repulsive. Thus the net energy involved in creating the ions and letting them come together to the equilibrium separation of 0.24 nm is $(-5.5 + 1.3)$ or -4.2 eV, and this is the binding energy of the molecule. To put it another way, 4.2 eV of energy would have to be added to dissociate the molecule into the separate atoms.

Ionic bonds can involve more than one electron per atom. The alkaline-earth elements form ionic compounds in which each atom loses *two*

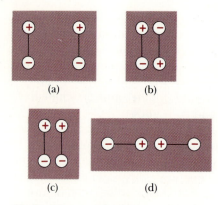

43–2 When two dipoles (a) are brought together, the interaction may be (b) attractive, or (c and d) repulsive.

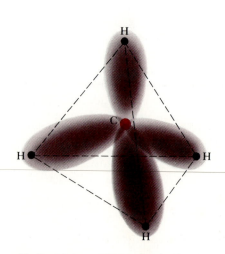

43–3 Methane (CH_4) molecule, showing four covalent bonds. The electron cloud between the central carbon atom and each of the four hydrogen nuclei represents the two electrons of a covalent bond. The hydrogen nuclei are at the corners of a regular tetrahedron.

electrons; an example is $Mg^{++}Cl_2^-$. Loss of more than two electrons is relatively rare, and instead a *different kind of bond* comes into operation.

The *covalent* or *homopolar* bond is characterized by a more nearly symmetric participation of the two atoms, as contrasted with the complete asymmetry involved in the electron-transfer process of the ionic bond. The simplest example of the covalent bond is the hydrogen molecule, a structure containing two protons and two electrons. As a preliminary to understanding this bond we consider first the interaction of two electric dipoles, as in Fig. 43–2. In (a) the dipoles are far apart; in (b) the like charges are farther apart than the unlike charges, and there is a net *attractive* force. In (c) and (d) the interaction is repulsive.

In a molecule the charges are, of course, not at rest, but Fig. 43–2b at least makes it seem plausible that if the electrons are in the region between the protons, the attractive force the electrons exert on each proton may more than counteract the repulsive interactions of the protons on each other and the electrons on each other. That is, in the covalent bond in the hydrogen molecules, the attractive interaction is supplied by a pair of electrons, one contributed by each atom, whose charge clouds are concentrated primarily in the region between the two atoms. Thus this bond may also be thought of as a shared-electron or electron-pair bond.

According to the exclusion principle, two electrons can occupy the same region of space only when they have opposite spin orientations. When the spins are parallel, the state that would be most favorable from energy considerations is forbidden by the exclusion principle, and the lowest-energy state permitted is one in which the electron clouds are concentrated *outside* the central region between atoms. The nuclei then repel each other and the interaction is repulsive rather than attractive. Thus, opposite spins are an essential requirement for an electron-pair bond, and no more than two electrons can participate in such a bond.

This is not to say, however, that an atom cannot have several electron-pair bonds. On the contrary, an atom having several electrons in its outermost shell can form covalent bonds with several other atoms. The bonding of carbon and hydrogen atoms, of central importance in organic chemistry, is an example. In the *methane* molecule (CH_4) the carbon atom is at the center of a regular tetrahedron, with a hydrogen atom at each corner. The carbon atom has four electrons in its L shell, and one of these electrons forms a covalent bond with each of the four hydrogen atoms, as shown in Fig. 43–3. Similar patterns occur in more complex organic molecules.

Ionic and covalent bonds represent two extremes in the nature of molecular bonds, but there is no sharp division between the two types. In many situations there is a *partial* transfer of one or more electrons (corresponding to a greater or smaller distortion of the electron wave functions) from one atom to another. As a result, many molecules having dissimilar atoms have electric dipole moments, that is, a preponderance of positive charge at one end and of negative at the other. Such molecules are said to be *polar*. Water molecules have large dipole moments that are responsible for the exceptionally large dielectric constant of liquid water.

The bonds discussed above typically have energies in the range of 1 eV to 5 eV; they are called *strong bonds* to distinguish them from several types of much *weaker* bonds having energies typically of the order of 0.1 eV or less. One of these, the *van der Waals* bond, is an interaction between the electric dipole moments of two atoms or molecules. The existence of dipole moments in molecules has been mentioned above; as an example, the van der Waals interaction is chiefly responsible for the liquefaction and solidification of water.

Even when an atom or molecule has no permanent dipole moment, fluctuating charge densities in the interior of the structure can lead to fluctuating dipole moments; these in turn can induce dipole moments in neighboring structures. The resulting dipole-dipole interaction can be attractive and can lead to weak bonding of atoms or molecules. For example, the low-temperature liquefaction and solidification of such homopolar molecules as H_2, O_2, and N_2 and of the inert gases is due to interaction of the induced-dipole type. Not much energy of thermal agitation is needed to break these weak bonds, so these substances exist in the liquid and solid states only at very low temperatures.

A second type of important weak bond is the *hydrogen bond*, which can be regarded as the opposite of the covalent bond. In the latter, an electron pair serves to bind two positively charged structures together. In the hydrogen bond an ionized hydrogen atom, which is simply a bare proton, acts as the glue to bond two negatively charged structures. Again the bond energy is only about 0.1 eV, but nevertheless the hydrogen bond plays an essential role in many organic molecules, including, for example, cross-link bonding of the two strands of the famous double helix of the DNA molecule.

All these bond types play roles in the structure of *solids* as well as of molecules. Indeed, a solid is in many respects a gigantic molecule. Still another type of bonding, the *metallic bond,* comes into play in the structure of metallic solids. We return to this subject in Section 43–5.

43–4

MOLECULAR SPECTRA

The energy levels of atoms are associated with the kinetic and potential energies of electrons with respect to the nucleus. Energy levels of molecules have additional features resulting from relative motion of the nuclei of the atoms, and there are characteristic spectra associated with these levels.

For the sake of simplicity we consider only diatomic molecules. Viewed as a rigid dumbbell, a diatomic molecule can *rotate* about an axis through its center of mass. According to classical mechanics, the kinetic energy of a rotating body may be expressed as $\frac{1}{2}I\omega^2$, as given by Eq. (9–13). This energy may also be expressed in terms of the magnitude L of angular momentum, which is given by $L = I\omega$. Combining this with the preceding expression, we find that the energy E is given by

$$E = \frac{L^2}{2I}.$$

(43–1)

Now in a quantum-mechanical discussion of this motion, it is reasonable to assume that the angular momentum is quantized in the same way as for electrons in an atom, as given by Eq. (42–19):

$$L^2 = l(l + 1)\hbar^2 \qquad (l = 0, 1, 2, 3, \ldots). \tag{43–2}$$

Combining Eqs. (43–1) and (43–2), we obtain the result

$$E = l(l + 1)\left(\frac{\hbar^2}{2I}\right) \qquad (l = 0, 1, 2, 3, \ldots). \tag{43–3}$$

A more detailed analysis, using the Schrödinger equation, confirms that the angular momentum is indeed given by Eq. (43–2) and the energy levels by Eq. (43–3).

Considering the magnitudes involved, we note that for an oxygen molecule $I = 5 \times 10^{-46}$ kg·m^2. Thus, the constant $\hbar^2/2I$ in Eq. (43–3) is approximately equal to

$$\frac{\hbar^2}{2I} = \frac{(1.05 \times 10^{-34} \text{ J·s})^2}{2(5 \times 10^{-46} \text{ kg·m}^2)} = 1.11 \times 10^{-23} \text{ J} = 0.69 \times 10^{-4} \text{ eV}.$$

We note that this energy is much *smaller* than typical atomic energy levels associated with optical spectra. The energies of photons emitted and absorbed in transitions among rotational levels are correspondingly small, and they fall in the far infrared region of the spectrum.

The rigid-dumbbell model of a diatomic molecule suggests that the distance between the two nuclei is fixed; in fact, a more realistic model would represent the connection as a *spring* rather than a rigid rod. The atoms can undergo *vibrational* motion relative to the center of mass, and there is additional kinetic and potential energy associated with this motion. Application of the Schrödinger equation shows that the corresponding energy levels are given by

$$E = \left(n + \frac{1}{2}\right)hf \qquad (n = 0, 1, 2, 3, \ldots), \tag{43–4}$$

where f is the frequency of vibration. For typical diatomic molecules this turns out to be of the order of 10^{13} Hz; thus the constant hf in Eq. (43–4) is of the order of

$$hf = (6.6 \times 10^{-34} \text{ J·s})(10^{13} \text{ s}^{-1}) = 6.6 \times 10^{-21} \text{ J}$$
$$= 0.041 \text{ eV}.$$

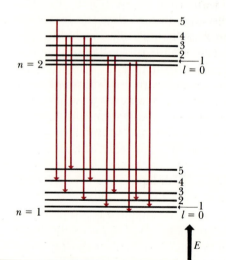

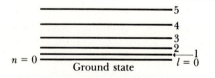

43–4 Energy-level diagram for vibrational and rotational energy levels of a diatomic molecule. For each vibrational level (*n*) there is a series of more closely spaced rotational levels (*l*). Several transitions corresponding to a single band in a band spectrum are shown.

Thus the vibrational energies, while still much smaller than those of the optical spectra, are typically considerably *larger* than the rotational energies.

An energy-level diagram for a diatomic molecule has the general appearance of Fig. 43–4. For each value of n, there are many values of l, leading to series of closely spaced levels. Transition between different pairs of n values gives different series of spectrum lines, and the resulting spectrum has the appearance of a series of *bands;* each band corresponds to a particular vibrational transition, and each individual line in a band, to a particular rotational transition. A typical band spectrum is shown in Fig. 43–5.

43–5   Typical band spectrum. (Courtesy of R. C. Herman.)

The same considerations can be applied to more complex molecules. A molecule with three or more atoms has several different kinds or *modes* of vibratory motion, each with its own set of energy levels related to the frequency by Eq. (43–4). The resulting energy-level scheme and associated spectra can be quite complex, but the general considerations discussed above still apply. In nearly all cases the associated radiation lies in the infrared region of the electromagnetic spectrum. Analysis of molecular spectra has proved to be an extremely valuable analytical tool, providing a great deal of information about the strength and rigidity of molecular bonds and the structure of complex molecules.

43–5

STRUCTURE OF SOLIDS

At ordinary temperatures and pressures most materials are in the solid state, a condensed state of matter characterized by interatomic interactions of sufficient strength to give the material a definite volume and shape, which change relatively little with applied stress. The separation of adjacent atoms in a solid is of the same order of magnitude as the diameter of the electron cloud around each atom.

A solid may be *amorphous* or *crystalline*. Crystalline solids have been discussed briefly in Section 20–7, and it would do no harm for the reader to review that discussion. A crystalline solid is characterized by *long-range order*, a recurring pattern in the arrangement of atoms; this pattern is called the *crystal structure* or the *lattice structure*. Amorphous solids have no long-range order but only short-range order, a state of affairs in which each atom has other atoms arranged around it in a more or less regular pattern, but without the recurring pattern characteristic of crystals. Amorphous solids have more in common with liquids than with crystalline solids; liquids also have short-range order but not long-range order.

The forces responsible for the regular arrangement of atoms in a crystal are, in some cases, the same as those involved in molecular binding. Corresponding to the classes of strong molecular bonds, there are ionic and covalent crystals. The alkali halides, of which ordinary salt (NaCl) is the most common variety, are the most familiar ionic crystals. The positive sodium ions and the negative chlorine ions occupy alternate positions in a cubic crystal lattice, as in Fig. 43–6. The forces are the familiar Coulomb's-law forces between charged particles; these forces are not directional, and the particular arrangement in which the material crystallizes is determined by the relative size of the two ions.

The simplest example of a *covalent* crystal is the *diamond structure*, a structure found in the diamond form of carbon and also in silicon, germanium, and tin, all elements in Group IV of the periodic table, with four electrons in the outermost shell. Each atom in this structure is situated at the center of a regular tetrahedron, with four nearest-neighbor

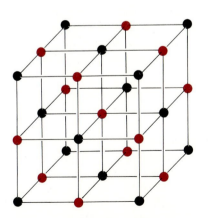

43–6   Symbolic representation of a sodium chloride crystal, with exaggerated distances between ions.

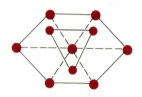

(a)

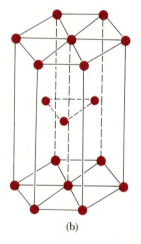

(b)

43–7 Close-packed crystal lattice structures. (a) Face-centered cubic; (b) hexagonal close-packed.

atoms at the corners, and it forms a covalent bond with each of these atoms. These bonds are strongly directional because of the asymmetical electron distribution, leading to the tetrahedral structure.

A third crystal type, which is less directly related to the chemical bond than ionic or covalent crystals, is the *metallic crystal*. In this structure the outermost electrons are not localized at individual lattice sites but are detached from their parent atoms and free to move through the crystal. The corresponding charge clouds (and their associated wave functions) extend over many atoms. Thus a metallic crystal can be visualized roughly as an array of positive ions from which one or more electrons have been removed, immersed in a sea of electrons whose attraction for the positive ions holds the crystal together. This sea has many of the properties of a gas, and indeed one speaks of the *electron-gas model* of metallic solids.

In a metallic crystal the situation is as though the atoms would like to form shared-electron bonds but there are not enough valence electrons. Instead, electrons are shared among *many* atoms. This bonding is not strongly directional in nature, and the shape of the crystal lattice is determined primarily by considerations of *close packing*, i.e., the maximum number of atoms that can fit into a given volume. The two most common metallic crystal lattices, the face-centered cubic and the hexagonal close-packed, are shown in Fig. 43–7. In each of these, each atom has 12 nearest neighbors.

As indicated in Section 43–3, van der Waals and hydrogen bonding also play a role in the structure of some solids. In polyethylene and similar polymers, there is covalent bonding of atoms to form long chain molecules, and hydrogen bonding to form cross-links between adjacent chains. In solid water both van der Waals and hydrogen bonds are significant, and together they determine the crystal structure of ice. Many other examples might be cited.

The discussion in this section has centered around *perfect crystals*, crystals in which the crystal lattice extends uninterrupted through the entire material. Real crystals exhibit a variety of departures from this idealized structure. Materials are often *polycrystalline*, composed of many small single crystals bonded together at *grain boundaries*. Within a single crystal there may be *interstitial* atoms in places where they do not belong, and *vacancies*, lattice sites that should be occupied by an atom but are not. An imperfection of particular interest in semiconductors, to be discussed

43–8 A dislocation. The concept is seen most easily by viewing the figure from various directions at a grazing angle with the page.

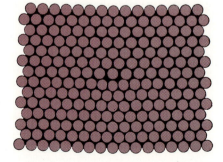

in Section 43–7, is the *impurity atom*, a foreign atom (e.g., arsenic in a silicon crystal) occupying a regular lattice site.

A more complex kind of imperfection is the *dislocation*, illustrated schematically in Fig. 43–8, in which one plane of atoms slips relative to another. The mechanical properties of metallic crystals are influenced strongly by the presence of dislocations; the ductility and malleability of some metals depends on the presence of dislocations that move through the lattice during plastic deformations.

43–6

PROPERTIES OF SOLIDS

Many *macroscopic* properties of solids, including mechanical, thermal, electrical, magnetic, and optical properties, can be understood by considering their relation to the *microscopic* structure of the material, and various aspects of the relation of structure to properties are part of the vigorous program of research in the physics of solids being carried out throughout the world. Although these topics cannot be discussed in detail in a book such as this one, a few examples will indicate the kinds of insights to be gained through study of the microscopic structure of solids.

We have already discussed, in Section 20–8, the subject of specific heat capacities of crystals, using the same principle of equipartition of energy as in the kinetic theory of gases. A simple analysis permits understanding of the empirical rule of Dulong and Petit on the basis of a microscopic model. This analysis has its limitations, to be sure; it does not include the energy of electron motion, which in metals makes a small additional contribution to specific heat, nor does it predict the temperature dependence of specific heats resulting from the quantization of the lattice-vibration energy discussed in Section 20–8. But these additional refinements *can* be included in the model to permit more detailed comparison of observed macroscopic properties with theoretical predictions.

Electrical conductivity is understood on the basis of the mobility or lack of mobility of electrons in the material. In a metallic crystal the valence electrons are not bound to individual lattice sites but are free to move through the crystal. Thus we expect metals to be good conductors, and they usually are. In a covalent crystal the valence electrons are involved in the bonds responsible for the crystal structure and are therefore *not* free to move. Thus there are no mobile charges available for conduction, and such materials are expected to be insulators. Similarly, an ionic crystal such as NaCl has no charges that are free to move, and solid NaCl is an insulator. However, when salt is melted, the ions are no longer locked to their individual lattice sites but are free to move, and *molten* NaCl is a good conductor.

There are, of course, no perfect conductors or insulators, although the resistivity of good insulators is greater than that of good conductors by an enormous factor, the order of at least 10^{15}. This great difference is one of the factors that make extremely precise electrical measurements possible. In addition, the resistivity of all materials depends on temperature; in general, the small conductivity of insulators *increases* with temperature, but that of good conductors usually *decreases* at increased temperatures.

Two competing effects are responsible for this difference. In metals the *number* of electrons available for conduction is nearly independent of temperature, and the resistivity is determined by the frequency of collisions of electrons with the lattice. Roughly speaking, lattice vibrations increase with increased temperature, and the ion cores present a larger target area for collisions with electrons. In insulators, what little conduction does take place is due to electrons that have gained enough energy from thermal motion of the lattice to break away from their "home" atoms and wander through the lattice. The number of electrons able to acquire the needed energy is very strongly temperature-dependent; a twofold increase for a 10° temperature rise is typical. There is also increased scattering at higher temperatures, as with metals, but the increased number of carriers is a far larger effect; thus insulators invariably become better conductors at higher temperatures.

A similar analysis can be made for *thermal* conductivity, which involves transport of microscopic mechanical energy rather than electric charge. The wave motion associated with lattice vibrations is one mechanism for energy transfer, and in metals the mobile electrons also carry kinetic energy from one region to another. This effect turns out to be much larger than that of the lattice vibrations, with the result that metals are usually much better thermal conductors than are electrical insulators, which have at most very few free electrons available to transport energy.

Optical properties are also related directly to microscopic structure. Good electrical conductors *cannot* be transparent to electromagnetic waves, for the electric fields of the waves induce currents in the material, and these dissipate the wave energy into heat as the electrons collide with the atoms in the lattice. All transparent materials are very good insulators. Metals are, however, good *reflectors* of radiation, and again the reason is the presence of free electrons at the surface of the material, which can move in response to the incident wave and generate a reflected wave. Reflection from the polished surface of an insulator is a somewhat more subtle phenomenon, dependent on polarization of the material, but again it is possible to relate macroscopic properties to microscopic structure.

43–7

BAND THEORY OF SOLIDS

The concept of *energy bands* in solids provides additional insight into some of the properties of solids discussed in Section 43–6 and also forms the basis for more detailed theoretical analysis of these properties on the basis of electron energy levels and wave functions.

To introduce the idea, let us suppose we have a large number N of identical atoms, far enough apart so their interactions are negligible. Then every atom has the same energy-level diagram, showing the possible quantum states for the electrons and their associated energy levels. We can draw an energy-level diagram for the entire system; it looks just

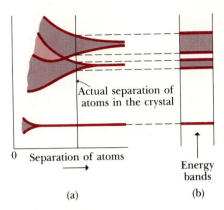

(a)   (b)

43–9 Origin of energy bands in a solid. (a) As the atoms are pushed together, the energy levels spread into bands. The vertical line shows the actual atomic spacing in the crystal lattice. (b) Symbolic representation of energy bands.

like the diagram for a single atom, except that the exclusion principle permits each state to be occupied by N electrons instead of just one.

Now we begin to push the atoms closer together. The wave functions begin to distort, especially those of the outer, or *valence*, electrons. The energy levels also shift somewhat, some upward, some downward, depending on the environment of each individual atom. Thus each valence electron state, formerly a sharp energy level that can accommodate N electrons, becomes spread out into a *band* containing N closely spaced levels, as shown in Fig. 43–9. Since N is ordinarily very large, of the order of Avogadro's number or of 10^{23}, the levels can be viewed as forming a continuous distribution of energies within a band. Between adjacent energy bands there are gaps or forbidden regions where there are *no* possible energy levels.

The inner electrons in an atom are affected much less than the valence electrons, and their energy levels remain relatively sharp. Ordinarily the behavior of an atom in a solid structure can be modeled as that of a rigid, unchanging *ion core* consisting of the positively charged nucleus and the inner electrons, surrounded by the valence electrons, whose states and energies *are* altered significantly by interactions with neighboring atoms.

Now let us see what this has to do with electrical conductivity. In insulators and semiconductors the valence electrons completely fill the highest occupied band, called the *valence band;* the next higher band, called the *conduction band,* is completely empty. The gap separating the two may be of the order of 1 eV to 5 eV. The situation is shown in Fig. 43–10b. The electrons in the valence band are not free to move in response to an applied electric field; to move, an electron would have to go to a different quantum-mechanical state with slightly different energy, but all the neighboring states are already occupied. The only way an electron can move is to jump into the conduction band, which would require an energy addition of a few electronvolts. Ordinarily that much energy is not available; for electric fields of reasonable magnitude the potential difference between two points spaced one atom apart is much

43–10 Three types of energy-band structure. (a) A conductor; there is a partially filled band, and electrons in this band are free to move when an electric field is applied. (b) An insulator; a completely full band is separated by a gap of several electronvolts from a completely empty band, and electrons in the full band cannot move. At finite temperatures a few electrons can reach the upper "conduction band." (c) A semiconductor; a completely filled band is separated by a small gap of 1 eV or less from an empty band; at finite temperatures substantial numbers of electrons can reach the upper "conduction band," where they are free to move.

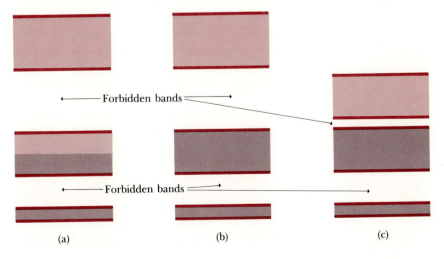

(a)   (b)   (c)

less than 1 eV, and there is no way such a field can do enough work on an electron to raise it across the gap into the conduction band. The situation is like a completely filled parking lot; none of the cars can move because there is no place to go. If a car could jump over the others, it could move!

However, at any finite temperature the crystal lattice has some vibrational motion, and there is some probability that an electron can gain enough energy from thermal motion to jump to the conduction band. Once in the conduction band, it is free to move in response to any applied electric field because there are plenty of nearby empty states available. At any finite temperature there are a few electrons in the conduction band, so no material is a perfect insulator. Furthermore, as the temperature increases, the population in the conduction band increases rapidly, and so the conductivity of an insulator always increases very rapidly with temperature. An increase of a factor of two for a temperature rise of 10 C° is typical.

With metals the situation is different because the valence band is only partially filled. The metal sodium is a simple example. The energy-level diagram for sodium in Fig. 41–5 shows that for an isolated atom the 3*p* "resonance-level" states for the valence electron are about 2.1 eV above the 3*s* ground state. The wavelength of the familiar yellow sodium doublet corresponds to this energy difference. But in the crystal lattice of solid sodium the 3*s* and 3*p* *bands* spread out enough so that they actually overlap to form a single band that is only $\frac{1}{4}$ filled. Electrons in states near the top of the filled portion of the band have many adjacent unoccupied states available, and they can easily gain or lose small amounts of energy in response to an applied electric field. Thus these electrons are mobile and can contribute to electrical and thermal conductivity. Partly filled bands are always characteristic of the metallic bond found in most metallic crystals. In the ionic NaCl crystal, on the other hand, there is no overlapping of bands; the valence band is completely filled, and solid sodium chloride is an insulator.

The band picture also adds insight to the phenomenon of *dielectric breakdown,* in which insulators subjected to a large enough electric field become conductors. If the electric field in a material is so large that there is a potential difference of a few volts over a distance comparable to atomic sizes (i.e., a field of the order of 10^{10} V·m^{-1}), then the field can do enough work on a valence electron to boost it over the forbidden region and into the conduction band. Real insulators usually have dielectric strengths much less than this because of structural imperfections that provide some energy states in the forbidden region.

The concept of energy bands is very useful in understanding the properties of semiconductors, to be discussed in the next section.

43–8
SEMICONDUCTORS

As the name implies, the electrical resistivity of a semiconductor is intermediate between that of good conductors and of good insulators. This is, however, only one aspect of the behavior of this important class of materials, so vital to present-day electronics. This discussion will be confined, for simplicity, to the elements germanium and silicon, the simplest semi-conductors, but these examples serve to illustrate the principal concepts.

Silicon and germanium, having four electrons in the outermost sub-shell, crystallize in the diamond structure described in Section 43–5; each atom lies at the center of a regular tetrahedron, with four nearest neighbors at the corners and a covalent bond with each. Thus all the electrons are involved in the bonding and the materials should be insulators. However, an unusually small amount of energy is needed to break one of the bonds and set an electron free to roam around the lattice, 1.1 eV for silicon and only 0.7 eV for germanium. This corresponds to the energy gap between the valence and conduction bands, discussed in Section 43–7 and shown in Fig. 43–10c. Thus even at room temperature a substantial number of electrons are dissociated from their parent atoms, and this number increases rapidly with temperature.

Furthermore, when an electron is removed from a covalent bond, it leaves a positively charged vacancy where there would ordinarily be an electron. An electron in a neighboring atom can drop into this vacancy, leaving the neighbor with the vacancy. In this way the vacancy, usually called a *hole*, can travel through the lattice and serve as an additional current-carrier. A hole behaves like a positively charged particle. In a pure semiconductor, holes and electrons are always present in equal numbers; the resulting conductivity is called *intrinsic conductivity* to distinguish it from conductivity due to impurities, to be discussed later.

The parking-lot analogy mentioned in Section 43–7 is useful in clarifying the mechanism of conduction in a semiconductor. A crystal with no bonds broken is like a completely filled floor of a parking garage. No cars (electrons) can move because there is nowhere for them to go. But if one car is removed to the empty floor above, it can move around freely, and the vacancy it leaves also permits cars to move on the nearly filled floor; this motion is most easily described in terms of *motion of the vacant space* from which the car has been removed. This corresponds to a vacancy in the normally filled valence band. Such a vacancy is called a *hole*, as mentioned above.

Now suppose we mix into melted germanium a small amount of arsenic, a Group V element having five electrons in its outermost sub-shell. When one of these electrons is removed, the remaining electron structure is essentially that of germanium; the only difference is that it is scaled down in size by the insignificant factor $\frac{32}{33}$ because the arsenic nucleus contains a charge $+33e$ rather than $+32e$. Thus an arsenic atom can take the place of a germanium atom in the lattice. Four of its five valence electrons form the necessary four covalent bonds with the nearest neighbors; the fifth is very loosely bound, with a binding energy of only 0.01 eV, corresponding in the band picture to an energy level 0.01 eV below the bottom of the conduction band. Even at ordinary temperature it can very easily escape and wander about the lattice. The corresponding positive charge is associated with the nuclear charge and is *not* free to move, in contrast to the situation with electrons and holes in pure germanium.

Because at ordinary temperatures only a very small fraction of the valence electrons are able to escape their sites and participate in the conduction process, a concentration of arsenic atoms as small as one part in 10^{10} can increase the conductivity so drastically that conduction due to impurities becomes by far the dominant mechanism. In such a case the conductivity is due almost entirely to *negative* charge motion; the material is called an *n-type* semiconductor, or is said to have *n*-type impurities.

Adding atoms of an element in Group III, with only three electrons in its outermost subshell, has an analogous effect. An example is gallium; placed in the lattice, the gallium atom would like to form four covalent bonds, but it has only three outer electrons. It can, however, steal an electron from a neighboring germanium atom to complete the bonding. This leaves the neighbor with a *hole* or missing electron, and this hole can then move through the lattice just as with intrinsic conductivity. In this case the corresponding negative charge is associated with the deficiency of positive charge of the gallium nucleus ($+31e$ instead of $+32e$), so it is not free to move. This state of affairs is characteristic of *p-type* semiconductors, materials with *p*-type impurities. The two types of impurities, *n* and *p*, are also called *donors* and *acceptors*, respectively, and the deliberate addition of these impurity elements is called *doping*.

The assertion that in *n* and *p* semiconductors the current *is* actually carried by electrons and holes, respectively, can be verified using the Hall effect, discussed in Section 30–11. The direction of the Hall emf is opposite in the two cases, and measurements of the Hall effect in various semiconductor materials confirm the above analysis of the conduction mechanisms.

43–9

SEMICONDUCTOR DEVICES

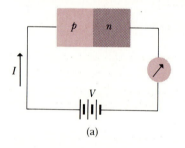

(a)

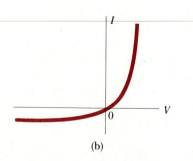

(b)

43–11   (a) A semiconductor *p-n* junction in a circuit; (b) graph showing the asymmetric voltage–current relationship given by Eq. (43–5).

The tremendous importance of semiconductor devices in practical electronic equipment results directly from the fact that the conductivity can be controlled within wide limits and changed from one region of a device to another, by varying impurity concentrations. The simplest example is the *p-n junction*, a crystal of germanium or silicon with *p*-type impurities in one region and *n*-type in the other, with the two separated by a boundary region called the *junction*. The details of how such a crystal is produced need not concern us; one way, not necessarily the most economical, is to grow a crystal by pulling a seed crystal very slowly away from the surface of a melted semiconductor. By varying the concentration of impurities in the melt while the crystal is grown, one can make a crystal with two or more regions of varying conductivity.

When a *p-n* junction is connected in an external circuit as shown in Fig. 43–11a and the potential V across the device is varied, the behavior of the current is as shown in Fig. 43–11b. The device conducts much more readily in the direction $p \rightarrow n$ than in the reverse, in striking contrast to the behavior of materials that obey Ohm's law. In the language of electronics, such a one-way device is called a *diode*.

The behavior of a *p-n* junction diode can be understood at least roughly on the basis of the conductivity mechanisms in the two regions. When the *p* region is at higher potential than the *n*, holes in the *p* region flow into the *n* region and electrons in the *n* region into the *p* region, so both contribute substantially to current. When the polarity is reversed,

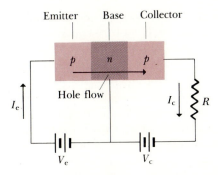

43–12 Schematic diagram of a *p-n-p* transistor and circuit. When $V_e = 0$ the current in the collector circuit is very small. When a potential V_e is applied between emitter and base, holes travel from emitter to base, as shown; when V_c is sufficiently large, most of them continue into the collector. The collector current I_c is controlled by the emitter current I_e.

the resulting electric fields tend to push electrons from *p* to *n* and holes from *n* to *p*. But there are very few electrons in the *p* region, only those associated with intrinsic conductivity and some that diffuse over from the *n* region. A similar condition prevails in the *n* region, and the current is much smaller than with the opposite polarity.

A more detailed analysis of this process, taking into account the effects of drift under the applied field and the diffusion that would take place even in the absence of a field, shows that the voltage–current relationship is given by

$$I = I_0(e^{eV/kT} - 1), \qquad (43-5)$$

where I_0 is a constant characteristic of the device, *e* is the electron charge, *k* is Boltzmann's constant, and *T* is absolute temperature.

The *transistor* includes two *p-n* junctions in a "sandwich" configuration, which may be either *p-n-p* or *n-p-n*. One of the former type is shown in Fig. 43–12. The three regions are usually called the emitter, base, and collector, as shown. In the absence of current in the left loop of the circuit, there is only a very small current through the resistor *R* because the voltage across the base–collector junction is in the "reverse" direction, i.e., the direction of small current flow, as in a simple *p-n* junction. But when a voltage is applied between emitter and base, as shown, the holes traveling from emitter to base can travel *through* the base to the second junction, where they come under the influence of the collector-to-base potential difference and thus flow on around through the resistor.

Thus the current in the collector circuit is *controlled* by the current in the emitter circuit. Furthermore, since V_c may be considerably larger than V_e, the power dissipated in *R* may be much larger than that supplied to the emitter circuit by the battery V_e. Thus the device functions as a power amplifier. If the potential drop across *R* is greater than V_e, it may also be a voltage amplifier.

In this configuration the *base* is the common element between the "input" and "output" sides of the circuit. Another widely used arrangement is the *common-emitter* circuit, shown in Fig. 43–13. In this circuit the current in the collector side of the circuit is much larger than in the base side, and the result is current amplification.

Transistors can perform many of the functions for which vacuum tubes have been used ever since the invention of the triode by de Forest in 1907; and where applicable they offer many advantages over vacuum tubes. These include mechanical ruggedness (since no vacuum container or fragile electrodes are involved), small size, long life (because they operate at relatively low temperatures and deteriorate very little with age), and efficiency (because no power is wasted heating a cathode for thermionic emission). Since their invention in 1948, semiconductor devices have completely revolutionized the electronics industry, including applications to communications, computer systems, control systems, and many other areas.

A further refinement in semiconductor technology is the *integrated circuit*. By successive depositing of layers of material and etching patterns to define current paths, one can combine the functions of several transistors, capacitors, and resistors on a single square of semiconductor mate-

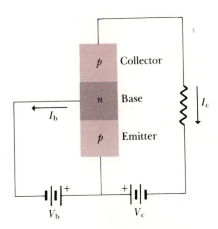

43–13 A common-emitter circuit. When $V_b = 0$, I_c is very small and most of the voltage V_c appears across the base–collector junction. As V_b increases, the base–collector potential decreases and more holes can diffuse into the collector; thus I_c increases. Ordinarily I_c is much larger than I_b.

rial that may be only a few millimeters on a side. A further elaboration of this idea leads to the *large-scale integrated circuit*. On a base consisting of a silicon chip, various layers are built up, including evaporated metal layers for conducting paths and silicon-dioxide layers for insulators and for dielectric layers in capacitors. Etching appropriate patterns into each layer by use of photosensitive etch-resistant materials and optically reduced patterns makes it possible to build up a circuit containing the functional equivalent of thousands of transistors, diodes, resistors, and capacitors on a single chip a few millimeters on a side. These so-called MOS (metal-oxide-semiconductor) chips are the heart of pocket calculators and microprocessors, as well as of many larger-scale computers. An example is shown in Fig. 43–14.

Semiconductors have many other practical applications. A thin slab of pure silicon or germanium can serve as a *photocell*. When the material is irradiated with light whose photons have at least as much energy as the gap between the valence and conduction bands, an electron in the valence band can absorb a photon and jump to the conduction band, where it contributes to the conductivity. Thus the conductivity increases when the material is exposed to light, and this change can be used to control the current in a circuit containing the device. Detectors for charged particles operate on the same principle. A high-energy charged particle passing through the material interacts with the electrons, and as it loses energy some of the electrons are excited from the valence to the conduction band. The conductivity increases momentarily, and a pulse of current in the external circuit is observed. Solid-state detectors are widely used in nuclear and high-energy physics research.

The inverse of the semiconductor photocell is the light-emitting diode (LED), which acts as a source of light. When a diode such as that shown in Fig. 43–9 is given a large *forward* voltage, many holes are injected across the junction into the *n*-region and electrons into the *p*-region. When these minority carriers recombine with majority carriers in the respective regions, they lose energy, which in some cases is radiated as photons of visible light. Light-emitting diodes are widely used for digital displays in clocks and electronic equipment.

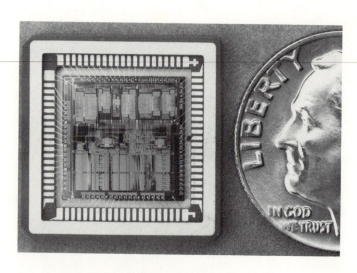

43–14 A large-scale integrated circuit. This dime-sized chip is a 32-bit microprocessor chip containing the equivalent of 150,000 transistors. (Courtesy of AT&T Bell Laboratories.)

43–10

SUPERCONDUCTIVITY

At very low temperatures some materials show a complete disappearance of all electrical resistance; such materials are called *superconductors*. Superconductivity was discovered in 1911 by the Dutch physicist H. Kamerlingh-Onnes during experiments on the temperature dependence of resistivity of materials. This area of research served at the time as a testing ground for competing theories of the behavior of electrons in solids. Three years earlier Kamerlingh-Onnes had succeeded in liquefying helium for the first time, thereby attaining lower temperatures than those available by any other means. Helium boils at 4.2 K at a pressure of 1 atm, and at lower temperatures under reduced pressure. Using liquid helium cooling, he found that when very pure solid mercury was cooled to 4.16 K its resistivity suddenly dropped to zero.

In later years many other superconducting metals and alloys have been found; each has a characteristic superconducting transition temperature called its *critical temperature*. The highest critical temperature found so far is 23 K for a niobium-germanium alloy, and several niobium alloys have critical temperatures in the range 15 to 20 K.

Below the superconducting transition the resistivity of a material is believed to be *exactly* zero. Experimental upper limits are of the order of 10^{-25} $\Omega \cdot$m, compared with typical values of 10^{-8} $\Omega \cdot$m for good conductors such as silver and copper at ordinary temperatures. When a current is magnetically induced in a superconducting ring, the current continues without measurable decrease for many months!

There can never be a magnetic field inside a superconducting material; any attempt to establish such a field results in induced eddy currents that exactly cancel the applied field everywhere inside the material. Related to this behavior is the fact that an applied magnetic field lowers the critical temperature, as shown in Fig. 43–15; a sufficiently strong field eliminates the superconducting transition completely. This transition can be regarded as a *phase transition* (although there is no associated heat of transition), and Fig. 43–15 is a *phase diagram*.

Although discovered in 1911, superconductivity was not well understood on a theoretical basis until 1957, when Bardeen, Cooper, and Schrieffer published the theory (now called the BCS theory) that was to earn them the Nobel Prize in 1972. It is difficult to describe the BCS theory in terms of elementary concepts. It is based on the existence of an *energy gap* in what would ordinarily be a continuum of electron energy states in a partly filled band. This gap is created through the collective motions of pairs of electrons with opposite spins and momenta. Once excited to energy states above the gap, electrons cannot decay back to their "normal" states below it by usual means, such as loss of energy by collisions with the ion cores in the crystal lattice. Thus these electrons become free to move through the lattice without any scattering by the ion cores.

Many important and exciting applications of superconductors are under development. For several years superconducting electromagnets have been used in research laboratories. Once a current is established in

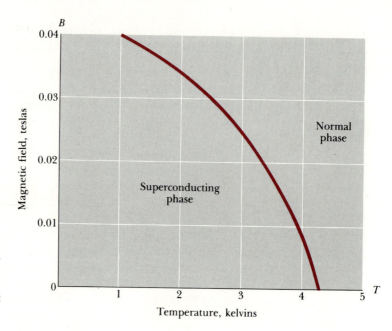

43–15 Phase diagram for pure mercury, showing the superconducting transition critical temperature and its dependence on magnetic field. Other superconducting materials have similar curves but with different scales.

the coil of such a magnet, no additional power input is required because there is no resistive energy loss. The coils can also be made more compact because there is no need to provide channels for the circulation of cooling fluids. Thus superconducting magnets can attain very large fields much more easily and economically than conventional magnets; fields of the order of 10 T are fairly routine. These considerations also make superconductors attractive for long-distance electric-power transmission, an active area of development work.

One of the most glamorous applications is in the field of magnetic levitation. Imagine a superconducting ring mounted on a railroad car that runs on a magnetized rail. The current induced in the ring leads to a repulsive interaction with the rail, and levitation is possible. Magnetically levitated trains have been running on an experimental basis in Japan for several years, and similar development projects are also under way in Germany.

QUESTIONS

43–1 In the ground state of the helium atom, the electrons must have opposite spins. Why?

43–2 The central-field approximation is more accurate for alkali metals than for transition metals (Group IV of the periodic table). Why?

43–3 The outermost electron in the potassium atom is in a $4s$ state. What does this tell you about the relative positions of the $3d$ and $4s$ states for this atom?

43–4 A student asserted that any filled shell (i.e., all the levels for a given n occupied by electrons) must have zero total angular momentum and hence must be spherically symmetric. Do you believe this? What about a filled *subshell* (all values of m for given values of n and l)?

43–5 What factors determine whether a material is a conductor of electricity or an insulator?

43–6 The nucleus of a gold atom contains 79 protons. How would you expect the energy required to remove a $1s$ electron completely from a gold atom to compare with the energy required to remove the electron from a hydrogen atom? In what region of the electromagnetic spectrum would a photon of the appropriate energy lie?

43–7 Elements can be identified by their visible spectra. Could analogous techniques be used to identify compounds from their molecular spectra? In what region of the electromagnetic spectrum would the appropriate radiation lie?

43–8 The ionization energies of the alkali metals (i.e., the energy required to remove an outer electron) are in the range from 4 eV to 6 eV, while those of the inert gases are in the range from 15 eV to 25 eV. Why the difference?

43–9 The energy required to remove the 3s electron from a sodium atom in its ground state is about 5 eV. Would you expect the energy required to remove an additional electron to be about the same, or more or less? Why?

43–10 Electrical conductivities of most metals decrease gradually with increasing temperature, but the intrinsic conductivity of semiconductors always *increases* rapidly with increasing temperature. Why the difference?

43–11 What are some advantages of transistors compared to vacuum tubes, for electronic devices such as amplifiers? What are some disadvantages? Are there any situations in which vacuum tubes *cannot* be replaced by solid-state devices?

PROBLEMS

43–1 Make a list of the four quantum numbers for each of the six electrons in the ground state of the carbon atom, and for each of the 14 electrons in the silicon atom.

43–2 Cobalt ($Z = 27$) has two electrons in 4s states. How many does it have in 3d states?

43–3 For germanium ($Z = 32$) make a list of the number of electrons in each state (1s, 2s, 2p, etc.).

43–4 Calculate the magnitude of the potential energy ($e^2/4\pi\epsilon_0 r$) of interaction of a charge $+e$ with a charge $-e$ (where e is the electron charge magnitude) at a distance of 2×10^{-10} m, roughly the magnitude of the lattice spacing in the NaCl crystal.

43–5 For the sodium chloride molecule discussed at the beginning of Section 43–3, what is the maximum separation of the ions for stability if they may be regarded as point charges? (The potential energy of interaction must be at least -1.3 eV.)

43–6 For magnesium, the first ionization potential is 7.6 eV; the second (additional energy required to remove a second electron) is almost exactly twice this, 15 eV, and the third ionization potential is much larger, about 80 eV. How can these numbers be understood?

43–7 The ionization potentials of the alkali metals are

Li	5.4 eV
Na	5.1
K	4.3
Rb	4.2
Cs	3.9

Which of these would you expect to be most active chemically?

43–8 The rotation spectrum of HCl contains the following wavelengths:

60.4 μm
69.0
80.4
96.4
120.4

a) Find the moment of inertia of the HCl molecule.

b) Sketch the first two *vibrational* levels of HCl, and show how the rotational levels modify the vibrational energies.

43–9 Show that the frequencies in a pure rotation spectrum (disregarding vibrational levels) are all integer multiples of the quantity $\hbar/2\pi I$.

43–10 If the distance between atoms in a diatomic oxygen molecule is 2×10^{-10} m, calculate the moment of inertia about an axis through the center of mass perpendicular to the line joining the atoms.

43–11 The dissociation energy of the hydrogen molecule (i.e., the energy required to separate the two atoms) is 4.72 eV. At what temperature is the average kinetic energy of a molecule equal to this energy?

43–12 The vibrational frequency of the hydrogen molecule is 1.29×10^{14} Hz.

a) What is the spacing of adjacent vibrational energy levels, in electronvolts?

b) Find the wavelength of radiation emitted in the transition from the $n = 2$ to $n = 1$ vibrational state, and from $n = 3$ to $n = 2$.

c) For what initial values of n do transitions to the ground state of vibrational motion yield radiation in the visible spectrum?

43–13 The vibration frequency for the molecule HCl is 8.6×10^{13} Hz.

a) If this molecule can be regarded as a simple harmonic oscillator with the Cl atom stationary (because it is much more massive than the H atom), what is the effective "spring constant" k corresponding to the interatomic force?

b) What is the spacing between adjacent vibrational energy levels, in joules? In electronvolts?

c) What is the wavelength of a photon emitted in a transition between two adjacent vibrational levels? In what region of the spectrum does it lie?

43–14 In Problem 43–13, suppose the hydrogen atom is replaced by an atom of deuterium, the isotope of hydrogen with atomic mass 2 g·mol^{-1}. The effective spring constant k is determined by the electron configuration, so it is the same as for the normal HCl molecule.

  a) What is the vibration frequency of this molecule?

  b) What is the wavelength of light emitted in the transition $n = 2$ to $n = 1$? In what region of the spectrum does it lie?

43–15 The distance between atoms in a hydrogen molecule is 0.074 nm.

  a) Calculate the moment of inertia of a hydrogen molecule about an axis perpendicular to the line joining the nuclei, at its center.

  b) Find the energies of the $l = 0$, $l = 1$, and $l = 2$ rotational states.

  c) Find the wavelength and frequency of the photon emitted in the transition from $l = 2$ to $l = 0$.

43–16 The spacing of adjacent atoms in a sodium-chloride crystal is 0.282 nm. Calculate the density of sodium chloride.

43–17 In the hexagonal close-packed crystal structure, regarding the atoms as rigid spheres, show that an atom can accommodate twelve nearest neighbors (all touching it) if six are placed around it in a plane hexagonal array, with three additional atoms above and three below the plane.

43–18 Suppose a piece of very pure germanium is to be used as a light detector by observing the increase in conductivity resulting from generation of electron–hole pairs by absorption of photons. If each pair requires 0.7 eV of energy, what is the maximum wavelength that can be detected? In what portion of the spectrum does it lie?

43–19 What are the answers for Problem 43–18 if the material is silicon, with an energy requirement of 1.1 eV per pair, corresponding to the gap between valence and conduction bands in silicon?

44

NUCLEAR AND HIGH-ENERGY PHYSICS

Every atom contains at its center a massive, positively charged *nucleus*, much smaller than the overall size of the atom but nevertheless containing most of its total mass. We review the experimental evidence for the existence of nuclei, and then describe their most important properties in terms of the protons and neutrons of which they are composed. The stability or instability of a particular nuclear structure is determined by the competition between the attractive nuclear force among the protons and neutrons and the repulsive electrical interactions among the protons. Unstable nuclei *decay* or transform themselves spontaneously into other structures by a variety of decay processes; structure-altering nuclear reactions can also be induced by bombardment of a nucleus with any of a variety of particles or other nuclei. Two classes of reactions of special interest are *fission* and *fusion*. Protons and neutrons are not among the most fundamental of particles, but instead they are believed to be composed of more basic entities called *quarks*, which also play a fundamental role in the understanding of the relationships among the four fundamental types of interactions: strong, electromagnetic, weak, and gravitational. Research into the nature and interactions of fundamental particles has required the construction of larger and larger experimental facilities such as particle accelerators, as scientists continue to probe more and more deeply into this most fundamental aspect of the nature of our physical universe.

44–1

THE NUCLEAR ATOM

The earliest experimental evidence for the existence of nuclei was found in the Rutherford scattering experiments, carried out in 1910–1911 by Sir Ernest Rutherford and two of his students, Hans Geiger and Ernest Marsden, at Cambridge, England.

The electron had been "discovered" in 1897 by Sir J. J. Thomson, and by 1910 its mass and charge were quite accurately known. It had also been well established that all atoms, with the sole exception of hydrogen, contain more than one electron. Thomson had proposed a model of the atom consisting of a relatively large sphere of positive charge (about 2 or 3×10^{-8} cm in diameter) within which were embedded, like chocolate chips in a cookie, the electrons.

What Rutherford and his co-workers did was to project other particles at the atoms under investigation; and, from observations of the way in which the projected particles were deflected or *scattered*, they drew conclusions about the distribution of charge within the "target" atoms.

At this time the high-energy particle accelerators now in common use for nuclear and high-energy physics research had not yet been developed, and Rutherford had to use as projectiles the particles produced in natural radioactivity, to be discussed later in this chapter. Some radioactive disintegrations result in the emission of *alpha particles;* these particles are now known to be identical with the nuclei of helium atoms, each consisting of two protons and two neutrons bound together, but without the two electrons normally present in a neutral helium atom. Alpha particles are ejected from unstable nuclei with speeds of the order of 10^7 m·s^{-1}, and they can travel several centimeters through air, or on the order of 0.1 mm through solid matter, before they are brought to rest by collisions.

The experimental setup is shown schematically in Fig. 44–1. A radioactive source at the left emits alpha particles. Thick lead screens stop all particles except those in a narrow *beam* defined by small holes. The beam then passes through a thin metal foil (gold, silver, and copper were used) and strikes a plate coated with zinc sulfide. A momentary flash or scintillation can be observed visually on the screen whenever it is struck by an alpha particle, and the number of particles that have been deflected through any angle from their original direction can therefore be determined.

According to the Thomson model, the atoms of a solid are packed together like marbles in a box. The experimental fact that an alpha particle can pass right through a sheet of metal foil forces one to conclude, if

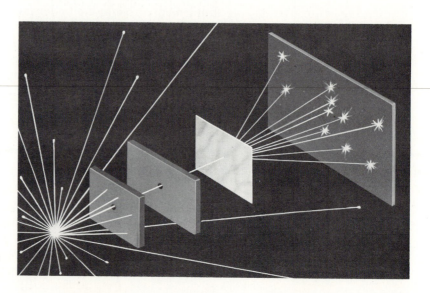

44–1   The scattering of alpha particles by a thin metal foil.

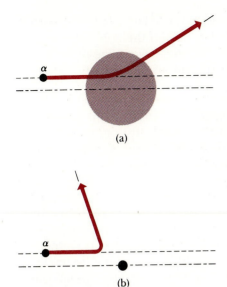

44–2 (a) Alpha particle scattered through a small angle by the Thomson atom. (b) Alpha particle scattered through a large angle by the Rutherford nuclear atom.

this model is correct, that the alpha particle is capable of actually penetrating the spheres of positive charge. Granted that this is possible, we can compute the deflection it would undergo. The Thomson atom is electrically neutral, so outside the atom no force would be exerted on the alpha particle. Within the atom, the electrical force would be due in part to the electrons and in part to the sphere of positive charge. However, the mass of an alpha particle is about 7400 times that of an electron, and from momentum considerations it follows that the alpha particle can suffer only a negligible scattering as a consequence of forces between it and the much less massive electrons. It is only interactions with the *positive* charge, which makes up most of the atomic mass, that can deviate the alpha particle.

When electric charge is distributed uniformly inside a spherical volume, the electric field at points inside the sphere is proportional to the distance from the center of the sphere; this can be proved using Gauss's law, as in Problem 25–23. Thus the positively charged alpha particle inside the Thomson atom would be *repelled* from the center of the sphere with a force proportional to its distance from the center, and its trajectory can be computed for any initial direction of approach such as that in Fig. 44–2a. On the basis of such calculations, Rutherford predicted the number of alpha particles that should be scattered at any angle with respect to the original direction.

The experimental results did not agree with the calculations based on the Thomson atom. In particular, many more particles were scattered through large angles than were predicted. To account for the observed large-angle scattering, Rutherford concluded that the positive charge, instead of being spread through a sphere of atomic dimensions (2 or 3×10^{-8} cm) was concentrated in a much *smaller* volume, which he called a *nucleus*. When an alpha particle approaches the nucleus, the entire nuclear charge exerts a repelling effect on it down to extremely small separations, with the consequence that much larger deviations can be produced. Figure 44–2b shows the trajectory of an alpha particle deflected by a Rutherford nuclear atom, for the same original path as that in part (a) of the figure.

Rutherford again computed the expected number of particles scattered through any angle, assuming an inverse-square law of force between the alpha particle and the nucleus of the scattering atom. Within the limits of experimental accuracy, the computed and observed results were in agreement down to distances of approach of about 10^{-12} cm. These experiments thus indicate that the size of the nucleus is no larger than the order of 10^{-12} cm.

44–2

PROPERTIES OF NUCLEI

As indicated in Section 44–1, the most obvious feature of the atomic nucleus is its size, of the order of 20,000 to 200,000 times smaller than the atom itself. Since Rutherford's initial experiments, many additional scattering experiments have been performed, using high-energy protons, electrons, and neutrons as well as alpha particles. Although the "surface" of a nucleus is not a sharp boundary, these experiments determine an approximate radius for each nucleus. The radius is found to depend on the mass, which in turn depends on the total number A of neutrons and protons, usually called the *mass number*. The radii of most nuclei are represented fairly well by the empirical equation

$$r = r_0 A^{1/3} \tag{44-1}$$

where r_0 is an empirical constant equal to 1.2×10^{-15} m, the same for all nuclei.

Since the volume of a sphere is proportional to r^3, Eq. (44-1) shows that the *volume* of a nucleus is proportional to A (i.e., to the total mass) and therefore that the mass per unit volume (proportional to A/r^3) is the same for all nuclei. That is, *all nuclei have approximately the same density.* This fact is of crucial importance in understanding nuclear structure.

Two additional important properties of nuclei are angular momentum and magnetic moment. The particles in the nucleus are in motion, just as the electrons in an atom are in motion; there is angular momentum associated with this motion, and, because circulating charge constitutes a current, there is also magnetic moment. Experimental evidence for the existence of nuclear angular momentum (often called *nuclear spin*) and magnetic moment came originally from spectroscopy. Some spectrum lines are found to be split into series of very closely spaced lines, called *hyperfine structure,* which can be understood on the basis of interactions between the electrons in an atom and the magnetic field produced by a nuclear magnetic moment. Detailed analysis of this phenomenon indicates that the nuclear angular momentum is *quantized,* just as it is for electrons and for molecular rotation; the component of angular momentum in a specified axis direction is a multiple of $h/2\pi$, but some nuclei seem to require *integer* multiples (as with orbital angular momentum of electrons) and some *half-integer* multiples, as with electron spin. As we shall see later, the nuclear spin can be understood on the basis of the particles making up the nucleus.

Although it was once believed that nuclei were made of protons and electrons, the discovery of the neutron (discussed in Section 44-11) and many other experiments have established that the basic building blocks of the nucleus are the proton and the neutron. The total number of protons, equal in a neutral atom to the number of electrons, is the *atomic number Z.* The total number of *nucleons* (protons and neutrons) is called the mass number A. The number of neutrons, denoted by N, is called the *neutron number.* For any nucleus these are related by

$$A = Z + N \tag{44-2}$$

Table 44-1 lists values of A, Z, and N for several nuclei. As the table shows, some nuclei have the same Z but different N. Since the electron structure, which determines the chemical properties, is determined by the charge of the nucleus, these are nuclei of the same element, but they have different masses and can be distinguished in precise experiments. The experimental investigation of *isotopes* through mass spectroscopy is discussed in Section 30-6, and the reader would do well to review this section. Nuclei of a given element having different mass numbers are called *isotopes* of the element, and a single nuclear species (single values of both Z and N) is called a *nuclide.*

TABLE 44–1   NUCLEAR PARTICLES

Nucleus	Mass number (total number of nuclear particles), A	Atomic number (number of protons), Z	Neutron number, $N = A - Z$
$^{1}_{1}\text{H}$	1	1	0
$^{2}_{1}\text{D}$	2	1	1
$^{4}_{2}\text{He}$	4	2	2
$^{6}_{3}\text{Li}$	6	3	3
$^{7}_{3}\text{Li}$	7	3	4
$^{9}_{4}\text{Be}$	9	4	5
$^{10}_{5}\text{B}$	10	5	5
$^{11}_{5}\text{B}$	11	5	6
$^{12}_{6}\text{C}$	12	6	6
$^{13}_{6}\text{C}$	13	6	7
$^{14}_{7}\text{N}$	14	7	7
$^{16}_{8}\text{O}$	16	8	8
$^{23}_{11}\text{Na}$	23	11	12
$^{65}_{29}\text{Cu}$	65	29	36
$^{200}_{80}\text{Hg}$	200	80	120
$^{235}_{92}\text{U}$	235	92	143
$^{238}_{92}\text{U}$	238	92	146

Table 44–1 also shows the notation usually used to denote individual nuclides; the symbol of the element is used, with a pre-subscript equal to the atomic number Z and a pre-superscript equal to the mass number A. This is of course redundant, since the element is determined by the atomic number, but the notation is a useful aid to memory.

The total mass of a nucleus is always *less* than the total mass of its constituent parts because of the mass equivalent ($E = mc^2$) of the negative potential energy associated with the attractive forces that hold the nucleus together. This mass difference is sometimes called the *mass defect*. In fact, the best way to determine the total potential energy, or *binding energy,* is to compare the mass of a nucleus with the masses of its constituents. The proton and neutron masses are

$$m_{\text{p}} = 1.6726 \times 10^{-27} \text{ kg,}$$

$$m_{\text{n}} = 1.6750 \times 10^{-27} \text{ kg.}$$

Since these are nearly equal, it is not surprising that many nuclear masses are approximately integer multiples of the proton or neutron mass.

This observation suggests defining a new mass unit equal to the proton or neutron mass. Instead, it has been found more convenient to define the new unit, called the *atomic mass unit* (u), as $\frac{1}{12}$ the mass of the neutral carbon atom having mass number $A = 12$. It is found that

$$1 \text{ u} = 1.660566 \times 10^{-27} \text{ kg.}$$

In atomic units, the masses of the proton, neutron, and electron are found to be:

$$m_p = 1.007276 \text{ u},$$

$$m_n = 1.008665 \text{ u},$$

$$m_e = 0.000549 \text{ u}.$$

The masses of some common atoms, including their electrons, are shown in Table 44–2. The masses of the bare nuclei are obtained by subtracting Z times the electron mass. The energy equivalent of 1 u is found from the relation $E = mc^2$:

$$E = (1.660566 \times 10^{-27} \text{ kg})(2.998 \times 10^8 \text{ m} \cdot \text{s}^{-1})^2$$
$$= 1.492 \times 10^{-10} \text{ J} = 931.5 \text{ MeV}.$$

TABLE 44–2   ATOMIC DATA*

Element	Atomic number, Z	Neutron number, N	Atomic mass, u	Mass number, A
Hydrogen H	1	0	1.00783	1
Deuterium H	1	1	2.01410	2
Helium He	2	1	3.01603	3
Helium He	2	2	4.00260	4
Lithium Li	3	3	6.01513	6
Lithium Li	3	4	7.01601	7
Beryllium Be	4	4	8.00508	8
Beryllium Be	4	5	9.01219	9
Boron B	5	5	10.01294	10
Boron B	5	6	11.00931	11
Carbon C	6	6	12.00000	12
Carbon C	6	7	13.00335	13
Nitrogen N	7	7	14.00307	14
Nitrogen N	7	8	15.00011	15
Oxygen O	8	8	15.99491	16
Oxygen O	8	9	16.99913	17
Oxygen O	8	10	17.99916	18

*American Institute of Physics Handbook, 1963

EXAMPLE Find the mass defect, the total binding energy, and the binding energy per nucleon for carbon (12).

Solution The mass of the neutral carbon atom, including the nucleus and the six electrons, is, according to Table 44–2, 12.00000 u. The mass of the bare nucleus is obtained by subtracting the mass of the six electrons:

$$m = 12.00000 \text{ u} - (6)(0.000549 \text{ u}) = 11.996706 \text{ u}.$$

The total mass of the six protons and six neutrons in the nucleus is

$$(6)(1.007276 \text{ u}) + (6)(1.008665 \text{ u}) = 12.095646 \text{ u}.$$

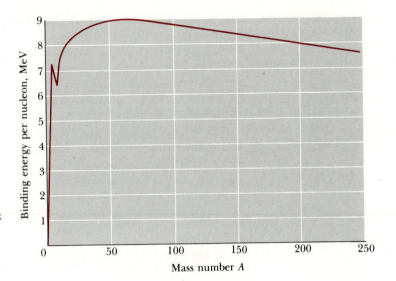

44-3 Binding energy per nucleon as a function of mass number A (the total number of nucleons). The curve reaches a peak of about 8.7 MeV at about A = 60, corresponding to the element iron. The spike at A = 4 shows the unusual stability of the α-particle structure.

The mass defect is therefore

$$12.095646 \text{ u} - 11.996706 \text{ u} = 0.09894 \text{ u}.$$

The energy equivalent of this mass is

$$(0.09894 \text{ u})(931.5 \text{ MeV} \cdot \text{u}^{-1}) = 92.16 \text{ MeV}.$$

Thus the total binding energy for the twelve nucleons is 92.16 MeV. To pull the carbon nucleus completely apart into twelve separate nucleons would require a minimum of 92.16 MeV. The binding energy *per nucleon* is $\frac{1}{12}$ of this, or 7.68 MeV per nucleon. Nearly all stable nuclei, from the lightest to the most massive, have binding energies in the range of 6 to 9 MeV per nucleon. Figure 44-3 is a graph of binding energy per nucleon, as a function of the mass number A. ◄

Because the nucleus is a composite structure, it can have internal motion, with corresponding energy levels. Thus each nucleus has a set of allowed energy levels, corresponding to a *ground state* (state of lowest energy) and several *excited states*. Because of the strength of nuclear interactions, excitation energies of nuclei are typically of the order of 1 MeV, compared with a few eV for atomic energy levels. In ordinary physical and chemical transformations the nucleus always remains in its ground state. When a nucleus is placed in an excited state, either by bombardment by high-energy particles or by a radioactive transformation, it can decay to the ground state by emission of one or more photons, called in this case *gamma rays* or *gamma photons*.

The forces that hold protons and neutrons together in the nucleus despite the electrical repulsion of the protons are not of any familiar sort but are unique to the nucleus. Some aspects of the nuclear force are still incompletely understood, but several qualitative features can be described. First, it does not depend on charge, since neutrons as well as protons must be bound. Second, it must be of short range; otherwise the nucleus would pull in additional protons and neutrons. But within the range it must be stronger than electrical forces; otherwise the nucleus could never be stable. Third, the nearly constant density of nuclear matter and the nearly constant binding energy per nucleon indicate that a given nucleon cannot interact simultaneously with all the other nucleons but only with those few in its immediate vicinity. This is again in contrast to the behavior of electrical forces, in which *every* proton in the nucleus repels every other one. This limitation on the maximum number of nucleons with which a nucleon can interact is called *saturation;* it is analogous in some respects to covalent bonding in solids. Finally, the nuclear forces appear to favor binding of pairs of particles (e.g., two protons with opposite spin, or two neutrons with opposite spin) and of *pairs of pairs,* a pair of protons and a pair of neutrons, with total spin zero. Thus the alpha particle is an exceptionally stable nuclear structure.

These qualitative features of the nuclear force are helpful in understanding the various kinds of nuclear instability, to be discussed in the following sections.

44–3

NATURAL RADIOACTIVITY

In studying the fluorescence and phosphorescence of compounds irradiated with visible light, Becquerel, in 1896, performed a crucial experiment that led to a deeper understanding of the properties of the nucleus of an atom. After illuminating some pieces of uranium–potassium sulfate with visible light, Becquerel wrapped them in black paper and separated the package from a photographic plate by a piece of silver. After several hours' exposure, the photographic plate was developed and showed a blackening due to something that must have been emitted from the compound and was able to penetrate both the black paper and the silver.

Rutherford showed later that the emanations given off by uranium sulfate were capable of ionizing the air in the space between two oppositely charged metallic plates (an ionization chamber). The current registered by a galvanometer in series with the circuit was taken to be a measure of the "activity" of the compound.

A systematic study of the activity of various elements and compounds led Mme. Curie to the conclusion that this activity was an atomic phenomenon; and by the methods of chemical analysis, she and her husband, Pierre Curie, found that "ionizing ability" or "activity" was associated not only with uranium but with two other elements that they discovered, radium and polonium. The activity of radium was found to be more than a million times that of uranium. Since the pioneer days of the Curies, many more radioactive substances have been discovered.

The activity of radioactive material may be easily shown to be the result of three different kinds of emanations. An early experiment is shown in Fig. 44–4. A small piece of radioactive material is placed at the bottom of a long groove in a lead block. At some distance above the lead block a photographic plate is placed, and the whole apparatus is highly evacuated. A strong magnetic field is applied at right angles to the plane of the diagram. When the plate is developed, three distinct spots are found, one in the direct line of the groove in the lead block, one de-

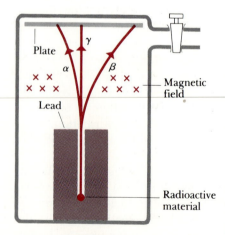

44–4   The three emanations from a radioactive material and their paths in a magnetic field perpendicular to the plane of the diagram.

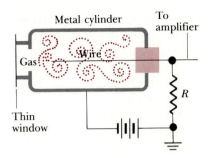

44-5 Schematic diagram of a Geiger counter.

flected to one side, and one to the other side. From a knowledge of the direction of the magnetic field, it is concluded that one of the emanations is positively charged (alpha particles), one is negatively charged (beta particles), and one is neutral (gamma rays).

Further investigation showed that not all three emanations are emitted simultaneously by all radioactive substances. Some elements emit alpha particles, others emit beta particles, while gamma rays sometimes accompany one and sometimes the other. Furthermore, no simple macroscopic physical or chemical process, such as raising or lowering the temperature, chemical combination with other nonradioactive substances, etc., could change or affect in any way the activity of a given sample. As a result, it was suspected from the beginning that radioactivity is a *nuclear* process and that the emission of a charged particle from the nucleus of an atom results in leaving behind a different atom, occupying a different place in the periodic table. In other words, radioactivity involves the *transmutation of elements*.

The first measurement of the charge of the alpha particle used a device called a *Geiger counter*, still an important instrument of modern physics.

As shown in Fig. 44-5, a Geiger counter consists of a metal cylinder and a wire along its axis. The cylinder contains a gas such as air or argon, at a pressure of from 50 to 100 mm of mercury. A difference of potential slightly less than that necessary to produce a discharge is maintained between the wire and the cylinder wall. Alpha particles (or, for that matter, any particles to be studied) can enter through a thin glass or mica window. The particle entering the counter produces ionization of the gas molecules. These ions are accelerated by the electric field and produce more ions by collisions, causing the ionization current to build up rapidly. The current, however, decays rapidly since the circuit has a small time constant. There is therefore a momentary surge of current or a momentary potential surge across R, which may be amplified and made to advance an electronic counter, or to activate a counting-rate meter.

Placing a known mass of radium at a known distance from the window of a Geiger counter, Rutherford and Geiger counted the number of alpha particles emitted in a known time interval. They found that 3.57×10^{10} alpha particles were emitted per second per gram of radium. They then allowed the alpha particles from the same source to fall upon a plate and measured its rate of increase of charge. Dividing the rate of increase of charge by the number emitted per second, Rutherford and Geiger determined the charge on an alpha particle to be 3.19×10^{-19} C, or practically twice the charge on an electron, but opposite in sign.

The next problem was to determine the mass of an alpha particle. This was accomplished by measuring the ratio of charge to mass by the electric and magnetic deflection method described in Chapter 30. The ratio was found by Rutherford and Robinson to be 4.82×10^7 C·kg^{-1}. When this result was combined with the charge of an alpha particle, the mass was found to be 6.62×10^{-27} kg, almost exactly four times the mass of a hydrogen atom.

Since a helium atom has a mass four times that of a hydrogen atom and, stripped of its two electrons (as a bare nucleus), has a charge equal in magnitude and opposite in sign to two electrons, it seemed certain that alpha particles were helium nuclei. To make the identification certain, however, Rutherford and Royds collected the alpha particles in a glass discharge tube over a period of about six days and then established an electric discharge in the tube. Examining the spectrum of the emitted light, they identified the characteristic helium spectrum and established without doubt that alpha particles *are* helium nuclei.

The speed of an alpha particle emitted from a given radioactive source such as radium may be measured by observing the radius of the circle traversed by the particle in a magnetic field perpendicular to the motion. Such experiments show that alpha particles are emitted with very high speeds. For example, the alpha particles emitted by radium, $^{226}_{88}$Ra, have speeds of about 1.5×10^7 m·s^{-1}. The corresponding kinetic energy is

$$K = \tfrac{1}{2}(6.62 \times 10^{-27} \text{ kg})(1.5 \times 10^7 \text{ m·s}^{-1})^2$$
$$= 7.4 \times 10^{-13} \text{ J} = 4.7 \times 10^6 \text{ eV}$$
$$= 4.7 \text{ MeV}.$$

We note that the speed, although large, is only five percent of the speed of light, so the nonrelativistic kinetic-energy expression may be used. We also note that the energy is larger than typical energies of atomic electrons by a factor of the order of a million. Because of these large energies, alpha particles are capable of traveling several centimeters in air, or a few tenths or hundredths of a millimeter through solids, before they are brought to rest by collisions.

Beta particles are *negatively* charged and are therefore deflected in an electric or magnetic field. Deflection experiments similar to those described in Chapter 30 prove conclusively that beta particles have the same charge and mass as electrons. They are emitted with tremendous speeds, some reaching a value of 0.9995 that of light. Thus relativistic relations must be used in analyzing their motion. Unlike alpha particles, which are emitted from a given nucleus with one speed or a few definite speeds, beta particles are emitted with a *continuous* range of speeds, from zero up to a maximum that depends on the nature of the emitting nucleus. If the principles of conservation of energy and of momentum are to hold in nuclear processes, it is necessary to assume that the emission of a beta particle is accompanied by the emission of another particle with no charge. This particle, called a *neutrino*, has zero rest mass and zero charge and therefore, even in traversing the densest matter, produces very little measurable effect. In spite of this, Reines and Cowan, in 1953 and again in 1956, succeeded in detecting its existence in a series of extraordinary experiments. Subsequent investigation has shown that in fact there are at least three varieties of neutrinos, the one associated with beta decay and others associated with the decay of unstable particles, the μ mesons and the τ particles.

Since gamma rays are not deflected by a magnetic field, they cannot consist of charged particles. They are, however, diffracted at the surface of a crystal in a manner similar to that of x-rays but with extremely small angles of diffraction. Experiments of this sort lead to the conclusion that gamma rays are actually electromagnetic waves of extremely short wavelength, about $\frac{1}{100}$ that of x-rays.

The gamma-ray spectrum of any one element is a line spectrum, suggesting that a gamma-ray photon is emitted when a nucleus proceeds from a state of higher to a state of lower energy. This view is substantiated in the case of radium by the following facts. When alpha particles are emitted from radium they are found to consist of two groups, those with a kinetic energy of 4.879 MeV and those with an energy of 4.695 MeV. When a radium atom emits an alpha particle of the smaller energy, the resulting nucleus (which corresponds to the element *radon*) has a *greater* amount of energy than if the higher-speed alpha particle had been emitted. This represents an excited state of the radon nucleus. If now the radon nucleus undergoes a transition from this excited state to the lower energy state, a gamma-ray photon of energy

$$(48.79 - 46.95) \times 10^5 = 1.84 \times 10^5 \text{ eV}$$

should be emitted. The *measured* energy of the gamma-ray photon emitted by radium is 1.89×10^5 eV, in excellent agreement.

Thus, by correlating alpha-particles energies and gamma-ray energies, it is possible in some cases to construct *nuclear energy-level diagrams* similar to atomic energy-level diagrams.

When a radioactive nucleus decays by alpha or beta emission, the resulting nucleus may also be unstable, and there may be a chain of successive decays until a stable configuration is reached. The most abundant radioactive nucleus found in nature is that of uranium $^{238}_{93}U$, which undergoes a series of 14 decays, including eight alpha emissions and six beta emissions, terminating at the stable isotope of lead, $^{206}_{82}Pb$. Decay series are discussed further in Section 44–5.

In alpha decay the neutron number N and the charge number Z each decrease by two, and the mass number A decreases by four, corresponding to the values $N = 2$, $Z = 2$, $A = 4$ for the alpha particle. The situation in beta decay is less obvious; one might well question how a nucleus composed of protons and neutrons can emit an *electron*. The answer is that, in beta decay, a neutron in the nucleus is transformed into a proton, an electron, and a neutrino. Such transformations of fundamental particles are discussed further in Section 44–11. The effect is to *increase* the charge number Z by one, decrease the neutron number N by one, and leave the mass number A unchanged. Finally, gamma emission leaves all three numbers unchanged. Both alpha and beta emission are often accompanied by gamma emission.

44–4

NUCLEAR STABILITY

Of about 1500 different nuclides now known, only about one fifth are stable. The others are radioactive, with lifetimes ranging from a small fraction of a second to many years. The stable nuclei are indicated by dots on the graph in Fig. 44–6, where the neutron number N is plotted against the charge number Z. Such a graph is called a *Segrè chart*, after its inventor.

Since the mass number A is the sum $N + Z$, a curve of constant A is a straight line perpendicular to the line $N = Z$. In general, lines of con-

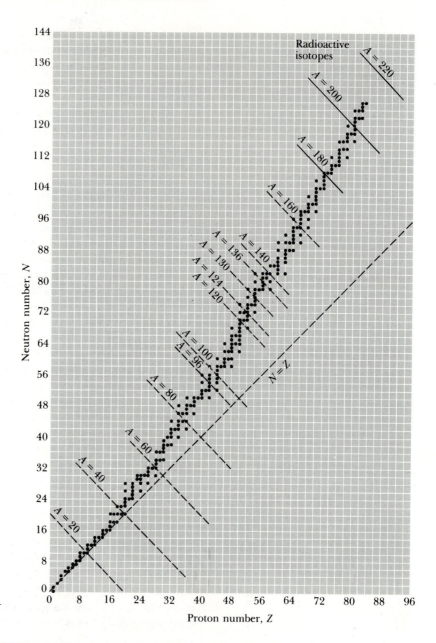

44–6 Segre chart, showing neutron number and proton number for stable nuclei.

stant A pass through only one or two stable nuclei (see $A = 20$, $A = 40$, $A = 60$, etc.), but there are four cases when such lines pass through three stable isotopes, namely at $A = 96$, 124, 130, and 136. Only four stable nuclei have both odd Z and odd N:

$$_{1}^{2}\text{H}, \quad _{3}^{6}\text{Li}, \quad _{5}^{10}\text{B}, \quad _{7}^{14}\text{N};$$

these are called *odd-odd nuclei*. Also, there is *no* stable nucleus with $A = 5$ or $A = 8$.

The points representing stable isotopes define a rather narrow stability region. For low mass numbers $N/Z = 1$. This ratio increases and becomes about 1.6 at large mass numbers. Points to the right of the stability region represent nuclei that have an excess of protons or a deficiency of neutrons; to the left of the stability region are the points representing nuclei with an excess of neutrons or a deficiency of protons. The graph also shows that there is a maximum value of A; no nucleus with A greater than 209 is stable.

The stability of nuclei can be understood qualitatively on the basis of the nature of the nuclear force and the competition between the attractive nuclear force and the repulsive electrical force. As pointed out in Section 44–2, the nuclear force favors pairs of nucleons, and pairs of pairs. In the absence of electrical interactions, the most stable nuclei would be those having equal numbers of neutrons and protons, $N = Z$. The electrical repulsion shifts the balance to favor greater numbers of neutrons, but a nucleus with *too many* neutrons is unstable because not enough of them are paired with protons. A nucleus with too many *protons* has too much repulsive electrical interaction, compared with the attractive nuclear interaction, to be stable.

Furthermore, as the number of nucleons increases, the total energy of electrical interaction increases faster than that of the nuclear interaction. To understand this, we recall the discussion of electrostatic energy in Section 27–4. The energy of a capacitor with a charge Q is proportional to Q^2. It can be shown that to place a total charge Q on the surface of a sphere of radius a requires a total energy $Q^2/8\pi\epsilon_0 a$.

Thus the (positive) electric potential energy in the nucleus increases approximately as Z^2, while the (negative) nuclear potential energy increases approximately as A, with corrections for pairing effects. Thus the competition of electric and nuclear forces accounts for the fact that the neutron–proton ratio in stable nuclei increases with Z, and also for the fact that there is a maximum A (and a maximum Z) for stability. At large A the electric energy *per nucleon* grows faster than the nuclear energy per nucleon, until the point is reached where stability is impossible. Unstable nuclei respond to these conditions in various ways; the next several sections discuss various types of decay of unstable nuclei.

44–5

RADIOACTIVE TRANSFORMATIONS

The number of radioactive nuclei in any sample of radioactive material decreases continuously as some of the nuclei disintegrate. The *rate* at which the number decreases, however, varies widely for different nuclei. Let N represent the number of radioactive nuclei in a sample at time t, and ΔN the number that undergo transformations in a short time interval Δt. Since every transformation results in a *decrease* in the number N, the corresponding change in N is $-\Delta N$ and the rate of change of N is $-\Delta N/\Delta t$. The larger the number of nuclei in the sample, the larger the number that will undergo transformations, so that the rate of change of N is proportional to N, or is equal to a constant λ multiplied by N. Thus,

$$\frac{\Delta N}{\Delta t} = -\lambda N. \qquad (44–3)$$

The constant λ is called the *decay constant*, and it has different values for different nuclides. Clearly, a large value of λ corresponds to rapid decay, and conversely.

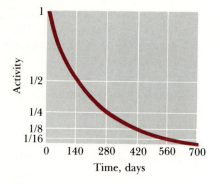

44–7 Decay curve for the radioactive element polonium. Polonium has a half-life of 140 days.

Because the number of decays per unit time is proportional to the number N of nuclei remaining (which decreases with time), the decay rate is not constant but decreases with time; thus a graph of N as a function of time is not a straight line but a curve, as shown in Fig. 44–7. By methods of calculus it can be shown that this curve is described by the function

$$N = N_0 e^{-\lambda t}, \qquad (44\text{--}4)$$

where N_0 is the number of undecayed nuclei at time $t = 0$.

The *half-life* $T_{1/2}$ of a radioactive sample is defined as the time at which the number of radioactive nuclei has decreased to one half the number at $t = 0$. At this time,

$$e^{-\lambda T_{1/2}} = \tfrac{1}{2}.$$

Taking natural logarithms of both sides and solving for $T_{1/2}$, we find

$$\lambda T_{1/2} = \ln 2,$$

$$T_{1/2} = \frac{\ln 2}{\lambda} = \frac{0.693}{\lambda}. \qquad (44\text{--}5)$$

Half of the original nuclei in a radioactive sample decay in a time interval $T_{1/2}$, half of those remaining at this time decay in a second interval $T_{1/2}$, and so on; thus the number remaining after successive intervals of $T_{1/2}$ is $\tfrac{1}{2}, \tfrac{1}{4}, \tfrac{1}{8}$, and so on.

The *mean lifetime* (or average lifetime) of a nucleus or of an unstable particle is related to the half-life $T_{1/2}$ as follows:

$$T_{\text{mean}} = \frac{1}{\lambda} = \frac{T_{1/2}}{\ln 2} = \frac{T_{1/2}}{0.693}. \qquad (44\text{--}6)$$

In particle physics, the life of an unstable particle is usually described in terms of the mean lifetime rather than the half-life.

The *activity* of a sample is defined to be the number of disintegrations per unit time. A commonly used unit is the *curie*, abbreviated C_i, defined to be 3.70×10^{10} decays per second. This is approximately equal to the activity of one gram of radium. Since the number of disintegrations is proportional to the number of radioactive nuclei in the sample, the activity decreases exponentially with time in the same way as the number N. Figure 44–7 is a graph of the activity of polonium, $^{210}_{84}\text{Po}$, which has a half-life of 140 days.

The SI unit of activity is the *becquerel*, abbreviated Bq. One becquerel is one distintegration per second. ($1 \text{ Bq} = 1 \text{ s}^{-1}$.)

In studying radioactivity the following questions are relevant:

1. What is the parent nucleus?

2. What particle is emitted from this nucleus?

3. What is the half-life of the parent nucleus?

4. What is the resulting nucleus (often called the *daughter nucleus*)?

5. Is the daughter nucleus radioactive and if so, what are the answers to questions 2, 3, and 4 for this nucleus?

Very extensive investigations have been carried on in the last 75 years, and these questions have been answered for many nuclides. The results are most conveniently presented on a Segre chart such as that shown in Fig. 44–8. The neutron number N is plotted along the vertical axis and the atomic number (or charge number) Z on the horizontal axis. Unit increase of Z with unit decrease of N indicates beta emission; decrease of two in both N and Z indicates alpha emission. The half-lives are given either in years (y), days (d), hours (h), minutes (m), or seconds (s).

Figure 44–8 shows the uranium decay series, which begins with the common uranium isotope ^{238}U. Each arrow represents a decay in which an alpha or beta particle is emitted. The decays can also be represented

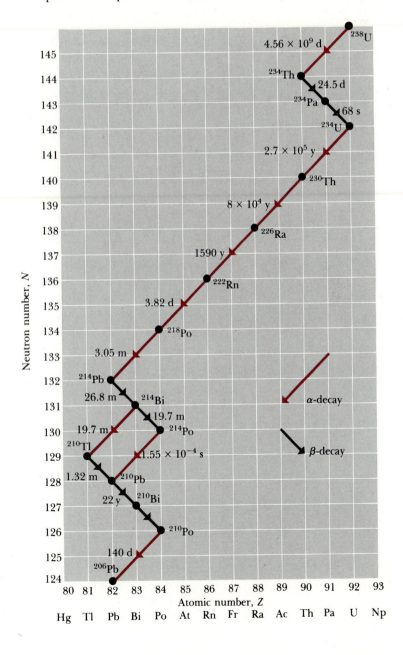

44–8 Segre chart showing the uranium ^{238}U decay series, terminating with the stable nuclide $^{206}_{82}$Pb.

in equation form; the first two decays in the series are written as

$$^{238}U \rightarrow {}^{234}Th + \alpha,$$

$$^{234}Th \rightarrow {}^{234}Pa + \beta,$$

or, even more briefly, as

$$^{238}U \xrightarrow{\alpha} {}^{234}Th,$$

$$^{235}Th \xrightarrow{\beta} {}^{234}Pa.$$

In the second decay a gamma photon is emitted following the beta; the reason for this is that the beta decay leaves the daughter nucleus not in its lowest energy state but in an excited state, from which it decays to the ground state by emitting a photon.

An interesting feature of the ^{238}U decay series is the branching that occurs at ^{214}Bi; this nuclide decays to ^{210}Pb by emission of an alpha and a beta, which can occur in either order. We also note that the series includes unstable isotopes of several elements that also have stable isotopes, including Tl, Pb, and Bi. The unstable isotopes of these elements that occur in the ^{238}U series all have too many neutrons to be stable, as discussed in Section 44–4.

Three other decay series are known; two of these occur in nature, one starting with the uncommon isotope ^{235}U, the other with thorium (^{232}Th). The last series starts with neptunium (^{237}Np), an element not found in nature but produced in nuclear reactors. In each case the series continues until a stable nucleus is reached; for these series the final members are ^{206}Pb, ^{207}Pb, ^{208}Pb, and ^{209}Bi, respectively.

An interesting application of radioactivity is the dating of archeological and geological specimens by measurement of the concentration of radioactive isotopes. The most familiar example is carbon dating; ^{14}C, an unstable isotope, is produced by nuclear reactions in the atmosphere caused by cosmic-ray bombardment, and there is a small portion of ^{14}C in the CO_2 in the atmosphere. Plants that obtain their carbon from this source contain the same proportion of ^{14}C as the atmosphere, but when a plant dies it stops taking in carbon, and the ^{14}C it has already taken in decays, with a half-life of 5,568 years. Thus by measuring the proportion of ^{14}C in the remains, one can determine how long ago the organism died. Similar techniques are used with other isotopes for dating of geologic specimens. A difficulty with carbon dating is that the ^{14}C concentration in the atmosphere changes with time, over long intervals.

*44–6

RADIATION AND THE LIFE SCIENCES

The interaction of radiation with living organisms is a topic that grows daily in interest and usefulness. In this context we construe *radiation* to include radiation emitted as a result of nuclear instability (alpha, beta, gamma, and neutrons) as well as electromagnetic radiation such as x-rays and γ rays. As these pass through matter they lose energy, breaking molecular bonds and creating ions, hence the term *ionizing radiation*. Charged particles interact through direct electrostatic forces on the electrons in the material. X-rays and γ rays can interact by photoelectric effect, where an electron absorbs a photon and is freed from its molecular bond, or by Compton scattering, discussed in Section 41–7. For sufficiently energetic photons a third possibility is pair production, discussed in Section 44–11. Neutrons cause ionization indirectly through absorption by nuclei with subsequent β or γ decay of the resulting unstable nucleus.

Ionizing radiation can change living tissue in various ways, as mentioned in Section 41–7 in connection with x-rays. The interactions of radiation with living tissue are extremely complex. It has been known for many years that excessive exposure to radiation, including sunlight, x-rays, and all the nuclear radiations, can cause destruction of tissues. In mild cases this destruction is manifested as a burn, as with common sunburn; greater exposure can cause very severe illness or death by a variety of mechanisms, one of which is the destruction of the components in bone marrow that produce red blood cells.

Radiation dosimetry is the quantitative description of the effect of radiation on living tissue. The *absorbed dose* of radiation is defined as the energy delivered to the tissue, per unit mass. The SI unit of absorbed dose, the joule per kilogram, is called the *gray* (Gy); that is, 1 Gy = $1\ \text{J} \cdot \text{kg}^{-1}$. Another unit, in more common use at present, is the *rad*, defined as $0.01\ \text{J} \cdot \text{kg}^{-1}$:

$$1\ \text{rad} = 0.01\ \text{J} \cdot \text{kg}^{-1} = 0.01\ \text{Gy}. \qquad (44\text{–}7)$$

Absorbed dose by itself is not an adequate measure of biological effect because equal amounts of energy of different kinds of radiation cause different extents of biological effect. This variation is described by a numerical factor called the *relative biological effectiveness* (RBE), also called the *quality factor* (QF), of the specific radiation under consideration. X-rays of energy 200 keV are defined to have an RBE of unity, and the effects of other radiations can be compared experimentally. The RBE also depends somewhat on the kind of tissue in which the radiation is absorbed. Table 44–3 shows approximate values of RBE for several radiations.

TABLE 44–3 RELATIVE BIOLOGICAL EFFECTIVENESS (RBE) FOR SEVERAL RADIATIONS	
Radiation	**RBE**
x-rays and γ rays	1
electrons	1.0–1.5
protons	10
α particles	20
heavy ions	20
slow neutrons	3–5

Thus the best measure of biological effect is the product of the absorbed dose and the RBE of the radiation; this quantity is called the *biologically equivalent dose*, or simply equivalent dose. The SI unit of equivalent dose for humans is the Sievert (Sv):

$$\text{equivalent dose (Sv)} = \text{RBE} \times \text{absorbed dose (Gy)}. \qquad (44\text{–}8)$$

A more common unit, corresponding to the rad, is the rem:

$$\text{equivalent dose (rem)} = \text{RBE} \times \text{absorbed dose (rad)}. \quad (44\text{--}9)$$

The unit millirem (1 mrem = 10^{-3} rem) is also in common use.

EXAMPLE   During a diagnostic x-ray, a broken leg with a mass of 5 kg receives an equivalent dose of 50 mrem. What total energy is absorbed, and how many x-ray photons are absorbed if the x-ray energy is 50 keV?

Solution   For x-rays, RBE = 1, so the absorbed dose is

$$50 \text{ mrad} = 0.050 \text{ rad} = (0.050)(0.01 \text{ J} \cdot \text{kg}^{-1}) = 5.0 \times 10^{-4} \text{ J} \cdot \text{kg}^{-1}.$$

The total energy absorbed is

$$(5.0 \times 10^{-4} \text{ J} \cdot \text{kg}^{-1})(5 \text{ kg}) = 2.5 \times 10^{-3} \text{ J} = 1.56 \times 10^{16} \text{ eV}.$$

The number of x-ray photons is

$$\frac{1.56 \times 10^{16} \text{ eV}}{5.0 \times 10^{4} \text{ eV}} = 3.12 \times 10^{11} \text{ photons}.$$

If the ionizing radiation had been a beam of α particles, for which RBE = 20, the absorbed dose needed for an equivalent dose of 50 mrem would be 2.5 mrad, corresonding to a total absorbed energy of 1.25×10^{-4} J.   ◀

Here are a few numbers for perspective. An ordinary chest x-ray delivers about 20 to 40 mrem to about 5 kg of tissue. A whole-body dose of up to about 20 rem causes no immediately detectable effect. A short-term whole-body dose of 500 rem or over usually causes death within a few days or weeks. A localized dose of 10,000 rem causes complete destruction of the exposed tissue. Radiation exposure from cosmic rays and natural radioactivity in soil, building materials, and so on, is of the order of 0.1 rem per year, considerably more at high altitudes.

The long-term hazards of radiation exposure in causing various cancers and genetic defects have been widely publicized, and the question whether there is any "safe" level of radiation exposure has been hotly debated. Various U. S. Government regulations are based on a maximum *yearly* exposure, from all except natural sources, of 0.2 to 0.5 rem. Workers having occupational exposure to radiation are permitted 5 rem per year. Recent studies suggest that these limits are too high, and that even extremely small exposures carry hazards. One study suggests that a single chest x-ray may induce leukemia in 1 of 10 million individuals. It has become clear that any use of x-rays for medical diagnosis should be preceded by a very careful consideration of the relation of risk to possible benefit.

Another sharply debated question is that of radiation hazards from nuclear power plants. It is certainly true that the radiation level from these plants is *not* zero. But to make a meaningful evaluation of hazards one must compare these levels with the alternatives, such as coal-powered plants. The health hazards of coal smoke are serious and well documented, and even the radioactivity in the smoke from a coal-fired power plant is believed to be greater than that from a properly operating nuclear plant of similar power capacity. It is clearly impossible to eliminate *all* hazards of health, and the next best alternative is an intelligent approach to the problem of *minimizing* hazards.

Radiation can also be used in beneficial ways; it is widely used for intentional selective destruction of tissue, as in treating tumors. The hazards are of course considerable, but if the disease would be fatal without treatment, then considerable hazard may be tolerable.

For sources of radiation for the treatment of cancers and related diseases, artificially produced isotopes are often used. One of the most commonly used is an isotope of cobalt, ^{60}Co. This is prepared by bombarding the stable isotope ^{59}Co with neutrons in a nuclear reactor. Neutron absorption leads to the unstable ^{60}Co; with $Z = 27$ and $N = 33$, this is an "odd-odd" nucleus, which decays to ^{60}Ni by beta and gamma emission, with a half-life of about 5 years. Such artificial sources have several advantages over naturally radioactive sources. Having shorter half-lives, they are more *intense* sources. They do not emit alpha particles, which are usually not wanted, and the electrons emitted are easily stopped by thin metal sheets without appreciably attenuating the intensity of the desired gamma radiation. Beams of π mesons produced with cyclotron beams have also been used. On a different line, work is currently under way on use of radiation to sterilize and preserve food.

Radioactive isotopes of an element have the same electron configurations as the stable isotopes and therefore have the same chemical behavior. But the location and concentration of radioactive isotopes can easily be detected by measurements of the emitted radiation. A familiar example is the use of an unstable isotope of iodine for thyroid studies. Nearly all the iodine ingested that is not eliminated eventually reaches the thyroid, and the body does not discriminate between the unstable isotope ^{131}I and the stable isotope ^{127}I. A minute quantity of ^{131}I is fed or injected, and the speed with which it becomes concentrated in the thyroid provides a measure of thyroid function. The half-life is about 8 days, so there are no long-lasting radiation hazards. By use of more sophisticated scanning detectors, one can also obtain a "picture" of the thyroid, which shows enlargement and other abnormalities. This procedure, a type of *autoradiography,* is comparable to photographing the glowing filament of an incandescent light bulb using the light emitted by the filament itself.

Similar techniques are used to visualize coronary arteries. A thin tube or *catheter* is threaded through a vein in the arm into the heart, and a radioactive material is injected. Narrowed or blocked arteries can actually be photographed by use of a scanning detector; such a picture is called an *angiogram* or *arteriogram.* A useful isotope for such purposes is an isotope of technicium, ^{99}Tc, which is formed in the decay of a molybdenum isotope, ^{99}Mo. It is formed in an excited state, from which it decays to the ground state by γ emission with a half-life of about 6 hours, unusually long for γ emission. Scanning detectors, often called *gamma cameras,* have been used for studies of the brain, kidneys, and numerous other organs.

More subtle applications of radioactive isotopes are found in *tracer* techniques. The use of radioactive iodine in the thyroid has been mentioned; tritium, a hydrogen isotope ^{3}H, is used to tag molecules in complex organic reactions. In the world of machinery, radioactive iron can be used to study piston ring wear; radioactive tags on pesticide molecules can be used to trace their passage through food chains. Laundry detergent manufacturers test the effectiveness of their products using radioactive dirt.

Thus ionizing radiation is a two-edged sword; it poses serious health hazards, yet it also provides a wide variety of benefits to mankind. As we have mentioned, these benefits include the diagnosis and treatment of disease and a wide variety of analytical techniques. Many direct effects of radiation are also useful, such as strengthening of polymers by cross-linking, sterilizing of food and surgical tools, and dispersion of unwanted static electricity by ionization of air.

44–7

NUCLEAR REACTIONS

The nuclear disintegrations described up to this point have consisted exclusively of a natural, uncontrolled emission of either an alpha or beta particle. Nothing was done to initiate the emission, and nothing could be done to stop it. It occurred to Rutherford in 1919 that it ought to be possible to penetrate a nucleus with a massive high-speed particle such as an alpha particle and thereby either produce a nucleus with greater atomic number and mass number or induce an artificial nuclear disintegration. Rutherford was successful in bombarding nitrogen with alpha particles and obtaining, as a result, an oxygen nucleus and a proton, according to the reaction

$$\ce{^4_2He} + \ce{^{14}_7N} \rightarrow \ce{^{17}_8O} + \ce{^1_1H}. \qquad (44\text{–}10)$$

Note that the sum of the initial atomic numbers is equal to the sum of the final atomic numbers, a condition imposed by conservation of charge. The sum of the initial mass numbers is also equal to the sum of the final mass numbers, but the initial rest mass is *not* equal to the final rest mass.

The difference between the rest masses corresponds to the *nuclear reaction energy*, according to the mass–energy relation $E = mc^2$. If the sum of the final rest masses *exceeds* the sum of the initial rest masses, energy is *absorbed* in the reaction. Conversely, if the final sum is less than the initial sum, energy is released in the form of kinetic energy of the final particles. (1 u = 931 MeV.)

For example, in the nuclear reaction represented by Eq. (44–10), the rest masses of the various particles, in u, are found from Table 44–2 to be

$$
\begin{array}{ll}
\ce{^4_2He} = 4.00260 \text{ u} \qquad & \ce{^{17}_8O} = 16.99913 \text{ u} \\
\ce{^{14}_7N} = 14.00307 \text{ u} \qquad & \ce{^1_1H} = 1.00783 \text{ u} \\
\hline
\phantom{\ce{^{14}_7N} = } 18.00567 \text{ u} \qquad & \phantom{\ce{^1_1H} = } 18.00696 \text{ u}
\end{array}
$$

(These values include nine electron masses in each case.) The total rest mass of the final products exceeds that of the initial particles by 0.00129 u, which is equivalent to 1.20 MeV. This amount of energy is *absorbed* in the reaction. If the initial particles do not have this much kinetic energy, the reaction cannot take place.

On the other hand, in the proton bombardment of lithium and consequent formation of two alpha particles,

$$\text{}^{1}_{1}\text{H} + \text{}^{7}_{3}\text{Li} \rightarrow \text{}^{4}_{2}\text{He} + \text{}^{4}_{2}\text{He},\qquad (44\text{–}11)$$

the sum of the final masses is *smaller* than the sum of the initial values, as shown by the following data:

$^{1}_{1}\text{He} = 1.00783$ u	$^{4}_{2}\text{He} = 4.00260$ u
$^{7}_{3}\text{Li} = 7.01601$ u	$^{4}_{2}\text{He} = 4.00260$ u
8.02384 u	8.00520 u

(Four electron masses on each side are included.) Since the decrease in mass is 0.01864 u, 17.4 MeV of energy is liberated and appears as kinetic energy of the two separating alpha particles. This computation may be verified by observing the distance the alpha particles travel in air at atmospheric pressure before being brought to rest by collisions with molecules. The distance is found to be 8.31 cm. A series of independent experiments is then performed in which the range of alpha particles of known energy is measured. These experiments show that, in order to travel 8.31 cm, an alpha particle must have an initial kinetic energy of 8.64 MeV. The energy of the two alphas together is therefore $2 \times 8.64 = 17.28$ MeV, in excellent agreement with the value 17.4 MeV obtained from the mass decrease.

Alpha particles and protons are not the only particles used to initiate artificial nuclear distintegration. The nucleus of a deuterium atom, known as the *deuteron* and represented by the symbol $^{2}_{1}\text{H}$, may be accelerated to high energy in one of a variety of particle accelerators. In order that positively charged particles such as the alpha particle, the proton, and the deuteron can be used to penetrate the nuclei of other atoms, they must travel with very high speeds to avoid being repelled or deflected by the positive charge of the nucleus they are approaching.

Absorption of neutrons by nuclei is an important class of nuclear reactions. Heavy nuclei bombarded by neutrons in a nuclear reactor may undergo a series of neutron absorptions alternating with beta decays, in which the mass number A increases by as much as 25. The *transuranic* elements, having Z larger than 92, are produced in this way; they do not occur in nature. Thirteen transuranic elements, having Z up to 105 and A up to about 260, have been identified.

The analytical technique of *neutron activation analysis* uses similar reactions. When stable nuclei are bombarded by neutrons, some absorb neutrons and then undergo beta decay. The energies of the beta and gamma emissions depend on the parent nucleus and hence provide a means of identifying it. The presence of quantities of elements far too small for conventional chemical analysis can be detected in this way.

44–8

NUCLEAR FISSION

Up to this point, all nuclear reactions considered have involved the ejection of relatively light particles, such as alpha particles, beta particles, protons, or neutrons. That this is not always the case was discovered by Hahn and Strassman in Germany in 1939. These scientists bombarded uranium ($Z = 92$) with neutrons, and after a careful chemical analysis discovered barium ($Z = 56$) and krypton ($Z = 36$) among the products. Cloud-chamber photographs showed the two heavy particles traveling in opposite directions with tremendous speed. The uranium nucleus is said to undergo *fission*. Measurement showed that an enormous amount of energy, 200 MeV, is released when uranium splits up in this way. The rest mass of a uranium atom exceeds the sum of the rest masses of the fission products. Thus it follows, from the Einstein mass–energy relation, that the extra energy released during fission is transformed into kinetic energy of the fission fragments. Uranium fission may be accomplished by either fast or slow neutrons. Of the two most abundant isotopes of uranium, $^{238}_{92}U$ and $^{235}_{92}U$, both may be split by a fast neutron, whereas only $^{235}_{92}U$ is split by a slow neutron.

When uranium undergoes fission, barium and krypton are not the only products. Over 100 different isotopes of more than 20 different elements have been detected among fission products. All of these atoms are, however, in the middle of the periodic table, with atomic numbers ranging from 34 to 58. Because the neutron–proton ratio needed for stability in this range is much *smaller* than that of the original uranium nucleus, the *fission fragments,* as the residual nuclei are called, always have too many neutrons for stability. A few free neutrons are liberated during fission, and the fission fragments undergo a series of beta decays (each of which increases Z by one and decreases N by one) until a stable nucleus is reached. During decay of the fission fragments, an average of 15 MeV of additional energy is liberated.

Discovery of the facts that 200 MeV of energy is released when uranium undergoes fission, and that other neutrons are liberated from the uranium nucleus during fission, suggested the possibility of a *chain reaction,* that is, a self-sustaining series of events that, once started, continues until much of the uranium in a given sample is used up (provided the sample stays together). In the case of a uranium chain reaction, a neutron causes one uranium atom to undergo fission, during which a large amount of energy and several neutrons are emitted. These neutrons then cause fission in neighboring uranium nuclei, which also give out energy and more neutrons. The chain reaction may be made to proceed slowly and in a controlled manner, and the device for accomplishing this is called a *nuclear reactor*. If the chain reaction is fast and uncontrolled, the device is a bomb.

In a nuclear reactor, the fissionable element is contained in *fuel elements,* whose configuration must be designed so that sufficient neutrons are slowed down in the surrounding material (often water) and cause further fissions, rather than escaping from the fuel region. On the average each fission produces about 2.5 free neutrons, so 40% of the neutrons are needed to sustain a chain reaction. The *rate* of the reaction is controlled by inserting or withdrawing *control rods* made of elements (often cadmium) whose nuclei *absorb* neutrons without undergoing any additional chain reaction.

The most familiar application of nuclear reactors is for the generation of electric power. To illustrate some of the numbers involved, we consider a hypothetical nuclear power plant with a generating capacity of 1000 MW; this figure is typical of large plants currently being built. As noted above, the fission energy appears as kinetic energy of the fission fragments, and its immediate result is to heat the fuel elements and the surrounding water. This heat generates steam to drive turbines, which in turn drive the electrical generators. The turbines, being heat engines, are subject to the efficiency limitations imposed by the second law of thermodynamics, as discussed in Chapter 19. In modern nuclear plants the overall efficiency is about one third, so 3000 MW of thermal power from the fission reaction are needed for 1000 MW of electrical power.

It is easy to calculate how much uranium must undergo fission per unit time to provide 3000 MW of thermal power. Each second we need 3000 MJ or 3000×10^6 J. Each fission provides 200 MeV, which is

$$200 \text{ MeV} = (200 \text{ MeV})(1.6 \times 10^{-13} \text{ J} \cdot \text{MeV}^{-1}) = 3.2 \times 10^{-11} \text{ J}.$$

Thus the number of fissions needed per second is

$$\frac{3000 \times 10^6 \text{ J}}{3.2 \times 10^{-11} \text{ J}} = 0.94 \times 10^{20}.$$

Each uranium atom has a mass of about $(235)(1.67 \times 10^{-27} \text{ kg}) = 3.9 \times 10^{-25}$ kg, so the mass of uranium needed per second is

$$(0.94 \times 10^{20})(3.9 \times 10^{-25} \text{ kg}) = 3.7 \times 10^{-5} \text{ kg} = 37 \text{ mg}.$$

In one day, (86,400 seconds) the consumption of uranium is

$$(3.7 \times 10^{-5} \text{ kg} \cdot \text{s}^{-1})(86,400 \text{ s} \cdot \text{d}^{-1}) = 3.2 \text{ kg} \cdot \text{d}^{-1}.$$

For comparison, we note that the 1000-MW coal-fired power plant discussed in Section 19–10 burns 10,600 tons (about 10^7 kg) of coal per day!

Nuclear fission reactors have several other practical uses; among these are the production of artificial radioactive isotopes for medical and other research; production of high-intensity neutron beams for research in nuclear structure, and production of fissionable transuranic elements such as plutonium from the common isotope ^{238}U. The last is the function of *breeder reactors*.

As noted above, about 15 MeV of the energy liberated as a result of fission of a ^{235}U nucleus comes from the subsequent beta decay of the fission fragments rather than the kinetic energy of the fragments themselves. This fact poses a serious problem with respect to control and safety of reactors. Even after the chain fission reaction has been completely stopped by insertion of control rods into the core, heat continues to be evolved by the beta decays, which cannot be stopped. For the 1000-MW reactor described above, this heat power amounts to about 200 MW, which, in the event of total loss of cooling water, is more than enough to cause a catastrophic "melt-down" of the reactor core and possible penetration of the containment vessel. The difficulty in achieving a "cold shut-down" following the accident at Three Mile Island in March 1979 resulted from the continued evolution of heat due to beta decays.

Fission appears to set an upper limit on the production of transuranic nuclei, discussed in Section 44–7. When a nucleus with $Z = 105$ is bombarded with neutrons, fission occurs essentially instantaneously; no $Z = 106$ nucleus is formed even for a short time. There are theoretical reasons to expect that nuclei in the vicinity of $Z = 114$, $N = 184$, might be stable with respect to spontaneous fission. These numbers correspond to *filled shells* in the nuclear energy-level structure, analogous to the filled shells of electrons in the inert gases, as discussed in Section 42–2. Such nuclei, called *superheavy nuclei,* would still be unstable with respect to alpha emission, but they might live long enough to be identified. Attempts to produce superheavy nuclei in the laboratory have not been successful; whether they exist in nature is still an open question.

44–9

NUCLEAR FUSION

There are two types of nuclear reactions in which large amounts of energy may be liberated. In both types, the rest mass of the products is less than the original rest mass. The fission or uranium, already described, is an example of one type. The other involves the combination of two light nuclei to form a nucleus that is more complex but whose rest mass is less than the sum of the rest masses of the original nuclei. Examples of such energy-liberating reactions are as follows:

$$_1^1H + {}_1^1H \rightarrow {}_1^2H + {}_1^0e,$$

$$_1^2H + {}_1^1H \rightarrow {}_2^3He + \gamma\text{-radiation},$$

$$_2^3He + {}_2^3He \rightarrow {}_2^4He + {}_1^1H + {}_1^1H.$$

In the first, two protons combine to form a deuteron and a positron (a positively charged electron, to be discussed in Section 44–11). In the second, a proton and a deuteron unite to form the light isotope of helium. For the third reaction to occur, the first two reactions must occur twice, in which case two nuclei of light helium unite to form ordinary helium. These reactions, known as the *proton–proton chain,* are believed to take place in the interior of the sun and also in many other stars that are known to be composed mainly of hydrogen.

The positrons produced during the first step of the proton–proton chain collide with electrons; annihilation takes place, and their energy is converted into gamma radiation. The net effect of the chain, therefore, is the combination of four hydrogen nuclei into a helium nucleus and gamma radiation. The net amount of energy released may be calculated from the mass balance as follows:

Mass of four hydrogen atoms (including electrons)	= 4.03132 u
Mass of one helium plus two additional electrons	= 4.00370 u
Difference in mass	= 0.02762 u
	= 25.7 MeV

In the case of the sun, 1 g of its mass contains about 2×10^{23} protons. Hence, if all of these protons were fused into helium, the energy released would be about 57,000 kWh. If the sun were to continue to radiate at its present rate, it would take about 30 billion years to exhaust its supply of protons.

For fusion to occur, the two nuclei must come together to within the range of the nuclear force, typically of the order of 2×10^{-15} m. To do this they must overcome the electrical repulsion of their positive charges; for two protons at this distance the corresponding potential energy is of the order of 1.1×10^{-13} J or 0.7 MeV, which thus represents the initial *kinetic* energy the fusion nuclei must have.

Such energies are available at extremely high temperatures. According to Section 20–4, the average translational kinetic energy of a gas molecule at temperature T is $3kT/2$, where k is Boltzmann's constant. For this to be equal to 1.1×10^{-13} J, the temperature must be of the order of 5×10^9 K. Of course, not all the nuclei have to have this energy, but this calculation shows that the temperature must be of the order of millions of kelvins if any appreciable fraction of the nuclei are to have enough kinetic energy to surmount the electrical repulsion and achieve fusion.

Such temperatures occur in stars as a result of gravitational contraction and its associated liberation of gravitational potential energy. When the temperature gets high enough, the reactions occur, more energy is liberated, and the pressure of the resulting radiation prevents further contraction. Only after most of the hydrogen has been converted into helium will further contraction and an accompanying increase of temperature result. Conditions are then suitable for the formation of heavier elements.

Temperatures and pressures similar to those in the interior of stars may be achieved on earth at the moment of explosion of a uranium or plutonium fission bomb. If the fission bomb is surrounded by proper proportions of the hydrogen isotopes, these may be caused to combine into helium and liberate still more energy. This combination of uranium and hydrogen is called a "hydrogen bomb."

Intensive efforts are underway in many laboratories to achieve controlled fusion reactions, which potentially represent an enormous new energy resource. In one kind of experiment, a plasma is heated to extremely high temperature by an electrical discharge, while being contained by appropriately shaped magnetic fields. In another, pellets of the material to be fused are heated by a high-intensity laser beam. One current laser experiment set-up is shown in Fig. 44–9. Reactions being studied include the following:

$$\begin{matrix} {}_1^2\text{H} + {}_1^2\text{H} \rightarrow {}_1^3\text{H} + {}_1^1\text{H} + 4 \text{ MeV}, & (1) \end{matrix}$$

$$\begin{matrix} {}_1^3\text{H} + {}_1^2\text{H} \rightarrow {}_2^4\text{He} + {}_0^1\text{n} + 17.6 \text{ MeV}, & (2) \end{matrix}$$

$$\begin{matrix} {}_1^2\text{H} + {}_1^2\text{H} \rightarrow {}_2^3\text{He} + {}_0^1\text{n} + 3.3 \text{ MeV}, & (3) \end{matrix}$$

$$\begin{matrix} {}_2^3\text{He} + {}_1^2\text{H} \rightarrow {}_2^4\text{He} + {}_1^1\text{H} + 18.3 \text{ MeV}. & (4) \end{matrix}$$

In the first, two deuterons combine to form tritium and a proton. In the second, the tritium nucleus combines with another deuteron to form helium and a neutron. The result of both of these reactions together is the conversion of three deuterons into a helium-4 nucleus, a proton, and a neutron, with the liberation of 21.6 MeV of energy. Reactions (3) and (4) together achieve the same conversion. In a plasma containing deuterium, the two pairs of reactions occur with roughly equal probability. As yet no one has succeeded in producing these reactions under controlled conditions in such a way as to yield a net surplus of usable energy, but the practical problems do not appear to be insurmountable.

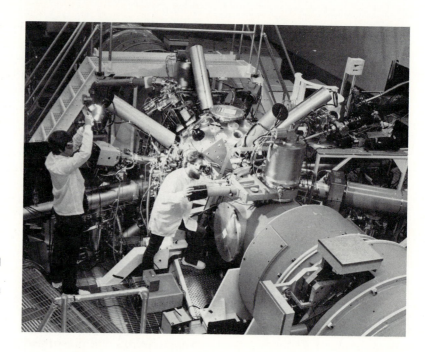

44–9 The Novette laser system at Lawrence Livermore National Laboratory, used for fusion research. This system went into operation in January 1983; it can deliver a power of 50×10^{12} W for a period of the order of 1 ns. (Courtesy Lawrence Livermore National Laboratory.)

44–10

PARTICLE ACCELERATORS

Many important experiments in nuclear and high-energy physics during the last 60 years or so have made use of beams of charged particles, such as protons or electrons, that have been accelerated to high speeds. Any device that uses electric and magnetic fields to guide and accelerate a beam of charged particles to high speed is called a *particle accelerator*. In a sense, the cathode-ray tubes of Thomson and his contemporaries were the first accelerators. In more recent times, accelerators have grown enormously in size, complexity, and energy range.

The *cyclotron,* developed in 1931 by Lawrence and Livingston at the University of California, is important historically because it was the first accelerator to use a magnetic field to guide particles in a nearly circular path so that they could be accelerated repeatedly by an electric field in a cyclic process. Lawrence was awarded the Nobel prize for physics in 1939 for his invention of the cyclotron.

The heart of the cyclotron is a pair of hollow metal chambers, labeled D_1 and D_2 in Fig. 44–10, resembling the halves of a hollow cylinder cut parallel to its axis. These are called *dees,* referring to their shape. A source of ions, often simply protons (ionized hydrogen atoms), is located near the midpoint of the gap between the dees, and the dees are connected to the terminals of a circuit generating an alternating voltage. The potential between the dees thus alternates rapidly, some millions of times per second; so the electric field in the gap between the dees is directed first toward one and then toward the other. But, because of the electrical shielding effect of the dees, the space *within* each is a region of nearly zero electric field.

The two dees are enclosed within, but insulated from, a somewhat larger cylindrical metal container from which the air is exhausted, and the whole apparatus is placed between the poles of a powerful electromagnet, which provides a uniform magnetic field perpendicular to the ends of the cylindrical container.

Consider an ion of charge $+q$ and mass m, emitted from the ion source S at an instant when D_1 in Fig. 44–10 is positive. The ion is accelerated by the electric field in the gap between the dees and then enters the region within D_2, where $E = 0$, with a speed v_1. Since its motion is at right angles to the magnetic field, the ion travels in a circular arc of radius

$$r_1 = \frac{mv_1}{Bq}.$$

If, now, in the time required for the ion to complete a half-circle, the *electric* field has reversed so that its direction is toward D_1, the ion is again accelerated as it crosses the gap between the dees and enters D_1 with a greater speed v_2. It therefore moves in a half circle of larger radius within D_1 to emerge again into the gap.

The *angular* velocity ω of the ion is

$$\omega = \frac{v}{r} = B\left(\frac{q}{m}\right). \tag{44–12}$$

Hence the angular velocity is *independent of the speed of the ion and of the radius of the circle* in which it travels, depending only on the magnetic field and the charge-to-mass ratio (q/m) of the ion. If, therefore, the electric field reverses at regular intervals, each equal to the time required for the ion to make a half-revolution, the field in the gap will always be in the proper direction to accelerate an ion each time the gap is crossed. It is this feature of the motion, that the time of rotation is independent of the radius, that makes the cyclotron feasible, since the regularly timed reversals are accomplished automatically by the oscillator circuit to which the dees are connected.

The path of an ion is a sort of spiral, composed of semicircular arcs of progressively larger radius connected by short segments along which

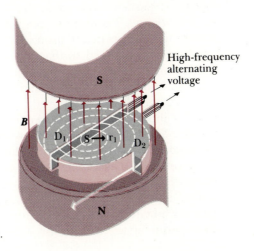

44–10   Schematic diagram of a cyclotron.

the radius is increasing. If R represents the outside radius of the dees, and v_{max} the speed of the ion when traveling in a path of this radius, then

$$v_{max} = BR\left(\frac{q}{m}\right),$$

and the corresponding kinetic energy of the ion is

$$K = \frac{1}{2}mv_{max}{}^2 = \frac{1}{2}m\left(\frac{q}{m}\right)^2 B^2 R^2. \qquad (44\text{--}13)$$

If the ions are protons,

$$\frac{q}{m} = \frac{1.60 \times 10^{-19}\text{ C}}{1.67 \times 10^{-27}\text{ kg}} = 9.58 \times 10^7\text{ C} \cdot \text{kg}^{-1}.$$

Taking values typical of early-day cyclotrons, we assume $B = 1.5$ T and $R = 0.5$ m. Then the maximum kinetic energy is, from Eq. (44–13),

$$K = \tfrac{1}{2}(1.67 \times 10^{-27}\text{ kg})(9.58 \times 10^7\text{ C} \cdot \text{kg}^{-1})^2(1.5\text{ T})^2(0.5\text{ m})^2$$
$$= 4.31 \times 10^{-12}\text{ J} = 2.69 \times 10^7\text{ eV} = 26.7\text{ MeV}.$$

This energy, considerably larger than the average binding energy per nucleon, is sufficient to cause a variety of interesting nuclear reactions.

The energy attainable with the cyclotron is limited by relativistic effects. For Eq. (44–12) to be relativistically correct, m should be replaced by $m/(1 - v^2/c^2)^{1/2}$. As the particles speed up, the angular velocity *decreases;* if the decrease is appreciable, the particle motion is no longer in the correct phase relative to the alternating dee voltage. In a variation called the *synchrocyclotron,* the particles are accelerated in bursts, and for each burst the frequency of the alternating voltage is decreased at just the right rate to maintain the correct phase relation with the particles' motion. A practical limitation of the cyclotron is the expense of building very large electromagnets. The largest synchrocyclotron ever built has a vacuum chamber about 8 m in diameter and accelerates protons to energies of about 600 MeV. The synchrocyclotron, incidentally, provides a very direct confirmation of relativistic mechanics.

To attain higher energies, another type of machine, called the *synchrotron,* is more practical. In a synchrotron, the vacuum chamber in which the particles move is in the form of a thin doughnut, called the *accelerating ring.* The particles are forced to move within this chamber by a series of magnets placed around it. As the particles speed up, the magnetic field is increased so that the particles retrace the same trajectory over and over. The synchrotron located at the Fermi National Accelerator Laboratory (Fermilab) in Batavia, Illinois, can accelerate protons to an energy of 500 GeV (500×10^9 eV), and modifications are under way which will permit a maximum energy of 1000 GeV. It uses an accelerating ring 2 km in diameter and cost about $400 million to build. In each machine cycle, of a few seconds' duration, it accelerates approximately 10^{13} protons. An aerial view of the Fermilab accelerator is shown in Fig. 44–11.

As higher and higher energies are sought in the attempt to investigate new phenomena in particle interactions, a new problem emerges. In an experiment where a beam of high-energy particles collides with a

44–11 An aerial view of the 500-GeV accelerator at the Fermi National Accelerator Laboratory, Batavia, Illinois. (Courtesy of Fermi National Accelerator Laboratory.)

stationary target (such as protons onto a liquid-hydrogen target), not all the kinetic energy of the incident particles is available to form new particle states. Because momentum must be conserved, the particles emerging from the collision must have some motion and thus some kinetic energy. The energy E_a available for creating new particles or particle configurations is the *difference* between initial and final kinetic energies.

In the extreme-relativistic range, where the kinetic energies of the particles are large compared to their rest energies, this is a severe limitation. When bombarding and target particles have equal mass, as with protons onto a hydrogen target, it can be shown from relativistic mechanics that the available energy E_a is related to the total energy E of the bombarding particle and to its mass m by

$$E_a = \sqrt{2mc^2E}. \tag{44–14}$$

For example, for the proton, $mc^2 = 931$ MeV $= 0.931$ GeV. If $E = 500$ GeV, as for the Fermilab accelerator, then

$$E_a = \sqrt{2(0.931 \text{ GeV})(500 \text{ GeV})} = 30.5 \text{ GeV},$$

and if $E = 1000$ GeV, $E_a = 43.1$ GeV. Thus, increasing the beam energy by 500 GeV increases the available energy by only 12.6 GeV.

This limitation may be circumvented in part by *colliding-beam* experiments, in which there is no stationary target but in which beams of particles circulate in opposite directions in arrangements called *storage rings*. In regions where the rings intersect, the beams are focused sharply onto one another, and collisions can occur. Because the total momentum in such a two-particle collision is zero, the available energy E_a is the total kinetic energy of the two particles.

An example is the storage ring facility at the Stanford Linear Accelerator Center (SLAC), where electron and positron beams collide with total available energy of up to 36 GeV. Some other storage-ring facilities are located at DESY (German Electron Synchrotron) in Hamburg, West Germany (E_a up to 38 GeV in e^+e^- collisions), and at the Cornell Electron Storage Ring Facility (CESR), where the maximum energy *per beam* is 8 GeV. At the CERN (European Council for Nuclear Research) laboratory in Geneva, Switzerland, construction has begun for a large electron–positron ($e^+ - e^-$) storage ring that will transport beams of particles with energies of 50 GeV or more. This facility is expected to have usable beams in 1989 and will provide total available energy of 100 GeV (in the center of mass system).

44–11

FUNDAMENTAL PARTICLES

The physics of fundamental particles has been a recognized field of research only in the past 40 years. The electron and proton were known by the turn of the century, but the existence of the neutron was not established definitely until 1930; its discovery is an interesting story and a useful illustration of nuclear reactions.

In 1930, Bothe and Becker in Germany observed that when beryllium, boron, or lithium was bombarded by fast alpha particles, the bombarded material emitted something, either particles or electromagnetic waves, of much greater penetrating power than the original alpha particles. Further experiments in 1932 by Curie and Joliot in Paris confirmed these results, but all attempts to explain them in terms of gamma rays were unsuccessful. Chadwick in England repeated the experiments and found that they could be satisfactorily interpreted on the assumption that *uncharged* particles of mass approximately equal to that of the proton were emitted from the nuclei of the bombarded material. He called the particles *neutrons*. The emission of a neutron from a beryllium nucleus takes place according to the reaction

$$\ce{^{4}_{2}He} + \ce{^{9}_{4}Be} \rightarrow \ce{^{12}_{6}C} + \ce{^{1}_{0}n}, \tag{44–15}$$

where $^{1}_{0}\text{n}$ is the symbol for a neutron.

Since neutrons have no charge, they produce no ionization in their passage through gases. They are not deflected by the electric field around a nucleus and can be stopped only by colliding with a nucleus in a direct hit, in which case they may either undergo an elastic impact or penetrate the nucleus. It was shown in Chapter 7 that if an elastic body strikes a motionless elastic body of the same mass, the first is stopped and the second moves off with the same speed as the first. Since the proton and neutron masses are almost the same, fast neutrons are slowed down most effectively by collisions with the hydrogen atoms in hydrogenous materials such as water or paraffin. The usual laboratory method of obtaining slow neutrons is to surround the fast neutron source with water or blocks of paraffin.

Once the neutrons are moving slowly, they may be detected by means of the alpha particles they eject from the nucleus of a boron atom, according to the reaction

$$\ce{^{1}_{0}n} + \ce{^{10}_{5}B} \rightarrow \ce{^{7}_{3}Li} + \ce{^{4}_{2}He}. \tag{44–16}$$

The ejected alpha particle then produces ionization that may be detected in a Geiger counter or an ionization chamber.

The discovery of the neutron gave the first real clue to the structure of the nucleus. Before 1930 it had been thought that the total mass of a nucleus was due to protons only. We now know that a nucleus consists of both protons and neutrons (except hydrogen, whose nucleus consists only of one proton) and that (1) the mass number A equals the total number of nuclear particles and (2) the atomic number Z equals the number of protons.

The study of *cosmic rays* has been a very fertile field of particle physics. It has been known since the early years of this century that air and other gases are slightly ionized, and therefore slightly conductive, at all times, even in the absence of obvious causes of ionization such as x-rays, ultraviolet light, or radioactivity. For example, if a charged electroscope is left standing, it will eventually lose its charge no matter how well it is insulated. Ionization of air inside a vessel is decreased slightly if the vessel is lowered into a lake, but it increases considerably if the vessel is transported in a balloon high into the stratosphere. Hess suggested that the ionization is due to some kind of penetrating waves or particles from outer space, and called them *cosmic rays*.

It is fairly certain that cosmic rays consists largely of high-speed protons with energies of the order of billions of electron volts. A collision between such a proton and the nucleus of a nitrogen or oxygen atom in the upper atmosphere gives rise to so many interesting secondary phenomena that the study of cosmic rays has become one of the richest sources of knowledge of the behavior and properties of fundamental particles.

An instrument called the *bubble chamber* has been used extensively not only to study cosmic rays but also to render visible the paths of particles involved in nuclear and fundamental-particle reactions; this instrument was described in Section 17–10. Bubble chambers are sometimes made several feet in diameter; particle tracks are photographed with two or three cameras to obtain stereoscopic pictures. Normally the chamber is placed in a magnetic field so that a charged particle will be deflected; by measuring the radius of curvature of the path we may determine the particle's momentum. The *cloud chamber*, also described in Section 17–10, was an important historical predecessor of the bubble chamber.

The positive electron or *positron* was first observed during the course of an investigation of cosmic rays by Dr. Carl D. Anderson in 1932, in a historic cloud-chamber photograph reproduced in Fig. 44–12. The photograph was made with the cloud chamber in a magnetic field perpendicular to the plane of the paper. A lead plate crosses the chamber and evidently the particle has passed through it. Since the curvature of the track is greater above the plate than below it, the velocity is less above than below; the inference is that the particle was moving upward, since it could not have *gained* energy going through the lead.

The *density* of droplets along the path is the same as would be expected if the particle were an electron. But the direction of the magnetic field and the direction of motion are consistent only with a particle of *positive* charge. Hence Anderson concluded that the track had been made by a positive electron or *positron*. Since the time of this discovery, the positron's existence has been definitely established. Its mass equals that of a negative electron and its charge is equal in magnitude but of opposite sign to that of the electron.

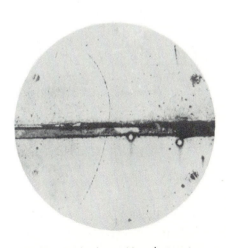

44–12　Track of a positive electron traversing a lead plate 6 mm thick. (Photograph by C. D. Anderson.)

Positrons have only a transitory existence and do not form a part of ordinary matter. They are produced in high-energy collisions of charged particles or gamma photons with matter in a process called *pair production,* in which an ordinary electron and a positron are produced simultaneously. Electric charge is conserved in this process, but enough energy must be available to account for the energy equivalent of the rest masses of the two particles, about 0.5 MeV each. The inverse process, e^+e^- *annihilation,* occurs when a positron collides with an ordinary electron. Both particles disappear and two or three gamma photons appear, with total energy $2mc^2$, where m is the electron rest mass. Decay into a *single* gamma photon is impossible because such a process cannot possibly conserve both energy and momentum.

Positrons also occur in the decay of some unstable nuclei. We recall that nuclei having too many neutrons for stability often emit a beta particle (electron), decreasing N by one and increasing Z by one. Similarly, a nucleus having *too few neutrons* for stability may respond by emitting a positron, increasing N by one and decreasing Z by one. Such nuclides do not occur in nature, but they can be produced artificially by neutron bombardment of stable nuclides in nuclear reactors. An example is the unstable nuclide $^{22}_{11}\text{Na}$, which has one less neutron than the stable $^{23}_{11}\text{Na}$. It emits a positron, leaving the stable nuclide $^{22}_{10}\text{Ne}$.

In 1935, the Japanese physicist Hideki Yukawa inferred, from theoretical considerations, the existence of a particle of mass intermediate between that of the electron and the proton. A particle of intermediate mass, but *not* identical with that predicted by Yukawa, was discovered one year later by Anderson and Neddermeyer as a component of cosmic radiation. This particle is now known as a μ *meson* (or *muon*). The μ^- has charge equal to that of the electron, and its antiparticle the μ^+ has a positive charge of equal magnitude. The two particles have equal mass, about 207 times the electron mass. The muons are unstable; each decays into an electron of the same sign, plus two neutrinos, with a lifetime of about 2.2×10^{-6} s.

Yukawa first proposed the mesons as a basis for nuclear forces; he suggested that nucleons could interact by emitting and absorbing unstable particles, just as two basketball players interact by tossing the ball back and forth, or by snatching it away from each other. It was established soon after the discovery of the muons that they could not be Yukawa's particles because their interactions with nuclei were far too weak. But in 1947 another family of mesons was discovered; called π *mesons* or *pions,* these can be positive, negative, and neutral. The charged pions have masses of about 273 times the electron mass and decay into muons with the same sign, plus a neutrino, with a lifetime of about 2.6×10^{-8} s. The neutral pion has a small mass, about 264 electron masses, and decays, with a very short lifetime of about 0.8×10^{-16} s, into two gamma-ray photons.

In the years since 1947, *high-energy physics* has emerged as a distinct branch of physics. These years have witnessed the attainment of higher and higher energies in particle accelerators, the discovery of a whole array of new particles, and intensive efforts to understand the properties of these new particles and their interactions.

Along with higher energies and the creation of new particles has come the need for new and more sophisticated detectors. Since energy and momentum must be conserved in any reaction or decay process, most large present day detectors are designed to handle the problems of mass, charge, and momentum identification. Early detectors such as scintillation counters, proportional tubes, cloud and bubble chambers have been replaced by vertex detectors, electromagnetic calorimeters, wire proportional chambers, large solenoidal magnets, and various kinds of Cerênkov counters. Modern electronics in the form of on-line computers and microprocessors has followed, along with higher accelerator duty cycles. Present-day electronic detectors can comfortably distinguish individual interactions (events) that occur 1 microsecond apart.

*44–12

HIGH-ENERGY PHYSICS

It was recognized, even in the early years of high-energy physics, that the fundamental particles are not *permanent* entities but can be created and destroyed in interactions with other particles. The earliest such interaction to be observed was that of creation and destruction of electron–positron pairs. Such pairs are *created* in collisions of high-energy cosmic-ray particles with stationary targets; when an electron and a positron collide, both *disappear* and two or three gamma-ray photons are created to carry away the energy. This transitory nature of the fundamental particles may seem disturbing, but in one sense it is a welcome development. We have seen that photons and electrons (and indeed all particles) share the dual wave-particle nature discussed in Section 42–2, and photons are known to be created and destroyed (or emitted and absorbed) in atomic transitions. Thus it seems natural that other particles can also be created and destroyed.

As an example, it was speculated as early as 1932 that there might be an *antiproton*, bearing the same relation to the ordinary proton as the positron does to the electron, that is, a particle with the same mass as the proton but negatively charged. Finally in 1955 proton–antiproton pairs were created by impact on a stationary target of a beam of protons with kinetic energy 6 GeV (6×10^9 eV) from the Bevatron at the University of California at Berkeley.

In the years after 1960, as higher-energy accelerators and more sophisticated detectors were developed, a veritable blizzard of new unstable particles occurred. Initially they were classified according to *mass*. The electrons, muons, and their associated neutrinos were called *leptons;* particles with masses between those of muons and nucleons were called *mesons,* and the nucleons and more massive particles were called *baryons*. Particles were further classified according to electric charge, spin, and two additional quantum numbers, *isospin* (the number determining the number of different charges for a particular type of particle) and *strangeness* (a number needed to account for the production and decay modes of certain particles). A partial list of some known particles is shown in Table 44–4. All particles having zero or integer spin (including photons and π and K mesons) are called *bosons,* and all particles having half-integer spin (including all leptons and baryons) are called *fermions*. These terms, derived from the names Bose and Fermi, refer to the different statistical energy distribution functions describing the behaviors of the two classes of particles.

During this period it became clear that particles could also be classified in terms of the types of *interactions* in which they participate, and in terms of the conservation laws associated with these interactions. There appear to be four classes of interactions; in order of decreasing strength, they are

1. Strong interactions,
2. Electromagnetic interactions,
3. Weak interactions,
4. Gravitational interactions.

Particles that experience strong interactions are called *hadrons;* these include all the mesons and baryons in Table 44–4. The strong interactions are responsible for the nuclear force and also for the creation of pions, heavy mesons, and hyperons in high-energy collisions. Electrons, muons, and neutrinos, have *no* strong interactions.

The *electromagnetic* interactions are those associated directly with electric charge; as noted previously, the electromagnetic interaction between two protons is weaker at distances of the order of nuclear di-

TABLE 44–4   SOME KNOWN PARTICLES AND THEIR PROPERTIES

	Particle	Mass (MeV/c^2)	Charge	Spin	Isopin	Strangeness	Mean lifetime (s)	Typical decay modes	Quark content
leptons	e^-	0.511	-1	$\frac{1}{2}$	—	0	stable	—	—
	ν_e	$0(<5\times10^{-5})$	0	$\frac{1}{2}$	—	0	stable	—	—
	μ^-	105.7	-1	$\frac{1}{2}$	—	0	2.2×10^{-6}	$e^-\bar{\nu}_e\nu_\mu$	—
	ν_μ	$0(<0.52)$	0	$\frac{1}{2}$	—	0	stable	—	—
	τ^-	1784	-1	$\frac{1}{2}$	—	0	5×10^{-13}	$\mu^-\bar{\nu}_\mu\nu_\tau$	—
	ν_τ	$0(<250)$	0	$\frac{1}{2}$	—	0	stable	—	—
mesons	π^0	135.0	0	0	1	0	0.83×10^{-16}	$\gamma\gamma$	$u\bar{u},\ d\bar{d}$
	π^+	139.6	$+1$	0	1	0	2.6×10^{-8}	$\mu^+\nu_\mu$	ud
	π^-	139.6	-1	0	1	0	2.6×10^{-8}	$\mu^-\bar{\nu}_\mu$	$\bar{u}d$
	K^+	493.7	$+1$	0	$\frac{1}{2}$	$+1$	1.24×10^{-8}	$\mu^+\nu_\mu$	$u\bar{s}$
	K^-	492.67	-1	0	$\frac{1}{2}$	-1	1.24×10^{-8}	$\mu^-\bar{\nu}_\mu$	$\bar{u}s$
	η^0	548.8	0	0	0	0	$\sim10^{-18}$	$\gamma\gamma$	$u\bar{u},\ d\bar{d},\ s\bar{s}$
baryons	p	938.3	$+1$	$\frac{1}{2}$	$\frac{1}{2}$	0	stable	—	uud
	n	939.6	0	$\frac{1}{2}$	$\frac{1}{2}$	0	917	$pe^-\bar{\nu}_e$	udd
	Λ	1115	0	$\frac{1}{2}$	0	-1	2.63×10^{-10}	$p\pi^-$ or $n\pi^0$	uds
	Σ^+	1189	$+1$	$\frac{1}{2}$	1	-1	0.80×10^{-10}	$p\pi^0$ or $n\pi^+$	uus
	Δ^{++}	1232	$+2$	$\frac{3}{2}$	$\frac{3}{2}$	0	$\sim10^{-23}$	$p\pi^+$	uuu
	Ξ^-	1321	-1	$\frac{1}{2}$	$\frac{1}{2}$	-2	1.64×10^{-10}	$\Lambda\pi^-$	dss
	Ω^-	1672	-1	$\frac{3}{2}$	0	-3	0.82×10^{-10}	ΛK^-	sss
	Λ_c^+	2273	1	$\frac{1}{2}$	0	0	$\sim7\times10^{-13}$	$\Lambda\pi\pi\pi$	udc

mensions than the strong interaction, but it has longer range. Neutral particles have no electromagnetic interactions, with the exception of effects due to the magnetic moments of neutral baryons; these magnetic moments are believed to be associated with the emission and absorption of charged pions and heavy mesons.

The *weak* interaction is responsible for beta decay such as the conversion of a neutron into a proton, an electron, and a neutrino. It is also responsible for the *decay* of many unstable particles (pions into muons, muons into electrons, Λ particles into protons, and so on). The *gravitational* interaction, although of central importance for the large-scale structure of celestial bodies, is not believed to be of significance in the analysis of fundamental-particle interactions. For example, the gravitational attraction of two electrons is smaller than their electrical repulsion by a factor of about 2.4×10^{-43}.

Several conservation laws are believed to be obeyed by *all* of the above interactions. These include the laws growing out of classical physics: energy, momentum, angular momentum, and electric charge. In addition, several new quantities having no classical analog have been introduced to help characterize the properties of particles. These include *baryon number* (the number of baryons minus the number of antibaryons), *isospin* (used also to describe the charge-independence of nuclear forces), *parity* (the comparative behavior of two systems that are mirror images of each other), *strangeness* (a quantum number used to classify particle production and decay reactions), and *lepton number* (the number of leptons minus the number of antileptons). Baryon number is conserved in *all* interactions; isospin is conserved in strong interactions but not in electromagnetic or weak interactions. Parity and strangeness are conserved in strong and electromagnetic but not in weak interactions. Lepton number is conserved in all interactions. Thus the new conservation laws are not absolute but instead serve as a means for *classifying* interactions.

The large number of supposedly fundamental particles discovered since 1960 (well over a hundred) suggests strongly that these particles *do not* represent the most fundamental level of the structure of matter, but that there is at least one additional level of structure. There is now fairly general agreement among physicists concerning the nature of this level; the theory is based on a proposal made initially in 1964 by Gell-Mann and his collaborators. We cannot discuss this theory in detail, but the following is a very brief sketch of some of its features.

Leptons are indeed fundamental particles. In addition to the electrons and muons and their associated neutrinos, a third massive lepton called the *tau* (τ), having spin $\frac{1}{2}$ and mass 1784 MeV was discovered at SLAC in 1974. The τ neutrino has thus far escaped detection, but experiments designed to establish its existence are now under way. In Table 44–4 upper limits on the three neutrino masses are given. Although zero-mass neutrinos are postulated in most theories, a small mass could be accommodated. If neutrinos are massive, oscillations in which one type of neutrino changes into another type are possible, and these oscillations allow for experimental detection of finite neutrino mass. Experiments designed to detect neutrino oscillations are under way, but there

44–13 Brookhaven National Laboratory's solar neutrino experiment, located 4900 feet underground in a gold mine in South Dakota in order to shield out cosmic rays and all other particles except neutrinos. The tank contains 100,000 gallons of perchloro-ethylene. Neutrinos from the interior of the sun are captured by ^{37}Cl nuclei, which then beta-decay into ^{37}A. The argon is then trapped and measured.

are no positive results at present. Meanwhile, neutrino research continues; an example is shown in Fig. 44–13.

It now appears that hadrons are *not* fundamental particles but are composite structures whose constituents are spin $\frac{1}{2}$ fermions called *quarks*. In fact, all known hadrons can be constructed as follows: baryons are composed of three quarks (*qqq*), and mesons are composed of quark–antiquark pairs (*q–q̄*). No other combinations seem to be necessary. This scheme requires that quarks have properties not previously allowed for fundamental particles. For example, quarks have fractional electric charge and have a strong affinity for each other through a new kind of charge known as "color" charge. Thus, color charge is responsible for strong interactions, and the force is known as the color force. The color force is mediated by color *gluons*, massless spin-one bosons that play the role in strong interactions that the photon plays in electromagnetic interactions.

Correspondingly, the weak and gravitational forces are mediated by weak bosons ($W^{\pm}$ and Z^0) and the graviton, respectively. The theory of strong forces is known as *quantum chromodynamics* (QCD). Most QCD theories require that phenomena associated with the creation of quark-antiquark pairs make it impossible to observe a free, isolated quark. The binding energy between quarks is thought to be so strong that any stray quark or antiquark in matter will always be reabsorbed and only baryons or mesons can emerge.

Needless to say, these strange new quark properties have challenged many experimentalists. To date, isolated free quarks have not been observed; the observation of fractional electric charge has been claimed by some experimenters (Section 26–6) and disputed by others. However, many indirect observations lead us to believe that the quark structure of hadrons is correct and that quantum chromodynamics may aid in the understanding of the strong force.

Early quark theory suggested the existence of three types (flavors) of quarks; these were labeled *u* (up), *d* (down) and *s* (strange). (See Table 44–5.) Protons, neutrons, π and K mesons, and so on, could all be constructed from these three quarks. For example, a proton has $Q/e = +1$, baryon number (B) = +1, strangeness (S) = 0. The proton quark content is *uud* if the *u* quark has $Q/e = \frac{2}{3}$ and $B = \frac{1}{3}$ and the *d* quark has $Q/e = -\frac{1}{3}$ and $B = \frac{1}{3}$. The neutron would then have quark content *udd*,

TABLE 44–5   PROPERTIES OF QUARKS

Symbol	Q/e	Spin	Baryon Number	Strange-ness	Charm	Bottom-ness	Top-ness
u	$\frac{2}{3}$	$\frac{1}{2}$	$\frac{1}{3}$	0	0	0	0
d	$-\frac{1}{3}$	$\frac{1}{2}$	$\frac{1}{3}$	0	0	0	0
s	$-\frac{1}{3}$	$\frac{1}{2}$	$\frac{1}{3}$	-1	0	0	0
c	$\frac{2}{3}$	$\frac{1}{2}$	$\frac{1}{3}$	0	$+1$	0	0
b	$-\frac{1}{3}$	$\frac{1}{2}$	$\frac{1}{3}$	0	0	$+1$	0
t	$\frac{2}{3}$	$\frac{1}{2}$	$\frac{1}{3}$	0	0	0	$+1$

the π^+ meson $\boldsymbol{u\bar{d}}$ and the K^+ meson $\boldsymbol{u\bar{s}}$. Antiparticles can easily be accommodated: $\bar{p} = \boldsymbol{\bar{u}\bar{u}\bar{d}}$, $\pi^- = \boldsymbol{\bar{u}d}$, and so on. Particles can then be arranged according to quark content, and families of particles can be classified according to intrinsic orbital angular momentum, spin, and parity. Thus a state $\boldsymbol{q\bar{q}}$, for example, could represent particles in different families depending on the spin configuration of the quarks and the orbital angular momentum. It was observed that because of the Pauli exclusion principle (Section 42–1) the quarks are required to have a property that distinguishes one quark from another of the same flavor. This new property was labeled color and thus each quark flavor has three colors. This "color charge" is then responsible for the strong force between quarks. Figure 44–14 shows how one can picture the decay process $K^0 \to \pi^- e^+ \nu_e$.

For symmetry and other compelling reasons, theorists later predicted the existence of a fourth quark flavor. This quark was labeled $\boldsymbol{c}$ (charm); it was required to have $Q/e = \frac{2}{3}$, $B = \frac{1}{3}$, $S = 0$, and a new quantum number $C = +1$. Charm was discovered in 1974 at SLAC and Brookhaven by the observation of a meson of mass 3100 MeV. This meson, named ψ at SLAC and J at Brookhaven, was found to decay into e^+e^-, $\mu^+\mu^-$, and into hardrons. The mean lifetime was found to be $\sim 10^{-20}$ s and that was consistent with J/ψ being the ground state of a bound $\boldsymbol{c\bar{c}}$ system much like the way in which the hydrogen atom is a bound p-e system. Immediately excited $\boldsymbol{c\bar{c}}$ states or energy levels were observed and finally (a few years later) isolated mesons having the charm quantum number were also observed. These mesons, $D^0(\boldsymbol{c\bar{u}})$ and $D^+(\boldsymbol{c\bar{d}})$, and their excited states are now firmly established and a charmed baryon, Λ_c^+, has been observed.

In 1977 a meson of mass 9460 MeV called upsilon (Υ) was discovered at Brookhaven. Because it had properties similar to J/ψ, it was conjectured that this meson was really the bound system of a new quark, $\boldsymbol{b}$, and its antiquark, $\boldsymbol{\bar{b}}$. Excited states of Υ were soon observed and $B^+(\boldsymbol{b\bar{u}})$, $B^0(\boldsymbol{b\bar{d}})$ mesons are now well established also.

Thus, five flavors of quarks ($\boldsymbol{u, d, s, c, b}$) are thought to exist along with six flavors of leptons, $e, \mu, \tau, \gamma_e, \gamma_\mu, \gamma_\tau$). If we presume that quarks and leptons are the basic particles of matter we can explain many of the strong and weak properties of hadrons and mesons and the weak properties of leptons. But it is easy to conjecture that nature would be symmetric in its building blocks and thus a sixth quark is thought to exist. This quark, labeled $\boldsymbol{t}$ (top), should have $Q/e = \frac{2}{3}$, $B = \frac{1}{3}$, and a new quan-

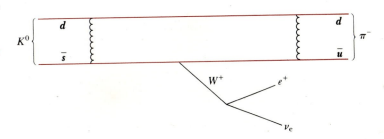

44–14 Diagram showing decay of the K^0 meson. The time sequence is shown from left to right. The initial K^0 meson consists of $\boldsymbol{d}$ and $\boldsymbol{\bar{s}}$ quarks, bound together by gluon exchange (wiggly line). The quark $\boldsymbol{\bar{s}}$ emits the weak boson W^+ and becomes quark $\boldsymbol{\bar{u}}$. This is bound by gluon exchange to the $\boldsymbol{d}$ quark, forming the final π^+ meson, while the W^+ decays into an electron and a neutrino.

tum number, $T = 1$. No experimental evidence requiring the existence of the *t*-quark has yet been found. Table 44–5 lists some properties of the six quarks.

Particle theorists have long tried to combine all four forces of nature into a single unified theory but with little success. Recently, the weak and electromagnetic forces were successfully unified by Weinberg, Salam, and Glashow. Thus, there is a more fundamental force in nature, the electro-weak force. The electro-weak theory was successfully verified in 1983 with the discovery of the weak force propagators, the Z^0 and $W^\pm$ bosons, by two experimental groups working at the $\bar{p}p$ collider at CERN, Geneva. Not only were these particles observed, but their masses are as predicted in the electro-weak theory. It is possible that quantum electro-dynamics, the theory of strong interaction and the electro-weak theory, when correctly unified, will give a valid theory of strong, weak, and electromagnetic forces. At present the photon, lepton, quarks, Z^0, and W particles have been established and provide the basis for attempts to construct this "grand-unified" theory. Although such schemes are still speculative in nature, the entire area is a very active field of present-day theoretical and experimental research.

QUESTIONS

44–1 How can one be sure that nuclei are not made of protons and electrons, rather than protons and neutrons?

44–2 In calculations of nuclear binding energies such as those in the examples of Sections 44–2 and 44–7, should the binding energies of the *electrons* in the atoms be included?

44–3 If different isotopes of the same element have the same chemical behavior, how can they be separated?

44–4 In beta decay a neutron becomes a proton, an electron, and a neutrino. This decay also occurs with free neutrons, with a half-life of about 15 minutes. Could a free *proton* undergo a similar decay?

44–5 Since lead is a stable element, why doesn't the ^{238}U decay series shown in Fig. 44–8 stop at lead, ^{214}Pb?

44–6 In the ^{238}U decay chain shown in Fig. 44–8, some nuclides in the chain are found much more abundantly in nature than others, despite the fact that every ^{238}U nucleus goes through every step in the chain before finally becoming ^{206}Pb. Why are the abundances of the intermediate nuclides not all the same?

44–7 Radium has a half-life of about 1600 years. If the universe was formed five billion or more years ago, why is there any radium left now?

44–8 Why is the decay of an unstable nucleus unaffected by the *chemical* situation of the atom, such as the nature of the molecule in which it is bound, and so on?

44–9 Fusion reactions, which liberate energy, occur only with light nuclei, and fission reactions only with heavy

nuclei. A student asserted that this shows that the binding energy per nucleon increases with A at small A but decreases at large A and hence must have a maximum somewhere in between. Do you agree?

44–10 Nuclear power plants use nuclear fission reactions to generate steam to run steam-turbine generators. How does the nuclear reaction produce heat?

44–11 There are cases where a nucleus having too few neutrons for stability can capture one of the electrons in the K shell of the atom. What is the effect of this process on N, A, and Z? Is this the same effect as that of β^+ emission? Might there be situations where K-capture is energetically possible while β^+ emission is not? Explain.

44–12 Is it possible that some parts of the universe contain antimatter whose atoms have nuclei made of antiprotons and antineutrons, surrounded by positrons? How could we detect this condition without actually going there? What problems might arise if we actually *did* go there?

44–13 When x-rays are used to diagnose stomach disorders such as ulcers, the patient first drinks a thick mixture of (insoluble) barium sulfate and water. What does this do? What is the significance of the choice of barium for this purpose?

44–14 Why are there so many health hazards associated with fission fragments that are produced during fission of heavy nuclei?

PROBLEMS

44–1 How many protons and neutrons are there in a nucleus of

a) hydrogen,

b) gold, and

c) uranium, ^{238}U?

44–2 Consider the three nuclei of Problem 44–1.

a) Estimate the radius of each nucleus.

b) Estimate the surface area of each.

c) Estimate the volume of each.

d) Determine the mass density (in $kg \cdot m^{-3}$) for each.

e) Determine the particle density (in $particles \cdot m^{-3}$) for each.

44–3 A beam of α particles is incident on gold nuclei. A particular α particle comes in "head-on" and stops 10^{-14} m away from the center of the nucleus.

a) Calculate the electrostatic potential of the α particle when it has stopped. Express your result in joules and in MeV.

b) What initial kinetic energy did the α particle have? Express in joules and in MeV.

44–4 A 4.7-MeV α particle from a radium (^{226}Ra) decay makes a head-on collision with a gold nucleus. What is the distance of closest approach of the alpha particle to the center of the nucleus?

44–5 What initial kinetic energy would an α particle need in order to "make contact" with the surface of a gold nucleus (radius 7.0×10^{-15} m). What does "make contact" mean?

44–6 Compute the approximate density of nuclear matter, and compute your result with typical densities of ordinary matter.

44–7 How much energy would be required to add a proton to a nucleus with $Z = 91$ and $A = 234$? Express your results in joules, and in MeV.

44–8 Using the data in Table 44–2, calculate the binding energy of deuterium.

44–9 Calculate the mass defect, the binding energy, and the binding energy per nucleon, for the common isotope of oxygen, ^{16}O.

44–10 Calculate the energy in MeV of an electron whose wavelength is of the order of nuclear dimensions, say 2×10^{-15} m. This calculation is one argument used to show that there cannot be any electrons in the nucleus.

44–11 Tritium is an unstable isotope of hydrogen, ^3_1H, its mass, including one electron, is 3.01647 u.

a) Show that it must be unstable with respect to beta decay because ^3_2He plus an emitted electron has less total mass.

b) Determine the total kinetic energy of the decay products, taking care to account for the electron masses correctly.

44–12 A carbon specimen found in a cave believed to have been inhabited by cavemen contained $\frac{1}{8}$ as much ^{14}C as an equal amount of carbon in living matter. Find the approximate age of the specimen.

44–13 Radium (^{226}Ra) undergoes alpha emission, leading to radon (^{222}Rn). The masses, including all electrons in each atom, are 226.0254 u and 222.0163 u, respectively. Find the maximum kinetic energy that the emitted alpha particle can have.

44–14 The common isotope of uranium, ^{238}U, has a half-life of 4.50×10^9 years, decaying by alpha emission.

a) What is the decay constant?

b) What mass of uranium would be required for an activity of one curie?

c) How many alpha particles are emitted per second by 1 g of uranium?

44–15 The unstable isotope ^{40}K is used for dating of rock samples. Its half-life is 2.4×10^8 years.

a) How many decays occur per second in a sample containing 2×10^{-6} g of ^{40}K?

b) What is the activity of the sample, in curies?

44–16 An unstable isotope of cobalt, ^{60}Co, has one more neutron in its nucleus than the stable ^{59}Co, and is a beta emitter with a half-life of 5.3 years. This isotope is widely used in medicine. A certain radiation source in a hospital contains 0.01 g of ^{60}Co.

a) What is the decay constant for this isotope?

b) How many atoms are in the source?

c) How many decays occur per second?

d) What is the activity of the source, in curies? How does this compare with the activity of an equal mass of radium?

44–17 A ^{60}Co source with activity 10 Ci is imbedded in a tumor having a mass of 0.5 kg. The Co source emits γ photons with average energy 1.25 MeV; half the photons are absorbed in the tumor and half escape.

a) What energy is delivered to the tumor per second?

b) What absorbed dose (in rad) is delivered per second?

c) What equivalent dose (in rem) is delivered per second, if the RBE for these γ rays is 0.7?

d) What exposure time is required for an equivalent dose of 200 rem?

44–18 A 70-kg person experiences whole-body exposure to α radiation of energy 1.5 MeV. A total of 10^{12} α particles are absorbed.

a) What is the absorbed dose, in rad, if all the α particles are absorbed?

b) What is the equivalent dose, in rem?

c) If the source is 0.01 g of ^{226}Ra (half-life 1600 years) somewhere in the body, what time is required for this dose to be delivered?

d) What is the activity of this source?

44–19 A free neutron decays into a proton, an electron and a neutrino, with a half-life of about 15 minutes. Calculate the total kinetic energy of the decay products.

44–20 The magnetic field in a cyclotron that is accelerating protons is 1.5 T.

a) How many times per second should the potential across the dees reverse?

b) The maximum radius of the cyclotron is 0.35 m. What is the maximum velocity of the protons?

c) Through what potential difference would the protons have to be accelerated to give them the maximum cyclotron velocity?

d) What is the energy of the protons when they emerge? Express your result in joules and in MeV.

44–21 Deuterons in a cyclotron describe a circle of radius 32.0 cm just before emerging from the dees. The frequency of the applied alternating voltage is 10 MHz. Find

a) the magnetic field, and

b) the energy and speed of the deuterons upon emergence.

44–22 In the fission of ^{238}U, 200 MeV of energy is released. Express this energy in joules per mole, and compare with typical heats of combustion.

44–23 Consider the fusion reaction.

$$^2H + {}^2H \rightarrow {}^4He + energy.$$

Compute the energy liberated in this reaction, in MeV and in joules. Compute the energy *per mole* of deuterium, re-

membering that the gas is diatomic, and compare with the heat of combustion of hydrogen, about 2.9×10^5 $J \cdot mol^{-1}$.

44–24 Why do elements with mass numbers of 210 and above decay be emission of alpha particles rather than single protons or neutrons?

44–25 If two gamma-ray photons are produced in e^+e^- annihilation, find the energy, frequency, and wavelength of each photon.

44–26 A neutral pion at rest decays into two gamma-ray photons. Find the energy, frequency, and wavelength of each photon.

44–27 Beams of π^- mesons are being used experimentally in the treatment of cancer. What is the minimum total energy a pion can release in a tumor?

44–28 Determine the electric charge, baryon number, strangeness, and charm quantum numbers for the following quark combinations:

a) *uus*,

b) *c$\bar{s}$*,

c) $\overline{ddu}$,

d) $\bar{u}s$.

44–29 A K^+ meson at rest decays into two π mesons. Find the kinetic energy of each π meson, assuming for simplicity that they have equal masses.

44–30 A Λ hyperon at rest decays into a proton and a π^-. Find the total kinetic energy of the decay products. What fraction of the energy is carried off by each particle?

44–31 The measured energy width of the ϕ meson is 4 MeV. Using the uncertainty principle, Eq. (42–16), estimate the lifetime of the ϕ meson. (Mass = 1020 MeV/c^2.)

44–32 A ϕ meson at rest decays via $\phi \rightarrow K^+K^-$.

a) Find the energy of the K^+ meson.

b) Suggest a reason why the decay $\phi \rightarrow K^+K^-\pi^0$ has not been observed.

c) Suggest reasons why the decays $\phi \rightarrow K^+\pi^-$ and $\phi \rightarrow K^+\mu^-$ have not been observed.

APPENDIX A

THE INTERNATIONAL SYSTEM OF UNITS

The Système International d'Unités, abbreviated SI, is the system developed by the General Conference on Weights and Measures and adopted by nearly all the industrial nations of the world. It is based on the mksa (meter-kilogram-second-ampere) system. The following material is adapted from NBS Special Publication 330 (1981 edition) of the National Bureau of Standards.

Quantity	Name of unit	Symbol	
SI Base Units			
length	meter	m	
mass	kilogram	kg	
time	second	s	
electric current	ampere	A	
thermodynamic temperature	kelvin	K	
luminous intensity	candela	cd	
amount of substance	mole	mol	
SI Derived Units			
area	square meter	m^2	
volume	cubic meter	m^3	
frequency	hertz	Hz	s^{-1}
mass density (density)	kilogram per cubic meter	$kg \cdot m^{-3}$	
speed, velocity	meter per second	$m \cdot s^{-1}$	
angular velocity	radian per second	$rad \cdot s^{-1}$	
acceleration	meter per second squared	$m \cdot s^{-2}$	
angular acceleration	radian per second squared	$rad \cdot s^{-2}$	
force	newton	N	$kg \cdot m \cdot s^{-2}$
pressure (mechanical stress)	pascal	Pa	$N \cdot m^{-2}$
kinematic viscosity	square meter per second	$m^2 \cdot s^{-1}$	
dynamic viscosity	newton-second per square meter	$N \cdot s \cdot m^{-2}$	
work, energy, quantity of heat	joule	J	$N \cdot m$
power	watt	W	$J \cdot s^{-1}$

(continued)

Quantity	Name of unit	Symbol	
quantity of electricity	coulomb	C	$A \cdot s$
potential difference, electromotive force	volt	V	$W \cdot A^{-1}, J \cdot C^{-1}$
electric field strength	volt per meter	$V \cdot m^{-1}$	$N \cdot C^{-1}$
electric resistance	ohm	Ω	$V \cdot A^{-1}$
capacitance	farad	F	$A \cdot s \cdot V^{-1}$
magnetic flux	weber	Wb	$V \cdot s$
inductance	henry	H	$V \cdot s \cdot A^{-1}$
magnetic flux density	tesla	T	$Wb \cdot m^{-2}$
magnetic field strength	ampere per meter	$A \cdot m^{-1}$	
magnetomotive force	ampere	A	
luminous flux	lumen	lm	$cd \cdot sr$
luminance	candela per square meter	$cd \cdot m^{-2}$	
illuminance	lux	lx	$lm \cdot m^{-2}$
wave number	1 per meter	m^{-1}	
entropy	joule per kelvin	$J \cdot K^{-1}$	
specific heat capacity	joule per kilogram kelvin	$J \cdot kg^{-1} \cdot K^{-1}$	
thermal conductivity	watt per meter kelvin	$W \cdot m^{-1} \cdot K^{-1}$	
radiant intensity	watt per steradian	$W \cdot sr^{-1}$	
activity (of a radioactive source)	becquerel	Bq	s^{-1}
radiation dose	gray	Gy	$J \cdot kg^{-1}$
radiation dose equivalent	sievert	Sv	$J \cdot kg^{-1}$

SI Supplementary Units

plane angle	radian	rad	
solid angle	steradian	sr	

Definitions of SI Units

meter (m)   The *meter* is the length equal to the distance traveled by light, in vacuum, in a time of 1/299,792,458 second.

kilogram (kg)   The *kilogram* is the unit of mass; it is equal to the mass of the international prototype of the kilogram. (The international prototype of the kilogram is a particular cylinder of platinum-iridium alloy that is preserved in a vault at Sèvres, France, by the International Bureau of Weights and Measures.)

second (s)   The *second* is the duration of 9,192,631,770 periods of the radiation corresponding to the transition between the two hyperfine levels of the ground state of the cesium-133 atom.

ampere (A)   The *ampere* is that constant current that, if maintained in two straight parallel conductors of infinite length, of negligible circular cross section, and placed 1 meter apart in vacuum, would produce between these conductors a force equal to 2×10^{-7} newton per meter of length.

kelvin (K)   The *kelvin*, unit of thermodynamic temperature, is the fraction 1/273.16 of the thermodynamic temperature of the triple point of water.

candela (cd)   The *candela* is the luminous intensity, in a given direction, of a source that emits monochromatic radiation of frequency 540×10^{12} hertz and that has a radiant intensity in that direction of 1/683 watt per steradian.

mole (mol)   The *mole* is the amount of substance of a system that contains as many elementary entities as there are carbon atoms in 0.012 kg of carbon 12. The elementary entities must be specified and may be atoms, molecules, ions, electrons, other particles, or specified groups of such particles.

newton (N)   The *newton* is that force that gives to a mass of 1 kilogram an acceleration of 1 meter per second per second.

joule (J)   The *joule* is the work done when the point of application of 1 newton is displaced a distance of 1 meter in the direction of the force.

watt (W)   The *watt* is the power that gives rise to the production of energy at the rate of 1 joule per second.

volt (V)   The *volt* is the difference of electric potential between two points of a conducting wire carrying a constant current of 1 ampere, when the power dissipated between these points is equal to 1 watt.

ohm (Ω)   The *ohm* is the electric resistance between two points of a conductor when a constant difference of potential of 1 volt, applied between these two points, produces in this conductor a current of 1 ampere, this conductor not being the source of any electromotive force.

coulomb (C)   The *coulomb* is the quantity of electricity transported in 1 second by a current of 1 ampere.

farad (F)   The *farad* is the capacitance of a capacitor between the plates of which there appears a difference of potential of 1 volt when it is charged by a quantity of electricity equal to 1 coulomb.

henry (H)   The *henry* is the inductance of a closed circuit in which an electromotive force of 1 volt is produced when the electric current in the circuit varies uniformly at a rate of 1 ampere per second.

weber (Wb)   The *weber* is the magnetic flux that, linking a circuit of one turn, produces in it an electromotive force of 1 volt as it is reduced to zero at a uniform rate in 1 second.

lumen (lm)   The *lumen* is the luminous flux emitted in a solid angle of 1 steradian by a uniform point source having an intensity of 1 candela.

radian (rad)   The *radian* is the plane angle between two radii of a circle that cut off on the circumference an arc equal in length to the radius.

steradian (sr)   The *steradian* is the solid angle that, having its vertex in the center of a sphere, cuts off an area of the surface of the sphere equal to that of a square with sides of length to the radius of the sphere.

SI Prefixes   The names of multiples and submultiples of SI units may be formed by application of the prefixes listed in Table 1–1, page 3.

APPENDIX B

USEFUL MATHEMATICAL RELATIONS

Algebra

$$a^{-x} = \frac{1}{a^x} \qquad\qquad a^{(x+y)} = a^x a^y \qquad\qquad a^{(x-y)} = \frac{a^x}{a^y}$$

Logarithms: If $\log a = x$, then $a = 10^x$. $\log a + \log b = \log (ab)$ $\log a - \log b = \log (a/b)$ $\log (a^n) = n \log a$

If $\ln a = x$, then $a = e^x$. $\ln a + \ln b = \ln (ab)$ $\ln a - \ln b = \ln (a/b)$ $\ln (a^n) = n \ln a$

Quadratic formula: If $ax^2 + bx + c = 0$, $x = \dfrac{-b \pm \sqrt{b^2 - 4ac}}{2a}$.

Binomial theorem

$$(a + b)^n = a^n + na^{n-1}b + \frac{n(n-1)a^{n-2}b^2}{2!} + \frac{n(n-1)(n-2)a^{n-3}b^3}{3!} + \cdots$$

Trigonometry

In the right triangle ABC, $x^2 + y^2 = r^2$.

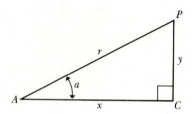

Definitions of the trigonometric functions: $\sin a = y/r$ $\cos a = x/r$ $\tan a = y/x$

Identities: $\sin^2 a + \cos^2 a = 1$

$\sin 2a = 2 \sin a \cos a$

$\sin \tfrac{1}{2}a = \sqrt{\dfrac{1 - \cos a}{2}}$

$\sin (-a) = -\sin a$

$\cos (-a) = \cos a$

$\sin (a \pm \pi/2) = \pm\cos a$

$\cos (a \pm \pi/2) = \mp\sin a$

$\tan a = \dfrac{\sin a}{\cos a}$

$\cos 2a = \cos^2 a - \sin^2 a = 2 \cos^2 a - 1$

$\cos \tfrac{1}{2}a = \sqrt{\dfrac{1 + \cos a}{2}}$

$\sin (a \pm b) = \sin a \cos b \pm \cos a \sin b$

$\cos (a \pm b) = \cos a \cos b \mp \sin a \sin b$

$\sin a + \sin b = 2 \sin \tfrac{1}{2}(a + b) \cos \tfrac{1}{2}(a - b)$

$\cos a + \cos b = 2 \cos \tfrac{1}{2}(a + b) \cos \tfrac{1}{2}(a - b)$

Geometry

Circumference of circle of radius r: $C = 2\pi r$

Area of circle of radius r: $A = \pi r^2$

Volume of sphere of radius r: $V = 4\pi r^3/3$

Surface area of sphere of radius r: $A = 4\pi r^2$

Volume of cylinder of radius r and height h: $V = \pi r^2 h$

APPENDIX C
THE GREEK ALPHABET

Name	Capital	Lowercase	Name	Capital	Lowercase
Alpha	A	α	Nu	N	ν
Beta	B	β	Xi	Ξ	ξ
Gamma	Γ	γ	Omicron	O	o
Delta	Δ	δ	Pi	Π	π
Epsilon	E	ϵ	Rho	P	ρ
Zeta	Z	ζ	Sigma	Σ	σ
Eta	H	η	Tau	T	τ
Theta	Θ	θ	Upsilon	Υ	υ
Iota	I	ι	Phi	Φ	ϕ
Kappa	K	κ	Chi	X	χ
Lambda	Λ	λ	Psi	Ψ	ψ
Mu	M	μ	Omega	Ω	ω

APPENDIX D

NATURAL TRIGONOMETRIC FUNCTIONS

Angle					Angle				
Degree	Radian	Sine	Cosine	Tangent	Degree	Radian	Sine	Cosine	Tangent
0°	.000	0.000	1.000	0.000					
1°	.017	.018	1.000	.018	46°	0.803	0.719	0.695	1.036
2°	.035	.035	0.999	.035	47°	.820	.731	.682	1.072
3°	.052	.052	.999	.052	48°	.838	.743	.669	1.111
4°	.070	.070	.998	.070	49°	.855	.755	.656	1.150
5°	.087	.087	.996	.088	50°	.873	.766	.643	1.192
6°	.105	.105	.995	.105	51°	.890	.777	.629	1.235
7°	.122	.122	.993	.123	52°	.908	.788	.616	1.280
8°	.140	.139	.990	.141	53°	.925	.799	.602	1.327
9°	.157	.156	.988	.158	54°	.942	.809	.588	1.376
10°	.175	.174	.985	.176	55°	.960	.819	.574	1.428
11°	.192	.191	.982	.194	56°	.977	.829	.559	1.483
12°	.209	.208	.978	.213	57°	.995	.839	.545	1.540
13°	.227	.225	.974	.231	58°	1.012	.848	.530	1.600
14°	.244	.242	.970	.249	59°	1.030	.857	.515	1.664
15°	.262	.259	.966	.268	60°	1.047	.866	.500	1.732
16°	.279	.276	.961	.287	61°	1.065	.875	.485	1.804
17°	.297	.292	.956	.306	62°	1.082	.883	.470	1.881
18°	.314	.309	.951	.325	63°	1.100	.891	.454	1.963
19°	.332	.326	.946	.344	64°	1.117	.899	.438	2.050
20°	.349	.342	.940	.364	65°	1.134	.906	.423	2.145
21°	.367	.358	.934	.384	66°	1.152	.914	.407	2.246
22°	.384	.375	.927	.404	67°	1.169	.921	.391	2.356
23°	.401	.391	.921	.425	68°	1.187	.927	.375	2.475
24°	.419	.407	.914	.445	69°	1.204	.934	.358	2.605
25°	.436	.423	.906	.466	70°	1.222	.940	.342	2.747
26°	.454	.438	.899	.488	71°	1.239	.946	.326	2.904
27°	.471	.454	.891	.510	72°	1.257	.951	.309	3.078
28°	.489	.470	.883	.532	73°	1.274	.956	.292	3.271
29°	.506	.485	.875	.554	74°	1.292	.961	.276	3.487
30°	.524	.500	.866	.577	75°	1.309	.966	.259	3.732
31°	.541	.515	.857	.601	76°	1.326	.970	.242	4.011
32°	.559	.530	.848	.625	77°	1.344	.974	.225	4.331
33°	.576	.545	.839	.649	78°	1.361	.978	.208	4.705
34°	.593	.559	.829	.675	79°	1.379	.982	.191	5.145
35°	.611	.574	.819	.700	80°	1.396	.985	.174	5.671
36°	.628	.588	.809	.727	81°	1.414	.988	.156	6.314
37°	.646	.602	.799	.754	82°	1.431	.990	.139	7.115
38°	.663	.616	.788	.781	83°	1.449	.993	.122	8.144
39°	.681	.629	.777	.810	84°	1.466	.995	.105	9.514
40°	.698	.643	.766	.839	85°	1.484	.996	.087	11.43
41°	.716	.656	.755	.869	86°	1.501	.998	.070	14.30
42°	.733	.669	.743	.900	87°	1.518	.999	.052	19.08
43°	.751	.682	.731	.933	88°	1.536	.999	.035	28.64
44°	.768	.695	.719	.966	89°	1.553	1.000	.018	57.29
45°	.785	.707	.707	1.000	90°	1.571	1.000	.000	

APPENDIX E

PERIODIC TABLE OF THE ELEMENTS

Period	IA	IIA	IIIB	IVB	VB	VIB	VIIB	VIIIB	VIIIB	IB	IIB	IIIA	IVA	VA	VIA	VIIA	Noble gases
1	1 **H** 1.008																2 **He** 4.003
2	3 **Li** 6.939	4 **Be** 9.012										5 **B** 10.811	6 **C** 12.011	7 **N** 14.007	8 **O** 15.999	9 **F** 18.998	10 **Ne** 20.183
3	11 **Na** 22.990	12 **Mg** 24.312										13 **Al** 26.982	14 **Si** 28.086	15 **P** 30.974	16 **S** 32.064	17 **Cl** 35.453	18 **Ar** 39.948
4	19 **K** 39.102	20 **Ca** 40.08	21 **Sc** 44.956	22 **Ti** 47.90	23 **V** 50.942	24 **Cr** 51.996	25 **Mn** 54.938	26 **Fe** 55.847	27 **Co** 58.933	28 **Ni** 58.71	29 **Cu** 63.54	30 **Zn** 65.37	31 **Ga** 69.72	32 **Ge** 72.59	33 **As** 74.922	34 **Se** 78.96	35 **Br** 79.909	36 **Kr** 83.80
5	37 **Rb** 85.47	38 **Sr** 87.62	39 **Y** 88.905	40 **Zr** 91.22	41 **Nb** 92.906	42 **Mo** 95.94	43 **Tc** (98)	44 **Ru** 101.07	45 **Rh** 102.91	46 **Pd** 106.4	47 **Ag** 107.87	48 **Cd** 112.40	49 **In** 114.82	50 **Sn** 118.69	51 **Sb** 121.75	52 **Te** 127.60	53 **I** 126.90	54 **Xe** 131.30
6	55 **Cs** 132.91	56 **Ba** 137.34	57 **La** 138.91	72 **Hf** 178.49	73 **Ta** 180.95	74 **W** 183.85	75 **Re** 186.2	76 **Os** 190.2	77 **Ir** 192.2	78 **Pt** 195.09	79 **Au** 196.97	80 **Hg** 200.59	81 **Tl** 204.37	82 **Pb** 207.19	83 **Bi** 208.98	84 **Po** (210)	85 **At** (210)	86 **Rn** (222)
7	87 **Fr** (223)	88 **Ra** (226)	89 **Ac** (227)	104 **Rf(?)** (261)	105 **Ha(?)** (262)	106 (263)	107 (261)											

58 **Ce** 140.12	59 **Pr** 140.91	60 **Nd** 144.24	61 **Pm** (145)	62 **Sm** 150.35	63 **Eu** 151.96	64 **Gd** 157.25	65 **Tb** 158.92	66 **Dy** 162.50	67 **Ho** 164.93	68 **Er** 167.26	69 **Tm** 168.93	70 **Yb** 173.04	71 **Lu** 174.97
90 **Th** (232)	91 **Pa** (231)	92 **U** (238)	93 **Np** (239)	94 **Pu** (239)	95 **Am** (243)	96 **Cm** (245)	97 **Bk** (247)	98 **Cf** (249)	99 **Es** (254)	100 **Fm** (253)	101 **Md** (255)	102 **No** (255)	103 **Lr** (257)

For each element the average atomic mass of the mixture of isotopes occurring in nature is shown. For elements having no stable isotope, the approximate atomic mass of the longest-lived isotope is shown in parentheses.

APPENDIX F

UNIT CONVERSION FACTORS

Length

1 m = 100 cm = 1000 mm = $10^6 \mu$m = 10^9 nm
1 km = 1000 m = 0.6214 mi
1 m = 3.281 ft = 39.37 in.
1 cm = 0.3937 in.
1 in. = 2.540 cm
1 ft = 30.48 cm
1 yd = 91.44 cm
1 mi = 5280 ft = 1.609 km
1 Å = 10^{-10} m = 10^{-8} cm = 10^{-1} nm
1 nautical mile = 6080 ft
1 light year = 9.461×10^{15} m

Area

1 cm^2 = 0.155 in^2
1 m^2 = 10^4 cm^2 = 10.76 ft^2
1 in^2 = 6.452 cm^2
1 ft^2 = 144 in^2 = 0.0929 m^2

Volume

1 liter = 1000 cm^3 = 10^{-3}m^3 = 0.0351 ft^3 = 61.02 in^3
1 ft^3 = 0.02832 m^3 = 28.32 liters = 7.477 gallons
1 gallon = 3.788 liters

Time

1 min = 60 s
1 hr = 3600 s
1 da = 86,400 s
1 yr = 365.24 da = 3.156×10^7 s

Angle

1 rad = 57.30° = 180°/π
1° = 0.01745 rad = π/180 rad
1 revolution = 360° = 2π rad
1 rev$\cdot$min^{-1} (rpm) = 0.1047 rad$\cdot$s^{-1}

Speed

1 m$\cdot$s^{-1} = 3.281 ft$\cdot$s^{-1}
1 ft$\cdot$s^{-1} = 0.3048 m$\cdot$s^{-1}
1 mi$\cdot$min^{-1} = 60 mi$\cdot$hr^{-1} = 88 ft$\cdot$s^{-1}
1 km$\cdot$hr^{-1} = 0.2778 m$\cdot$s^{-1} = 0.6214 mi$\cdot$hr^{-1}
1 mi$\cdot$hr^{-1} = 1.466 ft$\cdot$s^{-1} = 0.4470 m$\cdot$s^{-1} = 1.609 km$\cdot$hr^{-1}
1 furlong$\cdot$fortnight^{-1} = 1.662×10^{-4} m$\cdot$s^{-1}

Acceleration

1 m$\cdot$s^{-2} = 100 cm$\cdot$s^{-2} = 3.281 ft$\cdot$s^{-2}
1 cm$\cdot$s^{-2} = 0.01 m$\cdot$s^{-2} = 0.03281 ft$\cdot$s^{-2}
1 ft$\cdot$s^{-2} = 0.3048 m$\cdot$s^{-2} = 30.48 cm$\cdot$s^{-2}
1 mi$\cdot$hr$^{-1}\cdot$s^{-1} = 1.467 ft$\cdot$s^{-2}

Mass

1 kg = 10^3 g = 0.0685 slug
1 g = 6.85×10^{-5} slug
1 slug = 14.59 kg
1 u = 1.661×10^{-27} kg
1 kg has a weight of 2.205 lb when g = 9.80 m$\cdot$s^{-2}

Force

1 N = 10^5 dyn = 0.2248 lb
1 lb = 4.448 N = 4.448×10^5 dyn

Pressure

1 Pa = 1 N$\cdot$m^{-2} = 1.451×10^{-4} lb$\cdot$in^{-2} = 0.209 lb$\cdot$ft^{-2}
1 bar = 10^5 Pa
1 lb$\cdot$in^{-2} = 6891 Pa
1 lb$\cdot$ft^{-2} = 47.85 Pa
1 atm = 1.013×10^5 Pa = 1.013 bar
     = 14.7 lb$\cdot$in^{-2} = 2117 lb$\cdot$ft^{-2}
1 mm Hg = 1 torr = 133.3 Pa

Energy

1 J = 10^7 ergs = 0.239 cal
1 cal = 4.186 J (based on 15° calorie)
1 ft$\cdot$lb = 1.356 J
1 Btu = 1055 J = 252 cal = 778 ft$\cdot$lb
1 eV = 1.602×10^{-19} J
1 kWh = 3.600×10^6 J

Mass–Energy Equivalence

1 kg $\leftrightarrow$ 8.988×10^{16} J
1 u $\leftrightarrow$ 931.5 MeV
1 eV $\leftrightarrow$ 1.073×10^{-9} u

Power

1 W = 1 J$\cdot$s^{-1}
1 hp = 746 W = 550 ft$\cdot$lb$\cdot$s^{-1}
1 Btu$\cdot$hr^{-1} = 0.293 W

APPENDIX G

NUMERICAL CONSTANTS

Fundamental physical constants

Name	Symbol	Value
Speed of light	c	$2.9979 \times 10^8 \text{ m} \cdot \text{s}^{-1}$
Charge of electron	e	$1.602 \times 10^{-19} \text{ C}$
Gravitational constant	G	$6.673 \times 10^{-11} \text{ N} \cdot \text{m}^2 \cdot \text{kg}^{-2}$
Planck's constant	h	$6.626 \times 10^{-34} \text{ J} \cdot \text{s}$
Boltzmann's constant	k	$1.381 \times 10^{-23} \text{ J} \cdot \text{K}^{-1}$
Avogadro's number	N_0	$6.022 \times 10^{23} \text{ molecules} \cdot \text{mol}^{-1}$
Gas constant	R	$8.314 \text{ J} \cdot \text{mol}^{-1} \cdot \text{K}^{-1}$
Mass of electron	m_e	$9.110 \times 10^{-31} \text{ kg}$
Mass of neutron	m_n	$1.675 \times 10^{-27} \text{ kg}$
Mass of proton	m_p	$1.673 \times 10^{-27} \text{ kg}$
Permittivity of free space	ϵ_0	$8.854 \times 10^{-12} \text{ C}^2 \cdot \text{N}^{-1} \cdot \text{m}^{-2}$
	$1/4\pi\epsilon_0$	$8.987 \times 10^9 \text{ N} \cdot \text{m}^2 \cdot \text{C}^{-2}$
Permeability of free space	μ_0	$4\pi \times 10^{-7} \text{ Wb} \cdot \text{A}^{-1} \cdot \text{m}^{-1}$

Other useful constants

Mechanical equivalent of heat		$4.186 \text{ J} \cdot \text{cal}^{-1}$ (15° calorie)
Standard atmospheric pressure	1 atm	$1.013 \times 10^5 \text{ Pa}$
Absolute zero	0 K	$-273.15°\text{C}$
Electronvolt	1 eV	$1.602 \times 10^{-19} \text{ J}$
Atomic mass unit	1 u	$1.661 \times 10^{-27} \text{ kg}$
Electron rest energy	mc^2	0.511 MeV
Energy equivalent of 1 u	Mc^2	931.5 MeV
Volume of ideal gas (0°C and 1 atm)	V	$22.4 \text{ liter} \cdot \text{mol}^{-1}$
Acceleration due to gravity (sea level, at equator)	g	$9.78049 \text{ m} \cdot \text{s}^{-2}$

Astronomical data

Body	Mass, kg	Radius, m	Orbit radius, m	Orbit period
Sun	1.99×10^{30}	6.95×10^8	—	—
Moon	7.36×10^{22}	1.74×10^6	0.38×10^9	27.3 d
Mercury	3.28×10^{23}	2.57×10^6	5.8×10^{10}	88.0 d
Venus	4.82×10^{24}	6.31×10^6	1.08×10^{11}	224.7 d
Earth	5.98×10^{24}	6.38×10^6	1.49×10^{11}	365.3 d
Mars	6.34×10^{23}	3.43×10^6	2.28×10^{11}	687.0 d
Jupiter	1.88×10^{27}	7.18×10^7	7.78×10^{11}	11.86 y
Saturn	5.63×10^{26}	6.03×10^7	1.43×10^{12}	29.46 y
Uranus	8.61×10^{25}	2.67×10^7	2.87×10^{12}	84.02 y
Neptune	9.99×10^{25}	2.48×10^7	4.49×10^{12}	164.8 y
Pluto	5×10^{23}	4×10^5	5.90×10^{12}	247.7 y

ANSWERS TO
ODD-NUMBERED PROBLEMS

CHAPTER 1

1–1 1 mi = 1.6093 km
1–3 a) 88.0 ft·s^{-1} b) 27.8 m·s^{-1}
1–5 86,400; 3.15 × 10^7
1–7 40.2 mi·gal^{-1}
1–9 122 in^3
1–11 0.39%
1–13 5.50 × 10^3 kg·m^{-3}
1–15 10,000
1–17 Just barely
1–19 4000
1–21 3 km; 1 m
1–23 $10^{15}
1–25 $5 × 10^{12}
1–27 2 km
1–29 $[(x_2 - x_1)^2 + (y_2 - y_1)^2]^{1/2}$, arctan $\dfrac{y_2 - y_1}{x_2 - x_1}$

1–31 a) 5 cm, −53.1°   b) 13 m, −112°
    c) 3.61 km, 124°
1–33 6 in., at 45° to initial direction of line
1–35 230 N, 55° above negative x-axis; 1%
1–37 15.0 N, 53.3° above +x-axis; 25.9 N, 27.7° below
    −x-axis
1–39 a) 2 in., +x-direction
    b) 3.46 in., +y-direction, −y-direction
1–41 5 mi, 53.1° south of east
1–43 a) −1 cm, +1 cm   b) 1.41 cm, 135°
1–45 When the angle between A and B is 120°; when
    it is 60°

CHAPTER 2

2–1 25.7 N, 30.6 N
2–3 a) 18.5 N   b) 9.2 N
2–5 308 N, 25° above +x-axis
2–7 a) 7.01 N, 2.93 N   b) 7.60 N   c) 11.0 N
2–9 46.6 N in −y-direction
2–11 a) The earth   b) 4 N, the book   c) No
    d) 4 N, the earth, the book, upward   e) 4 N,
    the hand, the book, downward   f) First
    g) Third   h) No   i) No   j) Yes
    k) Yes   l) One   m) No   n) Nothing
2–13 a) 10 N   b) 20 N
2–15 a) T_1 = 80.0 N, T_2 = 46.2 N, T_3 = 92.3 N
    b) T_1 = 80.0 N, T_2 = 80.0 N, T_3 = 113.1 N
2–17 828 N, 214 N
2–19 a) 1545 N   b) 1.15°
2–21 14.1 N
2–25 a) $w/(2 \sin \theta)$ at angle θ below horizontal
    b) $w/(2 \tan \theta)$
2–27 22.3 N
2–29 a) 371.4 N   b) $\mu_k = 1$
2–31 a) Held back   b) 480 N
2–33 a) 3.00 N   b) 4.00 N   c) 5.00 N

CHAPTER 3

3–1 a) 15 mi·hr^{-1} b) 22.0 ft·s^{-1}
    c) 6.71 m·s^{-1}
3–3 61.0 cm·s^{-1}, 60.1 cm·s^{-1}, 60.01 cm·s^{-1};
    60.0 cm·s^{-1}

3–5 a) 0, 6.25 m·s^{-2}, −11.2 m·s^{-2} b) 100 m, 230 m, 320 m

3–9 a) 3.95 ft·s^{-2} b) 620 ft

3–11 a) 2.67 km·hr^{-1}·s^{-1} b) 7.50 s   c) 83.3 m, 104.2 m

3–13 a) 12.5 cm·s^{-2} b) 7840 cm

3–15 a) 100 m   b) 20 m·s^{-1}

3–17 a) 22.5 ft·s^{-1}, 27.5 ft·s^{-1}, 32.5 ft·s^{-1} b) 2.50 m·s^{-2} c) 20.0 m·s^{-1} d) 8.00 s   e) 21.25 ft   f) 9.06 s   g) 22.6 ft

3–19 a) 32 m·s^{-1} b) 2.83 s   c) 16 m   d) 22.6 m·s^{-1} e) 16 m·s^{-1}

3–21 a) 8.66 s   b) 37.5 m   c) 26.0 m·s^{-1}, 17.3 m·s^{-1}

3–23 a) 29.6 m·s^{-1} b) 39.6 m   c) 17.2 m·s^{-1} d) 50.0 m·s^{-2} e) 2.01 s   f) 29.7 m·s^{-1}

3–25 10.7 m

3–27 a) 48.0 ft·s^{-1} b) 36.0 ft   c) zero   d) 32.0 ft·s^{-2}, downward   e) 80 ft·s^{-1}

3–29 a) 10.2 m   b) 4.34 m·s^{-1}, −9.8 m·s^{-2} c) −15.3 m·s^{-1}, −9.8 m·s^{-2} d) 11.7 m·s^{-1}

3–31 a) 11.3 m   b) 0.51 s   c) 9.98 m·s^{-1}, −9.8 m·s^{-2}

3–33 a) 4 m·s^{-2} b) 6 m·s^{-1} c) 4.5 m

3–35 a) 46.8 s   b) 102 mi

3–37 a) 6000 m·s^{-1} b) 18.4 hr   c) 0.99

3–39 40 min, 30 min

CHAPTER 4

4–1 a) 0.102 kg   b) 0.00102 g

4–3 a) 5 kg   b) 200 m   c) 0.5 slug, 500 ft

4–5 a) 2 m·s^{-2} b) 100 m   c) 20 m·s^{-1}

4–7 a) 19.6 N   b) 2.55 s

4–9 a) 1.62 × 10^{-15} N   b) 3.33 × 10^{-9} s   c) 1.8 × 10^{15} m·s^{-2}

4–11 49.8 N, 2.54%

4–13 6.18 × 10^{24} kg

4–15 1.94 m·s^{-2}

4–17 21,600 N

4–19 4500 lb

4–21 a) 40 N   b) 2.45 m·s^{-2}, downward   c) Zero

4–23 a) 242 ft   b) 38.1 mi·hr^{-1}

4–25 3.13 s, 9.8 m

4–27 a) 2.5 × 10^4 N   b) 7.5 × 10^4 N

4–29 a) 36.9°   b) 1.95 m·s^{-2} c) 2.5 s

4–31 a) 36.8 N   b) 2.45 m·s^{-2}

4–33 a) 10.54 N   b) 10.0 N

4–35 a) 1.96 m·s^{-2} b) 3.14 N

4–37 $2m_2g/(4m_1 + m_2)$, $m_2g/(4m_1 + m_2)$

4–39 a) 2.70 m·s^{-2} b) 112.5 N   c) 87.5 N

4–41 g/μ

4–43 a) 4.0 m·s^{-2} b) 1.0 m·s^{-2} c) 3.6 m·s^{-1} d) 4.0 m·s^{-1}

4–45 a) Move up   b) Constant   c) Constant   d) Stop

CHAPTER 5

5–1 a) $x = (10 \text{ m·s}^{-1})t$, $y = -(4.9 \text{ m·s}^{-2})t^2$ b) (0, 0), (5 m, −1.22 m), (10 m, −4.9 m), (15 m, −11.0 m), (20 m, −19.6 m)   c) $v_x = 10 \text{ m·s}^{-1}$, $v_y = -(9.8 \text{ m·s}^{-2})t$ d) 14.0 m·s^{-1}, 44.4° below horizontal; yes

5–3 a) 0.452 s   b) 3.32 m·s^{-1} c) 5.53 m·s^{-1}, 53.1° below horizontal

5–5 a) 20.2 s   b) 2020 m   c) 100 m·s^{-1}, −198.0 m·s^{-1}

5–7 a) 30.6 m   b) 62.5 m   c) 25 m·s^{-1}, 24.5 m·s^{-1}; 35.0 m·s^{-1}, 44.4° below horizontal

5–9 a) 0.136 m   b) 1.225 m

5–11 a) 1.01 m   b) −3.91 m   c) Yes

5–13 17.0 ft·s^{-1}

5–15 a) 21.8 m·s^{-1} b) 13.5 m·s^{-1}, 36.2° below horizontal

5–19 Straight line, 27.0° to vertical

5–21 1.86 m·s^{-2}

5–23 a) 2.96 × 10^4 m·s^{-1} b) 0.0594 m·s^{-2}

5–25 6.00°

5–27 22.4 m·s^{-1}

5–29 a) 0.127   b) 2.19 in.

5–31 a) 6.73 s   b) No

5–33 a) 51.6°   b) No   c) The bead remains at the bottom.

5–35 a) 1537 ft   b) 1440 lb

5–37 a) 0.403w b) 4.49 s   c) 2w d) He would strike the ground 10.0 m from the center of the wheel.

5–39 1.98 hr

5–41 7560 m·s^{-1}

5–43 1.41 hr

5–45 6.55 knots

5–47 a) 100 km·hr^{-1}, 53.1° S of W   b) 30° N of W

5–49 a) 3.6 m·s^{-1}, 33.7° N of E   b) 667 m   c) 333 s

5–51 11.6 mi·hr^{-1}, 59° N of E

CHAPTER 6

6–1 6.00 × 10^7 J

6–3 600 J

6–5 a) 132.2 N   b) 573 J, −573 J

6–7 5.64 × 10^4 J

6–9 a) 1.85 × 10^4 J   b) 4 times

6–11 4.5 × 10^{-17} J

6–13 14.1 m·s^{-1}

6–15 4.0 m

6–17 a) 33 J   b) 13 J

6–19 a) 200 J   b) 8.12 m·s^{-1}, 200 J

6–21 a) 2400 J   b) 515 J   c) 1415 J
d) −470 J   e) $a + d = b + c$

6–23 a) 240 N   b) 238 J

6–25 a) 10 N, 20 N, 40 N   b) 0.5 J, 2 J, 8 J

6–27 a) 0.707 m·s^{-1} b) 0.612 m·s^{-1}

6–29 a) 11.2 m·s^{-1} b) 6.71 m·s^{-1}

6–31 a) 0.10 m

6–33 a) 0.272   b) 3.60 J

6–37 1.81 m·s^{-1}

6–39 8.02 m·s^{-1}

6–41 a) 5.51 m·s^{-1} b) 0.0182 m

6–43 a) 4.43 m·s^{-1} b) 14.7 N

6–45 1.12×10^4 m·s^{-1}

6–47 a) 38×10^6 m from the center of moon
b) 1.10×10^4 m·s^{-1} c) 2.24×10^3 m·s^{-1}

6–49 7.3 m·s^{-1}; yes, barely

6–51 0.63 hp, 470 W, 0.470 kW

6–53 148 hp

6–55 $1.07

6–57 a) 1070 N   b) 5 hp, 80 hp

6–59 a) 8232 J   b) 8232 J   c) 0.552 hp

6–61 a) 27,600 N   b) 10.9 m·s^{-1} c) 144×10^5 J
d) 3.53×10^5 J   e) 296 hp

6–63 a) 1070 N   b) 41.8 hp   c) 15.6 hp
d) 9.2%

6–65 227 km^2

CHAPTER 7

7–1 a) 2.0×10^5 kg·m·s^{-1} b) 40 m·s^{-1}
c) 28.3 m·s^{-1}

7–3 a) 8.0×10^5 m·s^{-2} b) 4.0×10^4 N
c) 5.0×10^{-4} s   d) 20 N·s

7–5 No; no; transferred to earth

7–7 180 N

7–9 4600 N, 1500 N

7–11 a) 0.667 m·s^{-1} b) 1.33×10^4 J
c) 1.0 m·s^{-1}

7–13 237 m·s^{-1}

7–15 a) 0.163   b) 240 J   c) 0.320 J

7–17 595 m·s^{-1}

7–19 24.0 km·h^{-1}, 33.7° S of E

7–21 a) 10.0 cm·s^{-1} b) 0.135 J
c) −70 cm·s^{-1}, 80 cm·s^{-1}

7–23 −25 cm·s^{-1}, 5 cm·s^{-1}

7–25 a) 50 cm·s^{-1}, 53.1° below x-axis   b) 0.037 J

7–27 d) $\rho A(v_J - v_B)$ e) $\rho A(v_J - v_B)^2$
f) $\rho A v_B(v_J - v_B)^2$

7–29 a) $v_A = 22.0$ m·s^{-1}, $v_B = 15.5$ m·s^{-1} b) No;
0.196

7–33 a) 21,208 m   b) 4.0×10^5 J

7–35 a) 7.20 ft·s^{-1} b) 375

7–37 0.91 m·s^{-1}

7–39 16 N

7–41 a) 20 m ahead of 1000-kg car
b) 50,000 kg·m·s^{-1} c) 16.7 m·s^{-1}
d) 50,000 kg·m·s^{-1}

7–43 On bisector of angle, 0.354 m from vertex

7–45 2.18 m

CHAPTER 8

8–1 1000 N, 0.8 m from 600-N end

8–3 350 N, 750 N

8–5 3 1/3 ft from other end

8–7 2400 N up, 1600 N down

8–9 2000 N up; 2000 N down, 200 N to left

8–11 b) 2.5 m   c) 2.89 m

8–13 125 N, 100 N to the right, 25 N up

8–15 a) 2.31 m   b) 213 N

8–17 0.3 m from the 1-kg end

8–19 0.0667 m, 0.0667 m; yes

8–21 0.646×10^{-10} m from carbon atom

8–23 1033 kg, 1167 kg

8–25 a) −30 N, 50 N   b) 5/3   c) 58.3 N
d) 1.0 m from right end

8–27 a) 270 N   b) 120 N

8–29 a) 326 N, 374 N   b) 224 N   c) 236 N
d) 864 N

8–31 a) 16.7°   b) 24 cm   c) 0.53

8–33 a) 66.7 N, 100 N   b) 120.2 N, 33.7° to
vertical

8–35 a) Tip at 26.6°, slip at 21.8°; it slips   b) Tip
before sliding; tip and slide at same angle.

8–37 a) 100 N, 700 N   b) 2.0 m

CHAPTER 9

9–1 a) 1.5 rad   b) 1.57 rad, 90°   c) 120 cm,
120 ft

9–3 5 rad·s^{-2}, 1000 rad

9–5 20 rad·s^{-2}

9–7 32.5 in.·s^{-1}, 6.98 in.·s^{-1}; 33.2 in.·s^{-1}, 12.1°
from radial line

9–9 a) At the bottom   b) 35.6 m·s^{-2} at 3° from
vertical

9–11 164,000 rev·min^{-1}

9–13 a) $a_\perp = 2\alpha\theta r$ b) 0.289 rad

9–15 a) 0.18 kg·m^2 b) 0.32 kg·m^2
c) 0.50 kg·m^2 (= 0.18 kg·m^2 + 0.32 kg·m^2)

9–17 a) 2.67 kg·m^2 b) 10.7 kg·m^2
c) 1.6×10^{-3} kg·m^2

9–19 0.748 kg·m^2

9–21 $2(g/3R)^{1/2}$

9–23 a) 247 rev·min^{-1} b) 1.07 hp

9–25 a) 19.6 N·m   b) 3084 J

9–27 a) 283 N·m   b) 1.33×10^5 J

9–29 a) 49.0 N b) 40.0 N c) 8.16 s

9–31 a) 6.25 rad$\cdot$s^{-2} b) 250 J c) 4.74 rad$\cdot$s^{-2}; tension is less than weight.

9–33 a) 65.3 N b) 16.2 m$\cdot$s^{-1} c) 2.47 s

9–35 a) 3.27 m$\cdot$s^{-2}, 0 rad$\cdot$s^{-2}, 26.1 N, 26.1 N
   b) 0.754 m$\cdot$s^{-2}, 7.54 rad$\cdot$s^{-2}, 21.1 N, 36.2 N

9–37 $(2/3)g \sin \theta$, $(1/3)mg \sin \theta$,
   $(1/3) \tan \theta$ (3.27 m$\cdot$s^{-2}, $mg/6$, 0.192)

9–39 $F/2m$, $F/2$

9–41 $2Ma^2/3$

9–43 1.98 N$\cdot$m

9–45 0.60 rad$\cdot$s^{-1}

9–47 a) 5.61 rad$\cdot$s^{-1} b) 0.0214 m
   c) 918 m$\cdot$s^{-1}

9–49 1.44 J

9–51 a) 12.0 rad$\cdot$s^{-1} b) 0.027 J

9–53 0.0789 rev$\cdot$s^{-1}

9–55 a) -0.05 rad$\cdot$s^{-1} b) 32.7° c) 36.0°

9–57 a) 4.0 kg$\cdot$m^2 b) 5260 J

9–59 a) 1.76 N b) 4290 rev$\cdot$min^{-1}

CHAPTER 10

10–1 a) 2.4×10^8 Pa b) 5.0×10^{-4}
   c) 4.8×10^{11} Pa

10–3 4.11×10^8 Pa, 0.054 mm

10–5 b) 1.02×10^{11} Pa c) 7.64×10^7 Pa

10–7 a) 62.8 N b) 5.71 mm c) 1.83×10^{-5} cm

10–9 a) 1.82 m b) 1.5×10^8 Pa, 3.0×10^8 Pa
   c) 1.36×10^{-3}, 1.50×10^{-3}

10–11 1.0×10^8 Pa, 0.5×10^{-3}, 2.5 mm, 9.5×10^{-5}

10–13 a) -0.0504 m^3 b) 1.085×10^3 kg$\cdot$m^{-3}

10–15 4.0×10^9 Pa, 0.25×10^{-9} Pa^{-1}

10–17 4.71×10^4 N

10–19 0.633×10^{-6} atm^{-1}, 49.6×10^{-6} atm^{-1}; water is 78 times more compressible.

10–21 0.17 m

10–25 1.66 cm

CHAPTER 11

11–1 10π s^{-1}, 0.2 s

11–3 493 N$\cdot$m^{-1}

11–5 $1/3$ s, 6π s^{-1}, 0.563 kg

11–7 a) 94.8 m$\cdot$s^{-2}, 3.77 m$\cdot$s^{-1} b) 56.9 m$\cdot$s^{-2}, 3.02 m$\cdot$s^{-1} c) 0.0369 s

11–9 a) 3.13 kg b) 4.33 cm to right of equilibrium, moving to left c) 36 N

11–11 a) 7106 m$\cdot$s^{-2} b) 3553 N
   c) 67.9 km$\cdot$hr^{-1}

11–13 22.5 kg

11–15 a) 7.88 Hz b) 9.8×10^{10} Pa

11–17 0.777 s

11–19 a) $k_1 + k_2$ b) $k_1 + k_2$ c) $k_1 k_2/(k_1 + k_2)$
   d) $\sqrt{2}$

11–21 a) 1.41 s, 3.53 cm b) Yes; friction c) No

11–23 0.248 m

11–25 4.80 s

11–27 a) 1.80 g$\cdot$cm^2 or 1.80×10^{-7} kg$\cdot$m^2
   b) 284 dyn$\cdot$cm$\cdot$rad^{-1} or
   0.284×10^{-4} N$\cdot$m$\cdot$rad^{-1}

11–29 a) $7ML^2/48$ b) $2\pi(7L/12g)^{1/2}$

11–31 a) 0.0402 m^2 b) 0.698 rad$\cdot$s^{-1}

11–33 24.8 m

CHAPTER 12

12–1 0.889×10^3 kg$\cdot$m^{-3}; yes

12–3 2.08×10^5 Pa, 2.05 atm

12–5 a) 1.077×10^5 Pa b) 1.037×10^5 Pa
   c) 1.037×10^5 Pa d) 5 cm Hg
   e) 68 cm H$_2$O

12–7 a) 128,000 lb$\cdot$ft^{-2} or 889 lb$\cdot$in^{-2}
   b) 25,132 lb

12–9 a) 1.483×10^{-3} m^3 b) 110.5 N

12–11 a) 1/4 b) 3/4

12–13 0.893

12–15 a) 98 N b) 1.95 N

12–17 2.41 m^2

12–19 8333 m^3, 9266 kg

12–21 a) 5 cm b) 490 Pa

12–23 a) 1667 kg$\cdot$m^{-3} b) E reads 2.5 kg, D reads 7.5 kg

12–27 Air bubbles do cause errors.

12–29 8.84×10^4 N$\cdot$m

12–31 a) 9.26 m b) Yes c) No effect

12–33 a) Yes b) For rubber, not for soap
   c) No

12–35 1372 Pa

12–37 72.8 Pa, 1.46×10^4 Pa

12–39 a) 70.7 cm Hg b) 71.3 cm Hg c) 10.8 cm

12–41 40 dyn$\cdot$cm^{-2} or 4.0 Pa

12–43 3.6×10^{-4} mm

CHAPTER 13

13–1 a) 14 m$\cdot$s^{-1} b) 4.4×10^{-3} m$^3\cdot$s^{-1}

13–3 10.0 cm

13–5 11.8 m$\cdot$s^{-1}

13–7 a) 1.22×10^{-3} m$^3\cdot$s^{-1} b) 0.90 m

13–9 a) 8 m$\cdot$s^{-1} b) 0.0628 m$^3\cdot$s^{-1}

13–11 4.95×10^5 Pa

13–13 a) 4.38 m$\cdot$s^{-1} b) 4.38 m$\cdot$s^{-1}, 0.39×10^5 Pa (gauge)

13–15 0.0268 m$^3\cdot$min^{-1}

13–17 0.438 m$^3\cdot$min^{-1}

13–19 0.816 m$^3\cdot$s^{-1}

13–21 38.8 m$\cdot$s^{-1}

13–23 4.0×10^4 N

13–25 8.04 Pa
13–29 a) 0.769 cm·s^{-1} b) 1.89 cm·s^{-1}
13–31 107 m·s^{-1}
13–33 a) Turbulent   b) 5.3 liter·s^{-1}
13–35 a) Yes   b) No

CHAPTER 14

14–1 600.4 K
14–3 a) 20°C   b) 35°C   c) −15°C
14–5 6.83 cm Hg
14–7 200.072 m
14–9 499.88 ml
14–11 a) −5 × 10^{-5} (C°)$^{-1}$ b) 1 × 10^{-5} (C°)$^{-1}$
       c) 7.50 × 10^{-5} (C°)$^{-1}$ d) 68.6 × 10^{-5} (C°)$^{-1}$
14–13 3.6 × 10^{-4}
14–15 1.0 × 10^{-5} (C°)$^{-1}$
14–17 0.2506 in.
14–19 20 cm, 10 cm
14–21 76.5 m
14–23 a) 1.16 C°; no   b) Lose
14–25 1.20 × 10^{6} N
14–27 1.04 × 10^{8} Pa
14–29 a) 0.036 ft   b) 1.74 × 10^{4} lb·in^{-2}
14–31 a) 9.01 × 10^{7} Pa   b) 36.2°C
14–33 5.04 × 10^{7} Pa

CHAPTER 15

15–1 a) Yes   b) No
15–3 4480 cal
15–5 371 years
15–7 a) 80.6
15–9 a) 6.70 × 10^{4} J   b) 670 s or 11 min 10 s
15–11 1.73 × 10^{5} J
15–13 29.8°C
15–15 a) 1:0.093:0.031   b) 1:0.83:0.35
15–17 0.1 Btu·lb^{-1}·(F°)$^{-1}$
15–19 0.091 cal·g^{-1}·(C°)$^{-1}$
15–21 a) 72.3°C   b) Temperature of apparatus does
       not change.
15–23 a) 0.806   b) 8.36 × 10^{4} cal
15–25 116 W, 13
15–27 357 m·s^{-1}
15–29 b) 3.01 J·g^{-1}·(C°)$^{-1}$
15–31 107 g
15–33 0°C with 0.2 g of ice remaining
15–35 539 cal·g^{-1}
15–37 23.9°C
15–39 39.9°C
15–41 1.84 kg
15–43 a) 1602 ft^{3} b) 310 ft^{3}
15–45 a) 12,000 Btu·hr^{-1} b) 3.51 kW
15–47 a) 8.05 × 10^{6} J   b) 805 N

CHAPTER 16

16–1 4.32 × 10^{5} J
16–3 a) 7.70 W   b) 19.6 cm
16–5 0.205 Btu·in.·hr^{-1}·ft^{-2}·(F°)$^{-1}$
16–7 111°C
16–9 a) −4°C   b) 8 W·m^{-2}
16–11 9.48°C, 82.2°C
16–13 a) 40°C   b) 2.4 cal·s^{-1} or 10.1 W
16–15 76.4 C°
16–17 a) 7.30 × 10^{5} cal   b) 7.83 × 10^{5} cal
16–19 a) 459 W·m^{-2} b) 4.59 × 10^{6} W·m^{-2}
16–21 1.78 W
16–23 20.027 K
16–25 6.89 W
16–27 72.2 W
16–29 111 m^{2}
16–31 34.4 kg

CHAPTER 17

17–1 a) 606°C   b) 6 l
17–3 a) 2493 mol   b) 79.8 kg
17–5 1.16 kg
17–7 3.0 × 10^{4} Pa
17–9 a) 203 mol   b) 0.812 kg
17–11 a) 760 × 10^{4} Pa   b) 1.95 g
17–13 $\rho = pM/RT$
17–15 3.21
17–17 a) 82.1 cm^{3} b) 0.333 g
17–19 When the piston has moved 13.1 in.
17–21 2.24 mg
17–25 Ice and vapor, pressure 610 Pa
17–27 321 kg·m^{-3}, 0.321
17–29 a) 37.3%   b) 868 Pa   c) 6.41 g·m^{-3}
17–31 a) 0.616 kg   b) 25.7 × 10^{-3} kg·m^{-3}
       c) 0.035 kg
17–33 10.8 kg·hr^{-1}

CHAPTER 18

18–1 a) Yes   b) No   c) Negative
18–3 a) No   b) Yes   c) Yes   d) Work done
       on the resistor equals heat transferred to water.
18–5 171 m
18–7 10.8 ft^{3}
18–9 $U_1 = U_2$
18–11 a) 0.167 × 10^{6} J   b) 2.03 × 10^{6} J
18–13 a) 20.64 mol   b) 2.142 × 10^{4} J
       c) 0.858 × 10^{4} J   d) 2.142 × 10^{4} J
18–15 a) 500 J, 0   b) 0, −500 J
18–17 a) 2.45 × 10^{-4} m^{3} b) −735 J
       c) 1.529 × 10^{6} J   d) 1.528 × 10^{6} J
18–19 b) 831 J   c) On the piston   d) 2110 J
       e) 2940 J   f) 831 J

18–21 a) 1247 J, 4411 J, 3165 J   b) 0, −3165 J, −3165 J   c) Zero
18–25 a) 314 atm, 189 K   b) 3.79 atm, 227 K
18–27 42.0 J, zero
18–29 254°C
18–31 a) When piston in 3.13 in. from bottom   b) 477 K

CHAPTER 19

19–1 a) 0.375   b) 5000 J   c) 0.16 g   d) 150 kW, 201 hp
19–3 a) 1250 MW   b) 50 kg, 4.32×10^6 kg   c) 750 MW   d) 35.8 m^3
19–5 a) 1 atm, 2.46×10^{-3} m^3; 2 atm, 2.46×10^{-3} m^3; 1 atm, 3.73×10^{-3} m^3 b) 52.0 J   c) 0.139
19–7 15.8%
19–9 a) 320 K   b) 20%
19–11 a) 900 cal   b) 1600 cal   c) 400 cal
19–13 a) 610 J·K^{-1} b) −568 J·K^{-1} c) 41.6 J·K^{-1}
19–15 a) −446 J·K^{-1} b) 610 J·K^{-1} c) Zero   d) 164 J·K^{-1}
19–19 0.0503

CHAPTER 20

20–1 a) 3.34×10^{-7} cm   b) About 10 times as great
20–3 1.95×10^{-4} cm
20–5 10^8 Pa (1000 atm)
20–7 2.82×10^{-10} m
20–9 a) 51.5 K   b) 100 K
20–11 1.00636
20–13 a) 517 m·s^{-1} b) 298 m·s^{-1}
20–15 a) 8650 m·s^{-1}, 1.25×10^{-19} J   b) 61,800 m·s^{-1}; no
20–17 a) 1.1 cm·s^{-1} b) No
20–19 5.21×10^{-25} J; 0.0251 K
20–21 Bi: 0.99 cm; Bi$_2$: 1.4 cm
20–23 10.4 J·g^{-1}·K^{-1}, 2.48 times as great
20–25 24.9 J·mol^{-1}·K^{-1}; 0.124, 0.205, 0.039
20–27 8.17×10^{-11} mol·s^{-1}
20–29 41.6 mol·m^{-3}
20–31 a) 1027 mol·m^{-3} b) 25.0×10^5 Pa (about 25 atm)   c) 25.0×10^5 Pa

CHAPTER 21

21–1 a) 1075 m   b) 282 Hz
21–3 4302 Hz
21–5 0.423 g·m^{-1}
21–7 a) 200 m·s^{-1}, 20 m   b) Increase by $2^{1/2}$
21–9 63.3 m·s^{-1}

21–11 a) 17.3 m, 0.0173 m   b) 72.5 m, 0.0725 m
21–13 4.76×10^{-10} Pa^{-1}
21–15 a) 323 m·s^{-1} b) 1316 m·s^{-1} c) 348 m·s^{-1}
21–17 a) 138 Hz   b) 1380 Hz   c) 13,800 Hz
21–19 0.683
21–21 a) 3.95 m·s^{-1} b) 2.21 m·s^{-1}
21–23 a) 10.3 m   b) 1.63 m
21–25 a) 2 cm, 4 cm, 100 Hz, 10^{-2} s, 400 cm·s^{-1} c) 8 N
21–29 a) 0.50 Hz   b) π s^{-1} c) $\pi/100$ cm^{-1} d) $y = 10 \sin[(\pi x/100) - \pi t]$ e) $y = -10 \sin \pi t$ f) $y = 10 \sin[(3\pi/2) - \pi t]$ g) 31.4 cm·s^{-1} h) $y = 7.07$ cm, $v = -22.2$ cm·s^{-1}

CHAPTER 22

22–1 a) 200 Hz   b) 49 (50th harmonic)
22–3 a) 36.0 m·s^{-1} b) 64.8 N
22–5 1.28
22–7 a) 4960 m·s^{-1} b) 342 m·s^{-1}
22–9 a) Diatomic
22–11 a) 1130, 2260, 3390, 4520, 5650 Hz   b) 565, 1695, 2825, 3955, 5085 Hz   c) 17, 35
22–13 a) 420 Hz   b) 415 Hz
22–15 12.2%
22–17 a) 0.655 m   b) 56°C

CHAPTER 23

23–1 a) 9   b) 4
23–3 a) 60 dB   b) 77 dB
23–5 b) No; $\beta = 160 + 10 \log I$
23–7 10^{-6} W
23–9 a) 0.252 μm   b) 0.290 Pa   c) 50 m
23–11 $BV^2/2c$
23–13 61 dB
23–15 a) π b) 3.98×10^{-6} W·m^{-2}, 11.9×10^{-6} W·m^{-2} c) 2.13×10^{-6} W·m^{-2}, 63.3 dB
23–17 a) 0.029 m   b) 0.21 m   c) 475 Hz
23–19 1.5000, 1.4983; just fifth is 0.11% larger
23–21 9.89 m·s^{-1}
23–23 a) 0.63 m   b) 0.75 m   c) 548 Hz   d) 460 Hz   e) 573 Hz   f) 440 Hz   g) 0.65 m, 0.73 m, 547, 458, 569, 439 Hz
23–25 a) 464.3 Hz   b) 468.8 Hz   c) 4.5 Hz
23–27 a) 0.345 m   b) 0.315 m   c) 1.476 wave   d) 345 m·s^{-1} e) 0.152 m

CHAPTER 24

24–1 0.899 N, repulsive
24–3 1.52×10^{-14} N
24–5 625

24–7 0.119 m

24–9 a) 8.34×10^{19} N   b) 8.34×10^5 N

24–11 a) 2.75×10^{26} b) 6.59×10^{15}
c) 2.40×10^{-11}

24–13 2.16×10^{-7} N

24–15 a) $q^2/2\pi\epsilon_0 a^2$, vertically upward
b) $q^2 a/2\pi\epsilon_0 (a^2 + x^2)^{3/2}$

24–17 a) 4.04×10^6 b) 4.93×10^{-17}

24–19 a) 5.09×10^5 N   b) 1.44×10^6 N

24–23 b) 2.79×10^{-5} N, 21.9^c below $+x$-axis

CHAPTER 25

25–1 7.19×10^4 N·C^{-1}

25–3 4.74 m

25–5 a) 4.0 N·C^{-1} upward   b) 6.4×10^{-19} N downward

25–7 a) 1010 N·C^{-1} b) 2670 km·s^{-1}

25–9 a) 1.42 cm   b) 9.86 cm

25–11 -24.9×10^{-9} C

25–13 a) 1.80×10^4 N·C^{-1}, $-x$-direction
b) 7.99×10^3 N·C^{-1}, $+x$-direction
c) 3.34×10^3 N·C^{-1}, 70° above $-x$-axis
d) 6.36×10^3 N·C^{-1}, $-x$-direction

25–15 40.8°

25–17 5.58×10^{-11} N·C^{-1}

25–19 8.85×10^{-10} C

25–23 b) $q/4\pi\epsilon_0 r^2$ or $\rho R^3/3\epsilon_0 r^2$

25–25 a) $Q/4\pi\epsilon_0 r^2$, 0, $Q/4\pi\epsilon_0 r^2$ b) $-Q/4\pi a^2$

CHAPTER 26

26–1 30 V; positively charged plate

26–3 a) 180 V   b) 180 V

26–5 a) -4.5×10^{-5} J   b) 3.0×10^5 N·C^{-1}
c) -1.5×10^4 V

26–7 b) 0.667 mm

26–9 a) 1798 V   b) Zero   c) -4.49×10^{-5} J

26–11 a) 3.00 m   b) 0.200×10^{-6} C

26–13 b) $q/2\pi\epsilon_0 a$ e) $\pm a\sqrt{3}$

26–15 a) It remains at rest.   b) It oscillates about the origin, along the y-axis.   c) It accelerates away from the origin along the x-axis.

26–17 b) Zero
c) $q/4\pi\epsilon_0[(x^2 + 2ax + 2a^2)^{-1/2} - (x^2 - 2ax + 2a^2)^{-1/2}]$

26–19 c) Zero

26–21 1.03×10^7 m·s^{-1}

26–23 a) $Q/4\pi\epsilon_0 (x^2 + a^2)^{1/2}$ c) Second curve is negative slope of first.

26–25 a) -539 V, $+59.9$ V   b) -2.40×10^{-6} J

26–27 a) 4.33×10^{-4} m·s^{-1} upward   b) 2.74×10^{-4} m·s^{-1}

26–29 938 MeV

26–31 2.27×10^{-14} m

26–33 a) 5114 V   b) $0.140c$ c) 9.38 MeV, $0.140c$

26–35 a) 0.704 cm   b) 19.4°   c) 4.92 cm

CHAPTER 27

27–1 a) 50 pF   b) 2.82×10^{-3} m^2 c) 400 V
d) 1.0×10^{-6} J

27–3 a) 1000 V   b) 2000 V   c) 0.5×10^{-3} J

27–5 a) 400 V   b) 1.13×10^{-2} m^2
c) 2.00×10^6 N·C^{-1} d) 1.77×10^{-5} C·m^{-2}

27–7 a) 1 μF   b) 900 μC   c) 100 V

27–9 a) 800 μC, 800 V; 800 μC, 400 V
b) 533 μC, 533 V; 1067 μC, 533 V

27–11 1.30×10^{-3} J

27–13 3.60 J

27–15 a) 0.226 m^2 b) 1250 V

27–17 a) 35.4 pF   b) 1.77 nC   c) 5000 V·m^{-1}
d) 4.42×10^{-8} J   e) 17.7 pF, 1.77 nC, 5000 V·m^{-1}, 8.84×10^{-8} J

27–19 a) $Q^2 x/2\epsilon_0 A$ b) $Q^2(x + \Delta x)/2\epsilon_0 A$

27–21 1770 pF

27–23 a) 26.6 μC·m^{-2} b) 17.7 μC·m^{-2}

27–25 a) 3.43   b) 0.708×10^{-5} C·m^{-2}

CHAPTER 28

28–1 a) 20 mA   b) 2.74×10^{-6} m·s^{-1}

28–3 1.30×10^5 mm^{-3}

28–5 a) 10^7 A·m^{-2} b) 0.172 V·m^{-1} c) 44 hr

28–7 a) 2.4×10^{-8} Ω·m   b) 20 A
c) 3.90×10^{-4} m·s^{-1}

28–9 45.6 m

28–11 a) No   c) About 1.0 Ω; smaller

28–13 0.735×10^{-6} Ω

28–15 a) 99.52 Ω b) 0.0148 Ω

28–17 28.5°C

28–19 284°C

28–21 60.9°C

28–23 2.47 mm

28–25 a) 0.50 Ω b) 10V

28–27 Open: a) 12 V   b) 0   c) 12 V
Closed: a) 11.1 V   b) 11.1 V   c) 0

28–29 a) 2.22 A   b) 1.4 A   c) 6.4 V

28–31 90 Ω

28–33 a) VI, $I^2 R$, V^2/R b) 90 Ω

28–35 a) 24 W   b) 4 W   c) 20 W

CHAPTER 29

29–1 a) 6 Ω b) 2 A   c) 2 A   d) 2 V, 4 V, 6 V   e) 4 W, 8 W, 12 W

29–3 9.6 W in 60-W lamp, 14.4 W in 40-W lamp

29–5 a) 141 V   b) 4.5 W   c) Two in series, connected in parallel with two others in series

29–7 a) Two in series, connected in parallel with two others in series. b) 1/2 W, if total power is 2 W.

29–9 a) $8\,\Omega$ b) 12 V

29–11 655 W

29–13 a) -12 V b) 3 A c) -12 V d) 12/7 A from b to a e) $4.5\,\Omega$ f) $4.2\,\Omega$

29–15 a) 18 V b) Point a c) 6 V d) $36\,\mu$C

29–17 18 V, 7 V, 13 V

29–19 6 A, 4 A, 2 A, 4 A, 6 A (top to bottom)

29–21 a) 0.8 A, 0.2 A, 0.6 A b) -3.2 V

29–23 a) $1200\,\Omega$ b) 30 V

29–25 10.9 V, 109.1 V

29–27 a) $5.00\,\text{m}\Omega$ b) $299{,}950\,\Omega$

29–29 $2985\,\Omega$, $12{,}000\,\Omega$, $135{,}000\,\Omega$; $3000\,\Omega$; $15{,}000\,\Omega$, $150{,}000\,\Omega$

29–31 a) $1450\,\Omega$ b) 4500, 1500, $500\,\Omega$ c) No

29–33 $55\,\Omega$

29–35 $2749\,\Omega$

29–37 a) 9.52 V b) No

29–39 $7.21\,\mu$F

29–41 a) $12\,\mu$s b) 2.53 V

29–43 a) $5.3\,\mu$C b) 2 mA c) 2 mA

29–47 a) 12.5 A, 10.0 A, 0.833 A b) Yes

29–49 a) 3.15×10^9 J, 876 kWh b) \$87.60

CHAPTER 30

30–1 a) qvB along $-z$-axis b) qvB along $+y$-axis c) Zero d) $qvB/\sqrt{2}$ parallel to $-y$-axis e) qvB in yz-plane at 45° to $-y$- and $-z$-axes f) $\sqrt{2}qvB/\sqrt{3}$ in yz-plane at 45° to $-y$- and $-z$-axes

30–3 3.27 T, perpendicular to direction of v

30–5 a) 1.14×10^{-3} T, into page b) 1.57×10^{-8} s

30–7 Negative

30–9 8 mm

30–11 2.15×10^{-7} Wb

30–13 a) 4.0×10^6 V·m^{-1} b) 20,000 V

30–15 2.13×10^{-2} m

30–17 21

30–19 1.20 N, $-z$-direction; 1.20 N, $-y$-direction; 1.70 N, 45° up, parallel to yz-plane; 120 N, $-y$-direction; zero

30–21 2.4 N

30–23 24 A

30–25 a) 0.16 N, 0.06 N, 0.0083 N·m b) 0.16 N, 0.104 N, 0.0048 N·m c) Same torque

30–27 3.6×10^{-6} N·m

30–29 a) 0.181 N·m b) When normal to coil is at 30° to field.

30–31 a) 0.5 A b) 4 A c) 108 V d) 60 W e) 48 W f) 540 W g) 71%

30–33 a) 8.45×10^{-4} m·s^{-1} b) 1.27 mV·m^{-1} c) $25.3\,\mu$V

CHAPTER 31

31–1 1.6×10^{-5} T; yes

31–3 a) 125 A b) 2.5×10^{-4} T, 1.25×10^{-4} T

31–5 b) $\mu_0 Ia/\pi(a^2 + x^2)$ d) Zero

31–7 a) $\mu_0 I/\pi a$ b) $\mu_0 I/3\pi a$ c) Zero d) $2\mu_0 I/3\pi a$

31–9 a) 1.92×10^{-4} N·m^{-1}, down b) 1.92×10^{-4} N·m^{-1}, up

31–11 1.02 cm

31–13 7.20×10^{-4} N to left

31–17 $1.91R$

31–19 16

31–21 a) $\mu_0 I/2\pi r$ b) Zero

31–23 6.0×10^{-4} T

31–25 a) 7958 m^{-1} b) 1250 m

31–27 a) Zero b) 1.41×10^{-6} T, 45° in xz-plane c) 5.00×10^{-6} T, $+x$-direction

31–29 397, 398

31–31 a) 0.250 V·m^{-1} b) 2.00×10^{-4} T c) 25.0 V·m^{-1}·s^{-1} d) 2.21×10^{-10} A·m^{-2} e) 3.54×10^{-21} T; no

CHAPTER 32

32–1 a) 5.0 m·s^{-1} b) 2.0 A c) 0.96 N, to left

32–3 a) 3 V b) 3 V c) Upper end

32–5 a) 1.0 V, from B to A b) 1.25 N c) 5.0 W

32–7 0.006 V

32–9 b) $FR/B^2 l^2$

32–11 a) 5.56 rad·s^{-1} b) Zero

32–15 a) 0.2 V b) Zero c) 0.2 V

32–17 a) Circles, clockwise b) 0.005 V·m^{-1}, 3.14 mV c) 1.57 mA d) Zero e) 3.14 mV

32–19 a) Right to left b) Right to left c) Left to right

32–21 a) 1.01×10^{-5} C

32–23 a) 0.737 rev·s^{-1} b) 4320 N·m

CHAPTER 33

33–1 0.790 mH

33–3 1.0 mH; 2.25 mH or 0.25 mH

33–5 a) 5.0×10^{-4} V; yes b) 5.0×10^{-4} V

33–7 a) 0.100 J b) 2.00 W

33–11 387

33–13 a) 0.05 A b) 1.0 A·s^{-1} c) 0.5 A·s^{-1} d) 0.23 s e) 0.0197, 0.0317, 0.0389, 0.0433 A

33–17 a) 2.39 mH   b) 4.41 pF
33–19 a) $4L$ b) L c) $\omega/2$

CHAPTER 34

34–1 a) 127 Hz   b) 7.96 Hz
34–3 a) 377 Ω b) 2.65 mH   c) 2650 Ω
    d) 2650 μF
34–5 a) 5 mA   b) 0.05 A   c) 0.5 A
34–7 a) 0.05 A   b) 5.0 mA   c) 0.50 mA
34–11 a) 583 Ω b) 0.0858 A   c) 25.7 V, 42.9 V
    d) 59°, lead
34–13 a) 626 Ω b) First larger, then smaller
    c) 61.4°, current leads
34–15 16.9, 25.4, 56.4, 31.0, 35.4 V (rms)
34–17 b) 745 rads·s^{-1} or 118 Hz   c) 1   d) 35.4,
    79.2, 79.2, 0, 35.4 V (rms)   e) 745 s^{-1}
    f) 0.354 A
34–19 a) 1592 Hz   b) 12.4 V (rms)
34–21 0.0184 H
34–23 a) 25 W   b) 25 W   c) Zero   d) Zero
    e) 0.127
34–25 a) 151 μF   b) 917 W
34–27 a) 31.6   b) 3.16 V
34–29 a) 1592 Hz, 10,000 s^{-1} b) 1.0 A
    c) 1000 V   d) 0.05 J
34–31 a) 150   b) 180 W   c) 1.5 A

CHAPTER 35

35–1 a) 3.0×10^8 m·s^{-1} b) 294 m
    c) 4.8×10^{-3} V·m^{-1}
35–3 a) 3.0×10^6 m·s^{-1}
    b) 3 cm
35–5 a) 1027 V·m^{-1}, 3.42×10^{-6} T
    b) 3.96×10^{26} W
35–7 About 6×10^4 V·m^{-1}, 2×10^{-4} T
35–9 a) $\rho I/\pi a^2$, parallel to wire   b) $Ir/2\pi a^2$,
    perpendicular to wire   c) $I^2 r\rho/2\pi^2 a^4$
    d) $2\pi r l S$
35–11 24.5 V·m^{-1}, 0.0650 A·m^{-1}
35–13 a) $(\mu_0 nr/2)(\Delta I/\Delta t)$, tangent
    b) $(\mu_0^2 n^2 r/2)(I \, \Delta I/\Delta t)$, radially inward
35–15 300 m
35–17 a) 10^{10} Hz   b) 1.0×10^{-7} T
    c) 1.19 W·m^{-2} d) 1.99×10^{-9} N

CHAPTER 36

36–1 a) 1.5×10^{-9} m, 1.5×10^{-3} μm, 1.50 nm,
    15 Å   b) 5.36×10^{-7} m, 0.536 μm, 536 nm,
    5360 Å
36–3 229 rad·s^{-1}
36–7 1.732

36–9 a) 32.1°   b) Independent of n
36–11 a) 2×10^8 m·s^{-1} b) 333 nm
36–13 1.87
36–15 1.88
36–17 1.02 cm
36–21 1.75°
36–23 7.07 m
36–25 a) 44   b) 44 kW   c) 354 hp
36–27 a) $I_0/2$, along first axis; $I_0/4$, along second axis;
    $I_0/8$, along third axis   b) No light transmitted
36–29 First filter at 45° to polarization of beam, second
    at 90° to beam; $I_0/4$
36–31 35.5°
36–33 a) 37°   b) Horizontal
36–35 a) 58.8°   b) 51.1°
36–37 2, linearly polarized at 45°; 3, elliptical; 4,
    linearly polarized at 60°
36–39 Left

CHAPTER 37

37–1 Half the observer's height
37–3 a) 9.88 cm   b) 5 cm
37–7 a) 3   b) 30 cm behind mirror
37–9 1.71 cm, 0.143
37–11 b) 10 cm behind mirror, 4 cm, erect, virtual
37–13 a) $|s| < 5$ cm   b) Erect
37–15 a) 45°   b) 1.41 cm
37–17 4.48 cm
37–19 1.35
37–21 30 cm, -1
37–23 0.667 cm

CHAPTER 38

38–1 60 cm, 5 cm, inverted
38–3 a) -5 cm, diverging   b) 0.4 cm, erect
38–5 a) 6 cm or 12 cm from object   b) -2, -0.5
38–7 a) 15, 20, 30, -10 cm   b) -0.5, -1, -2, -3
    c) Real, real, real, virtual   d) inverted,
    inverted, inverted, erect
38–9 a) 120 cm, -3 cm   b) 120 cm to right of
    second lens   c) 21.8 cm to right of second
    lens   d) 80 cm to left of second lens
38–11 a) 25 cm to the left   b) Virtual
38–13 -15 cm
38–15 a) The first image   b) 30 cm   c) Real
    d) At infinity
38–17 50 cm
38–19 a) 35 mm   b) 200 mm
38–21 2.0 cm
38–23 2 cm left of first lens, 2 cm right of second lens
38–25 0.714 cm
38–27 $f = 0.81$ cm in air, 2.4 cm in water

38–29 a) +3.2 diopters   b) −2 diopters
38–31 a) 50 cm   b) 2 m
38–33 a) 7.14 cm   b) 3.5 mm
38–35 1/400 s
38–37 a) 120 mm   b) 121.4 mm from lens
38–39 a) 842×   b) 50×
38–41 380×
38–43 20×
38–47 25×, 35 cm from mirror

CHAPTER 39

39–1 1.0 mm
39–3 1.0 mm
39–5 a) Perpendicular to line joining antennas
     b) $0.5I_0$, $0.197I_0$, $0.044I_0$, 0
39–7 a) 5.5   b) Zero
39–9 0.909 mm
39–11 36.7 fringes·cm^{-1}
39–13 143 nm
39–15 a) Yes   b) 89 nm
39–17 2.36 mm
39–19 17.5°, 36.9°, 64.1°
39–21 a) 570 nm   b) 43.2°
39–23 500 nm
39–25 2000 m
39–27 Smaller by factor of 5/6
39–29 No difference

CHAPTER 40

40–1 a) 4.3×10^{-8} s   b) 10.4 m
40–3 The bolt at A appears to come first.
40–5 $\Delta x' = [(\Delta x)^2 - c^2 (\Delta t)^2]^{1/2}$
40–9 5.77×10^{-9} s
40–11 $v/c = E/mc^2$
40–13 2.60×10^8 m·s^{-1}
40–15 a) 0.5×10^{-4} b) 15.5%   c) Factor of 2.29
40–17 1.18×10^{-18} m·s^{-1}
40–19 2.986×10^8 m·s^{-1}
40–21 1.64×10^{-13} J
40–23 a) 4.55×10^{-15} J, 4.92×10^{-15} J
     b) 18.2×10^{-15} J, 27.9×10^{-15} J
40–25 a) 1.8×10^{14} J   b) 1.8×10^{20} W
     c) 1.84×10^{10} kg
40–27 a) 1.8×10^5 eV   b) 6.91×10^5 eV
     c) 2.02×10^8 m·s^{-1} d) 2.52×10^8 m·s^{-1}

CHAPTER 41

41–1 1.78×10^{19} s^{-1}
41–3 a) 0.01 J, 6.24×10^{16} eV   b) 3.14×10^{-19} J,
     1.96 eV   c) 3.19×10^{16}
41–5 3.02×10^{19} s^{-1}, 1550 km

41–7 $hf_0 = \phi$
41–9 1.72×10^6 m·s^{-1}
41–11 a) 4.58×10^{14} Hz   b) 654 nm
     c) 1.90 eV   d) 6.62×10^{-34} J·s
41–13 0.34 V greater
41–15 a) 1.04 eV   b) 1190 nm
     c) 2.52×10^{14} Hz   d) 4.14×10^{-7} eV
     f) No
41–17 1.41×10^{19} s^{-1}
41–19 a) 6.16×10^{14} Hz   b) 486 nm
41–21 1.96 eV or 3.14×10^{-19} J
41–23 a) 2.48×10^4 V   b) 0.00062 mm or 0.62 pm
41–25 a) 355 cal·s^{-1} b) 33.8 C°·s^{-1} c) High
     melting temperature and good thermal
     conductivity
41–27 79,056 K
41–29 4.85×10^{-12} m
41–31 a) 2.481×10^{-11} m   b) 2.496×10^{-11} m
     c) 49.7 keV
41–33 a) 0.1248×10^{-9} m   b) 6.43×10^{-17} J or
     401.4 eV   c) 1.19×10^7 m·s^{-1}; no

CHAPTER 42

42–1 a) 13.6 eV   b) −26.2 eV   c) −13.6 eV
     d) 13.6 eV
42–3 a) Larger by factor of 4   b) Series ending at
     $n = 4$, $n = 5$, and one line of series ending at
     $n = 3$.
42–5 a) −2.80 keV   b) 2.57×10^{-13} m or 257 fm
     c) 0.59 nm
42–9 97.2 nm, 3.09×10^{15} Hz
42–11 636.3 nm, 364.6 nm
42–13 a) 12,400 eV   b) 150 eV
42–15 0.145 nm
42–17 8.67 pm
42–21 2.93×10^{42}
42–23 a) 2 eV, 621 nm   b) 3.29×10^{-10} eV,
     1.02×10^{-7} nm
42–25 2.34×10^{-23} s
42–27

l	0	1	1	1	2	2	2	2	2
m	0	−1	0	1	−2	−1	0	1	2

CHAPTER 43

43–1 $(1s)^2(2s)^2(2p)^2$, $(1s)^2(2s)^2(2p)^6(3s)^2(3p)^2$
43–3 $(1s)^2(2s)^2(2p)^6(3s)^2(3p)^6(3d)^{10}(4s)^2(4p)^2$
43–5 1.11 nm
43–7 Cesium
43–11 36,480 K
43–13 a) 488 N·m^{-1} b) 5.70×10^{-20} J, 0.356 eV
     c) 3.49 μm, infrared
43–15 a) 4.58×10^{-48} kg·m^2 b) 0, 2.43×10^{-21} J,
     7.28×10^{-21} J   c) 27.3 μm, 1.10×10^{13} Hz
43–19 1127 nm, near infrared

CHAPTER 44

44–1 a) 1, 0   b) 79, 118   c) 92, 146

44–3 a) 3.65×10^{-12} J, 22.7 MeV
b) 3.65×10^{-12} J, 22.7 MeV

44–5 5.21×10^{-12} J, 32.5 MeV

44–7 17.6 MeV, 2.81×10^{-12} J

44–9 0.137 u, 127.6 MeV, 7.97 MeV

44–11 $\Delta m = 0.000440$ u, 0.410 MeV

44–13 6.05 MeV

44–15 2.76×10^{6} s^{-1}, 7.46×10^{-5} curie

44–17 a) 3.70×10^{-2} J   b) 7.41 rad, 7.41×10^{-2} Gy
c) 5.19 rem, 5.19×10^{-2} Gy
d) 38.6 s

44–19 0.782 MeV

44–21 a) 1.32 T   b) 4.22 MeV, 2.01×10^{7} m·s^{-1}

44–23 23.8 MeV, 3.82×10^{-12} J, 2.3×10^{12} J·mol^{-1};
larger by 10^{7}

44–25 0.511 MeV (8.19×10^{-14} J), 1.24×10^{20} Hz,
2.43×10^{-12} m

44–27 56.5 MeV

44–29 107 MeV

44–31 1.65×10^{-22} s